Transfer Functions of Switching Converters

Fast Analytical Techniques at Work with Small-Signal Analysis

Christophe Basso

Transfer Functions of Switching Converters: Fast Analytical Techniques at Work with Small-Signal Analysis

Other books by Christophe Basso

Linear Circuit Transfer Functions: An Introduction to Fast Analytical Techniques (2016)
Switch-Mode Power Supplies, Second Edition: SPICE Simulations and Practical Designs (2014)
Designing Control Loops for Linear and Switching Power Supplies: A Tutorial Guide (2012)
Switch-Mode Power Supplies: Spice Simulations and Practical Designs (2008)
Switch-Mode Power Supply SPICE Cookbook (1996)

ISBN 978-1-949267-51-8

Cover Design by Guy D. Corp, www.GrafixCorp.com

Faraday Press
1000 West Apache Trail—Suite 126
Apache Junction, AZ 85120 USA

About the Author

CHRISTOPHE BASSO WAS a Technical Fellow at ON Semiconductor in Toulouse, France, where he led an application team dedicated to developing new offline PWM controller specifications. He originated numerous integrated circuits—among which the NCP120X series set new standards for low standby power converters.

Further to his 2012 book, *Designing Control Loops for Linear and Switching Power Supplies: a Tutorial Guide*, published by Artech House, he released a new title in 2016 with Wiley's IEEE Press imprint called *Linear Circuits Transfer Functions: An Introduction to Fast Analytical Techniques*. He holds 25 patents on power conversion and regularly teaches in IEEE-sponsored conferences like APEC in the U.S. He also publishes articles in trade magazines including How2Power and PET.

Christophe has over 25 years of power supply industry experience. Prior to joining ON Semiconductor in 1999, Christophe was an application engineer at Motorola Semiconductor in Toulouse. Before 1997, he worked at the European Synchrotron Radiation Facility in Grenoble, France for 10 years.

He holds a *diplome universitaire de technologie* (DUT) from the Montpellier University (France) and a MSEE from the Institut National Polytechnique of Toulouse (France). He is an IEEE Senior member.

When he is not writing, Christophe enjoys mountain biking in the Pyrenees.

Acknowledgements

A BOOK LIKE this one could not have been written and published without the help of many contributing friends. My warmest thanks and love first go to my sweet wife Anne who endured my ups and downs when determining the book's numerous transfer functions and ensuring compliance with simulations—often done late at night.

I was fortunate to share my work with my ON Semiconductor colleagues and friends Stéphanie Cannenterre and Joël Turchi in Toulouse who kindly reviewed portions of this work.

Three people did also accompany me from the beginning of this long writing process. Mon loyal ami Alain Laprade from ON Semiconductor in East Greenwich (US) who reviewed all chapters and exercised some of the early models. Monsieur Dennis Feucht from Innovatia (Belize) did also tremendous work in correcting my pages but also kindly polished my English in many occasions. Mister Tomas Gubek in Czech Republic carefully reviewed numerous equations and it was a lot of work – děkuji!

I want to warmly thank the following reviewers for their kind help in reading and correcting some of my pages: Gregory Mirsky (US), Dave Stewart (US) and Richard Redl (Switzerland).

Last but not least, I would like to thank Ken Coffman at Faraday Press for giving me the opportunity to publish my work.

Preface

I STARTED STUDYING the design of control loops with articles written by the legendary Lloyd Dixon. At that time, he was still contributing seminar papers and application notes for Unitrode. I remember successfully using his advices and formulas disclosed in *Closing the Feedback Loop*, a 31-page guide he published in the SEM-500 booklet. In this document, he taught how to arrange poles and zeroes to shape the compensation strategy decided after examining the control-to-output transfer function of the selected converter.

The formulas he used were reproduced as is, without explaining how they were derived. If I understood at that time that he could not detail all the calculations considering the adopted publication format, I remember feeling frustrated for not being able to figure out myself the derivation steps. Later on, in 1996, I signed up for a four-day course organized by Gordon (Ed) Bloom in Southampton (UK), another power electronics giant specialized in magnetics design and co-author with Rudy Severns of *Modern DC-to-DC Switchmode Power Converter Circuits*. Ed taught about integrated magnetics, Colonel McLyman detailed his approach on transformer design and Daniel Mitchell had matrixes covering all his transparent foils: it was my first confrontation with state-space averaging (SSA) and I remember having difficulty to keep up the pace.

In the months following the course, I tried to apply the newly-acquired knowledge, but I gave up when faced with the heavy mathematical manipulations inherent to SSA. At that time, I was in contact with Dr. Raymond Ridley and we were discussing by email some of the early SPICE models documented by Dr. Vincent Bello in several research papers. There was no Internet 25 years ago and I had to order all the publications through the local library of my company. To get me practically started with switching converters design, Ray Ridley graciously sent me an early version of his burgeoning design suite, POWER 4-5-6. I was amazed by the design capabilities around transient and small-signal responses but also by the automated compensation chain exactly meeting crossover frequency and phase margin goals. It was really the turning point for me: I had to understand how these equations were obtained and how to apply them to practical cases.

All the simulation models and equations implemented in this design suite were built around the *PWM switch model* described by Dr. Vatché Vorpérian in 1986. Ridley used the voltage-mode version of that model that he extended with a specific treatment around a 2nd-order polynomial to model and explain in his thesis the origins of sub-harmonic oscillations in current-mode-controlled converters. I then remember building ac SPICE models around his approach for the basic switching structures with automated parameters calculations. I later developed my own voltage- and current-mode models around the PWM switch which could work in ac, dc and transient while automatically selecting the right operating mode, DCM or CCM. These models have been refined over the years and ported to various simulation platforms with the help and contribution of many readers world-wide.

The PWM switch lends itself well to analyzing switching converters. In my opinion, it is the best approach for the basic switching cells that we know. One reason is the proximity with transistor-based circuit analysis that most engineers are familiar with: regardless of the configuration – common emitter, voltage follower or common base – replace the transistors symbol in the schematic diagram by its small-signal model and there you go, the entire circuit becomes linear, ready to be analyzed with Thévenin, superposition and Laplace. Same approach with the PWM switch which *averages* current and voltage waveforms of the commutating cell, describing switching events by time-continuous expressions. What is the difference with SSA? All governing equations remain the same regardless of the topology you study: the model is *invariant*. You no longer had to go through a canonical model before starting the analysis. And what is even better compared to SSA, the model can be simulated in its large- or small-signal version with SPICE, helping you to understand circuit operations and debug the node-mesh analyses.

This approach of SPICE-based analysis has been extensively applied in this book together with the fast analytical circuits techniques or FACTs. Introduced by Dr. Middlebrook with his extra-element theorem or EET and later formalized with a set of proven tools by Vorpérian, the FACTs are an invaluable tool for analyzing linearized converters. The ability to independently determine each polynomial coefficient of the numerator and the denominator with intermediate checks has helped me solve complicated configurations. Finally, a simulation program like SIMPLIS® was involved at the end of the calculations to ensure the frequency response of the switching circuit was in agreement with that obtained with the small-signal model.

This book includes five chapters.

The first one could be a small book in its own: it introduces the reader to the world of small-signal analysis and switching converters. I strived to write it in a smooth way, hopefully avoiding take-off turbulences that would discourage the reader. The PWM switch cell appears in this chapter and remains present throughout the whole book. The second chapter describes the most common converter, the buck structure. You can find it in several configurations, isolated or non-isolated, operated in different modes and various control types. Less common versions such as borderline operation or tapped versions were not forgotten. The boost converter occupies the third chapter and is analyzed in various configurations including power factor correction and multiphase. The fourth chapter tackles the buck-boost converter which is popular in its isolated version, the flyback converter. Quasi-resonance is part of the study with a particular emphasis on power factor correction applications. Finally, the fifth chapter deals with higher-order converters such as Ćuk, SEPIC or Zeta. The LLC is there as it was requested by many readers.

As you will quickly realize while reading, this book represents a tremendous amount of personal work and has been a challenge in many aspects: the number of drawings, calculations, equations editing, models to test and so on. I also wanted to detail all calculation steps so that you could follow the flow and understand how to get to the final result. I strived to cover the most popular switching converters and you should now be equipped with all the needed formulas to properly stabilize your power supply.

The Mathcad® files are available from my web page as well as many SIMPLIS® circuits which can be simulated by the free demonstration version Elements. For that purpose, a set of 60+ ready-to-simulate templates are freely downloadable from my webpage and most of the isolated or non-isolated converters are automatically compensated with specific macro instructions. The circuits cover many of the examples found in this book and a short operating manual is available from my website. As usual, feel free to send me your comments or any typos you may find at cbasso@wanadoo.fr. I will maintain an errata list in my personal web page as I did with the previous books.

Thank you and happy reading to you all.

Christophe Basso
June 22, 2020
https://cbasso.pagesperso-orange.fr/Spice.htm

1 Introduction to Small-Signal Modeling

SMALL-SIGNAL ANALYSIS IS a technique that aims to approximate nonlinear behaviors of passive or active components in a circuit by linear equations. You obtain this result by restricting signal perturbations to small enough levels leading to an approximately linear behavior of the network under study. Once a linear equivalent circuit is revealed, the Laplace transform or any of the classical theorems such as superposition, Thévenin or Norton can be invoked to determine a particular transfer function.

Switching circuits are highly nonlinear systems in essence and you must *linearize* them if you want to apply any of the aforementioned tools. Numerous techniques exist to study the frequency response of switching converters. However, we will not cover them here as articles and books abound on the subject. In my opinion, none of the proposed schemes rival in simplicity the PWM switch model introduced in the early 90's. The PWM switch model solely considers the switching elements and models them with a linearized 3-terminal cell similarly to what is done with the hybrid-pi model for transistors: identify the cell, replace it with its *invariant* small-signal model and start working on the newly-linearized circuit. Before delving into our topic, let's review some of the important aspects of linear systems.

1.1 Linear Systems

A system is said to be linear if it satisfies the *superposition principle*. The definition combines two important properties, *additivity* and *proportionality*. Additivity is mathematically described by:

$$f(u_1 + u_2) = f(u_1) + f(u_2) \tag{1.1}$$

Assume a box fed by an input signal u_1 delivering an output $y_1 = f(u_1)$. Now, change the input amplitude to u_2 and measure the new output variable, $y_2 = f(u_2)$. Then, bias the input with the sum of u_1 and u_2 to form $y_3 = f(u_1 + u_2)$. If the circuit inside the box is linear, then you should read $y_3 = y_1 + y_2$.

The second property is proportionality. It is described by the following expression:

$$k \cdot f(u_1) = f(k \cdot u_1) \tag{1.2}$$

Take the same box and bias its input with u_1 again. The output reads $y_1 = f(u_1)$. Now grow u_1's amplitude by a factor k and read the output $y_2 = f(k \cdot u_1)$. If proportionality is respected, then $k \cdot y_1 = y_2$.

In Figure 1.1, you see a straight line shifted by a positive offset. Assuming the x and y axes are labeled in volts, input u is first multiplied by 0.5 and a 2-V offset is added to form output y.

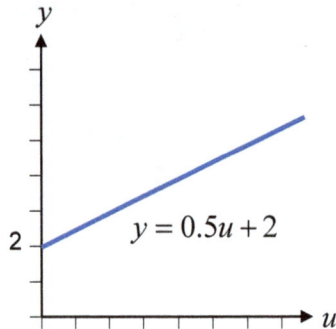

Figure 1.1 A function where the input stimulus u is divided by 2 and a fixed 2-V offset is added.

Is this a linear system per our definition? If you check the Figure 1.2 in which we have analyzed responses to various stimuli, the system described by (1.3) is nonlinear per our definitions. It is an *affine function*.

$$y = 0.5u + 2 \tag{1.3}$$

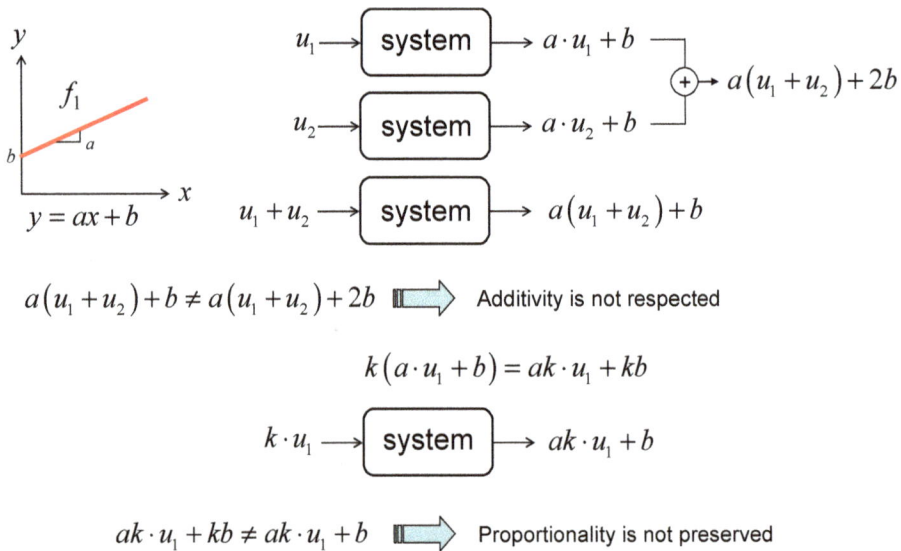

Figure 1.2 An affine function does not satisfy the superposition principle.

1.1.1 Linearization by Perturbation

If we want to analyze a given nonlinear circuit by applying our theorems and the Laplace transform, we must find a way to *linearize* it. Then, we will be able to determine the transfer function of our choice. A transfer function is a mathematical relationship linking a *response* (the output signal) to an *excitation* or a *stimulus* (the input signal). One widely adopted technique to linearize a given circuit is called *perturbation*. Figure 1.3 illustrates the principle. The variables that are part of the identified nonlinear equations are all perturbed by the addition of a small ac modulation around their operating bias point. The *bias* or *dc operating point* is the static state of a powered circuit in absence of stimulus. The voltage and current values you see displayed in the left-side of Figure 1.3 are the circuit *bias points* as you would measure them in the laboratory. Now, assume you inject a modulating stimulus of sufficiently-low amplitude to avoid distorting the response while the system is perturbed. This is the so-called *small-signal* excitation because the circuit is assumed to remain linear under this small-amplitude excursion. All variables now undergo a modulation around their static operating point. This is what the right-side of Figure 1.3 illustrates.

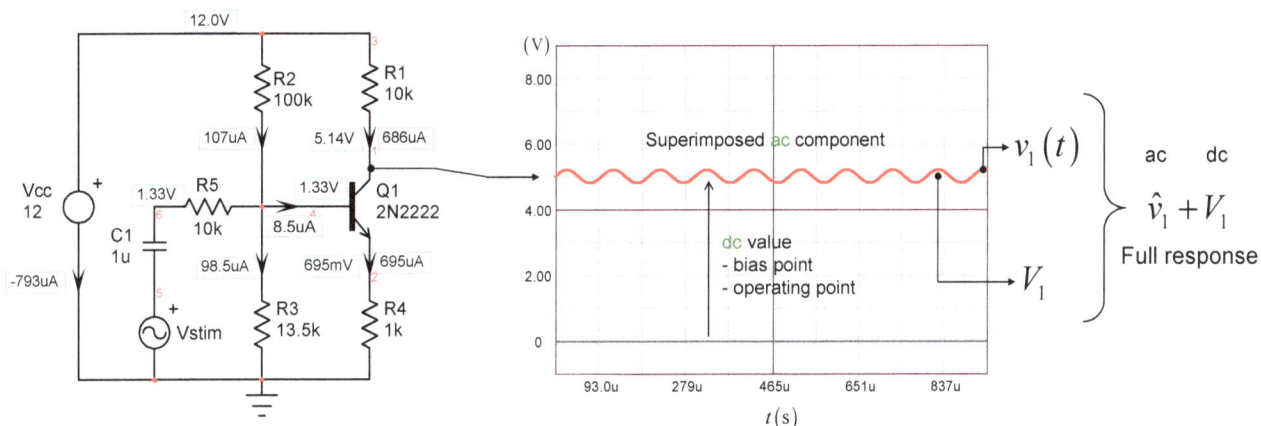

Figure 1.3 You perturbing a system by injecting of a sinusoidal variation around a bias point. The bias point is displayed as framed values in the electrical schematic.

In the adopted formalism, the varying component is designated by a small hat or caret (^) placed above the considered instantaneous variable written in lowercase like $\hat{v}_{out}$ or $\hat{i}_4$. The static component is usually upper case like V_1 or I_4. Sometimes, you can add a subscripted 0 to better identify the variable like I_{L0} for a steady-state inductor current for instance. If we come back to our expression in (1.3), we have two variables: the output y and the input u so let's perturb them by writing:

$$y = y_0 + \hat{y}$$
$$u = u_0 + \hat{u}$$

(1.4)

We wrote that the output y is now made of a dc term (y_0) to which the ac perturbation $\hat{y}$ is added. Similarly, the stimulus is now made of a dc bias u_0 merged with an *incremental* or *small-signal* perturbation, $\hat{u}$. Now, substitute these definitions into (1.3) and obtain:

$$\hat{y} + y_0 = 0.5(u_0 + \hat{u}) + 2 = 0.5u_0 + 0.5\hat{u} + 2$$

(1.5)

From this expression, collect the ac and dc terms: any dc variable associated with an ac variable becomes an ac term. A product of ac variables, if any, is neglected because a) it is a nonlinear term (multiplying two varying signals results in a nonlinear term not applicable to a linear analysis) and, most importantly, b) the product of two small terms will itself be nearly vanishingly small and may be safely ignored. From (1.5), you can sort the terms as follows:

AC: $\hat{y} = 0.5\hat{u}$ represents the small-signal or *incremental* transfer function linking the response y to the stimulus u.
DC: $y_0 = 0.5u_0 + 2$, this is the static operating or bias point value of the circuit when the ac excitation is removed.

Let's briefly consider another example.

Assume the voltage V_1 you observe is made of two waveforms V_2 and V_3 multiplied together:

$$V_1 = V_2 V_3 \qquad (1.6)$$

We have three variables, V_1, V_2 and V_3. To linearize this equation, perturb the variables with a small-signal component:

$$V_1 + \hat{v}_1 = \left(V_2 + \hat{v}_2\right)\left(V_3 + \hat{v}_3\right) \qquad (1.7)$$

Develop and collect the terms. Remember, the ac products are eliminated for the reasons mentioned above.

$$V_1 + \hat{v}_1 = V_2 V_3 + V_2 \hat{v}_3 + V_3 \hat{v}_2 + \hat{v}_2 \hat{v}_3 \qquad (1.8)$$

AC: $\hat{v}_1 = V_3 \hat{v}_2 + V_2 \hat{v}_3$ represents the small-signal or incremental transfer function linking the response $\hat{v}_1$ to the two stimuli $\hat{v}_2$ and $\hat{v}_3$. The dc values V_3 and V_2 are now coefficients affecting the circuit's *sensitivity* to $\hat{v}_2$ and $\hat{v}_3$.

DC: $V_1 = V_2 V_3$, this is the static operating or bias point value of the circuit when the ac excitation is removed.

The sum of the above definitions form the *large-signal* or *total-variable* response. We are interested by the dynamic response of our system, i.e. the way the output responds to a particular frequency-dependent stimulus, the sinusoidal input. The ratio of the response to the stimulus is the transfer function we want. In the separated ac and dc components, only the ac part depends on frequency excitation.

The dc component is a fixed offset, independent of frequency. Besides calculating operating points, it does not directly interest us for frequency analysis. What is important for us is the small-signal or *incremental* equation involving ac terms only. Since the dc term is of direct interest for our transfer function study, let's identify it as the offset $b = 2$ V in Figure 1.1 while the small-signal expression is simply $\hat{y} = 0.5\hat{u}$. We can redraw the curve by separating the two contributors as shown in Figure 1.4. As detailed in the drawing, the expression $\hat{y} = 0.5\hat{u}$ now satisfies the superposition principle and is linear.

We could now consider a real system and its instantaneous waveforms:

$$\hat{y}(t) = 0.5\hat{u}(t) \qquad (1.9)$$

Apply the Laplace transform to the above expression and obtain the transfer function linking the response y to the stimulus u:

$$L\{\hat{y}(t)\} = Y(s) = 0.5 \cdot U(s) \qquad (1.10)$$

leading to:

$$\frac{Y(s)}{U(s)} = 0.5 \qquad (1.11)$$

In this simple example, the gain, or more specifically the attenuation, is 0.5 and represents the slope a of the right-side curve in Figure 1.4.

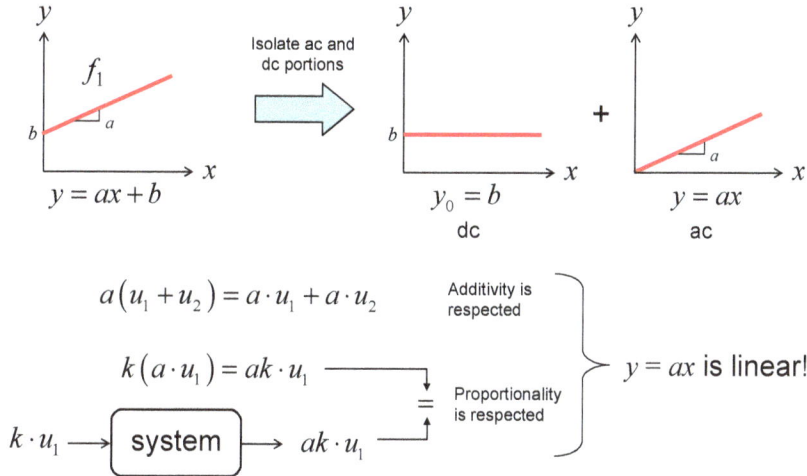

Figure 1.4 Linearization transforms the affine function into a linear function to which a static offset is added.

1.1.2 Gain and Slope

The slope of a function is determined at a given operating point and is defined as the ratio of the vertical displacement noted Δy to the horizontal displacement expressed as Δx. Figure 1.5 illustrates this simple definition. The slope can be positive as the picture shows or negative if Δy or Δx are negative.

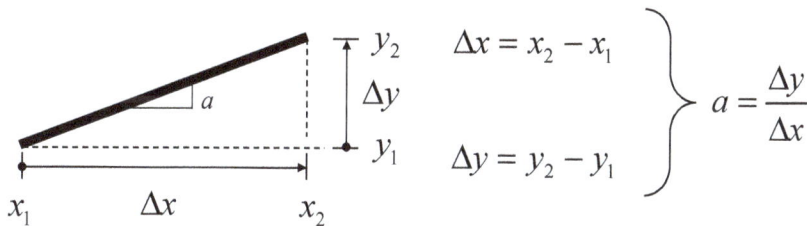

Figure 1.5 The slope of a curve a is defined as the ratio of the vertical displacement Δy to the horizontal distance Δx.

Knowing the slope of a function is an important piece of information if we want to predict the *state* of a system a short time after a known state, such as when power is first applied at $t = 0$. The initial value of the *state variables*, such as inductor current and capacitor voltage at $t = 0$ s – the initial conditions if you prefer –determine how the state of the system will evolve under excitation for $t > 0$. As shown in Figure 1.6, what is important for this purpose is the slope of the identified state variable.

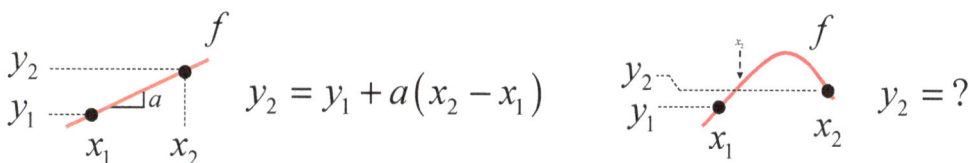

Figure 1.6 When the slope of the observed waveform remains constant along the observation interval, you can predict its future state.

In the left side of the figure, slope a is constant along the considered interval. If we know the value y_1 at x_1 then we are able to predict what y_2 will be after the time interval $x_2 - x_1$. On the contrary, in the right side, the curve is highly nonlinear and even if y_1 is known, you won't be able to predict y_2 value in a simple way. In this drawing, you see that if x_2 would come closer to x_1, you can identify a portion in which the slope can be considered constant.

This is the idea behind the representation in Figure 1.7. The curve is nonlinear along the horizontal axis. However, if you reduce the observation window by shortening the x-axis displacement, you can isolate a linear segment as shown on the right side of the picture. As the distance between x_1 and x_2 is decreased, you start observing a straight line. Electrically speaking, you would position the circuit bias point between y_1 and y_2 then ac-perturb with a small amplitude waveform around that point. The nonlinear function f_1 observed between x_1 and x_2 becomes a new affine function f_2.

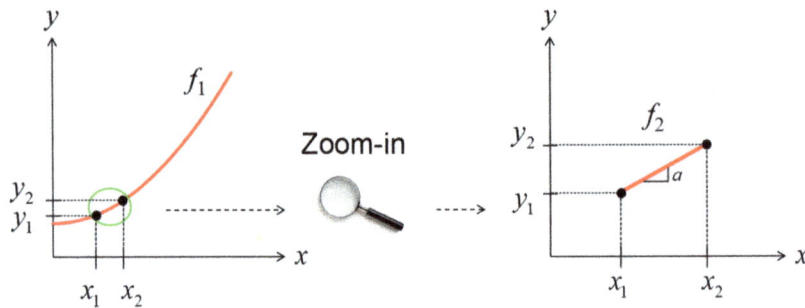

Figure 1.7 Rather than considering the curve entirely, only concentrate on a small part to isolate a linear segment.

To actually identify a linear zone, the change in x formerly designated by Δx must be a very small quantity. When x_2 approaches x_1, we will have calculated the slope at point x_1. In fact, this is the mathematical definition of the *derivative* of the function f at x_1 noted f prime:

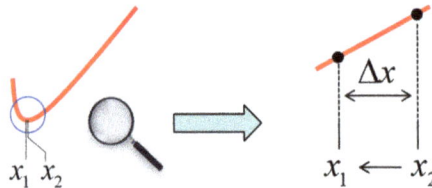

Figure 1.8 You must reduce the distance between x_2 and x_1 to isolate a linear portion.

$$f'(x_1) = \lim_{\Delta x \to 0} \frac{f(x_1 + \Delta x) - f(x_1)}{(x_1 + \Delta x) - x_1} \tag{1.12}$$

Considering an infinitesimal variation in x noted dx as proposed by Leibniz, we can reformulate (1.12) as

$$f'(x) = \frac{d}{dx} f(x) \tag{1.13}$$

A nonlinear characteristic can thus be approximated by a succession of linear segments whose slope is evaluated at selected points. The more linear segments you add, the better the approximation becomes. The resulting approximated waveform becomes a *Piece-Wise Linear* or PWL function. Figure 1.9 describes the concept for a given curve. The slopes of the obtained tangents are the slope of the curve differentiated at the considered point.

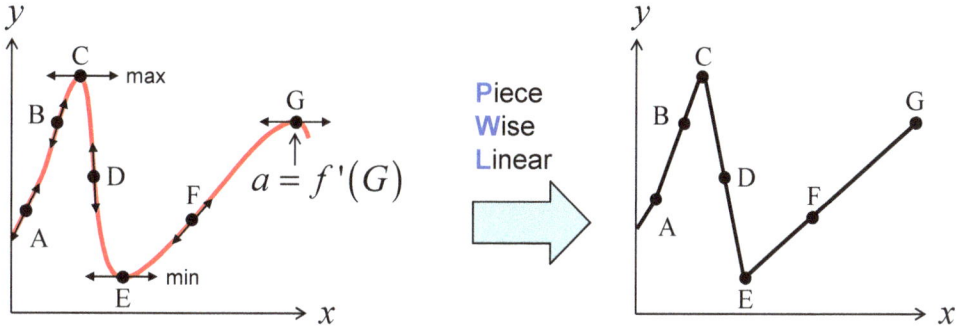

Figure 1.9 A PWL waveform is a possible way to approximate a nonlinear function.

Should you want to study a system described by the right-side curve in Figure 1.9, you could apply any of the cited theorems because the system remains linear along the x-axis as long as the slope is revaluated as operating points change.

The derivative of the function $y = f(x)$ determines the sensitivity of y to a given variation of x: "if x moves by a certain quantity, what is the change in y?" Sometimes, we can have a function depending on more than a single variable as in (1.6). In this case, we may be interested to determine the sensitivity of the output to each of the individual variables. This is called *partial differentiation* and we will come back to this important notion shortly.

To calculate slopes at various points on a given curve, differentiation must be possible. In other words, the operation described by (1.12) must return a finite number. A simple approximate definition for a linear system could be "a small input change brings a small output change". In the left side of Figure 1.10, you see a waveform smoothly toggling between two states. Differentiation is possible at any point and you could approximate this curve as shown in Figure 1.9.

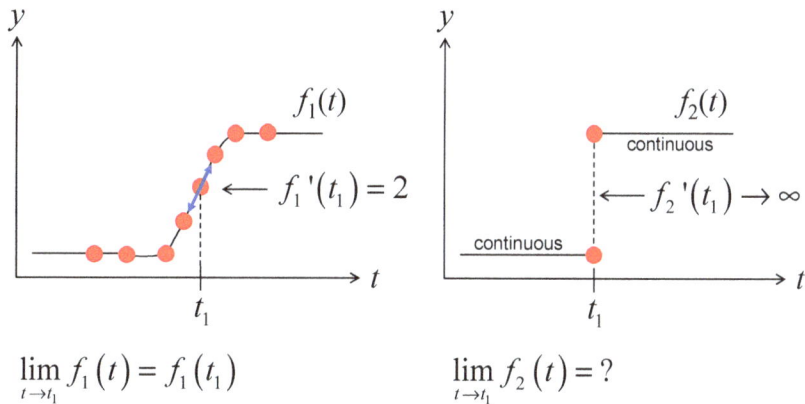

Figure 1.10 Function f_2 is discontinuous at point t_1: this is a singularity in the function.

On the opposite, the right side of Figure 1.10 shows a waveform instantaneously jumping from one state to the other with a zero time-span.

Even if the network is linear before and after the toggling point, the switching event at t_1 introduces a *singularity* which prevents differentiation at this point: the function is no longer continuous in time but becomes *time-discontinuous*: it is impossible to approximate it by a succession of segments as we did before.

We will need to identify a way to *smooth* the transition between the two considered states so that we can describe the new waveform by a time-continuous expression.

1.1.3 The Importance of the Bias Point

Figure 1.11 shows why it is important to maintain linearity if you perform frequency analysis on a given circuit.

Imagine you want to characterize the frequency response of an amplifier powering a load. You inject a sinusoidal waveform of amplitude u_1 into the amplifier input port and observe the response y_1 to this stimulus with an oscilloscope. Assume you want a Bode plot characterizing the input-to-output frequency response of this amplifier let's say from 10 Hz to 100 kHz.

The ratio of the two variables y_1 to u_1 captured at a given frequency gives the gain *magnitude G* while the delay between u_1 and y_1 characterizes the *phase* or *argument* of the transfer function. Now, sweep the frequency while keeping the amplitude of the excitation constant. As frequency increases, it is likely that the gain G starts reducing to approach 1 (0 dB) at the *crossover* or *transition* point.

However, imagine that at a particular frequency sample, the amplifier clips and the response is distorted. If you compute the ratio of y_1 to u_1 at this moment, you certainly obtain a new magnitude value. However, this number is meaningless because the system has lost its linearity at the measurement point.

In other words, the gain magnitude no longer depends on frequency but is affected by the clipping phenomenon due, in this instance, to the differential input stage temporarily saturating. The data points collection representing the frequency response now contains bad data and the Bode plot is corrupt. Fortunately, as you were observing the response with an oscilloscope, you realized that distortion occurred and you reduced the input amplitude to bring linearity back and proceed with the acquisition.

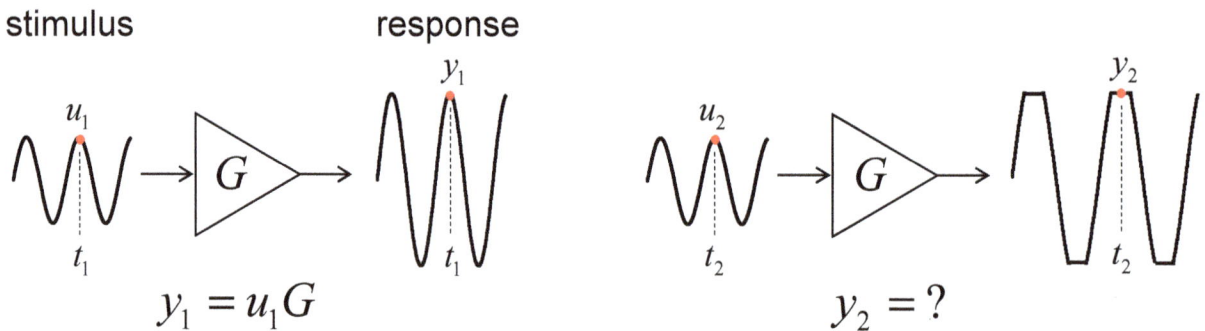

Figure 1.11 If distortion occurs, linearity is lost and induces wrong results.

In Figure 1.11, what could explain the clipping of the response? Consider the operational amplifier (op-amp) wired in a non-inverting configuration in Figure 1.12. Assume the op-amp is supplied from a 5-V dc source. What output bias point will you select to analyze the frequency response of that circuit?

The important things are that a) the stimulus is of a sufficient amplitude so that the response can be comfortably observed or measured (electrically distinguishable from noise for instance) b) the operating point is selected so that no distortion occurs during analysis.

As shown in the picture, if the op-amp hits the high or low rail, distortion occurs and corrupts the measurement.

Selecting a bias point in the middle of the total span is a favorable option and leads to a correct setup. You will thus need to adjust the source offset in V_{stim} to bring the output voltage to around 2 V and then start the modulation.

In an open-loop configuration like here, the bias point can drift with temperature and you will have to monitor the output signal during the analysis so that it does not approach the upper or lower stops as shown in Figure 1.12.

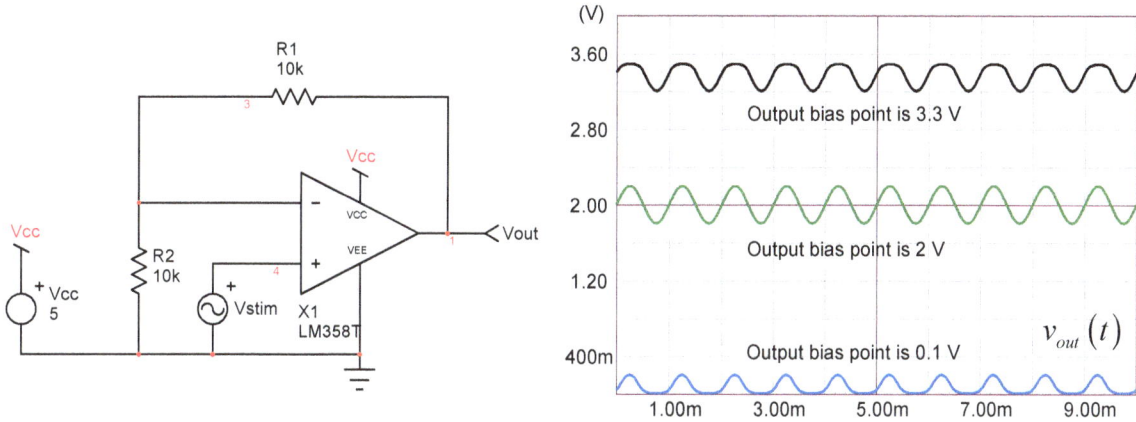

Figure 1.12 Bias point selection is important to ensure the system remains linear as you observe it. Source V_{stim} includes dc (bias) and ac components.

Something happens inside the op-amp as you select different biases. The op-amp, an LM358 in our example, is made of bipolar elements such as transistors and diodes. Junctions of these components endure different bias conditions as the operating environment changes (V_{cc} is adjusted to a new value, temperature changes and so on).

The typical I-V response of a silicon diode appears in Figure 1.13.

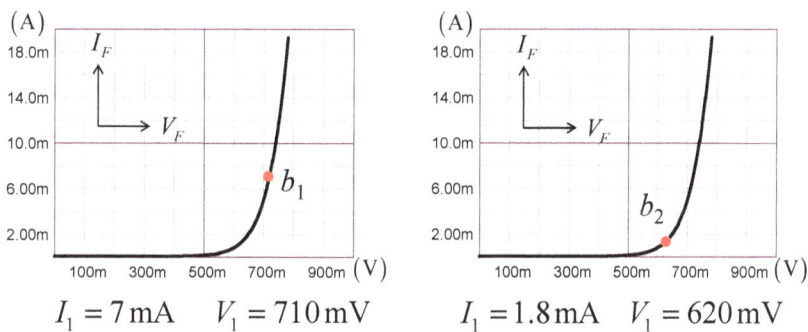

$$I_1 = 7\,\text{mA} \quad V_1 = 710\,\text{mV} \qquad I_1 = 1.8\,\text{mA} \quad V_1 = 620\,\text{mV}$$

Figure 1.13 As operating conditions change, point b moves along the I-V characteristic.

When we adjust the bias point in Figure 1.12, for instance by changing the dc offset in the stimulus, assume the operating point of one of the internal diodes slides along the I-V characteristic and moves from point b_1 to b_2. If you modulate the current of this diode, what voltage will you obtain across its terminals?

On the left side of Figure 1.13, the slope is steep and despite small variations of b_1 along the vertical axis, we can say with good approximation that the slope remains constant. If the diode current is now a sinusoid, b_1 will move around its operating point and the voltage drop across the diode terminals is also a sinusoid. For point b_2 in Figure 1.13, it is placed in a rather curved portion and the slope won't remain constant if b_2 moves around its position: a current modulation of sufficiently-large amplitude is likely to produce a distorted waveform.

If the amplitude increases further, b_2 can potentially touch the x-axis and clipping occurs. Figure 1.14 offers a graphical representation of the zoomed-in diode current and voltage when operating at these two different bias points. In the left side, the voltage response is undistorted while the right side of the picture shows a degraded response. At b_2, the option exists to reduce the modulation amplitude and bring the dynamic displacement within a smaller region where linearity is preserved. Another option is to change the operating point and find a more suitable region.

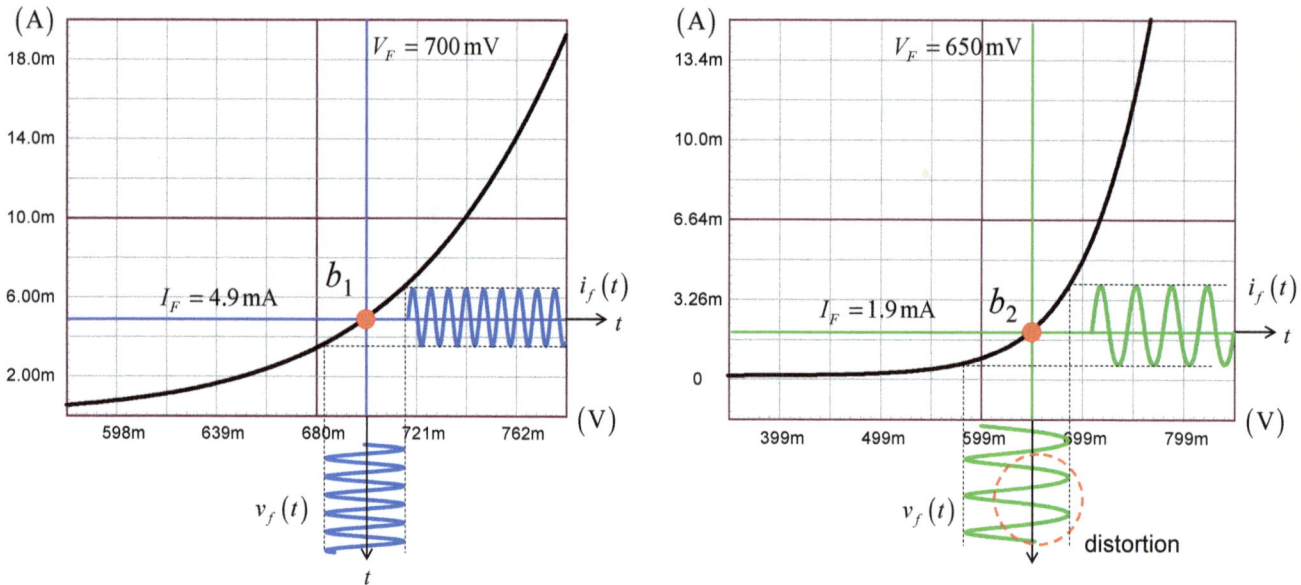

Figure 1.14 Depending on the selected bias or operating point, distortion or clipping effects can occur if the modulation is too strong.

1.1.4 A Small-Signal Model

To analyze a nonlinear circuit, we approximate its behavior by a set of linear equations derived around a given operating point. If the operating point changes, the model coefficients need updating to account for the modified operating conditions. A diode is a simple nonlinear active component to start with.

Figure 1.15 shows how the characteristic of the diode can be approximated by two straight segments to form the simplest PWL model of the component. The value of V_{T0} depends on the technology and is approximately 0.6 V for a silicon diode and 0.4 V for a Schottky. The slope at which the forward voltage drop grows as more current circulates is a function of the dynamic resistance r_d at the considered operating point.

In the PWL model, the dynamic resistance is idealized and constant regardless of the operating point.

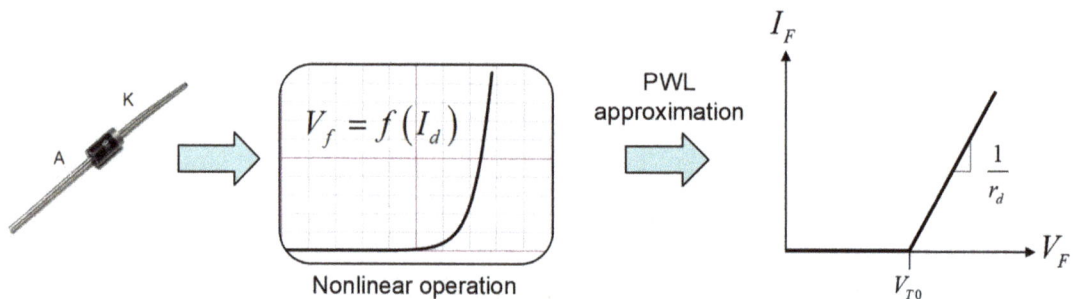

Figure 1.15 The diode dc characteristic can be approximated by a simple 2-segment PWL function.

The parameter V_{T0} is not frequency-dependent and can be seen as a fixed offset. Therefore, what matters for the low-frequency small-signal model of the diode is the resistive term r_d.

This is what Figure 1.16 illustrates by showing the possible simplified electrical representation of a diode.

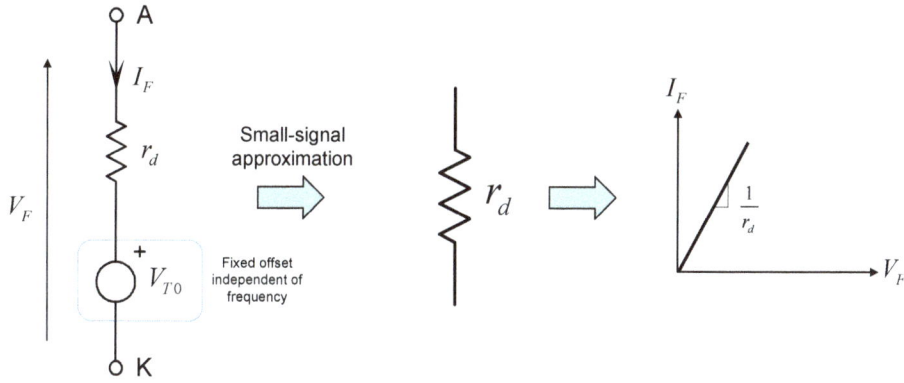

Figure 1.16 The small-signal model of a diode is a simple resistance evaluated at an operating point.

The forward current I_F in a diode obeys the well-known Shockley equation:

$$I_F = I_S \left(e^{\frac{V_F}{nV_T}} - 1 \right) \tag{1.14}$$

Where:

I_F, the diode forward current
V_F, the diode forward drop
V_T, the thermal voltage, ≈ 26 mV at a junction temperature of 25 °C
n, the emission coefficient which varies between 1 and 2
I_S, the reverse bias saturation current. It is a few nA for a 1N4148

The first thing we can do is further simplify (1.14) by considering the exponential term much larger than 1. Thus:

$$I_F = I_S \left(e^{\frac{V_F}{nV_T}} - 1 \right) \approx I_S e^{\frac{V_F}{nV_T}} \tag{1.15}$$

In this equation, I_S, n and V_T are constant and depend on the diode technology. I_F and V_F are the two variables we can perturb. It leads to:

$$I_F + \hat{i}_F = I_S e^{\frac{V_F + \hat{v}_F}{nV_T}} \tag{1.16}$$

Applying a simple manipulation around the exponential term, we have:

$$I_F + \hat{i}_F = I_S e^{\frac{V_F}{nV_T}} e^{\frac{\hat{v}_F}{nV_T}} \tag{1.17}$$

You recognize I_F in the first term expression (1.15) which leads to:

$$I_F + \hat{i}_F = I_F e^{\frac{\hat{v}_F}{nV_T}} \tag{1.18}$$

Considering a small-signal modulation, the exponent is of small magnitude. Therefore, the exponential term can be reduced to its first-order Padé approximant:

$$e^x \approx 1 + x \qquad (1.19)$$

and we obtain:

$$I_F + \hat{i}_F \approx I_F\left(1 + \frac{\hat{v}_F}{nV_T}\right) = I_F + I_F \frac{\hat{v}_F}{nV_T} \qquad (1.20)$$

This expression is made of continuous and alternating terms. Isolate the alternating component:

$$\hat{i}_F = I_F \frac{\hat{v}_F}{nV_T} \qquad (1.21)$$

The ratio of the small-signal voltage v_F by the modulating current i_F gives the diode dynamic resistance r_d evaluated at a given bias point, the forward current I_F:

$$r_d = \frac{\hat{v}_F}{\hat{i}_F} = \frac{nV_T}{I_F} \, [\Omega] \qquad (1.22)$$

This dynamic resistance changes as the operating point moves following the representation in Figure 1.17.

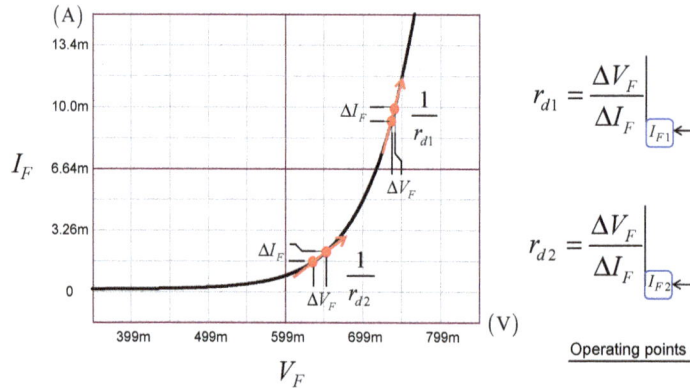

Figure 1.17 The diode dynamic resistance r_d decreases as more current flows in the component.

In a circuit with a diode, we can now approximate its behavior by a simple resistance computed at the considered operating condition. In Figure 1.18, you see a 2.3-V source forward-biasing a diode through a 1-kΩ resistance. The diode forward drop is 641 mV leading to a current I_F of 1.66 mA. From (1.22), considering an emission coefficient of 2, we find a dynamic resistance:

$$r_d = \frac{nV_T}{I_F} = \frac{2 \times 0.026}{0.00166} \approx 31 \, \Omega \qquad (1.23)$$

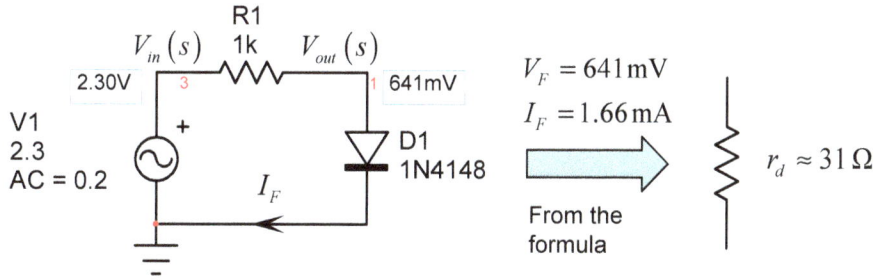

Figure 1.18 The diode dynamic resistance is evaluated at the operating point imposed by the circuit.

We can thus simply approximate the diode by a 31-Ω resistance and plug it in the circuit as a replacement for the diode symbol (Figure 1.19). It is now easy to calculate the output amplitude when a 200-mV alternating stimulus is applied as V_{in}. According to the resistance values, the response is a waveform of amplitude is:

$$V_{out} = V_{in} \frac{r_d}{r_d + R_1} = 200m \times \frac{31}{31 + 1k} = 6 \text{ mV} \tag{1.24}$$

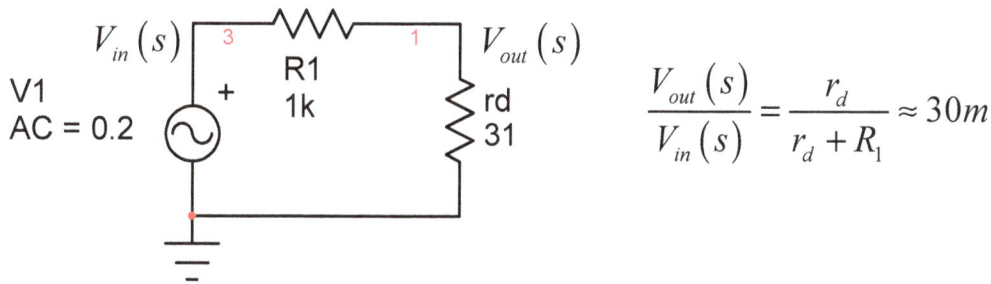

Figure 1.19 The circuit simplifies to a resistive divider.

Should you now modify the input bias and force more or less current in the diode, you will have to consider a new operating point and update the dynamic resistance value accordingly.

This is what Figure 1.21 shows when Figure 1.18 circuit is tested in Figure 1.20. The diode bias current is changed from 427 μA to 897 μA leading to a smaller r_d and, consequently, a reduced division ratio between v_{ac} and v_f.

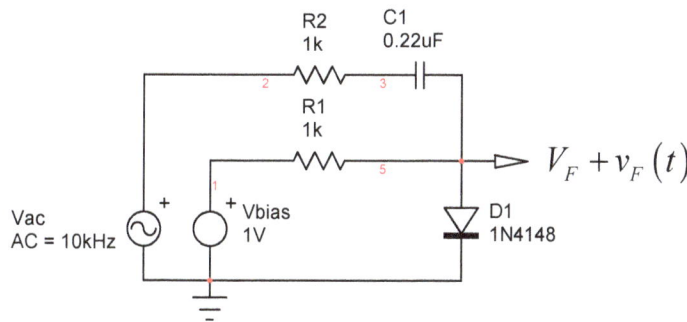

Figure 1.20 It is easy to adjust the bias value and modify the operating point of diode D_1.

13

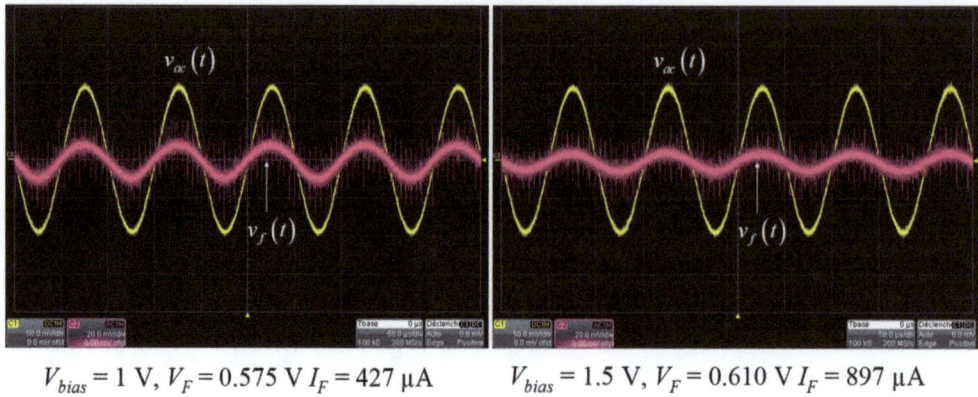

$V_{bias} = 1$ V, $V_F = 0.575$ V $I_F = 427$ µA $V_{bias} = 1.5$ V, $V_F = 0.610$ V $I_F = 897$ µA

Figure 1.21 If you change the bias condition, the diode dynamic resistance evolves and leads to new division ratio.

A simulator like SPICE exclusively solves linear equations. When facing a nonlinear behavior, SPICE reduces its internal time step – its sampling step if you will – until linearization around a bias point is possible. Sometimes, the nonlinearity is so strong that the time-step reduction algorithm hits the minimum time step and simulation fails. This is the frustrating nonconverging "time step too small" problem we have all experienced. SPICE can overcome this problem by temporarily offering a linear path across the nonlinear element via the *gmin* stepping routine until a solution is found. You realize how long it can be to simulate a switching circuit as SPICE continuously adjusts its time step and determine a linear solution along fast-moving waveforms. On the other hand, a simulator like SIMPLIS® uses perfect components (0-Ω resistance during the on-state, infinite resistance when open, zero time-span switching times) whose I-V characteristics are modeled with PWL segments. As such, the simulator always operates on a linear portion and saves the linearization time with all its potential convergence issues. As a result, simulation time is flashingly fast. One very attractive feature as we will see later on is that SIMPLIS® can extract the small-signal response of a switching circuit without resorting to an averaged model as SPICE requires. Another PWL simulator, PSIM® also offers the possibility to extract a dynamic response from a switching circuit.

1.1.5 Partial Differentiation

Perturbing equations is an easy task with a simple expression where few variables are involved. It can become a quite tedious exercise to perturb and sort out alternating and continuous terms when the equation is complicated with many contributing variables. Take the following expression for instance which defines a current as a function of two variables:

$$I_a = \frac{V_{ac} D_1^2}{2 F_{sw} L} \tag{1.25}$$

V_{ac} is the voltage between nodes a and c while D_1 represents the duty ratio. If you decide to apply the perturbation method, you must add a small modulation to I_a and the two identified variables. This is what you write:

$$I_a + \hat{i}_a = \frac{\left(V_{ac} + \hat{v}_{ac}\right)\left(D_1 + \hat{d}_1\right)^2}{2 F_{sw} L} \tag{1.26}$$

Expand this equation, collect continuous and alternating terms, get rid of products and obtain the two expressions you want. Nothing truly complicated here but you can easily make a mistake while expanding this equation and it is not easy to automate the process with a mathematical solver. As result, perturbing and collecting terms can quickly become a long and difficult exercise especially with large expressions containing many variables.

A faster and more efficient way is to determine the *sensitivity* or the gain of the output variable I_a to each of the individual variables: if V_{ac} moves while D_1 is held constant, how does it affect I_a? Then, if D_1 changes while V_{ac} remains unchanged, how does this modification of the control variable propagate to I_a? Rather than perturbing all variables at once (and sort out all contributions afterwards), you perturb one variable after the other and compute respective responses. At the end, you simply sum all responses to write the final small-signal equation. This implies that the system under study always remains linear while you perturb it. We could describe this operation in two steps:

$$k_1 = \left. \frac{dI_a(D_1)}{dD_1} \right|_{V_{ac} = constant} \tag{1.27}$$

and

$$k_2 = \left. \frac{dI_a(V_{ac})}{dV_{ac}} \right|_{D_1 = constant} \tag{1.28}$$

However, strictly speaking, the output current I_a is a function of two variables, V_{ac} and D_1. The correct notation for I_a should thus be

$$I_a = f(V_{ac}, D_1) \tag{1.29}$$

in which V_{ac} and D_1 are two *independent* variables while I_a is the dependent variable. The term independent means that there is no correlation between the considered inputs. In other words, there is no link between V_{ac} and D_1 in (1.29). We then need to decide if we differentiate with respect to V_{ac} or with respect to D_1. For that purpose, a specific notation is used, the symbol ∂ which designates the *partial derivative* of I_a:

$\frac{\partial I_a}{\partial V_{ac}}$ is the partial derivative of I_a with respect to V_{ac} while D_1 is held constant—the sensitivity of I_a to V_{ac}.

$\frac{\partial I_a}{\partial D_1}$ is the partial derivative of I_a with respect to D_1 while V_{ac} is held constant—the sensitivity of I_a to D_1.

If we apply these definitions to our converter, we could write that the total small current variation dI_a attributed to the small variations of D_1 and V_{ac} is defined by

$$dI_a = \frac{\partial I_a}{\partial D_1} dD_1 + \frac{\partial I_a}{\partial V_{ac}} dV_{ac} \tag{1.30}$$

Recognizing that dI_a is actually the result of a small-amplitude modulation of V_{ac} and D_1, the above expression can be similarly written using our small-signal notation

$$\hat{i}_a = \frac{\partial I_a}{\partial D_1} \hat{d}_1 + \frac{\partial I_a}{\partial V_{ac}} \hat{v}_{ac} \tag{1.31}$$

Please note that this is a dynamic equation. When we use partial differentiation, we lose the continuous expression.

We can now apply this concept to (1.25) directly and automate the calculation process with a mathematical solver like Mathcad®. Figure 1.22 shows a typical answer delivered by the program.

$$\frac{d}{dV_{ac}} \frac{V_{ac} \cdot D_1^{\,2}}{2 \cdot F_{sw} \cdot L} \rightarrow \frac{D_1^{\,2}}{2 \cdot F_{sw} \cdot L}$$

$$\frac{d}{dD_1} \frac{V_{ac} \cdot D_1^{\,2}}{2 \cdot F_{sw} \cdot L} \rightarrow \frac{D_1 \cdot V_{ac}}{F_{sw} \cdot L}$$

Figure 1.22 Mathcad® immediately computes the sensitivity of complex expressions to a specific variable.

The complete linearized version of I_a is now defined by the following expression:

$$\hat{i}_a = k_1 \hat{d}_1 + k_2 \hat{v}_{ac} \tag{1.32}$$

with:

$$k_1 = \frac{V_{ac} D_1}{F_{sw} L} \tag{1.33}$$

and:

$$k_2 = \frac{D_1^{\,2}}{2 F_{sw} L} \tag{1.34}$$

Equation (1.32) can also be expressed using the Laplace formalism and either one will be used indistinctively in this book:

$$I_a(s) = k_1 D_1(s) + k_2 V_{ac}(s) \tag{1.35}$$

In the above definitions, all parameters are either component or parameter values – L or F_{sw} for instance – or, if we are talking about a specific dc voltage (V_{ac}) or static operating point (D_1), they can be obtained from a measurement, a simulation or even derived from a dc model as we will shall later see.

1.2 Switching Converters—Basic Cells

This section is a brief introduction to switching power supplies and describes the basic blocks we will encounter in our analyses. For more details about operation and control of these converters, the reader will find in [1] and [2] a comprehensive coverage of the subject.

A basic switching circuit associates two energy-storing elements L and C with two semiconductors: a power switch *SW* and either a diode D or a second controlled switch that acts as a diode (we talk about *synchronous rectification* in this case). The converter provides a continuous voltage V_{out} to a load R by processing energy from an input source V_{in} at a switching frequency F_{sw}. The way the switch and the diode are connected together form a 3-terminal cell in which ports *a, p* and *c* respectively stand for *active, passive* and *common.*

Figure 1.23 shows how rotating the cell lets you cover the three basic non-isolated switching converters:

- The *buck* converter reduces the input voltage: $V_{out} \leq V_{in}$. The isolated version of the buck is the *forward* converter and can be extended to half- or full-bridge structures and push-pull (to name a few).

- The *buck-boost* converter reduces or increases the input voltage but reverses the output polarity with respect to ground: $|V_{out}| \leq V_{in}$ or $|V_{out}| \geq V_{in}$. The isolated version of the buck-boost is the *flyback* converter.

- The *boost* converter increases the input voltage: $V_{out} > V_{in}$.

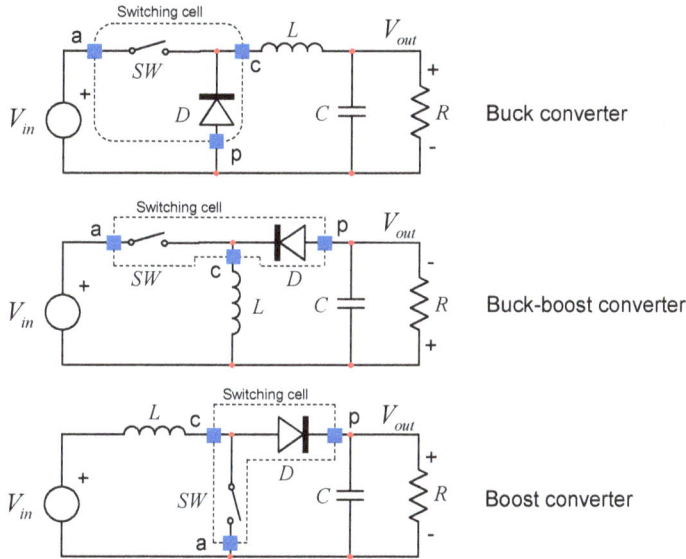

Figure 1.23 The same active and passive components can be rearranged in different ways to form the three basic switching converters. The switching cell remains the same despite the variations.

In all of these examples, we can show that the basic switching cell is actually made of an equivalent single-pole double-throw (SPDT) controlled-switch which dynamically steers the inductor current at the switching period pace. Figure 1.24 applies this principle to the three cells and illustrates the basis of the *PWM switch model* introduced by Dr. Vorpérian in 1986 [3]. You can apply this scheme to numerous switching cells from nonisolated to isolated topologies. The beauty lies in the fact that the cell is said to be *invariant*: its intrinsic properties (input and output variables relationship, internal architecture) remain the same regardless of its implementation in a converter.

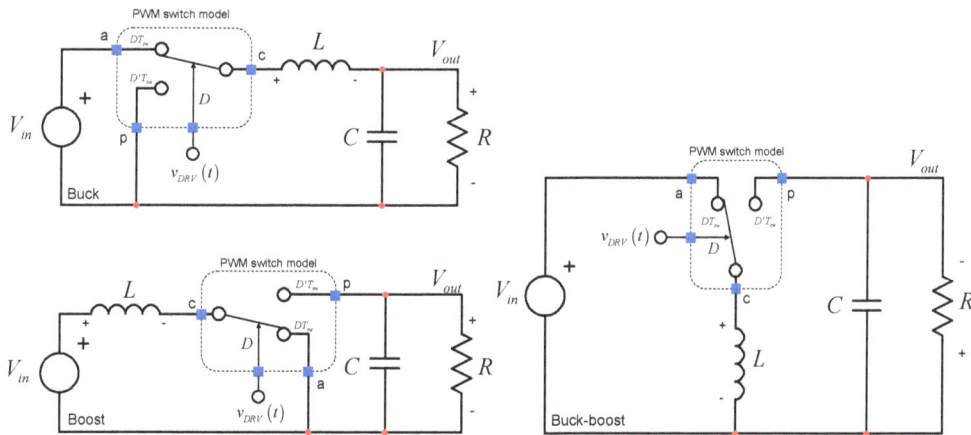

Figure 1.24 The switching cell is actually a single-pole double-throw (SPDT) power switch routing the inductor current to the output *RC* network at the switching frequency.

The modeling principle is similar to the hybrid-π model for a bipolar or MOSFET transistor: the model internals remain the same whether you wire the bipolar transistor in a common-base configuration or in a current mirror for instance. You just need to identify the terminals in the circuit and have them match those of the model.

1.2.1 Input and Output Ripple

To operate our converters, the SPDT switch will be actuated by a periodic square wave alternatively ensuring a temporary connection between terminals c and a then terminals c and p. The SPDT switch will occupy the first position c-a during a percentage D of the switching period while the second position c-p will be maintained during another percentage noted D' of the switching period. D is called the *duty ratio* and represents the *control variable* of all these converters. This square signal driving the SW switch appears in Figure 1.25.

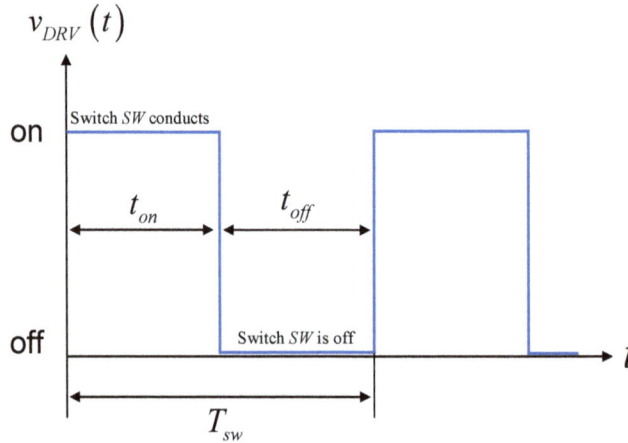

Figure 1.25 The duty ratio is the time during which the switch is closed – t_{on} – divided by the switching period T_{sw}.

With D, you control how long the switch SW remains on – this is t_{on} – and how long it stays off – this is t_{off} – before actuating it for the next cycle. V_{out} directly depends on D as we will later see. Based on these remarks, we can write the following expressions:

$$D = \frac{t_{on}}{T_{sw}} \rightarrow t_{on} = DT_{sw} \qquad (1.36)$$

and:

$$D' = \frac{t_{off}}{T_{sw}} \rightarrow t_{off} = D'T_{sw} \qquad (1.37)$$

Realizing that $t_{on} + t_{off} = T_{sw}$ and summing the above definitions, we obtain a relationship linking D and D':

$$
\begin{aligned}
t_{on} + t_{off} &= T_{sw} \\
T_{sw} &= DT_{sw} + D'T_{sw} \\
D' &= 1 - D
\end{aligned}
\qquad (1.38)
$$

Assuming a 10-μs switching period and a 30% duty ratio, the power switch SW in Figure 1.23 conducts current for 3 μs and remains off during 7 μs.

In the three configurations of Figure 1.24, the energy-storing element L undergoes different voltage biases as the SPDT switch toggles during t_{on} and t_{off}. If you consider the input source, it could be a battery or a rectified and filtered alternating source. None of these devices exhibit a zero output resistance. When they deliver energy to the switching converter, the voltage at terminal a in the buck converter for instance will not maintain V_{in} but is going to drop during the on-time. However, because a large-value front-end capacitor featuring a low equivalent series resistance (ESR, labeled r_C) is likely to decouple this node to ground, we will assume for the sake of simplifying the analysis that V_{in} remains constant during the switching events. A similar remark applies to the output voltage V_{out}. Indeed, the goal of a dc-dc converter is to deliver a continuous output voltage, precisely regulated at a target level. From the circuits we have drawn, you can see that the inductor current will flow in the output network made of R and C. As this current features continuous and alternating components, the capacitor is sized so that very little of the varying component flows in the load R and most of it is absorbed by capacitor C. Thus, a power supply is designed so that the output ripple remains of very small amplitude with respect to the regulated level. 1% of ripple or less is a typical value you can think of, e.g. 120 mV for a 12-V output. If we consider this ripple component to be of much smaller amplitude than the continuous voltage it affects or if:

$$\left| v_{ripple}(t) \right| << V_{out} \tag{1.39}$$

then, the instantaneous output voltage $v_{out}(t)$ can be approximated as a continuous value:

$$v_{out}(t) \approx V_{out} \tag{1.40}$$

This so-called *small-ripple approximation* helps in the analysis of switching converters by considering output and input voltages constant over a switching cycle. Figure 1.26 shows this principle in action.

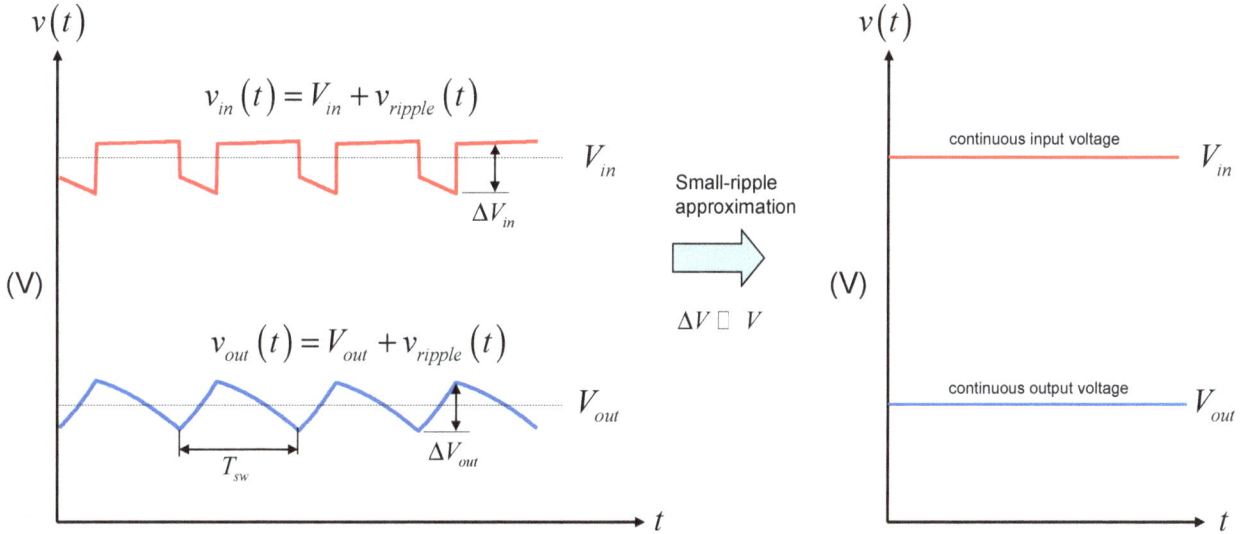

Figure 1.26 For the sake of simplifying the analysis, the small-ripple approximation considers the input and output voltages as continuous variables, without modulation over it.

1.2.2 Conduction Modes

As the SPDT switch continuously toggles between its two states, a fixed bias voltage is alternatively applied across the inductor terminals. For instance, in the buck converter, as the inductor right-terminal is connected to V_{out}, its left one at node c will move between V_{in} and ground, neglecting the diode forward drop V_f.

Respecting Figure 1.26 inductor polarity and considering the small-ripple approximation, we can write:

$$V_L\big|_{t_{on}} = V_{in} - V_{out} \qquad (1.41)$$

and:

$$V_L\big|_{t_{off}} = -V_{out} \qquad (1.42)$$

Applying a continuous bias V_L across a perfect inductor implies the linear growth of the current i_L as described by:

$$i_L(t) = \frac{V_L}{L} \cdot t \qquad (1.43)$$

The current increases if the applied voltage is positive as in (1.41), but decreases if the polarity reverses as shown in (1.42). The term V_L/L in (1.43) determines the inductor current slope S. It characterizes the inductor current rate of change per time and is expressed in ampere per second or A/s.

If the slope were always positive, the current would permanently rise in the inductor which would eventually saturate and impact operation.

For 'that reason, the inductor slope must change within a switching cycle to ensure the current remains in control and adjusts to what the load absorbs at the rated output voltage. Combining the above expressions lets us define the operating slopes S for the buck converters. When terminals a and c are connected during t_{on}, we have

$$S_{on} = \frac{V_{in} - V_{out}}{L} \qquad (1.44)$$

When terminals c and p are linked together during t_{off}, the slope changes to

$$S_{off} = -\frac{V_{out}}{L} \qquad (1.45)$$

The slope change occurs smoothly as the inductor current circulates in the same direction: when it can no longer enter port a, it instantaneously goes through port p and keeps flowing as long as the bias is maintained. This switch position describes the spontaneous diode conduction when the switch SW opens in Figure 1.23.

Figure 1.27 represents the two states we have described in a buck converter.

The right side of the figure depicts the bias changes and the corresponding inductor instantaneous voltage and current.

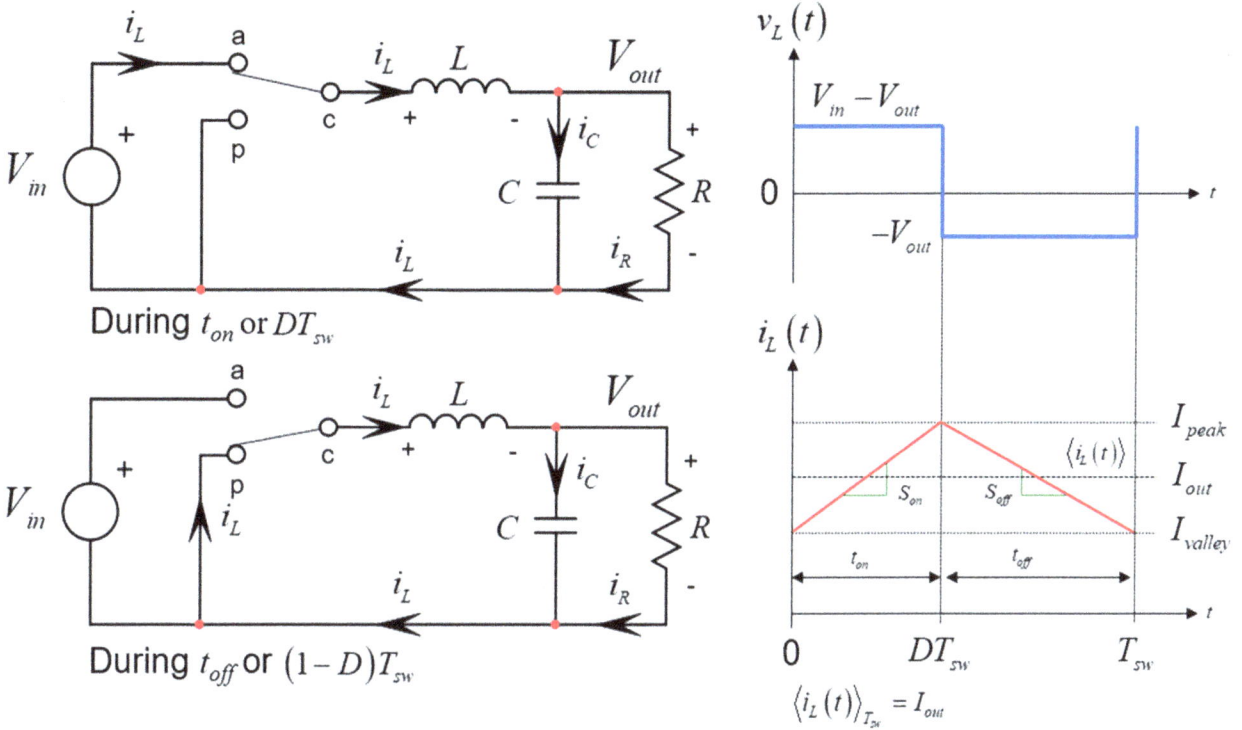

Figure 1.27 The SPDT switch imposes different bias voltages at the switching period pace (here, a buck converter operated in CCM).

As expected, the inductor current goes up and down when the power switch toggles between its two states. At the end of the on-time when the control waveform of Figure 1.25 instructs switch SW to open, the inductor current has reached a peak value I_{peak} defined using (1.43):

$$I_{peak} = I_L(0) + S_{on}t_{on} \tag{1.46}$$

The term $I_L(0)$ describes the initial current condition in the inductor when a new cycle takes place. If (1.45) tells you that the inductor linearly demagnetizes during the off-time, then, depending on the operating conditions, the demagnetization can be partial or complete within a switching cycle. If partial then there is still energy stored in L when a new cycle commences and $I_L(0)$ is the inductor current at the end of the off-time: it is I_{valley} shown in Figure 1.27. Thus, (1.46) updates to

$$I_{peak} = I_{valley} + S_{on}t_{on} \tag{1.47}$$

When the switch toggles and connects terminals c and p together, the bias across the inductor reverses and the current ramps down from its peak towards the valley. We can thus write:

$$I_{valley} = I_{peak} - S_{off}t_{off} \tag{1.48}$$

Depending on the component selection, switching frequency and operating conditions, it is very possible that the inductor completely depletes during the off-time. In this case, the new initial condition I_{valley} for the next switching cycle is 0 A and (1.46) becomes:

$$I_{peak} = S_{on}t_{on} \tag{1.49}$$

21

In the first case described by (1.47) and (1.48), the current in the inductor never returns to 0 within a switching cycle and I_{valley} is non-null value. It describes the *continuous conduction mode* or CCM. On the contrary, if the valley current is 0 A for a finite duration within a switching cycle, it indicates that the inductor has been fully depleted within that cycle and the converter has entered the *discontinuous conduction mode* or DCM. The conduction mode of the inductor current represents a fundamental characteristic of a switching converter and it affects its dynamic performance. Before we explain how the operating mode changes, we need to understand the link between the inductor current and the actual output current absorbed by the load.

If you look back at the inductor current i_L in Figure 1.27, you realize that it feeds the RC network during DT_{sw} but also during $(1-D)T_{sw}$. So, across a switching cycle, there is no *discontinuity* in the supply of the load current from the inductor. To determine what continuous current actually circulates in the load resistance R (even if it is obvious here with the buck converter) we need to introduce the concept of *averaging* by looking at Figure 1.28.

The idea behind averaging is to transform a time-discontinuous equation or waveform into a time-continuous expression that we can later differentiate or manipulate for further analysis. What does it mean? If you look at the current leaving the source V_{in} or flowing into terminal a in Figure 1.27, you see the instantaneous inductor current $i_L(t)$ as long as terminals c and a are bridged together during the on-time. When the SPDT moves to connect terminals c and p, the source current drops to zero and remains there during the entire t_{off} duration. As shown in the upper left corner of Figure 1.28 this waveform is highly discontinuous because of the sudden transition to 0 at $t = DT_{sw}$. To remove or *smooth* this discontinuity, one way is to observe the current activity no longer during t_{on} or t_{off}, but across the entire switching cycle T_{sw}: on *average*, how much current flows in terminal a during T_{sw}? Without resorting to an integral, we can graphically find the result by evaluating the area under the curve describing $i_a(t)$.

If we designate the average inductor current $\langle i_L(t) \rangle$ as the mid-point between the peak and valley currents:

$$\langle i_L(t) \rangle = \frac{I_{peak} + I_{valley}}{2} \tag{1.50}$$

we can move the small triangular area above $\langle i_L(t) \rangle$ and fill-in the empty space above the valley point. This is what is called the *flat-top approximation*, allowing us to transform a wavy current such as $i_a(t)$ into an equivalent square waveform whose flat amplitude is now $\langle i_L(t) \rangle$ during DT_{sw}. Once this is done, the area A_{on} of this approximated waveform is simply $\langle i_L(t) \rangle \cdot DT_{sw}$. If we now graphically stretch or average A_{on} over a switching period via a square box of similar area, it becomes the waveform in the upper right corner: during an entire cycle T_{sw}, on average, terminal a and the source V_{in} "sees" the average inductor current $\langle i_L(t) \rangle$ scaled by the duty ratio D. Mathematically, we can write:

$$I_a = \langle i_a(t) \rangle = \langle i_L(t) \rangle \cdot DT_{sw} \cdot \frac{1}{T_{sw}} = \langle i_L(t) \rangle \cdot D \tag{1.51}$$

The capital letter notation such as I_a designates an average value and will be often used later in the text. (1.51) represents a *time-continuous* nonlinear equation that we could, if necessary, differentiate.

If we now consider the current split between C and R during DT_{sw} and $(1-D)T_{sw}$, the lower-side drawing of Figure 1.28 appears. We can apply our flat-top approximation and sum the resulting shapes across a switching cycle keeping a similar total area. On *average*, across an entire switching cycle T_{sw}, we can say that the RC network "sees" the average inductor current $\langle i_L(t) \rangle$ flowing out of terminal c. As no direct current flows in capacitor C, we have:

$$I_c = I_{out} = \langle i_L(t) \rangle \tag{1.52}$$

Since the average inductor current is the current flowing out of terminal c, we can rewrite (1.51) by substituting $\langle i_L(t) \rangle$ by I_c. We obtain one of the invariant relationship of the PWM switch model:

$$I_a = DI_c \tag{1.53}$$

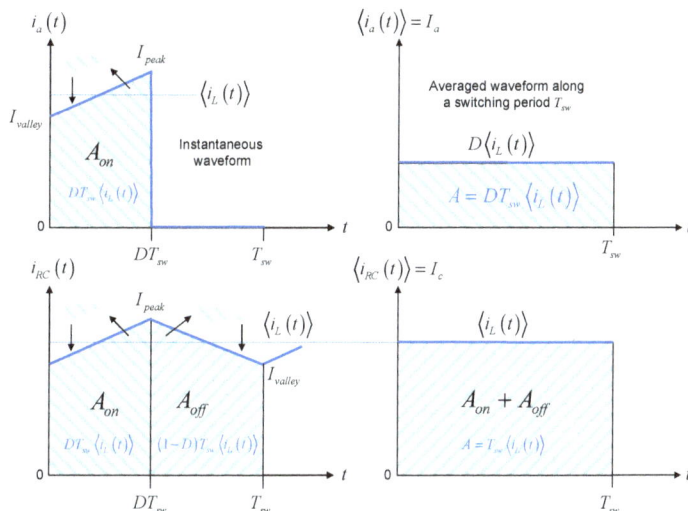

Figure 1.28 Waveform averaging is a fundamental principle of small-signal analysis as it allows the description of a time-discontinuous waveform by a time-continuous mathematical equation.

If the continuous inductor current is equal to the output current delivered by the buck converter to R, it means that this component varies with load conditions. Considering V_{in} and V_{out} constant (the converter regulates), the continuous component of the inductor current will reduce as the load gets lighter: the converter still operates in CCM but I_{peak} and I_{valley} are shifted down. At a certain moment, the valley current will reduce to zero and the converter enters DCM: at this time, the diode naturally or *spontaneously* turns off and both semiconductors are blocked. If the load current further reduces, a *dead time* noted DT appears. Figure 1.29 represents several operating cases when a buck converter delivers different output power levels P_{out}.

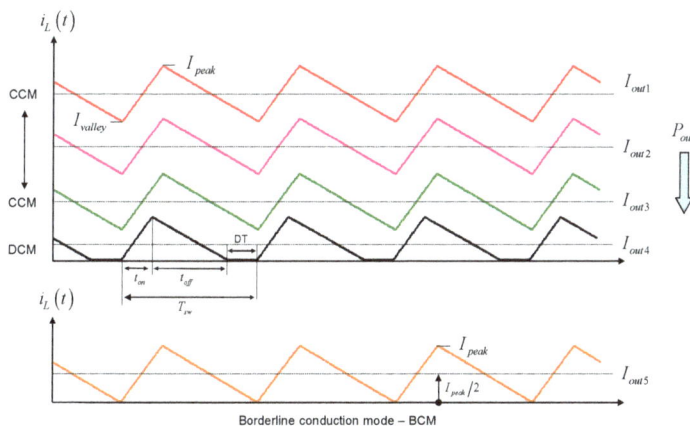

Figure 1.29 When the load current reduces, it shifts the instantaneous peak inductor currents down. The converter leaves CCM and enters DCM when the valley current reaches 0 A for a finite time. In BCM, the switch turns back on as soon as the inductor current hits 0 A.

23

From the 1st to the 3rd power level, the converter remains in CCM. As the output current drops from I_{out3} to I_{out4}, the inductor now fully demagnetizes within the switching cycle and the valley current is now 0 A : the converter has changed its mode from CCM to DCM. The inductor remains depleted until the next cycle. As the inductor current is 0, the capacitor remains alone during the dead time to power the load. As terminal c is now in a high-impedance state, the inductor voltage is 0 V during the dead time duration. Figure 1.30 shows how Figure 1.27 updates to account for this new event. A third theoretical position has been added to our original SPDT switch.

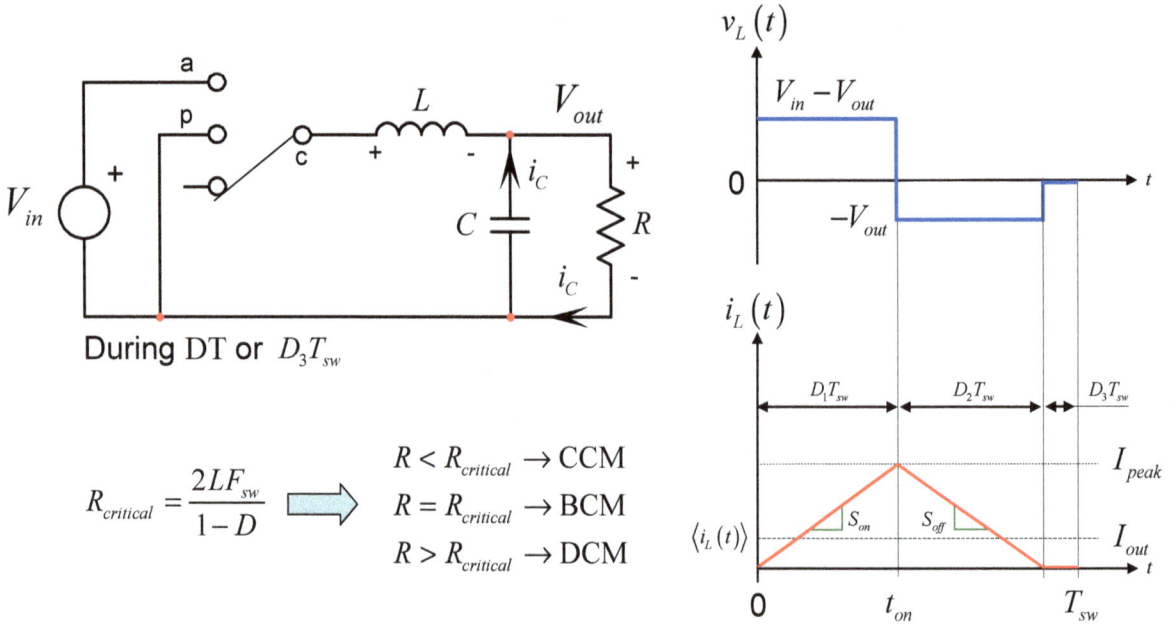

Figure 1.30 The dead time represents a third period of time during which both SW and D are blocked : capacitor C is alone to feed the load R.

If t_{on} still designates the time during which the SPDT switch bridges terminals c and a, t_{off} now includes the time during which terminals c and p are connected, plus the dead time DT. The new time periods describing these operations are conveniently renamed in D_1, D_2 and D_3:

- The duty ratio or the time during which SW conducts is $D_1 T_{sw} = t_{on}$
- The time during which diode D conducts is $D_2 T_{sw}$
- The dead time DT during which both SW and D are blocked is defined as $DT = D_3 T_{sw}$

(1.38) needs an update to become:

$$T_{sw} = D_1 T_{sw} + D_2 T_{sw} + D_3 T_{sw}$$
$$D_1 + D_2 + D_3 = 1$$
(1.54)

In the bottom of Figure 1.29, you can see a waveform in which the inductor current touches 0 A exactly at the point where a new switching period starts. The converter is thus at a boundary or at a transition point between CCM and DCM. We say the converter operates in *boundary conduction mode* (BCM) or in *borderline conduction mode* (also BCM) but also found in the literature as *critical conduction mode* (CrM). The term critical designates the particular value for R or L or even T_{sw} which induces a mode change: for a certain value of R, the converter operates in CCM and then enters BCM for the critical value $R_{critical}$. Increasing R above $R_{critical}$ implies DCM operation. Mathematically, we can determine the converter operating state by comparing the inductor average value with its peak value, $I_{L,peak}$. We can show that:

- $\left\langle i_L\left(t\right)\right\rangle > \dfrac{I_{L,peak}}{2} \rightarrow \text{CCM}$

- $\left\langle i_L\left(t\right)\right\rangle < \dfrac{I_{L,peak}}{2} \rightarrow \text{DCM}$

- $\left\langle i_L\left(t\right)\right\rangle = \dfrac{I_{L,peak}}{2} \rightarrow \text{BCM (no dead time)}$

Knowing the transition point is important when studying the dynamic response of a switching converter because you need to know if you deal with a converter operating in CCM, DCM or BCM. You can already intuitively sense the fact that a CCM-operated converter will need a larger inductance than the same converter purposely designed to operate in DCM while working in similar conditions. For CCM, you need a certain amount of inductance so that a bit of energy is always maintained in L cycle by cycle. However, we know that an inductor opposes sudden current changes. As the continuous inductor current is I_{out}, a buck purposely designed to maintain CCM across *line* (another word for input) and load conditions will be inherently slower than a DCM counterpart when a sudden current change is imposed by the load. Even if the control loop drastically extends the on-time, the inductor current won't move faster than what (1.43) allows and more switching cycles will be needed to match I_{out}. For the same available bias during t_{on} – we say for the same available volts-seconds – the current will grow faster cycle-by-cycle with a small inductor than with a large one. Obviously, the size of the inductance also affects other parameters such as switching ripple but it won't be discussed here. The key takeaway here is to realize that knowing the operating mode is fundamental before studying the small-signal response of a given converter. The critical load resistance value for the buck converter is derived in [1] and conveniently reproduced in Figure 1.30.

As a final note, a converter can transition from CCM to DCM (or the other way around) during normal operations, e.g. when V_{in} or I_{out} change. It is also very likely that mode crossing occurs during a fresh start-up sequence during which V_{out} is a small value for several cycles or during an overload or short circuit condition. In a well-designed converter, all these particular modes must be studied at an early design stage to make sure these natural (or faulty) transitions always happen smoothly and safely.

1.2.3 Indirect Energy Transfer

If you look at the upper sketch in Figure 1.27, it describes the buck converter configuration during the on-time. You can see how the source V_{in} delivers energy to magnetize the inductor and, at the same time, supplies the RC network. During the off-time, the input source is decoupled from the circuit and the inductor alone feeds the network. As such, the input current is said to be *pulsating* as it alternates between $i_L(t)$ during t_{on} and 0 for the rest of the time. As we have explained, the inductor current is permanently routed to the RC network without interruption when the converter operates in CCM. We say this output current is *non-pulsating* and if you increase the inductance value, its alternating component or ripple can be made small which is beneficial to all ohmic paths in the converter and the output capacitor lifetime in particular. However, increasing the inductor value poses other problems like response time and operating losses.

The arrangement typical to the buck structure shows that energy in CCM is permanently delivered to the load without an intermediate state. This is what is called a *direct energy transfer* configuration: as soon as the switch *SW* in Figure 1.23 closes, energy is immediately absorbed from the source and delivered to the load while the inductor magnetizes. Then, during the off-time, inductive energy keeps flowing without discontinuity in the output network. If the control loop decides to increase the duty ratio for a sudden power demand, as the on-time duration expands cycle by cycle, an increased energy flow – of course limited in amplitude by the available inductor volts-seconds – is provided to the load: considering perfect elements in the conversion chain, there is no delay in the execution of the control loop request. Dynamically speaking, the maximum theoretical crossover frequency for a buck converter can be as high to ½ F_{sw} as imposed by the Nyquist rate. Obviously, physical limits imposed by switching speed of semiconductors, delays in comparators and logic chains, noise susceptibility and so on will set a practical limit on the crossover choice. But beside the upper limit set by the Nyquist rate, there is no internal mechanism inherent to the buck which clamps down on the maximum crossover frequency. For example, low-frequency buck converters or their isolated forward counterparts used in telecom applications can have crossover frequencies up to several tens of kHz while switching at 200 kHz or more. In portable applications where tiny buck converters switch at several MHz, it is not unusual to see crossover frequencies at 100 kHz or more.

Contrary to the buck, the boost and buck-boost converters operate differently. The boost operating states appear in Figure 1.31. When the SPDT switch bridges nodes c and a, the inductor is connected across the input source V_{in}, decoupling the output from the input source. Following our previous notation and neglecting the switch voltage drop, the current builds up from the valley current (in CCM) to the peak with a slope equal to:

$$S_{on} = \frac{V_{in}}{L} \qquad (1.55)$$

During this time, the RC network does not receive any energy at all and the load current is supplied by the capacitor C. It is only when the switch toggles and connects ports c and p that the inductor becomes in series with the input source and transfers energy to the RC network. The current can only decay if the polarity across the inductor reverses from V_{in} to a negative value. As the inductor left side is connected to the source, this phenomenon only happens if V_{out} is greater than V_{in}. The new slope becomes:

$$S_{off} = -\frac{\left(V_{out} - V_{in}\right)}{L} \qquad (1.56)$$

and the inductor current linearly decays.

If you look back at the inductor current i_L in Figure 1.31, you see that it flows through the source during the on-time and the off-time: the input current of the boost converter is *nonpulsating*. However, it only feeds the RC network only during $(1-D)T_{sw}$. During DT_{sw}, the RC network does not see the inductor current and the output current of the boost converter across a switching cycle is pulsating. So here, we clearly have a discontinuity in the supply of the load current by the inductor. As we did with the buck converter and with the help of Figure 1.32, we can determine the relationship linking the average inductor current to the output current I_{out} in the case of the boost configuration. The left side of the picture shows the current split between C and R during t_{off} only as it is 0 during t_{on}. The area occupied by A_{off} during t_{off} is $\langle i_L(t) \rangle (1-D)T_{sw}$. If you now stretch this shape to occupy an entire switching cycle while keeping the area constant, you obtain a flat amplitude linking the inductor average current to the output current:

$$I_{out} = (1-D)T_{sw}\langle i_L(t)\rangle \cdot \frac{1}{T_{sw}} = (1-D)\langle i_L(t)\rangle \qquad (1.57)$$

The relationship given in (1.53) still holds here because terminal c sees the inductor current during t_{on} and t_{off}. (1.52) then updates to:

$$I_c = I_{in} = \langle i_L(t)\rangle \qquad (1.58)$$

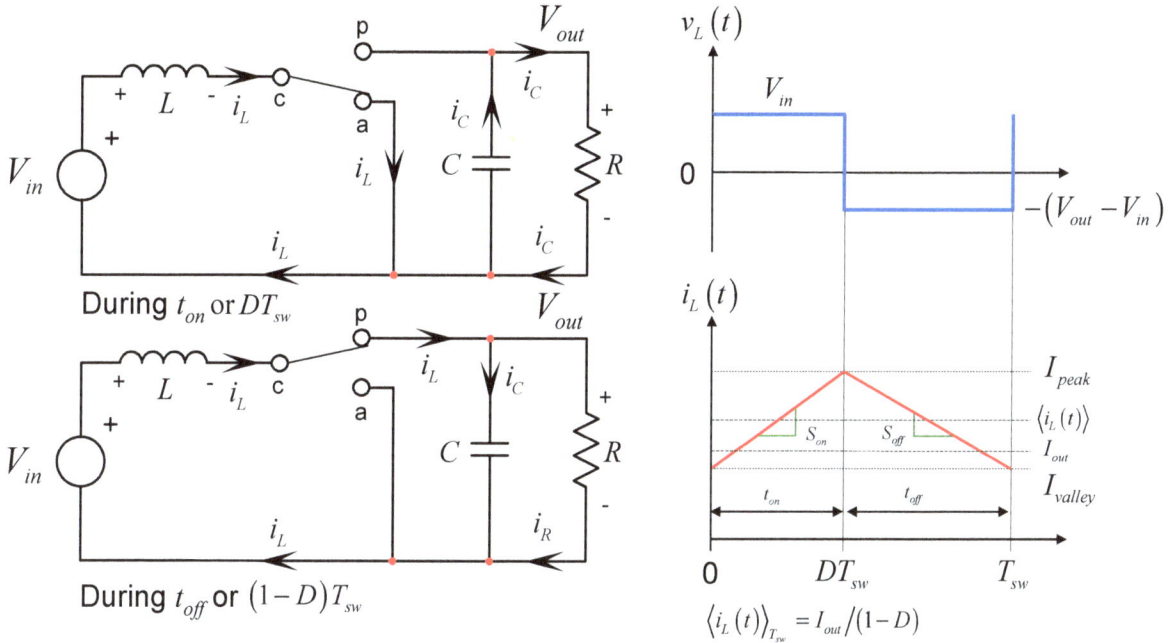

Figure 1.31 The boost converter needs to build up energy in the inductor during the on-time before it releases it to the load during the off-time.

while the current leaving terminal a is the inductor current during the on-time only. On average, we still have:

$$I_a = D \cdot \langle i_L(t) \rangle \tag{1.59}$$

which is a similar expression to that in (1.51). The link expressed in (1.53) is still valid.

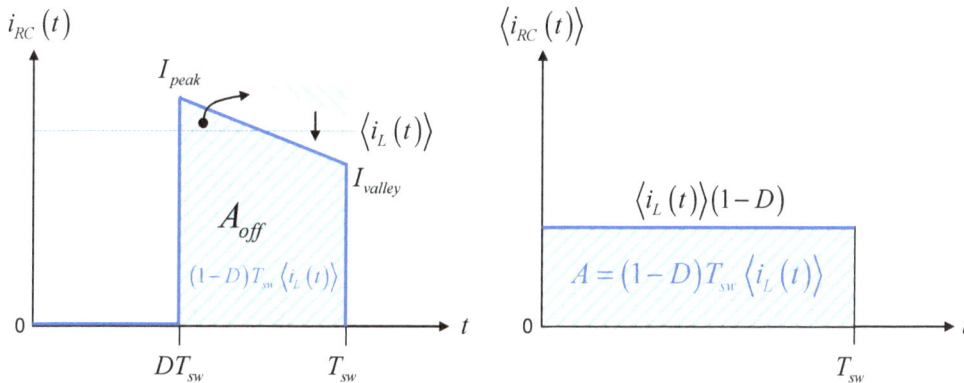

Figure 1.32 The RC network "sees" the inductor current during $(1-D)T_{sw}$ in the boost converter.

Exactly as in the buck converter, when the load current reduces, the average inductor current also drops and, at some point, the DCM mode is entered. Figure 1.31 updates to Figure 1.33 and shows that terminal c becomes a high-impedance point while the inductor voltage becomes 0 V. The figure includes the expression of the critical resistance value which sets the mode change.

One fundamental difference here is the way energy is processed and transferred from the source to the load.

Unlike the buck architecture, the boost converter operating scheme includes an intermediate state during which the RC network is decoupled from the inductor. This is during the on-time when the inductor is exclusively energized from the source to build up energy.

You can now realize that the conversion process is made of two states:

- t_{on}: energy is sourced to the inductor L
- t_{off}: energy stored in L is transferred in series with the input source to the RC network

If there is a sudden power demand, requiring more average current into L to satisfy (1.57), the control loop will certainly expand the on-time to supply more energy to L but given the decoupling of the RC network during that time, nothing will happen in the output. It is solely when the inductor feeds the RC network during t_{off} that the increase of energy supply propagates to the load. This intermediate storage phase induces an inherent delay in the conversion process and limits the maximum response speed theoretically obtainable from a boost converter. The boost architecture is an *indirect energy transfer* converter.

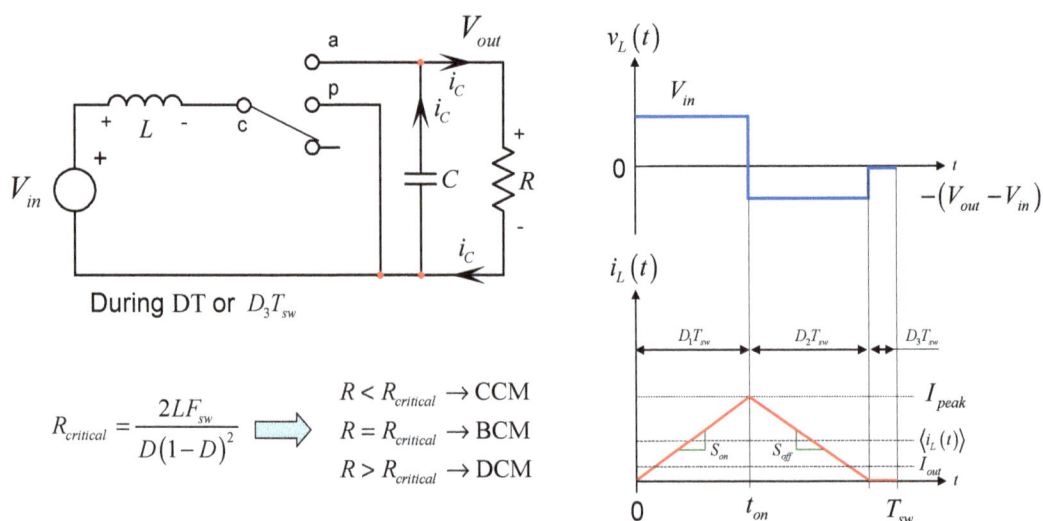

Figure 1.33 In DCM, both SW and D are blocked during the dead time.

From (1.57), we see that the output current depends on the average inductor current and the duty ratio. We said that our control system increases the duty ratio D in response of a sudden current demand. However, (1.57) equation tells us that if D increases, the $(1–D)$ factor reduces bringing I_{out} down, the inverse of what we want. The key here to is to make sure that the average inductor current $\langle i_L(t) \rangle$ always grows faster cycle by cycle than the decrease of $1-D$ when the loop reacts. Another way of looking at this is to purposely slow down the variation of duty ratio so that $\langle i_L(t) \rangle$ has always time to grow while the increase of D spreads across many switching cycles. The key is (1.55) which tells you that a small inductor will always let the current grow faster than a larger one but also that a small input voltage V_{in} in the boost case represents a worst-case situation for a fast increase of the inductive current.

For the sake of the illustration, we have simulated a simple *cycle-by-cycle* 100-kHz boost converter built around a 300-µH inductance. The narrow-pulse clock sets the latch on every 10 µs and resets occurs when the artificial 1-V peak ramp reaches the continuous value set by the control voltage V_{dc} (Figure 1.34). In a closed-loop converter, V_{dc} would be the error voltage delivered by the error amplifier or *compensator*. The comparator and the artificial ramp form the pulse width modulation (PWM) block. We will come back in detail on this important subcircuit.

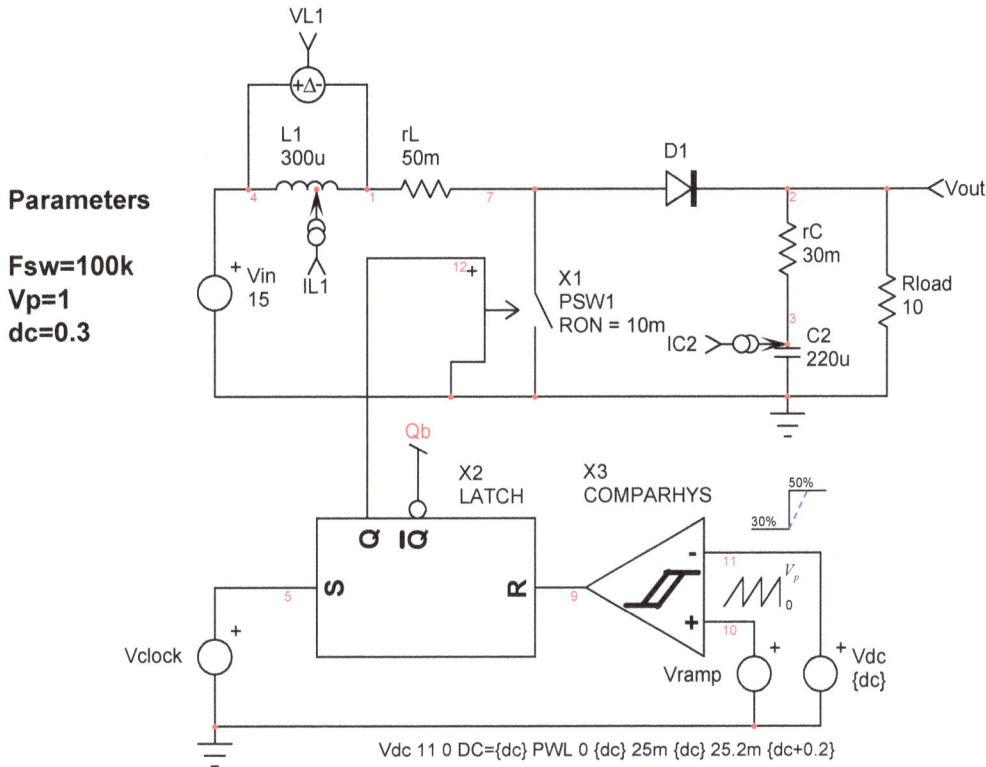

Figure 1.34 A simple open-loop boost converter illustrate how a sudden change in the control variable momentarily brings the output voltage down.

Please note that the circuit now includes some of the parasitic elements found in passive components. The small resistances r_L and r_C respectively model the ohmic losses in the inductor and the capacitor (their respective ESR). We have adopted fixed resistances but experience shows that some of these terms can be frequency-, temperature- and bias-dependent. You can refine the analysis by accounting for some of the dependencies and complicate models as you like. In this book, we will stick to fixed-value components.

In this configuration, the converter delivers 21 V in an open-loop configuration to a 10-Ω load from a 15-V input source. The duty ratio is initially set to 20% and stepped to 50% in different time spans while the load remained constant. In the upper graph of Figure 1.35, the variation occurs in 1 µs and the output voltage cannot instantaneously follow. It drops for a few cycles the time for the inductor energy to build up cycle by cycle to match the power requirement. In the middle graph, we slowed down the duty ratio change from 1 µs to 100 and 200 µs. The drop in V_{out} is less severe in both cases. Finally, in the lower side of the graph, we spread the duty ratio variation over 1 ms, giving time to the inductor current to build up cycle by cycle. As a result, V_{out} does not drop at all. Is it that important if V_{out} drops? After all, the loop's role is to compensate this, no? The phenomenon we observe here is the *open-loop control-to-output* response of a given converter: when its control variable D increases it should induce an output voltage increase. However, in the three upper graphs, we see that rather than increasing immediately, V_{out} first drops for a few cycles then eventually takes off as more current is building up in L. Control-wise, we have reversed the control law of the converter for several switching cycles because despite an increase in D, V_{out} drops: if you close the loop of such a system, oscillations are guaranteed.

This phenomenon is the representation of what is mathematically modeled as a *right half-plane zero* or RHPZ appearing in the control-to-output transfer function of the boost converter. The delay in the conversion process – store energy first in the inductance then deliver – explains its presence. Whether the converter operates in voltage-

or current-mode control, the zero occupies the same position. So here, you realize that for a given component selection, you have to make sure the loop is slow enough to let the current build up in the inductor and avoid the output drop.

As a refresh [2], the loop gain T is made of the cascaded gains of the power stage H and the compensator G. If you consider that the resulting crossover is not viable (your system is too slow), you must select a lower inductance value at the expense of a larger ripple. The theoretical crossover selection freedom we had with the buck disappears here with the boost converter.

In other words, you must know the lowest position of your RHP zero (lowest V_{in} and highest I_{out}) and limit the maximum crossover frequency to around 30% of this value. In this particular example, the RHP zero is calculated at 2.6 kHz naturally limiting a theoretical crossover to 780 Hz.

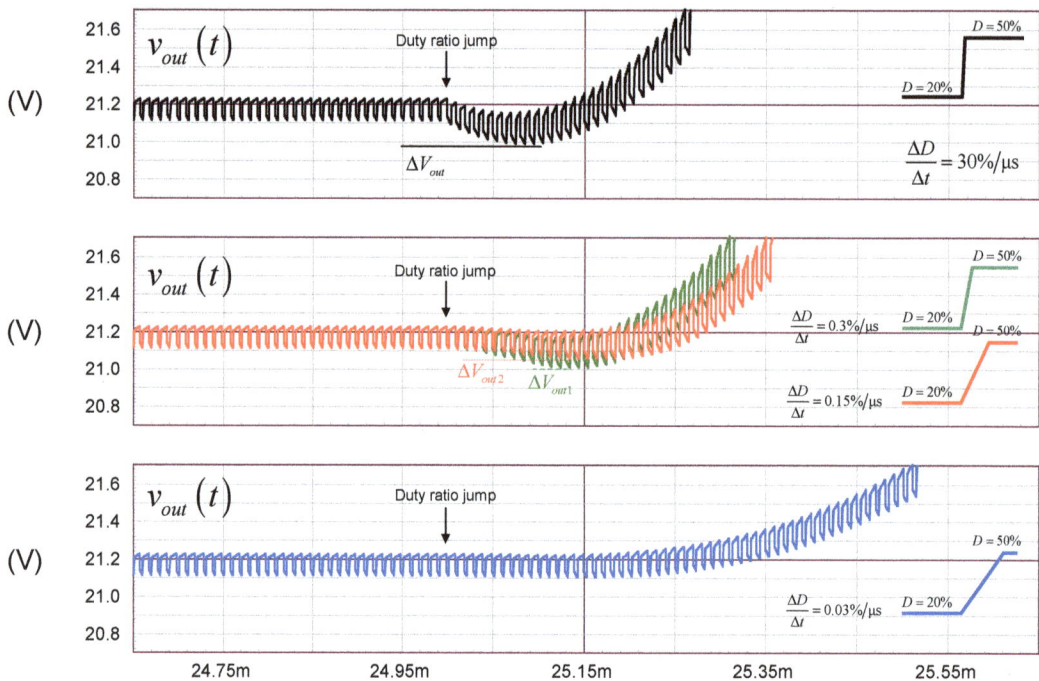

Figure 1.35 In CCM, if the duty ratio varies too quickly, the average inductor current takes time to build up and cannot supply the load: the output voltage momentarily drops until $\langle i_L(t) \rangle$ meets the demand.

If you try to push the crossover closer to the RHP zero, you will have difficulty in keeping enough phase margin and an oscillatory response is likely to appear. Remember, a zero in the left half-plane boosts the phase while a RHP zero lags it, as a pole would do.

A RHP zero in a transfer function is recognizable by the negative sign preceding the s: it is a positive root of the numerator while a left half-plane zero (LHPZ) is a negative one. In the following transfer function, ω_{z_1} is a LHPZ while ω_{z_2} is the RHPZ.

$$H(s) = H_0 \frac{\left(1 + \dfrac{s}{\omega_{z_1}}\right)\left(1 - \dfrac{s}{\omega_{z_2}}\right)}{D(s)} \tag{1.60}$$

1.2.4 Reversing Output Voltage Polarity

Both buck and boost converters ensure a positive output with respect to the input ground. The buck-boost converter, on the other hand, reverses the output polarity and delivers a negative voltage with respect to ground. Its operating states appear in Figure 1.36.

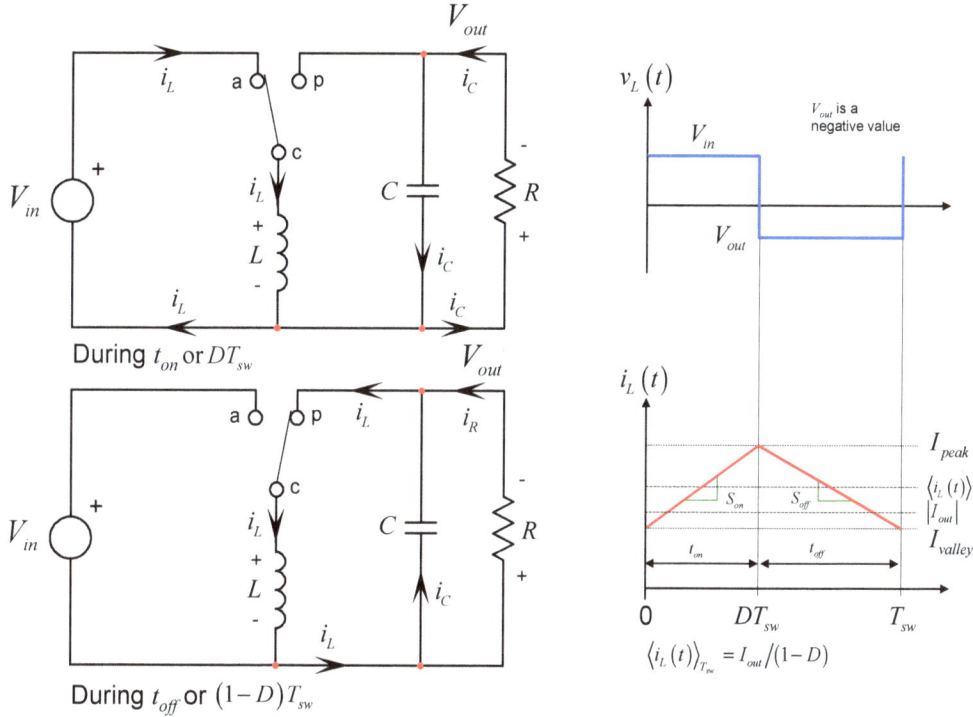

Figure 1.36 The buck-boost converter also needs to build up energy in the inductor during the on-time before it releases it to the load during the off-time.

Similar to the boost converter, the inductor connects across the input source V_{in} during the on-time. The current builds up in CCM from a valley value to a peak following a slope:

$$S_{on} = \frac{V_{in}}{L} \tag{1.61}$$

During this time, the RC network is on its own and capacitor C alone feeds the load resistance R. When the SPDT switch toggles and connects terminals p and c, the inductor current flows in the RC network. The current in the inductor depletes with a slope expressed as:

$$S_{off} = \frac{V_{out}}{L} \tag{1.62}$$

in which V_{out} is a negative value because the inductor current now circulates towards ground and biases the RC network from this point. As you can observe, the buck-boost converter also delivers energy to the load in a two-stage operation: energy is stored in L during t_{on} and is further delivered to the load during t_{off}. The source V_{in} "sees" the inductor current during the on-time while it flows in the RC network during the off-time only. As a result, both input and output currents are pulsating for the buck-boost converter. Because of the intermediate energy-storing phase in L when the power switch is closed, the buck-boost converter belongs to the indirect energy-transfer converter family and also exhibits a delay in the conversion process: the control-to-output transfer function hosts a RHP zero in CCM whose position is the same regardless of the control mode.

The output current signature of the buck-boost converter is similar in shape to that of the boost converter. Figure 1.32 is still valid for this structure and the relationship linking the output current to the inductor current remains the same for the buck-boost architecture:

$$I_{out} = \langle i_L(t) \rangle \cdot (1 - D) \tag{1.63}$$

From Figure 1.36, we see that the current leaving terminal c is the inductor current i_L. On average, we thus have:

$$I_c = \langle i_L(t) \rangle \tag{1.64}$$

Since during the on-time terminals c and a are connected, the current in a is that of c during DT_{sw} leading, again, to the expression already determined with (1.53). This property does not change by rotating the cell to accommodate with the different converters hence the qualifying term *invariant*.

When the load current decreases, the buck-boost converter enters DCM and a 3rd stage appears as shown in Figure 1.37. When both switches block, the inductor bias returns to zero and the capacitor feeds the load until a new cycle happens.

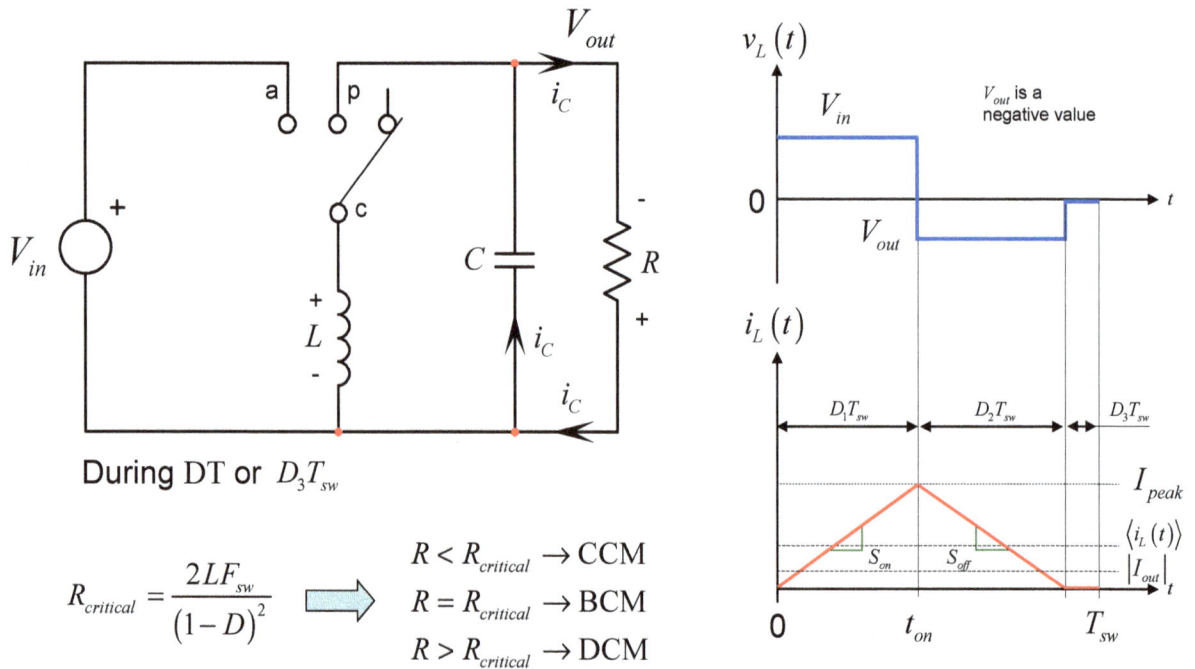

$$R_{critical} = \frac{2LF_{sw}}{(1-D)^2} \implies \begin{array}{l} R < R_{critical} \rightarrow \text{CCM} \\ R = R_{critical} \rightarrow \text{BCM} \\ R > R_{critical} \rightarrow \text{DCM} \end{array}$$

Figure 1.37 In DCM, both SW and D are blocked during the dead time and the capacitor alone feeds the load.

Looking at Figure 1.33 and Figure 1.37, a DCM-operated converter such as the boost or buck-boost still needs an intermediate storing phase before feeding the output RC network.

One legitimate question regards the disappearance of the RHP zero in this mode as often pointed out in the literature. In reality, the RHP zero is still present in DCM.

The physical explanation of its origin is given in Figure 1.38.

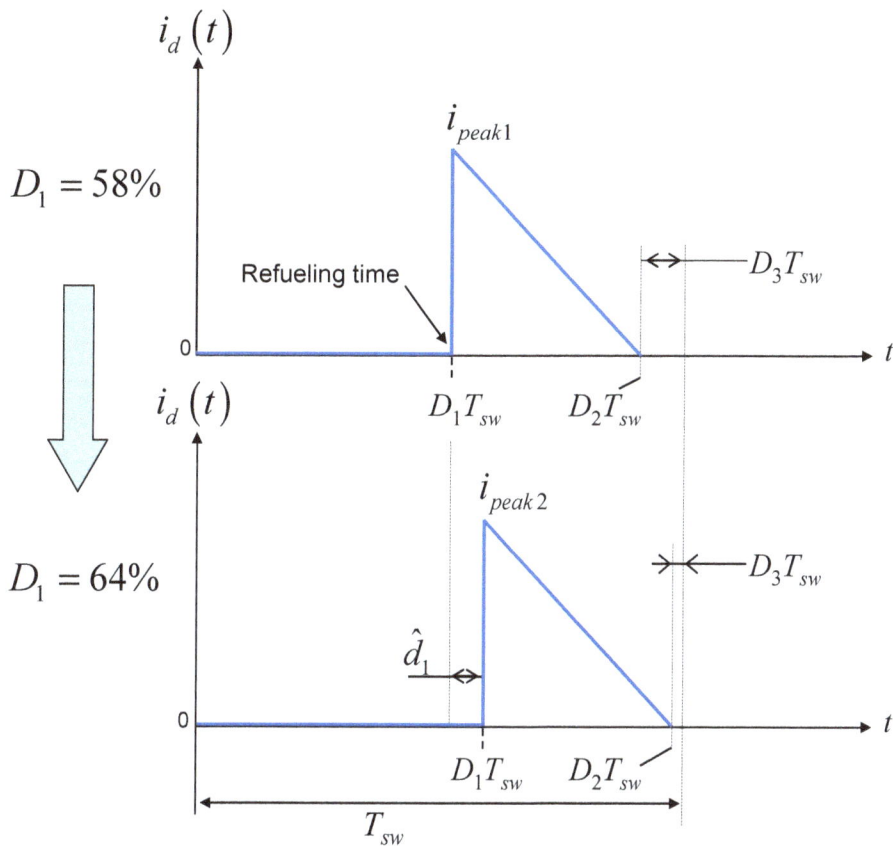

Figure 1.38 A RHP zero also exists in DCM but is constrained to a high frequency. The dead time reduces to accommodate the shift incurred by the duty ratio increase.

When the loop suddenly commands duty ratio increase, the on-time expands accordingly and the inductor current builds to a higher peak value.

Because of this event, the point at which the switch turns off is slightly shifted in time because the switch SW keeps on slightly longer than in the previous cycle. Unlike with the CCM operation, the inductor current builds up quickly but the moment where it transfers to the RC network is delayed and the capacitor remains on its own, feeding the load a little bit longer.

With a constant switching period, the dead time DT shrinks by the amount of duty ratio increase but V_{out} drops for a short period: again, the control law is momentarily reversed before V_{out} takes off. It is truly a high-frequency phenomenon and this is the manifestation of the RHP zero in DCM. Figure 1.39 shows the open-loop simulation of Figure 1.34 boost in which the inductor has been reduced to 100 µH and the load set to 500 Ω (C_{out} has decreased to 22 µF to reduce the settling time).

As you can observe, when the duty ratio abruptly changes from 30 to 40%, there is a small output voltage drop as expected but as the inductor current builds up quickly, the output voltage follows afterwards. Dr. Vorpérian [3] was the first to show the existence of the RHP zero in DCM-operated boost and buck-boost converters. He also proved with the help of the PWM switch model that despite DCM operation, all these converters remained 2nd-order systems however heavily damped.

We will see that in the following chapters. Before that, the state-space averaging (SSA) technique considered that a DCM converter was a 1st-order system, without RHPZ for the boost and the buck-boost versions.

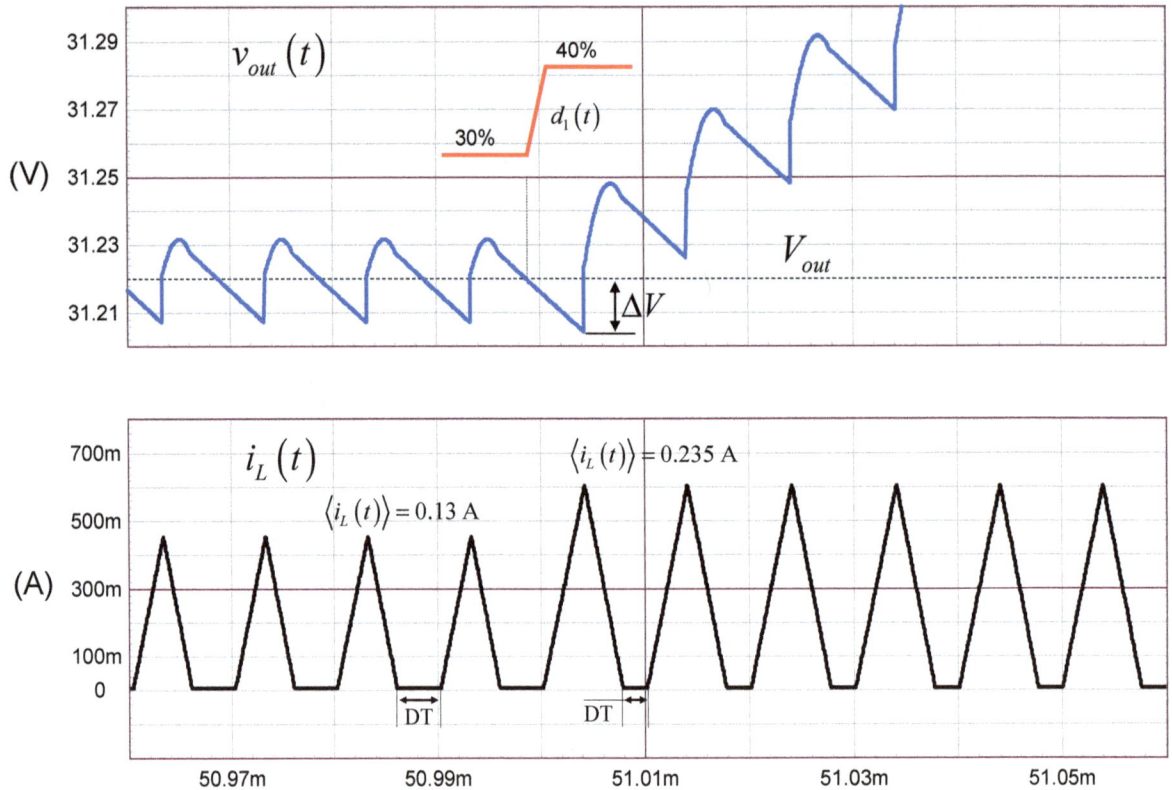

Figure 1.39 The inductor current increases immediately as the on-time expands but the small delay in the capacitor recharging induces a brief output voltage drop.

In Figure 1.40, we have gathered the definitions of the critical resistance values for the three basic dc-dc converters.

Converter	$R_{critical}$
Buck	$\dfrac{2LF_{sw}}{1-D}$
Boost	$\dfrac{2LF_{sw}}{D(1-D)^2}$
Buck-Boost	$\dfrac{2LF_{sw}}{(1-D)^2}$

R_{load} is below $R_{critical}$: CCM
R_{load} is above $R_{critical}$: DCM
R_{load} is equal to $R_{critical}$: BCM

Figure 1.40 Below a certain load current, the converter enters the discontinuous mode of operation.

1.2.5 Inductive and Capacitive Equilibrium

Building on the boost converter example from Figure 1.34, we can rearrange blocks and form a simple buck converter as shown in Figure 1.41. You recognize the voltage-controlled *SW* switch and the freewheel diode *D*. The SPICE diffusion parameter of the diode *N* is set to 0.01 so the forward drop V_f is almost 0 V making the component a perfect element.

The duty ratio is fixed to 30% and simulation runs for 1 ms.

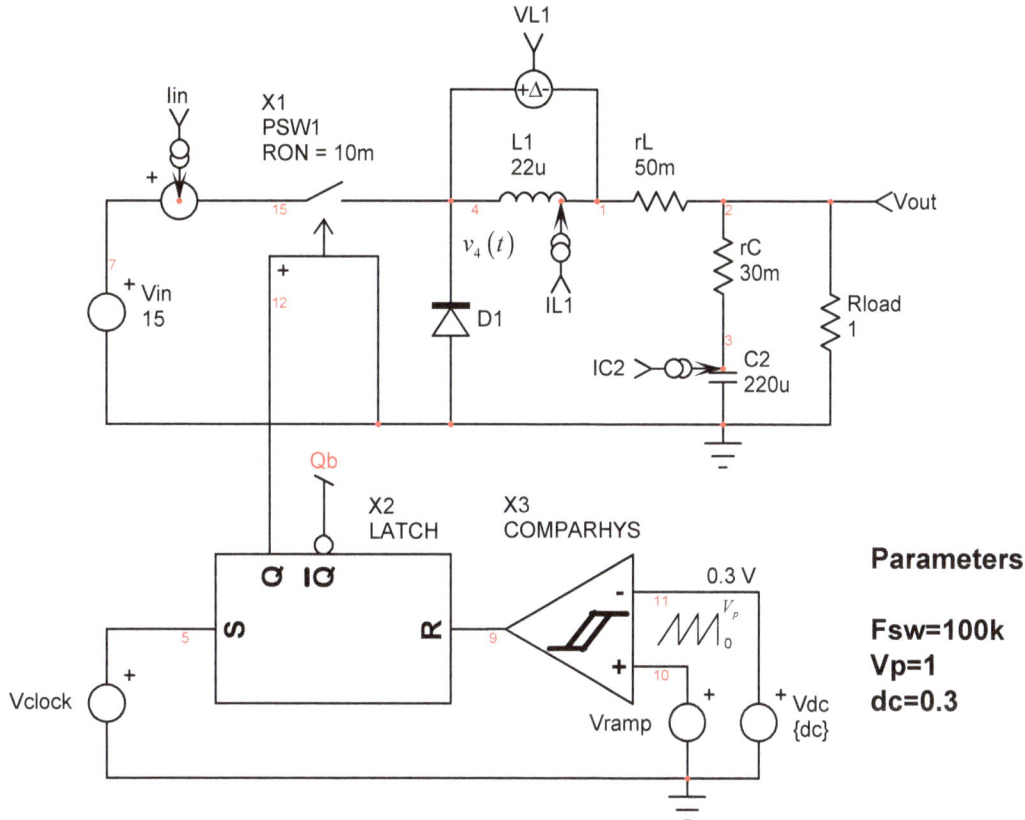

Figure 1.41 A SPICE simulation lets us look at the various voltages and currents present in the buck converter.

Figure 1.42 illustrates how the inductive and capacitive currents evolve with time. In the left side, at power on, V_{out} is 0 V and the slope calculated by (1.44) reaches its maximum while the off-slope described by (1.45) is zero. After a few switching cycles, the output voltage builds up, overshoots (2nd-order response) and converges towards a stable value around 4.4 V. At this point, the inductor current lands to its steady-state average value which is the output current I_{out}.

The capacitor current $i_C(t)$ follows $i_L(t)$ at the beginning and diverges as the converter approaches equilibrium in the right-side of the graph. Eventually, the average current in the capacitor becomes 0 A and confirms steady-state is reached. This is what is called capacitor *charge* balance: there is no net change in capacitor charge during a switching cycle. If we were to look at the average voltage across the inductor when the converter is at equilibrium, we would measure 0 V.

This is called inductor *volt-second* balance and implies no net change in the inductor current over a switching cycle.

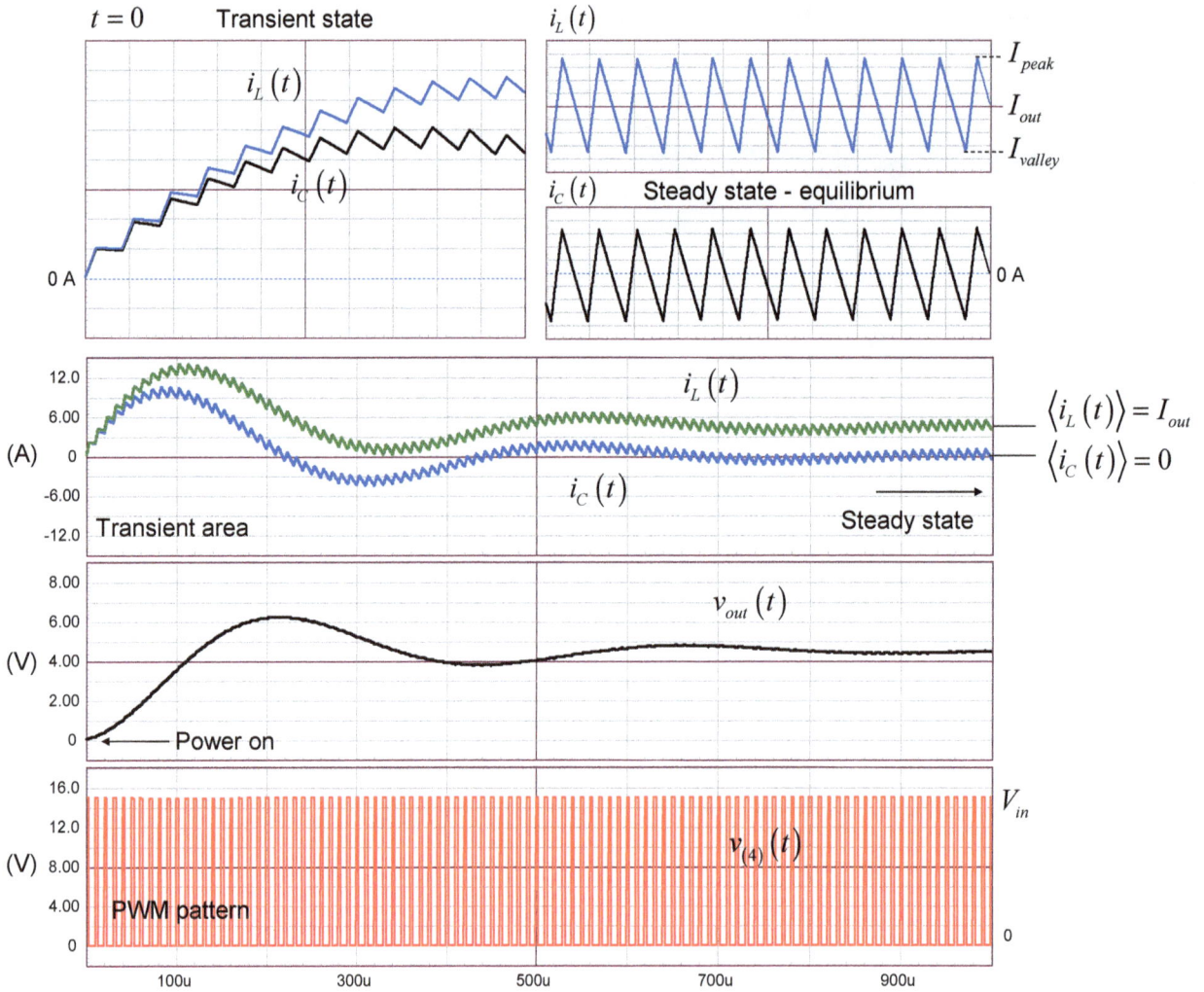

Figure 1.42 Inductor and capacitor current evolution at power on and during steady-state in the buck converter.

In Figure 1.43, we have represented the inductor current and its terminals voltage $v_L(t)$ in two operating phases of the buck converter: transient and steady state. During the transient, the net inductor current increases cycle by cycle. This can be inferred from the fact that the valley current I_{valley} after one switching cycle is larger than the initial current at the beginning of that cycle (see the upper left corner of Figure 1.42). We say there is a *net* increase of the inductor current cycle by cycle.

In the buck converter, the slopes defined by (1.44) and (1.45) depend on V_{out}, the output voltage. In the first cycle, V_{out} is 0 V and S_{on} is the steepest: the voltage applied across L is V_{in}. Because the output voltage is almost non-existing, S_{off} is flat and the inductor has not demagnetized at the end of the first cycle: $I_{peak} \approx I_{valley}$. As V_{out} builds up cycle by cycle, slopes change and eventually, the inductor current at the end of a switching cycle returns to its initial value at the beginning of that cycle: this is the so-called equilibrium in which there is no net inductor current increase per switching cycle. We can write:

$$i_L\left(nT_{sw}\right) = i_L\left(\left(n+1\right)T_{sw}\right) \tag{1.65}$$

36

In steady-state, the inductor current starts from the valley current, I_{valley}, and ramps up to the peak after DT_{sw} or t_{on}. The current swing from the valley to the peak represents the *inductor ripple current* ΔI_L.

From (1.47), we can write:

$$I_{peak} = I_{valley} + \frac{V_{L,on}}{L}t_{on} = I_{valley} + \Delta I_L \qquad (1.66)$$

When the switch turns off, the inductor current keeps circulating in the same direction and now biases the freewheeling diode D during $(1-D)T_{sw}$ or t_{off}. As the inductor voltage reverses to V_{out}, the current falls down from the peak to the valley according to (1.48):

$$I_{valley} = I_{peak} - \frac{V_{L,off}}{L}t_{off} = I_{peak} - \Delta I_L \qquad (1.67)$$

The ripple current ΔI_L is thus determined by two equivalent expressions:

$$\Delta I_L = I_{peak} - I_{valley} = \frac{V_{L,on}}{L}t_{on} = \frac{V_{L,off}}{L}t_{off} \qquad (1.68)$$

which implies:

$$V_{L,on}DT_{sw} = V_{L,off}(1-D)T_{sw} \qquad (1.69)$$

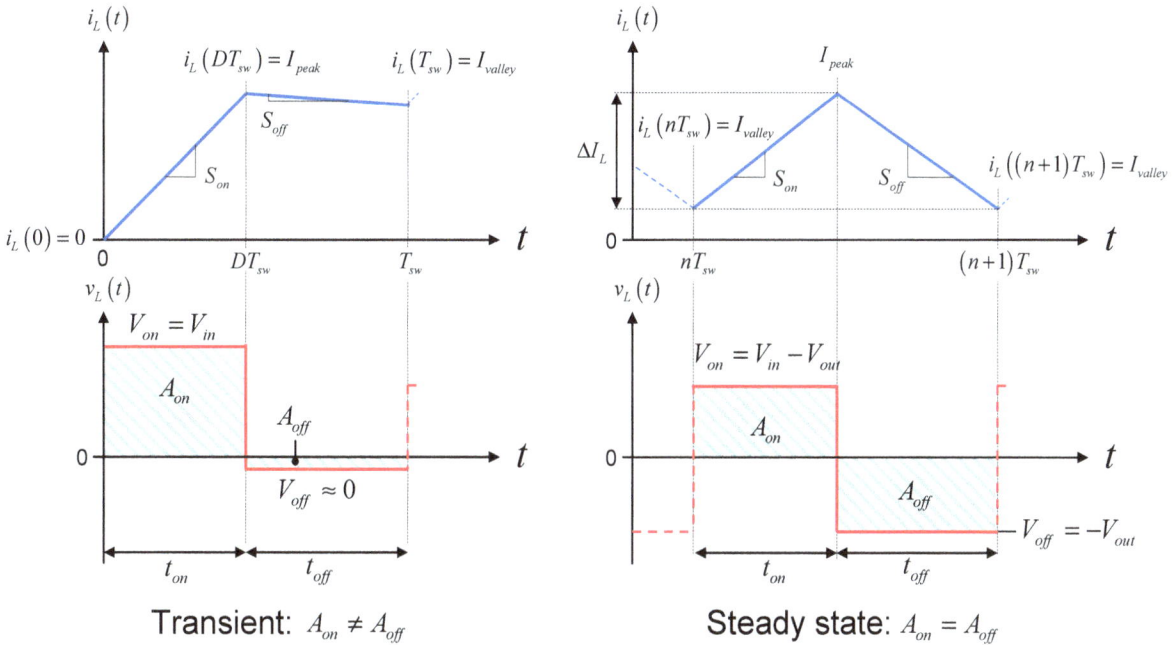

Figure 1.43 Currents ramping up and down in an inductor: transient and steady-state.

The inductor current slopes given in (1.68) derive from the *v-i* relationship for inductance which states:

$$v_L(t) = L\frac{di_L(t)}{dt} \qquad (1.70)$$

In our switching converter at steady-state, the current starts from the valley at the beginning of the cycle, $i_L(0) = I_{valley}$, and returns to the same level after a switching period: $i_L(T_{sw}) = I_{valley}$. If we integrate both parts of (1.70), we have:

$$\int_0^{T_{sw}} v_L(t) \cdot dt = L \int_0^{T_{sw}} \frac{di_L(t)}{dt} \cdot dt \qquad (1.71)$$

The left side in the above expression is proportional to the integral of the voltage applied across the inductor along a switching cycle T_{sw}. Its dimension is volt-seconds noted $V \cdot s$. The right part describes the inductor current change during the same switching event. As the current initially starts from the valley and returns to the valley when the converter operates at steady state, we have:

$$\int_0^{T_{sw}} v_L(t) \cdot dt = L\left[i_L(T_{sw}) - i_L(0)\right] = 0 \qquad (1.72)$$

At the converter's equilibrium, the net volt-seconds change of the inductor along a switching period must be 0: this is the so-called inductor *volt-second balance*. If we now divide both sides of (1.72) by the switching period, we have:

$$\langle v_L(t) \rangle = \frac{1}{T_{sw}} \int_0^{T_{sw}} v_L(t) \cdot dt = 0 \qquad (1.73)$$

which implies that the average voltage across the inductor at the equilibrium is 0 V. Looking at the lower right corner in Figure 1.43, we can determine the average inductor voltage by calculating the areas under the voltage waveform during t_{on} and t_{off}: these two areas must be equal to satisfy (1.73).

Using Faraday's law, we can also show that the inductor volt-seconds are the inductor internal flux activity during a switching cycle:

$$v_L(t) = N\frac{d\varphi_L(t)}{dt} \qquad (1.74)$$

with φ the internal core magnetic flux and N the number of turns on the core of the inductor. If we integrate both sides of this equation, we can link the inductor on- and off-time voltage-seconds with the internal core magnetic flux excursion incurred during the on-time:

$$V_L \cdot t = N \cdot \varphi_L \qquad (1.75)$$

At equilibrium, the flux in the inductor increases proportionally to the applied volt-seconds product during the on-time and swings back to its original point when the bias reverses during the off time: the core is said to be *reset*. To satisfy (1.73), we have:

$$\int_0^{T_{sw}} v_L(t) \cdot dt = N\left[\varphi_L(T_{sw}) - \varphi_L(0)\right] = 0 \qquad (1.76)$$

If core reset is not ensured at steady-state, a so-called *flux walk-away* situation may occur and risks of core material saturations exist with all the deleterious consequences (current runaway as the inductance L collapses).

As shown in Figure 1.42, the converter's equilibrium can also be inferred from the observation of the capacitor current $i_C(t)$. During the on-time, the inductor current rises from the valley to the peak and splits between capacitor C and the load R. We have:

$$i_C(t) = i_L(t) - i_R(t) \qquad (1.77)$$

If the capacitance is sufficiently large to minimize the ripple at the switching frequency and satisfies (1.40), we assume that the alternating component of the inductor current entirely flows in the capacitor while the direct portion feeds the load.

As such, (1.77) simplifies to:

$$i_C(t) \approx i_L(t) - \frac{V_{out}}{R}$$

(1.78)

The graphical representation of the capacitor current is thus the instantaneous inductor current minus by the direct current circulating in R. This is what Figure 1.44 represents.

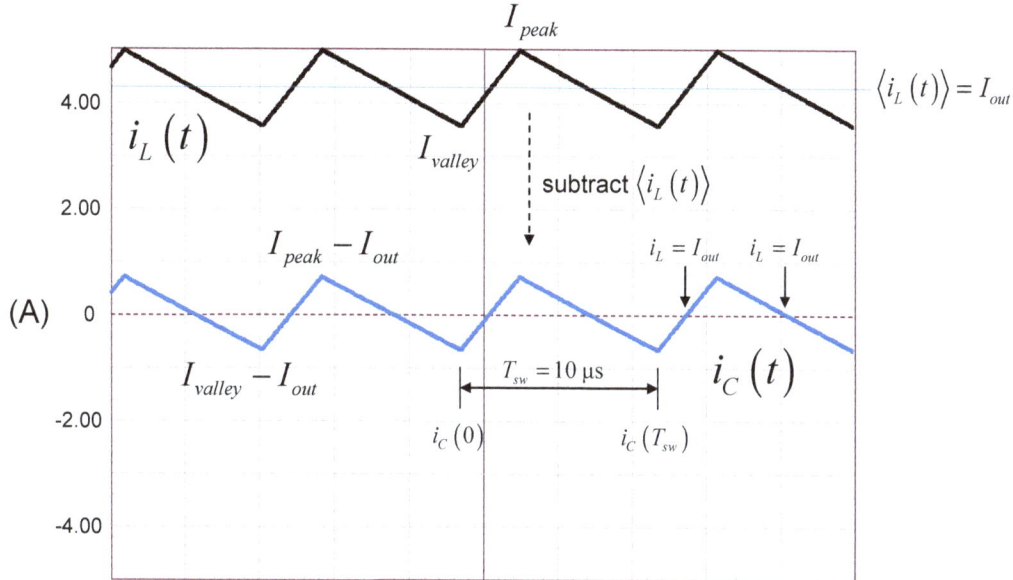

Figure 1.44 In the buck converter, the capacitor current is that of the inductor minus the dc component supplied to the load.

At the beginning of the cycle, when the inductor current is below the load current at I_{valley}, the capacitor current starts from a negative value equal to:

$$i_C(0) = I_{valley} - \frac{V_{out}}{R}$$

(1.79)

and increases with a slope dictated by (1.44). In this time interval, the capacitor discharges and supplies current to the load, compensating the energy deficit in the inductor. When the inductor current reaches the load current, the capacitor current is 0. As the inductor current keeps rising, the capacitor stores charges and the voltage across its terminals grows. The current keeps increasing with a constant slope until the switch turns off and the inductor current has reached I_{peak}. At that moment, the current in the capacitor is:

$$i_C(DT_{sw}) = I_{peak} - \frac{V_{out}}{R}$$

(1.80)

The voltage across the inductor has now reversed and is $-V_{out}$. The inductor current linearly decays but still charges the capacitor until i_L has reached the load current. At that moment, the capacitor starts supplying R and discharges as the inductor demagnetizes. At the end of the switching cycle, the inductor current has returned to I_{valley} and the capacitive current returned to the value given by (1.79).

The instantaneous current $i_C(t)$ in a capacitor depends on the voltage variation $v_C(t)$ across its terminals. Both variables are linked by the following expression:

$$i_C(t) = C \frac{dv_C(t)}{dt} \tag{1.81}$$

Similar to what we did with the inductor, we can integrate both sides of the above equation and obtain:

$$\int_0^{T_{sw}} i_C(t) \cdot dt = C \int_0^{T_{sw}} \frac{dv_C(t)}{dt} \cdot dt \tag{1.82}$$

Both sides of this expression are homogenous to a charge Q expressed in coulomb. As shown in Figure 1.45 lower right corner, when the converter regulates, the net change of the capacitor voltage over a switching cycle must be 0, naturally implying that the net capacitive charge change ΔQ is also zero. Otherwise stated:

$$\Delta Q = C\left[v_C(T_{sw}) - v_C(0) \right] = 0 \tag{1.83}$$

At the converter's equilibrium, the charge stored in the capacitor as its voltage increases equals the charge delivered during its discharge: this is the so-called capacitor *ampere-second* or charge *balance law*. It corresponds to the shaded areas below the capacitor instantaneous current in Figure 1.45. If we now divide both sides of (1.82) by the switching period, we have:

$$\frac{1}{T_{sw}} \int_0^{T_{sw}} i_C(t) \cdot dt = \langle i_C(t) \rangle = 0 \tag{1.84}$$

indicating a zero average current in the output capacitor when the converter has reached equilibrium.

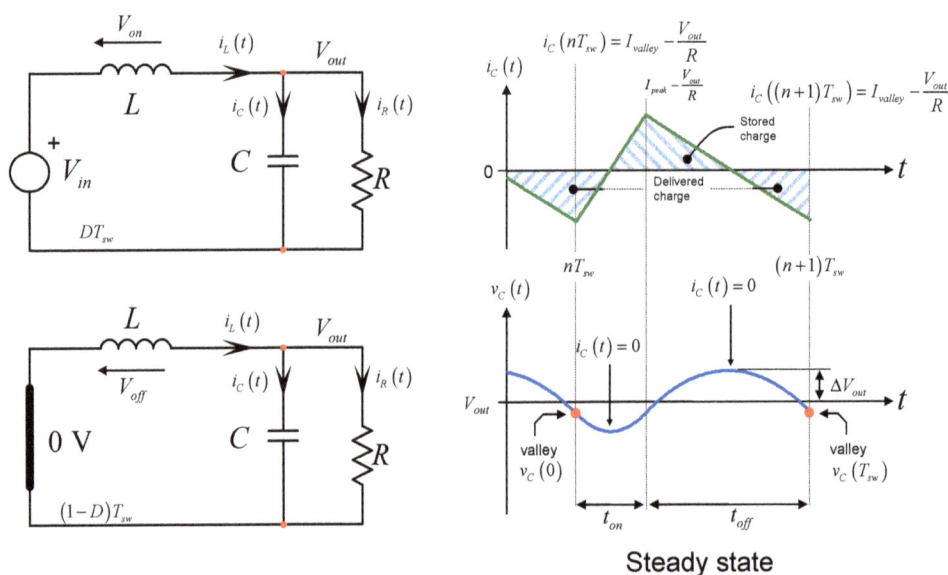

Figure 1.45 Currents ramping up and down in the output capacitor at steady-state.

These two laws, inductor volt-seconds balance and capacitor charge balance, are part of the converter analysis toolbox. In a simulation, looking at the inductor average voltage or considering the average capacitor current while confirming they are close to a minimum is a good indication that the converter operates in steady state.

1.2.6 Conversion Ratios of Switching Converters

One immediate application of the inductor volt-seconds balance is the derivation of the continuous conversion ratio noted M. It is the relationship linking V_{out} to V_{in} via the control variable D:

$$\frac{V_{out}}{V_{in}} = M \tag{1.85}$$

If we look back at Figure 1.27 for the buck converter, we know from (1.72) that the areas under $v_L(t)$ during the on- and off-times must be equal:

$$DT_{sw}\left(V_{in} - V_{out}\right) = V_{out}\left(1-D\right)T_{sw} \tag{1.86}$$

Solving for V_{out} gives:

$$V_{out} = DV_{in} \tag{1.87}$$

and factoring leads to the buck conversion ratio M in CCM:

$$M = D \tag{1.88}$$

If we follow these steps for the boost converter and look at Figure 1.31, we can write:

$$V_{in}DT_{sw} = \left(V_{out} - V_{in}\right)\left(1-D\right)T_{sw} \tag{1.89}$$

Solving for V_{out} and factoring gives:

$$M = \frac{1}{1-D} \tag{1.90}$$

Finally, for the buck-boost converter in Figure 1.36, the expression is quite simple and already ordered:

$$V_{in}DT_{sw} = \left(1-D\right)V_{out}T_{sw} \tag{1.91}$$

The conversion ratio is immediate:

$$M = \frac{D}{1-D} \tag{1.92}$$

In DCM, things complicate a little more given the presence of the dead time. For the buck converter, we now look at Figure 1.30.

The average inductor voltage is still 0 V during a switching cycle and (1.86) holds except that the off-time is now D_2T_{sw} (the voltage across the inductor is 0 V during the dead time DT):

$$D_1T_{sw}\left(V_{in} - V_{out}\right) = V_{out}D_2T_{sw} \tag{1.93}$$

From the discontinuous inductor current representation in Figure 1.30, we can see that the continuous component I_{out} is obtained by calculating the area under $i_L(t)$ during $D_1 T_{sw}$ and $D_2 T_{sw}$:

$$I_{out} = \langle i_L(t) \rangle = \frac{1}{2} I_{peak} (D_1 + D_2) \tag{1.94}$$

The peak current is defined by (1.49) and equals:

$$I_{peak} = \frac{V_{in} - V_{out}}{L} D_1 T_{sw} \tag{1.95}$$

If you substitute (1.95) in (1.94), replace I_{out} by V_{out}/R and solve for D_2, you should find:

$$D_2 = \frac{2L}{D_1 R T_{sw}} \frac{V_{out}}{V_{in} - V_{out}} - D_1 \tag{1.96}$$

If you consider a normalized time constant τ_L defined by:

$$\tau_L = \frac{L}{R T_{sw}} \tag{1.97}$$

Then (1.96) can be rewritten:

$$D_2 = \frac{2\tau_L}{D_1} \frac{V_{out}}{V_{in} - V_{out}} - D_1 \tag{1.98}$$

Now substitute (1.98) in (1.93), rearrange solve for V_{out} and you have the conversion ratio M for the buck converter operated in DCM:

$$M = \frac{D_1^2}{4\tau_L} \left[\sqrt{1 + \frac{8\tau_L}{D_1^2}} - 1 \right] = \frac{2}{1 + \sqrt{1 + \frac{8\tau_L}{D_1^2}}} \tag{1.99}$$

You can repeat the exercise for the boost and buck-boost converter as detailed in [1].

Figure 1.46 conveniently gathers these results for the three basic converters, including the duty ratio definitions for both operating modes.

Converter	M_{CCM}	M_{DCM}	$R_{critical}$	$L_{critical}$	D_{DCM}	D_{CCM}
Buck	D	$\dfrac{2}{1+\sqrt{1+\dfrac{8\tau_L}{D_1^2}}}$	$\dfrac{2LF_{sw}}{1-D}$	$\dfrac{(1-D)R}{2F_{sw}}$	$\sqrt{\dfrac{2\tau_L}{V_{in}(V_{in}-V_{out})}}\cdot V_{out}$	$\dfrac{V_{out}}{V_{in}}$
Boost	$\dfrac{1}{1-D}$	$\dfrac{1+\sqrt{1+\dfrac{2D_1^2}{\tau_L}}}{2}$	$\dfrac{2LF_{sw}}{D(1-D)^2}$	$\dfrac{RD(1-D)^2}{2F_{sw}}$	$\sqrt{\dfrac{2\tau_L V_{out}\left(\dfrac{V_{out}}{V_{in}}-1\right)}{V_{in}}}$	$1-\dfrac{V_{out}}{V_{in}}$
Buck-Boost	$-\dfrac{D}{1-D}$	$-D_1\sqrt{\dfrac{1}{2\tau_L}}$	$\dfrac{2LF_{sw}}{(1-D)^2}$	$\dfrac{(1-D)^2 R}{2F_{sw}}$	$-\dfrac{M}{\sqrt{\dfrac{1}{2\tau_L}}}$	$-\dfrac{V_{out}}{V_{in}-V_{out}}$

V_{out} is negative for the buck-boost, e.g. -12 V

Figure 1.46 The conversion ratio M changes depending on the operating mode. $\tau_L = \dfrac{L}{RT_{sw}}$ is the normalized time constant.

Please note that when operated in CCM and considering perfect elements (semiconductors and passive components), the converter output voltage is independent of loading conditions and switching frequency. Conversely, in DCM, the voltage becomes dependent on the absorbed current and operating frequency.

1.2.7 Considering Losses in Conversion Ratios

In all these examples, we have considered perfect elements only, neglecting all ohmic losses and voltage drops. Reality obviously differs and depending on input/output operating conditions, the duty ratio measured on the bench can notably diverge from the theoretically-calculated value. For a more precise analysis, we will now consider the switch $r_{DS(on)}$, the diode forward drop V_f and the inductor ohmic loss r_L. If we start with the CCM buck converter, the two stages from Figure 1.27 are detailed in Figure 1.47 now including the parasitic contributors we have listed. The principle remains the same, we determine the inductor voltage during the on- and off-times and apply the volt-seconds balance law already used. During the on-time, the positive inductor terminal is biased to the input source V_{in} minus two voltage drops while its negative terminal connects to V_{out}:

$$V_{L,on} = V_{in} - \frac{V_{out}}{R}\left[r_{DS(on)} + r_L\right] - V_{out} \tag{1.100}$$

When the diode conducts, it no longer brings the cathode to 0 V but to a negative voltage V_f, the diode forward drop. Observing the lower side of Figure 1.47, we can write:

$$V_{L,off} = V_{out} + V_f + \frac{V_{out}}{R}r_L \tag{1.101}$$

Applying the volt-second balance, at equilibrium, the following expression must be satisfied:

$$V_{L,on}DT_{sw} = V_{L,off}(1-D)T_{sw} \tag{1.102}$$

43

Solving for V_{out} and rearranging the expression, we have:

$$V_{out} = \left[DV_{in} - V_f \left(1 - D\right)\right] \frac{1}{1 + \dfrac{r_L + D \cdot r_{DS(on)}}{R}}$$

(1.103)

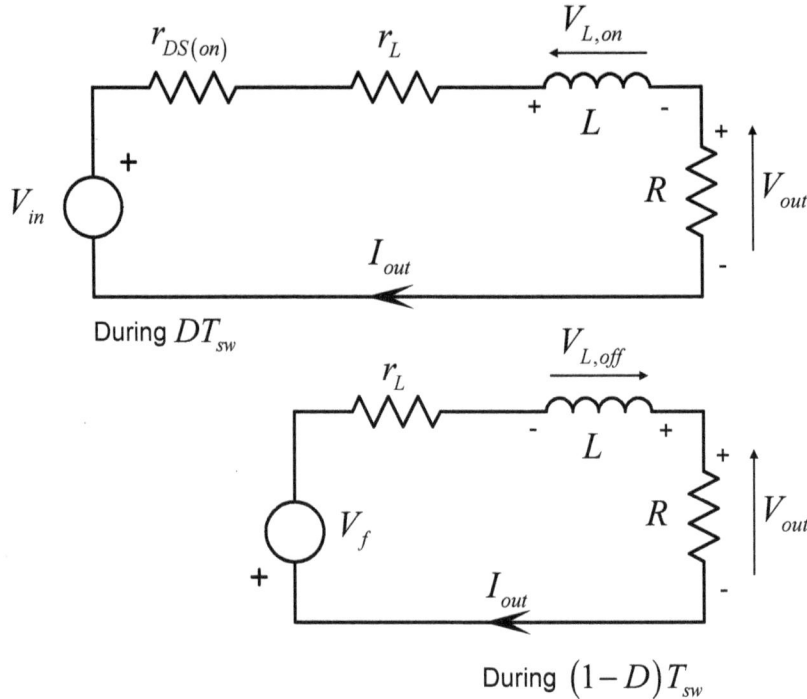

Figure 1.47 Adding parasitic contributors complicates the analysis somewhat.

We can rework this expression and have it fit (1.85) format in which V_f is a positive value:

$$M = D \left[\frac{1}{1 + \dfrac{r_{DS(on)}}{R} D + \dfrac{r_L}{R} + \dfrac{V_f}{V_{out}} D'} \right]$$

(1.104)

From this expression, you see that the switch $r_{DS(on)}$ contribution is weighted by the on-time while the diode drop V_f effect is averaged by the off-time duration. If all these terms reduce to 0, the expression returns to (1.88).

This exercise can be repeated with the boost converter whose two operating states now considering parasitic elements appear in Figure 1.48. During the on-time, the positive inductor terminal is biased to the input source V_{in} minus one drop while its negative terminal connects to ground via the switch $r_{DS(on)}$:

$$V_{L,on} = V_{in} - \frac{V_{out}}{R\left(1 - D\right)} \left[r_{DS(on)} + r_L \right]$$

(1.105)

When the diode conducts, it brings a forward drop in series with inductor while the cathode connects to the load resistance R.

Observing the lower side of Figure 1.48, we can write:

$$V_{L,off} = \left(V_f + V_{out}\right) - \left(V_{in} - r_L \frac{V_{out}}{R}\frac{1}{1-D}\right) \tag{1.106}$$

Applying the volt-seconds balance expression and solving for V_{out}, we find:

$$V_{out} = \left(\frac{V_{in}}{1-D} - V_f\right)\frac{1}{1 + \dfrac{r_L + D\cdot r_{DS(on)}}{R(1-D)^2}} \tag{1.107}$$

Rearranging in the form of a continuous transfer function, we have:

$$M = \frac{1}{D'}\left[\frac{1}{1 + \dfrac{r_L}{RD'} + \dfrac{D}{D'^2}\left(\dfrac{r_L + r_{DS(on)}}{R}\right) + \dfrac{V_f}{V_{out}}}\right] \tag{1.108}$$

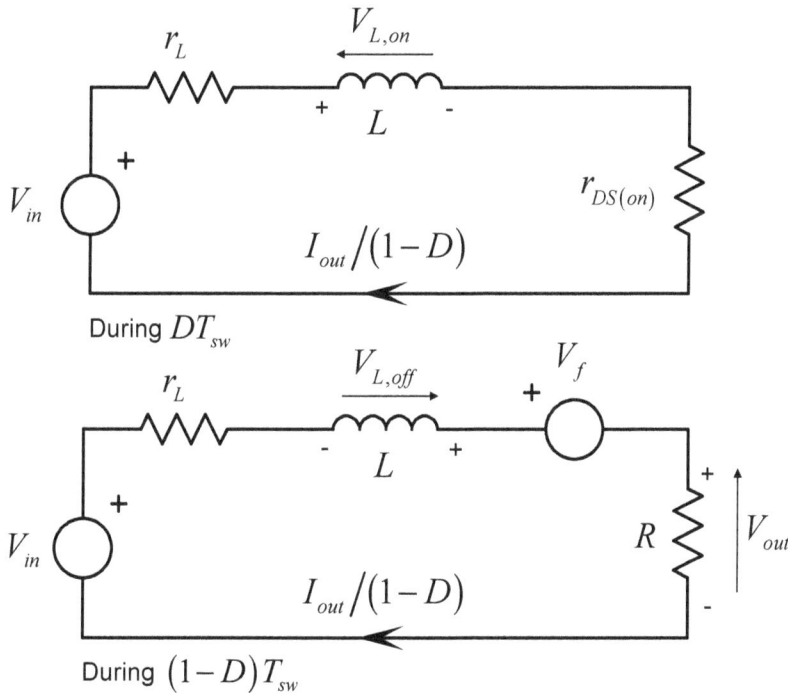

Figure 1.48 The boost converter can also be analyzed with the presence of parasitic terms.

If we plot the continuous transfer function of the boost converter featuring losses, we obtain the chart of Figure 1.49 in which you see the classical latch-up phenomenon.

If you try to push the duty ratio in an attempt to increase V_{OUT} while internal losses are too high, at a certain moment, the output voltage collapses.

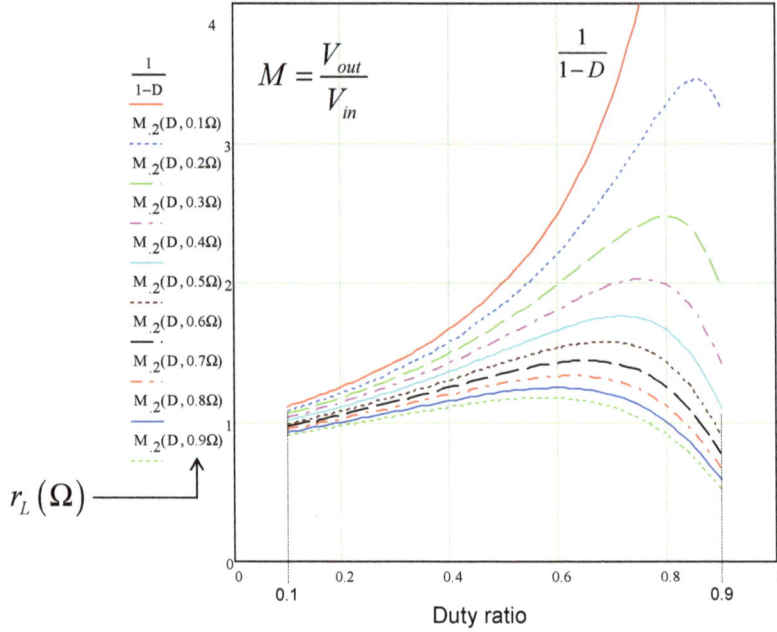

Figure 1.49 The boost converter latches if internal losses are too high.

The explanation is given in Figure 1.50. During the on-time, the switch closes and applies the input voltage through two resistances in series: $r_{DS(on)}$ and r_L. The current rises up linearly until it hits a maximum limit equal to:

$$I_{max} = \frac{V_{in}}{r_{DS(on)} + r_L}$$

(1.109)

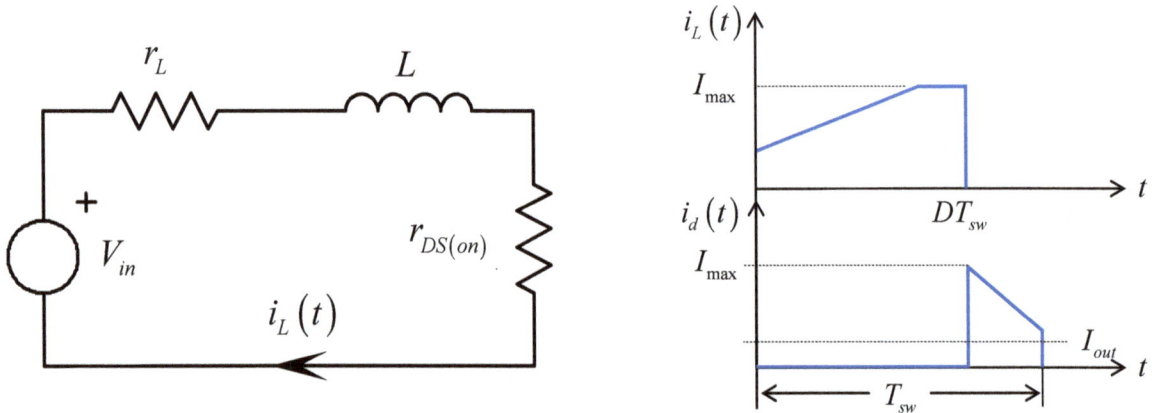

Figure 1.50 The current during the on-time is limited by the resistances in series with the inductor.

Beyond this point, the inductor cannot store more energy. If the load demand exceeds this upper limit while the loop tries to push the duty ratio higher, the output voltage cannot keep up and collapses. It explains the practical limit in the conversion ratio of the boost converter. If all parasitic contributors in (1.108) are made 0, the transfer function returns that of the perfect boost converter described by (1.90).

The buck-boost converter on- and off-sequences are drawn in Figure 1.51.

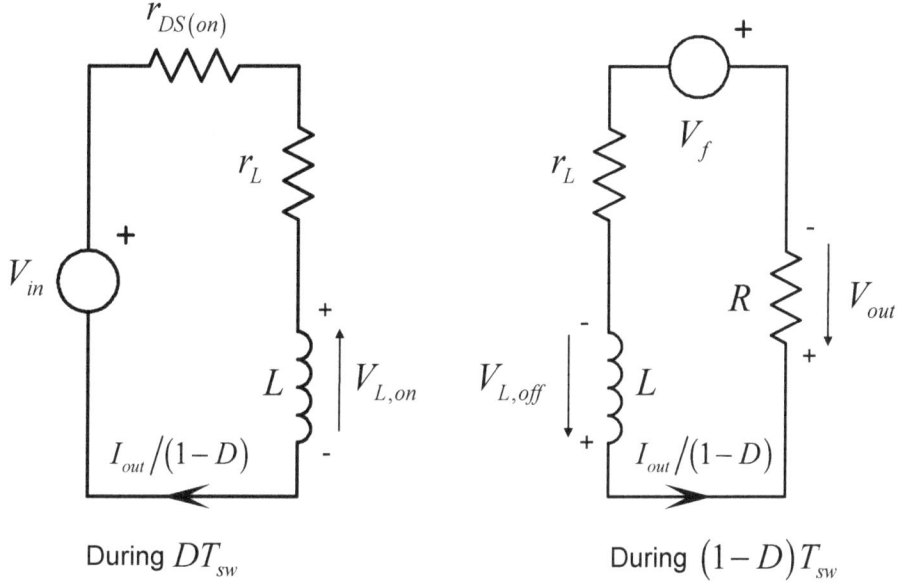

Figure 1.51 Two resistances also limit the maximum peak current in the buck-boost converter during the on-time.

During the on-time, the positive inductor terminal is biased to the input source V_{in} minus two drops while its negative terminal is grounded:

$$V_{L,on} = V_{in} - \frac{V_{out}}{R(1-D)}\left[r_L + r_{DS(on)}\right] \qquad (1.110)$$

When the switch opens, the upper terminal of r_L connects to the load R via the diode forward drop in series. Observing the lower side of Figure 1.51, we can write:

$$V_{L,off} = r_L \frac{V_{out}}{R(1-D)} + V_{out} + V_f \qquad (1.111)$$

Applying the volt-second balance expression and solving for V_{out}, we find immediately the definition of M:

$$M = -\frac{D}{1-D}\left[\frac{1 - \dfrac{V_f}{V_{in}}\dfrac{(1-D)}{D}}{\dfrac{r_L + r_{DS(on)}\dfrac{D}{R(1-D)^2} + 1}{}}\right] \qquad (1.112)$$

The buck-boost also suffers from a latch-up phenomenon if you try to boost the voltage too much (see Figure 1.52). The current in the inductor is also limited by the switch $r_{DS(on)}$ and the inductor ohmic loss as shown in Figure 1.50.

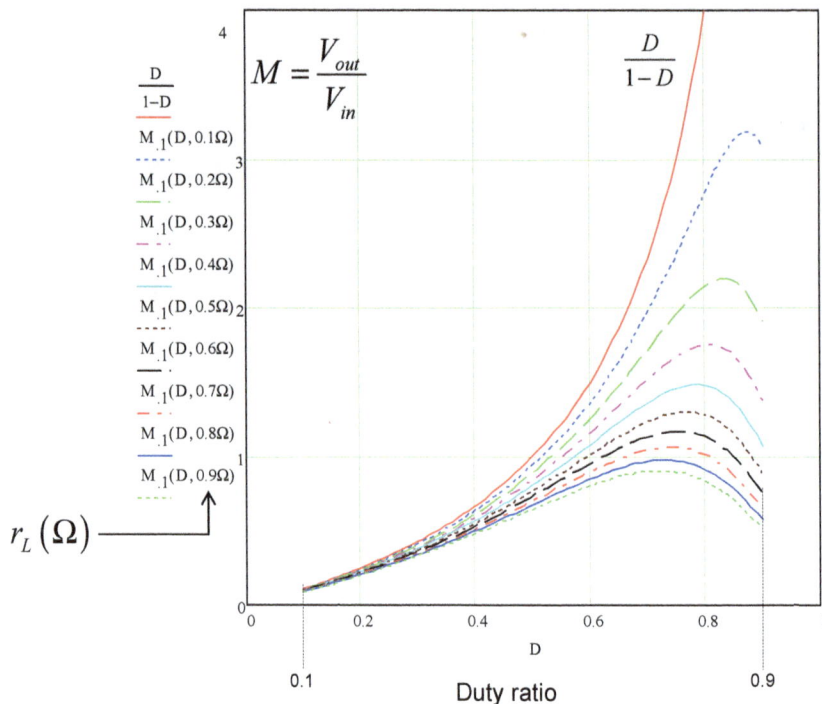

Figure 1.52 Similar to the boost converter, the buck-boost also latches if internal losses are too high.

The transfer functions of these lossy converters operated in CCM are combined in Figure 1.53.

Converter	M_{CCM}
Buck	$D\left[\dfrac{1}{1+\dfrac{r_{DS(on)}}{R}D+\dfrac{r_L}{R}+\dfrac{V_f}{V_{out}}D'}\right]$
Boost	$\dfrac{1}{D'}\left[\dfrac{1}{1+\dfrac{r_L}{RD'}+\dfrac{D}{D'^2}\left(\dfrac{r_L+r_{DS(on)}}{R}\right)+\dfrac{V_f}{V_{out}}}\right]$
Buck-Boost	$-\dfrac{D}{1-D}\left[\dfrac{1-\dfrac{V_f}{V_{in}}\dfrac{(1-D)}{D}}{\dfrac{r_L+r_{DS(on)}D}{R(1-D)^2}+1}\right]$

Figure 1.53 Transfer functions in CCM become more complicated when ohmic drops are considered.

1.2.8 Isolated Versions

The buck and buck-boost converters exist also in isolated versions. A transformer is added to provide galvanic isolation between the primary and secondary grounds but it also offers a convenient scaling factor via its turns ratio $N = N_s/N_p$. Figure 1.54 depicts a single-switch forward converter which is a buck-derived topology. You recognize the buck arrangement combining D_2, L, C and R in a familiar way. However, rather than having D_2 cathode swinging between 0 V and V_{in} as with the classical buck, its swings between 0 V and:

$$V_K\big|_{t_{on}} = N \cdot V_{in} \tag{1.113}$$

A third winding is present to ensure transformer demagnetization cycle by cycle. Multiple variations exist around this converter such as 2-switch forward, half- and full-bridge implementation, phase-shifted full-bridge, push-pull. All originally derive from the non-isolated buck converter.

Figure 1.55 represents a flyback converter. It is one of the most popular converters in the consumer world, particularly in ac-dc applications. It associates a power switch SW with a simple 2-winding transformer. A third winding is often added to provide an auxiliary self-supply for the switching controller. The flyback converter derives from the buck-boost structure and suffers from the same dynamic characteristics like the RHP zero.

When the switch closes, energy builds up in the primary inductance L_p as the secondary-side diode is blocked given the windings dots configuration. When the switch opens, the energy stored in the primary side transfers to the RC network via the diode D during the off-time. Considering a non-ideal coupling between the primary and secondary sides, a clamping network made of V_{clp} and D_{clp} is necessary to protect the switch SW from lethal voltages at turn off. Practically, the clamping source is often made by a resistor paralleled with a capacitor. Again, many variations exist with the flyback converter which can be designed in a 2-switch version, in an active clamp configuration and so on.

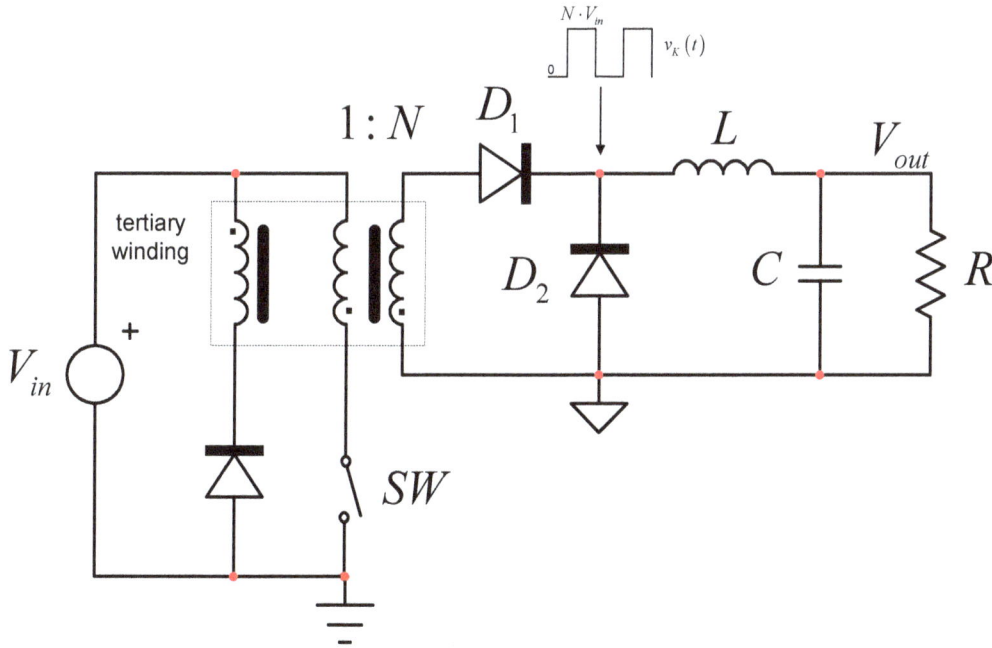

Figure 1.54 When a transformer is added, the buck becomes a forward converter.

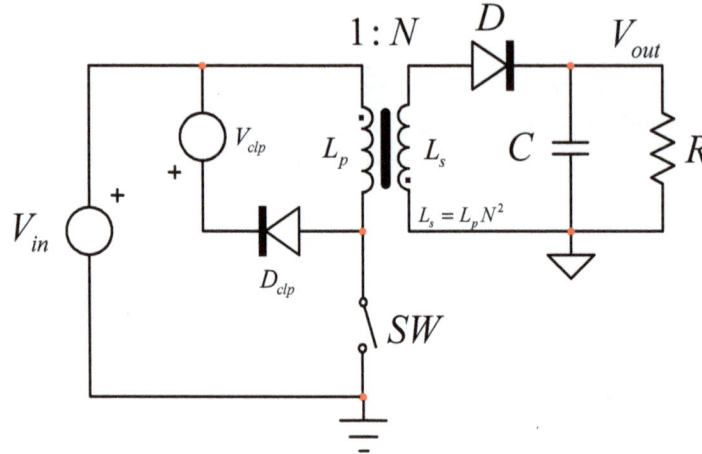

Figure 1.55 The flyback converter derives from an isolated version of the buck-boost converter.

Please note that adding isolation to the buck or the buck-boost increases the voltage stress of the power switch at turn-off. With a third winding featuring a 1:1 turns ratio (maximum duty ratio is smaller than 50%), the primary-side MOSFET must safely endure twice the maximum input voltage.

For a flyback converter, the turn-off voltage (neglecting the leakage inductance term) amounts to approximately $V_{in} + V_{out}/N$ where N is the transformer turns ratio (1:N). A flyback converter operated in an ac-dc adapter operated up to 265 V rms will typically use a 600- or 650-V transistor.

As we did with the isolated versions, we have gathered the static transfer functions in CCM and DCM of these isolated converters in Figure 1.56. Please note that we considered the primary inductance L_p in our flyback formulas. Should you want to consider the secondary inductance L_s instead, both are linked by

$$L_s = N^2 L_p \tag{1.114}$$

Operational and practical design details of these converters are given in [1].

Converter	V_{out}/V_{in} CCM	V_{out}/V_{in} DCM	$R_{critical}$	$L_{critical}$
Forward	ND	$\dfrac{2N}{1+\sqrt{1+\dfrac{8\tau_L}{D_1^2}}}$	$\dfrac{2LF_{sw}}{1-D}$	$\dfrac{(1-D)R}{2F_{sw}}$
Flyback	$\dfrac{ND}{1-D}$	$D_1\sqrt{\dfrac{R}{2L_pF_{sw}}}$	$\dfrac{2L_pN^2F_{sw}}{(1-D)^2}$	$\dfrac{(1-D)^2 R}{2N^2F_{sw}}$ L_p critical

Figure 1.56 Conversion ratios for isolated versions need to include the transformer turns ratio in the equation. Please note that we considered the primary inductance L_p for the flyback converter and the turns ratio disappears after a simplification. $\tau_L = \dfrac{L}{RT_{sw}}$ is the normalized time constant. The transformer turns ratio is $N = N_s/N_p$. L in the forward equations is the output inductor.

1.2.9 Generating the Duty Ratio—The PWM Block

The duty ratio D is a discrete value describing the on-time duration t_{on} with respect to the switching period T_{sw} as shown in Figure 1.25. By driving this variable, you have a means to adjust and control the output variable to your needs. To fulfill this goal, you need to build a square-wave generator whose output pulse width – our t_{on} – is externally adjustable. This is exactly what the *pulse-width modulator* or PWM block does. A simple version appears in Figure 1.57. You can see a comparator fed by an artificial ramp, the sawtooth, whose peak amplitude is V_p. The non-inverting pin of the comparator receives a continuous voltage V_{err}. This error voltage comes from the compensator which permanently observes the regulated variable and strives to minimize the error between target and output values. When the sawtooth is below V_{err}, the comparator output is high. When the sawtooth voltage meets the setpoint given by V_{err}, the comparator output toggles low and turns the power switch off, waiting for a new cycle to come up. Therefore, by adjusting the voltage setpoint, you have a means to shift the toggling point higher in the ramp, naturally increasing the on-time and the duty ratio: this is the principle of the *naturally-sampled pulse-width modulator* from Figure 1.57.

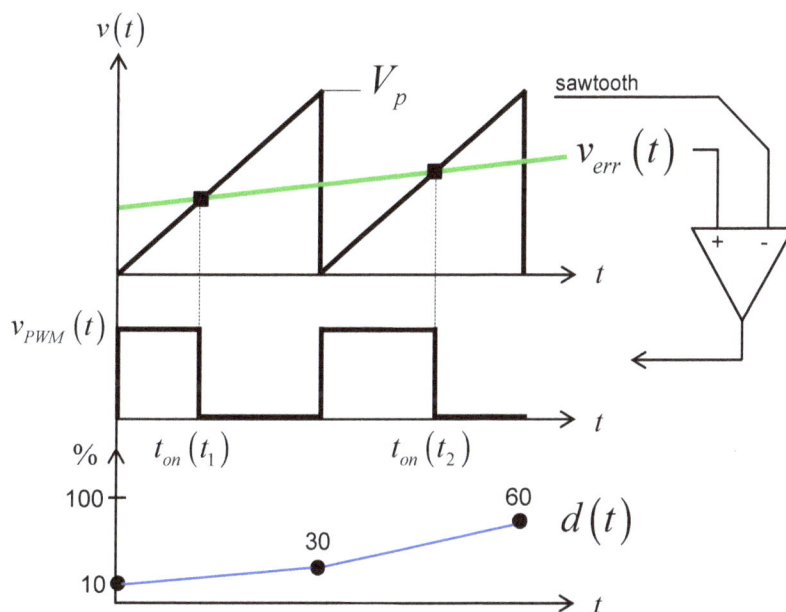

Figure 1.57 A PWM block is built around a simple comparator.

If the duty ratio D represents the control variable of the converter, the error voltage V_{err} *directly* adjusts it via the PWM block to meet the specifications. The sawtooth voltage can be described by the following equation:

$$v_{saw}(t) = V_p \frac{t}{T_{sw}}$$

(1.115)

When the sawtooth voltage meets the error voltage, the comparator toggles low and ends the on-time as illustrated in Figure 1.57. As this event occurs when $v_{err}(t_1) = v_{saw}(t_1)$, we can write:

$$V_{err@t=t_1} = V_p \frac{t_{on}}{T_{sw}}$$

(1.116)

As the right-side term is the duty ratio, the above equation becomes:

$$v_{err}(t) = V_p d(t) \qquad (1.117)$$

which can be rearranged as:

$$d(t) = \frac{v_{err}(t)}{V_p} \qquad (1.118)$$

In the above discrete expression, v_{err} modulates the duty ratio at a period much larger than that of the switching period. The numerous discrete duty ratio points which occur at each high-frequency switching cycle when observed at the low-frequency modulation pace look almost contiguous and can be approximated as a time-continuous ripple-free function. This is a mathematical abstraction which allows us to obtain a time-continuous equation valid up to half of the switching frequency.

Figure 1.58 illustrates this principle.

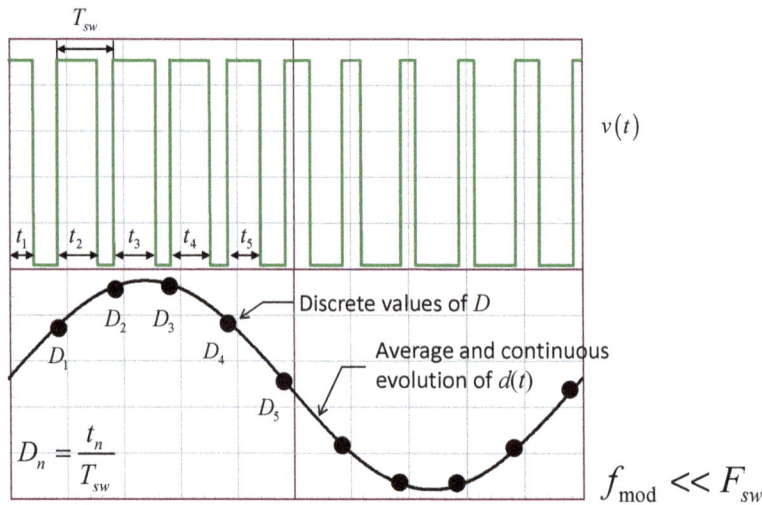

Figure 1.58 When the modulation frequency is much smaller than the switching period, the discrete points look contiguous and the average behavior is a time-continuous function.

We can now perturb (1.118) and obtain:

$$D_0 + \hat{d}(t) = \frac{V_{err0} + \hat{v}_{err}(t)}{V_p} \qquad (1.119)$$

In which D_0 and V_{err0} are static values. Sorting out the ac equation alone immediately gives us the small-signal gain of the PWM block:

$$G_{PWM} = \frac{\hat{d}(t)}{\hat{v}_{err}(t)} = \frac{1}{V_p} \qquad (1.120)$$

As an example, if the peak of the sawtooth is 2 V, the gain is 0.5 or -6 dB. In simulations or in analytical analysis, the gain block comes right after the compensator and ensures the conversion of the error voltage into a duty ratio driving the model. In Figure 1.59, you would multiply V_{err} by 0.5 before driving the D input of the PWM switch model. In our simulation, a duty ratio of 100% will lead to a 1-V bias value for the D node. If further to a bias point calculation you read 545 mV for node D, then the duty ratio is 54.5%.

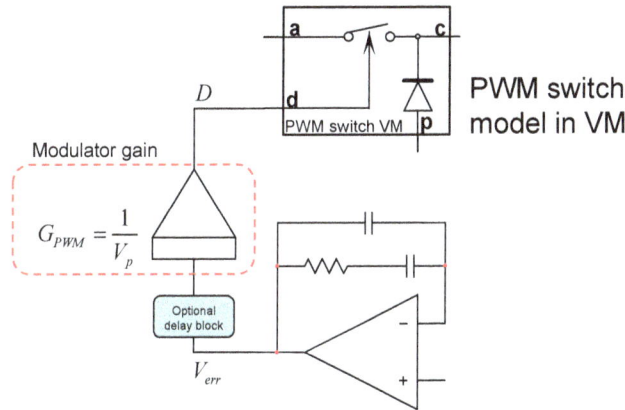

Figure 1.59 The gain block is inserted right after the compensator.

1.2.10 Frequency Response of the PWM Block

The frequency response of the naturally-sampled pulse width modulator is flat in magnitude and in phase as demonstrated in [3]. Measuring this dynamic response would require a specific equipment that most of us do not have. A simulator like SIMPLIS® can perform such analysis. SIMPLIS® features a *Piece-Wise Linear* (PWL) simulation engine which breaks a nonlinear curve into linear segments as shown in Figure 1.9. It can thus extract the dynamic response of a switching circuit without resorting to an average model. Figure 1.60 shows the circuit diagram and the various probes we have installed. The static duty ratio D_0 is 25% and the dynamic response is obtained by ac-sweeping the error voltage via the V_{mod} source.

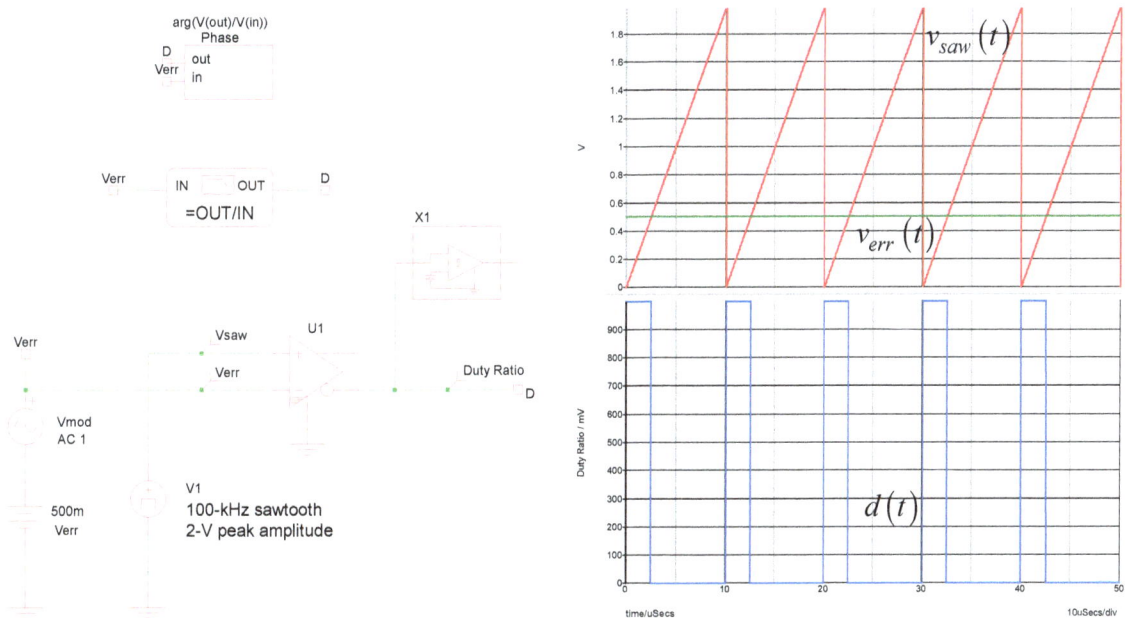

Figure 1.60 A simple PWM block is captured in SIMPLIS® which lets you extract its dynamic response.

The dynamic response appears in Figure 1.61 and confirms the flat frequency response with a 6-dB attenuation.

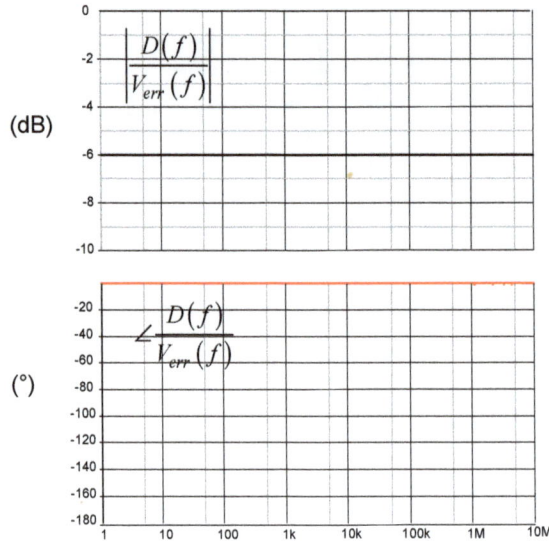

Figure 1.61 The response of the naturally-sampled PWM block is flat in magnitude and phase.

In this simulation, the comparator features a 1-ps propagation delay t_p which is unrealistic. A typical fast comparator will exhibit a delay of 100 ns or so. However, this delay is highly dependent on the internal bias currents. In integrated circuits where standby or light-load efficiency matter, it is very possible that these bias currents be scaled as the delivered power changes.

For instance, in heavy-load condition, small biases do not count in the consumption budget and efficiency remains unaffected. As the converter operates in light- or no-load conditions, a mechanism could certainly adjust the bias currents to save some extra mW, especially if the switching frequency has been folded back as in a lot of modern controllers. What happens if the reduced bias currents incur a larger propagation delay, say 1 µs or so? Figure 1.62 is eloquent and shows how the flat phase response now exhibits a lag which increases as we slide along the frequency axis.

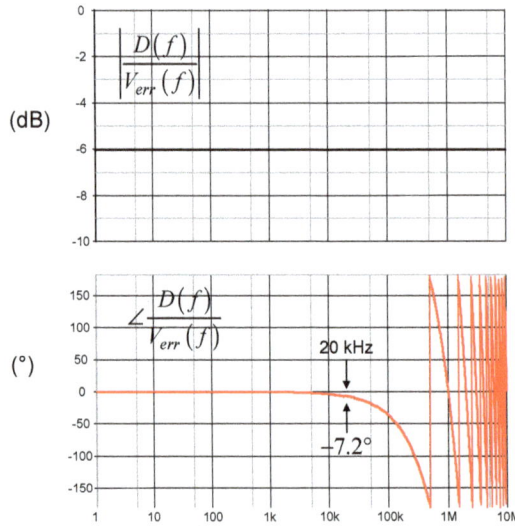

Figure 1.62 If the propagation delay increases, the phase response is no longer flat and lags in high frequencies.

Assume you have selected a 20-kHz crossover frequency and compensated your circuit for $60°$ phase margin, then accounting for this extra delay may reduce the margin by $7.2°$. Of course, this example deals with a 100-kHz switching frequency and conversion delays of hundred of ns are unlikely to affect your design and you can neglect them.

It is no longer the case for switching frequency of a few MHz and crossover points of 100 kHz or more. These are common figures found in cell phones dc-dc converters for instance. In these applications, it is important to account for the pure delay brought by the conversion chain. In this example, we have shown a PWM block but the delay could also come from a digital section featuring a conversion time. A pure delay can be modeled by a delay line as shown in Figure 1.63. The transfer function of such a block is:

$$H_1(s) = e^{-s \cdot \tau} \tag{1.121}$$

in which τ represents the delay inherent to the considered conversion chain. As shown in [2], the magnitude of this expression is 1 or 0 dB and the delay is equal to:

$$\varphi = -\omega \tau \tag{1.122}$$

What is the phase lag at 20 kHz for a 1-µs delay?

$$\varphi_{20kHz} = -2\pi \times 20k \times 1u = 0.1256 \times \frac{180}{\pi} = -7.2° \tag{1.123}$$

Such delay must be included in the loop gain but the exponential term is impractical to manipulate in Laplace analysis. One way to approximate a pure delay in the Laplace domain is to use a 1st-order Padé approximant.

We can show that combining a right half-plane zero and the left half-plane pole placed at $\omega_z = \omega_p = \dfrac{2}{\tau}$ is a good way to model a delay [2]:

$$H_2(s) = \frac{1 - \dfrac{s}{\omega_z}}{1 + \dfrac{s}{\omega_p}} \tag{1.124}$$

With a 1-µs delay, the pole/zero pair is placed at:

$$f_z = f_p = \frac{1}{\pi \cdot \tau} = \frac{1}{3.14159 \times 1u} = 318.3 \text{ kHz} \tag{1.125}$$

The phase of (1.124) is computed at 20 kHz and gives:

$$\varphi = -\tan^{-1}\left(\frac{f}{f_z}\right) - \tan^{-1}\left(\frac{f}{f_p}\right) = -\tan^{-1}\left(\frac{20k}{318.3k}\right) - \tan^{-1}\left(\frac{20k}{318.3k}\right) = -7.19° \tag{1.126}$$

which is very close to our $-7.2°$ measured in Figure 1.62.

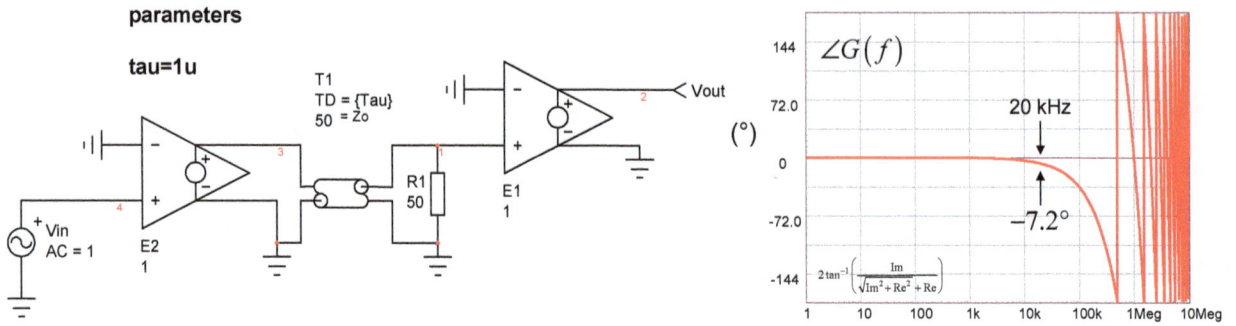

Figure 1.63 A delay line is the perfect SPICE primitive to model a pure delay in the conversion chain.

When more precision is needed, for instance in higher frequencies, the following expression gives good results by combining a second-order polynomial form in the denominator with a zero at the origin:

$$H_3(s) = 1 - \frac{s \cdot \tau}{1 + \dfrac{s}{Q\omega} + \left(\dfrac{s}{\omega}\right)^2} \tag{1.127}$$

with $Q = \dfrac{2}{\pi}$ and $\omega = \dfrac{\pi}{\tau}$. If your SPICE simulator does not handle delay line properly, you can still emulate (1.127) with a few passive elements as shown in Figure 1.64 which compares the response of the delay line (bottom) to that of the op-amp-based structure (top).

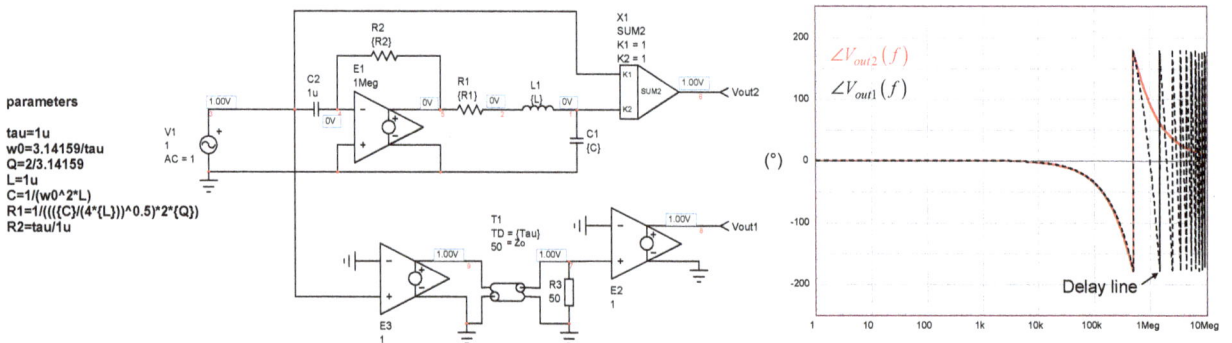

Figure 1.64 In lack of delay line in your simulator, combining an op-amp with a zero at the origin and a 2nd-order network lets you approximate a pure delay quite well.

The block you have selected to model the delay will be inserted in series with the PWM block in Figure 1.59.

1.2.11 From a Static to a Quasi-Static Transfer Function

Figure 1.56 details expressions linking the output voltage V_{out} to the control variable, the duty ratio D. These are the dc transfer functions of the considered perfect converter when operating in CCM or in DCM. If you now build a prototype and power it, by adjusting the duty ratio via a PWM block and a dc source, you have a means to set V_{out} to the desired value. Assume you have assembled a CCM-operated boost converter powered from a 10-V source as illustrated in Figure 1.65. This is a simplified open-loop converter delivering 15 W to a resistive load.

The duty ratio is adjusted so that V_{out} is 15 V. According to (1.90), the duty ratio should be set to:

$$D = \frac{V_{out} - V_{in}}{V_{out}} = \frac{15 - 10}{15} = 33.3\% \tag{1.128}$$

In practice, the duty ratio is adjusted to 36% as we must compensate the various losses already mentioned in this chapter. To get this value, the error voltage (a simple dc source) is set to 720 mV considering a 2-V peak sawtooth. With a 10-µs switching period, the MOSFET t_{on} is thus 3.6 µs. If we would now ac-modulate the error voltage V_{err} around its 720-mV dc bias, we would be able to extract the dynamic response of the power stage, also called the *control-to-output transfer function*. This transfer function is made of a *quasi-static gain* H_0 (determined for $s = 0$) followed by a nth-order polynomial form:

$$H(s) = \frac{V_{out}(s)}{V_{err}(s)} = H_0 \frac{N(s)}{D(s)} \tag{1.129}$$

The term quasi-static is employed here because it differs from the static transfer function linking V_{out} to V_{in} which is D for a buck converter. Here, we talk about a dynamic law relating a small variation of the output voltage to a small variation of the control input, V_{err}:

$$H_0 = \frac{dV_{out}(V_{err})}{dV_{err}} \tag{1.130}$$

Before going through a complete small-signal analysis, we can already determine this quasi-static gain in a very simple way:

1. Bias the error voltage to a level V_{err1}, e.g. 720 mV, and measure V_{out1}. We read 14.95 V.
2. Slightly change the error voltage to V_{err2} so that a measurable change occurs in V_{out}. In this example, $V_{err2} = 700$ mV led to an output voltage V_{out2} of 14.73 V.
3. The quasi-static gain is simply $H_0 = \dfrac{V_{out1} - V_{out2}}{V_{err1} - V_{err2}} = \dfrac{14.95 - 14.73}{720m - 700m} = 11$ or ≈ 20.8 dB.

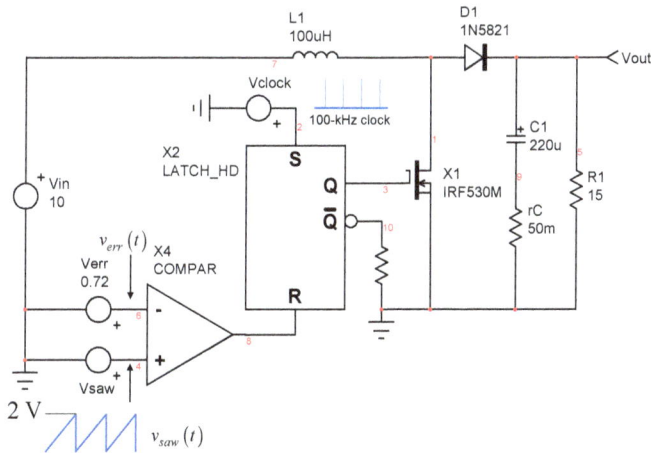

Figure 1.65 This simple boost converter delivers a 15-V output when the duty ratio is adjusted to 36%.

By running the above exercise, you have performed a differentiation around the operating point as already described in Figure 1.8. By building a simple prototype and without resorting to a frequency response analyzer (FRA), you can already measure the quasi-static gain of your converter. This is extremely useful to compare the first results delivered by the FRA and check if the dc part lies in the ballpark of what is expected.

The important lesson of this experiment is that you can determine the quasi-static gain of a given power stage if you know its dc conversion ratio or *static transfer characteristic*. With the CCM boost converter, the convertion ratio M is given by:

$$M = \frac{V_{out}}{V_{in}} = \frac{1}{1-D} \tag{1.131}$$

If you apply (1.130) to this expression, you will determine the power stage quasi-static gain H_{PS} linking V_{out} to D:

$$H_{PS} = \frac{dV_{out}(D)}{dD} = \frac{d}{dD}\left(\frac{V_{in}}{1-D}\right) = \frac{V_{in}}{(1-D)^2} \tag{1.132}$$

If we now include the pulse-width modulator gain which is 0.5 ($V_p = 2$ V), the above formula leads to a theoretical control-to-output gain of:

$$H_0 = \frac{dV_{out}(V_{err})}{dV_{err}} = G_{PWM}H_{PS} = \frac{1}{V_p}\frac{V_{in}}{(1-D)^2} = \frac{1}{2}\times\frac{10}{(1-0.333)^2} = 11.2 \tag{1.133}$$

which is not far from the measured value of 11. The formula given in (1.131) does not include the contribution of the various losses and explains the small discrepancy you will always have between the theoretical calculation and the practical measurement.

Building on this observation, we can now determine the quasi-static gain for the buck converter:

$$H_{PS} = \frac{dV_{out}(D)}{dD} = \frac{d}{dD}DV_{in} = V_{in} \tag{1.134}$$

which is scaled by modulator gain to form the control-to-output transfer function in CCM:

$$H_0 = \frac{dV_{out}(V_{err})}{dV_{err}} = G_{PWM}H_{PS} = \frac{V_{in}}{V_p} \tag{1.135}$$

For the buck-boost, repeat the operation:

$$H_{PS} = \frac{dV_{out}(D)}{dD} = \frac{d}{dD}\frac{DV_{in}}{1-D} = \frac{V_{in}}{(1-D)^2} \tag{1.136}$$

which is similar to the boost converter gain. When we scale the result by the PWM gain, we have:

$$H_0 = \frac{dV_{out}(V_{err})}{dV_{err}} = G_{PWM}H_{PS} = \frac{V_{in}}{V_p(1-D)^2} \tag{1.137}$$

Figure 1.66 gathers all the quasi-static control-to-output gains H_0 of the main switching converters operated in CCM. The DCM expressions will be derived in the dedicated converter section later in the book.

Converter	H_0
Buck	$\dfrac{V_{in}}{V_p}$
Boost	$\dfrac{V_{in}}{V_p(1-D)^2}$
Buck-Boost	$\dfrac{V_{in}}{V_p(1-D)^2}$
Forward	$\dfrac{NV_{in}}{V_p}$
Flyback	$\dfrac{NV_{in}}{V_p(1-D)^2}$

Figure 1.66 These expressions are the quasi-static control-to-output gains of the main converters when operated in CCM.

1.2.12 Adding Feedforward to the PWM Block

From these expressions, you see how the dc gain depends on the input voltage V_{in}. Changes in the input voltage will shift the power plant ac response up and down, affecting the crossover frequency and most likely stability. One way to make this input voltage dependency disappear is to make the artificial ramp amplitude of the PWM block dependent on V_{in} also:

$$V_p = k_{FF} \cdot V_{in} \tag{1.138}$$

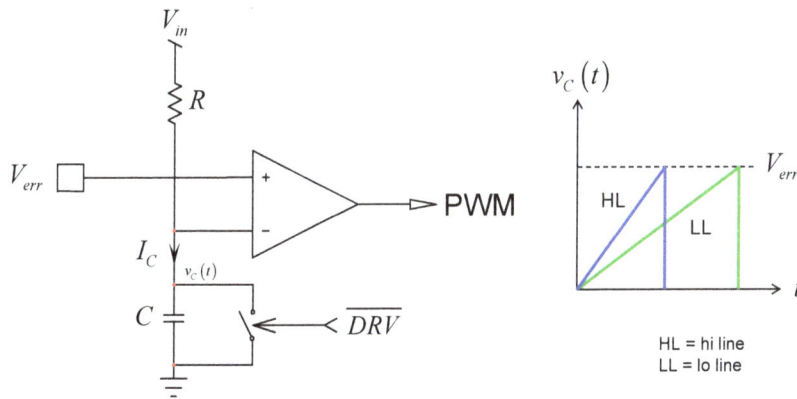

Figure 1.67 If the peak amplitude now depends on the input voltage, you implement feedforward control.

59

In this case, the dc gain of the CCM buck converter operated in voltage mode becomes:

$$H_0 = \frac{V_{in}}{k_{FF} \cdot V_{in}} = \frac{1}{k_{FF}} \tag{1.139}$$

and the dependency to V_{in} is gone. This is called *feedforward* and it is a known technique to improve the input line rejection capability of voltage-mode converters: before the line change effects propagate through the control system and impact V_{out}, the correction is immediate as the gain instantaneously changes via the V_p modulation. From Figure 1.67, assuming a fairly low ramp amplitude compared to V_{in}, we can say:

$$I_C \approx \frac{V_{in}}{R} \tag{1.140}$$

The peak value V_p is linked to the circuit time constant τ by:

$$V_p = \frac{I_C}{C_{ramp}} T_{sw} = \frac{V_{in}}{C_{ramp} R_{ramp}} T_{sw} = \frac{V_{in}}{\tau F_{sw}} \tag{1.141}$$

In the time domain, the ramp peak voltage change is described by:

$$v_{ramp}(t) = V_p \frac{t}{T_{sw}} = \frac{V_{in}}{\tau F_{sw}} \frac{t}{T_{sw}} \tag{1.142}$$

For $t = t_{on}$, the error voltage equals V_{ramp}. Therefore:

$$V_{err} = \frac{V_{in}}{\tau F_{sw}} \frac{t_{on}}{T_{sw}} = \frac{V_{in}}{\tau F_{sw}} D \tag{1.143}$$

Which implies:

$$D(V_{err}) = V_{err} \frac{\tau F_{sw}}{V_{in}} \tag{1.144}$$

If we differentiate this expression with respect to V_{err}, we obtain the new small-signal gain of this block:

$$G_{PWM} = \frac{\partial D(V_{err})}{\partial V_{err}} = \frac{\tau F_{sw}}{V_{in}} = \frac{1}{k_{FF} V_{in}} \tag{1.145}$$

in which:

$$k_{FF} = \frac{1}{F_{sw}\tau} \tag{1.146}$$

The implementation is very similar to that of Figure 1.59 and requires a simple analog behavioral model (ABM) source as illustrated in Figure 1.68.

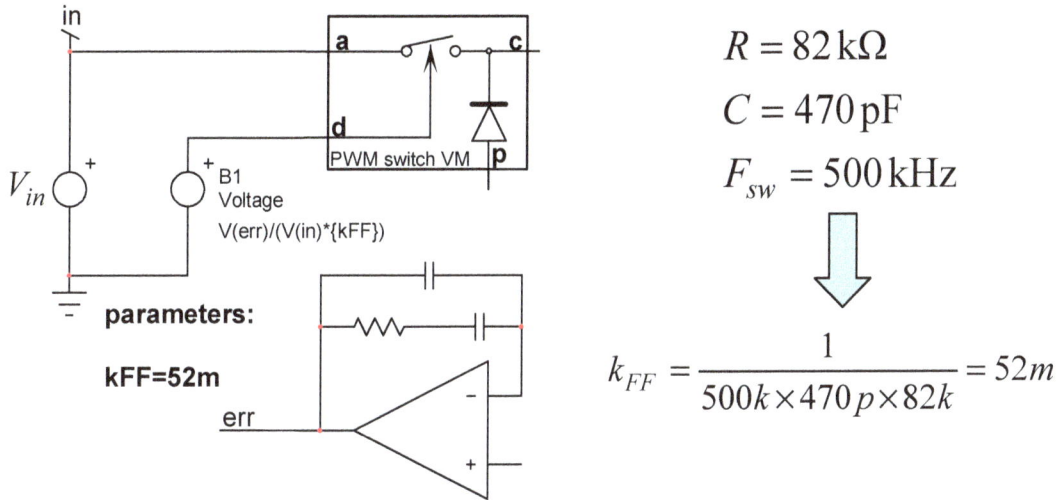

Figure 1.68 A simple ABM source lets you model the feedforward control contribution to your average model.

1.2.13 Voltage-Mode and Current-Mode Control

The switching converters we have reviewed are controlled by the duty ratio D. When the error voltage V_{err} directly drives the duty ratio via the pulse-width modulator, we talk about *direct duty ratio* control or *voltage-mode* control. The two terms designate the same control architecture: an error voltage representative of the output variable deviation from its setpoint is permanently compared to an artificial ramp to elaborate a duty ratio D. The typical internals of a voltage-mode or VM controller are given in Figure 1.69.

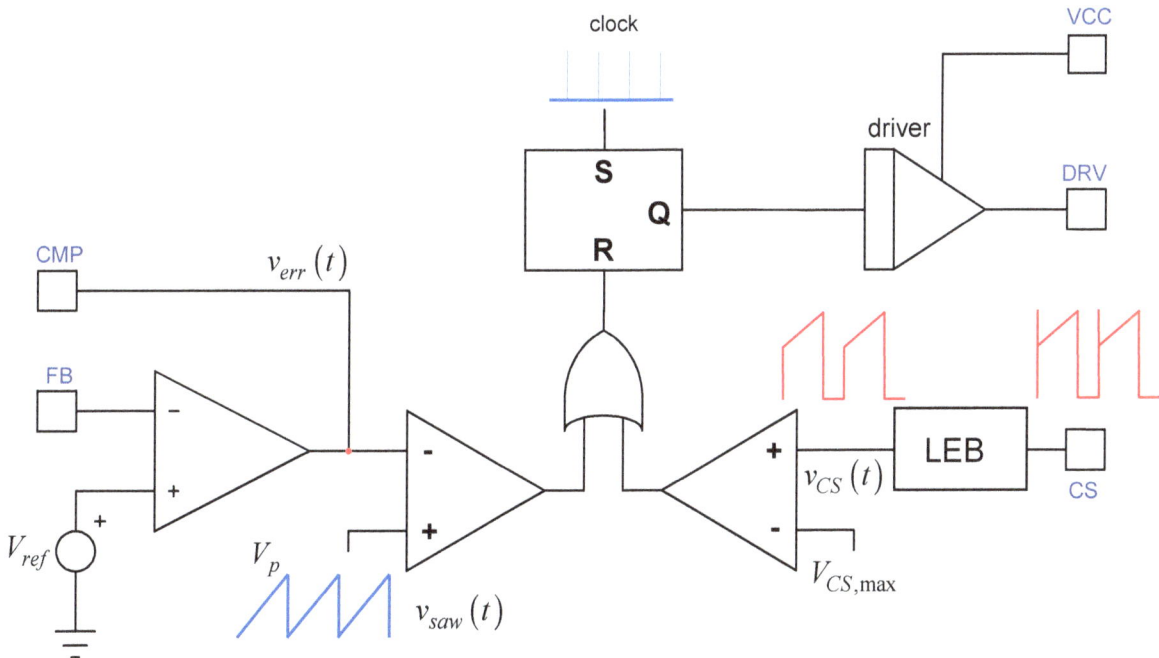

Figure 1.69 In voltage-mode control, the error amplifier output is compared to the artificial sawtooth.

Rather than directly driving a power stage, the PWM comparator resets a latch set by a clock signal.

That way, when the clock instructs the driver to go high and drives the transistor (a MOSFET or a bipolar), the information is latched: even if a glitch or spurious event occurs at turn-on, the output pulse is not affected and the drive output remains high.

This is a so-called *double-pulse suppression* scheme which prevents extra transitions and additional switching losses during a switching cycle.

The circuitry is supplemented with a maximum current comparator as voltage-mode control is blind to the inductor current. Indeed, the error compensator instructs the modulator to deliver a given duty ratio to meet the output target.

However, if a fault like a short circuit or an overload occurs, the loop is very likely to request a large duty ratio which may bring an over-current condition, potentially destroying the converter.

To prevent this dangerous situation, a comparator permanently monitors the cycle-by-cycle current via the current sense pin (CS) and immediately turns the power switch off in case of too high a current. If the $V_{CS,max}$ source is set to 1 V while you sense the current via a 0.5-Ω resistance, the comparator will trip when the current exceeds 2 A.

During normal operation, this comparator is silent.

You can observe a block labeled LEB in series with the current sense pin. It stands for *leading-edge blanking*. Its role is to blind the comparator input for a small period of time after the MOSFET has been turned on — a few hundreds of ns or less — so that spurious spikes incurred to parasitics or diode t_{rr} do not false-trip the comparator.

This is a very common block found in the vast majority of modern switching controllers. You can see the spike in the current waveform shown in the right-side of Figure 1.69 and it is gone because of the LEB action when it appears at the current sense comparator input.

In high-current applications, a current-sense transformer is used instead of a resistance as power dissipation and efficiency can be at stake with a resistive element. Figure 1.70 shows the principle of operating a current sense transformer.

The current flows in the 1-turn primary and the secondary current is scaled by the turns ratio. Considering as an example a 1:100 turns ratio, the secondary current will be that of the primary divided by 100.

If a 10-A current circulates in the primary and the secondary-side resistance R_1 is 10 Ω, then the secondary-side voltage applied at the current sense comparator input is:

$$V_s = \frac{I_L}{N} R_1 = \frac{10}{100} \times 10 = 1 \text{ V} \tag{1.147}$$

which is equivalent to a sense resistance of:

$$R_{sense} = \frac{V_{sense}}{I_L} = \frac{1 \text{ V}}{10 \text{ A}} = 0.1 \, \Omega \tag{1.148}$$

You will pass this value as a parameter to the current-mode model during the small-signal analysis of a power stage featuring a current sense transformer.

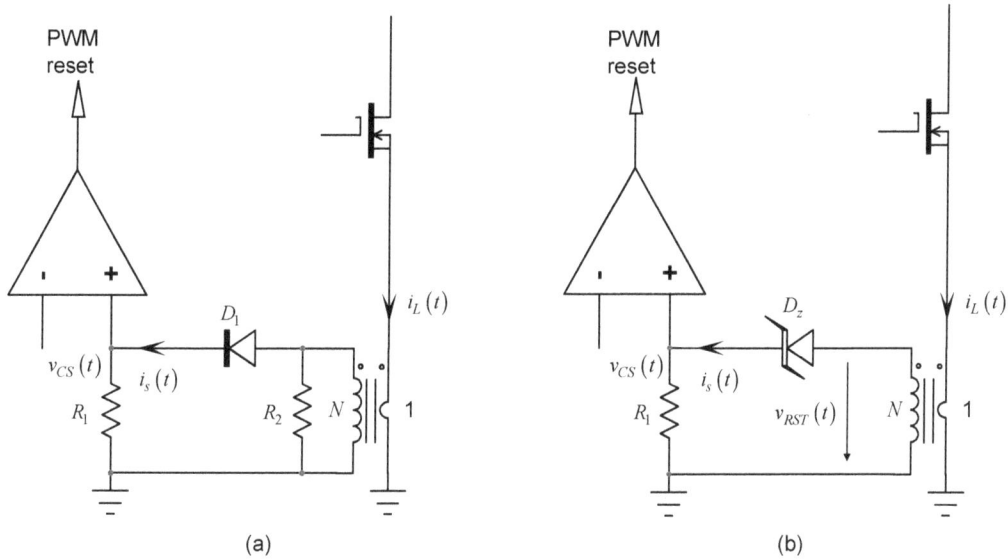

Figure 1.70 In high-current applications, a sense resistor is impractical so designers resort to a current sense transformer. Two possible reset schemes exist: with a simple resistance (R_2) or with a Zener diode.

In sketch (a), R_2 and D_1 in the circuit ensures the transformer demagnetization cycle by cycle by letting the secondary-side voltage swing negative when the power transistor turns off. Adding a Zener diode to ensure transformer reset is also a popular approach and is proposed in sketch (b).

Similarly, some integrated circuits include a current amplifier as illustrated in Figure 1.71 where current sensing requires a differential measurement. The UC1846, for instance, features such amplifier with a gain of 3.

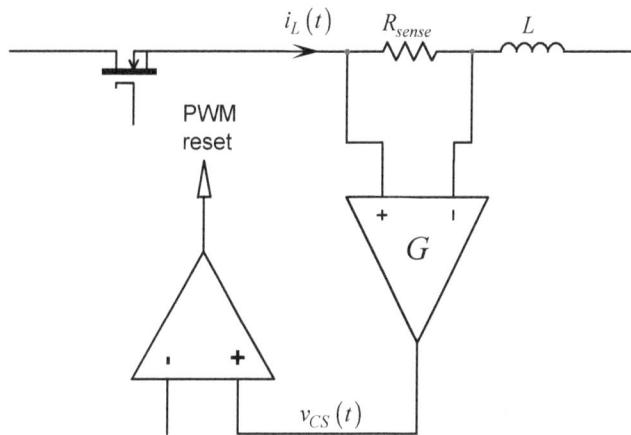

Figure 1.71 In some parts, the current sense information is amplified. This is often found in differential sensing.

The current read by the controller is the sense voltage multiplied by the gain G:

$$v_{CS}(t) = i_L(t) R_{sense} G \qquad (1.149)$$

The equivalent current sense resistance is therefore defined as:

$$R_{eq} = \frac{v_{CS}(t)}{i_L(s)} = R_{sense}G \tag{1.150}$$

This is the value you will use in the small-signal analysis of a circuit featuring an amplifier in the current loop.

Current-Mode Control, sometimes abbreviated CMC or simply CM, uses a power stage similar to that of a voltage-mode converter. Actually, should you observe switching waveforms coming from a given converter, you could not tell if this is a voltage- or a current-mode control type of scheme. And if you could change the control technique by the flip of a switch, you would observe a similar duty ratio at a given operating point whether voltage- or current-mode control is at work. What is modified though, is the way the duty ratio is elaborated. In voltage mode, the error voltage is compared to an artificial ramp as illustrated in Figure 1.57. In current mode, the error voltage is compared to the instantaneous inductor current $i_L(t)$ and sets the peak inductor value at which the power switch turns off. Figure 1.72 shows the basic architecture of a current-mode control in which you recognize the same latch as before but the reset of this element now occurs when the peak inductor current matches the setpoint imposed by the error voltage. As such, the peak current is equal to the error voltage divided by the current sense resistor labeled R_{sense} (also found as R_i in the literature):

$$i_{peak}(t) = \frac{v_{err}(t)}{R_{sense}} \tag{1.151}$$

This fact is illustrated by Figure 1.73 in which you see how the control voltage v_{err} or v_c *directly* modulates the inductor peak current. As such, the duty ratio is *indirectly* set by the control voltage, unlike in voltage-mode control. In DCM, equation (1.49) relates the peak current with the on-time duration:

$$i_{peak}(t) = S_{on}t_{on}(t) = S_{on}d(t)T_{sw} \tag{1.152}$$

The instantaneous duty ratio in a DCM current mode converter is thus defined as:

$$d_{DCM}(t) = \frac{v_{err}(t)}{S_{on}R_{sense}T_{sw}} \tag{1.153}$$

In CCM, the valley current I_v should be considered to define the duty ratio:

$$i_{peak}(t) = I_v + S_{on}t_{on}(t) = I_v + S_{on}d(t)T_{sw} \tag{1.154}$$

leading to:

$$d_{CCM}(t) = \frac{1}{S_{on}T_{sw}}\left(\frac{v_{err}(t)}{R_{sense}} - I_v\right) \tag{1.155}$$

In practical implementations, R_{sense} can take on very low values for efficiency reasons. Thus, setting the peak current setpoint directly via the error voltage would be rather impractical. Assume a $0.1\text{-}\Omega$ sense resistance value and a current setpoint of 3 A. To meet this target, V_{err} should be as low as 300 mV, bringing the op-amp output to almost its minimum level. For this reason, designers often insert a division block k_{div} between the op-amp output and the current sense comparator. If a ratio of 0.25 is implemented, the previous 300 mV become 1.2 V which is more comfortable for the op-amp and less susceptible to noise pollution. This division ratio will have to be included in the control-to-output transfer function of a converter in which the control section includes a division block. For instance, the classical UC384x controller features a division ratio of 0.33.

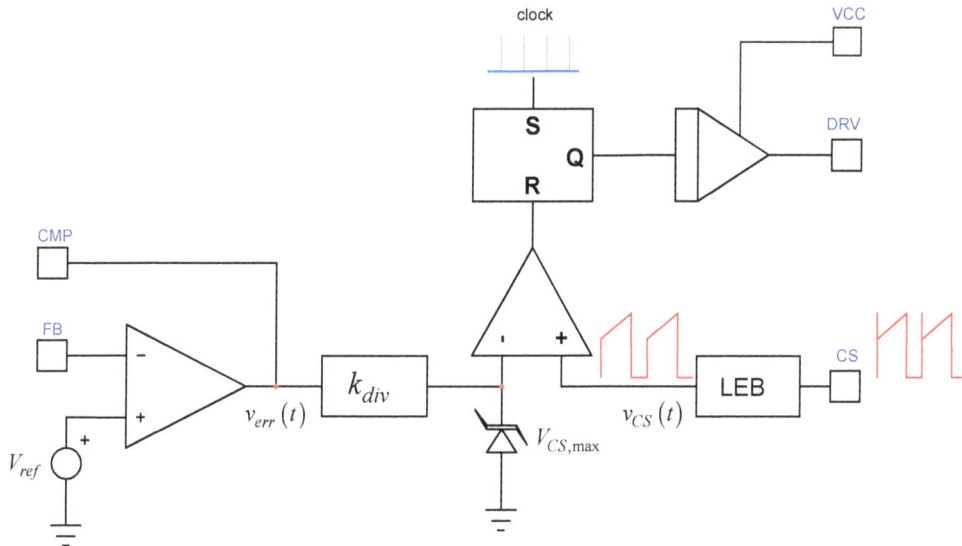

Figure 1.72 In a current-mode control circuit, the error voltage sets the inductor peak current cycle by cyle.

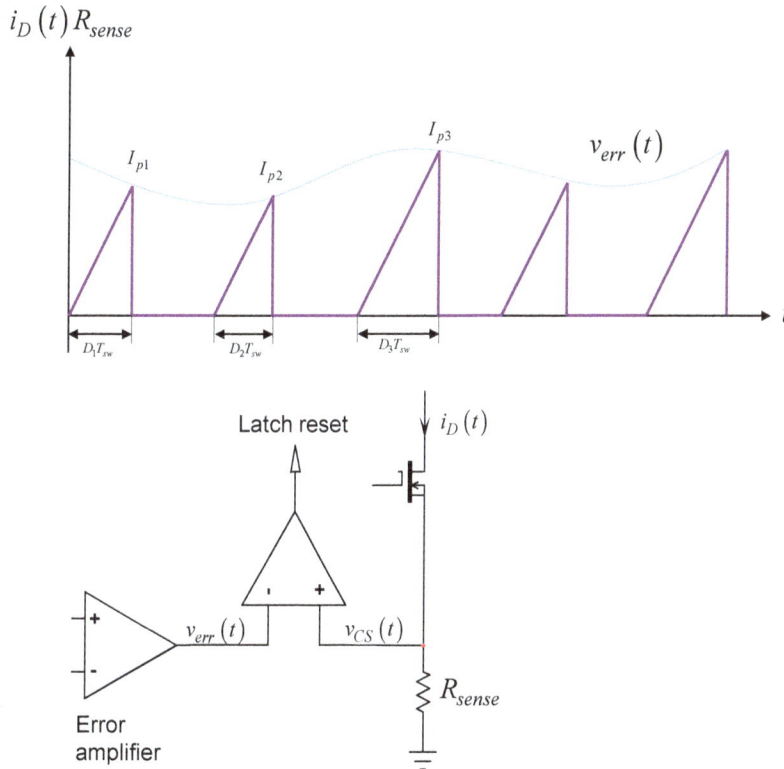

Figure 1.73 In current-mode control, the error voltage no longer *directly* controls the duty ratio but the inductor peak current.

In Figure 1.72, you see a clamp placed after the divider.

This clamp limits the maximum acceptable peak current when the loop is open in a fault condition or simply during the start-up sequence. When the feedback information is missing, the op-amp will rail up trying to push the current setpoint to its maximum value.

The maximum output the op-amp can deliver is not guaranteed with precision: from 5 to 6.2 V for the UC3845. The maximum current limit, on the other hand, must be set with a great precision. It represents an important data-sheet parameter and its precision is typically ±10% or even better in some cases. In a UC384x, the maximum authorized peak sense voltage is 1.1 V (the maximum of the data-sheet specifications).

You will thus select an inductor or a transformer knowing that the highest current it will endure will be limited to 1.1 V across the current sense element to which you must add all the delay affecting the final peak value [1].

It is well-known that current-mode converters are subject to subharmonic instabilities when they operate in continuous conduction mode with a duty ratio approaching 50% [5]. The upper-side of Figure 1.74 shows how a small perturbation in the inductor current propagates through switching cycles. When the duty ratio is less than 50%, the perturbation naturally dies out after a few switching events and the response is asymptotically stable.

When the same current glitch occurs in a CCM converter operated with a duty ratio greater than 50%, the perturbation grows and engenders a completely erratic operation leading to a chaotic behavior known as *subharmonic oscillation*.

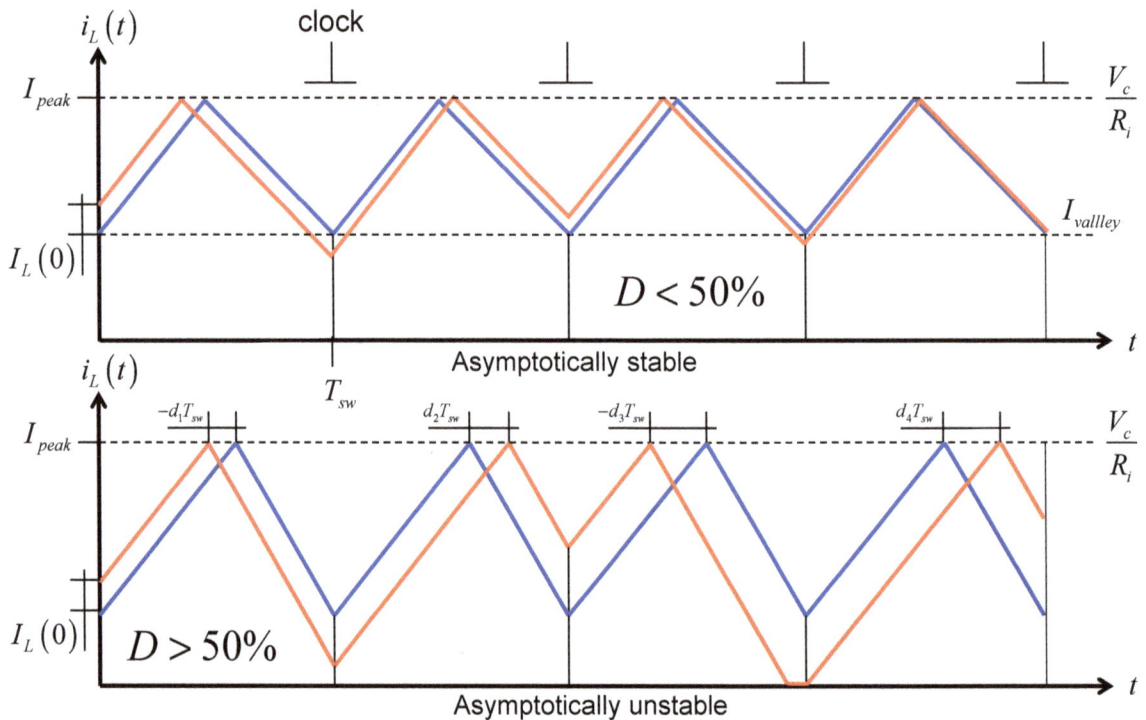

Figure 1.74 Instabilities can occur when a perturbation occurs in a converter operating near or beyond 50% of duty ratio.

It has been shown after Dr. Ridley's work who applied sampled data analysis to the current loop that this instability was usually approximated by a pair of double RHP zeroes located at $F_{sw}/2$. When the outer voltage loop closes around the current loop, these RHP zeros turn into a pair of complex poles tuned at $F_{sw}/2$. Mathematically, this behavior can be approximated by a second-order polynomial form taking place in the denominator of the control-to-output transfer function describing a CCM-operated current-mode control converter.

It is in the form of:

$$D(s) = 1 + \frac{s}{\omega_n Q} + \left(\frac{s}{\omega_n}\right)^2 \tag{1.156}$$

with:

$$\omega_n = \frac{\pi}{T_{sw}} \tag{1.157}$$

$$Q = \frac{1}{\pi(m_c D' - 0.5)} \tag{1.158}$$

$$m_c = 1 + \frac{S_e}{S_n} \tag{1.159}$$

In this expression, T_{sw} is the switching period, S_n or S_{on} is the on-slope – see (1.44) for a buck converter for instance – while S_e (also sometimes found as S_a in the literature) is the external or artificial stabilizing ramp as we will see in a few lines. If S_e is an artificial voltage ramp, then S_n which is expressed in A/s must be scaled by the sense resistance R_{sense} or R_i to ensure homogeneity. The relationship linking S_e and S_n is given by m_c: $m_c = 1$ means no stabilization ramp ($S_e = 0$) whereas $m_c = 1.5$ for instance, implies $S_e = 0.5 S_n$. Depending on the operating conditions and the duty ratio in particular, the quality factor Q can grow beyond 1 provoking an oscillatory response as the duty ratio approaches 50%.

To damp the poles by reducing the quality factor to 1, one must inject an artificial ramp labeled S_e whose slope is adjusted to damp the poles. The ramp can be subtracted from the feedback voltage or added to the current sense information as shown in Figure 1.75. In both cases, the new peak current setpoint is affected by the presence of the ramp S_e as illustrated in Figure 1.76. Equation (1.151) must be updated to account for the duty ratio reduction brought by the addition of the ramp:

$$i_{peak}(t) = \frac{v_{err}(t)}{R_{sense}} - \frac{S_e}{R_{sense}} D T_{sw} \tag{1.160}$$

You can see S_e divided by R_{sense} in this equation and it is so because S_e is a voltage ramp expressed in V/s. This ramp can come from the buffered oscillator ramp as drawn in Figure 1.77 or can be externally constructed as described in [6]. The value in V/s is passed to the model as a parameter when dealing with current-mode control analyses.

If you now average (1.152) and (1.160), equate them and solve for D, you obtain the current-mode pulse-width modulator expression involving the two ramps as defined in [5]:

$$\frac{D}{V_c} = \frac{1}{(S_e + S_n R_{sense}) T_{sw}} \tag{1.161}$$

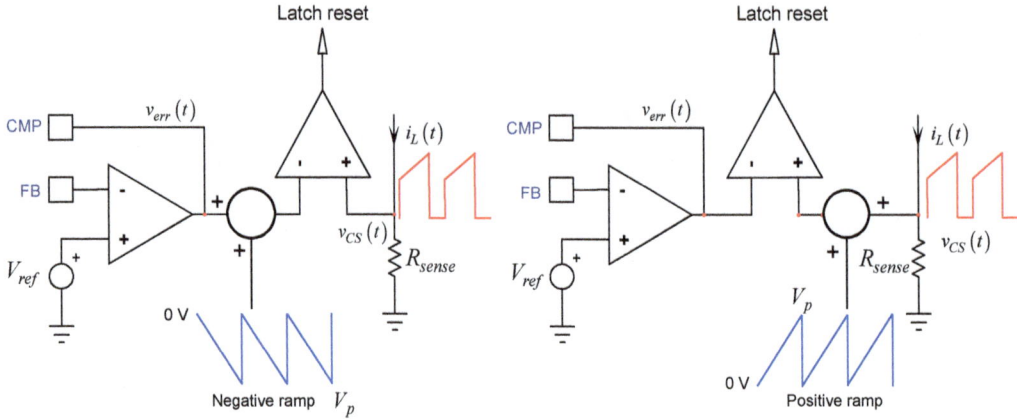

Figure 1.75 An artificial ramp is either subtracted from the error voltage or added to the current sense information to damp the subharmonic poles.

The calculation of this ramp amplitude is quite simple. Assuming a Q of 1 is the target, from the definition below (1.156) you extract the value of m_c:

$$m_c = \frac{\frac{1}{\pi}+0.5}{1-D} \qquad (1.162)$$

Assume we want to stabilize a 1-MHz 5-V CCM current-mode boost converter supplied by a 2.7-V input voltage. The inductance is 5 μH and the sense resistance 100 mΩ. The static duty ratio is calculated using (1.90) and is evaluated to:

$$D = 1-\frac{1}{M} = 1-\frac{2.7}{5} = 46\% \qquad (1.163)$$

If you substitute this value in (1.162), you have:

$$m_c = \frac{\frac{1}{\pi}+0.5}{1-0.46} \approx 1.52 \qquad (1.164)$$

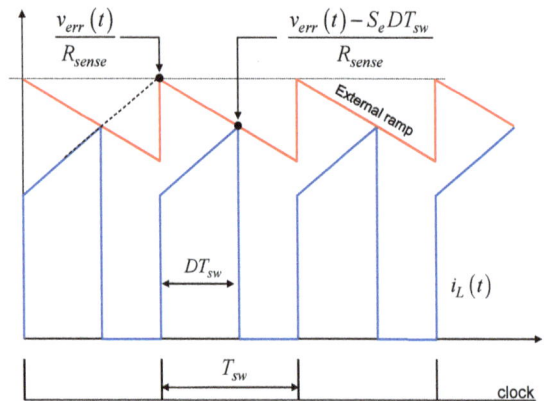

Figure 1.76 With a compensation ramp, the new peak current setpoint is reduced compared to the original value.

From (1.159), the needed external ramp amplitude is thus:

$$S_e = S_n(m_c - 1) R_{sense} = \frac{2.7}{5u} \times (1.52 - 1) \times 0.1 \approx 281 \text{ kV/s or } 28 \text{ mV/}\mu s \qquad (1.165)$$

In modern PWM controllers, the internal oscillator sawtooth is buffered and applied to the current-sense pin (CS pin) via a resistance. In a circuit like the NCP1250 from ON Semiconductor, the internal sawtooth peaks to 2 V peak while the resistance R_1 is 19 kΩ as shown in Figure 1. 77. Considering the needed 28 mV/μs indicated by the above expression and the 2-V peak ramp (over a 1-μs switching period), a scaling factor must be introduced to meet (1.165) recommendation. This factor is equal to:

$$\frac{S_e}{S_{ramp}} = \frac{28m}{2} = 14m \qquad (1.166)$$

With this value in hand and as detailed in [1], it is possible to compute the ramp resistance value as follows:

$$R_{ramp} = R_1 \cdot \frac{S_e}{S_{ramp}} = 19k \times 14m = 266 \ \Omega \qquad (1.167)$$

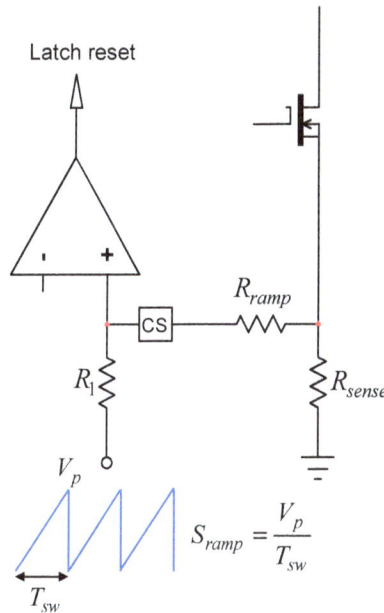

Figure 1.77 The new peak current setpoint is reduced compared to the original value.

You will probably need to a add a small 100-pF capacitor between the CS pin and the controller ground to filter out any additional noise. R_{ramp} and this extra capacitor must be placed very close to the controller.

1.2.14 Quasi-Square-Wave Resonant Converters

In the above lines, we have considered a fixed switching frequency with a converter operating in CCM at heavy load and entering DCM as the load current lowers. In DCM, the diode spontaneously blocks as the inductor current reaches 0 A. The parasitic capacitor at the switch node remains on its own, charged to a peak value present before turn-off. This element lumps all the stray capacitances incurred to the MOSFET itself, the inductor, the diode and, in isolated versions, the transformer inter-winding capacitances.

Figure 1.78 shows a simple buck-boost converter whose series switch is driven by a pulse generator. When the switch closes, node V_K is biased to the input source and the current in inductor L_1 linearly grows. When the power switch turns off, the diode conducts and V_K drops to the output voltage which is set by source V_{out}: the inductor current decays and if the switch remains off long enough, it eventually reaches zero at which point the diode blocks. The capacitance C_{lump} is now charged with a negative voltage and has some stored energy. If we were to turn the switch back on at this moment, the energy stored in the capacitor would be entirely lost through a sharp current spike combined with the instantaneous voltage across the switch. The dissipated loss can be approximated to

$$P_{loss} \approx \frac{1}{2} C_{lump} V_k^2 F_{sw} \qquad (1.168)$$

I said approximated because of a lot of the capacitances forming the lumped value are nonlinear in essence. However, you see the term F_{sw} which indicates that the higher the operating frequency, the higher the loss. Decreasing the switching cycle in an attempt to shrink the magnetics quickly finds its limit with switching losses.

Instead of turning the transistor on at the maximum voltage, we can wait for the oscillation mechanism to take place at turn off: the energy stored in the capacitor will transfer back and forth to the inductor, naturally creating peaks and valleys. This is shown in the right side of Figure 1.78. When the peak exceeds the input voltage, the body diode of the transistor conducts. If we turn the MOSFET on exactly at this moment, the capacitor is entirely discharged and the turn-on mechanism is loss-free. The switch operates in a zero voltage switching (ZVS) mode.

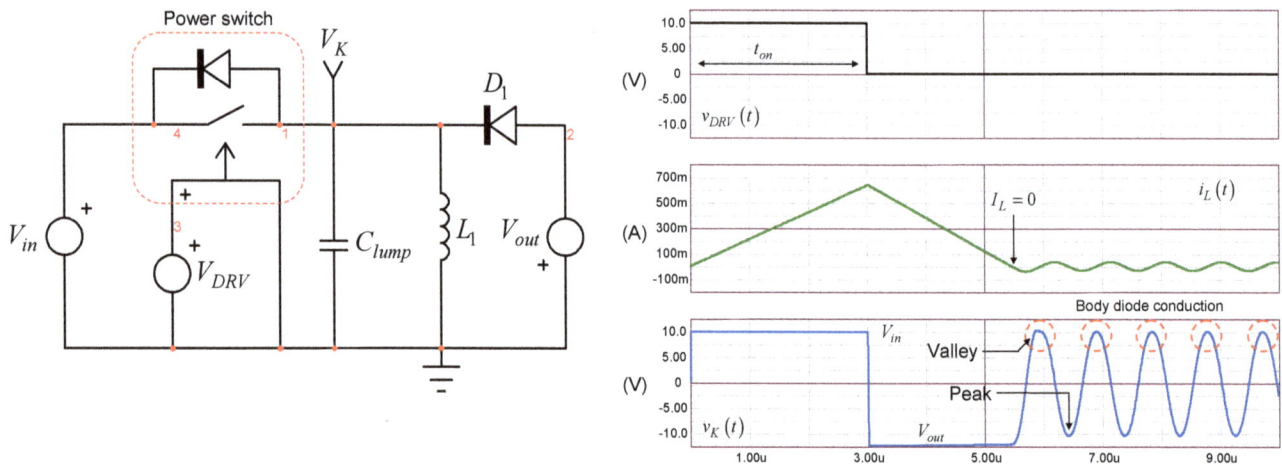

Figure 1.78 The resonance between the inductor and the lumped capacitance induces peaks and valleys.

In a quasi-square-wave resonant converter, often abbreviated QR, the switch turn-on event is synchronized with the valley of the resonating voltage. A circuit monitors the inductor current and when it reaches 0 A, it sets the PWM latch to initiate a new cycle. A small delay is usually inserted to perfectly set the turn-on pulse in the resonating valley. The moment at which the inductor current reaches 0 A depends on the programmed peak value and, of course, on the downslope linked to the output voltage in a buck-boost converter. As such, a QR converter is often referred to as a *self-relaxing* circuit. In the absence of an internal clock, the operating frequency depends on external conditions: V_{in}, V_{out} and I_{out}. In heavy-load conditions, the switching frequency is low while it increases significantly as the converter operates in lighter loads. Modern controllers clamp down on the frequency excursion and fold the frequency back as the converter enters in standby. The turn-on event thus jumps in different valleys as frequency reduces.

QR converters are popular with high-voltage flyback converters and can be found in ac-dc notebook adapters. They are particularly well suited to operate with a secondary-side synchronous rectifier. Figure 1.79 details how a dedicated winding observes the core magnetic activity and instructs the controller when the primary inductance is fully depleted. A delay ensures the controller issues the next drive pulse exactly in the valley where turn-on occurs at minimum loss.

Figure 1.80 shows the typical waveform that you can observe on the drain of the switching MOSFET of such QR flyback.

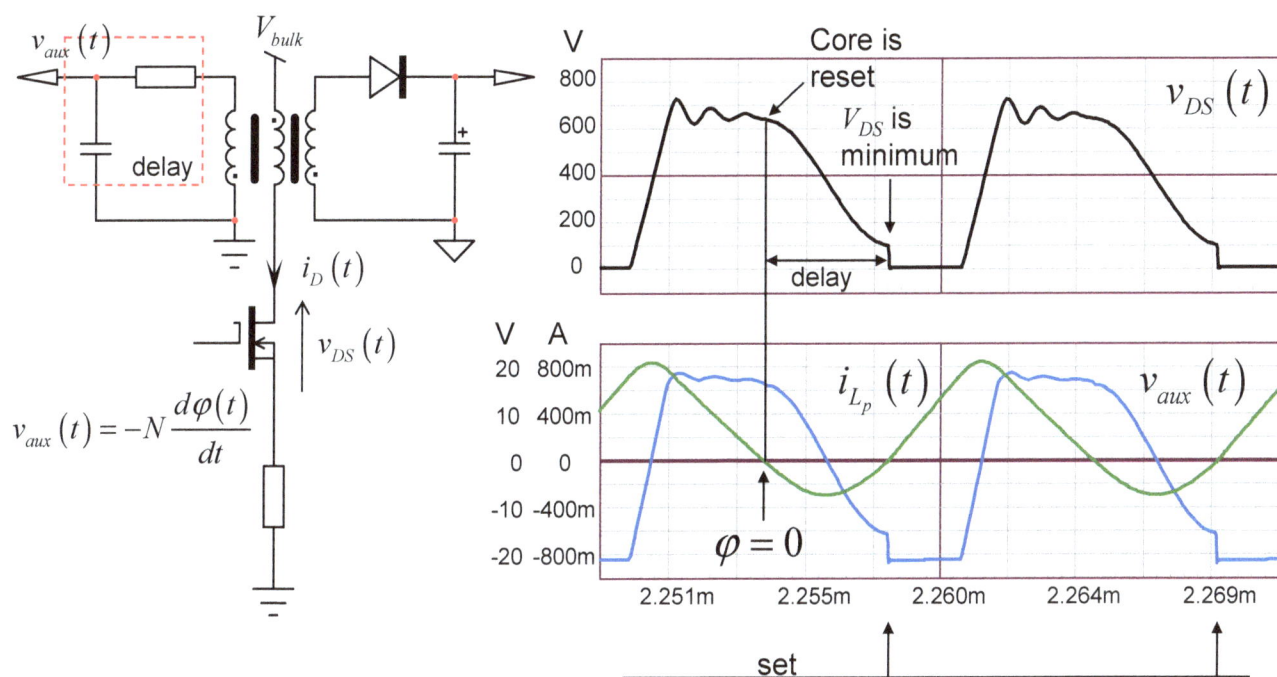

Figure 1.79 A QR flyback converter uses an auxiliary winding to monitor the magnetic activity in the core. $I_{Lp}(t)$ represents the primary-side magnetizing current.

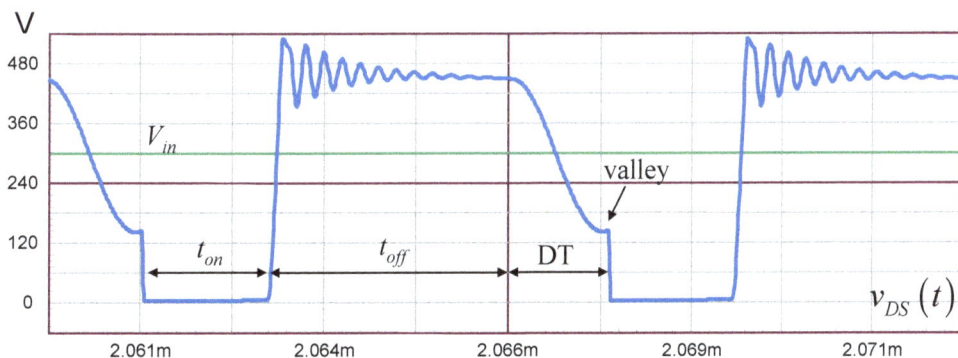

Figure 1.80 The typical drain-voltage voltage of a QR-operated flyback converter when turn-on occurs in the first valley.

As these converters never enter CCM, they perfectly lend themselves to the implementation of a synchronous rectifier. As the secondary-side rectifier naturally turns off, there is no shoot-through as you can observe in a CCM converter when the MOSFET brutally interrupts the secondary-side diode.

QR operation is also found under the name of *boundary* or *borderline* conduction mode (BCM) but also *critical conduction mode* (CrM) as the converter operates at the border between DCM and CCM.

A lot of power factor correction control circuits built around a boost converter operate in this scheme. You can find them in voltage- or current-mode control [1].

1.2.15 Hysteretic Converters

A hysteretic regulator uses the output voltage ripple and compares it to a reference voltage to form a simple pulse-width modulator. The block is made of a comparator affected by a hysteresis band and toggles the power switch on and off when the output voltage is respectively below or above the target. There is not internal clock for the simplest version as shown in Figure 1.81.

At power on, the power switch closes as long as V_{out} is below the 5-V target. In this simplified circuitry, there is no current limit and the output will certainly overshoot.

As soon as V_{out} enters the regulation band, the comparator actuates the switch on and off to maintain the output around 5 V.

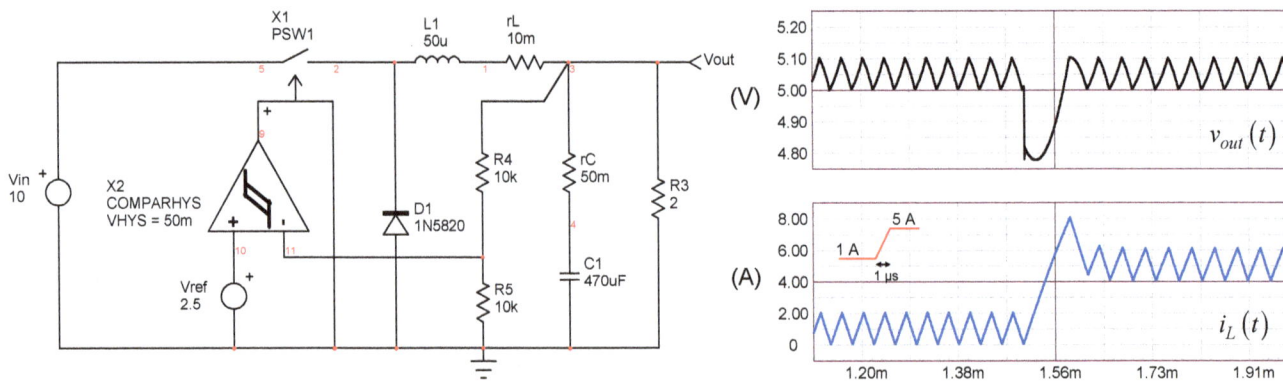

Figure 1.81 In its simplest form, the hysteretic converter is a self-relaxing system, without a clock.

If a load change occurs and V_{out} drops below the target, the power switch remains closed until the voltage meets the 5-V threshold. In Figure 1.81, the output current has been stepped from 1 to 5 A in 1 μs. The inductor current is immediately pushed to answer the sudden power change.

The response time is extremely short making these *ripple-based* circuits the fastest converters within the family of switching circuits. The frequency depends on numerous parameters among which the capacitor stray elements (ESR and ESL) play a significant role, making switching frequency prediction a difficult exercise.

You also realize that this simple structure could not be applied to a boost or buck-boost converter as the possible long on-time could quickly lead to the inductor or transformer saturation in an isolated structure. The circuit can be modified by including a clock and a peak current protection.

The Fairchild's μA78S40, released in the 80's, quickly followed by the MC34063 from Motorola, implemented this improvement and a simplified version of these switchers appears in Figure 1.82.

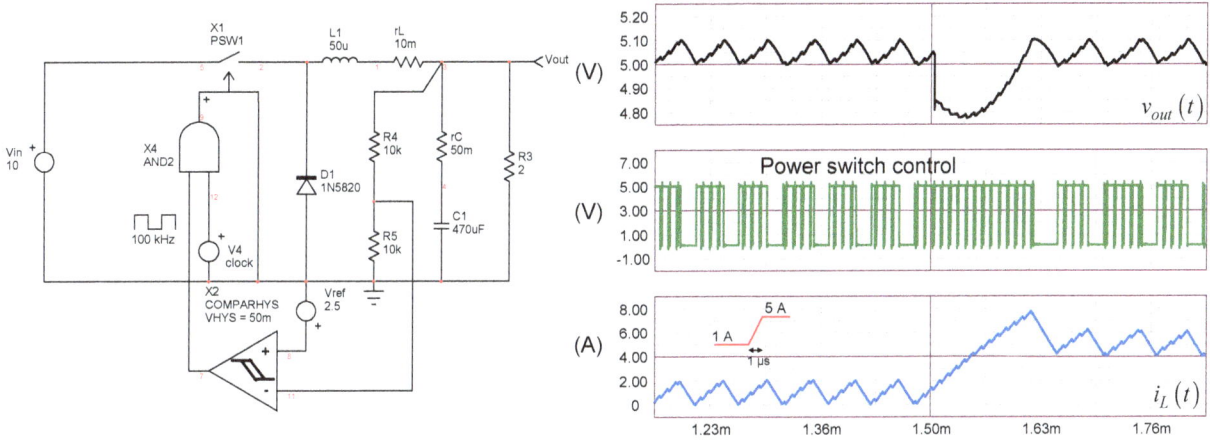

Figure 1.82 The circuit can be improved by adding a clock and a current limit as in the μA78S40.

The control is still hysteretic but the power switch is now activated at a 100-kHz pace. A current limit nicely supplements this implementation. These circuits have been very popular and the 34063 still sells by millions per year. One major drawback is the acoustic noise generated by the on-off type of operation and the large frequency spectrum covered by the hysteretic operation. A lot of possibilities exist to stabilize the frequency and [7] offers a nice overview of what is available on the subject.

Improvements over the pure hysteretic converter shown in Figure 1.82 are the constant on-time (COT) and fixed off-time (FOT) structures. Rather than turning the power switch on for a possibly long time (as long as V_{out} is below the target), the COT controller will calibrate the turn-on time t_{on} and force operations at a frequency now depending on the operating conditions.

One nice advantage of the COT converter lies in its natural *frequency-foldback* characteristic when the load current goes down and forces DCM operation. As the output capacitor takes longer time to discharge now that the load is lighter, the power switch remains off until the low comparator threshold is reached. The operating frequency is thus reduced and scales down all switching losses. The efficiency benefits from this operation and COT converter excel in standby power performance.

Figure 1.83 shows a simple simulated version featuring a minimum off-time generator which limits the maximum duty ratio in fault and the maximum switching frequency. Please note the constant peak current operation in DCM.

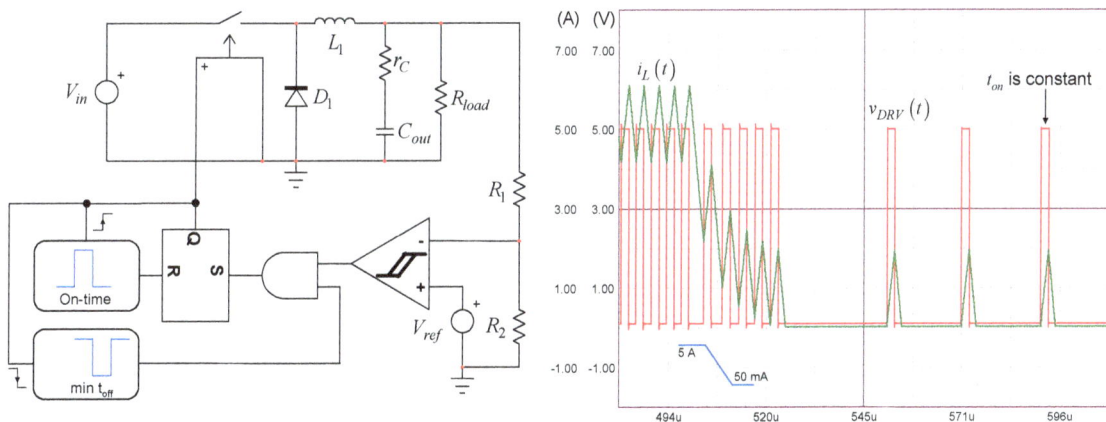

Figure 1.83 A hysteretic constant on–time converter in which the operating frequency is easier to predict than with its pure hysteretic counterpart.

In the fixed off-time version, the hysteretic comparator turns the MOSFET on and off but a timer keeps it off for a predetermined amount of time. The resulting switching period also varies with operating conditions when the converter enters DCM. However, unlike with the COT converter, the FOT version reduces the on-time (the inductor peak current diminishes) while the off-time remains constant: the switching frequency increases as the load current gets lighter in DCM which is not good for efficiency.

Figure 1.84 illustrates a simplified FOT structure and the corresponding simulated waveforms as the load current reduces. This time, the frequency increases in DCM and the peak inductor current is no longer constant.

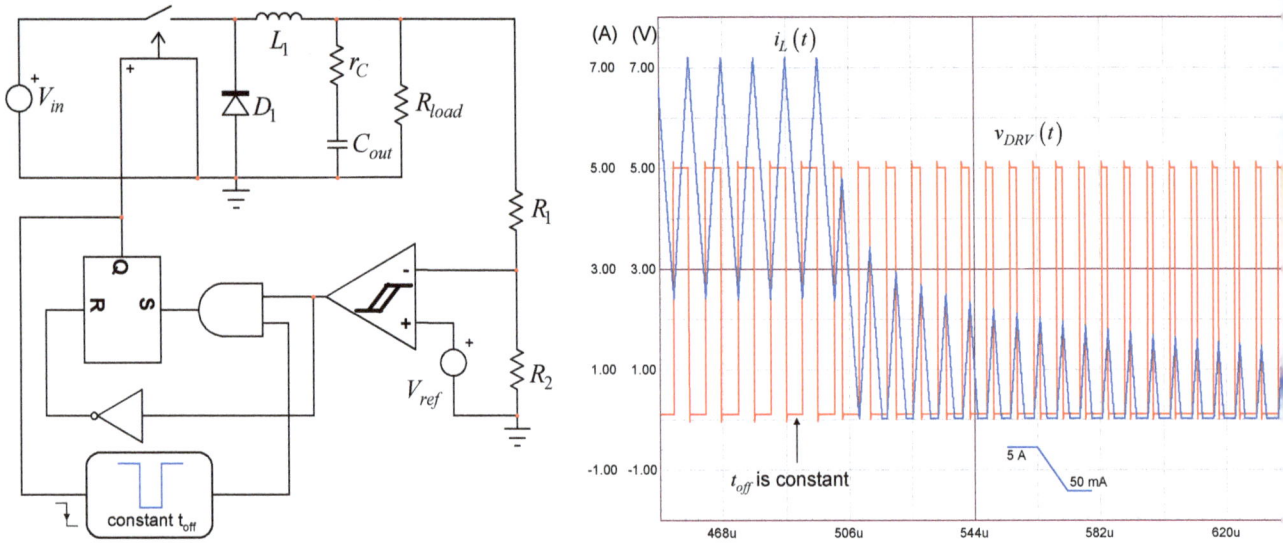

Figure 1.84 A hysteretic constant off-time converter in which the operating frequency increases as the converter enters DCM.

Please note that both of these converters can become unstable in CCM operation.

A simple stability condition as highlighted in [8] is defined as follows for the FOT converter:

$$r_C C > \frac{t_{off}}{2} \tag{1.169}$$

and:

$$r_C C > \frac{t_{on}}{2} \tag{1.170}$$

for the COT converter.

Within the hysteretic converters, the valley current control and fixed off-time peak current control converters have found applications a few decades ago in high-performance dc-dc power supplies designed for motherboards.

As the name implies, either the on- or off-time is kept constant while the off- and on-time is respectively adjusted to maintain regulation. However, regulation is done by adjusting the inductor valley current for the COT while the peak is driven for the FOT. Figure 1.85 shows the typical waveforms of these specific converters.

In the COT converter, the loop adjusts the inductor current valley point at which a new constant on-time cycle occurs. With the FOT, the peak current is adjusted and it triggers the off-time period. Unlike fixed-frequency operation, neither the FOT or the COT are subject to subharmonic instabilities and no artificial ramp is needed.

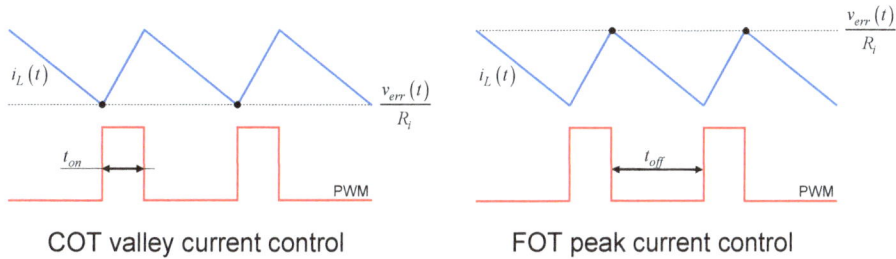

COT valley current control   FOT peak current control

Figure 1.85 The constant on- and off-time modified to operate in current mode.

Figure 1.86 shows a possible implementation of a COT converter. The error voltage sets the valley point at which the power switch turns on again for a programmed duration. When the load current goes down and the converter enters DCM, the error voltage blocks the converter re-start and the off-time naturally expands as shown in the figure. This brings excellent efficiency performance in light-load operation.

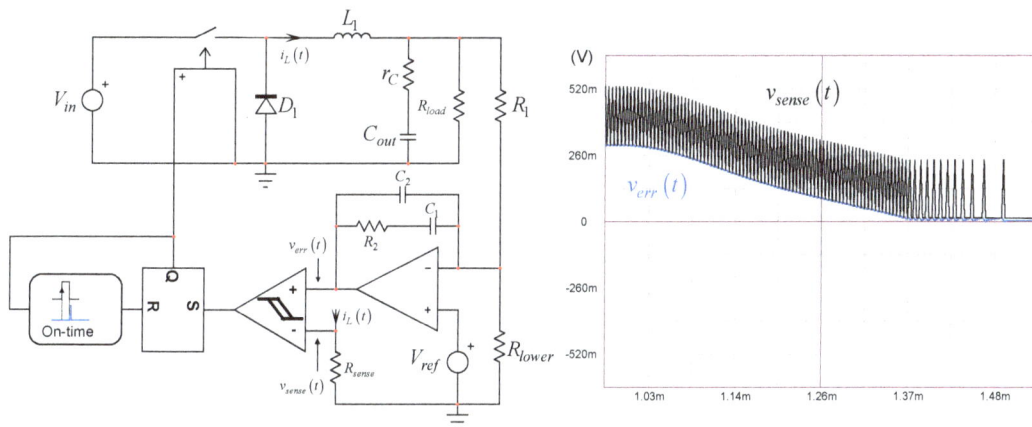

Figure 1.86 A possible implementation of the COT valley current.

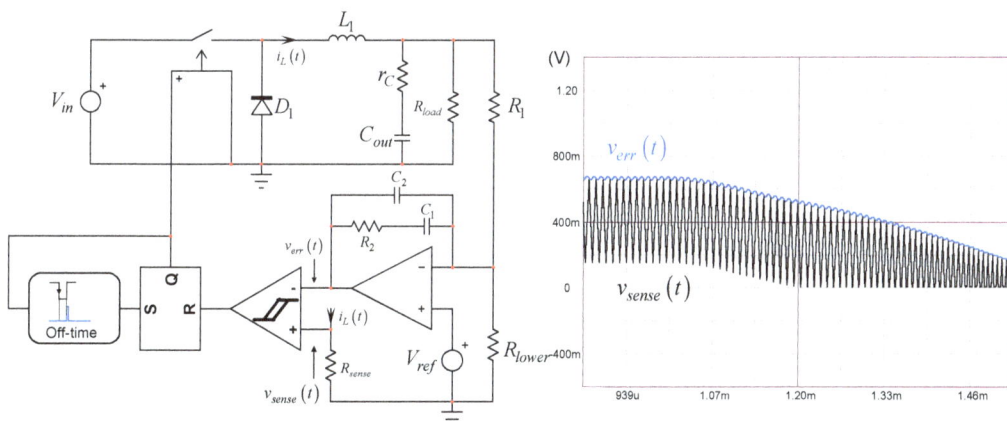

Figure 1.87 A peak current fixed off-time converter observes the inductor peak current set by the loop.

Figure 1.87 shows how a FOT converter could be built. It does not differ much from a COT converter except that the peak current is now controlled by the loop and the converter transitions in DCM nicely as shown in the right side of the picture.

As its COT counterpart, the FOT converter does not require the addition of an external ramp to ensure stability as with fixed-frequency peak current mode control. However, the FOT is less popular than the COT probably because its switching frequency increases as the load current gets lighter (constant off-time but smaller t_{on}) which negatively impacts efficiency in light-load operations.

1.3 The PWM Switch Model in Voltage Mode Control

In 1986, Dr. Vatché Vorpérian developed the concept of the PWM switch model [3]. At about the same time, Larry Meares from Intusoft also presented a paper in which the approach of the PWM switch was also explored, although in a less comprehensive manner since CCM was the only case Meares covered [9].

As the diode and the power switch were guilty for introducing the non-linearity, these gentlemen considered modeling the switching network alone, to finally replace it with an equivalent small-signal three-terminal model as we have seen in Figure 1.23. The analysis of a switching converter was considerably simplified considering the switching cell alone, leaving the rest of the elements (R, C and L) untouched: in the switching converter, identify the commutating cell and replace it with a linearized version to solve a linear circuit.

Figure 1.88 illustrates the process.

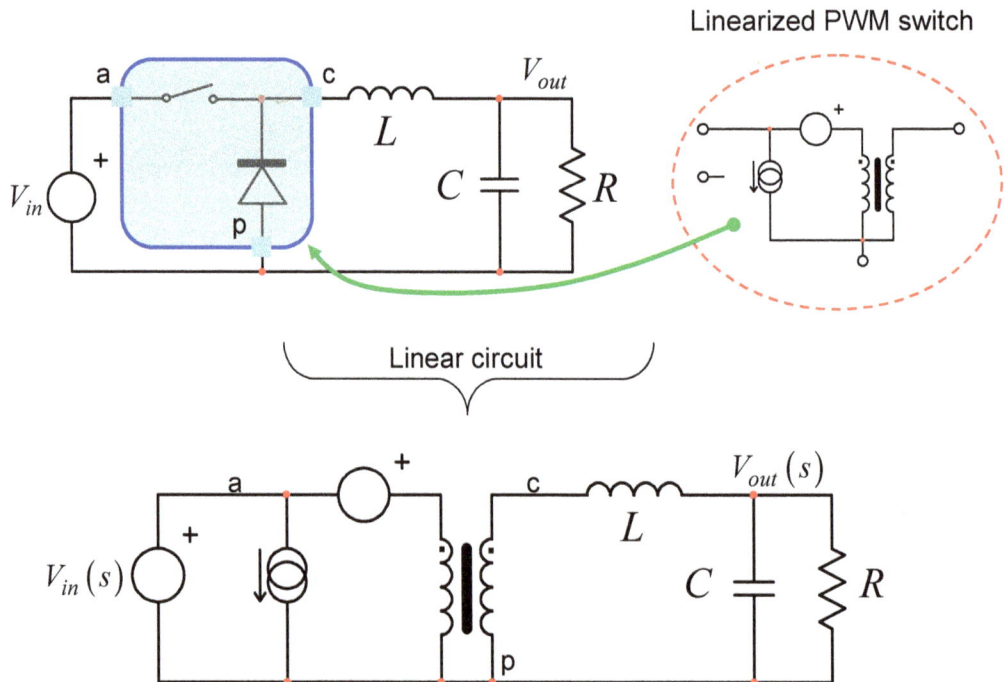

Figure 1.88 Replace the switching cell by the PWM switch model and work on a linear circuit.

Should you later add parasitic elements or another energy-storing device like those found in a front-end filter, the PWM cell internals remain the same and you only work on an updated linear circuit.

The model is *invariant* meaning that once derived, it fits all topologies simply through different model rotations (see Figure 1.89).

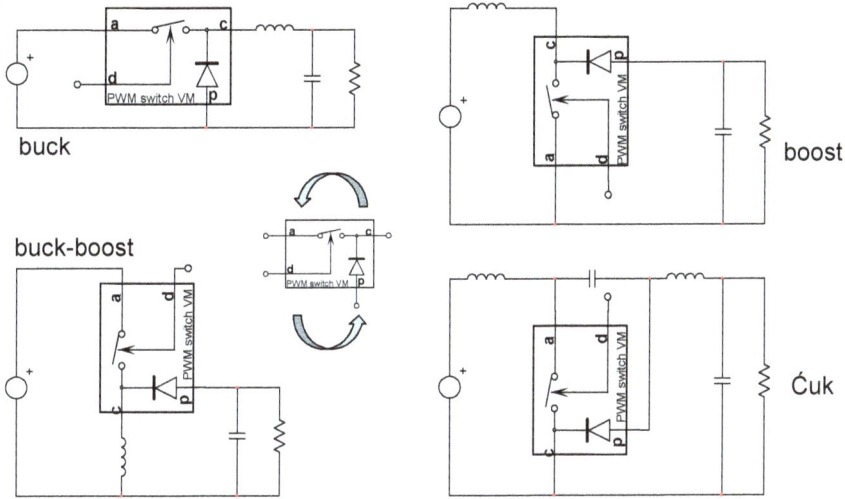

Figure 1.89 By rotating the original structure, you can explore a variety of dc-dc switching converters.

There are several types of PWM switch models: voltage mode (VM), current mode (CM), quasi-square wave (QR) and hysteretic with valley and peak current control. We will see all these models and their small-signal versions that we will later use in switching converters to determine the transfer functions of our choice.

A word often used in small-signal modeling is *averaging* as already illustrated with Figure 1.58. It consists of describing a time-discontinuous waveform by a time-continuous equation that we can later linearize. Averaging has been introduced graphically in Figure 1.28. We will repeat this process with the PWM switch when plugged in a classical buck converter: identify and draw input/output currents and voltages then describe their average value by an equation. Figure 1.90 represents the PWM switch when dropped into a buck converter.

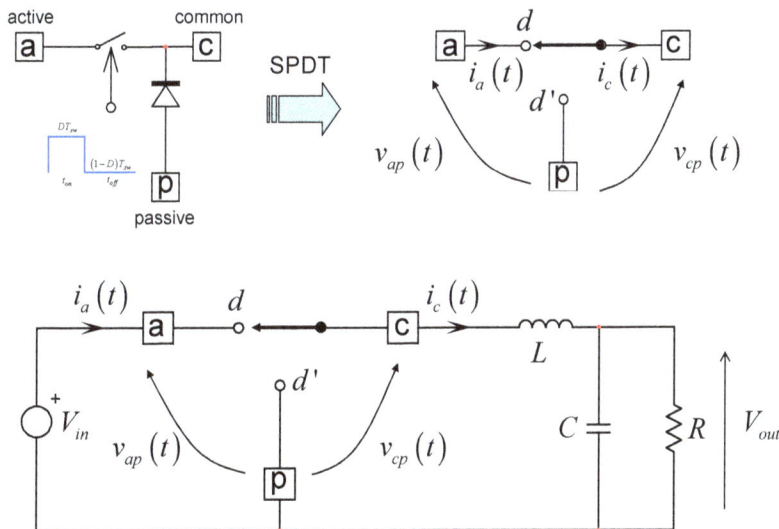

Figure 1.90 The PWM switch is revealed by inserting it in a buck converter. This is the so-called *common passive* configuration.

Assuming fixed-frequency operation and CCM, you can now draw currents and voltages pertinent to this arrangement. You obtain Figure 1.91. The upper side represents the current leaving terminal c. Just below, this is the current entering terminal a. This instantaneous current equals the inductor current when the switch turns on and equals 0 A the rest of the time. We can say terminal a "sees" the current leaving terminal c during DT_{sw}. On average, across a switching period, terminal a "sees" DI_c. Mathematically, we can write:

$$\langle i_a(t) \rangle = I_a = \frac{1}{T_{sw}} \int_0^{dT_{sw}} i_a(t)\,dt = D\langle i_c(t) \rangle = DI_c \qquad (1.171)$$

and this is the formula already obtained in (1.53). Please note (1.171) combines the averaged quantities I_a and I_c in a time-continuous equation.

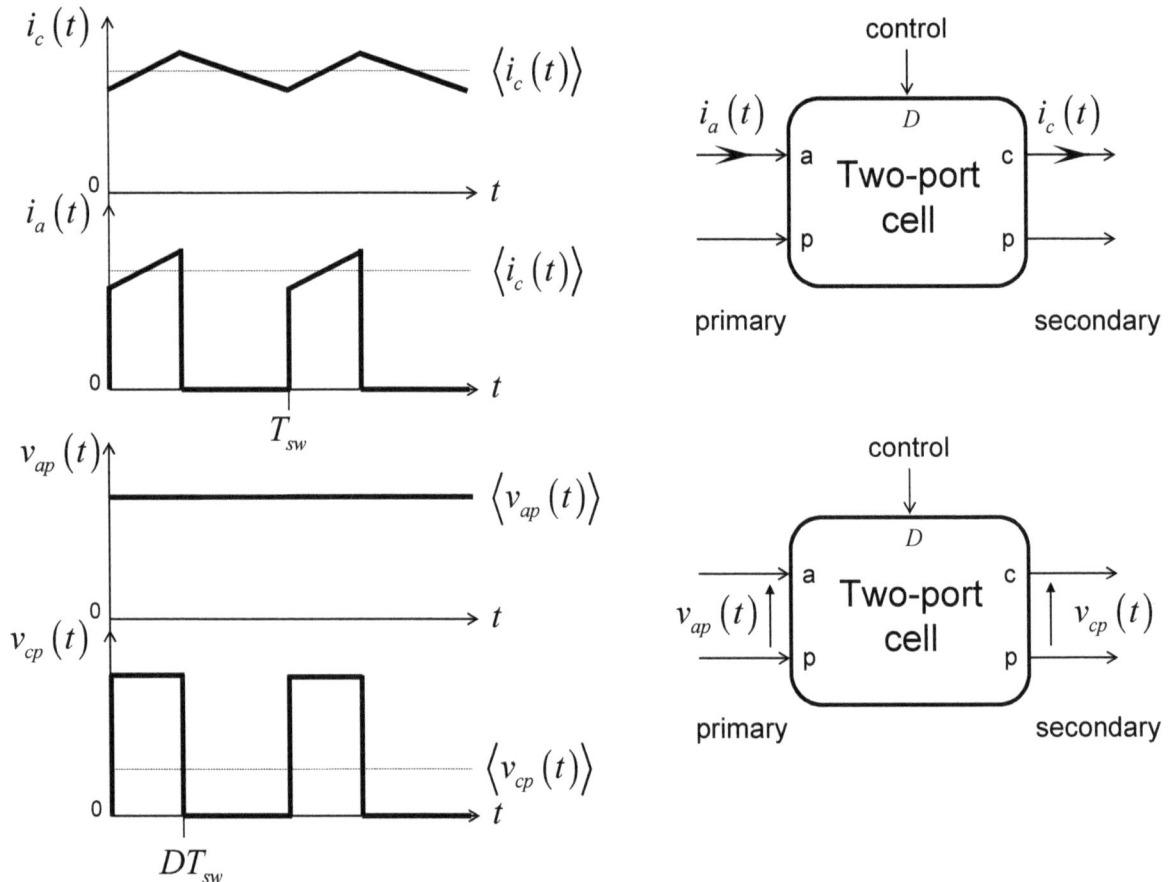

Figure 1.91 The buck converter lends itself well to revealing the key waveforms of the PWM switch.

If we now look at the voltages, the instantaneous level measured between terminals c and p is the voltage between a and p present during DT_{sw} while it is zero-volts the rest of the time. Mathematically, we can write:

$$\langle v_{cp}(t) \rangle = V_{cp} = \frac{1}{T_{sw}} \int_0^{dT_{sw}} v_{cp}(t)\,dt = D\langle v_{ap}(t) \rangle = DV_{ap} \qquad (1.172)$$

It is the second invariant relationship of the PWM switch model. Please note that V_{cp} and V_{ap} are now time-continuous averaged voltages values.

We now have a cell made of a primary and secondary side. The primary absorbs a secondary current described by (1.171) while the secondary is biased to a primary voltage following (1.172). When you look at Figure 1.92, you realize that the two cited equations actually describe a transformer whose turns ratio is D, the control variable. This is of course a theoretical transformer as it can accept continuous currents and voltages but it is extremely practical to visualize the large-signal VM-PWM switch model.

I say large signal because (1.171) and (1.172) are obviously nonlinear considering the product of two variables. If we want to study the small-signal response of a converter featuring the PWM switch model, we will have to linearize these two equations.

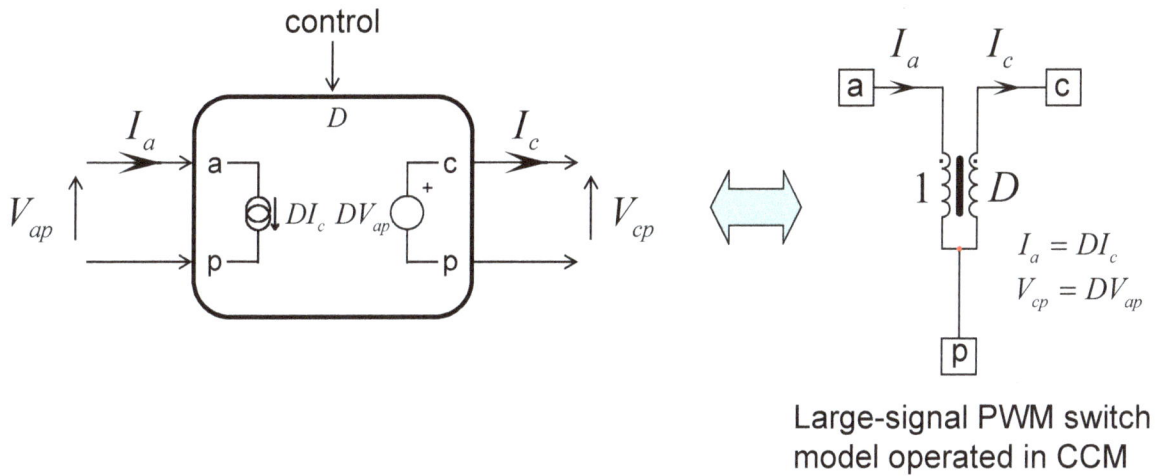

Large-signal PWM switch
model operated in CCM

Figure 1.92 The large-signal PWM switch model is a perfect dc transformer whose turns ratio is the duty ratio D.

Before going through this process, we can already drop this equivalent transformer into a boost converter as shown in Figure 1.93.

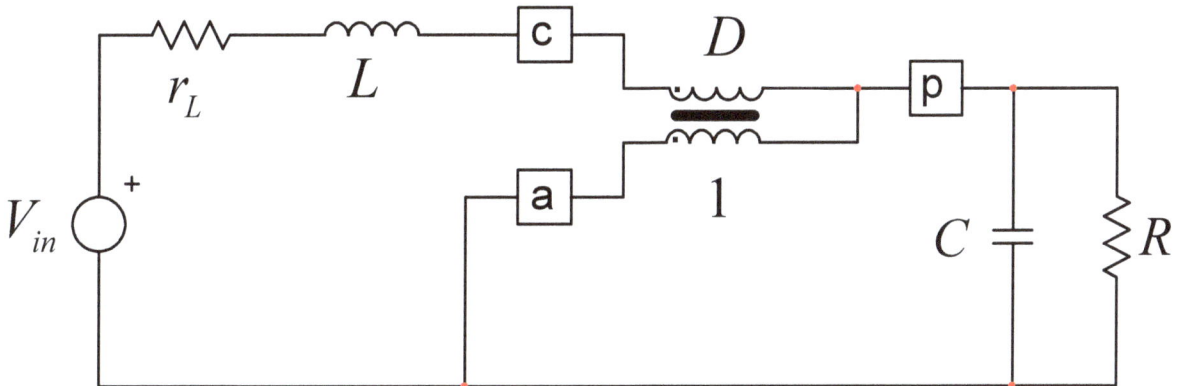

Figure 1.93 You can drop the large-signal model into any converter for dc and ac analyses.

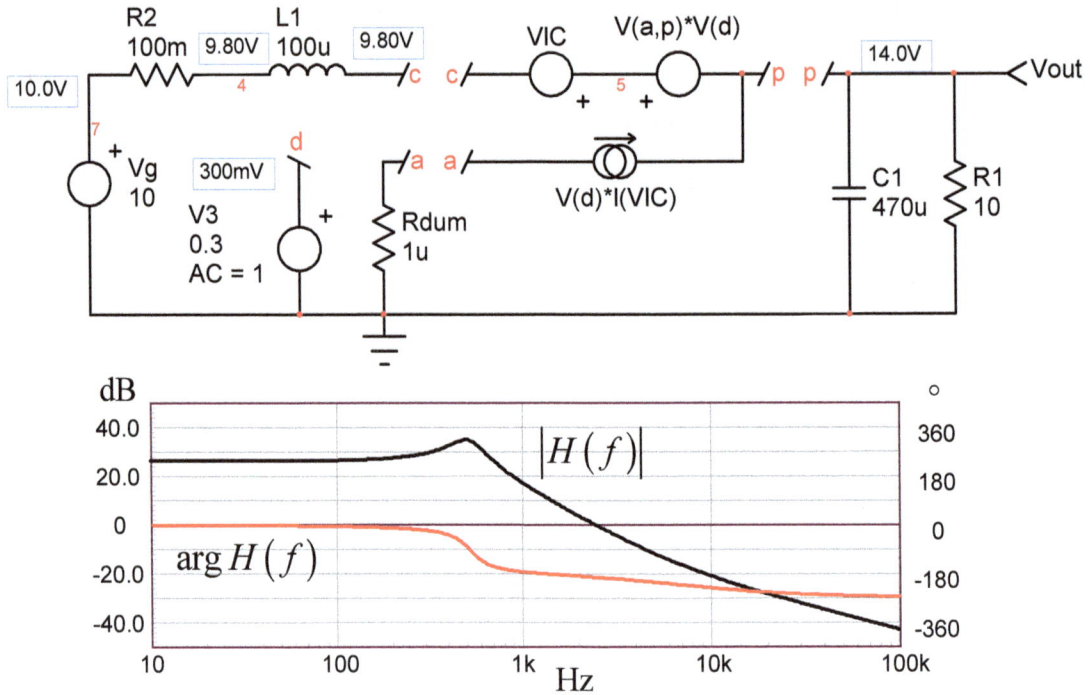

Figure 1.94 The dynamic response is immediately obtained. Please note the dc operating points computed by the simulator.

If you replace the transformer by controlled sources, you can run an ac simulation with SPICE and immediately obtain the converter dynamic response as illustrated in Figure 1.94. The framed numbers represent the operating point computed by SPICE prior to launching the ac simulation. It is extremely important to verify these bias points and make sure they are what you expect. For instance, source V_3 sets the duty ratio to 30% (1 V is 100%). With a 10-V input source, we expect an output voltage to be in the ballpark of:

$$V_{out} = V_{in} \frac{1}{1-D} = 10 \times \frac{1}{1-0.3} = 14.3 \text{ V} \tag{1.173}$$

which is roughly what we have. A bit less actually considering the inductor ohmic loss but how much? From this simple circuit, we can already work out the dc transfer function of the CCM-operated boost converter. See Figure 1.95 for the updated circuit.

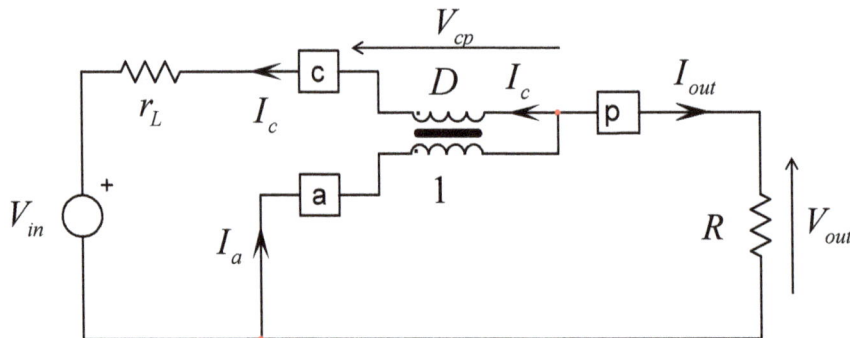

Figure 1.95 Short the inductor and open the capacitor to determine the dc transfer function.

The set of equations is quite simple actually considering the dc transformer we have installed. Let's start with the obvious relationships:

$$V_{out} = I_{out} R \tag{1.174}$$

The output current I_{out} is $I_a - I_c$ therefore:

$$V_{out} = (I_a - I_c) R \tag{1.175}$$

Knowing the link between I_a and I_c via (1.171), we can update (1.175):

$$V_{out} = I_c (D-1) R \tag{1.176}$$

Now if we consider the mesh involving V_{in} and the ohmic drop across r_L, we have:

$$V_{in} + r_L I_c - V_{cp} = V_{out} \tag{1.177}$$

From the schematic and the transformer connection, you see that V_{out} appears across terminals p and a. Using (1.172), the above expression updates to:

$$V_{in} + r_L I_c + D V_{out} = V_{out} \tag{1.178}$$

from which we determine that current I_c is equal to

$$I_c = \frac{V_{out}(1-D) - V_{in}}{r_L} \tag{1.179}$$

Substituting (1.179) in (1.176) and solving the transfer function, we have:

$$\frac{V_{out}}{V_{in}} = \frac{1}{(1-D) - \dfrac{r_L}{(D-1)R}} = \frac{1}{D'} \cdot \frac{1}{1 + \dfrac{r_L}{RD'^2}} \tag{1.180}$$

which links the output and input voltages including the inductor ohmic loss. We can now capture the schematic values and precisely determine the output voltage:

$$V_{out} = 10 \frac{1}{(1-0.3)} \frac{1}{1 + \dfrac{100m}{10 \times (1-0.3)^2}} = 14 \text{ V} \tag{1.181}$$

This is exactly what the operating point computation gives in Figure 1.94. If for any reason the computed dc points are wrong, it won't prevent SPICE from delivering an answer but it is likely to be wrong. Engineering judgment is important via a first bias point analysis. By the way, why can we get a small-signal response from a nonlinear model? This is because SPICE is a linear solver. When it encounters a nonlinearity, it seeks an operating point at which the linearization around that point is possible.

Therefore, the large-signal model is automatically transformed into a linear network by SPICE when it starts the ac analysis. Why bother with a small-signal model then? Because without it, we could not analytically determine the transfer function of our choice and see how the various components impact the dynamic response. The compensation circuit shall then be designed to neutralize the impact of these components and ensures stability is maintained in all operating conditions.

1.3.1 Small-Signal Model in CCM

If we want to determine the transfer function of our choice, we need a linear circuit to work with and the large-signal model of the VM-PWM switch needs to be linearized. We have two sources each featuring two variables. We can either perturb the original definitions from (1.171) and (1.172) or apply partial differentiation as shown below:

$$\hat{i}_a = \frac{\partial f(D,I_c)}{\partial D}\hat{d} + \frac{\partial f(D,I_c)}{\partial I_c}\hat{i}_c = I_c\hat{d} + D\hat{i}_c \tag{1.182}$$

$$\hat{v}_{cp} = \frac{\partial f(D,V_{ap})}{\partial D}\hat{d} + \frac{\partial f(D,V_{ap})}{\partial V_{ap}}\hat{v}_{ap} = V_{ap}\hat{d} + D\hat{v}_{ap} \tag{1.183}$$

In these expressions, I_c and D are dc coefficients whose values derive from the bias point analysis as carried around Figure 1.95. Sometimes, I will add a subscripted 0 for readability like I_{c0} and D_0. Any transfer function determination starts with an operating point assessment. It can be done by equations, simulations or even hardware measurements if needed. V_{ap} in Figure 1.95, for instance, is $-V_{out}$. From the two small-signal equations, we can draw the PWM switch model operated in CCM.

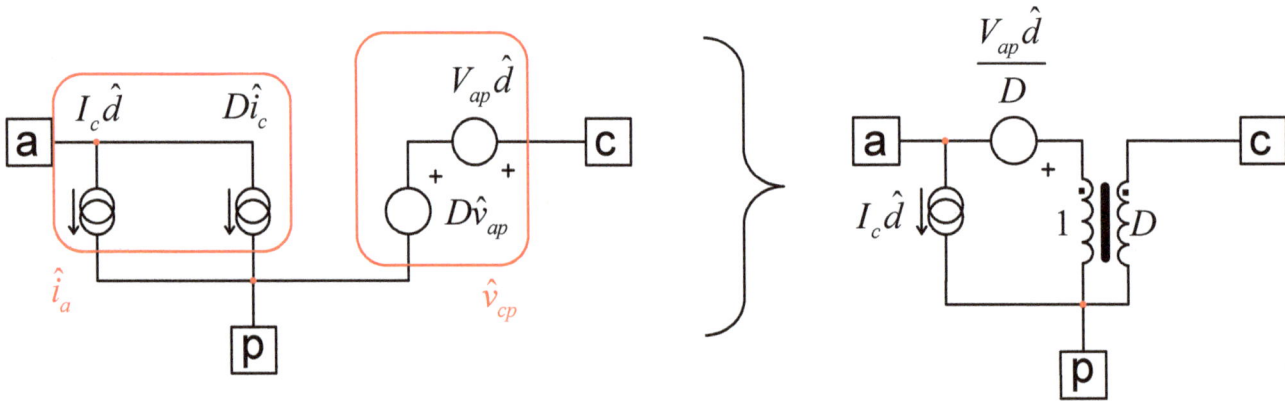

Figure 1.96 The small-signal version combines the sources determined in (1.182) and (1.183).

For the sake of revealing dc and ac sections, source $V_{ap}\hat{d}$ can be reflected to the primary side of the transformer simply by dividing it by D. In this drawing, D, V_{ap} and I_c are bias points. This model will extensively be exercised in various converters as we shall later see.

1.3.2 The PWM Switch Operated in DCM

The founding paper from [3] described a DCM version of the PWM switch model however, wired in common-common configuration which did not lend itself to deriving an auto-toggling CCM-DCM SPICE model as described in [1]. For this reason, I kept the original common-passive configuration and applied the same methodology as for the CCM version: draw the input/output ports waveforms and average them across a switching period. From the obtained large-signal expressions, linearize them to derive a small-signal model.

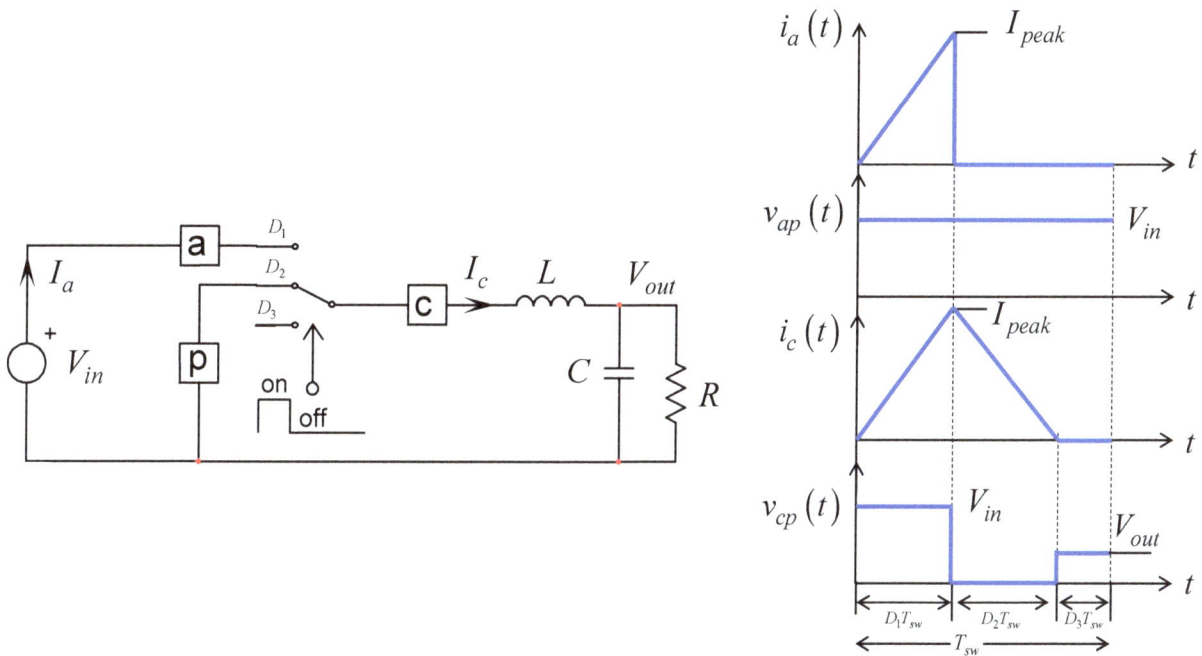

Figure 1.97 In DCM, a third state exists and must be accounted for when determining average values.

The DCM circuit appears in Figure 1.97 and highlights the presence of the third state $D_3 T_{sw}$ (or deadtime) which occurs when the inductor is demagnetized and both switches are open. The average current I_c is the sum of the areas occupied by two triangles peaking at I_{peak}:

$$I_c = \frac{I_{peak}}{2} D_1 + \frac{I_{peak}}{2} D_2 = \frac{I_{peak}}{2}(D_1 + D_2) \qquad (1.184)$$

The average current flowing into terminal a is the area of the triangle peaking at I_{peak}:

$$I_a = \frac{I_{peak}}{2} D_1 \qquad (1.185)$$

Extracting $I_{peak}/2$ from (1.185) and substituting it in (1.184) leads to the relationship linking I_c to I_a:

$$I_c = \frac{2I_a}{D_1} \frac{D_1 + D_2}{2} = \frac{I_a}{N} \qquad (1.186)$$

in which:

$$N = \frac{D_1}{D_1 + D_2} \qquad (1.187)$$

The same approach is applied to the instantaneous voltage observed between terminals c and p. The average value requires the calculation of the two rectangles involving D_1 and D_3.

We can write:

$$V_{cp} = V_{ap}D_1 + V_{cp}D_3 \tag{1.188}$$

The term D_3 can be extracted from (1.54) and substituted in (1.188). Developing and rearranging, we have:

$$V_{cp} = V_{ap}D_1 + V_{cp}(1 - D_1 - D_2) \rightarrow V_{cp} = NV_{ap} \tag{1.189}$$

with N defined by (1.187).

We now have the large-signal relationships linking the input/output currents and voltages.

The DCM model appears in Figure 1.98 and also operates around an equivalent dc transformer whose turns ratio now depends on both D_1 and D_2.

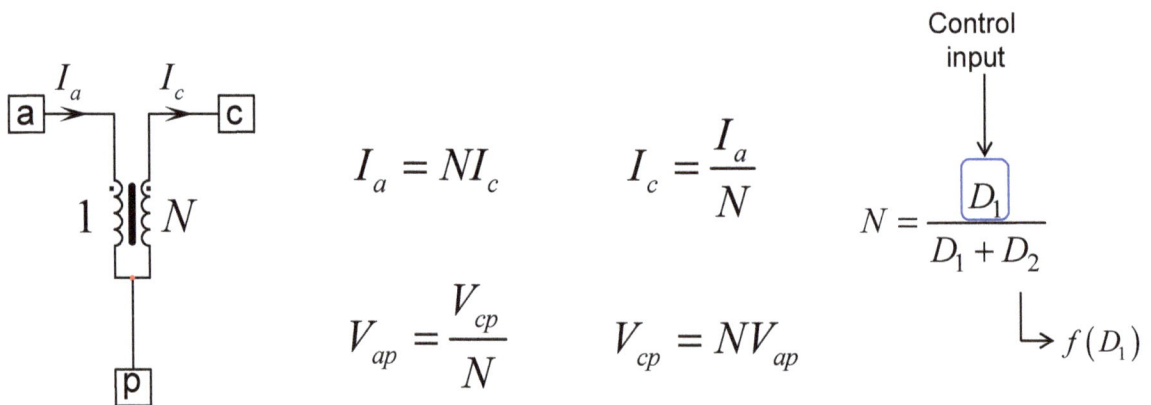

$$I_a = NI_c \qquad I_c = \frac{I_a}{N} \qquad N = \frac{D_1}{D_1 + D_2}$$

$$V_{ap} = \frac{V_{cp}}{N} \qquad V_{cp} = NV_{ap} \qquad \rightarrow f(D_1)$$

Figure 1.98 The large-signal model in DCM can still be modeled by transformer whose turns ratio N depends on D_1 and D_2.

In DCM, the error voltage sets D_1, implying an inductor current peaking at $D_1 T_{sw}$. The inductor demagnetizes according to the voltage applied across its terminals during the freewheel period.

At $D_2 T_{sw}$, the inductor current reaches 0 A and the inductor is demagnetized. We need to determine the variable D_2 and finalize the model definition. We know that the average voltage across an inductor at steady-state is 0 V. In DCM, this is true even under an ac excitation. If, on average along a switching cycle, the inductor voltage is zero, then, in Figure 1.97:

$$V_{cp} = V_{out} \tag{1.190}$$

Now, under an ac modulation, the peak inductor current takes on different discrete I_{peak} values depending on the modulated setpoint.

Considering a modulating waveform with a period much larger than the switching frequency (see Figure 1.58), the average function links together all the individual discrete events under a time-continuous function $i_{peak}(t)$ shown in Figure 1.99.

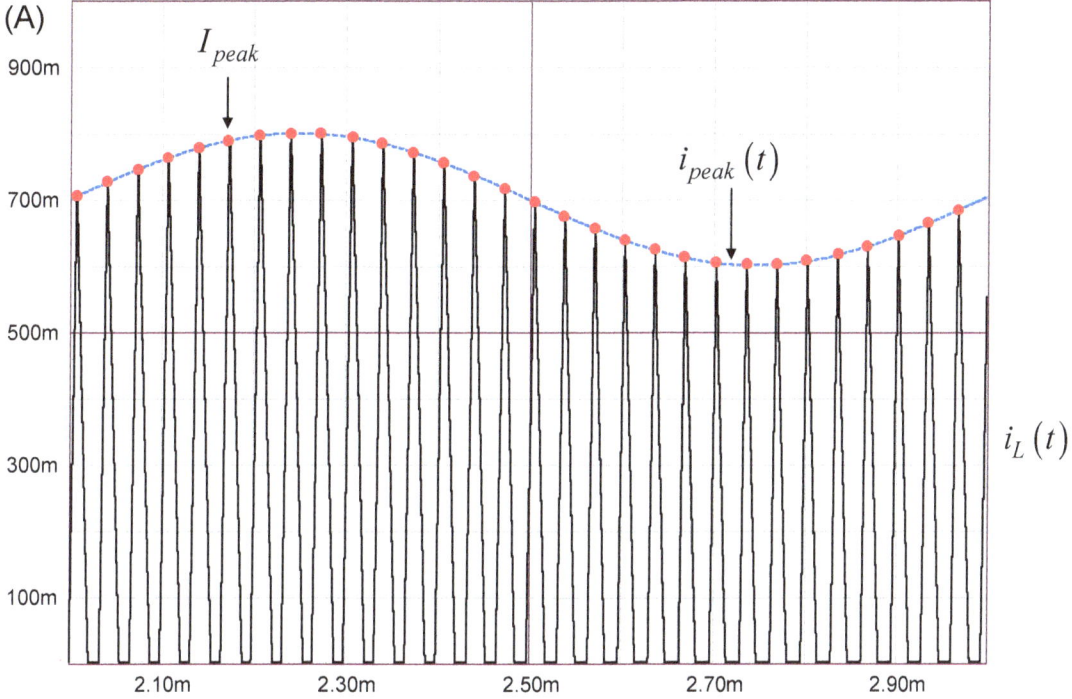

Figure 1.99 $i_{peak}(t)$ **is a time-continuous mathematical abstraction linking all discrete peak currents happening at the switching frequency pace.**

In this figure, $i_{peak}(t)$ is an averaged time-continuous representation of the peak current evolution as you ac-modulate the converter. It has no physical meaning (the peak current I_{peak} is a fixed value reached at the end of the on-time) but it is a convenient way, a mathematical abstraction, to link all these discrete values into a time-continuous waveform. We conveniently did a similar operation with the duty ratio in Figure 1.58.

The average of this function depends on the inductor voltage applied during the on-time or $D_1 T_{sw}$:

$$I_{peak} = \frac{\langle v_L(t) \rangle_{D_1 T_{sw}}}{L} D_1 T_{sw} \tag{1.191}$$

In our buck representation, regardless of its operating mode, the value applied across the inductor during t_{on} is $V_{out} - V_{in}$. It is not an invariant relationship as it characterizes the buck converter only. Capitalizing on the fact that the inductor average voltage across a switching cycle is 0 V, then V_{out} appears at terminal c while terminal a is biased at V_{in}. The voltage applied across the inductor during t_{on} can thus be described by the following relationship:

$$\langle v_L(t) \rangle_{D_1 T_{sw}} = V_{ac} \tag{1.192}$$

Therefore, substituting (1.192) in (1.191) leads to the *invariant* relationship:

$$V_{ac} = L \frac{I_{peak}}{D_1 T_{sw}} \tag{1.193}$$

in which I_{peak} is the averaged time-continuous peak current function described in Figure 1.99. If we now extract

I_{peak} from (1.184) and substitute it in (1.193), we can determine D_2 easily:

$$D_2 = \frac{2LF_{sw}}{D_1}\frac{I_c}{V_{ac}} - D_1 \tag{1.194}$$

Now substituting this expression in (1.187), we have a new expression for N:

$$N = \frac{V_{ac}D_1^2}{2F_{sw}I_c L} \tag{1.195}$$

We now have everything we need to assemble the large-signal DCM PWM switch model and use it in a boost converter as shown in Figure 1.100. Source B_{mode} calculates if the converter operates in DCM or CCM. A 0-V *mode* node indicates DCM [1].

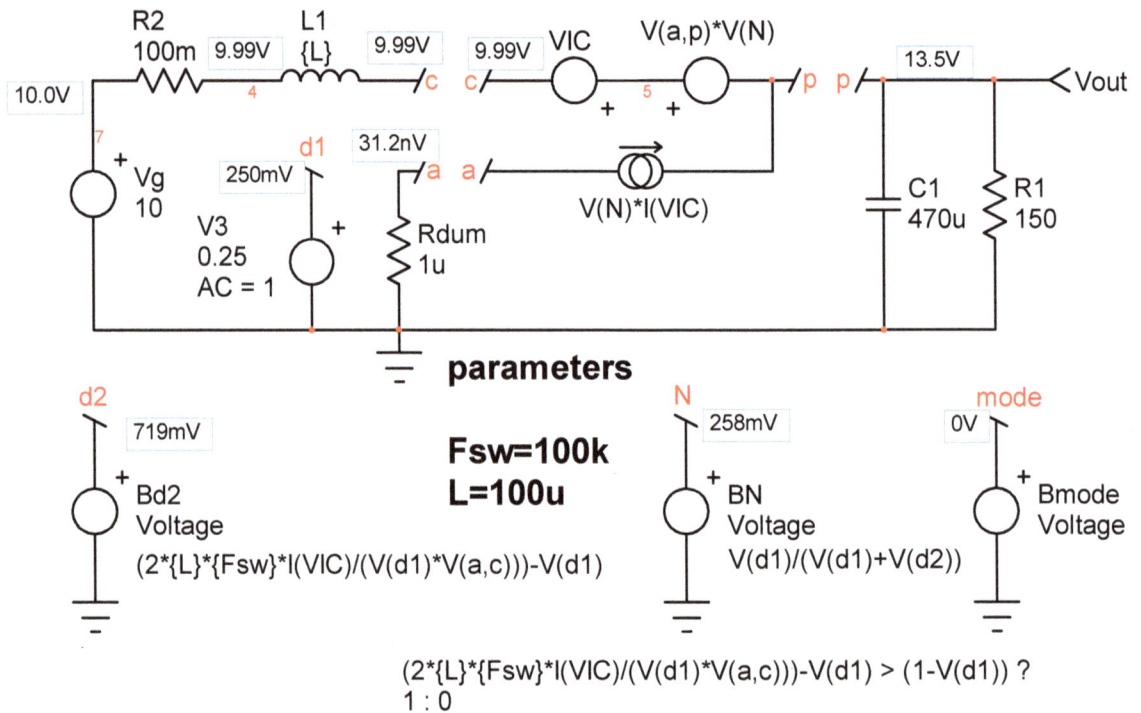

Figure 1.100 The DCM model requires the computation of the new turns ratio and the duration of the demagnetization phase, $D_2 T_{sw}$.

The simulation results appear in Figure 1.101 and show that the dynamic response of the DCM-operated boost converter is still of second-order and features a right-half-plane zero as illustrated in Figure 1.38. None of these results were predicted by the SSA and Dr. Vorpérian was the first to show these effects with the PWM switch model. How do we know, by the way, that a RHPZ hides in the Bode plot? Because if the -2-slope breaks into -1-slope, it implies the presence of zero. The zero should bring the phase up at this point — we say a zero *boosts* the phase. However, observing the phase response, you see that it lags even more after the inflection point.

This is the typical action of a RHP zero: the dynamic responses of the boost and buck-boost converters operated in CCM or DCM are described by non-minimum-phase functions.

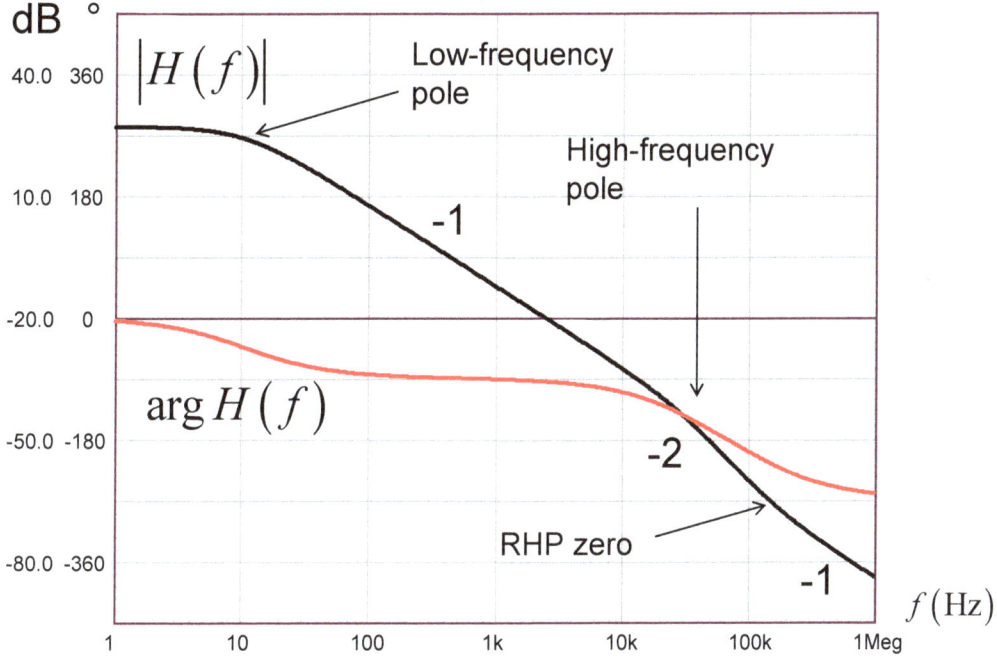

Figure 1.101 Small-signal response of the boost converter using the DCM PWM switch model.

Before looking at the small-signal model of the DCM PWM switch, we need the complete large-signal equations. The first one is the current flowing into terminal a. I_a is defined by (1.186) in which we have substituted (1.195):

$$I_a = NI_c = \frac{V_{ac}D_1^2}{2F_{sw}I_cL}I_c = \frac{V_{ac}D_1^2}{2F_{sw}L} \tag{1.196}$$

A similar operation is performed over (1.189):

$$V_{cp} = NV_{ap} = V_{ap}\frac{V_{ac}D_1^2}{2F_{sw}I_cL} \tag{1.197}$$

We have the two detailed large-signal equations governing the DCM model, we can now proceed with the linearization process.

1.3.3   Small-Signal Model in DCM

We now need to linearize the sources previously determined. The first source, I_a, depends on two variables, D_1 and the voltage V_{ac}. We can write:

$$\hat{i}_a = \frac{\partial I_a\left(D_1, I_c, V_{ac}\right)}{\partial D_1}\hat{d}_1 + \frac{\partial I_a\left(D_1, I_c, V_{ac}\right)}{\partial V_{ac}}\hat{v}_{ac} \tag{1.198}$$

The differentiation leads to:

$$\hat{i}_a = \frac{V_{ac}D_1}{F_{sw}L}\hat{d}_1 + \frac{D_1^2}{2F_{sw}L}\hat{v}_{ac} = k_1\hat{d}_1 + k_2\hat{v}_{ac} \tag{1.199}$$

with:

$$k_1 = \frac{V_{ac} D_1}{F_{sw} L} \tag{1.200}$$

and:

$$k_2 = \frac{D_1^2}{2 F_{sw} L} \tag{1.201}$$

The same operation is performed on (1.197):

$$\hat{v}_{cp} = \frac{\partial V_{cp}\left(D_1, I_c, V_{ac}, V_{ap}\right)}{\partial D_1} \hat{d}_1 + \frac{\partial V_{cp}\left(D_1, I_c, V_{ac}, V_{ap}\right)}{\partial V_{ap}} \hat{v}_{ap} + \frac{\partial V_{cp}\left(D_1, I_c, V_{ac}, V_{ap}\right)}{\partial I_c} \hat{i}_c + \frac{\partial V_{cp}\left(D_1, I_c, V_{ac}, V_{ap}\right)}{\partial V_{ac}} \hat{v}_{ac} \tag{1.202}$$

The differentiation leads to

$$\hat{v}_{cp} = \frac{V_{ap} V_{ac} D_1}{F_{sw} I_c L} \hat{d}_1 + \frac{V_{ac} D_1^2}{2 F_{sw} I_c L} \hat{v}_{ap} - \frac{V_{ap} V_{ac} D_1^2}{2 F_{sw} I_c^2 L} \hat{i}_c + \frac{V_{ap} D_1^2}{2 F_{sw} I_c L} \hat{v}_{ac} = k_3 \hat{d}_1 + k_4 \hat{v}_{ap} + k_5 \hat{i}_c + k_6 \hat{v}_{ac} \tag{1.203}$$

with

$$k_3 = \frac{V_{ap} V_{ac} D_1}{F_{sw} I_c L} \tag{1.204}$$

$$k_4 = \frac{V_{ac} D_1^2}{2 F_{sw} I_c L} \tag{1.205}$$

$$k_5 = -\frac{V_{ap} V_{ac} D_1^2}{2 F_{sw} I_c^2 L} \tag{1.206}$$

$$k_6 = \frac{V_{ap} D_1^2}{2 F_{sw} I_c L} \tag{1.207}$$

With all these definitions on hand, we can compare the dynamic response obtained with the linearized model with those obtained from the model in Figure 1.100.

As confirmed by Figure 1.102, the results are identical.

Please note the automation of the k coefficients in the left side of the picture.

parameters

Fsw=100k
L=100u
d1=250m
Vac=-9.99
Vap=-13.5
Ia=-31.21m
Ic=-120.956m

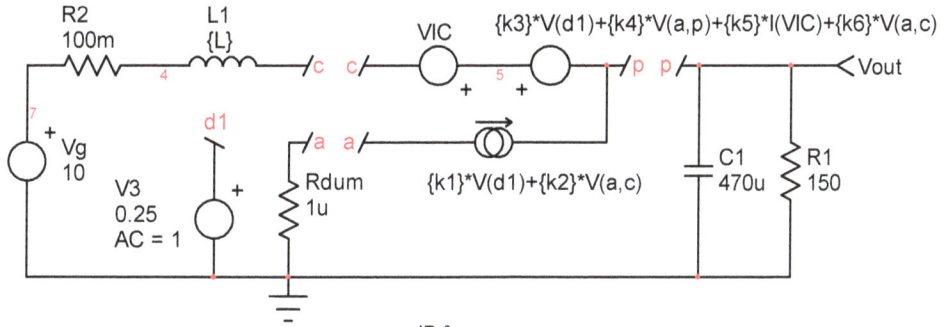

k1=Vac*d1/(Fsw*L)
k2=d1^2/(2*Fsw*L)
k3=Vac*Vap*d1/(Fsw*L*Ic)
k4=Vac*d1^2/(2*Fsw*L*Ic)
k5=-Vac*Vap*d1^2/(2*Fsw*Ic^2*L)
k6=Vap*d1^2/(2*Fsw*Ic*L)

Automated calculations

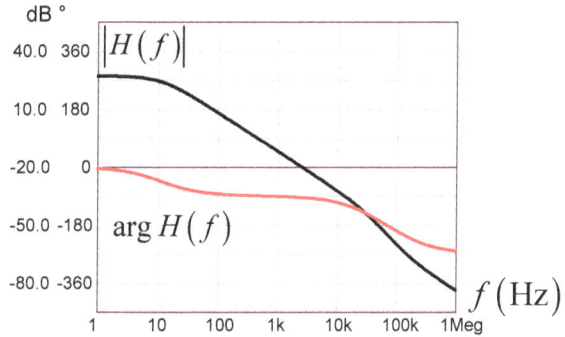

Figure 1.102 The linearized model of the DCM PWM switch model gives results similar to those of Figure 1.101.

Figure 1.103 summarizes both large- and small-signal models of the PWM switch which can be used when studying a converter operated in discontinuous conduction mode.

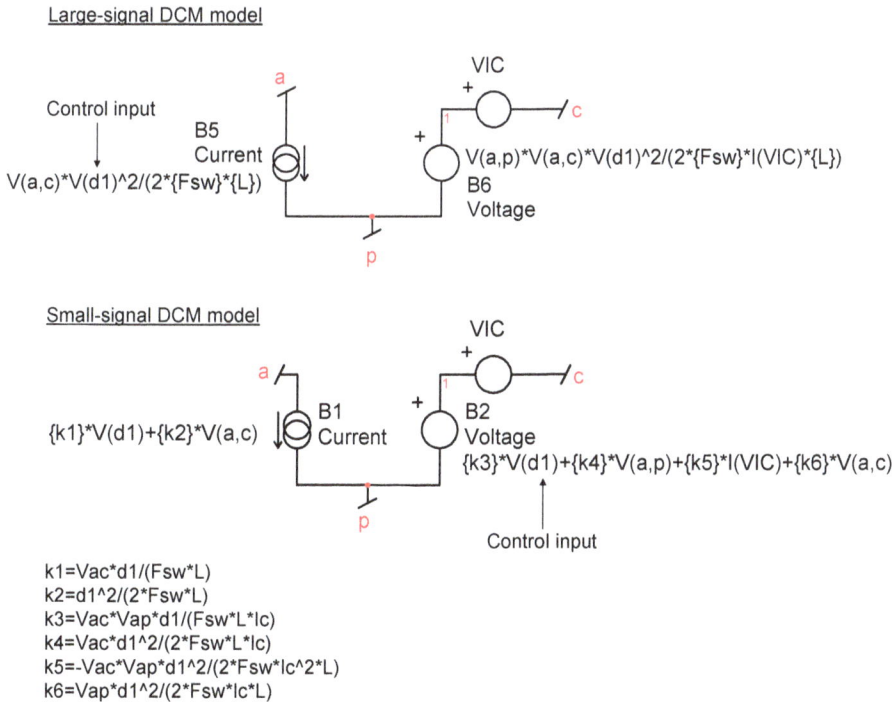

Figure 1.103 The large- and small-signal models of the PWM switch operated in DCM.

1.3.4 DC Operating Points with the PWM Switch Model

Whenever we talk about a small-signal or linear model, we need to specify the operating point at which the linearization is performed. For instance, if the converter delivers 2 A from a 10-V input, then regardless of the operating mode, the PWM switch model needs to be fed with dc values indicative of I_c, I_a, V_{ap} and V_{cp}.

   These values can be obtained by simulation – you can use the large-signal model and collect the values – or by calculation. For the determination of these variables in dc analysis, you will open-circuit the capacitors and short-circuit the inductors. We can start with the simplest configuration (no loss considered), the buck converter shown in Figure 1.104.

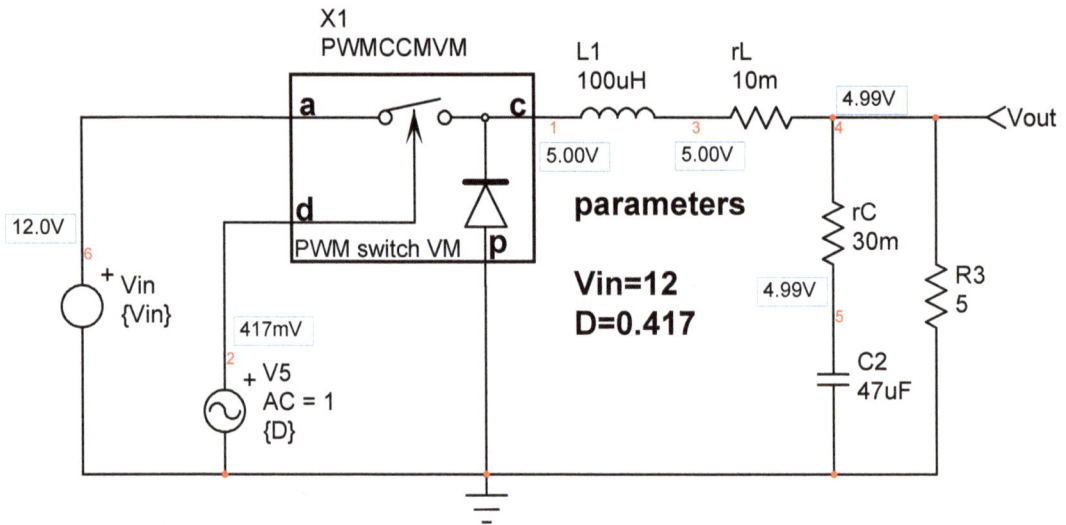

Figure 1.104 The large-signal CCM PWM switch installed in a buck converter.

The CCM values are obvious since the current in terminal c is the continuous current supplied to the load. As we short the inductor, the voltage at terminal c is V_{OUT} (neglecting r_L). Therefore, we have:

CCM:

$$I_c = \frac{V_{out}}{R} \tag{1.208}$$

$$I_a = DI_c \tag{1.209}$$

$$V_{ap} = V_{in} \tag{1.210}$$

$$V_{ac} = V_{in} - V_{out} \tag{1.211}$$

DCM: For the DCM operation, voltage bias points won't change but the current definitions and duty ratio will. Figure 1.105 shows how the model defined in the above lines takes place in the DCM-operated buck converter.

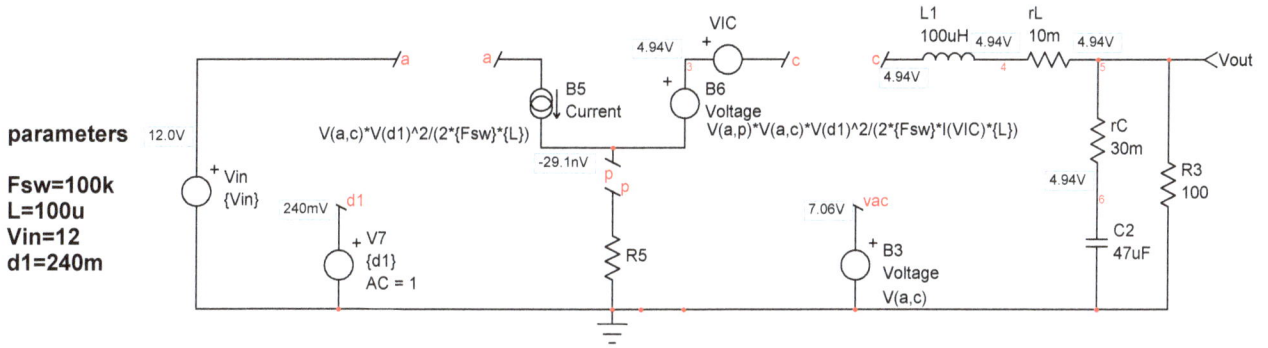

Figure 1.105 The DCM PWM switch is installed in the buck converter while it operates in the discontinuous mode.

We can easily determine the current in terminal c which is still the output voltage divided by the load resistance. Using Figure 1.46 value for V_{out} while considering $r_L \ll R$, we have:

$$I_c = \frac{V_{out}}{R} \approx \frac{V_{in}D_1^2\left(\sqrt{\dfrac{8F_{sw}L}{D_1^2 R}+1}-1\right)}{4F_{sw}L} \tag{1.212}$$

$$I_a = NI_c = \frac{V_{ac}D_1^2}{2F_{sw}L} \tag{1.213}$$

$$V_{ap} = V_{in} \tag{1.214}$$

$$V_{ac} = V_{in} - V_{out} \tag{1.215}$$

For the CCM boost converter (see Figure 1.106), the current in terminal c has already been determined with (1.57) and (1.58) when extracted from a perfect converter. Please note the negative sign respecting the original polarity of I_c entering the model. A similar comment applies to the voltages. As the inductor is a short circuit in dc analysis, the voltage at terminal c is V_{in} (ignoring r_L). We have:

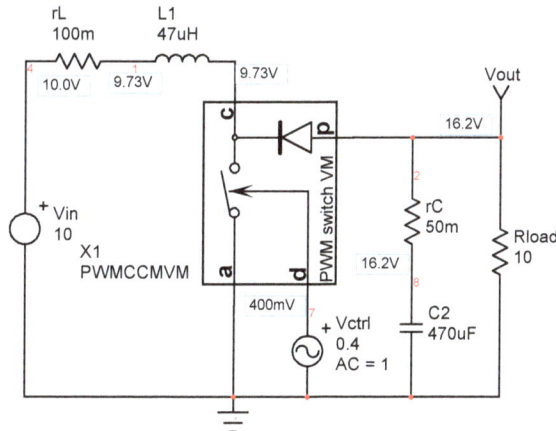

Figure 1.106 The large-signal CCM PWM switch installed in a boost converter.

CCM:

$$I_c = -\frac{I_{out}}{1-D} = -\frac{V_{out}}{R(1-D)}$$

(1.216)

$$I_a = DI_c$$

(1.217)

$$V_{ap} = -V_{out}$$

(1.218)

$$V_{ac} = -V_{in}$$

(1.219)

DCM: Determining the current in terminal c requires a bit of work for the DCM mode because of the dead time presence. The DCM model is inserted in the boost converter according to what is reproduced in Figure 1.107.

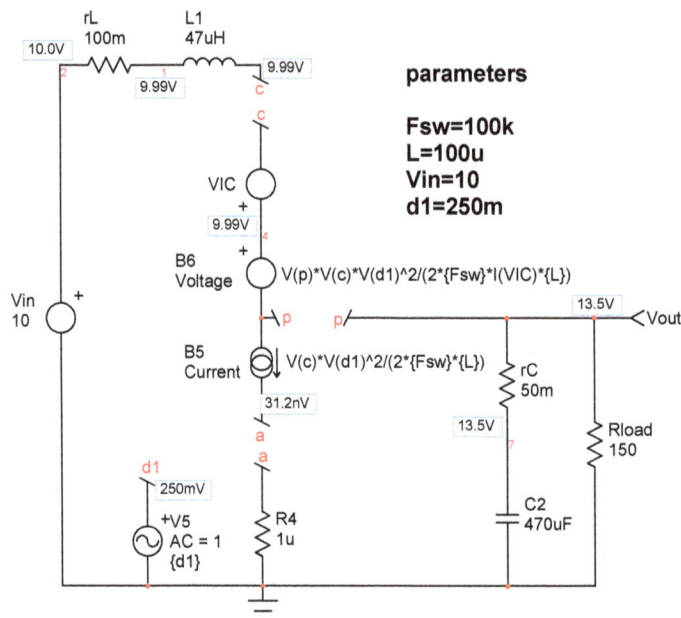

Figure 1.107 The large-signal DCM PWM switch is installed in the boost converter run in the discontinuous mode.

For the dc operating point, we will short the inductor and rearrange the sources with node a grounded. This is what we did in Figure 1.108. It is important to verify that the new schematic once all sources are rearranged delivers bias points identical to the original circuit from Figure 1.107. The new circuit is similar to that of Figure 1.100, but with N determined by its definition in (1.187) and D_2 is substituted with (1.194):

$$N = \frac{V_{ac}D_1^2}{2F_{sw}I_cL}$$

(1.220)

However, you have noticed that terminal a was grounded in the boost converter. Therefore, (1.220) can be rewritten as:

$$N = \frac{\left(0 - V_{(c)}\right)D_1^2}{2F_{sw}I_cL} = \frac{-V_{(c)}D_1^2}{2F_{sw}I_cL}$$

(1.221)

Similarly, the expression in analog behavioral model (ABM) B_2 features also V_{ap} multiplied by N. Considering $V_{(a)} = 0$, B_2 can be rewritten using (1.221) as:

$$V_{B_2} = \frac{V_{(p)}V_{(c)}D_1^2}{2F_{sw}I_cL} \tag{1.222}$$

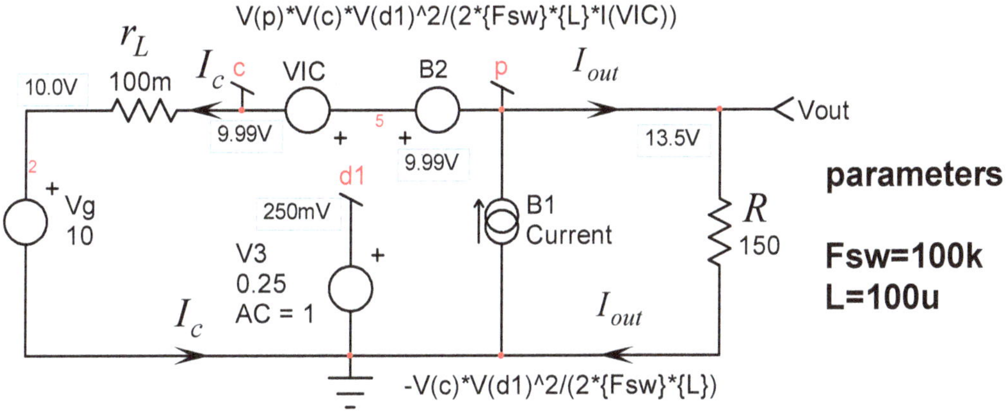

Figure 1.108 The large-signal DCM model is used to compute the dc operating point and the current in terminal *c*.

From Figure 1.108, we can write 3 equations:

$$V_{(c)} = V_{in} + r_L I_c \tag{1.223}$$

$$-\frac{V_{(c)}D_1^2}{2F_{sw}L} = \frac{V_{out}}{R} + I_c \tag{1.224}$$

$$V_{(c)} = V_{out} + \frac{V_{out}V_{(c)}D_1^2}{2LF_{sw}I_c} \tag{1.225}$$

The voltage at node *c* is determined as:

$$V_{(c)} = \frac{V_{in} - \dfrac{V_{out}r_L}{R}}{\dfrac{D_1^2 r_L}{2LF_{sw}} + 1} \tag{1.226}$$

Which, if r_L is very small becomes:

$$V_{(c)} \approx V_{in} \tag{1.227}$$

Substituting (1.227) in (1.224) and solving for I_c, we have:

$$I_c = -\left(\frac{V_{out}}{R} + \frac{V_{in}D_1^2}{2LF_{sw}} \right) \tag{1.228}$$

The current in terminal a, I_a, is obtained by developing (1.186) and plugging D_2's definition from (1.194):

$$I_a = \frac{V_{ac}D_1^2}{2F_{sw}L} \tag{1.229}$$

The rest of the variables comes easily:

$$V_{ap} = -V_{out} \tag{1.230}$$

$$V_{ac} = -V_{in} \tag{1.231}$$

The large-signal model of the CCM-operated buck-boost converter using the PWM switch model appears in Figure 1.109 where the voltage at terminal c is 0 this time.

The converter's I_c parameter is derived from (1.63) and (1.64). Considering a negative output voltage, V_{out} in the below equations, we have:

CCM:

$$I_c = -\frac{I_{out}}{1-D} = -\frac{V_{out}}{R(1-D)} \tag{1.232}$$

The rest of the variables come easily by observing the voltages at the model in Figure 1.109.

$$I_a = DI_c \tag{1.233}$$

$$V_{ap} = V_{in} - V_{out} \tag{1.234}$$

$$V_{ac} = V_{in} \tag{1.235}$$

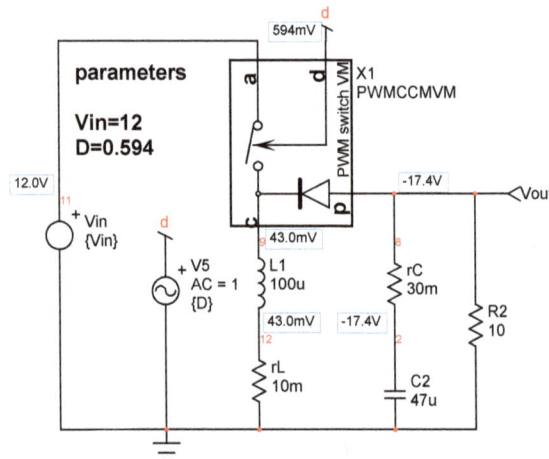

Figure 1.109 The large-signal CCM model in a buck-boost converter.

DCM: To determine the variables in this mode, we will install the large-signal DCM PWM switch model in the buck-boost converter as shown in Figure 1.110. For the dc analysis, we will short the inductor and open the capacitor then rearrange sources while checking the new schematic matches the original one in bias points.

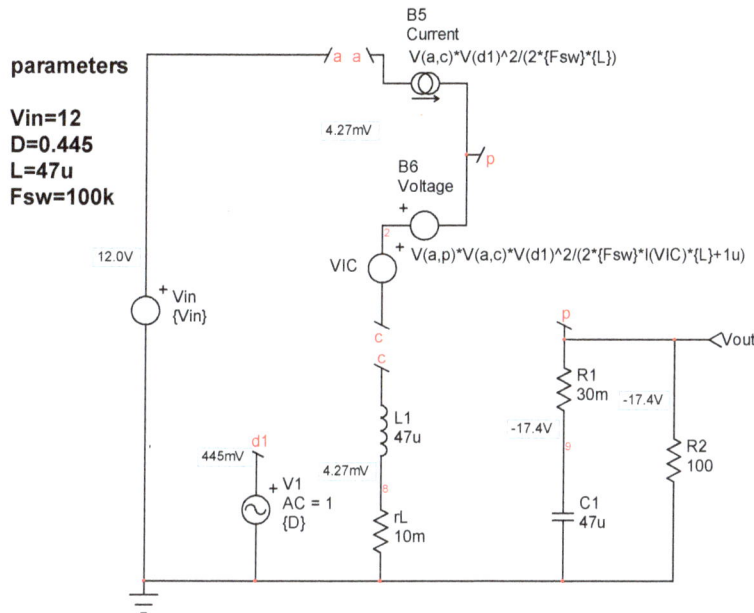

Figure 1.110 The large-signal DCM model is installed in the buck-boost converter and lets us compute the dc operating point and the current in terminal c.

The new circuit appears in Figure 1.111 and is now simpler to study in dc.

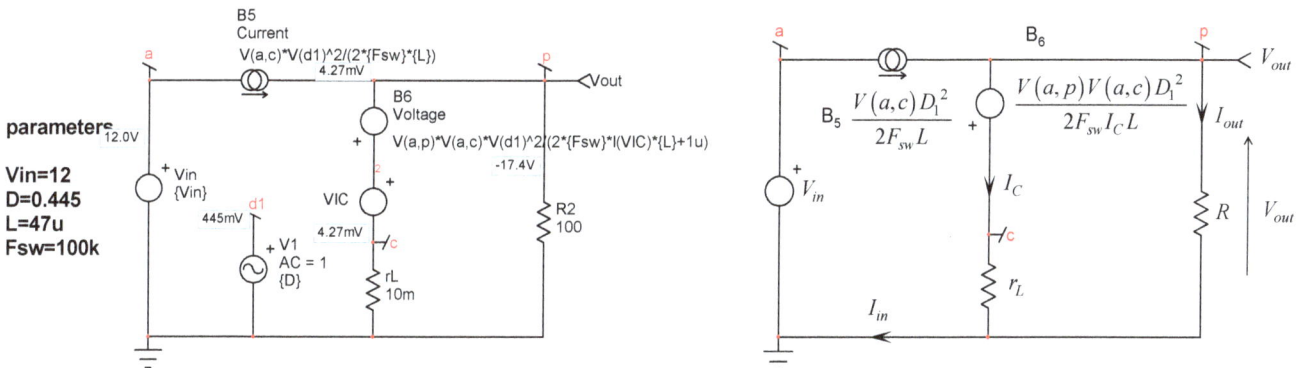

Figure 1.111 The key is to redraw a complex circuit into a simpler one by rearranging sources. Always check that ac and dc responses are not altered when doing so.

We can start with a few equations, for instance, the sum of I_c and I_{out} equal to the input current:

$$I_c + I_{out} = I_{in} = \frac{D_1^2 \left(V_{in} - V_{(c)} \right)}{2 F_{sw} L} \tag{1.236}$$

From which, recognizing that $I_{out} = V_{out}/R$:

$$I_c = \frac{D_1^2 \left(V_{in} - V_{(c)} \right)}{2 \tau_L R} - \frac{V_{out}}{R} \tag{1.237}$$

In which τ_L is the normalized time constant defined in (1.97). The current in terminal c is also:

$$I_c = \frac{V_{(c)}}{r_L} \qquad (1.238)$$

Equating (1.237) and (1.238) then solving for $V_{(c)}$, we have:

$$V_{(c)} = \frac{V_{in}D_1^2 r_L - 2V_{out}r_L\tau_L}{r_L D_1^2 + 2\tau_L R} \qquad (1.239)$$

The voltage at node c is also equal to the output voltage V_{out} plus source B$_6$:

$$V_{(c)} = V_{out} + \frac{(V_{in}-V_{out})(V_{in}-V_{(c)})D_1^2}{2\tau_L R \dfrac{V_{(c)}}{r_L}} \qquad (1.240)$$

If we now extract V_{out} from (1.240) then plug (1.239) in the result and rearrange, we should find:

$$M = \frac{V_{out}}{V_{in}} = -D_1 \frac{\left(r_L D_1^2 + 2\tau_L R\right)\sqrt{8\tau_L R(R+r_L)+D_1^2 r_L^2} - r_L D_1\left(D_1^2 r_L + 6\tau_L R\right)}{8\left(R^2\tau_L^2 + RD_1^2 r_L\tau_L + Rr_L\tau_L^2 + \dfrac{D_1^4 r_L^2}{4}\right)} \qquad (1.241)$$

which simplifies to:

$$M = -\frac{D_1}{\sqrt{2\tau_L}} \qquad (1.242)$$

when r_L is negligible. This is what is given in Figure 1.46. Finally, the current in terminal c is obtained by combining (1.242), (1.239) and (1.238) while neglecting r_L:

$$I_c = \frac{V_{in}\left(D_1^2 - 2M\tau_L\right)}{2\tau_L R} = \frac{V_{in}D_1^2}{2F_{sw}L} - \frac{V_{out}}{R} \qquad (1.243)$$

The rest of the voltages are similar to those of the CCM arrangement:

$$I_a = DI_c \qquad (1.244)$$

$$V_{ap} = V_{in} - V_{out} \qquad (1.245)$$

$$V_{ac} = V_{in} \qquad (1.246)$$

These dc points calculations will be useful when determining the transfer functions. We have gathered them all below in the arrays of Figure 1.112 and Figure 1.113 respectively for the CCM and DCM cases.

CCM:

Converter	I_c	I_a	V_{ap}	V_{ac}
Buck	$\dfrac{V_{out}}{R}$	DI_c	V_{in}	$V_{in} - V_{out}$
Boost	$-\dfrac{V_{out}}{R(1-D)}$	DI_c	$-V_{out}$	$-V_{in}$
Buck-Boost	$-\dfrac{V_{out}}{R(1-D)}$	DI_c	$V_{in} - V_{out}$	V_{in}

V_{out} is a negative value for the buck-boost converter

Figure 1.112 Summary of the dc points for the large-signal PWM switch model operated in CCM.

DCM:

Converter	I_c	I_a	V_{ap}	V_{ac}
Buck	$\dfrac{V_{in}D_1^2\left(\sqrt{\dfrac{8F_{sw}L}{D_1^2 R}+1}-1\right)}{4F_{sw}L}$	$\dfrac{V_{ac}D_1^2}{2F_{sw}L}$	V_{in}	$V_{in} - V_{out}$
Boost	$-\left(\dfrac{V_{out}}{R}+\dfrac{V_{in}D_1^2 R}{2LV_{out}F_{sw}}\right)$	$\dfrac{V_{ac}D_1^2}{2F_{sw}L}$	$-V_{out}$	$-V_{in}$
Buck-Boost	$\dfrac{V_{in}D_1^2}{2F_{sw}L}-\dfrac{V_{out}}{R}$	DI_c	$V_{in} - V_{out}$	V_{in}

V_{out} is a negative value for the buck-boost converter

Figure 1.113 Summary of the dc points for the large-signal PWM switch model operated in DCM.

1.3.5 Quasi-Resonant Operation

A model of the PWM switch can be derived when operated in the so-called quasi-square-wave resonant mode often abbreviated QR. The principle does not change: draw the voltage/currents waveforms and determine the average values for I_a, I_c, V_{ap} and V_{cp}. The drawing from Figure 1.114 shows the currents and voltages in borderline mode of a buck converter wired in the common-passive configuration. We have purposely ignored the dead time for the sake of simplicity. You can see the current reaching a peak and returning to zero at the end of the demagnetization time. The controller detects the zero current condition and issues a new restart pulse at this moment. Please note that QR converters are self-relaxing systems operating without an internal clock. The frequency varies with operating conditions as shown in [1] and needs to be determined. Voltage-mode control QR is very popular in constant-on-time power factor correction (PFC) control circuits.

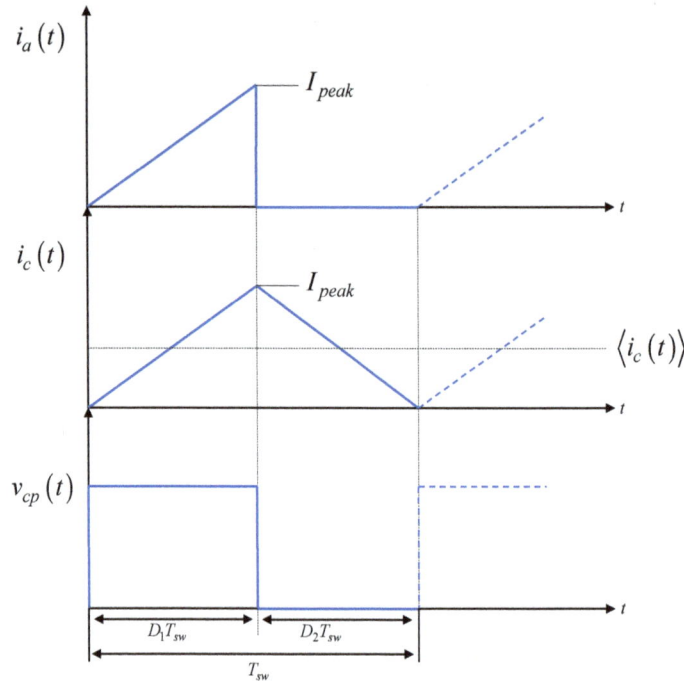

Figure 1.114 Waveforms of the PWM switch model operated in borderline mode.

In voltage mode control, the control variable is the on-time t_{on}. There is no PWM circuit like in fixed-frequency operation but rather a capacitor-based modulator driven by the error voltage. Equations are not really complicated mainly because in absence of dead time. We can show that the average current in terminal c is exactly the peak current divided by 2:

$$I_c = \frac{I_{peak}}{2} \tag{1.247}$$

If we still consider a common-passive configuration like what we showed in Figure 1.90, the inductor downslope is the voltage applied across the inductor during the off-time t_{off} or D_2T_{sw}. In a buck converter, the voltage is $-V_{out}$ as the freewheel diode conducts.

In an invariant form, meaning we write the average equations in relationship with terminals a, c and p, the off-time is defined as:

$$D_2T_{sw} = t_{off} = \frac{L}{V_{cp}}I_{peak} \tag{1.248}$$

If we now extract I_{peak} from (1.247) and substitute it in (1.248), we have:

$$t_{off} = \frac{2LI_c}{V_{cp}} \tag{1.249}$$

This expression (modeled as a source) will determine the off-time duration based on the inductor peak current indirectly set by the error voltage. Considering t_{on} as the input parameter, without dead time, we have:

$$T_{sw} = t_{on} + t_{off} \qquad (1.250)$$

The duty ratio D_1 comes easily then:

$$D_1 = \frac{t_{on}}{t_{on}+t_{off}} = \frac{t_{on}}{t_{on}+\frac{2LI_c}{V_{cp}}} = \frac{1}{1+\frac{2LI_c}{V_{cp}t_{on}}} \qquad (1.251)$$

This is it, we have the large-signal model of the voltage-mode PWM switch model operated in QR. As shown in Figure 1.115, it is the voltage-mode model to which a duty ratio generator is added.

Please note that the error voltage controlling t_{on} is scaled up by 1 million so that 1 V represents 1 µs.

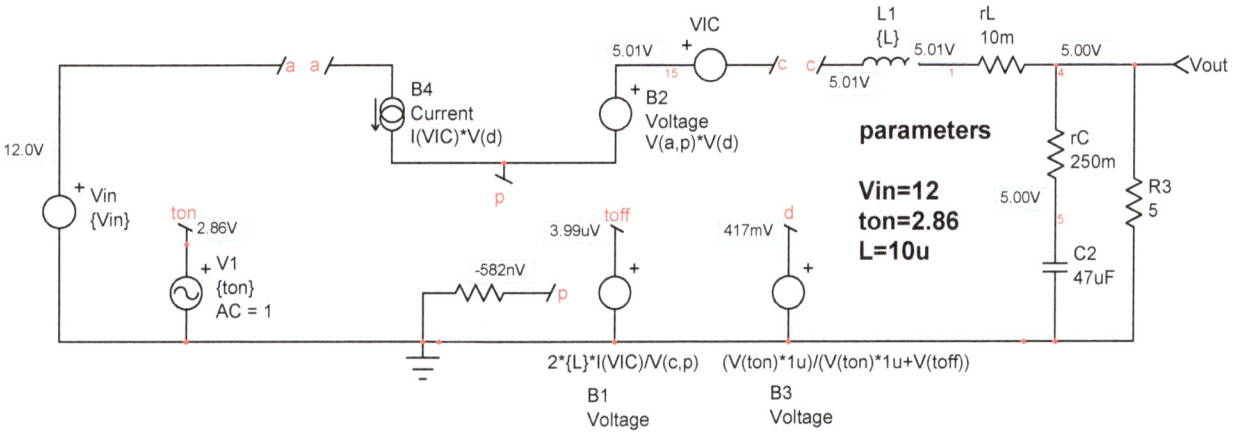

Figure 1.115 The large-signal BCM PWM switch model is a voltage-mode model to which a duty ratio generator is added.

In this QR buck converter, the frequency is obtained by combining the t_{off} source and the t_{on} setpoint:

$$F_{sw} = \frac{1}{t_{on}+t_{off}} = \frac{1}{3.99u+2.86u} \approx 146 \text{ kHz} \qquad (1.252)$$

The first source I_a is classically defined as:

$$I_a = D_1 I_c \qquad (1.253)$$

which, using (1.251), becomes:

$$I_a = I_c \frac{t_{on}}{t_{on}+\frac{2LI_c}{V_{cp}}} = \frac{I_c V_{cp} t_{on}}{2I_c L+V_{cp}t_{on}} \qquad (1.254)$$

The voltage source V_{cp} is defined as:

$$V_{cp} = D_1 V_{ap} = V_{ap}\frac{t_{on}}{t_{on}+\frac{2LI_c}{V_{cp}}} \qquad (1.255)$$

which if you solve for V_{cp} becomes:

$$V_{cp} = \frac{V_{ap}t_{on} - 2I_c L}{t_{on}} \qquad (1.256)$$

Figure 1.116 shows how to assemble the final large-signal model with only two sources as in the original PWM switch model. We can now linearize these I_a and V_{cp} generators to obtain a small-signal model. Please note that the dc bias points are identical between the models in Figure 1.115 and Figure 1.116. We also made sure both dynamic responses were also similar.

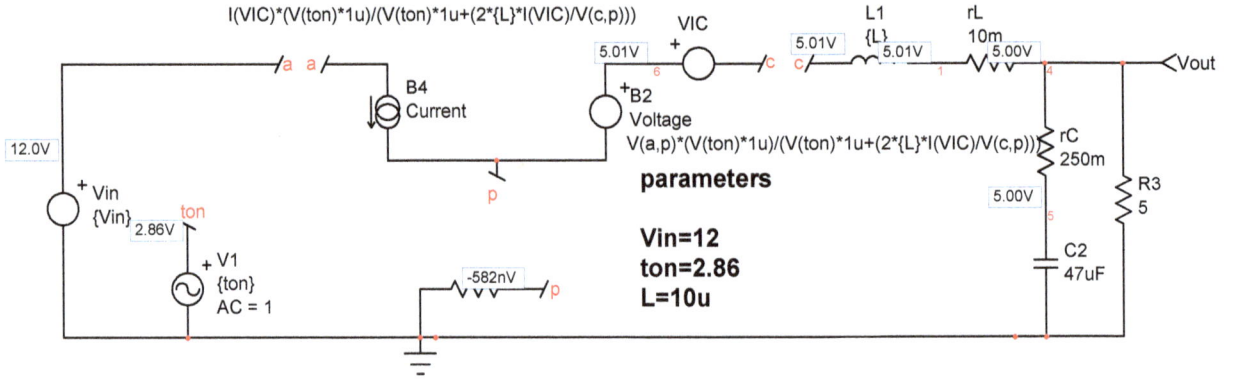

Figure 1.116 Once all sources are combined together, the model gains in compactness.

The V_{cp} source depends on three variables: V_{ap}, t_{on} and I_c. Let's determine their small-signal coefficients.

$$k_{ton} = \frac{d}{dt_{on}} \frac{V_{ap}t_{on} - 2I_c L}{t_{on}} = \frac{2I_c L}{t_{on}^2} \qquad (1.257)$$

$$k_{ic} = \frac{d}{dI_c} \frac{V_{ap}t_{on} - 2I_c L}{t_{on}} = -\frac{2L}{t_{on}} \qquad (1.258)$$

$$k_{ap} = \frac{d}{dV_{ap}} \frac{V_{ap}t_{on} - 2I_c L}{t_{on}} = 1 \qquad (1.259)$$

Combining these results, we have:

$$\hat{v}_{cp} = k_{ton}\hat{t}_{on} + k_{ic}\hat{i}_c + k_{ap}\hat{v}_{ap} \qquad (1.260)$$

We can proceed with the I_a source defined by (1.254).

This expression depends on three variables and will thus need three small-signal coefficients:

$$k_{tonia} = \frac{d}{dt_{on}} \frac{I_c V_{cp}t_{on}}{2I_c L + V_{cp}t_{on}} = \frac{2I_c^2 L V_{cp}}{\left(2I_c L + V_{cp}t_{on}\right)^2} \qquad (1.261)$$

$$k_{ica} = \frac{d}{dI_c} \frac{I_c V_{cp} t_{on}}{2I_c L + V_{cp} t_{on}} = \frac{V_{cp}^2 t_{on}^2}{\left(2I_c L + V_{cp} t_{on}\right)^2} \tag{1.262}$$

$$k_{vcpia} = \frac{d}{dI_c} \frac{I_c V_{cp} t_{on}}{2I_c L + V_{cp} t_{on}} = \frac{2t_{on} I_c^2 L}{\left(2I_c L + V_{cp} t_{on}\right)^2} \tag{1.263}$$

The small-signal equation of the current source becomes:

$$\hat{i}_a = k_{tonia}\hat{i}_{on} + k_{ica}\hat{i}_c + k_{cpia}\hat{v}_{cp} \tag{1.264}$$

We have tested the dynamic responses delivered by the large-signal model from Figure 1.115 and the linearized model whose expressions were determined in the above lines. The coefficients are calculated in the left side of the picture. As Figure 1.117 confirms, respective dynamic responses perfectly superimpose.

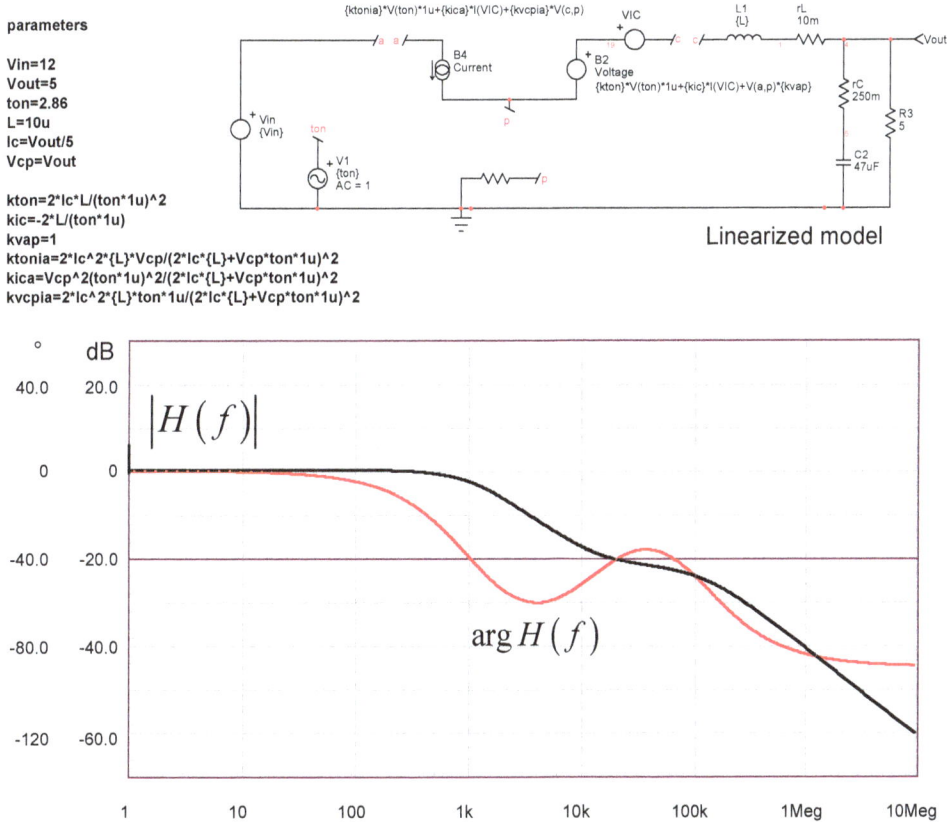

Figure 1.117 This is a small–signal model with coefficients automatically computed in the left side of the picture.

1.4 The PWM Switch Model in Current Mode Control

The CM-PWM switch model has been derived by Dr. Vorpérian in 1990 [11].

The CCM case alone was documented. The current-mode model departs from the original voltage-mode type by associating two current sources. The principle behind this model remains the same: we will average the instantaneous current waveforms $i_a(t)$ and $i_c(t)$ observed in a buck converter.

This last one appears in Figure 1.118. The peak current is determined by the setpoint voltage V_c – the control variable – and the sense resistance R_{sense} or R_i. However, because of the artificial ramp added to fix subharmonic oscillations, the real setpoint is lower. The current at point x in Figure 1.118 is given as:

$$I_{c(x)} = \frac{V_c}{R_i} - \frac{S_e DT_{sw}}{R_i} \qquad (1.265)$$

What we want, however, is the average current flowing out of terminal c.

This value, designated at point y in Figure 1.118, lies in the middle of the current downslope S_2 which, by the way, designates half of the inductor ripple ΔI_L. The final expression is thus given by:

$$\langle i_c(t) \rangle = I_{c(x)} - \frac{\Delta I_L}{2} = I_{c(x)} - \frac{S_2 D'T_{sw}}{2} = \frac{V_c}{R_i} - \frac{S_e}{R_i}DT_{sw} - \frac{S_2 D'T_{sw}}{2} \qquad (1.266)$$

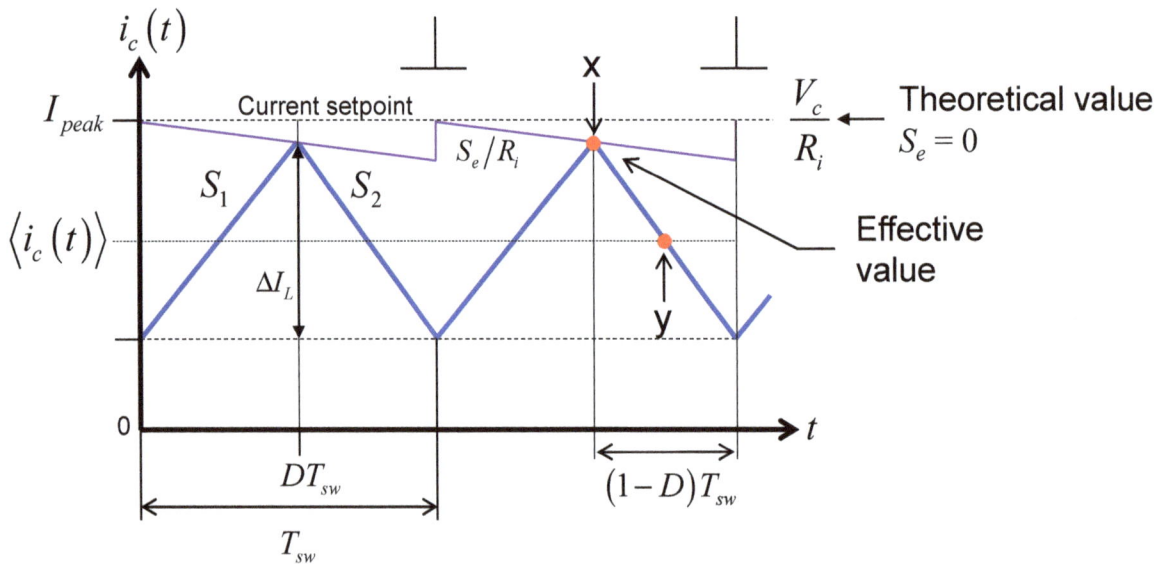

Figure 1.118 The waveform describing the instantaneous current $i_c(t)$.

S_2 in this expression represents the inductor current downslope which links points a and b during $(1-D)T_{sw}$. In Figure 1.119, we see that the downslope current in this buck converter depends on V_{out}:

$$S_2 = \frac{V_{out}}{L} \qquad (1.267)$$

102

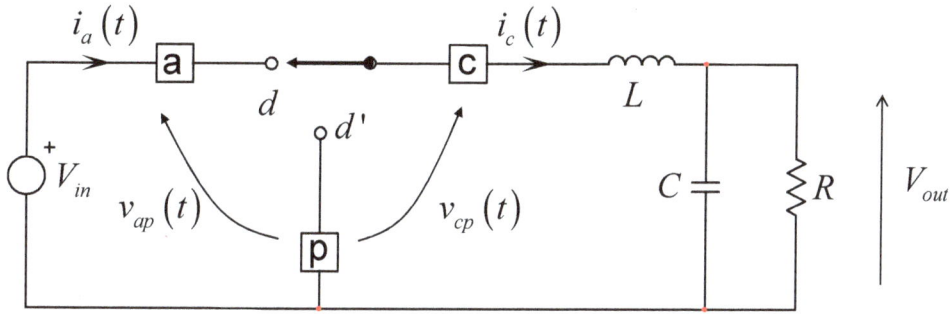

Figure 1.119 The inductor downslope in this buck converter depends on the output voltage.

However, we also know that the average value across the inductor at steady-state is 0 V. Therefore, we can update (1.267) by an invariant relationship:

$$S_2 = \frac{V_{cp}}{L} \qquad (1.268)$$

We can now substitute this value into (1.266) and obtain:

$$I_c = \frac{V_c}{R_i} - V_{cp}(1-D)\frac{T_{sw}}{2L} - \frac{S_e}{R_i}DT_{sw} \qquad (1.269)$$

In which:

$\dfrac{V_c}{R_i}$ is the peak inductor setpoint

$V_{cp}(1-D)\dfrac{T_{sw}}{2L}$ represents half of the inductor current ripple, $\dfrac{\Delta I_L}{2}$

$\dfrac{S_e}{R_i}DT_{sw}$ is the artificial ramp effect

If we group the second and third terms together in a source designated as:

$$I_\mu = V_{cp}(1-D)\frac{T_{sw}}{2L} + \frac{S_e}{R_i}DT_{sw} \qquad (1.270)$$

We can draw the right-side of the PWM switch model in current mode as shown in Figure 1.120.

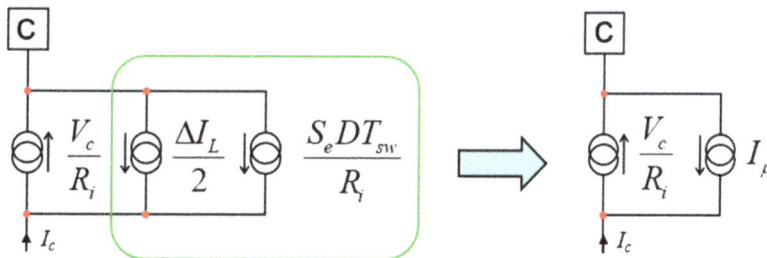

Figure 1.120 The complete invariant equation describing the current leaving terminal *c*.

103

It simply shows that without external ramp ($S_e = 0$), the average inductor current in a CCM peak-current-mode-operated converter is the setpoint V_c/R_i minus half the inductor ripple current ΔI_L.

Now, the current in terminal a is similar to the current observed in the PWM switch model operated in voltage mode: when the switch closes during DT_{sw}, I_a "sees" I_c and returns to 0 for the rest of the time:

$$I_a = DI_c \tag{1.271}$$

This is what Figure 1.121 illustrates.

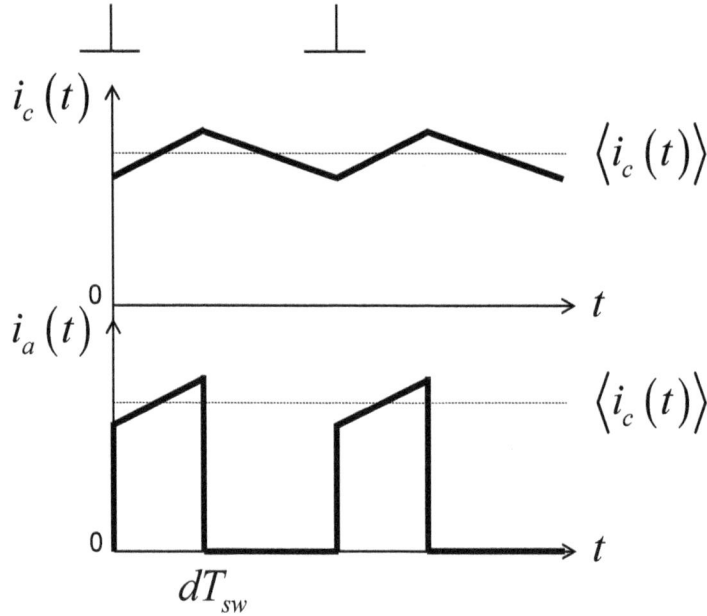

Figure 1.121 The average current entering terminal a is similar to that determined in the voltage-mode PWM switch model.

The relationship linking I_a and I_c is thus that already determined with (1.171), $I_a = DI_c$.

However, in this expression, D is not the control variable but I_{peak} is. The duty ratio is thus *indirectly* controlled by the peak current setpoint. In the perfect buck converter of Figure 1.119, the duty ratio is defined as:

$$D = \frac{V_{out}}{V_{in}} \tag{1.272}$$

Observing that V_{in} is applied across V_{ap} and V_{out} appears across V_{cp} (average inductor voltage is 0 V), then (1.272) can be rewritten in an invariant form as:

$$D = \frac{V_{cp}}{V_{ap}} \tag{1.273}$$

And (1.271) is updated as:

$$I_a = \frac{V_{cp}}{V_{ap}} I_c \tag{1.274}$$

This is it, the PWM switch model in current mode is now entirely defined in Figure 1.122's drawing.

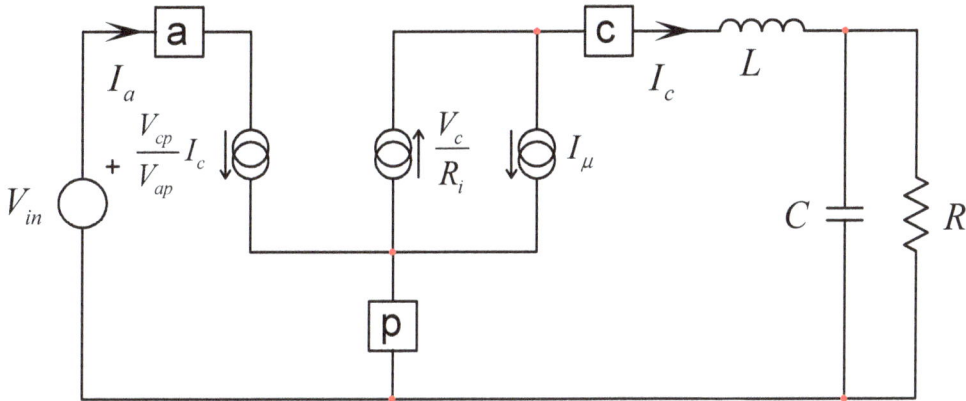

Figure 1.122 The large-signal PWM switch model in current mode is the simplest arrangement you can think of. V_c is the control variable which sets the peak current setpoint via the sense resistance R_i.

This model is truly a marvel as it requires three simple sources to mimic a converter operated in peak current mode control.

And now, by adding a simple capacitor between terminals c and p, sub-harmonic oscillations are modeled exactly as detailed in [1]. The interesting point is that Vorpérian with his simple model obtained identical results to those obtained by Ridley in [5] using sampled-data analysis.

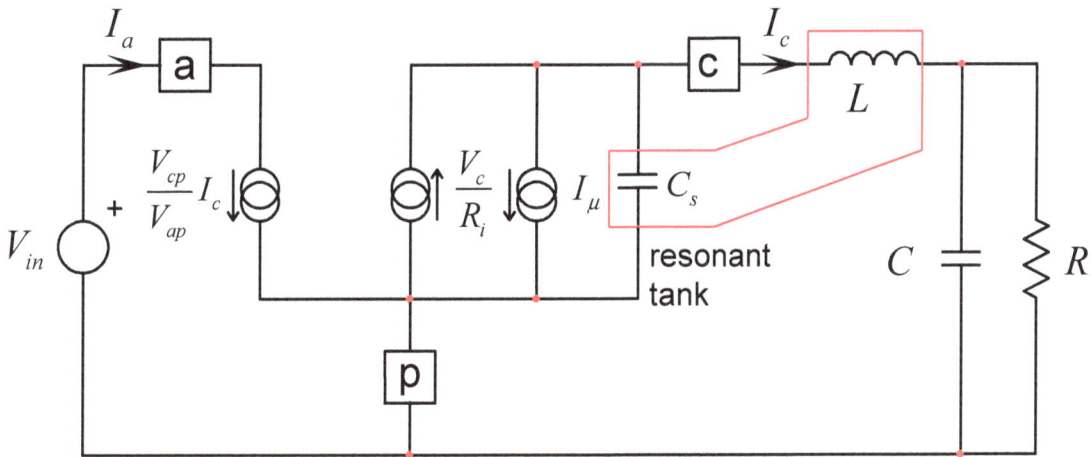

Figure 1.123 The model with this added capacitor predicts sub-harmonic instabilities as Ridley did in his thesis.

This extra capacitor is calculated as follows:

$$C_s = \frac{1}{L\left(F_{sw}\,\pi\right)^2} \tag{1.275}$$

in which F_{sw} is the switching frequency.

$$\{Se\}*V(D)/(\{Ri\}*\{Fsw\}) + v(c,p)*(1-V(D))*(\{1/Fsw\}/(2*\{L\}))$$

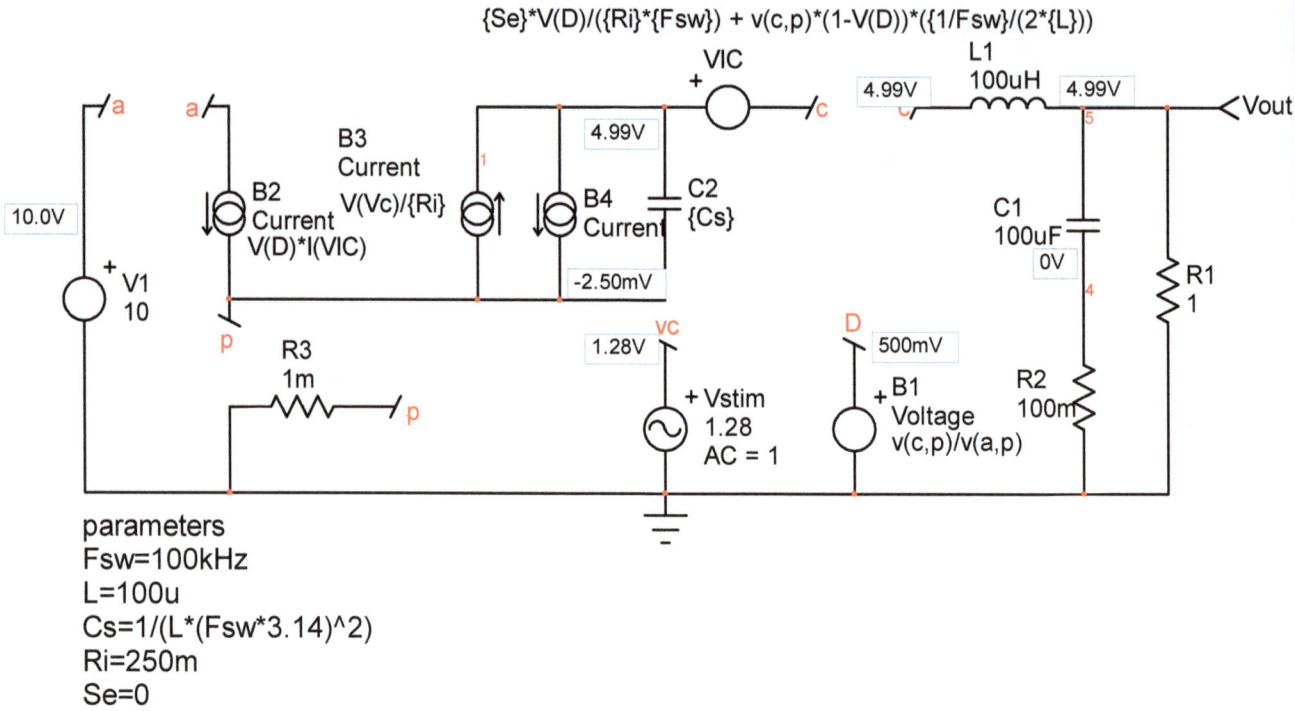

Figure 1.124 The model is tested in a buck converter delivering 5 V/5 A.

parameters
Fsw=100kHz
L=100u
Cs=1/(L*(Fsw*3.14)^2)
Ri=250m
Se=0

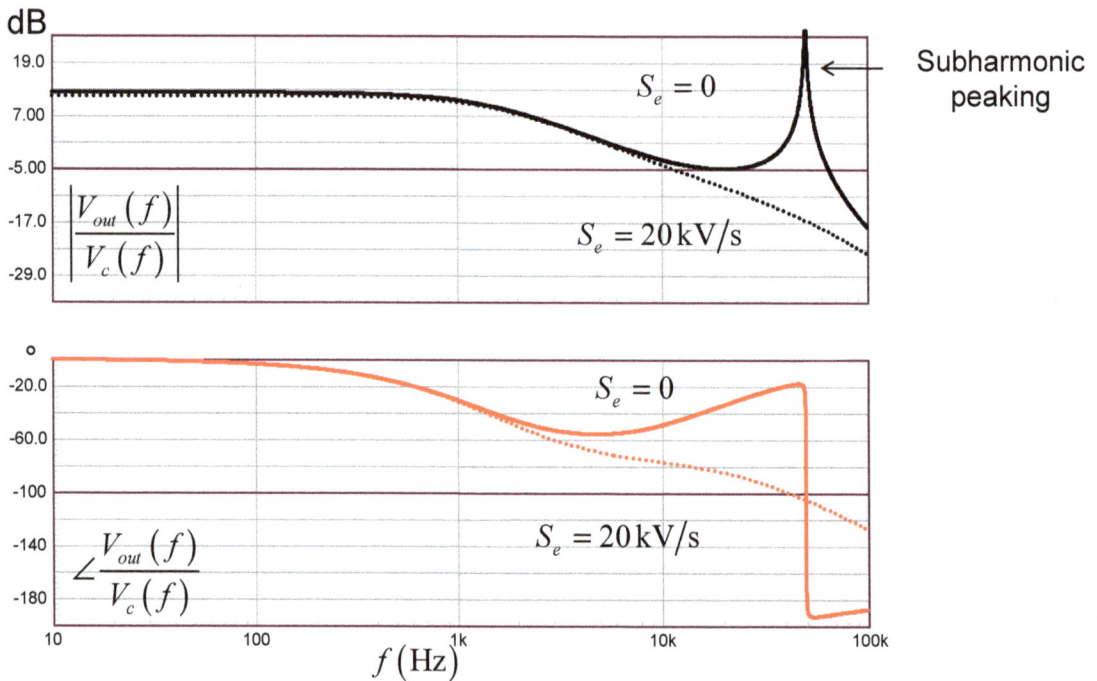

Figure 1.125 Without slope compensation ($S_e = 0$), the magnitude severely peaks as expected. When slope compensation is added, the poles are well damped.

Figure 1.124 shows the model wired in a current-mode buck converter. Source B_1 calculates the duty ratio while the two other sources implement the current absorbed by terminal a and that delivered by terminal c.

The simulation results from Figure 1.125 acknowledge for the sub-harmonic peaking and show how injecting slope compensation tames the quality factor Q. The control voltage V_c has been tweaked to get a 5-V output. To determine this value precisely, we used (1.269) from which we simply extracted V_c realizing that $I_c = I_{out} = 5$ A, $V_{cp} = V_{out} = 5$ V and D_0 is 50%:

$$V_c = R_i \left(I_c + \frac{T_{sw}V_{cp}(1-D_0)}{2L} + \frac{D_0 T_{sw} S_e}{R_i} \right) = 0.25 \times \left(5 + \frac{10u \times 5 \times (1-0.5)}{2 \times 100u} \right) = 1.28 \text{ V} \qquad (1.276)$$

Once last comment though: when you configure the CM-PWM switch model for a boost converter following Figure 1.89, you can see that the current enters terminal c rather than leaving it as for the buck configuration. To make the CM-PWM model work in this configuration, R_i must take on a negative value.

Another way exists to model a current-mode converter using the VM-PWM switch model.

Between a converter operated in fixed-frequency current-mode control and the same converter operated in voltage mode, nothing, when observing the waveforms with an oscilloscope, tells you which one of the two you look at.

For instance, with a CCM buck, $V_{out} = DV_{in}$ is always true whether you operate in voltage- or current-mode control. This is because the power stage is similar; it is the way the duty ratio is elaborated that changes the picture.

Take a look at Figure 1.126 which shows how two different pulse-width modulators drive the power stage. In the left side, in voltage mode, the duty ratio depends on the error voltage and the artificial ramp. In the right side, the duty ratio depends on the artificial (stabilizing) ramp and the inductor current information transformed into voltage by the sense resistance R_i.

The duty ratio now depends on both the artificial and the current sense signal. And you see that if you exaggeratedly increase the artificial ramp in the current-mode case – in other words if you overcompensate the converter – the artificial ramp S_e dominates over the current sense information S_n and you degenerate your current-mode converter into a voltage-mode converter.

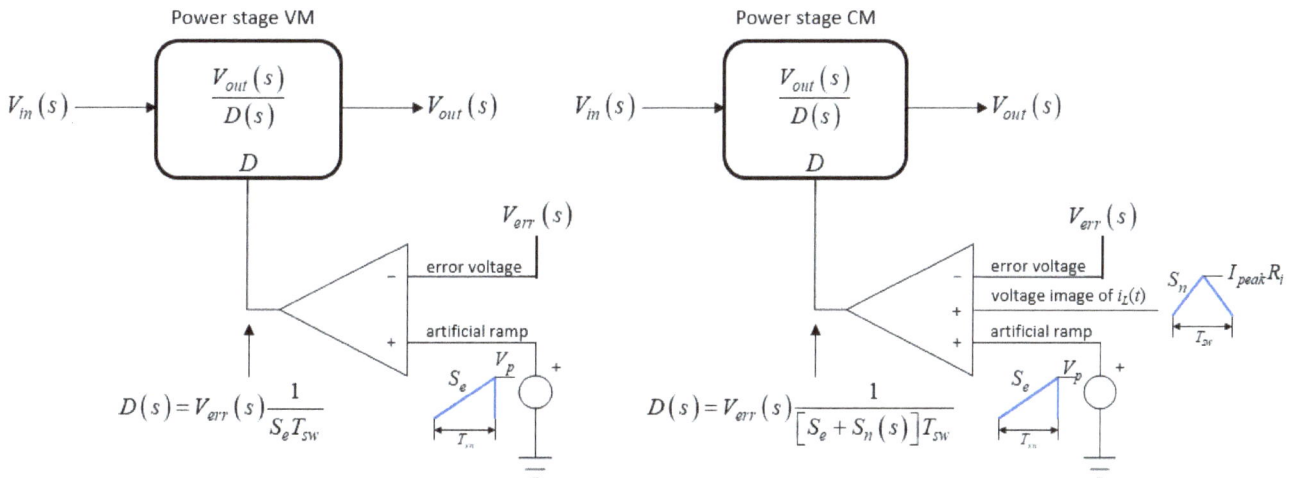

Figure 1.126 In voltage-mode control, the duty ratio depends on the artificial ramp and the error voltage. In current-mode control, the duty ratio depends on the artificial ramp, the inductor current and the error voltage.

Using this approach, we can think of modeling the current-mode converter by associating the PWM switch model operated in voltage mode with a block generating the duty ratio D.

Rewriting (1.269) now considering V_{ac} to express the ripple, we have the following alternate expression:

$$I_c = \frac{V_c}{R_i} - DT_{sw}\frac{V_{ac}}{2L} - \frac{S_e}{R_i}DT_{sw} \tag{1.277}$$

We can extract the duty ratio as follows:

$$D = \frac{F_{sw}\left(V_c - R_i I_c\right)}{S_e + \dfrac{R_i V_{ac}}{2L}} \tag{1.278}$$

We can install this generator together with the CCM PWM switch model we have already described. The application circuit appears in Figure 1.127.

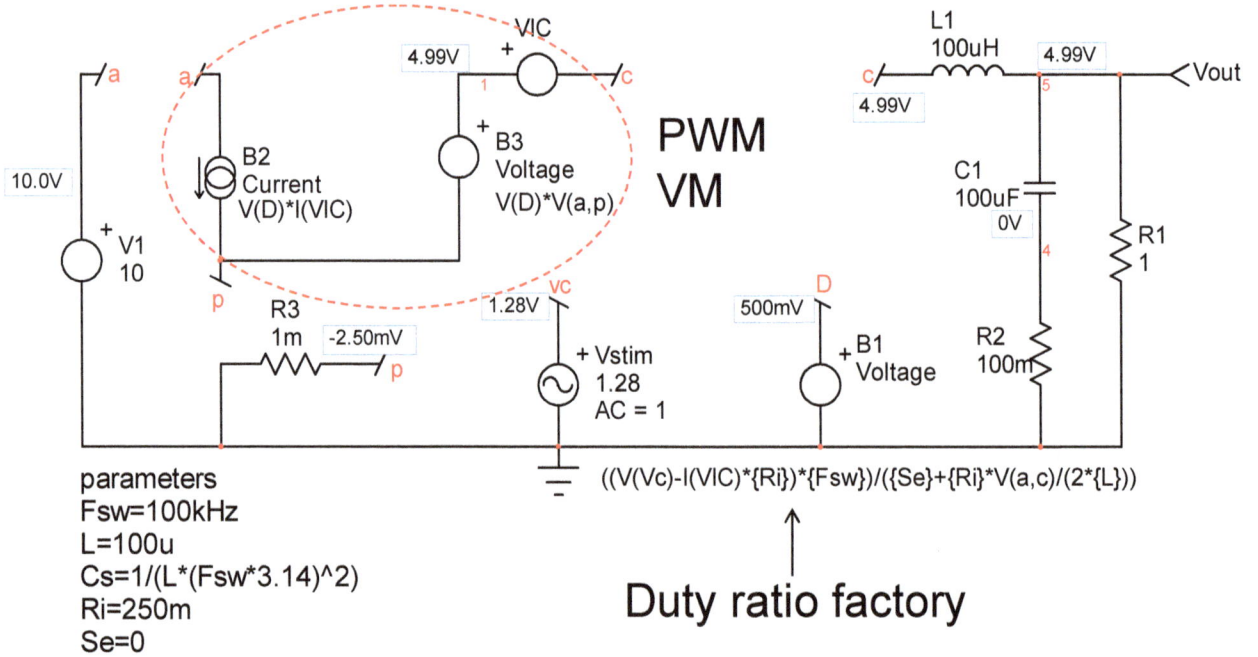

Figure 1.127 A large-signal independent source models the duty ratio and feeds the D input of the PWM switch model.

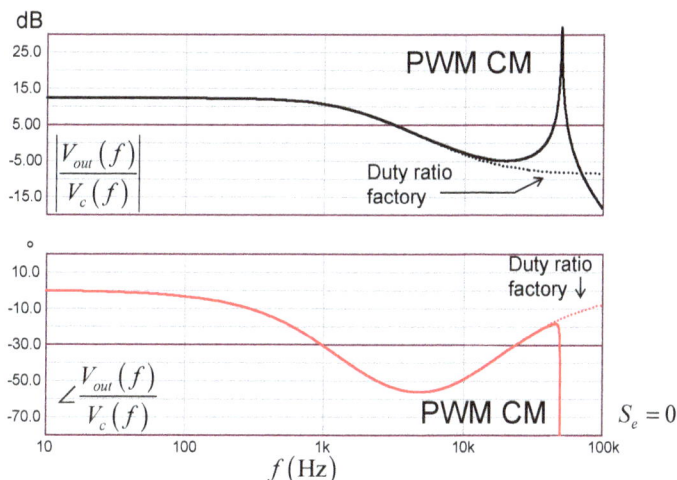

Figure 1.128 The simple VM-PWM switch model associated with duty ratio generator predicts the low-frequency response of the current-mode converter.

The dynamic response of this circuit was compared with that of the CM-PWM switch model and curves appear in Figure 1.128: the low-frequency response of the VM-PWM switch plus the duty ratio generator matches that of the CM-PWM switch but diverges as sub-harmonic poles start to kick-in. Nevertheless, this model is good enough as long as the crossover frequency is far from the $F_{sw}/2$ poles, which is often the case.

If you want to see the effects of the subharmonic poles with the VM-PWM switch and the duty ratio generator, it is possible to place an extra capacitor C_s – value determined with (1.275) – and approximate the response of the CM-PWM switch. This is just for the sake of documenting this circuit (Figure 1.129) as we won't use it in the book but I have found it to be an interesting simple way to predict the subharmonic poles without resorting to the CM-PWM switch. The phase response slightly diverges beyond F_{sw} but it is acceptable as shown in Figure 1.30.

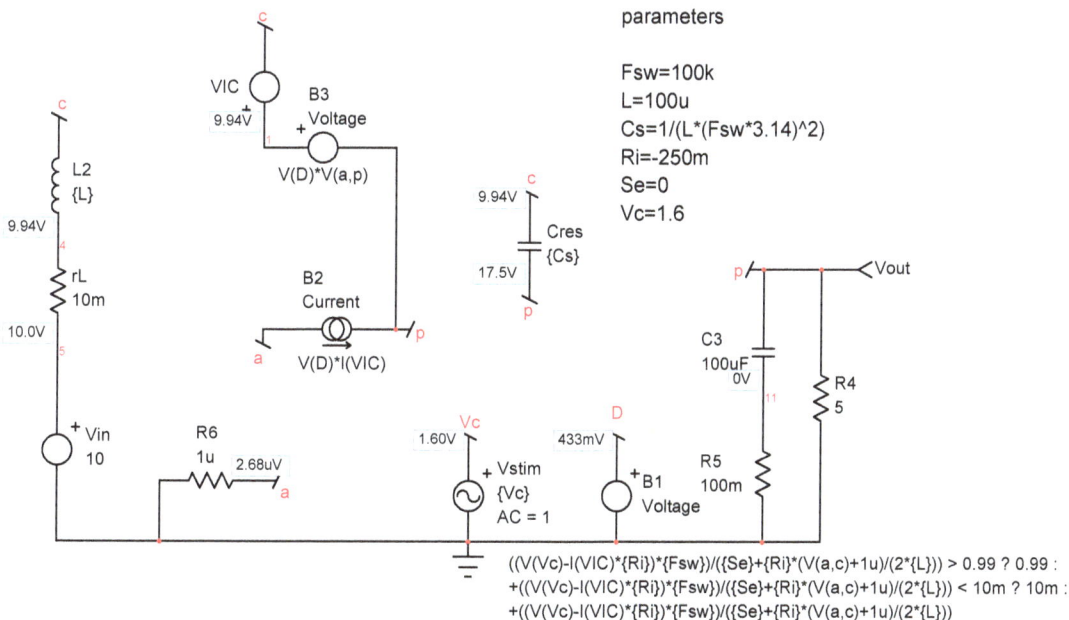

Figure 1.129 By adding an extra capacitor, it becomes possible to predict sub-harmonic peaking with the VM-PWM switch. Here in a boost converter.

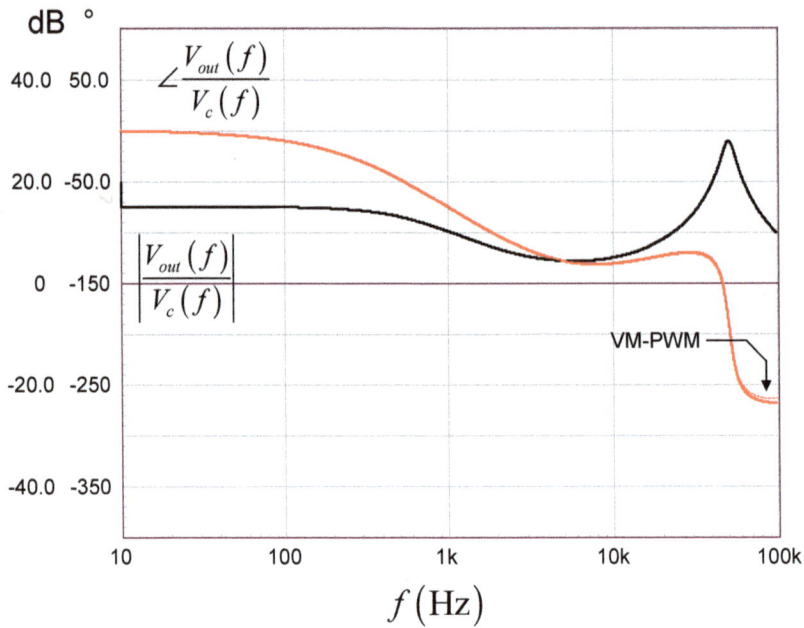

Figure 1.130 If you compare the dynamic responses between the CM-PWM switch and the VM-PWM switch with added capacitor, they are very close up to the switching frequency.

In 2009, Dr. Jian Li from CPES worked on a new scheme, aiming to unify all possible variations around current-mode control like peak-current control, COT, FOT, charge control and valley-current control in a single model [12]. He came up with an updated 3-terminal model based on the observation that intermodulation products contributed by the switching frequency and the modulating input are coupled back to the modulator via the current loop. Using describing functions, the author derived a new model whose components values can be adjusted based on the adopted switching strategy. Figure 1.131 shows the model for the determination of the control-to-output transfer function of a current-mode control CCM buck converter. The circuit shines by its simplicity despite the complexity of the mathematical derivation. In this large-signal canonical model, the component definitions depend on the type of operations. For current-mode control, the authors determined the values for R_e and C_e as follows:

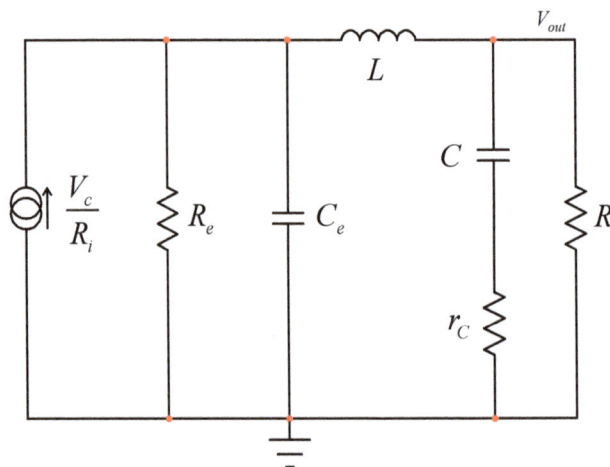

Figure 1.131 Jian Li's large-signal CM CCM model is quite simple and allows the modeling of various current mode topologies.

$$R_e = \frac{L}{T_{sw}\left(\dfrac{S_n + S_e}{S_n + S_f} - 0.5\right)}$$

(1.279)

$$C_e = \frac{T_{sw}^2}{\pi^2 L}$$

(1.280)

This last definition is similar to that given in (1.275) and tunes the LC network to half the switching frequency. In (1.279), S_n and S_f are respectively the inductor current on- and off-slopes. Please note that they should be scaled by the sense resistance R_i as the S_e, the compensation ramp, is expressed in volts per second. The SPICE implementation of the model appears in Figure 1.132.

parameters

Vin=12
Vout=5
Fsw=100k
Tsw=1/Fsw
L=100u
pi=3.14159
Cs=1/(L*(Fsw*pi)^2)
Ri=250m
Vac=Vin-Vout
Vcp=Vout
Se=7.049k
Sn=(Vac/L)*Ri
Sf=(Vcp/L)*Ri
Vc=1.28
Re=L/(((((Sn+Se)/(Sn+Sf))-0.5)*Tsw)
Ce=Tsw^2/(L*pi^2)

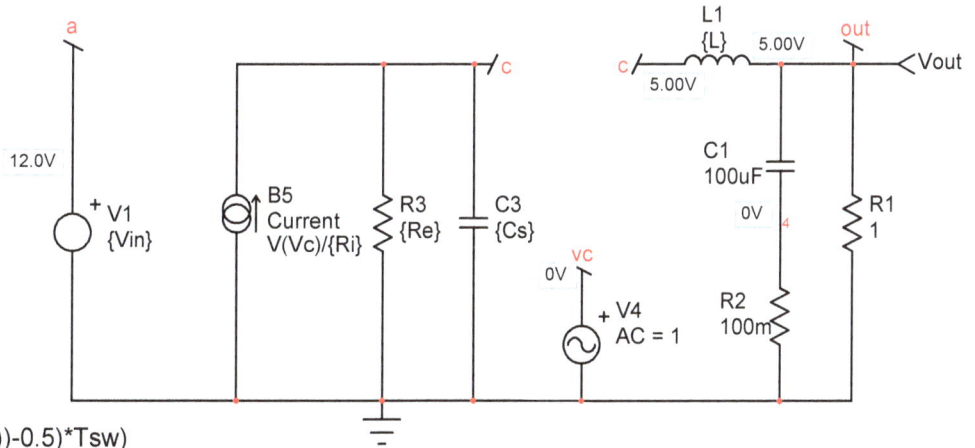

Figure 1.132 The SPICE implementation is not complicated and requires a few calculations with the inductor slopes.

We can calculate the double poles quality factor affecting this buck converter which delivers 5 V from a 12-V source. The duty ratio is:

$$D = \frac{V_{out}}{V_{in}} = \frac{5}{12} = 41.7\%$$

(1.281)

The double poles peak with a Q of:

$$Q = \frac{1}{\pi\left[m_c\left(1-D\right)0.5\right]} = 3.82$$

(1.282)

with $m_c = 1 + \dfrac{S_e}{S_n} = 1$ in absence of external ramp S_e. If we want to reduce Q to 1, the amount of injected ramp must amount to:

$$m_c = \frac{\dfrac{1}{\pi} + 0.5}{1-D} = 1.403$$

(1.283)

111

It implies an external ramp S_e of amplitude:

$$S_e = S_n(m_c - 1) = \frac{V_{in} - V_{out}}{L} R_i(m_c - 1) = \frac{12 - 5}{100u} \times 0.25 \times 0.403 = 7.05\,\text{kV/s} \tag{1.284}$$

We have passed this value to Figure 1.132 circuit and compared its response to the CM-PWM switch from Figure 1.124 with the same amount of ramp. As confirmed by Figure 1.133, dynamic responses are identical and you see a properly-damped circuit.

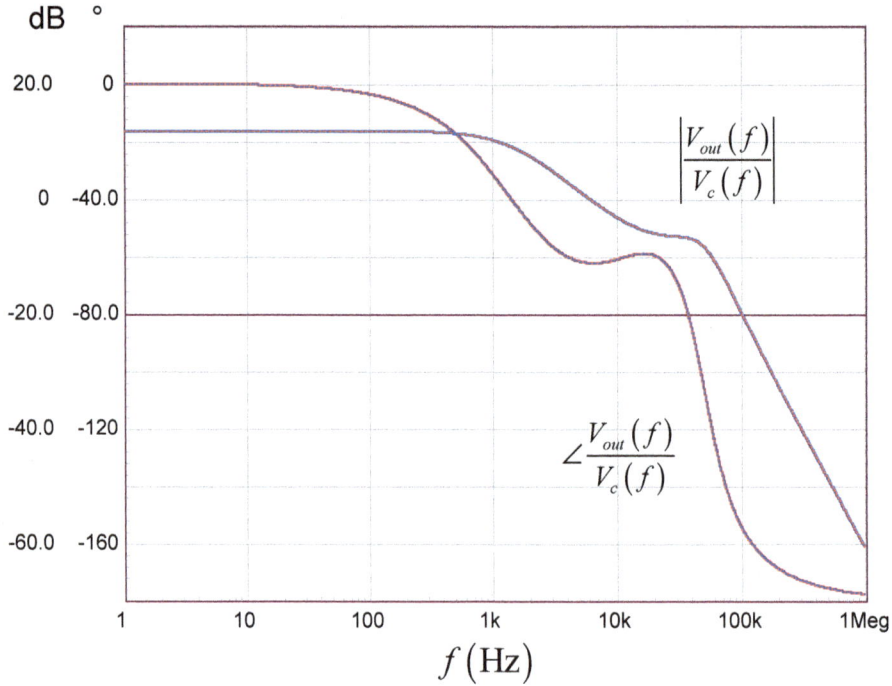

Figure 1.133 Simulation results are identical between the CM-PWM switch model and Jian Li's model.

1.4.1 Small-Signal Model in CCM

The principle here is similar to that of the VM-PWM switch in which we have differentiated the large-signal sources the model is made of (I_a and I_c).

Both are made of three variables as indicated by (1.274) and (1.269). For the small-signal version of I_c, we have:

$$\hat{i}_c = \frac{\partial I_c(V_c, V_{ap}, V_{cp})}{\partial V_c}\hat{v}_c + \frac{\partial I_c(V_c, V_{ap}, V_{cp})}{\partial V_{ap}}\hat{v}_{ap} + \frac{\partial I_c(V_c, V_{ap}, V_{cp})}{\partial V_{cp}}\hat{v}_{cp} \tag{1.285}$$

featuring the following coefficient names:

$$\hat{i}_c = k_o\hat{v}_c + g_f\hat{v}_{ap} + g_o\hat{v}_{cp} \tag{1.286}$$

For I_a, we need to determine:

$$\hat{i}_a = \frac{\partial I_a\left(V_c,V_{ap},V_{cp}\right)}{\partial V_c}\hat{v}_c + \frac{\partial I_a\left(V_c,V_{ap},V_{cp}\right)}{\partial V_{ap}}\hat{v}_{ap} + \frac{\partial I_a\left(V_c,V_{ap},V_{cp}\right)}{\partial V_{cp}}\hat{v}_{cp} \tag{1.287}$$

with the following coefficients:

$$\hat{i}_a = k_i\hat{v}_c + g_i\hat{v}_{ap} + g_r\hat{v}_{cp} \tag{1.288}$$

The definitions of these coefficients appear below:

$$k_o = \frac{1}{R_i} \tag{1.289}$$

$$g_f = \frac{S_e T_{sw} D}{R_i V_{ap}} - \frac{T_{sw} D^2}{2L} \tag{1.290}$$

$$g_o = -\left(\frac{T_{sw}(D-1)}{2L} + \frac{T_{sw}D}{2L} - \frac{S_e T_{sw} D}{R_i V_{cp}}\right) \tag{1.291}$$

$$k_i = \frac{D}{R_i} \tag{1.292}$$

$$g_i = D\left(g_f - \frac{I_c}{V_{ap}}\right) \tag{1.293}$$

$$g_r = \frac{I_c}{V_{ap}} - \frac{V_{cp} g_o}{V_{ap}} \tag{1.294}$$

In [10], Vorpérian has adopted a slightly different format considering the negative sign in (1.291). The definition of terminal c current updates to:

$$\hat{i}_c = k_o\hat{v}_c + g_f\hat{v}_{ap} - g_o\hat{v}_{cp} \tag{1.295}$$

Further to some rearrangement, the above expressions become:

$$k_o = \frac{1}{R_i} \tag{1.296}$$

$$g_f = Dg_o - \frac{DD'T_{sw}}{2L} \tag{1.297}$$

$$g_o = \frac{T_{sw}}{L}\left(D'\frac{S_e}{S_n} + \frac{1}{2} - D\right) \tag{1.298}$$

$$k_i = \frac{D}{R_i} \tag{1.299}$$

$$g_i = D\left(g_f - \frac{I_c}{V_{ap}}\right) \tag{1.300}$$

$$g_r = \frac{I_c}{V_{ap}} - g_o D \qquad\qquad (1.301)$$

In these formulas, S_n is the inductor on-slope, S_f is the inductor off-slope and S_e represents the external ramp in volts per seconds. The final small-signal model appears in Figure 1.134. To validate the model, we have captured in the same sheet a buck using the large-signal model of the CM-PWM switch and the small-signal version presented in Figure 1.134. The schematic appears in Figure 1.135.

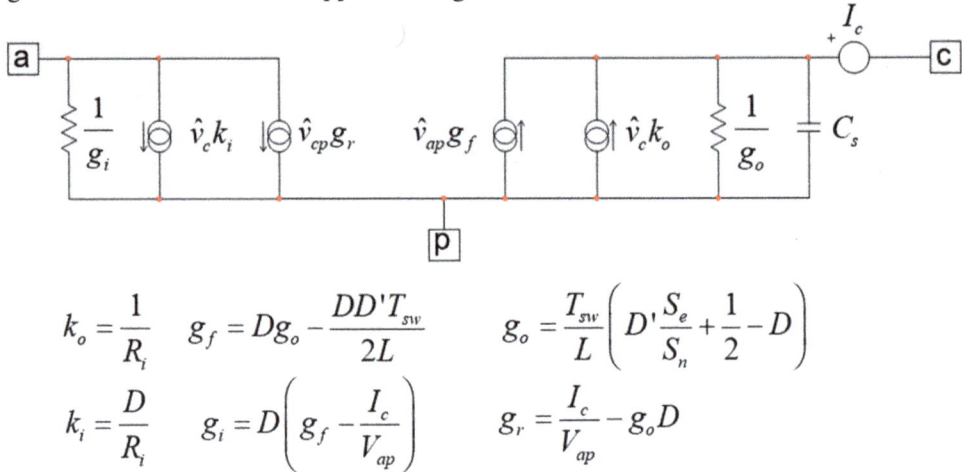

$$k_o = \frac{1}{R_i} \qquad g_f = Dg_o - \frac{DD'T_{sw}}{2L} \qquad g_o = \frac{T_{sw}}{L}\left(D'\frac{S_e}{S_n} + \frac{1}{2} - D\right)$$

$$k_i = \frac{D}{R_i} \qquad g_i = D\left(g_f - \frac{I_c}{V_{ap}}\right) \qquad g_r = \frac{I_c}{V_{ap}} - g_o D$$

Figure 1.134 The small-signal model of the CM-PWM switch model operated in CCM associates various linear current- and voltage-sources.

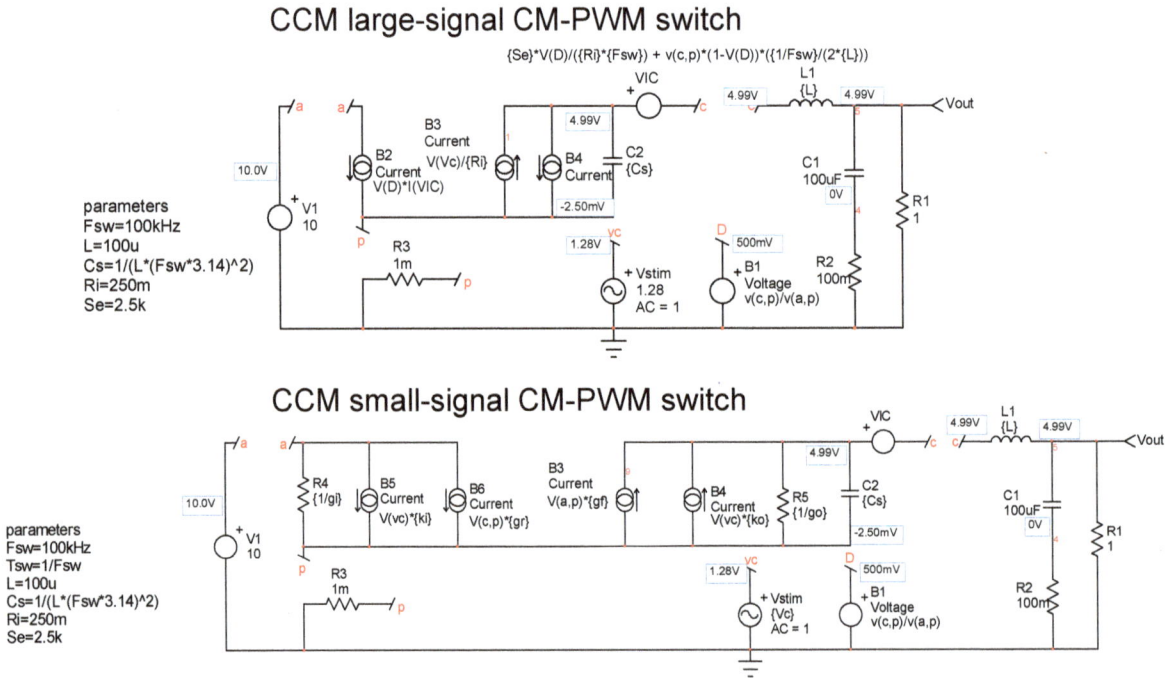

Figure 1.135 To validate the small-signal version, we can compare its dynamic response with that of the large-signal version. V_c, the control voltage, is 1.28 V for both schematics.

The simulation results appear in Figure 1.136 and are rigorously identical in magnitude/phase.

We have tested other transfer functions like the input and output impedances and also match the large-signal version.

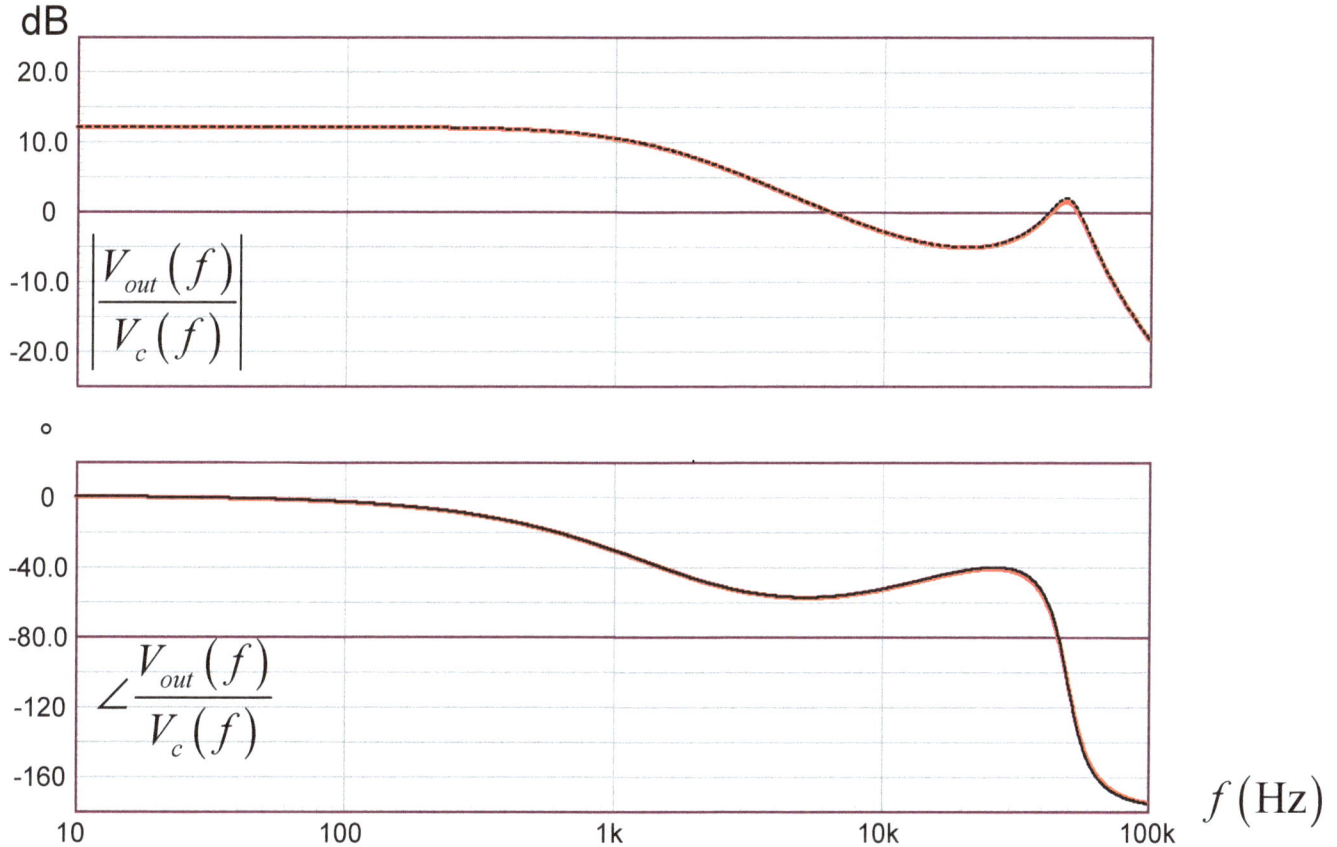

Figure 1.136 The dynamic responses between the large-signal CM-PWM switch and its small-signal counterpart are identical.

1.4.2 Small-Signal Model in DCM

Dr. Vorpérian did not publish his CM-PWM model operated in DCM. I have derived a version in [1] and encapsulated it in an auto-toggling 3-terminal CCM-DCM SPICE model.

However, the high-frequency dynamic response is not exactly that of the other models operated in DCM such as described in [5] for instance. This 2nd-order effect exists because the duty ratio calculation in DCM needs to sense the exact inductor voltage rather than reusing the voltage across the PWM switch terminals as in the CCM mode.

For the purpose of this book, I have assembled a new small-signal current-mode model exclusively operated in DCM where I sense the inductor voltage for deriving the duty ratio. Nothing truly complicated actually as it builds on the VM-PWM switch model operated in discontinuous mode associated with a duty ratio generator as in Figure 1.127.

To obtain this expression, we can look at the inductor current waveform shown in Figure 1.137.

This is a discontinuous mode waveform to which an external ramp is associated.

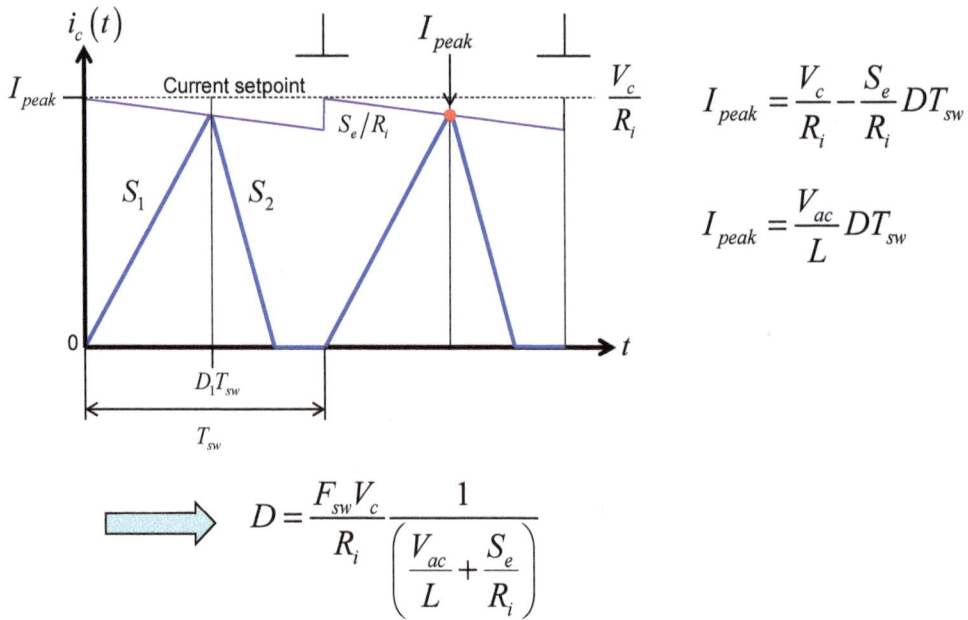

$$I_{peak} = \frac{V_c}{R_i} - \frac{S_e}{R_i} DT_{sw}$$

$$I_{peak} = \frac{V_{ac}}{L} DT_{sw}$$

$$\implies \quad D = \frac{F_{sw} V_c}{R_i} \frac{1}{\left(\dfrac{V_{ac}}{L} + \dfrac{S_e}{R_i} \right)}$$

Figure 1.137 The inductor current returns to zero within the switching cycle in DCM.

We could get rid of the ramp because it is not needed in DCM most of the time. With the forward converter, the magnetizing current forms a natural ramp which affects the peak current and, in many, cases, it can stabilize the converter. In some very specific cases, an external ramp might be needed: the DCM buck can be instable when operated at a particular M ratio and a stabilization ramp fixes the problem.

With this model, you can include the ramp damping effect and see how it impacts the power stage dynamic response. From the inductor current drawing, we can quickly determine the duty ratio expression via two equations:

$$I_{peak} = \frac{V_c}{R_i} - \frac{S_e}{R_i} D_1 T_{sw} \tag{1.302}$$

But the peak current is also equal to inductor slope times the on-time. This is where the difference is with the large-signal CCM model: we need the actual inductor on-time voltage V_L for this calculation:

$$I_{peak} = \frac{V_L}{L} D_1 T_{sw} \tag{1.303}$$

Extracting D_1 from these two equations gives us:

$$D_1 = \frac{F_{sw} V_c}{R_i} \frac{1}{\left(\dfrac{V_L}{L} + \dfrac{S_e}{R_i} \right)} \tag{1.304}$$

Assembling all of these sources together with those of the DCM VM-PWM switch, we have the template of Figure 1.138.

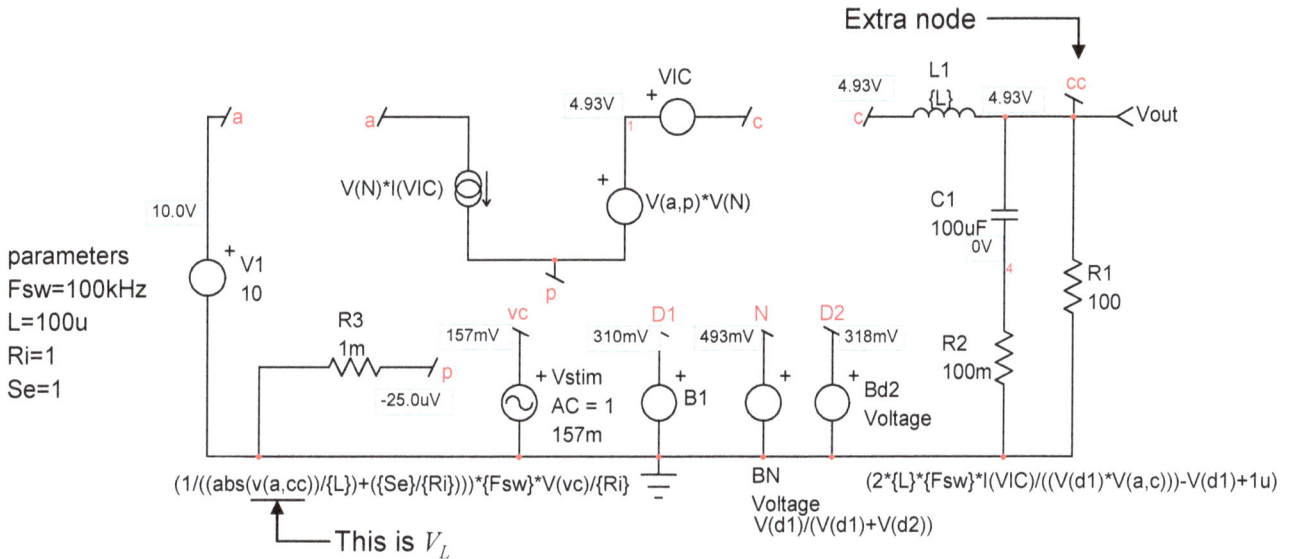

Figure 1.138 This is a DCM-operated buck converter delivering 5 V from a 10-V source.

You can recognize (1.304) coded in source B_1 generating D_1. Please note that the inductor voltage is no longer $V_{(a,c)}$ as in the CCM model but $V_{(a,cc)}$ in which node cc is the output voltage. The dynamic response appears in Figure 1.139 and confirms that the DCM-operated buck converter is still a 2nd-order system.

We now need to linearize these sources and build a small-signal version of the DCM CM-PWM switch model. By merging D_1's definition with D_2's into N, a simpler representation appears in Figure 1.140. The dynamic response of this new iteration is tested against that of Figure 1.138 and they are identical.

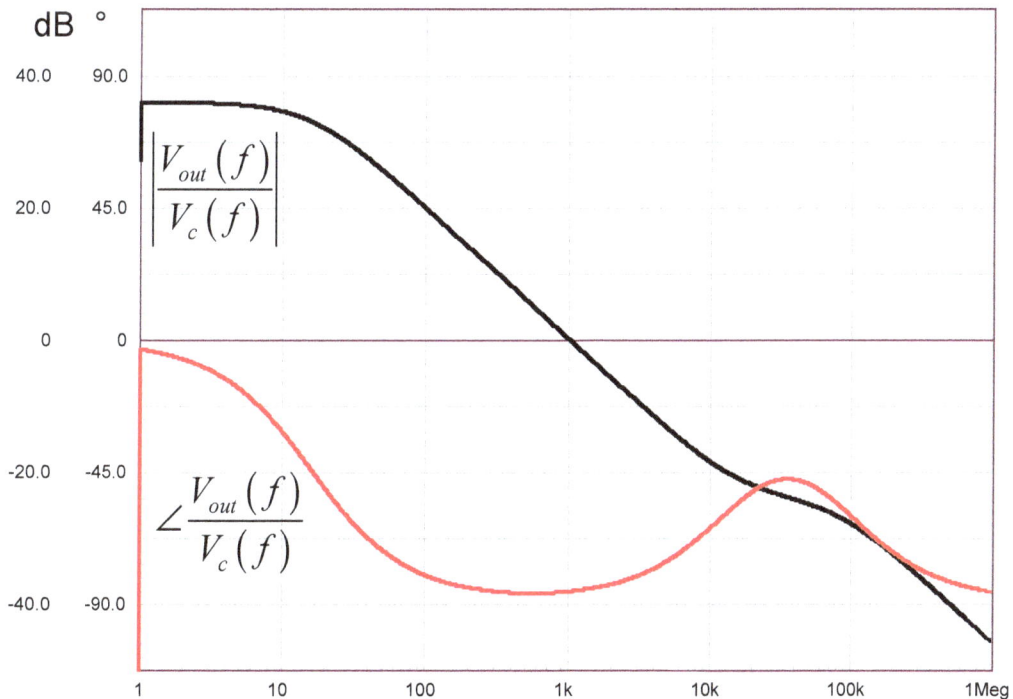

Figure 1.139 The DCM buck in current mode exhibits a 2nd-order response.

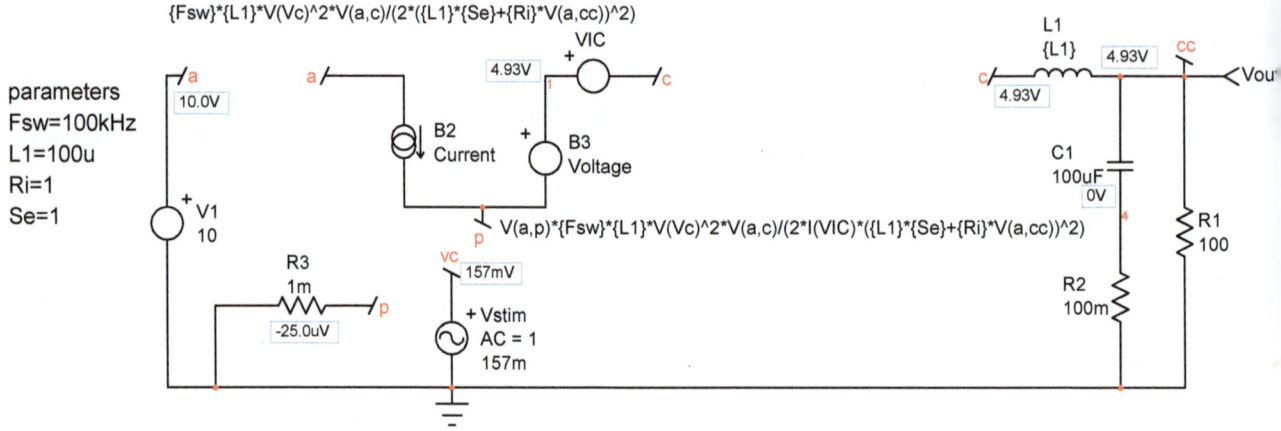

Figure 1.140 Once merged into two separate sources, the model looks simpler to analyze.

Having these two generators, we can determine the small-signal coefficients using partial differentiation. Here we go for source B_3:

$$V_{cp} = V_{ap} \frac{F_{sw}LV_c^2 V_{ac}}{2I_c\left(LS_e + R_i V_{acc}\right)^2} \tag{1.305}$$

$$k_1 = \frac{d}{dV_c} \frac{F_{sw}LV_c^2 V_{ac}V_{ap}}{2I_c\left(LS_e + R_i V_{acc}\right)^2} = \frac{F_{sw}LV_c V_{ap}V_{ac}}{I_c\left(LS_e + R_i V_{acc}\right)^2} \tag{1.306}$$

$$k_2 = \frac{d}{dV_{ac}} V_{ap} \frac{F_{sw}LV_c^2 V_{ac}}{2I_c\left(LS_e + R_i V_{acc}\right)^2} = \frac{F_{sw}LV_c^2 V_{ap}}{2I_c\left(LS_e + R_i V_{acc}\right)^2} \tag{1.307}$$

$$k_3 = \frac{d}{dI_c} \frac{F_{sw}LV_c^2 V_{ac}V_{ap}}{2I_c\left(LS_e + R_i V_{acc}\right)^2} = -\frac{F_{sw}LV_c^2 V_{ap}V_{ac}}{2I_c^2\left(LS_e + R_i V_{acc}\right)^2} \tag{1.308}$$

$$k_4 = \frac{d}{dV_{ap}} \frac{F_{sw}LV_c^2 V_{ac}V_{ap}}{2I_c\left(LS_e + R_i V_{acc}\right)^2} = \frac{F_{sw}LV_c^2 V_{ac}}{2I_c\left(LS_e + R_i V_{acc}\right)^2} \tag{1.309}$$

$$k_5 = \frac{d}{dV_{acc}} \frac{F_{sw}LV_c^2 V_{ac}V_{ap}}{2I_c\left(LS_e + R_i V_{acc}\right)^2} = -\frac{F_{sw}LR_i V_c^2 V_{ap}V_{ac}}{I_c\left(LS_e + R_i V_{acc}\right)^3} \tag{1.310}$$

Combining these equations gives the small-signal source $\hat{v}_{cp}$:

$$\hat{v}_{cp} = k_1\hat{v}_c + k_2\hat{v}_{ac} + k_3\hat{i}_c + k_4\hat{v}_{ap} + k_5\hat{v}_{acc} \tag{1.311}$$

In the above static coefficient definitions, considering 0 V across the inductor, we obviously have $V_{acc} = V_{ac}$.

The second equation is:

$$I_a = I_c \frac{F_{sw}LV_c^2 V_{ac}}{2I_c \left(LS_e + R_i V_{acc}\right)^2} = \frac{F_{sw}LV_c^2 V_{ac}}{2\left(LS_e + R_i V_{acc}\right)^2} \qquad (1.312)$$

Applying partial differentiation to the above formula gives:

$$k_6 = \frac{d}{dV_c} \frac{F_{sw}LV_c^2 V_{ac}}{2\left(LS_e + R_i V_{acc}\right)^2} = \frac{F_{sw}LV_c V_{ac}}{\left(LS_e + R_i V_{acc}\right)^2} \qquad (1.313)$$

$$k_7 = \frac{d}{dV_{ac}} \frac{F_{sw}LV_c^2 V_{ac}}{2\left(LS_e + R_i V_{acc}\right)^2} = \frac{F_{sw}LV_c^2}{2\left(LS_e + R_i V_{acc}\right)^2} \qquad (1.314)$$

$$k_8 = \frac{d}{dV_{acc}} \frac{F_{sw}LV_c^2 V_{ac}}{2\left(LS_e + R_i V_{acc}\right)^2} = -\frac{F_{sw}LR_i V_c^2 V_{ac}}{\left(LS_e + R_i V_{acc}\right)^3} \qquad (1.315)$$

Combining these equations gives the small-signal source $\hat{i}_a$:

$$\hat{i}_a = k_6 \hat{v}_c + k_7 \hat{v}_{ac} + k_8 \hat{v}_{acc} \qquad (1.316)$$

The above coefficient definitions have been included in the application circuit of Figure 1.141 which shows a current-mode buck converter operated in DCM.

To verify its dynamic response is correct, I have added a third example built with the auto-toggling DCM-CCM large-signal voltage-mode from CoPEC and described in [11]. The circuit appears in Figure 1.142 while the simulation results are given in Figure 1.143: curves perfectly superimpose.

Please note that we have tested the same model in the boost and buck-boost configuration also as the I_a source in the buck configuration has no effect in the control-to-output transfer function.

All tests confirm the validity of this small-signal model and we will use it to determine some of the transfer functions in the coming chapters.

{k6}*V(Vc)+{k7}*V(a,c)+{k8}*V(a,cc)

{k1}*V(Vc)+{k2}*V(a,c)+{k3}*I(VIC)+{k4}*V(a,p)+{k5}*V(a,cc)

parameters
Fsw=100kHz
L1=100u
Ri=1
Se=1
RL=100
Vout=4.93054
Vin=10
Vac=Vin-Vout
Vacc=Vac
Vap=Vin
Ic=Vout/RL
D1=0.3097
Vc=157m

B2 Current

VIC

B3 Voltage

L2 {L1}

C1 100uF

R1 100

R3 1m

VC

R2 100m

V1 10

+ Vstim
AC = 1
157m

Vout

k1=Fsw*L1*Vc*Vap*Vac/(Ic*(L1*Se+Ri*Vacc)^2)
k2=Fsw*L1*Vc^2*Vap/(2*Ic*(L1*Se+Ri*Vacc)^2)
k3=-Fsw*L1*Vc^2*Vap*Vac/(2*Ic^2*(L1*Se+Ri*Vacc)^2)
k4=Fsw*L1*Vc^2*Vac/(2*Ic*(L1*Se+Ri*Vacc)^2)
k5=-Fsw*L1*Vc^2*Vac*Vap*Ri/(Ic*(L1*Se+Ri*Vacc)^3)
k6=Fsw*L1*Vc*Vac/(L1*Se+Ri*Vacc)^2
k7=Fsw*L1*Vc^2/(2*(L1*Se+Ri*Vacc)^2)
k8=-Fsw*L1*Vc^2*Ri*Vac/(L1*Se+Ri*Vacc)^3

K1 = 6.28e+001
K2 = 9.73e-001
K3 = -1.00e+002
K4 = 4.93e-001
K5 = -1.95e+000
K6 = 3.10e-001
K7 = 4.80e-003
K8 = -9.59e-003

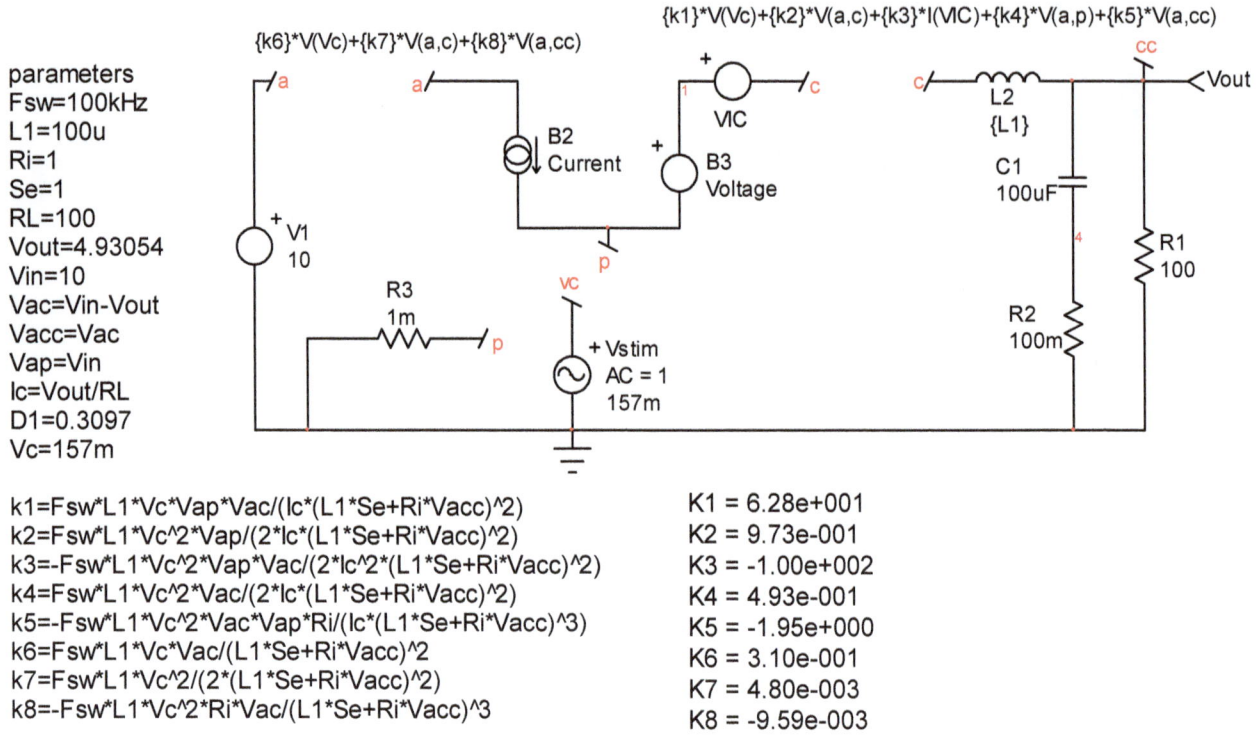

Figure 1.141 The small-signal coefficients appear in the left side of the circuit while I pasted the calculated values in the right side.

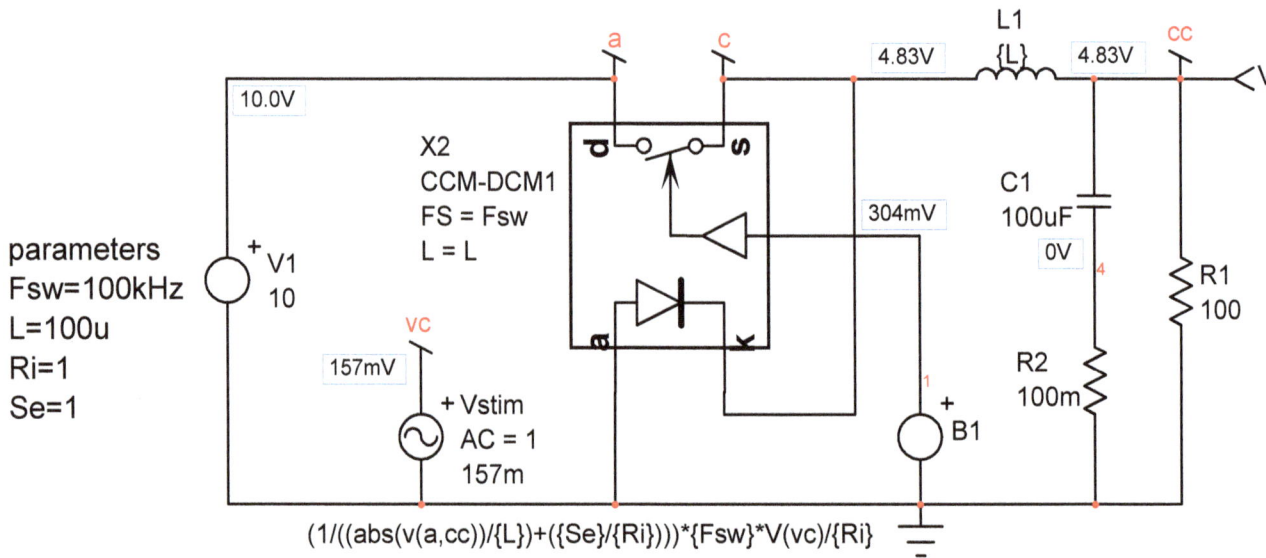

L1 {L}

4.83V   4.83V   CC

10.0V

X2
CCM-DCM1
FS = Fsw
L = L

304mV

C1 100uF

0V

R1 100

parameters
Fsw=100kHz
L=100u
Ri=1
Se=1

V1 10

VC

157mV

+ Vstim
AC = 1
157m

R2 100m

B1

(1/((abs(v(a,cc))/{L})+({Se}/{Ri})))*{Fsw}*V(vc)/{Ri}

Figure 1.142 The large-signal DCM current-mode buck built with the model from Colorado University, CoPEC.

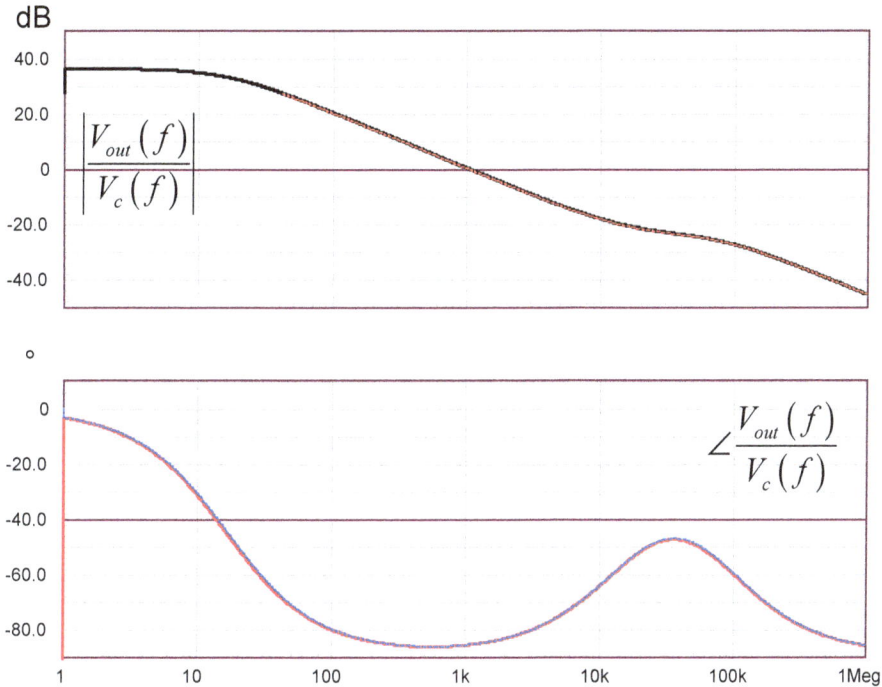

Figure 1.143 The dynamic response from the linearized circuit and the large-signal model from CoPEC are identical.

1.4.3 Small-Signal Model in QR

In the previous lines, we have determined a QR model operated in voltage-mode control. To operate the new circuit in current-mode, we could also derive a duty ratio generator and reuse the voltage-mode model. Rather, we will assemble two current sources following the CM-PWM architecture. We start with the currents flowing in terminal a and leaving terminal c. They appear in Figure 1.144.

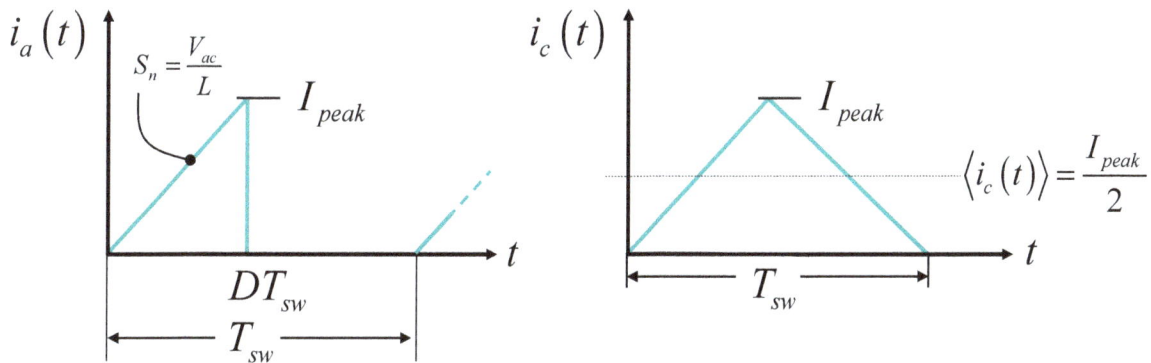

Figure 1.144 Determining the average value is simple in borderline mode (no dead time).

The current in the left side is I_c "seen" during DT_{sw}. On average, we have:

$$I_a = DI_c \qquad (1.317)$$

In the right side, the average value is easily found by calculating the total area under the curve and averaging it over the switching period. It is the sum of two triangles areas:

$$I_c = \frac{\frac{1}{2}I_{peak}DT_{sw} + \frac{1}{2}I_{peak}(1-D)T_{sw}}{T_{sw}} = \frac{I_{peak}}{2} \qquad (1.318)$$

In a current mode controller, the peak current is controlled by the control voltage V_c and the sense resistance R_{sense} or R_i:

$$I_{peak} = \frac{V_c}{R_i} \qquad (1.319)$$

Substituting (1.319) in (1.318) leads to:

$$I_c = \frac{V_c}{2R_i} \qquad (1.320)$$

and this is it, we have our QR CM-PWM switch model in which we did not consider the dead time – the added delay to switch right in the valley, see Figure 1.79 – for the sake of simplicity. The 3-terminal model appears in Figure 1.145 and shines by its simplicity when inserted in a buck converter.

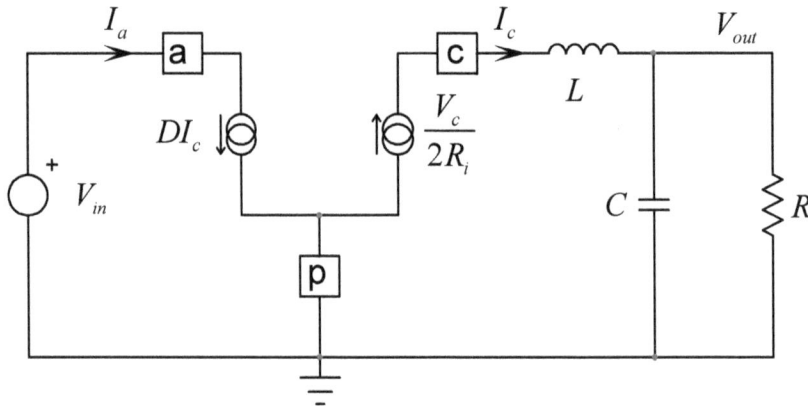

Figure 1.145 The lare-signal current-mode QR model requires two simple current sources to operate.

In a self-relaxing QR, the duty ratio and the frequency will depend on the operating conditions and change as the latter move. Considering the invariant inductor on-slope defined as:

$$S_n = \frac{V_{ac}}{L} \qquad (1.321)$$

the duty ratio can be computed knowing that:

$$I_{peak} = \frac{V_c}{R_i} \qquad (1.322)$$

and:

$$I_{peak} = S_n t_{on} = \frac{V_{ac}}{L}DT_{sw} \qquad (1.323)$$

from which:

$$D = \frac{V_c}{R_i} \frac{L}{V_{ac} T_{sw}}$$

(1.324)

The switching period now needs to be determined. We can calculate t_{on} and t_{off} easily via the respective inductor slopes:

$$t_{on} = \frac{V_c}{R_i} \frac{L}{V_{ac}}$$

(1.325)

$$t_{off} = \frac{V_c}{R_i} \frac{L}{V_{cp}}$$

(1.326)

Summing these two variables gives us the switching period:

$$T_{sw} = t_{on} + t_{off} = \frac{V_c L}{R_i} \left(\frac{1}{V_{ac}} + \frac{1}{V_{cp}} \right)$$

(1.327)

The complete large-signal model appears in Figure 1.146 with the expressions we have derived.

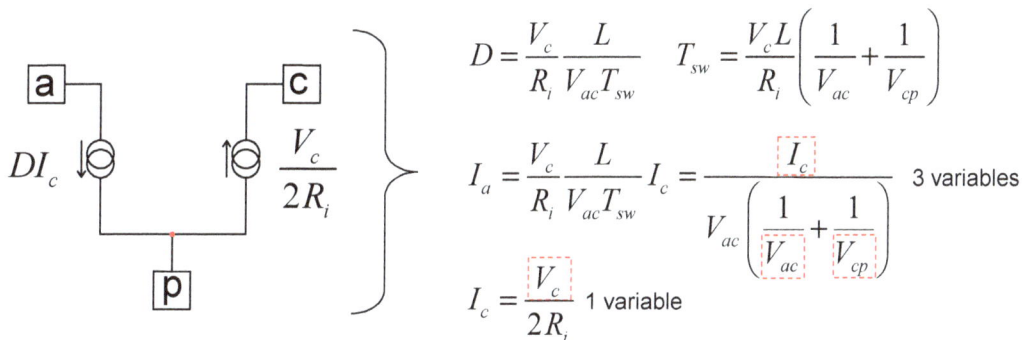

Figure 1.146 The expressions defining the model internals are quite simple.

We can exercise this model in a QR buck converter as shown in Figure 1.147.

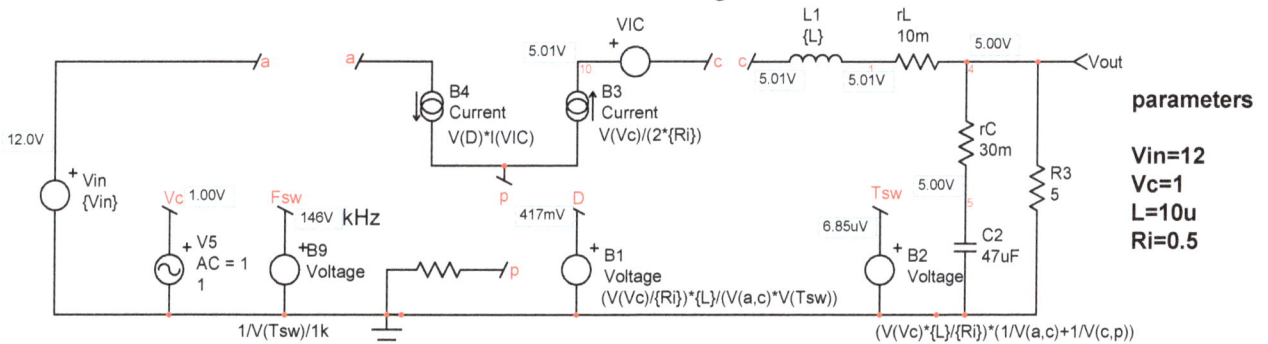

Figure 1.147 The current-mode model requires two simple current sources to operate.

The circuit delivers 5 V from a 12-V source and the duty ratio is, no surprise, equal to V_{out}/V_{in}. The on-time is calculated to 6.85 μs and leads to switching frequency of 146 kHz. If you ac-sweep this converter, you obtain the

Bode plot of Figure 1.148. You can see a pole in the low-frequency portion of the spectrum followed by a zero, most probably contributed by the output capacitor ESR.

The sources in Figure 1.146 need to be linearized to let us later determine transfer functions of our choice. Linearizing the I_c generator is easy and immediate:

$$\hat{i}_c = \frac{\partial I_c(V_c)}{\partial V_c}\hat{v}_c = \hat{v}_c\left(\frac{1}{2R_i}\right) = \hat{v}_c k_c \tag{1.328}$$

with $k_c = \dfrac{1}{2R_i}$

Source I_a depends on three variables, V_{cp}, I_c and V_{ac}. We have:

$$\hat{i}_a = \frac{\partial I_a(V_{cp}, V_{ac}, I_c)}{\partial V_{cp}}\hat{v}_{cp} + \frac{\partial I_a(V_{cp}, V_{ac}, I_c)}{\partial I_c}\hat{i}_c + \frac{\partial I_a(V_{cp}, V_{ac}, I_c)}{\partial V_{ac}}\hat{v}_{ac} \tag{1.329}$$

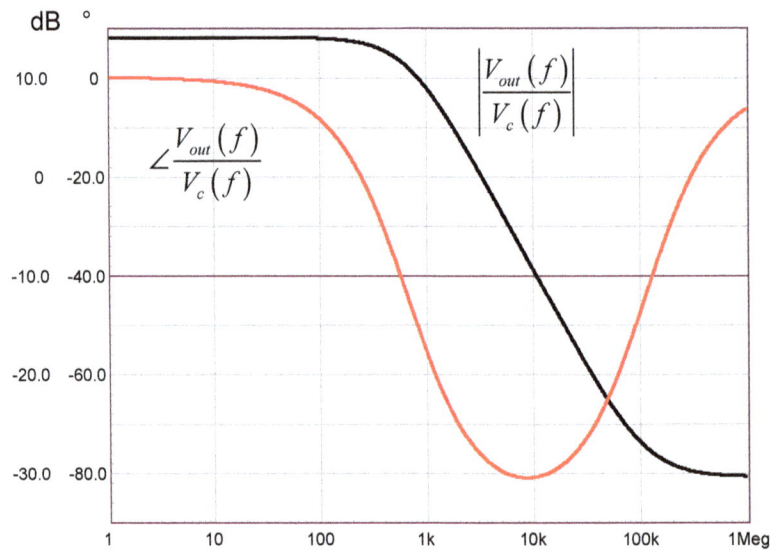

Figure 1.148 The dynamic response of this QR-operated buck converter is of 1ˢᵗ-order type.

Which leads to:

$$\hat{i}_a = \hat{v}_{cp}k_{cp} + \hat{i}_c k_{ic} - \hat{v}_{ac}k_{ac} \tag{1.330}$$

Where:

$$k_{cp} = \frac{I_c V_{cp}}{\left(V_{ac} + V_{cp}\right)^2} \tag{1.331}$$

$$k_{ic} = \frac{V_{cp}}{V_{cp} + V_{ac}} \tag{1.332}$$

$$k_{ac} = \frac{V_{cp} I_c}{\left(V_{ac} + V_{cp} \right)^2} \tag{1.333}$$

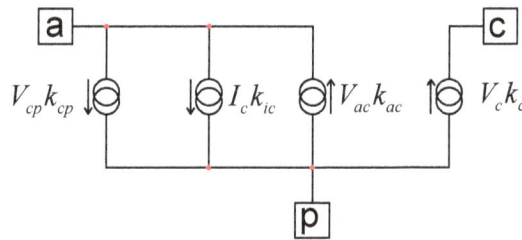

Figure 1.149 This is the small-signal model of the CM-PWM switch operated in QR.

Once all these sources are put together, you obtain the small-signal model in Figure 1.149. Install this model in the buck converter from Figure 1.145 and you have the circuit from Figure 1.150. If we plot the ac response of this circuit, it is identical to that of Figure 1.148, confirming the validity of this small-signal version.

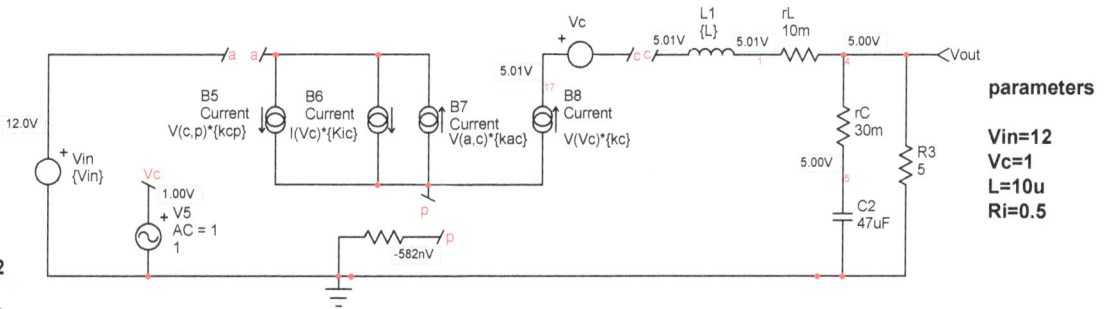

parameters

R=5
Vin=12
Vout=5
M=Vout/Vin
Ic=Vout/R
Vap=Vin
Vac=Vin-Vout
Vcp=Vout
kcp=Ic*Vac/(Vac+Vcp)^2
kic=Vcp/(Vcp+Vac)
kac=Vcp*Ic/(Vac+Vcp)^2
kc=1/(2*Ri)

parameters

Vin=12
Vc=1
L=10u
Ri=0.5

Figure 1.150 You simply install the small-signal in the original buck converter and compare its dynamic response with the large-signal version.

1.4.4 Constant On-Time Valley Current Control

A model for the constant on-time valley control is necessary to determine the transfer function of your choice. The principle remains the same, we draw the voltage and current of the 3-terminal PWM switch and we average them across a switching cycle. The inductor current appears in Figure 1.151. The on-time is a constant parameter (set by the designer) while the loop controls the valley current via V_c.

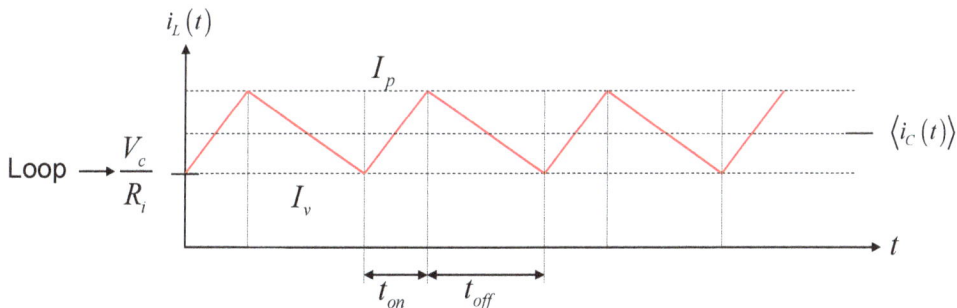

Figure 1.151 The loop controls the valley current I_v in this operating mode while t_{on} is constant.

The average current I_c can be obtained by starting from the valley current controlled by the loop via V_c. You can see that the average value lies in the middle of the inductor current ripple. Therefore:

$$I_c = I_v + \frac{\Delta I_L}{2} = \frac{V_c}{R_i} + \frac{V_{ac}}{2L} t_{on} \qquad (1.334)$$

in which V_{ac} is the on-time inductor voltage. The off-time is determined also using the valley current and the peak value I_p:

$$t_{off} = \Delta I_L \frac{L}{V_{cp}} = \left(I_p - \frac{V_c}{R_i} \right) \frac{L}{V_{cp}} \qquad (1.335)$$

I_p, on the other hand is defined as:

$$I_p = \frac{V_c}{R_i} + \frac{V_{ac}}{L} t_{on} \qquad (1.336)$$

Now substitute (1.336) in (1.335) and solve for t_{off}:

$$t_{off} = \frac{V_{ac}}{V_{cp}} t_{on} \qquad (1.337)$$

The switching frequency is thus equal to:

$$T_{sw} = t_{on} + t_{off} = t_{on} \left(1 + \frac{V_{ac}}{V_{cp}} \right) \qquad (1.338)$$

The duty ratio D is computed by dividing t_{on} by T_{sw} which leads to:

$$D = \frac{t_{on}}{T_{sw}} = \frac{1}{1 + \frac{V_{ac}}{V_{cp}}} \qquad (1.339)$$

We have our large-signal model: the I_a current source which is classically equal to DI_c involves (1.339) and I_c while the I_c generator is defined by (1.334). The complete model appears in Figure 1.152.

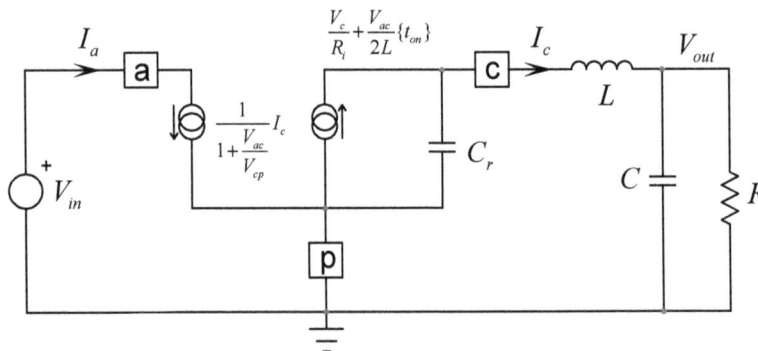

Figure 1.152 The large-signal COT model is truly simple and requires two easy-to-determine current sources.

Even if the COT valley current control is not subject to subharmonic instabilities, there is a need for a resonating capacitor whose value is [10]:

$$C_r = \frac{1}{L}\left(\frac{t_{on}}{\pi}\right)^2 \tag{1.340}$$

The complete SPICE model appears in Figure 1.153 and you recognize the sources we have just identified.

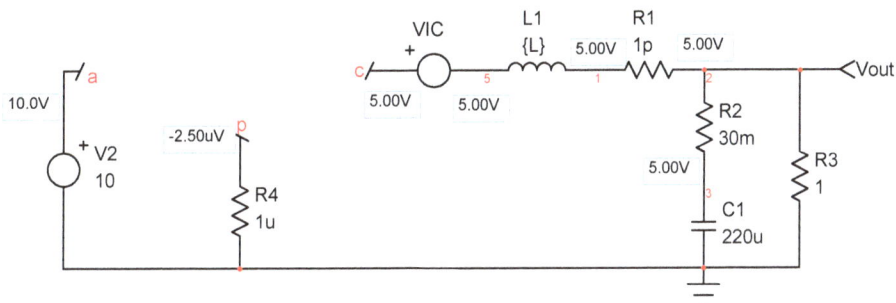

Parameters

Ri=100m
L=5u
pi=3.14159
W=pi/ton
Cr=1/(L*W^2)
ton=1.5u

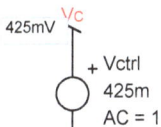

Control Voltage
Valley Current Setpoint

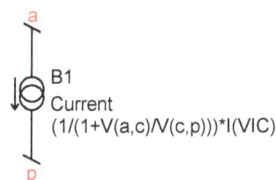

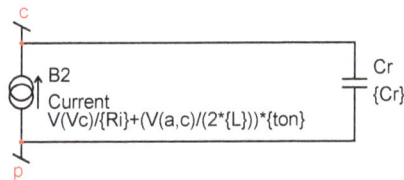

Figure 1.153 The SPICE implementation implies the selection of the on-time (1.5 μs in this example) and the calculation of the resonating capacitor.

To test our model, we have assembled the model presented in Figure 1.131 and derived by Dr. Li in [12]. It is shown in Figure 1.154 and required the selection of different values for R_e and C_e:

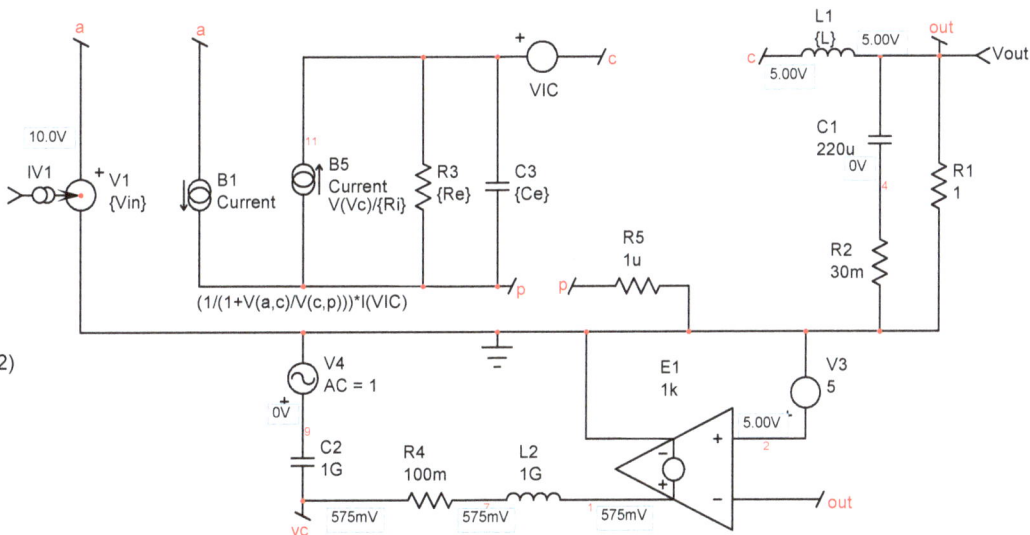

Figure 1.154 The large-signal model derived by Dr. Jian Li can also be used for checking the control-to-output response of the model we derived.

$$R_e = \frac{2L}{t_{on}} \tag{1.341}$$

$$C_e = \frac{1}{L}\left(\frac{t_{on}}{\pi}\right)^2 \tag{1.342}$$

In Figure 1.154, the additional op-amp E_1 automatically computes the setpoint to meet the 5-V target. During the bias point calculation, SPICE shorts the inductors and open the caps. L_2 is a short circuit and closes the loop so that E_1 adjusts the control voltage V_c to deliver 5 V to the load.

When the ac sweep begins, C_2 becomes a short circuit while L_2 blocks the signal propagation and opens the loop in ac. That way, if you change the input voltage or the load, the operating point will automatically be recalculated to the right value by E_1. This is a very useful trick when simulating average models and we will reuse it through the remaining examples of this book. After the simulation is run with both models, their control-to-output dynamic responses appear in Figure 1.155 and are similar.

We compared dynamic input currents and output impedances of both models, they were also similar in phase and magnitude.

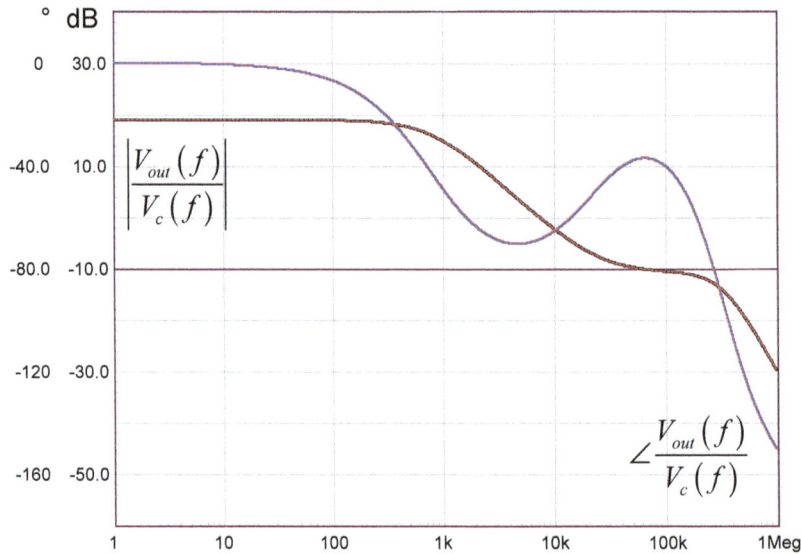

Figure 1.155 Dynamic responses from both models are rigorously equivalent.

To derive a small-signal version of Figure 1.152 large-signal circuit, we must determine the small-signal coefficients for both current sources. We will invoke partial differentiation with I_a and obtain three coefficients:

$$I_a = DI_c = \frac{1}{1+\dfrac{V_{ac}}{V_{cp}}} I_c \tag{1.343}$$

$$k_{ic} = \frac{d}{dI_c}\left(I_c \frac{1}{1+\dfrac{V_{ac}}{V_{cp}}}\right) = \frac{1}{1+\dfrac{V_{ac}}{V_{cp}}} \tag{1.344}$$

$$k_{ac} = \frac{d}{dV_{ac}}\left(I_c \frac{1}{1 + \frac{V_{ac}}{V_{cp}}} \right) = -\frac{I_c}{V_{cp}\left(1 + \frac{V_{ac}}{V_{cp}}\right)^2} \qquad (1.345)$$

$$k_{cp} = \frac{d}{dV_{cp}}\left(I_c \frac{1}{1 + \frac{V_{ac}}{V_{cp}}} \right) = \frac{I_c V_{ac}}{V_{cp}^{\;2}\left(1 + \frac{V_{ac}}{V_{cp}}\right)^2} \qquad (1.346)$$

leading to:

$$\hat{i}_a = k_{ic}\hat{i}_c + k_{ac}\hat{v}_{ac} + k_{cp}\hat{v}_{cp} \qquad (1.347)$$

We repeat the same operation with I_c which contains two variables:

$$I_c = \frac{V_c}{R_i} + \frac{V_{ac}}{2L}t_{on} \qquad (1.348)$$

$$k_{vc} = \frac{d}{dV_c}\left(\frac{V_c}{R_i} + \frac{V_{ac}}{2L}t_{on} \right) = \frac{1}{R_i} \qquad (1.349)$$

$$k_{vac} = \frac{d}{dV_{ac}}\left(\frac{V_c}{R_i} + \frac{V_{ac}}{2L}t_{on} \right) = \frac{t_{on}}{2L} \qquad (1.350)$$

It leads to the following definition:

$$\hat{i}_c = k_{vc}\hat{v}_c + k_{vac}\hat{v}_{ac} \qquad (1.351)$$

The linearized circuit appears in Figure 1.156 and its dynamic response is exactly that of Figure 1.155.

Parameters

Vout=5V
Vin=10V
R=1
Ri=100m
L=5u
pi=3.14159
W=pi/ton
Cr=1/(L*W^2)
ton=1.5u
Ic=Vout/R
Vac=Vin-Vout
Vcp=Vout

kic=1/(1+Vac/Vcp)
kac=-Ic/(Vcp*(Vac/Vcp+1)^2)
kcp=Ic*Vac/(Vcp^2*(Vac/Vcp+1)^2)
kvc=1/Ri
kvac=ton/(2*L)

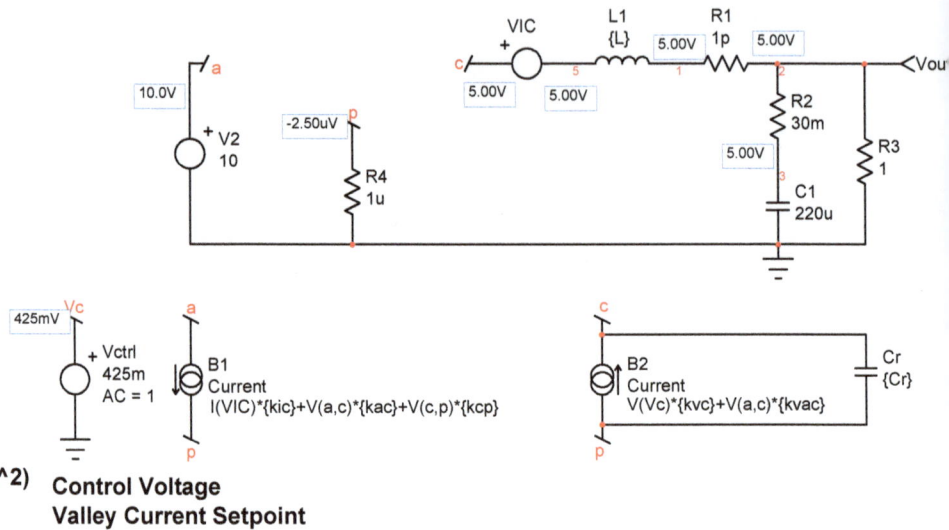

VIC
L1 {L}
R1
5.00V 1p 5.00V
10.0V 5.00V 5.00V
V2 10
-2.50uV p
R4 1u
R2 30m
5.00V
R3 1
C1 220u
Vou

425mV Vc
+ Vctrl 425m AC = 1
a
B1 Current
I(VIC)*{kic}+V(a,c)*{kac}+V(c,p)*{kcp}
p
Control Voltage
Valley Current Setpoint

c
B2 Current
V(Vc)*{kvc}+V(a,c)*{kvac}
p
Cr {Cr}

Figure 1.156 The small-signal of the PWM switch in a COT converter is quite simple.

1.4.5 Constant Off-Time Peak Current Control

To determine the CCM FOT large-signal PWM switch model, we will follow the steps adopted for the COT model. The inductor current waveform appears in Figure 1.157. The off-time is a constant parameter (set by the designer) while the loop controls the peak current via V_c.

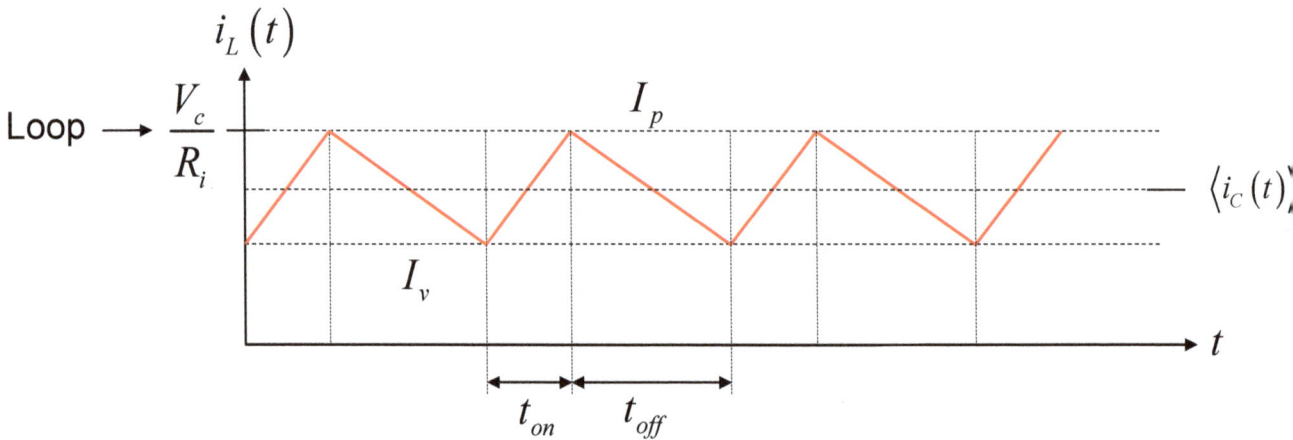

Figure 1.157 In the FOT converter, the loop controls the peak current in relationship with operating conditions.

The average inductor current is the peak set by the loop minus half the inductor current ripple:

$$I_c = \frac{V_c}{R_i} - \frac{\Delta I_L}{2} = \frac{V_c}{R_i} - \frac{V_{ac}}{2L}t_{on} \qquad (1.352)$$

From this expression, we can define the on-time value as:

$$t_{on} = \frac{\left(\dfrac{V_c}{R_i} - I_c\right)2L}{V_{ac}}$$

(1.353)

The duty ratio D is immediately determined as:

$$D = \frac{t_{on}}{t_{on} + t_{off}} = \frac{1}{1 + \dfrac{t_{off}V_{ac}}{\left(\dfrac{V_c}{R_i} - I_c\right)2L}}$$

(1.354)

(1.352) can be rewritten now considering the downslope during t_{off}:

$$I_c = \frac{V_c}{R_i} - \frac{\Delta I_L}{2} = \frac{V_c}{R_i} - \frac{V_{cp}}{2L}t_{off}$$

(1.355)

in which t_{off} is a fixed value programmed by the converter designer. The source I_a is classically defined as:

$$I_a = DI_c$$

(1.356)

with D calculated using (1.354). Finally, a resonating capacitor defined as:

$$C_r = \frac{1}{L}\left(\frac{t_{off}}{\pi}\right)^2$$

(1.357)

will take place between terminals c and p as explained in [10] to finally form the FOT PWM switch model shown in Figure 1.158.

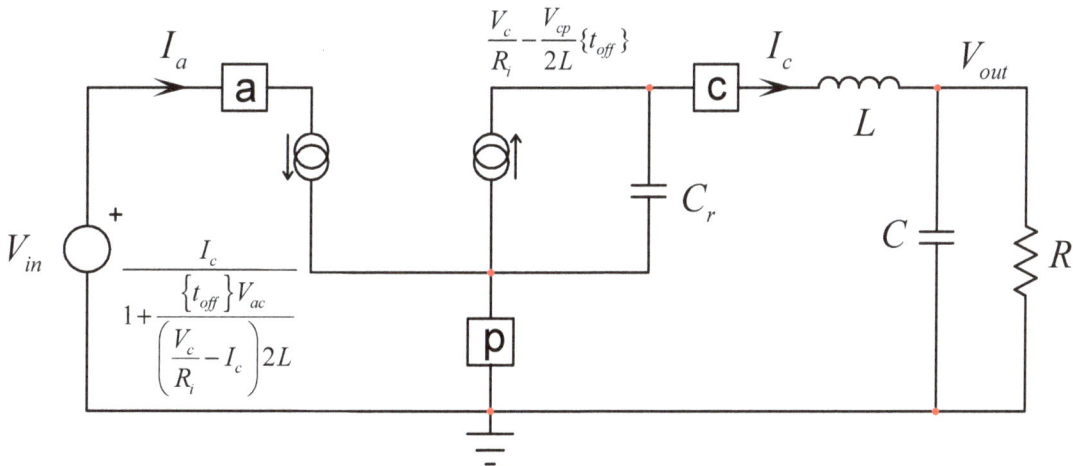

Figure 1.158 The large-signal FOT model is easy to derive and uses two current sources. t_{off} is the fixed off-time value set by designer.

The SPICE implementation of this model appears in Figure 1.159.

The left-side panel computes the different parameter values as t_{off} is now fixed by the designer.

The 1.25-V setpoint voltage delivers 5 V across the 0.5-Ω load.

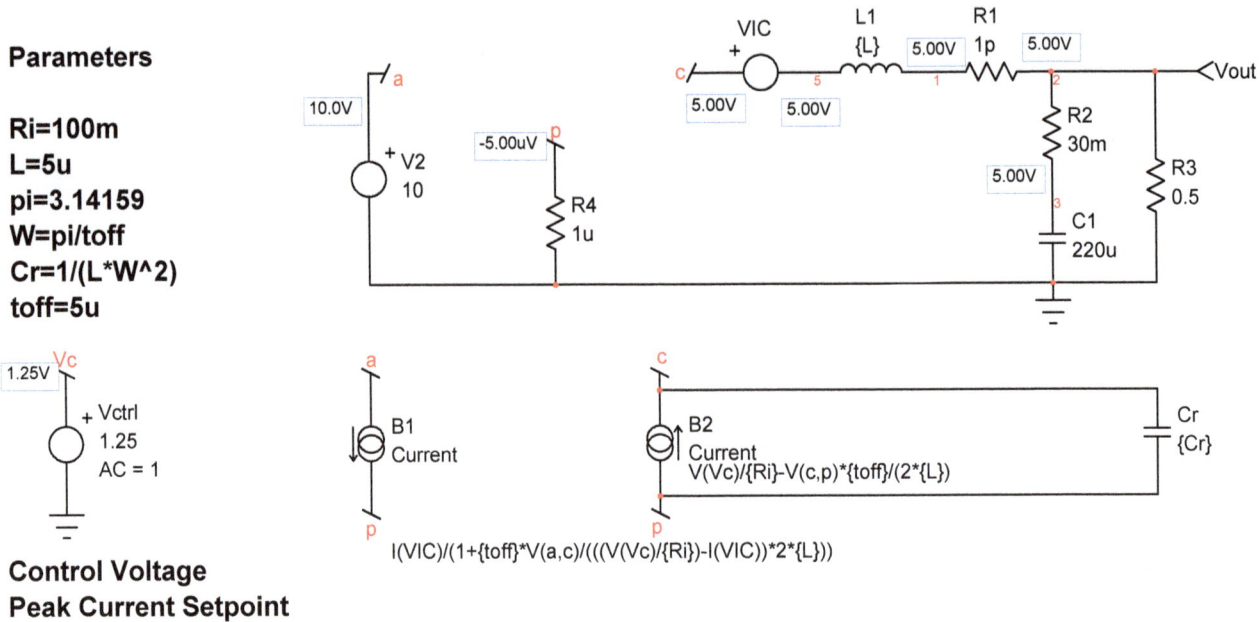

Parameters

Ri=100m
L=5u
pi=3.14159
W=pi/toff
Cr=1/(L*W^2)
toff=5u

Control Voltage
Peak Current Setpoint

Figure 1.159 The large-signal FOT model is easy to derive and uses two current sources.

To check if the small-signal response of the converter is correct, we have tested Dr. Li's model already presented in Figure 1.154.

This model is designed to cover a lot of different control modes by changing the values for R_e and C_e.

For a FOT converter, the new definitions are:

$$R_e = \frac{2L}{t_{off}} \tag{1.358}$$

$$C_e = \frac{1}{L}\left(\frac{t_{off}}{\pi}\right)^2 \tag{1.359}$$

The updated model appears in Figure 1.160 while the dynamic responses of both circuits are given in Figure 1.161. They are perfectly similar in magnitude and phase.

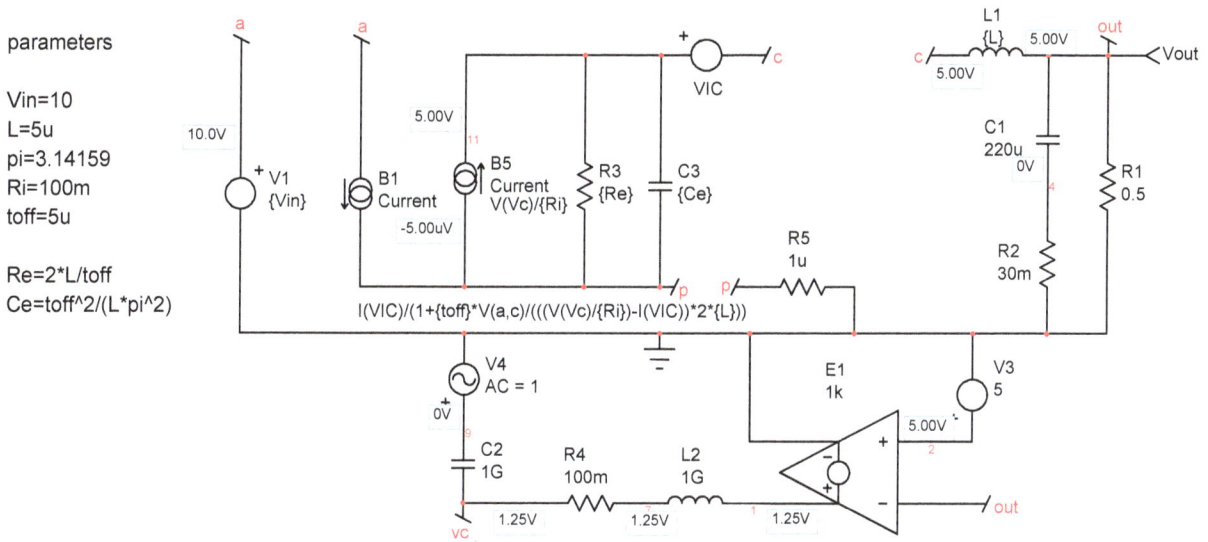

Figure 1.160 Jian Li's circuit can be revised to model a FOT converter by changing values for R_e and C_e.

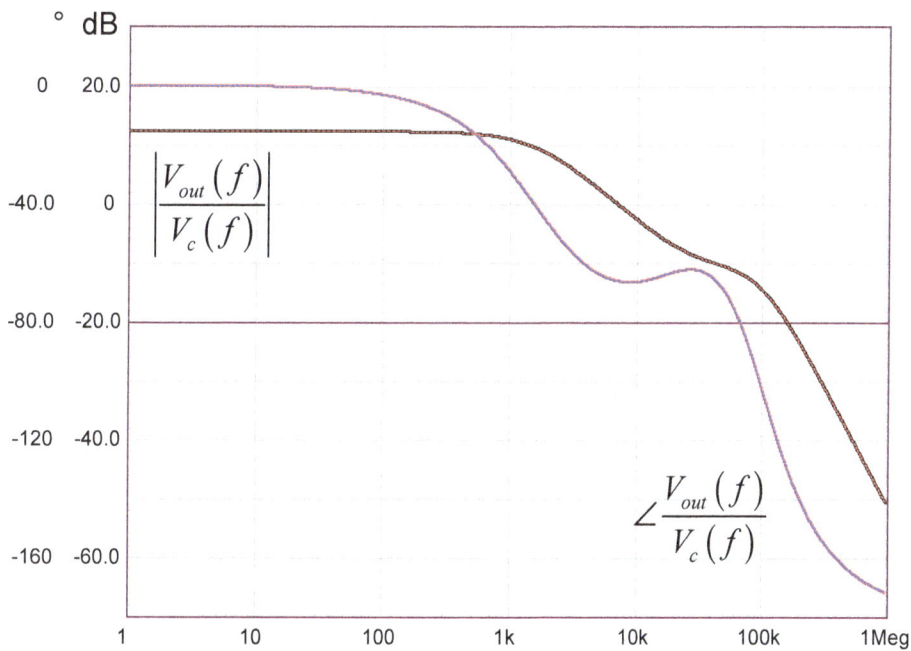

Figure 1.161 Dynamic responses between the newly-derived FOT-PWM switch and Jian Li's model are similar.

To derive a small-signal model, we need to determine the coefficients for the I_a and I_c sources. We have:

$$I_a = \frac{I_c}{1 + \dfrac{t_{off} V_{ac}}{\left(\dfrac{V_c}{R_i} - I_c\right) 2L}} \qquad (1.360)$$

and:

$$I_c = \frac{V_c}{R_i} - \frac{V_{cp}}{2L} t_{off} \qquad (1.361)$$

Let's start partial differentiation to linearize I_a:

$$k_{ic} = \frac{d}{dI_c} \frac{I_c}{1 + \dfrac{t_{off} V_{ac}}{\left(\dfrac{V_c}{R_i} - I_c\right) 2L}} = \frac{2L\left(2LI_c^2 R_i^2 - 2V_{ac} t_{off} I_c R_i^2 - 4LI_c R_i V_c + V_{ac} t_{off} R_i V_c + 2LV_c^2\right)}{\left(2LV_c + R_i V_{ac} t_{off} - 2I_c LR_i\right)^2} \qquad (1.362)$$

$$k_{vac} = \frac{d}{dV_{ac}} \frac{I_c}{1 + \dfrac{t_{off} V_{ac}}{\left(\dfrac{V_c}{R_i} - I_c\right) 2L}} = -\frac{2LI_c R_i t_{off} \left(V_c - I_c R_i\right)}{\left(2LV_c + R_i V_{ac} t_{off} - 2I_c LR_i\right)^2} \qquad (1.363)$$

$$k_{vc} = \frac{d}{dV_c} \frac{I_c}{1 + \dfrac{t_{off} V_{ac}}{\left(\dfrac{V_c}{R_i} - I_c\right) 2L}} = \frac{2LI_c R_i V_{ac} t_{off}}{\left(2LV_c + R_i V_{ac} t_{off} - 2I_c LR_i\right)^2} \qquad (1.364)$$

which leads to the following definition:

$$\hat{i}_a = k_{ic} \hat{i}_c + k_{vac} \hat{v}_{ac} + k_{vc} \hat{v}_c \qquad (1.365)$$

We can repeat the operation for I_c:

$$k_{ivc} = \frac{d}{dV_c} \left(\frac{V_c}{R_i} - \frac{V_{cp}}{2L} t_{off}\right) = \frac{1}{R_i} \qquad (1.366)$$

$$k_{vcp} = \frac{d}{dV_{cp}} \left(\frac{V_c}{R_i} - \frac{V_{cp}}{2L} t_{off}\right) = -\frac{t_{off}}{2L} \qquad (1.367)$$

Assembling these results leads to the linearized current source describing I_c:

$$\hat{i}_c = k_{ivc} \hat{v}_c + k_{vcp} \hat{v}_{cp} \qquad (1.368)$$

The complete linearized model appears in Figure 1.162.
Its dynamic response is similar to that of the large-signal model response from Figure 1.159.

Parameters

Ri=100m
RL=0.5
L=5u
pi=3.14159
W=pi/toff
Cr=1/(L*W^2)
toff=5u

Vc=1.25
Vin=10
Vout=5
Ic=Vout/RL
Vac=Vin-Vout
Vcp=Vout

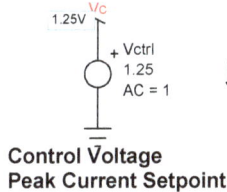

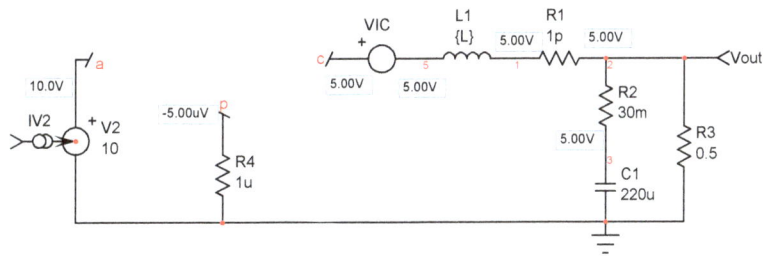

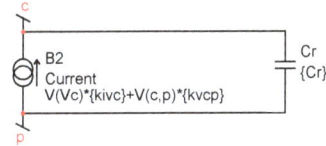

kica=2*L*(2*L*Ic^2*Ri^2-2*Vac*toff*Ic*Ri^2-4*L*Ic*Ri*Vc+Vac*toff*Ri*Vc+2*L*Vc^2)
kicb=(2*L*Vc+Ri*Vac*toff-2*Ic*L*Ri)^2
kic=kica/kicb
kvaca=2*Ic*L*Ri*toff*(Vc-Ic*Ri)
kvacb=(2*L*Vc+Ri*Vac*toff-2*Ic*L*Ri)^2
kvac=-kvaca/kvacb
kvca=2*Ic*L*Ri*Vac*toff
kvcb=(2*L*Vc+Ri*Vac*toff-2*Ic*L*Ri)^2
kvc=kvca/kvcb
kivc=1/Ri
kvcp=-toff/(2*L)

Figure 1.162 This is the FOT model in its small–signal version.

This last model terminates this first chapter introducing switching converters and some of their large- and small-signal models. We can now apply these models to some of the switching cells we are familiar with and obtain the transfer functions of our choice.

However, before delving into the subject, we need to master the tool we will use and known as fast analytical circuits techniques or FACTs. For this purpose, I invite you to discover these techniques by reading Appendix A and exercising your skills by working through some of the examples.

1.5   What Should I Retain from this Chapter?

In this first chapter, we have learned key information that is summarized below:

1. A linear circuit must satisfy the superposition principle: additivity and proportionality. If a circuit is nonlinear, we must restrict the analysis to small signals to linearize it around an operating bias point so that we can apply the Laplace transform to it.

2. A nonlinear model can be made linear by superimposing a small alternating signal across its operating point. The reduced-amplitude modulation imposes small variations around the bias point where the system is linear. This is a small-signal observation.

3. A circuit can be linearized by perturbing the large-signal equation describing the network which leads to ac and dc equations. Sorting out these equations lets you build dc and small-signal models. However, with complicated expressions, perturbation leads to numerous mixed terms that can be difficult to sort. Partial differentiation, on the other hand, calculates the sensitivity of the large-signal expression to each of its

variables. You end up with coefficients whose calculation can be automated and once assembled together by the application of superposition, will form a complete small-signal equation.

4. Switching converter are efficient circuits associating two energy-storage elements – an inductor and a capacitor – to form the basic cells: buck, buck-boost and boost converters. Operated on and off, the two switches these circuits are made can be grouped into a common switching cell, the PWM switch.

5. The switches are operated at the switching frequency and introduce a nonlinearity when turning on and off. The PWM switch model solely considers the nonlinear system formed by the two switches and replaces them with an averaged time-continuous model.

6. This new large-signal model also needs to be linearized so that you can study the circuit using Laplace transform and determine the transfer function of your choice. Note that once the power converter switches have been replaced with an averaged model, circuit simulators may be used to calculate both the large-signal time-domain response and the small-signal frequency-domain response.

7. We have derived the PWM switch model in several configurations such as fixed-frequency voltage- and current-mode control, variable frequency quasi-resonant voltage- and current-mode models, constant on- and off-time models.

1.6 References

1. C. Basso, *Switch-Mode Power Supplies: Spice Simulations and Practical Designs*, 2nd edition, McGraw-Hill, New-York 2014
2. C. Basso, *Designing Control Loops for Linear and Switching Power Supplies – A Tutorial Guide*, Artech House, Boston 2012
3. V. Vorpérian, *Simplified Analysis of PWM Converters using Model of PWM Switch, parts I and II*, IEEE Transactions on Aerospace and Electronic Systems, Vol. 26, NO. 3, 1990
4. R. D. Middlebrook, *Predicting modulator phase lag in PWM converter feedback loops*, Proceedings of Powercon 8, pp. H4.1-H4.6, 1981
5. R. B. Ridley, *A New Small-Signal Model for Current-Mode Control*, PhD Dissertation, Virginia Polytechnic Institute and State University, November, 1990.
6. C. Basso, *Compensation Ramp for the NCP1200*, Application Note AND8029/D, ON Semiconductor, https://www.onsemi.com/pub/Collateral/AND8029-D.PDF
7. R. Redl, J. Sun, *Ripple-Based Control of Switching Regulators – An Overview*, IEEE Transactions on Power Electronics, Vol. 24, Issue 12, Dec. 2009
8. R. Redl, *Feedforward Control of Switching Regulators*, Professional Education Seminars, Proceedings of the IEEE Applied Power Electronics Conference, 2009.
9. L.G Meares, *New Simulation techniques Using Spice*, Proceedings of the IEEE Applied Power Electronics Conference, pp. 198-205, 1986
10. V. Vorpérian, *Analysis of Current-Controlled Converters using the Model of the Current-Controlled PWM Switch*, Power Conversion and Intelligent Motion Conference, pp. 183-195, 1990
11. D. Maksimovic, R. Erickson, *Advances in Averaged Switch Modeling and Simulation*, Power Electronics Specialists Conference, PESC 1999 - https://ecee.colorado.edu/~rwe/references/pesc99-seminar.pdf
12. Jian Li, Fred C. Lee, *New modeling approach and equivalent circuit representation for current mode control*, IEEE Transaction on Power Electronics, vol. 25, no. 5, pp.1218–1230, May. 2010

2 The Buck Converter and its Derivatives

IN THE PREVIOUS chapter, we have introduced the 3 basic switching dc-dc converters and several small-signal models built around the PWM switch.

In this second chapter, we will cover the derivation of the four transfer functions describing a voltage-mode CCM- and DCM-operated buck converter: control to output, input to output, input and output impedances. To let you progress at your own pace, I have detailed each step of the derivation process and verified the results at the end with SPICE and SIMPLIS® simulations. I will detail the steps also for the current-mode CCM- and DCM-operated converter.

The other buck-derived topologies such as the tapped buck, the COT- and FOT-operated buck, the forward, half- and full-bridge converters, push-pull and active-clamp forward will not be forgotten and are also covered but at a lower level of detail.

Numerous tools exist to determine transfer functions of a linear network. I personally favor the fast analytical circuits techniques or FACTs briefly introduced in Appendix A and object of comprehensive coverage in [1] and [2]. By slicing a complex circuit into several smaller sketches that you individually solve, the FACTs let you determine transfer functions of any order in a swift and efficient manner.

They represent a set of truly powerful tools that I encourage you to acquire whether you are a seasoned engineer or a student.

2.1 Buck Transfer Functions in CCM – Fixed Frequency Voltage-Mode Control

We will start this chapter with the control-to-output transfer function. It describes the way a stimulus applied to the control input of a given converter – the error voltage V_{err} or the duty ratio D – propagates through the network and creates the output response.

To determine this transfer function with a converter operated in voltage mode and CCM, we simply plug the CCM VM-PWM switch model derived in Chapter 1 into the buck converter shown in Figure 2.1.

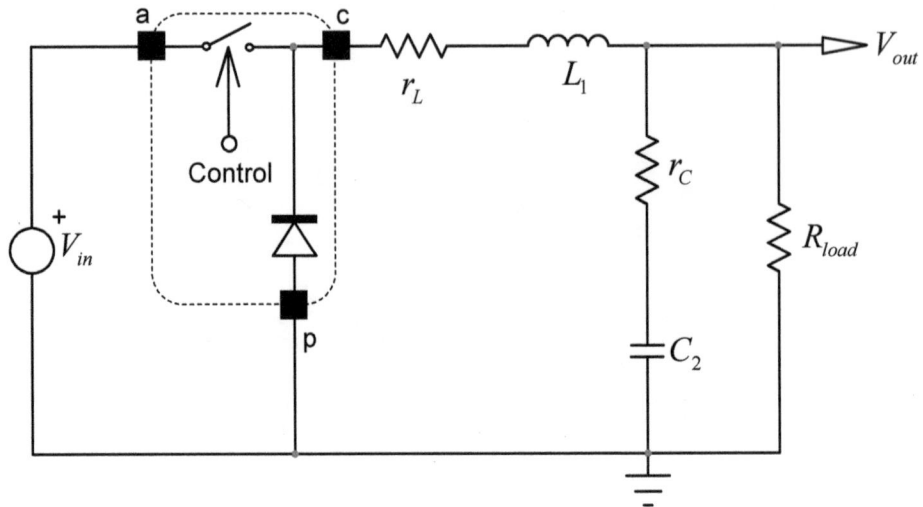

Figure 2.1 You identify the PWM switch in the selected structure and replace the active switch + diode with the model correctly connected.

Rather than directly going to the small-signal model, I first run a simulation with the large-signal model in place.

The Bode plot of this configuration (5 V/1 A from a 12-V source) represents the *reference* response. The duty ratio source V_1 is automatically tweaked so that the output delivers 5 V as confirmed by Figure 2.2 bias points. You recognize the extra circuit built around E_1 which closes the loop in dc via *LoL* and lets you ac-sweep in open loop with the low-pass filter formed by *LoL* and *CoL*.

This method allows you to change the load or the input voltage without having to manually adjust the duty ratio for maintaining a 5-V output. The resulting duty ratio voltage image here is 417.15 mV which corresponds to 41.7% Then, I capture a new circuit diagram in which I replace the large-signal model by its linear counterpart. This is what I have done in Figure 2.1. The dummy resistor connecting node *p* to ground is there to let me label that node which otherwise would be grounded without label.

Once the simulation is over, I verify that the bias points are correct (the 1-mV difference in the output comes from the rounded duty ratio). I also make sure the dynamic response exactly matches the reference one. When this work is finished, the electrical integrity is confirmed and you can start deriving the transfer function of your choice with the linear model.

Why going through these steps by the way? To simply ensure that you are going to spend time on a circuit that is electrically sound. There is nothing more infuriating to realize at the end of the work that the model was wrong from the beginning. Comparing bias points and dynamic response of the two models is thus an essential exercise.

Also, a method I recommend when deriving transfer functions consists of rearranging the original network in a simpler circuit in which components and sources are assembled in a way that you naturally can see how things flow. Very often, you start with a raw, complicated circuit and step-by-step simplify it by rearranging sources, reflecting components on the other side of a transformer and so on.

Again, any change in the circuit must be backed up by a quick simulation ensuring the hypothesis you made or the simplification you brought (neglecting a term for instance) does not significantly alter the response.

SPICE, to that respect, is an invaluable assistant and will be used extensively for that purpose.

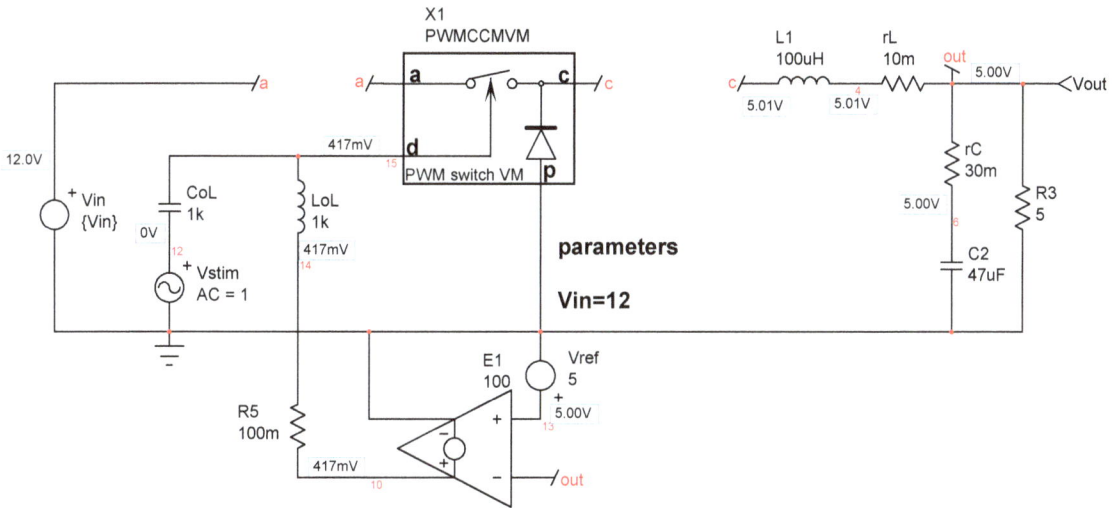

Figure 2.2 The large-signal model delivers the reference transfer function obtained from a 12-V source.

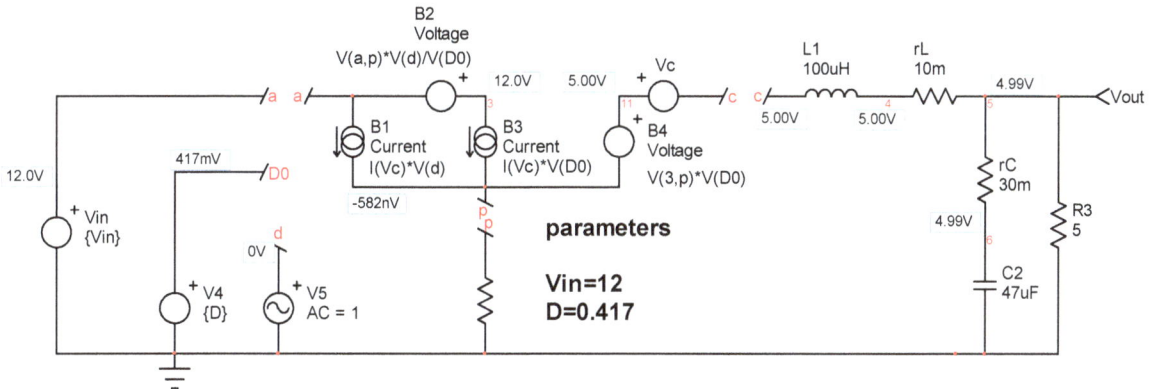

Figure 2.3 Plug the small-signal VM-PWM switch model in the converter under study and you have a linear network.

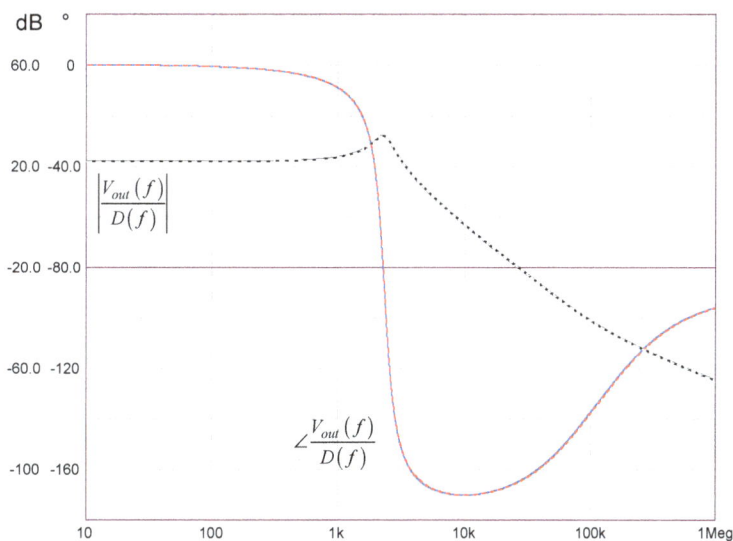

Figure 2.4 Dynamic responses of the two circuits are absolutely identical and we are good to go.

In Figure 2.1, the D_0 parameter represents the static duty ratio while the small-signal modulation applied at node d designates the stimulus $\hat{d}$ or $D(s)$. For this first transfer function, we want the relationship linking $V_{out}(s)$ to $D(s)$:

$$\left.\frac{V_{out}(s)}{D(s)}\right|_{V_{in}(s)=0} \tag{2.1}$$

In this configuration, the input voltage source V_{in} is held constant while node d is modulated for the analysis. If V_{in} does not change while d is swept in frequency, then the ac value of V_{in} is 0 V and node a can be grounded.

The expression $V(a,p)$ in source B_2 becomes a dc bias, V_{in}. Node 3 bias is thus $D(s)V_{in}/D_0$ and appears in source B_4 where a simplification by D_0 occurs. The circuit is immediately simplified for the purpose of this analysis as illustrated in Figure 2.5. Ac-sweeping this network gives the exact same response as in Figure 2.4.

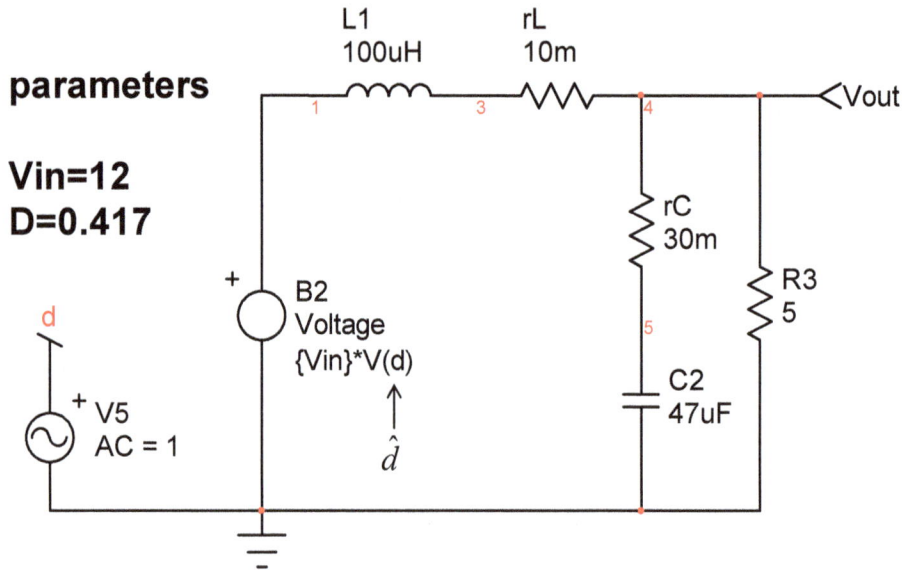

Figure 2.5 By grounding terminal a, the circuit greatly simplifies to a source B_2 feeding a low-pass filter.

The control-to-output transfer function H of a switching converter, also called the *plant* or *power stage* response, combines a leading term (the quasi-static gain often noted H_0) and a quotient combining a numerator $N(s)$ and a denominator $D(s)$:

$$H(s) = H_0 \frac{N(s)}{D(s)} \tag{2.2}$$

The roots of the numerator are the *zeroes* of the transfer function while the denominator roots are called the *poles*. The poles are related to the natural time constants of the network studied under zero excitation as explained in Appendix A: reduce the stimulus to zero and determine the time constants involving one or several energy-storing elements (the inductors and the capacitors of the circuit).

In this first transfer function determination, the stimulus is the small-signal duty ratio d while the response is the output voltage V_{out}. In Figure 2.5, you can see that setting d to 0 in source B_2 will let you simplify the circuit to that of Figure 2.6.

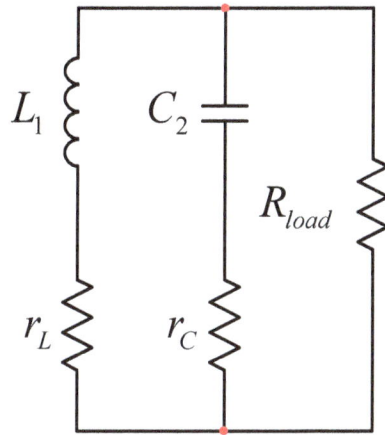

Figure 2.6 Setting the excitation to 0 reveals the network structure.

The exercise will now consist of determining the denominator of this 2nd-order circuit which obeys the following normalized form [1]:

$$D(s) = 1 + b_1 s + b_2 s^2 \qquad (2.3)$$

The first term b_1 is obtained by summing the two time constants involving L_1 and C_2. They are determined by expressing the resistance "seen" from one energy-storing element connecting terminals while the other is left in its dc state (a capacitor is open circuited, an inductor short circuited).

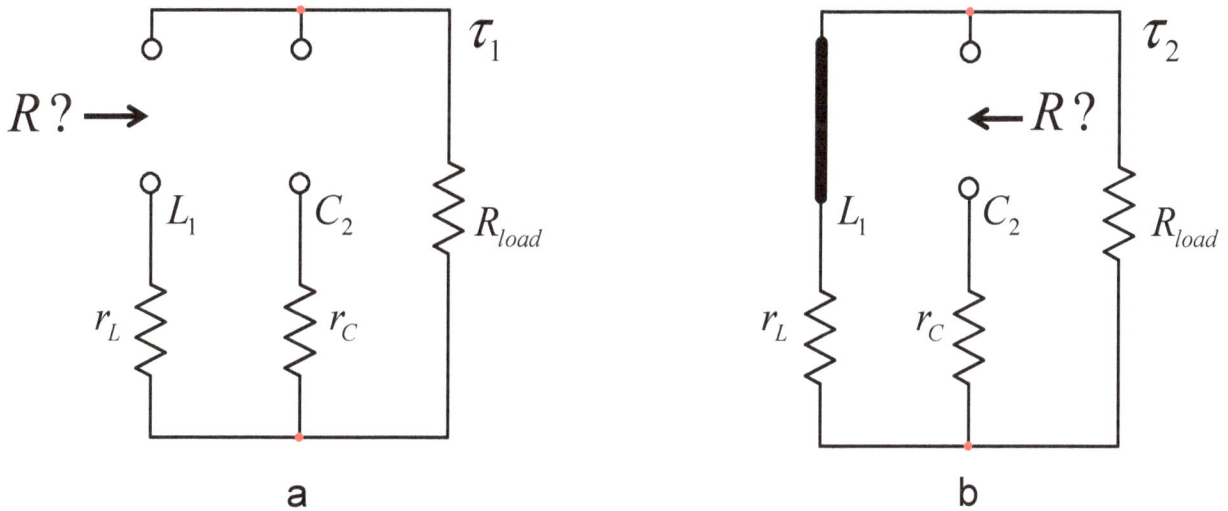

Figure 2.7 The coefficient for b_2 is quickly obtained by inspecting two simple sketches.

Figure 2.7 shows the steps. In sketch (a), the resistance you "see" through L_1's connecting terminals is the series combination of r_L and R_{load}, leading to the first time constant:

$$\tau_1 = \frac{L_1}{r_L + R_{load}} \qquad (2.4)$$

The second time constants involving C_2 is immediate with sketch (b):

$$\tau_2 = C_2 \left(r_L \parallel R_{load} + r_C \right)$$ (2.5)

We can form b_1 without further efforts:

$$b_1 = \tau_1 + \tau_2 = \frac{L_1}{r_L + R_{load}} + C_2 \left[r_L \parallel R_{load} + r_C \right]$$ (2.6)

The second term, b_2, is obtained by reusing one of the first time constants – τ_1 or τ_2 – and associating it with another time constant determined according to the sketches in Figure 2.8.

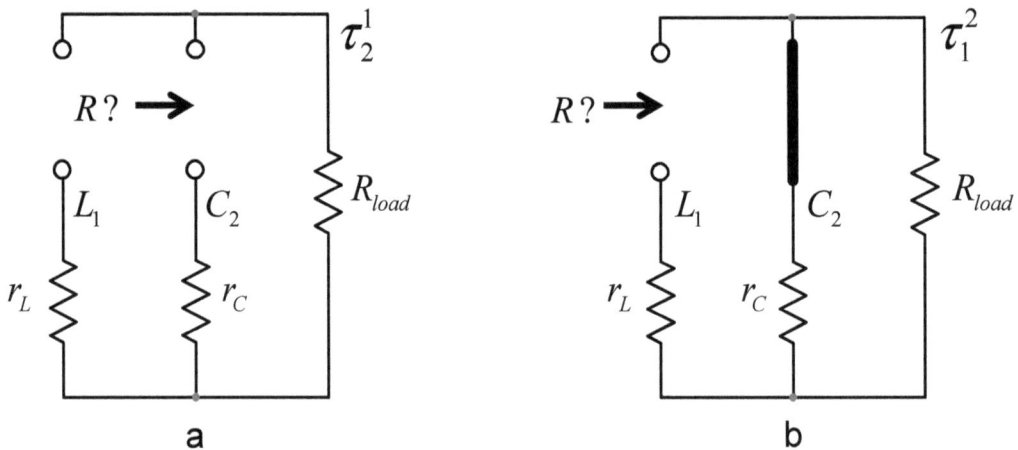

Figure 2.8 You now set one of the energy-storing elements in its high-frequency state and determine the resistance "seen" from the second element connecting terminals.

In sketch (a), you place inductor L_1 in its high-frequency state (open circuit it) while you determine the resistance "seen" from the capacitor terminals:

$$\tau_2^1 = C_2 \left(r_c + R_{load} \right)$$ (2.7)

In sketch (b), place capacitor C_2 in its high-frequency state (a short circuit) and look at the resistance offered from inductor L_1's connecting terminals:

$$\tau_1^2 = \frac{L_1}{r_L + R_{load} \parallel r_C}$$ (2.8)

The term b_2 is now defined by either combining τ_1 with (2.7) or τ_2 with (2.8). Both expressions are equal and you are free to pick the simplest one:

$$b_2 = \tau_1 \tau_2^1 = \frac{L_1}{r_L + R_{load}} C_2 \left(r_C + R_{load} \right)$$ (2.9)

or:

$$b_2 = \tau_2 \tau_1^2 = C_2 \left(r_L \parallel R_{load} + r_C \right) \frac{L_1}{r_L + R_{load} \parallel r_C} \tag{2.10}$$

The above expressions are identical but (2.10) looks more complicated than (2.9). This redundancy is interesting when one of the time constants association leads to an indeterminacy or a fairly complicated expression needing more simplification work. We will use (2.9) to complete the denominator determination:

$$D(s) = 1 + s \left(\frac{L_1}{r_L + R_{load}} + C_2 \left[r_L \parallel R_{load} + r_C \right] \right) + s^2 \left(L_1 C_2 \frac{r_C + R_{load}}{r_L + R_{load}} \right) \tag{2.11}$$

Now that we have the poles, we can check the presence of zeroes. Figure 2.9 shows the simplified buck network in which two impedances are circled: Z_a and Z_b. What condition in this circuit would prevent the stimulus $V_{in} \cdot D(s)$ from producing a response across the load resistance? Could Z_a become a transformed open circuit for a particular value of s?

$$Z_a(s) = r_L + sL_1 \rightarrow \infty \tag{2.12}$$

Certainly not so L_1 does not contribute a zero. For Z_b, can it become a transformed short circuit, effectively nulling the response?

$$Z_b(s) = r_C + \frac{1}{sC_2} = \frac{1 + sr_C C_2}{sC_2} = 0 \tag{2.13}$$

Yes, and it implies a root s_z equal to:

$$s_z = -\frac{1}{r_C C_2} \tag{2.14}$$

with a zero located at:

$$\omega_z = |s_z| = \frac{1}{r_C C_2} \tag{2.15}$$

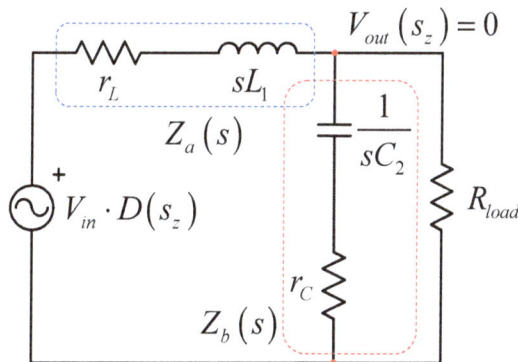

Figure 2.9 The zero is determined by inspecting the network with a nulled response.

The numerator $N(s)$ is immediately expressed as:

$$N(s) = 1 + \frac{s}{\omega_z} \tag{2.16}$$

What we miss now is the dc gain, H_0. To obtain it, we open the capacitor and short circuit the inductor (dc state). This is the drawing from Figure 2.10 illustrating the circuit for $s = 0$.

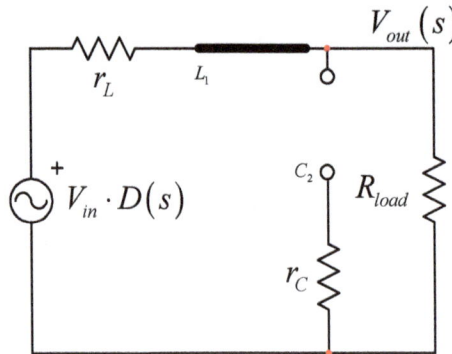

Figure 2.10 Shorting the inductor reveals a resistive divider involving the inductor series resistance r_L and the load.

The gain is immediate and equal to:

$$H_0 = V_{in} \frac{R_{load}}{R_{load} + r_L} \tag{2.17}$$

We can now assemble the pieces and form the complete control-to-output transfer function to which we add the PWM modulator gain as suggested by Figure 2.11.

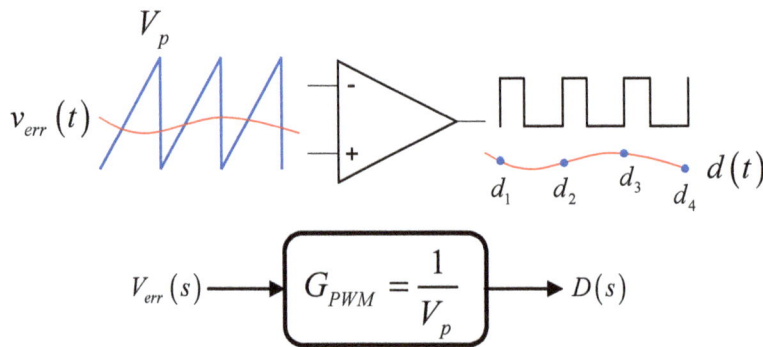

Figure 2.11 The duty ratio is generated by comparing the error signal to a sawtooth of peak value V_p.

$$\frac{V_{out}(s)}{V_{err}(s)} = \frac{V_{in}}{V_p} \frac{R_{load}}{r_L + R_{load}} \frac{1 + sr_C C_2}{1 + s\left(\frac{L_1}{r_L + R_{load}} + C_2\left[r_L \parallel R_{load} + r_C\right]\right) + s^2\left(L_1 C_2 \frac{r_C + R_{load}}{r_L + R_{load}}\right)} \tag{2.18}$$

This expression can advantageously be rewritten under a 2nd-order canonical form such as:

$$H(s) = H_0 \frac{1 + s/\omega_{z_1}}{1 + \dfrac{s}{\omega_0 Q} + \left(\dfrac{s}{\omega_0}\right)^2} \tag{2.19}$$

in which:

$$H_0 = \frac{V_{in}}{V_p} \frac{R_{load}}{r_L + R_{load}} \tag{2.20}$$

$$\omega_0 = \frac{1}{\sqrt{b_2}} = \frac{1}{\sqrt{L_1 C_2}} \sqrt{\frac{r_L + R_{load}}{r_C + R_{load}}} \tag{2.21}$$

$$Q = \frac{\sqrt{b_2}}{b_1} = \frac{L_1 C_2 \omega_0 \left(r_C + R_{load}\right)}{L_1 + C_2 \left(r_L r_C + r_L R_{load} + r_C R_{load}\right)} \tag{2.22}$$

$$\omega_z = \frac{1}{r_C C_2} \quad (2.23)$$

Please note that this expression was obtained without writing a single line of algebra, just inspecting simple individual drawings.

These equations have been captured in a Mathcad® sheet and the dynamic response appears in Figure 2.12. We have superimposed the SPICE response from Figure 2.4 and there is no difference.

The final check consists of simulating a SIMPLIS® schematic shown in Figure 2.13. Subcircuit U_2 builds the pulse-width modulator comparing a fixed 417-mV bias to a 1-V peak sawtooth, leading to a 41.7% duty ratio. The comparator output resets a flip-flop whose output is paced by the 100-kHz clock. The buck converter uses ideal components but can be modified to implement a MOSFET model affected by a $r_{DS(on)}$ or a diode featuring some t_{rr} (recovery time) for instance.

As these non-ideal component induce extra losses, the quality factor Q will show a significant reduction.

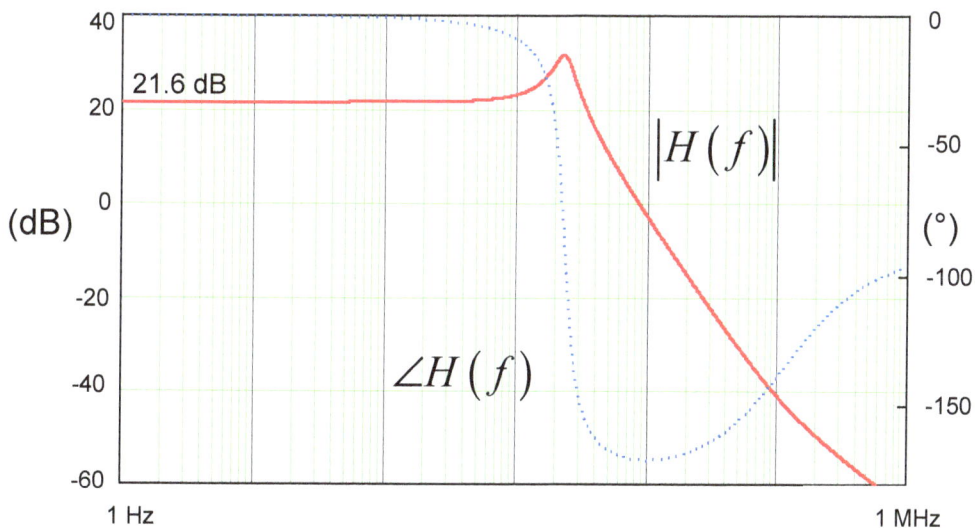

Figure 2.12 Mathcad® immediately plots the buck control-to-output dynamic response.

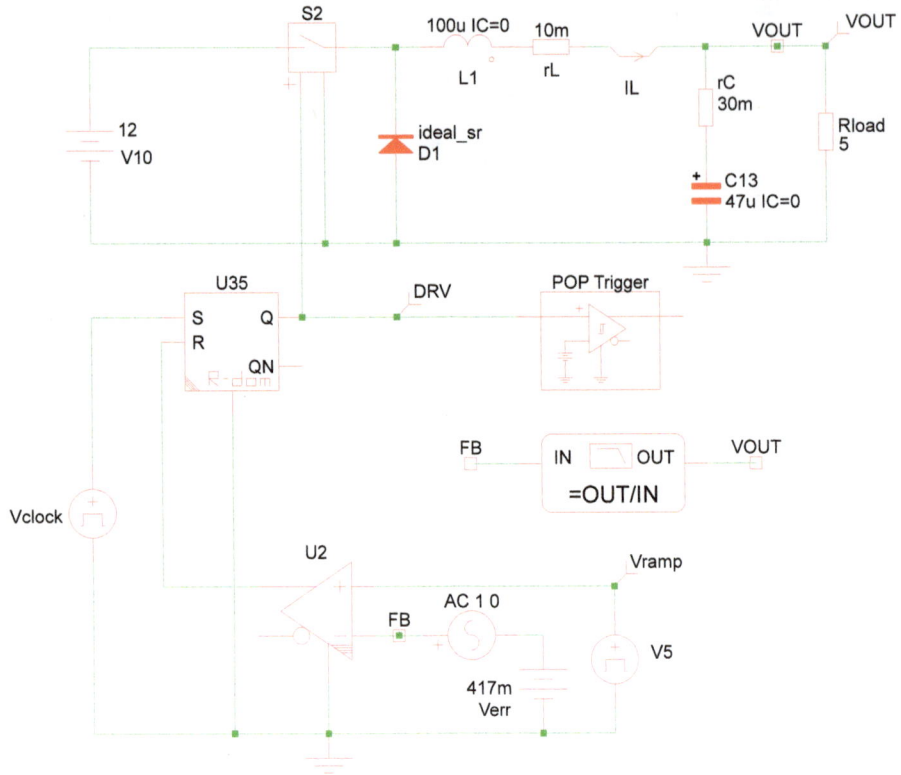

Figure 2.13 The SIMPLIS® schematic is quite simple for this open-loop buck converter.

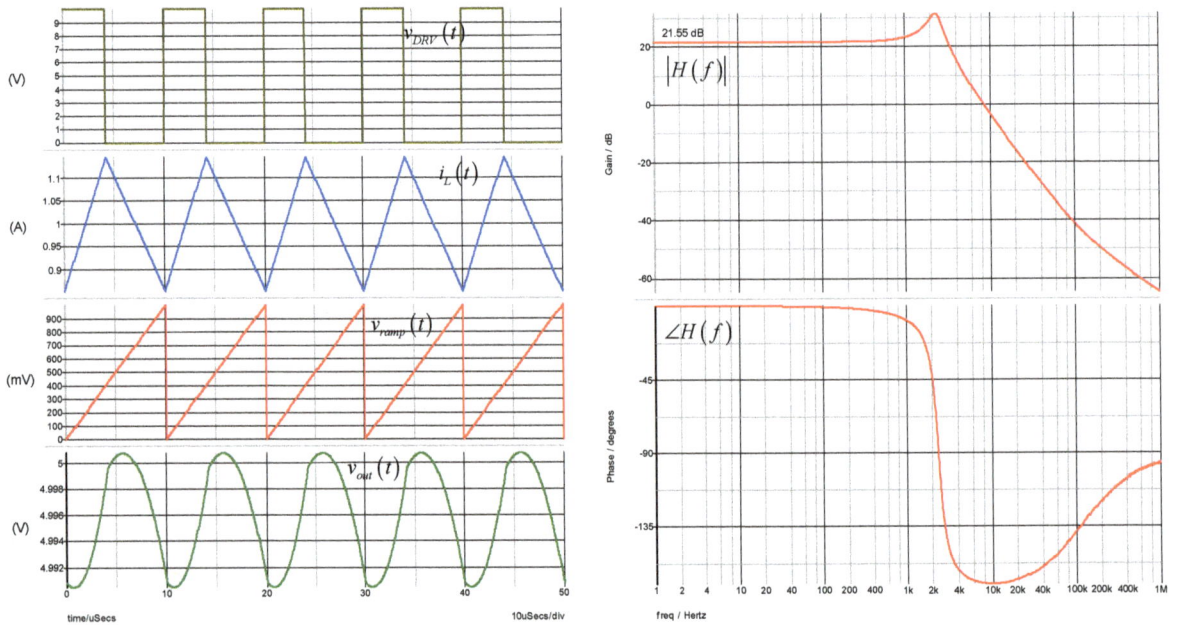

Figure 2.14 On top of switching waveforms, the dynamic response appears after a few seconds of simulation time.

The results appear in Figure 2.14 and confirm our analysis.

2.1.1 Input to Output

The input-to-output transfer function characterizes the ability of the converter to reject incoming perturbations from the input source. In this definition, the converter operates in open-loop conditions meaning that $D(s) = 0$ but the static duty ratio D_0 ensures the correct bias point ($V_{out} = 5$ V). The response is still V_{out} but the stimulus has become V_{in}.

$$\left.\frac{V_{out}(s)}{V_{in}(s)}\right|_{D(s)=0} \qquad (2.24)$$

Set node d to 0 in Figure 1.1 and source B_2 becomes 0 V (a short circuit). Node 3 is now biased to $V_{in}(s)$ which appears in source B_4. The simplified circuit updates to that of Figure 2.15, sketch (a).

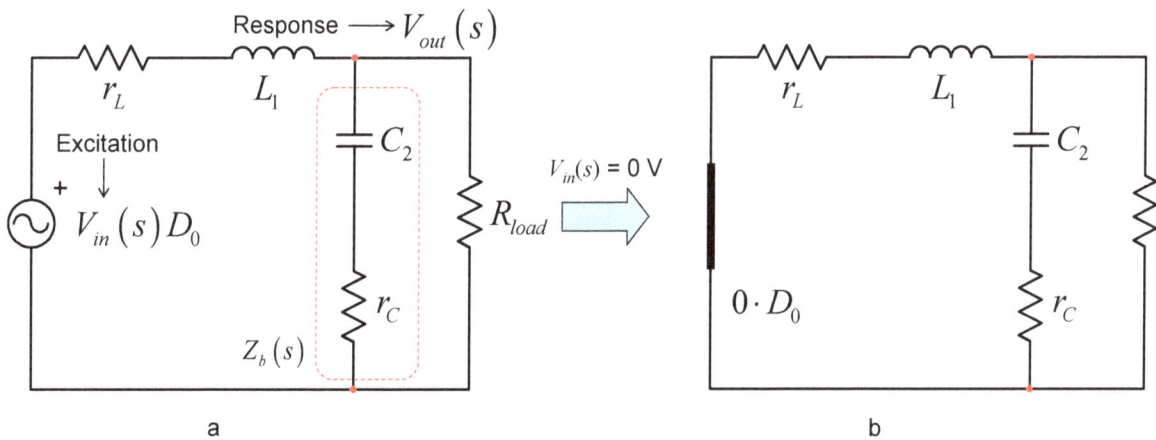

Figure 2.15 The input-to-output transfer function is almost instantaneously determined by inspection.

To determine the denominator, reduce the excitation V_{in} to 0 V and obtain the structure shown in sketch (b). This is exactly the same as in Figure 2.6: we can reuse the denominator $D(s)$ determined in (2.11). Do we have a zero in this new circuit? Yes, when the series connection of C_2 and r_C can become a transformed short circuit:

$$Z_b(s) = r_C + \frac{1}{sC_2} = 0 \qquad (2.25)$$

implying a zero positioned at:

$$\omega_z = \frac{1}{r_C C_2} \qquad (2.26)$$

For the dc gain at $s = 0$, in your head, short-circuit L_1 and open-circuit C_2. You are left with a resistive divider involving R_{load} and r_L:

$$H_0 = D_0 \frac{R_{load}}{R_{load} + r_L} \qquad (2.27)$$

This expression tells you that a narrow duty ratio will bring a better rejection than with a larger one (in open loop). By the way, is there a faster way to obtain this result rather going through this analysis? Yes, we know that V_{out} and V_{in} are linked by the static conversion ratio M:

$$V_{out}(V_{in}, D) = M(D)V_{in} \tag{2.28}$$

The small-signal dc gain linking the output and input variables is simply obtained by differentiating V_{out} with respect to V_{in} while the duty ratio is fixed:

$$H_0 = \left.\frac{\partial V_{out}(V_{in}, D)}{\partial V_{in}}\right|_{D_0} = M \tag{2.29}$$

We have determined the complete transfer function in a few simple steps, without solving complicated equations:

$$\frac{V_{out}(s)}{V_{in}(s)} = H_0 \frac{1 + \dfrac{s}{\omega_z}}{1 + \dfrac{s}{\omega_0 Q} + \left(\dfrac{s}{\omega_0}\right)^2} \tag{2.30}$$

Q and ω_0 are respectively defined by (2.22) and (2.21). With the component values shown in Figure 2.2, the quasi-static gain is computed to -7.6 dB, as confirmed by Figure 2.16 which gathers the Mathcad® response.

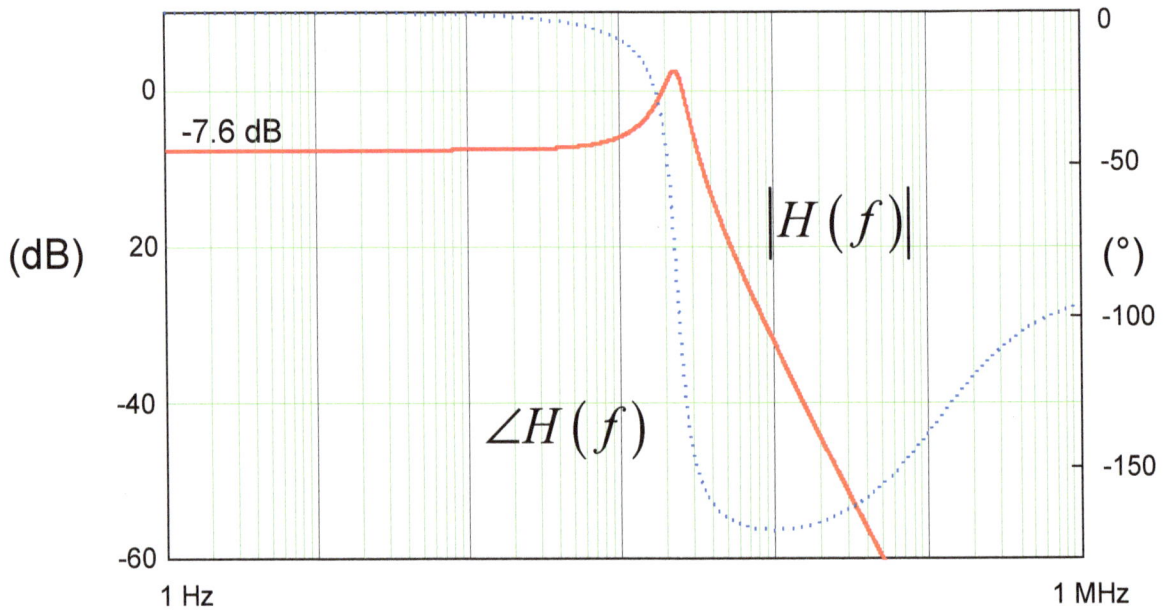

Figure 2.16 The input voltage attenuation or rejection capability of the open-loop CCM buck is only 7.6 dB in dc.

The simple SPICE template of Figure 2.17 lets us check this response using the VM-PWM model. As confirmed by the Bode plot in Figure 2.18, magnitude and phase responses agree with those of Figure 2.16.

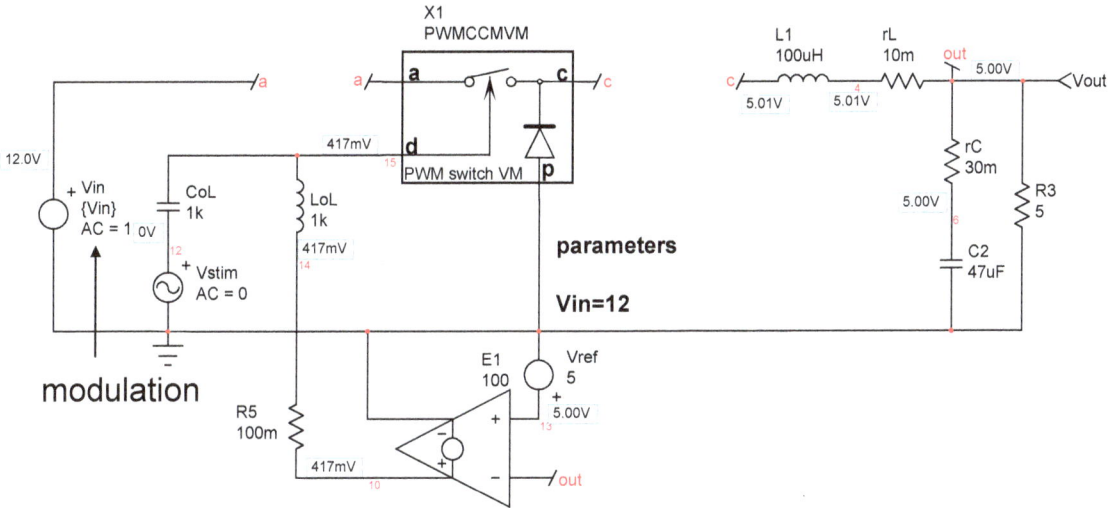

Figure 2.17 Sweeping the input source while the duty ratio is set to a static value is the way to reveal the input-to-output response.

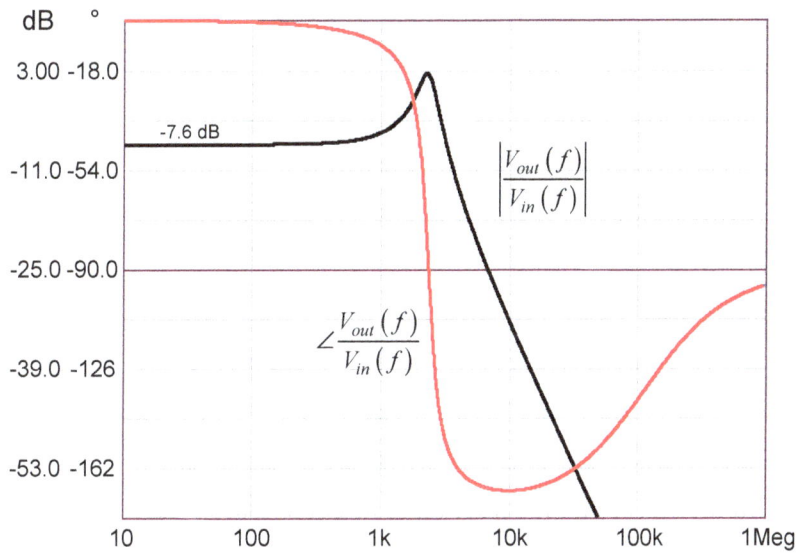

Figure 2.18 The SPICE simulation confirms the response obtained from Mathcad®.

2.1.2 Output Impedance

The output impedance is a transfer function obtained by installing a current test generator across the output node while the load is present. The voltage generated across its terminals (the response) divided by the injected current (the stimulus) leads to the transfer function we want. During this analysis, the small-signal duty ratio node d is kept to 0 V, implying an open-loop configuration. Also, the input voltage source V_{in} is perfect (0-ohm output resistance) and set to 0 V ac as in any output impedance determination exercise:

$$Z_{out}(s) = \frac{V_{out}(s)}{I_{out}(s)}\bigg|_{V_{in}(s)=0,\, D(s)=0} \tag{2.31}$$

149

The circuit reduces to sketch (a) in Figure 2.19.

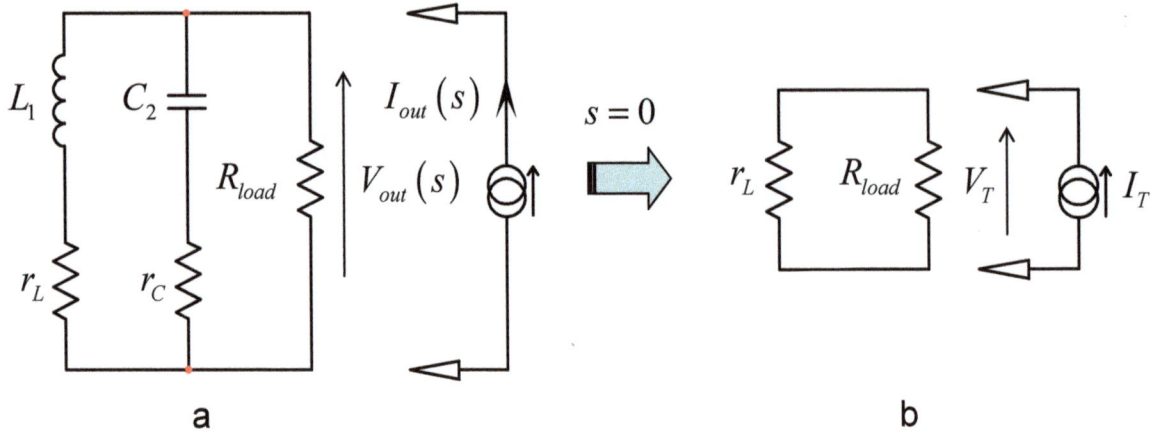

Figure 2.19 A current source is installed across the output node to determine the output impedance.

Setting $s = 0$ (shorting the inductor and open circuiting the capacitor) immediately gives the dc resistance R_0 from sketch (b):

$$R_0 = \frac{V_T}{I_T} = r_L \parallel R_{load} \tag{2.32}$$

What impedance combination in sketch (a) could lead to a null in the response? As shown in Figure 2.20, the null in the response is obtained when a transformed short circuit is created by L_1 and r_L but also more classically by C_2 and r_C.

The position of the zeroes is immediately determined:

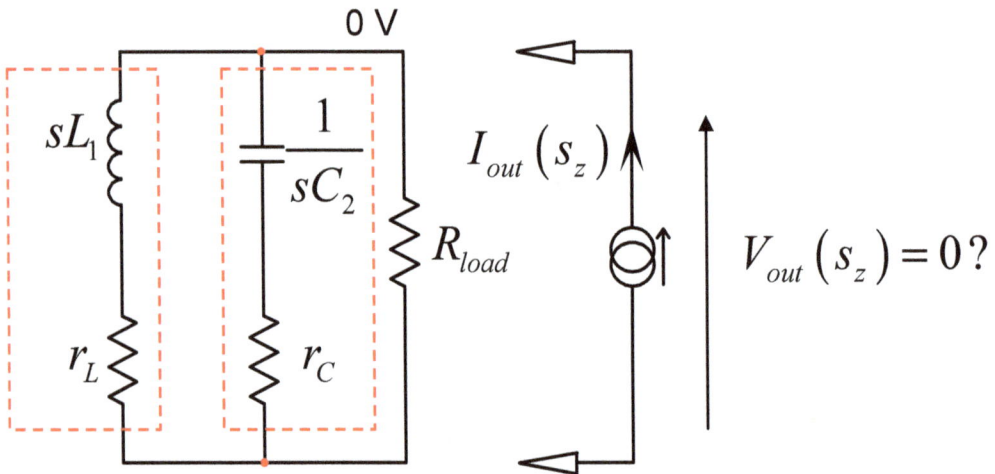

Figure 2.20 The two zeroes are easily determined by inspection.

$$sL_1 + r_L = 0 \tag{2.33}$$

leading to:

$$\omega_{z_1} = \frac{r_L}{L_1} \qquad (2.34)$$

The second zero is determined solving:

$$1 + s r_C C_2 = 0 \qquad (2.35)$$

which leads to:

$$\omega_{z_2} = \frac{1}{r_C C_2} \qquad (2.36)$$

To obtain the denominator, simply turn the excitation off or open circuit the current source I_{out} in Figure 2.19. Observing the illustration reveals that you have returned to the circuit of Figure 2.6 for which we have already determined the denominator $D(s)$.

The output impedance expression is thus:

$$Z_{out}(s) = R_0 \frac{\left(1 + \frac{s}{\omega_{z_1}}\right)\left(1 + \frac{s}{\omega_{z_2}}\right)}{1 + \frac{s}{Q\omega_0} + \left(\frac{s}{\omega_0}\right)^2} \qquad (2.37)$$

in which the two zeroes are defined by (2.34) and (2.36) while Q and ω_0 are respectively defined by (2.22) and (2.21). R_0 is obtained with (2.32) and carries the unit of Ω. Plotting (2.37) shows some peaking around 2.3 kHz (Figure 2.21).

When closing the loop, the crossover frequency f_c must be selected high enough so that sufficient gain exists at 2.3 kHz to damp the oscillations brought by $L_1 C_2$. Designers usually pick a crossover greater than 3 times the resonant frequency or above 7 kHz in this example.

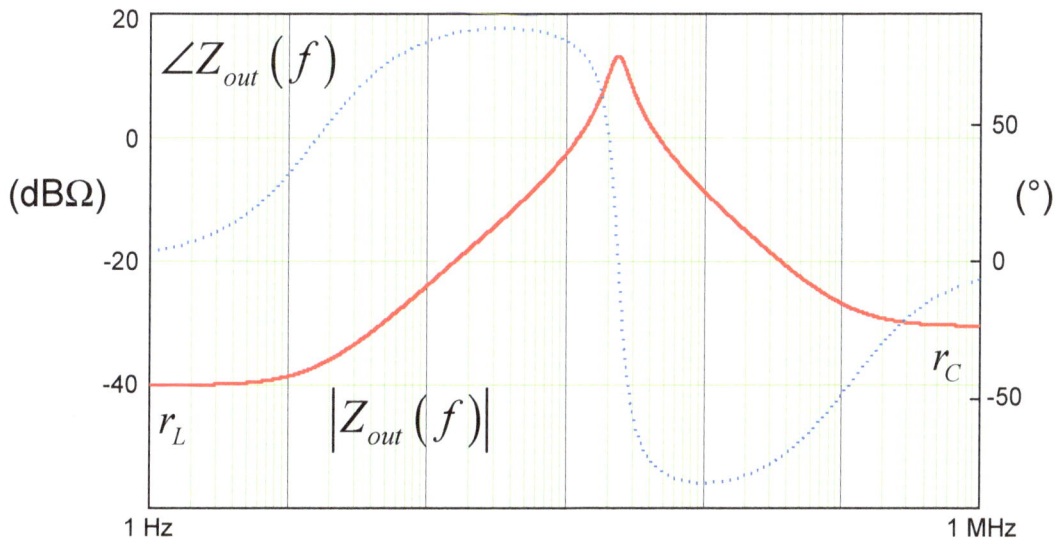

Figure 2.21 The buck output impedance is dictated by the inductor ohmic loss at dc.

A simple simulation template is proposed in Figure 2.22. To sweep the open-loop output impedance, install a 1-A ac current source across the load and monitor the voltage across its terminals, V_{out} in this case.

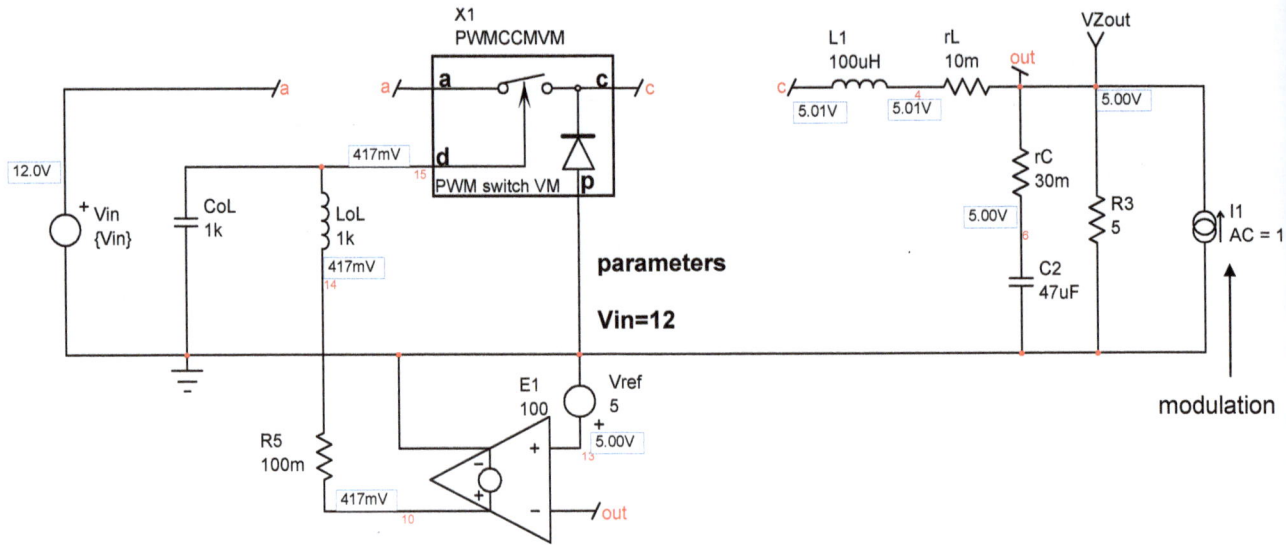

Figure 2.22 Installing a 1-A ac source across the loaded output of the buck converter and maintaining a static duty ratio is the way to unveil the output impedance.

The resulting response is shown in Figure 2.23 and confirms the shape from Figure 2.21: the inductor ohmic loss r_L dominates in dc then the impedance becomes inductive.

Resonance occurs and the capacitor now fixes the response until r_C dominates in the upper portion of the spectrum.

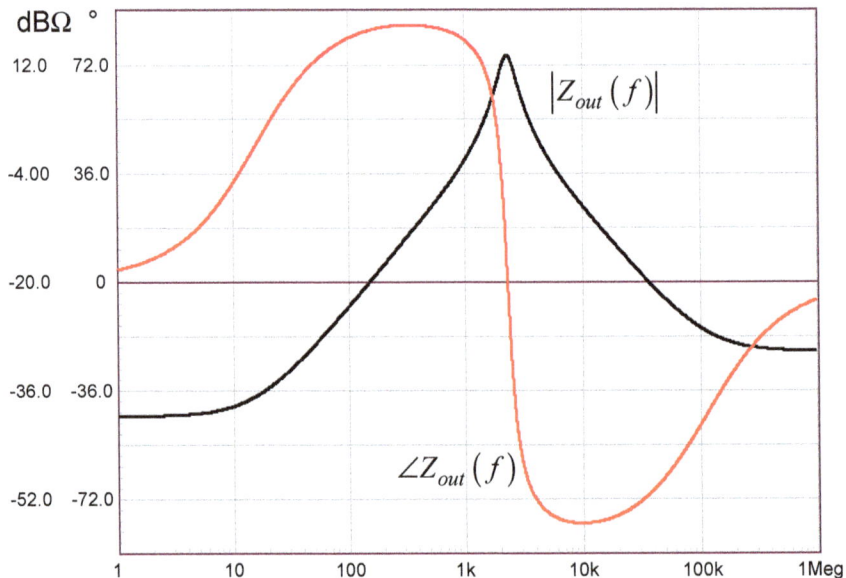

Figure 2.23 The buck output impedance is resistive, then inductive. After the peak, it becomes capacitive and again resistive as the ESR dominates the response.

2.1.3 Input Impedance

The input impedance represents the fourth transfer function we want to determine. To do so, we will connect a current generator sweeping the input node of the buck converter. In this mode, the injected current plays the stimulus role while the voltage across the source terminals represents the response:

$$Z_{in}(s) = \frac{V_{in}(s)}{I_{in}(s)}\bigg|_{D(s)=0} \tag{2.38}$$

In SPICE, we will install an inductor *LoL* in series with the input source V_{in} while a 1-A ac current generator biases node *a* in Figure 2.24. Displaying the voltage at terminal *a* will let us plot the input impedance image.

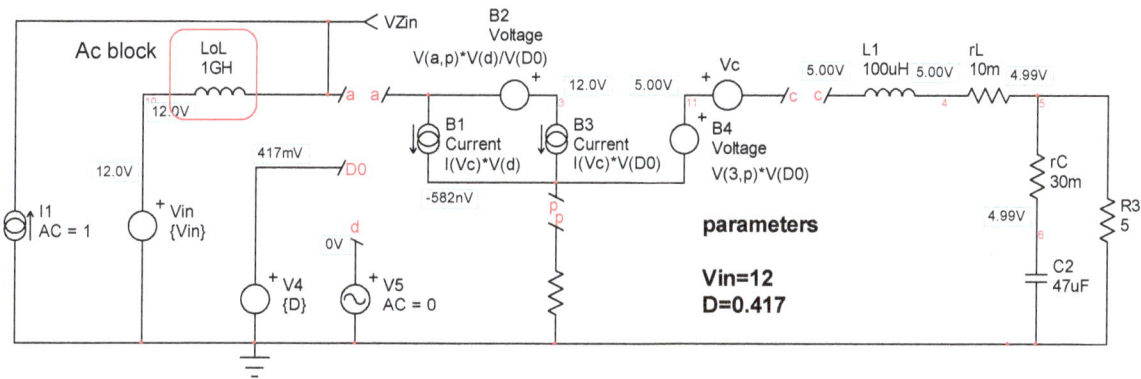

Figure 2.24 The circuit finds its correct dc operating point owing to the presence of *LoL*.

As we did before, we can run simplifications considering node *d* equal to 0 V: sources B_1 and B_2 are turned off leading to the circuit of Figure 2.25. This circuit can be further simplified by realizing that sources B_3 and B_4 are actually the dc transformer modeling the PWM switch (see Chapter 1).

This transformer features a $1:D_0$ "turns" ratio and will help us reflect all the secondary-side elements to the primary side, where the input impedance is measured. The updated circuit before reflection appears in Figure 2.26. You could very well directly determine the input impedance from Figure 2.24 but going through the transformer steps adds in simplification.

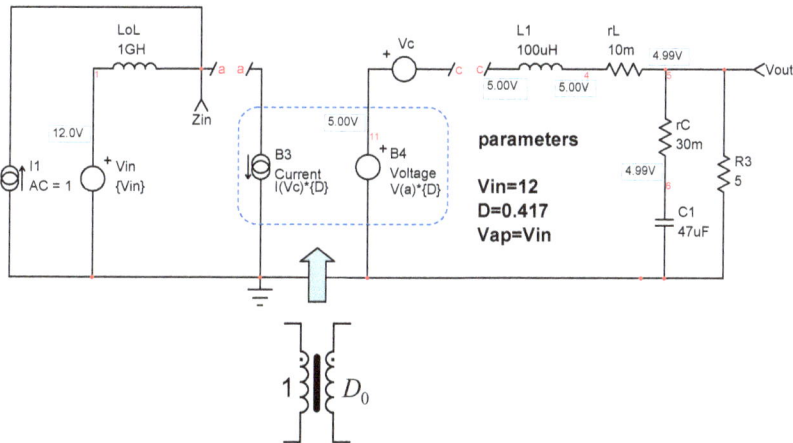

Figure 2.25 The simplification consists of plugging the dc transformer representation of the PWM switch.

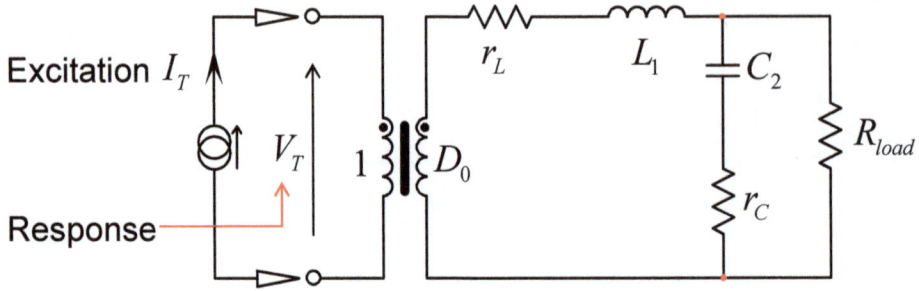

Figure 2.26 Once the dc transformer is plugged, it is possible to reflect all secondary-side components to the left side of the transformer.

We can now refer the inductor and the capacitor together with their stray elements and the load to the left side of the transformer. We simply divide all impedances/resistances by the turns ratio squared or D_0^2.

This is what we have done in Figure 2.27.

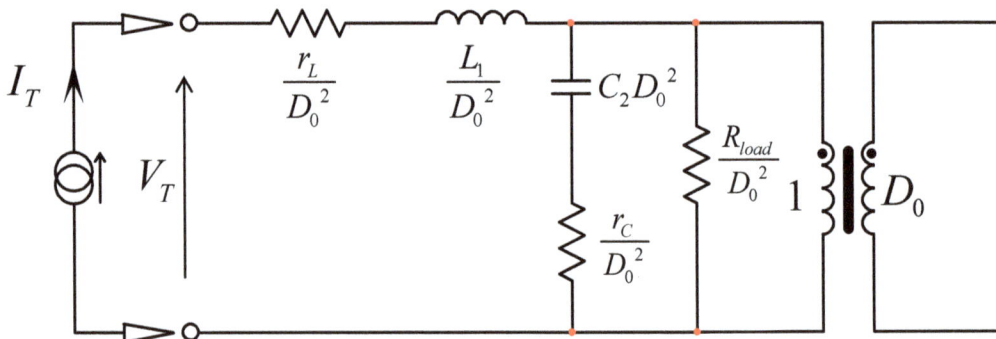

Figure 2.27 Components are all referred to the left side of the transformer where the input measurement takes place.

We first start with the resistance offered by this circuit for $s = 0$: short L_1 and open circuit C_2.

Figure 2.28 shows the updated drawing from which you determine the input resistance equal to the two reflected resistances in series, R_{load} and r_L:

$$R_0 = \frac{r_L + R_{load}}{D_0^2} \tag{2.39}$$

The denominator is determined by setting the excitation to 0 A i.e. open circuiting the current source I_T. The first time constant is determined according to sketch (a) of Figure 2.29.

The capacitor is open circuited and you express the resistance "seen" through L_1's terminals. As the inductor left side is floating, we have:

$$\tau_1 = \frac{L_1}{D_0^2 \cdot \infty} = 0 \tag{2.40}$$

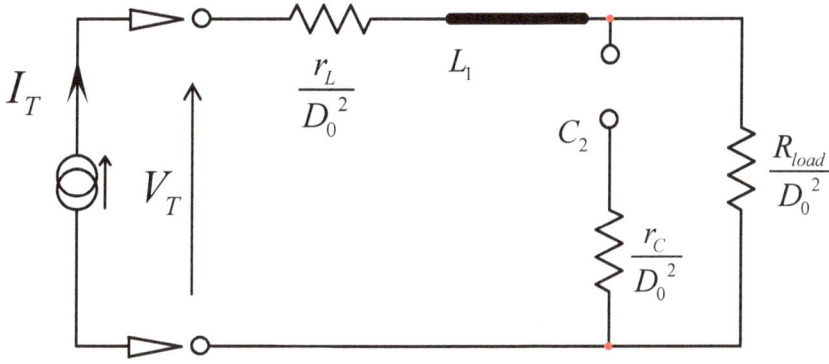

Figure 2.28 The dc input resistance determination is immediate when the inductor is shorted.

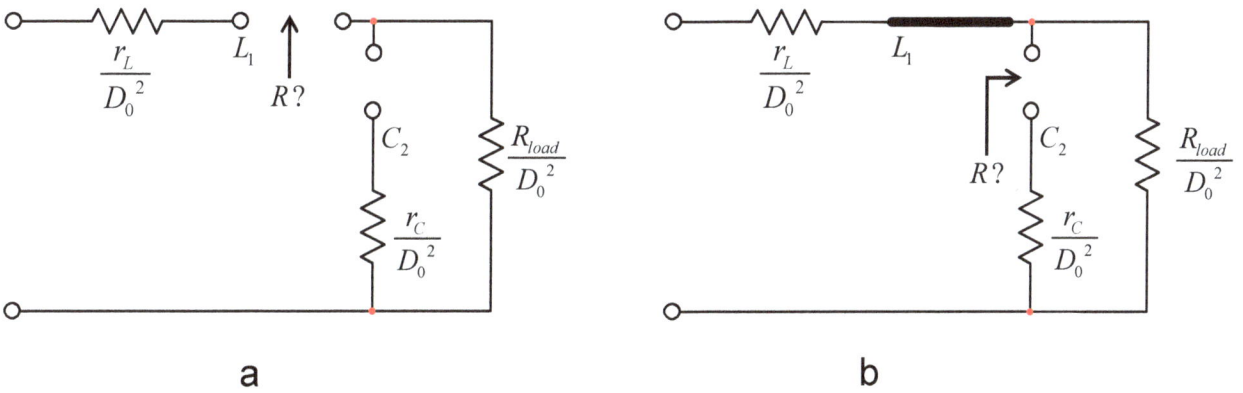

a   b

Figure 2.29 The two first time constants are found quite quickly by inspection.

In sketch (b), we can also immediately express the time constant involving capacitor C_2 by:

$$\tau_2 = C_2 D_0^2 \left(\frac{r_C + R_{Load}}{D_0^2} \right) \tag{2.41}$$

We can assemble the first term of the 2nd-order denominator:

$$b_1 = \tau_1 + \tau_2 = 0 + C_2 D_0^2 \left(\frac{r_C + R_{Load}}{D_0^2} \right) = C_2 \left(r_C + R_{load} \right) \tag{2.42}$$

For the second term b_2, we pick τ_2 and "look" through L_1's connections to express τ_1^2 with the help of Figure 2.30. Considering the dangling left connection, we have:

$$\tau_1^2 = \frac{L_1}{D_0^2 \cdot \infty} = 0 \tag{2.43}$$

which multiplied by (2.41) leads to:

$$b_2 = \tau_2 \tau_1^2 = C_2 D_0^2 \left(\frac{r_C + R_{Load}}{D_0^2} \right) \frac{L_1}{D_0^2 \cdot \infty} = 0 \tag{2.44}$$

The denominator is now complete and equal to:

$$D(s) = 1 + sC_2(r_C + R_{load}) = 1 + \frac{s}{\omega_p} \tag{2.45}$$

with:

$$\omega_p = \frac{1}{C_2(r_L + R_{load})} \tag{2.46}$$

This is a first-order expression despite the presence of two energy-storing elements. This is a degenerate case because the inductor state variable is dependent upon the current source and thus plays no role in the dynamic response.

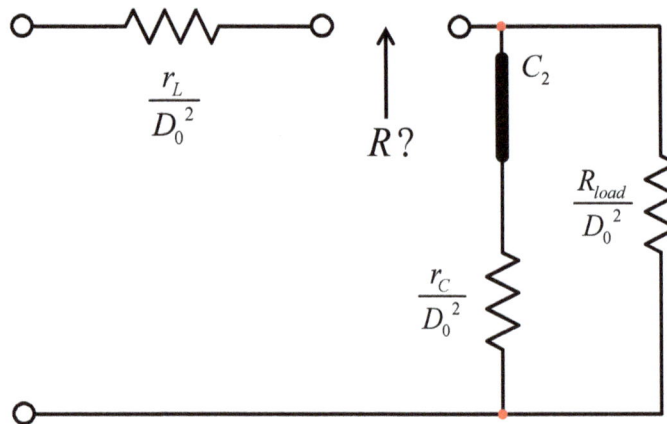

Figure 2.30 With a dangling connection, resistance R is infinite.

The numerator is obtained by finding the circuit's time constants when the response is nulled. A null across a current source (0 V between its terminals) represents a degenerate case and the current source can be replaced by a short circuit [1, 2].

The circuit redrawn with the nulled response appears in Figure 2.31. You recognize the circuit already shown in Figure 2.6 meaning the numerator $N(s)$ for Z_{in} is the denominator $D(s)$ already determined in (2.11).

Once more, what a gain in time!

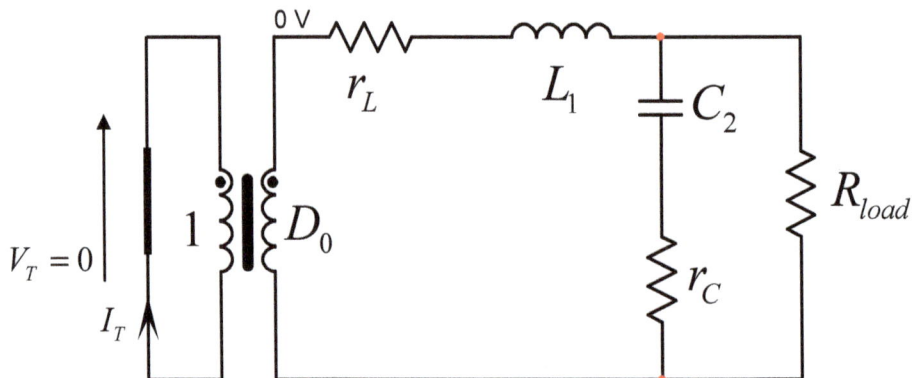

Figure 2.31 The null in the response brings the circuit back into its original structure from Figure 2.6 and you can reuse the expression derived in (2.11).

The complete transfer function is thus expressed as:

$$Z_{in}(s) = R_0 \frac{1 + \dfrac{s}{\omega_0 Q} + \left(\dfrac{s}{\omega_0}\right)^2}{1 + \dfrac{s}{\omega_p}} \qquad (2.47)$$

with R_0 defined in (2.39), Q and ω_0 are respectively defined by (2.22) and (2.21) while ω_p is calculated using (2.46). We have plotted this equation with Mathcad® and the response appears in Figure 2.32.

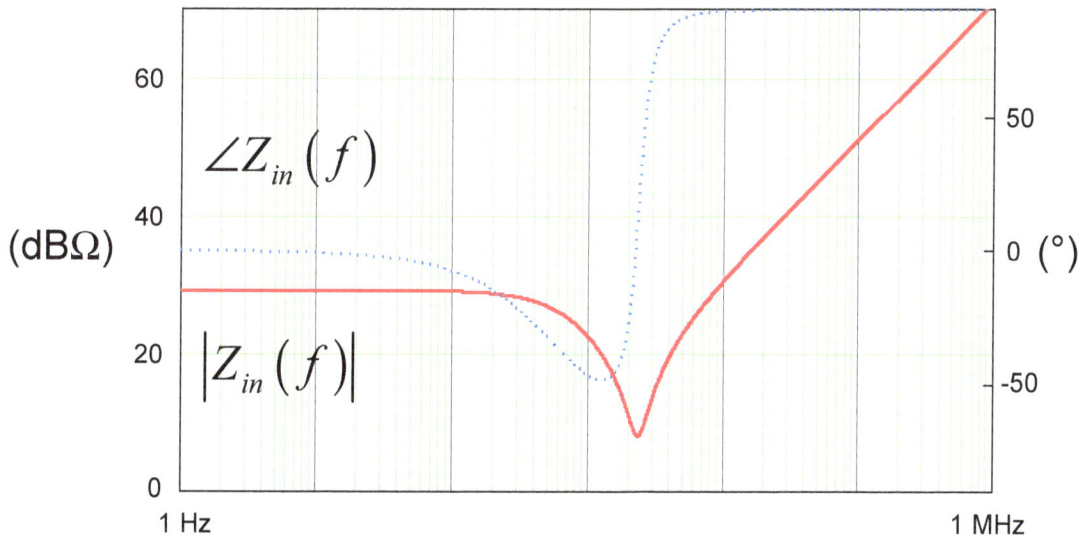

Figure 2.32 The input impedance is flat then dips before going up as *s* approaches infinity.

SPICE can also display the open-loop input impedance by adding an extra high-value inductance in series with the input node *a*. By sweeping this junction by a 1-A ac source, the collected voltage is the direct image of the input impedance.

During this operation, the duty ratio is fixed. The template is shown in Figure 2.33 while results appear in Figure 2.34.

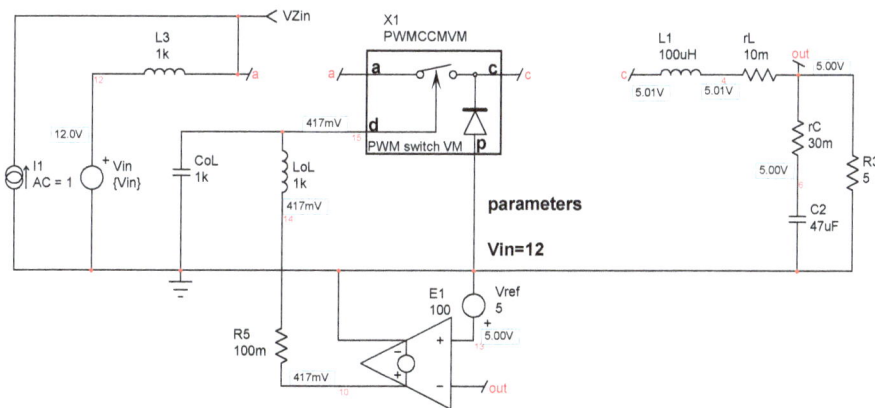

Figure 2.33 Current source I_1 sweeps node *a* while *D* is maintained to a fixed value.

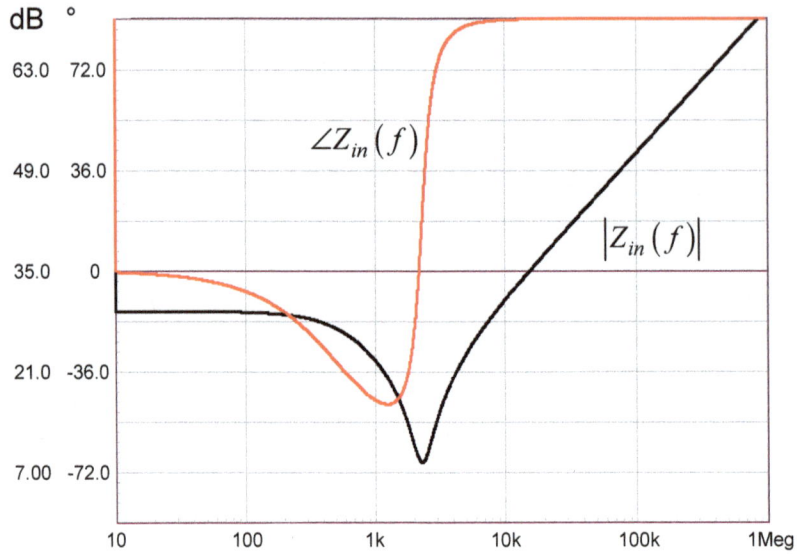

Figure 2.34 As expected, the input impedance is flat then dips before going up as *s* approaches infinity.

We have determined the 4 transfer functions of the buck in the simplest possible manner.

Forget about the complicated matrix manipulations with state-space averaging techniques: fast analytical techniques are the way to go!

We have gathered all these expressions in the convenient table below:

$\dfrac{V_{out}(s)}{V_{err}(s)}$ Control to Output	$H_0 \dfrac{1+\dfrac{s}{\omega_z}}{1+\dfrac{s}{\omega_0 Q}+\left(\dfrac{s}{\omega_0}\right)^2}$	$\omega_z = \dfrac{1}{r_C C_2}$		$H_0 = \dfrac{V_{in}}{V_p}\dfrac{R_{load}}{R_{load}+r_L}$
$\dfrac{V_{out}(s)}{V_{in}(s)}$ Input to Output	$H_0 \dfrac{1+\dfrac{s}{\omega_z}}{1+\dfrac{s}{\omega_0 Q}+\left(\dfrac{s}{\omega_0}\right)^2}$	$\omega_z = \dfrac{1}{r_C C_2}$		$H_0 = D\dfrac{R_{load}}{R_{load}+r_L}$
$Z_{in}(s)$ Input impedance	$R_0 \dfrac{1+\dfrac{s}{\omega_0 Q}+\left(\dfrac{s}{\omega_0}\right)^2}{1+\dfrac{s}{\omega_p}}$		$\omega_p = \dfrac{1}{(r_C+R_{Load})C_2}$	$R_0 = \dfrac{r_L+R_{load}}{D^2}$
$Z_{out}(s)$ Output impedance	$R_0 \dfrac{(1+s/\omega_{z_1})(1+s/\omega_{z_2})}{1+\dfrac{s}{\omega_0 Q}+\left(\dfrac{s}{\omega_0}\right)^2}$	$\omega_{z_1}=\dfrac{r_L}{L_1}$ $\omega_{z_2}=\dfrac{1}{r_C C_2}$		$R_0 = r_L \parallel R_{load}$

D is the duty ratio $\qquad Q = \dfrac{L_1 C_2 \omega_0 (r_C+R_{load})}{L_1+C_2(r_L r_C+r_L R_{load}+r_C R_{load})} \qquad \omega_0 = \dfrac{1}{\sqrt{L_1 C_2}}\sqrt{\dfrac{r_L+R_{load}}{r_C+R_{load}}}$

Figure 2.35 These transfer functions have been determined by looking at small sketches to determine time constants in the simplest possible way (who can beat that?).

2.2 Buck Transfer Functions in DCM – Fixed-Frequency Voltage-Mode Control

We will now repeat the exercise with the buck converter operating in the discontinuous conduction mode or DCM. From Chapter 1, we can evaluate the load resistance value at which the 5-V converter enters borderline conduction mode or BCM when supplied from a 12-V input:

$$R_{crit} = \frac{2L_1 F_{sw}}{1 - \frac{V_{out}}{V_{in}}} = \frac{2 \times 100u \times 100k}{1 - \frac{5}{12}} \approx 34\ \Omega \qquad (2.48)$$

We start by simulating this buck converter with the DCM large-signal model we have derived in Chapter 1.3.2. This is what Figure 2.36 illustrates. The converter delivers 5 V to a load whose value is lower than that recommended by (2.48) to ensure DCM. We have adopted a 100-Ω resistance (50 mA) with a duty ratio tweaked to 24.4% via the *LoL/CoL* components.

Considering all the intermediate sources in this circuit, I will lump them into two independent sources, further simplifying the circuit. This is what Figure 2.37 shows, confirming the operating points. Finally, I replaced the large-signal model by the small-signal version obtained in Chapter 1.3.3. For this next step, we need the operating points defining the duty ratio d_1, voltages V_{ac} and V_{ap} with currents I_a and I_c. Fortunately, all these parameters were determined in chapter 1.3.4 and we have translated equations in parameters expressions as shown in Figure 2.38.

You could also very well extract these values from Figure 2.36 simulation dc bias points result and feed the parameters list.

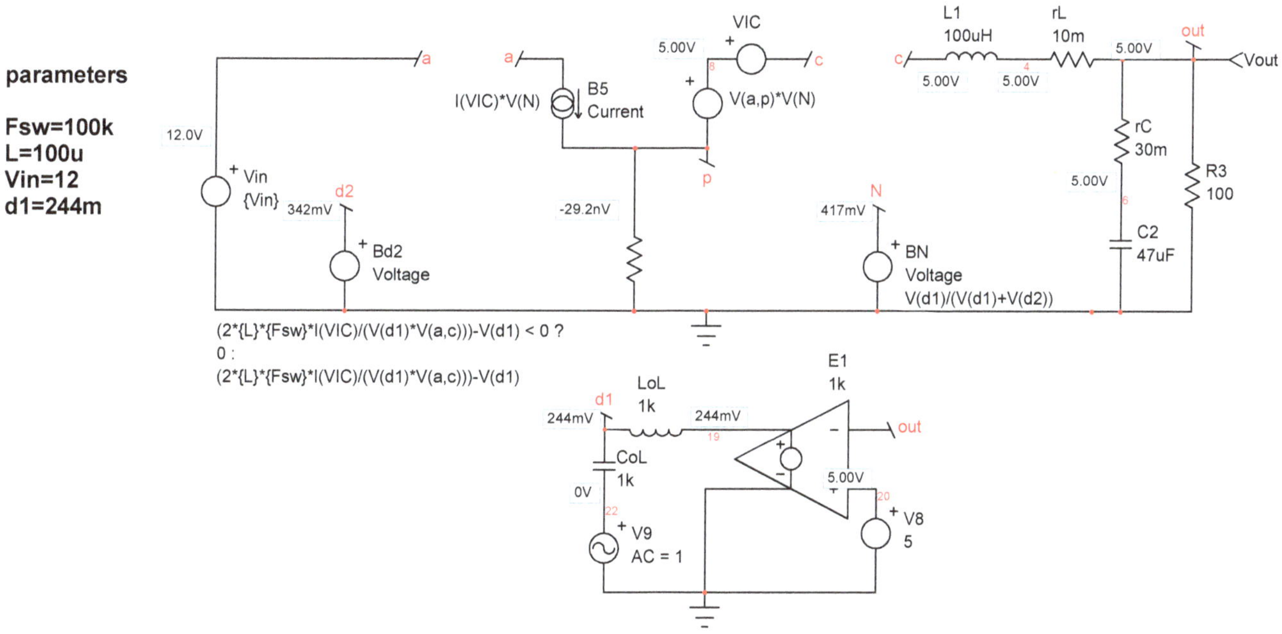

Figure 2.36 The large-signal DCM model delivers 5 V to the 100-Ω load.

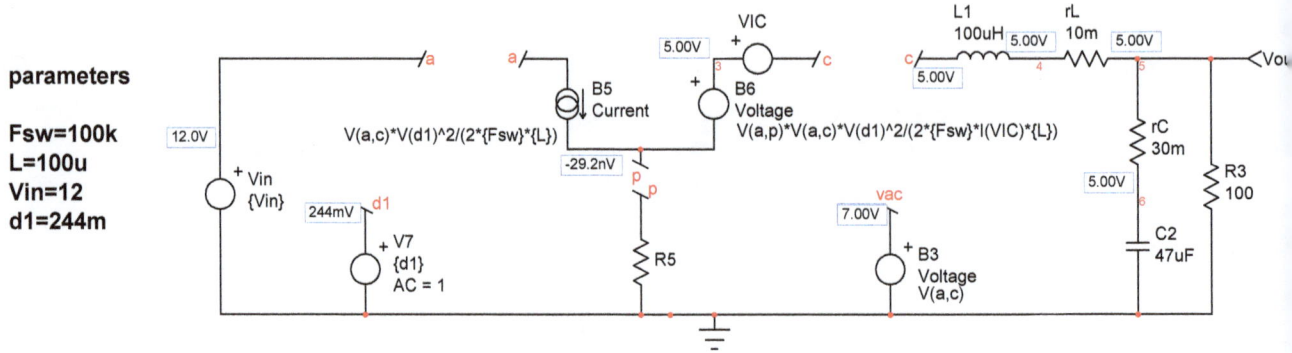

Figure 2.37 The intermediate model confirms the operating point.

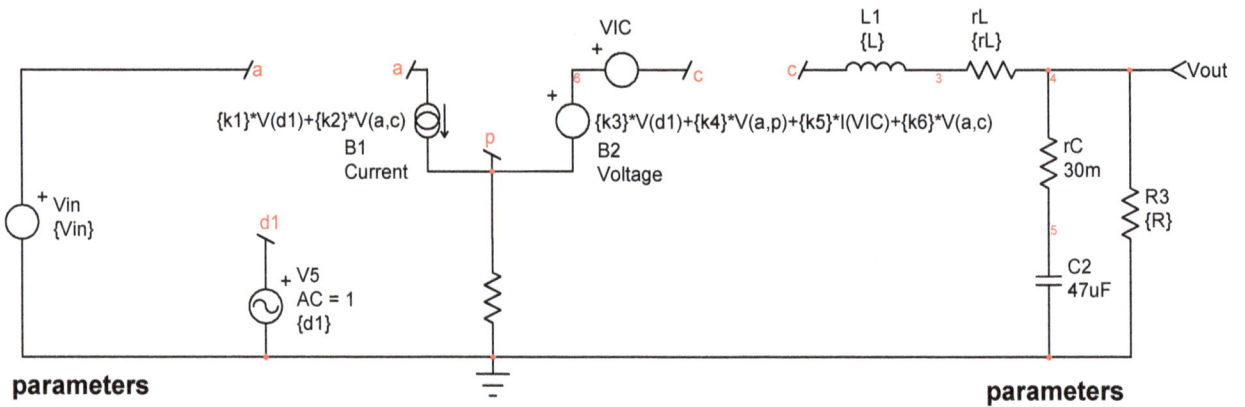

parameters

Fsw=100k
L=100u
Vin=12
d1=244m

(Figure 2.37 circuit diagram with labels: VIC, L1 100uH, rL 10m, B5 Current V(a,c)*V(d1)^2/(2*{Fsw}*{L}), B6 Voltage V(a,p)*V(a,c)*V(d1)^2/(2*{Fsw}*I(VIC)*{L}), Vin {Vin}, V7 {d1} AC = 1, R5, B3 Voltage V(a,c), rC 30m, R3 100, C2 47uF. Node voltages: 5.00V, 12.0V, -29.2nV, 7.00V, 244mV, 5.00V, Vou)

Figure 2.38 The small-signal DCM requires the computation of the operating points before assessing the *k* coefficients.

(Figure 2.38 circuit diagram with labels: VIC, L1 {L}, rL {rL}, B1 Current {k1}*V(d1)+{k2}*V(a,c), B2 Voltage {k3}*V(d1)+{k4}*V(a,p)+{k5}*I(VIC)+{k6}*V(a,c), Vin {Vin}, V5 AC = 1 {d1}, rC 30m, R3 {R}, C2 47uF, Vout)

parameters

M=Vout/Vin
tauL=L*Fsw/R
Vac=7
Vap=12
d1=M*tauL*sqrt(2/(Fsw*L/(R+rL)-M*tauL))

Ia=Vac*d1^2/(2*Fsw*L)
Ic=Vin*d1^2*(sqrt(1+8*Fsw*L/(d1^2*(rL+R)))-1)/(4*tauL*R)

k1=Vac*d1/(Fsw*L)
k2=d1^2/(2*Fsw*L)
k3=Vac*Vap*d1/(Fsw*L*Ic)
k4=Vac*d1^2/(2*Fsw*L*Ic)
k5=-Vac*Vap*d1^2/(2*Fsw*Ic^2*L)
k6=Vap*d1^2/(2*Fsw*Ic*L)

parameters

Fsw=100k
L=100u
Vin=12
Vout=5
R=100
rL=10m

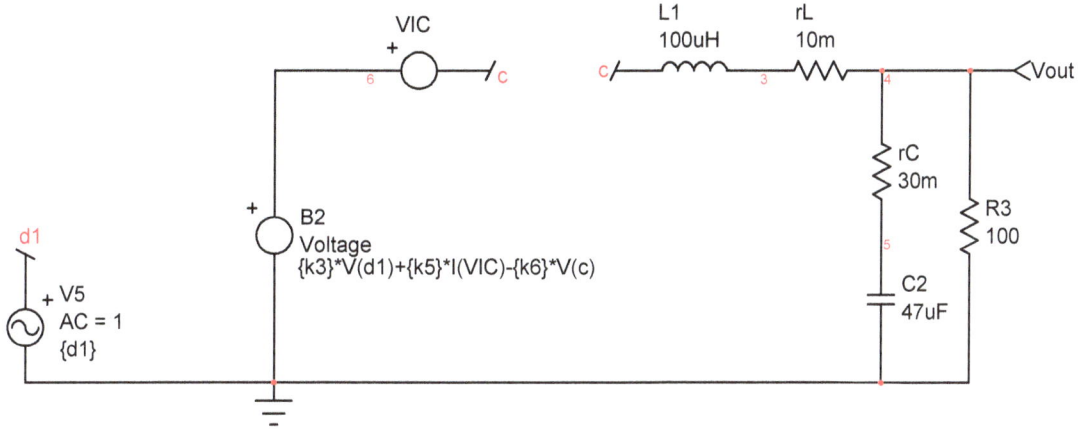

Figure 2.39 Once the control-to-output section is identified, the circuit simplifies significantly.

In the control-to-output transfer function, the stimulus is the duty ratio $\hat{d}$ (node d_1) while the small-signal input source $\hat{v}_{in}$ is kept to zero: $V_{in}(s) = 0$. In Figure 2.38, set node a to zero in all equations and simplify the expression in B_2. Source B_1 is of no interest in this transfer function determination and can be ignored. The complete circuit appears in Figure 2.39 and still represents a 2nd-order circuit. We now plot the control-to-output responses of these 4 circuits and make sure all curves perfectly superimpose as confirmed in Figure 2.40.

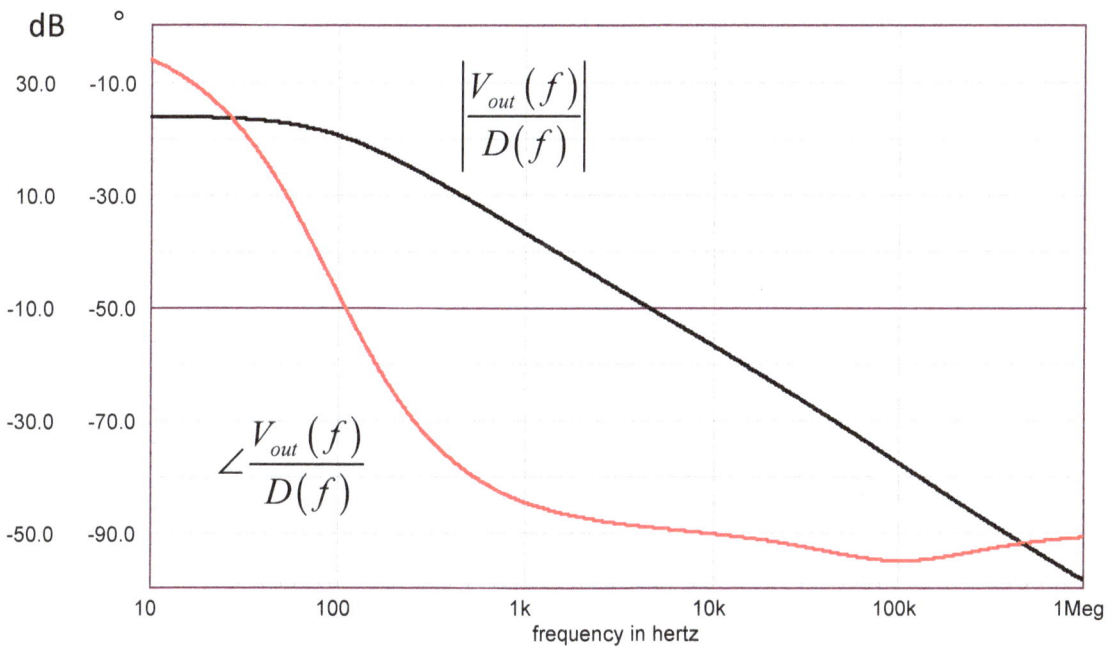

Figure 2.40 The four circuits deliver the exact same frequency response confirming all the simplification steps we took were correct.

I know we have determined the dc transfer function of the DCM buck converter in Chapter 1 but for the sake of the illustration, we can redo the exercise with the PWM switch in place. Using Figure 2.37 electrical schematic and simplifying it further, we see that the output voltage is actually source B_6 minus the voltage drop across r_L as shown in Figure 2.41.

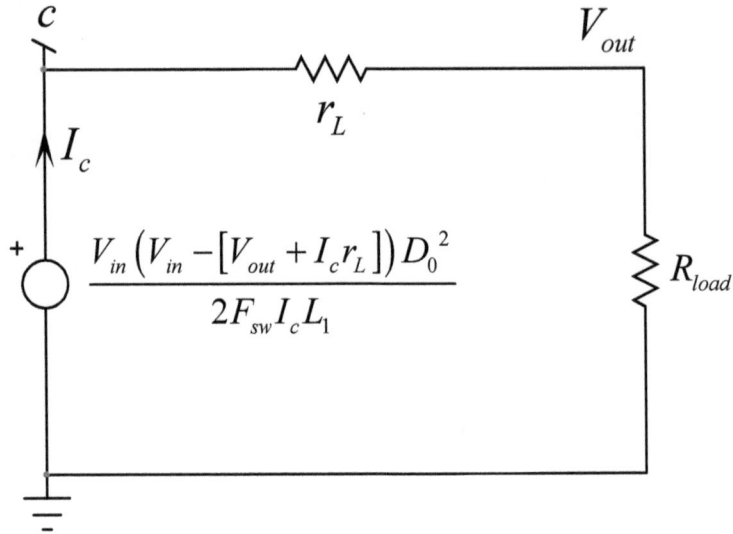

Figure 2.41 The PWM switch can be used to determine the dc transfer function of the DCM-operated buck converter.

From this picture, we can write:

$$V_{out} = \frac{V_{in}\left(V_{in} - [V_{out} + I_c r_L]\right) D_0^{\,2}}{2F_{sw} I_c L_1} - I_c r_L \tag{2.49}$$

In a buck converter, the current leaving terminal c is the load current defined as:

$$I_c = \frac{V_{out}}{R_{load}} \tag{2.50}$$

which lets us update (2.49) as:

$$V_{out} = \frac{V_{in}\left(V_{in} - \left[V_{out} + \dfrac{V_{out}}{R_{load}} r_L\right]\right) D_0^{\,2}}{2F_{sw} \dfrac{V_{out}}{R_{load}} L_1} - \frac{V_{out}}{R_{load}} r_L \tag{2.51}$$

Solving for V_{out} leads to:

$$M = \frac{V_{out}}{V_{in}} = \frac{D_0^{\,2}\left(\sqrt{1 + 8\dfrac{F_{sw} L_1}{D_0^{\,2}\left(r_L + R_{load}\right)}}\right)}{4\tau_L} \tag{2.52}$$

162

in which:

$$\tau_L = \frac{L_1}{R_{load} T_{sw}} \tag{2.53}$$

When r_L is negligible and after a few manipulations, (2.52) simplifies to:

$$M = \frac{2}{1 + \sqrt{1 + \frac{8\tau_L}{D_0^2}}} \tag{2.54}$$

which is the classical formula found in the literature and in Chapter 1.2.6.

We now have to determine the dc gain, H_0. In Figure 2.39, we short the inductor and open the capacitor. We obtain the circuit drawn in Figure 2.42. The voltage at terminal c is defined as:

$$V_{(c)} = k_3 D + k_5 I_c - k_6 V_{(c)} \tag{2.55}$$

Solving for $V_{(c)}$ leads to:

$$V_{(c)} = \frac{D k_3 + I_c k_5}{1 + k_6} \tag{2.56}$$

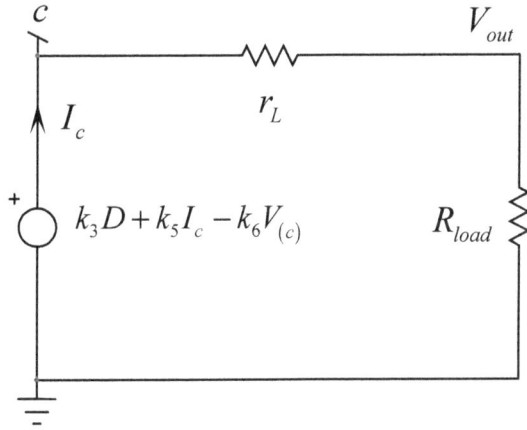

Figure 2.42 The quasi-static gain is obtained by shorting inductor L_1 and open circuiting capacitor C_2.

The output voltage is obtained by applying a simple resistive divider giving:

$$V_{out} = V_{(c)} \frac{R_{load}}{r_L + R_{load}} \tag{2.57}$$

If we extract $V_{(c)}$ from (2.57), equate it with (2.56) and solve for V_{out}, we have:

$$\frac{V_{out}}{D} = \frac{R_{load} k_3}{R_{load} - k_5 + r_L + k_6 (R_{load} + r_L)} \tag{2.58}$$

The exercise now consists of replacing each k coefficient by its respective value as defined in Figure 2.43 and rearranging the result in an intelligible format.

$$k_1 = \frac{(V_{in} - V_{out})D_0}{F_{sw}L_1} \quad k_2 = \frac{D_0{}^2}{2F_{sw}L_1} \quad k_3 = \frac{D_0(V_{in} - V_{out})V_{in}}{F_{sw}I_cL_1} \quad M = \frac{V_{out}}{V_{in}}$$

$$k_4 = \frac{(V_{in} - V_{out})D_0{}^2}{2F_{sw}I_cL_1} \quad k_5 = -\frac{V_{in}(V_{in} - V_{out})D_0{}^2}{2F_{sw}I_c{}^2L_1} \quad k_6 = \frac{V_{in}D_0{}^2}{2F_{sw}I_cL_1}$$

Figure 2.43 Coefficients k pertaining to the small-signal DCM PWM switch model used in a buck converter.

If you do the algebra correctly, you should find the following result:

$$H_0 = \frac{2V_{in}(1 - M)}{\dfrac{D_0}{M} + \dfrac{r_L}{R_{load}}D_0 + 2F_{sw}L_1 M \dfrac{R_{load} + r_L}{R_{load}{}^2 D_0}} \tag{2.59}$$

When $r_L \approx 0$, it simplifies to:

$$H_0 \approx \frac{2V_{in}(1 - M)}{\dfrac{D_0}{M} + 2\tau_L \dfrac{M}{D_0}} \tag{2.60}$$

This result can be further rearranged as suggested by [3]:

$$H_0 \approx \frac{2V_{out}}{D_0} \frac{1 - M}{2 - M} \tag{2.61}$$

Now that we have the dc gain, we can start determining the circuit time constants by reducing the excitation $D(s)$ to zero. The circuit from Figure 2.39 simplifies to that of Figure 2.44 as all sources referring to node d_1 are now set to zero and capacitor C_2 set in its dc state (open circuited). To determine the time constant, we have to determine the resistance R "seen" from the inductor's terminals. To do so, we install a test generator I_T developing a voltage V_T across its connections. Once in place, we can start by defining the voltage at node c:

$$V_{(c)} = k_5 I_c - k_6 V_{(c)} \tag{2.62}$$

Then, node 2 voltage comes easily:

$$V_{(2)} = -I_T(r_L + R_{load}) \tag{2.63}$$

Finally, we have:

$$V_T = V_{(c)} - V_{(2)} \tag{2.64}$$

If we now substitute (2.62) and (2.63) into (2.64) then solve for V_T, we have:

$$V_T = I_T \left[(r_L + R_{load}) - \frac{k_5}{k_6 + 1} \right] \qquad (2.65)$$

The resistance we want is equal to:

$$R = \frac{V_T}{I_T} \qquad (2.66)$$

Combining (2.66) with L_1, we have determined our first time constant τ_1:

$$\tau_1 = \frac{L_1}{R} = \frac{L_1}{r_L + R_{load} - \dfrac{k_5}{k_6 + 1}} \qquad (2.67)$$

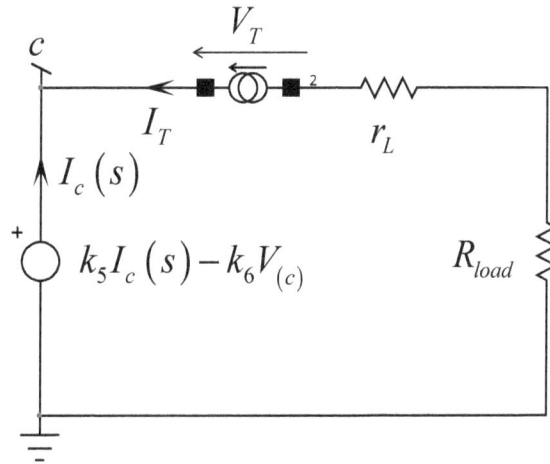

Figure 2.44 The first time constant involving inductor L_1 is determined using this sketch.

The second time constant involves C_2 and is determined by setting L_1 in its dc state, a short circuit. The new circuit appears in Figure 2.45 and we want to determine the resistance "seen" from C_2's connecting terminals.

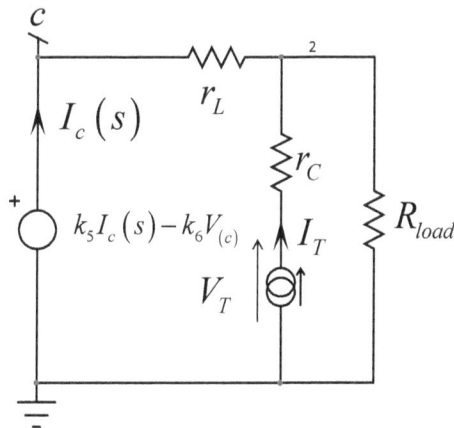

Figure 2.45 The second time constant involves capacitor C_2.

The voltage at node c and 2 are defined as:

$$V_{(c)} = k_5 I_c - k_6 V_{(c)} \tag{2.68}$$

$$V_{(2)} = V_{(c)} - I_c r_L \tag{2.69}$$

Extracting I_c leads to:

$$I_c = \frac{V_{(c)} - V_{(2)}}{r_L} \tag{2.70}$$

From (2.68), we extract $V_{(c)}$ and substitute it in (2.70):

$$I_C = -\frac{V_{(2)} - I_c k_5 + V_{(2)} k_6}{r_L (1 + k_6)} \tag{2.71}$$

The current flowing into R_{load} is the sum of I_T and I_c:

$$I_T + I_c = \frac{V_{(2)}}{R_{load}} \tag{2.72}$$

Substitute (2.71) in the above expression and determine the voltage at node 2:

$$V_{(2)} = \frac{I_T R_{load} (r_L - k_5 + k_6 r_L)}{R_{load} - k_5 + r_L + k_6 (r_L + R_{load})} \tag{2.73}$$

Now, we can see that the voltage V_T across the test generator is determined by:

$$V_T = V_{(2)} + r_C I_T \tag{2.74}$$

We now substitute (2.73) in (2.74) and obtain the resistance R following (2.66). From this result, we can express the second time constant τ_2:

$$\tau_2 = RC_2 = C_2 \left[r_C + \frac{R_{load} (r_L - k_5 + k_6 r_L)}{R_{load} - k_5 + r_L + k_6 (r_L + R_{load})} \right] \tag{2.75}$$

The remaining time constant is determined while inductor L_1 is set in its high-frequency state or open circuited. The new circuit appears in Figure 2.46 and we determine the resistance looking into C_1's connections. Considering an open-circuited inductor, current I_c is 0 A, leaving the generator biasing the two resistances in series:

$$\tau_2^1 = C_2 (r_C + R_{load}) \tag{2.76}$$

We are all set and can combine all the time constants together to form the 2nd-order polynomial form of the denominator we want:

$$D(s) = 1 + s b_1 + s^2 b_2 = 1 + s (\tau_1 + \tau_2) + s^2 \tau_1 \tau_2^1 \tag{2.77}$$

166

Associating the time constants we have determined in (2.67), (2.75) and (2.76) leads to:

$$D(s)=1+\left[\frac{L_1}{r_L+R_{load}-\dfrac{k_5}{k_6+1}}+C_2\left[r_C+\frac{R_{load}\left(r_L-k_5+k_6 r_L\right)}{R_{load}-k_5+r_L+k_6\left(r_L+R_{load}\right)}\right]\right]s+\left(\frac{L_1}{r_L+R_{load}-\dfrac{k_5}{k_6+1}}C_2\left(r_C+R_{load}\right)\right)s^2 \quad (2.78)$$

This expression can be rearranged under the classical 2nd-order normalized form such as:

$$D(s)=1+\frac{s}{\omega_0 Q}+\left(\frac{s}{\omega_0}\right)^2 \tag{2.79}$$

From (2.77), we can show [1] that:

$$\omega_0=\frac{1}{\sqrt{b_2}} \tag{2.80}$$

and:

$$Q=\frac{\sqrt{b_2}}{b_1}=\frac{1}{\omega_0 b_1} \tag{2.81}$$

In a heavily-damped converter like with this DCM buck converter, we can apply the low-Q approximation which considers a dominant pole at low frequency associated with a higher-frequency pole:

$$D(s)\approx\left(1+\frac{s}{\omega_{p_1}}\right)\left(1+\frac{s}{\omega_{p_2}}\right) \tag{2.82}$$

where:

$$\omega_{p_1}=\omega_0 Q \tag{2.83}$$

$$\omega_{p_2}=\frac{\omega_0}{Q} \tag{2.84}$$

The exercise now consists of rearranging the terms while the k coefficients are replaced by their respective values from
Figure 2.43. If you neglect the inductor ohmic loss – $r_L \ll R_{load}$ – then you should find:

$$\omega_{p_1}\approx\frac{1+\dfrac{2M^2\tau_L}{D_0^2}}{C_2 R_{load}\left(1-M\right)} \tag{2.85}$$

also expressed as in the literature:

$$\omega_{p_1}\approx\frac{2-M}{1-M}\frac{1}{C_2 R_{load}} \tag{2.86}$$

The second high-frequency pole is determined using:

$$\omega_{p_2} \approx \frac{L_1 + C_2 R_{load}{}^2 \left(1 - M\right)}{L_1 C_2 R_{load}}$$

(2.87)

The zero is obtained very easily by bringing the excitation back as illustrated in Figure 2.47. In this transformed circuit, the condition to null the output voltage is observed when the series connection of r_C and C_2 form a transformed short circuit:

$$Z_1(s) = r_C + \frac{1}{sC_2} = 0$$

(2.88)

It leads to a negative root expressed as:

$$s_z = -\frac{1}{r_C C_2}$$

(2.89)

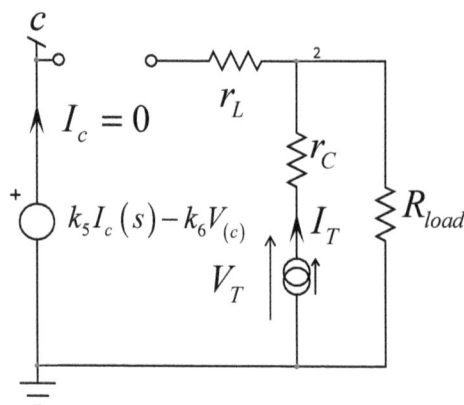

Figure 2.46 Open circuit inductor L_1 and determine the resistance looking into C_1's terminals.

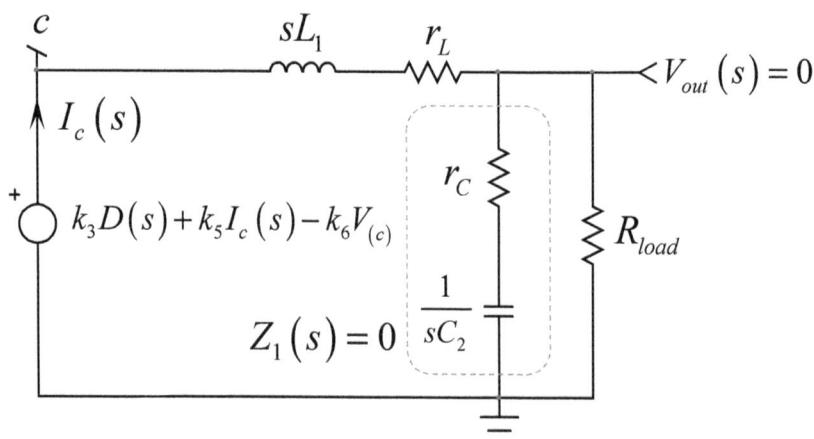

Figure 2.47 The zero is determined by nulling the response.

and a zero located at:

$$\omega_z = \frac{1}{r_C C_2}$$

(2.90)

The numerator is thus defined as:

$$N(s) = 1 + \frac{s}{\omega_z} \tag{2.91}$$

We have determined all the needed elements and the control-to-output transfer function can be obtained by combining (2.59), (2.78) and (2.91):

$$H(s) = \frac{V_{out}(s)}{D(s)} = H_0 \frac{1 + \dfrac{s}{\omega_z}}{\left(1 + \dfrac{s}{\omega_{p_1}}\right)\left(1 + \dfrac{s}{\omega_{p_2}}\right)} \tag{2.92}$$

We can now plot this expression with Mathcad® and superimpose the magnitude-phase response with that of SPICE displayed in Figure 2.40. As confirmed by Figure 2.48, the curves perfectly match with each other.

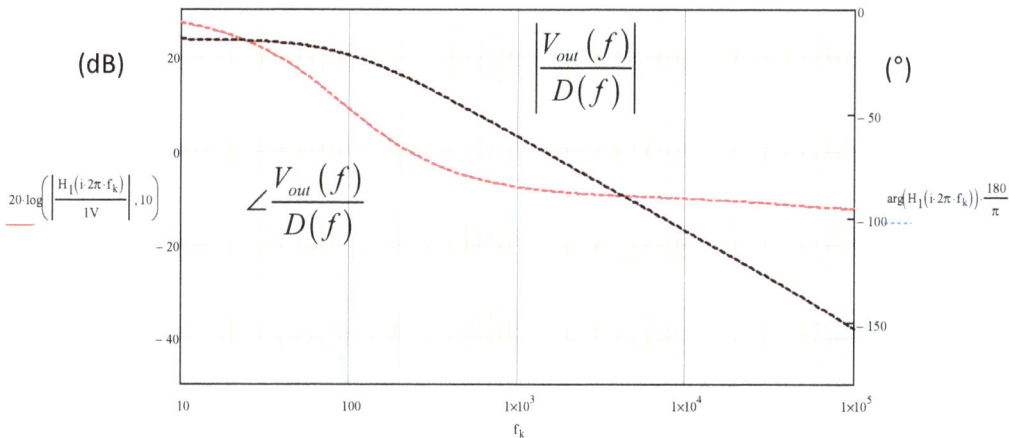

Figure 2.48 Mathcad® and SPICE responses perfectly match.

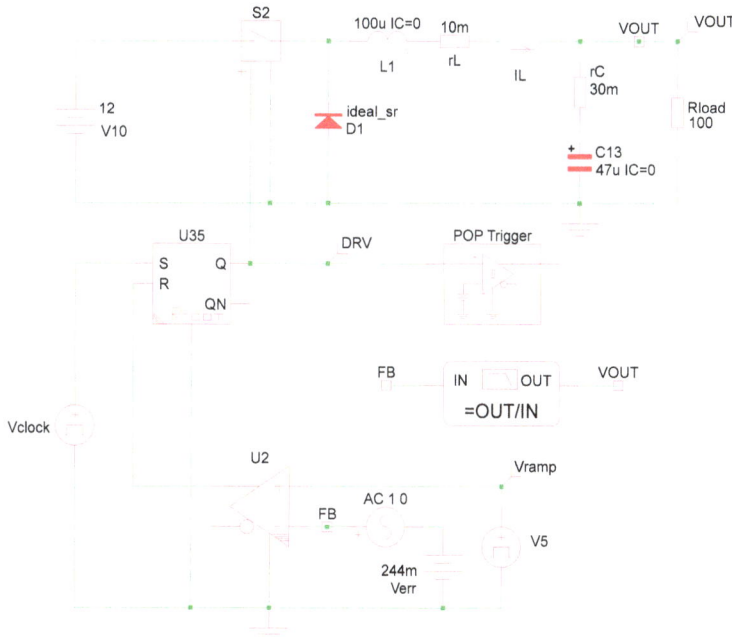

Figure 2.49 The SIMPLIS® simulation circuit does not change from the CCM example. The load is reduced to 100 Ω and the duty ratio set to 24.4%.

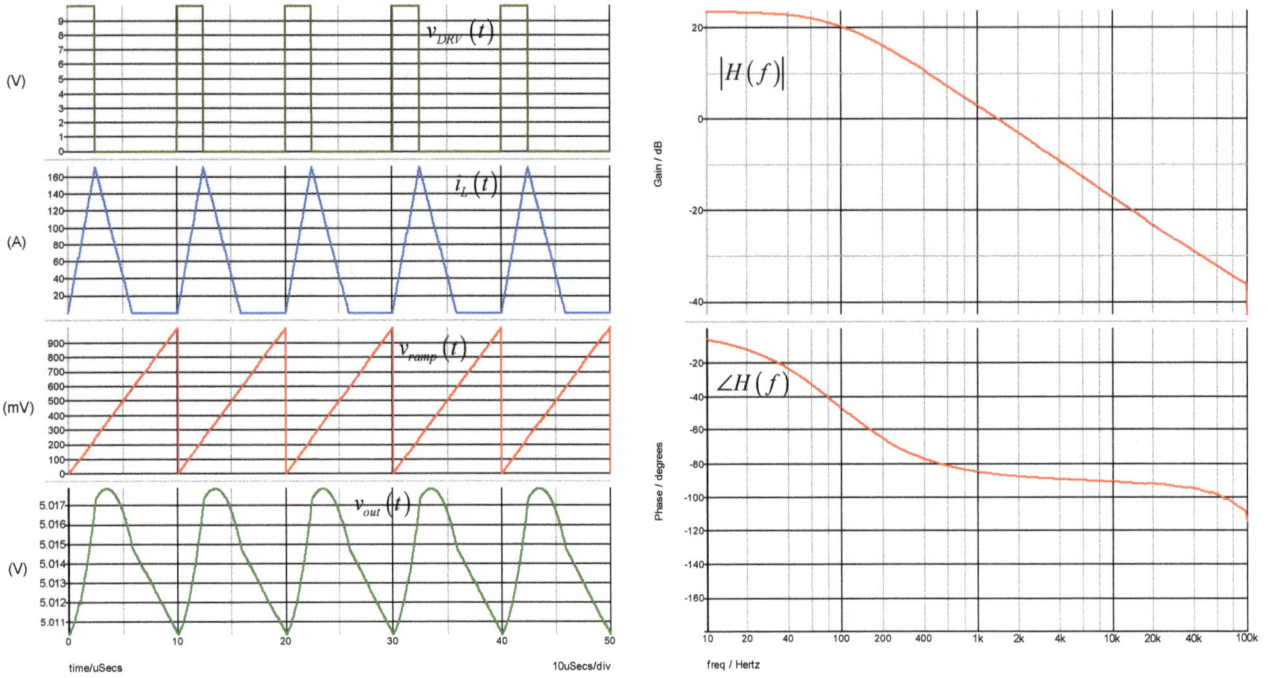

Figure 2.50 The inductor current shows a DCM operation and the small-signal response from SIMPLIS® confirms the analysis carried with the DCM PWM switch model.

We have now updated the original buck converter SIMPLIS circuit described in Figure 2.13 to that of Figure 2.49. The load is reduced to $100\,\Omega$ and source V_{err} sets the duty ratio to 24.4% as in the SPICE simulation. Figure 2.50 confirms the DCM operation and shows a Bode plot very close to that of Figure 2.48.

Finally, for the sake of illustrating the coefficients calculations, I have included their values in Figure 2.51 sheet. It is a convenient way for check your own approach in case you would use another solver like Excel® for instance.

$L_1 := 100\mu H \qquad F_{sw} := 100kHz \qquad\qquad C_2 := 47\mu F \qquad\qquad d_1 := 0.24 \qquad\qquad r_L := 0.01\Omega$

$$\tau_L := \frac{L_1}{R_L} \cdot F_{sw} \qquad \|(x,y) := \frac{x \cdot y}{x + y} \qquad R_L := 100\Omega \qquad V_{in} := 12V \qquad r_C := 0.03\Omega$$

Static operating points determination:

$$V_{out} := V_{in} \cdot \frac{2}{1 + \sqrt{1 + \dfrac{8\tau_L}{d_1{}^2}}} = 4.9396821V \qquad M := \frac{V_{out}}{V_{in}}$$

santity check on duty ratio

$$V_{ac} := V_{in} - V_{out} = 7.0603179V \qquad V_{ap} := V_{in} = 12V \qquad d_{11} := \sqrt{\frac{2 \cdot \tau_L}{V_{in} \cdot (V_{in} - V_{out})}} \cdot V_{out} = 0.24$$

si rL est proche de 0 $\qquad M \cdot \sqrt{\dfrac{2 \cdot \tau_L}{1 - M}} = 0.24 \qquad$ duty ratio definition for the buck in dcm. This is the literature definition

$$R_{crit} := \frac{2 \cdot L_1 \cdot F_{sw}}{1 - \dfrac{V_{out}}{V_{in}}} = 33.9928034\Omega$$

$$I_c := \frac{V_{in} \cdot d_1{}^2 \cdot \left[\sqrt{1 + 8 \cdot \dfrac{F_{sw} \cdot L_1}{d_1{}^2 \cdot (r_L + R_L)}} - 1\right]}{4 \cdot \tau_L \cdot R_L} = 0.0493937A \qquad I_{c1} := \frac{M \cdot V_{in}}{R_L} = 0.0493968A$$

Small-signal parameters calculations:

$$k_1 := \frac{V_{ac} \cdot d_1}{F_{sw} \cdot L_1} = 0.1694476A \qquad\qquad k_2 := \frac{d_1{}^2}{2 \cdot F_{sw} \cdot L_1} = 2.88 \times 10^{-3} \frac{1}{\Omega}$$

$$k_3 := \frac{V_{ac} \cdot V_{ap} \cdot d_1}{F_{sw} \cdot I_c \cdot L_1} = 41.1666087V \qquad k_4 := \frac{V_{ac} \cdot d_1{}^2}{2 \cdot F_{sw} \cdot I_c \cdot L_1} = 0.4116661 \qquad k_5 := -\frac{V_{ac} \cdot V_{ap} \cdot d_1{}^2}{2 \cdot F_{sw} \cdot I_c{}^2 \cdot L_1} = -100.0125918\Omega$$

$$k_6 := \frac{V_{ap} \cdot d_1{}^2}{2 \cdot F_{sw} \cdot I_c \cdot L_1} = 0.6996842 \qquad d_2 := \frac{2 \cdot L_1 \cdot F_{sw} \cdot I_c}{d_1 \cdot V_{ac}} - d_1 = 0.3429968$$

Figure 2.51 This sheet gives you the coefficients computed by Mathcad® for this example.

2.2.1 Input to Output

We will now operate the converter in open-loop conditions meaning that $D(s) = 0$. The static duty ratio D_0 however still ensures the correct bias point ($V_{out} = 5$ V). The response is still V_{out} but the stimulus is the input voltage V_{in}. From Figure 2.38, all sources containing a reference to node d_1 are zeroed and node a previously set to 0 V is now back in the equations. The circuit updates to that of Figure 2.52.

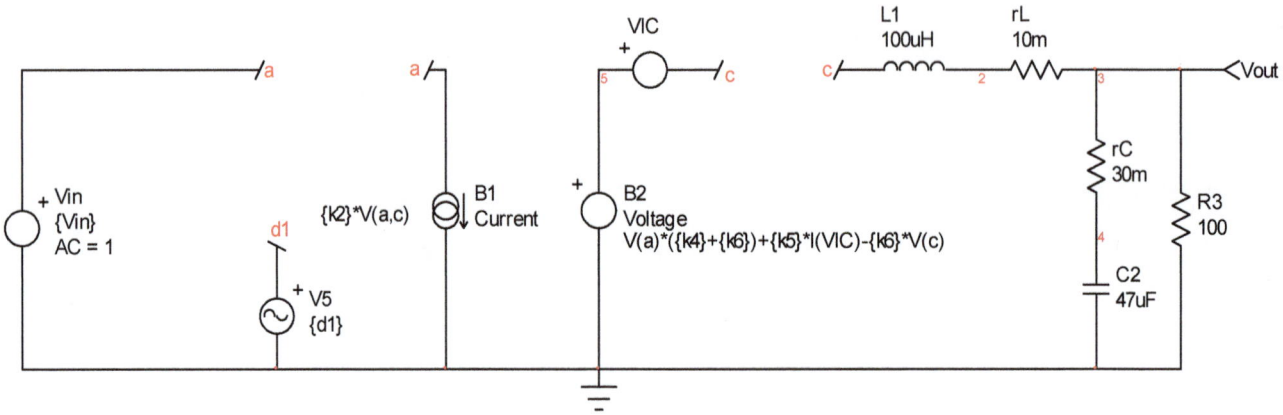

Figure 2.52 The input-to-output transfer function is determined with a constant duty ratio.

The quasi-static gain H_0 is determined by shorting inductor L_1 and opening capacitor C_2. The circuit simplifies to the sketch of Figure 2.53. A bunch of equations will get us straight to the result we want:

$$V_{(c)} = V_{in}\left(k_4 + k_6\right) + k_5 I_c - k_6 V_{(c)} \tag{2.93}$$

Solving for $V_{(c)}$ from this expression gives:

$$V_{(c)} = \frac{V_{in}\left(k_4 + k_6\right) + k_5 I_c}{1 + k_6} \tag{2.94}$$

The output current is given by:

$$I_c = \frac{V_{(c)}}{r_L + R_{load}} \tag{2.95}$$

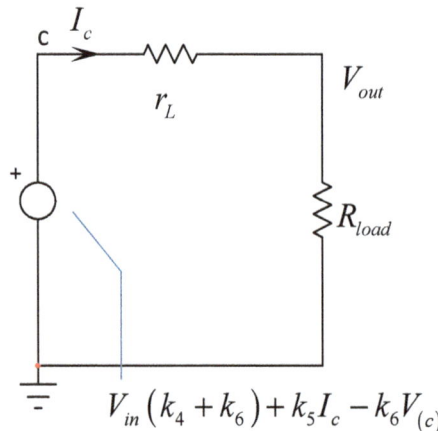

Figure 2.53 Shorting the inductor and opening the capacitor gives the quasi-static gain we need.

If we substitute (2.95) in (2.94) and solve for $V_{(c)}$, we have:

$$V_{(c)} = \frac{V_{in}\left(r_L + R_{load}\right)\left(k_4 + k_6\right)}{R_{load} - k_5 + r_L + k_6\left(r_L + R_{load}\right)} \tag{2.96}$$

The output voltage is obtained by applying a simple impedance divider:

$$V_{out} = V_{(c)} \frac{R_{load}}{r_L + R_{load}}$$

(2.97)

Replacing $V_{(c)}$ by its definition from (2.96), we obtain:

$$H_0 = \frac{(r_L + R_{load})(k_4 + k_6)}{R_{load} - k_5 + r_L + k_6(r_L + R_{load}) R_{load} + r_L} \frac{R_{load}}{R_{load} + r_L}$$

(2.98)

If we neglect the r_L term and replace the k coefficients by their values from Figure 2.43, this expression simplifies to:

$$H_0 \approx \frac{k_4 + k_6}{1 + k_6 - \dfrac{k_5}{R_{load}}} \approx M$$

(2.99)

Now, to determine the circuit time constants, we can reduce the excitation to 0 V: zero all sources featuring node a in the in-line equations and the new circuit is that of Figure 2.54. If you compare this circuit with that of Figure 2.44, the circuit is similar: the time constants are unchanged and we can reuse the denominator $D(s)$ already determined in (2.82).

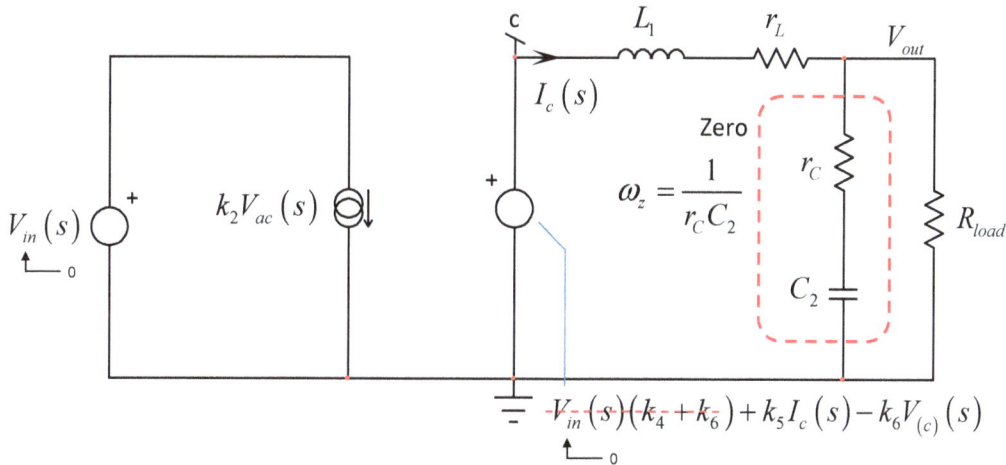

Figure 2.54 When the stimulus reduces to 0 V, the circuit returns to its natural structure already studied for the control-to-output transfer function: the denominator $D(s)$ is the same.

From Figure 2.44, we can see that the series combination of r_C and C_2 classically contributes the zero of the transfer function already identified with (2.90):

$$\omega_z = \frac{1}{r_C C_2}$$

(2.100)

We now have all the terms we need to determine the input-to-output transfer function of the DCM-operated buck converter:

$$\frac{V_{out}(s)}{V_{in}(s)} = H_0 \frac{\left(1 + \dfrac{s}{\omega_z}\right)}{\left(1 + \dfrac{s}{\omega_{p_1}}\right)\left(1 + \dfrac{s}{\omega_{p_2}}\right)}$$

(2.101)

with:

$$H_0 = \frac{2}{1 + \sqrt{1 + \dfrac{8\tau_L}{D_0^2}}}$$

(2.102)

and the poles-zeroes defined by (2.86), (2.87) and (2.100).

Figure 2.55 gathers the plots obtained from the Mathcad® sheet and the SPICE model. As you can see, they perfectly superimpose.

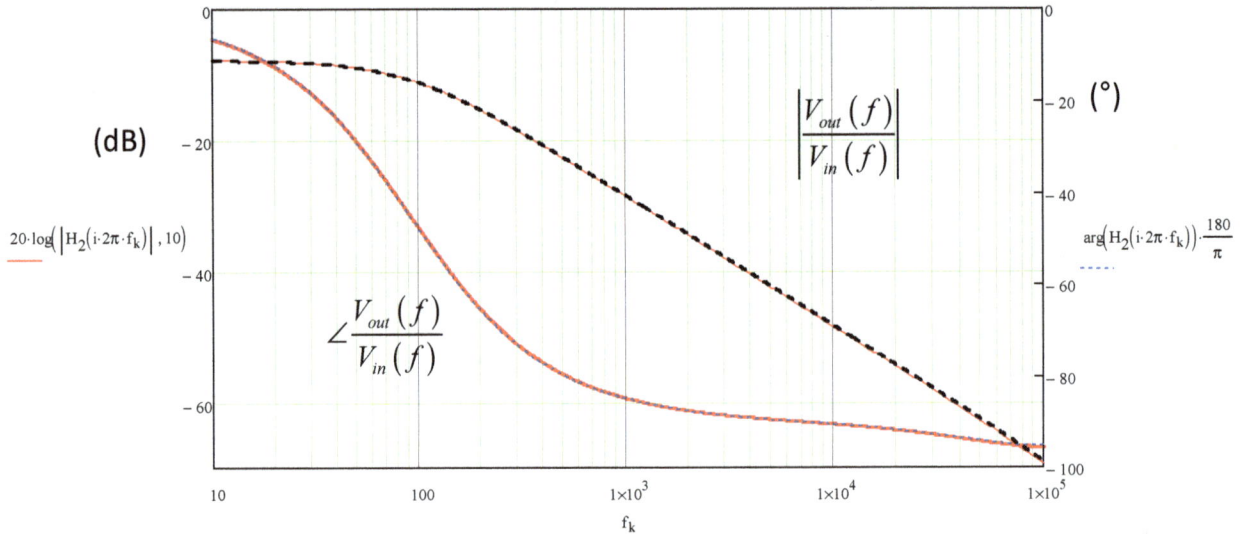

Figure 2.55 The input-to-output transfer function shows a 7.7-dB attenuation as a quasi-static gain.

A SPICE simulation is run from Figure 2.56 schematic and the response appears in Figure 2.57.

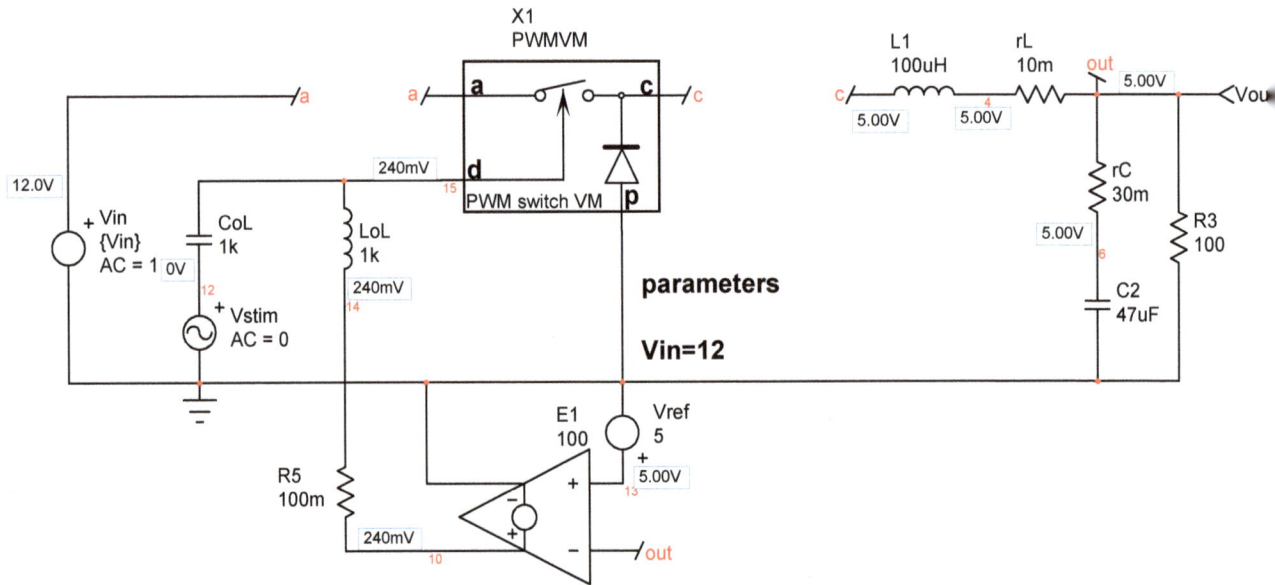

Figure 2.56 The SPICE template remains the same as in the previous example but the load current reduction forces the DCM operation.

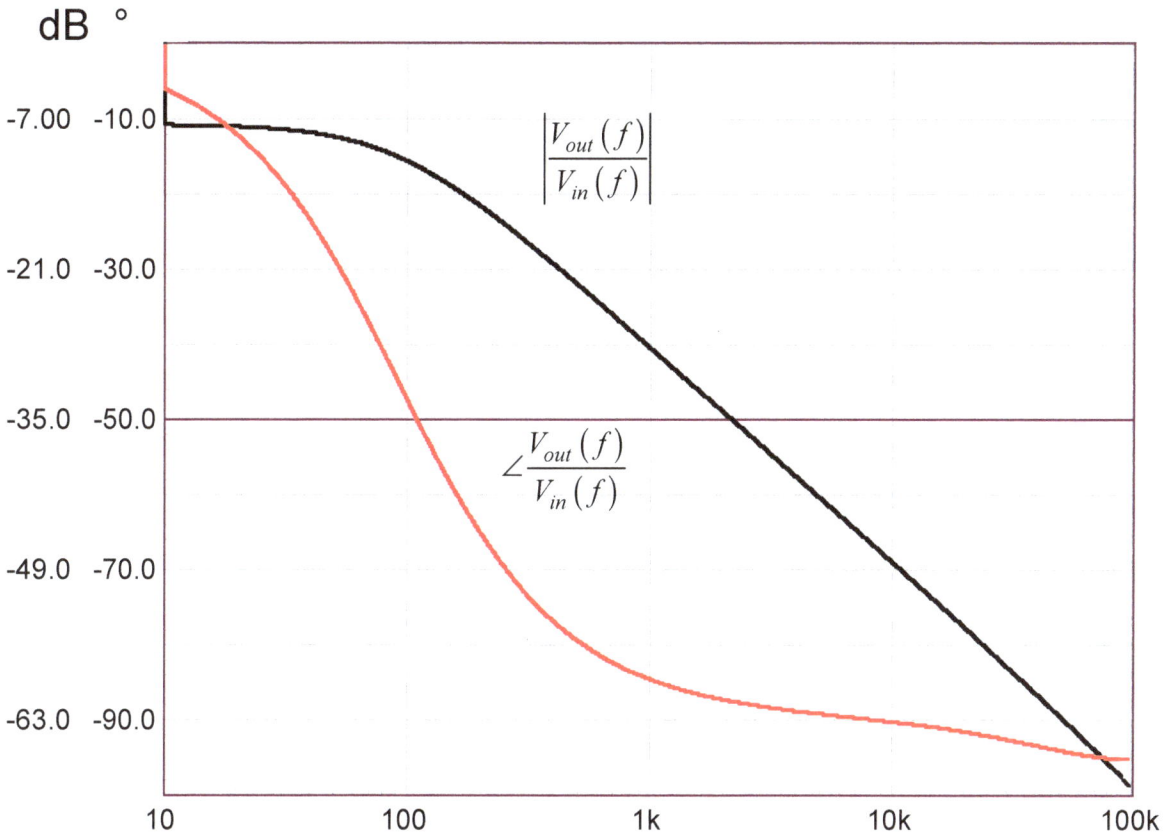

Figure 2.57 The simulated response is similar to that of the Mathcad® sheet.

2.2.2 Output Impedance

To determine the output impedance, we will install a generator across the output, in parallel with R_{load}. The new circuit appears in Figure 2.58. In this circuit, the small-signal duty ratio is turned off as well as the small-signal input voltage. As such, all sources involving d_1 and node a are set to zero.

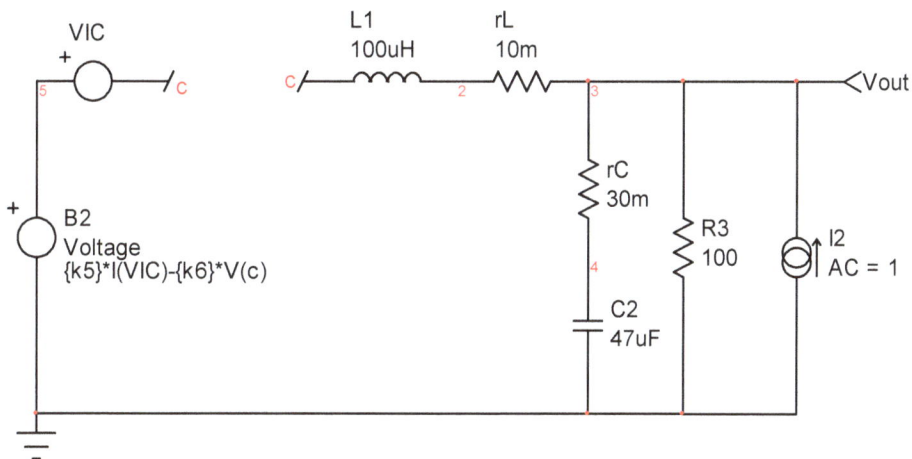

Figure 2.58 The output impedance is determined by stimulating the output of the loaded buck converter.

The resistance R_0 is determined for $s = 0$: short L_1 and open circuit C_2. This is what the circuit from Figure 2.59 illustrates. The circuit implies the test generator I_T (the stimulus) generating a response V_T across its terminals. To simplify the analysis, we can temporarily disconnect R_{load} and apply it later in parallel with the result. A few equations will get us there. The test current I_T and I_c are in opposite directions:

$$I_T = -I_c \tag{2.103}$$

The voltage V_T depends on the left-side source minus the voltage drop across r_L:

$$V_T = k_5 I_c - k_6 V_{(c)} - I_c r_L \tag{2.104}$$

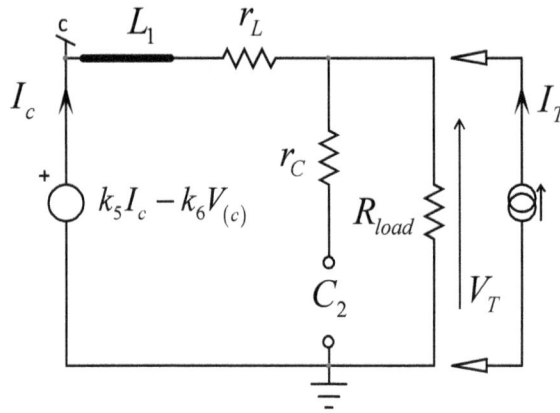

Figure 2.59 The dc output resistance is determined by shorting the inductor and open-circuiting the capacitor.

The voltage at node c is V_T plus the voltage drop across r_L:

$$V_{(c)} = V_T + I_c r_L \tag{2.105}$$

Combining the above equations and substituting I_c by I_T leads to:

$$V_T = I_T r_L - I_T k_5 - V_T k_6 + I_T k_6 r_L \tag{2.106}$$

Rearranging this expression gives:

$$\frac{V_T}{I_T} = \frac{r_L - k_5 + k_6 r_L}{k_6 + 1} \tag{2.107}$$

Now paralleling the load resistance leads to the dc resistance we want:

$$R_0 = \left(\frac{r_L - k_5 + k_6 r_L}{k_6 + 1} \right) \parallel R_{load} \tag{2.108}$$

If you now replace the k coefficients with their respective values from Figure 2.43 and run simplifications, you obtain:

$$R_0 = \left[r_L + R_{load}(1 - M) \right] \parallel R_{load} \approx \left[R_{load}(1 - M) \right] \parallel R_{load} \tag{2.109}$$

The denominator is studied by reducing the excitation I_T to zero: the current source disappears. Without surprise, the circuit reduces to the structure already examined in the previous DCM transfer functions and we can reuse the same denominator defined with (2.82). The new circuit appears in Figure 2.60.

To determine the zero, we have to realize that the response is the voltage V_T across the current source (Figure 2.61). The first zero is immediate and involves r_C and C_2 which induce a transformed short circuit and null the response at the following zero location:

$$\omega_{z_1} = \frac{1}{r_C C_2} \tag{2.110}$$

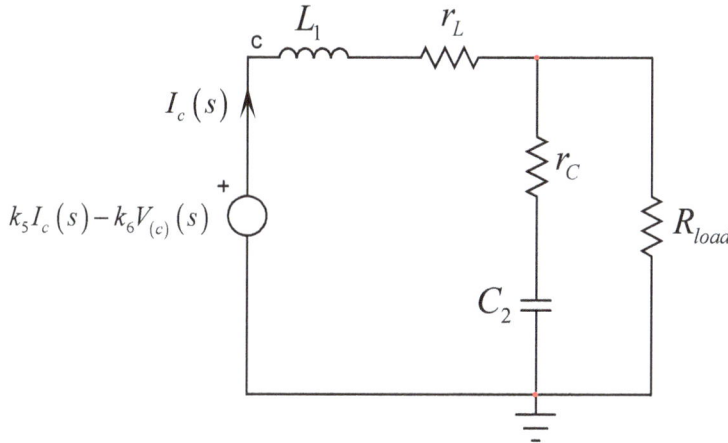

Figure 2.60 When the excitation reduces to 0 A, the circuit reduces to a known structure whose time constants have already been determined.

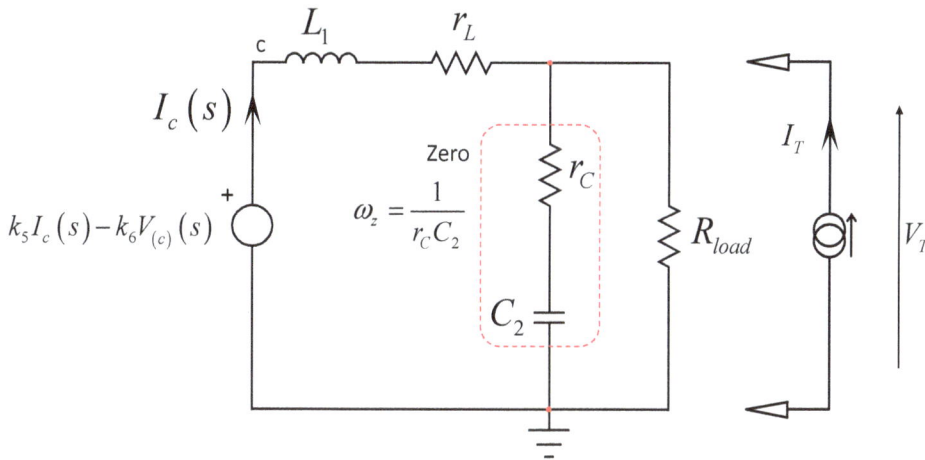

Figure 2.61 The response is nulled to determine the zeroes positions. A 0-V condition across a current source is a degenerate case: replace the current source by a short circuit.

The second zero considers observing the circuit for a null in the output. In this mode, the response V_T is nulled across the test generator I_T. This is a degenerate case [1] and the current source can be replaced by a short circuit installed across R_{load}, grounding the right side of r_L. What we need is the resistance R seen from L_1's connecting terminals to determine the time constant involving L_1.

Figure 2.62 illustrates the exercise and we have to resort to a test generator I_T to determine R.

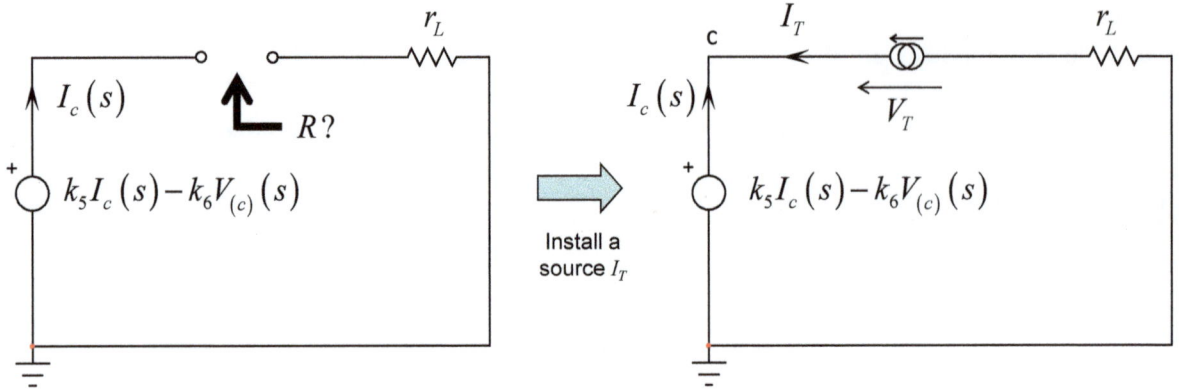

Figure 2.62 When the response is nulled, you can determine the time constant involving L_1.

The voltage V_T is defined by considering the drop across resistance r_L:

$$V_T = k_5 I_c - k_6 V_{(c)} + I_T r_L \qquad (2.111)$$

The voltage at terminal c is:

$$V_{(c)} = V_T - I_T r_L \qquad (2.112)$$

Combining the two above expressions and realizing that I_c and I_T run in opposite directions, we can write:

$$V_T = -I_T k_5 + I_T r_L - V_T k_6 + I_T k_6 r_L \qquad (2.113)$$

Factoring and rearranging, we obtain the resistance we need:

$$R = \frac{r_L(1+k_6) - k_5}{1+k_6} \qquad (2.114)$$

It leads to the second zero:

$$\omega_{z_2} = \frac{\dfrac{r_L(1+k_6) - k_5}{1+k_6}}{L_1} \qquad (2.115)$$

If you neglect the inductor ohmic loss r_L and replace the k coefficients by their respective values, then the above formula simplifies to:

$$\omega_{z_2} = \frac{R_{load}(1-M)}{L_1} \qquad (2.116)$$

178

The complete transfer function describing the output impedance of the DCM-operated buck converter is therefore:

$$Z_{out}(s) = R_0 \frac{\left(1 + \dfrac{s}{\omega_{z_1}}\right)\left(1 + \dfrac{s}{\omega_{z_2}}\right)}{\left(1 + \dfrac{s}{\omega_{p_1}}\right)\left(1 + \dfrac{s}{\omega_{p_2}}\right)} \qquad (2.117)$$

Where R_0 is defined by (2.109), the two zeroes by (2.110) and (2.116) while the denominator poles are obtained using (2.86) and (2.87). We have checked the response delivered by (2.117) and that obtained from the SPICE simulation of Figure 2.58. The curves perfectly match in magnitude and phase as confirmed by Figure 2.63.

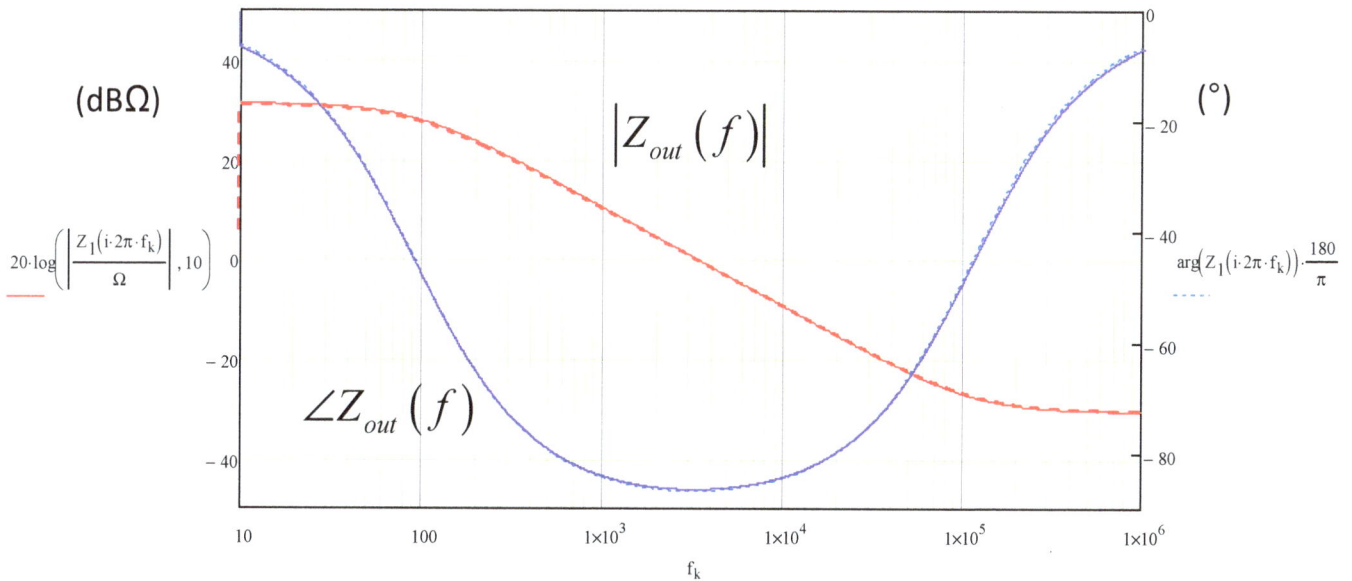

Figure 2.63 Mathcad® and SPICE curves perfectly superimpose: our derivation is correct.

A SPICE simulation circuit for the output impedance appears in Figure 2.64 while results from Figure 2.65 confirm our analytic results.

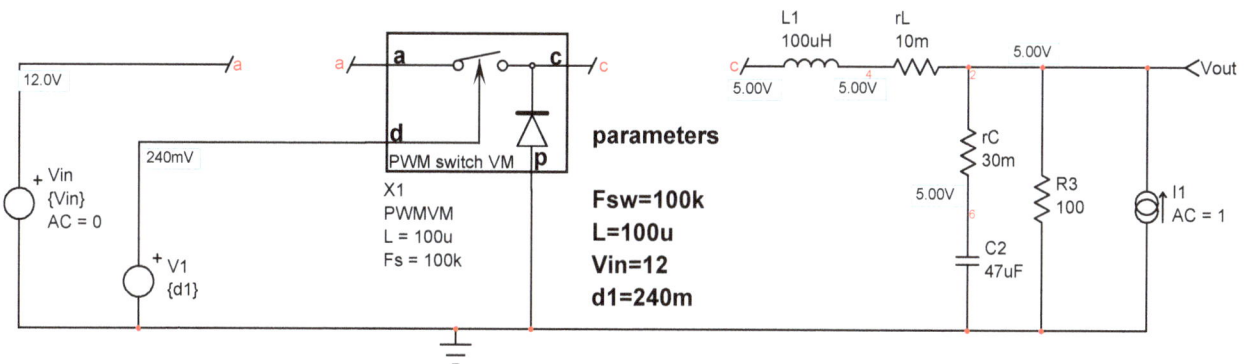

Figure 2.64 Installing a 1-A ac source across the output load reveals the converter output impedance.

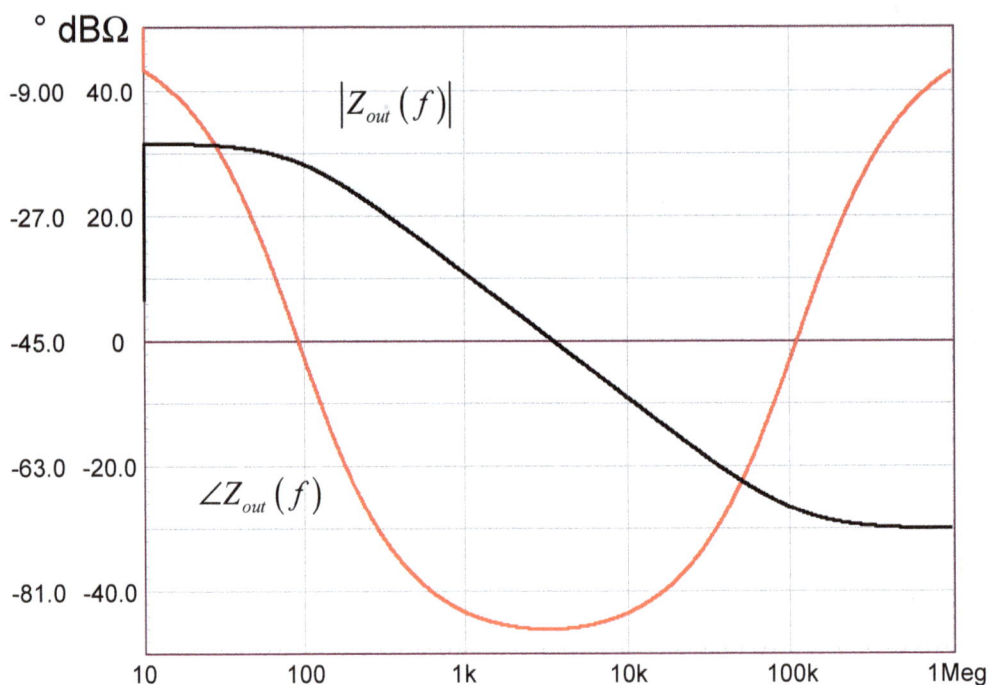

Figure 2.65 The large-signal SPICE model delivers curves in good agreement with those obtained from the small-signal circuit.

2.2.3 Input Impedance

The determination of the input impedance by simulation requires the connection of an inductor LoL to maintain the right input operating point and a current source I_1 to ac-sweep the circuit. The voltage collected at node a divided by the current source value represents the input impedance we want to determine. This is what Figure 2.66 shows.

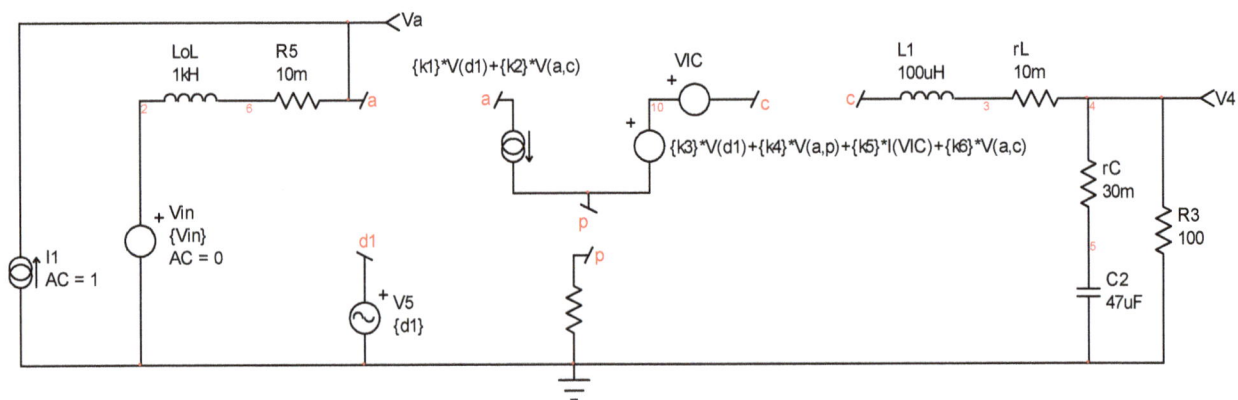

Figure 2.66 The inductor biases the circuit during the bias point analysis and decouples the circuit when the ac sweep starts.

In this mode, the dynamic duty ratio observed at node d_1 is equal to zero and a simplification can occur as illustrated by Figure 2.67 circuit.

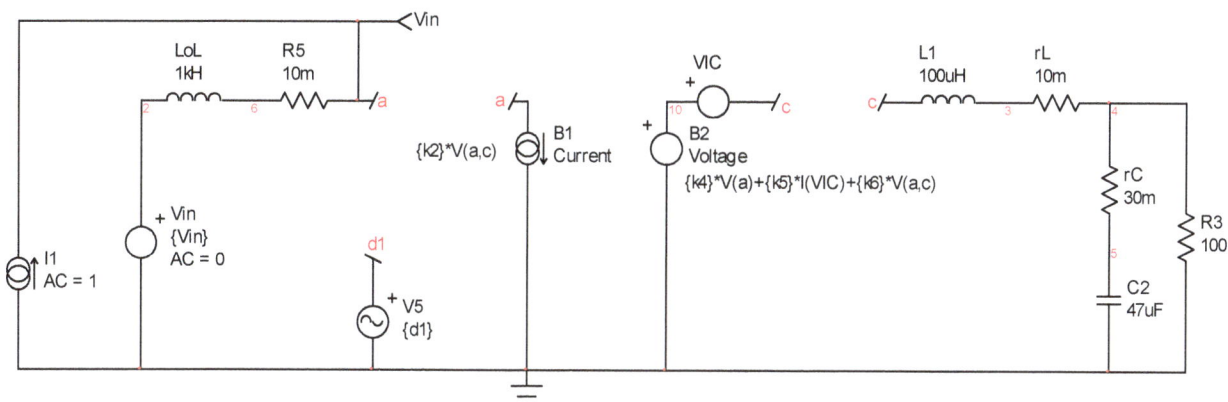

Figure 2.67 Reducing d_1 to 0 V helps simplifying the circuit under analysis.

For the analysis, the circuit is redrawn in a more convenient way in Figure 2.68. The test generator I_T produces the response V_T across its terminals.

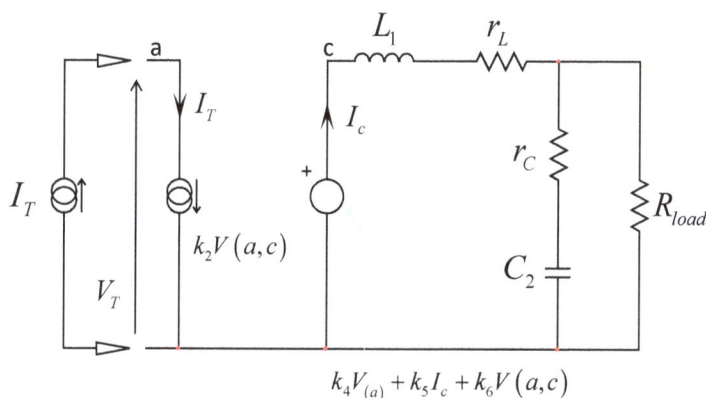

Figure 2.68 The current source I_T biases the input of the DCM converter.

We start with the dc analysis in which L_1 is shorted and C_2 is open. The circuit updates to that of Figure 2.69 and a few equations are necessary to proceed. You can see that the voltage at terminal a is the voltage V_T we want. Thus, current source $k_2 V(a,c)$ can be advantageously replaced by:

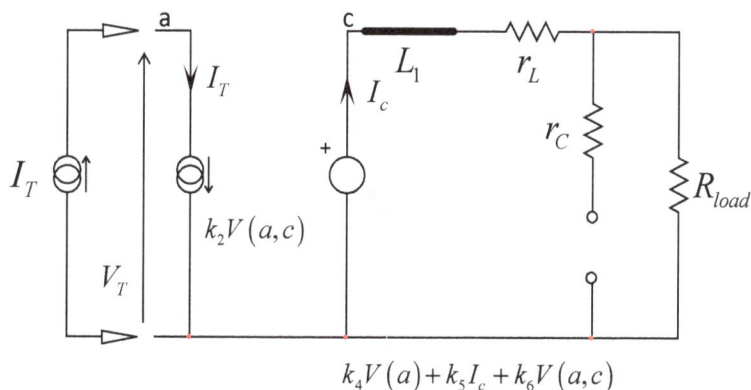

Figure 2.69 You must find the link between V_T and I_T to get the input resistance in dc.

$$I_T = k_2 V_T - k_2 V_{(c)} \tag{2.118}$$

The voltage at node c is defined as:

$$V_{(c)} = k_4 V_T + k_5 I_c + k_6 V_T - k_6 V_{(c)} \tag{2.119}$$

From this expression, you can extract $V_{(c)}$:

$$V_{(c)} = \frac{V_T (k_4 + k_6) + k_5 I_c}{1 + k_6} \tag{2.120}$$

The output current I_c is the output voltage V_{out} divided by the load resistance R_{load}. As the output and input (here V_T) voltages are linked by M, the above formula updates to:

$$V_{(c)} = \frac{V_T (k_4 + k_6) + k_5 \dfrac{M V_T}{R_{load}}}{(1 + k_6)} \tag{2.121}$$

Now, substitute (2.121) in (2.118) and factor V_T / I_T. You should get the following expression for the input resistance when $s = 0$.

$$R_0 = \frac{R_{load} (1 + k_6)}{k_2 \left[R_{load} (1 - k_4) - M \cdot k_5 \right]} \tag{2.122}$$

We can now determine the first time constant involving inductor L_1. The excitation is turned off and node a now becomes floating. To avoid any indeterminacy, we connect a dummy finite resistance of value R_{inf} from terminal a to ground as shown in Figure 2.70.

A current source I_T is installed across the inductor connections which develops a voltage V_T.

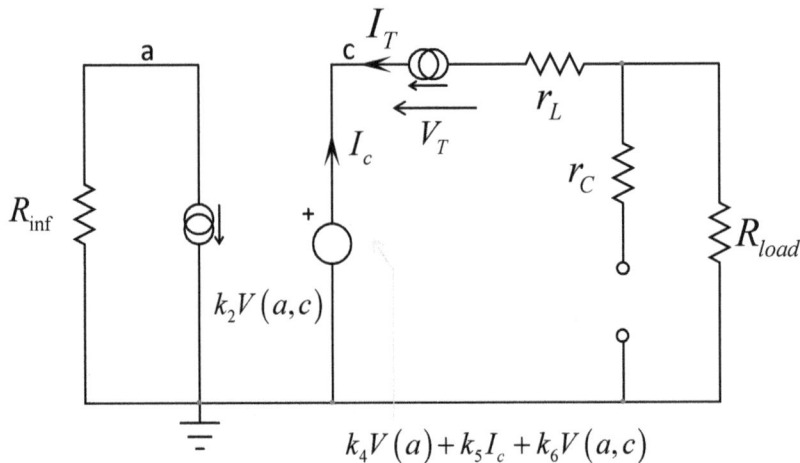

Figure 2.70 The excitation is now turned off to determine the various time constants.

The voltage at node a is simply the current source times $R_{\inf}$:

$$V_{(a)} = R_{\inf} \left(k_2 V_{(c)} - k_2 V_{(a)} \right)$$

(2.123)

Factoring $V_{(a)}$ leads to:

$$V_{(a)} = \frac{k_2 V_{(c)} R_{\inf}}{1 + k_2 R_{\inf}} \approx V_{(c)}$$

(2.124)

Voltage V_T is the voltage at node c minus the voltage across r_L and R_{load}:

$$V_T = V_{(c)} - I_c \left(r_L + R_{load} \right)$$

(2.125)

Observing the picture, we see that I_c and I_T are flowing in opposite directions. Therefore, (2.125) updates to:

$$V_T = V_{(c)} + I_T \left(r_L + R_{load} \right)$$

(2.126)

The voltage at node c follows the source expression:

$$V_{(c)} = k_4 V_{(c)} + k_5 I_c + k_6 V(a,c)$$

(2.127)

From (2.124), since nodes a and c are the same potential, (2.127) simplifies to:

$$V_{(c)} = \frac{I_T k_5}{k_4 - 1}$$

(2.128)

We can now substitute (2.128) in (2.126) and express V_T:

$$V_T = I_T \frac{\left[k_5 - R_{load} - r_L + k_4 \left(R_{load} + r_L \right) \right]}{k_4 - 1}$$

(2.129)

This leads us to the time constant definition involving L_1:

$$\tau_1 = \frac{L_1}{\dfrac{k_5 - R_{load} - r_L + k_4 \left(R_{load} + r_L \right)}{k_4 - 1}}$$

(2.130)

The second time constant involves capacitor C_2. The inductor is placed in its dc state (a short circuit) and the test generator I_T is now installed across C_2's connections.
Figure 2.71 shows the arrangement.

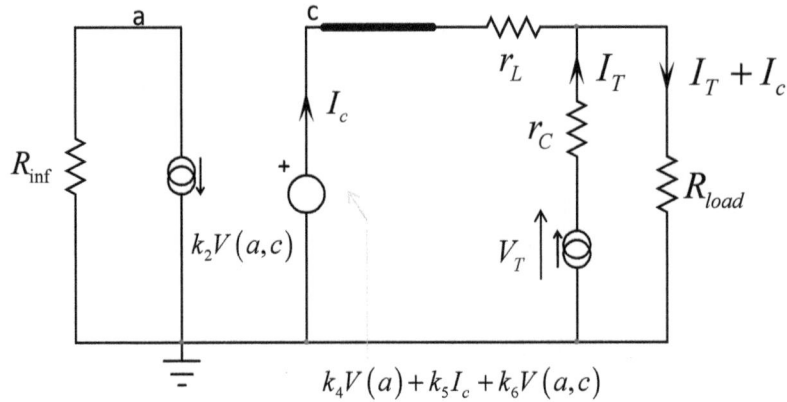

Figure 2.71 The source I_T now biases capacitor C_2's connections.

The relationship between node a and c is preserved and (2.127) leads to:

$$V_{(c)} = \frac{I_c k_5}{1-k_4} \qquad (2.131)$$

The voltage at terminal c is also determined by:

$$V_{(c)} = (I_T + I_c) R_{load} + I_c r_L \qquad (2.132)$$

Equating the above equations and solving for I_c gives:

$$I_c = -\frac{I_T R_{load}}{R_{load} + r_L + \dfrac{k_5}{k_4 - 1}} \qquad (2.133)$$

Finally, the voltage V_T across the current source accounts for the voltage drop across the resistance r_C:

$$V_T = R_{load}(I_T + I_c) + I_T r_C \qquad (2.134)$$

Substituting (2.133) in (2.134) lets us determine the ratio V_T over I_T and thus time constant τ_2:

$$\tau_2 = C_2 \left[r_L + R_{load} + \frac{R_{load}^2}{\dfrac{k_5}{1-k_4} - (R_{load} + r_L)} \right] \qquad (2.135)$$

The first term of the second-order expression is the sum of τ_1 and τ_2:

$$b_1 = \tau_1 + \tau_2 = \frac{L_1}{\dfrac{k_5 - R_{load} - r_L + k_4(R_{load} + r_L)}{k_4 - 1}} + C_2 \left(r_L + R_{load} + \frac{R_{load}^2}{\dfrac{k_5}{1-k_4} - (R_{load} + r_L)} \right) \qquad (2.136)$$

The last term b_2 requires one of the energy-storing element to be set in its high-frequency state. We conveniently select L_1 as, in this case, we open-circuit this element and current I_c drops to 0 A (Figure 2.72).

The only place for the circulating current remains the series connection of r_C and R_{load}:

$$\tau_2^1 = C_2 \left[r_C + R_{load} \right] \tag{2.137}$$

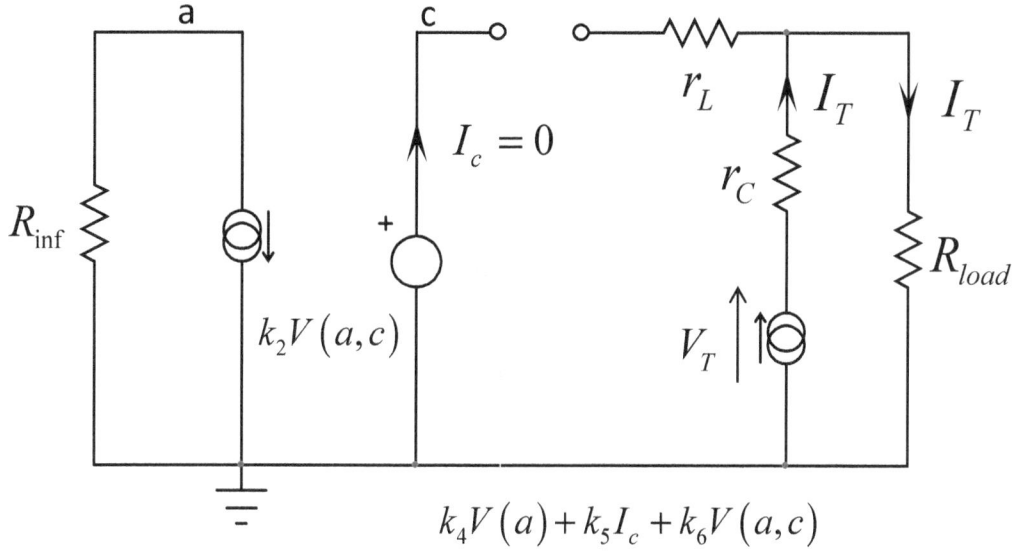

Figure 2.72 Opening L_1 brings the current to zero and simplifies the analysis.

And consequently:

$$b_2 = \tau_1 \tau_2^1 = \frac{L_1}{\dfrac{k_5 - R_{load} - r_L + k_4 \left(R_{load} + r_L \right)}{k_4 - 1}} C_2 \left[r_C + R_{load} \right] \tag{2.138}$$

Assembling (2.136) and (2.138) leads to the denominator determination:

$$D(s) = 1 + s \left[\frac{L_1}{\dfrac{k_5 - R_{load} - r_L + k_4 \left(R_{load} + r_L \right)}{k_4 - 1}} C_2 \left(r_L + R_{load} + \frac{R_{load}^2}{\dfrac{k_5}{1 - k_4} - \left(R_{load} + r_L \right)} \right) \right]$$

$$+ s^2 \frac{L_1}{\dfrac{k_5 - R_{load} - r_L + k_4 \left(R_{load} + r_L \right)}{k_4 - 1}} C_2 \left[r_L + R_{load} \right] \tag{2.139}$$

Now that we have the poles, where are the zeroes located?

If we null the response V_T while the main excitation I_T is back, we know that a 0-V voltage across a current source represents a degenerate case. As shown in Figure 2.72, shorting the current generator brings the circuit back into its natural structure already introduced in Figure 2.54: the time constants we have already determined in the previous transfer functions (control to output, input to output and output impedance) now become the s coefficients of the input impedance numerator:

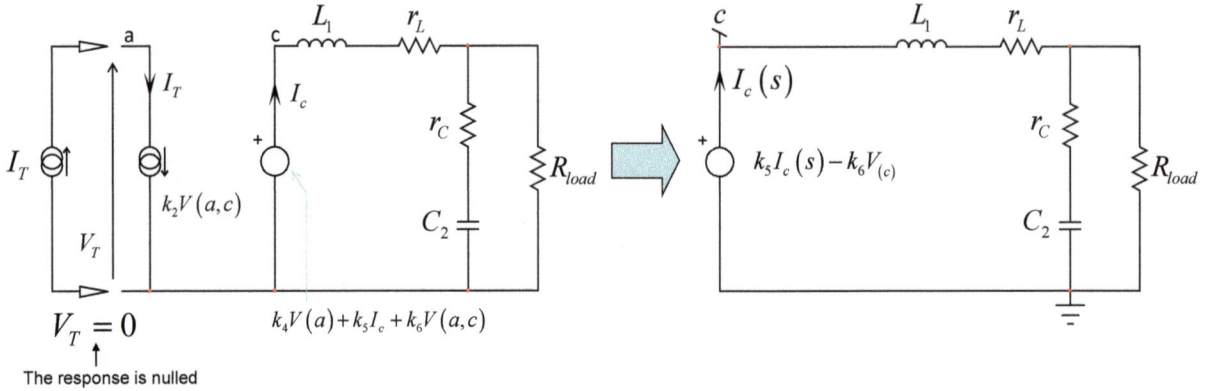

$V_T = 0$

The response is nulled

Figure 2.73 Nulling the response implies a short circuit across the current generator which returns the circuit into its original structure (right side).

$$N(s) = 1 + \left[\frac{L_1}{r_L + R_{load} - \frac{k_5}{k_6+1}} + C_2 \left[r_C + \frac{R_{load}(r_L - k_5 + k_6 r_L)}{R_{load} - k_5 + r_L + k_6(r_L + R_{load})} \right] \right] s + \left(\frac{L_1}{r_L + R_{load} - \frac{k_5}{k_6+1}} C_2(r_C + R_{load}) \right) s^2$$

$$(2.140)$$

Applying the low-Q approximation, the complete expression is thus:

$$Z_{in}(s) = R_0 \frac{\left(1 + \frac{s}{\omega_{z_1}}\right)\left(1 + \frac{s}{\omega_{z_2}}\right)}{\left(1 + \frac{s}{\omega_{p_1}}\right)\left(1 + \frac{s}{\omega_{p_2}}\right)} \qquad (2.141)$$

In which R_0 is defined by (2.122), ω_{z_1} obeys (2.86) while ω_{z_2} is determined with (2.87).

$$\omega_{z_1} \approx \frac{2-M}{1-M} \frac{1}{C_2 R_{load}} \qquad (2.142)$$

$$\omega_{z_2} \approx \frac{L_1 + C_2 R_{load}^2 (1-M)}{L_1 C_2 R_{load}} \qquad (2.143)$$

The poles are simplified versions derived from (2.139) in which r_L is neglected and the low-Q approximation described by (2.83) and (2.84) applies:

$$\omega_{p_1} \approx \frac{R_{load}\left[\left(\frac{2\tau_L}{D_0^2}+1\right)M^2 - 2M + 1\right]}{L_1 M^2\left(1 + \frac{2\tau_L}{D_0^2}\right) - M\left(C_2 R_{load}^2 + L_1\right) + C_2 R_{load}^2} \qquad (2.144)$$

$$\omega_{p_2} \approx \frac{L_1 - \left(C_2 R_{load}^2 + L_1\right)\frac{1}{M} + C_2\left(\frac{R_{load}}{M}\right)^2 + \frac{2L_1 \tau_L}{D_0^2}}{C_2 L_1 R_{load}\left(1 - \frac{1}{M} + \frac{2\tau_L}{D_0^2}\right)}$$

(2.145)

We have plotted the various expressions (the most complete and its simplified version) and all curves perfectly match as shown in the impedance plot from Figure 2.74. The SPICE simulation of Figure 2.75 confirms that our approach is correct as confirmed by Figure 2.76 results.

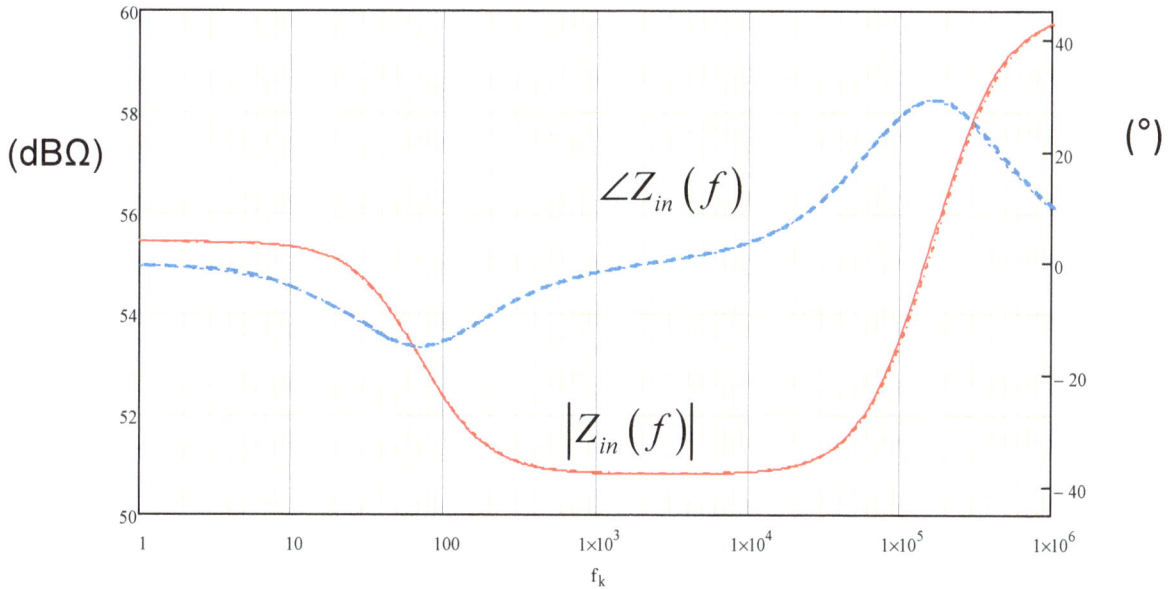

Figure 2.74 The full-blown expression and the simplified version produce the same set of curves.

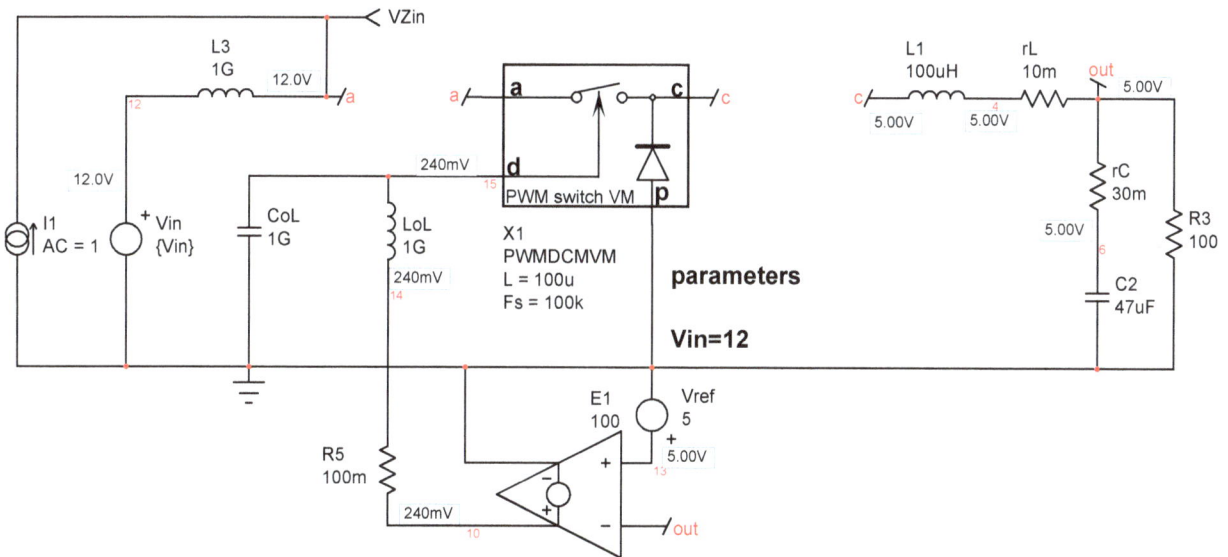

Figure 2.75 The large-signal DCM VM-PWM switch lets us also simulate the input impedance magnitude and phase.

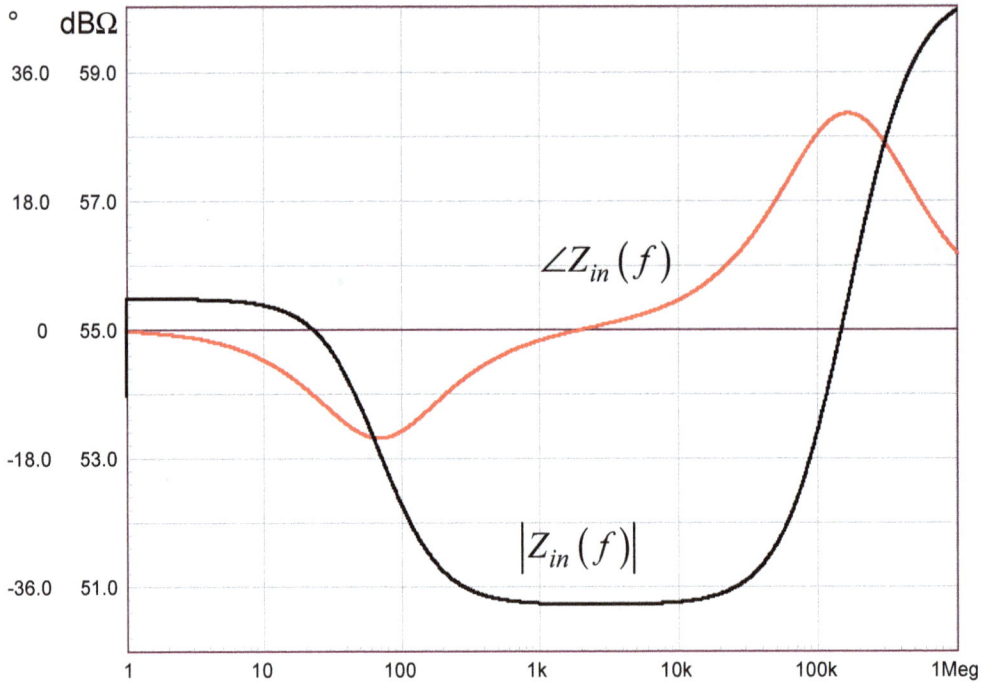

Figure 2.76 The SPICE results perfectly match the Mathcad® derivations.

Now that we have the four transfer functions of the voltage-mode buck converter operated in DCM, I have gathered all of them in a convenient table shown in Figure 2.77.

$\dfrac{V_{out}(s)}{V_{err}(s)}$ Control to Output	$H_0\dfrac{1+\dfrac{s}{\omega_z}}{\left(1+\dfrac{s}{\omega_{p1}}\right)\left(1+\dfrac{s}{\omega_{p2}}\right)}$	$\omega_z=\dfrac{1}{r_C C_2}$	$\omega_{p_1}\approx\dfrac{2-M}{1-M}\dfrac{1}{C_2 R_{load}}$ $\omega_{p_2}\approx\dfrac{L_1+C_2 R_{load}{}^2(1-M)}{L_1 C_2 R_{load}}$	$H_0\approx\dfrac{2V_{out}}{D}\dfrac{1-M}{2-M}$
$\dfrac{V_{out}(s)}{V_{in}(s)}$ Input to Output	$H_0\dfrac{1+\dfrac{s}{\omega_z}}{\left(1+\dfrac{s}{\omega_{p1}}\right)\left(1+\dfrac{s}{\omega_{p2}}\right)}$	$\omega_z=\dfrac{1}{r_C C_2}$	$\omega_{p_1}\approx\dfrac{2-M}{1-M}\dfrac{1}{C_2 R_{load}}$ $\omega_{p_2}\approx\dfrac{L_1+C_2 R_{load}{}^2(1-M)}{L_1 C_2 R_{load}}$	$H_0=M$
$Z_{in}(s)$ Input impedance	$R_0\dfrac{\left(1+\dfrac{s}{\omega_{z1}}\right)\left(1+\dfrac{s}{\omega_{z2}}\right)}{\left(1+\dfrac{s}{\omega_{p1}}\right)\left(1+\dfrac{s}{\omega_{p2}}\right)}$	$\omega_{z_1}\approx\dfrac{2-M}{1-M}\dfrac{1}{C_2 R_{load}}$ $\omega_{z_2}\approx\dfrac{L_1+C_2 R_{load}{}^2(1-M)}{L_1 C_2 R_{load}}$	$\omega_{p_1}\approx\dfrac{R_{load}\left[\left(\dfrac{2\tau_L}{D_1^2}+1\right)M^2-2M+1\right]}{L_1 M^2\left(1+\dfrac{2\tau_L}{D_1^2}\right)-M\left(C_2 R_{load}{}^2+L_1\right)+C_2 R_{load}{}^2}$ $\omega_{p_1}\approx\dfrac{L_1-\left(C_2 R_{load}{}^2+L_1\right)\dfrac{1}{M}+C_2\left(\dfrac{R_{load}}{M}\right)^2+\dfrac{2L_1\tau_L}{D_1^2}}{C_2 L_1 R_{load}\left(1-\dfrac{1}{M}+\dfrac{2\tau_L}{D_1^2}\right)}$	$R_0\approx\dfrac{R_{load}}{M^2}$
$Z_{out}(s)$ Output impedance	$R_0\dfrac{(1+s/\omega_{z_1})(1+s/\omega_{z_2})}{\left(1+\dfrac{s}{\omega_{p1}}\right)\left(1+\dfrac{s}{\omega_{p2}}\right)}$	$\omega_{z_1}=\dfrac{1}{r_C C_2}$ $\omega_{z_2}=\dfrac{R_{load}(1-M)}{L_1}$	$\omega_{p_1}\approx\dfrac{2-M}{1-M}\dfrac{1}{C_2 R_{load}}$ $\omega_{p_2}\approx\dfrac{L_1+C_2 R_{load}{}^2(1-M)}{L_1 C_2 R_{load}}$	$R_0\approx\left[R_{load}(1-M)\right]\|R_{load}$

$$M\approx\dfrac{2}{1+\sqrt{1+\dfrac{8\tau_L}{D_0^2}}}\qquad \tau_L=\dfrac{L_1}{R_{load}}F_{sw}$$

Figure 2.77 The four transfer functions of the DCM-operated voltage-mode buck converter are gathered in this array.

2.3 Buck Transfer Functions in CCM — Peak-Current-Mode Control

The buck converter in current-mode control is studied in a similar way as we did with the voltage-mode case: identify the switch/diode couple and plug the small-signal model as illustrated in Figure 2.78. This model has been described in Chapter 1.4.1 and once inserted in the buck converter, the circuit to solve is that of Figure 2.79. We have neglected the inductor series resistance r_L for the sake of simplicity but with the fast analytical techniques, it can easily be brought back in the time constants definitions.

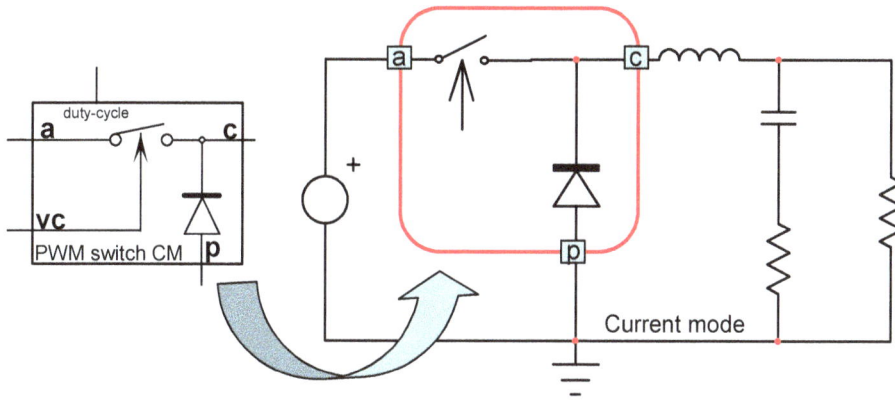

Figure 2.78 To determine the control-to-output transfer function of the current-mode buck converter, plug the small-signal model of the PWM switch.

As usual, the keyword is simplification. In the control-to-output transfer function, the input voltage V_{in} is ac-silent meaning that $\hat{v}_{in} = 0$. As such, sources including a reference to terminal a which is connected to V_{in} can undergo a simplification.

This is the case for the term $\hat{v}_{ap} g_f$ which naturally disappears from the circuit. The input contribution (all sources connected between terminals a and p) can also go away as we do not have interest in the input impedance in this example. The circuit is now reduced to that of Figure 2.80. You can count the number of energy-storing elements and there are three of them: this is a 3rd-oder system whose denominator will obey the below expression:

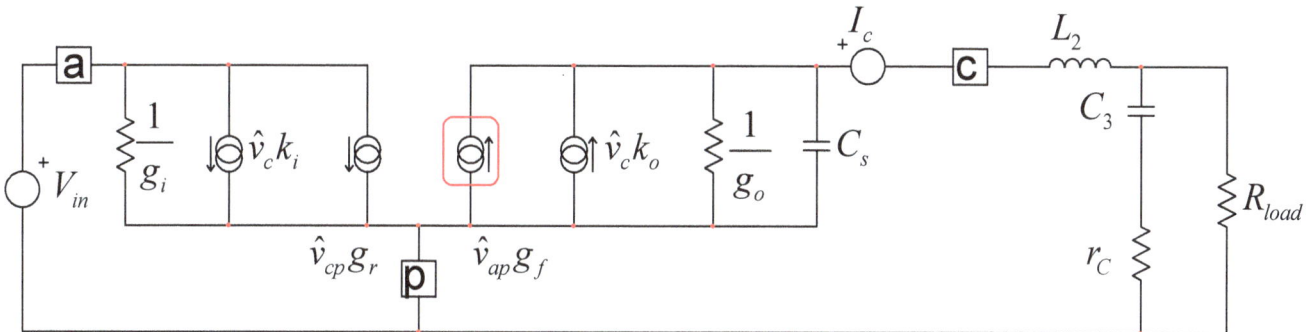

Figure 2.79 The circuit host three energy-storing components and is thus a 3rd-order system.

$$D(s) = 1 + sb_1 + s^2 b_2 + s^3 b_3 \qquad (2.146)$$

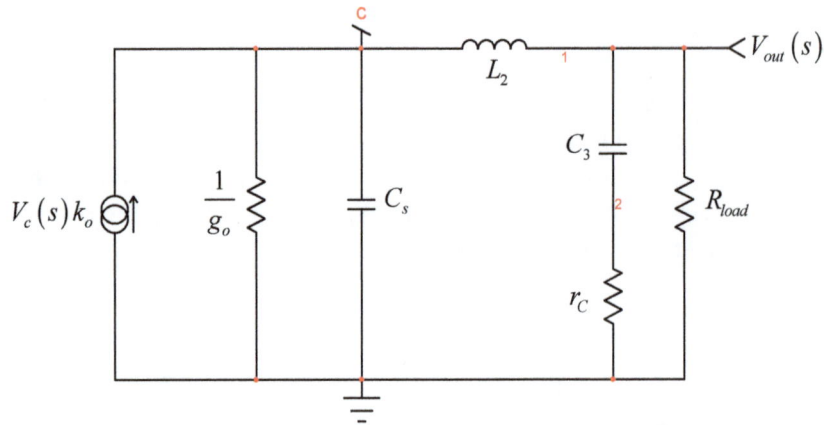

Figure 2.80 It is now easier to determine the transfer function of the converter as simplifications were carried out over the original circuit.

Setting the inductor in its dc state while opening all capacitors leads to the schematic diagram of Figure 2.81. In this circuit, the excitation source is the control voltage, V_c while the response is V_{out}. From this circuit, we can immediately determine the dc gain H_0:

$$H_0 = k_o \left(\frac{1}{g_o} \| R_{load} \right) \tag{2.147}$$

You can use this expression as is but it is better to rearrange it in a more readable form as proposed in [4]:

$$H_0 = \frac{R_{load}}{R_i} \frac{1}{1 + \dfrac{R_{load} T_{sw}}{L_2} \left[m_c (1-D) - 0.5 \right]} \tag{2.148}$$

We reduce the stimulus V_c to zero for determining the time constants and the left-side current source open-circuits. We can now determine the time constants for C_s, L_2 and C_3 in a simple way by inspecting the schematic. Figure 2.82 shows the steps.

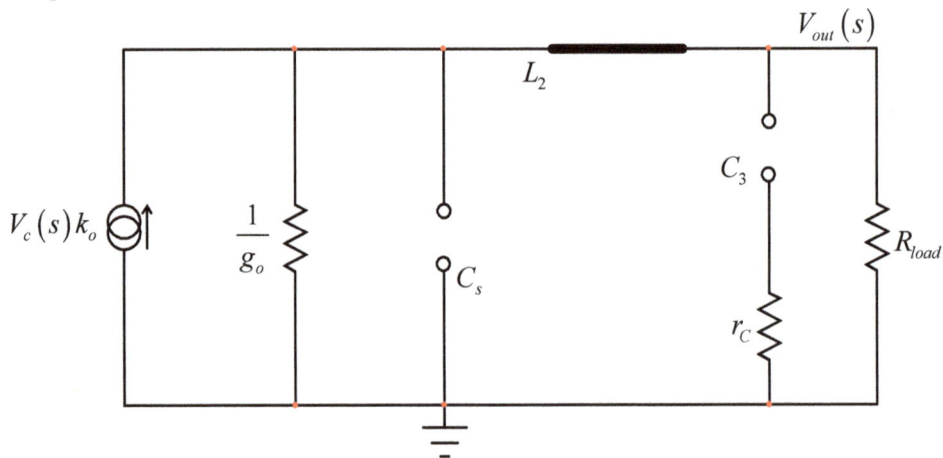

Figure 2.81 The quasi-static gain is obtained immediately by combining the transconductor g_o and the load resistance.

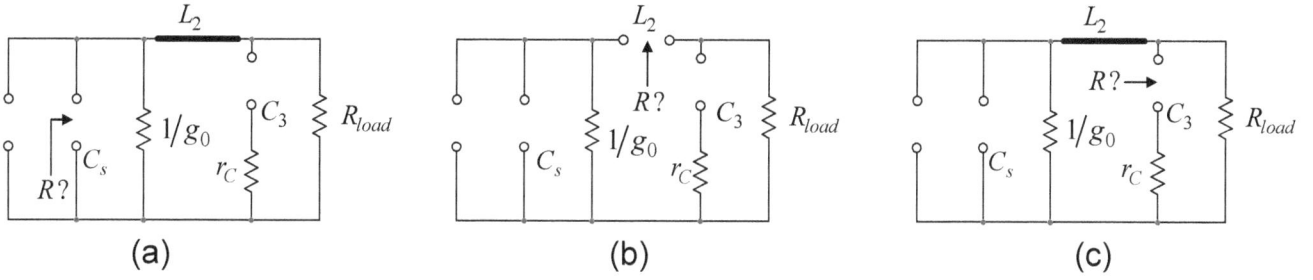

Figure 2.82 Three simple sketches are necessary to determine the time constants.

For these expressions, the time constants are determined immediately. For sketch (a) we have:

$$\tau_1 = C_s \left(\frac{1}{g_o} \| R_{load} \right) \tag{2.149}$$

Then for (b):

$$\tau_2 = \frac{L_2}{\dfrac{1}{g_0} + R_{load}} \tag{2.150}$$

And finally:

$$\tau_3 = C_3 \left[r_C + \left(\frac{1}{g_0} \| R_{load} \right) \right] \tag{2.151}$$

We can form the first coefficient b_1 by summing all these time constants:

$$b_1 = \tau_1 + \tau_2 + \tau_3 = C_s \left(\frac{1}{g_o} \| R_{load} \right) + \frac{L_2}{\dfrac{1}{g_0} + R_{load}} + C_3 \left[r_C + \left(\frac{1}{g_0} \| R_{load} \right) \right] \tag{2.152}$$

The second coefficient b_2 is obtained by combining time constants products:

$$b_2 = \tau_1 \tau_2^1 + \tau_1 \tau_3^1 + \tau_2 \tau_3^2 \tag{2.153}$$

As explained in appendix A, b_2 is built by associating one of the time constants determined in the above lines with a second time constant obtained as indicated in Figure 2.83: in τ_2^1 the element involved in time constant number 1 (C_s) is set in a high-frequency state (a short circuit) while you determine the time constant involving element 2 (L_2). During this exercise, the third element, C_3, is set in its dc state (open circuit).
Figure 2.83 summarizes these recommendations.

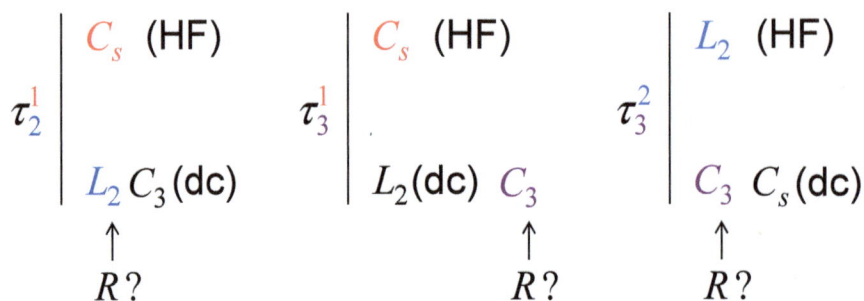

$$\tau_2^1 \begin{vmatrix} C_s \text{ (HF)} \\ \\ L_2 C_3 \text{(dc)} \\ \uparrow \\ R? \end{vmatrix} \qquad \tau_3^1 \begin{vmatrix} C_s \text{ (HF)} \\ \\ L_2 \text{(dc)} \ C_3 \\ \uparrow \\ R? \end{vmatrix} \qquad \tau_3^2 \begin{vmatrix} L_2 \text{ (HF)} \\ \\ C_3 \ C_s \text{(dc)} \\ \uparrow \\ R? \end{vmatrix}$$

Figure 2.83 This time, one of the energy-storing elements is set in its high-frequency state while you "look" at the resistance offered by the second one's terminals. The third element is in its dc state.

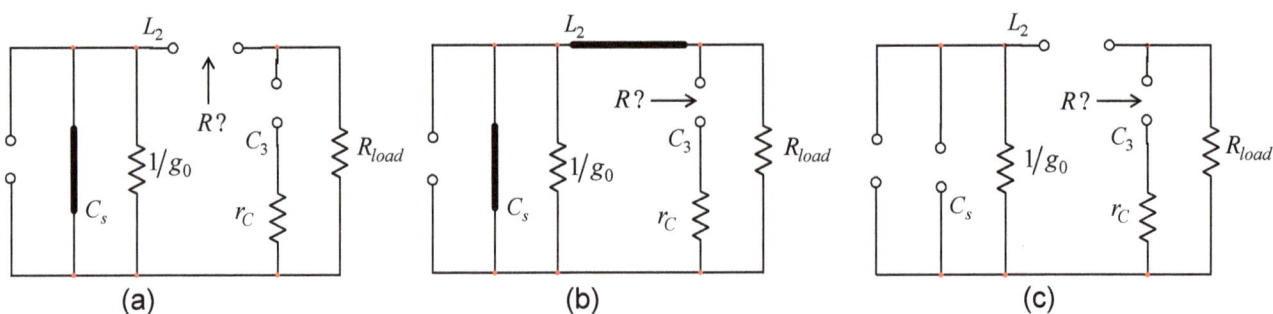

Figure 2.84 The time constants are obtained in a snap shot with these simple sketches.

With three terms in (2.153), there are three drawings as shown in Figure 2.84.

Given the simple electrical diagrams, you can quickly infer the time constants by inspection. In sketch (a), we have:

$$\tau_2^1 = \frac{L_2}{R_{load}} \tag{2.154}$$

In sketch (b):

$$\tau_3^1 = r_C C_3 \tag{2.155}$$

And, finally, in (c):

$$\tau_3^2 = (r_C + R_{load}) C_3 \tag{2.156}$$

All these results can now be assembled for b_2:

$$b_2 = C_s \left(\frac{1}{g_o} \| R_{load} \right) \frac{L_2}{R_{load}} + C_s \left(\frac{1}{g_o} \| R_{load} \right) r_C C_3 + \frac{L_2}{\frac{1}{g_o} + R_{load}} (r_C + R_{load}) C_3 \tag{2.157}$$

We are almost there and the third-order coefficient is composed as in the following expression:

$$b_3 = \tau_1 \tau_2^1 \tau_3^{12}$$

(2.158)

This term is composed by time constants previously defined multiplied by a third time constant in which two elements are set in their high-frequency state (C_s is shorted and L_2 is open-circuited). With the simple corresponding sketch of Figure 2.85, the third time constant is immediately determined:

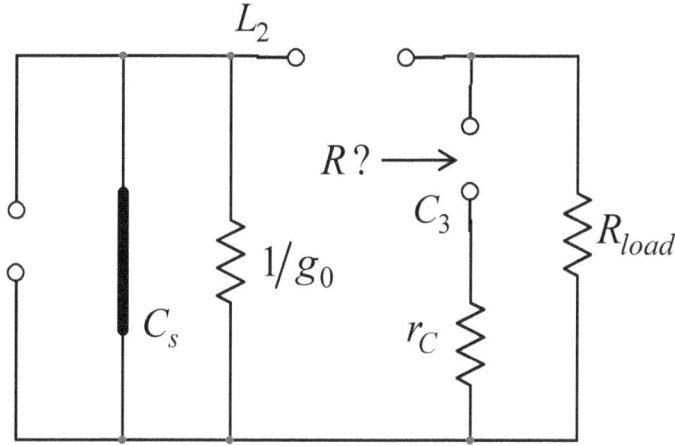

Figure 2.85 The third time constant τ_3^{12} is easily determined with a simple intermediate sketch.

$$\tau_3^{12} = \left(r_C + R_{load} \right) C_3$$

(2.159)

Leading to:

$$b_3 = C_s \left(\frac{1}{g_o} \| R_{load} \right) \frac{L_2}{R_{load}} \left(r_C + R_{load} \right) C_3$$

(2.160)

We now have everything to form the denominator following (2.146). The numerator is determined looking at Figure 2.80 and realizing that a null in the output can only be obtained if r_C and C_3 form a transformed short circuit. As such, we classically have:

$$\omega_z = \frac{1}{r_C C_3}$$

(2.161)

The control-to-output transfer function of the CM buck converter is completely determined and is equal to:

$$\frac{V_{out}(s)}{V_c(s)} = H_0 \frac{1 + \dfrac{s}{\omega_z}}{1 + s b_1 + s^2 b_2 + s^3 b_3}$$

(2.162)

The static control voltage V_c can be evaluated using two equations based on the inductor current observation (Figure 2.86). From this waveform, we can determine the output voltage as a function of the control voltage V_c. The peak current is defined as V_c divided by the sense resistance R_i but it is reduced by the compensation slope S_e:

$$I_{peak} = \frac{V_c}{R_i} - \frac{S_e DT_{sw}}{R_i}$$ (2.163)

The peak current is also equal to the average inductor current plus half of the inductor ripple current:

$$I_{peak} = \frac{V_{out}}{R_{load}} + \frac{V_{out}}{2L_2}(1-D)T_{sw}$$ (2.164)

Equating (2.163) with (2.164) and solving for V_c gives:

$$V_c = \left[V_{out}\left(\frac{1}{R_{load}} + \frac{1-D}{2L_1}T_{sw} \right) + \frac{S_e DT_{sw}}{R_i} \right] R_i$$ (2.165)

The control voltage is extracted and equals 1.28 V for a 4.95-V output.

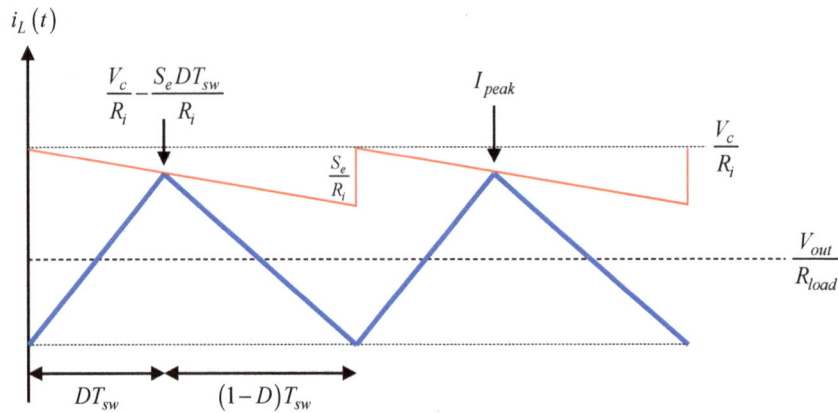

Figure 2.86 The inductor peak current lets us calculate the needed control voltage to deliver V_{out}.

All the small-signal parameters for a 5-V/5-A buck converter (biased from a 10-V input voltage) are evaluated in Figure 2.87 and use 1.28 V as the control voltage.

We will now plot (2.162) magnitude/phase response versus that of the corresponding large-signal SPICE model from Figure 2.88.

$$H(s) = H_0 \frac{1 + \dfrac{s}{\omega_z}}{1 + b_1 s + b_2 s^2 + b_3 s^3}$$

$$H_0 = k_0 \left(R_{load} \left\| \frac{1}{g_o} \right. \right)$$

$$\omega_z = \frac{1}{r_C C_3}$$

$$b_1 = C_s \left(\frac{1}{g_o} \| R_{load} \right) + \frac{L_2}{\dfrac{1}{g_o} + R_{load}} + C_3 \left(r_C + \left(\frac{1}{g_o} \| R_{load} \right) \right)$$

$$b_2 = C_s \left(\frac{1}{g_o} \| R_{load} \right) \frac{L_2}{R_{load}} + C_s \left(\frac{1}{g_o} \| R_{load} \right) r_C C_3 + \frac{L_2}{\dfrac{1}{g_o} + R_{load}} (r_C + R_{load}) C_3$$

$$b_3 = C_s \left(\frac{1}{g_o} \| R_{load} \right) \frac{L_2}{R_{load}} (r_C + R_{load}) C_3$$

$$\|(x,y) := \frac{x \cdot y}{x + y} \qquad F_{sw} := 100\text{kHz} \qquad T_{sw} := \frac{1}{F_{sw}} \qquad L_2 := 100\mu H \qquad R_1 := 1\Omega$$

$$C_s := \frac{1}{L_2 \cdot (\pi \cdot F_{sw})^2} = 101.32118\text{nF} \qquad R_i := 0.25\Omega \qquad S_e := 2.5 \frac{kV}{s} \qquad r_C := 0.1\Omega$$

$$C_3 := 100\mu F$$

$$V_c := 1.28V \qquad V_{ac} := 5V \qquad V_{cp} := 5V \qquad V_{ap} := 10V \qquad D := \frac{V_{cp}}{V_{ap}} = 50\%$$

$$S_n := \frac{V_{ac} \cdot R_i}{L_2} = 12.5 \frac{kV}{s} \qquad S_f := \frac{V_{cp}}{L_2} \cdot R_i = 12.5 \frac{kV}{s}$$

$$I_c := \frac{V_c}{R_i} - D \cdot T_{sw} \cdot \frac{S_e}{R_i} - V_{cp} \cdot (1-D) \cdot T_{sw} \cdot \frac{1}{2 \cdot L_2} = 4.945A \qquad m_c := 1 + \frac{S_e}{S_n} = 1.2$$

Small-Signal Coefficients:

$$k_i := \frac{D}{R_i} = 2\frac{1}{\Omega} \qquad g_o := \frac{T_{sw}}{L_2} \cdot \left[(1-D) \cdot \frac{S_e}{S_n} + 0.5 - D \right] = 0.01\frac{1}{\Omega} \qquad k_o := \frac{1}{R_i} = 4\frac{1}{\Omega}$$

$$g_f := D \cdot g_o - D \cdot (1-D) \cdot T_{sw} \cdot \frac{1}{2 \cdot L_2} = -7.5 \times 10^{-3} \frac{1}{\Omega} \qquad g_r := \frac{I_c}{V_{ap}} - g_o \cdot D = 0.4895\frac{1}{\Omega}$$

$$g_i := D \left(g_f - \frac{I_c}{V_{ap}} \right) = -0.251\frac{1}{\Omega}$$

$$H_0 := k_o \left[R_1 \| \left(\frac{1}{g_o} \right) \right] = 3.9604 \qquad G_{dB} := 20 \cdot \log(H_0) = 11.95477$$

Figure 2.87 This is the complete expression and the computed small-signal coefficients for a 5-V/5-A buck converter.

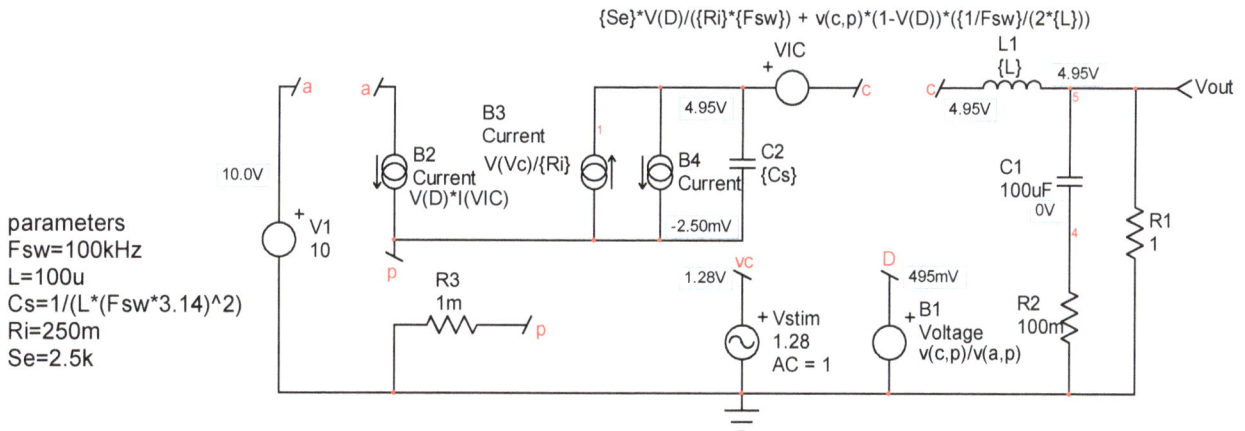

Figure 2.88 The simulation template uses the large-signal PWM switch model which confirms the operating point when a 1.28-V level biases the control pin.

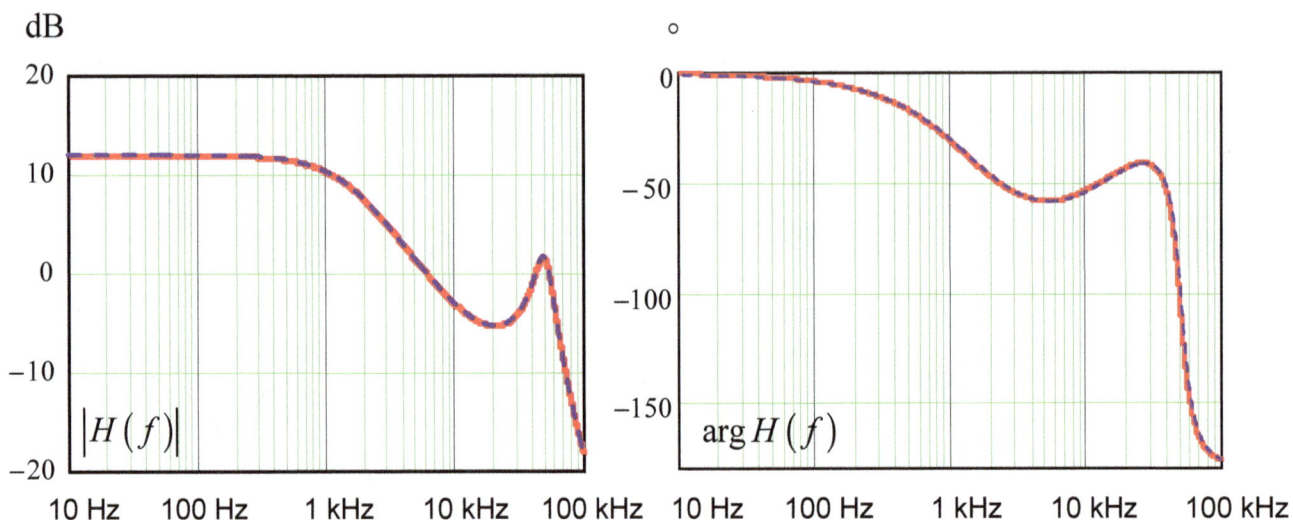

Figure 2.89 Mathcad® and SPICE responses are in excellent agreement, confirming our derivation.

The magnitude/phase plots of Figure 2.89 show a perfect matching between the derived expression and the simulation results. Observing the plots reveals the domination of a pole in the low-frequency spectrum followed by double poles located at half the switching frequency.

The exercise now consists of rewriting (2.162) to unveil and rearrange meaningful terms following the guidelines described in Figure 2.90: a low-frequency term dominates the frequency response before the double poles peak as we approach half the switching frequency. We can thus rewrite the transfer function as a pole cascaded with a double pole.

This is the solution detailed in [4] which leads to the following definitions:

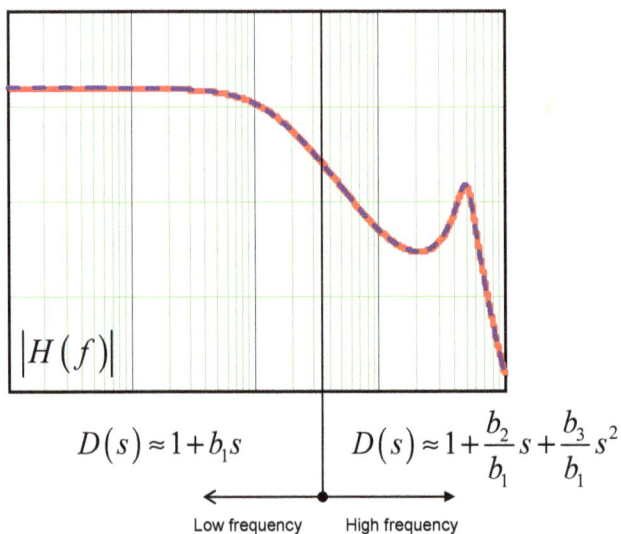

$$D(s) \approx 1 + b_1 s \qquad D(s) \approx 1 + \frac{b_2}{b_1} s + \frac{b_3}{b_1} s^2$$

Low frequency   High frequency

Figure 2.90 We can simplify the denominator observing a pole dominating the response in low frequency while two poles peak at $F_{sw}/2$.

$$\frac{V_{out}(s)}{V_c(s)} = H_0 \frac{1+\dfrac{s}{\omega_z}}{1+sb_1+s^2b_2+s^3b_3} \approx H_0 \frac{1+\dfrac{s}{\omega_z}}{1+\dfrac{s}{\omega_p}}\frac{1}{1+\dfrac{s}{\omega_nQ}+\left(\dfrac{s}{\omega_n}\right)^2} \qquad (2.166)$$

with:

$$H_0 = \frac{R_{load}}{R_i}\frac{1}{1+\dfrac{R_{load}T_{sw}}{L_2}\Big[m_c(1-D)-0.5\Big]} \qquad (2.167)$$

$$\omega_p = \frac{1}{R_{load}C_3}+\frac{T_{sw}}{L_2C_3}\Big[m_c(1-D)-0.5\Big] \qquad (2.168)$$

$$\omega_z = \frac{1}{r_C C_3} \qquad (2.169)$$

$$\omega_n = \frac{\pi}{T_{sw}} \qquad (2.170)$$

$$Q = \frac{1}{\pi\Big[m_c(1-D)-0.5\Big]} \qquad (2.171)$$

$$m_c = 1+\frac{S_e}{S_n} \qquad (2.172)$$

We have captured this expression in Mathcad® and checked its response versus the full-blown formula from (2.162). As Figure 2.91 confirms, the two expressions give very close magnitude responses. Please note that this example assumes an arbitrarily-selected 2.5-kV/s slope for the compensation ramp.

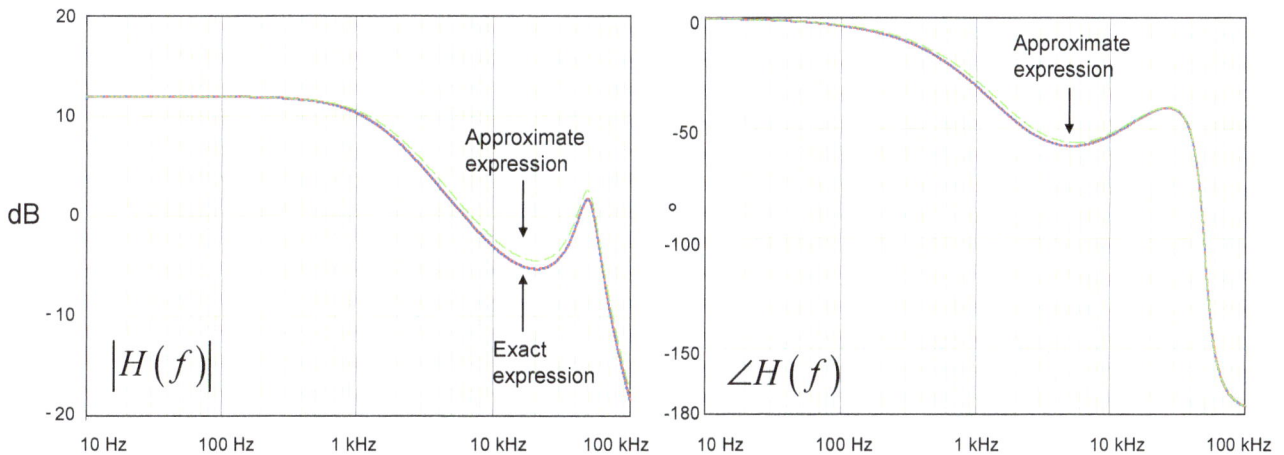

Figure 2.91 The factored expression gives excellent results compared to the raw formula.

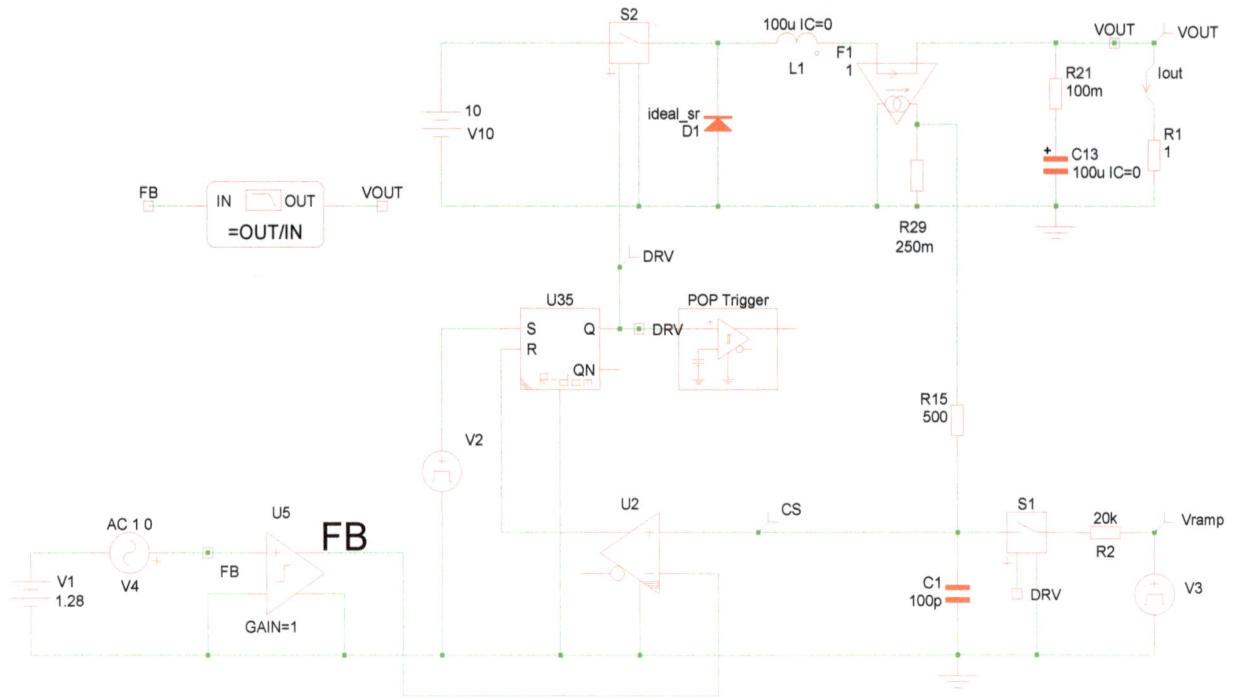

Figure 2.92 A simple buck operated in current mode is captured in SIMPLIS®.

SIMPLIS® lends itself well to testing the ac response of a switching converter. Figure 2.92 represents the CM buck converter operated at a 100-kHz switching frequency. Slope compensation is implemented with the 1-V peak-amplitude sawtooth source V_3 which, together with R_{15} and R_2, provide a 2.5-kV/s compensation level.

The magnitude/phase response is obtained in a few seconds with Elements, the demonstration version. The graphs are given in Figure 2.93 and confirm our derivation.

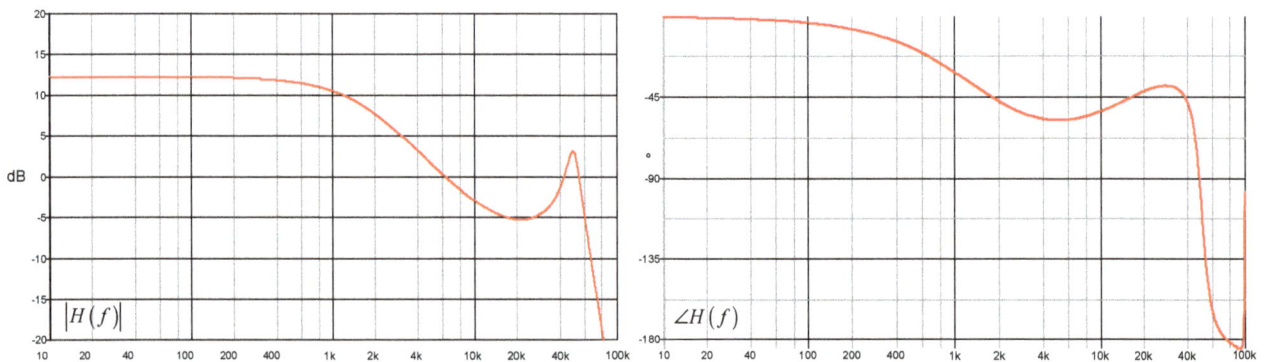

Figure 2.93 The simulation results confirm the small–signal response analytically obtained with the average model.

2.3.1   Input to Output

The input-to-output transfer function of the CCM buck can be derived reusing Figure 2.79 in which the stimulus becomes V_{in}. We know from the previous lines that for the four transfer functions we want, only Z_{in} does not share the denominator determined with (2.146). So, for the audio susceptibility of this converter, we already have $D(s)$.

What is the quasi-static gain of this circuit? We can determine the transfer function by opening the capacitors and shorting inductor L_2 as shown in Figure 2.94. The definition of the output voltage is immediate and equals the controlled source $\hat{v}_{ap}g_f$ times the parallel arrangement of the two resistances. The dc gain linking V_{out} to V_{in} is thus:

$$H_0 = g_f \left(\frac{1}{g_o} \| R_{load} \right) \tag{2.173}$$

Substituting all the small-signal coefficients from the PWM switch model and rearranging this expression leads to [±]:

$$H_0 = \frac{D\left[m_c (1-D) - \left(1 - \frac{D}{2} \right) \right]}{\frac{L_2}{R_{load}T_{sw}} + \left[m_c (1-D) - 0.5 \right]} \tag{2.174}$$

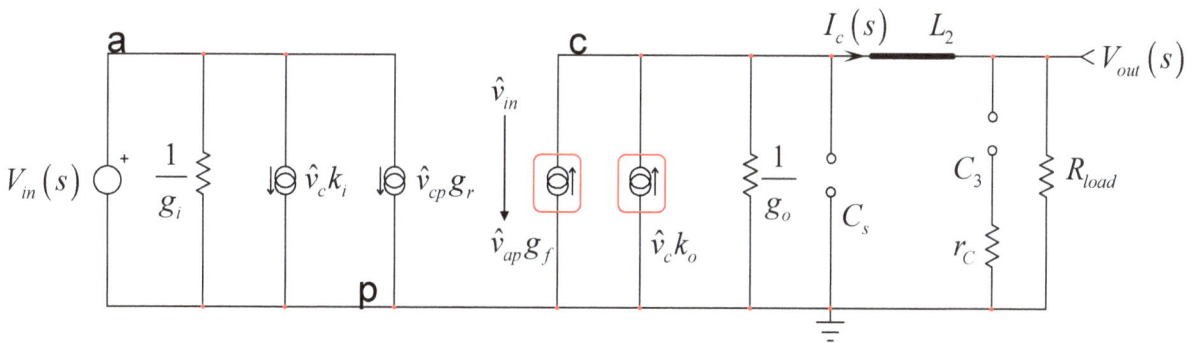

Figure 2.94 The circuit simplifies to the right-side section as we have no interest in the input current for that particular exercise.

To check the presence of zeroes, simply set the energy-storing elements in their high-frequency state and check that if the stimulus (V_{in}) propagates to form a response (V_{out}), then a zero is associated with that element. In Figure 2.94, if you open-circuit L_2 or short C_s, the response disappears if you modulate V_{in}.

On the opposite, shorting C_3 (high-frequency state), V_{out} is still present. There is a classical zero involving r_C and C_3:

$$\omega_z = \frac{1}{r_C C_3} \tag{2.175}$$

The complete input-to-output transfer function is thus expressed as:

$$H(s) = \frac{V_{out}(s)}{V_{in}(s)} = H_0 \frac{1 + \frac{s}{\omega_z}}{1 + \frac{s}{\omega_p}} \frac{1}{1 + \frac{s}{\omega_n Q} + \left(\frac{s}{\omega_n} \right)^2} \tag{2.176}$$

With the pole, zero and resonant parameters defined from (2.168) to (2.172). This equation has been plotted in a Mathcad® sheet and the results appear in Figure 2.95. As expected, some peaking occurs in the magnitude response.

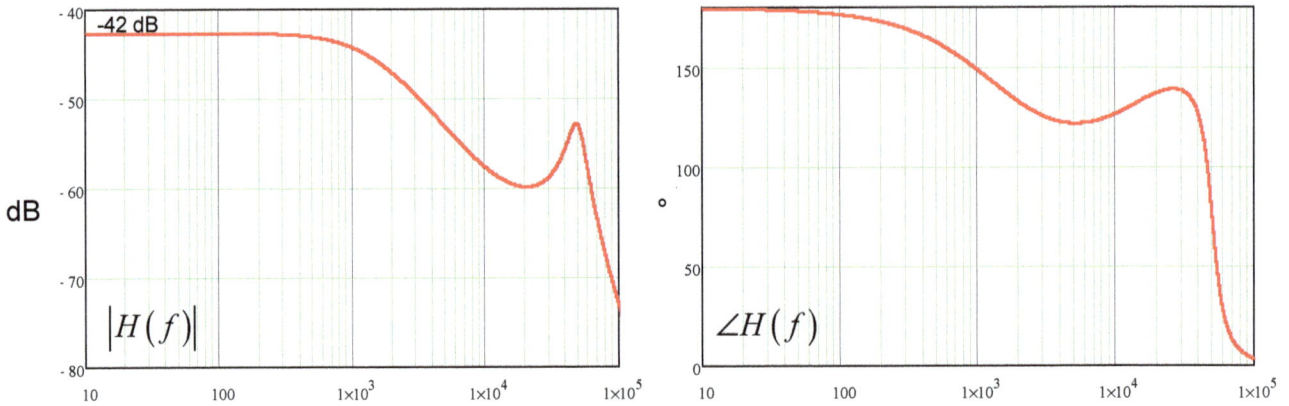

Figure 2.95 The magnitude/phase response given by (2.176) also show sub-harmonic peaking.

A SPICE template has been assembled. The control voltage modulation is now 0 V while the input source is ac-modulated (Figure 2.96). Results appear in Figure 2.97 and confirm our derivation.

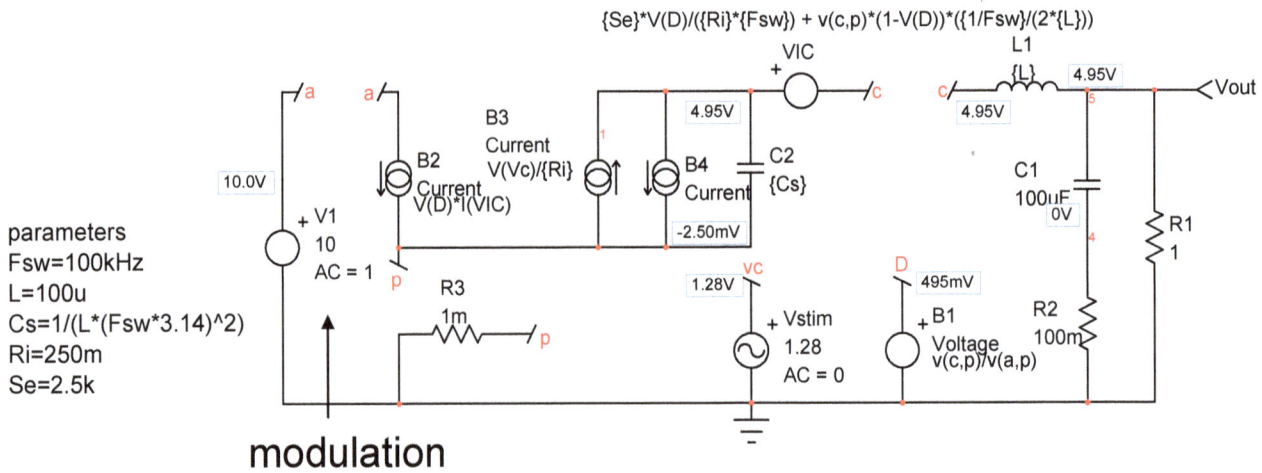

Figure 2.96 The SPICE simulation consists of modulating V_{in} while V_c remains constant at 1.28 V.

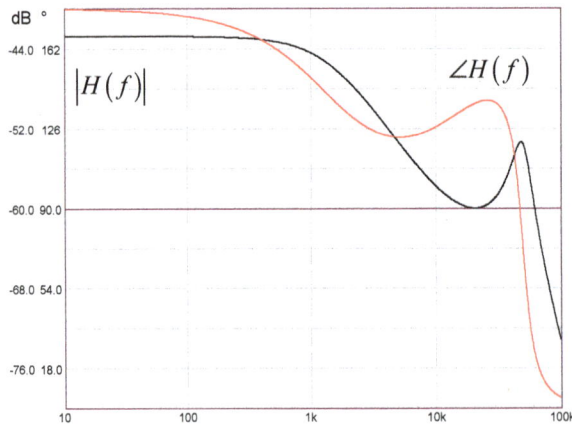

Figure 2.97 This plot is similar to that obtained with the Mathcad® sheet.

It is now interesting to look at the expression defining the dc gain H_0 in (2.174). If we cancel the numerator, the input rejection becomes theoretically infinite. To meet this condition, we must solve the following simple expression:

$$D\left[m_c\left(1-D\right)-\left(1-\frac{D}{2}\right)\right]=0 \tag{2.177}$$

The root leads to the optimum compensation ramp needed to cancel the quasi-static gain:

$$m_c=\frac{D-2}{2D-2} \tag{2.178}$$

For a 50% duty ratio, the expression returns $m_c = 1.5$ which implies $S_e = 0.5 \cdot S_n$ or 50% of the inductor on-slope scaled by the sense resistor. We have tested this formula in a 5-V/1-A CCM-operated CM buck converter with a 56.6% duty ratio. The optimum ramp level is 15.6 kV/s. As confirmed by the simulation results in Figure 2.98, when the ramp level reaches the right value, the rejection goes from -31 dB in an uncompensated version ($S_e = 0$) to -55 dB with the optimum ramp value.

$$D := 56.6\% \qquad V_{in} := 9V \qquad V_{out} := 5V \qquad L_2 := 100\mu H \qquad R_L := 5\Omega \qquad F_{sw} := 100kHz \qquad R_i := 0.6\Omega$$

$$T_{sw} := \frac{1}{100kHz} \qquad S_n := \frac{V_{in}-V_{out}}{L_2}\cdot R_i = 24\frac{kV}{s} \qquad S_e := 15.65\frac{kV}{s} \qquad m_c := \frac{S_e}{S_n}+1 = 1.652$$

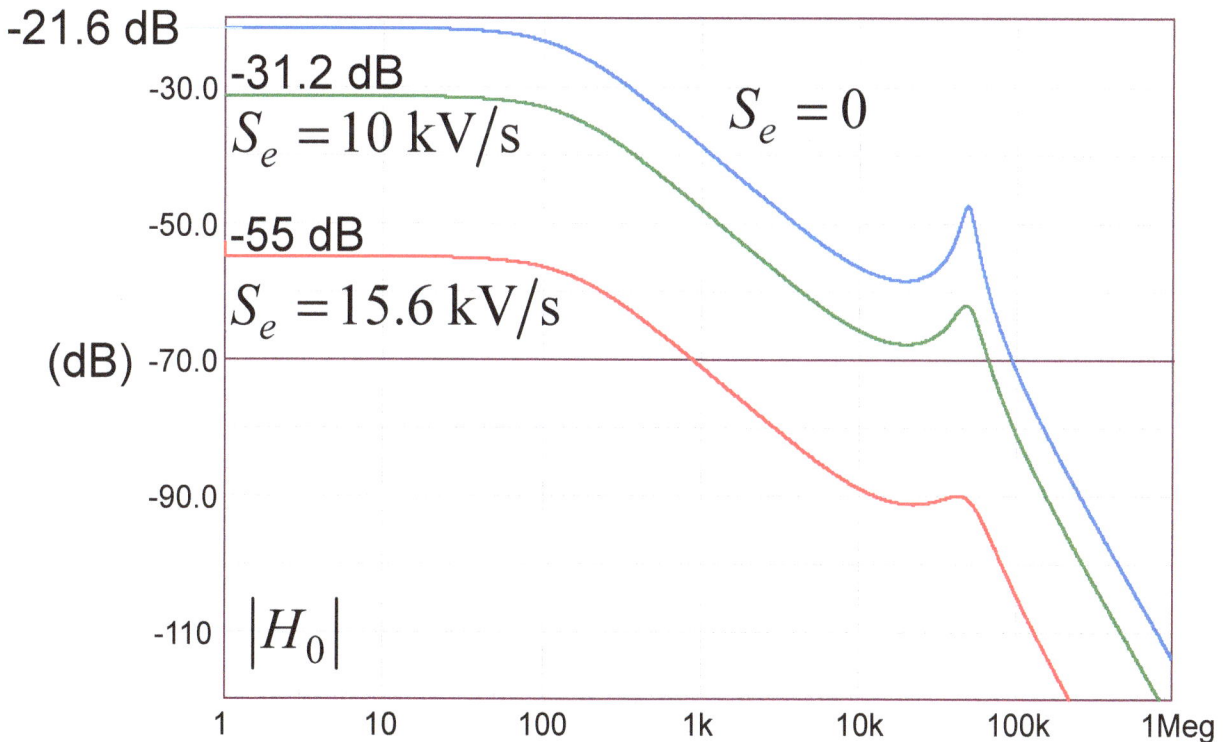

Figure 2.98 When the optimum ramp level is obtained, the input rejection becomes very high.

2.3.2 Output Impedance

The output impedance is determined by installing a test generator across the output load as shown in Figure 2.99.

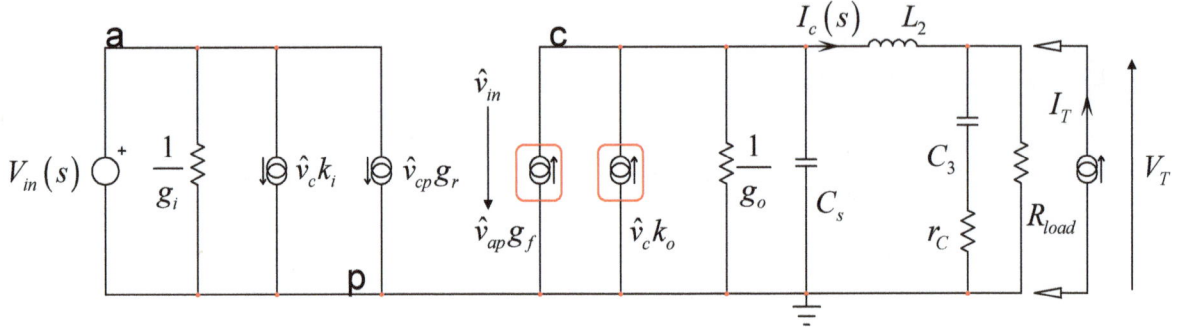

Figure 2.99 A current source is installed across the load resistance to determine the output impedance of the CM buck converter. The circled sources are reduced to 0 A and open-circuited.

As usual, it is important to simplify the circuit considering that a) the input voltage is 0 V in ac and b) the control voltage V_c is also 0 V (open-loop operation). The stimulus is the current I_T while the response V_T is the voltage across the current source terminals. The circuit simplifies to that of Figure 2.100.

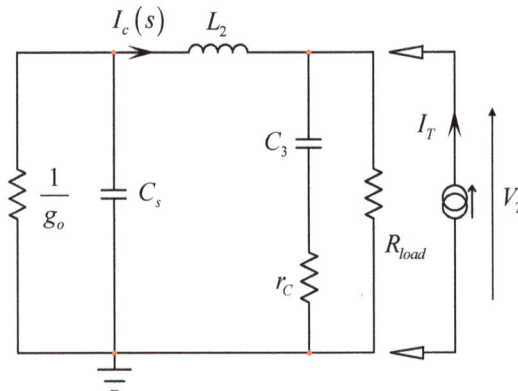

Figure 2.100 The circuit becomes simpler to analyze when 0-V sources are removed from the network.

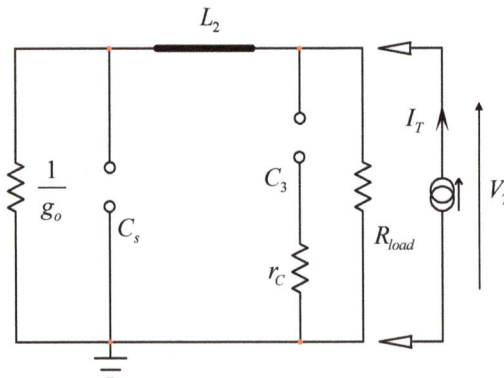

Figure 2.101 In dc, the inductor is shorted while the two capacitors are open circuited.

The first thing is to determine the output resistance R_0: short the inductor and open-circuit all capacitors as shown in Figure 2.101. The result is immediate by inspection:

$$R_0 = \frac{1}{g_o} \| R_{load} \qquad (2.179)$$

If you now reduce the excitation to 0 A, the circuit reduces to that of Figure 2.80 and we can reuse the denominator $D(s)$ from (2.146). What remains to determine now are the zeroes of this transfer function. Zeroes manifest themselves by nulling the response V_T despite the presence of the excitation stimulus. The first zero is classically defined by the series connection of r_C and C_3 which form a transformed short circuit at the first zero frequency:

$$\omega_z = \frac{1}{r_C C_3} \qquad (2.180)$$

A null in the response also implies that the injected current does not cross the load resistance R_{load} but entirely flows through the series connection of L_2 and the parallel connection of C_s with $1/g_o$. This is what Figure 2.102 shows. This happens in case this arrangement becomes a transformed short circuit. The impedance of such network is written as

$$Z(s) = sL_2 + \frac{1}{g_o} \| \frac{1}{sC_s} = sL_2 + \frac{1}{g_o + sC_s} \qquad (2.181)$$

It can be rearranged under the form:

$$Z(s) = \frac{1 + sg_o L_2 + s^2 L_2 C_s}{D(s)} \qquad (2.182)$$

In this expression, $D(s)$ is the denominator determined in (2.162).

This impedance becomes a transformed short if the numerator in (2.182) cancels to zero.

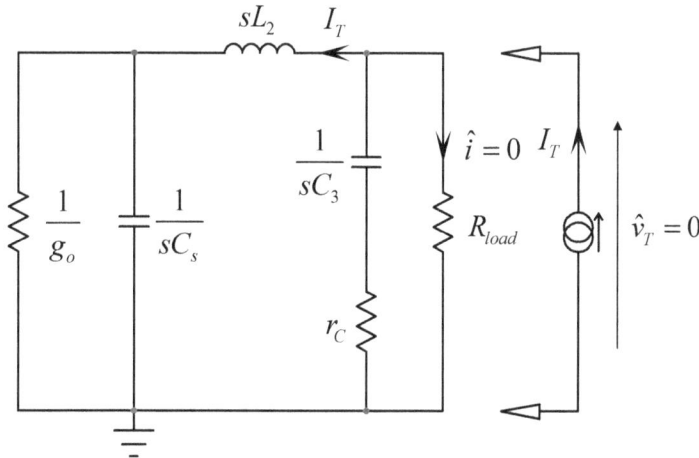

Figure 2.102 If the network formed by L_2 and the parallel connection of C_s and $1/g_o$ becomes a transformed short, the response is nulled: no current flows in R_{load}.

We recognize a second-order polynomial form that we can factor as:

$$N(s) = 1 + \frac{s}{\omega_N Q_N} + \left(\frac{s}{\omega_N}\right)^2 \tag{2.183}$$

In which:

$$\omega_N = \frac{1}{\sqrt{L_2 C_s}} \tag{2.184}$$

$$Q_N = \frac{1}{g_o}\sqrt{\frac{C_s}{L_2}} \tag{2.185}$$

and this is the numerator of the output impedance transfer function we want. The final expression is then defined as:

$$Z_{out}(s) = R_0 \frac{\left(1 + \frac{s}{\omega_z}\right)\left(1 + \frac{s}{\omega_N Q_N} + \left(\frac{s}{\omega_N}\right)^2\right)}{1 + sb_1 + s^2 b_2 + s^3 b_3} \tag{2.186}$$

The expression proposed in [4] offers a simpler format, neglecting the high-frequency zeroes:

$$Z_{out}(s) \approx \frac{R_{load}}{1 + \frac{R_{load}}{L_2}T_{sw}\left[m_c(1-D) - 0.5\right]} \cdot \frac{1 + \frac{s}{\omega_z}}{1 + \frac{s}{\omega_p}} \tag{2.187}$$

With the pole and the zero respectively defined with (2.168) and (2.180). With a large inductance or a small switching period, the incremental resistance R_0 reduces to the output load value.

This equation together with (2.186) have been plotted in a Mathcad® sheet and the results appear in Figure 2.103. There is no peaking in the magnitude response.

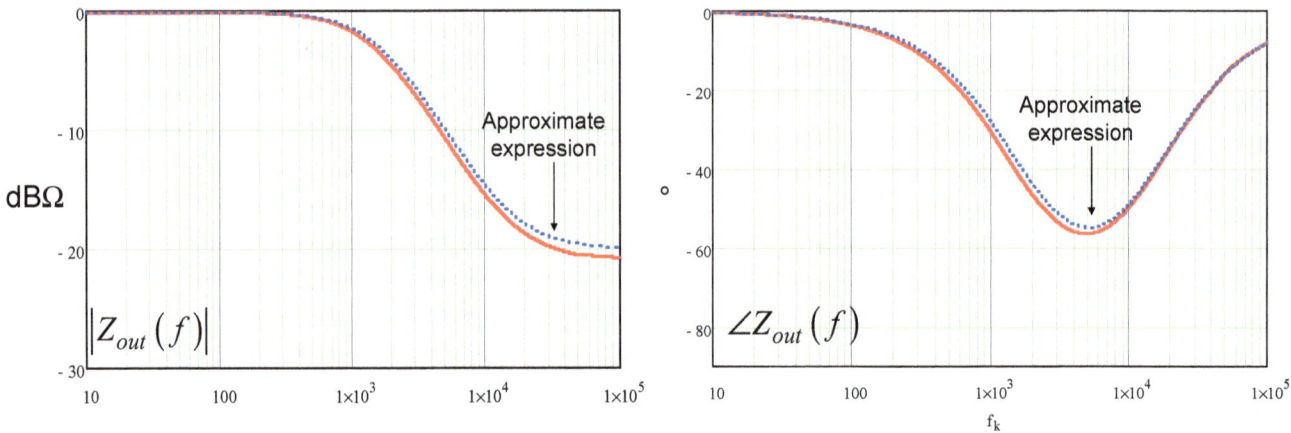

Figure 2.103 The output impedance is dominated by the load resistance in low frequency.

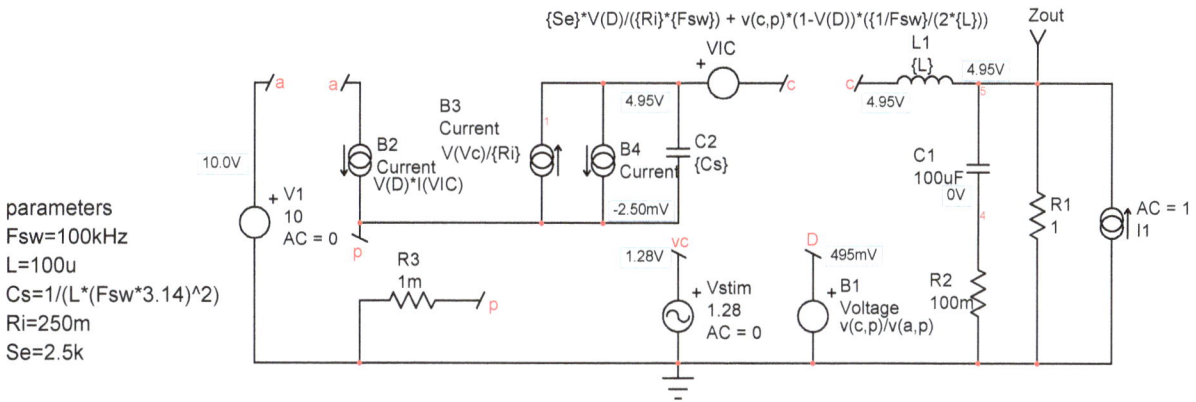

Figure 2.104 A simple ac current source connected across the load unveils the output impedance of the CCM CM buck converter.

We have reused the previous template and added a 1-A ac current source across the load. The collected voltage across R_1 represents the direct image of the output impedance.

The simulated circuit is given in Figure 2.104 while the results appear in Figure 2.105. They are identical with that of the Mathcad® sheet.

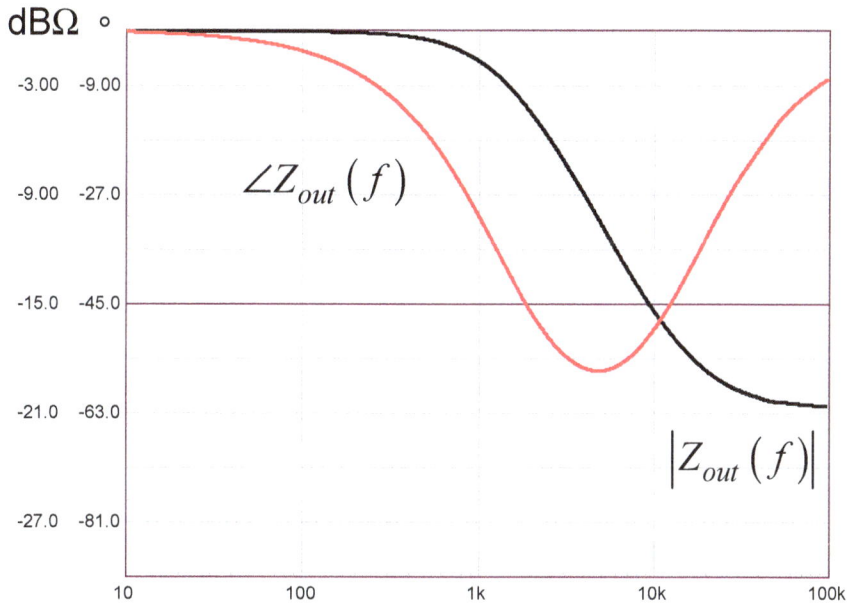

Figure 2.105 The response matches that of the Mathcad® sheet.

2.3.3   Input Impedance

The determination of the input impedance requires the connection of a stimulus current source I_T to the input side of the circuit. It develops a response V_T across its terminals. Figure 2.106 shows the configuration you are now familiar with.

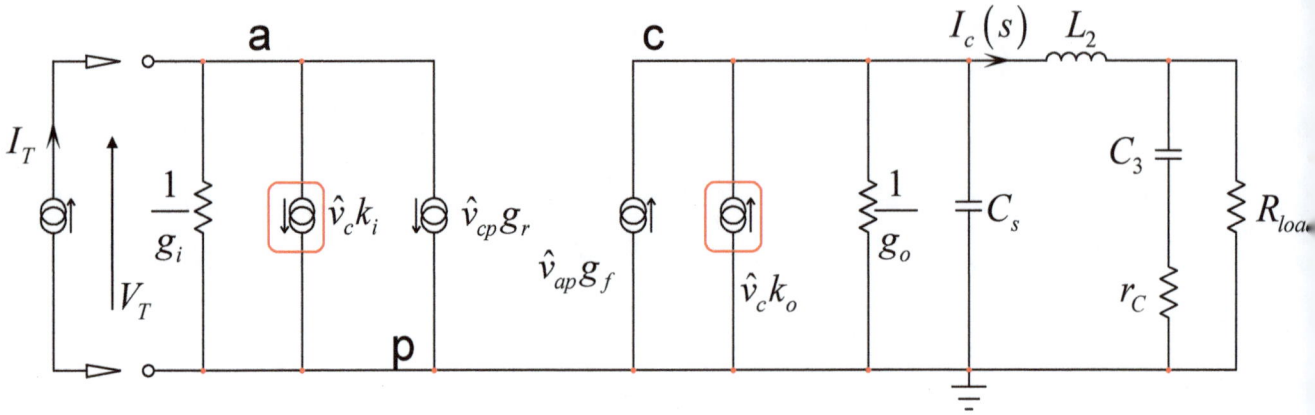

Figure 2.106 The input impedance requires the connection of a current source at the input source terminals.

Since we want the open-loop input impedance, the control voltage V_c is ac-silent and $\hat{v}_c = 0$.

The circled current sources are thus open-circuited further simplifying the network as drawn in Figure 2.107.

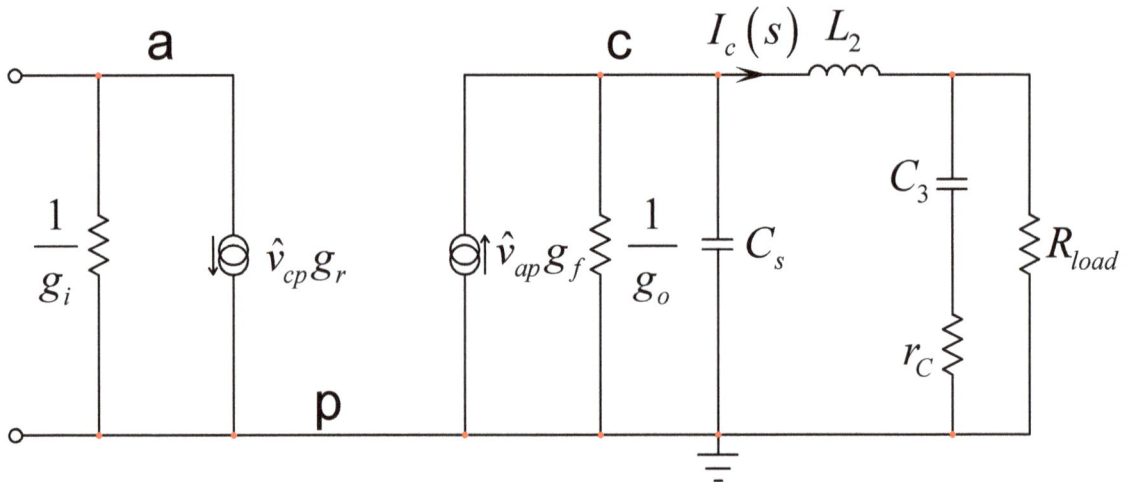

Figure 2.107 The circuit is simplified by turning off the excitation but also removing the zeroed current sources.

If we now open all capacitors and short the inductor, what is the dc input resistance R_0? First, in Figure 2.108, we see that $1/g_i$ can be temporarily removed from the schematic and placed back later in parallel with the intermediate result we will determine.

Without this resistance, the input current is:

$$I_T = \hat{v}_{cp} g_r \tag{2.188}$$

206

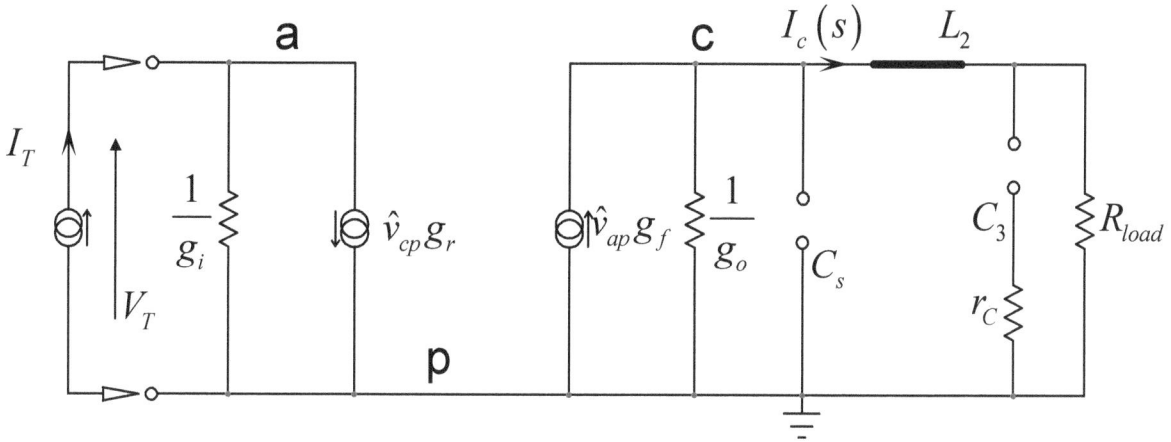

Figure 2.108 Short the inductor and open-circuit the capacitor to determine the input resistance at $s = 0$.

The voltage across terminals c and p is the source $\hat{v}_{ap}g_f$ flowing into the parallel arrangement of R_{load} and resistance $1/g_o$:

$$\hat{v}_{cp} = \hat{v}_{ap}g_f \left(\frac{1}{g_o} \| R_{load} \right) \tag{2.189}$$

Substituting (2.189) in (2.188) gives:

$$I_T = \hat{v}_{ap}g_f \left(\frac{1}{g_o} \| R_{load} \right) g_r = \frac{\hat{v}_{ap}}{r_{ap}} \tag{2.190}$$

$\hat{v}_{ap}$ is actually the response voltage V_T while the resistance r_{ap} is expressed by:

$$r_{ap} = \frac{1}{g_f g_r \left(\dfrac{1}{g_o} \| R_{load} \right)} \tag{2.191}$$

Finally, the input resistance R_0 is defined as:

$$R_0 = \frac{1}{g_i} \| \frac{1}{g_f g_r \left(\dfrac{1}{g_o} \| R_{load} \right)} \tag{2.192}$$

If you now replace all these definitions by their respective small-signal coefficients and rearrange the result, you will find:

$$R_0 = -\frac{R_{load}\left[\left(\dfrac{L_2}{R_{load}T_{sw}} + 0.5 - D \right) S_n + S_e (1-D) \right]}{\left[\left(S_e + \dfrac{S_n}{2} \right) D^3 - \left(S_e + \dfrac{M}{2}S_n \right) D^2 + \dfrac{M}{2}S_n D \right] + MDS_n \dfrac{L_2}{R_{load}T_{sw}}} \tag{2.193}$$

What you see is an interesting characteristic of the buck converter operated in current mode: the incremental input resistance is naturally negative, despite an open-loop configuration. This changes from the buck in voltage-mode whose incremental input resistance is positive. It is only when you close the loop that it becomes negative at low frequencies. It is worth spending a bit of time to understand where the negative value in the case of current-mode comes from.

Figure 2.110 represents the inductor current in low- (LL) and high-line (HL) conditions. In both conditions, the peak current is always constant (we neglect any propagation delay) and dictated by the sense resistance R_i and control voltage V_c. The valley current I_v depends on the off-time or $(1-D)T_{sw}$ and V_{out}. When the input voltage increases, the time to meet the peak current imposed by V_c and R_i reduces as more volt-seconds are available to magnetize the inductor. Therefore, the on-time and thus the duty ratio become smaller. More time is allocated for the inductor demagnetization time t_{off} and the valley current I_v reduces. If I_v reduces, so does I_{out} and the power delivered to the load also goes down: despite a voltage increase at the converter's input, the absorbed current becomes smaller, considering a constant efficiency between the two input voltages.

This is the effect of a negative resistance hence the negative sign in (2.193).

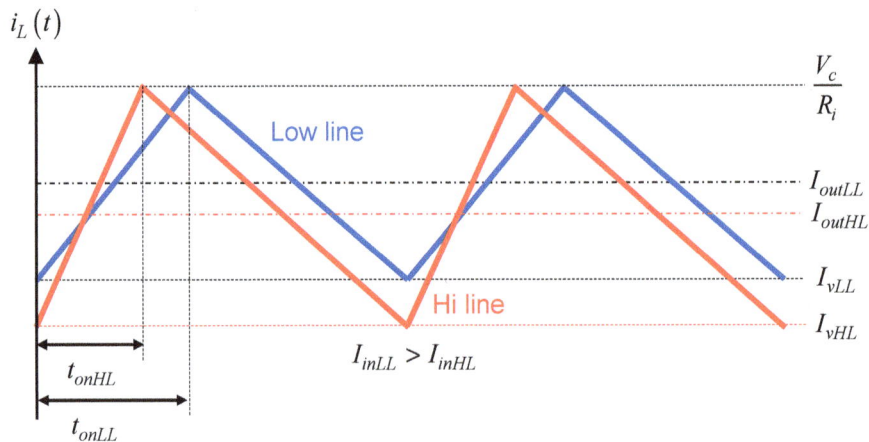

Figure 2.109 With a constant peak current imposed by the current loop, the valley current reduces at a higher input voltage.

In voltage-mode control, this phenomenon does not exist. As D the duty ratio is the controlled variable (VM is also named *direct duty ratio control*), it remains constant to fix the operating point in open-loop operations. When V_{in} increases, as V_{out} equals DV_{in}, the delivered power goes up. This is reflected on the input current, mimicking a positive resistance. In current mode, the controlled variable is the peak current while D is a consequence of I_{peak} hence the term coined for CM operations, *indirect duty ratio control*.

We have our dc input resistance, what about the time constants of Figure 2.107? To ease our work, this circuit can be further rearranged if you notice that the voltage at terminal a is equal to:

$$V_{(a)} = -\frac{1}{g_i}V(c,p)g_r \qquad (2.194)$$

This voltage is turned into a current via the source $\hat{v}_{ap}g_f$. This source can be updated with (2.194) to become:

$$\hat{v}_{ap}g_f = -\hat{v}_{cp}g_r\frac{g_f}{g_i} = -\frac{\hat{v}_{cp}}{r_{cp}} \qquad (2.195)$$

208

It can be seen as a current source placed across terminals c and p pointing towards ground or more advantageously by a simple resistance of value:

$$r_{cp} = \frac{g_i}{g_r g_f} \tag{2.196}$$

This truly simplifies our circuit which becomes an assembly of passive elements without control sources (Figure 2.110). The determination of time constants will be straightforward.

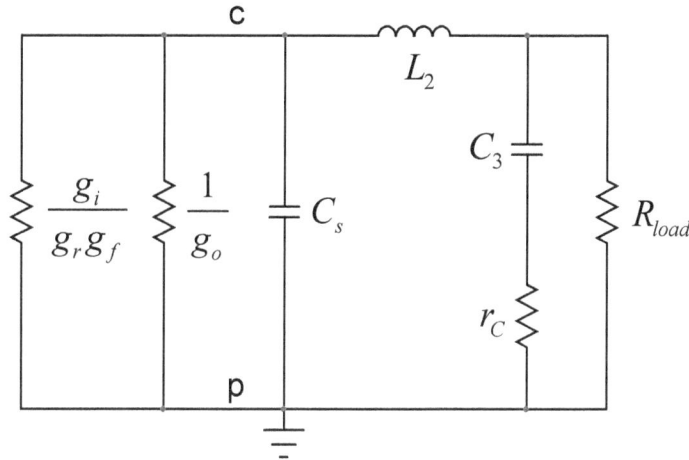

Figure 2.110 After simple manipulations, the circuit transforms into a fully passive circuit in which all controlled sources have disappeared.

The time constants associated with each energy-storing element can be found mentally without intermediate drawings. For the first one involving C_s, short L_2 and open C_3:

$$\tau_1 = C_s \left(\frac{1}{g_o} \parallel R_{load} \parallel \frac{g_i}{g_r g_f} \right) \tag{2.197}$$

For the second time constant, remove L_2 and "look" through its connections for C_s and C_3 open-circuited:

$$\tau_2 = \frac{L_2}{\dfrac{1}{g_o} \parallel \dfrac{g_i}{g_r g_f} + R_{load}} \tag{2.198}$$

Finally, short L_2, open-circuit C_s and "look" through C_3's connections:

$$\tau_3 = C_3 \left(r_C + \frac{1}{g_o} \parallel R_{load} \parallel \frac{g_i}{g_r g_f} \right) \tag{2.199}$$

We have:

$$b_1 = \tau_1 + \tau_2 + \tau_3 \tag{2.200}$$

For the first higher-order time constant, set C_s in its high-frequency state (a short circuit) and "look" through L_2's connections while C_3 is open-circuited:

$$\tau_2^1 = \frac{L_2}{R_{load}} \tag{2.201}$$

Repeat the same operation now with L_2 short-circuited (dc state) and C_s shorted:

$$\tau_3^1 = C_3 r_C \tag{2.202}$$

Finally, determine the resistance "seen" from C_3's connections while L_2 and C_s are open-circuited:

$$\tau_3^2 = C_3 \left(r_C + R_{load} \right) \tag{2.203}$$

Combining the above lines leads to the definition of b_2:

$$b_2 = \tau_1 \tau_2^1 + \tau_1 \tau_3^1 + \tau_2 \tau_3^2 \tag{2.204}$$

Last step, determine the resistance "seen" from C_3's terminals with L_2 open-circuited and C_s shorted:

$$\tau_3^{12} = C_3 \left(r_C + R_{load} \right) \tag{2.205}$$

We have b_3 defined as:

$$b_3 = \tau_1 \tau_2^1 \tau_3^{12} \tag{2.206}$$

Our denominator is now complete and equal to:

$$D(s) = 1 + s b_1 + s^2 b_2 + s^3 b_3 \tag{2.207}$$

With:

$$b_1 = C_s \left(\frac{1}{g_o} \parallel R_{load} \parallel \frac{g_i}{g_r g_f} \right) + \frac{L_2}{\dfrac{1}{g_o} \parallel \dfrac{g_i}{g_r g_f}} + C_1 \left(r_C + \frac{1}{g_o} \parallel R_{load} \parallel \frac{g_i}{g_r g_f} \right) \tag{2.208}$$

$$b_2 = C_s \left(\frac{1}{g_o} \parallel R_{load} \parallel \frac{g_i}{g_r g_f} \right) \frac{L_2}{R_{load}} + C_s \left(\frac{1}{g_o} \parallel R_{load} \parallel \frac{g_i}{g_r g_f} \right) C_3 r_C + \frac{L_2}{\dfrac{1}{g_o} \parallel \dfrac{g_i}{g_r g_f}} C_3 \left(r_C + R_{load} \right) \tag{2.209}$$

$$b_3 = C_s \left(\frac{1}{g_o} \parallel R_{load} \parallel \frac{g_i}{g_r g_f} \right) \frac{L_2}{R_{load}} C_3 \left(r_C + R_{load} \right) \tag{2.210}$$

Applying what we learned with the VM cases to determine the zeroes, the degenerate case of the nulled response V_T with a current source is similar to replacing the source with a short circuit. In this case, if you short the I_T generator in Figure 2.107, the circuit returns to the natural structure of Figure 2.80. It implies that the numerator we want for Z_{in} is already determined by (2.146). Our input impedance is thus equal to:

$$Z_{in}(s) = R_0 \frac{N(s)}{D(s)} = R_0 \frac{1 + a_1 s + a_2 s^2 + a_3 s^3}{1 + b_1 s + b_2 s^2 + b_3 s^3} \tag{2.211}$$

With $D(s)$ defined by (2.208) to (2.210) and $N(s)$ by (2.152), (2.157) and (2.160). We have captured all these data into Mathcad® (Figure 2.111) and the magnitude/phase responses from 100 Hz to 1 MHz appear in Figure 2.112. The 180° phase indicates a negative resistance value.

To check these results, we have modified the large-signal current-mode template as shown in Figure 2.113 to plot the input impedance. The dc operating point is automatically provided by the E_1 "op-amp" but ac-isolated form the circuit through the low-pass filter made of L_3-C_3. The plots from Figure 2.114 confirm our equations are correct.

$$R_0 := -\frac{R_1 \cdot \left[\left(\frac{L_1}{R_1 \cdot T_{sw}} + \frac{1}{2} - D \right) \cdot S_n + S_e \cdot (1 - D) \right]}{\left[\left(S_e + \frac{S_n}{2} \right) \cdot D^3 - \left(S_e + \frac{M}{2} \cdot S_n \right) \cdot D^2 + \frac{M}{2} \cdot S_n \cdot D \right] + D \cdot \frac{L_1}{R_1 \cdot T_{sw}} \cdot M \cdot S_n}$$

$$20 \cdot \log\left(-\frac{R_0}{\Omega} \right) = 11.78696 \quad \text{dBohms}$$

$$\tau_1 := C_s \cdot \left[\left(\frac{1}{g_o} \right) \| R_1 \| \left(\frac{g_i}{g_r \cdot g_f} \right) \right] = 0.09889 \mu s$$

$$\tau_2 := \frac{L_1}{\left(\frac{1}{g_o} \right) \| \left(\frac{g_i}{g_r \cdot g_f} \right) + R_1} = 2.40346 \mu s$$

$$\tau_3 := C_1 \cdot \left[r_C + \left(\frac{1}{g_o} \right) \| R_1 \| \left(\frac{g_i}{g_r \cdot g_f} \right) \right] = 0.1076 \, ms$$

$$\tau_{12} := \frac{L_1}{R_1} \qquad \tau_{13} := C_1 \cdot r_C \qquad \tau_{23} := C_1 \cdot (r_C + R_1) \qquad \tau_{123} := C_1 \cdot (r_C + R_1)$$

$$D_1(s) := 1 + s \cdot (\tau_1 + \tau_2 + \tau_3) + s^2 \cdot (\tau_1 \cdot \tau_{12} + \tau_1 \cdot \tau_{13} + \tau_2 \cdot \tau_{23}) + s^3 \cdot \tau_1 \cdot \tau_{12} \cdot \tau_{123}$$

$$R_0 := \frac{1}{g_i + g_f \cdot g_r \cdot \left[\left(\frac{1}{g_o} \right) \| R_1 \right]} = -3.92719 \Omega \qquad \left(\frac{1}{g_i} \right) \| \left[\frac{1}{\left[\left(\frac{1}{g_o} \right) \| R_1 \right] \cdot g_f \cdot g_r} \right] = -3.92719 \Omega$$

$$Z_{in}(s) := R_0 \cdot \frac{1 + a_1 \cdot s + a_2 \cdot s^2 + a_3 \cdot s^3}{D_1(s)}$$

Figure 2.111 A Mathcad® sheet helps plotting the response from these numerous time constants. The numerator in $Z_{in}(s)$ is that determined in the previous CM transfer functions.

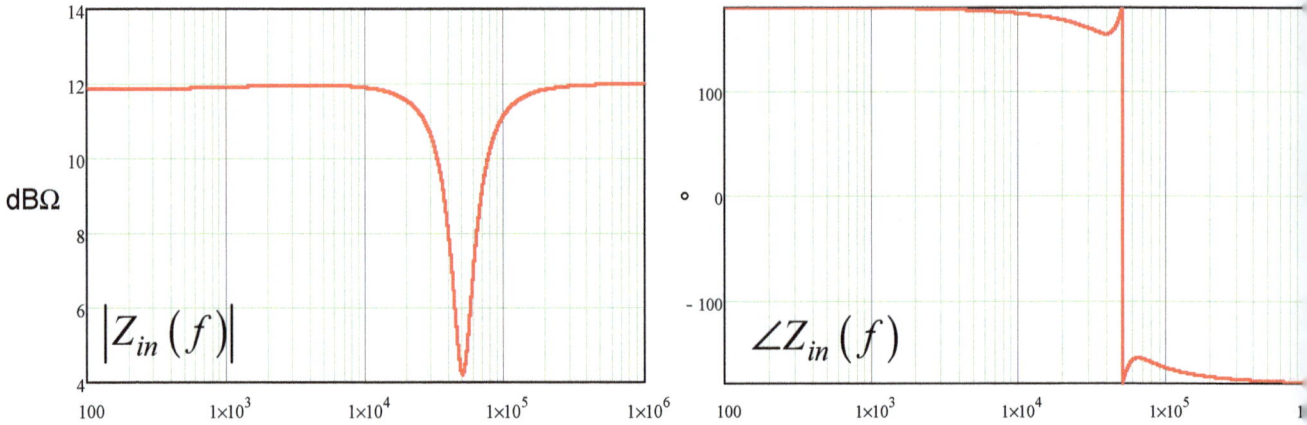

Figure 2.112 The magnitude/phase plots show a flat impedance until a dip occurs while the phase is negative at low frequencies.

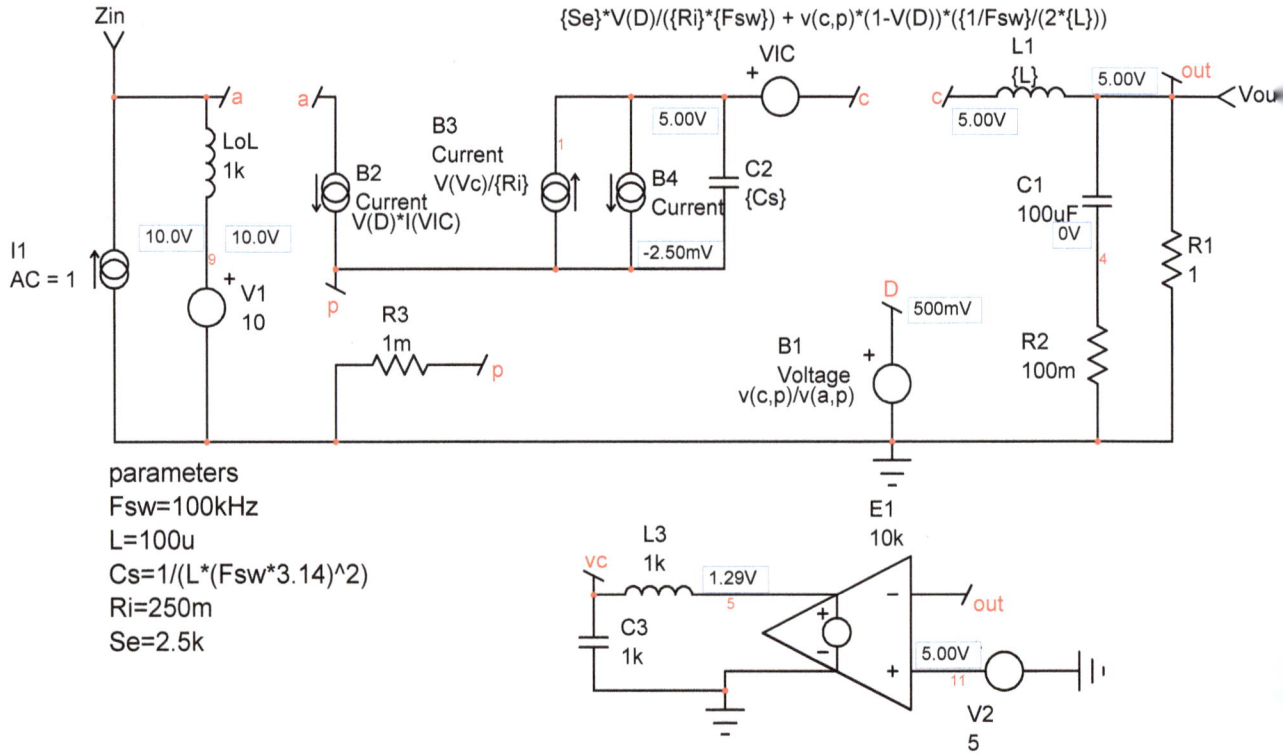

Figure 2.113 The SPICE template built around the large-signal model will tell us if our derivation for Z_{in} is correct.

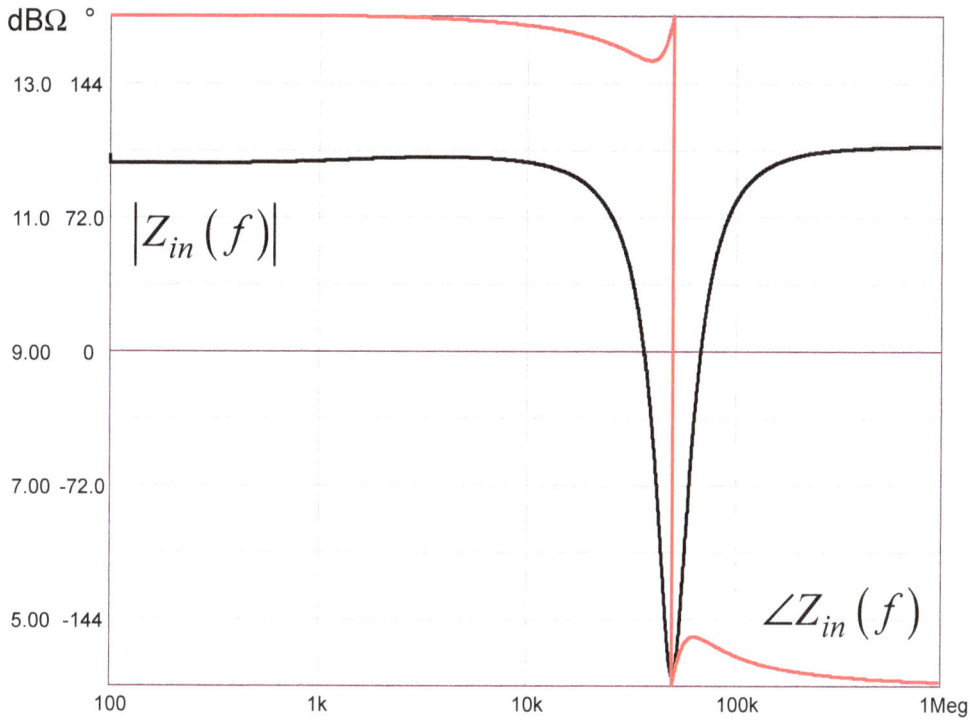

Figure 2.114 A Mathcad® helps plotting the response from these numerous time constants.

$\dfrac{V_{out}(s)}{V_{err}(s)}$ Control to Output	$H_0 \dfrac{1+\dfrac{s}{\omega_z}}{1+\dfrac{s}{\omega_p}} \dfrac{1}{1+\dfrac{s}{\omega_n Q}+\left(\dfrac{s}{\omega_n}\right)^2}$	$\omega_z=\dfrac{1}{r_C C_3}$ $\omega_p \approx \dfrac{1}{R_{load}C_3}+\dfrac{T_{sw}}{L_2 C_3}\left[m_c(1-D)-0.5\right]$	$H_0=\dfrac{R_{load}}{R_i}\dfrac{1}{1+\dfrac{R_{load}T_{sw}}{L_2}\left[m_c(1-D)-0.5\right]}$
$\dfrac{V_{out}(s)}{V_{in}(s)}$ Input to Output	$H_0 \dfrac{1+\dfrac{s}{\omega_z}}{1+\dfrac{s}{\omega_p}} \dfrac{1}{1+\dfrac{s}{\omega_n Q}+\left(\dfrac{s}{\omega_n}\right)^2}$	$\omega_z=\dfrac{1}{r_C C_3}$ $\omega_p \approx \dfrac{1}{R_{load}C_3}+\dfrac{T_{sw}}{L_2 C_3}\left[m_c(1-D)-0.5\right]$	$H_0=\dfrac{D\left[m_c(1-D)-\left(1-\dfrac{D}{2}\right)\right]}{\dfrac{L_2}{R_{load}T_{sw}}+\left[m_c(1-D)-0.5\right]}$
$Z_{in}(s)$ Input impedance	$R_0 \dfrac{1+a_1 s+a_2 s^2+a_3 s^3}{1+s b_1+s^2 b_2+s^3 b_3}$	3rd order polynomial form – see text 3rd order polynomial form – see text	$R_0=-\dfrac{R_{load}\left[\left(\dfrac{L_2}{R_{load}T_{sw}}+0.5-D\right)S_n+S_e(1-D)\right]}{\left[\left(S_e+\dfrac{S_n}{2}\right)D^3-\left(S_e+\dfrac{M}{2}S_n\right)D^2+\dfrac{M}{2}S_n D\right]+MDS_n\dfrac{L_2}{R_{load}T_{sw}}}$
$Z_{out}(s)$ Output impedance	$R_0 \dfrac{1+\dfrac{s}{\omega_z}}{1+\dfrac{s}{\omega_p}}$	$\omega_z=\dfrac{1}{r_C C_3}$ $\omega_p \approx \dfrac{1}{R_{load}C_3}+\dfrac{T_{sw}}{L_2 C_3}\left[m_c(1-D)-0.5\right]$	$R_0=\dfrac{R_{load}}{1+\dfrac{R_{load}}{L_2}T_{sw}\left[m_c(1-D)-0.5\right]}$

C_s is the resonating capacitor, C_3 is the output capacitor with r_C its ESR, L_2 is the inductor and m_c is the slope compensation. $\omega_n=\dfrac{\pi}{T_{sw}}$ $Q=\dfrac{1}{\pi\left[m_c(1-D)-0.5\right]}$ $m_c=1+\dfrac{S_e}{S_n}$

Figure 2.115 The four transfer functions of the CCM CM buck converter are conveniently assembled in this table.

213

2.4 Buck Transfer Functions in DCM – Peak-Current-Mode Control

The buck converter load is now increased to 100 Ω as suggested with (2.48) and the converter enters DCM. We replace CCM CM-PWM switch by its DCM the small-signal version derived in chapter 1.

   The new circuit appears in Figure 2.116 with all the computed small-signal coefficients k_1 to k_8.

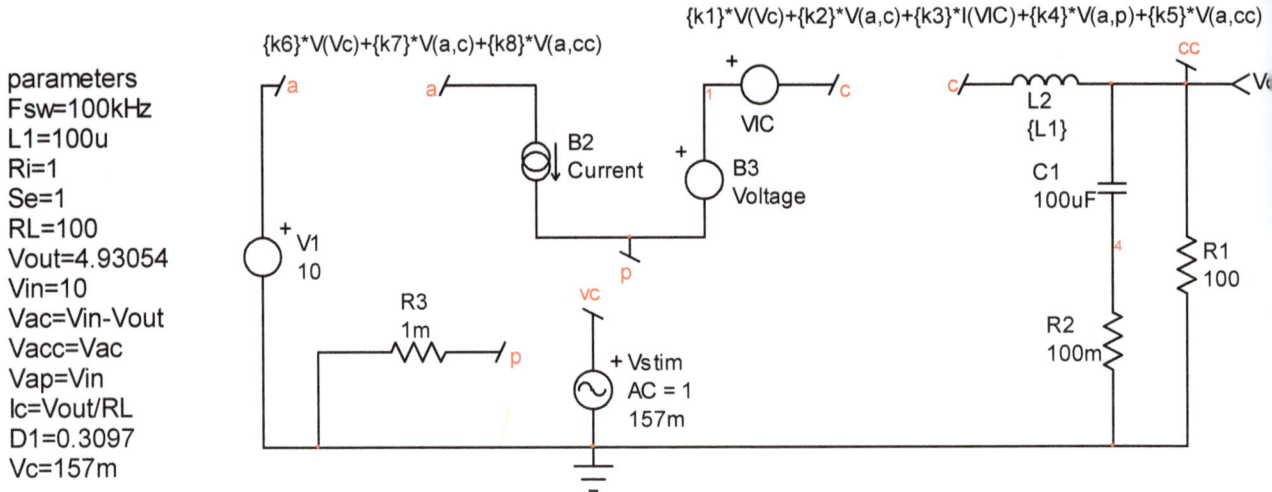

parameters
Fsw=100kHz
L1=100u
Ri=1
Se=1
RL=100
Vout=4.93054
Vin=10
Vac=Vin-Vout
Vacc=Vac
Vap=Vin
Ic=Vout/RL
D1=0.3097
Vc=157m

{k6}*V(Vc)+{k7}*V(a,c)+{k8}*V(a,cc)

{k1}*V(Vc)+{k2}*V(a,c)+{k3}*I(VIC)+{k4}*V(a,p)+{k5}*V(a,cc)

k1=Fsw*L1*Vc*Vap*Vac/(Ic*(L1*Se+Ri*Vacc)^2)
k2=Fsw*L1*Vc^2*Vap/(2*Ic*(L1*Se+Ri*Vacc)^2)
k3=-Fsw*L1*Vc^2*Vap*Vac/(2*Ic^2*(L1*Se+Ri*Vacc)^2)
k4=Fsw*L1*Vc^2*Vac/(2*Ic*(L1*Se+Ri*Vacc)^2)
k5=-Fsw*L1*Vc^2*Vac*Vap*Ri/(Ic*(L1*Se+Ri*Vacc)^3)
k6=Fsw*L1*Vc*Vac/(L1*Se+Ri*Vacc)^2
k7=Fsw*L1*Vc^2/(2*(L1*Se+Ri*Vacc)^2)
k8=-Fsw*L1*Vc^2*Ri*Vac/(L1*Se+Ri*Vacc)^3

Figure 2.116 The buck converter is now operated in DCM and uses the small-signal model derived in chapter 1.

Simplifications immediately take place considering a 0-V input voltage in ac. You can first ignore the left-side current source B_2 for the control-to-output transfer function. Then all sources containing a reference to node a undergo a simplification. The final circuit appears in Figure 2.117. We have checked the magnitude/phase responses of both circuits and they are identical: we can now proceed with this network.

   We start with the quasi-static gain in which the inductor is shorted while the capacitor is open-circuited. Node c and cc are now similar and biased to V_{out}, our desired response.

   Current I_c is equal to the output current:

$$I_c = \frac{V_{out}}{R_{load}} \qquad (2.212)$$

parameters
Fsw=100kHz
L1=100u
Ri=1
Se=1
RL=100
Vout=4.93054
Vin=10
Vac=Vin-Vout
Vap=Vin
Vacc=Vac
Ic=Vout/RL
D1=0.3097
Vc=157m

VIC

L2
{L1}

C1
100uF

R1
100

B3
Voltage

VC

+ Vstim
AC = 1
157m

R2
100m

CC

Vout

{k1}*V(Vc)-{k2}*V(c)+{k3}*I(VIC)-{k5}*V(cc)

k1=Fsw*L1*Vc*Vap*Vac/(Ic*(L1*Se+Ri*Vacc)^2)
k2=Fsw*L1*Vc^2*Vap/(2*Ic*(L1*Se+Ri*Vacc)^2)
k3=-Fsw*L1*Vc^2*Vap*Vac/(2*Ic^2*(L1*Se+Ri*Vacc)^2)
k4=Fsw*L1*Vc^2*Vac/(2*Ic*(L1*Se+Ri*Vacc)^2)
k5=-Fsw*L1*Vc^2*Vac*Vap*Ri/(Ic*(L1*Se+Ri*Vacc)^3)
k6=Fsw*L1*Vc*Vac/(L1*Se+Ri*Vacc)^2
k7=Fsw*L1*Vc^2/(2*(L1*Se+Ri*Vacc)^2)
k8=-Fsw*L1*Vc^2*Ri*Vac/(L1*Se+Ri*Vacc)^3

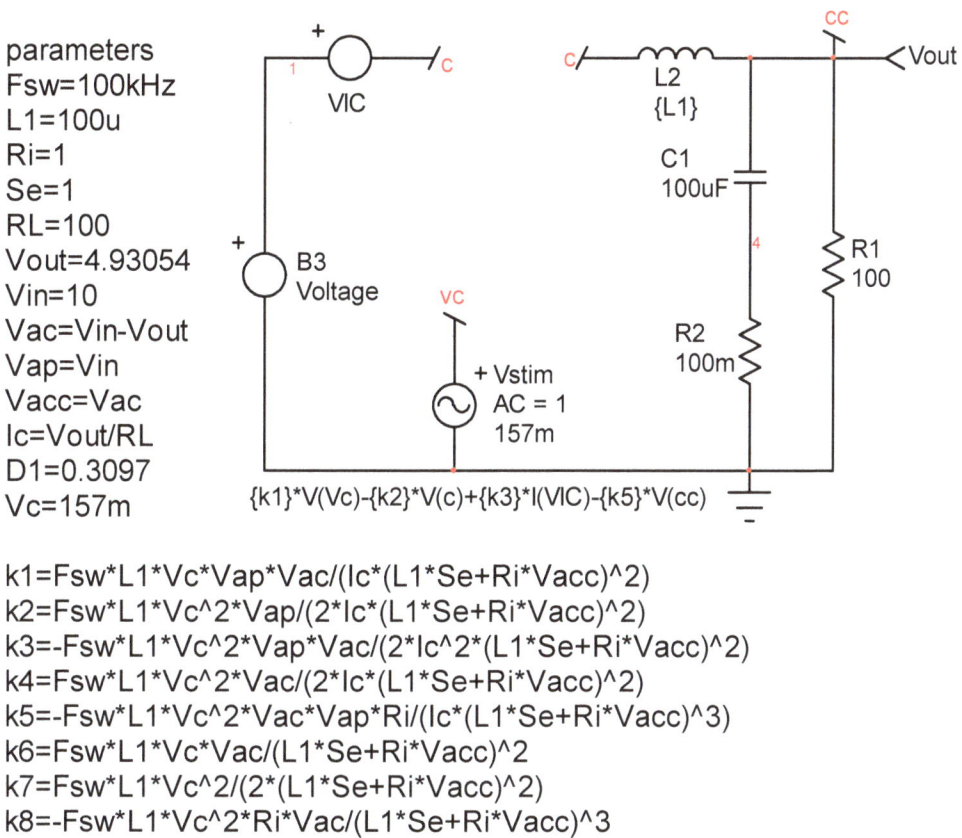

Figure 2.117 The DCM CM buck converter small-signal model is simplified when node *a* is grounded.

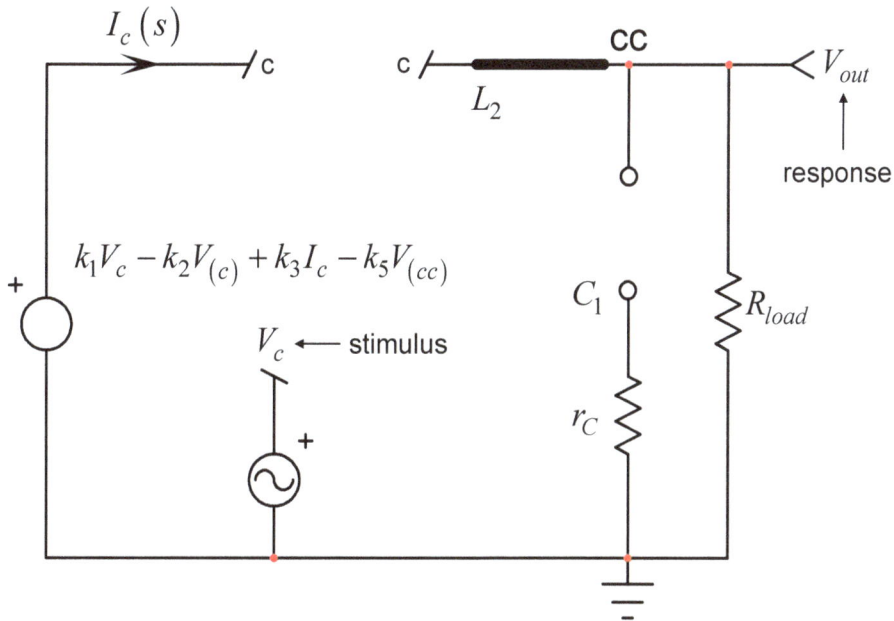

$I_c(s)$

c

c

CC

V_{out}

L_2

$k_1 V_c - k_2 V_{(c)} + k_3 I_c - k_5 V_{(cc)}$

$V_c \leftarrow$ stimulus

C_1

r_C

R_{load}

response

Figure 2.118 Shorting the inductor and open-circuiting the capacitor is the way to go for determining the quasi-static gain.

Simply write the expression determining V_{out} in which I_c is replaced by (2.212):

$$V_{out} = k_1 V_c - k_2 V_{out} + k_3 \frac{V_{out}}{R_{load}} - k_5 V_{out} \tag{2.213}$$

Solving for V_{out} and rearranging gives:

$$H_0 = \frac{V_{out}}{V_c} = \frac{k_1}{k_2 + k_5 - \dfrac{k_3}{R_{load}} + 1} \tag{2.214}$$

In the literature [4], this expression has been approximately factored as:

$$H_0 \approx \frac{2 m_c M V_{in}}{D} \frac{1 - M}{2 m_c - (2 + mc) M} \frac{1}{S_n m_c T_{sw}} \tag{2.215}$$

In this expression, S_n is the inductor upslope and m_c is the amount of compensation ramp while M is classically V_{out}/V_{in} .

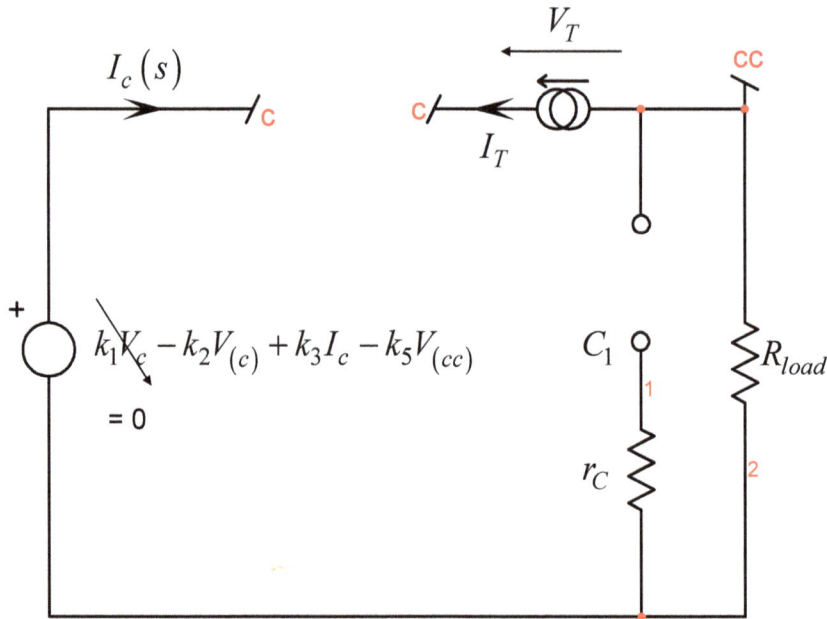

Figure 2.119 Installing a test generator I_T across the inductor lets us find the resistance "seen" from its terminals.

Now that the dc gain has been determined, the resistance driving inductor L_2 is obtained by installing a test generator I_T as shown in Figure 2.119. In this sketch, the original excitation source V_c has been reduced to zero, bringing some simplifications in the equations. Node cc is biased to:

$$V_{(cc)} = -R_{load} I_T \tag{2.216}$$

While the voltage at node c is that of node cc plus the voltage across the current source, V_T:

$$V_{(c)} = V_T - I_T R_{load} \qquad (2.217)$$

Finally, observing that current I_c is actually $-I_T$, we can write:

$$-I_T k_3 + k_5 R_{load} I_T - k_2 (V_T - R_{load} I_T) + I_T R_{load} = V_T \qquad (2.218)$$

Solving for V_T and rearranging to unveil a resistance leads us to the time constant involving L_2:

$$\tau_2 = \frac{L_2}{\dfrac{R_L (1 + k_2 + k_5) - k_3}{1 + k_2}} \qquad (2.219)$$

We are now interested in the time constant involving capacitor C_1 and the sketch evolves to that of Figure 2.120. From the circuit, we can see that r_C is in series with C_1. We can thus temporarily exclude it from the circuit and add it back later when we will find the intermediate result.

Doing this and with L_2 in its dc state (a short circuit), node c and cc share a common potential which is V_T. We thus have:

$$V_T = V_{(c)} = V_{(cc)} = (I_T + I_c) R_{load} \qquad (2.220)$$

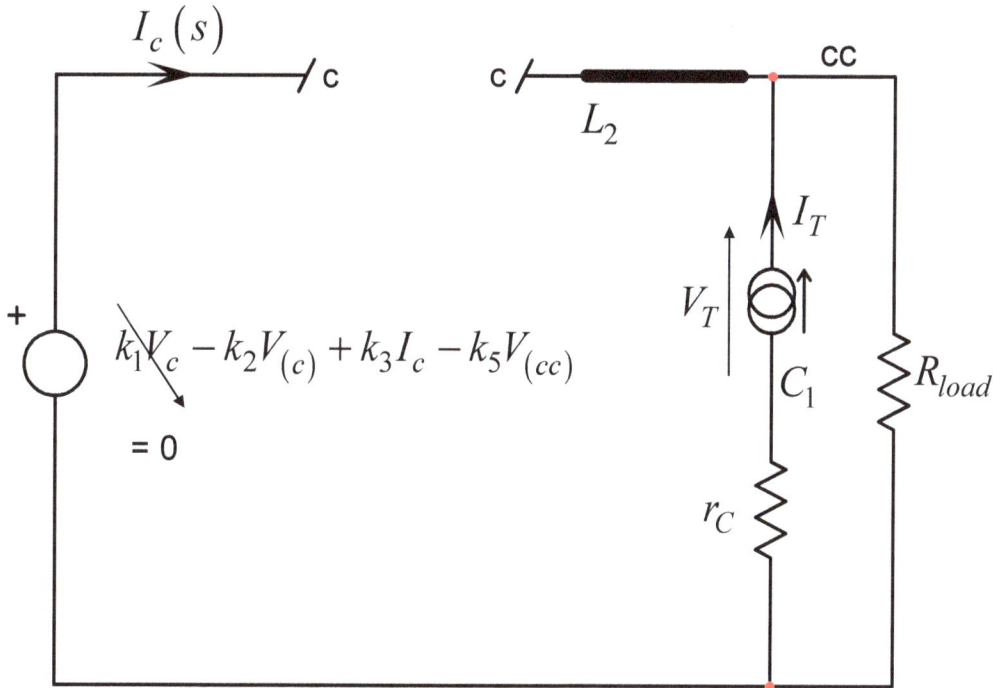

Figure 2.120 In this mode, the inductor is replaced by a short circuit and we want the resistance driving capacitor C_1.

Current I_c is nothing else than:

$$I_c = \frac{V_T}{R_{load}} - I_T \qquad (2.221)$$

It leads us to:

$$V_T = k_3\left(\frac{V_T}{R_{load}} - I_T\right) - k_5 V_T - k_2 V_T \qquad (2.222)$$

Solving for V_T, rearranging the result in the form of a resistance to which we can add r_C, we have:

$$\tau_1 = C_1\left(\frac{k_3}{\dfrac{k_3}{R_{load}} - k_5 - k_2 - 1} + r_C\right) \qquad (2.223)$$

We have b_1 expressed as:

$$b_1 = \tau_1 + \tau_2 \qquad (2.224)$$

For the final time constant, we will set L_2 in its high-frequency state (it simplifies the circuit as I_c equals 0 A) and from Figure 2.120 in which L_2 is open circuited, by inspection, the time constant is equal to:

$$\tau_1^2 = C_1\left(r_C + R_{load}\right) \qquad (2.225)$$

We are done and can express the denominator of our transfer function as:

$$D(s) = 1 + s(\tau_1 + \tau_2) + s^2 \tau_2 \tau_1^2$$

$$= 1 + s\left[C_1\left(\frac{k_3}{\dfrac{k_3}{R_{load}} - k_5 - k_2 - 1} + r_C\right) + \frac{L_2}{\dfrac{R_L(1 + k_2 + k_5) - k_3}{1 + k_2}}\right] + s^2 \frac{L_2}{\dfrac{R_L(1 + k_2 + k_5) - k_3}{1 + k_2}} C_1\left(r_C + R_{load}\right) \qquad (2.226)$$

For the zero, we can see from Figure 2.117 that if L_2 is set in its high-frequency state (open-circuited), the stimulus given by V_c cannot propagate through the circuit and give a response: there is no zero associated with L_2. The only zero classically occurs with the transformed short given by the series connection of r_C and C_1:

$$\omega_z = \frac{1}{r_C C_1} \qquad (2.227)$$

The complete control-to-output transfer function of the DCM CM buck converter is expressed as:

$$H(s) = H_0 \frac{1 + \dfrac{s}{\omega_z}}{D(s)} \qquad (2.228)$$

With H_0 given by (2.214) and the denominator $D(s)$ determined via (2.226).

In the literature [4], the author applied the low-Q approximation to the second-order denominator splitting the poles into one dominant at low frequency while the second lies in the upper portion of the spectrum:

$$H(s) \approx H_0 \frac{1 + \dfrac{s}{\omega_z}}{\left(1 + \dfrac{s}{\omega_{p_1}}\right)\left(1 + \dfrac{s}{\omega_{p_2}}\right)} \tag{2.229}$$

With H_0 defined by (2.215) and the poles as:

$$\omega_{p_1} = \frac{1}{R_{load}C_1} \frac{2m_c - (2 + m_c)M}{m_c(1 - M)} \tag{2.230}$$

$$\omega_{p_2} = 2F_{sw}\left(\frac{M}{D}\right)^2 \tag{2.231}$$

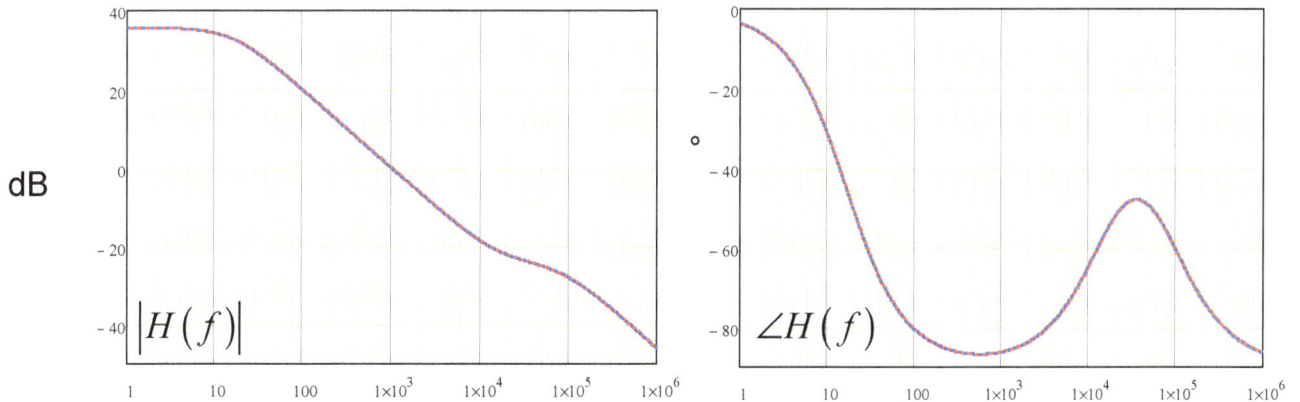

Figure 2.121 The control-to-output transfer function of the CM DCM buck converter is that of a damped second-order system.

The frequency response appears in Figure 2.121 and shows identical responses between the two expressions.

Despite DCM operation, the buck converter operated in current mode remains a second-order system heavily damped.

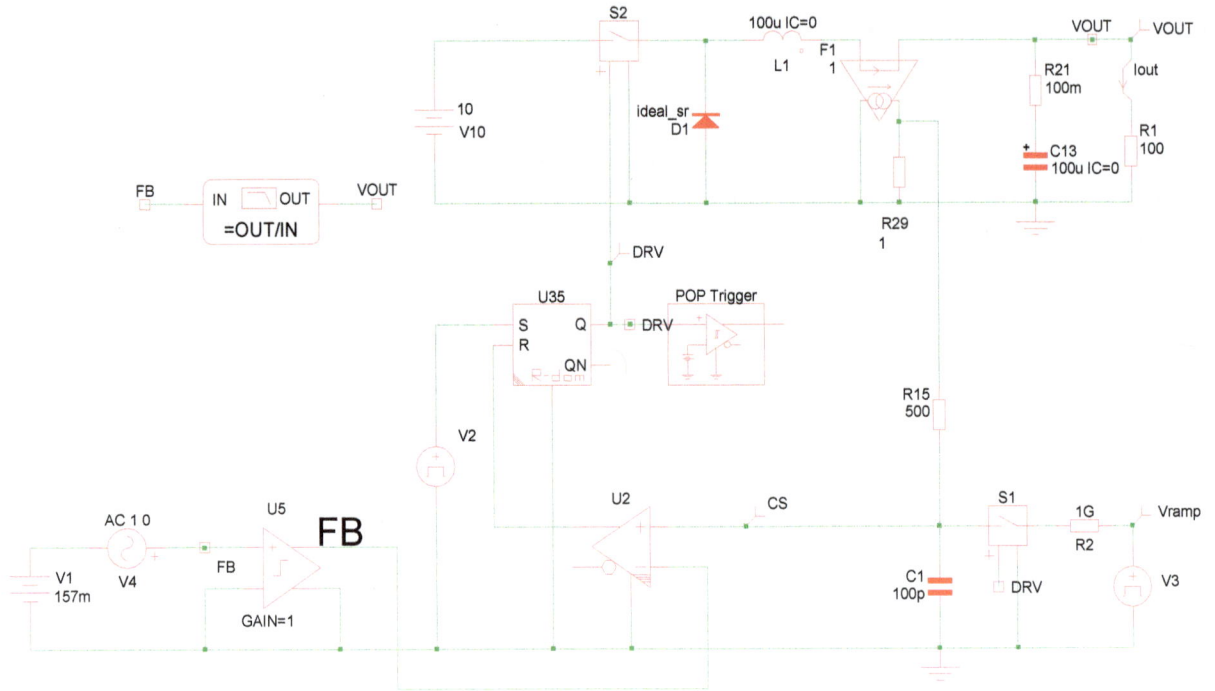

Figure 2.122 We have captured a simulation circuit in SIMPLIS® to test our equations.

We have run the SIMPLIS® template from Figure 2.122 whose load forces DCM operation. The frequency response is given in Figure 2.123 and confirms our Mathcad® results.

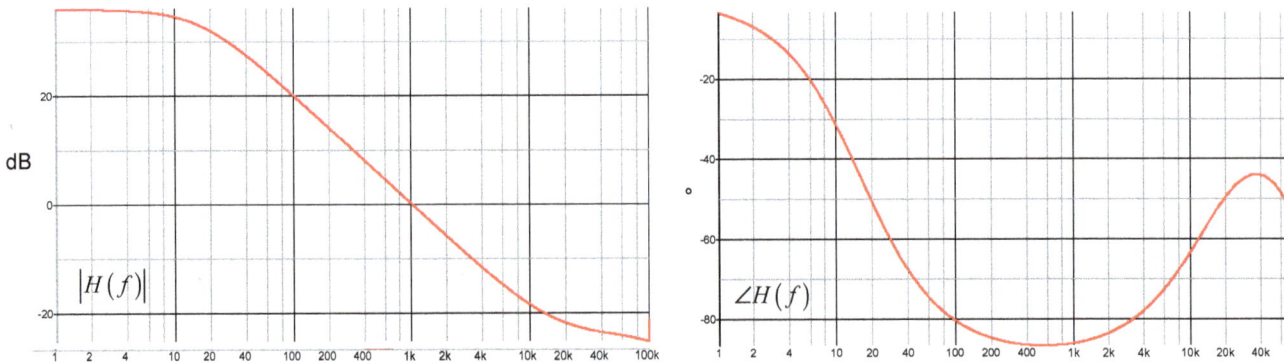

Figure 2.123 The simulation results confirm our analysis with a dc gain of around 36 dB.

It is now interesting to look at (2.230). We see a pole determined by the load and the output capacitor affected by a coefficient depending on M and m_c. In absence of stabilization ramp ($m_c = 1$), the term reduces to:

$$\omega_{P_1} = \frac{1}{R_{load}C_1} \frac{2-3M}{1-M} \qquad (2.232)$$

As you can see, this pole will move up and down in frequency depending on the adopted dc transfer ratio M. We can plot the right-side term of (2.232) and see how M affects the pole set by R_{load} and C_1. This is what you can see in Figure 2.124. For $M = 0.666$, the pole hits the origin. This is confirmed by the SIMPLIS® template of Figure 2.122 when you tweak the control voltage V_1 to 169.4 mV. In this mode, the gain is very high and the converter becomes very sensitive to perturbations (Figure 2.125).

If you keep increasing M (by increasing the control voltage V_c), the coefficient turns negative and the pole jumps into the right half-plane; the converter becomes unstable. Consequently, you still need to inject slope compensation in the buck converter operated in current mode even when it transitions to the discontinuous mode.

For a given M, you will compute m_c so that the right term in (2.230) remains positive and different from zero.

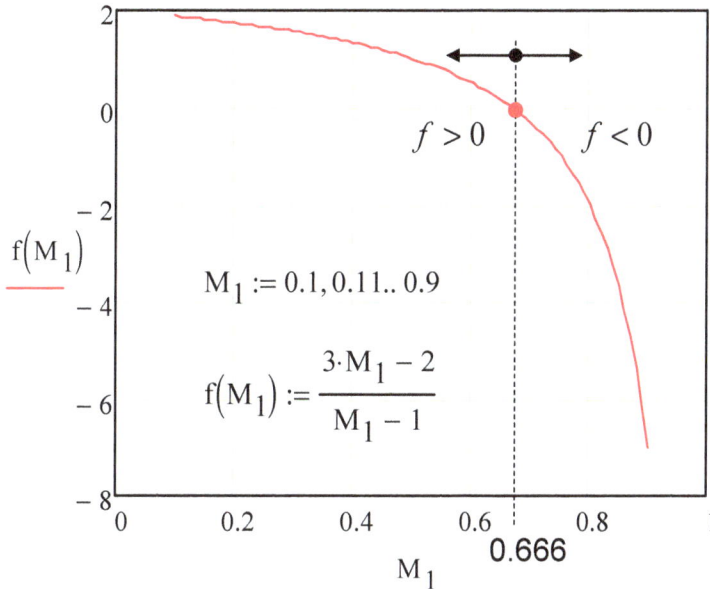

Figure 2.124 The pole moves to the origin for $M = 0.666$ and jumps into the right half-plane above this value.

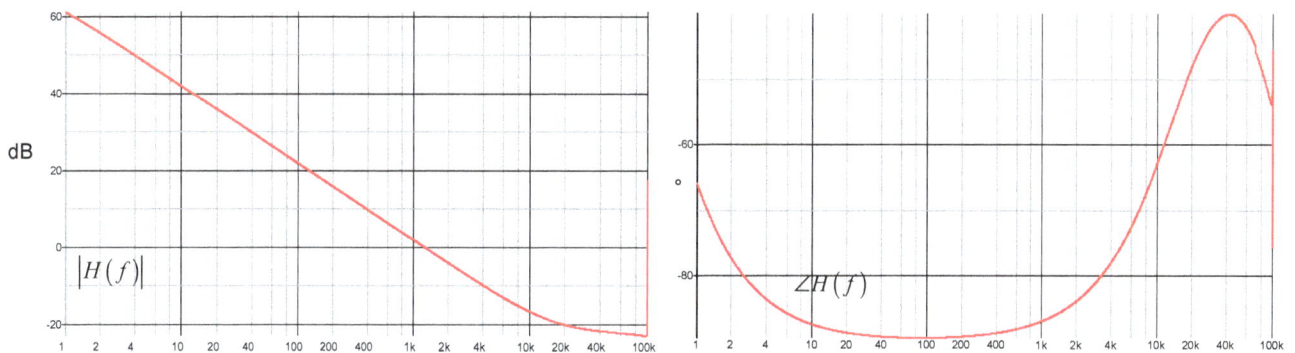

Figure 2.125 The simulation results confirms the pole moving to the origin for M approaching 0.666.

Like we did for the DCM voltage-mode case, I have gathered all the coefficients calculated by Mathcad® in Figure 2.126 so that you can check your own calculations.

$$F_{sw} := 100 \text{kHz} \qquad L_1 := 100 \mu\text{H} \qquad R_L := 100\Omega \qquad R_i := 1\Omega \qquad S_e := 1\frac{V}{s} \qquad V_c := 157\text{mV}$$

$$V_{in} := 10V \qquad V_{out} := 5V \qquad I_c := \frac{V_{out}}{R_L} = 0.05A \qquad V_{ap} := V_{in} \qquad T_{sw} := \frac{1}{F_{sw}}$$

$$V_{ac} := V_{in} - V_{out} = 5V \qquad V_{acc} := V_{ac} \qquad m_c := 1 + \frac{S_e}{S_n} = 1.00002 \qquad S_n := \frac{V_{in} - V_{out}}{L_1} \cdot R_i = 50 \cdot \frac{kV}{s}$$

$$M := \frac{V_{out}}{V_{in}} \qquad F_m := \frac{1}{S_n \cdot m_c \cdot T_{sw}} = 1.99996\frac{1}{V} \qquad r_C := 0.1\Omega \qquad C_2 := 100\mu\text{F} \qquad S_{off} := \frac{V_{out}}{L_1} \cdot R_i = 50 \cdot \frac{kV}{s}$$

$$D_1 := \frac{F_{sw} \cdot V_c}{R_i} \cdot \frac{1}{\frac{V_{ac}}{L_1} + \frac{S_e}{R_i}} = 0.31399 \qquad D_2 := \frac{2 \cdot L_1 \cdot F_{sw} \cdot I_c}{D_1 \cdot V_{ac}} - D_1 = 0.32296$$

$$N_1 := \frac{D_1}{D_1 + D_2} = 0.49296 \qquad \frac{F_{sw} \cdot L_1 \cdot V_c^2 \cdot V_{ac}}{2 \cdot I_c \cdot (L_1 \cdot S_e + R_i \cdot V_{ac})^2} = 0.49296 \qquad \|(x,y) := \frac{x \cdot y}{x + y}$$

$$k_1 := \frac{F_{sw} \cdot L_1 \cdot V_c \cdot V_{ac} \cdot V_{ap}}{I_c \cdot (L_1 \cdot S_e + R_i \cdot V_{acc})^2} = 62.79749 \qquad I_a := I_c \cdot \frac{F_{sw} \cdot L_1 \cdot V_c^2 \cdot V_{ac}}{2 \cdot I_c \cdot (L_1 \cdot S_e + R_i \cdot V_{acc})^2} = 0.02465A$$

$$k_2 := \frac{F_{sw} \cdot L_1 \cdot V_c^2 \cdot V_{ap}}{2 \cdot I_c \cdot (L_1 \cdot S_e + R_i \cdot V_{acc})^2} = 0.98592 \qquad k_6 := \frac{F_{sw} \cdot L_1 \cdot V_c \cdot V_{ac}}{(L_1 \cdot S_e + R_i \cdot V_{acc})^2} = 0.31399\frac{1}{\Omega}$$

$$k_3 := -\frac{F_{sw} \cdot L_1 \cdot V_c^2 \cdot V_{ac} \cdot V_{ap}}{2 \cdot I_c^2 \cdot (L_1 \cdot S_e + R_i \cdot V_{acc})^2} = -98.59206\Omega \qquad k_7 := \frac{F_{sw} \cdot L_1 \cdot V_c^2}{2 \cdot (L_1 \cdot S_e + R_i \cdot V_{acc})^2} = 4.9296 \times 10^{-3}\frac{1}{\Omega}$$

$$k_4 := \frac{F_{sw} \cdot L_1 \cdot V_c^2 \cdot V_{ac}}{2 \cdot I_c \cdot (L_1 \cdot S_e + R_i \cdot V_{acc})^2} = 0.49296 \qquad k_8 := -\frac{F_{sw} \cdot L_1 \cdot R_i \cdot V_c^2 \cdot V_{ac}}{(L_1 \cdot S_e + R_i \cdot V_{acc})^3} = -9.85901 \times 10^{-3}\frac{1}{\Omega}$$

$$k_5 := -\frac{F_{sw} \cdot L_1 \cdot R_i \cdot V_c^2 \cdot V_{ac} \cdot V_{ap}}{I_c \cdot (L_1 \cdot S_e + R_i \cdot V_{acc})^3} = -1.9718$$

Figure 2.126 This sheet shows how all the DCM coefficients were calculated.

2.4.1 Input to Output

The input-to-output transfer function of the DCM current-mode buck converter can be derived reusing the model shown in Figure 2.116 in which the stimulus becomes V_{in}. The control voltage V_c is constant (0 V in ac) and, again, the circuit can be simplified (Figure 2.127) considering L_2 shorted and C_1 open-circuited.

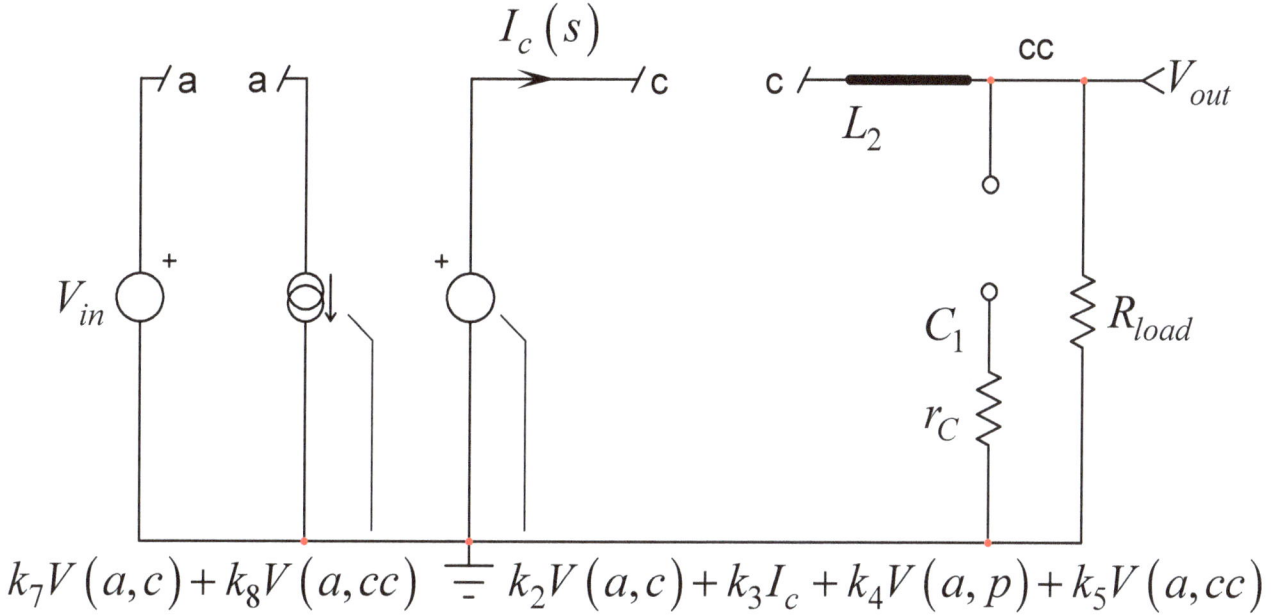

$$k_7 V(a,c) + k_8 V(a,cc) \doteq k_2 V(a,c) + k_3 I_c + k_4 V(a,p) + k_5 V(a,cc)$$

Figure 2.127 Shorting the inductor and opening the capacitor while V_{in} becomes the stimulus lets us determine the input-to-output transfer function quasi-static gain.

Observing the network, we can see that node a is connected to V_{in} while nodes c and cc are both V_{out}. The output voltage is thus defined as:

$$V_{out} = k_2 V_{in} - k_2 V_{out} + k_3 \frac{V_{out}}{R_{load}} + k_4 V_{in} - k_5 V_{in} - k_5 V_{out} \tag{2.233}$$

Solving for V_{out} and rearranging gives:

$$H_0 = \frac{k_2 + k_4 + k_5}{k_2 + k_5 - \dfrac{k_3}{R_{load}} + 1} \tag{2.234}$$

If we now reduce V_{in} to 0 V, the circuit returns to its natural structure and we can reuse the denominator from (2.226).

The only zero in this circuit is, again, given by the series connection of r_C and C_1:

$$\omega_z = \frac{1}{r_C C_1} \tag{2.235}$$

The input-to-output transfer function is determined as:

$$H(s) = H_0 \frac{1 + \dfrac{s}{\omega_z}}{D(s)} \tag{2.236}$$

In this expression, H_0 is defined by (2.234) while $D(s)$ is obtained using (2.226) or its simplified form in (2.229). The dynamic response appears in Figure 2.128 and shows a poor input rejection compared to the CCM version of this buck CM converter. The phase is still inverted, implying a decreasing output voltage when the input voltage increases.

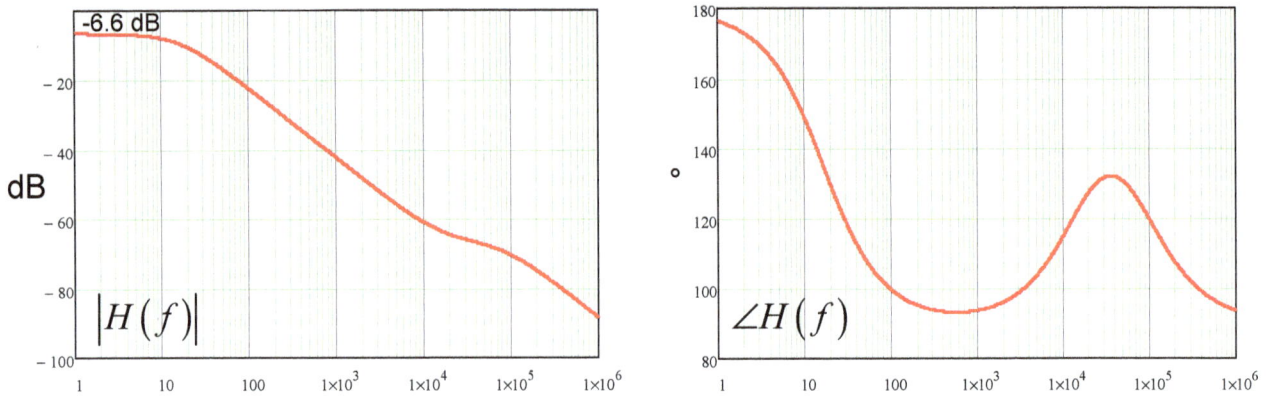

Figure 2.128 The response shows a poor input rejection compared to the CCM version.

Finally, we have run a SPICE simulation with Figure 2.129 template and the results from Figure 2.130 confirm our derivation exercise.

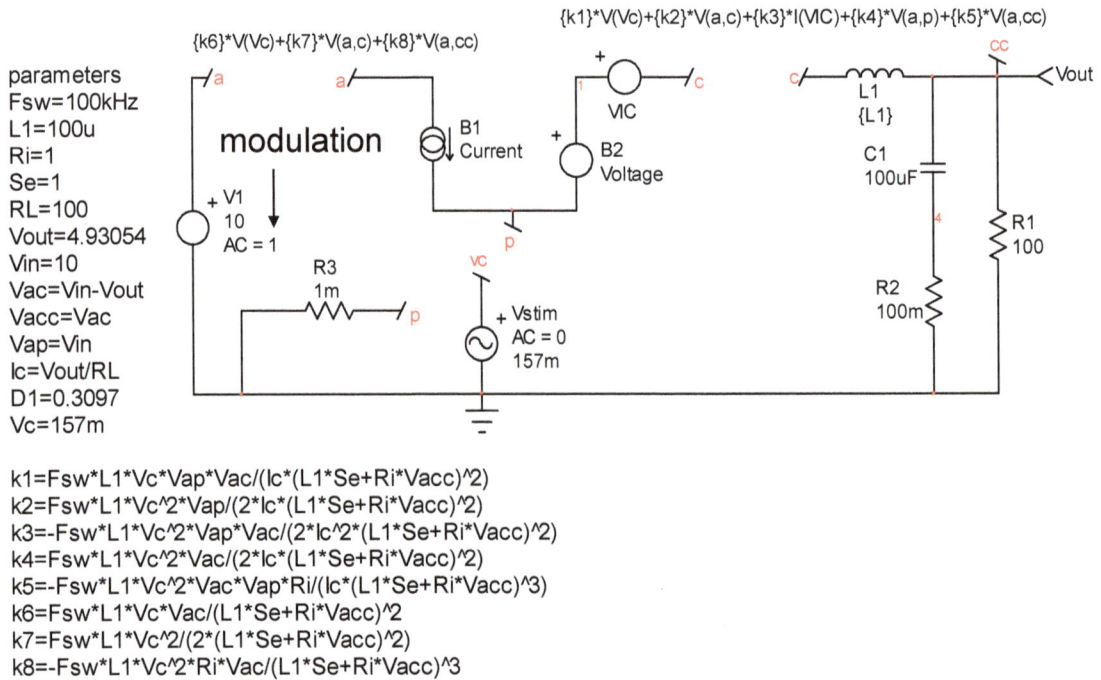

```
k1=Fsw*L1*Vc*Vap*Vac/(Ic*(L1*Se+Ri*Vacc)^2)
k2=Fsw*L1*Vc^2*Vap/(2*Ic*(L1*Se+Ri*Vacc)^2)
k3=-Fsw*L1*Vc^2*Vap*Vac/(2*Ic^2*(L1*Se+Ri*Vacc)^2)
k4=Fsw*L1*Vc^2*Vac/(2*Ic*(L1*Se+Ri*Vacc)^2)
k5=-Fsw*L1*Vc^2*Vac*Vap*Ri/(Ic*(L1*Se+Ri*Vacc)^3)
k6=Fsw*L1*Vc*Vac/(L1*Se+Ri*Vacc)^2
k7=Fsw*L1*Vc^2/(2*(L1*Se+Ri*Vacc)^2)
k8=-Fsw*L1*Vc^2*Ri*Vac/(L1*Se+Ri*Vacc)^3
```

Figure 2.129 The SPICE template ac-modulates the input voltage while the control voltage V_c remains constant.

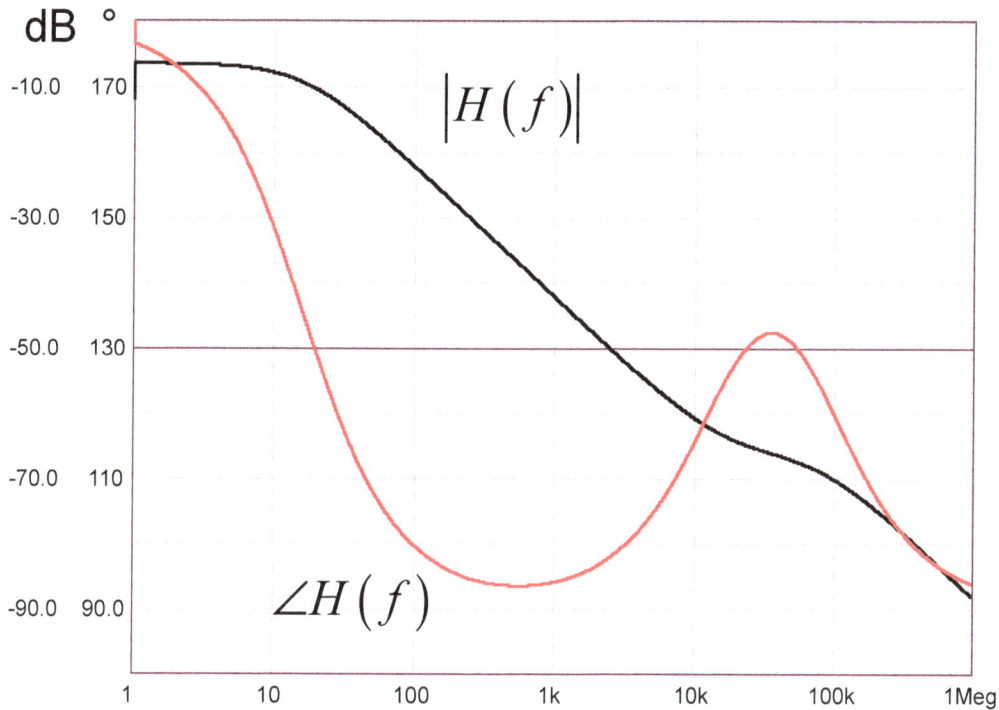

Figure 2.130 The resulting curves confirm the Mathcad® plots.

2.4.2 Output Impedance

The output impedance is determined by reusing Figure 2.117 circuit (in which $V_{in} = 0$) and setting V_c to 0 V also. A test generator is installed across the output load as shown in Figure 2.131 in which equations have already been simplified considering the zeroed node a and V_c.

The dc resistance R_0 is determined by shorting inductor L_2 and open-circuiting capacitor C_1 as shown in Figure 2.132. In this configuration, owing to the short circuit of L_2, nodes c and cc are at V_T, the variable we want to determine. Also, we can see that considering R_{load} connected across the test generator, implies that the final result will include this resistance in a parallel arrangement with an intermediate expression.

We can thus temporarily disconnect R_{load} and work on a simpler circuit in which I_T and I_c are equal but flow in opposite directions:

$$-V_T k_2 - k_3 I_T - k_5 V_T = V_T \qquad (2.237)$$

Solving for V_T and rearranging with I_T and R_{load}, we find:

$$R_0 = \left(-\frac{k_3}{k_2 + k_5 + 1} \right) \| R_{load} \qquad (2.238)$$

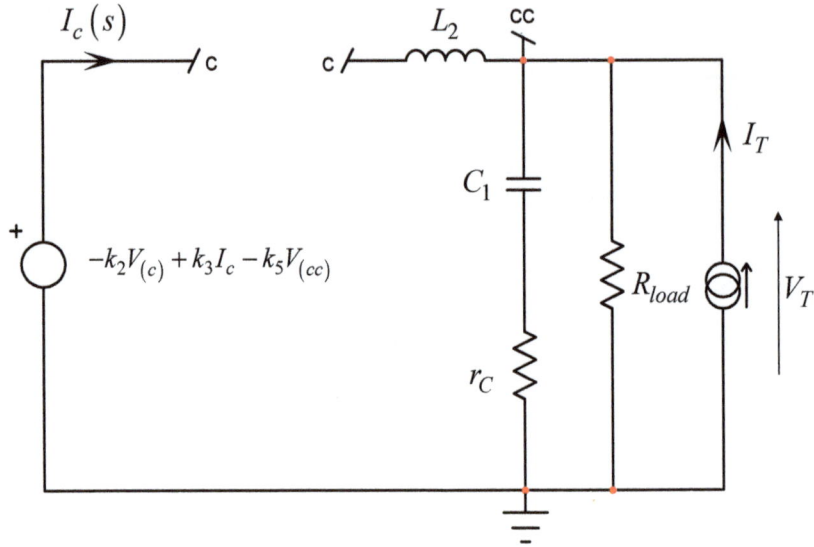

Figure 2.131 The study of the output impedance requires the installation of a test generator I_T across the load.

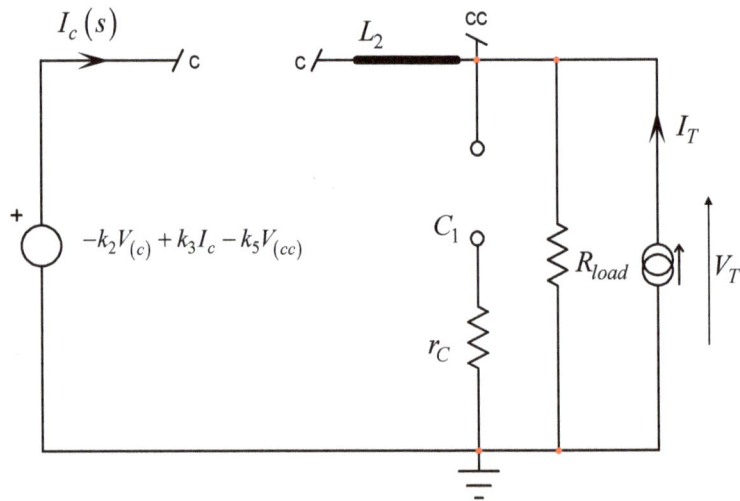

Figure 2.132 The dc resistance R_0 is quickly determined when shorting inductor L_2 and open-circuiting capacitor C_2.

We are familiar with what happens if we suppress the excitation (the current source I_T): the circuit returns to its natural structure and we can reuse the denominator $D(s)$ already determined in (2.226) or in (2.229) for a simplified version.

The zeroes are determined observing Figure 2.133. What conditions in this circuit could null the response across current source I_T? The obvious one is the transformed short circuit made by r_C and C_1.

We immediately have:

$$\omega_{z_1} = \frac{1}{r_C C_1} \tag{2.239}$$

To check if a second zero exists, we turn L_2 in its high-frequency state and check if the response still exists in this mode while I_T injects current. This is the case for L_2 alone but also when both L_2-C_1 are in their high-frequency state: we have a double zero in the numerator. To determine the second zero involving L_2, what condition implies a null at node cc? If the voltage developed across sL_2 equals the voltage generated by the left-side source.

In other words:

$$-k_2 V_{(c)} + k_3 I_c - k_5 V_{(cc)} - I_c sL_2 = 0 \tag{2.240}$$

Node cc is 0 V so the associated term disappears from the equation. The voltage at node c is:

$$V_{(c)} = I_c sL_2 \tag{2.241}$$

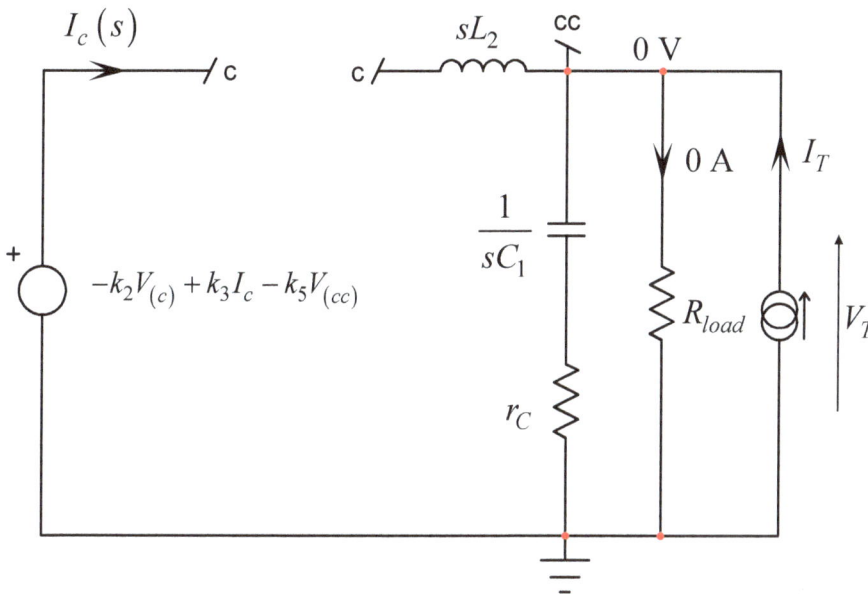

Figure 2.133 The transformed circuit shows two possible zeroes.

After factoring I_c, the equation updates to:

$$I_c \left[k_3 - sL_2 (1 + k_2) \right] = 0 \tag{2.242}$$

The root of this equation is immediate and equals:

$$s_{z_2} = \frac{k_3}{(1 + k_2) L_2} \tag{2.243}$$

k_3 is a negative value so the root is also negative, leading to a zero determined as:

$$\omega_{z_2} = -\frac{k_3}{(1 + k_2) L_2} \tag{2.244}$$

Replacing k_3 and k_2 by their definitions then rearranging leads to:

$$\omega_{z_2} = \frac{F_{sw} R_{load}{}^2 V_c{}^2 (1-M)}{\left(2\left[L_2 S_e + R_i \left(V_{in} - M V_{in}\right)\right]^2 M + F_{sw} L_2 R_{load} V_c{}^2\right) M} \qquad (2.245)$$

We have the complete definition for the output impedance of the DCM-operated current-mode buck defined as:

$$Z_{out}(s) = R_0 \frac{\left(1 + \dfrac{s}{\omega_{z_1}}\right)\left(1 + \dfrac{s}{\omega_{z_2}}\right)}{D(s)} \qquad (2.246)$$

With R_0, the two zeroes and the denominator respectively determined by (2.238), (2.239), (2.245) and (2.226). If we consider that the second zero and the second pole determined in (2.231) only affect the high-frequency response, we can advantageously approximate (2.246) as:

$$Z_{out}(s) \approx R_0 \frac{1 + \dfrac{s}{\omega_z}}{1 + \dfrac{s}{\omega_p}} \qquad (2.247)$$

With ω_z determined by (2.239) and ω_p by (2.230). Figure 2.134 shows dynamic response delivered by (2.246) and (2.247). There is no significant difference.

Figure 2.135 represents the SPICE template built around the large-signal model while Figure 2.136 confirms our calculations.

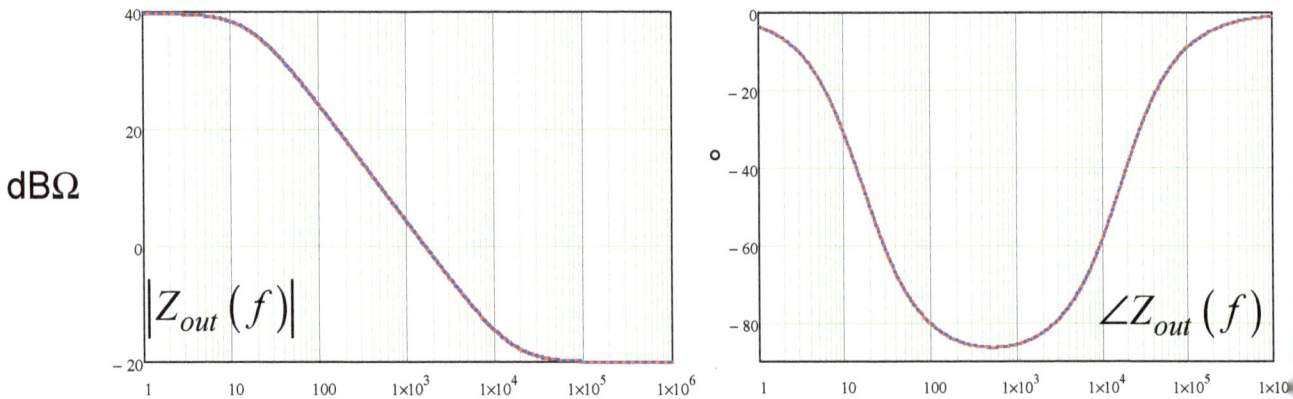

Figure 2.134 The output impedance of the DCM CM-buck exhibits two zeroes in the transfer function.

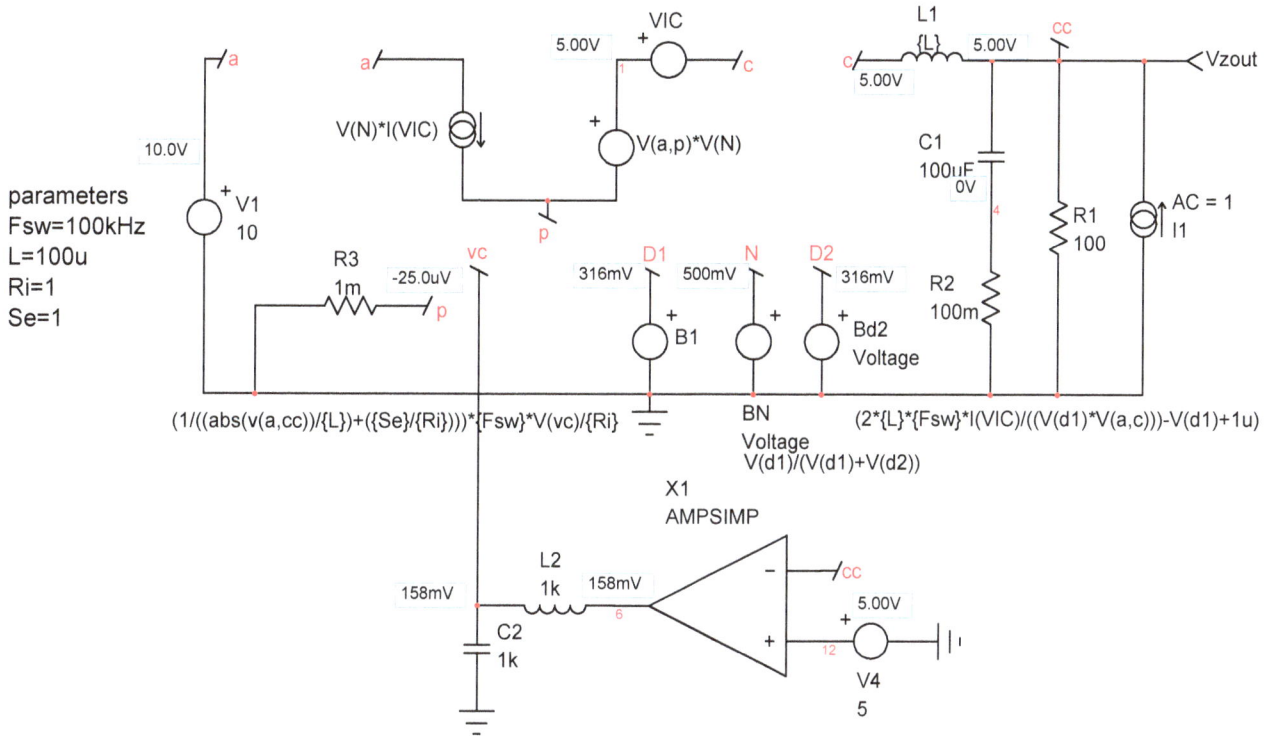

Figure 2.135 The large-signal model uses an op-amp to stabilize the operating point at exactly 5 V.

The parameters shown in the figure:

parameters
Fsw=100kHz
L=100u
Ri=1
Se=1

$(1/((abs(v(a,cc))/\{L\})+(\{Se\}/\{Ri\})))*\{Fsw\}*V(vc)/\{Ri\}$

$V(N)*I(VIC)$

$V(a,p)*V(N)$

BN
Voltage
$V(d1)/(V(d1)+V(d2))$

$(2*\{L\}*\{Fsw\}*I(VIC)/((V(d1)*V(a,c)))-V(d1)+1u)$

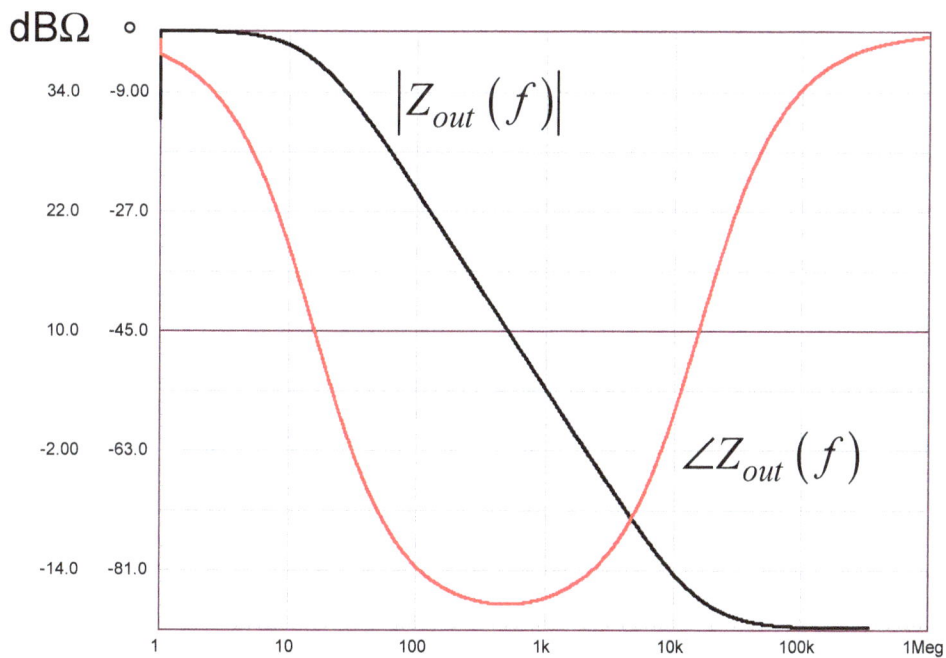

$|Z_{out}(f)|$

$\angle Z_{out}(f)$

Figure 2.136 The magnitude and phase responses are identical to those plotted by Mathcad®.

2.4.3 Input Impedance

The input impedance is classically determined by biasing the input with a test generator I_T producing a response V_T. This is what is shown in Figure 2.137 where simplifications can occur considering the zeroed control voltage V_c. Also, the voltage at terminal a is V_T while the injected current I_T is absorbed by the current source. In dc, the inductor is shorted and the capacitor open-circuited. It leads to the Figure 2.138 circuit.

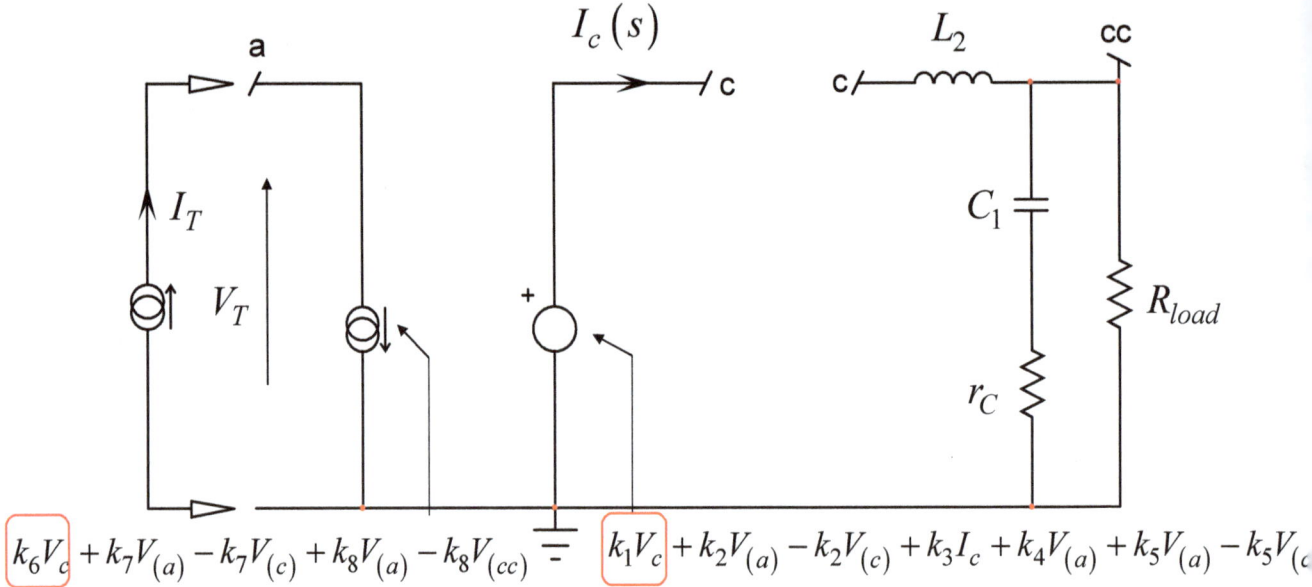

$$k_6 V_c + k_7 V_{(a)} - k_7 V_{(c)} + k_8 V_{(a)} - k_8 V_{(cc)} \qquad k_1 V_c + k_2 V_{(a)} - k_2 V_{(c)} + k_3 I_c + k_4 V_{(a)} + k_5 V_{(a)} - k_5 V_{(c)}$$

Figure 2.137 A test generator is installed to bias the input voltage connecting port.

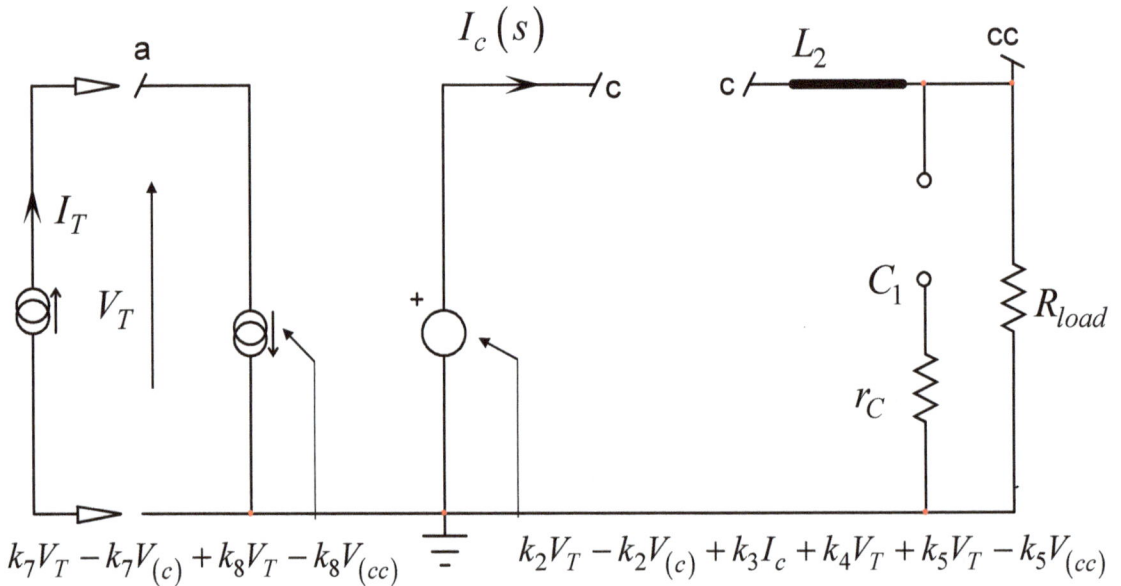

$$k_7 V_T - k_7 V_{(c)} + k_8 V_T - k_8 V_{(cc)} \qquad k_2 V_T - k_2 V_{(c)} + k_3 I_c + k_4 V_T + k_5 V_T - k_5 V_{(cc)}$$

Figure 2.138 For $s = 0$, the inductor is shorted while the capacitor is removed. Node c and cc are now at the same potential.

Current I_c is defined as:

$$I_c = \frac{V_{(c)}}{R_{load}}$$
(2.248)

The voltage at node c is given by the voltage source:

$$V_{(c)} = k_2 V_T - k_2 V_{(c)} + k_3 I_c + k_4 V_T + k_5 V_T - k_5 V_{(c)}$$
(2.249)

Substitute (2.248) and solve for V_c:

$$V_{(c)} = \frac{V_T \left(k_2 + k_4 + k_5\right)}{k_2 + k_5 - \dfrac{k_3}{R_{load}} + 1}$$
(2.250)

Now express the input current I_T using the input current source connected between node a and ground:

$$I_T = k_7 V_T - k_7 V_{(c)} + k_8 V_T - k_8 V_{(c)}$$
(2.251)

Substitute (2.250) in the above equation, solve for V_T and rearrange to define R_0:

$$R_0 = \frac{V_T}{I_T} = -\frac{\left(k_2 + k_5 + 1\right) R_{load} - k_3}{R_{load} \left(k_4 k_7 - k_8 - k_7 + k_4 k_8\right) + k_3 \left(k_7 + k_8\right)}$$
(2.252)

To determine the denominator, we turn the excitation off (zero I_T or open-circuit it) and look at the time constants involving L_2 and C_1. However, doing this operation leaves node a open with a current source attached to it. To close the current path, we add an extra resistance labeled R_{inf} which will approach infinity later on. A test generator is placed across L_2's terminals while C_1 is open-circuited.

The circuit to study is given in Figure 2.139. We start by defining the voltage at node a:

$$V_{(a)} = -R_{inf} \left[k_7 V_{(a)} - k_7 V_{(c)} + k_8 V_{(a)} - k_8 V_{(cc)}\right]$$
(2.253)

Solving for $V_{(a)}$ leads to:

$$V_{(a)} = \frac{R_{inf} \left(V_{(c)} k_7 + V_{(cc)} k_8\right)}{R_{inf} \left(k_7 + k_8\right) + 1}$$
(2.254)

231

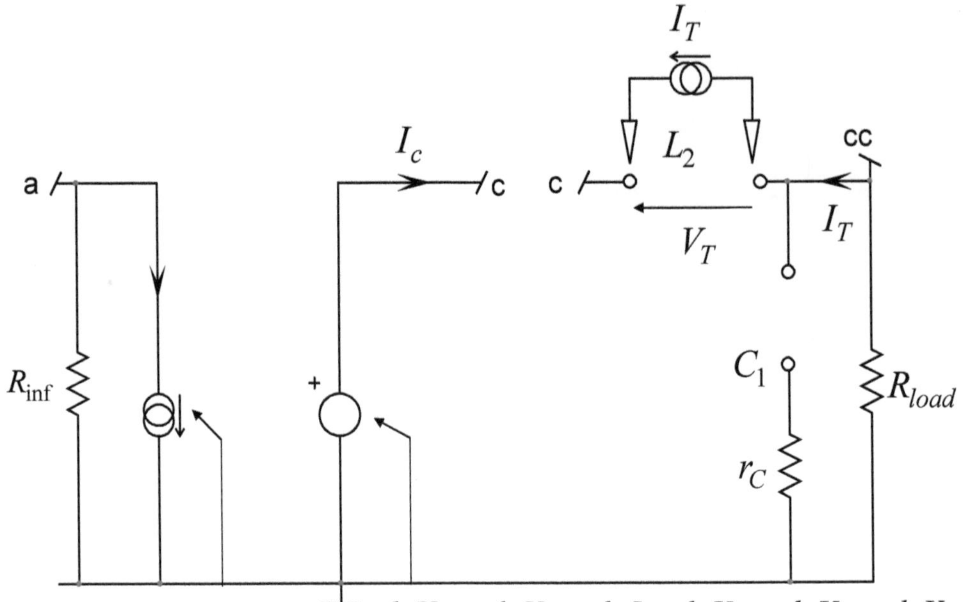

$$k_7 V_{(a)} - k_7 V_{(c)} + k_8 V_{(a)} - k_8 V_{(cc)} \;\doteq\; k_2 V_{(a)} - k_2 V_{(c)} + k_3 I_c + k_4 V_{(a)} + k_5 V_{(a)} - k_5 V_{(cc)}$$

Figure 2.139 The excitation is shut off and a resistance is added to fix the potential at node *a*.

If we consider R_{inf} much larger than 1, this expression simplifies to:

$$V_{(a)} \approx \frac{V_{(c)} k_7 + V_{(cc)} k_8}{k_7 + k_8} \tag{2.255}$$

The voltage at node *c* is expressed as:

$$V_{(c)} = k_2 V_{(a)} - k_2 V_{(c)} + k_3 I_c + k_4 V_{(a)} + k_5 V_{(a)} - k_5 V_{(cc)} \tag{2.256}$$

The voltage at node *cc*, depends on the injected current I_T:

$$V_{(cc)} = -I_T R_{load} \tag{2.257}$$

Now combining (2.257) and (2.255) into (2.256) then rearranging in the form of a resistance gives:

$$\frac{V_T}{I_T} = R_{load} + \frac{R_{load}\left(k_5 k_7 - k_2 k_8 - k_4 k_8\right) - k_3\left(k_7 + k_8\right)}{k_7 + k_8\left(1 + k_2\right) - k_7\left(k_4 + k_5\right)} \tag{2.258}$$

The numerical application shows that the second term is negligible compared to R_{load}. The time constant involving L_2 can thus be approximated by:

$$\tau_2 \approx \frac{L_2}{R_{load}} \tag{2.259}$$

For the time constant associated with C_1, we can see r_C in series with the capacitor. We can temporarily remove this resistance and put it back in series with the intermediate result. It is easier as V_T is now the voltage at c or cc since both are connected together. The current flowing in R_{load} is the sum of I_c and I_T:

$$V_T = (I_T + I_c) R_{load} \tag{2.260}$$

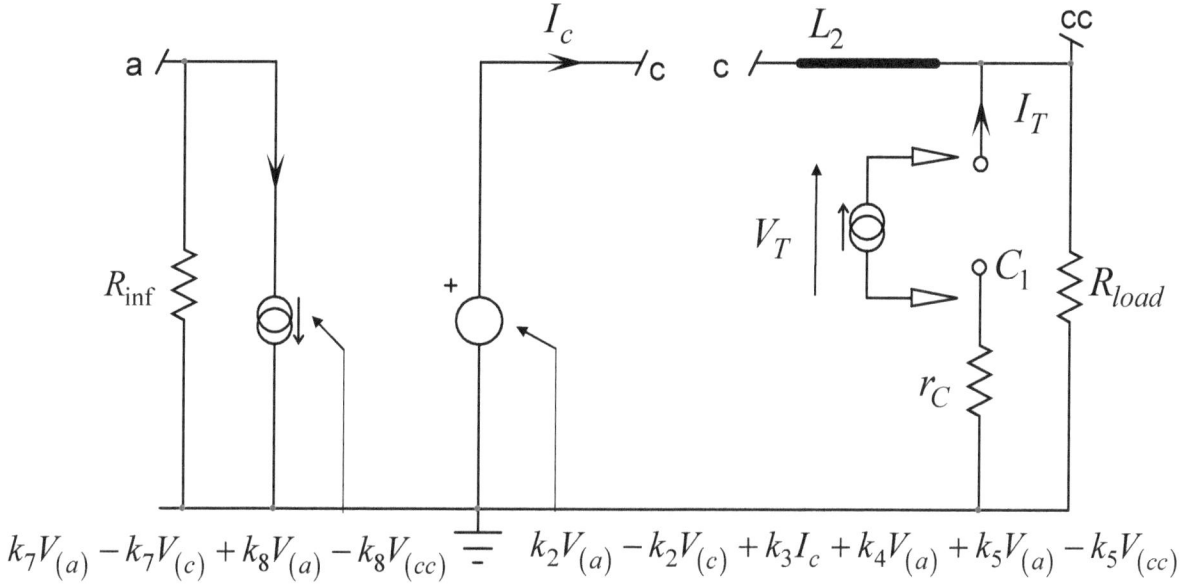

$$k_7 V_{(a)} - k_7 V_{(c)} + k_8 V_{(a)} - k_8 V_{(cc)} \quad\quad k_2 V_{(a)} - k_2 V_{(c)} + k_3 I_c + k_4 V_{(a)} + k_5 V_{(a)} - k_5 V_{(cc)}$$

Figure 2.140 Inductor L_2 is shorted to determine the resistance across the C_1's terminals.

From this we extract I_c as:

$$I_c = \frac{V_T - I_T R_{load}}{R_{load}} \tag{2.261}$$

We can reuse (2.255) in which $V_{(c)}$ and $V_{(cc)}$ are equal to V_T. In this case, the equation simplifies to:

$$V_a = V_T \tag{2.262}$$

The voltage at node c is V_T and expressed as:

$$V_T = k_2 V_T - k_2 V_T + k_3 I_c + k_4 V_T + k_5 V_T - k_5 V_T \tag{2.263}$$

Solving for V_T after substituting (2.261) in the above equation and rearranging in the form of a resistance and bringing back r_C leads to:

$$\frac{V_T}{I_T} = \frac{k_3}{k_4 + \dfrac{k_3}{R_{load}} - 1} + r_C \tag{2.264}$$

233

The time constant associated with C_1 is thus:

$$\tau_1 = C_1 \left(\frac{k_3}{k_4 + \dfrac{k_3}{R_{load}} - 1} + r_C \right) \tag{2.265}$$

For the last time constant, we can open-circuit L_2 to isolate the right-side network involving C_1 ($I_c = 0$). In this case, by inspection, we see that only r_C and R_{load} are involved in the resistive path:

$$\tau_1^2 = C_2 \left(r_C + R_{load} \right) \tag{2.266}$$

The complete denominator is obtained by combining the above expressions as follows:

$$D(s) = 1 + s\left(\tau_1 + \tau_2\right) + s^2 \left(\tau_2 \tau_1^2\right) \tag{2.267}$$

Combining the derived results, we have:

$$D(s) = 1 + s\left[\frac{L_2}{R_{load}} + C_1 \left(\frac{k_3}{k_4 + \dfrac{k_3}{R_{load}} - 1} + r_C \right) \right] + s^2 \frac{L_2}{R_{load}} C_1 \left(r_C + R_{load} \right) \tag{2.268}$$

If we consider r_C very small compared to R_{load}, then this expression simplifies to:

$$D(s) \approx 1 + s\left[\frac{L_2}{R_{load}} + C_1 \frac{k_3}{k_4 + \dfrac{k_3}{R_{load}} - 1} \right] + s^2 L_2 C_1 \tag{2.269}$$

The zeroes are obtained when the response is nulled, meaning that the voltage across the generator in Figure 2.137 is nulled. As a current source with zero volts across its connections can be replaced by a short circuit (degenerate case), the network returns to its natural structure and we can reuse the denominator $D(s)$ already determined in (2.226) or in (2.229) for a simplified version. The complete input impedance transfer function is thus:

$$Z_{in}(s) = R_0 \frac{N(s)}{1 + s\left[\dfrac{L_2}{R_{load}} + C_1 \dfrac{k_3}{k_4 + \dfrac{k_3}{R_{load}} - 1} \right] + s^2 L_2 C_1} \tag{2.270}$$

With R_0 and the numerator N respectively determined by (2.252) and (2.226).

We have plotted the complete response in Figure 2.141 which shows the magnitude and phase response.

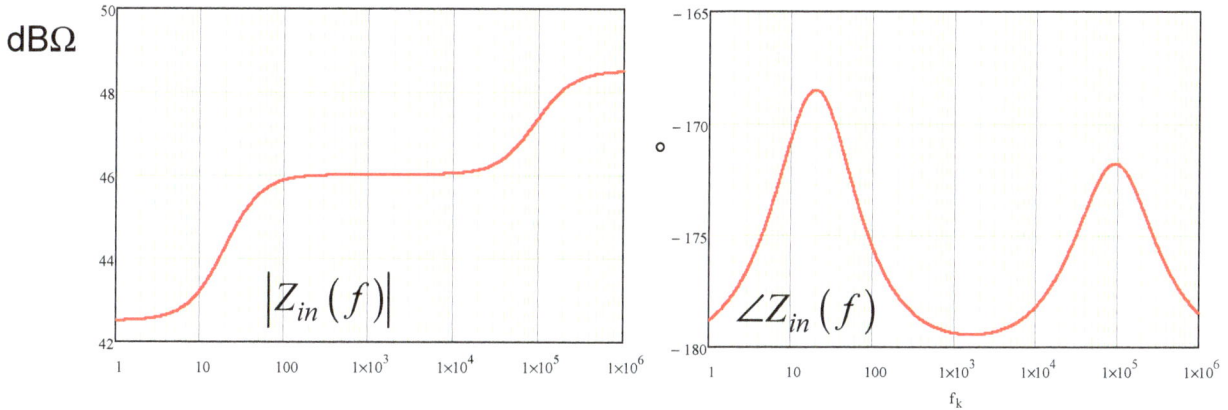

Figure 2.141 The input impedance plot reveals a negative incremental resistance.

To check our derivation, we have simulated the large-signal template from Figure 2.142 whose results are given in Figure 2.143. They perfectly match the results we have obtained with Mathcad®. Please note that operating points from the simulator must be perfectly reflected in the sheet if you want an exact match between the curves.

We have finished the study of the transfer functions describing the current-mode buck converter when operated in the discontinuous conduction mode.

All transfer functions are conveniently gathered in the table of Figure 2.144.

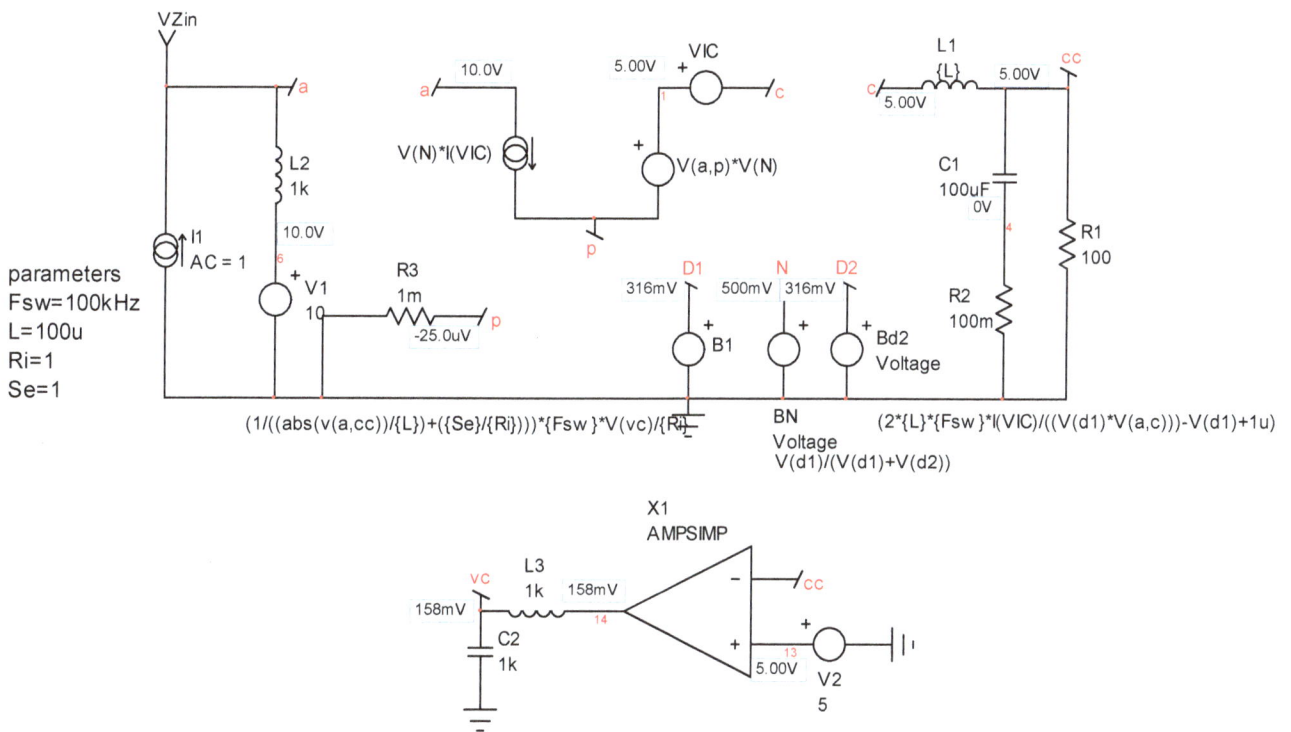

Figure 2.142 The large-signal model lets us verify our derivation results.

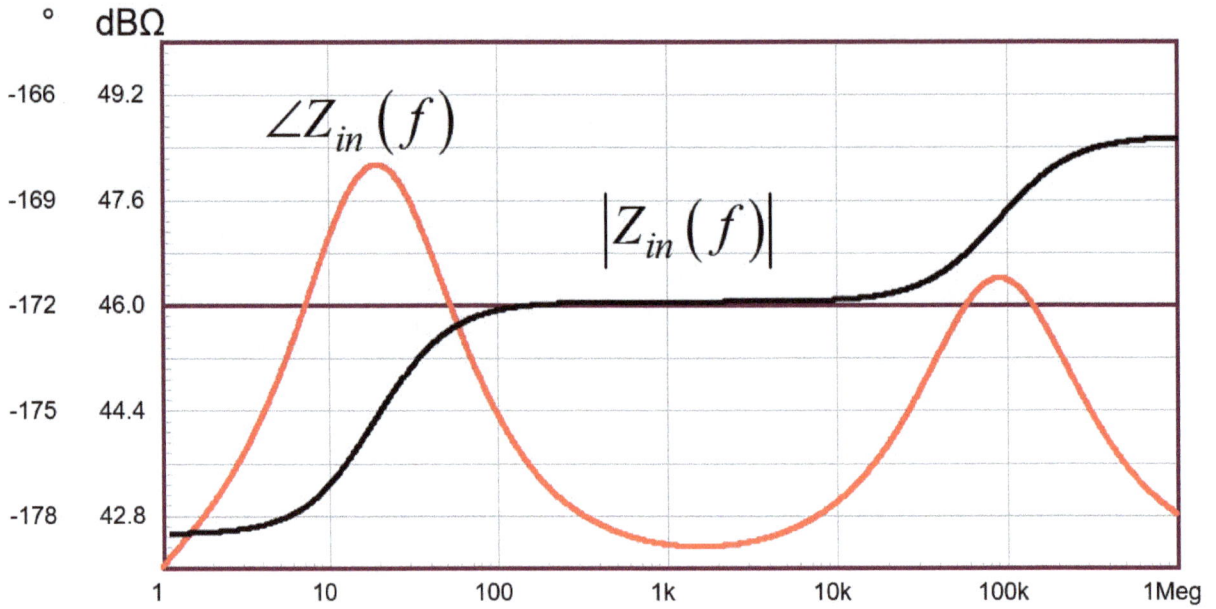

Figure 2.143 The response is identical to that of the Mathcad® sheet.

$\dfrac{V_{out}(s)}{V_{err}(s)}$ Control to Output	$H_0 \dfrac{1+\dfrac{s}{\omega_z}}{\left(1+\dfrac{s}{\omega_{p_1}}\right)\left(1+\dfrac{s}{\omega_{p_2}}\right)}$	$\omega_z = \dfrac{1}{r_C C_1}$	$\omega_p \approx \dfrac{1}{R_{load}C_1}\dfrac{2m_c-(2+m_c)M}{m_c(1-M)}$ see text for second pole	$H_0 \approx \dfrac{2m_c M V_{in}}{D}\dfrac{1-M}{2m_c-(2+mc)M}\dfrac{1}{S_n m_c T_{sw}}$
$\dfrac{V_{out}(s)}{V_{in}(s)}$ Input to Output	$H_0 \dfrac{1+\dfrac{s}{\omega_z}}{\left(1+\dfrac{s}{\omega_{p_1}}\right)\left(1+\dfrac{s}{\omega_{p_2}}\right)}$	$\omega_z = \dfrac{1}{r_C C_1}$	$\omega_p \approx \dfrac{1}{R_{load}C_1}\dfrac{2m_c-(2+m_c)M}{m_c(1-M)}$ see text for second pole	$H_0 = \dfrac{k_2+k_4+k_5}{k_2+k_5-\dfrac{k_3}{R_{load}}+1}$ see text for coefficients values
$Z_{in}(s)$ Input impedance	$R_0 \dfrac{1+\dfrac{s}{\omega_z}}{1+\dfrac{s}{\omega_p}}$	$\omega_z = \dfrac{1}{r_C C_1}$	$\omega_p \approx \dfrac{1}{R_{load}C_1}\dfrac{2m_c-(2+m_c)M}{m_c(1-M)}$	$R_0 = \left(-\dfrac{k_3}{k_2+k_5+1}\right)\| R_{load}$
$Z_{out}(s)$ Output impedance	$R_0 \dfrac{1+\dfrac{s}{\omega_z}}{1+\dfrac{s}{\omega_p}}$	$\omega_z = \dfrac{1}{r_C C_1}$	$\omega_p \approx \dfrac{1}{R_{load}C_3}+\dfrac{T_{sw}}{L_2 C_3}\left[m_c(1-D)-0.5\right]$	$R_0 = -\dfrac{(k_2+k_5+1)R_{load}-k_3}{R_{load}(k_4 k_7-k_5-k_7+k_4 k_8)+k_3(k_7+k_8)}$ see text for coefficients values

C_1 is the output capacitor with r_C its ESR, L_2 is the inductor and m_c is the slope compensation.

Figure 2.144 This table summarizes the four transfer functions of the current-mode buck converter operated in DCM.

2.5 Tapped Buck Transfer Function in CCM – Voltage-Mode Control

The tapped buck converter is an interesting structure used in applications where the difference between input and output is large. For instance, should you want to reduce a 300-V input to a 5-V regulated voltage, then the duty ratio – provided the converter operates in CCM – is 1.6% which is extremely small.

A way to improve this value consists of tapping the buck inductor and connecting either the diode (passive tap) or the switch (active tap) to the tap. Figure 2.145 shows the two possibilities.

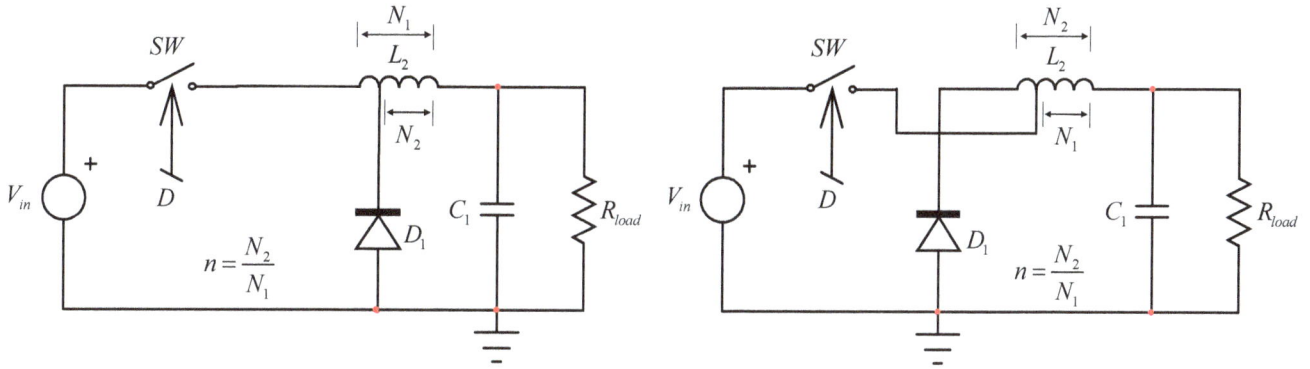

Figure 2.145 The inductor is tapped to connect the freewheel diode.

Dr. Vopérian has shown in [5] how the PWM switch could be revealed, suiting both configurations. This is what Figure 2.146 illustrates.

In this arrangement, the current and voltage of the diode in the PWM switch model are different than those of the original configuration: the diode is normally grounded but now sits before the transformer hence the scaling action in voltage and current.

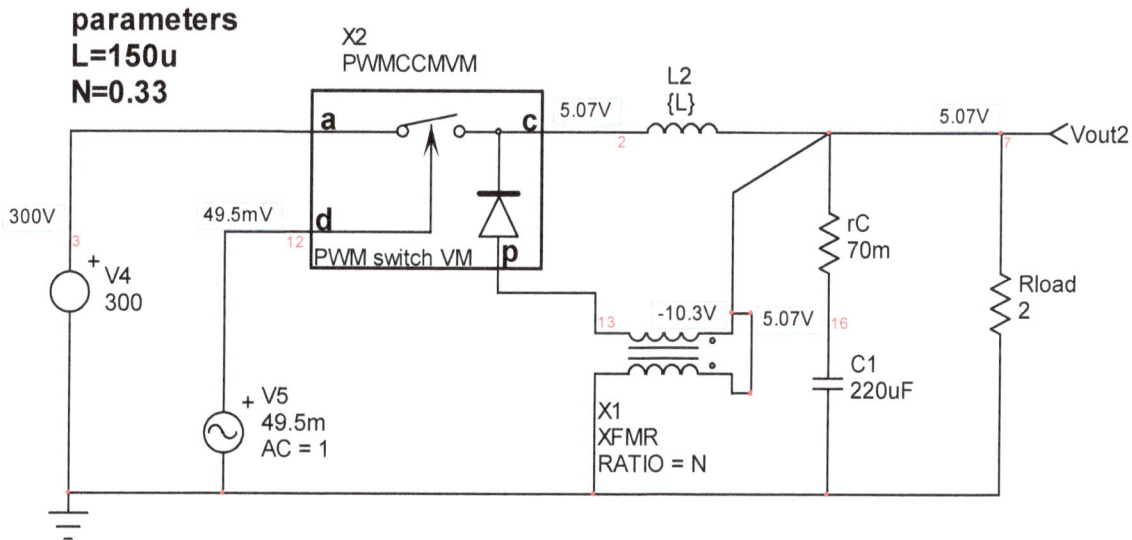

Figure 2.146 The PWM switch is associated with a transformer to model the tapped buck.

The dc conversion ratio is obtained by shorting the inductor while open-circuiting the capacitor as indicated in Figure 2.147. The mesh involving the input source with the voltage between terminals a-p, the reflected voltage between terminals p-c and V_{out} leads to writing:

$$V_{in} = \frac{V_{out}}{nD} - \frac{V_{out}}{n} + V_{out} = V_{out}\left(\frac{1}{nD} - \frac{1}{n} + 1\right) \qquad (2.271)$$

Rearranging gives:

$$M = \frac{V_{out}}{V_{in}} = \frac{D}{D + \frac{1-D}{n}} \qquad (2.272)$$

Having the dc conversion ratio on hand, we can immediately calculate the quasi-static gain H_0:

$$H_0 = \frac{dV_{out}(D)}{dD} = \frac{nV_{in}}{\left[D(n-1)+1\right]^2} \tag{2.273}$$

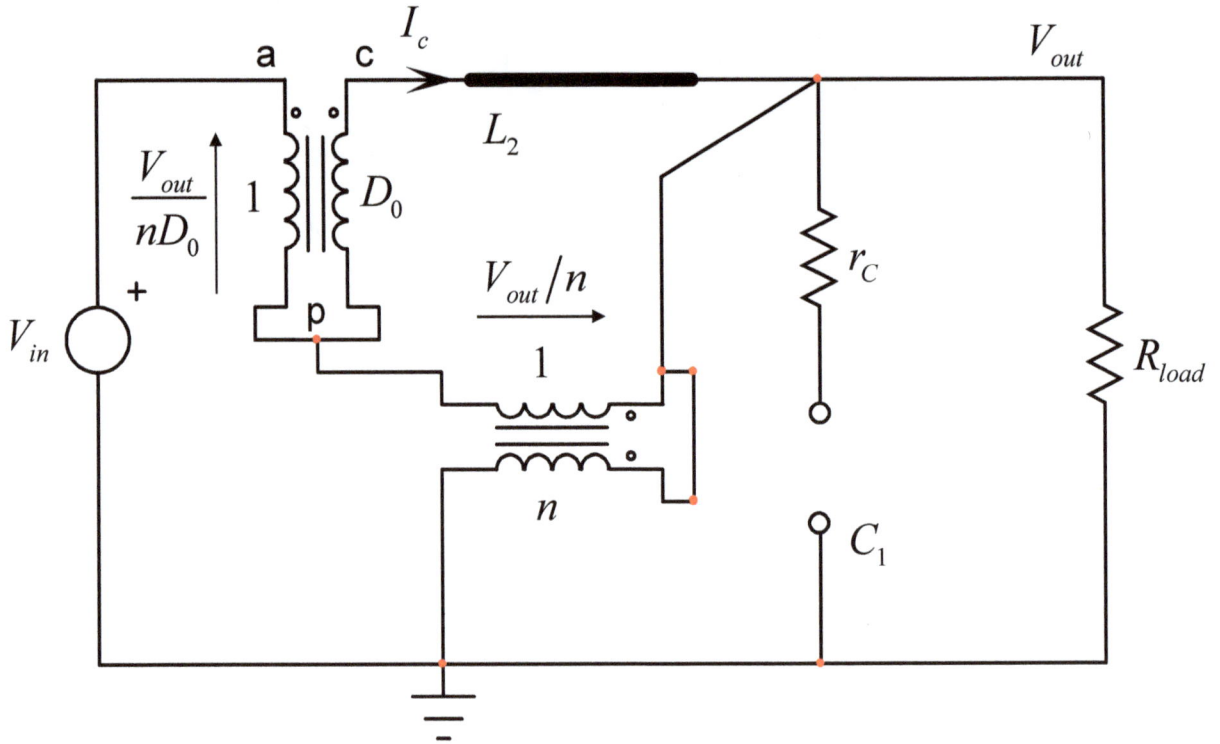

Figure 2.147 You can determine the conversion ratio by setting s to 0.

The determination of the time constants requires the replacement of the large-signal model by its small-signal version. Figure 2.148 shows the updated circuit with, on the left side, the bias points V_{ap} and I_c calculated according to [5].

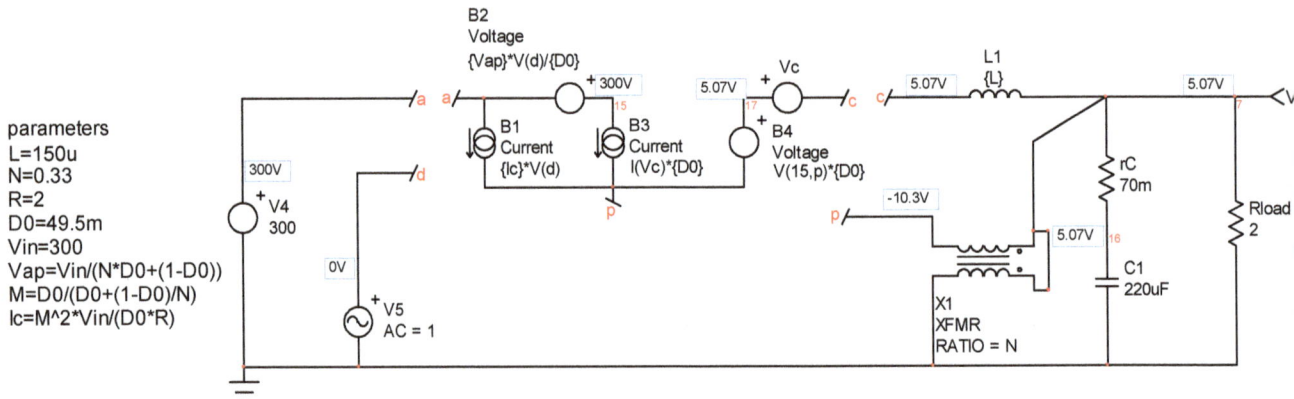

Figure 2.148 The small-signal model requires the calculation of operating points.

If you do the math around the time constants involving L_1 and C_2, you should obtain:

$$\tau_1 = r_C C_1 \tag{2.274}$$

$$\tau_2 = \frac{L_2}{R_{load}\left(D + \frac{1-D}{n}\right)^2} \tag{2.275}$$

$$\tau_1^2 = (r_C + R_{load})C_1 \tag{2.276}$$

The denominator follows the form:

$$D(s) = 1 + sb_1 + s^2 b_2 = 1 + s(\tau_1 + \tau_2) + s^2 \tau_2 \tau_1^2 \tag{2.277}$$

from which we can determine a resonant angular frequency ω_0 and a quality factor Q [1]:

$$\omega_0 = \frac{1}{\sqrt{b_2}} = \frac{D + \frac{1-D}{n}}{\sqrt{\frac{r_C + R_{load}}{R_{load}}}} \frac{1}{\sqrt{L_2 C_1}} \tag{2.278}$$

$$Q = \frac{\sqrt{b_2}}{b_1} = \frac{1}{\left[\frac{L_2}{R_{load}} \frac{1}{\left(D + \frac{1-D}{n}\right)^2} + r_C C_1\right]\omega_0} \tag{2.279}$$

The numerator is made of two terms: a zero classically given by the series combination of r_C and C_1 plus a right-half-plane zero:

$$\omega_{z_1} = \frac{1}{r_C C_1} \tag{2.280}$$

$$\omega_{z_2} = \frac{DR_{load}}{M^2(1-n)L_2} \tag{2.281}$$

Associating these terms leads to the following transfer function:

$$H(s) = \frac{V_{out}(s)}{D(s)} = H_0 \frac{\left(1 + \frac{s}{\omega_{z_1}}\right)\left(1 - \frac{s}{\omega_{z_2}}\right)}{1 + \frac{s}{\omega_0 Q} + \left(\frac{s}{\omega_0}\right)^2} \tag{2.282}$$

We can use SPICE to verify these equations, following the guidelines given in [1]. The natural time constants are evaluated by reducing the excitation to zero and installing 1-A current sources across the energy-storing elements's terminals. The voltage across these sources is the image of the resistance we want.

Figure 2.149 represents the experiments run for the first two time constants.

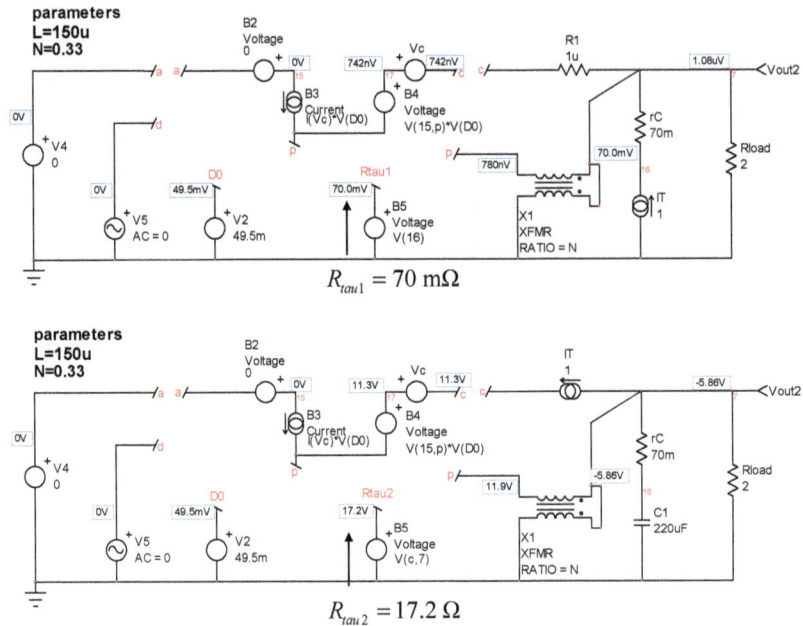

$$R_{tau1} = 70 \ m\Omega$$

$$R_{tau2} = 17.2 \ \Omega$$

Figure 2.149 SPICE is an extremely useful resource to verify the results given by equations.

The bottom of Figure 2.150 describes the nulling of the output via source G_1 which adjusts the injected current to maintain a 0-V V_{out}. Please note that the excitation is back with source V_5 arbiratrily set to 0.5 V. The measured resistance is negative, L_2 contributes a RHPZ.

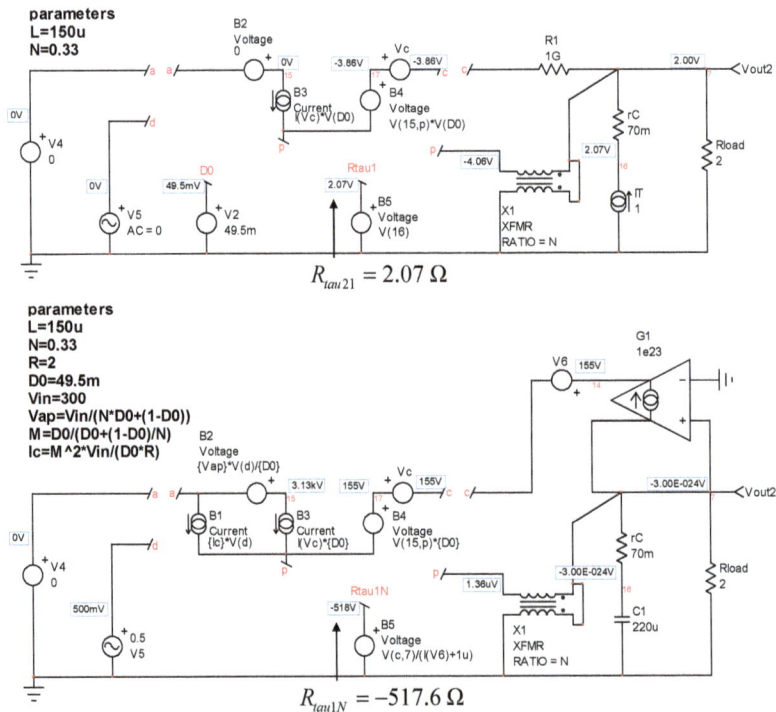

$$R_{tau21} = 2.07 \ \Omega$$

$$R_{tau1N} = -517.6 \ \Omega$$

Figure 2.150 The resistance determined by the output null is negative: this is a RHP zero.

$$V_{in} := 300V \qquad R_L := 2\Omega \qquad C_1 := 220\mu F \qquad r_C := 0.07\Omega$$
$$L_2 := 150\mu H \qquad D := 4.95\% \qquad D_p := 1 - D = 95.05\% \qquad N_1 := 0.33$$

$$M := \frac{D}{D + \frac{1-D}{N_1}} = 0.017 \qquad V_{out} := M \cdot V_{in} = 5.069V \qquad I_c := \frac{M}{D} \cdot \frac{V_{out}}{R_L} = 0.865A$$

Dc gain H0:

$$H_0 := \frac{d}{dD} V_{out}(D) = \frac{N_1 \cdot V_{in}}{(D \cdot N_1 - D + 1)^2} \qquad H_0 := \frac{N_1 \cdot V_{in}}{(D \cdot N_1 - D + 1)^2} = 105.908V$$

$$20 \cdot \log\left(\frac{H_0}{V}\right) = 40.499$$

Numerator coefficients:

$$\omega_{z1} := \frac{1}{r_C \cdot C_1} \qquad\qquad f_{z1} := \frac{\omega_{z1}}{2\pi} = 10.335\,kHz$$

$$\omega_{z2} := \frac{D \cdot R_L}{M^2 \cdot (1 - N_1) \cdot L_2} \qquad f_{z2} := \frac{\omega_{z2}}{2\pi} = 0.549\,MHz$$

Denominator coefficients:

$$\tau_2 := \frac{L_2}{\left(D + \frac{D_p}{N_1}\right)^2 \cdot R_L} = 8.737\,\mu s \qquad \tau_1 := C_1 \cdot r_C = 1.54 \times 10^{-5}\,s \qquad \tau_{21} := C_1 \cdot (R_L + r_C) = 455.4\,\mu s$$

$$b_1 := \tau_1 + \tau_2 \qquad b_2 := \tau_2 \cdot \tau_{21} \qquad \omega_0 := \frac{1}{\sqrt{b_2}} = 1.585 \times 10^4\,\frac{1}{s} \qquad Q := \frac{\sqrt{b_2}}{b_1} = 2.613$$

$$\omega_0 := \frac{D + \frac{D_p}{N_1}}{\sqrt{L_2 \cdot C_1} \cdot \sqrt{\frac{r_C + R_L}{R_L}}} = 1.585 \times 10^4\,\frac{1}{s} \qquad Q := \frac{1}{\left[\frac{L_2}{R_L} \cdot \frac{1}{\left(D + \frac{D_p}{N_1}\right)^2} + r_C \cdot C_1\right] \cdot \omega_0} = 2.613$$

$$D_1(s) := 1 + s \cdot \left[\frac{L_2}{R_L} \cdot \frac{1}{\left(D + \frac{D_p}{N_1}\right)^2} + r_C \cdot C_1\right] + s^2 \cdot \left[\frac{L_2}{R_L} \cdot \frac{1}{\left(D + \frac{D_p}{N_1}\right)^2} \cdot (r_C + R_L) \cdot C_1\right]$$

$$H_1(s) := H_0 \frac{\left(1 + \frac{s}{\omega_{z1}}\right) \cdot \left(1 - \frac{s}{\omega_{z2}}\right)}{1 + \frac{s}{\omega_0 \cdot Q} + \left(\frac{s}{\omega_0}\right)^2} \qquad\qquad H_2(s) := H_0 \frac{\left(1 + \frac{s}{\omega_{z1}}\right) \cdot \left(1 - \frac{s}{\omega_{z2}}\right)}{D_1(s)}$$

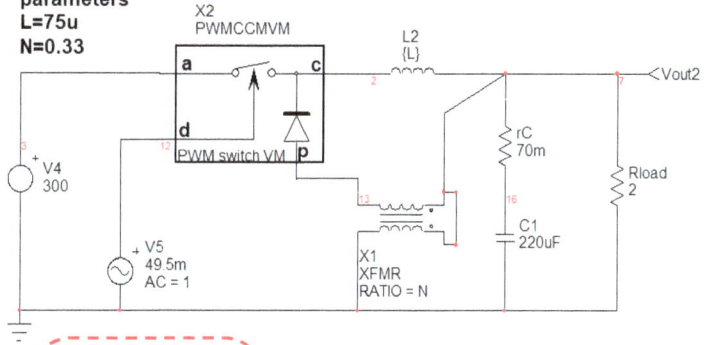

parameters
L=75u
N=0.33

X2
PWMCCMVM

L2 {L}

<Vout2

a — c

d — p

PWM switch VM

+ V4 300

+ V5 49.5m AC = 1

X1 XFMR RATIO = N

rC 70m

C1 220uF

Rload 2

$$R_{tau1} := \left(D + \frac{D_p}{N_1}\right)^2 \cdot R_L = 17.167\Omega$$

$$R_{tau2} := r_C = 0.07\Omega$$

$$R_{tau12} := r_C + R_L = 2.07\Omega$$

$$R_{tau1N} := \frac{D \cdot R_L}{M^2 \cdot (1 - N_1)} = 517.639\Omega$$

$$f_0 := \frac{\omega_0}{2\pi} = 2.523\,kHz$$

Figure 2.151 Mathcad® and SPICE deliver the exact same resistive results.

As confirmed by Figure 2.151, these results are similar to those computed by Mathcad®, confirming the equations. Finally, Figure 2.152 plots the control-to-output transfer function of the tapped buck.

The LHP zero brings the phase up when it kicks in but the RHP zero brings it back down.

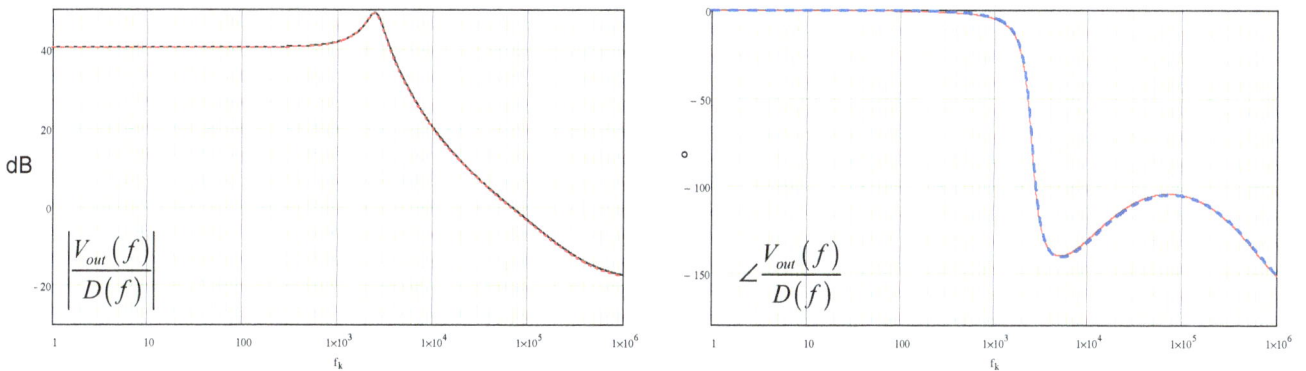

Figure 2.152 The magnitude/phase frequency responses between SPICE and Mathcad® are identical.

A quick check with a SIMPLIS® simulation (Figure 2.153) confirms that our calculations are correct as shown in Figure 2.154.

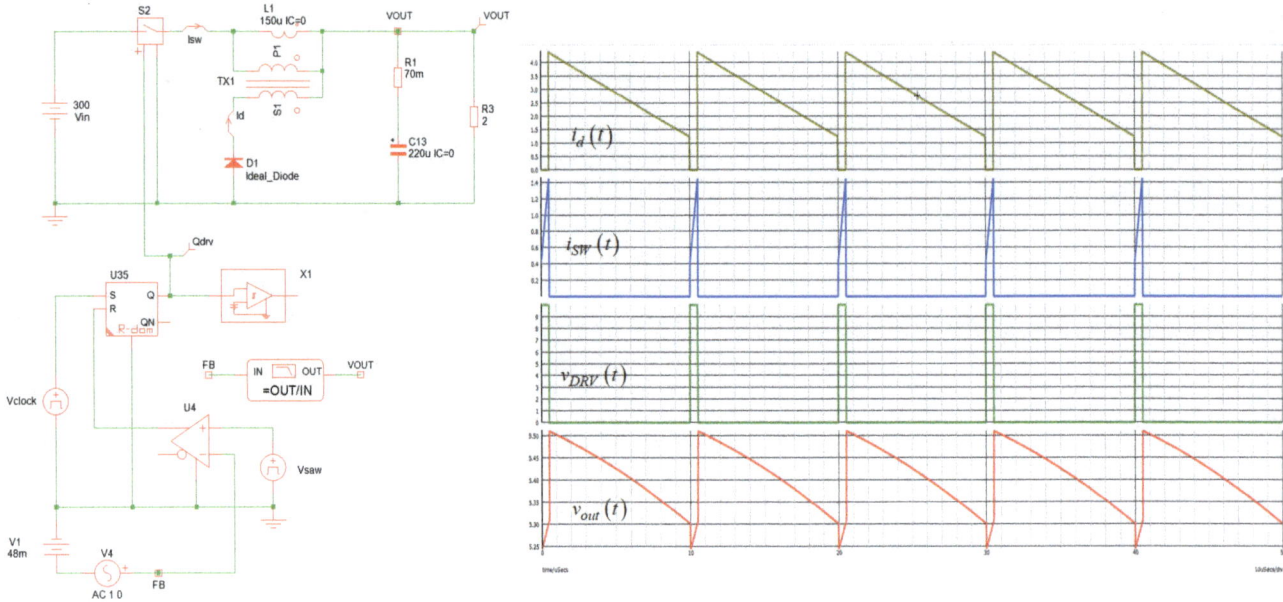

Figure 2.153 SIMPLIS® can simulate the tapped-buck in an easy with a transformer.

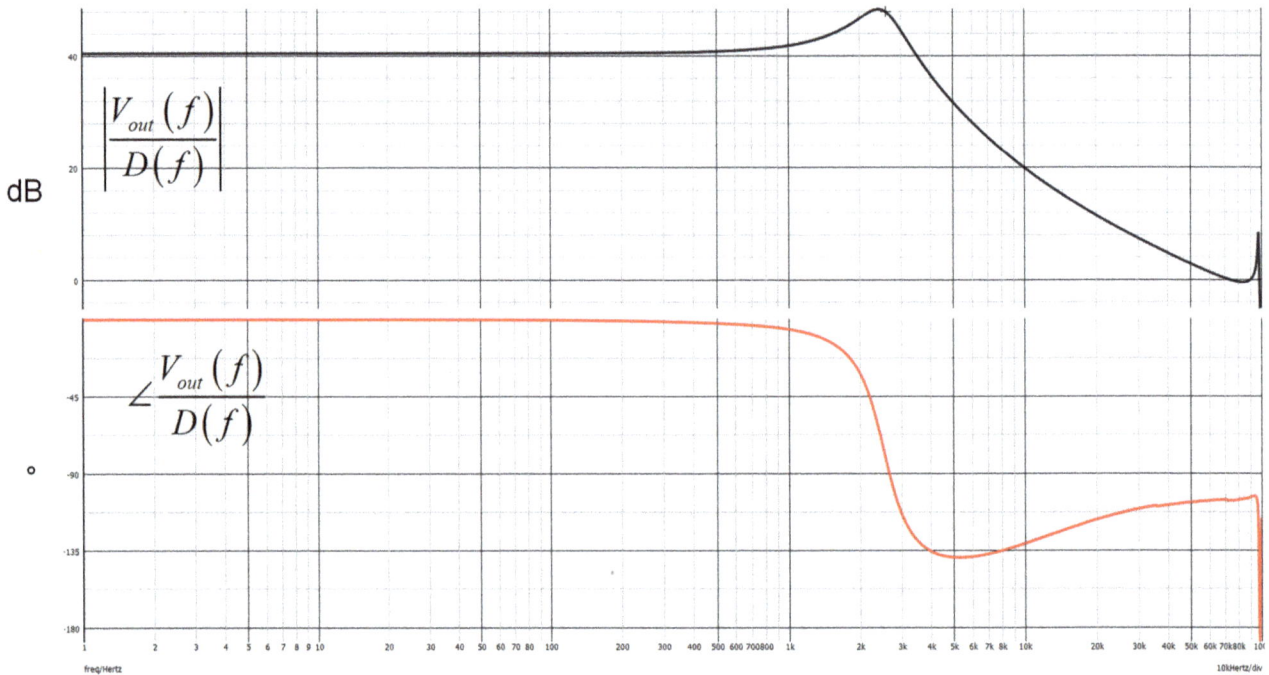

Figure 2.154 The magnitude/phase frequency responses between SPICE and SIMPLIS® are very close.

2.6 Quasi-Resonant Buck Transfer Function – Current-Mode Control

The buck operated in quasi-resonance is well suited to reduce switching losses but suffers from wide frequency variations in relationship to input and output conditions.

To determine the control-to-output transfer function of the QR buck operated in current mode, we will use the PWM switch model specifically developed in [6] and insert it in a buck configuration as shown in Figure 2.155.

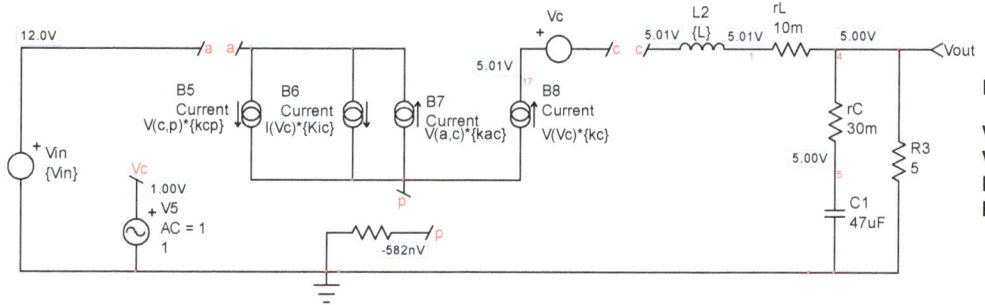

parameters

R=5
Vin=12
Vout=5
M=Vout/Vin
Ic=Vout/R
Vap=Vin
Vac=Vin-Vout
Vcp=Vout
kcp=Ic*Vac/(Vac+Vcp)^2
kic=Vcp/(Vcp+Vac)
kac=Vcp*Ic/(Vac+Vcp)^2
kc=1/(2*Ri)

parameters

Vin=12
Vc=1
L=10u
Ri=0.5

Figure 2.155 The modified PWM switch now operates in quasi-resonant mode.

The circuit reduces to that of Figure 2.156 considering the zeroed input source in ac (the voltage at node *a* is 0 V).

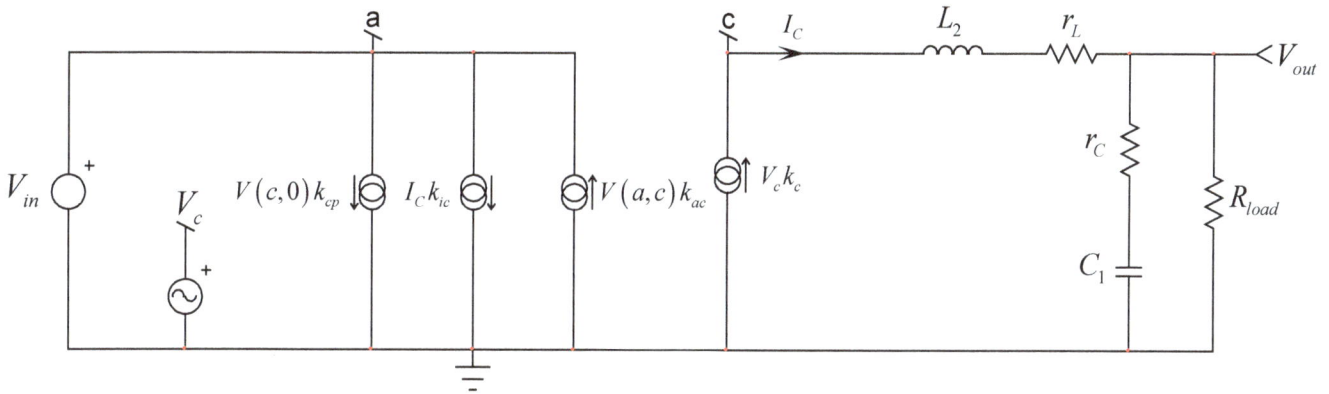

Figure 2.156 The left-side sources can be ignored for the control-to-output transfer function.

The left-side part affects the input current and is no use to determine the control-to-output transfer function.

To determine the dc gain, the circuit is updated in Figure 2.157 in which inductor L_2 is shorted while capacitor C_1 is open-circuited.

243

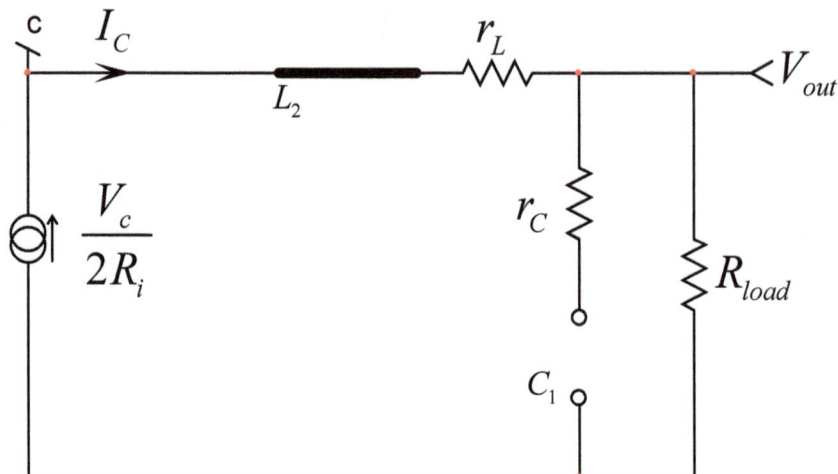

Figure 2.157 The quasi-static gain is quickly determined with L_2 shorted and C_1 open-circuited.

The voltage at node c is given by:

$$V_{(c)} = \frac{V_c}{2R_i}(r_L + R_{load})$$

(2.283)

The output voltage is obtained using a simple resistive divider involving r_L and R_{load}:

$$V_{out} = V_{(c)} \frac{R_{load}}{r_L + R_{load}} = \frac{V_c}{2R_i} R_{load}$$

(2.284)

Implying that the quasi-static gain H_0 equals:

$$H_0 = \frac{dV_{out}(V_c)}{dV_c} = \frac{R_{load}}{2R_i}$$

(2.285)

The time constants are obtained in a quick way noticing that the inductor in series with the current source plays no role in ac. If you determine its time constant when V_c the excitation is reduced to 0 V (Figure 2.158), the left node is open-circuited and:

$$\tau_2 = \frac{L_2}{\infty} = 0$$

(2.286)

The time constant involving capacitor C_1 is obtained by shorting L_2 (but its left terminal is floating) as shown in sketch (b) of Figure 2.158.

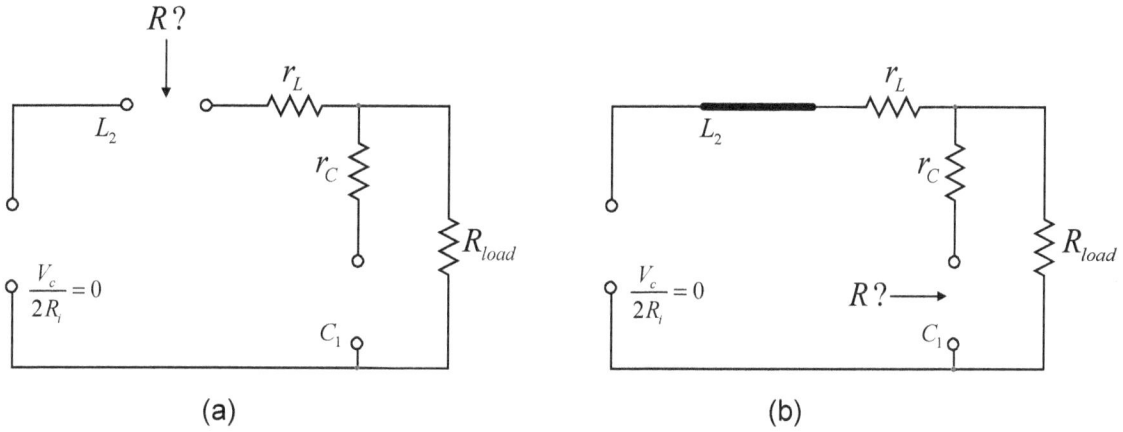

Figure 2.158 The inductor is driven by a current source and plays no role in ac.

$$\tau_1 = C_1 \left(R_{load} + r_C \right) \tag{2.287}$$

If you try to determine τ_1^2 or τ_2^1, they will respectfully return either a finite value or zero. When multiplied by τ_2 for the first one or τ_1 for the second one, it will return zero: there is no second-order term in the denominator. It makes sense as L_2 being driven by a current source, it then has no role in ac. The denominator is thus defined as:

$$D(s) = 1 + s\left(\tau_1 + \tau_2\right) + s^2 \tau_1 \tau_2^1 = 1 + \frac{s}{\omega_p} \tag{2.288}$$

with:

$$\omega_p = \frac{1}{C_1 \left(r_C + R_{load} \right)} \tag{2.289}$$

The zero is classically given by the series connection of r_C and C_1 as illustrated by Figure 2.159 and henceforth:

$$\omega_z = \frac{1}{r_C C_1} \tag{2.290}$$

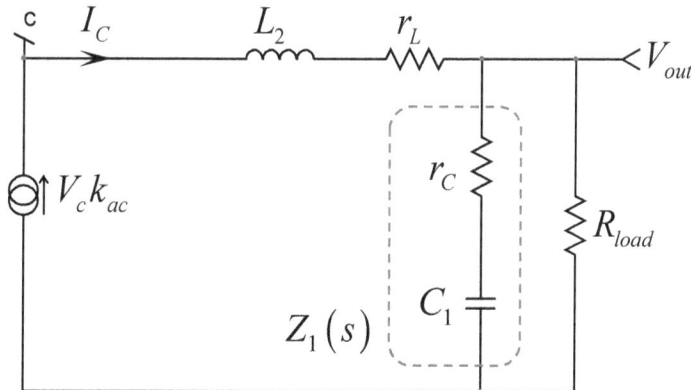

Figure 2.159 The series connection of r_C and C_1 contributes the zero of the transfer function.

The transfer function is assembled as follows:

$$H(s) = \frac{V_{out}(s)}{D(s)} = H_0 \frac{1 + \dfrac{s}{\omega_z}}{1 + \dfrac{s}{\omega_p}} \qquad (2.291)$$

The plot from Figure 2.160 confirms a first-order response.

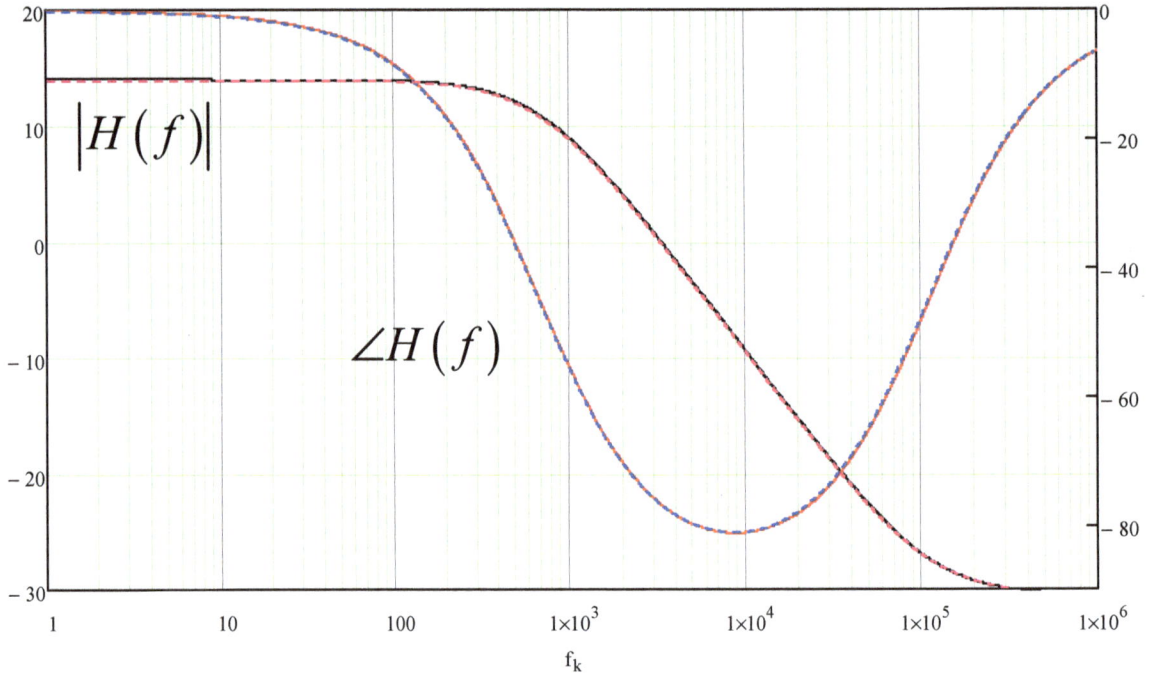

Figure 2.160 The QR-operated buck response in current-mode is that of a 1st-order system.

To verify our analysis, we have simulated a quasi-resonant buck converter with SIMPLIS®. It appears in Figure 2.161.

The inductor needs an extra winding to sense the core flux activity which instructs the control circuit to restart operations when the core is reset. A small 15-pF capacitor creates a delay so that the high-side switch turns on right in the drain-source valley voltage. This delay impacts the switching frequency but has almost no influence on the transfer function beside slightly altering the dc gain (especially if the controller forces operation in 2nd or the following valleys).

Figure 2.162 and Figure 2.163 confirm the correct operation of the circuit.

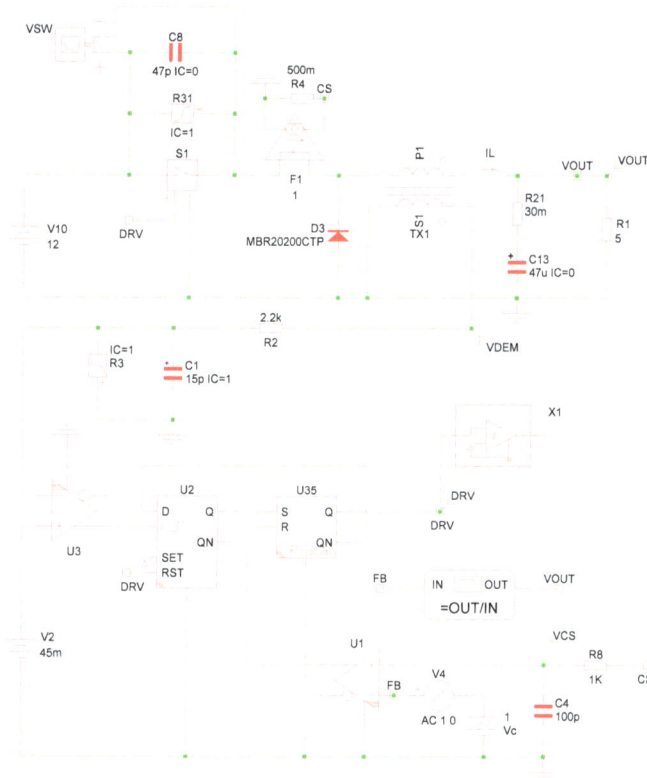

Figure 2.161 An extra winding senses the inductor state and lets the control section know when the current is zero.

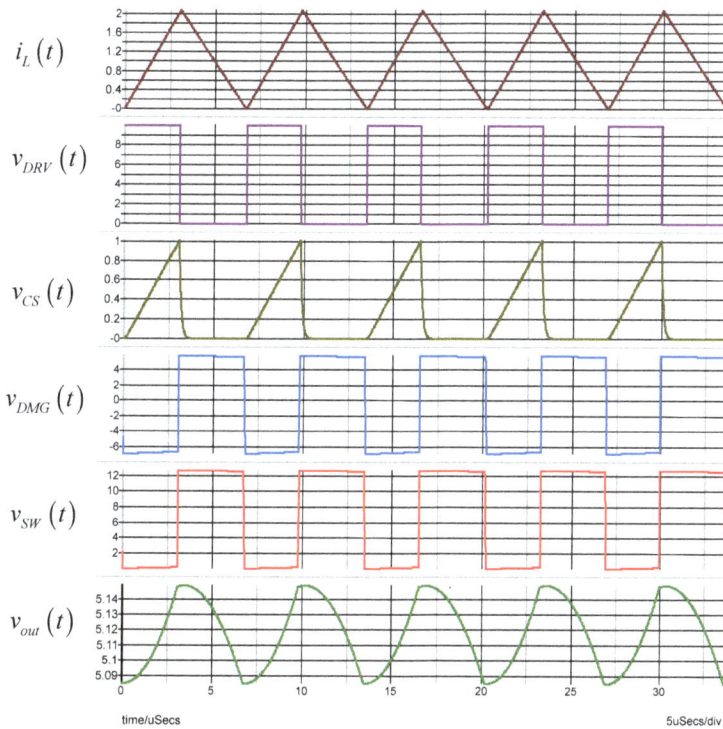

Figure 2.162 The inductor current touches zero and a cycle immediately restarts.

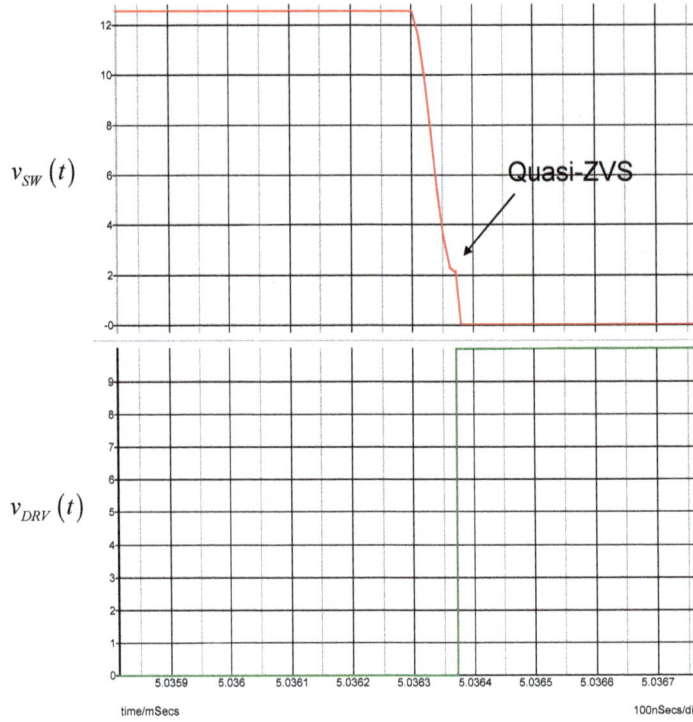

Figure 2.163 With an adequate delay, turn-on losses can virtually be eliminated.

SIMPLIS® can extract the small-signal dynamic response and confirms the pole-zero shape we expect in Figure 2.164.

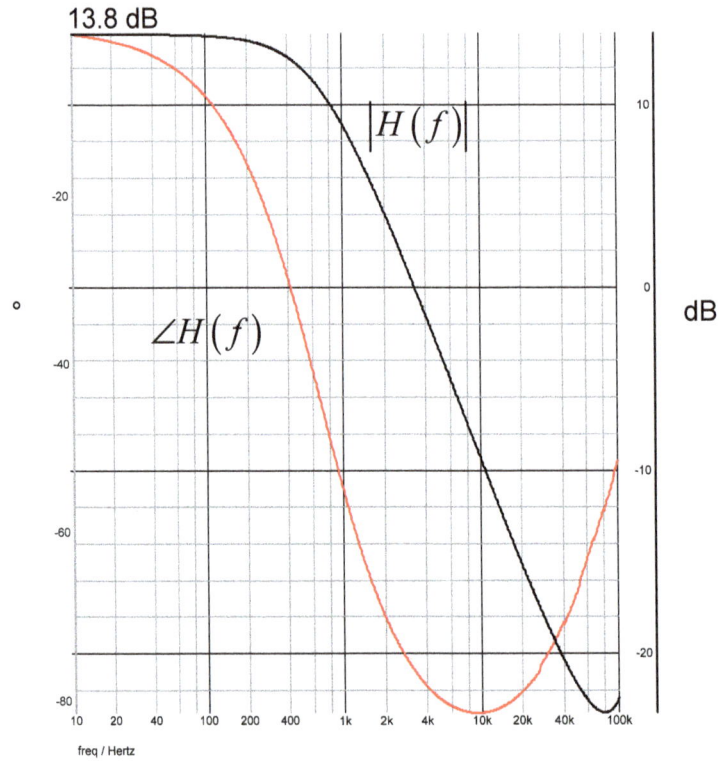

Figure 2.164 The ac response delivered by the simulator confirms the 1st-order response of the QR CM buck converter.

As a final test, we have compared the step response of a cycle-by-cycle circuit with that of the averaged CM-PWM switch specifically developed for QR operations in [6]. As shown in Figure 2.165, both models deliver the exact same transient response.

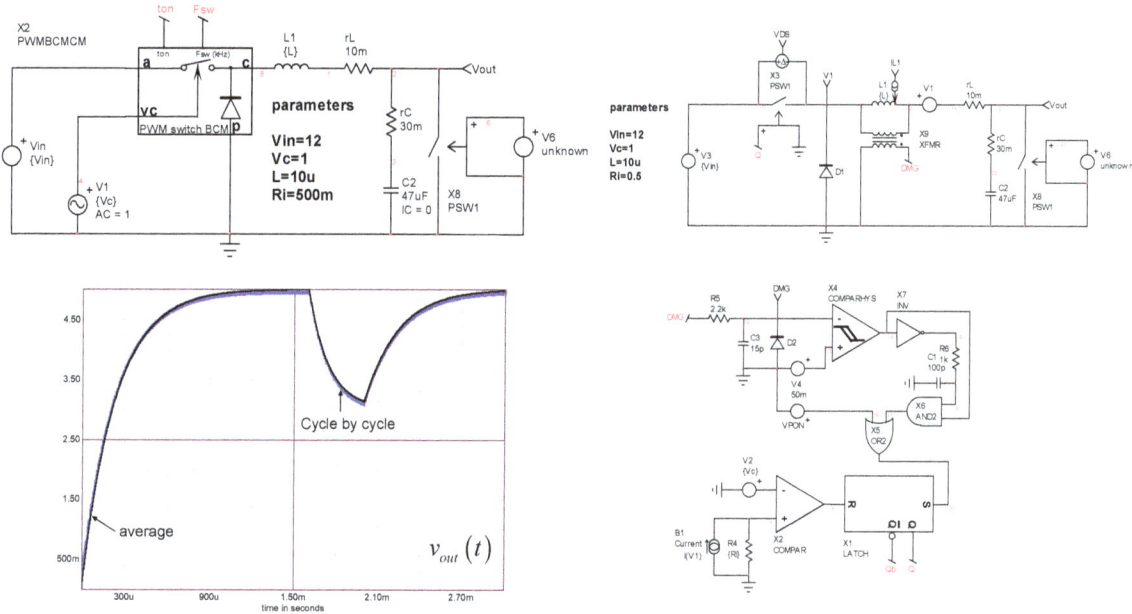

Figure 2.165 The step-load responses between the averaged model and its cycle-by-cycle counterpart are identical.

2.7 Constant On-Time Buck Transfer Function – Current-Mode Control

A COT converter works by varying the off-time while the on-time is held constant. In a current-mode version, this is achieved by adjusting the valley current setpoint. The small-signal model of the buck operated in COT shows up in Figure 2.166. It is based on the subcircuit introduced in Chapter 1.4.4.

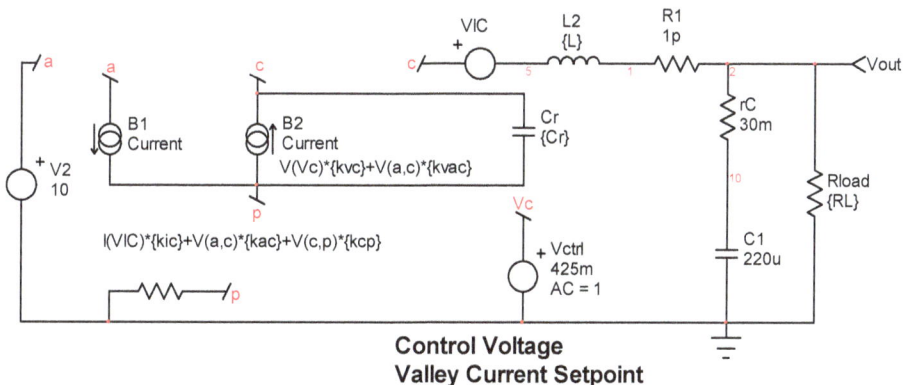

Parameters

Vout=5V
Vin=10V
RL=1
Ri=100m
L=5u
pi=3.14159
W=pi/ton
Cr=1/(L*W^2)
ton=1.5u
Ic=Vout/RL
Vac=Vin-Vout
Vcp=Vout

kic=1/(1+Vac/Vcp)
kac=-Ic/(Vcp*(Vac/Vcp+1)^2)
kcp=Ic*Vac/(Vcp^2*(Vac/Vcp+1)^2)
kvc=1/Ri
kvac=ton/(2*L)

Figure 2.166 The buck operated in COT requires a specific model introduced in Chapter 1.4.4.

We will not go through the details of the time constant determination with a zeroed input or a nulled output. You should now be familiar with the methodology and we will reproduce the control-to-output expression derived in [7]:

$$\frac{V_{out}(s)}{V_c(s)} = H_0 \frac{1 + \dfrac{s}{\omega_z}}{1 + \dfrac{s}{\omega_p}} F_h(s) \qquad (2.292)$$

With:

$$\omega_p = \frac{1 - \dfrac{R_{load} k_2}{R_i}}{\left(R_{load} + r_C\right)C_1 - \dfrac{R_{load} r_C C_1 k_2}{R_i}} \qquad (2.293)$$

$$k_1 = \frac{t_{on} R_i}{2L_2} \qquad (2.294)$$

$$k_2 = -\frac{t_{on} R_i}{2L_2} \qquad (2.295)$$

$$\omega_z = \frac{1}{r_C C_1} \qquad (2.296)$$

$$H_0 = \frac{R_{load}}{R_i \left(1 - \dfrac{k_2}{R_i} R_{load}\right)} \qquad (2.297)$$

$$F_h(s) = \frac{1}{1 + \dfrac{s}{\omega_1 Q} + \left(\dfrac{s}{\omega_1}\right)^2} \qquad (2.298)$$

$$\omega_1 = \frac{\pi}{t_{on}} \qquad (2.299)$$

$$Q = \frac{2}{\pi} \qquad (2.300)$$

In this expression, the double poles are still located at a position depending on the selected t_{on} but the quality factor Q is fixed and no longer depends on operating conditions as with peak-current mode control.

We have plotted the Mathcad® response together with that of the PWM switch from Figure 2.166 and results appear in Figure 2.167.

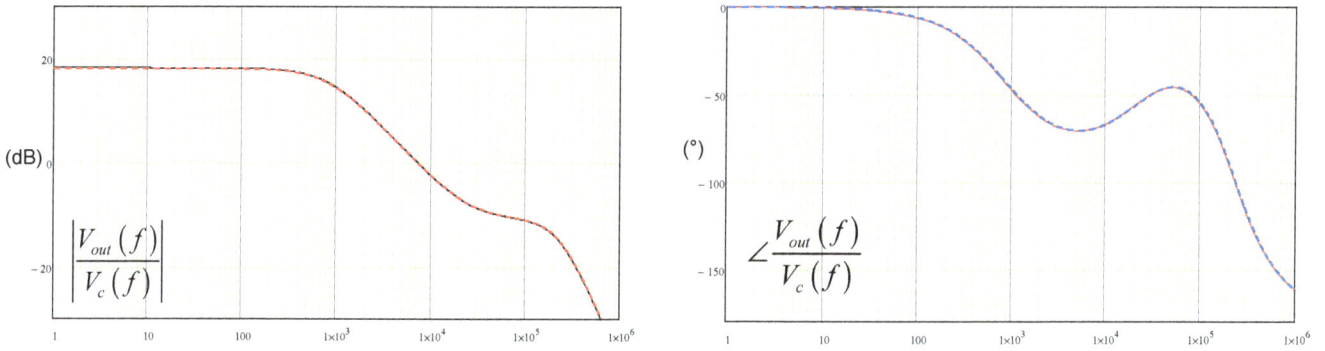

Figure 2.167 The magnitude/phase responses from the simulated small-signal model and the equations are identical.

We have also built a COT buck in SIMPLIS® and it is represented in Figure 2.168. The on-time generator involves the 10-μA source I_1 and the 23-pF capacitor.

With a 1-V reference voltage, the on-time is calibrated to 2.3 μs. Comparator U_2 restarts the main latch when the inductor current passes below the setpoint imposed by V_1.

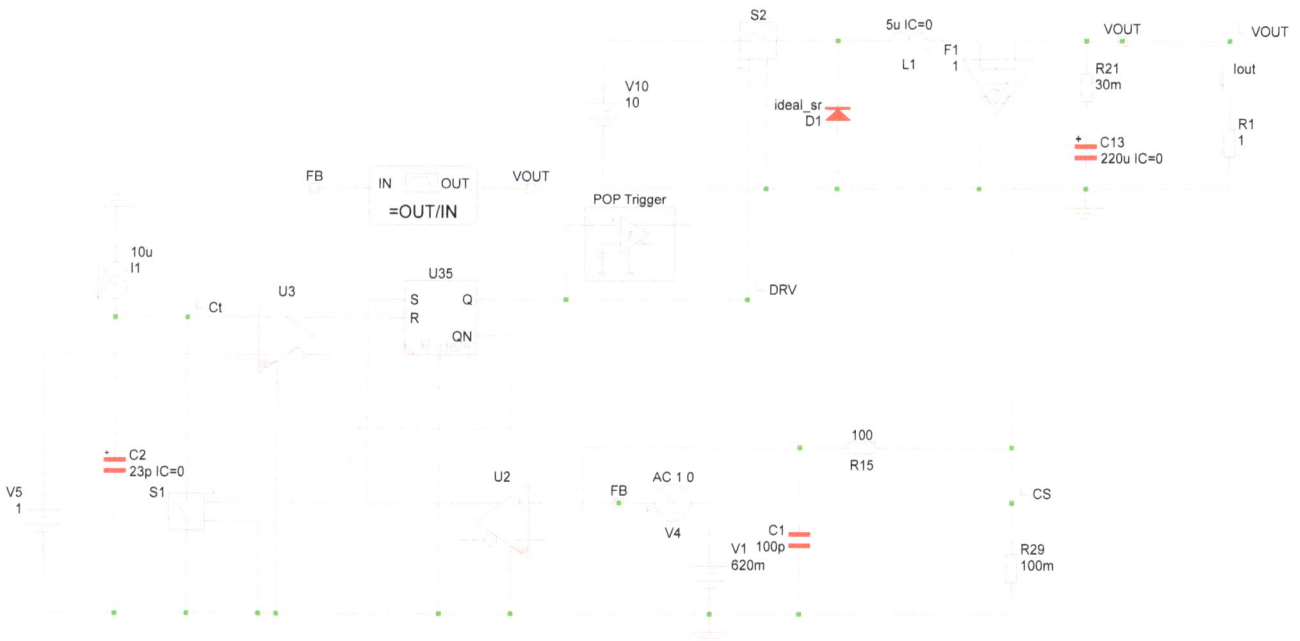

Figure 2.168 SIMPLIS® easily simulates a COT buck working in current mode.

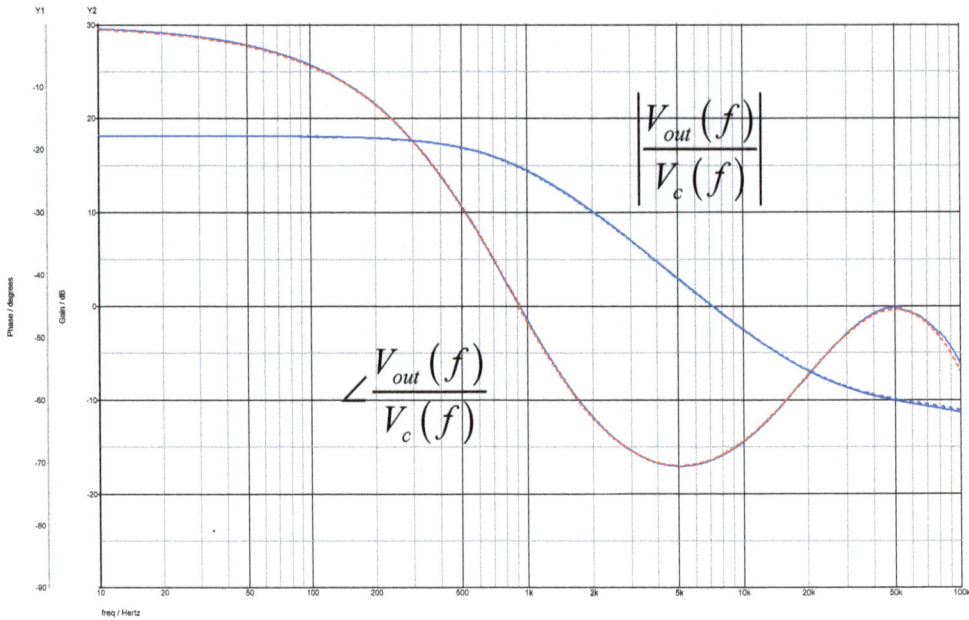

Figure 2.169 The SIMPLIS® simulation confirms the response obtained with both SPICE and Mathcad® (SPICE results are in the dashed lines).

As confirmed by Figure 2.169 in which we have pasted the SPICE curves, simulation results are in excellent agreement with the small-signal circuit's response.

2.8 Fixed Off-Time Buck Transfer Function – Current-Mode Control

A FOT converter works by varying the on-time while the off-time is kept constant. In a current-mode version, this is achieved by adjusting the peak current setpoint. The small-signal model of the buck operated in FOT is given in Figure 2.170. It is based on the subcircuit introduced in Chapter 1.4.5.

Like we did with the COT converter, we have reproduced the control-to-output transfer function expression derived in [7] which perfectly matches the model presented in Figure 2.170.

$$\frac{V_{out}(s)}{V_c(s)} = H_0 \frac{1 + \dfrac{s}{\omega_z}}{1 + \dfrac{s}{\omega_p}} F_h(s) \qquad (2.301)$$

with:

$$\omega_p = \frac{1 - \dfrac{R_{load} k_2}{R_i}}{\left(R_{load} + r_C\right) C_1 - \dfrac{R_{load} r_C C_1 k_2}{R_i}} \qquad (2.302)$$

$$k_1 \approx 0 \qquad (2.303)$$

$$k_2 = -\frac{t_{off} R_i}{2 L_2} \qquad (2.304)$$

252

$$\omega_z = \frac{1}{r_C C_1} \tag{2.305}$$

$$H_0 = \frac{R_{load}}{R_i \left(1 - \frac{k_2}{R_i} R_{load}\right)} \tag{2.306}$$

$$F_h(s) = \frac{1}{1 + \frac{s}{\omega_1 Q} + \left(\frac{s}{\omega_1}\right)^2} \tag{2.307}$$

$$\omega_1 = \frac{\pi}{t_{off}} \tag{2.308}$$

$$Q = \frac{2}{\pi} \tag{2.309}$$

Parameters

Ri=100m
RL=0.5
L=5u
pi=3.14159
W=pi/toff
Cr=1/(L*W^2)
toff=5u

Vc=1.25
Vin=10
Vout=5
Ic=Vout/RL
Vac=Vin-Vout
Vcp=Vout

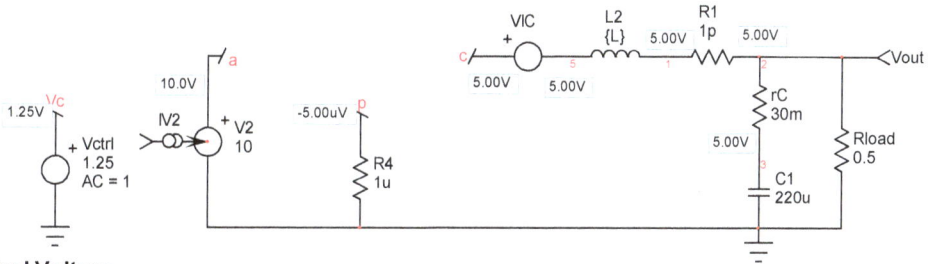

Control Voltage
Peak Current Setpoint

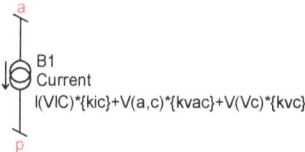

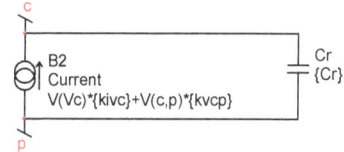

kica=2*L*(2*L*Ic^2*Ri^2-2*Vac*toff*Ic*Ri^2-4*L*Ic*Ri*Vc+Vac*toff*Ri*Vc+2*L*Vc^2)
kicb=(2*L*Vc+Ri*Vac*toff-2*Ic*L*Ri)^2
kic=kica/kicb
kvaca=2*Ic*L*Ri*toff*(Vc-Ic*Ri)
kvacb=(2*L*Vc+Ri*Vac*toff-2*Ic*L*Ri)^2
kvac=-kvaca/kvacb
kvca=2*Ic*L*Ri*Vac*toff
kvcb=(2*L*Vc+Ri*Vac*toff-2*Ic*L*Ri)^2
kvc=kvca/kvcb
kivc=1/Ri
kvcp=-toff/(2*L)

Figure 2.170 The FOT buck requires several small-signal parameters to be computed before launching an ac simulation. The t_{off} is fixed to 5 μs in this example.

The small-signal model SPICE response has been tested against that of Mathcad® obtained with the above equations.

Figure 2.171confirms the perfect match between these tools.

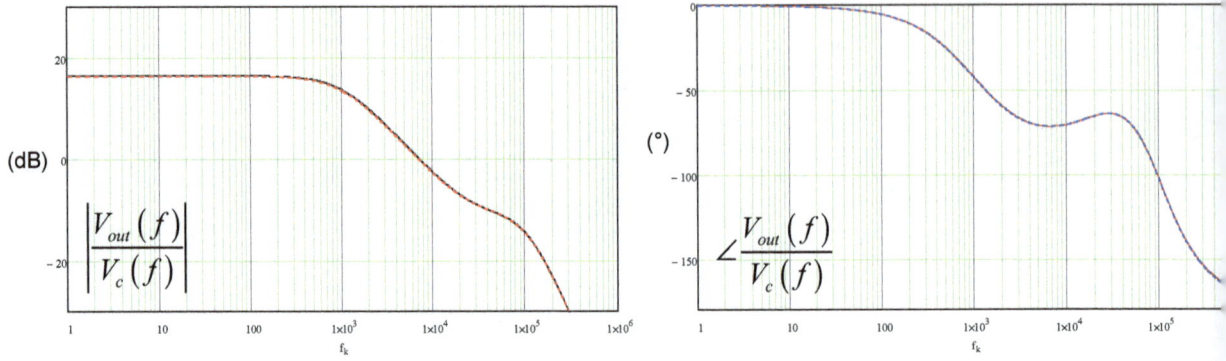

Figure 2.171 SPICE and Mathcad® deliver the exact same response.

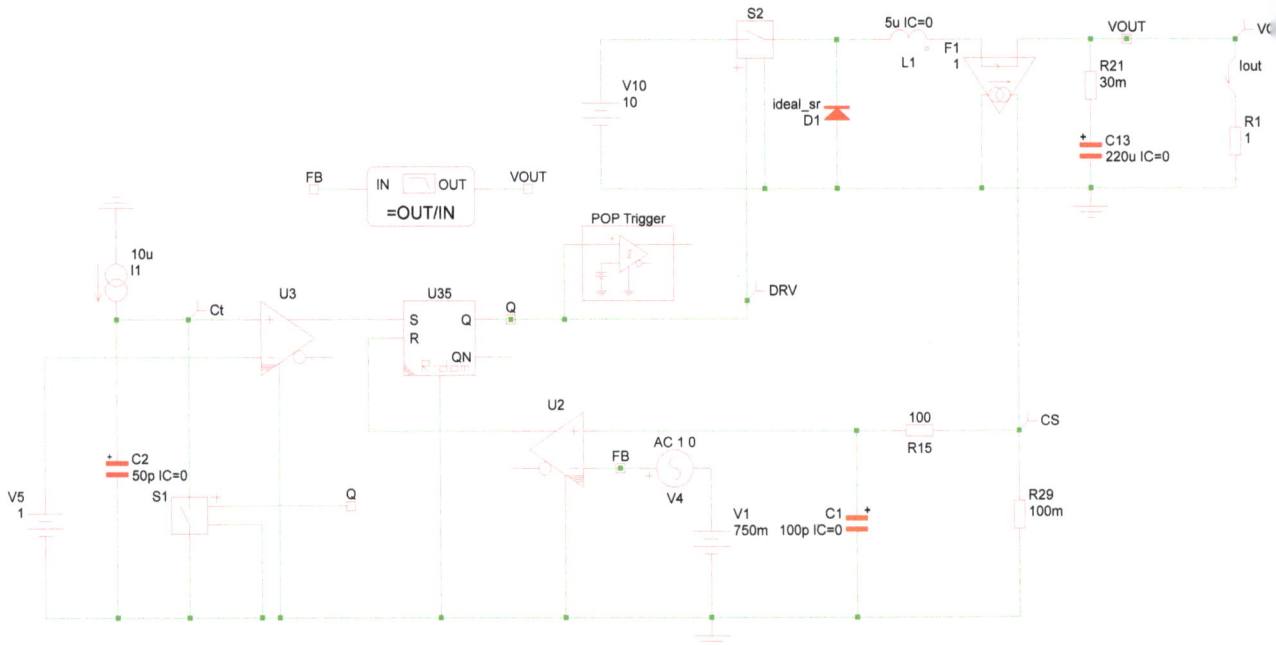

Figure 2.172 SIMPLIS® easily simulates a FOT buck operated in current mode.

A SIMPLIS® simulation file is presented in Figure 2.172 and will let us test our SPICE response versus that of the PWL simulator.

The graphs are plotted in Figure 2.172 and show a good agreement between the simulation engines.

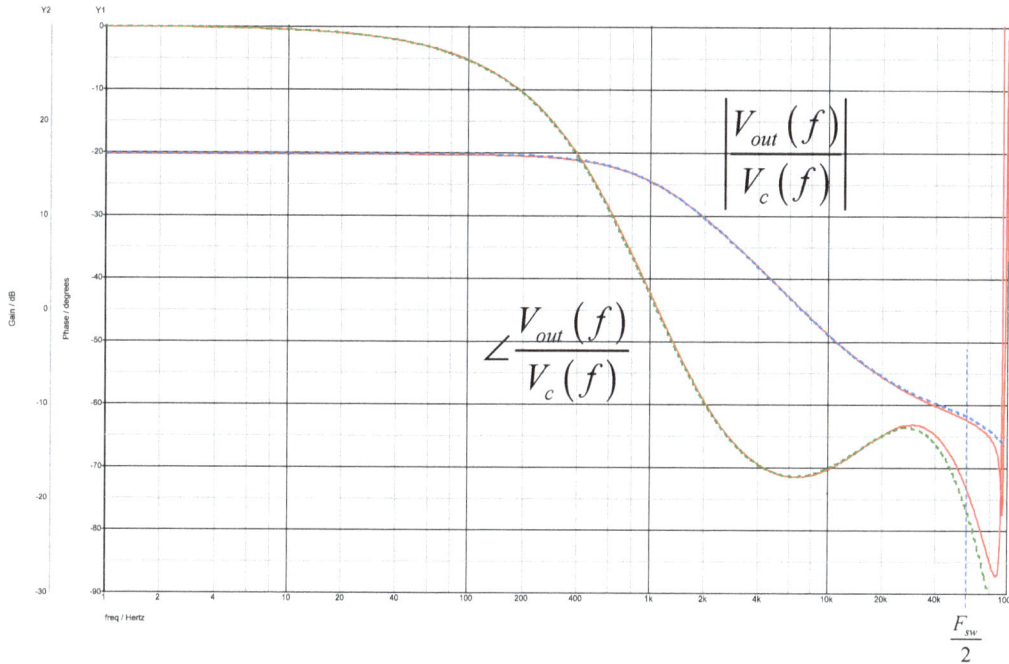

Figure 2.173 Very good match between SPICE curves (dashed) and SIMPLIS® results.

2.9 Fixed Frequency Multiphase Buck Transfer Function – Voltage- and Current-Mode Control

When very high output currents are required, it can be interested to add more switching phases to the buck converter which becomes a multiphase type. These high-current dc-dc converters are extremely popular in point-of-load converters for low-voltage high-speed processor units which can absorb peak currents up to 1 A/ns in extreme cases.

Small-signal wise, although the ripple frequency is increased, increasing the number of phases does not increase the equivalent control frequency which would have otherwise helped adopting a higher crossover frequency. You only enhance the control frequency by raising the frequency of each of the phases. However, considering the reduced power carried by each branch, we can say that the multiphase topology facilitates raising of the clock frequency.

The equivalent single-inductance power stage small-signal model features an inductance L_{eq} equal to the inductance L per phase divided by the number of phases n:

$$L_{eq} = \frac{L}{n} \tag{2.310}$$

The rest of the elements in the small-signal formulas from the CCM voltage-mode fixed-frequency analysis remain of identical values.

A 1.5-V 200-kHz two-phase voltage-mode buck converter appears in Figure 2.174 and will be simulated with SIMPLIS®. This is a simplified version of a real converter as for a high output current and a low output voltage, synchronous rectification would be implemented. Fortunately, the freewheel diodes are perfect and won't drop any voltage in this example. The duty ratio is limited to 50% but suppressing the crossed clock inputs to the OR gates removes this limit if necessary. The converter features two 5-μH inductors leading to an equivalent value of 2.5 μH.

255

With a 1000-µF output capacitor, the resonant frequency of the equivalent stage is thus:

$$f_0 = \frac{1}{2\pi\sqrt{2.5u \times 1m}} \approx 3.2 \text{ kHz} \tag{2.311}$$

The transfer function derived in the CCM voltage-mode section can be applied with the updated inductance value. The switching waveforms are given in Figure 2.175. You can see how the 200-kHz driving signals are interleaved and create a 400-kHz capacitor ripple current.

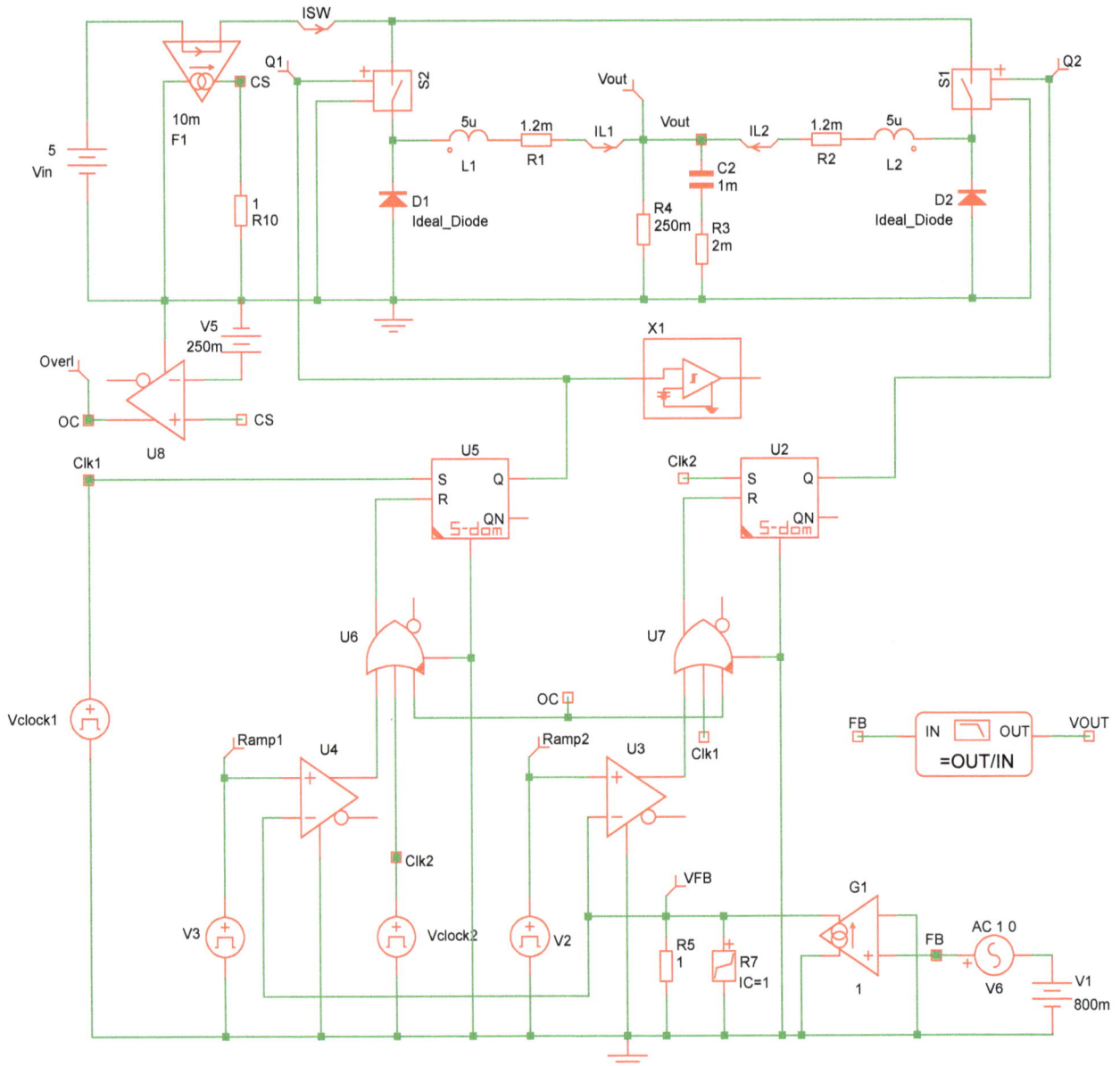

Figure 2.174 A multiphase buck converter can be simulated using more than one switching branch.

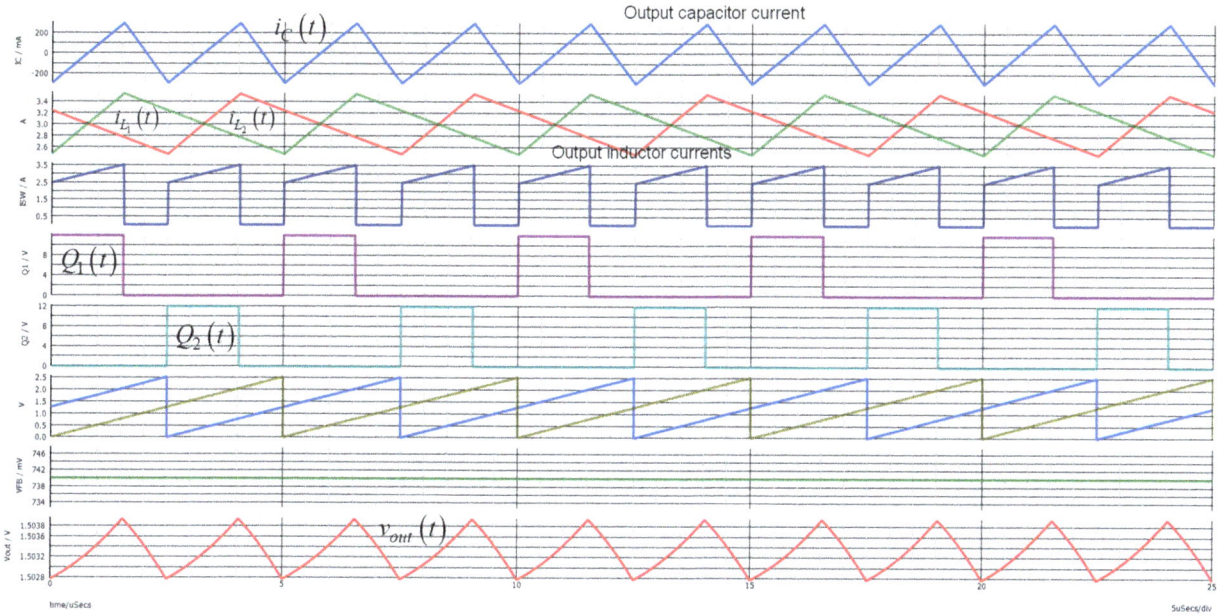

Figure 2.175 The switching waveforms show a capacitor ripple current frequency doubled compared to that of the individual swiches.

The frequency response appears in Figure 2.176 and confirms the resonance at 3.2 kHz. The 30-A version of this converter (50-mΩ load) features two 470-nH inductors with a 6-mF output capacitor. I purposely changed these values to observe a meaningful peaky response in voltage mode which would have otherwise been damped with the 50-mΩ load.

These are the values I used for the current-mode version that is shown in Figure 2.177. Each current loop receives some external ramp ($m_c = 1.2$) for slope compensation purposes. The duty ratio is well below 50% considering the present dc transfer characteristic. The cycle-by-cycle waveforms are given in Figure 2.178 and are very similar to those of the voltage-mode case at steady state.

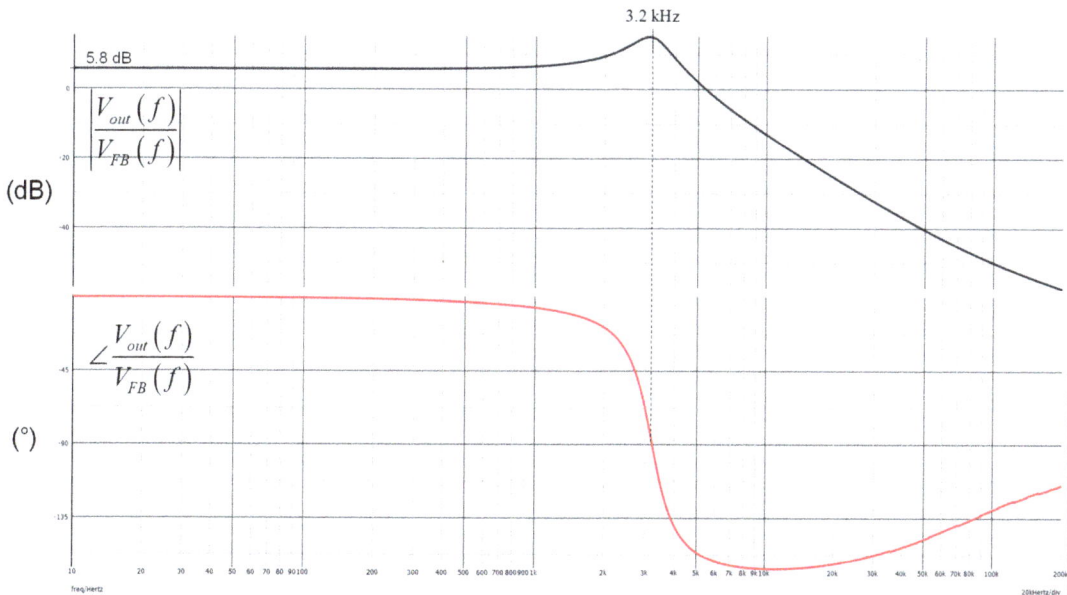

Figure 2.176 Very good match between SPICE curves (dashed) and SIMPLIS® results.

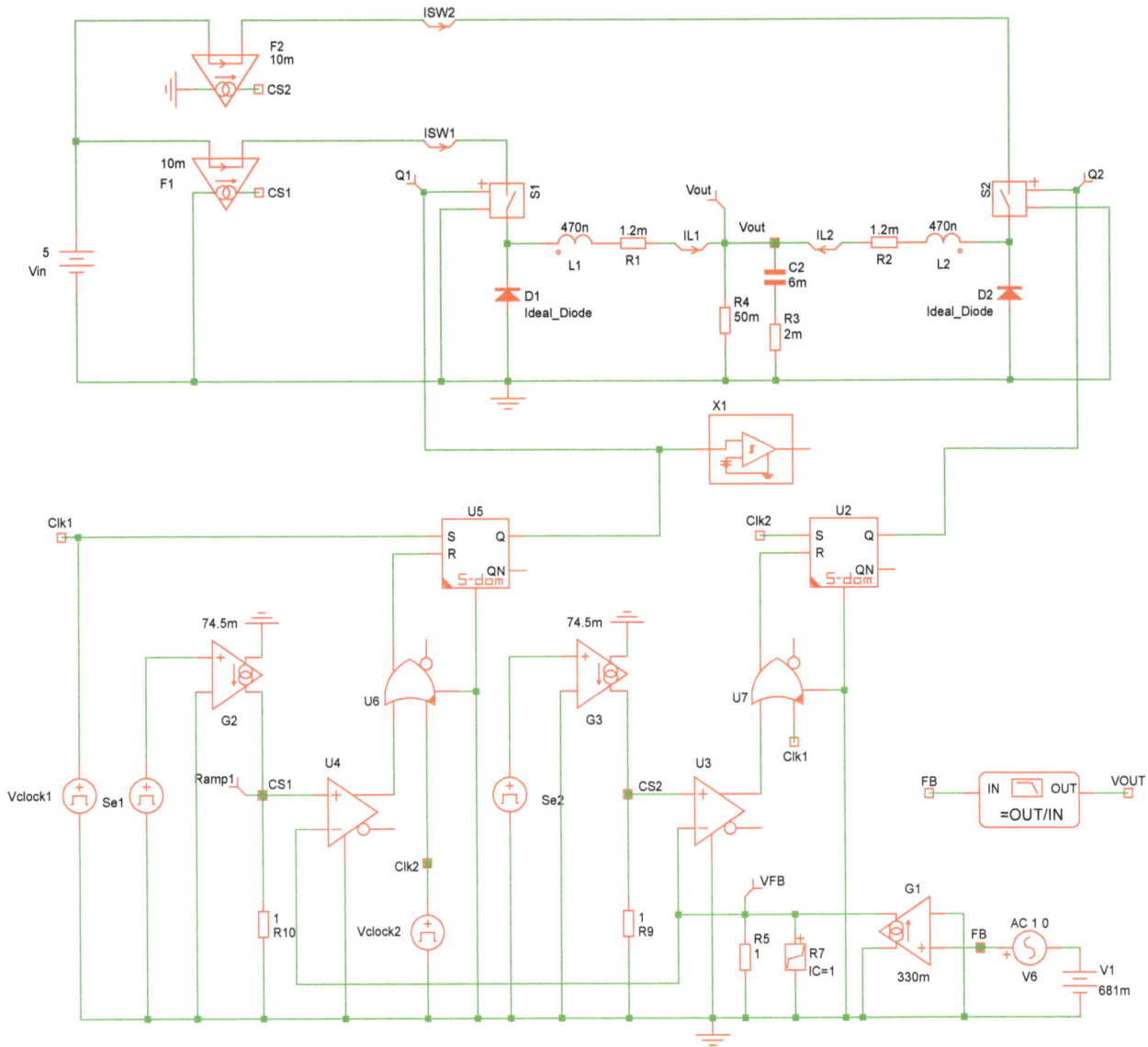

Figure 2.177 This 200-kHz current-mode converter delivers 1.5 V/30 A.

In current-mode control, the component values feeding the original buck formulas need an update as follows:

$$L_{eq} = \frac{L}{n} = \frac{470\ \text{nH}}{2} = 235\ \text{nH} \tag{2.312}$$

Which means the equivalent inductance for the single-inductor model is the same as in the multiphase voltage-mode version. The sense resistance also needs an update: the single-inductance equivalent small-signal model has a current loop gain amplified by a ratio depending on the number of phases n:

$$R_{iequ} = \frac{R_i}{n} = \frac{10\ \text{m}\Omega}{2} = 5\ \text{m}\Omega \tag{2.313}$$

The rest of the contributors in the small-signal formulas previously derived for the buck converter operated in CCM and current-mode control remain the same.

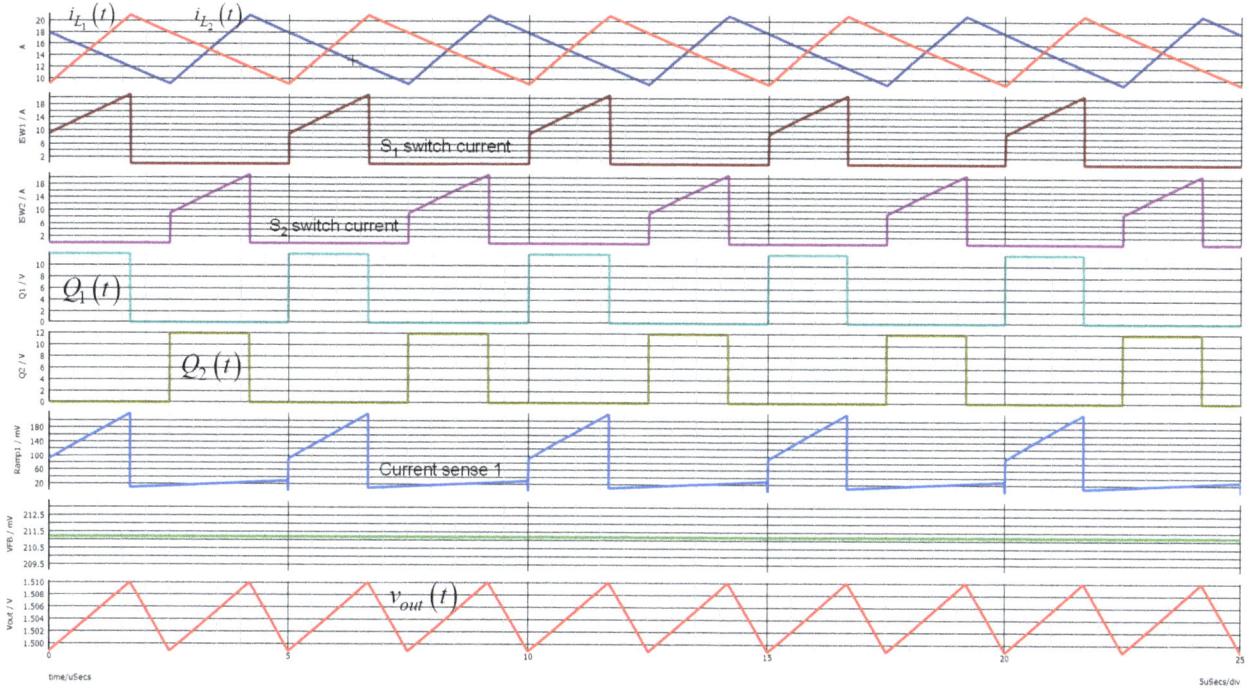

Figure 2.178 The switching waveforms are not different from the voltage-mode version in steady-state.

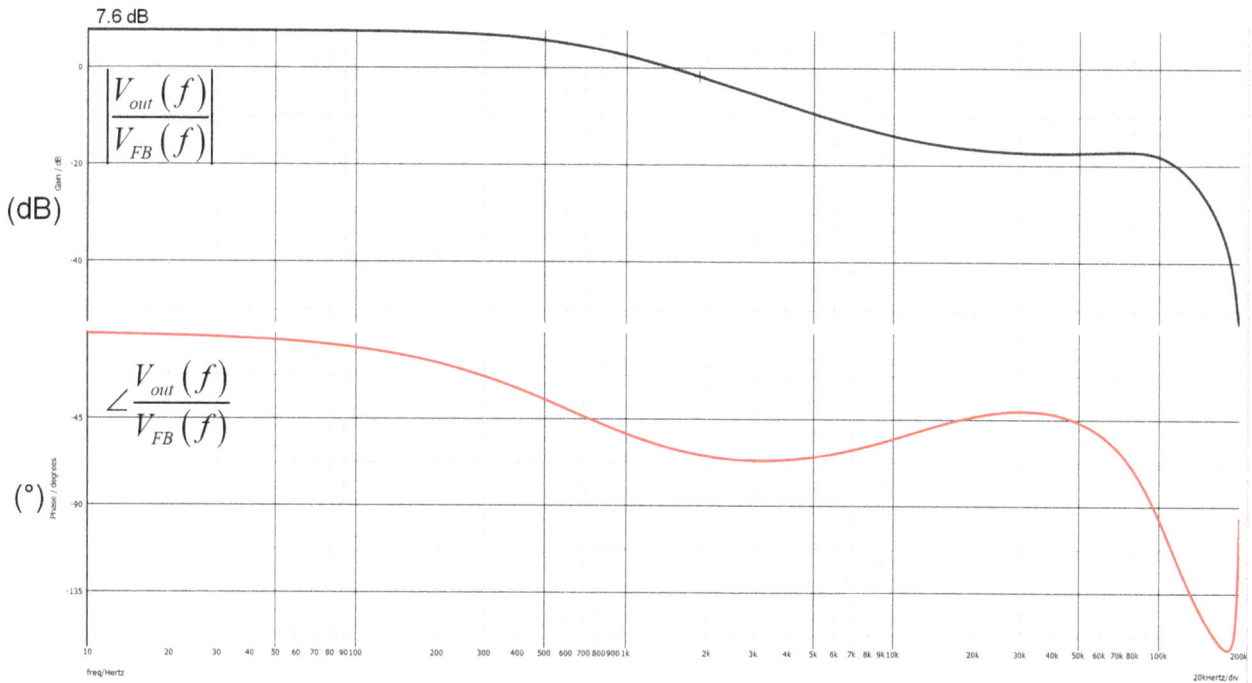

Figure 2.179 The dynamic response is that of a classical current mode converter.

In this application example, the quasi-static gain H_0 of the 200-kHz 2-phase buck converter operated in current-mode control is determined as:

$$H_0 = \frac{R_{load}}{Div \cdot R_{ieq}} \frac{1}{1 + \frac{R_{load} T_{sw}}{L_{eq}} \left[m_c (1-D) - 0.5 \right]} = \frac{50m}{3 \times \frac{10m}{2}} \frac{1}{1 + \frac{50m \times 5u}{470n} \left[1.2 \times (1-0.33) - 0.5 \right]} = 2.423 \text{ or} \approx 7.7 \text{ dB}$$

$$(2.314)$$

In this expression, Div represents the divide-by-3 voltage-controlled current-source G_1 setting the peak current setpoint from the op-amp output. The simulation results from Figure 2.179 confirm this value.

2.10 Isolated Buck-Derived Converters: The Forward Converter in Voltage-Mode Control

The forward converter has been introduced in Chapter 1. It is an isolated version of the buck converter. A simplified version of this converter is given in Figure 2.180. The transformer scales V_{in} by its turns ratio N. In voltage-mode control, the control-to-output transfer functions of the isolated buck in CCM or in DCM still hold but V_{in} and M need to include the transformer action:

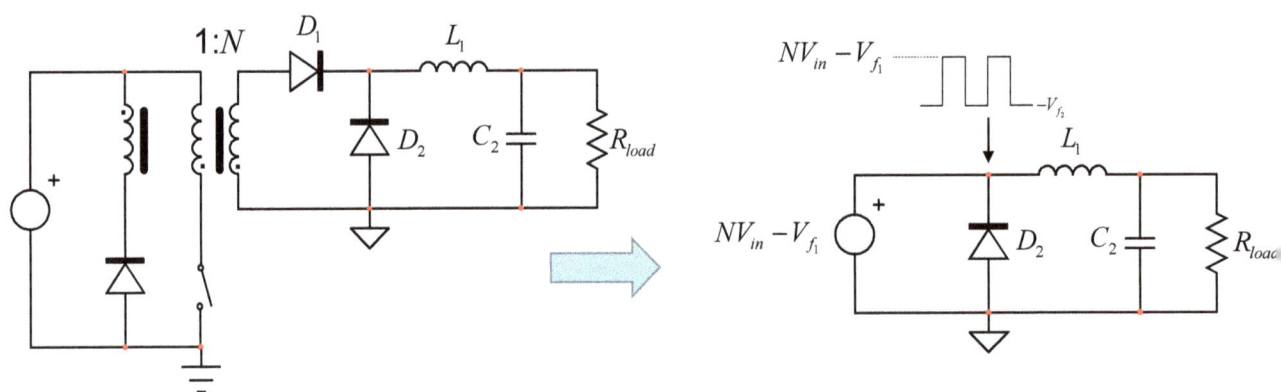

Figure 2.180 The transformer scales the input voltage during the on-time and the secondary-side buck converter remains the same.

$$V_{in} \rightarrow N \cdot V_{in} \tag{2.315}$$

$$M = \frac{V_{out}}{N \cdot V_{in}} \tag{2.316}$$

The summary of the control-to-output transfer functions of the forward operated in voltage-mode control appears in Figure 2.181.

To test our formulas versus simulation results, we have captured the schematic diagram of a VM forward converter operated in CCM and DCM. The circuit appears in Figure 2.182.

$\dfrac{V_{out}(s)}{V_{err}(s)}$ CCM	$H_0 \dfrac{1+\dfrac{s}{\omega_z}}{1+\dfrac{s}{\omega_0 Q}+\left(\dfrac{s}{\omega_0}\right)^2}$	$\omega_z = \dfrac{1}{r_C C_2}$	Double poles	$H_0 = \dfrac{NV_{in}}{V_p}\dfrac{R_{load}}{R_{load}+r_L}$

$$Q = \frac{L_1 C_2 \omega_0 (r_C + R_{load})}{L_1 + C_2(r_L r_C + r_L R_{load} + r_C R_{load})} \quad \omega_0 = \frac{1}{\sqrt{L_1 C_2}}\sqrt{\frac{r_L + R_{load}}{r_C + R_{load}}} \quad V_{out} = D \cdot N \cdot V_{in}\frac{R_{load}}{R_{load}+r_L} \quad M = \frac{V_{out}}{NV_{in}} \quad \text{Turns ratio } 1{:}N$$

$\dfrac{V_{out}(s)}{V_{err}(s)}$ DCM	$H_0 \dfrac{1+\dfrac{s}{\omega_z}}{\left(1+\dfrac{s}{\omega_{p1}}\right)\left(1+\dfrac{s}{\omega_{p2}}\right)}$	$\omega_z = \dfrac{1}{r_C C_2}$	$\omega_{p_1} \approx \dfrac{2-M}{1-M}\dfrac{1}{C_2 R_{load}}$ $\omega_{p_2} \approx \dfrac{L_1 + C_2 R_{load}^2 (1-M)}{L_1 C_2 R_{load}}$	$H_0 \approx \dfrac{2V_{out}}{DV_p}\dfrac{1-M}{2-M}$

$$M \approx \frac{2}{1+\sqrt{1+\dfrac{8\tau_L}{D_0^2}}} \quad \tau_L = \frac{L_1}{R_{load}}F_{sw} \quad D = \sqrt{\frac{2\tau_L}{NV_{in}(NV_{in}-V_{out})}}V_{out}$$

Figure 2.181 The forward converter transfer functions now need to account for the input voltage scaled by the transformer turns ratio. r_L and r_C are respectively the inductor and output capacitor ESRs.

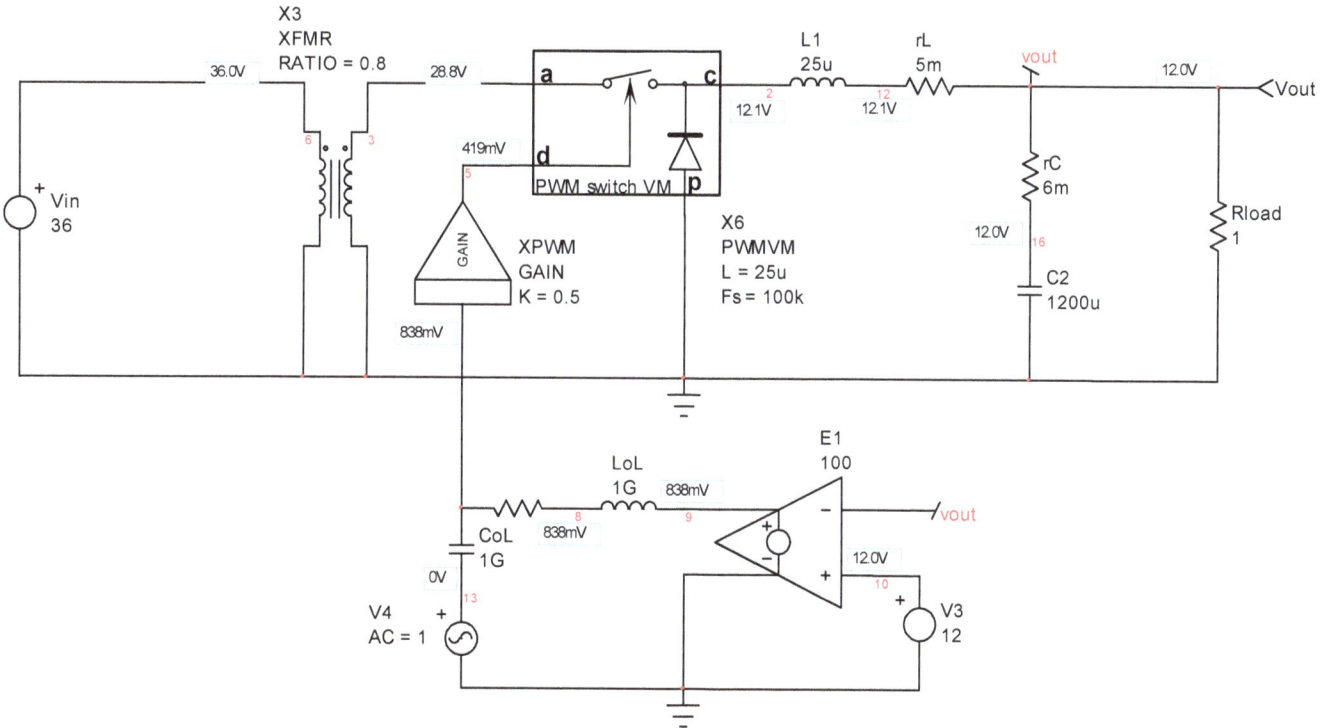

Figure 2.182 The VM-PWM switch model lends itself well to simulating a forward converter.

Updating (2.48) with the transformer turns ratio, we can find the load resistance value beyond which the converter transitions from CCM to DCM:

$$R_{crit} = \frac{2L_1 F_{sw}}{1 - \frac{V_{out}}{N V_{in}}} = \frac{2 \times 25u \times 100k}{1 - \frac{12}{0.8 \times 36}} \approx 8.6\ \Omega \tag{2.317}$$

With a 1-Ω load, the simulated buck converter operates in CCM. The Mathcad® sheet given in Figure 2.183 details the calculation steps and the obtained dynamic response for CCM ($R_{load} = 1\ \Omega$) and DCM ($R_{load} = 10\ \Omega$).

As confirmed by Figure 2.184, the SPICE model delivers the exact same dynamic responses, automatically toggling between CCM and DCM.

$L_1 := 25\mu H \qquad r_L := 0.005\Omega \qquad C_2 := 1200\mu F \qquad r_C := 0.006\Omega \qquad N_{ps} := 0.8$

$V_p := 2V \qquad V_{in} := 36V \qquad R_{load} := 1\Omega \qquad F_{sw} := 100kHz$

$\omega_z := \dfrac{1}{r_C \cdot C_2} \qquad \omega_0 := \dfrac{1}{\sqrt{L_1 \cdot C_2}} \sqrt{\dfrac{r_L + R_{load}}{r_C + R_{load}}}$

$Q := \dfrac{L_1 \cdot C_2 \cdot \omega_0 \cdot (r_C + R_{load})}{L_1 + C_2 \left[r_L \cdot r_C + R_{load} \cdot (r_L + r_C) \right]} = 4.555 \qquad 20 \cdot \log(Q) = 13.169$ dB

$H_0 := \dfrac{N_{ps} \cdot V_{in}}{V_p} \cdot \dfrac{R_{load}}{R_{load} + r_L} \qquad 20 \cdot \log(H_0) = 23.124$

$H_1(s) := H_0 \cdot \dfrac{1 + \dfrac{s}{\omega_z}}{1 + \dfrac{s}{Q \cdot \omega_0} + \left(\dfrac{s}{\omega_0}\right)^2}$

$L_1 := 25\mu H \qquad r_L := 0.005\Omega \qquad C_2 := 1200\mu F \qquad r_C := 0.006\Omega \qquad N_{ps} := 0.8$

$V_p := 2V \qquad V_{in} := 36V \qquad R_{load} := 10\Omega \qquad F_{sw} := 100kHz \qquad V_{out} := 12V$

$M := \dfrac{V_{out}}{N_{ps} \cdot V_{in}} = 0.417 \qquad \tau_L := \dfrac{L_1}{R_{load}} \cdot F_{sw} \qquad R_{crit} := \dfrac{2 \cdot L_1 \cdot F_{sw}}{1 - \dfrac{V_{out}}{N_{ps} \cdot V_{in}}} = 8.571\Omega$

$d_1 := \sqrt{\dfrac{2 \cdot \tau_L}{N_{ps} \cdot V_{in} \cdot (N_{ps} \cdot V_{in} - V_{out})}} \cdot V_{out} = 38.576\%$

$H_0 := \dfrac{2 \cdot V_{out}}{d_1 \cdot V_p} \cdot \dfrac{1 - M}{2 - M} = 11.461 \qquad 20 \cdot \log(H_0) = 21.184$ dB

$\omega_z := \dfrac{1}{r_C \cdot C_2} \qquad \omega_{p1} := \dfrac{2 - M}{1 - M} \cdot \dfrac{1}{R_{load} \cdot C_2} \qquad \omega_{p2} := \dfrac{L_1 + C_2 \cdot R_{load}^2 \cdot (1 - M)}{C_2 \cdot L_1 \cdot R_{load}}$

$H_1(s) := H_0 \cdot \dfrac{1 + \dfrac{s}{\omega_z}}{\left(1 + \dfrac{s}{\omega_{p1}}\right)\left(1 + \dfrac{s}{\omega_{p2}}\right)}$

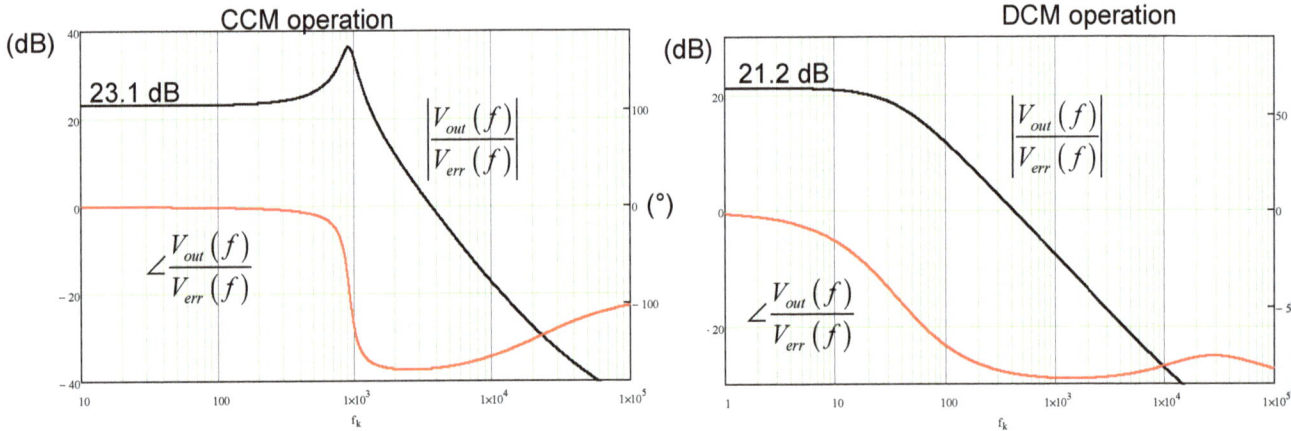

Figure 2.183 You need to account for the transformer turns ratio in the CCM and DCM control-to-output transfer functions.

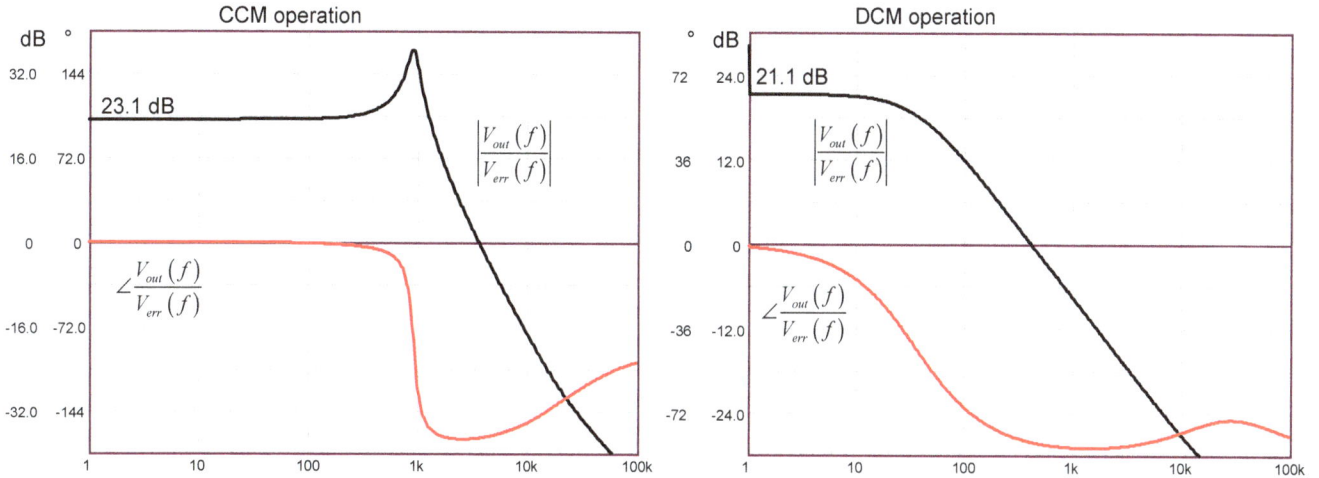

Figure 2.184 SPICE simulations lead to the exact same results with the auto-toggling averaged model.

2.11 Isolated Buck-Derived Converters: The Forward Converter in Current-Mode Control

In current-mode control, the control-to-output transfer functions of the isolated buck in CCM or in DCM still hold as in the voltage-mode case.

However, the transformer presence changes some of the original parameters. We will stick to the circuit of Figure 2.182 as shown in Figure 2.185. It is the simplest way to include the transformer as the PWM switch remains configured as a buck converter. Another option consists of inserting the transformer between the PWM switch and the output inductor as described in Figure 2.186 [6]. The operating points and dynamic responses are rigorously similar.

Using the Figure 2.185 circuit, some of the PWM switch static parameters change as follows:

$$V_{ac} = NV_{in} - V_{out} \tag{2.318}$$

$$V_{ap} = NV_{in} \tag{2.319}$$

The peak current in the output inductor is dictated by the primary-side current by:

$$I_{peak,sec} = \frac{V_c}{R_{sense}N} \tag{2.320}$$

In which R_{sense} is the sense resistance of your forward converter and N the transformer turns ratio, $N_p:N_s$.

263

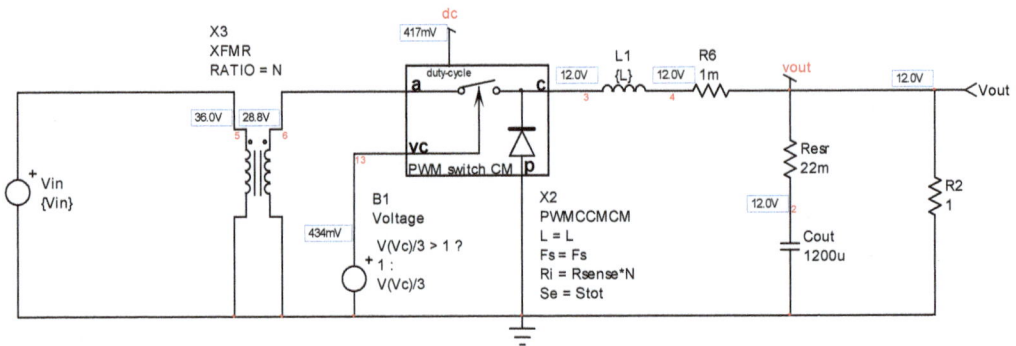

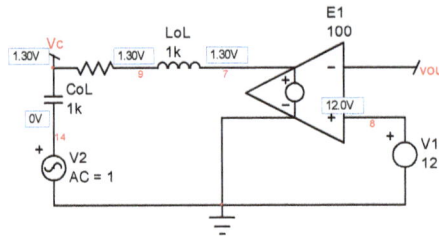

Figure 2.185 The transformer is inserted in series with V_{in} but the sense resistance needs to be scaled by the turns ratio.

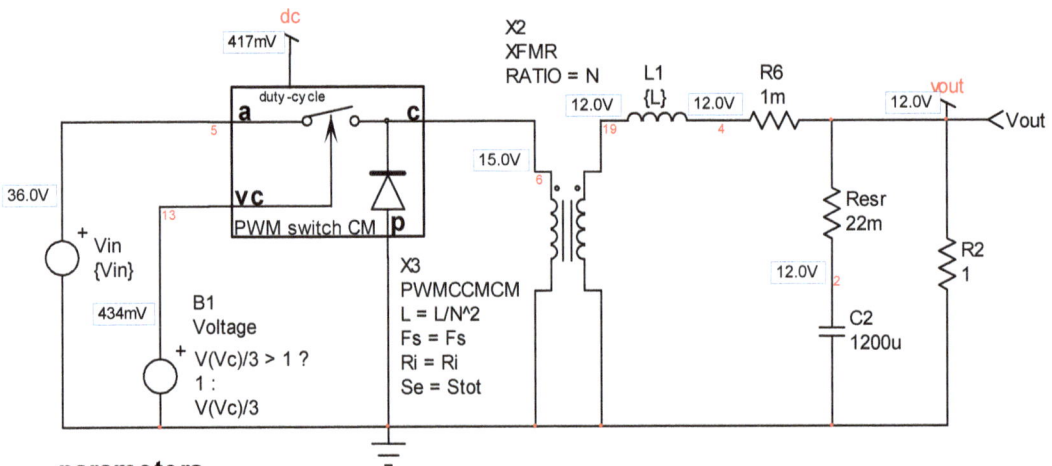

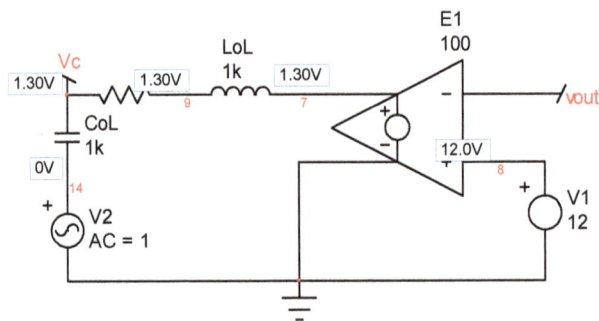

Figure 2.186 The transformer can also be inserted in series with the PWM switch but the output inductance needs to be scaled back to the primary side by the squared turns ratio.

However, as the transformer is inserted before the PWM switch, you must apply a scaling factor to R_i – the sense resistance parameter passed to the model – so that (2.320) is respected:

$$R_i = R_{sense} N \qquad (2.321)$$

You must also account for the natural ramp provided by the transformer magnetizing inductance L_{mag}. This slope is defined as:

$$S_{mag} = \frac{V_{in}}{L_{mag}} R_{sense} \qquad (2.322)$$

This magnetization current plays the role of the compensation ramp and naturally compensates the current-mode forward converter. It can be insufficient to adequately damp the subharmonic poles and, in this case, it must be supplemented with an external artificial ramp S_x.

In this configuration, the total external ramp parameter S_e passed to the model is:

$$S_e = S_{mag} + S_x \qquad (2.323)$$

Depending on the magnetizing current, the converter can be overcompensated (L_{mag} is small) and there is nothing you can do.

Accounting for these changes, the dc gain is as follows, including an internal feedback-to-current sense ratio – Div – as found in a variety of PWM controllers (e.g. 3 in a UC384x):

$$H_0 = \frac{R_{load}}{NR_{sense} Div} \frac{1}{1 + \frac{R_{load} T_{sw}}{L_1}\left[m_c\left(1-D\right)-0.5\right]} \qquad (2.324)$$

The rest of the transfer function valid for the CM buck (poles and zeros) is untouched. Figure 2.187 details the steps calculated by the Mathcad® sheet.

It is important to verify the average output current I_c is correct and corresponds to the 12-A we expect in this example (12 V, 1 Ω).

Design parameters:

$L_1 := 25\mu H$ $r_L := 0.001\Omega$ $C_1 := 1200\mu F$ $r_C := 0.022\Omega$ $L_{mag} := 1mH$

$N_{ps} := 0.8$ $V_{in} := 36V$ $R_1 := 1\Omega$ $F_{sw} := 100kHz$ $R_{sense} := 0.04\Omega$

$V_{out} := 12V$ $D := \dfrac{V_{out}}{N_{ps} \cdot V_{in}} = 41.66667\%$ $D_p := 1 - D = 58.33333\%$ $T_{sw} := \dfrac{1}{F_{sw}} = 10\mu s$

$||(x,y) := \dfrac{x \cdot y}{x + y}$ $R_i := R_{sense} \cdot N_{ps} = 0.032\Omega$ $M := \dfrac{V_{out}}{N_{ps} \cdot V_{in}} = 0.41667$

$C_s := \dfrac{1}{L_1 \cdot (\pi \cdot F_{sw})^2} = 405.28473nF$

$Div := 3$ UC384x division ratio

$S_{mag} := \dfrac{V_{in}}{L_{mag}} \cdot R_{sense} = 1.44\dfrac{kV}{s}$ artificial ramp brought by magnetizing current

$S_x := 0\dfrac{kV}{s}$ added artificial ramp

$S_e := S_x + S_{mag} = 1.44\dfrac{kV}{s}$

$V_c := 434mV$

$V_{ac} := N_{ps} \cdot V_{in} - V_{out} = 16.8V$ $V_{cp} := V_{out} = 12V$ $V_{ap} := V_{in} \cdot N_{ps}$

$S_n := \dfrac{V_{ac}}{L_1} \cdot R_i = 21.504\dfrac{kV}{s}$ $m_c := 1 + \dfrac{S_e}{S_n} = 1.06696$

$I_c := \dfrac{V_c}{R_i} - D \cdot T_{sw} \cdot \dfrac{S_e}{R_i} - V_{cp} \cdot (1 - D) \cdot T_{sw} \cdot \dfrac{1}{2 \cdot L_1} = 11.975A$

Small-signal parameters:

$k_i := \dfrac{D}{R_i} = 13.02083\dfrac{1}{\Omega}$ $g_o := \dfrac{T_{sw}}{L_1} \cdot \left[(1 - D) \cdot \dfrac{S_e}{S_n} + 0.5 - D \right] = 0.04896\dfrac{1}{\Omega}$

$k_o := \dfrac{1}{R_i} = 31.25\dfrac{1}{\Omega}$

$g_f := D \cdot g_o - D \cdot (1 - D) \cdot T_{sw} \cdot \dfrac{1}{2 \cdot L_1} = -0.02821\dfrac{1}{\Omega}$

$g_r := \dfrac{I_c}{V_{ap}} - g_o \cdot D = 0.3954\dfrac{1}{\Omega}$

$g_i := D \cdot \left(g_f - \dfrac{I_c}{V_{ap}} \right) = -0.185\dfrac{1}{\Omega}$

$G_0 := k_o \cdot \left[R_1 \,\middle\|\, \left(\dfrac{1}{g_o} \right) \right] \cdot \dfrac{1}{Div} = 9.93049$ $G_{dB} := 20 \cdot \log(G_0) = 19.93941$

Approximate gain expression

$H_0 := \dfrac{R_{load}}{R_i \cdot Div} \cdot \dfrac{1}{1 + \dfrac{R_{load} \cdot T_{sw}}{L_{out}} \cdot (m_c \cdot D_p - 0.5)} = 9.93$

$20 \cdot \log(H_0) = 19.939$ dc gain in dB

Figure 2.187 The small-signal parameters need a small tweak to account for the isolated forward converter.

In DCM, the path is not different and we will adopt similar relationship between parameters. The SPICE model is given in Figure 2.188 and will let us assess all the operating points.

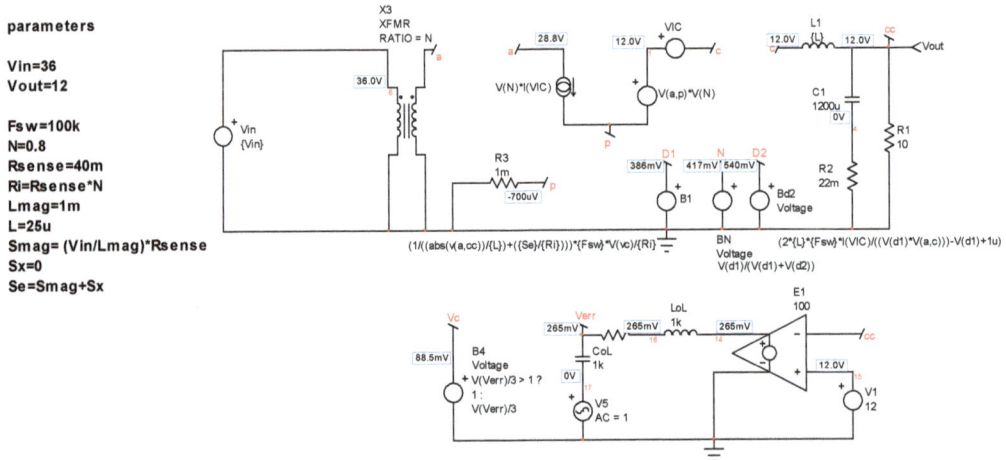

Figure 2.188 The DCM model uses the duty ratio factory approach described in the DCM CM section.

The duty ratio is computed to 38.6%, very close to what is calculed in the DCM VM version of the same forward converter. All calculations are gathered in Figure 2.189 and confirm the simulated bias points. The approximate dc gain H_0 for the DCM-operated forward converter is that of the DCM CM buck converter to which the transformer turns ratio N is added:

$$H_0 \approx \frac{2m_c V_{out}}{D \cdot Div} \frac{1 - M}{2m_c - (2 + mc) M} \frac{1}{S_n m_c T_{sw}} \tag{2.325}$$

In this expression, D is computed using the expression given in Chapter 1 and updated with the turns ratio:

$$D = \sqrt{\frac{2\tau_L}{NV_{in}(NV_{in} - V_{out})}} \cdot V_{out} \tag{2.326}$$

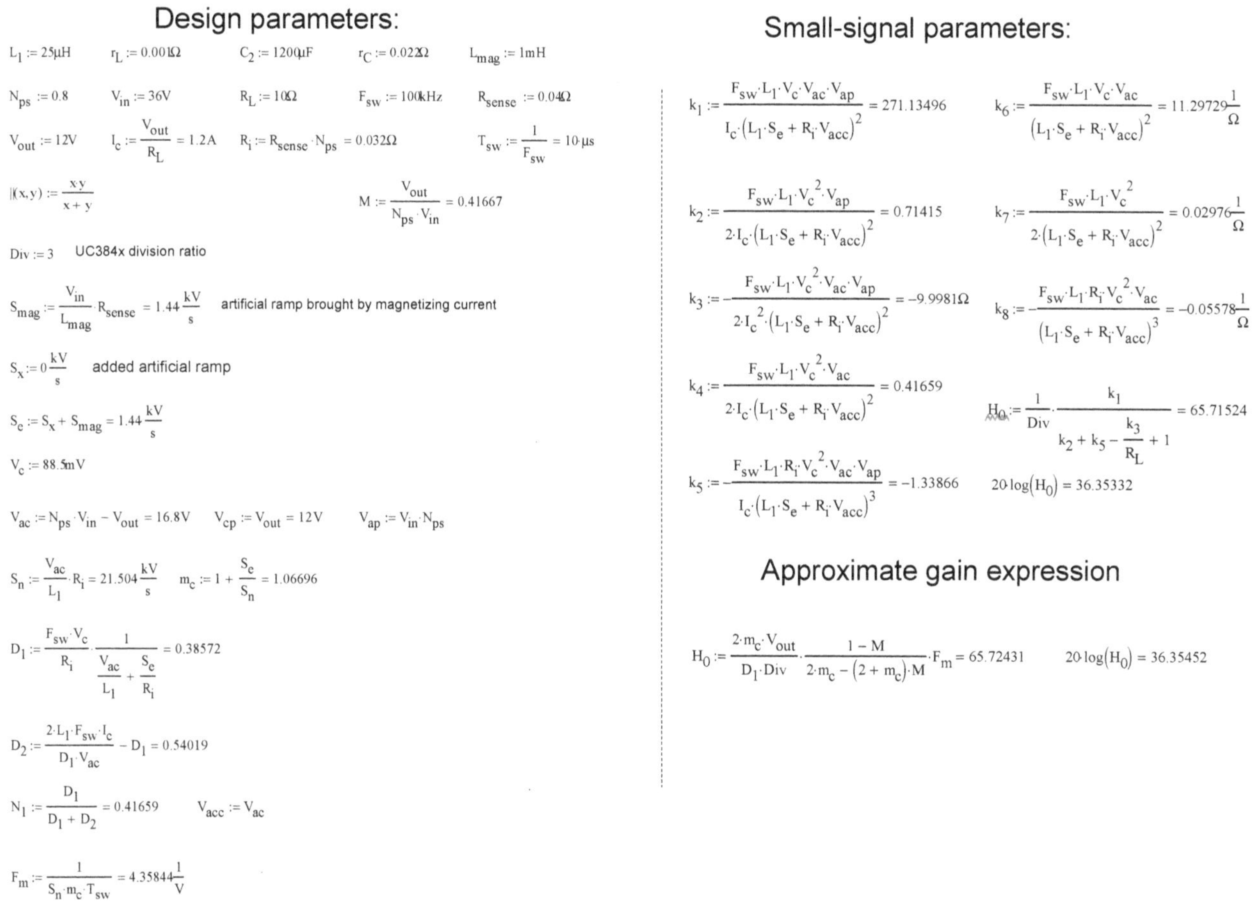

Design parameters:

$L_1 := 25\mu H$ $r_L := 0.001\Omega$ $C_2 := 1200\mu F$ $r_C := 0.022\Omega$ $L_{mag} := 1mH$

$N_{ps} := 0.8$ $V_{in} := 36V$ $R_L := 10\Omega$ $F_{sw} := 100kHz$ $R_{sense} := 0.04\Omega$

$V_{out} := 12V$ $I_c := \dfrac{V_{out}}{R_L} = 1.2A$ $R_i := R_{sense} \cdot N_{ps} = 0.032\Omega$ $T_{sw} := \dfrac{1}{F_{sw}} = 10\,\mu s$

$l(x,y) := \dfrac{x \cdot y}{x + y}$ $M := \dfrac{V_{out}}{N_{ps} \cdot V_{in}} = 0.41667$

$Div := 3$ UC384x division ratio

$S_{mag} := \dfrac{V_{in}}{L_{mag}} \cdot R_{sense} = 1.44 \dfrac{kV}{s}$ artificial ramp brought by magnetizing current

$S_x := 0 \dfrac{kV}{s}$ added artificial ramp

$S_e := S_x + S_{mag} = 1.44 \dfrac{kV}{s}$

$V_c := 88.5mV$

$V_{ac} := N_{ps} \cdot V_{in} - V_{out} = 16.8V$ $V_{cp} := V_{out} = 12V$ $V_{ap} := V_{in} \cdot N_{ps}$

$S_n := \dfrac{V_{ac}}{L_1} \cdot R_i = 21.504 \dfrac{kV}{s}$ $m_c := 1 + \dfrac{S_e}{S_n} = 1.06696$

$D_1 := \dfrac{F_{sw} \cdot V_c}{R_i} \cdot \dfrac{1}{\dfrac{V_{ac}}{L_1} + \dfrac{S_e}{R_i}} = 0.38572$

$D_2 := \dfrac{2 \cdot L_1 \cdot F_{sw} \cdot I_c}{D_1 \cdot V_{ac}} - D_1 = 0.54019$

$N_1 := \dfrac{D_1}{D_1 + D_2} = 0.41659$ $V_{acc} := V_{ac}$

$F_m := \dfrac{1}{S_n \cdot m_c \cdot T_{sw}} = 4.35844 \dfrac{1}{V}$

Small-signal parameters:

$k_1 := \dfrac{F_{sw} \cdot L_1 \cdot V_c \cdot V_{ac} \cdot V_{ap}}{I_c \cdot (L_1 \cdot S_e + R_i \cdot V_{acc})^2} = 271.13496$

$k_6 := \dfrac{F_{sw} \cdot L_1 \cdot V_c \cdot V_{ac}}{(L_1 \cdot S_e + R_i \cdot V_{acc})^2} = 11.29729 \dfrac{1}{\Omega}$

$k_2 := \dfrac{F_{sw} \cdot L_1 \cdot V_c^2 \cdot V_{ap}}{2 \cdot I_c \cdot (L_1 \cdot S_e + R_i \cdot V_{acc})^2} = 0.71415$

$k_7 := \dfrac{F_{sw} \cdot L_1 \cdot V_c^2}{2 \cdot (L_1 \cdot S_e + R_i \cdot V_{acc})^2} = 0.02976 \dfrac{1}{\Omega}$

$k_3 := -\dfrac{F_{sw} \cdot L_1 \cdot V_c^2 \cdot V_{ac} \cdot V_{ap}}{2 \cdot I_c^2 \cdot (L_1 \cdot S_e + R_i \cdot V_{acc})^2} = -9.9981\Omega$

$k_8 := \dfrac{F_{sw} \cdot L_1 \cdot R_i \cdot V_c^2 \cdot V_{ac}}{(L_1 \cdot S_e + R_i \cdot V_{acc})^3} = -0.05578 \dfrac{1}{\Omega}$

$k_4 := \dfrac{F_{sw} \cdot L_1 \cdot V_c^2 \cdot V_{ac}}{2 \cdot I_c \cdot (L_1 \cdot S_e + R_i \cdot V_{acc})^2} = 0.41659$

$H_0 := \dfrac{1}{Div} \cdot \dfrac{k_1}{k_2 + k_5 - \dfrac{k_3}{R_L} + 1} = 65.71524$

$k_5 := \dfrac{F_{sw} \cdot L_1 \cdot R_i \cdot V_c^2 \cdot V_{ac} \cdot V_{ap}}{I_c \cdot (L_1 \cdot S_e + R_i \cdot V_{acc})^3} = -1.33866$

$20 \cdot \log(H_0) = 36.35332$

Approximate gain expression

$H_0 := \dfrac{2 \cdot m_c \cdot V_{out}}{D_1 \cdot Div} \cdot \dfrac{1 - M}{2 \cdot m_c - (2 + m_c) \cdot M} \cdot F_m = 65.72431$ $20 \cdot \log(H_0) = 36.35452$

Figure 2.189 The original DCM CM Mathcad® sheet just needs an update on the PWM switch parameters to reflect the transformer insertion.

We now have everything to launch simulations of the CM forward converter operated in CCM and DCM. The computed results are given in Figure 2.190.

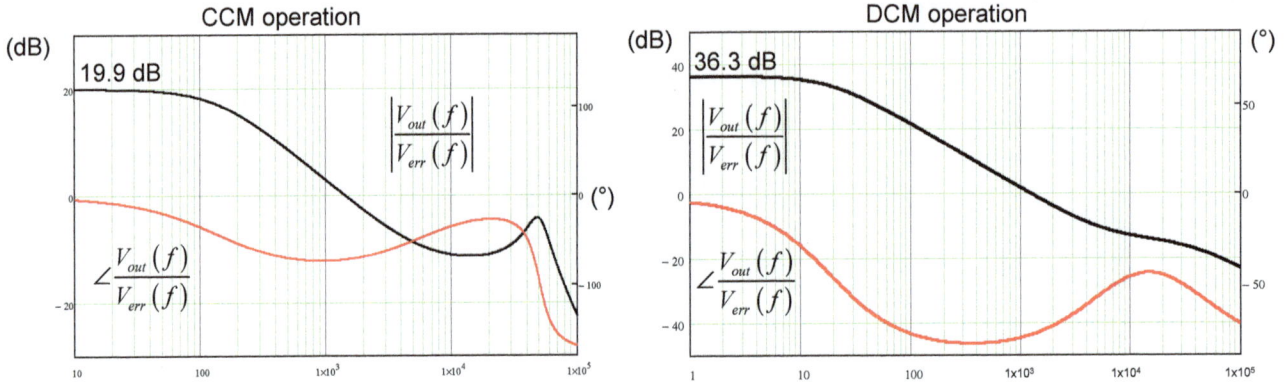

Figure 2.190 Dynamic responses of the CM forward are those of the CM buck converter scaled by the transformer ratio.

The simulation results of the CCM- and DCM-operated converters appear in Figure 2.191. They confirm the curves obtained with Mathcad®.

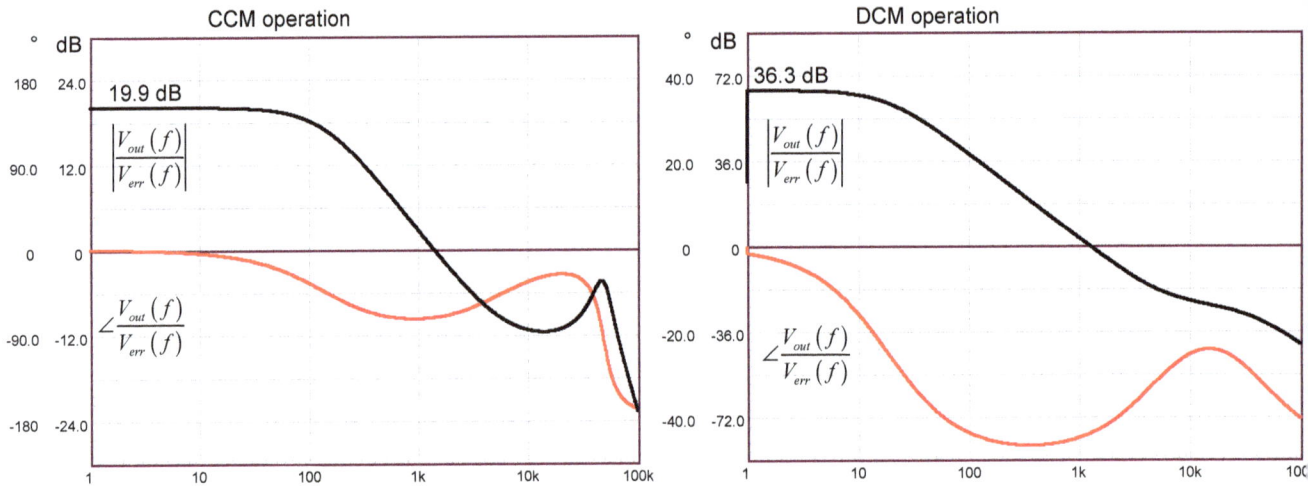

Figure 2.191 SPICE models confirm the curves obtained using the mathematical solver.

The summary of the control-to-output transfer functions of the forward converter operated in current mode in Figure 2.182.

$$\frac{V_{out}(s)}{V_c(s)}\bigg|_{CCM} \quad H_0 \frac{1+\dfrac{s}{\omega_z}}{1+\dfrac{s}{\omega_p}}\frac{1}{1+\dfrac{s}{\omega_n Q}+\left(\dfrac{s}{\omega_n}\right)^2} \quad\Bigg| \omega_z = \frac{1}{r_C C_2} \quad\Bigg| \omega_p = \frac{1}{R_{load}C_2}+\frac{T_{sw}}{L_1 C_2}\left[m_c(1-D)-0.5\right] \quad\Bigg| H_0 = \frac{R_{load}}{NR_{sense}Div}\frac{1}{1+\dfrac{R_{load}T_{sw}}{L_1}\left[m_c(1-D)-0.5\right]}$$

$$Q = \frac{1}{\pi\left[m_c(1-D)-0.5\right]} \quad \omega_n = \frac{\pi}{T_{sw}} \quad M = \frac{V_{out}}{NV_{in}} \quad \text{Turns ratio } 1{:}N \quad L_1 \text{ is the output inductance, } C_2 \text{ the output capacitance, } Div \text{ is the internal feedback-to-current setpoint divider.}$$

$$\frac{V_{out}(s)}{V_c(s)}\bigg|_{DCM} \quad H_0 \frac{1+\dfrac{s}{\omega_z}}{\left(1+\dfrac{s}{\omega_{p_1}}\right)\left(1+\dfrac{s}{\omega_{p_2}}\right)} \quad\Bigg| \omega_z = \frac{1}{r_C C_2} \quad\Bigg| \begin{aligned}\omega_{p_1} &= \frac{1}{R_{load}C_2}\frac{2m_c-(2+m_c)M}{m_c(1-M)}\\ \omega_{p_2} &= 2F_{sw}\left(\frac{M}{D}\right)^2\end{aligned} \quad\Bigg| H_0 \approx \frac{2m_c V_{out}}{D\cdot Div}\frac{1-M}{2m_c-(2+mc)M}\frac{1}{S_n m_c T_{sw}}$$

$$M = \frac{V_{out}}{NV_{in}} \quad \tau_L = \frac{L_1}{R_{load}}F_{sw} \quad D = \sqrt{\frac{2\tau_L}{NV_{in}(NV_{in}-V_{out})}}V_{out}$$

Figure 2.192 The transfer functions of the CM forward converter account for the turns ratio.

A final word on the two-swich forward whose transfer functions are similar to those of the single-switch version in VM and CM. The two-switch forward does not need a reset winding and recycles the magnetizing current via the input source, positively impacting efficiency. It requires a high-side drive (the upper transistor source is not ground-referenced) and two extra diodes but it is a rugged converter used in ATX power supplies some years ago.

There's no need to care about cross-conduction of the two MOSFETs as they turn on and off simultaneously. Figure 2.193 represents a voltage-mode implementation.

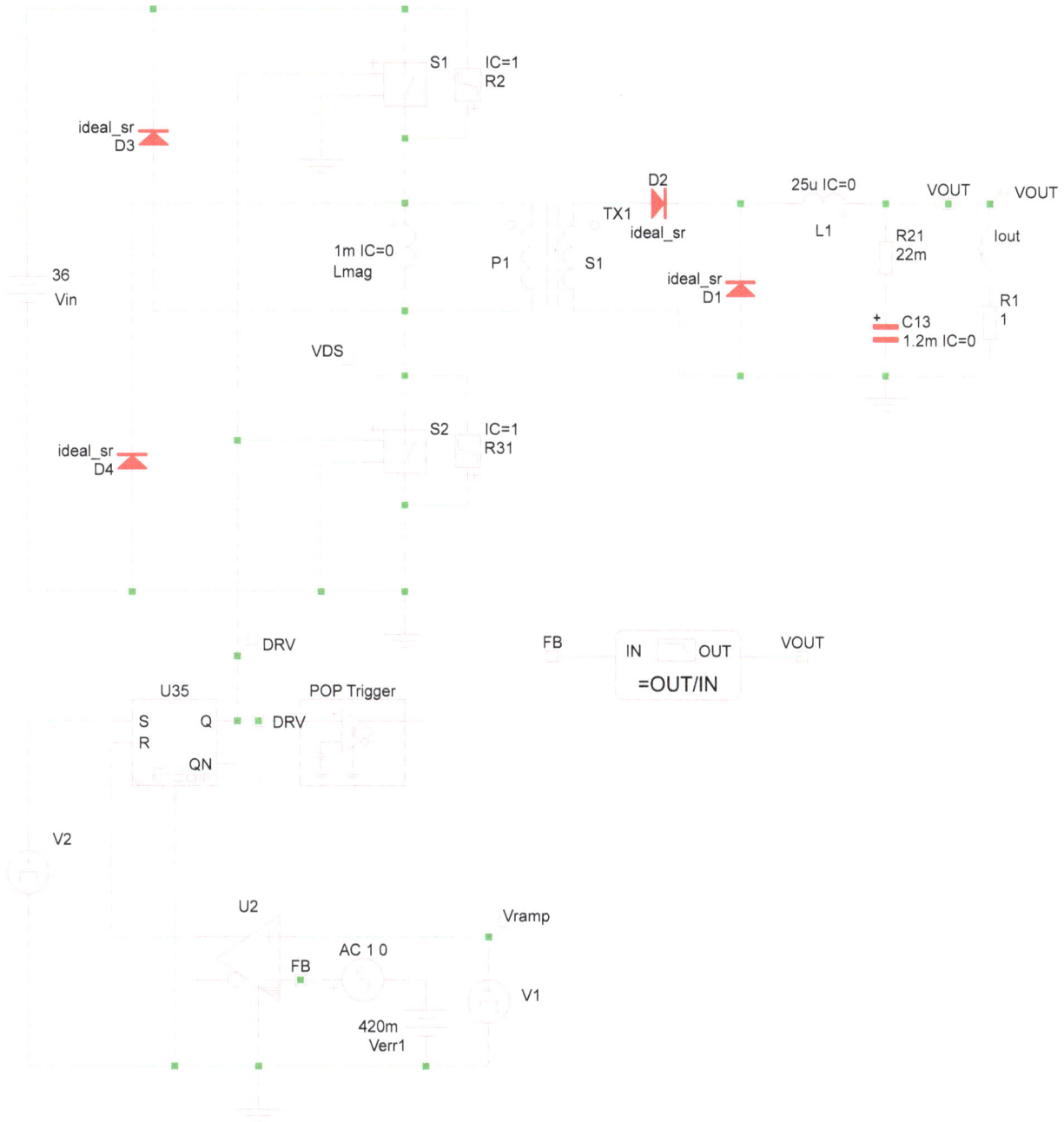

Figure 2.193 The 2-switch forward does not require a tertiary winding.

2.12 Isolated Buck-Derived Converters: The Forward Converter with Active-Clamp in Voltage-Mode Control

The active-clamp forward converter recycles the energy stored in the transformer leakage inductance to discharge the capacitance lumped at the drain node in a single-switch configuration. This technique helps reducing switching losses while increasing the operating frequency and offering a wider duty ratio dynamic (beyond 50%) compared to a classical forward converter (below 50% with a 1:1 demagnetization winding ratio).

The structure lends itself very well to self-driven synchronous rectification considering the longer demagnetization time compared to a classical forward converter.

To perform near-zero-volt switching (ZVS), the structure requires an extra switch which can be referenced to to either ground (P-channel) or the high-voltage rail (N-channel). The adequate timing control of this switch routes the leakage current via the input source and makes it circulate in the drain lumped capacitance, providing an efficient means to discharge it prior to turning the main MOSFET on again.

The complete control-to-output transfer function in voltage-mode control has been derived and documented in a 3-part article published in the on-line newsletter, How2Power.com [8]. The paper shows how transfer equations were derived together with an associated SPICE model.

The frequency response of the converter is described by the below equations:

$$\frac{V_{out}(s)}{D(s)} = F_0 \frac{1 + \dfrac{s}{\omega_{zF}}}{1 + \dfrac{s}{\omega_{0F}Q_F} + \left(\dfrac{s}{\omega_{0F}}\right)^2} \cdot N \left(V_{in} - D_0 r_{on1} M_0 \frac{sC_{clp}}{1 + \dfrac{s}{\omega_{0M}Q_M} + \left(\dfrac{s}{\omega_{0M}}\right)^2} \right) \tag{2.327}$$

In which:

$$F_0 = \frac{R_{Load}}{R_{load} + r_L} \tag{2.328}$$

$$\omega_{zF} = \frac{1}{r_C C_{out}} \tag{2.329}$$

$$\omega_{0F} = \frac{1}{\sqrt{L_{out}C_{out}}}\sqrt{\frac{r_L + R_{load}}{r_C + R_{load}}} \tag{2.330}$$

$$Q_F = \frac{L_{out}C_{out}\omega_{0F}(r_C + R_{Load})}{L_{out} + C_{out}\left[r_L r_C + R_{load}(r_L + r_C)\right]} \tag{2.331}$$

$$M_0 = \frac{V_{clamp}}{(1-D)^2} = \frac{V_{in}}{(1-D)^3} \tag{2.332}$$

$$\omega_{0M} = \frac{1-D}{\sqrt{L_{mag}C_{clp}}} \tag{2.333}$$

$$Q_M = \sqrt{\frac{L_{mag}}{C_{clp}}} \frac{1-D}{r_{on2}(1-D)+Dr_{on1}} \qquad (2.334)$$

In these expressions, r_{on1} represents the $r_{DS(on)}$ of the primary-side MOSFET while r_{on2} corresponds to the $r_{DS(on)}$ of the active-clamp MOSFET. This latter can lump the MOSFET on-resistance value with a damping resistance if any. L_{mag} is the transformer primary inductance, C_{clp} designates the clamp capacitor and L_{out}/C_{out} are respectively the output inductance and filter capacitor.

The right-side of the 4th-order polynomial form in (2.327) shows a term subtracting from the input voltage V_{in}. This term expresses the loss across the primary-side power MOSFET created by the resonant magnetizing current.

This current peaks and invokes the notch observed in the frequency response of Figure 2.194.

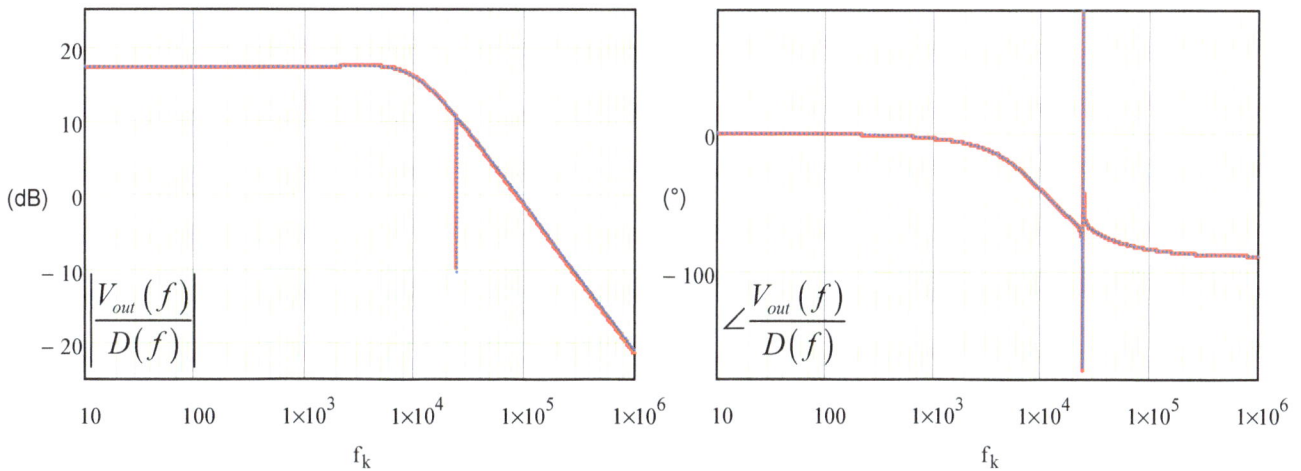

Figure 2.194 The notch in the frequency response is due to the resonating magnetizing current involving the active-clamp capacitor. [8] shows how the SPICE model responses matches these results perfectly.

Unlike in the other examples, we have built a simple open-loop prototype to verify our equations and the SPICE model. This 50-kHz 5-V/5-A converter associates a few logic gates to create the proper control timings for both power MOSFETs.

The circuit appears in Figure 2.195.

We have used a P-channel referenced to ground which together with C_6 represent the active-clamp circuitry. A deadtime is inserted between A and M outputs so that near-ZVS is obtained. Typical operating waveforms are given in Figure 2.196. The left-side picture describes the delay effect in the control of both MOSFETs.

The right side illustrates the magnetizing current peaking to a negative value prior to opening the active-clamp P-channel MOSFET.

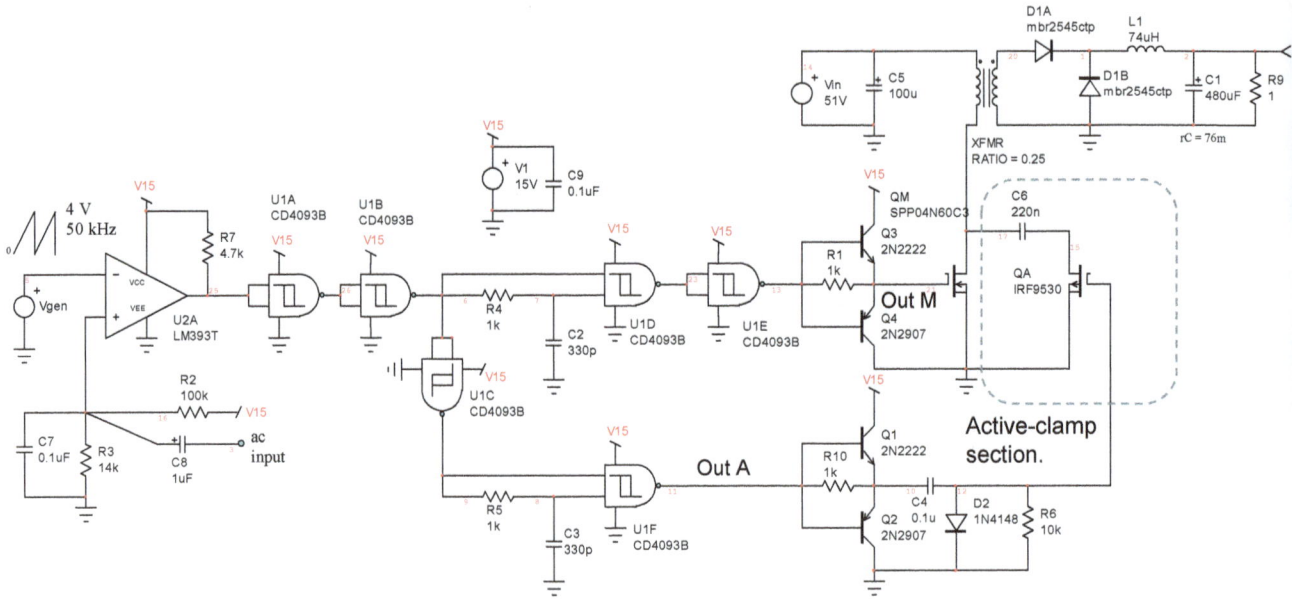

Figure 2.195 A prototype has been assembled to check if its dynamic response matches what our model predicts.

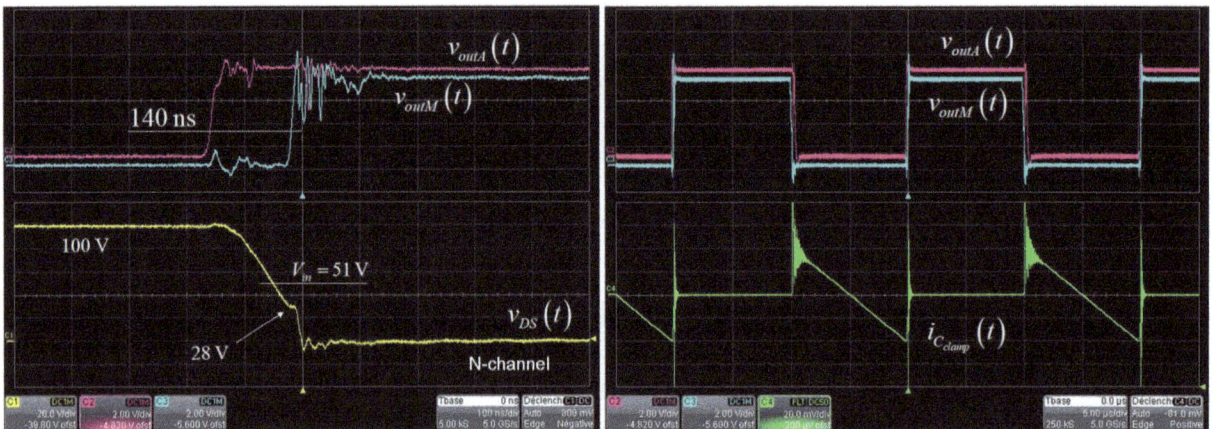

Figure 2.196 Proper deadtime adjustment leads to a near-ZVS operation at full load.

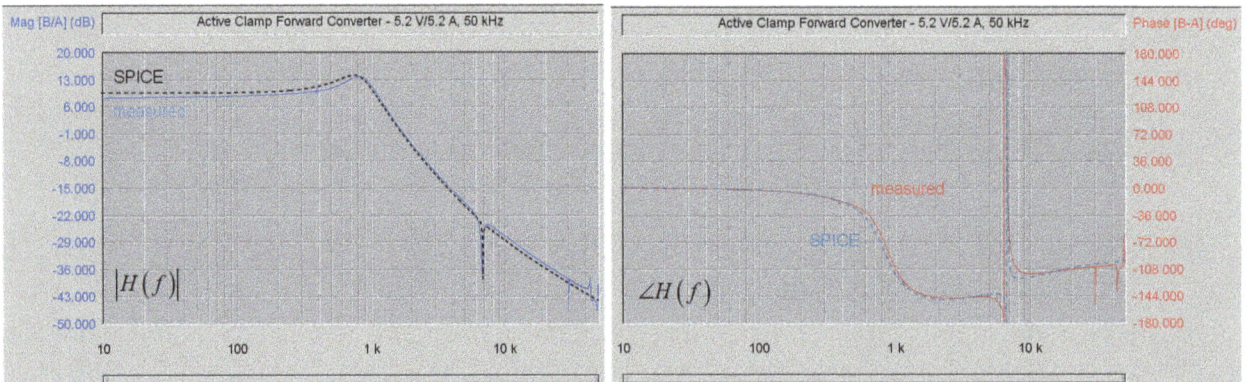

Figure 2.197 Magnitude and phase responses from the prototype are identical to that of the SPICE model.

The measurements results match the SPICE response very well as confirmed by Figure 2.197. The magnitude and phase responses are those of a classical forward converter until the notch resonance affects the curves. Without proper damping action of the clamping circuit, it is difficult to enjoy a wide bandwith as crossover frequency would have to be selected well before the worst-case notch occurrence. Fortunately, adding a small resistance (10 Ω in this example) in series with the clamping capacitor helps damping the whole circuit and the dynamic response now resembles that of a classical forward converter as Figure 2.198 confirms. It is now possible to select a crossover frequency beyond the notch occurrence, enjoying a wider bandwidth.

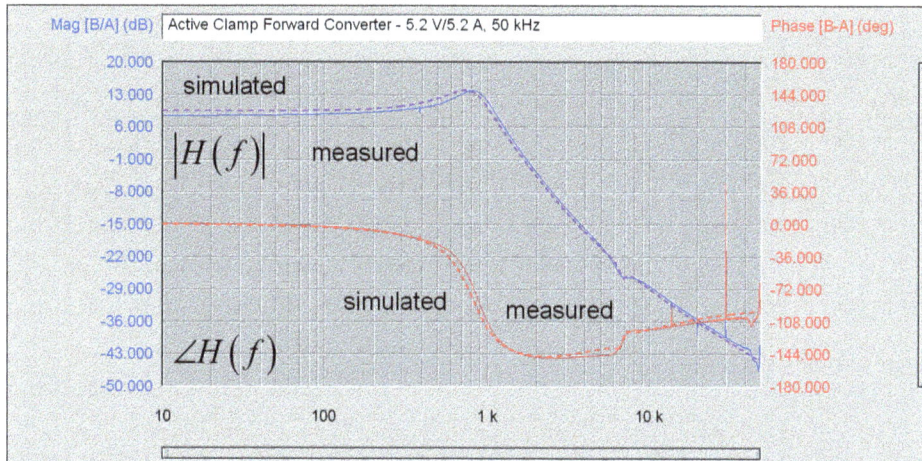

Figure 2.198 Adding a small resistance in series with the active-clamp capacitor nicely damps the circuit and makes its frequency response look closer to that of a classical voltage-mode forward converter.

I will not present the control-to-output transfer function of the current-mode control version in this edition.

However, when determing the control-to-output transfer function, a resonance occurs in the magnitude response rather than a notch as in the above curves. The phase distortion at this point compromises the crossover frequency selection and you must keep it well below this point.

Options such as damping or reducing the clamp capacitor value exist but make the exercise more complicated than with the voltage-mode version in my opinion.

2.13 Isolated Buck-Derived Converters: The Push-Pull Converter in Current- and Voltage-Mode Control

The push-pull converter is well suited for high-power 200-300-W dc-dc converters usually operated from a low input voltage. The transformer usage is excellent since the flux swings in both directions across zero. Proper balancing must thus be ensured and that is the Achille's heel of these buck-derived converters (push-pull, half- and full-bridge).

Transfer functions in voltage- or current-mode of the push-pull converter are those of the forward converter in which the transformer turns ratio must be accounted for. A SIMPLIS® simulation template for a voltage-mode version appears in Figure 2.199.

The PWM circuit is built around U_4 while the D flip-flop generates the correct drive signals via two NOR gates.

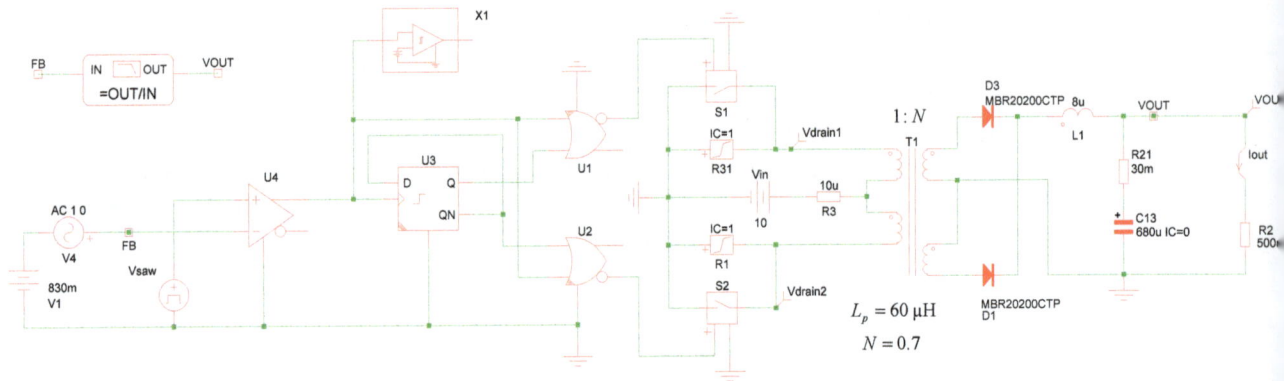

Figure 2.199 The push-pull in voltage-mode control is simulated with a simple D flip-flop.

The simulation results are given in Figure 2.200 and show the 2nd-order response of the buck converter scaled by the transformer turns ratio.

The transfer function describing the control-to-output transfer function of the push-pull converter operated in voltage-mode is that of the forward converter working in similar conditions.

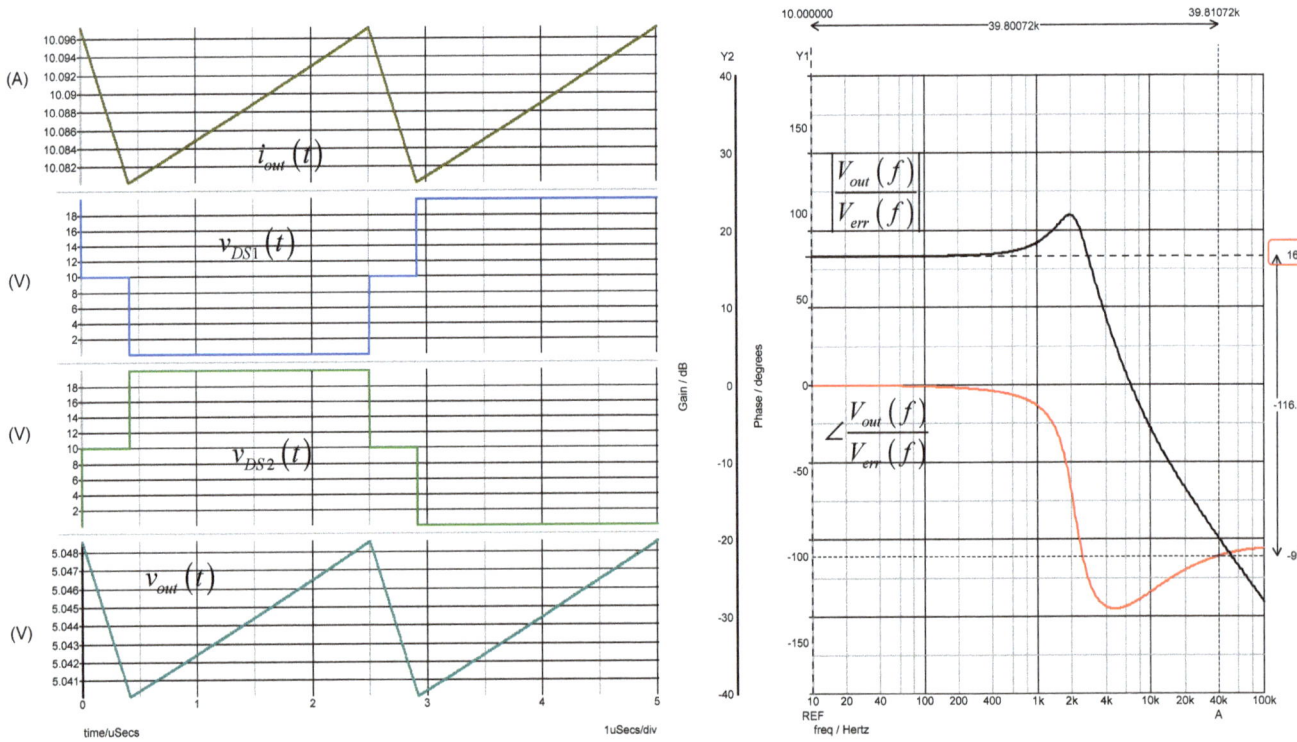

Figure 2.200 The response is similar to that of the CCM forward converter operated in voltage-mode.

In current-mode control, the logic is a bit different as current sensing is required (it is also mandatory in VM but was not represented for simplicity reasons). The corresponding template appears in Figure 2.201.

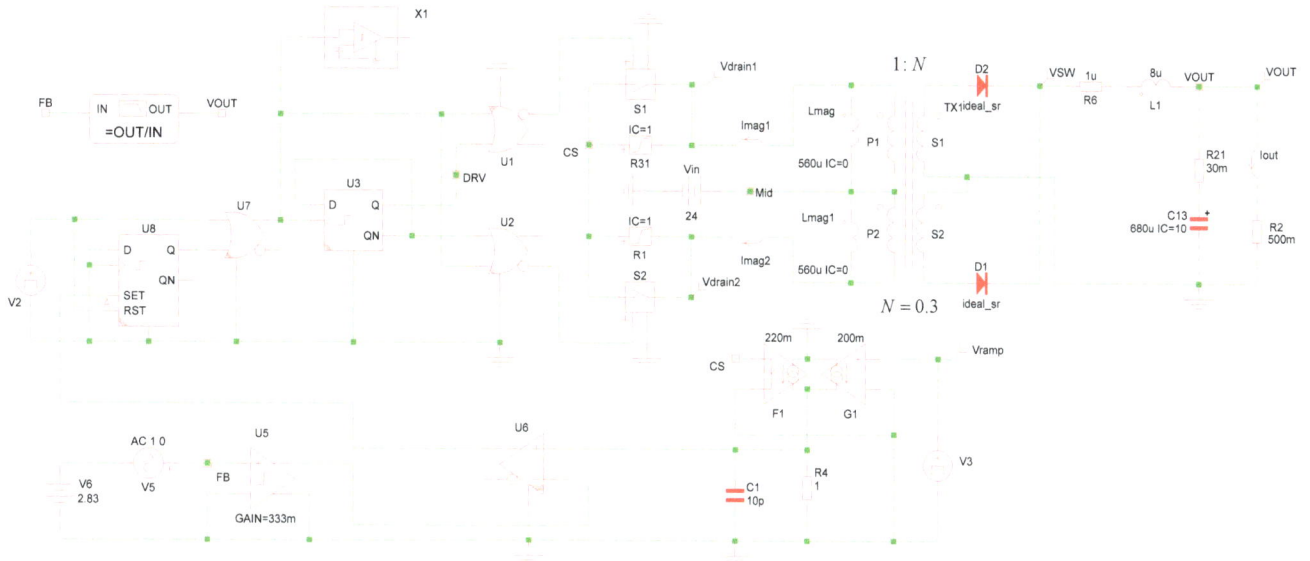

Figure 2.201 The logic blocks are slightly more complex for a current-mode version.

The primary inductance is 1.12 mH and you have to split this value in two for the calculation of the magnetizing current contribution. In this example, despite its presence, it is supplemented with an artificial ramp produced by V_3. Since it delivers a 1-V/10-μs ramp or a slope of 100 mV/μs, we need 20 mV/μs as extra ramp that we sum up with the current sense information via two current-controlled (F_1) and voltage-controlled (G_1) elements in a 1-Ω resistance (R_4). We could have used a resistive arrangement but going through these sources is easier since you see how much ramp you add. The converter delivers 5 V to 0.5-Ω load and is supplied from a 24-V source. The simulation results are given in Figure 2.202.

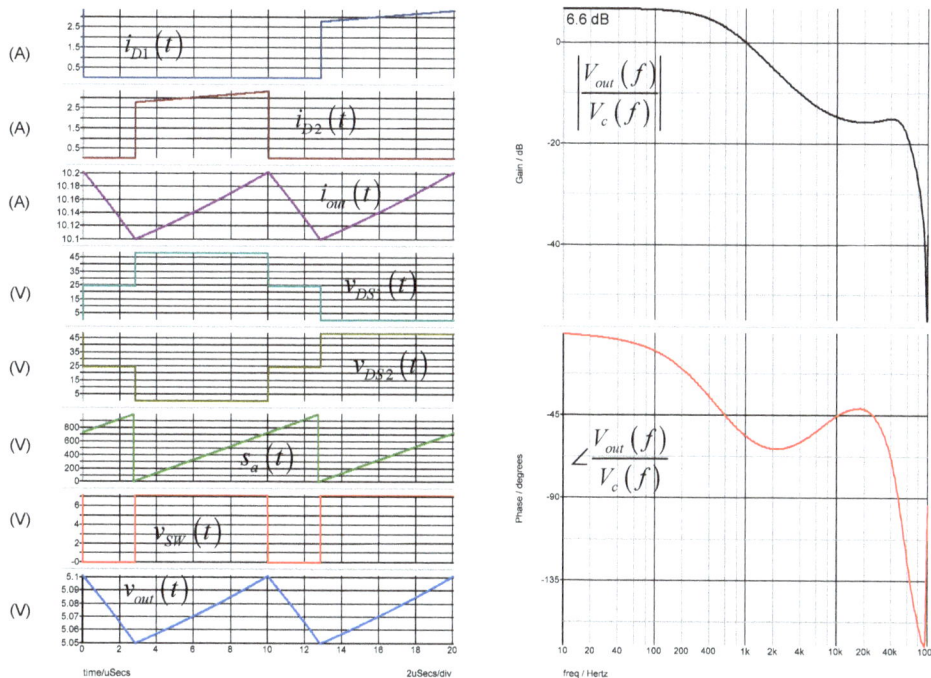

Figure 2.202 The response is similar to the CCM forward converter operated in current-mode.

2.14 Isolated Buck-Derived Converters: The Half- and Full-Bridge Converters

These converters are used in high-power applications. The half-bridge converter usually operates in voltage-mode only as current-mode control can potentially bring flux imbalance via the depletion of one of the capacitors. Figure 2.203 represents a typical half-bridge converter. Please note the presence of R_3, a 100-$\mu\Omega$ resistance, necessary to break the capacitive loop: without it, the simulation engine fails to converge. In this application, the capacitive node provides a bias of half the input voltage. The output voltage is thus determined as:

$$V_{out} = D\frac{NV_{in}}{2}\frac{R_{load}}{R_{load}+r_L} = 0.72\times\frac{1.55\times10}{2\times1}\frac{500m}{500m+10m} = 5.47\ \text{V} \tag{2.335}$$

The quasi-static gain accounts also for this division by 2:

$$H_0 = \frac{dV_{out}(D)}{dD} = \frac{NV_{in}}{2V_p}\frac{R_{load}}{R_{load}+r_L} = \frac{1.55\times10}{2\times1}\frac{500m}{500m+10m} \approx 7.6\ \text{or}\ 17.6\ \text{dB} \tag{2.336}$$

The rest of the expressions for the forward converter in voltage-mode control still hold.

Simulations results are given in Figure 2.204 and confirm our calculations.

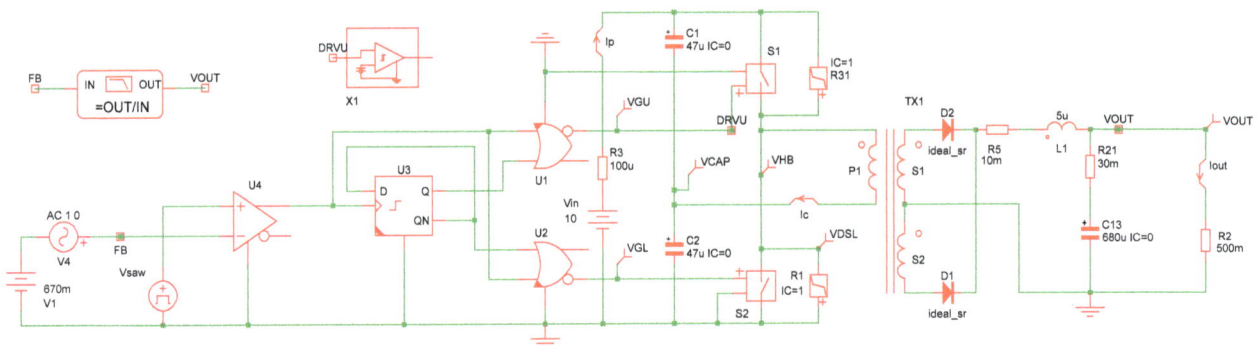

Figure 2.203 The half-bridge converter is usually operated in voltage-mode control.

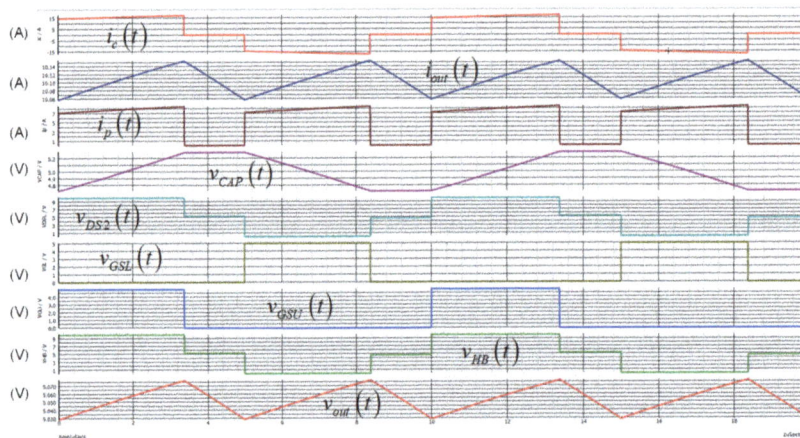

Figure 2.204 The voltage at the capacitive node must be kept constant for proper operation and can require large capacitor values.

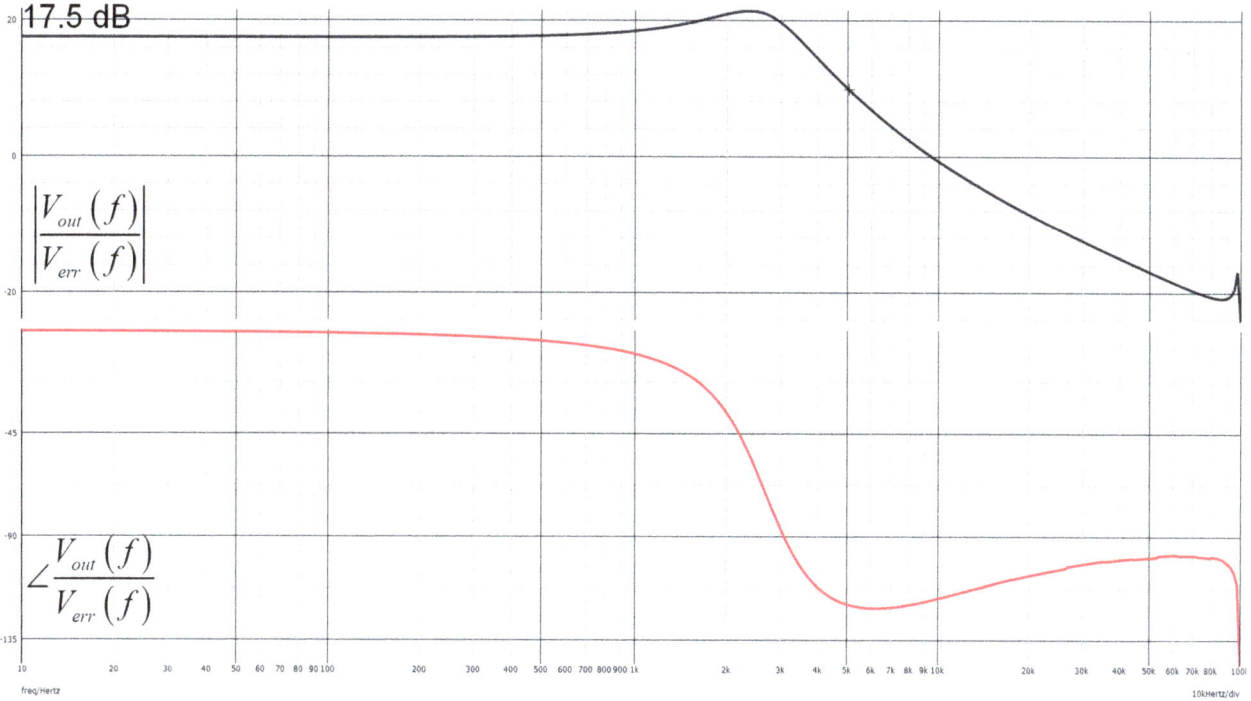

Figure 2.205 The quasi-static gain of the half-bridge requires a division by two compared to its equivalent in forward mode.

The full-bridge converter is well suited for large output power levels, beyond 500 W and over several kW. Its schematic diagram appears in Figure 2.206. It resembles the 2-switch forward in which the primary-side freewheel diodes have been replaced by a pair of controlled switches. In the proposed circuits, the diagonal legs are alternatively switched on and off: S_1 and S_4 turn on then S_2 and S_3.

The transfer functions of the forward converter operated in voltage-mode hold for the full-bridge converter. Transient and frequency responses are given in Figure 2.207.

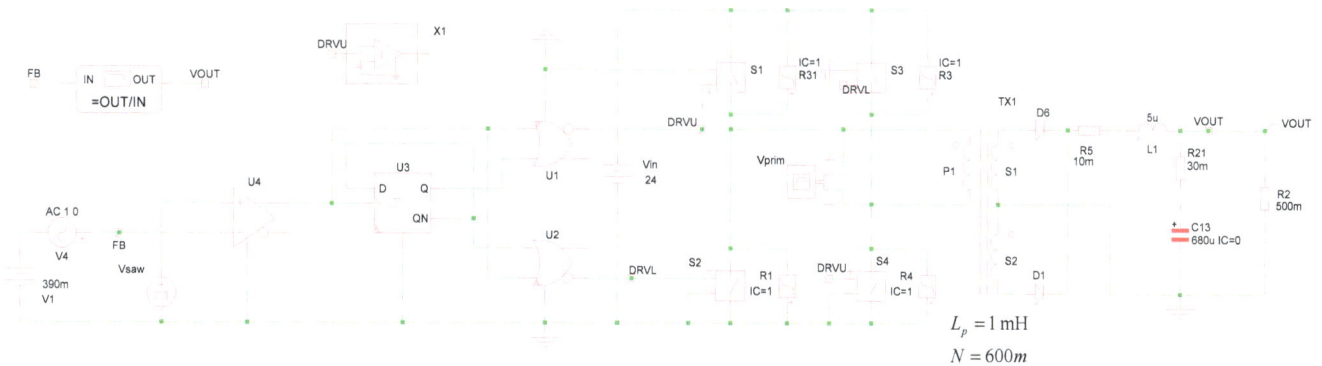

Figure 2.206 The full-bridge in voltage-mode control can deliver a large amount of power, over 5 kW.

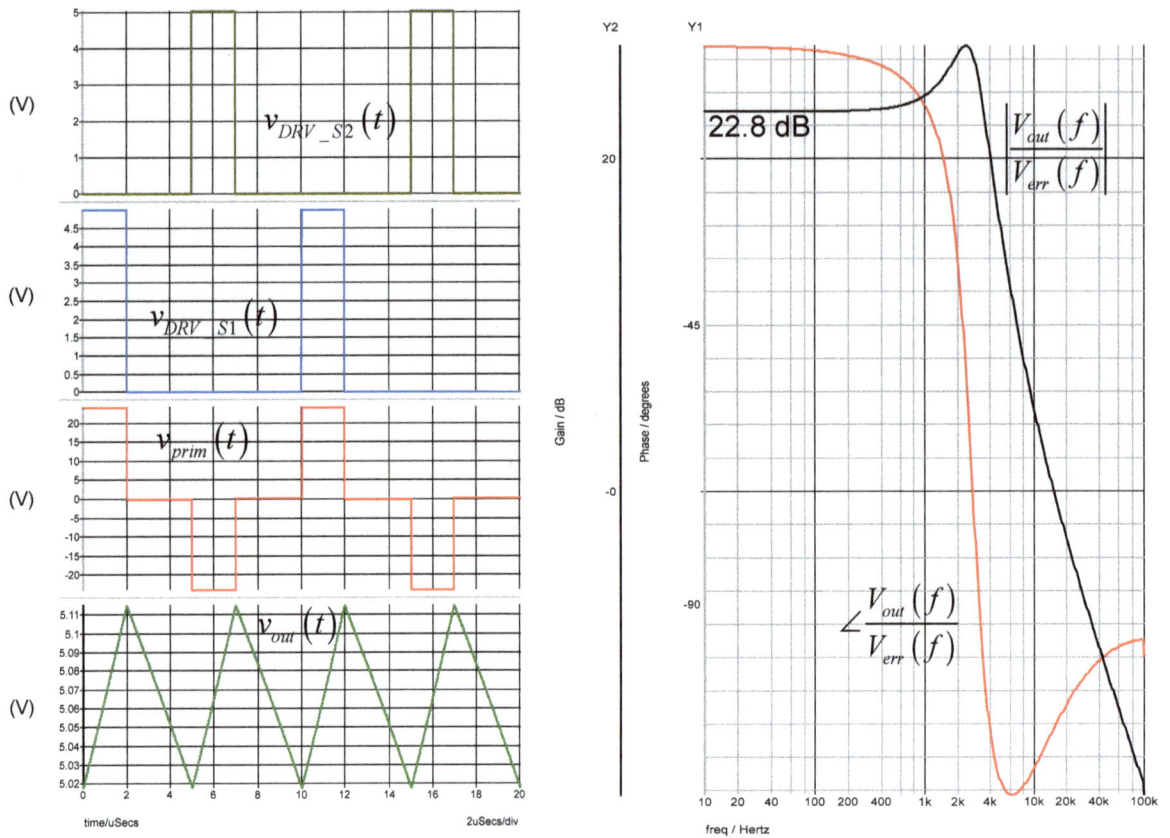

Figure 2.207 The VM full-bridge dynamic response is that of the voltage-mode forward converter.

The current-mode control version of the full-bridge converter appears in Figure 2.208. The current is sensed in the input rail via the current-controlled current source F_1 and mimics a current-sense transformer as implemented in a real implementation.

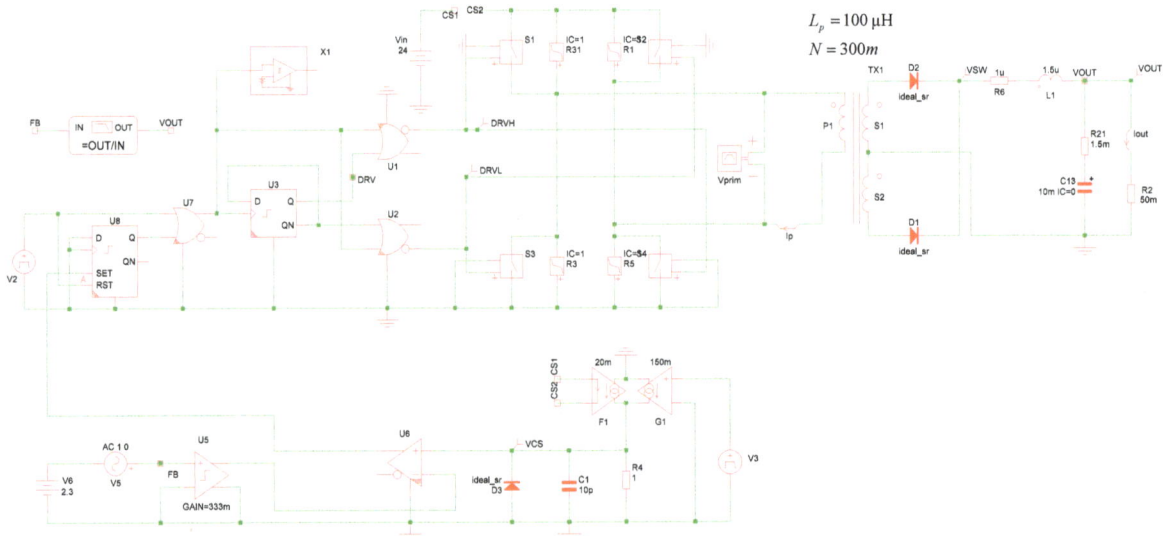

Figure 2.208 This full-bridge application operated in current mode delivers 5 V / 100 A to the load in this example.

The dynamic response is that of the current-mode forward converter and appears in Figure 2.209.

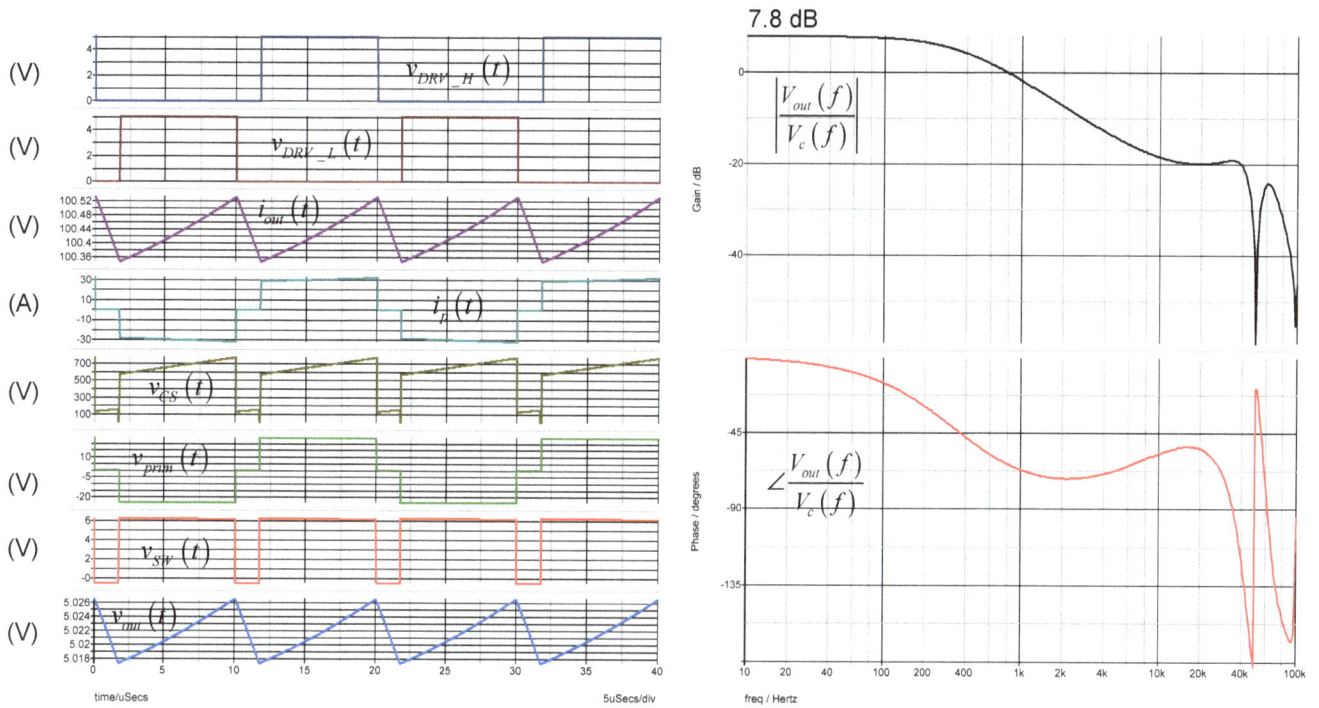

Figure 2.209 The full-bridge in current mode exhibits the dynamic response of the isolated forward converter.

2.15 Isolated Buck-Derived Converters: The Phase-Shifted Full-Bridge Converter in Voltage-Mode Control

This is an extension of the full-bridge converter in which the two legs are now operated at a 50% duty ratio. The control loop adjusts the overlap between them to bias the transformer primary side and deliver energy to the load. The leakage inductance combined with the various lumped capacitances resonates and offers a means to turn the MOSFETs on while their drain-source voltage is close to 0 V. This zero-volt switching (ZVS) operation lets the designer increase the operating frequency without the penalty of associated switching losses.

The control-to-output transfer function has been originally derived in [9] while additional information on the operating principle can be found in [10]. The leakage inductance L_r damps the whole system and the peaking present in the CCM-operated voltage-mode-controlled buck converter disappears in the phase-shifted version.

From [9], the control-to-output transfer function for the voltage-mode control is defined as:

$$H(s) = H_0 \frac{1 + \dfrac{s}{\omega_z}}{1 + \dfrac{s}{\omega_0 Q} + \left(\dfrac{s}{\omega_0}\right)^2} \tag{2.337}$$

in which:

$$H_0 = \frac{N R_{load} V_{in}}{R_{load} + r_d} \tag{2.338}$$

$$r_d = 2 N^2 L_r F_{sw} \tag{2.339}$$

279

$$\omega_z = \frac{1}{r_C C_1} \tag{2.340}$$

$$Q = \frac{\left(R_{load} + R_d\right)\sqrt{\dfrac{C_1 L_2 R_{load}}{R_{load} + R_d}}}{L_2 + C_1 R_{load} R_d} \tag{2.341}$$

$$\omega_0 = \frac{1}{\sqrt{\dfrac{C_1 L_2 R_{load}}{R_{load} + R_d}}} \tag{2.342}$$

The conversion ratio M is defined as:

$$M \approx \frac{ND}{F_{sw} L_r \left(\dfrac{2}{R_{load}} - \dfrac{1-D}{L_2 F_{sw}}\right) N^2 + 1} \tag{2.343}$$

The so-called *effective* duty ratio D_{eff} is actually the duty ratio D imposed by the modulator but reduced in reality because of the resonating inductor.

It is possible to show that this effective duty ratio – measureable at the rectifying diodes cathodes – depends on a reduction brought by the leakage inductance and defined as:

$$\Delta D = \frac{NF_{sw} L_r \left[2I_{out} - \dfrac{1}{F_{sw}} \dfrac{V_{out}}{L_2}(1-D)\right]}{V_{in}} \tag{2.344}$$

With this definition on hand, the effective duty ratio is expressed as:

$$D_{eff} = D - \Delta D \tag{2.345}$$

In these expressions, F_{sw} is the clock frequency, L_r the resonating inductor, L_2 the output inductor, N the transformer turns ratio $1:N$ and D the duty ratio set by the pulse-width modulator.

To verify this expression, I have derived a specific auto-toggling PWM switch model for the phase-shifted converter adapted from [11]. This model is described in [12] and includes a dedicated expression for computing the effective duty ratio at node dL in Figure 2.210.

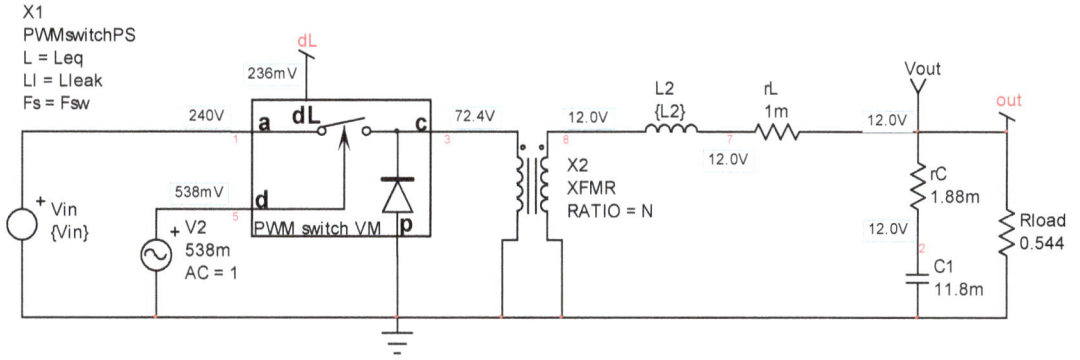

Figure 2.210 The averaged SPICE model computes the operating point and delivers the dynamic response of the phase-shifted converter.

We have compared the frequency response of the given equations with that of the averaged model. Results appear in Figure 2.211 and are identical.

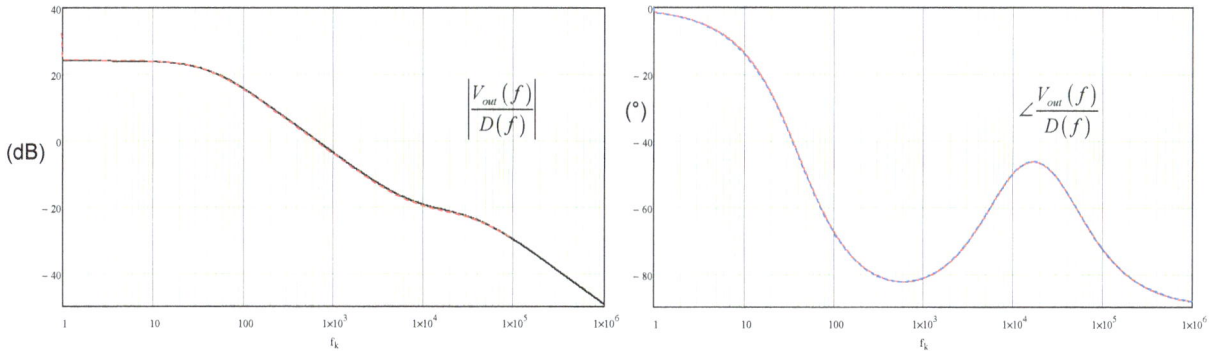

Figure 2.211 The average model delivers the exact same dynamic response as that of the equation-based Mathcad® sheet.

A SIMPLIS® simulation template has been built to check these results with a cycle-by-cycle model. The circuit is shown in Figure 2.212. Both legs are operated from a D flip-flop but I did not insert a deadtime circuit for the sake of simplicity. The converter delivers 12 V to 544-mΩ load (264 W) from a 240-V source. The leakage inductance L_r amounts to 15.5 µH in this example. The dynamic response and some of the key operating waveforms are reproduced in Figure 2.213. The quasi-static gain amounts to 26.6 dB, very close to what (2.338) gives. Please note that the PWM sawtooth peak voltage is 1 V.

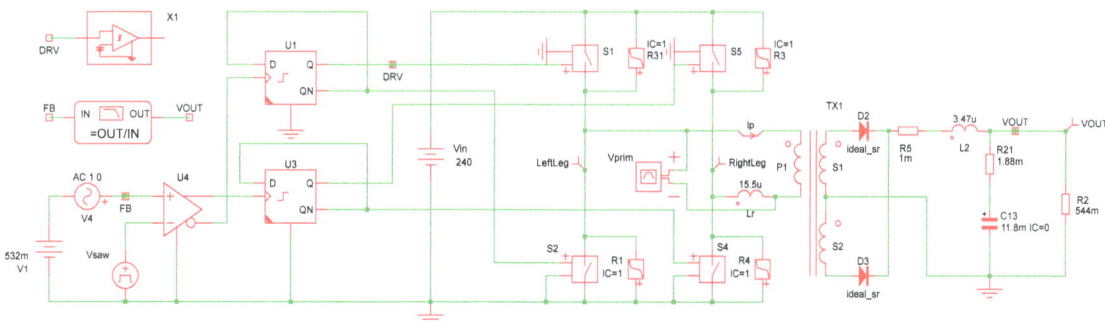

Figure 2.212 The leakage inductance appears as the series element L_r in the electrical diagram.

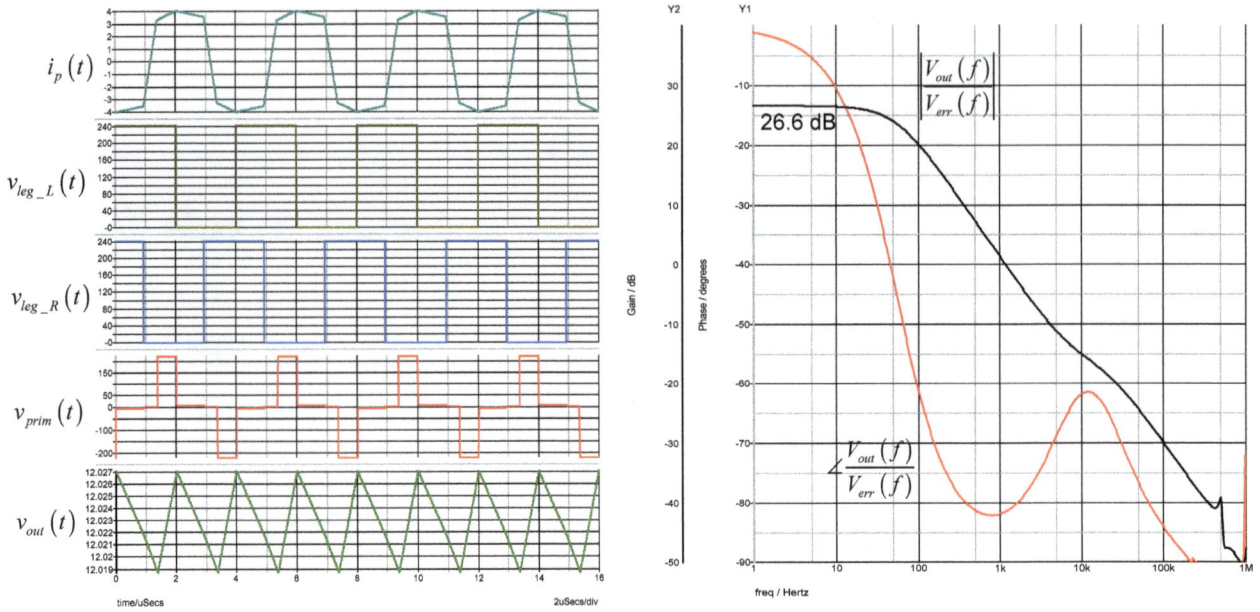

Figure 2.213 The phase-shifted full-bridge response no longer shows the *LC* peaking inherent to voltage-mode control.

The ac response shows the absence of peaking that would otherwise appear in the classical voltage-mode version.

2.16 Isolated Buck-Derived Converters: The Phase-Shifted Full-Bridge Converter in Current-Mode Control

The phase-shifted full-bridge converter can be operated in current-mode control. The duty ratio computation remains similar to what has been presented in the voltage-mode version. However, despite the duty ratio reduction induced by resonating inductor, the control-to-output transfer function is very close to that of the normal current-mode full-bridge converter without the leakage term. For that reason, it is not necessary to develop a new current-mode small-signal model.

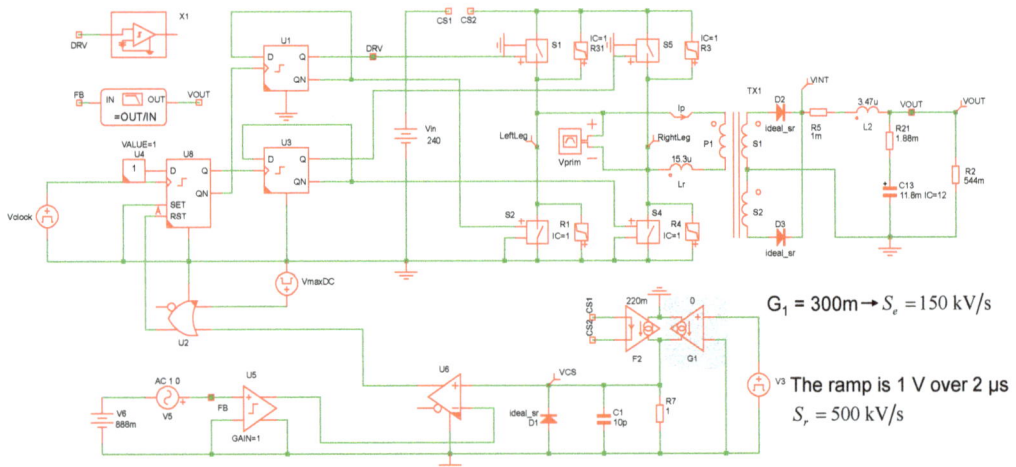

Figure 2.214 The phase-shifted full-bridge operated in current mode hosts the same series inductor L_2.

Figure 2.214 shows the SIMPLIS® simulation template for the current-mode version. We have run simulations with and without the resonating inductor L_2. The cycle-by-cycle waveforms appear in Figure 2.215 with L_2 in place. The converter delivers 22 A in this configuration. After the ac simulation was run with and without L_2, the superimposed responses appear in Figure 2.216: there is almost no difference between the curves meaning the extra inductor does not change the dynamic response of the regular full-bridge converter.

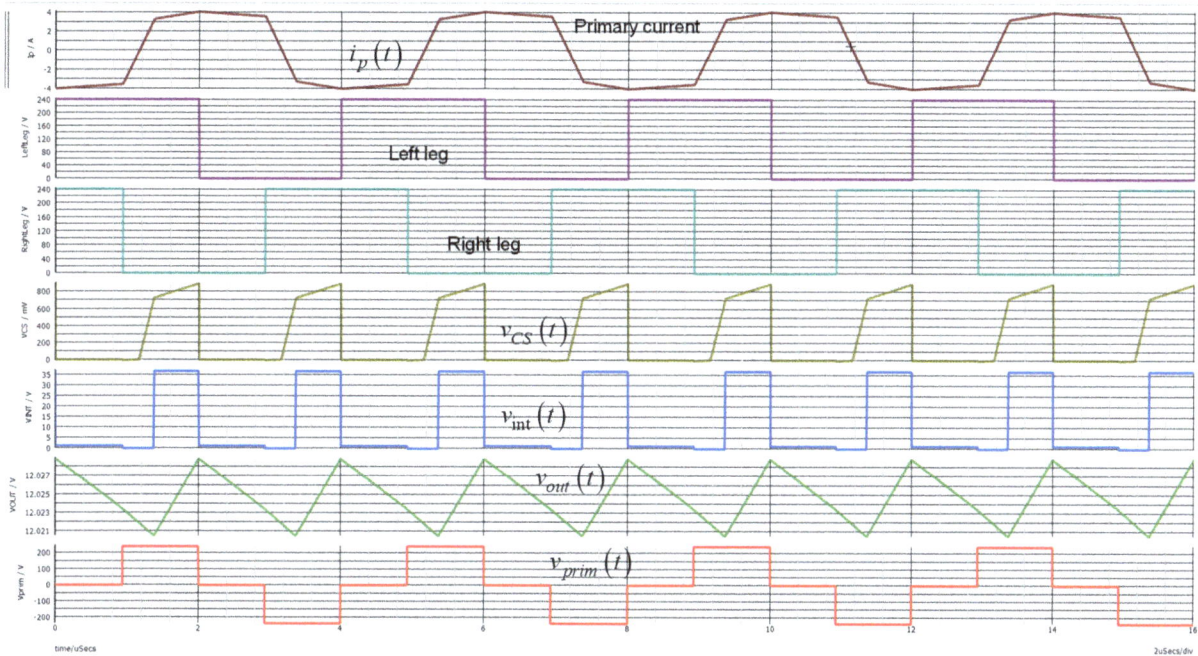

Figure 2.215 The operating curves show the effect of the extra inductor affecting the primary-side current.

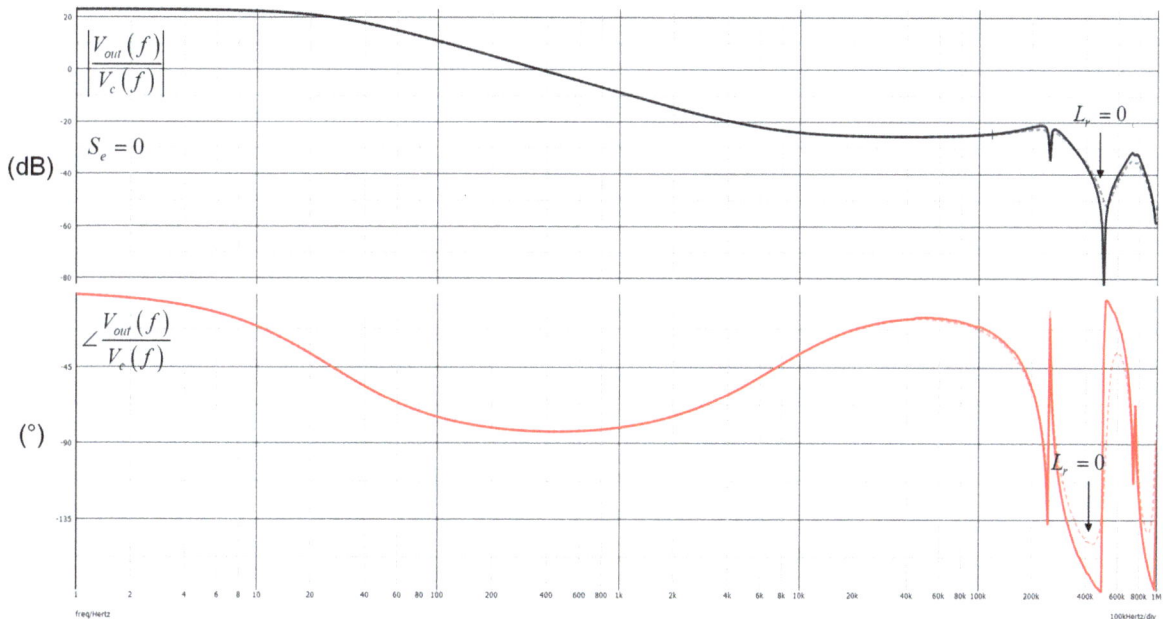

Figure 2.216 The ac response remains very close to that of the regular full-bridge converter operated in current-mode control without a resonating inductor.

As such, the expression given in the forward converter operated in current-mode control will be fine to analyze this converter. In this example, without a compensation ramp – $S_e = 0$ – the dc gain H_0 is determined as:

$$H_0 = \frac{R_{load}}{NR_{sense}} \frac{1}{1 + \frac{R_{load}T_{sw}}{L_2}\left[m_c(1-D)-0.5\right]} = \frac{544m}{0.166 \times 220m} \frac{1}{1 + \frac{544m \times 2u}{3.47u}\left[1 \times (1-0.538)-0.5\right]} = 13.5 \text{ or } 22.6 \text{ dB} \quad (2.346)$$

It corresponds to what Figure 2.216 shows.

2.17 What Should I Retain from this Chapter?

In this second chapter, we have learned key information that is summarized below:

1. A CCM buck converter operated in voltage-mode control is a second-order system. When transitioning to DCM, it still exhibits a second-order dynamic response but well damped.

2. In fixed-frequency current-mode control, the CCM buck converter becomes a third-order converter with a pole dominating the low-frequency response and two subharmonic poles located at half the switching frequency. These poles must be damped by some additional ramp as the duty ratio approaches 50%.

3. In DCM current-mode, some external ramp is still needed as an instability can appear when M equals 0.666 or above.

4. The incremental input resistance of an open-loop buck converter operated in voltage-mode is positive. On the other hand, when the same converter is operated in current-mode, the incremental resistance is negative: if the input voltage increases, the output current reduces and so does the input current.

5. Beside fixed-frequency operation, COT, FOT and QR control were explored, each offering different small-signal characteristics. COT is popular in high-power converters as it naturally improves efficiency in light-load conditions. QR control is interesting to provide ZVS operations at the expense of an increased switching frequency as the load gets lighter. Frequency response of these three structures were covered in this chapter.

6. There are many buck-derived topologies such as the tapped buck which is useful in high-voltage applications where operated at too low a duty ratio could cause problems. The tapped buck provides a solution by tapping the inductor via an extra connection. Small-signal analysis shows the appearance of a right-half-plane zero in this particular configuration.

7. Within the family of isolated converters, the forward structure is certainly the most popular one. Easy to operate in a single-switch version, it requires a 1:1 demagnetization winding to properly reset the core cycle by cycle. It naturally limits the duty ratio below 50%. Its dynamic response is that of the buck now accounting for the transformer turns ratio.

8. Several variations of the forward help push the delivered power to higher levels: push-pull, full-bridge and so on. Care must be taken to always ensure transformer flux balance in worst-case conditions.

9. The active-clamp forward elegantly solves this issue by providing a lossless demagnetization cycle via the addition of a controlled switch. The duty ratio can now exceed 50% and these converters are very popular in dc-dc converters used in telecommunication applications. The voltage-mode response now includes a notch in the magnitude curve and it must be damped if you want a high crossover. The current-mode version includes resonant poles which hamper the control loop design as phase distortion at resonance makes compensation a difficult exercise.

10. Finally, the phase-shifted converter was characterized and shows a damped second-order response owing to the added leakage inductance. This extra element resonates with the lumped capacitances and provides ZVS, naturally improving efficiency.

2.18 References

1. C. Basso, *Linear Circuit Transfer Functions – An Introduction to Fast Analytical Techniques*, Wiley, 2016.
2. V. Vorpérian, *Fast Analytical Techniques for Electrical and Electronic Circuits*, Cambridge University Press, 2002.
3. V. Vorpérian, *Simplified Analysis of PWM Converters using Model of PWM Switch, parts I and II*, IEEE Transactions on Aerospace and Electronic Systems, Vol. 26, NO. 3, 1990
4. R. B. Ridley, *A New Continuous-Time Model for Current-Mode Control*, IEEE Transactions on Power Electronics, Vol. 6, April 1991.
   http://www.ridleyengineering.com/images/current_mode_book/CurrentModeControl.pdf
5. V. Vorpérian, *Analytical Methods in Power Electronics,* in-house course, Toulouse 2000
6. C. Basso, *Switch-Mode Power Supplies: SPICE Simulations and Practical Designs*, 2nd edition, McGraw-Hill, 2014
7. J. Li, *Current-Mode Control: Modeling and its Digital Application*, MSEE Thesis, Virginia Tech, 2009
8. C. Basso, *The Small-Signal Model of an Active-Clamp Forward Converter*, How2Power.com, 2014, http://www.how2power.com/pdf_view.php?url=/newsletters/1402/articles/H2PToday1402_design_ON%20Semiconductor.pdf
9. V. Vlatcović et al., *Small-Signal Analysis of the Phase-Shifted PWM Converter*, IEEE Transactions on Power Electronic, Vol. 7, NO. 1, 1992
10. M. Schutten, D. Torrey, *Improved Small-Signal Analysis for the Phase-Shifted PWM Power Converter,* IEEE Transactions on Power Electronic, Vol. 18, NO. 2, 2003
11. F.S. Tsai, *Small-Signal and Transient Analysis of a Zero-Voltage-Switched, Phase-Controlled PWM Converter Using Averaged Switch Model*, IEEE Transactions, Vol. 29, NO. 3, May/June 1993
12. C. Basso, *A Phase-Shifted Averaged Model*, cbasso.pagesperso-orange.fr/Downloads/Papers/A%20phase%20shifted%20average%20model.pdf

3 The Boost Converter and its Derivatives

THIS CHAPTER WILL detail the derivation of the four boost converter transfer functions when operated in fixed-frequency CCM/DCM voltage- and current-mode control. A quasi-resonant version will also be covered as this structure is popular in power factor correction circuits. The methodology follows what has already been exposed for the buck converter: FACTs are at work together with SPICE to swiftly determine the various time constants while final expressions are tested in a Mathcad® sheet and later confronted with a SIMPLIS® simulation. Going step by step and verifying intermediate results with these tools is the recipe to success considering the complexity of some of the studied structures. Let's begin with the popular voltage-mode control.

3.1 Boost Transfer Functions in CCM – Fixed-Frequency Voltage-Mode Control

We will start this chapter with the control-to-output transfer function linking $V_{out}(s)$ to $D(s)$. To determine this transfer function with a converter operated in voltage mode and CCM, we simply plug the CCM VM-PWM switch model derived in Chapter 1 section 1.3 in the boost converter shown in Figure 3.1.

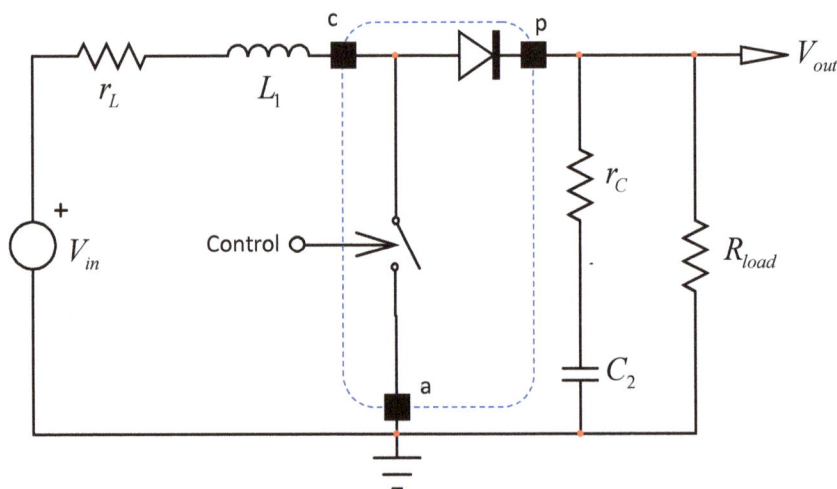

Figure 3.1 The boost converter increases the input voltage with an inductor placed in series with the input source.

The insertion of the large-signal switch model is straightforward and will allow us to immediately obtain the reference dynamic response. Figure 3.2 depicts the simulation templates showing a 16.2-V output obtained from a 10-V source. The duty ratio is set by the V_{ctrl} source and indicates 400 mV or 40%.

The dynamic response appears in Figure 3.3 and shows a flat quasi-static gain until the resonance appears. After the peak, the magnitude falls with a -2-slope (-40 dB per decade) considering the second-order LC filter. The phase should asymptotically hit -180° but does not, indicating the presence of a left-half-plane zero willing to bring it back to -90° (slope breaks to -1 and the phase goes up).

However, you can see that very quickly in higher frequencies, the magnitude breaks to a flat slope, showing that a second zero is at work here. The phase should then head towards 0° but lags further: this is the indication of a right-half-plane zero presence, degrading the phase response of this power stage. This RHPZ is typical of the boost converter and models the delay brought by the 2-step conversion process: store energy in the inductor during the on-time first then transfer it to the capacitor during the off-time as a second step.

If a sudden power demand occurs, you must first increase the energy stored in the inductor before answering the demand. We have illustrated this phenomenon in Chapter 1.

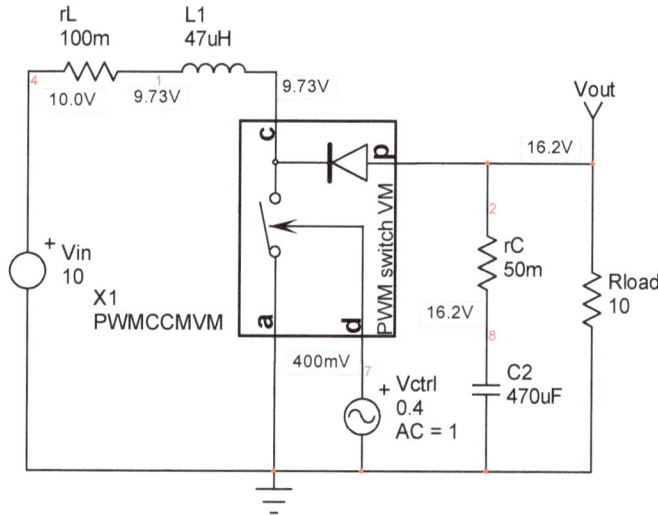

Figure 3.2 The boost converter transfer functions are easily obtained with the VM-PWM switch model.

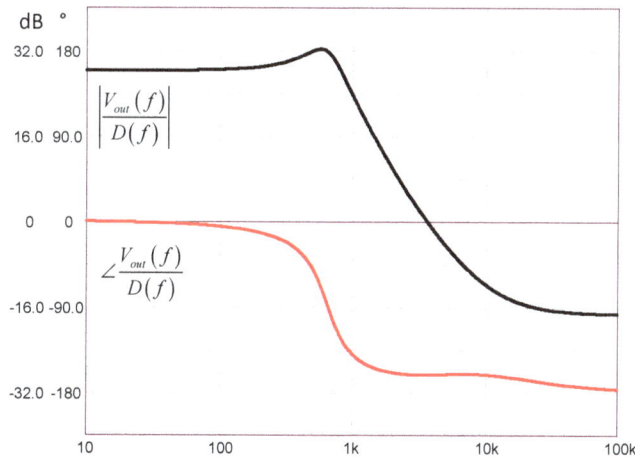

Figure 3.3 The gain is flat at low frequency, peaks at the resonance and falls until the RHPZ kicks-in.

Now that we have a reference curve, we can install the small-signal model of the PWM switch model. This is what Figure 3.4 shows and the included graph confirms the approach: magnitude and phase responses between the large- and small-signal models perfectly superimpose. We can proceed with the simplification of the circuit as started with Figure 3.5. As we want the control-to-output transfer function, the input source is ac-silent and $\hat{v}_{in} = 0$. Source B_2 is back on the other side of the transformer as in the original small-signal model and combines with B_4 in a single ground-referenced source as represented in Figure 3.5. The intermediate sanity check confirms the dynamic response is untouched compared to the previous results. What is the quasi-static gain?

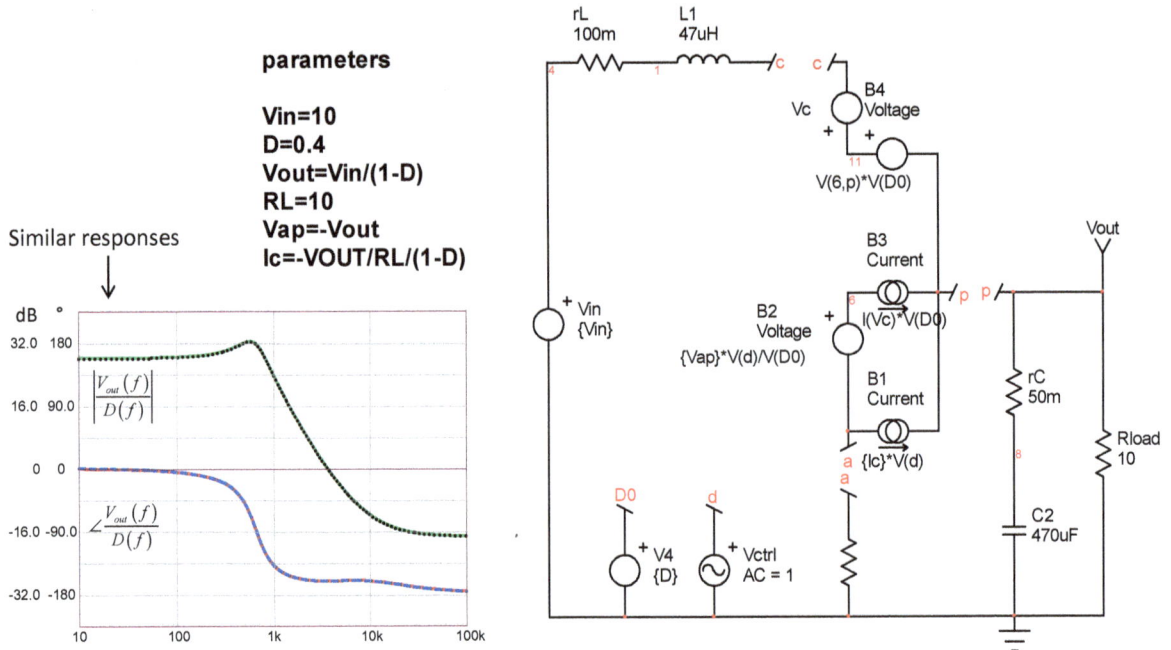

Figure 3.4 The response in magnitude and phase is similar to that obtained from the large-signal model.

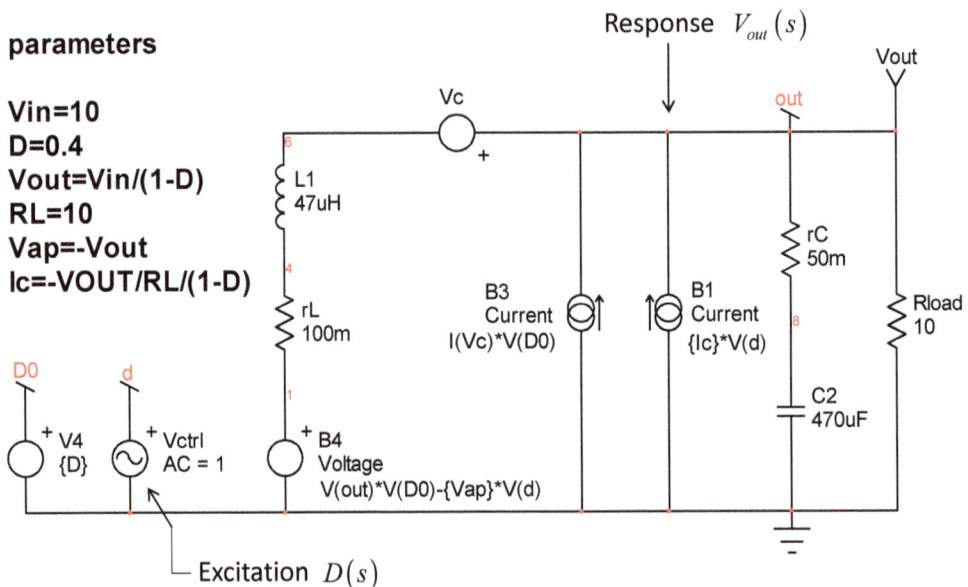

Figure 3.5 The circuit is now considerably simpler and is arranged in an easier-to-read form.

To perform this analysis, open the capacitor and short the inductor as depicted in Figure 3.6.

We can define the current flowing in terminal c as follows, with the 0-subscripted notation for static values:

$$I_c(s) = \frac{V_{out}(s) - V_{out}(s)D_0 + V_{ap0}D(s)}{r_L} \tag{3.1}$$

The current flowing in the load resistance R_{load} is the difference between currents I_1 and I_C:

$$I_2(s) = I_{c0}D(s) - I_c(s)(1 - D_0) \tag{3.2}$$

We also know that the response V_{out} is the load resistance multiplied by I_2:

$$V_{out}(s) = I_2(s)R_{load} \tag{3.3}$$

Finally, from Figure 3.4, we can see that the voltage between terminals a and p is the inverse of the output voltage:

$$V_{ap0} = -V_{out} \tag{3.4}$$

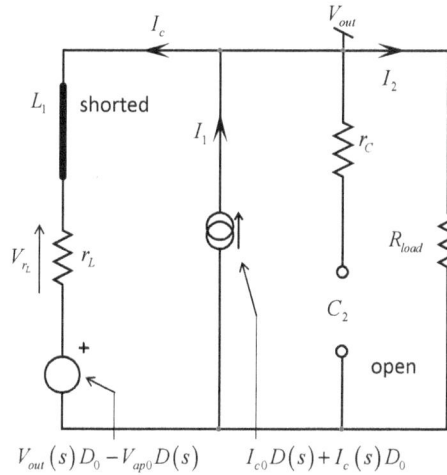

$$V_{out}(s)D_0 - V_{ap0}D(s) \qquad I_{c0}D(s) + I_c(s)D_0$$

Figure 3.6 The inductor is shorted while the capacitor is open in a dc analysis.

Combining all the above equations leads to:

$$H_0 = \frac{V_{out}(s)}{D(s)} = V_{in}R_{load}\frac{\left[(1-D_0)^2 R_{load} - r_L\right]}{\left[(1-D_0)^2 R_{load} + r_L\right]^2} \tag{3.5}$$

If we now neglect the inductor ohmic loss ($r_L = 0$), this expression simplifies to:

$$H_0 \approx \frac{V_{in}}{(1-D_0)^2} \tag{3.6}$$

D_0 is the static duty ratio (40% in our example).

Now that the dc gain is obtained, we can reduce the excitation $-D(s) = 0$ — and determine the time constants involving C_2 and L_1. This is what Figure 3.8 and Figure 3.7 respectively show for L_1 and C_2. The test generator I_T forces a current which is flowing in reverse compared to I_c:

$$I_c(s) = -I_T(s) \tag{3.7}$$

The first equation is:

$$V_T(s) = V_{out}(s) + I_T(s)r_L - V_{out}(s)D_0 \tag{3.8}$$

Factoring and rearranging leads to:

$$V_T(s) = I_T(s)r_L + V_{out}(s)(1 - D_0) \tag{3.9}$$

The output voltage depends on the load resistance and the current flowing through it:

$$V_{out}(s) = \left[I_c(s)D_0 - I_c(s)\right]R_{load} \tag{3.10}$$

Factoring, rearranging and using (3.7), we have:

$$V_{out}(s) = I_T(s)(1 - D_0)R_{load} \tag{3.11}$$

Finally, inserting (3.11) in (3.9), factoring and rearranging to get V_T/I_T leads to our first time constant:

$$\tau_1 = \frac{L_1}{r_L + R_{load}(1 - D_0)^2} \tag{3.12}$$

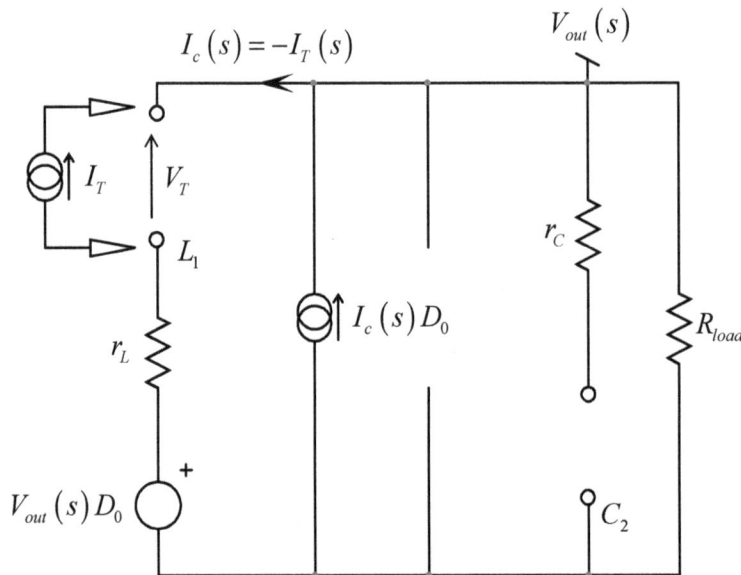

Figure 3.7 The excitation is now reduced to zero and we can determine the natural time constants involving L_1.

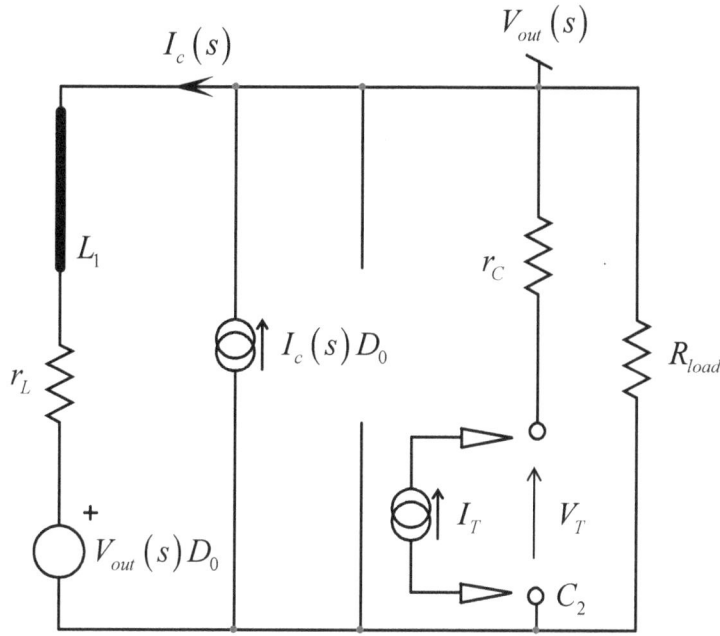

Figure 3.8 The inductor is now set in its dc state (a short circuit) and we determine the time constant involving C_2.

The current in terminal c is defined as:

$$I_c(s) = \frac{V_{out}(s) - V_{out}(s)D_0}{r_L} = \frac{V_{out}(s)(1-D_0)}{r_L}$$ (3.13)

The output voltage depends on the load resistance R_{load}:

$$V_{out}(s) = R_{load}\left[I_c(s)(D_0 - 1) + I_T(s)\right]$$ (3.14)

but can also be determined with:

$$V_{out}(s) = V_T(s) - I_T(s)r_C$$ (3.15)

Inserting (3.14) into (3.13) to eliminate I_c and finally substituting V_{out} by the above equation gives the definition the second time constant:

$$\tau_2 = \left[r_C + \frac{r_L}{\left(1-D_0\right)^2 + \dfrac{r_L}{R_{load}}}\right]C_2$$ (3.16)

We can form the first term b_1 in the denominator:

$$b_1 = \tau_1 + \tau_2 = \frac{L_1}{r_L + R_{load}\left(1-D_0\right)^2} + \left[r_C + \frac{r_L}{\left(1-D_0\right)^2 + \dfrac{r_L}{R_{load}}}\right]C_2$$ (3.17)

291

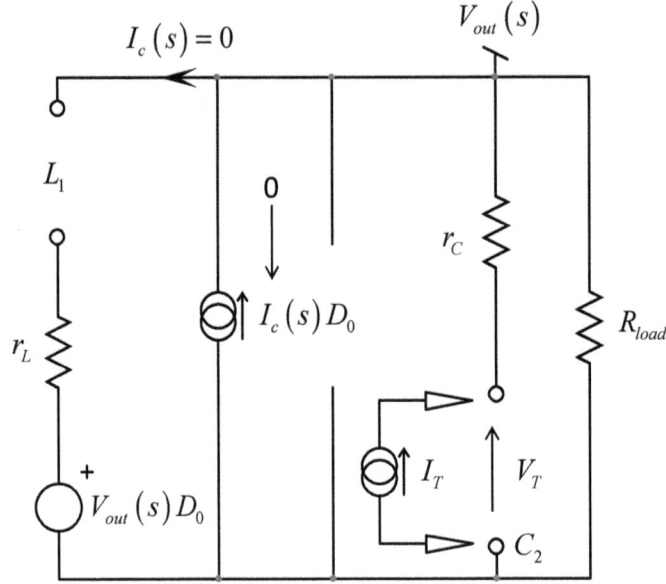

Figure 3.9 The higher-order coefficient is determined by setting the inductor in its high-frequency state (an open circuit).

The time constant needed to form b_2 — τ_2^1 for instance — is determined by setting the inductor in its high-frequency state as shown in Figure 3.9. The resistance seen from C_2's connecting terminals is obtained by inspection as $I_c(s) = 0$:

$$\tau_2^1 = C_2\left(r_C + R_{load}\right) \tag{3.18}$$

Combined with τ_1, we have determined the second-order coefficient b_2:

$$b_2 = \tau_1\tau_2^1 = \frac{L_1}{r_L + R_{load}\left(1 - D_0\right)^2}\left(r_C + R_{load}\right)C_2 \tag{3.19}$$

Assembling all these elements leads to the denominator expression:

$$D(s) = 1 + \left[\frac{L_1}{r_L + \left(1 - D_0\right)^2 R_{load}} + C_2\left(r_C + \frac{R_{load}\,r_L}{R_{load}\left(1 - D_0\right)^2 + r_L}\right)\right]s + \left[L_1 C_2 \frac{r_C + R_{load}}{r_L + \left(1 - D_0\right)^2 R_{load}}\right]s^2 \tag{3.20}$$

It can be put under the classical canonical form:

$$D(s) = 1 + \frac{s}{\omega_0 Q} + \left(\frac{s}{\omega_0}\right)^2 \tag{3.21}$$

in which we have:

$$\omega_0 = \frac{1}{\sqrt{L_1 C_2}}\sqrt{\frac{r_L + R_{load}\left(1 - D_0\right)^2}{r_C + R_{load}}} \approx \frac{1 - D_0}{\sqrt{L_1 C_2}} \tag{3.22}$$

$$Q \approx \frac{\omega_0}{\dfrac{r_L}{L_1} + \dfrac{1}{C_2 \left(r_C + R_{load} \right)}}$$

(3.23)

To determine the zeroes, we bring the excitation back and null the output. This is what Figure 3.10 shows.

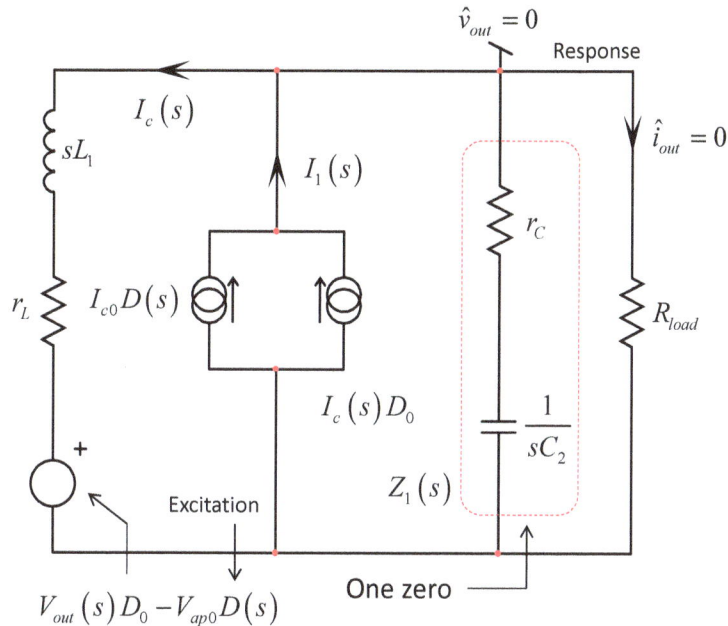

Figure 3.10 The zeroes are obtained by nulling the output voltage when the excitation is back in place.

The first zero is classically brought by the series connection of r_C and C_2 forming a transformed short circuit:

$$Z_1 \left(s_z \right) = r_C + \frac{1}{sC_2} = 0$$

(3.24)

Implying that the root is:

$$s_{z_1} = -\frac{1}{r_C C_2}$$

(3.25)

and gives a zero positioned at:

$$\omega_{z_1} = \frac{1}{r_C C_2}$$

(3.26)

The other zero is obtained by determining the second condition for which no current circulates in the load (Figure 3.11).

One of the sources simplifies considering the output null.

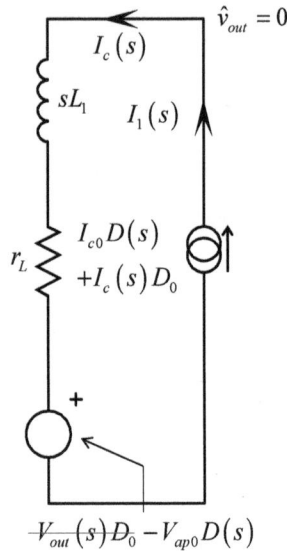

Figure 3.11 When both I_1 and I_c are equal, there is no current in R_{load}, creating an output null.

This condition is obtained if:

$$I_1(s) - I_c(s) = 0 \tag{3.27}$$

We have the definition for each of these generators:

$$I_1(s) = I_{c0}D(s) + I_c(s)D_0 \tag{3.28}$$

while $I_c(s)$ is defined as:

$$I_c(s) = \frac{V_{ap0}D(s)}{sL_1 + r_L} \tag{3.29}$$

Substitute these expressions in (3.27) and you have:

$$I_{c0}D(s) + \frac{V_{ap0}D(s)}{r_L + sL_1}D_0 - \frac{V_{ap0}D(s)}{r_L + sL_1} = 0 \tag{3.30}$$

The root of this expression is the zero we are looking for. Solving for it gives:

$$s_{z_2} = \frac{(1-D_0)^2 R_{load} - r_L}{L_1} \tag{3.31}$$

It is a positive root located in the right half-plane, also called a RHPZ.
If we neglect the inductor ohmic loss r_L, the zero is defined as:

$$\omega_{z_2} \approx \frac{(1-D_0)^2 R_{load}}{L_1} \tag{3.32}$$

We can now assemble the pieces to form the control-to-output transfer function of the boost converter operated in CCM:

$$\frac{V_{out}(s)}{V_{err}(s)} = \frac{V_{out}(s)}{V_p D(s)} = H_0 \frac{\left(1 + \frac{s}{\omega_{z_1}}\right)\left(1 - \frac{s}{\omega_{z_2}}\right)}{1 + \frac{s}{\omega_0 Q} + \left(\frac{s}{\omega_0}\right)^2} \tag{3.33}$$

$$H_0 \approx \frac{V_{in}}{V_p (1 - D_0)^2} \tag{3.34}$$

In which V_p represents the peak voltage of the pulse width modulator sawtooth waveform while D_0 designates the static duty ratio at the considered operating point. Other terms are defined as follows:

$$\omega_{z_1} = \frac{1}{r_C C_2} \tag{3.35}$$

$$\omega_{z_2} = \frac{R_{load} (1 - D_0)^2}{L_1} \tag{3.36}$$

$$\omega_0 = \frac{1}{\sqrt{L_1 C_2}} \sqrt{\frac{r_L + R_{load} (1 - D_0)^2}{r_C + R_{load}}} \tag{3.37}$$

$$Q \approx \frac{\omega_0}{\frac{r_L}{L_1} + \frac{1}{C_2 (r_C + R_{load})}} \tag{3.38}$$

If we neglect the ohmic contributions of r_C and r_L, the two above expressions simplify to:

$$Q \approx (1 - D_0) R_{load} \sqrt{\frac{C_2}{L_1}}$$

$$\omega_0 \approx \frac{1 - D_0}{\sqrt{L_1 C_2}} \tag{3.39}$$

We have captured these equations – the simplified versions but also the complete expressions – in a Mathcad® sheet (Figure 3.12). Without surprise, the ohmic losses in the inductor but also in the capacitor damp the response and affect the quasi-static gain H_0.

Other elements will damp the circuit such as the MOSFET $r_{DS(on)}$ but also the diode dynamic resistance r_d and even its recovery loss. For the sake of simplicity, they have been ignored here.

$r_L := 0.1\Omega$ $r_C := 0.05\Omega$ $C_2 := 470\mu F$ $L_1 := 47\mu H$ $R_L := 10\Omega$ $\|(x,y) := \dfrac{x \cdot y}{x + y}$

$V_{in} := 10V$ $V_p := 1V$ $D_0 := 40\%$

$V_{out1} := V_{in} \cdot \dfrac{1}{1 - D_0} = 16.6666667V$ theoretical value

$V_{out2} := \dfrac{R_L \cdot V_{in} \cdot (1 - D_0)}{r_L \left[\dfrac{R_L \cdot (1 - D_0)^2}{r_L} + 1 \right]} = 16.2162162V$ value accounting for losses

$H_{01} := \dfrac{V_{in}}{V_p \cdot (1 - D_0)^2} = 27.7777778$ $20 \cdot \log(H_{01}) = 28.87395$ dB

$H_{02} := \dfrac{V_{in} \cdot R_L}{V_p} \cdot \dfrac{(1 - D_0)^2 \cdot R_L - r_L}{\left[(1 - D_0)^2 \cdot R_L + r_L \right]^2} = 25.5661066$ $20 \cdot \log(H_{02}) = 28.1532919$ dB

$\omega_{z1} := \dfrac{1}{r_C \cdot C_2}$ $f_{z1} := \dfrac{\omega_{z1}}{2 \cdot \pi} = 6.7725508kHz$ LHP zero

$\omega_{z2} := \dfrac{R_L \cdot (1 - D_0)^2}{L_1}$ $f_{z2} := \dfrac{\omega_{z2}}{2 \cdot \pi} = 12.1905914kHz$ RHP zero

$Q_1 := (1 - D_0) \cdot R_L \cdot \sqrt{\dfrac{C_2}{L_1}} = 18.973666$ $\omega_{01} := \dfrac{1 - D_0}{\sqrt{L_1 \cdot C_2}}$ $f_{01} := \dfrac{\omega_{01}}{2 \cdot \pi} = 642.5005801Hz$

$\omega_{02} := \dfrac{1}{\sqrt{L_1 \cdot C_2}} \cdot \sqrt{\dfrac{r_L + R_L \cdot (1 - D_0)^2}{r_C + R_L}}$ $Q_2 := \dfrac{\omega_{02}}{\dfrac{r_L}{L_1} + \dfrac{1}{C_2 \cdot (r_C + R_L)}} = 1.7451052$

$f_{02} := \dfrac{\omega_{02}}{2 \cdot \pi} = 649.7407492Hz$

$H_1(s) := H_{01} \cdot \dfrac{\left(1 + \dfrac{s}{\omega_{z1}}\right) \cdot \left(1 - \dfrac{s}{\omega_{z2}}\right)}{1 + \dfrac{s}{\omega_{01} \cdot Q_1} + \left(\dfrac{s}{\omega_{01}}\right)^2}$ $H_2(s) := H_{02} \cdot \dfrac{\left(1 + \dfrac{s}{\omega_{z1}}\right) \cdot \left(1 - \dfrac{s}{\omega_{z2}}\right)}{1 + \dfrac{s}{\omega_{02} \cdot Q_2} + \left(\dfrac{s}{\omega_{02}}\right)^2}$

Figure 3.12 Mathcad® gathers the equations we have derived and will plot both transfer functions.

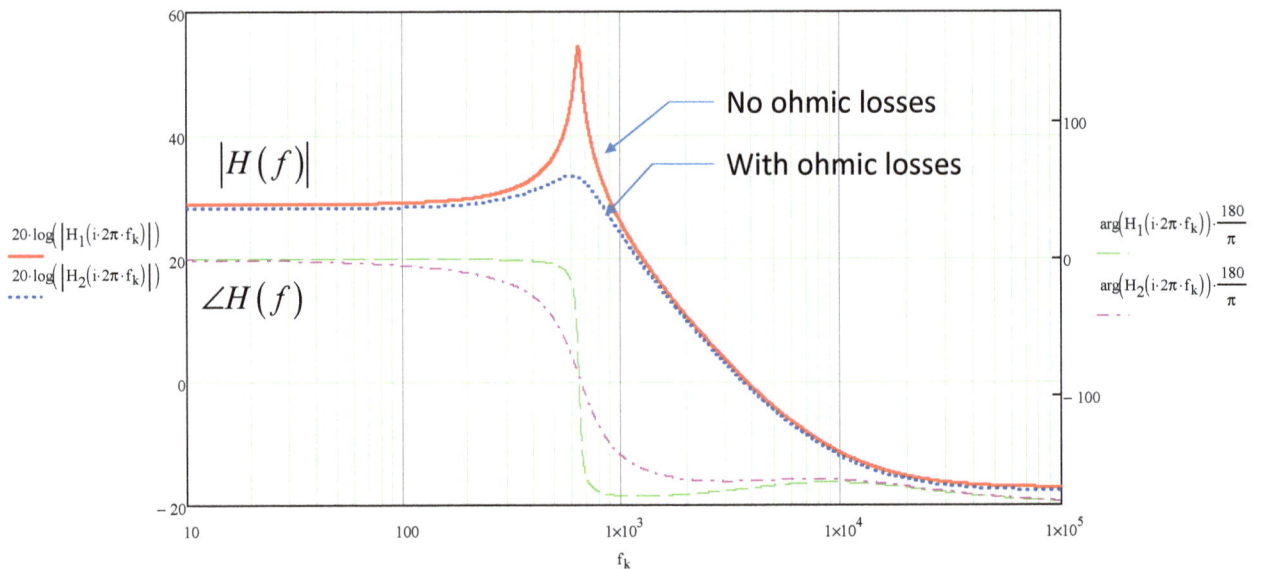

Figure 3.13 Neglecting ohmic losses masks the damping effect brought by these resistances.

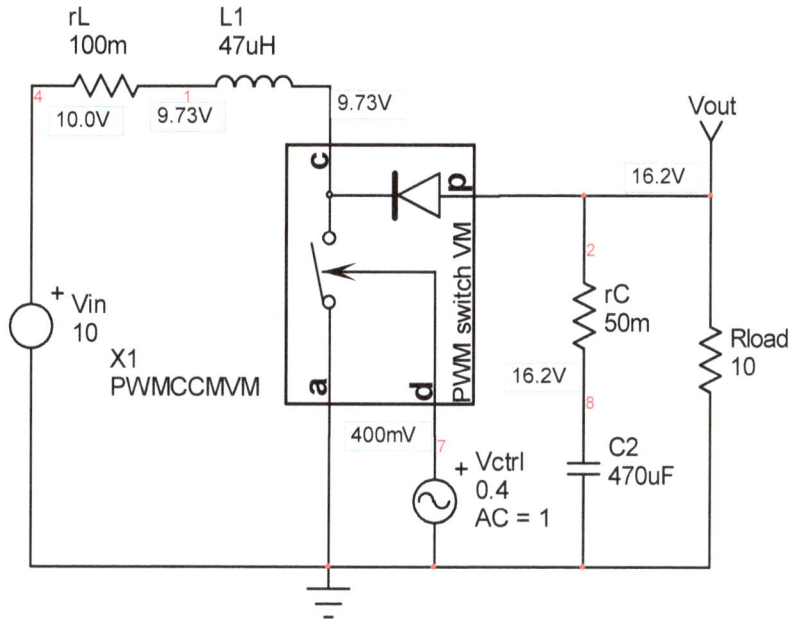

Figure 3.14 The large-signal VM-PWM switch model delivers the small-signal response of the CCM-operated boost converter. A G_{PWM} gain of 1 is assumed here.

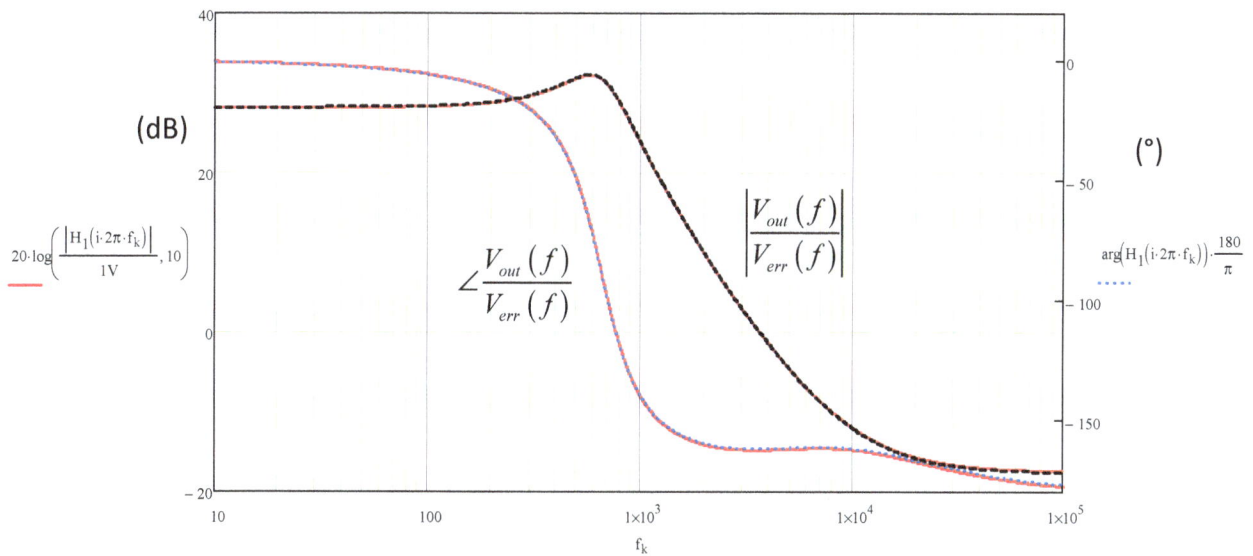

Figure 3.15 The dynamic response delivered by the SPICE model and that of Mathcad® (full expression) are identical.

To confirm all these analyses, we have run a simple boost circuit in SIMPLIS® as shown in Figure 3.16.

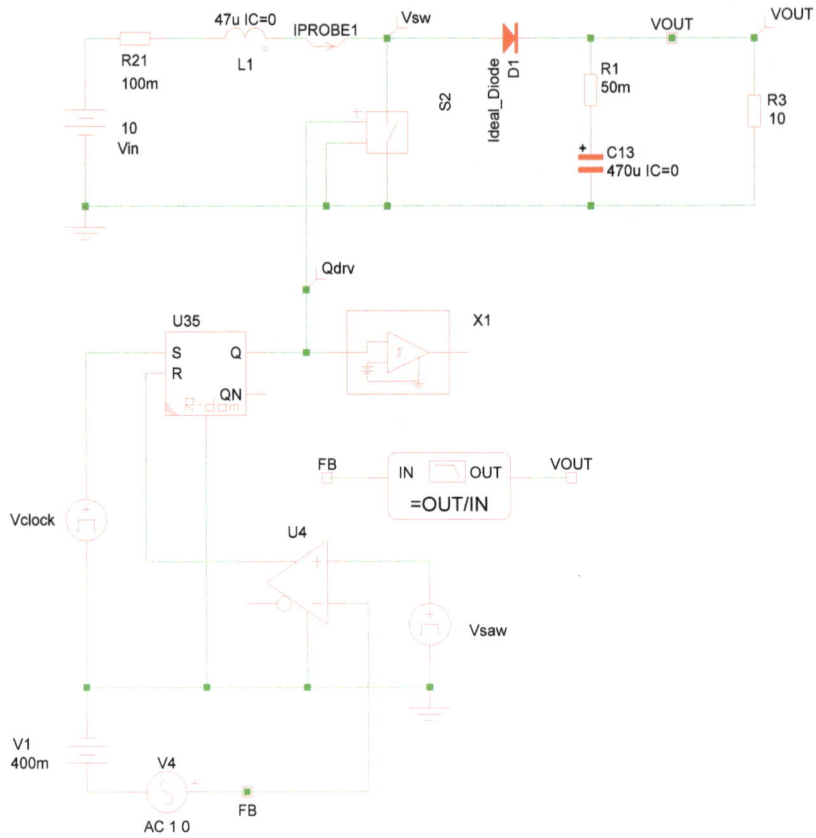

Figure 3.16 SIMPLIS® will give us the switching waveform and the dynamic response in a few seconds.

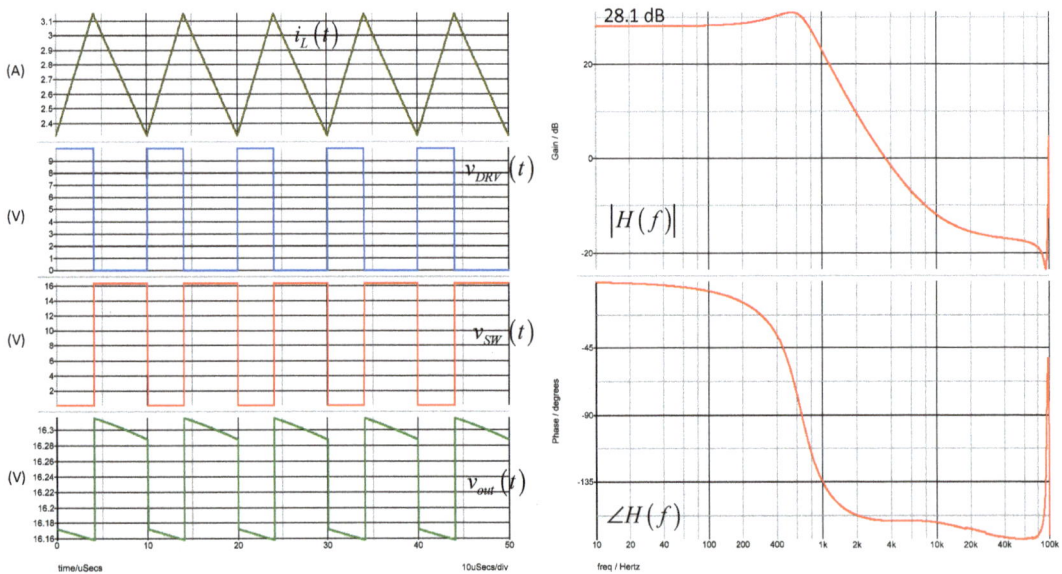

Figure 3.17 The delivered results agree quite well with the SPICE simulations.

The dynamic response shown in Figure 3.17 confirm the SPICE simulation results.

3.1.1 Input to Output

The input-to-output transfer function characterizes the ability of the boost converter to reject perturbations present in the input source. In this analysis, the converter operates in open-loop conditions meaning that $\hat{d}$ is zero but the static duty ratio D_0 ensures the correct bias point ($V_{out} = 16$ V). The response is still $V_{out}(s)$ but the stimulus has become $V_{in}(s)$. The new circuit to analyze appears in Figure 3.18. It is derived from Figure 3.4 in which the sources referring to $D(s)$ have been reduced to zero and only one excitation source involving V_{in} remains. When this source is turned off (replaced by a short circuit), the circuit returns to its natural structure and the time constants we have determined for the denominator are the same: we can reuse the denominator $D(s)$ defined by (3.20).

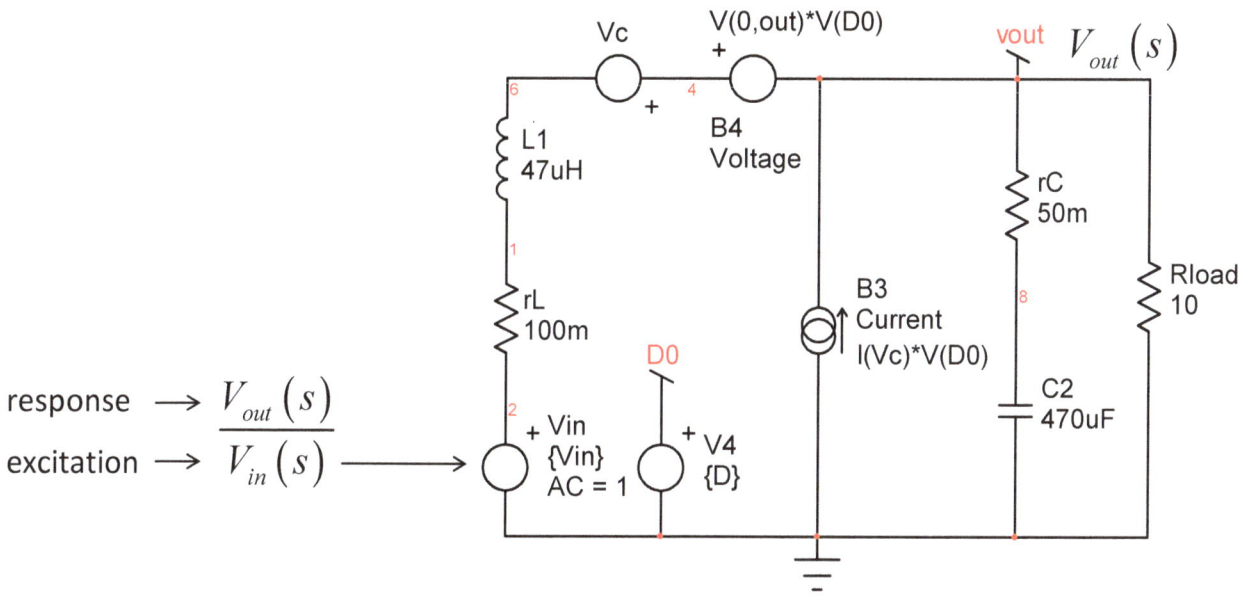

Figure 3.18 As the excitation V_{in} is reduced to 0 V, the circuit returns to its natural structure.

Since the denominator is known, we can concentrate our efforts on the quasi-static gain H_0 with the help of Figure 3.19. We can define the current flowing out of terminal c as:

$$I_c(s) = \frac{V_{out}(s)(1-D_0)-V_{in}(s)}{r_L} \tag{3.40}$$

I_1 is this current scaled by the static duty ratio D_0:

$$I_1 = \left(\frac{V_{out}(s)(1-D_0)-V_{in}(s)}{r_L}\right)D_0 \tag{3.41}$$

Finally, we define the output voltage then substitute the above definitions in this expression and rearrange the result:

$$V_{out} = (I_1 - I_c)R_{load} \tag{3.42}$$

$$H_0 = \frac{V_{out}(s)}{V_{in}(s)} = \frac{(1-D_0)R_{load}}{(1-D_0)^2 R_{load} + r_L} \approx \frac{1}{1-D_0} \qquad (3.43)$$

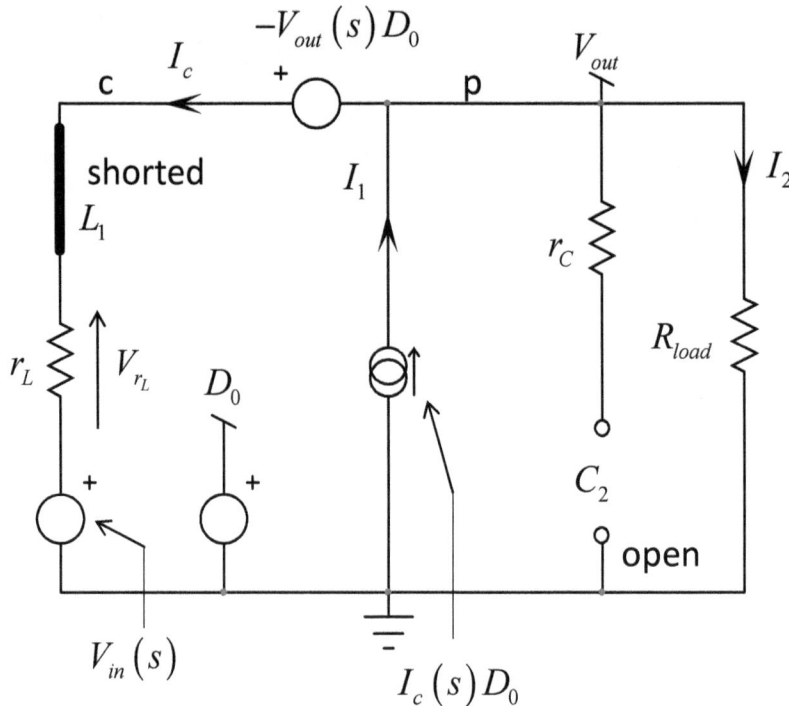

Figure 3.19 In this figure, we determine the dc gain linking V_{out} to V_{in}.

Now that we have the denominator and the leading term H_0, what conditions in the circuit would prevent a stimulus to propagate in the network and create a response? In other words, where are the zeroes? We can use the circuit from Figure 3.20 and immediately realize that the series combination of r_C and C_2 forms a transformed short circuit:

$$Z_1(s) = r_C + \frac{1}{sC_2} = 0 \qquad (3.44)$$

It brings a zero located at:

$$\omega_z = \frac{1}{r_C C_2} \qquad (3.45)$$

Is there a second zero? To check this, we set L_1 in its high-frequency state (an open circuit), and we check if a stimulus in this mode could generate a response?

In other words, if V_{in} biases the input, do you obtain a voltage at node V_{out}? No, as L_1 is open, the current in terminal c is zero and you have no response: there is no zero associated with L_1 and we have one single zero defined by (3.45).

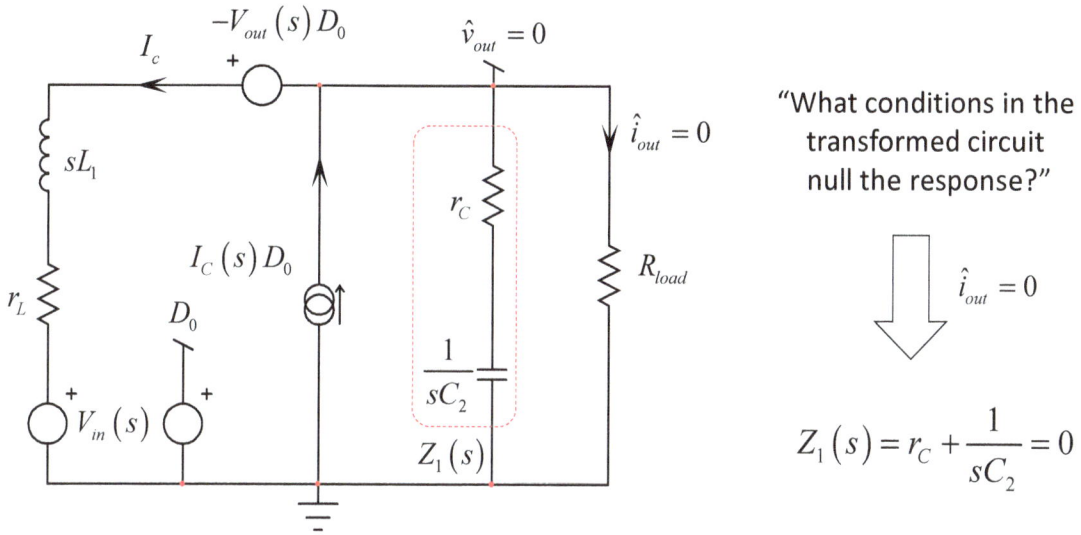

Figure 3.20 In this figure, we determine the dc gain linking V_{out} to V_{in}.

We now have all the elements we need to assemble the input-to-output transfer function of the CCM-operated boost converter:

$$\frac{V_{out}(s)}{V_{in}(s)} \approx \frac{1}{1-D_0} \frac{1+\dfrac{s}{\omega_z}}{1+\dfrac{s}{\omega_0 Q}+\left(\dfrac{s}{\omega_0}\right)^2}$$

(3.46)

with terms defined by (3.43), (3.45), (3.37) and (3.38).

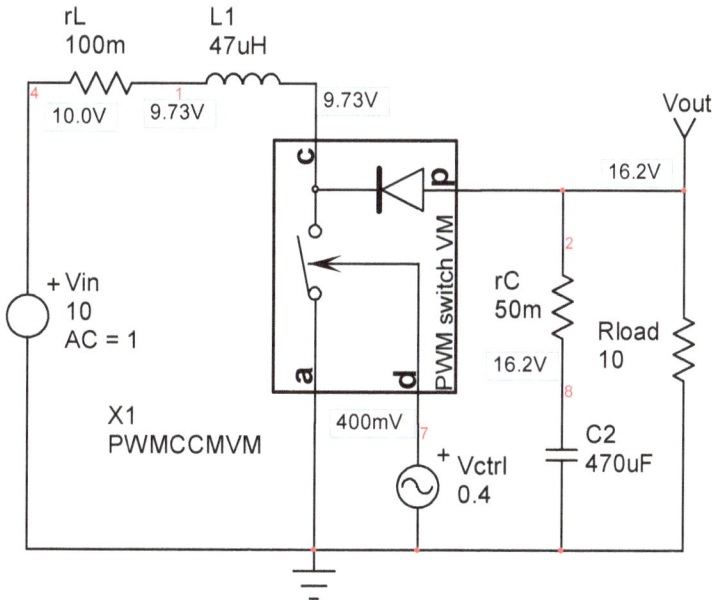

Figure 3.21 The large-signal model is used to unveil the input-to-output transfer function.

We can now test the response from the SPICE circuit of Figure 3.21 and compare the magnitude and phase curves to those obtained with (3.46) via a Mathcad® sheet. As confirmed by Figure 3.22, our analysis is correct.

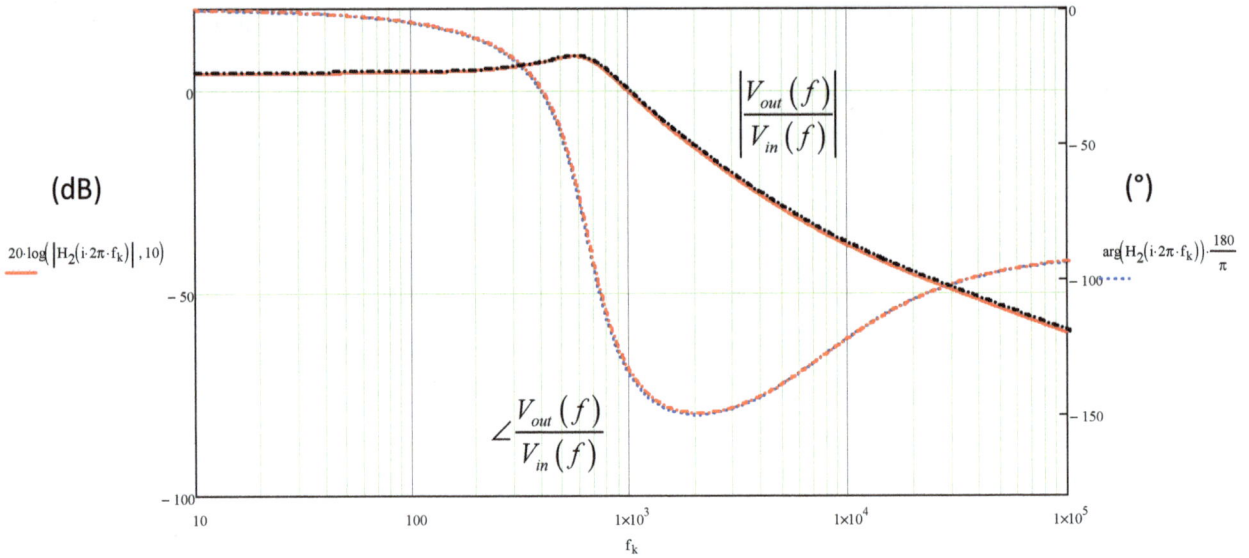

Figure 3.22 The perfectly-superimposed plots confirm our analysis is correct.

3.1.2 Output Impedance

The output impedance is obtained by installing a current test generator across the loaded output node of Figure 3.18 in which the input source V_{in} is turned off. The voltage generated across its terminals (the response) is divided by the injected current (the stimulus) and leads to the transfer function we want. In this mode, the duty ratio $\hat{d}$ is reduced to zero, implying an open-loop configuration. Also, the input voltage source V_{in} is perfect (0-ohm output resistance) and set to 0 V ac as in any output impedance determination exercise.

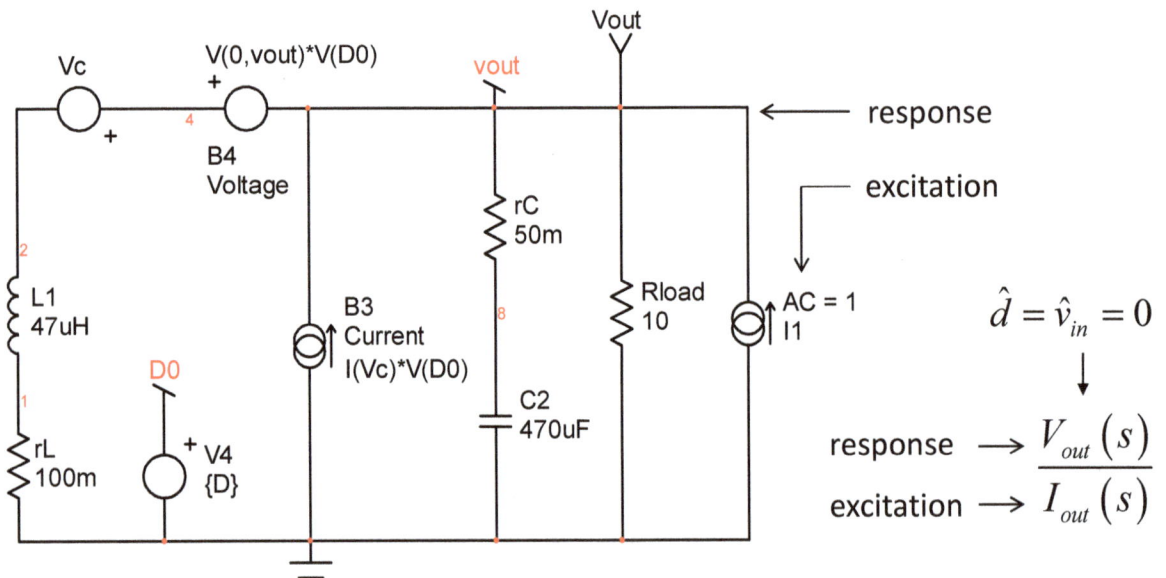

Figure 3.23 The output impedance is obtained by installing a test generator across the load. When this stimulus is set to 0 A, the circuit returns to its natural structure.

If we turn the excitation source off – the right-side current generator is open circuited – the circuit returns to its natural structure and we can reuse the denominator already defined by (3.20). The leading term in an impedance definition is the quasi-static or incremental resistance R_0 determined when the inductor is shorted and the output capacitor open.

This is what Figure 3.24 illustrates.

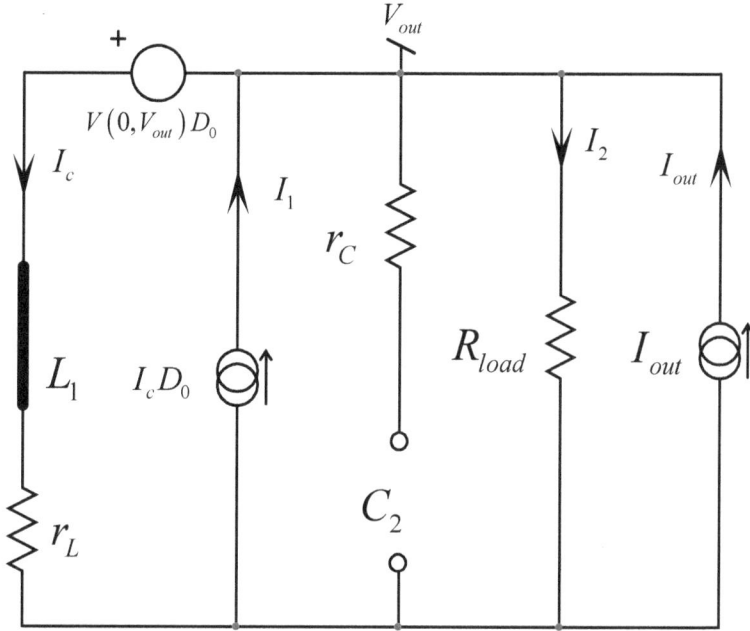

Figure 3.24 A few equations are necessary to determine the incremental resistance R_0.

We are just a few equations away from the result we want. Current I_c is defined as:

$$I_c = \frac{V_{out} - V_{out}D_0}{r_L} = \frac{V_{out}(1-D_0)}{r_L} \tag{3.47}$$

Current I_1 is simply:

$$I_1 = I_c D_0 = \left(\frac{V_{out}(1-D_0)}{r_L} \right) D_0 \tag{3.48}$$

While current I_2 is the current circulating in the load resistance:

$$I_2 = \frac{V_{out}}{R_{load}} \tag{3.49}$$

Finally, KCL applied at node V_{out} tells us:

$$I_{out} = I_c + I_2 - I_1 \tag{3.50}$$

Substituting the current definitions in (3.50) then factoring V_{out} and rearranging the result we obtain:

$$R_0 = \frac{V_{out}}{I_{out}} = \frac{r_L}{\left(1-D_0\right)^2 + \dfrac{r_L}{R_{load}}}$$

(3.51)

To determine the zeroes, we look at our circuit considering a nulled response or $V_{out}(s) = 0$ V while the excitation is back in place. This is what the circuit in Figure 3.25 illustrates.

The response nulls if we have two transformed short circuits brought by Z_1 and Z_2:

$$Z_1(s) = r_C + \frac{1}{sC_2} = 0$$

(3.52)

It implies a first zero located at:

$$\omega_{z_1} = \frac{1}{r_C C_2}$$

(3.53)

while:

$$Z_2(s) = r_L + sL_1 = 0$$

(3.54)

leads to a second zero placed at:

$$\omega_{z_2} = \frac{r_L}{L_1}$$

(3.55)

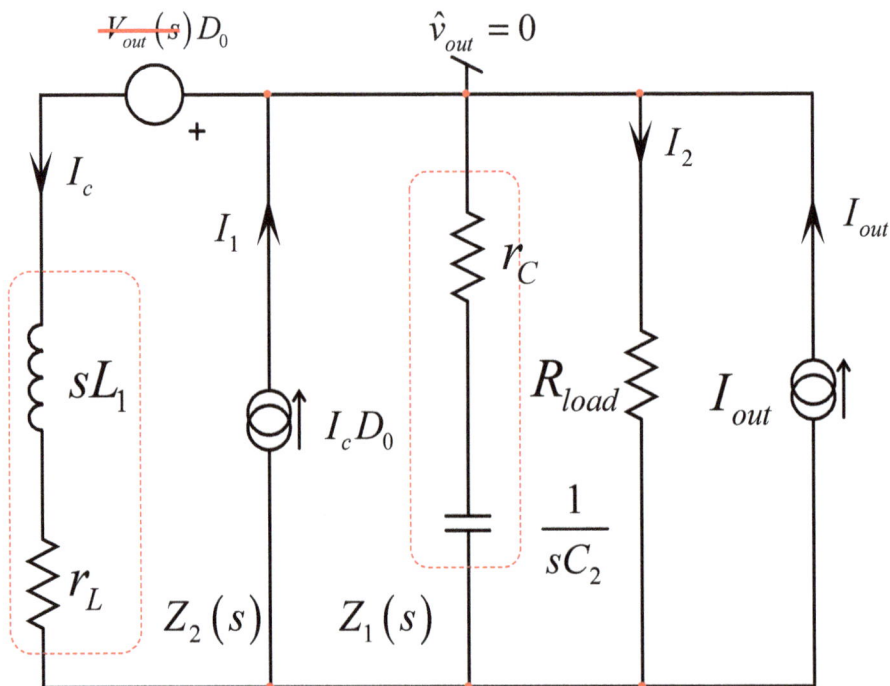

Figure 3.25 The excitation is back in place: what conditions null the response?

Combining these expressions gives us the denominator $N(s)$ properly factored:

$$N(s) = \left(1 + \frac{s}{\omega_{z_1}}\right)\left(1 + \frac{s}{\omega_{z_2}}\right)$$

(3.56)

We can form the complete transfer function expressed as:

$$Z_{out}(s) \approx R_0 \frac{\left(1 + \frac{s}{\omega_{z_1}}\right)\left(1 + \frac{s}{\omega_{z_2}}\right)}{1 + \frac{s}{\omega_0 Q} + \left(\frac{s}{\omega_0}\right)^2}$$

(3.57)

R_0 is defined by (3.51), the two zeroes by (3.53) and (3.55) while the denominator uses terms defined by (3.37) and (3.38). We have captured this formula in a Mathcad® sheet and drawn its response together with that of the SPICE circuit from Figure 3.26. As Figure 3.27 confirms, both responses are identical, confirming that our analysis is correct.

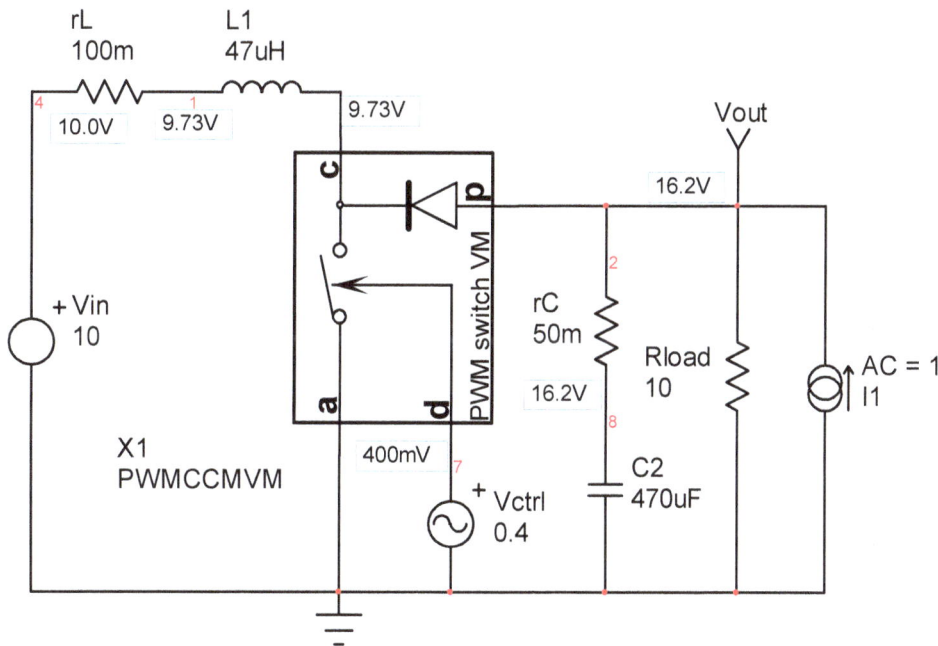

Figure 3.26 The large-signal model helps us verify if our expression is correct.

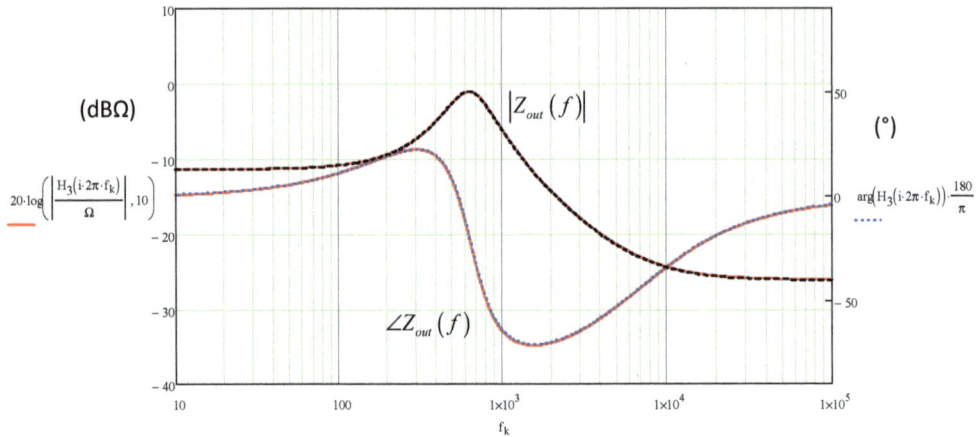

Figure 3.27 Magnitude and phase responses are identical confirming our analysis is good.

3.1.3 Input Impedance

Determining the input impedance requires a change in the stimulus connections. This time, the current source will sweep the converter input nodes. The SPICE simulation template involving a large inductor *LoL* ac-isolating the input source V_{in} is shown in Figure 3.28 and represents the circuit to simulate and analyze.

We start with the incremental resistance value R_0 for which the power inductor is shorted and the output capacitor open. The circuit to analyze is that of Figure 3.29. The current in terminal c is actually the I_T current flowing in reverse direction. The output voltage is thus determined as:

$$V_{out} = V_T - I_T r_L + V_{out} D_0 \qquad (3.58)$$

If we rearrange this expression, we have:

$$V_{out} = \frac{V_T - I_T r_L}{1 - D_0} \qquad (3.59)$$

Applying KCL, we have:

$$I_T = I_2 + I_T D_0 \qquad (3.60)$$

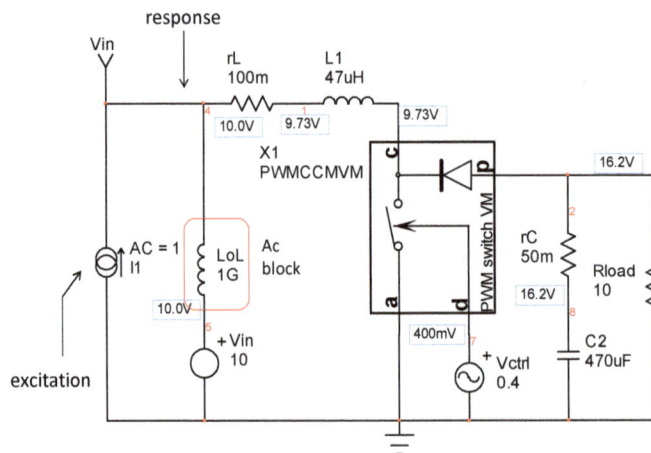

Figure 3.28 Adding a large-value inductor isolates the input source in this ac analysis.

Factoring and rearranging, we have a definition for I_T:

$$I_T = \frac{I_2}{1 - D_0} = \frac{V_{out}}{R_{load}(1 - D_0)} \tag{3.61}$$

Now substituting (3.59) in the above expression and solving for R_0, we have:

$$\frac{V_T}{I_T} = R_0 = (1 - D_0)^2 R_{load} + r_L \tag{3.62}$$

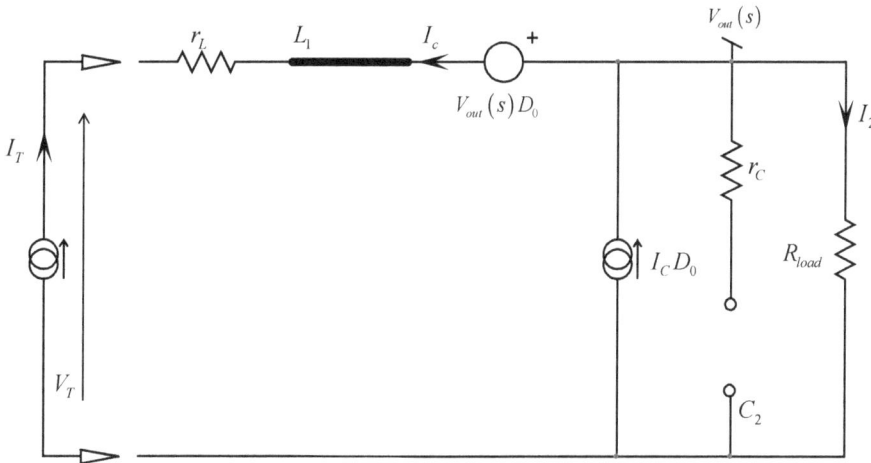

Figure 3.29 The incremental resistance requires a few equations to determine its value.

In this exercise, the stimulus is the current source I_T. Turning it off means open-circuiting the generator as shown in Figure 3.30. You can see the left-side terminal of L_1 is now dangling: it is no longer the previously-analyzed natural structure and we cannot reuse the denominator determined in (3.20). By inspection, we can infer that the resistance "seen" from the inductor connecting terminal approaches infinity.

As such, we have:

$$\tau_1 = \frac{L_1}{\infty} = 0 \tag{3.63}$$

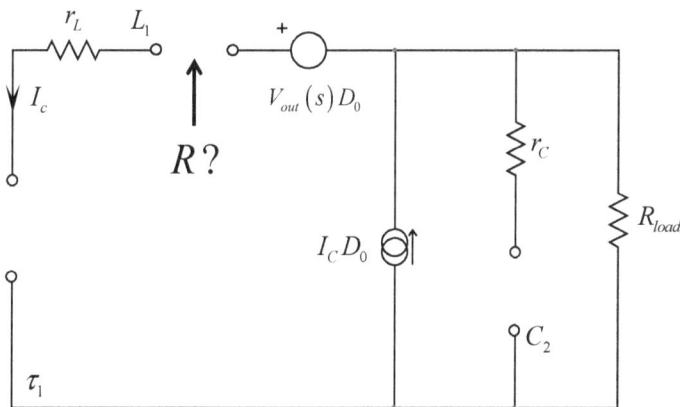

Figure 3.30 Turning the current source off does not return the circuit into its natural state: $D(s)$ **previously determined cannot be reused.**

For the time constant involving C_2, look at Figure 3.31 in which the inductor is set in its dc state and replaced by a short circuit. As no current flows out of terminal c, the current source $I_c D_0$ is 0 A and the generator disappears from the circuit. Inspecting the circuit shows a time constant defined as:

$$\tau_2 = C_2 \left(r_C + R_{load} \right) \tag{3.64}$$

The 1st-order term b_1 is obtained by summing τ_1 and τ_2:

$$b_1 = \tau_1 + \tau_2 = \frac{L_2}{\infty} + C_2 \left(r_C + R_{load} \right) = C_2 \left(r_C + R_{load} \right) \tag{3.65}$$

Now, we are going to set L_1 in its high-frequency state (an open circuit) and look at the resistance driving capacitor C_2. The circuit is shown in Figure 3.32 and leads to the result obtained in the expression above:

$$\tau_2^1 = C_2 \left(r_C + R_{load} \right) \tag{3.66}$$

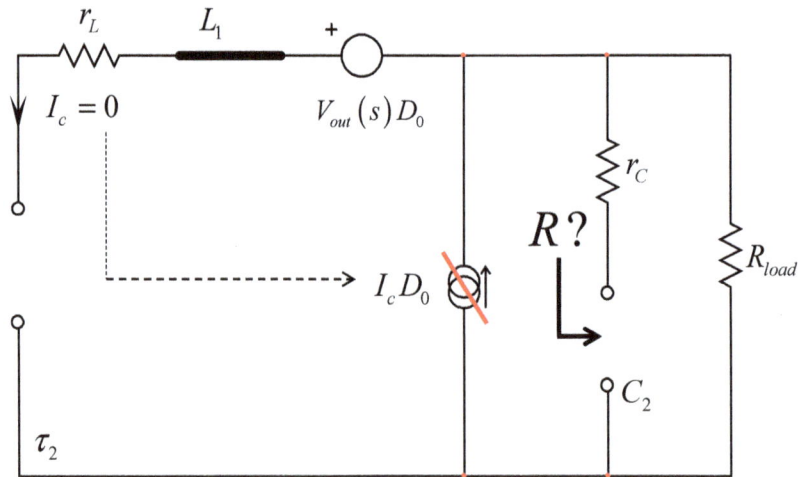

Figure 3.31 The circuit can be simplified and the time constant determined via inspection.

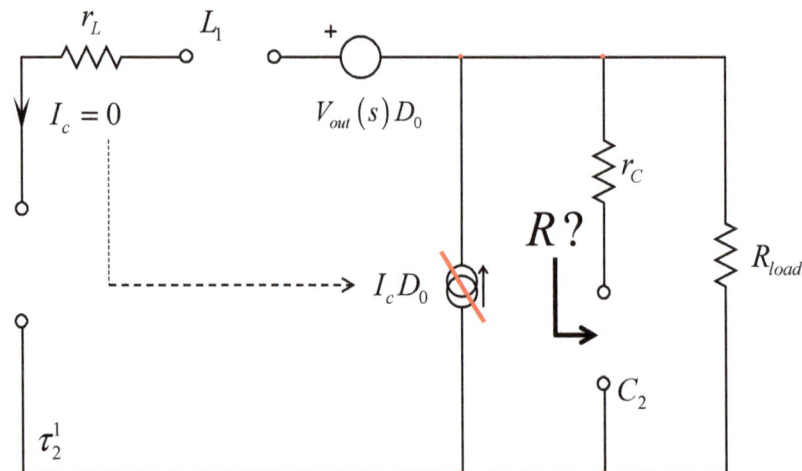

Figure 3.32 In this mode, the current flowing out of terminal c is still 0 A.

We can now define b_2 as the product of (3.63) by (3.66):

$$b_2 = \tau_1 \tau_2^1 = \frac{L_1}{\infty} \cdot C_2 (r_C + R_{load}) = 0 \tag{3.67}$$

We have our denominator equal to:

$$D(s) = 1 + b_1 s + b_2 s^2 = 1 + s(r_C + R_{load}) C_2 \tag{3.68}$$

You can see this is a first-order type despite the presence of two energy-storing elements. This is because the inductor – actually its state variable – is driven by a current source in this input impedance configuration: this is a degenerate case.

For the zeroes, we are going to bring the excitation back and null the response. The response is V_T, the voltage across the current source. A 0-V condition across a current source is similar to replacing the generator by a short circuit as shown in Figure 3.33.

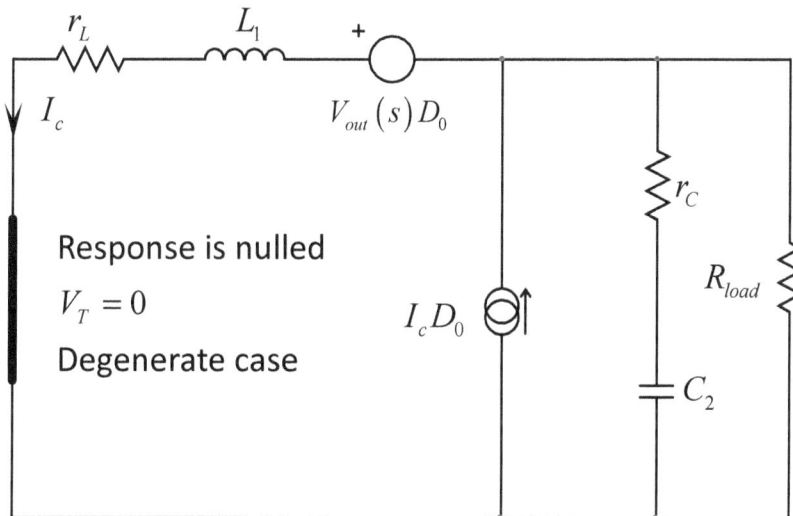

Figure 3.33 Having 0 V across the current source returns the circuit into its natural state.

This configuration is similar to that already studied to determine the denominator in the previous transfer functions. We can reuse the expression determined in (3.20) for the numerator of $Z_{in}(s)$. The final transfer function is thus defined as

$$Z_{in}(s) \approx R_0 \frac{1 + \frac{s}{\omega_0 Q} + \left(\frac{s}{\omega_0}\right)^2}{1 + \frac{s}{\omega_p}} \tag{3.69}$$

In which we have:

$$R_0 = (1 - D_0)^2 R_{load} + r_L \tag{3.70}$$

$$\omega_0 \approx \frac{1-D_0}{\sqrt{L_1 C_2}} \qquad (3.71)$$

$$Q \approx (1-D_0) R_{load} \sqrt{\frac{C_2}{L_1}} \qquad (3.72)$$

$$\omega_p = \frac{1}{(r_C + R_{load}) C_2} \qquad (3.73)$$

We have run a SPICE simulation of the input impedance whose circuit appears in Figure 3.34. The duty ratio is set to 40% and ensures a 16.2-V output voltage for the bias point. We have superimposed the SPICE dynamic response in magnitude and phase with those of the Mathcad® sheet and, as shown in Figure 3.35, they perfectly superimpose.

We have determined the four transfer functions of the voltage-mode boost converter operated in CCM and gathered all its transfer functions in the table proposed in Figure 3.36.

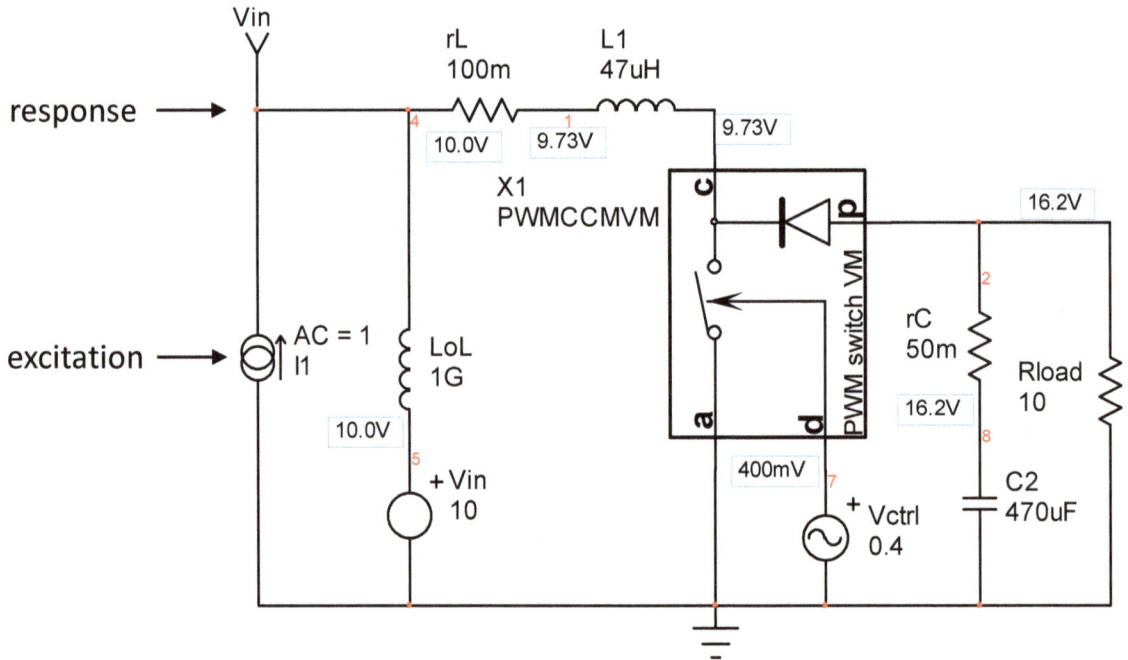

Figure 3.34 The input impedance is obtained by inserting a large-value inductor between the input voltage source and the circuit input.

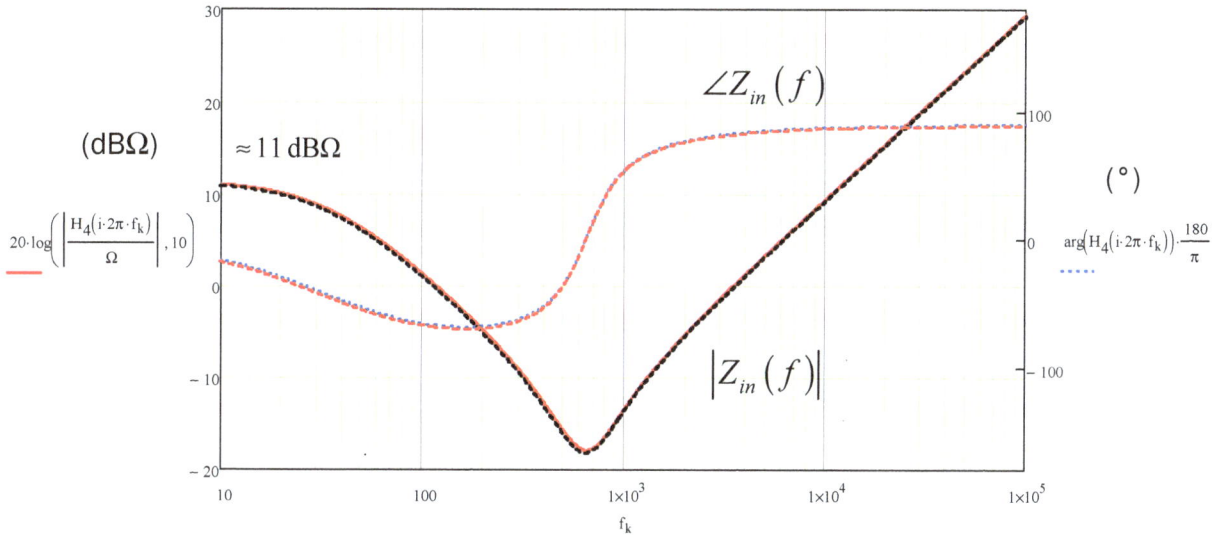

Figure 3.35 Response from SPICE and results from the Mathcad® sheet perfectly superimpose.

$\dfrac{V_{out}(s)}{V_{err}(s)}$ Control to Output	$H_0 \dfrac{\left(1+\dfrac{s}{\omega_{z_1}}\right)\left(1-\dfrac{s}{\omega_{z_2}}\right)}{1+\dfrac{s}{\omega_0 Q}+\left(\dfrac{s}{\omega_0}\right)^2}$	$\omega_{z_1} = \dfrac{1}{r_C C_2}$ $\omega_{z_2} = \dfrac{(1-D)^2 R_{load}}{L_1}$	$\omega_0 \approx \dfrac{1-D}{\sqrt{L_1 C_2}}$ $Q \approx (1-D) R_{load}\sqrt{\dfrac{C_2}{L_1}}$	$H_0 \approx \dfrac{V_{in}}{V_p(1-D)^2}$
$\dfrac{V_{out}(s)}{V_{in}(s)}$ Input to Output	$H_0 \dfrac{1+\dfrac{s}{\omega_z}}{1+\dfrac{s}{\omega_0 Q}+\left(\dfrac{s}{\omega_0}\right)^2}$	$\omega_z = \dfrac{1}{r_C C_2}$	$\omega_0 \approx \dfrac{1-D}{\sqrt{L_1 C_2}}$ $Q \approx (1-D) R_{load}\sqrt{\dfrac{C_2}{L_1}}$	$H_0 \approx \dfrac{1}{1-D}$
$Z_{in}(s)$ Input impedance	$R_0 \dfrac{1+\dfrac{s}{\omega_0 Q}+\left(\dfrac{s}{\omega_0}\right)^2}{1+\dfrac{s}{\omega_p}}$	$\omega_0 \approx \dfrac{1-D}{\sqrt{L_1 C_2}}$ $Q \approx (1-D) R_{load}\sqrt{\dfrac{C_2}{L_1}}$	$\omega_p = \dfrac{1}{(r_C + R_{Load})C_2}$	$R_0 \approx (1-D)^2 R_{load}$
$Z_{out}(s)$ Output impedance	$R_0 \dfrac{(1+s/\omega_{z_1})(1+s/\omega_{z_2})}{1+\dfrac{s}{\omega_0 Q}+\left(\dfrac{s}{\omega_0}\right)^2}$	$\omega_{z_1} = \dfrac{r_L}{L_1}$ $\omega_{z_2} = \dfrac{1}{r_C C_2}$	$\omega_0 \approx \dfrac{1-D}{\sqrt{L_1 C_2}}$ $Q \approx (1-D) R_{load}\sqrt{\dfrac{C_2}{L_1}}$	$R_0 \approx \dfrac{r_L}{(1-D)^2}$

Figure 3.36 The four transfer functions of the VM boost converter operated in CCM are given.

3.2 Boost Transfer Functions in DCM – Fixed-Frequency Voltage-Mode Control

We will now reduce the load current and force the converter to enter the discontinuous conduction mode. To determine the load value leading to this mode, we can calculate the critical resistance using the expression given in Chapter 1 section 1.2.4:

$$R_{critical} = \frac{2F_{sw}L}{D(1-D)^2} = \frac{2 \times 47u \times 100k}{0.4 \times (1-0.4)^2} \approx 63 \ \Omega \tag{3.74}$$

By selecting a 100-Ω resistance, we know the converter from Figure 3.37 will operate in DCM.

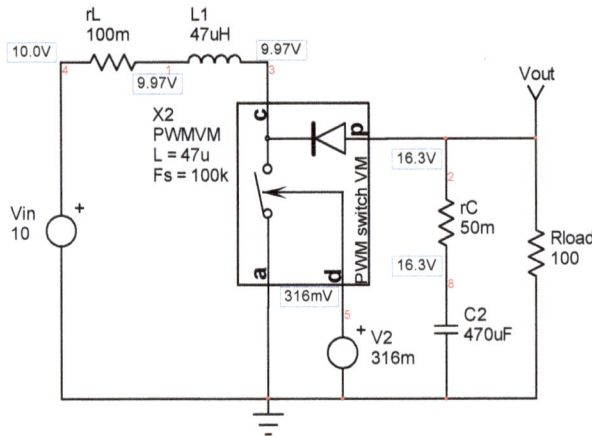

Figure 3.37 The load current is reduced and forces a transition to the discontinuous mode.

We can now replace the PWM switch symbol by its small-signal version as depicted in Figure 3.38. You can see all the small-signal coefficients computed in the left side of the schematic diagram.

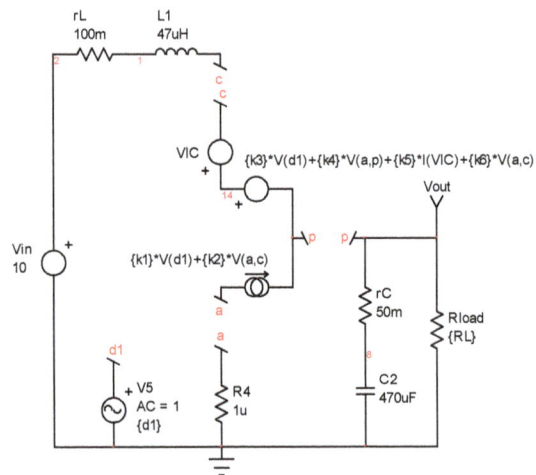

Figure 3.38 The small-signal model is now part of the circuit and we can start analyzing the converter.

312

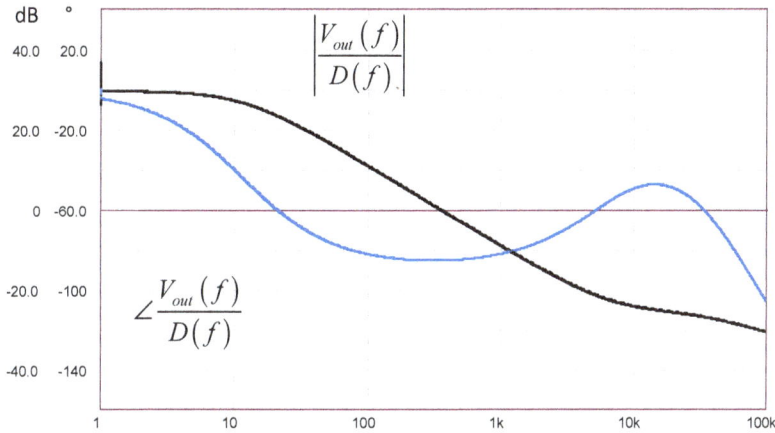

Figure 3.39 Dynamic responses between the large- and small-signal models are identical.

Before proceeding with the analysis, it is important to run a sanity check and make sure the electrical diagram we are about to study is sound. As confirmed by Figure 3.39, both circuits deliver the exact same response: we are good to go.

We will start with the bias point calculations for which we short the inductor and open the capacitor. Figure 3.40 confirms the bias point is correct and we can rearrange sources as in Figure 3.41 for a more convenient analysis. The current flowing out of terminal c is determined as follows:

$$I_c = \frac{V_{(c)} - V_{in}}{r_L} \tag{3.75}$$

and:

$$I_c = -\left(\frac{V_{out}}{R_{load}} + \frac{V_{(c)}D^2}{2R_{load}\tau_L}\right) \tag{3.76}$$

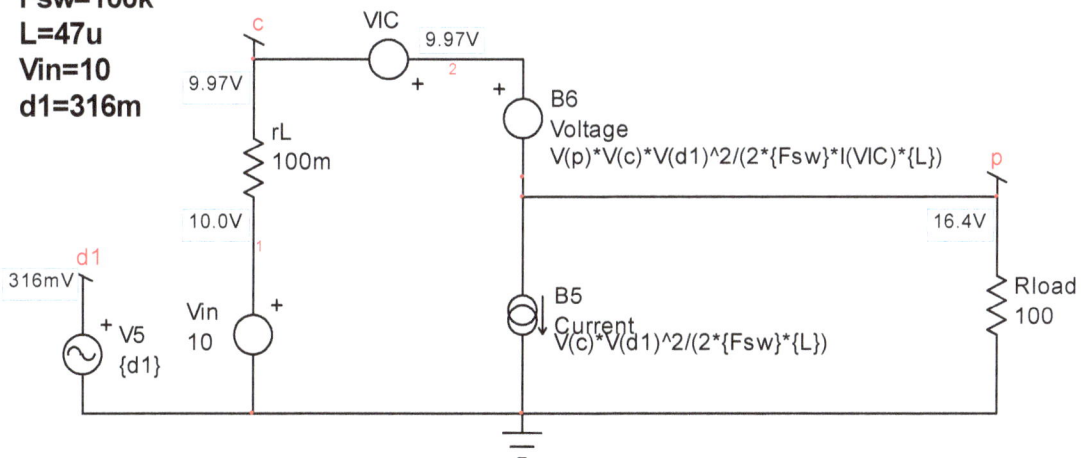

Figure 3.40 For the dc analysis, replace the inductor by a short circuit and open the capacitor.

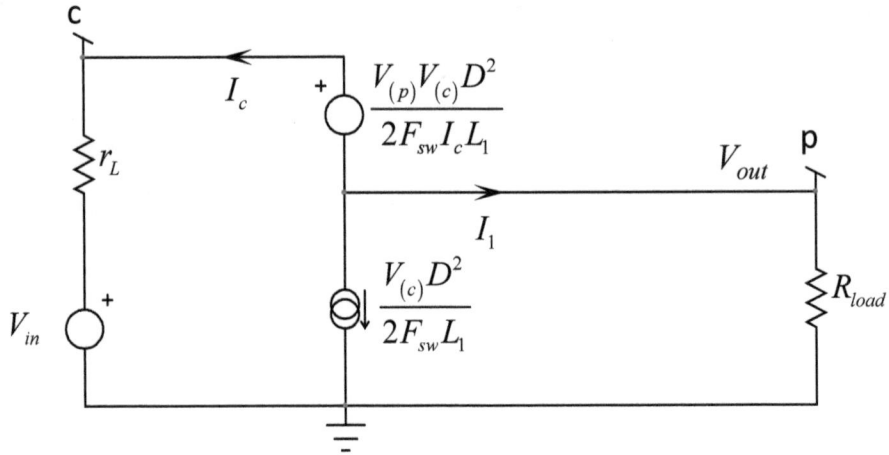

Figure 3.41 The updated circuit with rearranged sources.

In this expression, we use the normalized time constant τ_L defined by:

$$\tau_L = \frac{L_1}{R_{load}} F_{sw} \tag{3.77}$$

From (3.76), we extract $V_{(c)}$:

$$V_{(c)} = -\frac{2R_{load}\tau_L\left(I_c + \dfrac{V_{out}}{R_{load}}\right)}{D_0^{\,2}} \tag{3.78}$$

We replace I_c by its definition from (3.75) and we solve again for $V_{(c)}$:

$$V_{(c)} = \frac{2R_{load}V_{in}\tau_L - 2V_{out}r_L\tau_L}{r_L D_0^{\,2} + 2R_{load}\tau_L} \tag{3.79}$$

We can substitute it back in (3.75) and solving for I_c, we obtain:

$$I_c = \frac{V_{(c)} - V_{in}}{r_L} = -\frac{V_{in}D_0^{\,2} + 2V_{out}\tau_L}{r_L D_0^{\,2} + 2R_{load}\tau_L} \tag{3.80}$$

From mesh analysis, we can express the input voltage V_{in} as:

$$V_{in} = -r_L I_c + \frac{V_{out}V_{(c)}D_0^{\,2}}{2I_c R_{load}\tau_L} + V_{out} \tag{3.81}$$

Solving for V_{out} gives:

$$V_{out} = \frac{V_{in} + I_c r_L}{\dfrac{V_{(c)}D_0^{\,2}}{2I_c R_{load}\tau_L} + 1} \tag{3.82}$$

If we substitute (3.79) and (3.80) in (3.82), we obtain a definition for V_{out} equal to:

$$V_{out} = V_{in} \frac{\sqrt{2D_0^2 \tau_L + \tau_L^2} \left(\dfrac{2R_{load}\tau_L^2}{\sqrt{2D_0^2 \tau_L + \tau_L^2}} + 2R_{load}\tau_L - \dfrac{D_0^2 r_L \tau_L}{\sqrt{2D_0^2 \tau_L + \tau_L^2}} + D_0^2 r_L \right)}{4 \left(R_{load}\tau_L^2 + D_0^2 r_L \tau_L + r_L \tau_L^2 + D_0^4 \dfrac{r_L^2}{4R_{load}} \right)} \tag{3.83}$$

Considering a small ohmic loss r_L, this expression fortunately simplifies to:

$$V_{out} = V_{in} \frac{\tau_L + \sqrt{2D_0^2 \tau_L + \tau_L^2}}{2\tau_L} = V_{in} \frac{1}{2} \left(1 + \sqrt{1 + \frac{2D_0^2}{\tau_L}} \right) \tag{3.84}$$

This is the dc relationship linking V_{out} to V_{in} for a boost converter operated in the discontinuous conduction mode. To obtain the quasi-static gain H_0, we can either determine the control-to-output transfer function in the same configuration but since we have determined the link between the output and input variables, we can perform a differentiation on (3.84) and obtain the gain immediately:

$$H_0 = \frac{d}{dD} V_{out}(D) = \frac{d}{dD} V_{in} \frac{\tau_L + \sqrt{2D^2 \tau_L + \tau_L^2}}{2\tau_L} = \frac{V_{in} D_0}{\sqrt{2D_0^2 \tau_L + \tau_L^2}} \tag{3.85}$$

The numerical application returns a gain of 29.3 dB confirming what you can read in Figure 3.39. With the dc gain on hand, we now need to determine the rest of the control-to-output transfer functions. The excitation in this approach is the duty ratio (d_1 or D_0 in the figures) and turning it off to zero brings us to Figure 3.42.

For the first time constant involving the inductor, open the capacitor and determine the resistance "seen" from its connecting terminals.

We have installed a test generator I_T as shown in Figure 3.43. I_T and I_c flow in opposite directions:

$$I_T = -I_c \tag{3.86}$$

The voltage at node c is that of the current source minus the drop across r_L:

$$V_{(c)} = V_T - I_T r_L \tag{3.87}$$

Observing that the voltage at terminal p is actually the output voltage, we have:

$$V_{(p)} = \left(I_T - k_2 V_{(c)} \right) R_{load} \tag{3.88}$$

Writing the main mesh involving V_T gives:

$$V_T = I_T r_L + k_5 I_c - k_6 V_{(c)} - k_4 V_{(p)} + V_{(p)} \tag{3.89}$$

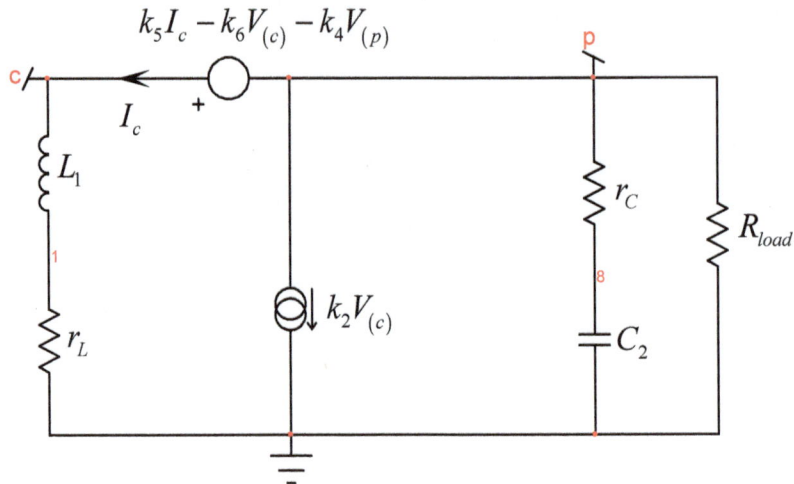

Figure 3.42 This is the natural structure of the DCM-operated VM boost converter.

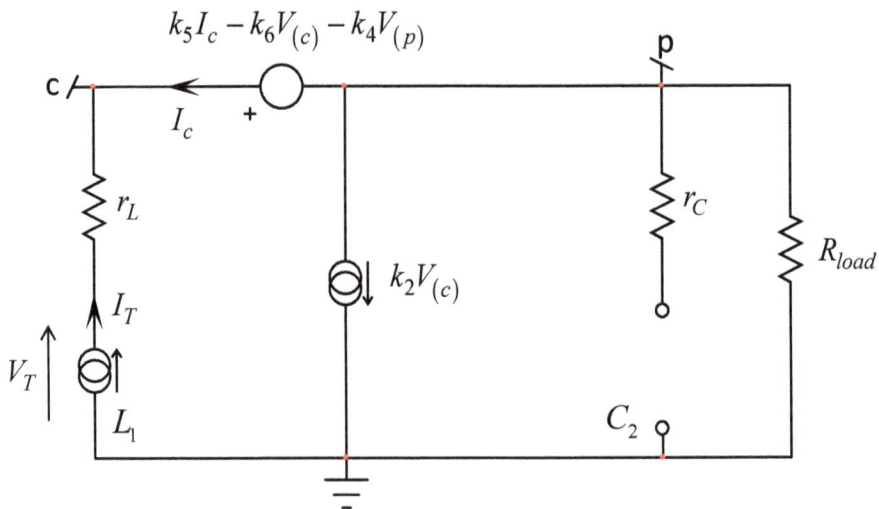

Figure 3.43 Determine the first time constant by setting the capacitor is in its dc state.

Finally, substituting the above definitions in (3.89) and determining:

$$R = \frac{V_T}{I_T} \tag{3.90}$$

we have:

$$\tau_1 = \frac{L_1}{R} = \frac{L_1}{r_L + \dfrac{R_{load}(1 - k_4) - k_5}{k_6 + R_{load}k_2(1 - k_4) + 1}} \tag{3.91}$$

For the second time constant involving capacitor C_2, we look at Figure 3.44 in which the inductor is replaced by a short circuit.

We first determine the voltage at terminal p:

$$V_{(p)} = \left[I_T - I_c - k_2 V_{(c)} \right] R_{load} \tag{3.92}$$

The voltage across the current generator I_T is:

$$V_T = r_C I_T + V_{(p)} \tag{3.93}$$

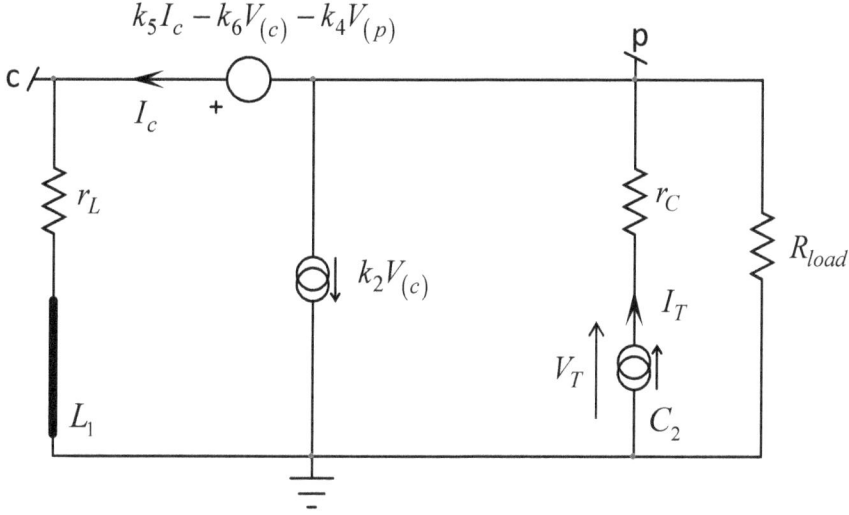

Figure 3.44 The inductor is now replaced by a short circuit.

while the voltage at terminal c is simply:

$$V_{(c)} = I_c r_L \tag{3.94}$$

We can define the voltage at terminal p in another way considering the left-side mesh:

$$V_{(p)} = I_c r_L - k_5 I_c + k_6 V_{(c)} + k_4 V_{(p)} \tag{3.95}$$

Once rearranged, we have:

$$V_{(p)} = -\frac{I_c r_L - I_c k_5 + V_c k_6}{k_4 - 1} \tag{3.96}$$

Equating (3.96) and (3.92) then solving for $V_{(c)}$ leads to:

$$V_{(c)} = \frac{I_T R_{load} - I_c R_{load} + I_c k_5 - I_c r_L - I_T R_{load} k_4 + I_c R_{load} k_4}{k_6 + R_{load} k_2 - R_{load} k_2 k_4} \tag{3.97}$$

Substitute (3.96) in (3.93), then plug I_c obtained via (3.94) and (3.97), factor I_T and express the second time constant such as:

$$\tau_2 = \frac{V_T}{I_T} C_2 = C_2 \left[\frac{r_C (1 + k_2 R_{load})}{1 + k_2 r_L} + \frac{r_L (1 + k_6) - k_5}{1 - \frac{k_5}{R_{load}} + \frac{r_L}{R_{load}}(1 + k_6) - k_4 + k_2 r_L (1 - k_4)} \right] \tag{3.98}$$

We can form the first denominator term by adding τ_1 and τ_2:

$$b_1 = \tau_1 + \tau_2 = \frac{L_1}{r_L + \dfrac{R_{load}(1-k_4)-k_5}{k_6 + R_{load}k_2(1-k_4)+1}} + C_2 \left[\frac{r_C(1+k_2 R_{load})}{1+k_2 r_L} + \frac{r_L(1+k_6)-k_5}{1 - \dfrac{k_5}{R_{load}} + \dfrac{r_L}{R_{load}}(1+k_6)-k_4+k_2 r_L(1-k_4)} \right] \quad (3.99)$$

For the second-order term, we set C_2 in its high-frequency state (a short circuit) and look at the resistance driving L_1. This is what Figure 3.45 suggests.

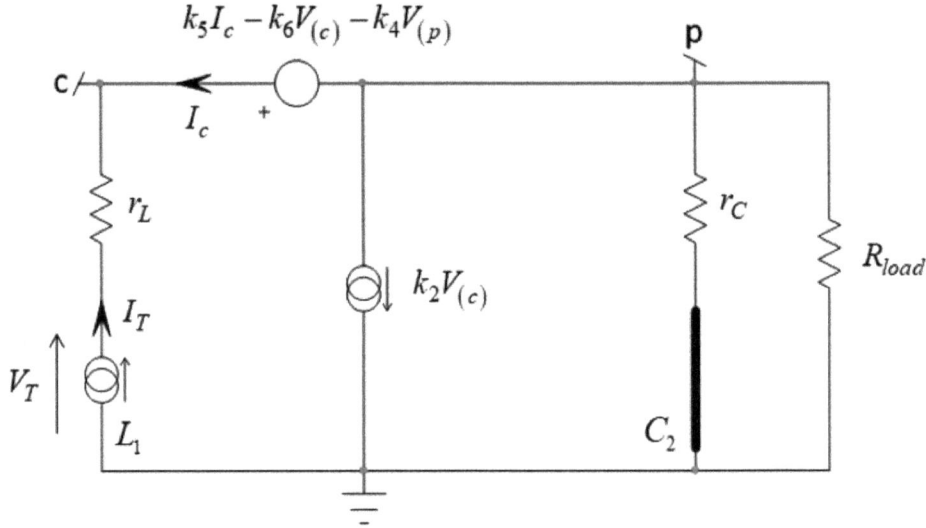

Figure 3.45 The capacitor is set in its high-frequency state and we determine the resistance driving L_1.

This configuration is similar to that of Figure 3.43 except that r_C now comes in parallel with R_{load}. We can thus immediately reuse (3.91) with a slight update:

$$\tau_1^2 = \frac{L_1}{r_L + \dfrac{(R_{load} \| r_C)(1-k_4)-k_5}{k_6 + (R_{load} \| r_C)k_2(1-k_4)+1}} \quad (3.100)$$

We have our second-order term b_2 defined as:

$$b_2 = \tau_2 \tau_1^2 = C_2 \left[\frac{r_C(1+k_2 R_{load})}{1+k_2 r_L} + \frac{r_L(1+k_6)-k_5}{1 - \dfrac{k_5}{R_{load}} + \dfrac{r_L}{R_{load}}(1+k_6)-k_4+k_2 r_L(1-k_4)} \right] \frac{L_1}{r_L + \dfrac{(R_{load} \| r_C)(1-k_4)-k_5}{k_6 + (R_{load} \| r_C)k_2(1-k_4)+1}}$$

$$(3.101)$$

The denominator is obtained by combining b_1 and b_2 as:

$$D(s) = 1 + sb_1 + s^2 b_2 \quad (3.102)$$

If you now calculate the quality factor Q defined as [1] with numerical values:

$$Q = \frac{\sqrt{b_2}}{b_1} \tag{3.103}$$

you will find a very low value indicating that the boost converter operated in DCM is still a second-order system but heavily damped. Considering the low-Q approximation, (3.102) can be advantageously rewritten as:

$$D(s) \approx \left(1 + \frac{s}{\omega_{p_1}}\right)\left(1 + \frac{s}{\omega_{p_2}}\right) \tag{3.104}$$

In this expression, a pole dominates the low-frequency response:

$$\omega_{p_1} = \frac{1}{b_1} \tag{3.105}$$

A second pole appears in higher frequencies:

$$\omega_{p_2} = \frac{b_1}{b_2} \tag{3.106}$$

As shown in [2], these poles can advantageously be rearranged as:

$$\omega_{p_1} \approx \frac{2M - 1}{M - 1}\frac{1}{R_{load}C_2} \tag{3.107}$$

and:

$$\omega_{p_2} \approx 2F_{sw}\left(\frac{1 - \frac{1}{M}}{D_0}\right)^2 \tag{3.108}$$

In which M is V_{out}/V_{in}.

For the zeroes, we need to identify impedance conditions in the circuit which would prevent the excitation $- D(s) -$ to propagate and form a response V_{out}. In other words, what conditions null the output voltage despite the presence of a stimulus? Figure 3.46 shows the configuration to study. We have the usual suspects with r_C and C_2 forming a transformed short circuit and bringing a zero:

$$\omega_{z_1} = \frac{1}{r_C C_2} \tag{3.109}$$

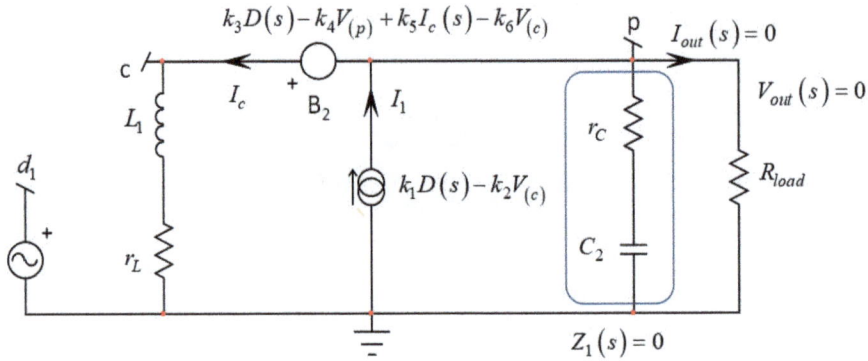

Figure 3.46 The zeroes are identified by looking at the circuit when the response is nulled.

Then, the other conditions for which no current circulates in R_{load} is when all the current delivered in I_1 is flowing into terminal c. When this phenomenon happens, we have an output null. The current I_c is defined as the voltage at terminal c divided by the impedance made of L_1 and r_L in series:

$$k_1 D(s) - k_2 V_{(c)} = \frac{V_{(c)}}{r_L + sL_1}$$

(3.110)

From this expression, we extract the voltage at terminal c:

$$V_{(c)} = \frac{D(s)k_3}{k_6 - \dfrac{k_5}{r_L + sL_1} + 1}$$

(3.111)

However, the voltage at this terminal is given by source B_2 since V_{out} is 0 V:

$$V_{(c)} = k_3 D(s) - k_4 V_{(p)} + k_5 I_c(s) - k_6 V_{(c)}$$

(3.112)

In this expression, the voltage at terminal p is also 0 V because it is connected to V_{out}. Equating (3.111) and (3.112) then solving for the root leads to the second zero:

$$\omega_{z_2} = \frac{\dfrac{k_3 + k_1 k_5}{k_1 - k_2 k_3 + k_1 k_6} - r_L}{L_1}$$

(3.113)

The numerical application shows that this second zero is a positive root and is thus located in the right half-plane. Chapter 1 explains why a RHPZ occurs in a DCM boost or buck-boost converter. We now have everything we need for this first control-to-output transfer function of the boost converter operated in discontinuous mode. Adopting notations from [2], we have:

$$H(s) = \frac{V_{out}(s)}{D(s)} = H_0 \frac{\left(1 + \dfrac{s}{\omega_{z_1}}\right)\left(1 - \dfrac{s}{\omega_{z_2}}\right)}{\left(1 + \dfrac{s}{\omega_{p_1}}\right)\left(1 + \dfrac{s}{\omega_{p_2}}\right)}$$

(3.114)

with:

$$H_0 = \frac{V_{in}D_0}{\sqrt{2D_0^2\tau_L + \tau_L^2}} \approx \frac{2V_{in}}{2M-1}\sqrt{\frac{M(M-1)}{2\tau_L}} \qquad (3.115)$$

in which:

$$M = \frac{V_{out}}{V_{in}} \qquad (3.116)$$

and:

$$\tau_L = \frac{L_1}{R_{load}T_{sw}} \qquad (3.117)$$

The poles are defined as:

$$\omega_{p_1} \approx \frac{2M-1}{M-1}\frac{1}{R_{load}C_2} \qquad (3.118)$$

$$\omega_{p_2} \approx 2F_{sw}\left(\frac{1-\frac{1}{M}}{D}\right)^2 \qquad (3.119)$$

While the zeroes are given by:

$$\omega_{z_1} = \frac{1}{r_C C_2} \qquad (3.120)$$

$$\omega_{z_2} \approx \frac{R_{load}}{M^2 L_1} \qquad (3.121)$$

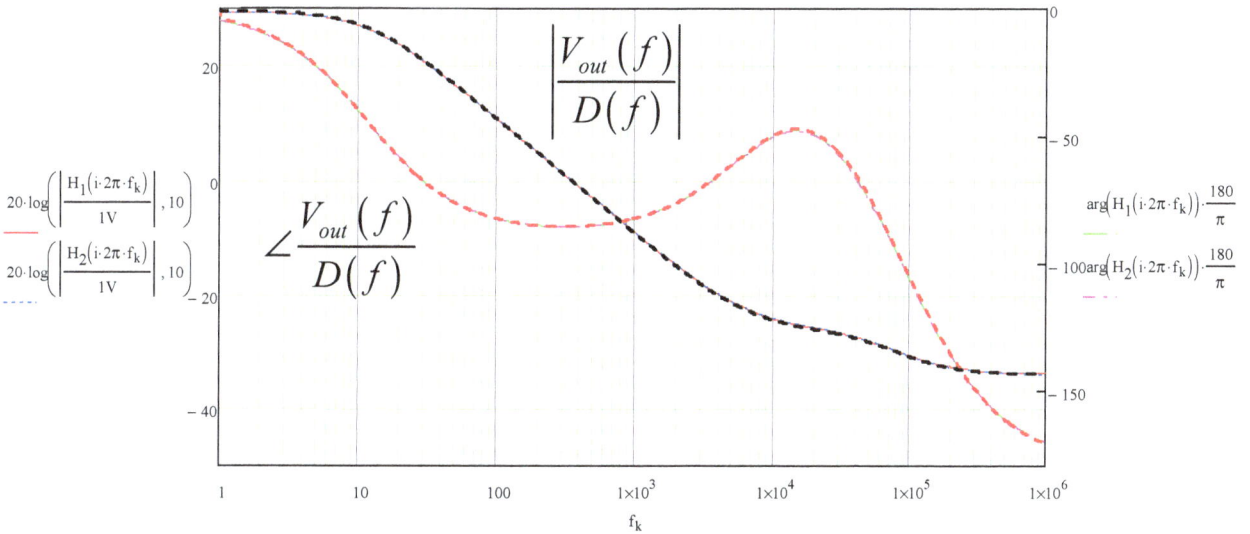

**Figure 3.47 Superimposing the SPICE simulation results and the response from the Mathcad®
shows no difference.**

We have reused the SIMPLIS® template implemented for the CCM version and reduced the load current with a 100-Ω load (Figure 3.48). The inductor current is discontinuous as confirmed by Figure 3.49 which also shows the dynamic response. The phase is very close to that of the averaged model but the magnitude curves deviate a little as we approach half the switching frequency.

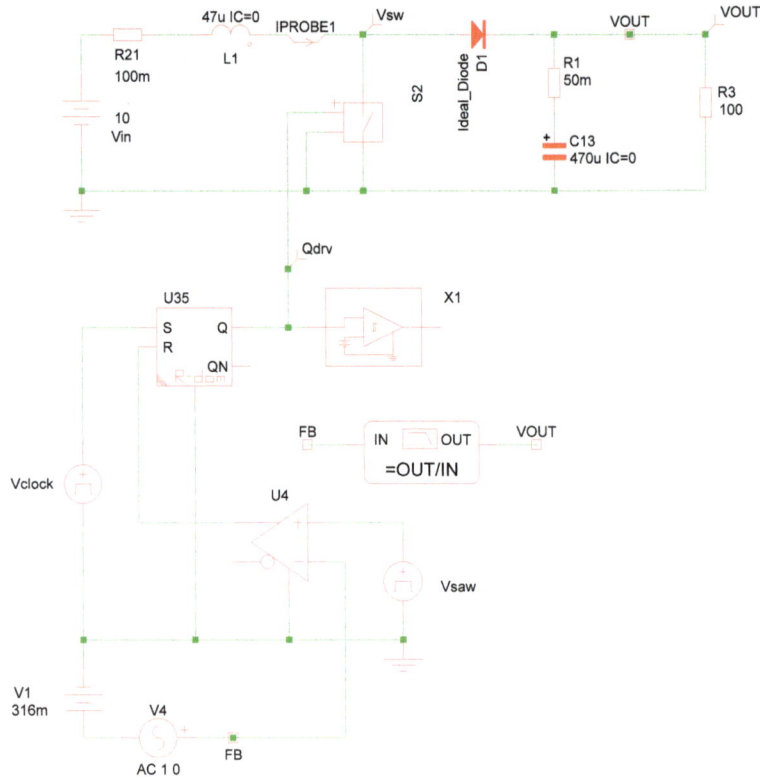

Figure 3.48 Reducing the load current in the SIMPLIS® template forces DCM operation.

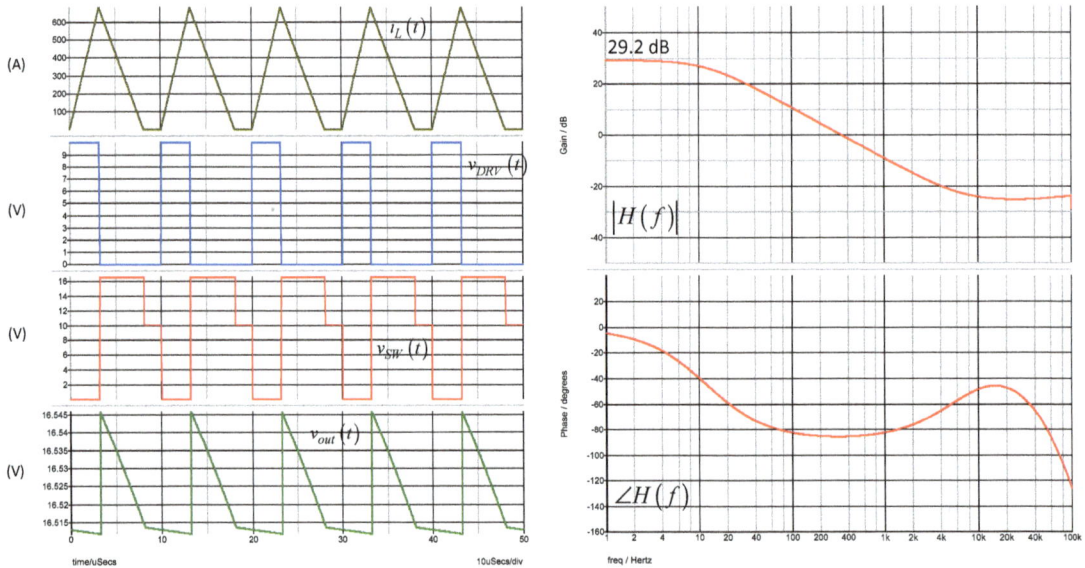

Figure 3.49 The dynamic response is close to what SPICE delivers with a slight deviation at higher frequencies.

For the sake of illustrating the coefficients calculations, I have included their values in Figure 3.50 sheet.

It is a convenient way for checking your own approach in case you would use another solver like Excel® for instance.

$$L_1 := 47\mu H \qquad F_{sw} := 100kHz \qquad C_2 := 470\mu F \qquad V_{in} := 10V \qquad R_L := 100\Omega$$

$$\tau_L := \frac{L_1}{R_L} \cdot F_{sw} \qquad d_1 := 0.316 \qquad r_L := 0.1\Omega \qquad r_C := 0.05\Omega$$

$$V_{out} := V_{in} \cdot 0.5 \cdot \left(1 + \sqrt{1 + \frac{2 \cdot d_1^2}{\tau_L}}\right) = 16.45556V \qquad M := \frac{V_{out}}{V_{in}} = 1.64556$$

$$I_c := -\frac{V_{in} \cdot d_1^2 + 2 \cdot V_{out} \cdot \tau_L}{r_L \cdot d_1^2 + 2 \cdot R_L \cdot \tau_L} = -0.2705A$$

Ic is negative for the PWM switch model because original model assumed Ic leaves terminal c whereas it enters it in a boost application.

$$D := \sqrt{\frac{2 \cdot L_1 \cdot V_{out} \cdot \left(\frac{V_{out}}{V_{in}} - 1\right) \cdot F_{sw}}{V_{in} \cdot R_L}} = 0.316$$

dc gain for small inductive loss:

$$H_0 := \frac{V_{in} \cdot d_1}{\sqrt{2 \cdot d_1^2 \cdot \tau_L + \tau_L^2}} = 29.3456V \qquad 20 \cdot \log\left(\frac{H_0}{V}\right) = 29.35086$$

Small signal coefficients calculations:

$$V_{ac} := -V_{in} = -10V \qquad V_{ap} := -V_{out} = -16.45556V$$

$$k_1 := \frac{V_{ac} \cdot d_1}{F_{sw} \cdot L_1} = -0.67234A \qquad k_2 := \frac{d_1^2}{2 \cdot F_{sw} \cdot L_1} = 0.01062\frac{1}{\Omega}$$

$$k_3 := \frac{V_{ac} \cdot V_{ap} \cdot d_1}{F_{sw} \cdot I_c \cdot L_1} = -40.90136V \qquad k_4 := \frac{V_{ac} \cdot d_1^2}{2 \cdot F_{sw} \cdot I_c \cdot L_1} = 0.39272 \qquad k_5 := -\frac{V_{ac} \cdot V_{ap} \cdot d_1^2}{2 \cdot F_{sw} \cdot I_c^2 \cdot L_1} = -23.89008\Omega$$

$$k_6 := \frac{V_{ap} \cdot d_1^2}{2 \cdot F_{sw} \cdot I_c \cdot L_1} = 0.64624 \qquad d_2 := \frac{2 \cdot L_1 \cdot F_{sw} \cdot I_c}{d_1 \cdot V_{ac}} - d_1 = 0.48865$$

Figure 3.50 This sheet gives you the coefficients computed by Mathcad® for this DCM boost example.

3.2.1 Input to Output

In this configuration, the stimulus is now the input voltage V_{in} while the duty ratio is 0 V in ac. The large-signal configuration we have used to compare dynamic responses is that of Figure 3.51. The duty ratio now fixes the operating point while the input source is ac-modulated.

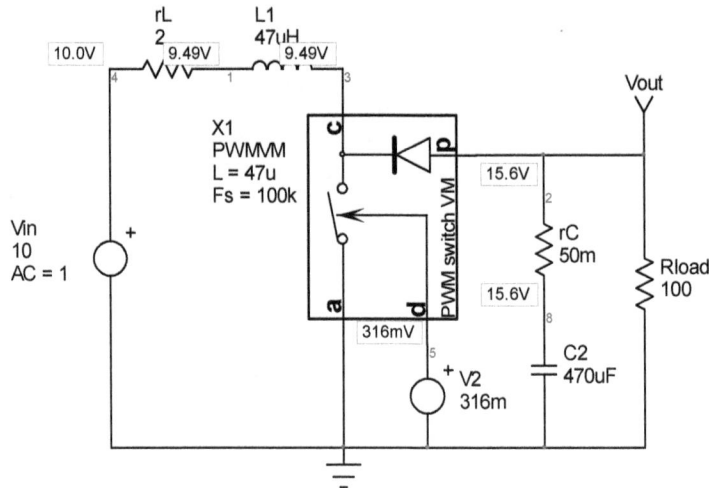

Figure 3.51 The large-signal model remains unchanged except the input source which is now ac-modulated. The duty ratio generator is a simple dc source ensuring the correct bias point.

Figure 3.52 depicts the adopted circuit in which we have already inserted the small-signal model. The left-side parameter list includes I_a and I_c determined from the large-signal simulation. The quasi-static gain H_0 is determined by differentiating V_{out} with respect to V_{in}:

$$V_{out}\left(V_{in}\right) = M \cdot V_{in} \tag{3.122}$$

$$H_0 = \frac{dV_{out}\left(V_{in}\right)}{dV_{in}} = \frac{1}{2}\left(1+\sqrt{1+\frac{2D^2}{\tau_L}}\right) \tag{3.123}$$

To determine the time constants, we reduce the stimulus to zero as illustrated in Figure 3.53. The circuit is similar to that of Figure 3.42 implying that we can reuse the denominator already determined in (3.104).

parameters

d1=316m
Vin=10
Vout=16.3216
Vac=-Vin
Vap=-Vout
Fsw=100k
L=47u
RL=100

Ia=-103.892m
Ic=-267.108m

Ica=(Vout/RL)*(-1-Vin*d1^2*RL/(2*L*Vout*Fsw))
N=Vac*d1^2/(2*Fsw*Ica*L)
Iaa=N*Ica

k1=Vac*d1/(Fsw*L)
k2=d1^2/(2*Fsw*L)
k3=Vac*Vap*d1/(Fsw*L*Ic)
k4=Vac*d1^2/(2*Fsw*L*Ic)
k5=-Vac*Vap*d1^2/(2*Fsw*Ic^2*L)
k6=Vap*d1^2/(2*Fsw*Ic*L)

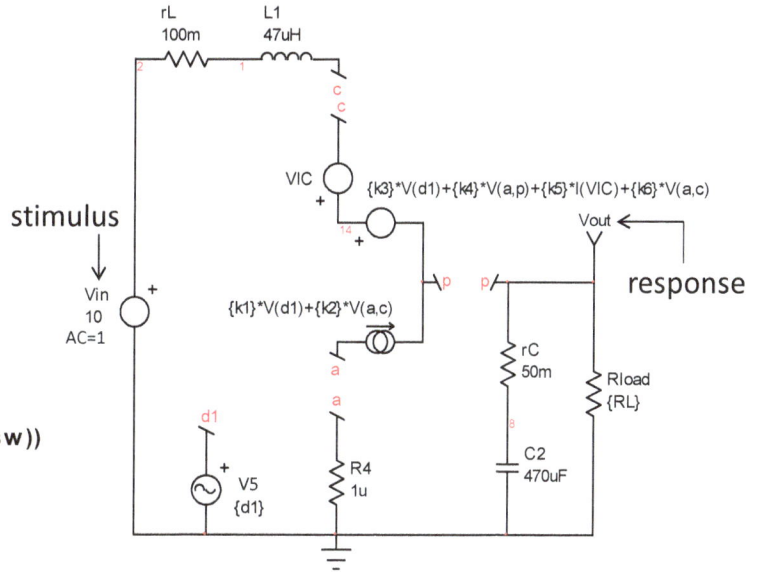

stimulus

response

Excitation is now the
input voltage and d_1 is a
fixed value (0 in ac).

Figure 3.52 The duty ratio is set to 0 V and the stimulus is now V_{in}.

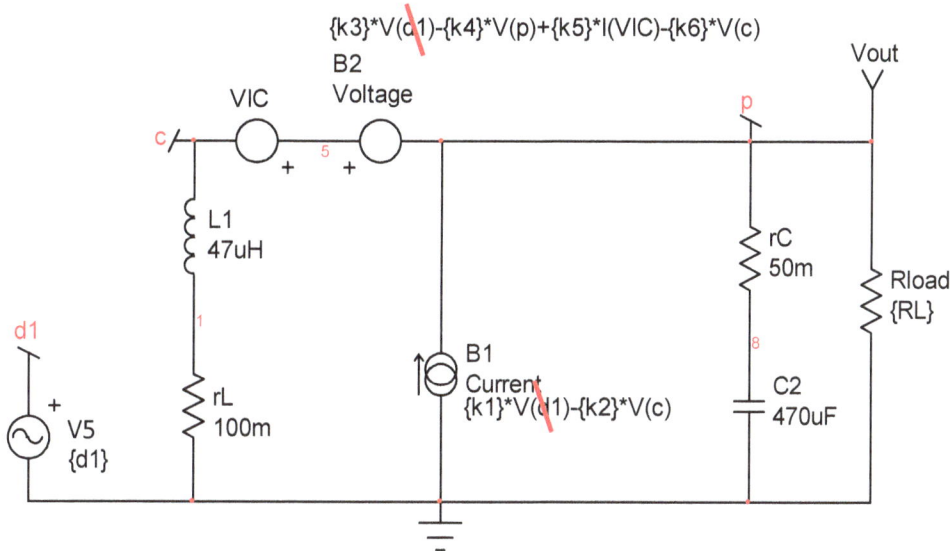

Figure 3.53 Reduce the stimulus – V_{in} – to 0 V and you bring the circuit back to its natural structure.

What we now need to do is find the zeroes of the transfer function. What conditions in Figure 3.51 would prevent the stimulus (V_{in}) to produce a response V_{out}? The usual suspects are r_C and C_2 which can form a transformed short circuit at the zero angular frequency:

$$Z_1(s) = r_C + \frac{1}{sC_2} = 0 \tag{3.124}$$

Implying a zero located at:

$$\omega_z = \frac{1}{r_C C_2} \qquad (3.125)$$

parameters

d1=316m
Vin=10
Vout=16.3216
Vac=-Vin
Vap=-Vout
Fsw=100k
L=47u
RL=100

la=-103.892m
lc=-267.108m

lca=(Vout/RL)*(-1-Vin*d1^2*RL/(2*L*Vout*Fsw))
N=Vac*d1^2/(2*Fsw*lca*L)
laa=N*lca

k1=Vac*d1/(Fsw*L)
k2=d1^2/(2*Fsw*L)
k3=Vac*Vap*d1/(Fsw*L*lc)
k4=Vac*d1^2/(2*Fsw*L*lc)
k5=-Vac*Vap*d1^2/(2*Fsw*lc^2*L)
k6=Vap*d1^2/(2*Fsw*lc*L)

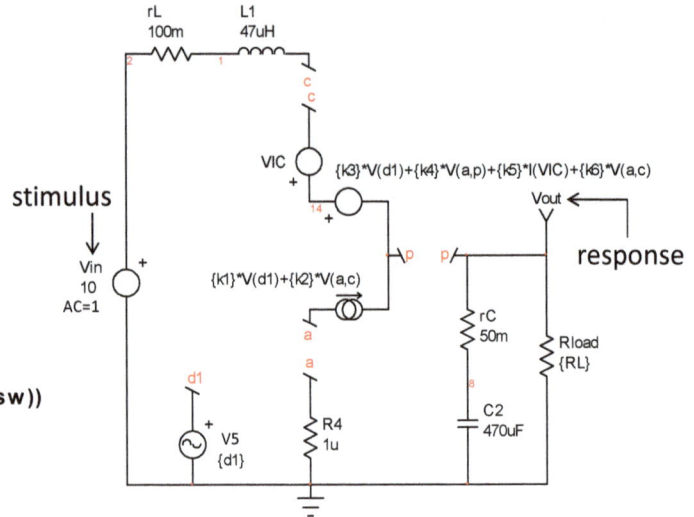

Excitation is now the input voltage and d_1 is a fixed value (0 in ac).

Figure 3.54 There is a single zero in this configuration.

To check whether a zero is associated with L_1, set it in its high-frequency state (open-circuit it) and check if V_{in} propagates to form a response. It does not since r_L no longer connects to terminal c: there is no zero associated with L_1 and the complete transfer function is:

$$H(s) = H_0 \frac{1 + \dfrac{s}{\omega_z}}{\left(1 + \dfrac{s}{\omega_{P_1}}\right)\left(1 + \dfrac{s}{\omega_{P_2}}\right)} \qquad (3.126)$$

with:

$$H_0 = \frac{1}{2}\left(1 + \sqrt{1 + \frac{2D^2}{\tau_L}}\right) \qquad (3.127)$$

With τ_L defined by (3.117). The poles are given by:

$$\omega_{P_1} \approx \frac{2M-1}{M-1}\frac{1}{R_{load}C_2} \qquad (3.128)$$

326

$$\omega_{p_2} \approx 2F_{sw} \left(\frac{1 - \frac{1}{M}}{D_0} \right)^2 \tag{3.129}$$

The single zero is:

$$\omega_z = \frac{1}{r_C C_2} \tag{3.130}$$

These expressions have been tested against a SPICE simulation and Figure 3.55 confirms that the analysis is correct.

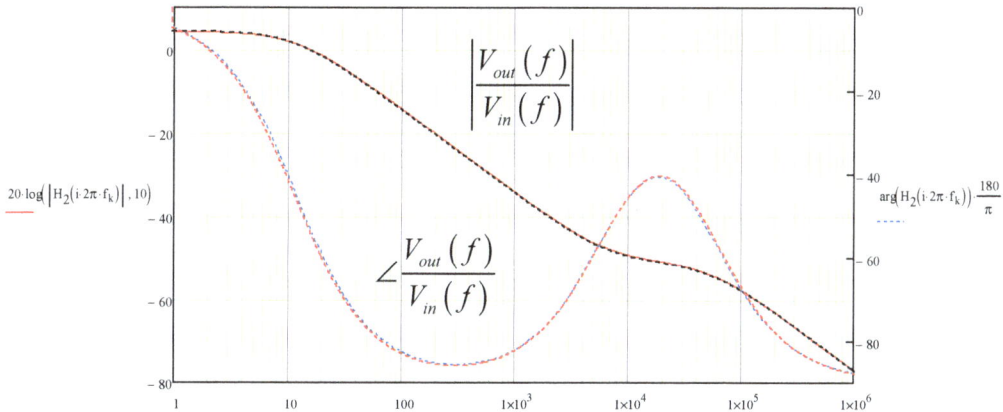

Figure 3.55 The input-to-output transfer function obtained with Mathcad® matches the SPICE response.

3.2.2 Output Impedance

To determine the open-loop output impedance, we will install a generator across the load as portrayed by Figure 3.56. From the small-signal model of Figure 3.57 you can see that turning the test generator I_1 off (open-circuit it) brings the circuit back to its natural structure: time constants are similar to those already determined and we can reuse the denominator expressed in (3.104).

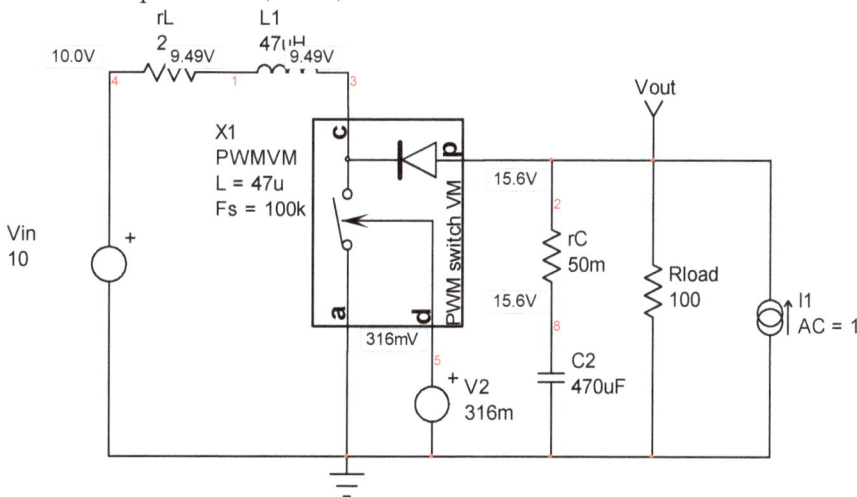

Figure 3.56 A 1-A current source installed across the load sweeps the output impedance.

parameters

d1=316m
Vin=10
Vout=16.3216
Vac=-Vin
Vap=-Vout
Fsw=100k
L=47u
RL=100

Ia=-103.892m
Ic=-267.108m

Ica=(Vout/RL)*(-1-Vin*d1^2*RL/(2*L*Vout*Fsw))
N=Vac*d1^2/(2*Fsw*Ica*L)
Iaa=N*Ica

k1=Vac*d1/(Fsw*L)
k2=d1^2/(2*Fsw*L)
k3=Vac*Vap*d1/(Fsw*L*Ic)
k4=Vac*d1^2/(2*Fsw*L*Ic)
k5=-Vac*Vap*d1^2/(2*Fsw*Ic^2*L)
k6=Vap*d1^2/(2*Fsw*Ic*L)

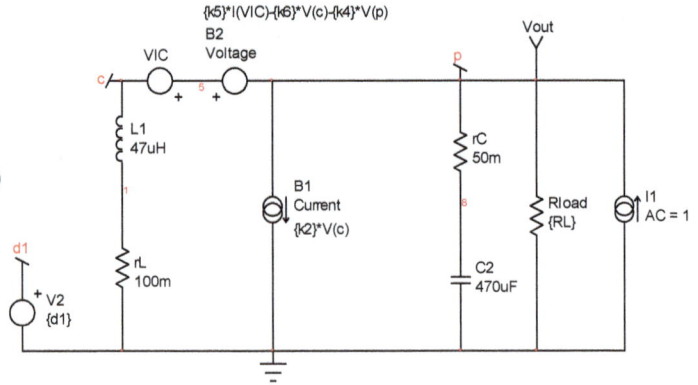

Figure 3.57 Turning the excitation off will bring the circuit back to its natural structure.

To determine the incremental output resistance R_0, we short the inductor and open the capacitor (Figure 3.58). From this drawing, you can see R_{load} coming in parallel with the current source I_T. We can temporarily disconnect R_{load}, work on a simpler circuit without it and express the final result by bringing it back in parallel.

Let's start with the current flowing in terminal c:

$$I_c = I_T - k_2 V_{(c)} \tag{3.131}$$

The voltage at this terminal is given by:

$$V_{(c)} = V_T + k_5 I_c - k_6 V_{(c)} - k_4 V_T \tag{3.132}$$

Inserting (3.131) into (3.132) then solving for $V_{(c)}$ gives:

$$V_{(c)} = \frac{V_T + I_T k_5 - V_T k_4}{k_6 + k_2 k_5 + 1} \tag{3.133}$$

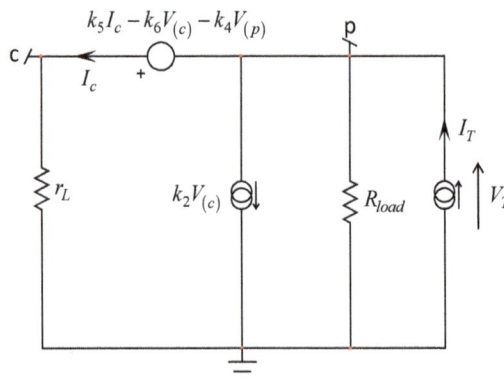

Figure 3.58 The incremental output resistance requires a few lines of algebra to determine it.

From (3.131), we can extract I_T in which I_c is replaced by $V_{(c)}/r_L$ and have:

$$I_T = \frac{V_{(c)}}{r_L} + k_2 V_{(c)}$$ (3.134)

From the above equation, solve for $V_{(c)}$ and equate the result with (3.133). Factor V_T and you obtain:

$$R_0 = \frac{V_T}{I_T} \| R_{load} = \frac{r_L(1+k_6)-k_5}{(1+k_2 r_L)(1-k_4)} \| R_{load}$$ (3.135)

If we neglect r_L and rearrange this expression, we have:

$$R_0 \approx \frac{R_{load} V_{in} D^2}{V_{in} D^2 + 2V_{out} R_{load}} \| R_{load}$$ (3.136)

The zeroes are found by nulling the response which is the voltage V_T across the current source. The configuration to study is described in Figure 3.59. Because the response is nulled, there is no current flowing in R_{load} and node p is at a 0-V potential. The test current I_T is thus the sum of two contributors:

$$I_T(s) = I_c(s) + k_2 V_{(c)}(s)$$ (3.137)

The voltage at node c depends on the impedance offered by the series connection of L_1 and r_L:

$$V_{(c)}(s) = I_c(s)(r_L + sL_1)$$ (3.138)

Extract I_c from (3.137) and substitute it in (3.138):

$$V_{(c)}(s) = \left(I_T(s) - V_{(c)}(s)k_2 \right)(r_L + sL_1)$$ (3.139)

Solve for $V_{(c)}$ and you have:

$$V_{(c)}(s) = \frac{I_T(s)(r_L + sL_1)}{1 + k_2(r_L + sL_1)}$$ (3.140)

We have another expression to define the voltage at node c:

$$V_{(c)}(s) = k_5 I_c(s) - k_6 V_{(c)}$$ (3.141)

After an update with I_c extracted from (3.137), this expression becomes:

$$V_{(c)}(s) = I_T(s)k_5 - V_{(c)}k_6 - V_{(c)}k_2 k_5$$ (3.142)

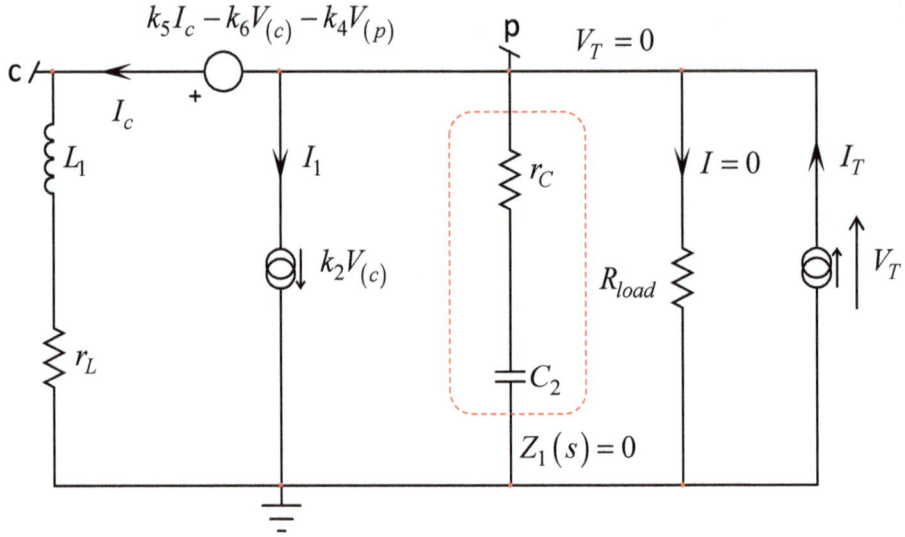

Figure 3.59 The response is V_T and you must find the conditions in the circuit which null it.

Solving for $V_{(c)}$ gives:

$$V_{(c)}(s) = \frac{I_T(s)k_5}{1+k_6+k_2k_5} \quad (3.143)$$

Now equate (3.143) with (3.140) which eliminates I_T.

Rearrange the result to obtain:

$$\omega_{z_2} = \frac{r_L - \dfrac{k_5}{1+k_6}}{L_1} \quad (3.144)$$

If we apply and simplify all the k coefficients from Figure 3.50 while neglecting r_L, we have:

$$\omega_{z_2} \approx \frac{2F_{sw}}{\left(D_0^2 + 2M\tau_L\right)\left(1+\dfrac{1}{M}+2\dfrac{\tau_L}{D_0^2}\right)} \quad (3.145)$$

The other zero is classically given by the series combination of r_C and C_2:

$$\omega_{z_1} = \frac{1}{r_C C_2} \quad (3.146)$$

The complete transfer function can now be assembled:

$$Z_{out}(s) = R_0 \frac{\left(1+\dfrac{s}{\omega_{z_1}}\right)\left(1+\dfrac{s}{\omega_{z_2}}\right)}{\left(1+\dfrac{s}{\omega_{p_1}}\right)\left(1+\dfrac{s}{\omega_{p_2}}\right)} \quad (3.147)$$

In which we have:

$$R_0 \approx \frac{R_{load} V_{in} D_0^2}{V_{in} D_0^2 + 2V_{out} R_{load}} \| R_{load} \qquad (3.148)$$

$$\omega_{z_2} \approx \frac{2F_{sw}}{\left(D_0^2 + 2M\tau_L\right)\left(1 + \frac{1}{M} + 2\frac{\tau_L}{D_0^2}\right)} \qquad (3.149)$$

$$\omega_{z_1} = \frac{1}{r_C C_2} \qquad (3.150)$$

$$\omega_{p_1} = \frac{2M-1}{M-1}\frac{1}{R_{load}C_2} \qquad (3.151)$$

$$\omega_{p_2} = 2F_{sw}\left(\frac{1-\frac{1}{M}}{D_0}\right)^2 \qquad (3.152)$$

If we consider that the second zero and pole occur in the high-frequency range, this expression can be approximated as:

$$Z_{out}(s) \approx R_0 \frac{1+\frac{s}{\omega_{z_1}}}{1+\frac{s}{\omega_{p_1}}} \qquad (3.153)$$

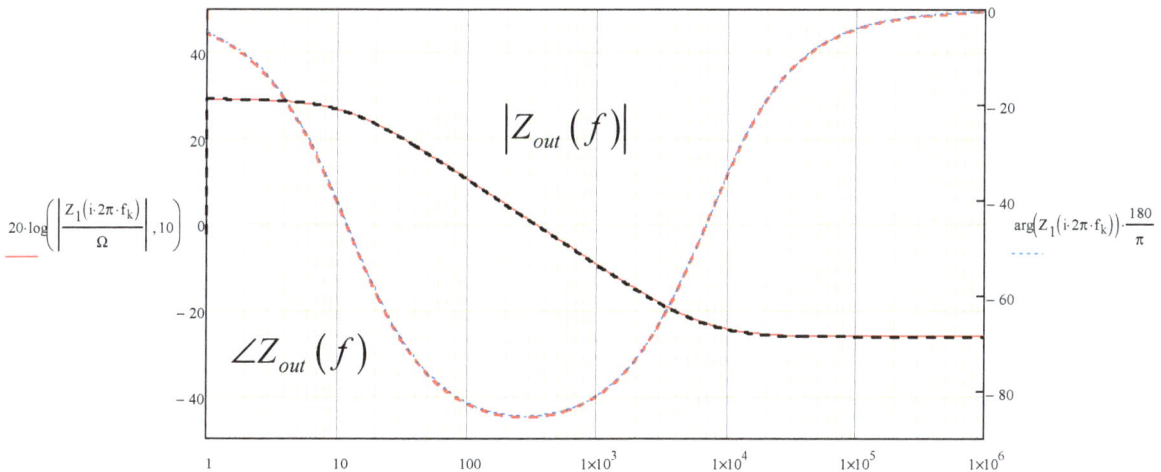

Figure 3.60 The output impedance shows a response dominated by a pole and a zero in the low frequency range.

We compared the dynamic response obtained with a SPICE simulation to Mathcad® and the result appears in Figure 3.60. It confirms the analysis is correct.

3.2.3 Input Impedance

The input impedance is the last of the four transfer functions in DCM. To determine its expression, we will connect a test generator I_T biasing the input port. In a SPICE environment, we insert a high-value inductor LoL to let the simulator calculate the correct bias point while the ac modulation is injected at the input port. LoL actually ac-isolates the circuit from the 0-Ω output impedance of input source V_{in}. The recommended circuit is shown in Figure 3.61.

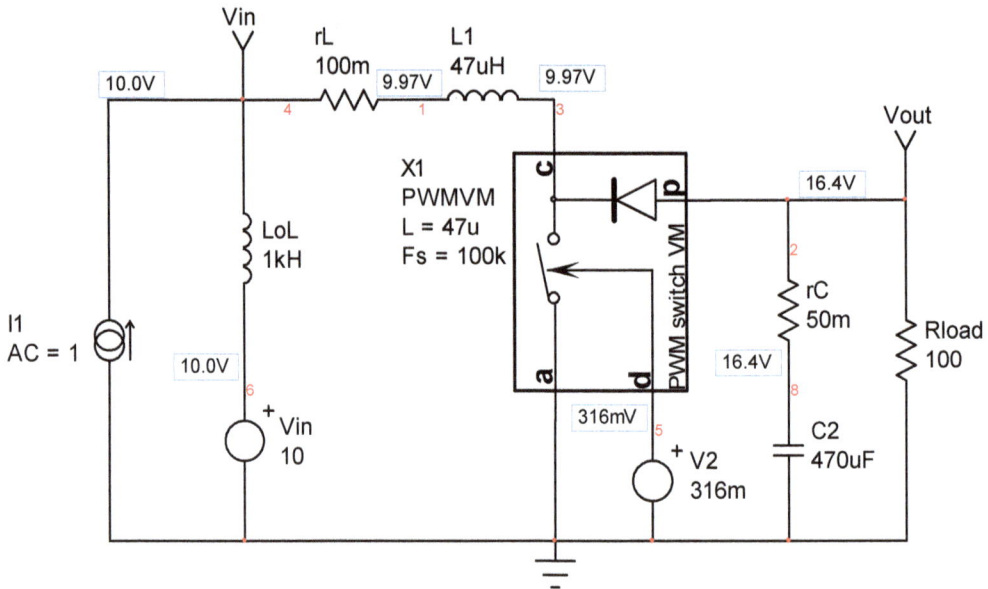

Figure 3.61 The input impedance is obtained by ac-sweeping the input port.

Unlike in the other exercises, we can immediately run an ac analysis and check what this impedance looks like. Figure 3.62 shows how the magnitude and phase evolve along the frequency axis.

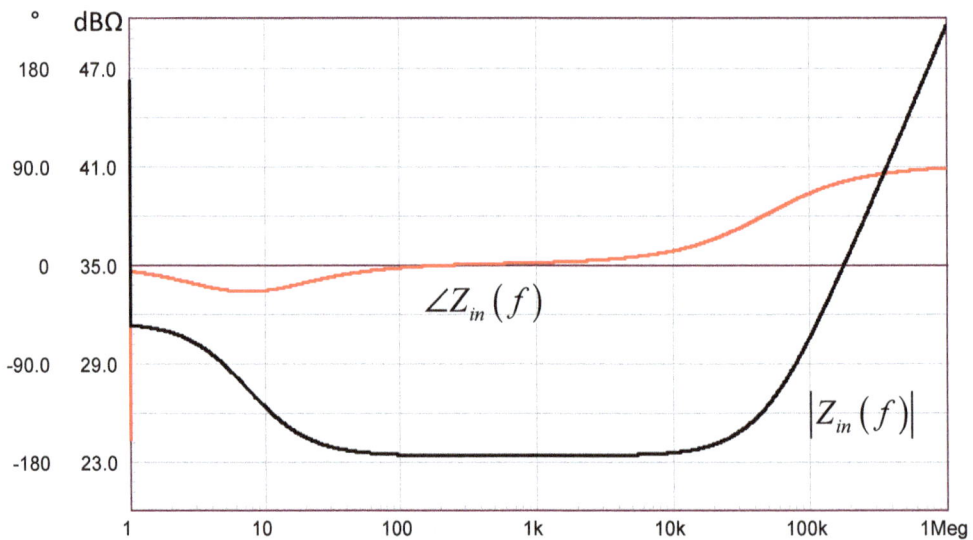

Figure 3.62 The impedance is flat in the middle and increases in magnitude at higher frequencies.

We can now plug the small-signal version of the PWM switch model and update the figure to that of Figure 3.63. Running this circuit helps us confirm the electrical diagram is sound by comparing its response to that of Figure 3.62.

In this analysis, the duty ratio d_1 is 0 V (the circuit operates in open loop). One comment though: you see that the inductor current is driven by I_T, the test generator. It implies that its state variable is no longer independent and despite the presence of two energy-storing elements, you lose an order: the denominator will host a single pole.

We start with the dc analysis in which we short the inductor and open the capacitor. This is what Figure 3.64 shows.

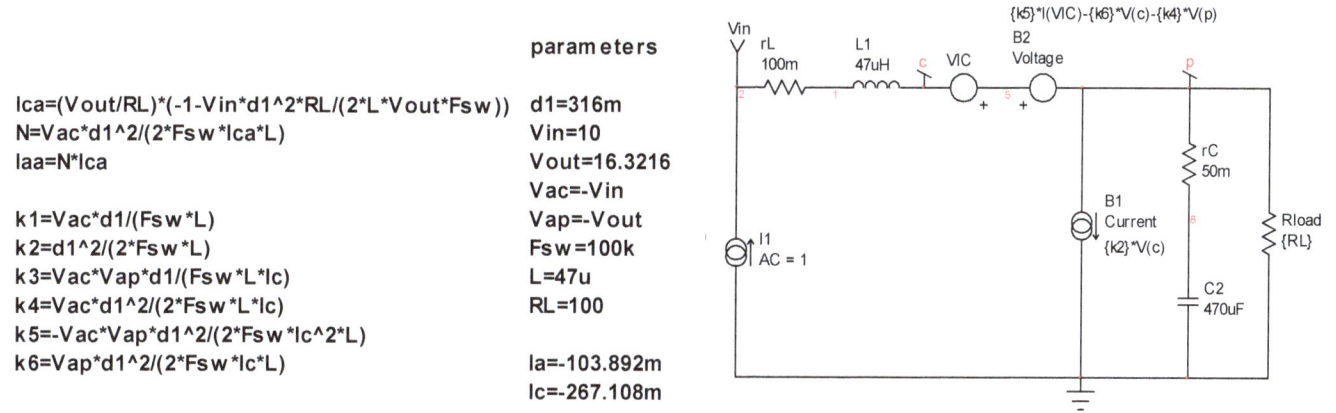

Figure 3.63 The inductor is now driven by the current source I_T while the duty ratio d_1 is set to zero.

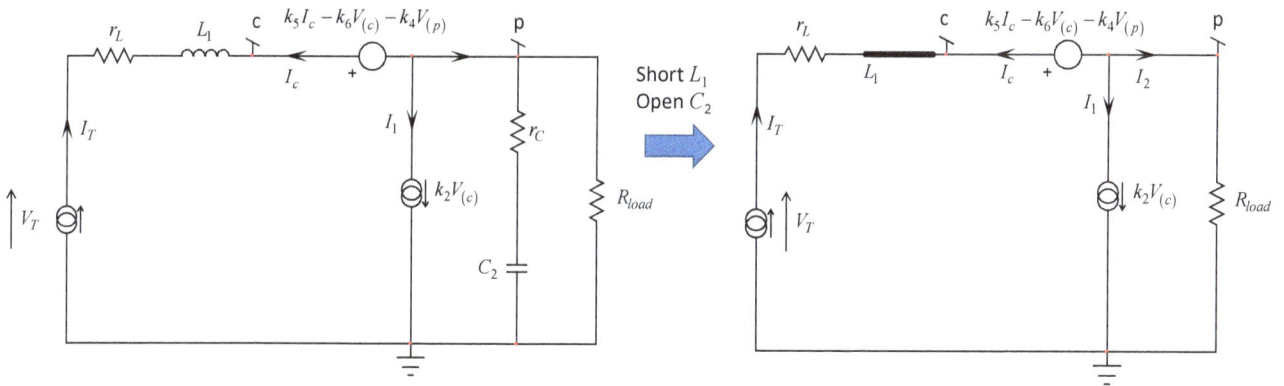

Figure 3.64 For dc analysis, the inductor is replaced by a short circuit while the capacitor is open circuited.

To help with the determination of R_0, the incremental resistance, we can temporarily disconnect r_L (replace it by a short) and bring it back later in series with the intermediate result (Figure 3.65).

With this condition in mind, the voltage at terminal c is expressed as:

$$V_{(c)} = V_T = k_5 I_c - k_6 V_T - k_4 V_{(p)} + V_{(p)} \qquad (3.154)$$

333

The test current I_T is the combination of I_1 and I_2:

$$I_T = -I_c = I_1 + I_2 = k_2 V_T + \frac{V_{(p)}}{R_{load}}$$

(3.155)

The voltage at terminal p depends on the load resistance:

$$V_{(p)} = (I_T - k_2 V_T) R_{load}$$

(3.156)

Substituting the two above expressions in (3.154) and solving for V_T while factoring I_T gives the final result (after bringing r_L back):

$$R_0 = r_L + \frac{R_{load}(1-k_4) - k_5}{k_6 + R_{load} k_2 (1-k_4) + 1}$$

(3.157)

Neglecting r_L and substituting the k coefficient values from chapter 1, this expression becomes:

$$R_0 \approx \frac{R_{load} \tau_L \left(D_0^2 + M\tau_L\right)}{D_0^2 \left[\tau_L (M+1) + 0.5 D_0^2 + \frac{D_0^2}{4M}\right] + M\tau_L^2}$$

(3.158)

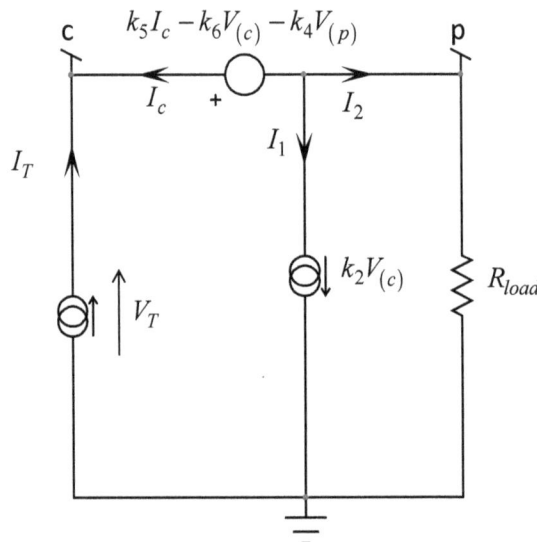

Figure 3.65 Removing r_L simplifies the analysis.

We can now look at the natural time constants of this circuit. When turning the excitation off (open-circuit the current source), the left side of the inductor is left unconnected. The circuit no longer returns to the natural state encountered when studying the three transfer functions: we cannot reuse the denominator we have determined before. Figure 3.66 shows the new configuration in which C_2 is set in its dc state.

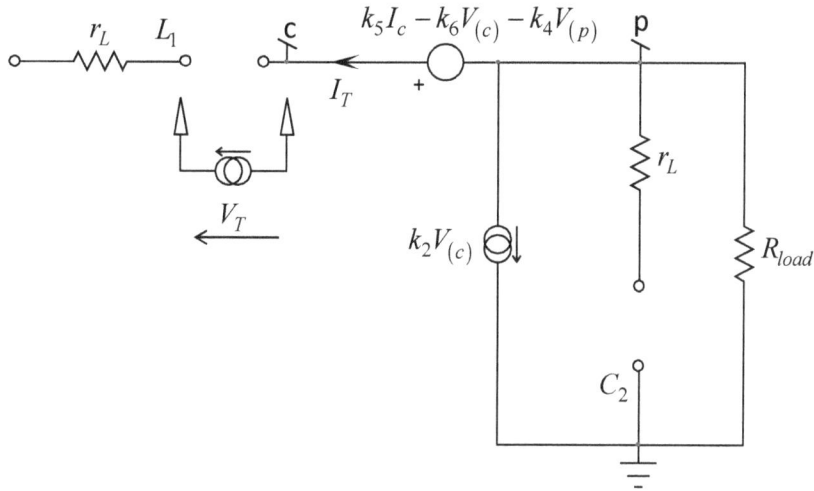

Figure 3.66 The left side of the inductor is left dangling as the current source is open-circuited.

The first time constant τ_1 is simply:

$$\tau_1 = \frac{L_1}{\infty} = 0 \qquad (3.159)$$

For the second time constant, L_1 is set in its dc state (a short circuit) and we determine the resistance seen from the capacitor terminals. Figure 3.67 shows you the circuit.

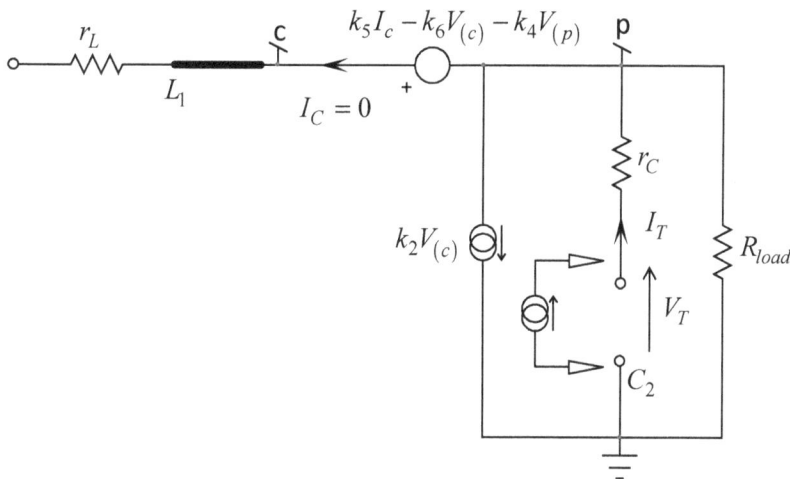

Figure 3.67 The inductor is replaced by a short circuit for this second time constant determination.

As we did before, we can ignore resistance r_C in series with C_2 and bring it back in the final expression. Therefore, the expression for $V_{(c)}$ is:

$$V_{(c)} = k_5 I_c - k_6 V_{(c)} - k_4 V_{(p)} + V_{(p)} \qquad (3.160)$$

In this configuration, considering a current I_c equal to 0 A and the voltage at p which is V_T:

$$V_{(c)} = -k_6 V_{(c)} - k_4 V_T + V_T \tag{3.161}$$

The voltage at node p can also be expressed as the drop across the load resistance:

$$V_{(p)} = V_T = -R_{load}\left(k_2 V_{(c)} - I_T\right) \tag{3.162}$$

If you extract $V_{(c)}$ from (3.161) and (3.162), equate the results and solve for V_T while factoring I_T, you should find the following final time constant after bringing r_C back:

$$\tau_2 = C_2\left[r_C + \frac{R_{load}\left(1+k_6\right)}{k_6 + 1 + R_{load} k_2\left(1-k_4\right)}\right] \tag{3.163}$$

The first-order term b_1 is equal to:

$$b_1 = \tau_1 + \tau_2 = C_2\left[r_C + \frac{R_{load}\left(1+k_6\right)}{k_6 + 1 + R_{load} k_2\left(1-k_4\right)}\right] \tag{3.164}$$

For the second-order time constant, the inductor is set in its high-frequency state (an open circuit) while we determine the resistance seen from C_2's terminals. The configuration is that of Figure 3.68 and you can see that it is equivalent to the circuit of Figure 3.67 since the I_c current is 0 A in both cases. Therefore, we can reuse the result we have previously derived:

$$\tau_2^1 = C_2\left[r_C + \frac{R_{load}\left(1+k_6\right)}{k_6 + 1 + R_{load} k_2\left(1-k_4\right)}\right] \tag{3.165}$$

The second-order term is defined as:

$$b_2 = \tau_1 \tau_2^1 = 0 \cdot C_2\left[r_C + \frac{R_{load}\left(1+k_6\right)}{k_6 + 1 + R_{load} k_2\left(1-k_4\right)}\right] = 0 \tag{3.166}$$

The denominator follows a 1st-order form:

$$D(s) = 1 + s C_2\left[r_C + \frac{R_{load}\left(1+k_6\right)}{k_6 + 1 + R_{load} k_2\left(1-k_4\right)}\right] \tag{3.167}$$

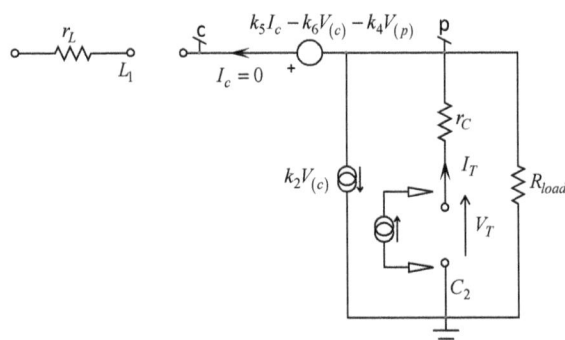

Figure 3.68 The inductor is replaced by an open circuit for the second-order time constant determination.

Regarding the zeroes, you have to find the time constants involving L_1 and C_2 when the response is nulled. The response in this impedance determination exercise is the voltage V_T across the current source I_T. If this voltage is 0 V, it is a degenerate case and we can replace the current generator by a short circuit. Doing so places the circuit back in its natural structure already studied in Figure 3.42: the denominator obtained in (3.104) now becomes our numerator:

$$Z_{in}(s) = R_0 \frac{\left(1 + \dfrac{s}{\omega_{z_1}}\right)\left(1 + \dfrac{s}{\omega_{z_2}}\right)}{1 + \dfrac{s}{\omega_p}} \tag{3.168}$$

In this expression, we have:

$$R_0 \approx \frac{R_{load}\tau_L\left(D_0^2 + M\tau_L\right)}{D_0^2\left[\tau_L(M+1) + 0.5D_0^2 + \dfrac{D_0^2}{4M}\right] + M\tau_L^2} \tag{3.169}$$

$$\omega_{z_1} \approx \frac{2M-1}{M-1}\frac{1}{R_{load}C_2} \tag{3.170}$$

$$\omega_{z_2} \approx 2F_{sw}\left(\frac{1-\dfrac{1}{M}}{D_0}\right)^2 \tag{3.171}$$

$$\omega_p = \frac{1}{C_2\left[r_C + \dfrac{R_{load}(1+k_6)}{k_6 + 1 + R_{load}k_2(1-k_4)}\right]} \tag{3.172}$$

We have compared the dynamic responses delivered by Mathcad® with those of the SPICE simulation and Figure 3.69 shows that they perfectly agree.

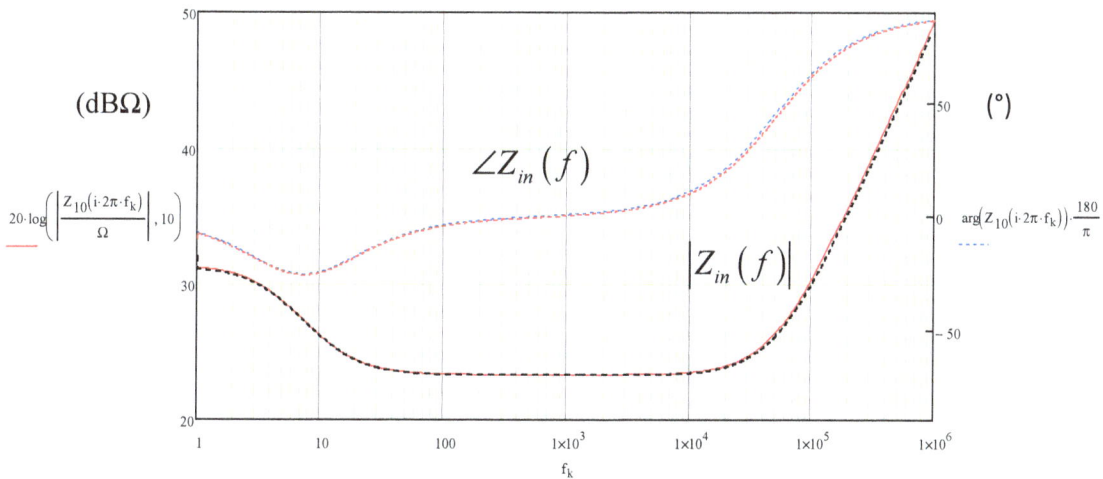

Figure 3.69 The agreement between the SPICE curves and those from the Mathcad® sheet is excellent.

We have determined the four transfer functions of the voltage-mode boost converter operated in DCM and all transfer functions are conveniently given in the table of Figure 3.70.

$\dfrac{V_{out}(s)}{V_{err}(s)}$ Control to Output	$H_0 \dfrac{\left(1+\dfrac{s}{\omega_{z_1}}\right)\left(1+\dfrac{s}{\omega_{z_2}}\right)}{\left(1+\dfrac{s}{\omega_{p_1}}\right)\left(1+\dfrac{s}{\omega_{p_2}}\right)}$	$\omega_{z_1}=\dfrac{1}{r_C C_2}$ $\omega_{z_2}\approx\dfrac{R_{load}}{M^2 L_1}$	$\omega_{p_1}\approx\dfrac{2M-1}{M-1}\dfrac{1}{R_{load}C_2}$ $\omega_{p_2}\approx 2F_{sw}\left(\dfrac{1-\dfrac{1}{M}}{D}\right)^2$	$H_0\approx\dfrac{1}{V_p}\dfrac{2V_{in}}{2M-1}\sqrt{\dfrac{M(M-1)}{2\tau_L}}$
$\dfrac{V_{out}(s)}{V_{in}(s)}$ Input to Output	$H_0\dfrac{1+\dfrac{s}{\omega_z}}{\left(1+\dfrac{s}{\omega_{p_1}}\right)\left(1+\dfrac{s}{\omega_{p_2}}\right)}$	$\omega_z=\dfrac{1}{r_C C_2}$	$\omega_{p_1}\approx\dfrac{2M-1}{M-1}\dfrac{1}{R_{load}C_2}$ $\omega_{p_2}\approx 2F_{sw}\left(\dfrac{1-\dfrac{1}{M}}{D}\right)^2$	$H_0=\dfrac{1}{2}\left(1+\sqrt{1+\dfrac{2D^2}{\tau_L}}\right)$
$Z_{in}(s)$ Input impedance	$R_0\dfrac{\left(1+\dfrac{s}{\omega_{z_1}}\right)\left(1+\dfrac{s}{\omega_{z_2}}\right)}{1+\dfrac{s}{\omega_p}}$	$\omega_{z_1}\approx\dfrac{2M-1}{M-1}\dfrac{1}{R_{load}C_2}$ $\omega_{z_2}\approx 2F_{sw}\left(\dfrac{1-\dfrac{1}{M}}{D}\right)^2$	$\omega_p=\dfrac{1}{C_2\left[r_C+\dfrac{R_{load}(1+k_6)}{k_6+1+R_{load}k_2(1-k_4)}\right]}$ See Chap. 1 for coefficients values	$R_0\approx\dfrac{R_{load}\tau_L\left(D^2+M\tau_L\right)}{D^2\left[\tau_L(M+1)+0.5D^2+\dfrac{D^2}{4M}\right]+M\tau_L^2}$
$Z_{out}(s)$ Output impedance	$R_0\dfrac{\left(1+s/\omega_{z_1}\right)\left(1+s/\omega_{z_2}\right)}{1+\dfrac{s}{\omega_0 Q}+\left(\dfrac{s}{\omega_0}\right)^2}$	$\omega_{z_1}=\dfrac{1}{r_C C_2}$ $\omega_{z_2}=\dfrac{r_L-\dfrac{k_5}{1+k_6}}{L_1}$	$\omega_{p_1}\approx\dfrac{2M-1}{M-1}\dfrac{1}{R_{load}C_2}$ $\omega_{p_2}\approx 2F_{sw}\left(\dfrac{1-\dfrac{1}{M}}{D}\right)^2$	$R_0\approx\dfrac{R_{load}V_{in}D_0^2}{V_{in}D_0^2+2V_{out}R_{load}}\ \|\ R_{load}$

D is the duty ratio, M is V_{out}/V_{in} and $\tau_L=\dfrac{L_1}{R_{load}T_{sw}}$

Figure 3.70 The four transfer functions of the DCM boost converter operated in voltage-mode control appear in this array.

3.3 Boost Transfer Functions in CCM – Fixed-Frequency Current-Mode Control

The boost converter operated in current-mode control is analyzed in the same way as with the voltage-mode case: identify the switch/diode couple and plug the small-signal model already used in the case of the current-mode buck converter. The large-signal model used for the reference curves is shown in Figure 3.71. This is a 5-V converter operated at a 1-MHz switching frequency. Please note the negative value for the sense resistance R_i already explained in chapter 1.4.

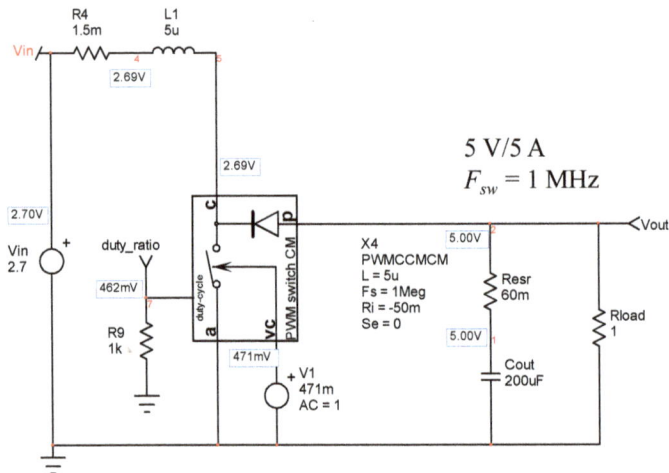

Figure 3.71 The current mode boost converter uses the large-signal CM PWM switch model.

Once the small-signal model is installed as in Figure 3.72, we must verify the control-to-output frequency response of this circuit before starting the study. Please note that L_1 became L_2 because the resonating capacitor (internal to the model) is labeled C_1. The output capacitor is now C_3. Dynamic responses between the two models (large- and small-signal versions) must be similar otherwise a flaw in the electric wiring or in the computed coefficients is likely to be present. As confirmed by Figure 3.73, responses are identical and we are good to go.

Before working on this quite complex schematic, I will go through a few simplifications (neglect the inductor ohmic loss r_l) and rearrangements as shown successively in Figure 3.74 and Figure 3.75. Once the inductor is shorted and the output capacitor open, we are ready to look at the circuit for $s = 0$ as shown in Figure 3.76. Starting from the left of the picture, we can express the output current:

$$V_{out} g_f - V_c k_o - V_{out}\left(g_o + g_i\right) - V_{out} g_r + V_c k_i = \frac{V_{out}}{R_{load}} \tag{3.173}$$

Rearranging this expression by factoring V_{out} and V_c, we have:

$$H_0 = \frac{V_{out}(s)}{V_c(s)} = \frac{k_0 - k_i}{g_f - \left(g_o + g_i\right) - g_r - \dfrac{1}{R_{load}}} \tag{3.174}$$

A simpler equivalent expression is given in the literature [3]:

$$H_0 = \frac{R_{load}}{R_i} \frac{1}{2M + \dfrac{1}{\tau_L M^2}\left(\dfrac{1}{2} + \dfrac{S_e}{S_n}\right)} \tag{3.175}$$

It can also be expressed as:

$$H_0 = \frac{R_{load}}{2R_i} \frac{1}{\left(1-D\right)^2} \frac{1}{\dfrac{1}{\left(1-D\right)^3} + \dfrac{R_{load} T_{sw}}{2L_2}\left(\dfrac{1}{2} + \dfrac{S_e}{S_n}\right)} \tag{3.176}$$

with:

$$M = \frac{V_{out}}{V_{in}} \tag{3.177}$$

$$\tau_L = \frac{L_2}{R_{load} T_{sw}} \tag{3.178}$$

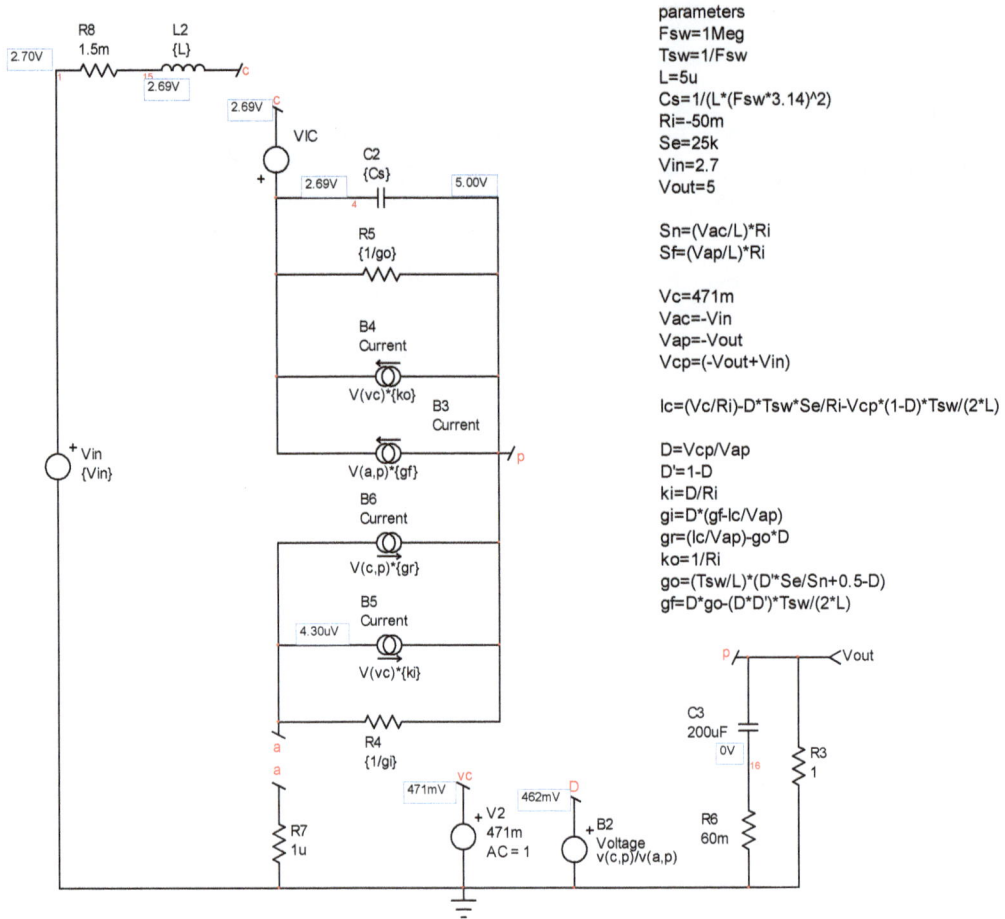

Figure 3.72 The small–signal version requires the computation of some of the coefficients.

parameters
Fsw=1Meg
Tsw=1/Fsw
L=5u
Cs=1/(L*(Fsw*3.14)^2)
Ri=-50m
Se=25k
Vin=2.7
Vout=5

Sn=(Vac/L)*Ri
Sf=(Vap/L)*Ri

Vc=471m
Vac=-Vin
Vap=-Vout
Vcp=(-Vout+Vin)

Ic=(Vc/Ri)-D*Tsw*Se/Ri-Vcp*(1-D)*Tsw/(2*L)

D=Vcp/Vap
D'=1-D
ki=D/Ri
gi=D*(gf-Ic/Vap)
gr=(Ic/Vap)-go*D
ko=1/Ri
go=(Tsw/L)*(D*Se/Sn+0.5-D)
gf=D*go-(D*D')*Tsw/(2*L)

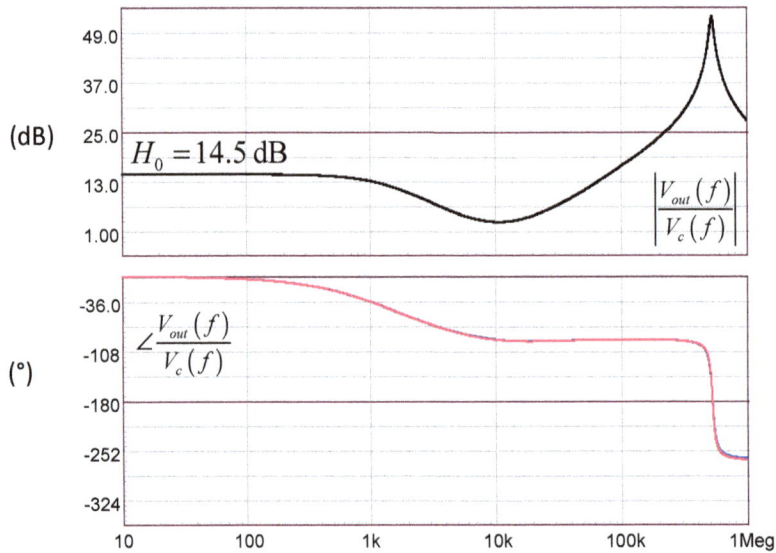

$H_0 = 14.5 \text{ dB}$

$\dfrac{|V_{out}(f)|}{|V_c(f)|}$

$\angle \dfrac{V_{out}(f)}{V_c(f)}$

Figure 3.73 The dynamic responses between the two models are rigorously identical.

$$S_n = \frac{V_{in}}{L_2} R_i \qquad\qquad (3.179)$$

and S_e is the external stabilization ramp expressed in volts per seconds.

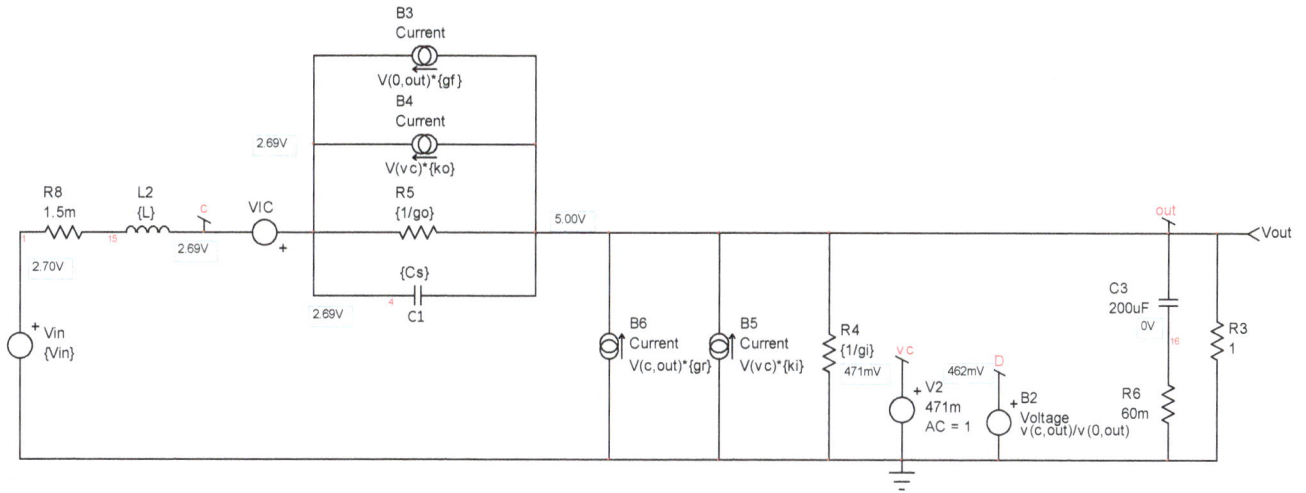

Figure 3.74 This is the first step in rearranging the circuit.

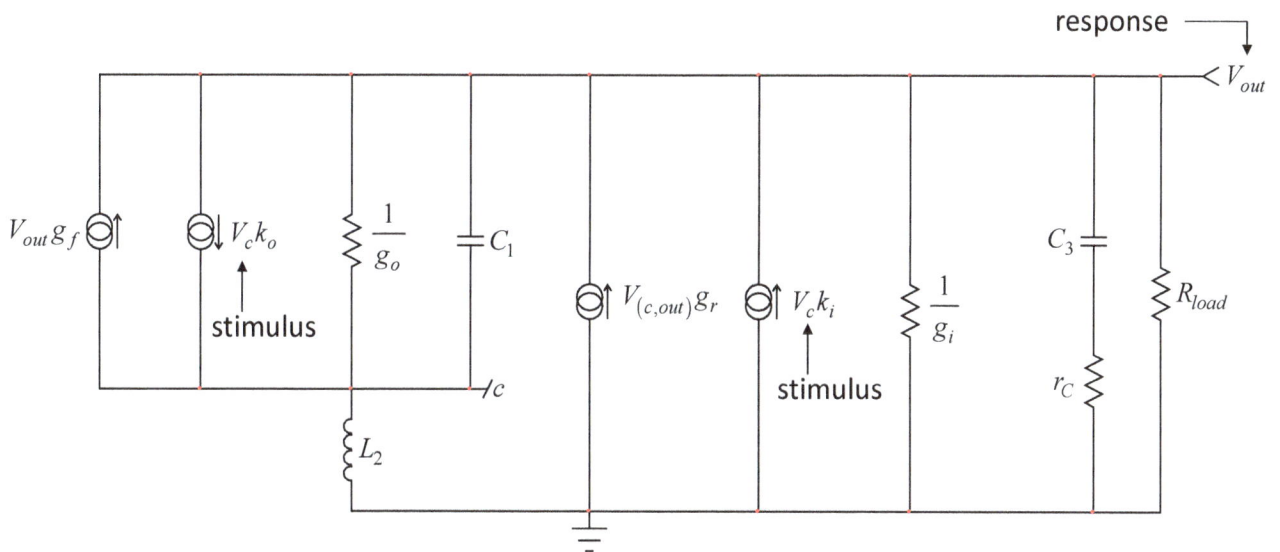

Figure 3.75 Neglecting the inductor series resistance r_L grounds L_2's lower terminal and further simplifies some of the expressions.

To determine the natural time constants of the circuit, we are going to reduce the excitation to zero after assigning new labels to the three energy-storing elements. This is what Figure 3.77 shows.

This stage is important so that a consistent and clear notation is adopted when manipulating these elements.

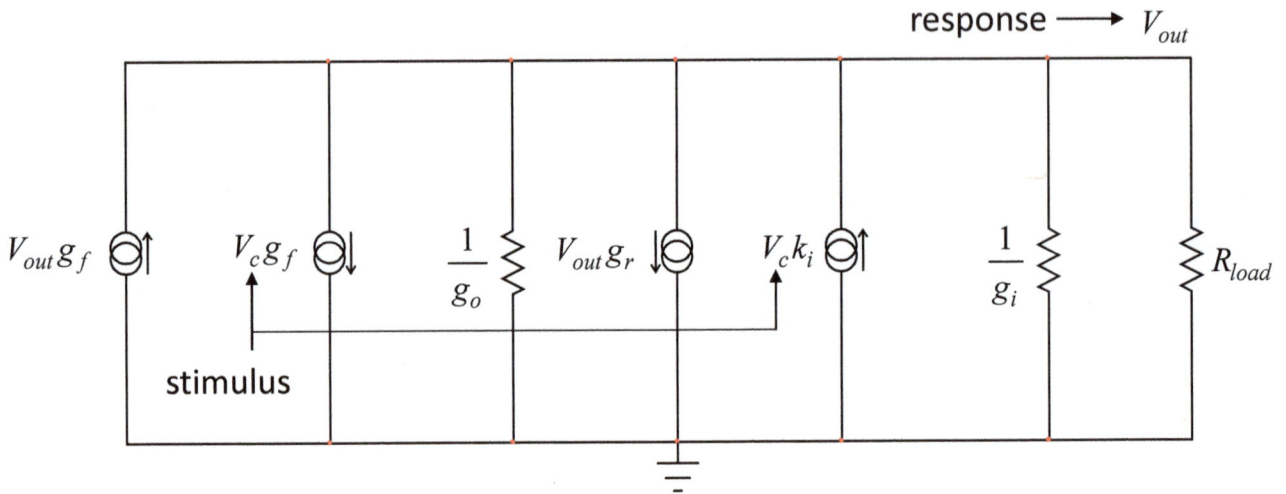

Figure 3.76 The circuit is now ready to be studied for the case $s = 0$.

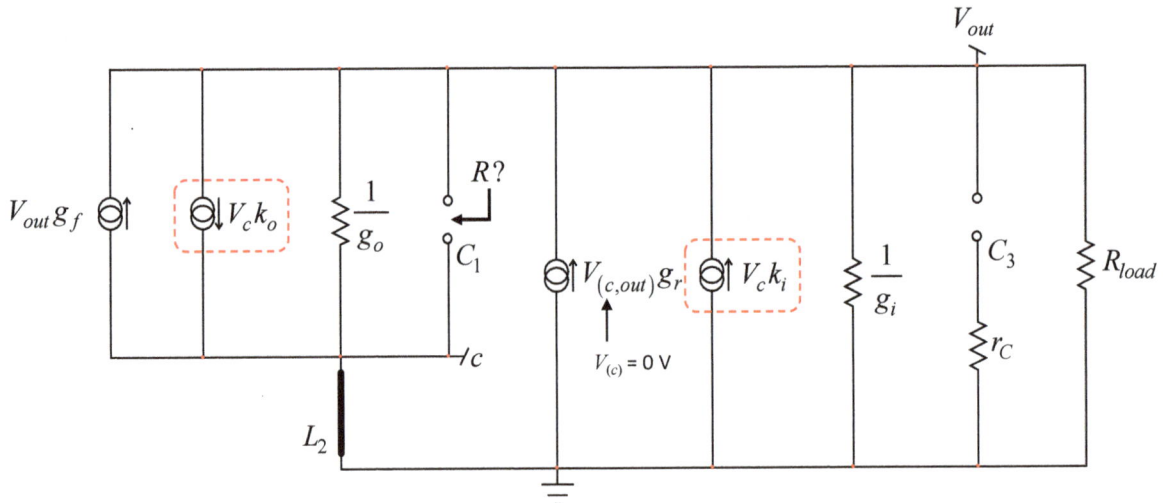

Figure 3.77 During the first time constant analysis, inductor L_2 is replaced by a short circuit while V_c is made zero.

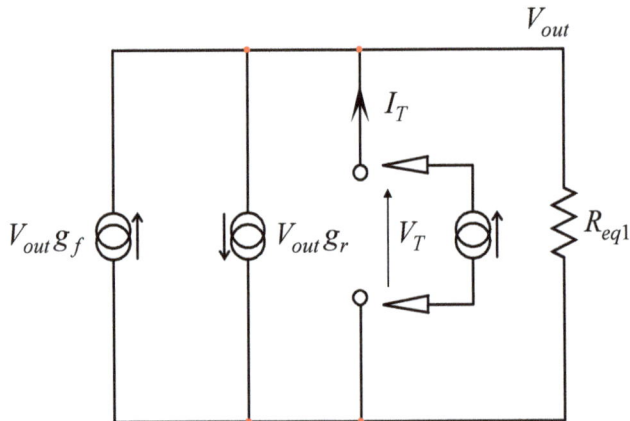

Figure 3.78 We can rearrange the circuit in a form easier to analyze.

To determine the time constants, the excitation source V_c is reduced to zero naturally open-circuiting some of the current sources. The two conductances end up in parallel with the load resistance and I have lumped them in an equivalent term R_{eq1} defined as:

$$R_{eq1} = R_{load} \parallel \frac{1}{g_i} \parallel \frac{1}{g_o} \tag{3.180}$$

The final circuit is given in Figure 3.78 in which a test generator I_T has been installed across C_1's connecting terminals. From this electrical diagram, current I_T is defined by:

$$I_T = V_{out}\left(g_r - g_f\right) + \frac{V_{out}}{R_{eq1}} \tag{3.181}$$

Realizing that the voltage V_T is the output voltage, we immediately have:

$$R = \frac{V_T}{I_T} = \frac{V_{out}}{I_T} = \frac{1}{g_r - g_f + \dfrac{1}{R_{eq1}}} \tag{3.182}$$

The first time constant is defined by:

$$\tau_1 = C_1 \left[\frac{1}{g_r - g_f + \dfrac{1}{R_{eq1}}} \right] \tag{3.183}$$

The second time constant involves inductor L_2. To determine the resistance driving this element, all capacitors are set in their dc state (open circuited) and a test generator I_T is placed across L_2's connecting terminals. The first step is shown in Figure 3.79. In this figure, the voltage at terminal c is no longer 0 V and the current source $V_{(c,out)}g_r$ can be split in two elements: a current source $V_{(c)}g_r$ keeping the original polarity and another one, $-V_{out}g_r$ pointing towards ground. This new source can advantageously be replaced by a resistance $1/g_r$ installed from V_{out} to ground. This is shown in the updated schematic of Figure 3.80.

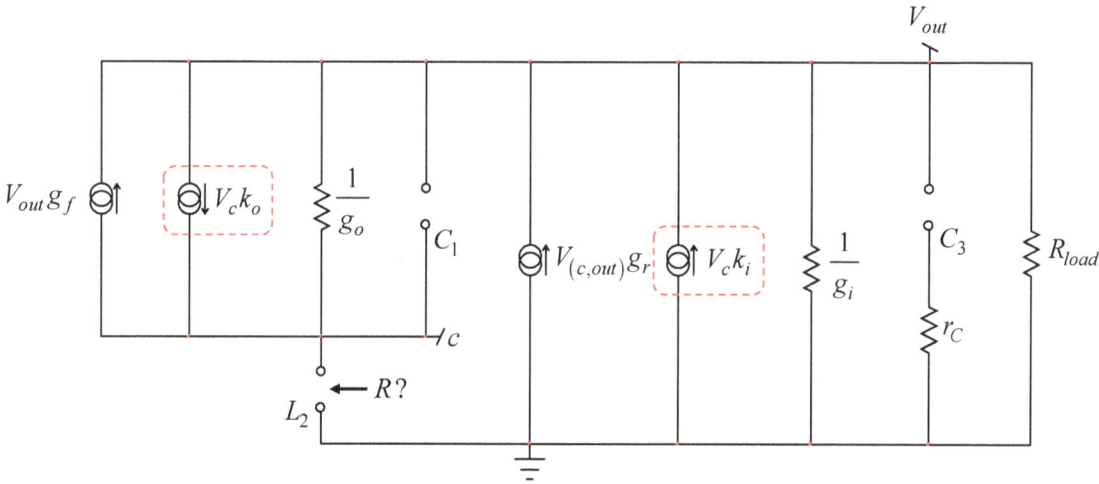

Figure 3.79 We are now looking at the resistance driving L_2. The sources circled in red are open-circuited.

Christophe Basso

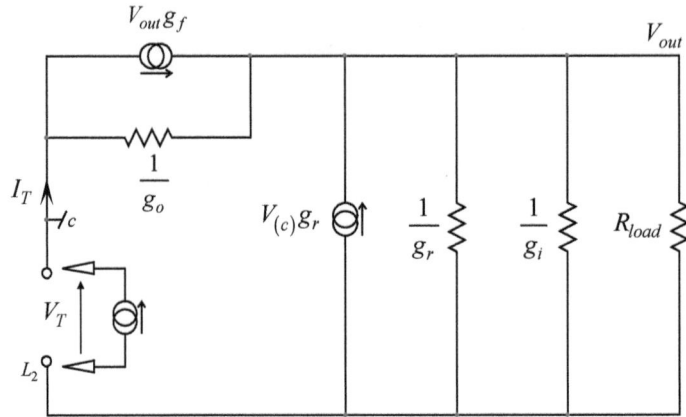

Figure 3.80 The circuit can be rearranged in a simpler form.

The current I_T is defined as:

$$I_T = V_{out} g_f + \frac{V_T - V_{out}}{\frac{1}{g_0}} \tag{3.184}$$

We know also that:

$$V_{out} = \left(I_T + V_T g_r\right) R_{eq2} \tag{3.185}$$

In which R_{eq2} is defined as:

$$R_{eq2} = R_{load} \parallel \frac{1}{g_i} \parallel \frac{1}{g_r} \tag{3.186}$$

Manipulating the above equations to factor V_T and I_T leads to:

$$R = \frac{V_T}{I_T} = \frac{1 - R_{eq2}\left(g_f - g_o\right)}{g_o + R_{eq2} g_r \left(g_f - g_o\right)} \tag{3.187}$$

The time constant is obtained by associating this resistance with L_2:

$$\tau_2 = \frac{L_2}{\frac{1 - R_{eq2}\left(g_f - g_o\right)}{g_o + R_{eq2} g_r \left(g_f - g_o\right)}} \tag{3.188}$$

For the third time constant, we are going back to Figure 3.77 and determine the resistance driving C_3 while L_2 is a short circuit and C_2 is open-circuited.

Running a few arrangements on this diagram leads to the circuit of Figure 3.81 in which a test generator has been installed across C_2's connecting ports.

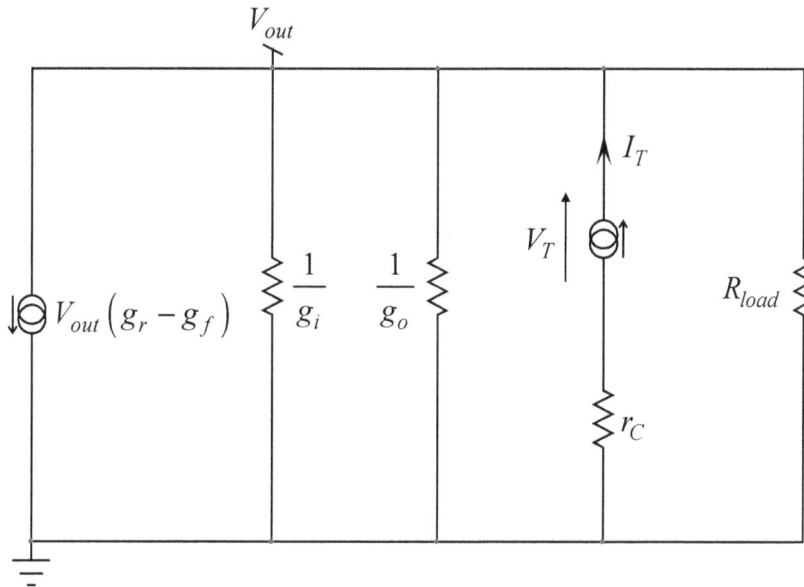

Figure 3.81 We can rearrange the circuit in a form easier to analyze.

Actually, if you look closer to this new circuit, it is similar to that already studied for τ_1 in Figure 3.77 except that a resistance r_C is now in series with the result.

Thus, the third time constant is defined as:

$$\tau_3 = \left(\frac{1}{g_r - g_f + \dfrac{1}{R_{eq1}}} + r_C \right) C_3 \tag{3.189}$$

With R_{eq1} defined in (3.180).

We can form the first term b_1 of the denominator:

$$b_1 = \tau_1 + \tau_2 + \tau_3 = C_1 \left[\frac{1}{g_r - g_f + \dfrac{1}{R_{eq1}}} \right] + \frac{L_2}{\dfrac{1 - R_{eq2}\left(g_f - g_o\right)}{g_o + R_{eq2}g_r\left(g_f - g_o\right)}} + \left(\frac{1}{g_r - g_f + \dfrac{1}{R_{eq1}}} + r_C \right) C_3 \tag{3.190}$$

For the second-order terms, we will apply the strategy described in Figure 3.82: we determine the time constant involving a selected energy-storing element while the second element is set in its high-frequency state and the rest remains in its dc state.

For instance, to determine τ_2^1 involving inductor L_2, capacitor C_1 is replaced by a short circuit and C_3 is open circuited.

345

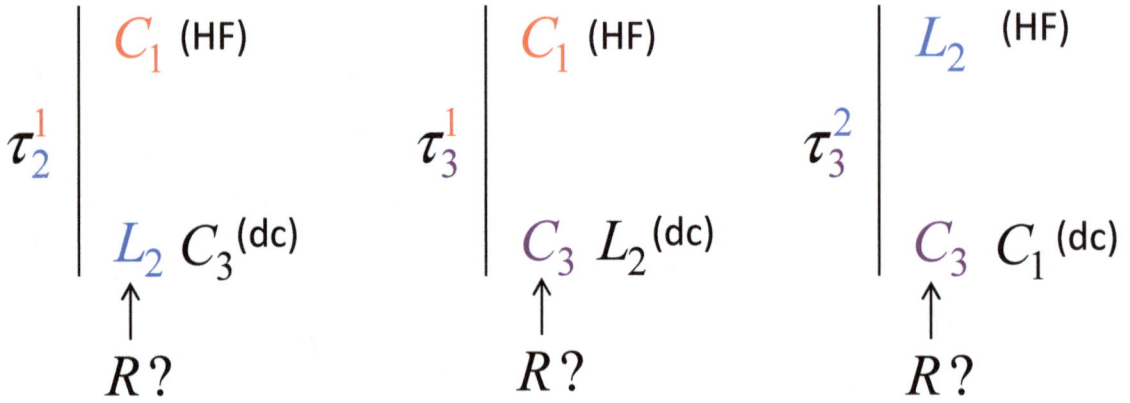

Figure 3.82 Second-order calculations require a different impedance arrangement.

The goal is to assemble these new terms as follows:

$$b_2 = \tau_1 \tau_2^1 + \tau_1 \tau_3^1 + \tau_2 \tau_3^2 \tag{3.191}$$

The first circuit is shown in Figure 3.83 and will considerably simplify considering the short circuit brought by C_1. If we install a test generator I_T across L_2's connecting terminals, the circuit simplifies further to that of Figure 3.84 where we see that V_T is actually V_{out}.

The time constant is immediate:

$$\tau_2^1 = \frac{L_2}{\dfrac{1}{\dfrac{1}{R_{eq2}} - g_r}} \tag{3.192}$$

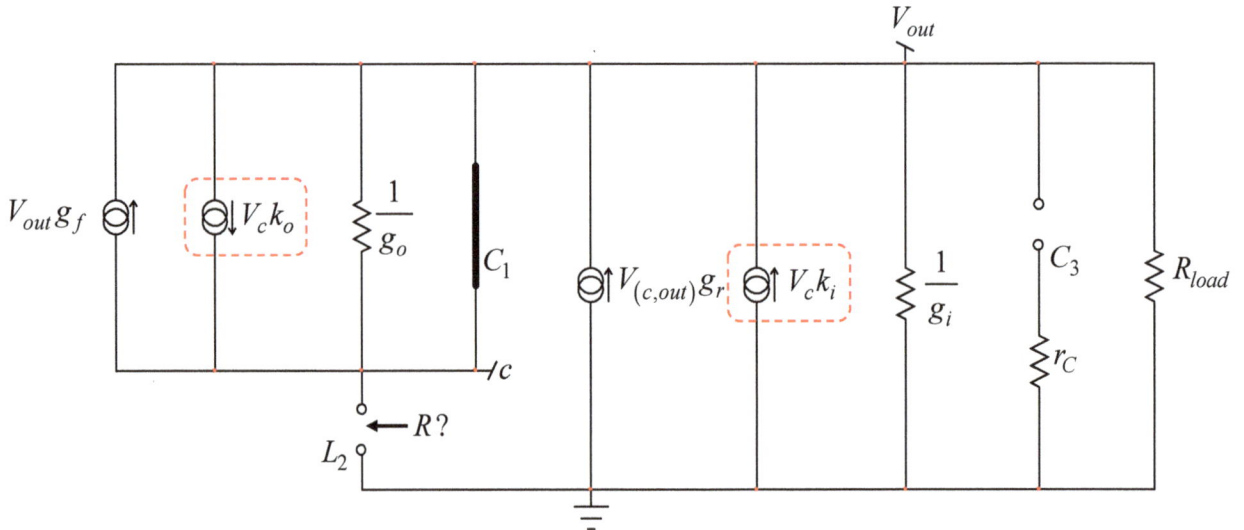

Figure 3.83 As C_1 is a short circuit, many elements disappear from the picture.

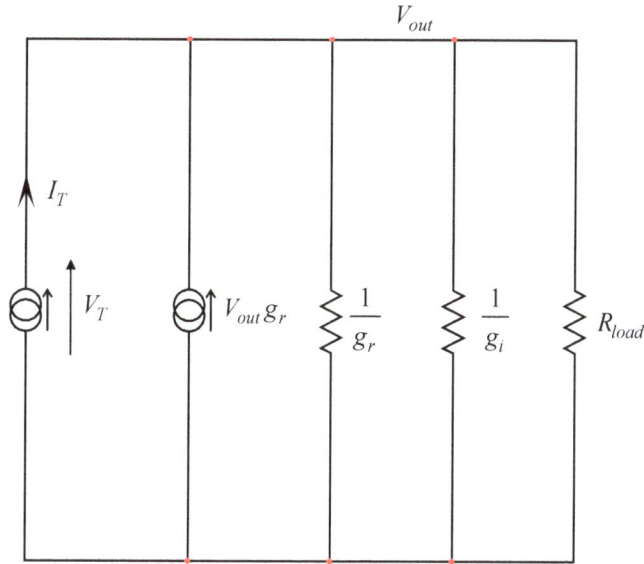

Figure 3.84 The circuit reduces to a simpler form.

For the next round, we will look at the configuration τ_3^1 in which C_1 is a short circuit, L_2 is set in its dc state and we determine the resistance offered by C_3's connecting terminals. The circuit is shown in Figure 3.85.

By inspection, you can see that the only non-shorted resistance is r_C leading to:

$$\tau_3^1 = r_C C_3 \tag{3.193}$$

For the last term, τ_3^2, the circuit is that of Figure 3.86. The dangling node c complexifies the analysis and I recommend to reshuffle the time constants as follows [1].

$$\tau_2 \tau_3^2 = \tau_3 \tau_2^3 \tag{3.194}$$

The new circuit is that of Figure 3.87, further simplified in Figure 3.88.

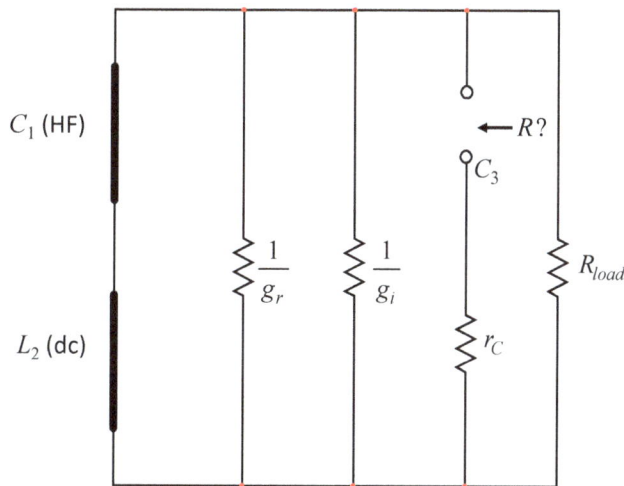

Figure 3.85 The circuit reduces to a simpler form.

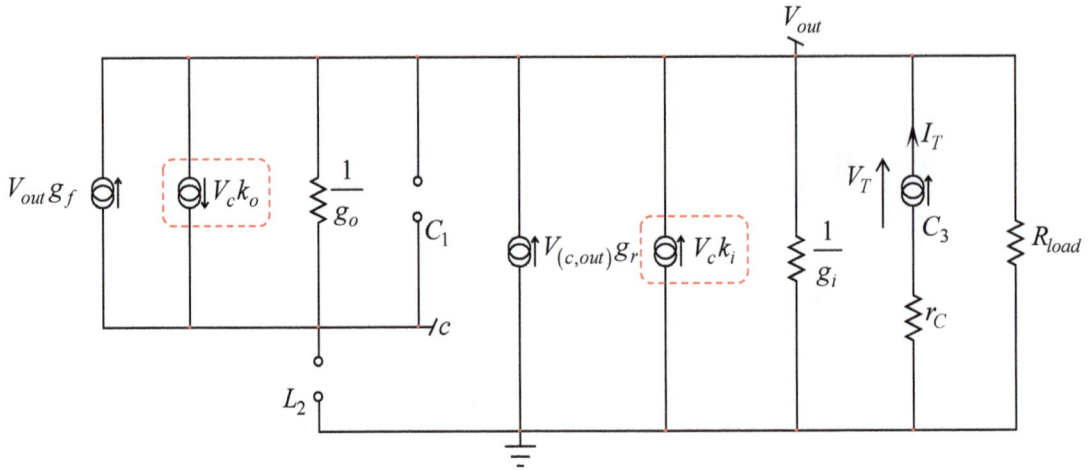

Figure 3.86 The circuit is not obvious at first glance and I recommend a different combination.

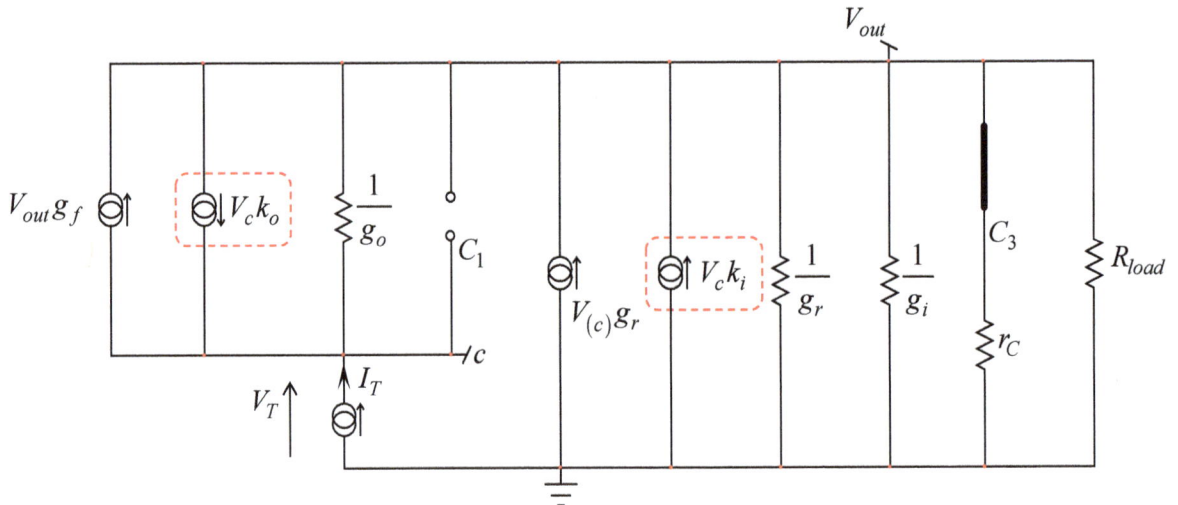

Figure 3.87 You can explore different combinations (C_3 is a short circuit) until a simpler circuit appears as in here.

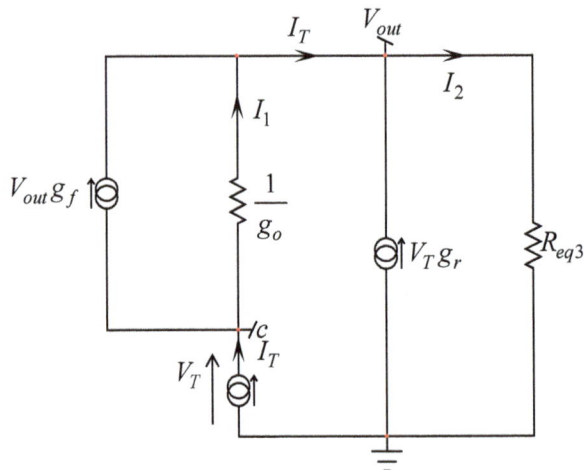

Figure 3.88 This new combination greatly simplifies the analysis.

In this approach, all conductances and resistances are lumped into a single resistance defined as:

$$R_{eq3} = r_C \parallel R_{load} \parallel \frac{1}{g_i} \parallel \frac{1}{g_r} \tag{3.195}$$

The first current I_1 is determined as:

$$I_1 = \frac{V_T - V_{out}}{\frac{1}{g_o}} \tag{3.196}$$

The second current, I_2, is defined by:

$$I_2 = I_T + V_T g_r \tag{3.197}$$

We also know that:

$$I_T = I_1 + V_{out} g_f \tag{3.198}$$

The output voltage V_{out} is:

$$V_{out} = R_{eq3} I_2 = R_{eq3} \left(I_T + V_T g_r \right) \tag{3.199}$$

Now using (3.198) in which you substitute (3.196) and (3.199) then rearranging the result, you should find:

$$R = \frac{V_T}{I_T} = \frac{1 - R_{eq3} \left(g_f - g_o \right)}{g_0 + R_{eq3} g_r \left(g_f - g_o \right)} \tag{3.200}$$

Leading to the time constant we want:

$$\tau_2^3 = \frac{L_2}{\dfrac{1 - R_{eq3} \left(g_f - g_o \right)}{g_0 + R_{eq3} g_r \left(g_f - g_o \right)}} \tag{3.201}$$

We can form the second-order coefficient now that we have all the time constants we needed:

$$b_2 = \tau_1 \tau_2^1 + \tau_1 \tau_3^1 + \tau_3 \tau_2^3 \tag{3.202}$$

Replacing with the expressions we have determined so far, it gives:

$$
b_2 = \left(\frac{1}{g_r - g_f + \dfrac{1}{R_{eq1}}} \right) C_1 \frac{L_2}{\dfrac{1}{R_{load} \parallel \dfrac{1}{g_r} \parallel \dfrac{1}{g_i}} - g_r} + \left(\frac{1}{g_r - g_f + \dfrac{1}{R_{eq1}}} \right) C_1 r_C C_3
$$

$$
+ \left(\frac{1}{g_r - g_f + \dfrac{1}{R_{eq1}}} + r_C \right) C_3 \frac{L_2}{\dfrac{1 - R_{eq3} \left(g_f - g_o \right)}{g_0 + R_{eq3} g_r \left(g_f - g_o \right)}} \tag{3.203}
$$

With R_{eq1}, R_{eq2} and R_{eq3} respectively defined in (3.180), (3.186) and (3.195).

For the final time constant, we analyze the circuit in which C_1 and L_2 are in their high-frequency state while we want the time constant associated with C_3:

$$b_3 = \tau_1 \tau_2^1 \tau_3^{12} \tag{3.204}$$

Figure 3.89 represents the circuit to analyze. Since node c and V_{out} are at the same potential, the source $V_{(c)}$ and the resistance $1/g_r$ respectively generate and absorb the same current thus cancel each other: they can disappear from the figure. The reduced circuit is given in Figure 3.90 and lets you determine the time constant immediately by inspection:

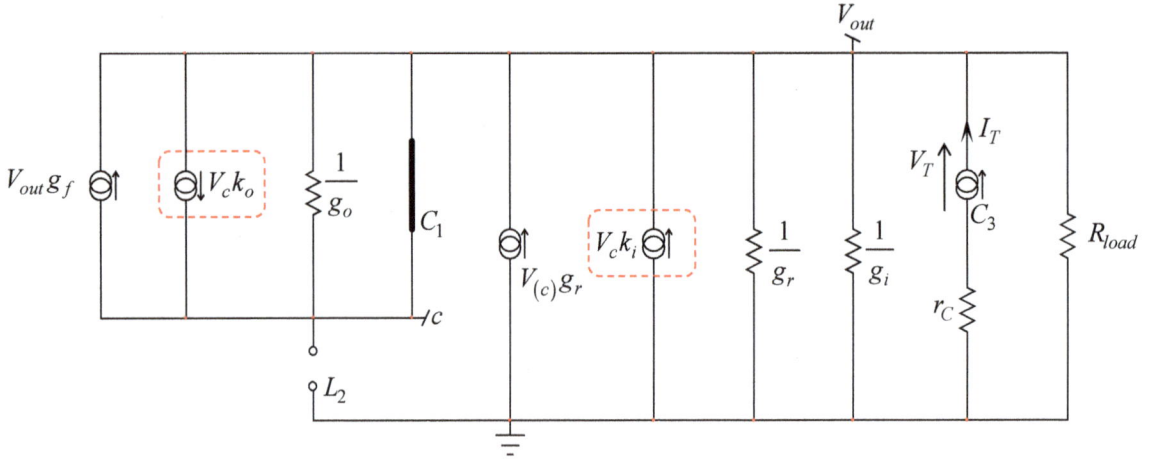

Figure 3.89 This final time constant analysis implies the opening of L_2 and the shortening of C_1.

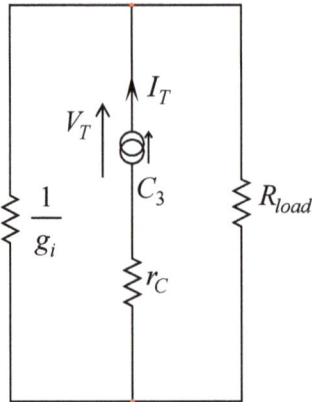

Figure 3.90 The circuit is really simple to solve by inspection.

$$\tau_3^{12} = \left(\frac{1}{g_i} \| r_L + r_C \right) C_3 \tag{3.205}$$

The last coefficient is thus defined by:

$$b_3 = \left(\frac{1}{g_r - g_f + \frac{1}{R_{eq1}}} \right) C_1 \frac{\frac{L_2}{1}}{\frac{1}{R_{eq2}} - g_r} \left(\frac{1}{g_i} \| r_L + r_C \right) C_3 \tag{3.206}$$

350

We can now form the denominator which is of 3rd order:

$$D(s) = 1 + b_1 s + b_2 s^2 + b_3 s^3 \qquad (3.207)$$

This 3rd-order polynomial form can be rearranged considering a dominating low-frequency pole and two high-frequency poles close to each other:

$$1 + b_1 s + b_2 s^2 + b_3 s^3 \approx \left(1 + b_1 s\right)\left(1 + \frac{b_2}{b_1} s + \frac{b_3}{b_1} s^2\right) \qquad (3.208)$$

It can be shown that this denominator can be put under the form:

$$\left(1 + b_1 s\right)\left(1 + \frac{b_2}{b_1} s + \frac{b_3}{b_1} s^2\right) \approx \left(1 + \frac{s}{\omega_{p_1}}\right) \frac{1}{1 + \frac{s}{\omega_n Q_p} + \left(\frac{s}{\omega_n}\right)^2} \qquad (3.209)$$

with:

$$\omega_{p_1} \approx \frac{\dfrac{2}{R_{load}} + \dfrac{T_{sw}}{L_2 M^3}\left(1 + \dfrac{S_e}{S_n}\right)}{C_3} \qquad (3.210)$$

The quality factor and the double poles position have already been defined in the previous chapters:

$$Q_p = \frac{1}{\pi \left(m_c D' - 0.5\right)} \qquad (3.211)$$

$$\omega_n = \frac{\pi}{T_{sw}} \qquad (3.212)$$

with:

$$m_c = 1 + \frac{S_e}{S_n} \qquad (3.213)$$

In this expression, S_e is the external stabilization ramp in volts per seconds. S_n represents the boost converter on-slope scaled in volts per seconds via the sense resistance R_i and defined in (3.179).

Now that we have determined the poles of this circuit, let's have a look at the zeroes. Zeroes are salient angular frequency points for which the response is nulled despite the presence of the excitation. The circled current sources which were turned off because of a zeroed control voltage V_c are back in the game. The exercise now consists of finding the electrical conditions for which the response is 0 V. The circuit to study is that of Figure 3.91. Current sources referring entirely or partially to V_{out} can be turned off as this variable is nulled.

The transconductance $1/g_i$ also goes away considering the absence of voltage across its terminals.

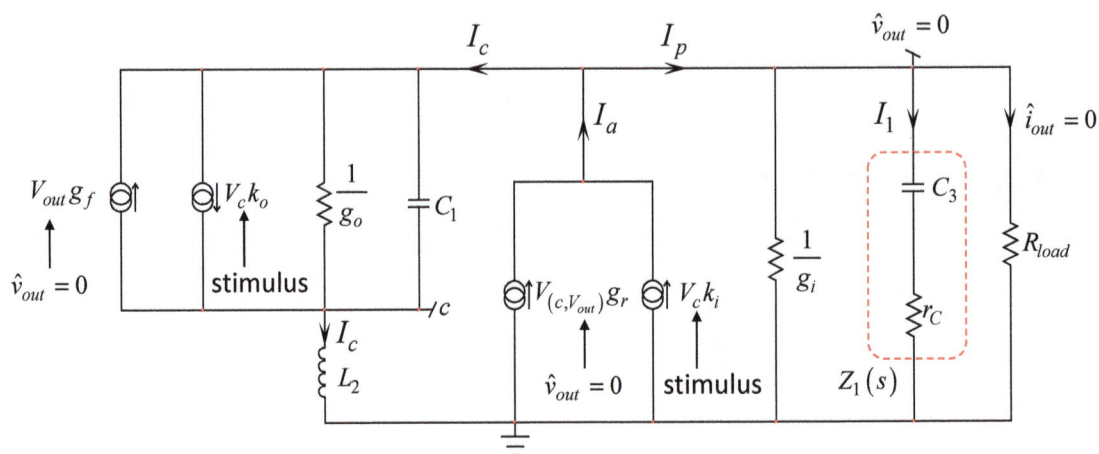

Figure 3.91 A null in the response means that no current flows in R_{load}.

The first obvious condition for which an output null is obtained is when the series combination of C_3 and r_C forms a transformed short circuit: $I_p = I_1$.

In this case, the zero is immediate and obtained by solving:

$$Z_1(s) = r_C + \frac{1}{sC_3} = 0 \tag{3.214}$$

It gives the classical location:

$$\omega_{z_1} = \frac{1}{r_C C_3} \tag{3.215}$$

For the second zero, the null is obtained when $I_a = I_c$ meaning that all the current generated by B_5 and B_6 is absorbed by the left-side network:

$$I_a = V_{(c)} g_r + V_c k_i \tag{3.216}$$

Current I_c is defined by:

$$I_c = V_c k_o - V_{(c)} g_o - V_{(c)} sC_1 \tag{3.217}$$

Homogeneity is respected as g_o and sC_1 have the dimension of a conductance. Now equating (3.216) and (3.217) gives:

$$V_{(c)} g_r + V_c k_i = V_c k_o - V_{(c)} g_o - V_{(c)} sC_1 \tag{3.218}$$

In this expression, the unknown is the voltage at terminal c.

We can obtain it by observing that current I_c also depends on the voltage at terminal c applied across inductor L_2:

$$\frac{V_{(c)}}{sL_2} = V_c k_o - V_{(c)} (g_o - sC_1) \tag{3.219}$$

From this expression, we can extract the voltage at terminal c:

$$V_{(c)} = \frac{V_c k_o}{\dfrac{1}{sL_2} + g_o + sC_1} \tag{3.220}$$

If we now substitute this definition in (3.218) and rearrange the result, we have:

$$\frac{V_c \left(k_i + L_2 g_o k_i s + L_2 g_r k_o s + C_1 L_2 k_i s^2 \right)}{1 + sL_2 g_o + s^2 C_1 L_2} = \frac{V_c k_o}{1 + sL_2 g_o + s^2 C_1 L_2} \tag{3.221}$$

As denominators are equal, to satisfy this equation, we must equate numerators. After moving k_o to the left side and properly factoring, we end up solving:

$$\left(k_i - k_o \right)\left(1 + sL_2 \left(\frac{g_o k_i + g_r k_o}{k_i - k_o} \right) + s^2 \frac{C_1 L_2 k_i}{k_i - k_o} \right) = 0 \tag{3.222}$$

The left side can be advantageously written under the following polynomial form:

$$1 + sL_2 \left(\frac{g_o k_i + g_r k_o}{k_i - k_o} \right) + s^2 \frac{C_1 L_2 k_i}{k_i - k_o} = 0 \tag{3.223}$$

If we consider spread zeroes [1], we can approximate this expression as:

$$1 + a_1 s + a_2 s^2 \approx \left(1 + a_1 s \right)\left(1 + \frac{a_2}{a_1} s \right) \tag{3.224}$$

The factored expression equals zero if one of its members is also zero. The low-frequency zero set by a_1 is defined as:

$$1 + sL_2 \left(\frac{g_o k_i + g_r k_o}{k_i - k_o} \right) = 1 - \frac{s}{s_{z_2}} \tag{3.225}$$

Numerical application shows that L_2's factor is negative; we have a right-half-plane zero:

$$\omega_{z_2} = \frac{k_i - k_o}{L_2 \left(g_o k_i + g_r k_o \right)} \tag{3.226}$$

Replacing the coefficients by their value and simplifying the expression leads to a simpler approximate formula:

$$\omega_{z_2} \approx \frac{\left(1 - D \right)^2 R_{load}}{L_2} \tag{3.227}$$

Looking at the right-side member in (3.224) sheds light on a third zero:

$$1 + s \frac{C_1 k_i}{g_o k_i + g_r k_o} = 1 + \frac{s}{s_{z_3}} \tag{3.228}$$

However, it is relegated to higher frequencies and will be neglected:

$$\omega_{z_3} = \frac{g_o k_i + g_r k_o}{C_1 k_i} \tag{3.229}$$

Gathering the quasi-static gain H_0, the poles and the zeroes, we have the complete control-to-output transfer function of the current-mode boost converter:

$$\frac{V_{out}(s)}{V_c(s)} \approx H_0 \frac{\left(1 + \dfrac{s}{\omega_{z_1}}\right)\left(1 - \dfrac{s}{\omega_{z_2}}\right)}{1 + \dfrac{s}{\omega_{p_1}}} \frac{1}{1 + \dfrac{s}{\omega_n Q_p} + \left(\dfrac{s}{\omega_n}\right)^2} \tag{3.230}$$

For which we have previously determined all the terms:

$$\frac{V_{out}(s)}{V_c(s)} \approx \frac{R_{load}}{R_i} \frac{1}{2M + \dfrac{R_{load} T_{sw}}{L_2 M^2}\left(\dfrac{1}{2} + \dfrac{S_e}{S_n}\right)} \frac{\left(1 - s\dfrac{L_2}{(1-D)^2 R_{load}}\right)(1 + sr_C C_3)}{1 + s\dfrac{C_3}{\dfrac{2}{R_{load}} + \dfrac{T_{sw}}{LM^3}\left(1 + \dfrac{S_e}{S_n}\right)}} \frac{1}{1 + \dfrac{s}{\omega_n Q_p} + \left(\dfrac{s}{\omega_n}\right)^2} \tag{3.231}$$

To test the validity of these equations, I have compared the response between the Mathcad® plots generated with the most comprehensive version with all the raw k coefficients and SPICE. Then I confronted the simplified version of the expressions with SPICE.

All these results are gathered in Figure 3.92 and confirm the approach is correct.

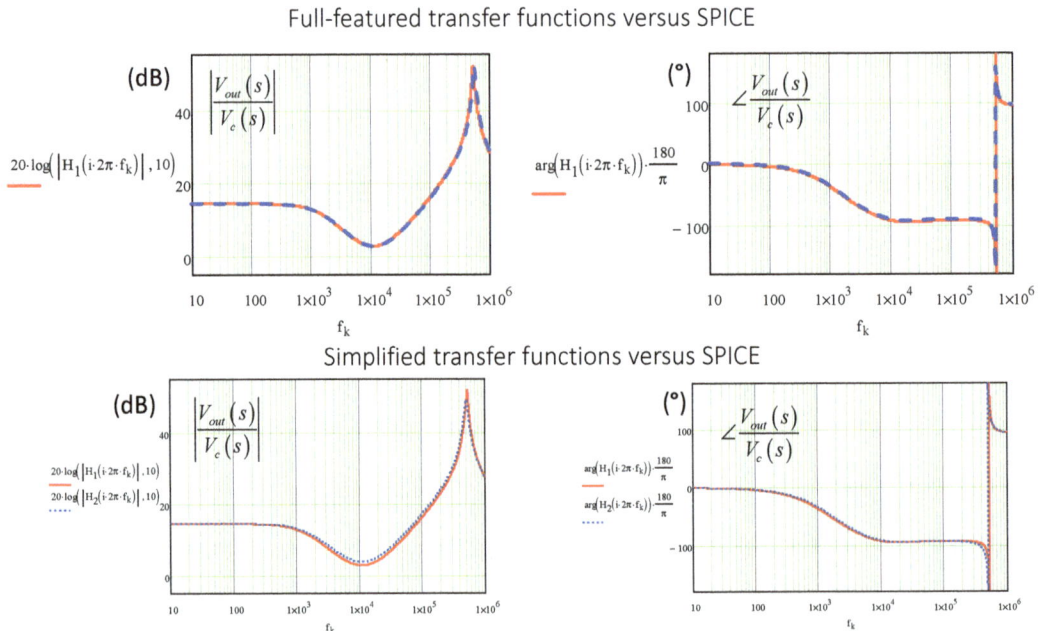

Figure 3.92 The curves given by Mathcad® with different complexity perfectly match the SPICE simulations.

As we did in the previous examples, a SIMPLIS® simulation is run to check the SPICE model correctly reflects reality. The simulated schematic diagram appears in Figure 3.93. A very small amount of slope compensation helps for stabilizing the ac analysis considering the high Q_p of the circuit. The sense resistor is purposely excluded from the turn-on path like with the model.

In reality, the on-time will be affected by the series connection of r_L, the $r_{DS(on)}$ and the sense resistor, impacting the converter. The simulation results are given in Figure 3.94 and show a good overall shape with a small deviation on the dc gain.

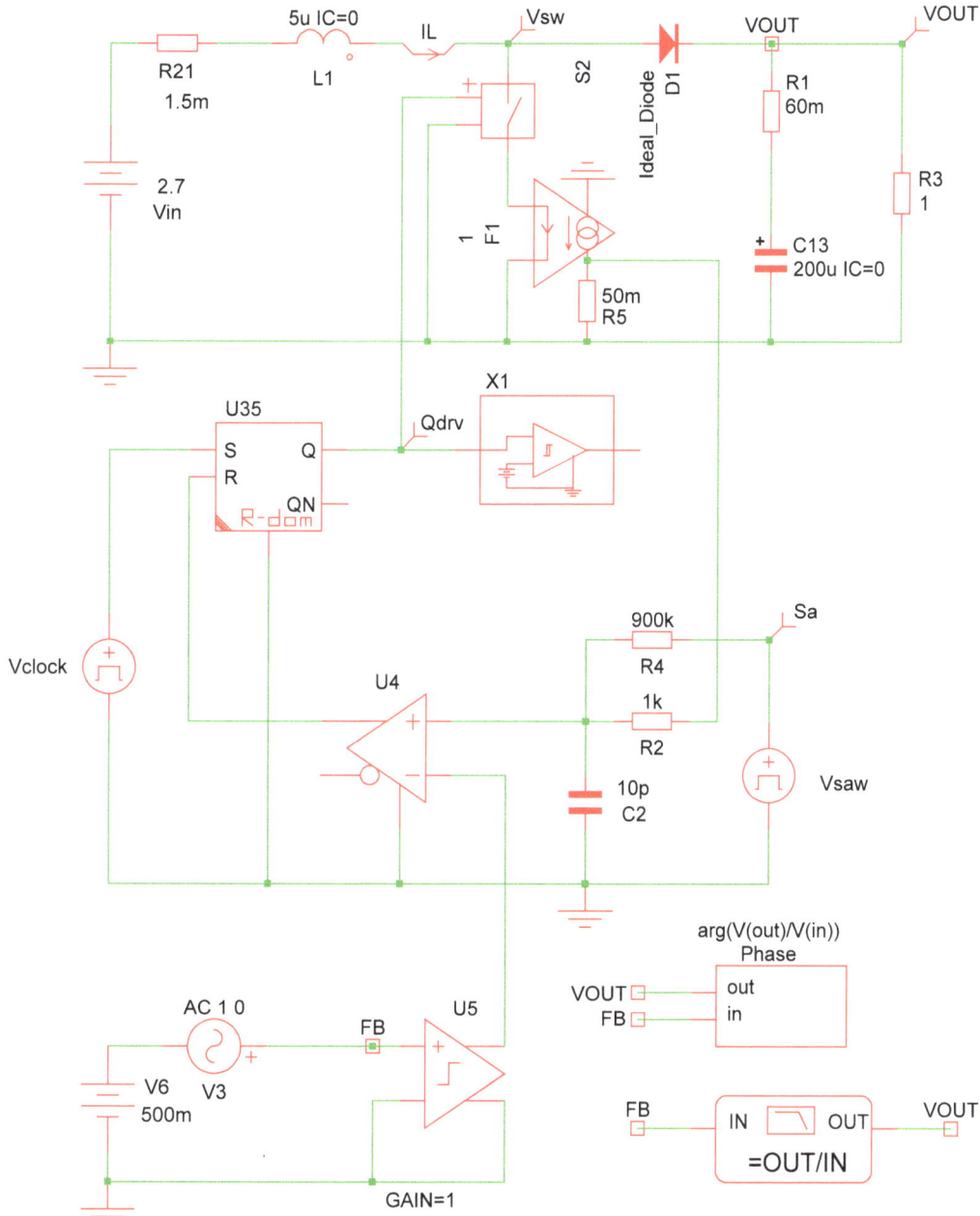

Figure 3.93 The simulated boost converter includes a little bit of slope compensation to stabilize the ac analysis.

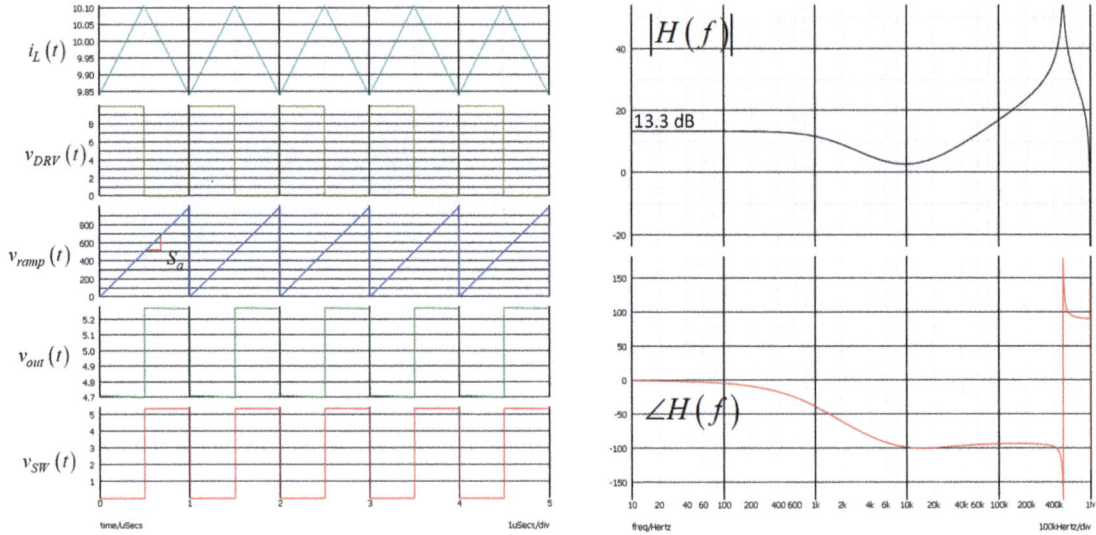

Figure 3.94 The dynamic response is very close to what SPICE has predicted.

For a sanity check on your side, I have included in Figure 3.95 the small-signal coefficients computed by Mathcad®
and used to plot all the CCM transfer functions of the CM boost converter.

$$V_{in} := 2.7V \qquad V_{out} := 5.01V \qquad R_i := -0.05\Omega \qquad C_{out} := 200\mu F \qquad r_C := 0.06\Omega \qquad R_{inf} := 10^{10}\Omega$$

$$F_{sw} := 1000kHz \qquad T_{sw} := \frac{1}{F_{sw}} \qquad r_L := 0.0015\Omega \qquad L := 5 \cdot \mu H \qquad R_L := 1\Omega$$

$$V_{ap} := -V_{out} \qquad V_{cp} := -V_{out} + V_{in} \qquad V_{ac} := -V_{in} \qquad D := \frac{V_{cp}}{V_{ap}} = 0.461$$

$$V_c := 471mV \qquad S_a := 0\frac{kV}{s} \qquad D_p := 1 - D = 0.539 \qquad D = 0.461 \qquad M := \frac{V_{out}}{V_{in}} = 1.856$$

$$S_n := \frac{V_{ac}}{L} \cdot R_i = 27 \cdot \frac{kV}{s} \qquad S_f := \frac{V_{ap}}{L} \cdot R_i = 50.1 \cdot \frac{kV}{s} \qquad C_s := \frac{1}{L \cdot (F_{sw} \cdot \pi)^2} = 20.264nF$$

$$I_c := \frac{V_c}{R_i} - V_{cp} \cdot \left(1 - \frac{V_{cp}}{V_{ap}}\right) \cdot \frac{T_{sw}}{2 \cdot L} - \frac{S_a}{R_i} \cdot \frac{V_{cp}}{V_{ap}} \cdot T_{sw} = -9.296A \qquad \tau_L := \frac{L}{R_L \cdot T_{sw}} = 5$$

$$I_a := \frac{V_{cp}}{V_{ap}} \cdot \left[\frac{V_c}{R_i} - V_{cp} \cdot (1 - D) \cdot \frac{T_{sw}}{2 \cdot L} - \frac{S_a}{R_i} \cdot D \cdot T_{sw}\right] = -4.286A$$

$$g_o := \left[(1 - D) \cdot \frac{S_a}{S_n} + \frac{1}{2} - D\right] \cdot \frac{T_{sw}}{L} = 7.784 \times 10^{-3} \frac{1}{\Omega} \qquad k_o := \frac{1}{R_i} = -20\frac{1}{\Omega}$$

$$g_f := D \cdot g_o - \frac{D \cdot D_p \cdot T_{sw}}{2 \cdot L} = -0.021\frac{1}{\Omega} \qquad k_i := \frac{D}{R_i} = -9.222\frac{1}{\Omega}$$

$$g_i := D \cdot \left(g_f - \frac{I_c}{V_{ap}}\right) = -0.865\frac{1}{\Omega} \qquad g_r := \frac{I_c}{V_{ap}} - g_o \cdot D = 1.852\frac{1}{\Omega}$$

**Figure 3.95 These are the small-signal coefficients for the current-mode model computed by
Mathcad®.**

3.3.1 Input to Output

The input-to-output transfer function is obtained by changing the stimulus from V_c to V_{in}. The relationship linking V_{in} to V_{out} is now the wanted expression. The large-signal simulation template is given in Figure 3.96.

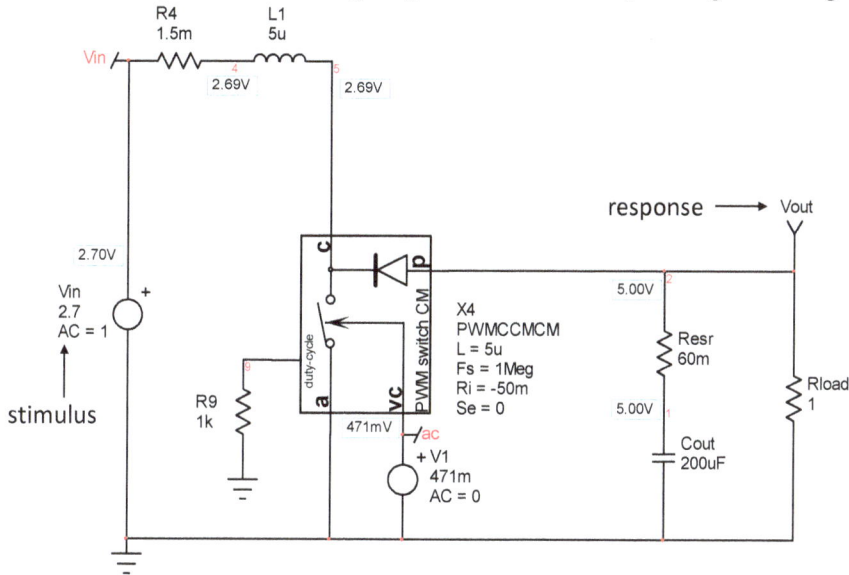

Figure 3.96 The stimulus is now the input voltage.

We can now plug the small-signal version of the CM PWM-switch model as shown in Figure 3.97. The response from this circuit is identical to that given by Figure 3.96 model. This is an important sanity check to confirm you will work on an electrically-sound circuit.

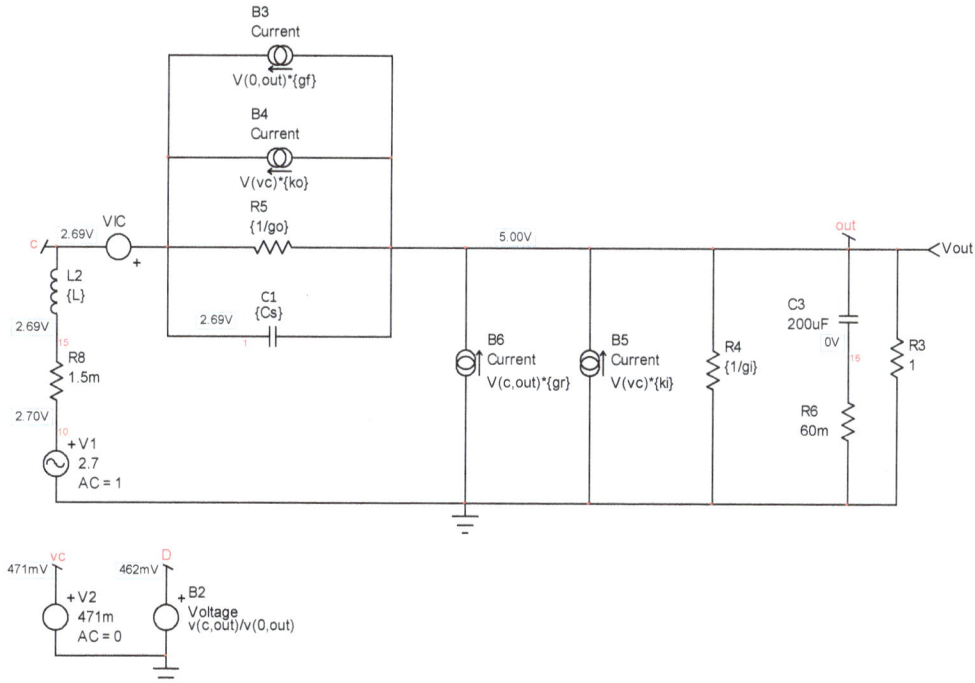

Figure 3.97 The circuit is very close to that already used to study the control-to-output transfer function.

Here also we can neglect the inductor ohmic loss r_L and rearrange the circuit considering zeroed current sources such as those containing the control voltage V_c. The updated diagram is given in Figure 3.98.

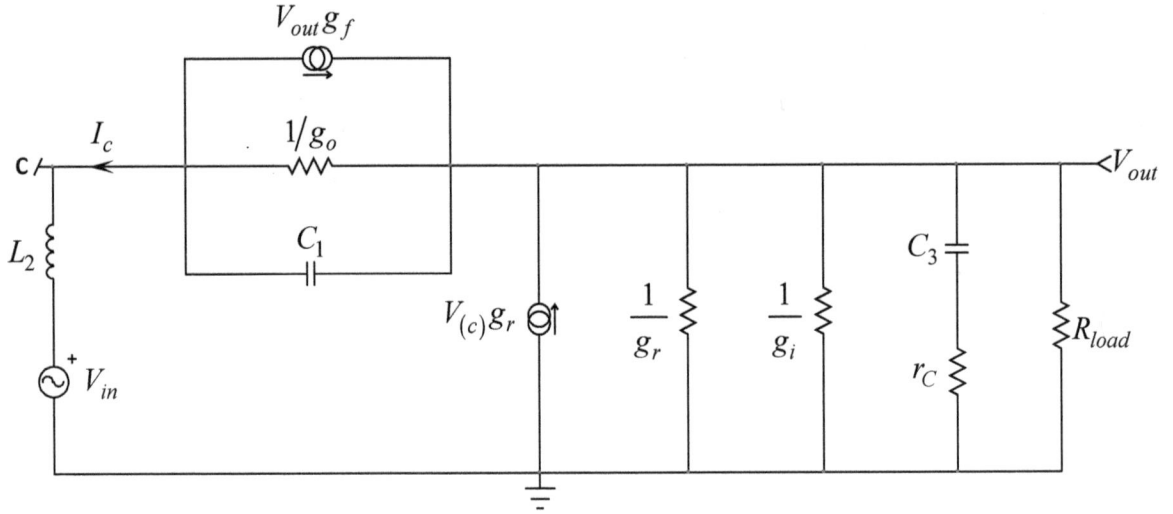

Figure 3.98 This is the final small-signal drawing for the input-to-output transfer function determination.

For $s = 0$, we short the inductor and open all the capacitors to end up with the circuit of Figure 3.99.

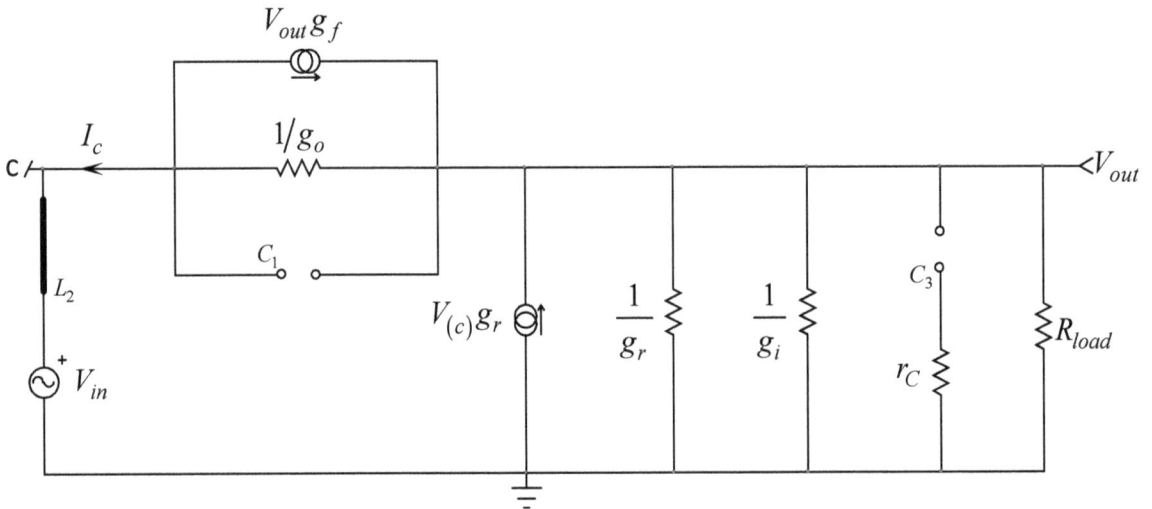

Figure 3.99 The inductor is shorted and all capacitors are open-circuited for the dc gain determination.

We can lump all right-side elements into a single resistance R_{eq2} already defined in (3.186). The new circuit appears in Figure 3.100 and looks simpler to solve.

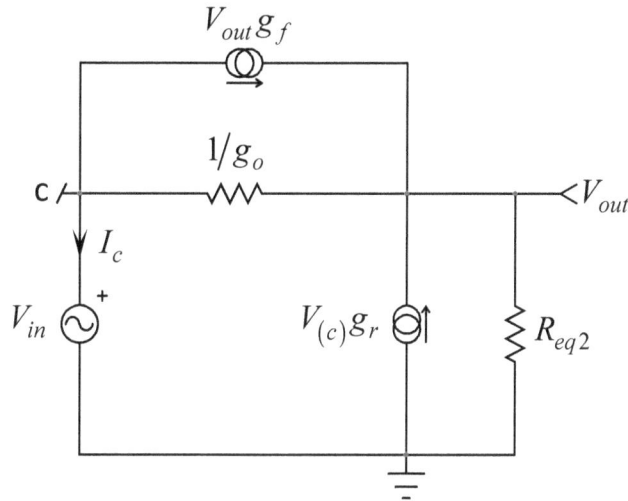

Figure 3.100 Reducing the circuit to a simpler arrangement is the way to go.

We can write:

$$V_{out} = R_{eq2}\left[V_{out}g_f + V_{(c)}g_r + \left(V_{(c)} - V_{out}\right)g_o\right] \qquad (3.232)$$

Solve for V_{out} observing that $V_{in} = V_{(c)}$. Rearrange the result with V_{in} as a factor and you obtain H_0:

$$H_0 = \frac{g_o + g_r}{g_o - g_f + \dfrac{1}{R_{eq2}}} \qquad (3.233)$$

Now, if you reduce the excitation to 0 V, i.e. short the input source V_{in}, the circuit from Figure 3.98 turns to that of Figure 3.75: time constants are similar and we can reuse the denominator already found in (3.209).

The zeroes need now to be determined. In Figure 3.98, we need to know what particular condition would bring the response to 0 V despite the presence of the input source? The first solution is easy as C_3 classically introduces a zero when teaming with r_C, its ESR:

$$\omega_{z_1} = \frac{1}{r_C C_3} \qquad (3.234)$$

For the second condition, if the output is nulled, then no current flows in R_{load} and the two conductances g_r and g_i disappear (no voltage across their terminals). All the current generated by $V_{(c)}g_r$ is absorbed by the left-side circuit involving L_1, C_1 and the g_o conductance.

Fortunately, the voltage-controlled current source goes away as V_{out} the control variable is nulled. The circuit reduces to that of Figure 3.101.

359

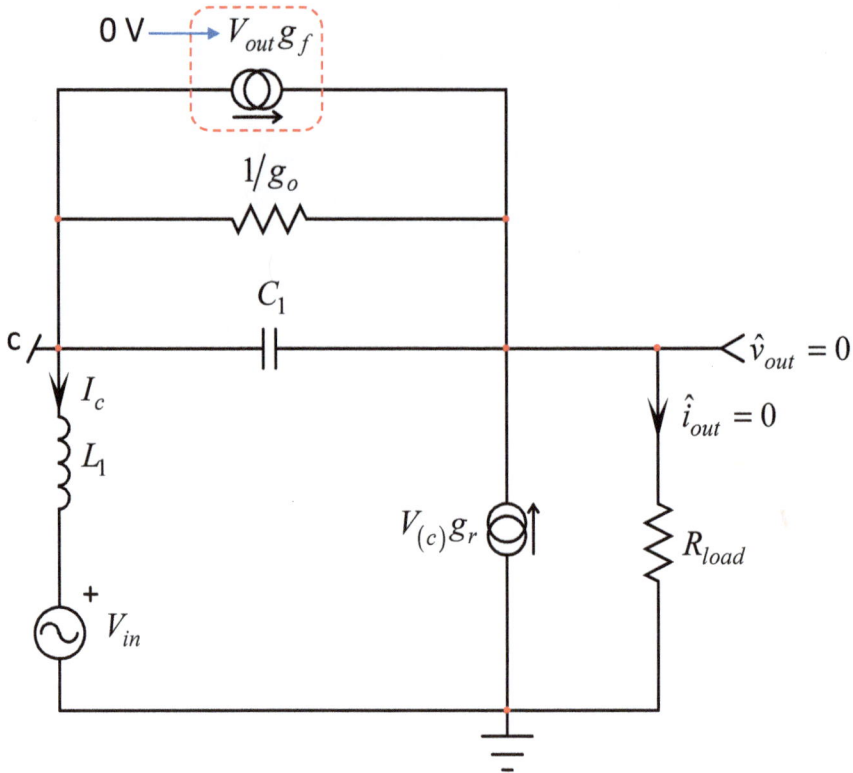

Figure 3.101 There is no current flowing in the load when the output null is observed.

We can write a first equation to define the voltage at node c:

$$V_{(c)} = V_{in} + I_c sL_2 \tag{3.235}$$

We extract the current I_c:

$$I_c = \frac{V_{(c)} - V_{in}}{sL_2} \tag{3.236}$$

The voltage at node c is also the current I_c flowing in the parallel arrangement made of C_1 and the conductance g_o

$$V_{(c)} = -I_c \left(\frac{1}{g_o} \parallel \frac{1}{sC_1} \right) \tag{3.237}$$

If you substitute (3.236) in the above expression and solve for $V_{(c)}$, you have:

$$V_{(c)} = \frac{V_{in}}{1 + sL_2 g_o + s^2 L_2 C_1} \tag{3.238}$$

Now, we know that the current provided by $V_{(c)}g_r$ must equate I_c to generate the output null. From (3.236), we can write:

$$\frac{V_{(c)} - V_{in}}{sL_2} = V_{(c)}g_r \tag{3.239}$$

Substituting in this formula the definition of $V_{(c)}$ from (3.238), we have:

$$-\frac{V_{in}\left(g_o + sC_1\right)}{1 + sL_2g_o + s^2L_2C_1} = \frac{V_{in}g_r}{1 + sL_2g_o + s^2L_2C_1} \quad (3.240)$$

Equating numerators accounting for the sign and solving for s leads to the second zero position:

$$\omega_{z_2} = \frac{g_o + g_r}{C_1} \quad (3.241)$$

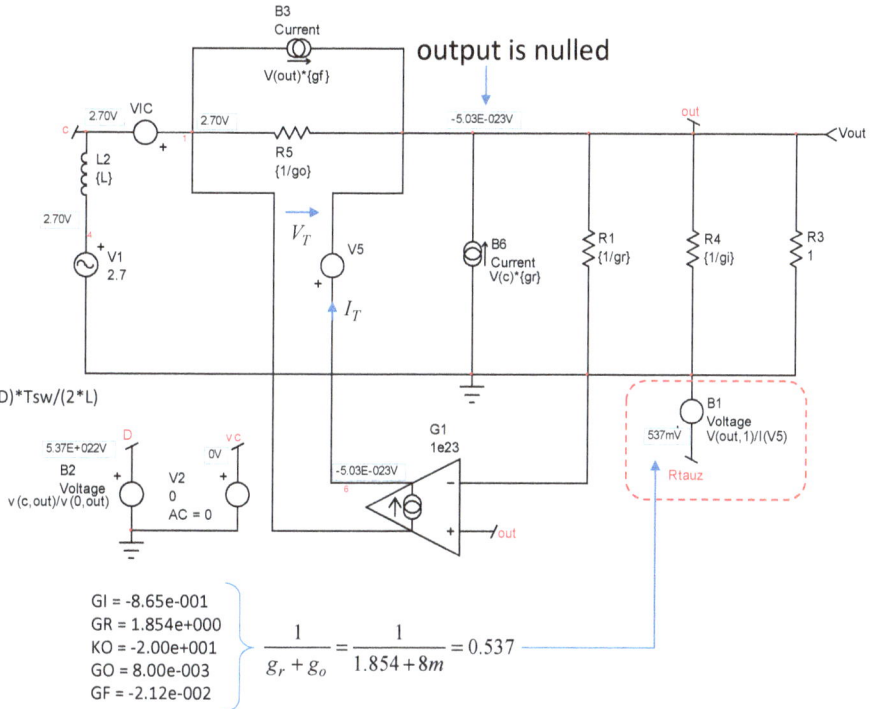

parameters
Fsw =1Meg
Tsw =1/Fsw
L=5u
Cs=1/(L*(Fsw *3.14)^2)
Ri=-50m
Se=0
Vin=2.7
Vout=5

Sn=(Vac/L)*Ri
Sf=(Vap/L)*Ri

Vc=471m
Vac=-Vin
Vap=-Vout
Vcp=(-Vout+Vin)

Ic=(Vc/Ri)-D*Tsw*Se/Ri-Vcp*(1-D)*Tsw/(2*L)

D=Vcp/Vap
D'=1-D
ki=D/Ri
gi=D*(gf-Ic/Vap)
gr=(Ic/Vap)-go*D
ko=1/Ri
go=(Tsw /L)*(D*Se/Sn+0.5-D)
gf=D*go-(D*D')*Tsw /(2*L)

GI = -8.65e-001
GR = 1.854e+000
KO = -2.00e+001
GO = 8.00e-003
GF = -2.12e-002

$$\frac{1}{g_r + g_o} = \frac{1}{1.854 + 8m} = 0.537$$

Figure 3.102 The SPICE simulation nulls the output and tells us the resistance seen from C_1's connections in this mode.

How do we know this is the correct value? We can use SPICE to check what resistance is offered from the capacitor's connecting terminals when the output is nulled. This configuration appears in Figure 3.101 and was described in [1]. Source G_1 injects a current and adjusts its value to exactly null the output. The resistance computed by B_1 is the resistance offered by C_1's connections in this mode and matches what (3.241) suggested. This technique is extremely useful when a doubt exists in the determination of a zero in a complicated circuit diagram. The SPICE simulation will immediately tell you if your result is correct or not.

Numerical application shows that this LHP zero is located at a high frequency and its action can be neglected in the final result. The complete input-to-output transfer function is then defined as follows:

$$\frac{V_{out}(s)}{V_{in}(s)} \approx H_0 \frac{1 + \dfrac{s}{\omega_{z_1}}}{1 + \dfrac{s}{\omega_{p_1}}} \frac{1}{1 + \dfrac{s}{\omega_n Q_p} + \left(\dfrac{s}{\omega_n}\right)^2} \quad (3.242)$$

In which we have previously determined all the terms. The magnitude-phase curves from Mathcad® and SPICE are shown in Figure 3.103. The curves are very close but there is a small mismatch in the peaking. Small rounding differences in the calculation of the small-signal parameters from SPICE and Mathcad® can explain this mismatch.

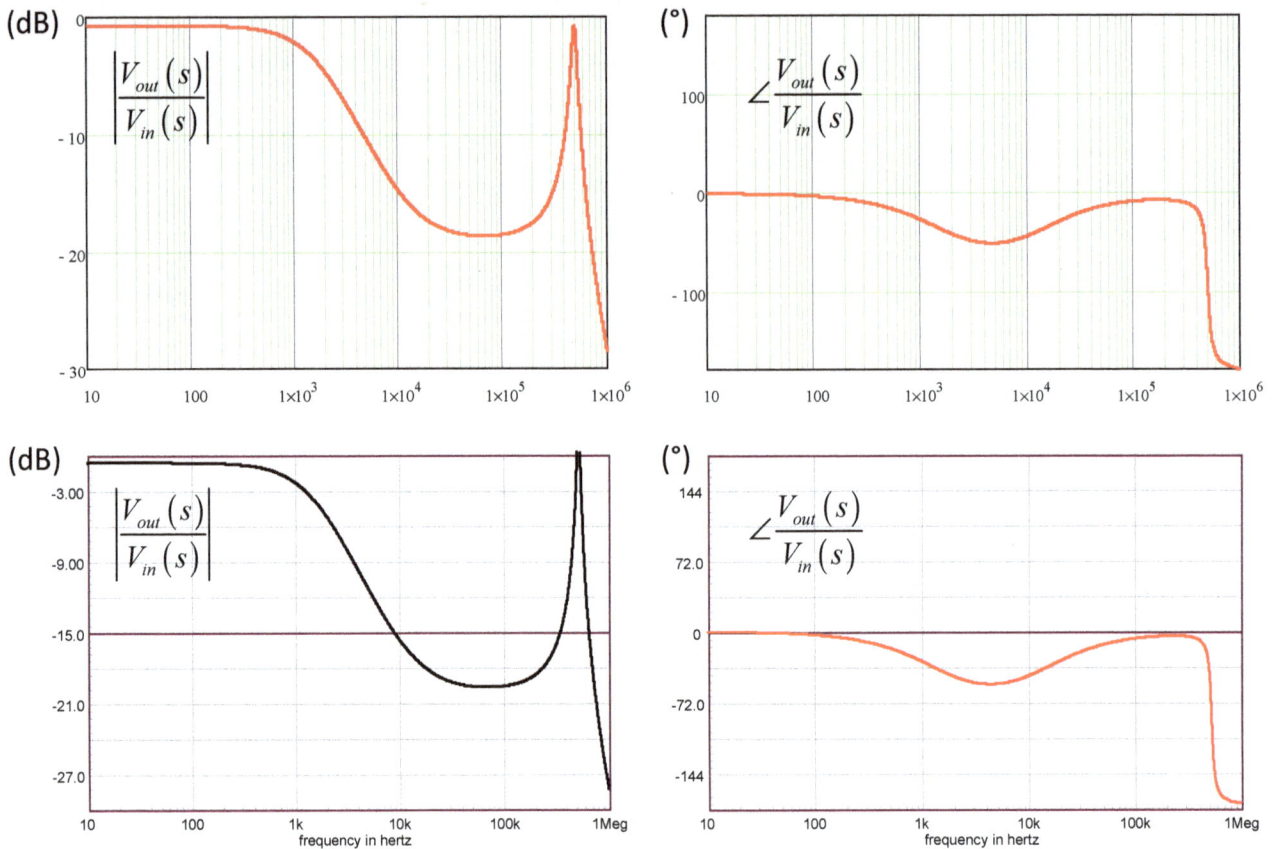

Figure 3.103 Mathcad® and SPICE instantaneously plot the input-to-output transfer function dynamic response: they are very close to each other.

3.3.2 Output Impedance

The output impedance is obtained by installing a current source across the load as Figure 3.104 details. The small-signal circuit is simply updated by adding this extra generator while the input source is now 0 V ac. The circuit to solve is that of Figure 3.105. The circled sources are turned off and disappear as the control voltage V_c is also 0 V ac during this open-loop analysis.

We will start with R_0, the output impedance determined for $s = 0$: open the capacitors and short the inductor neglecting its ESR r_L. The updated circuit diagram appears in Figure 3.106. Because node c is grounded, the controlled source $V_{(c)}g_f$ is now turned off also and further simplifies the diagram.

The three resistances are lumped into an equivalent resistance R_{eq2} already defined in (3.186).

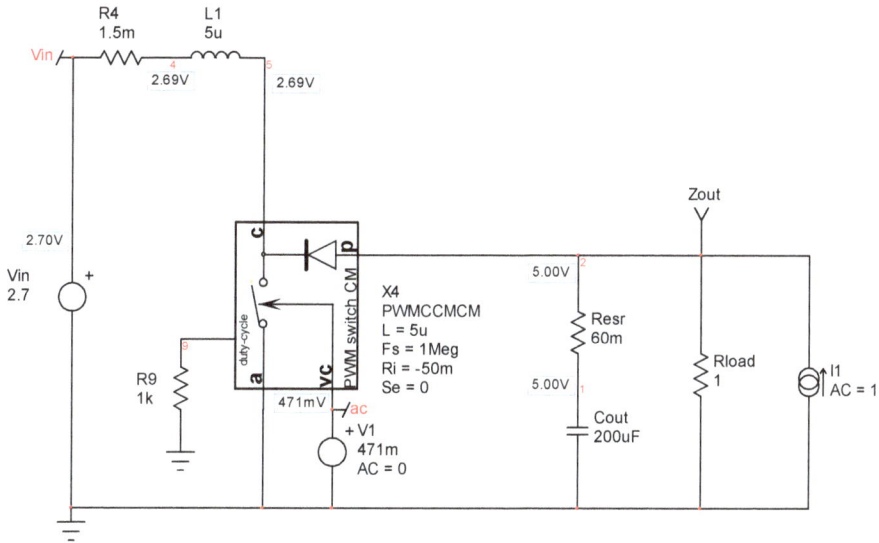

Figure 3.104 An ac current source loads the converter output to unveil the output impedance.

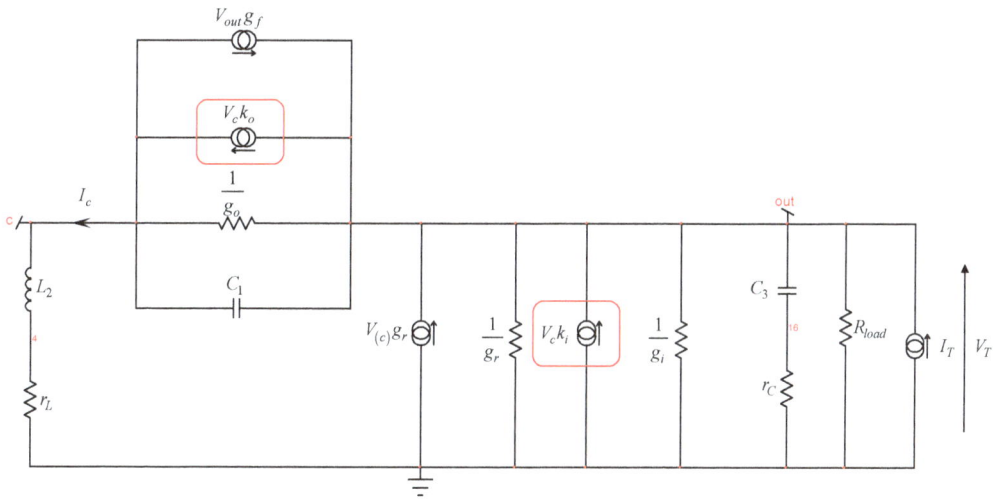

Figure 3.105 This is the small-signal circuit once rearranged for determining Z_{out}.

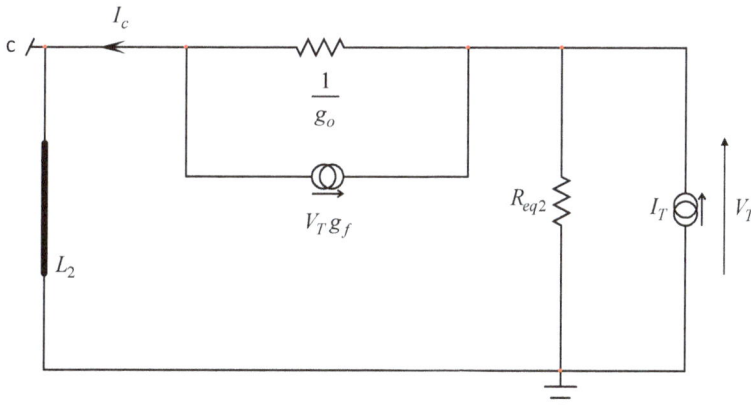

Figure 3.106 This circuit is used to determine the quasi-static output resistance R_0.

To further ease the analysis, we can temporarily remove R_{eq2} and the conductance $1/g_o$ to bring them back in parallel with the intermediate result we will find. Looking at the picture, we see the current source imposing a current opposite to I_T:

$$I_T = -V_T g_f \tag{3.243}$$

From which we immediately determine an intermediate resistance equal to:

$$\frac{V_T}{I_T} = -\frac{1}{g_f} \tag{3.244}$$

This leads us to definition of R_0 given by:

$$R_0 = -\frac{1}{g_f} \left\| \frac{1}{g_o} \right\| \frac{1}{g_r} \left\| \frac{1}{g_i} \right\| \frac{1}{R_{load}} = \frac{1}{g_i - g_f + g_o + g_r + \dfrac{1}{R_{load}}} \tag{3.245}$$

The time constants are determined by turning the stimulus off. Doing so brings the circuit back into its natural state already depicted by Figure 3.75 with zeroed V_c and V_{in}. As such, the denominator determined in (3.209) can be reused for the output impedance expression.

The zeroes need now to be determined considering the response as the voltage V_T developed across the current source I_T. In Figure 3.105, we need to know what particular condition would bring the response to 0 V despite the presence of the stimulus I_T? The first solution is easy as C_3 classically introduces a zero when associated with r_C, its parasitic resistance:

$$\omega_{z_1} = \frac{1}{r_C C_3} \tag{3.246}$$

To determine the other zeroes, we must null the response and determine the time constants involving C_1 and L_2 in this configuration. Zeroing the response across the current source I_T implies that we can replace it by a short circuit [1], naturally removing many elements from the circuit. The updated schematic diagram is given in Figure 3.107 while the final circuit to study appears in Figure 3.108: difficult to make it simpler.

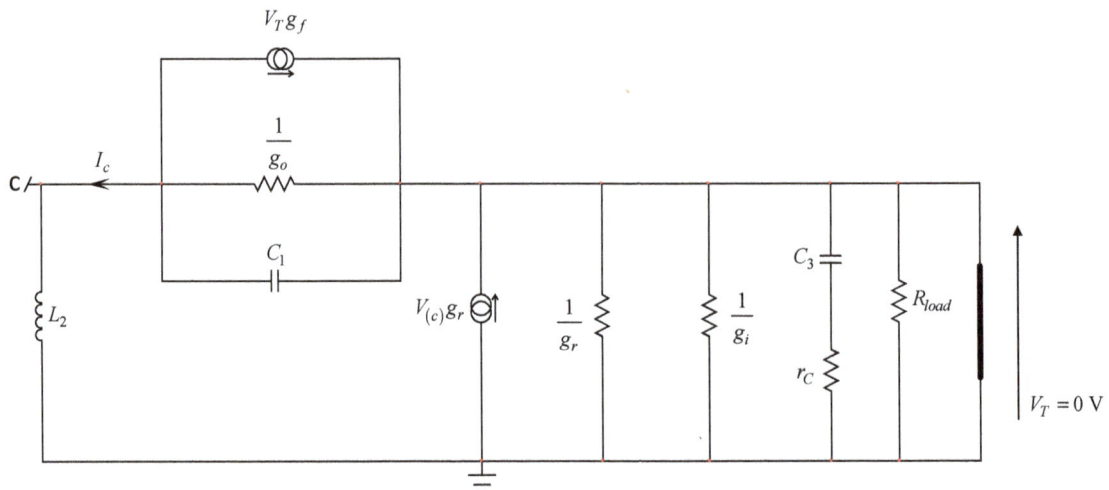

Figure 3.107 Zeroing the response across the current source is similar to replacing it by a short circuit.

The two zeroes involving L_2 and C_1 will fill out the numerator under the following form:

$$N(s)=1+s\left(\tau_{1N}+\tau_{2N}\right)+s^2\tau_{2N}\tau_{1N}^2 \tag{3.247}$$

The time constants are obtained by inspecting the intermediate sketches from Figure 3.108 and the final result is as follows:

$$N(s)=1+s\left(0+L_2g_o\right)+s^2L_2g_o\frac{C_1}{g_o}=1+sL_2g_o+s^2L_2C_1 \tag{3.248}$$

It can be factored under the classical 2nd-order form:

$$N(s)=1+\frac{s}{\omega_{0N}Q_z}+\left(\frac{s}{\omega_{0N}}\right)^2 \tag{3.249}$$

with:

$$\omega_{0N}=\frac{1}{\sqrt{L_2C_1}} \tag{3.250}$$

$$Q_z=\sqrt{\frac{C_1}{L_2}}\frac{1}{g_o} \tag{3.251}$$

Finally, the output impedance of the CCM boost CM converter obeys the following expression:

$$Z_{out}(s)=R_0\frac{\left(1+\frac{s}{\omega_{z_1}}\right)\left[1+\frac{s}{\omega_{0N}Q_z}+\left(\frac{s}{\omega_{0N}}\right)^2\right]}{D(s)} \tag{3.252}$$

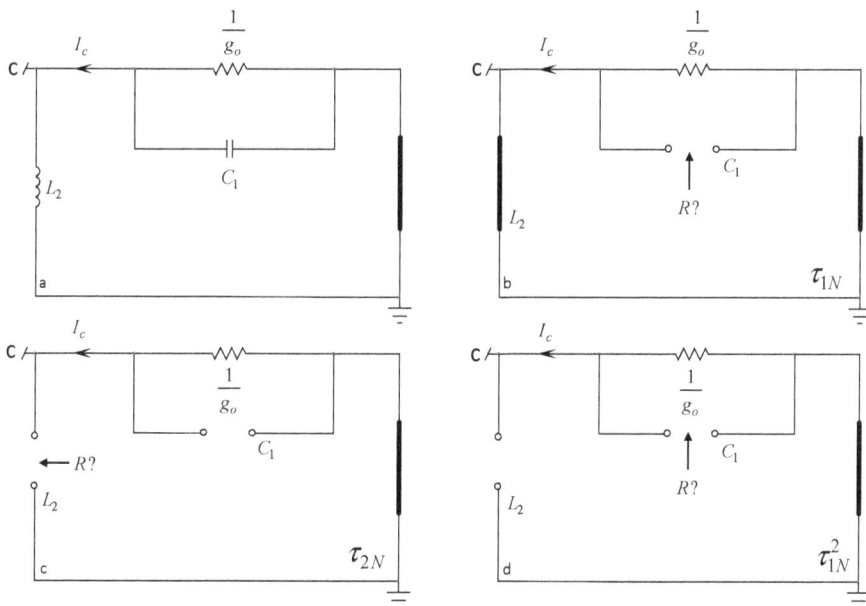

Figure 3.108 Determining the remaining zeroes is a simple exercise.

In this expression, $D(s)$ is the third order denominator defined in (3.207). We thus further rearrange Z_{out} by bringing in the simplified denominator given in (3.209):

$$Z_{out}(s) \approx R_0 \frac{1+\dfrac{s}{\omega_{z_1}} \quad 1+\dfrac{s}{\omega_{0N}Q_z}+\left(\dfrac{s}{\omega_{0N}}\right)^2}{1+\dfrac{s}{\omega_{p_1}} \quad 1+\dfrac{s}{\omega_n Q_p}+\left(\dfrac{s}{\omega_n}\right)^2} \qquad (3.253)$$

The right-side term is actually equal to 1 as both numerator and denominator are equal but written in a different format. Capitalizing on this fact, we end-up with fairly simple equation for Z_{out}:

$$Z_{out}(s) \approx R_0 \frac{1+\dfrac{s}{\omega_{z_1}}}{1+\dfrac{s}{\omega_{p_1}}} \qquad (3.254)$$

With ω_{p_1} defined by (3.210).

The Mathcad® plot of these transfer functions appear in Figure 3.109 and compares the complete expression from (3.252) with (3.254) and shows a good agreement between the curves.

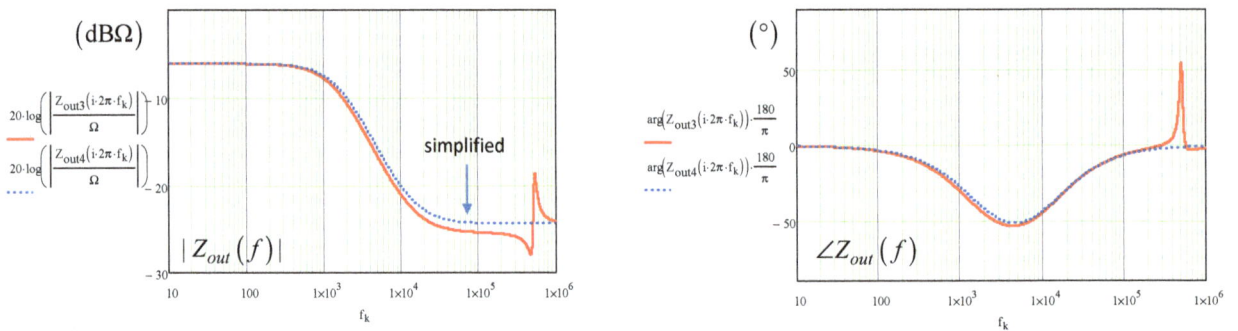

Figure 3.109 The two transfer functions – full-blown formula and the simplified one – deliver responses close to each other.

Finally, the simulation of Figure 3.104 gives a set of Z_{out} curves reproduced in Figure 3.110: they are in perfect agreement with equation (3.252) we have derived.

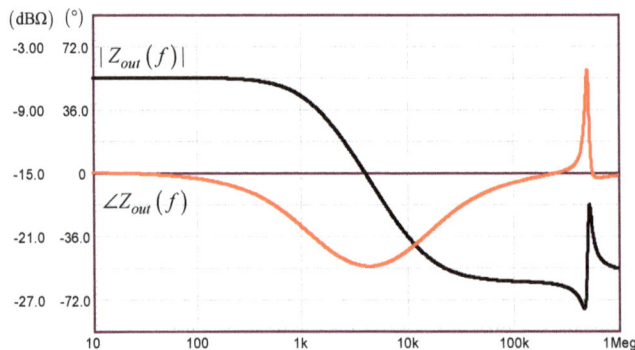

Figure 3.110 The output impedance response of the CCM-operated CM boost converter combines a zero and a pole.

3.3.3 Input Impedance

The input impedance is determined by installing the current source I_T as depicted by Figure 3.111: the big inductor LoL authorizes the bias point calculation (it is a short circuit during the operating point calculation) and later blocks the ac current once the sweep has started. As a result, the voltage observed across the 1-A current source is the image – the response – of the open-loop input impedance.

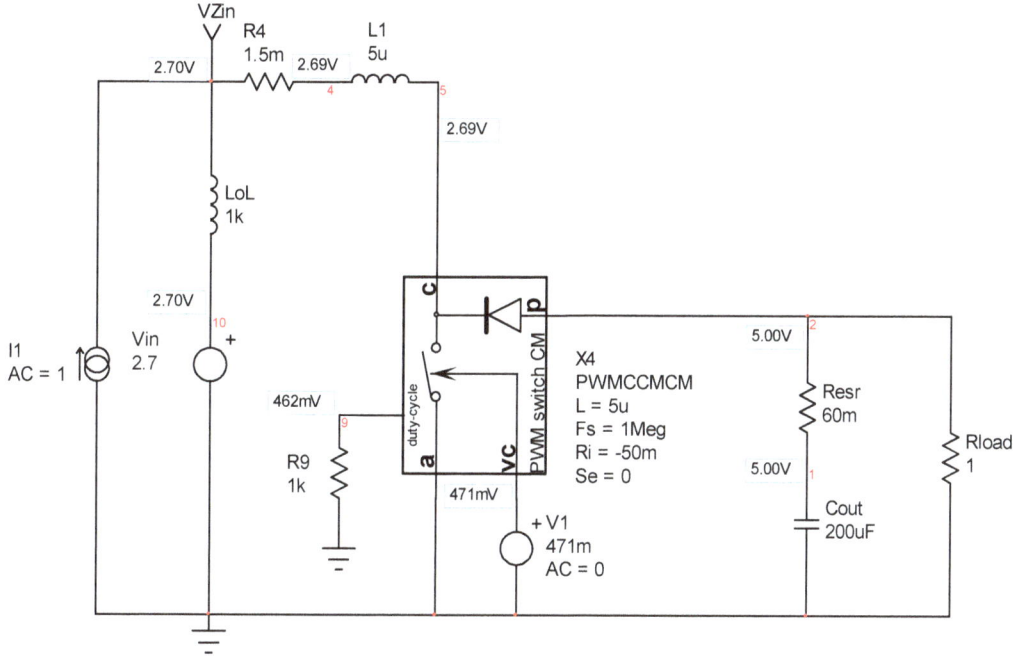

Figure 3.111 The input impedance is obtained by installing an ac-blocking high-value inductor in series with the input source.

The rearranged small-signal version of this configuration appears in Figure 3.112 with all zeroed sources already removed. As you can see, L_2 is driven by the stimulus I_T and its state variable is no longer independent: this is a degenerate case and L_2 won't participate in the denominator expression.

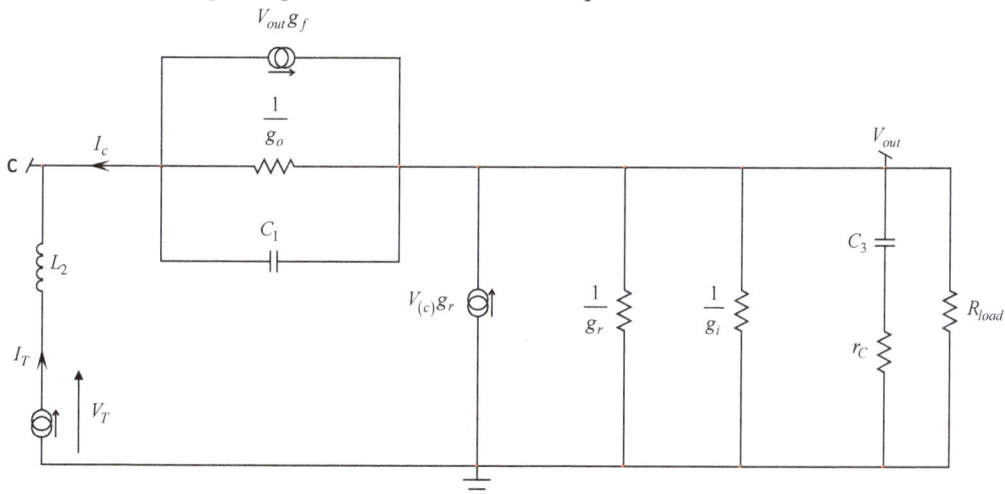

Figure 3.112 The small-signal circuit shows that inductor L_2 appears in series with the current source: the denominator loses one degree and will be of second order.

We can start by shorting the inductor and opening all capacitors to obtain the incremental input resistance R_0. The schematic diagram appears in Figure 3.113. We can define the test current I_T as follows:

$$I_T = V_{out} g_f + (V_T - V_{out}) g_o \tag{3.255}$$

I_T is also the output current flowing in the load:

$$\frac{V_{out}}{R_{eq2}} = I_T + V_T g_r \tag{3.256}$$

Solving for V_{out} gives:

$$V_{out} = R_{eq2} (I_T + V_T g_r) \tag{3.257}$$

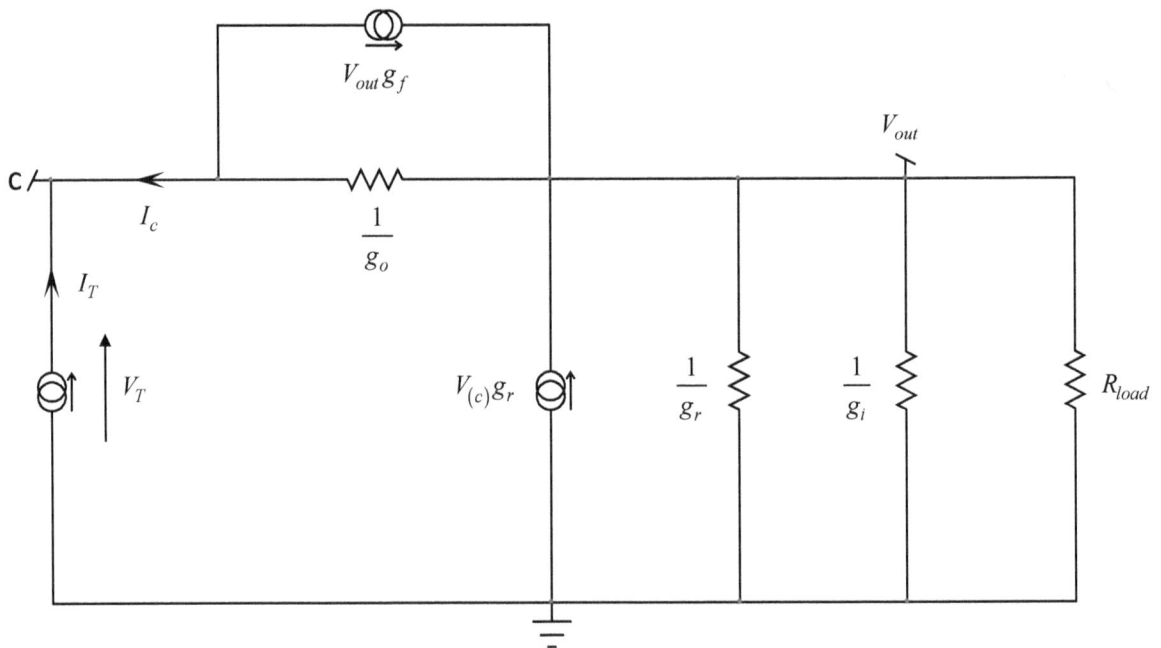

Figure 3.113 The incremental input resistance is determined after shorting the inductor and opening the capacitors.

Substitute (3.257) in (3.255) then solve for V_T, rearrange and obtain:

$$R_0 = \frac{V_T}{I_T} = \frac{1 + R_{eq2}(g_o - g_f)}{g_o + R_{eq2} g_r (g_f - g_o)} \tag{3.258}$$

Where R_{eq2} has been defined in (3.186). If you now compute this resistance with a zeroed compensation ramp, you actually find a negative value: the open-loop incremental resistance of the current-mode CCM-operated boost converter is negative. We have found the same result with the buck operated in similar open-loop conditions. The phenomenon is similar: the input voltage increase reduces the on-time for the same peak current setpoint (the slope V_{in}/L is steeper). Therefore, the off-time duration naturally increases in a converter operated at a fixed switching frequency. With a longer demagnetization time, the inductor valley current is lower and so is the average current. As the input current is the inductor current in a boost converter, the input current goes down. Figure 3.114 illustrates this fact.

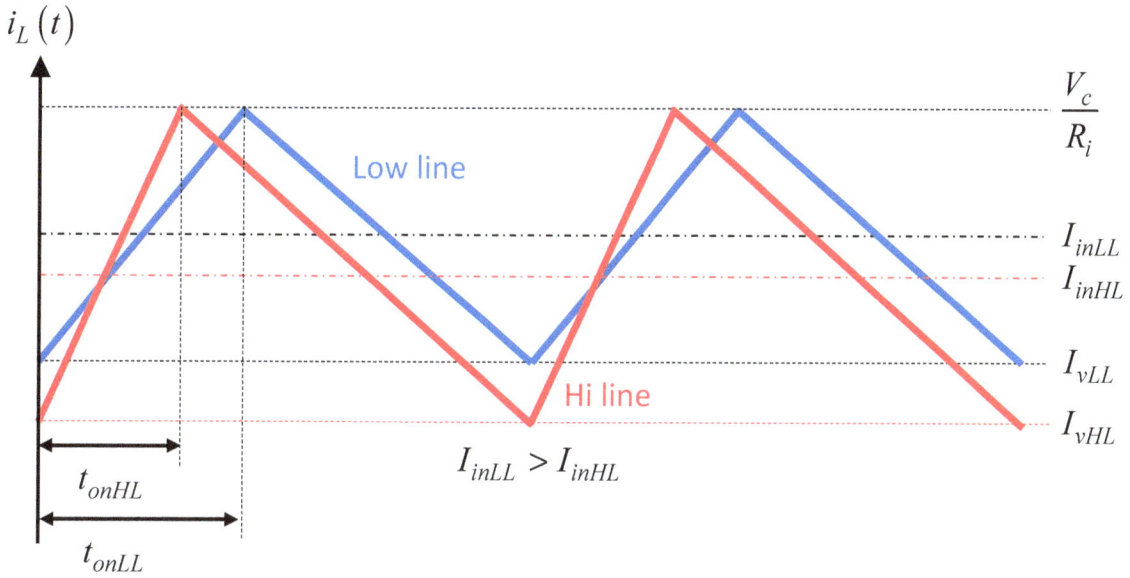

Figure 3.114 The input current decreases when the input voltage increases. This is typical of a negative incremental resistance.

To determine the time constants, we turn the excitation off which is similar to open-circuiting the test generator in Figure 3.112. The natural structure is modified and we cannot reuse the denominator already determined. The updated diagram to determine the time constants appears in Figure 3.115. If we start with the resistance looking into L_2's ports, it is infinite as the left node is floating. Therefore:

$$\tau_2 = \frac{L_2}{\infty} = 0 \tag{3.259}$$

If we now look at the resistance driving capacitor C_1, we can redraw the circuit to make it simpler to analyze. This is what Figure 3.116 proposes in sketch (a). To further simplify the circuit, we can remove the conductance g_o which will come later in parallel with the intermediate result. Finally, as a first approach, I will install a finite resistance from the low-side connection of L_2 to close the current path to ground (sketch b). This resistance will then approach infinity once factored in the final expression.

Observing this circuit, we have a few equations to describe it:

$$V_{out} = R_{eq2}\left(V_{(c)}g_r + I_T + V_{out}g_f\right) \tag{3.260}$$

$$V_T = V_{out} - V_c \tag{3.261}$$

$$I_c = -\left(V_{out}g_f + I_T\right) \tag{3.262}$$

$$V_{(c)} = I_c R_{\inf} \tag{3.263}$$

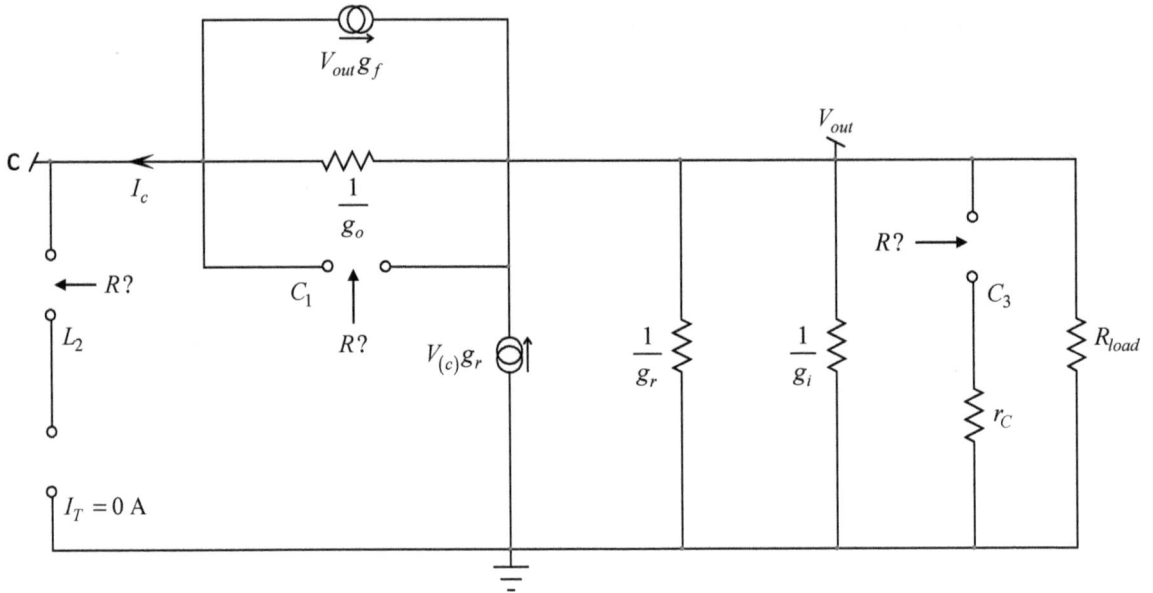

Figure 3.115 Open-circuiting the current source leaves the inductor floating.

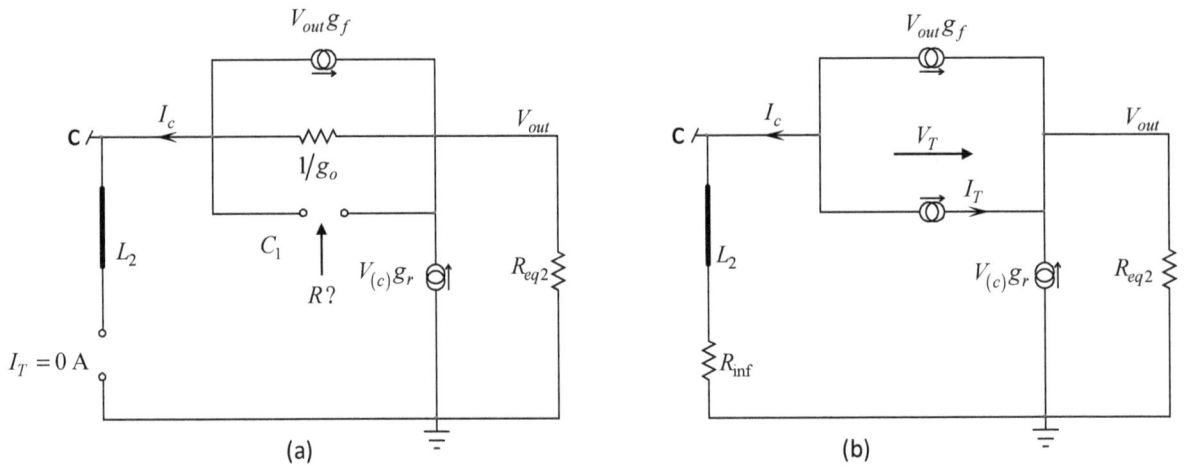

Figure 3.116 We can close the current to ground via R_{inf} and make the analysis simpler.

Combining (3.262) and (3.263), we have:

$$V_{(c)} = -\left(V_{out}g_f + I_T\right)R_{inf} \tag{3.264}$$

Now substituting this definition in (3.260) and solving for V_{out} gives:

$$V_{out} = -\frac{R_{eq2}\left(I_T - I_T R_{inf}g_r\right)}{R_{eq2}\left(g_f - R_{inf}g_f g_r\right) - 1} \tag{3.265}$$

Then plug V_{out} in (3.264) and rearrange to obtain:

$$V_{(c)} = -\frac{I_T R_{inf}}{R_{eq2}R_{inf}g_f g_r - R_{eq2}\,g_f + 1} \tag{3.266}$$

Finally, substitute both expressions in (3.261) and find:

$$V_T = \frac{R_{eq2} + R_{inf} - R_{eq2} R_{inf} g_r}{R_{eq2} R_{inf} g_f g_r - R_{eq2} g_f + 1} \qquad (3.267)$$

Now if R_{inf} approaches infinity and you bring g_o back in the picture, the time constant associated with C_1 is:

$$\tau_1 = C_1 \left(\frac{1}{g_o} \, \| \, \frac{1 - R_{eq2} g_r}{R_{eq2} g_f g_r} \right) \qquad (3.268)$$

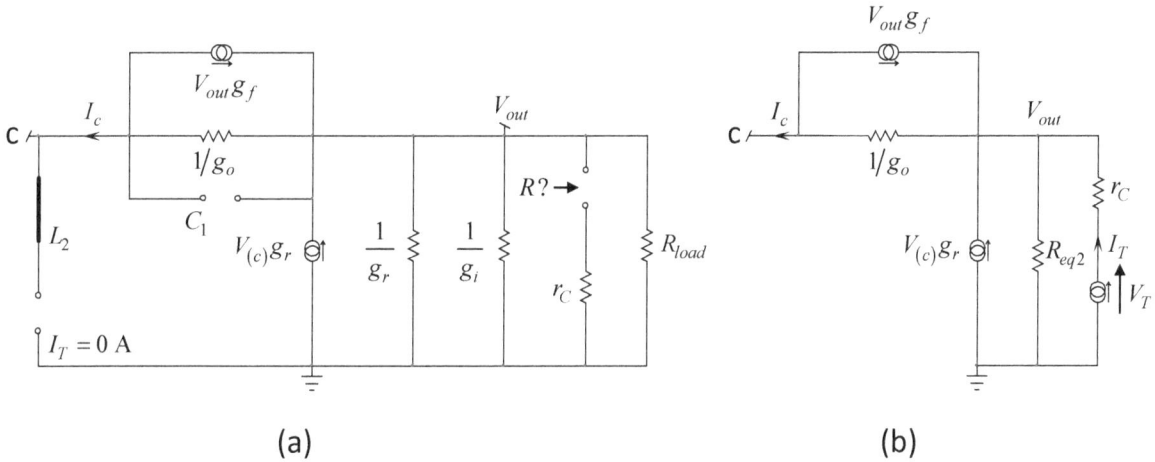

(a)   (b)

Figure 3.117 To determine the resistance "seen" from C_3's connections, you can further simplify the circuit by temporarily removing some elements such as r_C and R_{eq2}.

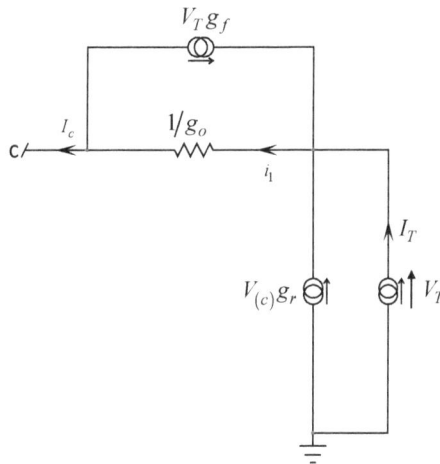

Figure 3.118 The remaining circuit is really simple to solve.

To determine the time constant involving C_3, the circuit evolves into that of Figure 3.117. To simplify the work, we see that the resistance to determine will be in series with r_C. We can thus remove r_C and sum it up with the intermediate result later. Also, be removing r_C, we see that the resistance R_{eq2} comes in parallel with the intermediate resistance. Again, we can temporarily disconnect it and bring it back at the end. Finally, the circuit we will study is that of Figure 3.118. We can first observe that the current i_1 flowing in the conductance g_o is that delivered by the current source $V_T g_f$. The voltage at node c is thus defined as:

$$V_{(c)} = V_T - i_1 \frac{1}{g_o} = V_T \left(1 - g_f \frac{1}{g_o} \right)$$

(3.269)

Current I_T is set by the second current source $V_{(c)} g_r$:

$$I_T = -V_{(c)} g_r$$

(3.270)

Substituting (3.269) into the above equation and solving for V_T leads to:

$$\frac{V_T}{I_T} = \frac{g_o}{g_r \left(g_f - g_o \right)}$$

(3.271)

Bringing the other elements back, we have the final time constant we want:

$$\tau_3 = C_3 \left[r_C + R_{eq2} \parallel \frac{g_o}{g_f \left(g_r - g_o \right)} \right]$$

(3.272)

We can form the first coefficient of our denominator expression:

$$b_1 = \tau_1 + \tau_2 + \tau_3 = C_1 \left(\frac{1}{g_o} \parallel \frac{1 - R_{eq2} g_r}{R_{eq2} g_f g_r} \right) + \frac{L_2}{\infty} + C_3 \left[r_C + R_{eq2} \parallel \frac{g_o}{g_f \left(g_r - g_o \right)} \right]$$

$$= C_1 \left(\frac{1}{g_o} \parallel \frac{1 - R_{eq2} g_r}{R_{eq2} g_f g_r} \right) + C_3 \left[r_C + R_{eq2} \parallel \frac{g_o}{g_f \left(g_r - g_o \right)} \right]$$

(3.273)

The second term, b_2, is defined as follows:

$$b_2 = \tau_1 \tau_2^1 + \tau_1 \tau_3^1 + \tau_2 \tau_3^2$$

(3.274)

The beauty of the fast analytical circuits techniques is that you can reshuffle the time constant combinations in the above expressions to obtain easier-to-solve circuits. As shown in [1], you can rewrite this expression as follows:

$$b_2 = \tau_2 \tau_1^2 + \tau_1 \tau_3^1 + \tau_2 \tau_3^2$$

(3.275)

We see that τ_2 is actually factoring two of the terms, τ_1^2 and τ_3^2. The first one corresponds to the circuit observed when L_2 is in its high-frequency state while we look at the resistance offered by C_1. The result is already found and it is (3.268) because L_2 is floating and plays no role regardless of its state. Now, because τ_2 is equal to zero and factors a finite term, then the product is also zero. For the second term, τ_3^2, it is already determined in (3.272). Same result, its product by τ_2 is also equal to zero. As a matter of fact, the term b_2 simplifies to:

$$b_2 = \tau_1 \tau_3^1$$

(3.276)

What resistance is offered from C_3's connections while C_1 is replaced by a short circuit? This is the circuit proposed in Figure 3.119. Again, we can remove r_C and bring it back later on. The left-side conductance and current generator are disabled by the short circuit of C_1 and the voltage at terminal c is equal to V_T. The current generator $V_{(c)}g_r$ is actually equal to $V_T g_r$ and its current is absorbed entirely by the conductance g_i: these two elements go off the circuit and we are left with conductance g_i in parallel with the load resistance. The time constant we want is thus:

$$\tau_3^1 = C_3 \left(\frac{1}{g_i} \| R_{load} + r_C \right) \tag{3.277}$$

The term b_2 is thus defined as:

$$b_2 = \tau_1 \tau_3^1 = C_1 \left(\frac{1}{g_o} \| \frac{1 - R_{eq2}g_r}{R_{eq2}g_f g_r} \right) C_3 \left(\frac{1}{g_i} \| R_{load} + r_C \right) \tag{3.278}$$

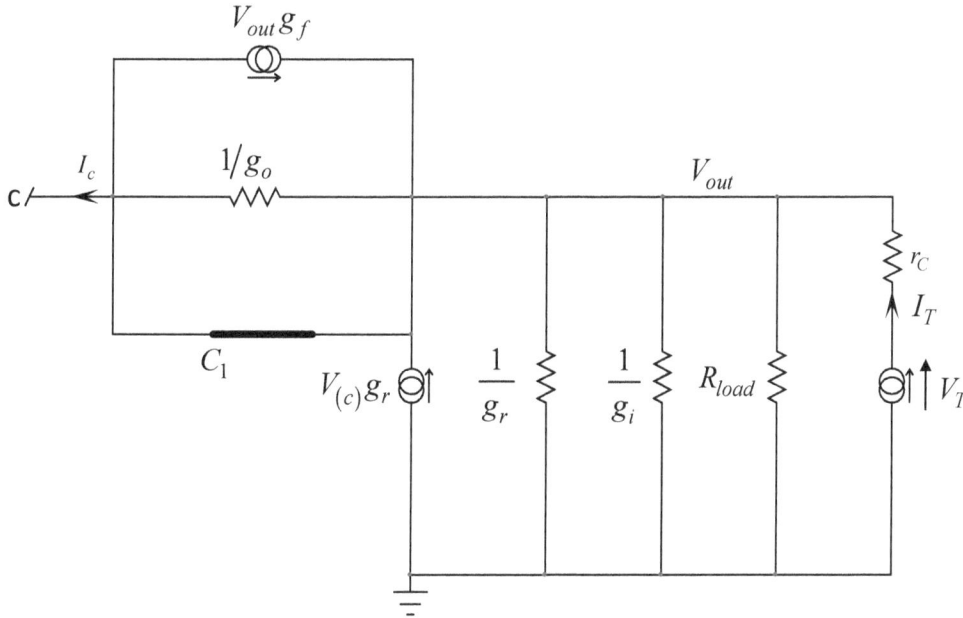

Figure 3.119 One single configuration is necessary to determine b_2.

The denominator determination is completed and equal to:

$$D(s) = 1 + b_1 s + b_2 s^2 = 1 + s \left[C_1 \left(\frac{1}{g_o} \| \frac{1 - R_{eq2}g_r}{R_{eq2}g_f g_r} \right) + C_3 \left(r_C + R_{eq2} \| \frac{g_o}{g_f (g_r - g_o)} \right) \right]$$
$$+ s^2 C_1 \left(\frac{1}{g_o} \| \frac{1 - R_{eq2}g_r}{R_{eq2}g_f g_r} \right) C_3 \left(\frac{1}{g_i} \| R_{load} + r_C \right) \tag{3.279}$$

Considering a low quality factor, this expression can be advantageously factored by the product of two poles:

$$D(s) \approx \left(1 + \frac{s}{\omega_{p_1}} \right) \left(1 + \frac{s}{\omega_{p_2}} \right) \tag{3.280}$$

with:

$$\omega_{p_1} = \frac{1}{b_1} \tag{3.281}$$

$$\omega_{p_2} = \frac{b_1}{b_2} \tag{3.282}$$

To determine the zeroes, we null the response in Figure 3.112. Nulling the response across the current source is similar to replacing it by a short circuit. By doing so, the circuit returns in its natural state as already analyzed in the previous transfer functions determination. The denominator obtained in (3.209) is thus our numerator in this particular configuration. The input impedance transfer function is given by:

$$Z_{in}(s) = R_0 \frac{1 + \dfrac{s}{\omega_{z_1}} \left(1 + \dfrac{s}{\omega_{0N} Q_Z} + \left(\dfrac{s}{\omega_{0N}} \right)^2 \right)}{1 + \dfrac{s}{\omega_{p_1}} \left(1 + \dfrac{s}{\omega_{p_2}} \right)} \tag{3.283}$$

with:

$$R_0 = \frac{1 + R_{eq2}(g_o - g_f)}{g_o + R_{eq2} g_r (g_f - g_o)} \tag{3.284}$$

$$\omega_{z_1} \approx \frac{\dfrac{2}{R_{load}} + \dfrac{T_{sw}}{L_2 M^3} \left(1 + \dfrac{S_e}{S_n} \right)}{C_3} \tag{3.285}$$

$$Q_Z = \frac{1}{\pi (m_c D' - 0.5)} \tag{3.286}$$

$$\omega_{0N} = \frac{\pi}{T_{sw}} \tag{3.287}$$

$$m_c = 1 + \frac{S_e}{S_n} \tag{3.288}$$

$$\omega_{p_1} = \frac{1}{C_1 \left(\dfrac{1}{g_o} \| \dfrac{1 - R_{eq2} g_r}{R_{eq2} g_f g_r} \right) + C_3 \left(r_C + R_{eq2} \| \dfrac{g_o}{g_f (g_r - g_o)} \right)} \tag{3.289}$$

$$\omega_{p_2} = \frac{C_1 \left(\dfrac{1}{g_o} \| \dfrac{1 - R_{eq2} g_r}{R_{eq2} g_f g_r} \right) + C_3 \left(r_C + R_{eq2} \| \dfrac{g_o}{g_f (g_r - g_o)} \right)}{C_1 \left(\dfrac{1}{g_o} \| \dfrac{1 - R_{eq2} g_r}{R_{eq2} g_f g_r} \right) C_3 \left(\dfrac{1}{g_i} \| R_{load} + r_C \right)} \tag{3.290}$$

$$R_{eq2} = R_{load} \parallel \frac{1}{g_i} \parallel \frac{1}{g_r} \tag{3.291}$$

We have plotted the response of equation (3.283) and it appears in Figure 3.120. The phase of the transfer function is negative at low frequencies as explained in the above lines. There are two curves: one is plotting the complete expression using (3.279) while the dashed lines are those corresponding to the transfer function using the low-Q approximation of (3.280). They are very similar in shape. To verify these curves, we have graphed the results obtained by simulating Figure 3.111.

They appear in Figure 3.121 and confirm our analyses.

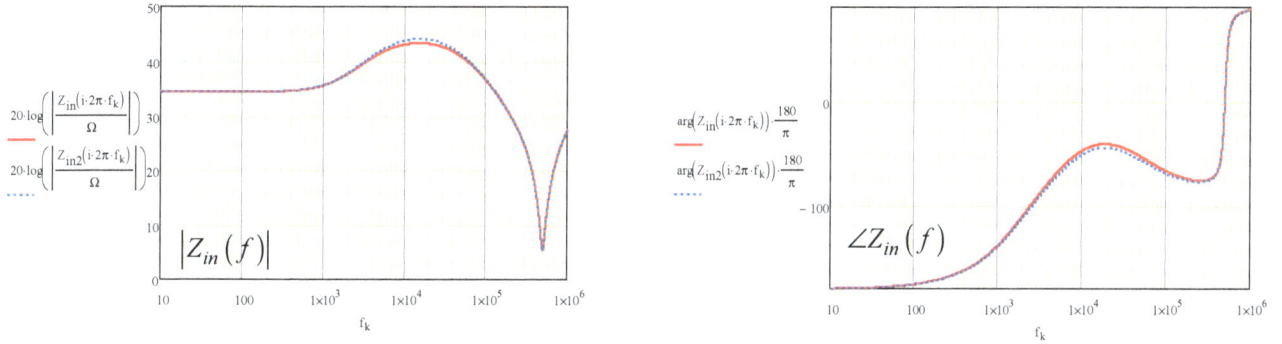

Figure 3.120 The input impedance shows a negative argument at low frequencies.

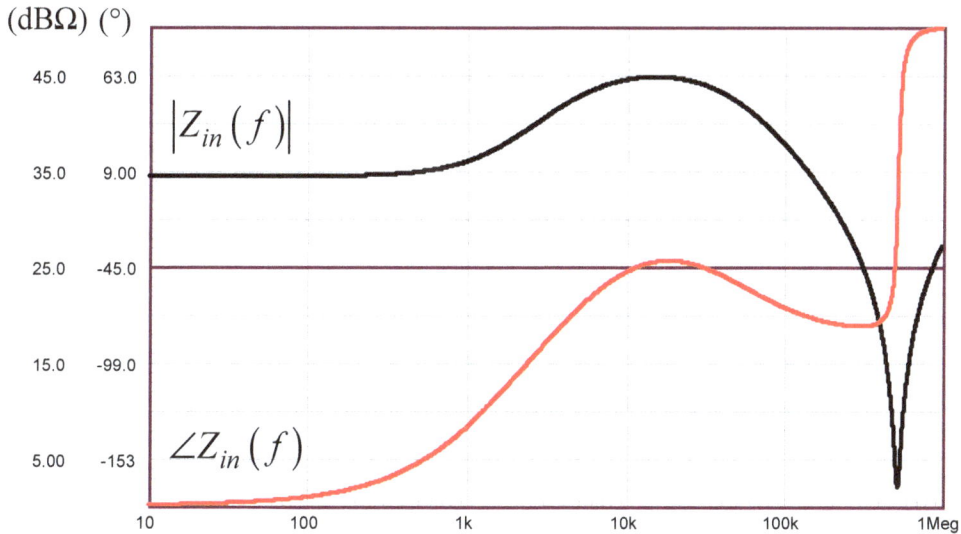

Figure 3.121 The SPICE simulation confirms the results obtained with Mathcad®.

We have determined the four transfer functions of the current-mode-controlled boost converter operated in CCM. These transfer functions are summarized in Figure 3.122.

$\dfrac{V_{out}(s)}{V_{err}(s)}$ Control to Output	$H_0\dfrac{\left(1+\frac{s}{\omega_{z_1}}\right)\left(1-\frac{s}{\omega_{z_2}}\right)}{1+\frac{s}{\omega_{p_1}}}\dfrac{1}{1+\frac{s}{\omega_n Q_p}+\left(\frac{s}{\omega_n}\right)^2}$	$\omega_{z_1}=\dfrac{1}{r_C C_3}$ $\omega_{z_2}\approx\dfrac{(1-D)^2 R_{load}}{L_2}$	$\omega_{p_1}\approx\dfrac{\frac{2}{R_{load}}+\frac{T_{sw}}{L_2 M^3}\left(1+\frac{S_e}{S_n}\right)}{C_3}$ $Q_p=\dfrac{1}{\pi(m_c D'-0.5)}$ $\omega_n=\dfrac{\pi}{T_{sw}}$	$H_0\approx\dfrac{R_{load}}{R_i}\dfrac{1}{2M+\frac{1}{\tau_L M^2}\left(\frac{1}{2}+\frac{S_e}{S_n}\right)}$
$\dfrac{V_{out}(s)}{V_{in}(s)}$ Input to Output	$H_0\dfrac{1+\frac{s}{\omega_{z_1}}}{1+\frac{s}{\omega_{p_1}}}\dfrac{1}{1+\frac{s}{\omega_n Q_p}+\left(\frac{s}{\omega_n}\right)^2}$	$\omega_{z_1}=\dfrac{1}{r_C C_3}$	$\omega_{p_1}\approx\dfrac{\frac{2}{R_{load}}+\frac{T_{sw}}{L_2 M^3}\left(1+\frac{S_e}{S_n}\right)}{C_3}$ $Q_p=\dfrac{1}{\pi(m_c D'-0.5)}$ $\omega_n=\dfrac{\pi}{T_{sw}}$	$H_0=\dfrac{g_o+g_r}{g_o-g_f+\frac{1}{R_{eq2}}}$
$Z_{in}(s)$ Input impedance	$R_0\dfrac{\left(1+\frac{s}{\omega_{z_1}}\right)\left(1+\frac{s}{\omega_{0N}Q_Z}+\left(\frac{s}{\omega_{0N}}\right)^2\right)}{\left(1+\frac{s}{\omega_{p_1}}\right)\left(1+\frac{s}{\omega_{p_2}}\right)}$	$\omega_{z_1}\approx\dfrac{\frac{2}{R_{load}}+\frac{T_{sw}}{L_2 M^3}\left(1+\frac{S_e}{S_n}\right)}{C_3}$ $Q_Z=\dfrac{1}{\pi(m_c D'-0.5)}$ $\omega_{0N}=\dfrac{\pi}{T_{sw}}$	$\omega_{p_1}\approx\dfrac{1}{C_1\left(\frac{1}{g_o}\|\frac{1-R_{eq2}g_r}{R_{eq2}g_f g_r}\right)+C_3\left(r_C+R_{eq2}\|\frac{g_o}{g_f(g_r-g_o)}\right)}$ $\omega_{p_2}\approx\dfrac{C_1\left(\frac{1}{g_o}\|\frac{1-R_{eq2}g_r}{R_{eq2}g_f g_r}\right)+C_3\left(r_C+R_{eq2}\|\frac{g_o}{g_f(g_r-g_o)}\right)}{C_1\left(\frac{1}{g_o}\|\frac{1-R_{eq2}g_r}{R_{eq2}g_f g_r}\right)C_3\left(\frac{1}{g_i}\|R_{load}+r_C\right)}$	$R_0=\dfrac{1+R_{eq2}(g_o-g_f)}{g_o+R_{eq2}g_r(g_f-g_o)}$
$Z_{out}(s)$ Output impedance	$R_0\dfrac{1+\frac{s}{\omega_{z_1}}}{1+\frac{s}{\omega_{p_1}}}$	$\omega_{z_1}=\dfrac{1}{r_C C_3}$	$\omega_{p_1}\approx\dfrac{\frac{2}{R_{load}}+\frac{T_{sw}}{L_2 M^3}\left(1+\frac{S_e}{S_n}\right)}{C_3}$	$R_0=\dfrac{1}{g_i-g_f+g_o+g_r+\frac{1}{R_{load}}}$

C_3 is the output capacitor, L_2 the inductor. C_1 is the resonating capacitor: $C_1=\dfrac{1}{L_2(\pi F_{sw})^2}$ $m_c=1+\dfrac{S_e}{S_n}$, M is V_{out}/V_{in} and $\tau_L=\dfrac{L_2}{R_{load}T_{sw}}$

Figure 3.122 The four transfer functions of the CCM CM boost converter operated in current-mode control are conveniently gathered in this picture.

3.4 Boost Transfer Functions in DCM – Fixed-Frequency Current-Mode Control

The DCM boost chosen for the analysis now delivers 15 V from a 10-V source and its large-signal schematic diagram derived in Chapter 1.4.2 appears in Figure 3.123. The resonating capacitor has disappeared because DCM operation is not subject to subharmonic oscillations. We have two energy-storing elements, this is a 2nd-order system.

The various sources compute the duty ratios d_1 and d_2 obtained with a control level V_c of 250 mV. We can now install the linearized sources determined in Chapter 1 for the buck converter. Please note that in the boost configuration of the DCM PWM CM switch model, the voltage applied across the inductor $V_{(a,cc)}$ during the on-time is positive and equal to V_{in}. The large-signal equation in B_1 which computes d_1 needs a negative sign as node a is grounded. The small-signal version of this circuit is given in Figure 3.124. Before proceeding with the analysis, we want to check that the dynamic responses of Figure 3.123 and Figure 3.124 lead to the exact same set of magnitude/phase curves. Figure 3.125 confirms simulation results between the large- and small-signal models are identical and we can start analyzing the small-signal circuit. First of all, we need to simplify and rearrange the structure into a friendlier shape. The input voltage in ac is 0 V hence the voltage at terminal cc is 0 V. Terminal a is grounded and sources including a reference to this node can be advantageously rewritten. Finally, after all these manipulations, the final circuit used to determine the control-to-output transfer function appears in Figure 3.126 and redrawn in Figure 3.127. We have verified that its dynamic response perfectly matches that already graphed in Figure 3.125. We can now start analyzing the circuit of Figure 3.127.

We start with the quasi-static gain H_0 obtained by shorting the inductor and opening the capacitor as shown in Figure 3.128. The voltage at terminal c is now zero volt as we have neglected the inductor ohmic drop.

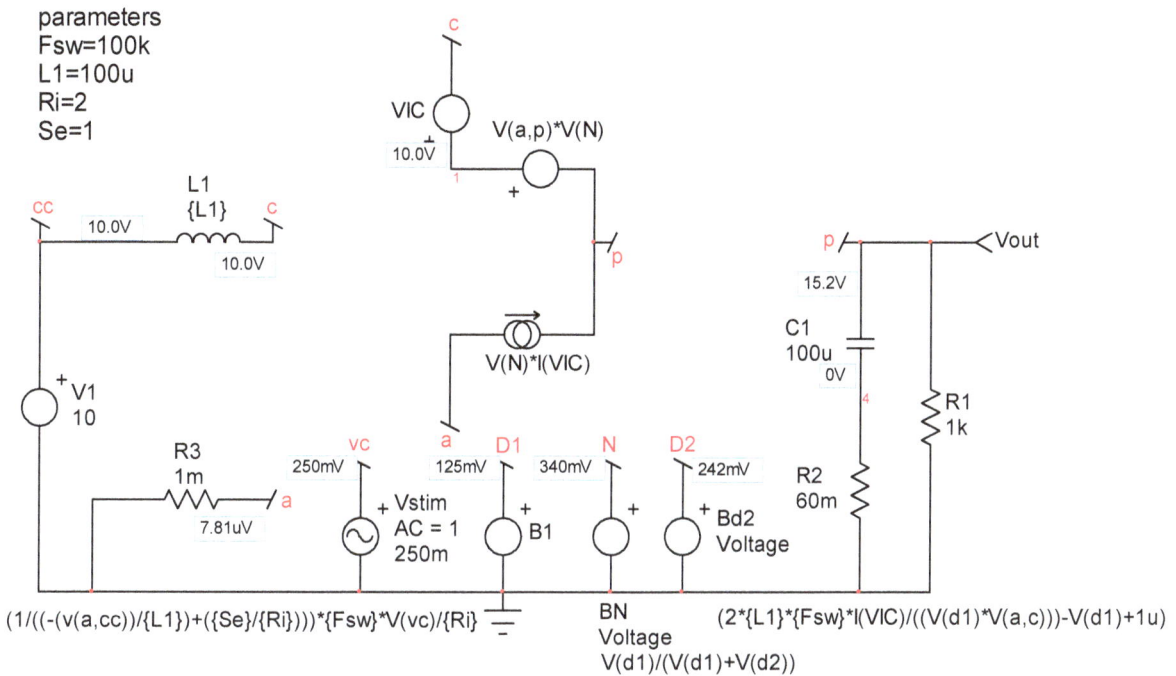

Figure 3.123 The current-mode-controlled large-signal boost converter now operates in discontinuous conduction mode.

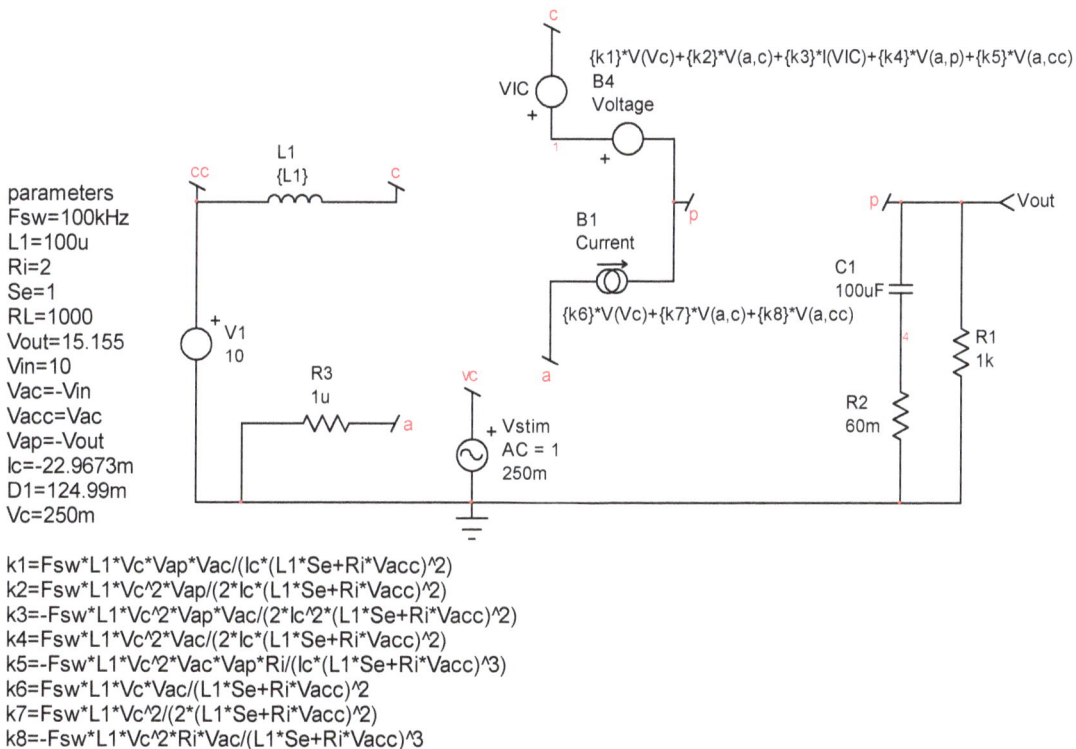

k1=Fsw*L1*Vc*Vap*Vac/(Ic*(L1*Se+Ri*Vacc)^2)
k2=Fsw*L1*Vc^2*Vap/(2*Ic*(L1*Se+Ri*Vacc)^2)
k3=-Fsw*L1*Vc^2*Vap*Vac/(2*Ic^2*(L1*Se+Ri*Vacc)^2)
k4=Fsw*L1*Vc^2*Vac/(2*Ic*(L1*Se+Ri*Vacc)^2)
k5=-Fsw*L1*Vc^2*Vac*Vap*Ri/(Ic*(L1*Se+Ri*Vacc)^3)
k6=Fsw*L1*Vc*Vac/(L1*Se+Ri*Vacc)^2
k7=Fsw*L1*Vc^2/(2*(L1*Se+Ri*Vacc)^2)
k8=-Fsw*L1*Vc^2*Ri*Vac/(L1*Se+Ri*Vacc)^3

Figure 3.124 This is the small-signal version of the DCM CM boost converter with all its coefficients.

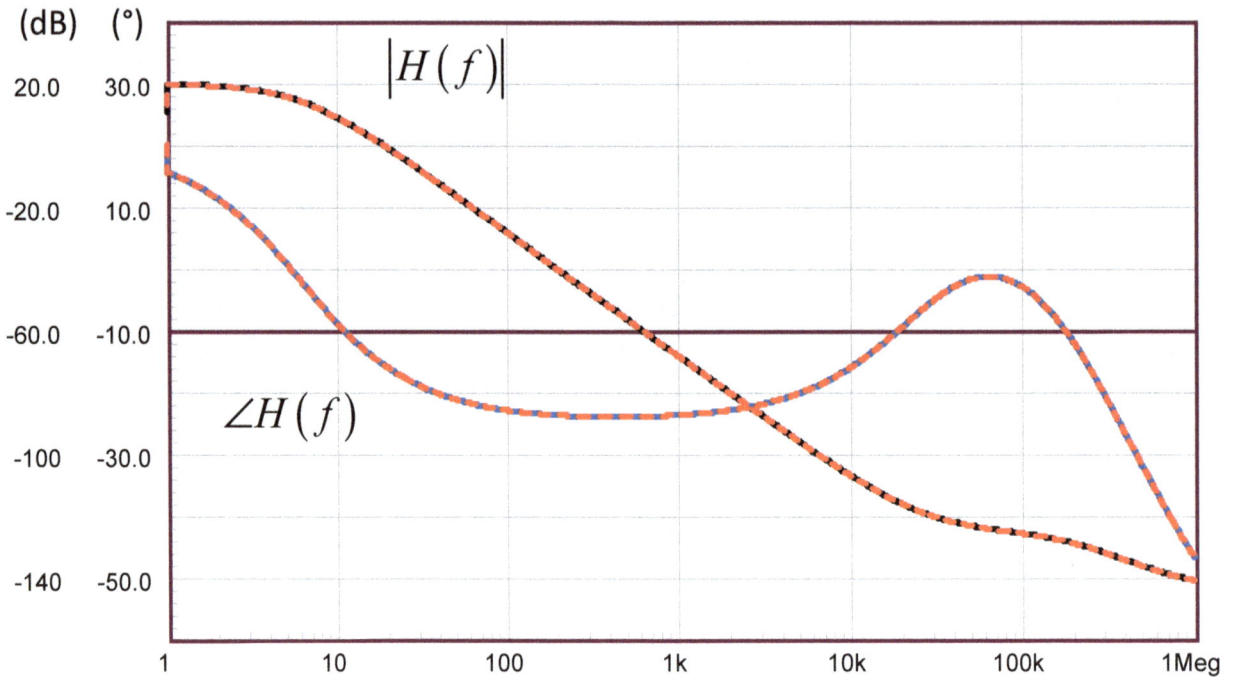

Figure 3.125 The dynamic responses of the large-signal model and that of the linearized version are identical.

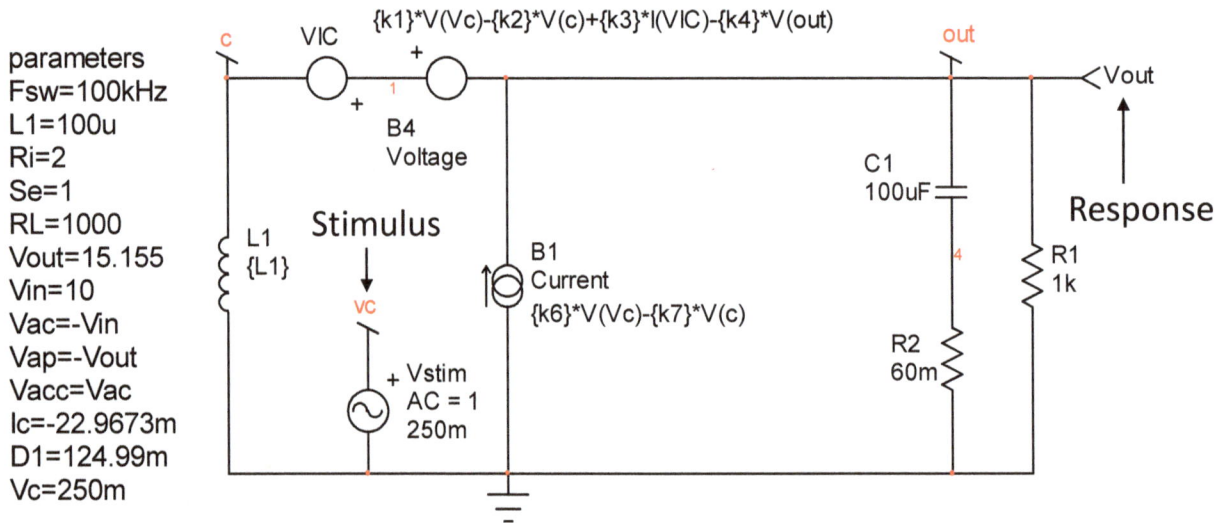

k1=Fsw*L1*Vc*Vap*Vac/(Ic*(L1*Se+Ri*Vacc)^2)
k2=Fsw*L1*Vc^2*Vap/(2*Ic*(L1*Se+Ri*Vacc)^2)
k3=-Fsw*L1*Vc^2*Vap*Vac/(2*Ic^2*(L1*Se+Ri*Vacc)^2)
k4=Fsw*L1*Vc^2*Vac/(2*Ic*(L1*Se+Ri*Vacc)^2)
k5=-Fsw*L1*Vc^2*Vac*Vap*Ri/(Ic*(L1*Se+Ri*Vacc)^3)
k6=Fsw*L1*Vc*Vac/(L1*Se+Ri*Vacc)^2
k7=Fsw*L1*Vc^2/(2*(L1*Se+Ri*Vacc)^2)
k8=-Fsw*L1*Vc^2*Ri*Vac/(L1*Se+Ri*Vacc)^3

Figure 3.126 Once all sources are rearranged, the circuit looks less intimidating to analyze.

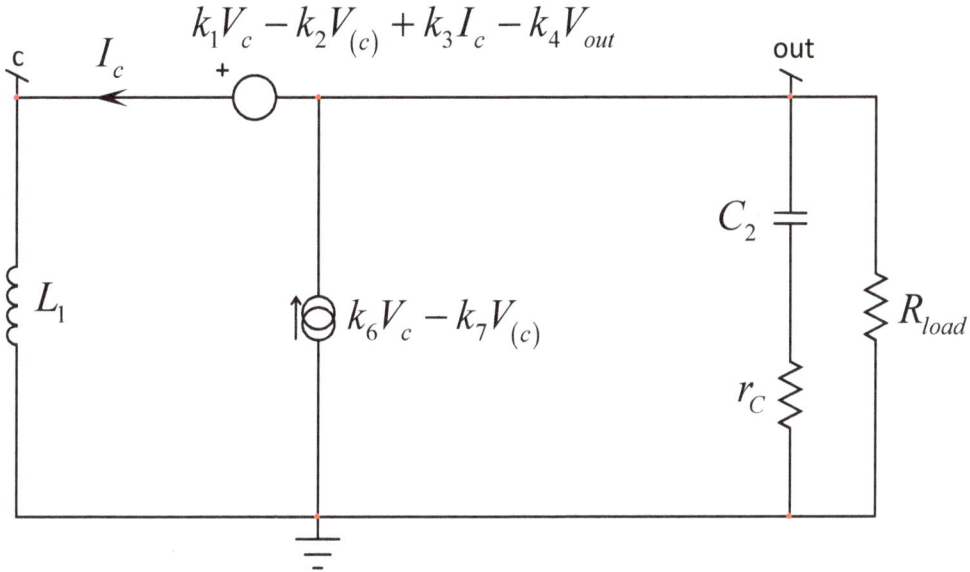

Figure 3.127 There are two energy-storing elements in this 2nd-order linear system.

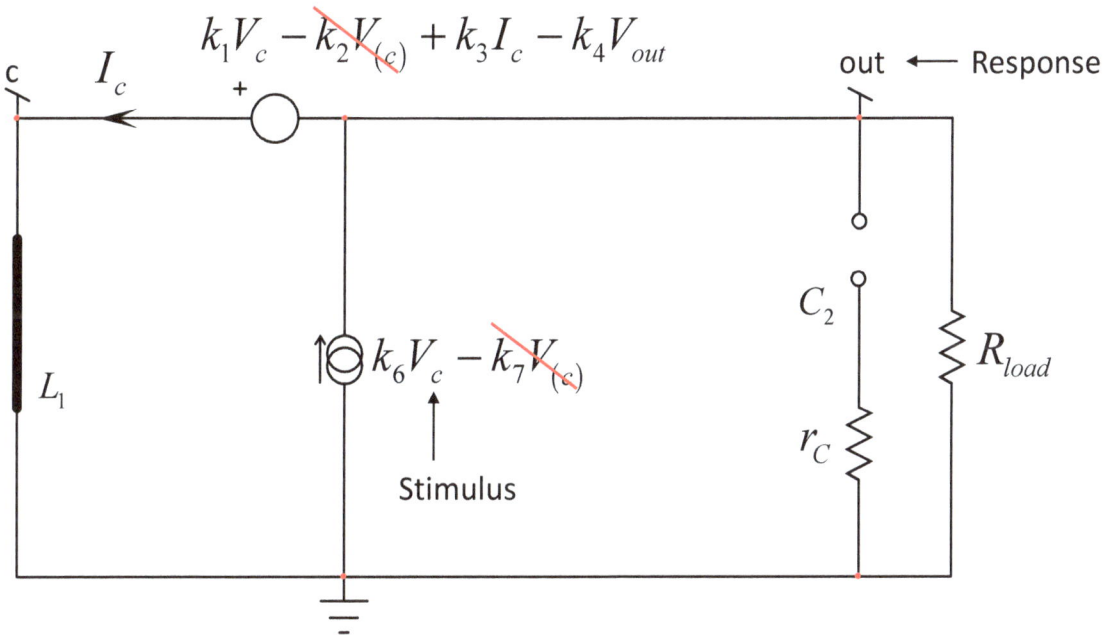

Figure 3.128 The inductor is shorted and the capacitor open-circuited to determine H_0.

We can write a first equation describing V_{out}:

$$V_{out} = -\left(k_1 V_c + k_3 I_c - k_4 V_{out}\right) \tag{3.292}$$

From which we extract the current flowing out of terminal c:

$$I_c = -\frac{V_{out} + V_c k_1 - V_{out} k_4}{k_3} \tag{3.293}$$

The current circulating in the load resistance is defined as:

$$\frac{V_{out}}{R_{load}} = k_6 V_c - I_c \tag{3.294}$$

If we substitute the definition of I_c in the above and solve for V_{out}, then, after rearranging, we find:

$$H_0 = \frac{V_{out}}{V_c} = \frac{R_{load}\left(k_1 + k_3 k_6\right)}{k_3 - R_{load}\left(1 - k_4\right)} \tag{3.295}$$

In the literature [2], this gain is presented as follows and is identical to the above expression:

$$H_0 = \frac{2V_{out}}{D}\frac{M-1}{2M-1}\frac{1}{S_n m_c T_{sw}} \tag{3.296}$$

Further rearranging this formula leads to the following different form:

$$H_0 = V_{in}\sqrt{\frac{2M}{\tau_L\left(M-1\right)}}\frac{M-1}{2M-1}\frac{1}{S_n m_c T_{sw}} \tag{3.297}$$

with:

$$M = \frac{1+\sqrt{1+\dfrac{2D^2}{\tau_L}}}{2} \tag{3.298}$$

$$D = \frac{F_{sw}V_c}{R_i}\frac{1}{\dfrac{S_e}{R_i}-\dfrac{V_{in}}{L_1}} \tag{3.299}$$

$$\tau_L = \frac{L_1}{R_{load}T_{sw}} \tag{3.300}$$

$$m_c = 1 + \frac{S_e}{S_n} \tag{3.301}$$

V_c is the control voltage setting the peak current setpoint, 250 mV in this example. S_e is the external stabilizing ramp expressed in volts per seconds while the on-slope scaled by the sense resistance R_i is given by:

$$S_n = \frac{V_{in}}{L_1}R_i \tag{3.302}$$

Now that we have the quasi-static gain, we can look at the various time constants involving L_1 and C_2. First, we turn the excitation off: V_c is made 0 V and all sources containing a reference to it are simplified accordingly. Starting with L_1, we have to analyze the circuit from Figure 3.129. In this figure, considering a zeroed excitation V_c, expressions are simplified.

The current source is reversed to keep a positive current definition. In this mode, the test current I_T flows in the opposite direction of I_c:

$$I_T = -I_c \tag{3.303}$$

The voltage V_T is defined by:

$$V_T = k_3 I_c - k_4 V_{out} - k_2 V_T + V_{out} \tag{3.304}$$

The output voltage involves the load resistance R_{load}:

$$V_{out} = -R_{load}\left(k_7 V_T - I_T\right) \tag{3.305}$$

If you now substitute the above expression in (3.304) while I_c is replaced by $-I_T$, then after factoring and rearranging, you should find the resistance "seen" from the inductor port. The time constant involving L_1 is thus:

$$\tau_1 = \frac{L_1}{\dfrac{R_{load}\left(1-k_4\right)-k_3}{R_{load}k_7\left(1-k_4\right)+1+k_2}} \tag{3.306}$$

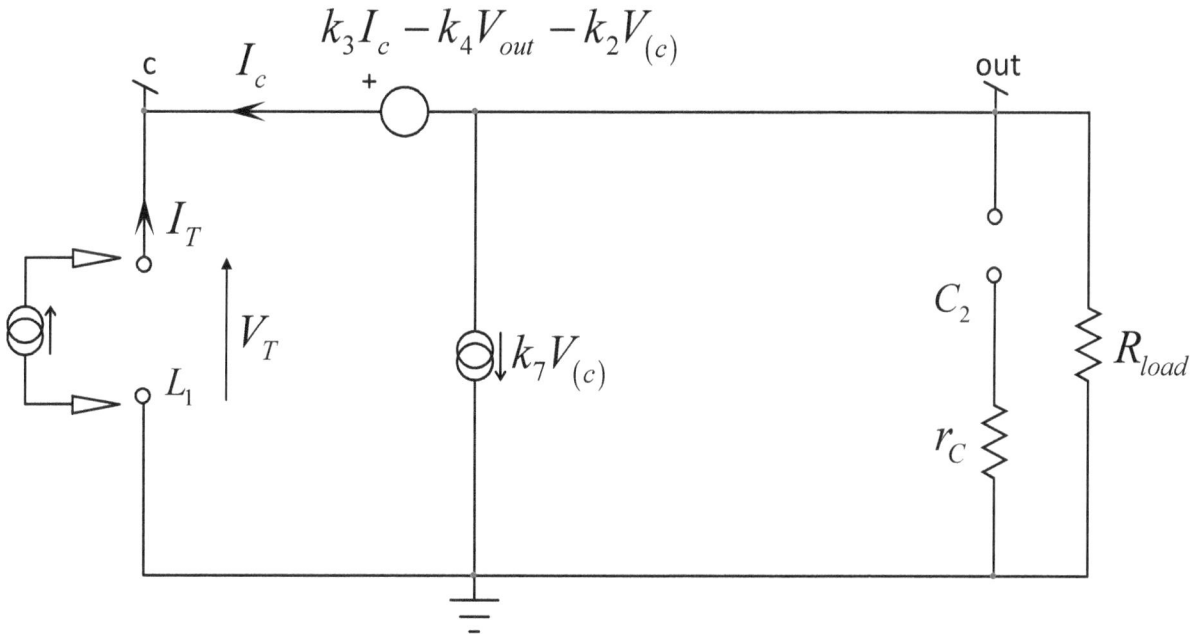

Figure 3.129 The circuit to determine the first time constant simplifies considering a zeroed excitation.

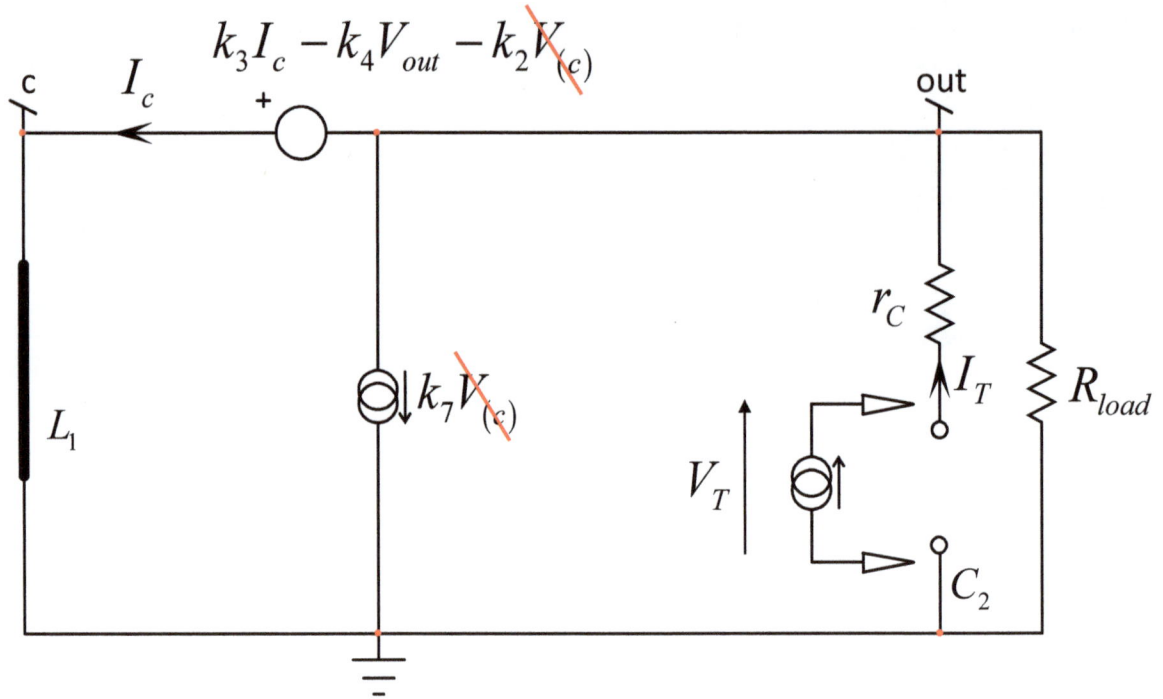

Figure 3.130 The inductor is now shorted to determine the time constant involving C_2.

Figure 3.130 shows the circuitry to determine the second time constant. As we did in the previous examples, r_C will be temporarily disconnected and brought back in series with the intermediate result determined without it. By doing so, V_T, the voltage across the current source, becomes the voltage at node *out*. Also, as inductor L_1 is shorted for the analysis, the voltage at terminal c is 0 V, further simplifying expressions. We can write:

$$V_T = -\left(k_3 I_c - k_4 V_{out}\right) \tag{3.307}$$

and:

$$V_T = -\left(I_c - I_T\right) R_{load} \tag{3.308}$$

Now extract I_c from the above expression, plug it into (3.307), solve for V_T, factor I_T. Then add r_C the capacitor ESR and you have:

$$\tau_2 = C_2 \left(\frac{k_3}{\dfrac{k_3}{R_{load}} - 1 + k_4} + r_C \right) \tag{3.309}$$

We have the first low-frequency term in the definition of the denominator:

$$b_1 = \tau_1 + \tau_2 = \frac{L_1}{\dfrac{R_{load}\left(1 - k_4\right) - k_3}{R_{load} k_7 \left(1 - k_4\right) + 1 + k_2}} + C_2 \left(\frac{k_3}{\dfrac{k_3}{R_{load}} - 1 + k_4} + r_C \right) \tag{3.310}$$

For the high-frequency term, we will calculate τ_2^1 in which L_1 is set in its high-frequency state (an open circuit) while we look at the resistance offered by C_2's connections. This is the circuit given in Figure 3.131.

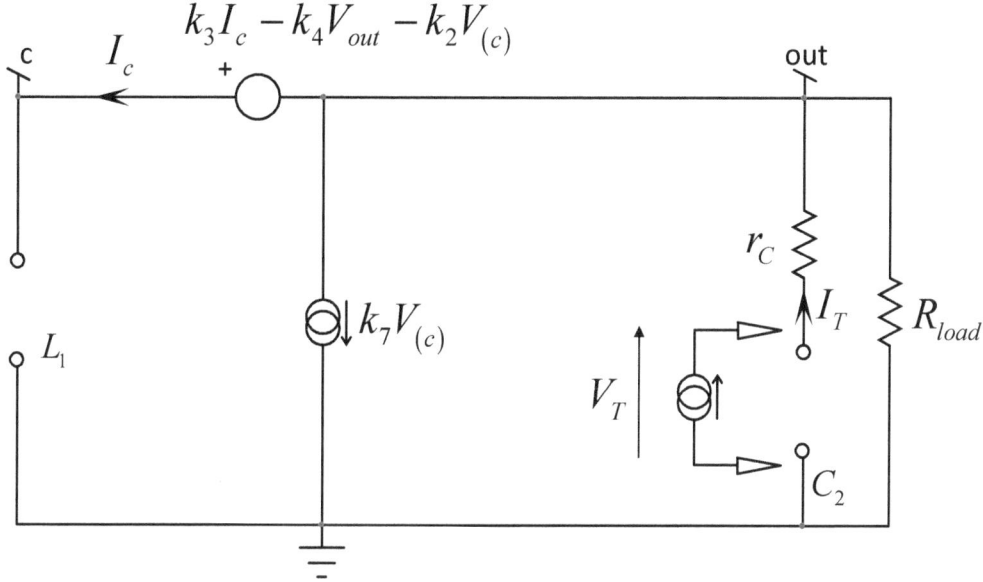

Figure 3.131 The inductor is now open-circuited to determine the 2nd-order term.

Because the open-circuited inductor interrupts the return path at node c, the current I_c is zero. We can again temporarily disconnect r_C and bring it back later. The voltage at node c is defined as follows:

$$V_{(c)} = V_T + 0 \cdot k_3 - k_4 V_T - k_2 V_{(c)} \tag{3.311}$$

The voltage at node *out* is V_T and defined by the following expression:

$$V_T = -\left(k_7 V_{(c)} - I_T\right) R_{load} \tag{3.312}$$

We extract $V_{(c)}$ from (3.311) and substitute it in the above formula. We then solve for V_T and rearrange the result. Bringing r_C back in series, we have:

$$\tau_2^1 = C_2 \left(\frac{R_{load}}{1 - \dfrac{R_{load} k_7 (k_4 - 1)}{1 + k_2}} + r_C \right) \tag{3.313}$$

We can now form the second high-frequency term b_2 as:

$$b_2 = \tau_1 \tau_2^1 = \frac{L_1}{\dfrac{R_{load}(1 - k_4) - k_3}{R_{load} k_7 (1 - k_4) + 1 + k_2}} C_2 \left(\frac{R_{load}}{1 - \dfrac{R_{load} k_7 (k_4 - 1)}{1 + k_2}} + r_C \right) \tag{3.314}$$

The denominator is now complete and expressed as:

$$D(s) = 1 + s(\tau_1 + \tau_2) + s^2 \tau_1 \tau_2' = 1 + s \left[\frac{\dfrac{L_1}{R_{load}(1-k_4) - k_3}}{R_{load} k_7 (1-k_4) + 1 + k_2} + C_2 \left(\frac{k_3}{\dfrac{k_3}{R_{load}} - 1 + k_4} + r_C \right) \right]$$

(3.315)

$$+ s^2 \frac{\dfrac{L_1}{R_{load}(1-k_4) - k_3}}{R_{load} k_7 (1-k_4) + 1 + k_2} C_2 \left(\frac{R_{load}}{1 - \dfrac{R_{load} k_7 (k_4 - 1)}{1 + k_2}} + r_C \right)$$

This expression can be put under the classical canonical form

$$D(s) = 1 + \frac{s}{\omega_0 Q} + \left(\frac{s}{\omega_0} \right)^2$$

(3.316)

where:

$$Q = \frac{\sqrt{b_2}}{b_1}$$

(3.317)

and:

$$\omega_0 = \frac{1}{\sqrt{b_2}}$$

(3.318)

Applying the low-Q approximation, the denominator can be rewritten in the following way:

$$D(s) \approx \left(1 + \frac{s}{\omega_{p_1}} \right) \left(1 + \frac{s}{\omega_{p_2}} \right)$$

(3.319)

In which:

$$\omega_{p_1} = Q \omega_0$$

(3.320)

and:

$$\omega_{p_2} = \frac{\omega_0}{Q}$$

(3.321)

Combining all these elements together will give you the poles position but in a complex formulation. Fortunately, you can find a simplified form in [2] which gives results very close to the above poles definitions:

$$\omega_{p_1} \approx \frac{1}{R_{load} C_2} \frac{2M - 1}{M - 1}$$

(3.322)

$$\omega_{p_2} \approx 2F_{sw} \left(\frac{1 - \frac{1}{M}}{D} \right)^2 \tag{3.323}$$

Now that the denominator is known, we can take a look at the numerator where zeroes are located. The excitation V_c is back in the circuit and some impedance combination prevents the stimulus to generate an output response. This is what Figure 3.132 shows. The first obvious contributor is when impedance $Z_1(s)$ becomes a transformed short circuit at an angular frequency defined as:

$$\omega_{z_1} = \frac{1}{r_C C_2} \tag{3.324}$$

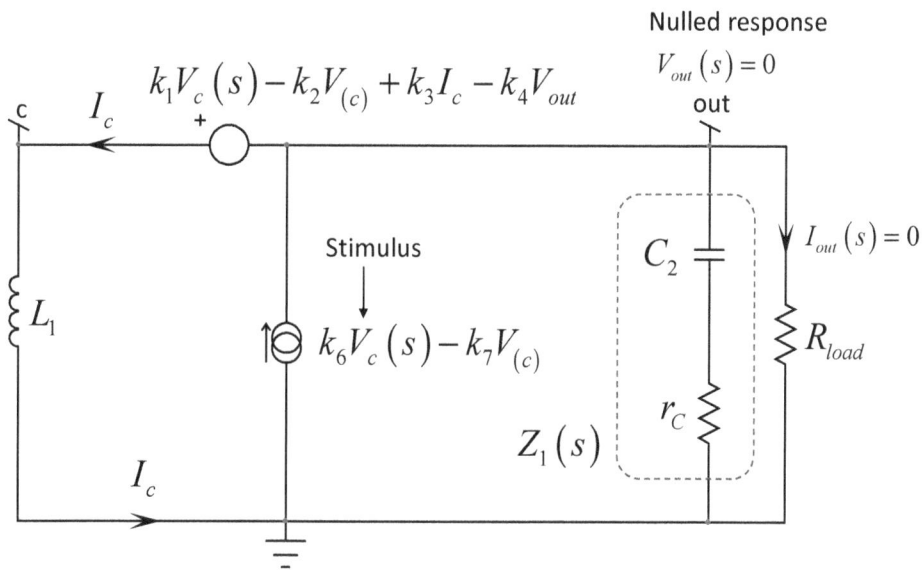

Figure 3.132 To determine the zeroes, bring the stimulus back in and determine the conditions for which the output is nulled.

The second zero is realized when all the current flowing out of terminal c is absorbed by the current source. When this happens, there is no current flowing in the load and V_{out} is zero volt. We can write a few equations describing the circuit:

$$V_{(c)} = k_1 V_c - k_2 V_{(c)} + k_3 I_c \tag{3.325}$$

From which we can extract the voltage at terminal c:

$$V_{(c)} = \frac{I_c k_3 + V_c k_1}{1 + k_2} \tag{3.326}$$

We know that:

$$I_c = \frac{V_{(c)}}{s L_1} \tag{3.327}$$

385

Substituting (3.327) in (3.326) and solving for $V_{(c)}$ leads to:

$$V_{(c)} = \frac{k_1 V_c}{k_2 + 1 - \dfrac{k_3}{sL_1}} \tag{3.328}$$

The second equation defines the current I_c:

$$\frac{V_{(c)}}{sL_1} = k_6 V_c - k_7 V_{(c)} \tag{3.329}$$

From this expression, we can also extract the voltage at terminal c:

$$V_{(c)} = \frac{L_1 V_c k_6 s}{1 + sL_1 k_7} \tag{3.330}$$

Now equating (3.330) with (3.328) and solving for s gives us the second zero position:

$$s_{z_2} = \frac{k_1 + k_3 k_6}{L_1 (k_6 - k_1 k_7 + k_2 k_6)} \tag{3.331}$$

It is a positive root making this second zero a root located in the right half-plane. Reference [2] gives a simpler equivalent form defined as:

$$\omega_{z_2} \approx \frac{R_{load}}{L_1 M^2} \tag{3.332}$$

The denominator is immediately determined and leads us to the complete control-to-output transfer function of the DCM boost converter operated in current-mode control:

$$H(s) = H_0 \frac{\left(1 + \dfrac{s}{\omega_{z_1}}\right)\left(1 - \dfrac{s}{\omega_{z_2}}\right)}{\left(1 + \dfrac{s}{\omega_{p_1}}\right)\left(1 + \dfrac{s}{\omega_{p_2}}\right)} \tag{3.333}$$

We can plot the frequency response of the above equation using Mathcad® and the results appear in Figure 3.133. To check this graph, we captured a switching circuit under SIMPLIS® (Figure 3.134) and ran an ac analysis. After a few seconds, the small-signal response is delivered together with the switching waveforms. The results, appearing in Figure 3.135, confirm our approach is correct.

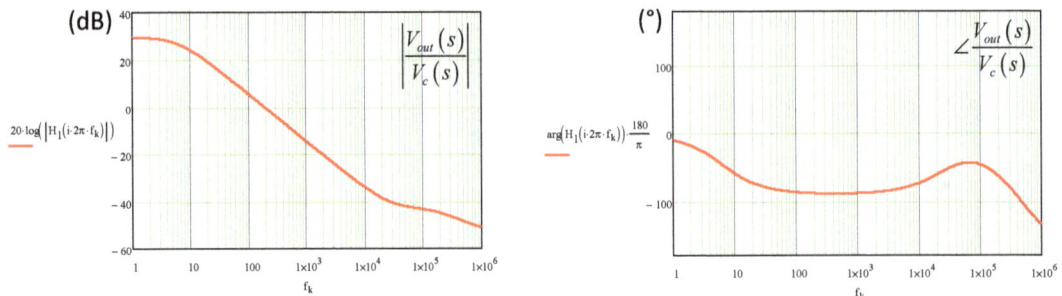

Figure 3.133 The dynamic response is that of a damped second-order system.

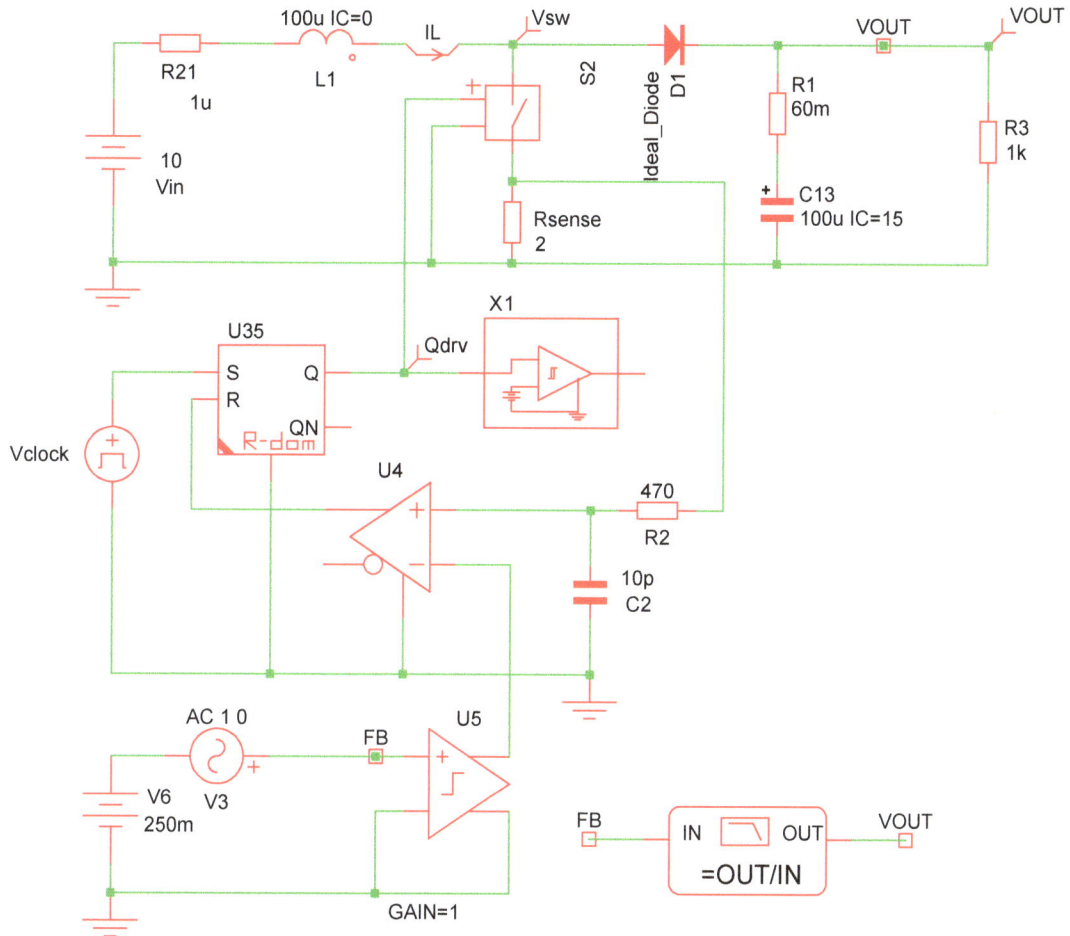

Figure 3.134 SIMPLIS® quickly delivers the cycle-by-cycle simulation results and the ac response.

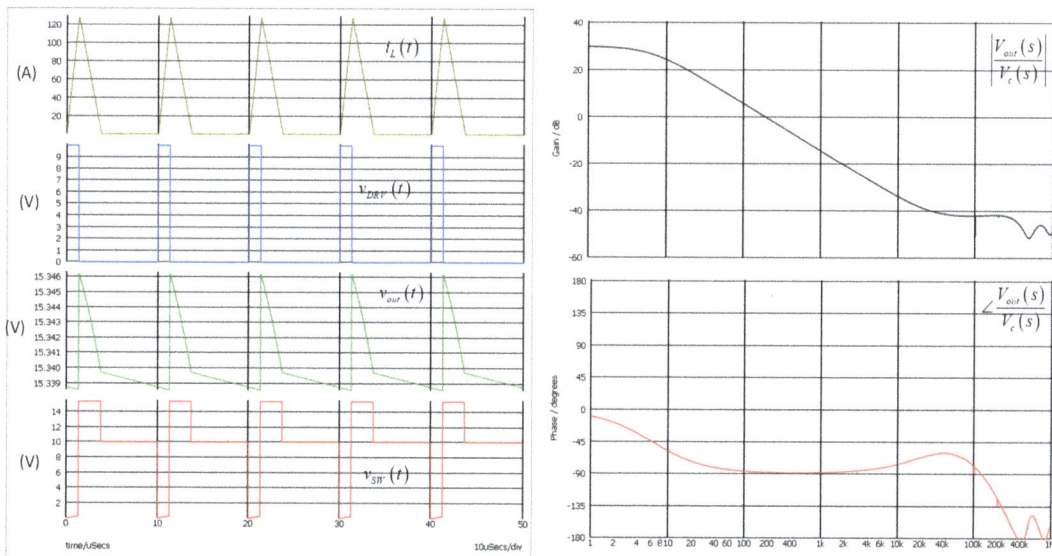

Figure 3.135 The magnitude and phase response are in a good agreement with the equations-based plots.

To let you calculate the small-signal coefficients yourself and check your results versus what Mathcad® gives, I have included the DCM CM values in Figure 3.136.

$$V_{in} := 10V \qquad V_{out} := 15.155V \quad R_i := -2\Omega \qquad C_{out} := 100\mu F \qquad r_C := 0.06\Omega \qquad R_{inf} := 10^{10}\Omega$$

$$F_{sw} := 100kHz \qquad T_{sw} := \frac{1}{F_{sw}} \qquad r_L := 0.0015\Omega \qquad L_1 := 100\,\mu H \qquad R_L := 1k\Omega$$

$$V_{ap} := -V_{out} \quad V_{cp} := -V_{out} + V_{in} \qquad V_{ac} := -V_{in} \qquad V_{acc} := V_{ac} \qquad \text{positive on-slope} \atop \text{always}$$

$$V_c := 250mV \qquad S_e := 0\frac{kV}{s} \qquad M := \frac{V_{out}}{V_{in}} = 1.5155$$

$$S_n := \frac{V_{ac}}{L_1}\cdot R_i = 200\frac{kV}{s} \qquad S_f := \frac{V_{ap}}{L_1}\cdot R_i = 303.1\cdot\frac{kV}{s}$$

$$D_1 := \frac{F_{sw}\cdot V_c}{R_i}\cdot\frac{1}{\frac{V_{ac}}{L_1} + \frac{S_e}{R_i}} = 12.5\% \qquad \tau_L := \frac{L_1}{R_L\cdot T_{sw}} = 0.01 \qquad m_c := 1 + \frac{S_e}{S_n} = 1$$

$$I_c := -\left(\frac{V_{out}}{R_L} + \frac{V_{in}\cdot D_1^{\,2}}{2\cdot L_1\cdot F_{sw}}\right) = -22.9675mA \qquad I_a := \frac{V_{ac}\cdot D_1^{\,2}}{2\cdot F_{sw}\cdot L_1} = -7.8125mA$$

$$M_1 := \frac{1 + \sqrt{1 + \frac{2\cdot D_1^{\,2}}{\tau_L}}}{2} = 1.5155 \qquad\qquad \text{Ia coefficients:}$$

$$k_1 := \frac{F_{sw}\cdot L_1\cdot V_c\cdot V_{ac}\cdot V_{ap}}{I_c\cdot\left(L_1\cdot S_e + R_i\cdot V_{acc}\right)^2} = -41.24034 \qquad k_6 := \frac{F_{sw}\cdot L_1\cdot V_c\cdot V_{ac}}{\left(L_1\cdot S_e + R_i\cdot V_{acc}\right)^2} = -0.0625\frac{1}{\Omega}$$

$$k_2 := \frac{F_{sw}\cdot L_1\cdot V_c^{\,2}\cdot V_{ap}}{2\cdot I_c\cdot\left(L_1\cdot S_e + R_i\cdot V_{acc}\right)^2} = 0.5155 \qquad k_7 := \frac{F_{sw}\cdot L_1\cdot V_c^{\,2}}{2\cdot\left(L_1\cdot S_e + R_i\cdot V_{acc}\right)^2} = 7.8125\times 10^{-4}\frac{1}{\Omega}$$

$$k_3 := -\frac{F_{sw}\cdot L_1\cdot V_c^{\,2}\cdot V_{ac}\cdot V_{ap}}{2\cdot I_c^{\,2}\cdot\left(L_1\cdot S_e + R_i\cdot V_{acc}\right)^2} = -224.44944\Omega \qquad k_8 := -\frac{F_{sw}\cdot L_1\cdot R_i\cdot V_c^{\,2}\cdot V_{ac}}{\left(L_1\cdot S_e + R_i\cdot V_{acc}\right)^3} = -1.5625\times 10^{-3}\frac{1}{\Omega}$$

$$k_4 := \frac{F_{sw}\cdot L_1\cdot V_c^{\,2}\cdot V_{ac}}{2\cdot I_c\cdot\left(L_1\cdot S_e + R_i\cdot V_{acc}\right)^2} = 0.34015$$

$$k_5 := -\frac{F_{sw}\cdot L_1\cdot R_i\cdot V_c^{\,2}\cdot V_{ac}\cdot V_{ap}}{I_c\cdot\left(L_1\cdot S_e + R_i\cdot V_{acc}\right)^3} = -1.03101$$

Figure 3.136 The computed small-signal coefficients of DCM CM boost converters are given.

3.4.1 Input to Output

Now that we have determined the denominator, we can proceed with the input source rejection capability of the CM boost converter operated in DCM.

In this configuration, the control voltage fixes the static operating conditions (15 V across the 1-kΩ load resistance) and the input source is ac-swept. This is what Figure 3.137 represents with the large-signal model response used as a reference. The small-signal version is given in Figure 3.138 and we have checked that both circuits give the exact same dynamic response in magnitude and phase.

We can now rework this diagram and rearrange sources noting that node a is grounded and the control voltage V_c is 0 V in this ac analysis. The simplified circuit appears in Figure 3.139. We start with the quasi-static gain H_0 that we determine by shorting inductor L_1 and opening output capacitor C_2. The electrical diagram appears in Figure 3.140.

We can see that nodes cc, c and V_{in} are all connected and we can thus write a first expression:

$$V_{in} = k_3 I_c - k_2 V_{in} - k_4 V_{out} - k_5 V_{in} + V_{out} \qquad (3.334)$$

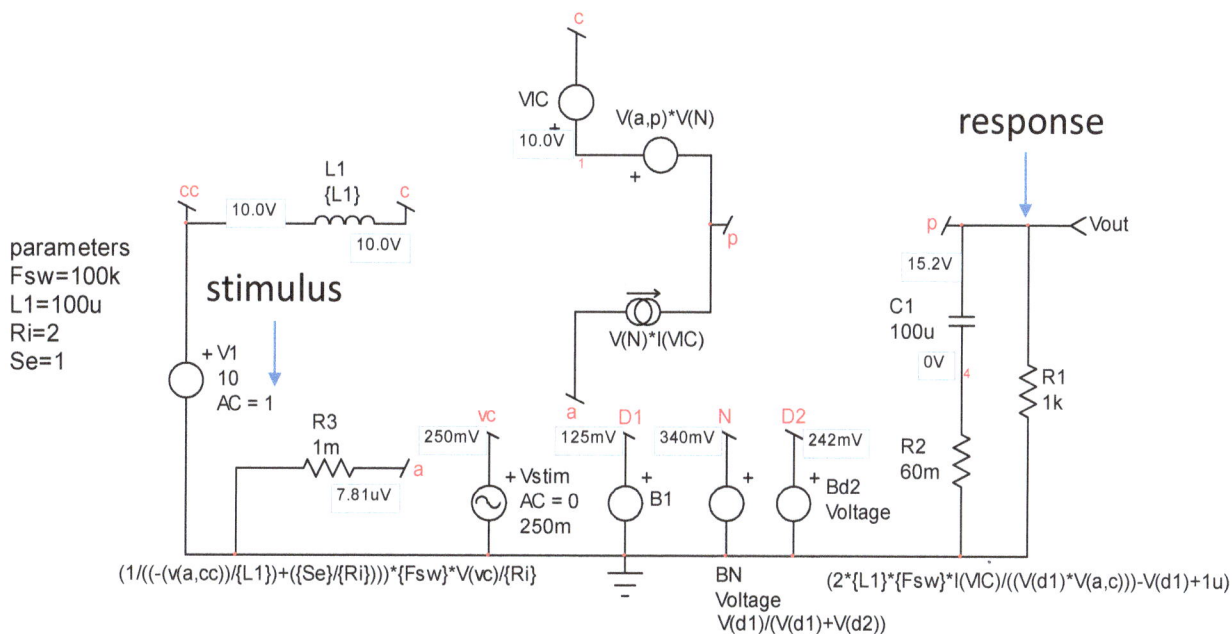

Figure 3.137 The source is now the stimulus while the control voltage V_c is made 0 V in ac.

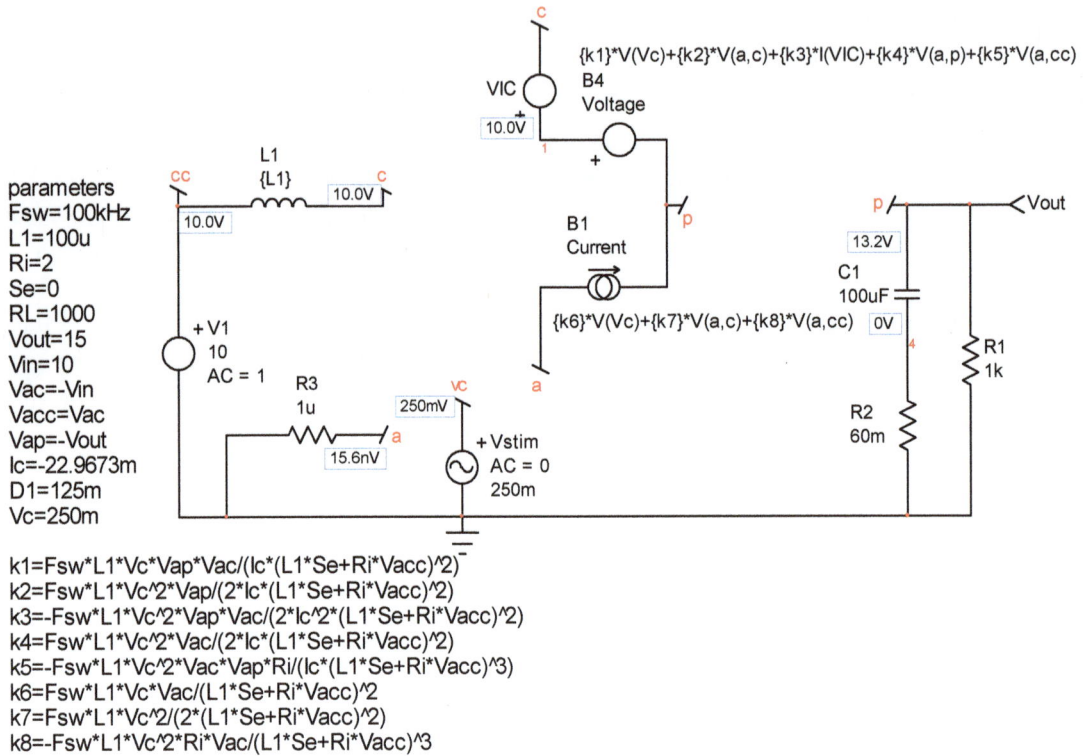

parameters
Fsw=100kHz
L1=100u
Ri=2
Se=0
RL=1000
Vout=15
Vin=10
Vac=-Vin
Vacc=Vac
Vap=-Vout
Ic=-22.9673m
D1=125m
Vc=250m

k1=Fsw*L1*Vc*Vap*Vac/(Ic*(L1*Se+Ri*Vacc)^2)
k2=Fsw*L1*Vc^2*Vap/(2*Ic*(L1*Se+Ri*Vacc)^2)
k3=-Fsw*L1*Vc^2*Vap*Vac/(2*Ic^2*(L1*Se+Ri*Vacc)^2)
k4=Fsw*L1*Vc^2*Vac/(2*Ic*(L1*Se+Ri*Vacc)^2)
k5=-Fsw*L1*Vc^2*Vac*Vap*Ri/(Ic*(L1*Se+Ri*Vacc)^3)
k6=Fsw*L1*Vc*Vac/(L1*Se+Ri*Vacc)^2
k7=Fsw*L1*Vc^2/(2*(L1*Se+Ri*Vacc)^2)
k8=-Fsw*L1*Vc^2*Ri*Vac/(L1*Se+Ri*Vacc)^3

Figure 3.138 The small-signal model needs simplification considering a zeroed V_c.

The output voltage depends on the current I_c and the controlled current source:

$$V_{out} = -R_{load}\left[I_c + V_{in}\left(k_7 + k_8 \right)\right] \tag{3.335}$$

If you now extract I_c from the above, substitute it in (3.334) then solve for V_{out}, you should find the following expression for the gain:

$$H_0 = \frac{V_{out}}{V_{in}} = \frac{1 + k_3\left(k_7 + k_8 \right) + k_5 + k_2}{1 - k_4 - \dfrac{k_3}{R_{load}}} \tag{3.336}$$

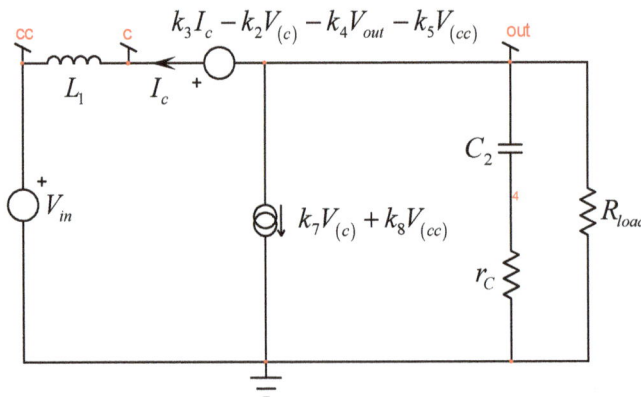

Figure 3.139 The input-to-output transfer function will be determined using this circuit.

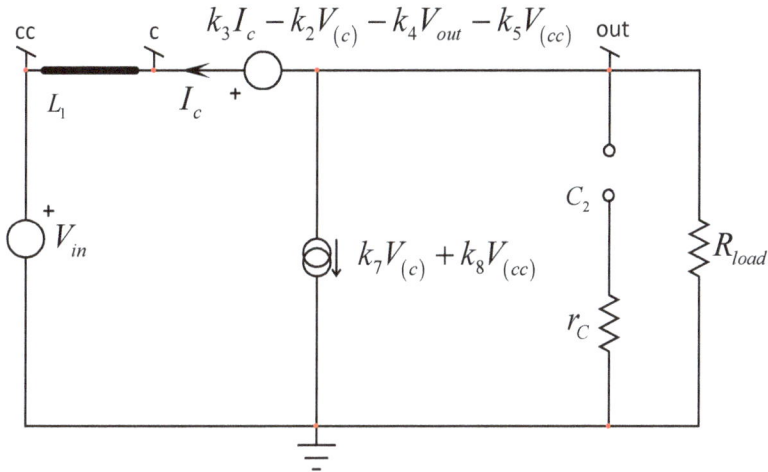

Figure 3.140 The dc gain is obtained by shorting the inductor and opening the capacitor.

If we now reduce the excitation to zero – $V_{in} = 0$ V – then the circuit returns to the state already studied in the control-to-output transfer function determination: we can reuse $D(s)$ defined by (3.319) as we did in the other examples.

Regarding the zeroes, the first one is classically brought by r_C and C_2:

$$\omega_{z_1} = \frac{1}{r_C C_2} \tag{3.337}$$

The second zero requires an update of the electrical circuit by bringing the excitation back while nulling the output. This is what Figure 3.141 details. Considering an output null and an I_T current equal to I_c, we can write a first equation to determine the voltage at node c:

$$V_{(c)} = k_3 I_T - k_2 V_{(c)} - k_5 V_{(cc)} \tag{3.338}$$

We also know that:

$$V_{(cc)} = V_{(c)} + V_T \tag{3.339}$$

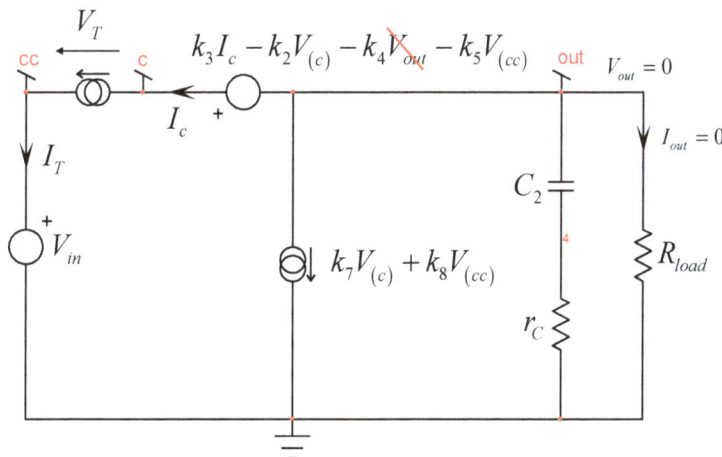

Figure 3.141 The inductor contributes a second zero that must be unveiled.

Christophe Basso

Substituting this expression in (3.338) and solving for $V_{(c)}$ gives:

$$V_{(c)} = \frac{I_T k_3 - V_T k_5}{1 + k_2 + k_5} \qquad (3.340)$$

To null the current in the load, the current source imposes the current in the left-side mesh:

$$I_T = -\left(k_7 V_{(c)} + k_8 V_{(cc)}\right) \qquad (3.341)$$

If we substitute (3.340) in the above and solve for $V_{(cc)}$, we have:

$$V_{(cc)} = -\frac{I_T + \dfrac{I_T k_3 - V_T k_5}{1 + k_2 + k_5} k_7}{k_8} \qquad (3.342)$$

The voltage across the test generator I_T is the difference between nodes cc and c:

$$V_T = V_{(cc)} - V_{(c)} = -\frac{I_T + \dfrac{I_T k_3 - V_T k_5}{1 + k_2 + k_5} k_7}{k_8} - \frac{I_T k_3 - V_T k_5}{1 + k_2 + k_5} \qquad (3.343)$$

Solving for V_T and factoring I_T leads to a resistance driving L_1 equal to:

$$R_{tau} = \frac{1 + k_3\left(k_8 + k_7\right) + k_2 + k_5}{k_5 k_7 - k_8\left(1 + k_2\right)} \qquad (3.344)$$

The numerical application with Mathcad® gives a resistance of −1.29 kΩ: we have a right-half-plane zero. To verify our calculations, the SPICE circuit shown in Figure 3.142 confirms our calculations are correct. This second zero is defined as follows:

$$\omega_{z_2} = \frac{1 + k_3\left(k_8 + k_7\right) + k_2 + k_5}{L_1\left[k_8\left(1 + k_2\right) - k_5 k_7\right]} \qquad (3.345)$$

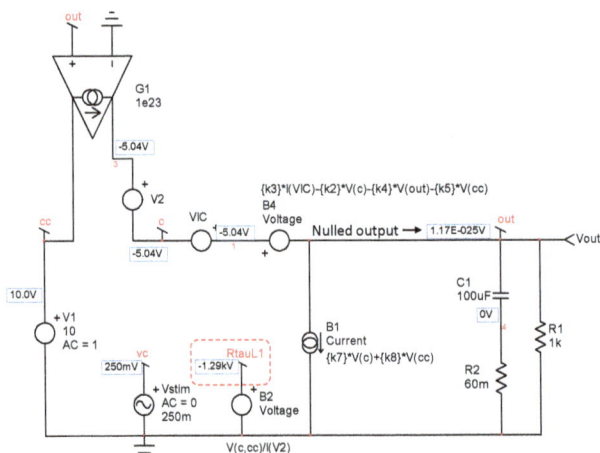

Figure 3.142 SPICE confirms the resistance value by nulling the output through an infinite-gain voltage-controlled current source.

392

We now have completed the input-to-output transfer function which is expressed as:

$$H(s) = H_0 \frac{\left(1 + \dfrac{s}{\omega_{z_1}}\right)\left(1 - \dfrac{s}{\omega_{z_2}}\right)}{\left(1 + \dfrac{s}{\omega_{p_1}}\right)\left(1 + \dfrac{s}{\omega_{p_2}}\right)} \tag{3.346}$$

With H_0 defined by (3.336), the two poles by (3.322) and (3.323) while the zeroes appear in the above lines.

It is time to compare the dynamic response delivered by the large-signal circuit of Figure 3.137 with that of the above equation plotted with Mathcad® in Figure 3.143.

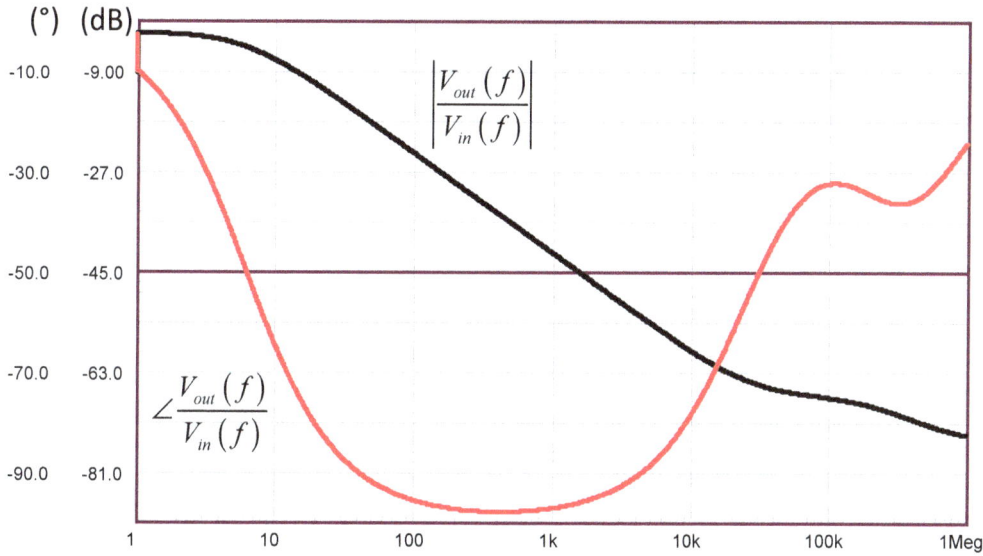

Figure 3.143 SPICE shows a poor static rejection of the CM boost converter operated in deep DCM.

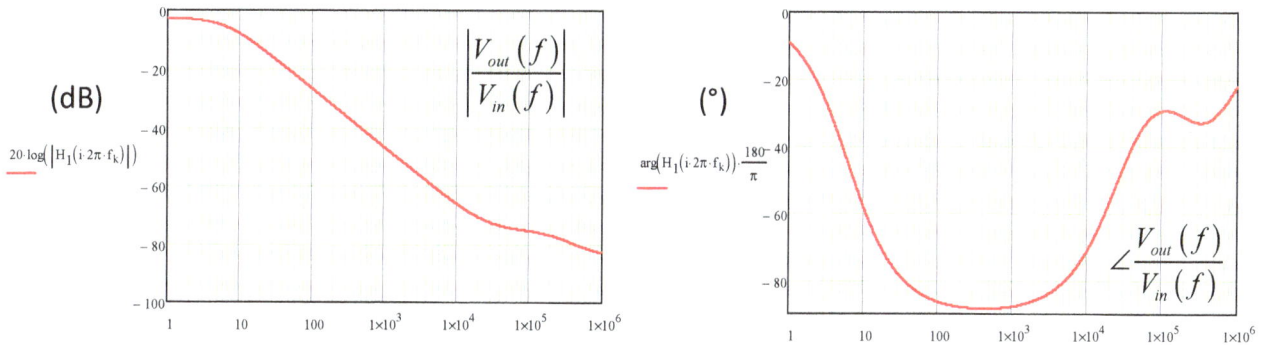

Figure 3.144 The Mathcad® sheet results perfectly match the SPICE simulation.

3.4.2 Output Impedance

The output impedance is obtained by installing a test generator I_T across the load resistance. This is what Figure 3.145 shows with the large-signal model. The stimulus is the test current while the voltage across the current source represents the response.

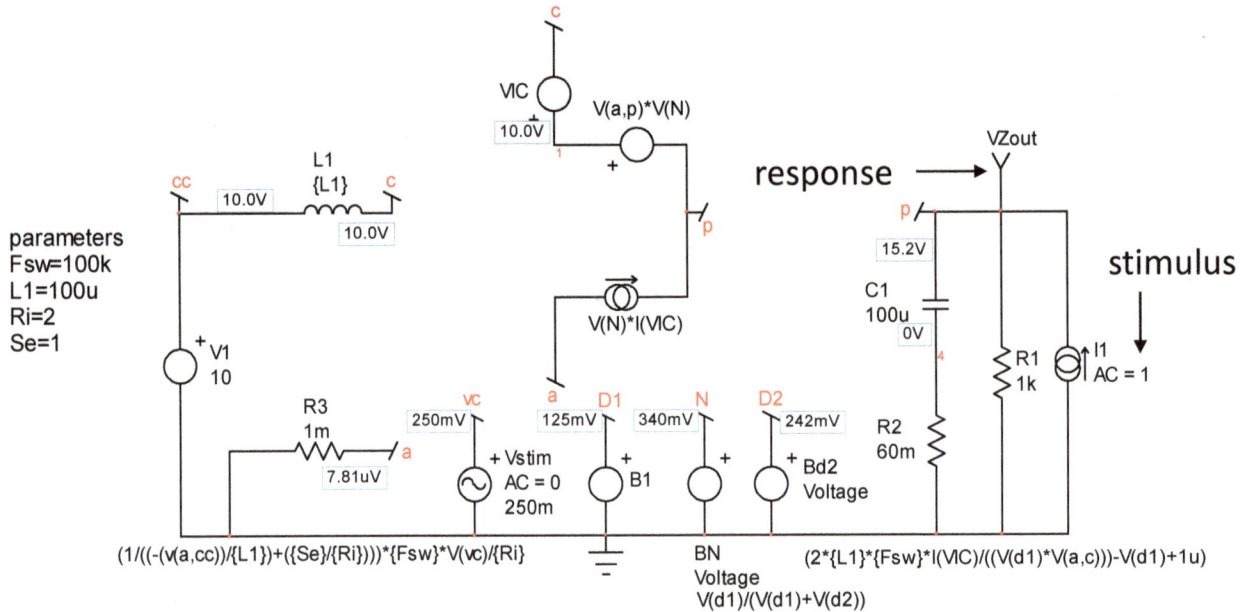

Figure 3.145 The large-signal model lets us plot the output impedance by installing a current generator across the load resistance.

To study the small-signal output impedance, we will consider the linear model shown in Figure 3.146. It is very close to that of Figure 3.127 with the added current source. In this mode, the input contribution is zeroed and the left-side of inductor L_1 is grounded.

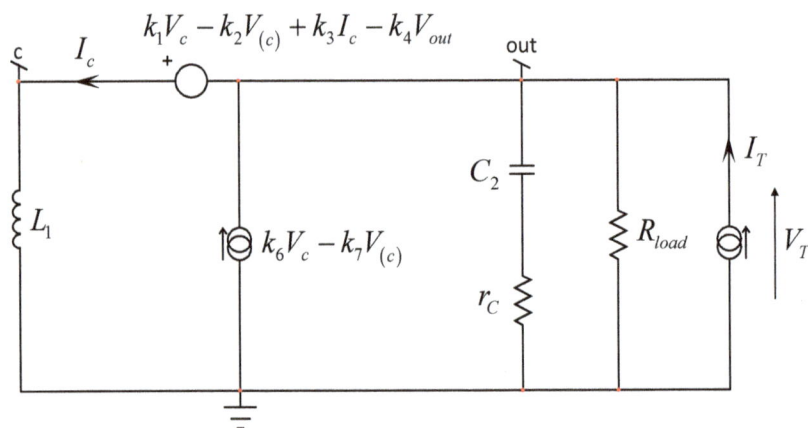

Figure 3.146 The input contribution V_{in} is zeroed in ac when studying the output impedance.

Zeroing the excitation I_T (open-circuit the source) and the control voltage V_c (0 V in this ac analysis) brings the circuit back to its natural structure that we studied in the control-to-output transfer function: we can reuse the denominator $D(s)$ determined with (3.319).

We can now start determining the incremental output resistance R_0 with the help of Figure 3.147. The load resistance has been temporarily disconnected and will be brought back in parallel with the intermediate result. In this mode, I_T and I_c currents are similar while the voltage at node c is 0 V owing to the shorting of L_1: the analysis gains in simplicity.

We can write a single equation describing this circuit:

$$V_T = -(k_3 I_T + k_4 V_T) \tag{3.347}$$

Solving for V_T, rearranging and bringing R_{load} back gives:

$$R_0 = \frac{k_3}{k_4 - 1} \| R_{load} \tag{3.348}$$

To determine the zeroes, we have to null the response (V_T) and determine the time constants to form the numerator $N(s)$:

$$N(s) = 1 + s(\tau_{1N} + \tau_{2N}) + s^2 \tau_{1N} \tau_{2N}^1 \tag{3.349}$$

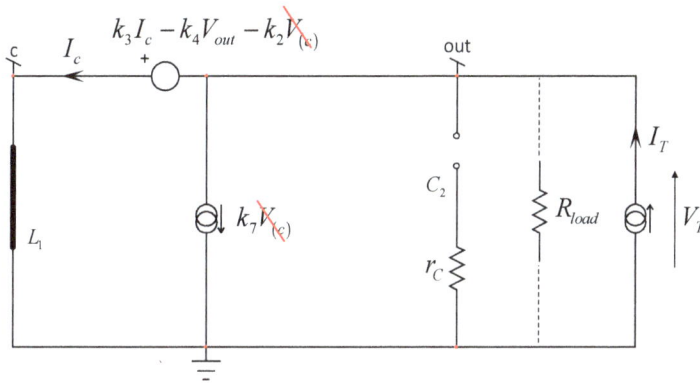

Figure 3.147 To determine the output impedance, the load can be temporarily disconnected and be brought back later.

The first time constant is obtained by installing a test generator across the inductor's connections as shown in Figure 3.148. The excitation is back in place while the response at node *out* is nulled. The test current I_T is equal to the current I_c. This helps simplify the main expression which leads us to a single equation (3.350) to solve.

$$V_T = -k_3 I_T - k_2 V_T \tag{3.350}$$

After factoring V_T, the time constant involving L_1 is immediate and equal to:

$$\tau_1 = \frac{L_1}{-\dfrac{k_3}{1 + k_2}} \tag{3.351}$$

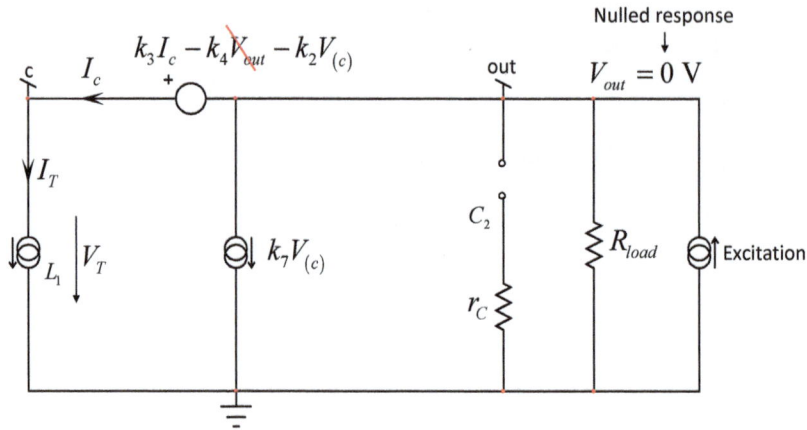

Figure 3.148 The time constant involving L_1 is determined with the excitation back in place while the response is nulled.

The second time constant involves C_2 and the circuit to determine it appears in Figure 3.149 in which L_1 is set in its dc state which is a short circuit.

Considering a nulled output, the time constant is immediate as it involves r_C alone:

$$\tau_2 = r_C C_2 \tag{3.352}$$

The high-frequency term τ_2^1 implies an inductor set in its high-frequency state (an open-circuit) while the output is nulled. The circuit is the same as in Figure 3.149 except that L_1 is open-circuited and we want the resistance driving C_2. Considering the nulled output, the result previously obtained with τ_2 does not change and we have:

$$\tau_2^1 = r_C C_2 \tag{3.353}$$

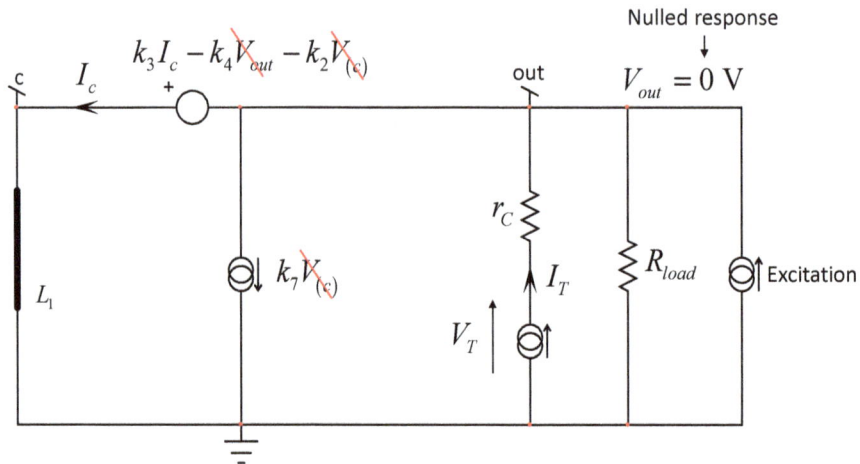

Figure 3.149 Inspection quickly shows a time constant built around C_2 and its ESR r_C.

We can form the numerator as follows:

$$N(s) = 1 + s\left(\frac{L_1}{-\dfrac{k_3}{1+k_2}} + r_C C_2\right) + s^2 \frac{L_1}{-\dfrac{k_3}{1+k_2}} r_C C_2 \tag{3.354}$$

This expression can be advantageously factored knowing that:

$$1 + s(a+b) + s^2 ab = (1+as)(1+bs) \tag{3.355}$$

Rearranging (3.354), we can write:

$$N(s) = \left(1 + \frac{s}{\omega_{z_1}}\right)\left(1 + \frac{s}{\omega_{z_2}}\right) \tag{3.356}$$

with:

$$\omega_{z_1} = \frac{1}{r_C C_2} \tag{3.357}$$

and:

$$\omega_{z_2} = -\frac{k_3}{(1+k_2)L_1} \tag{3.358}$$

The second zero occurs at a high frequency as the second pole defined in (3.323) does. Therefore, the output impedance can be approximated by the following transfer function:

$$Z_{out}(s) \approx R_0 \frac{1 + \dfrac{s}{\omega_{z_1}}}{1 + \dfrac{s}{\omega_{p_1}}} \tag{3.359}$$

With R_0 defined by (3.348), the zero defined in the above lines and the pole by (3.322). Using this formula, we can plot the output impedance using Mathcad® as shown in Figure 3.150. The agreement between the complete formula and its simplified expression from (3.359) is excellent.

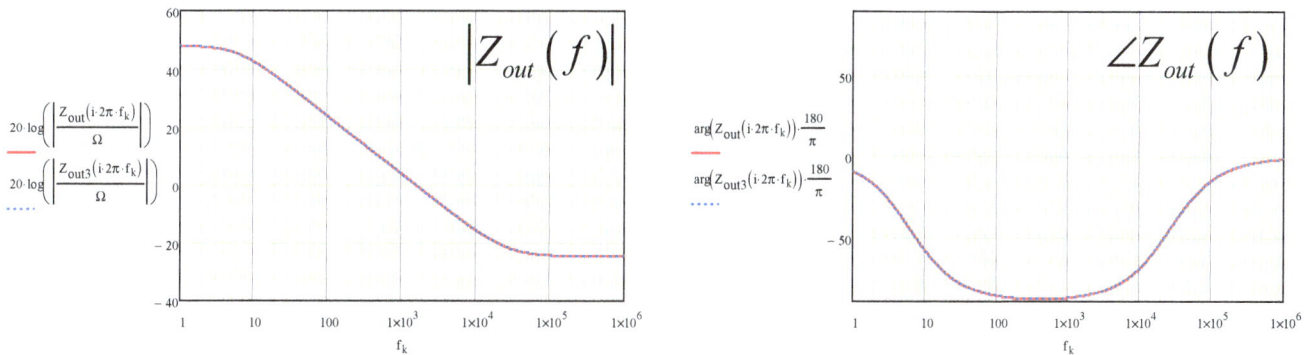

Figure 3.150 The output impedance hosts one pole and one zero in its simplified form.

We have also plotted the output impedance from the circuit given in Figure 3.145 and results agree well with the mathematical analysis.

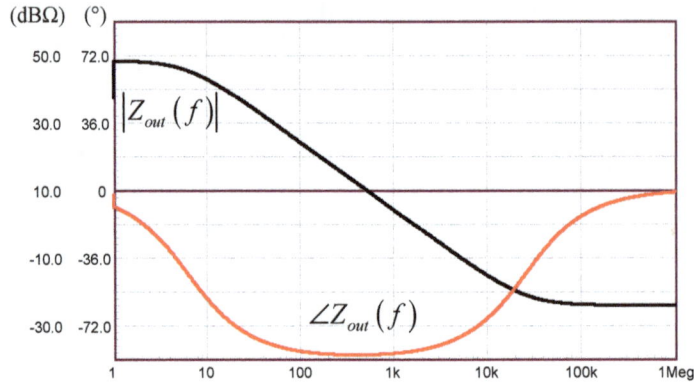

Figure 3.151 SPICE can also deliver the output impedance quickly.

3.4.3 Input Impedance

For the input impedance, we will connect the 1-A ac current source across the input node of the DCM-operated large-signal CM boost converter as shown in Figure 3.152.

The 1-kH inductor is a short circuit during the bias point calculation and will later block the ac modulation. Vector VZ_{In} becomes the voltage image of the impedance we want.

After plugging the small-signal model and rearranging all sources considering 0-V input and control modulations, we come up with the reduced simulation setup presented in Figure 3.153. As usual, we check that responses in phase and magnitude from both approaches – large- and small-signal schematic diagrams – are identical.

We are good to go and can now work on the graphical representation of Figure 3.154. As we did in the previous examples, we start with the incremental input resistance R_0 determined with L_1 short-circuited and C_2 open-circuited (Figure 3.155).

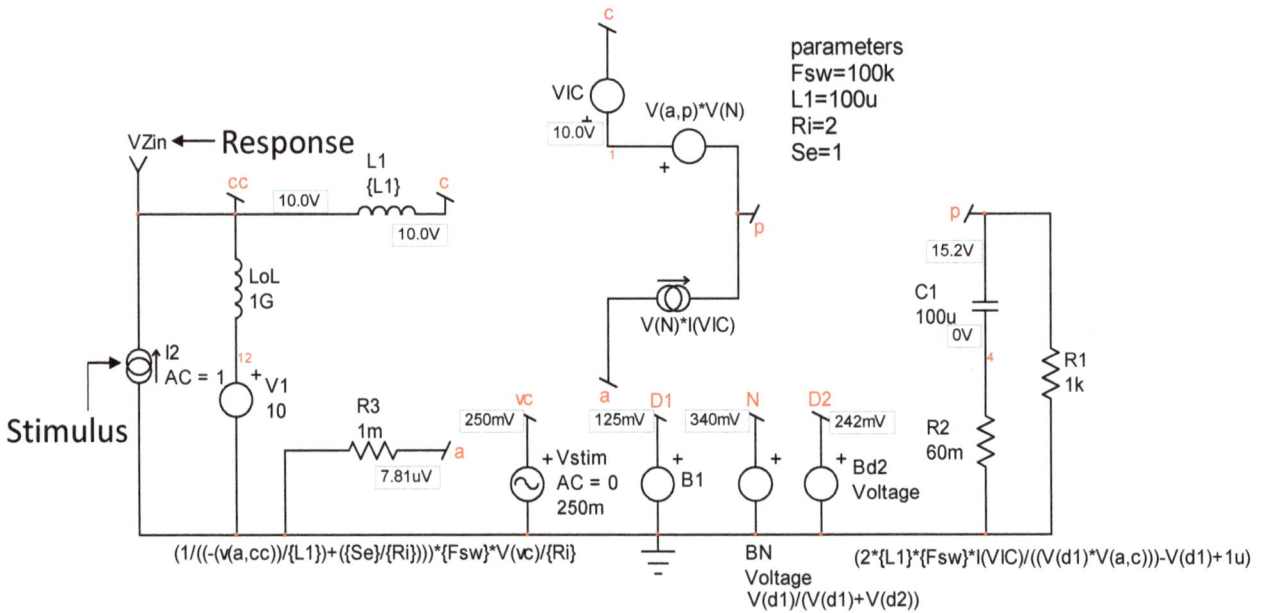

Figure 3.152 The input impedance is swept using a 1-A ac current source isolated from the dc source via a high-value inductor.

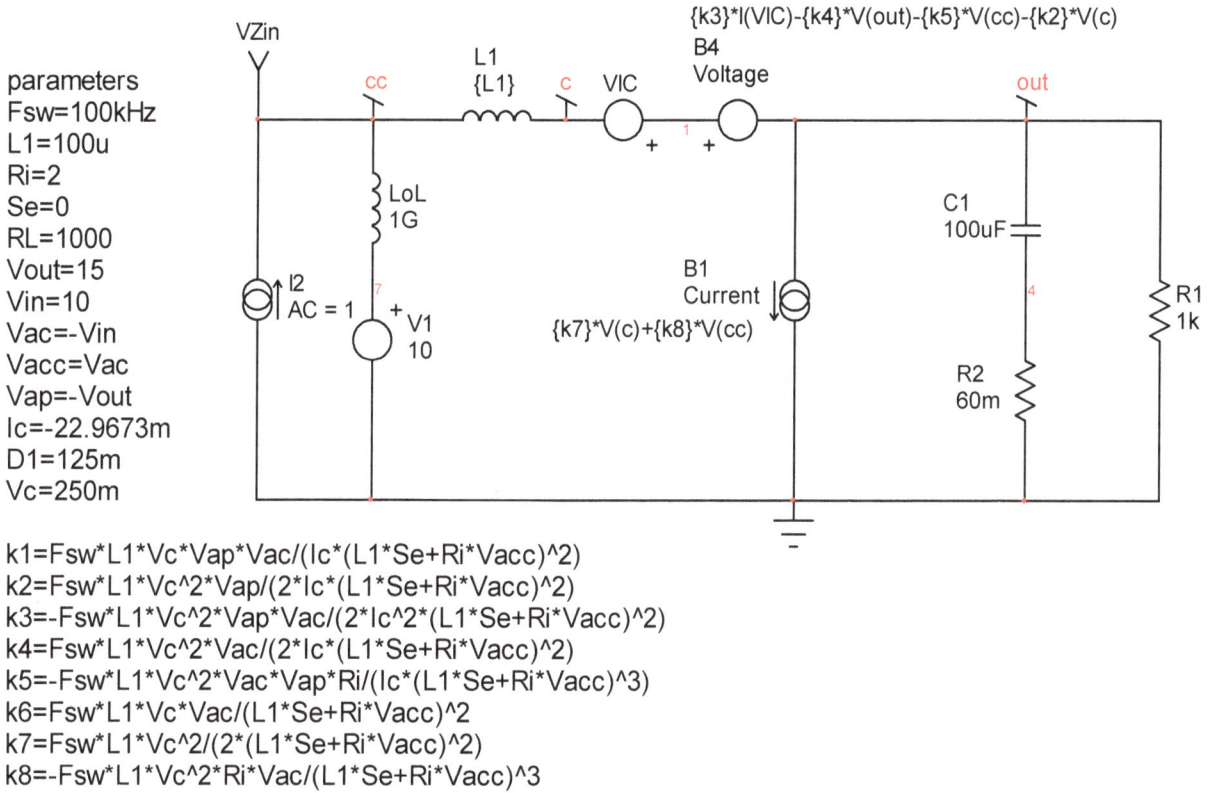

$\{k3\}*I(VIC)-\{k4\}*V(out)-\{k5\}*V(cc)-\{k2\}*V(c)$

parameters
Fsw=100kHz
L1=100u
Ri=2
Se=0
RL=1000
Vout=15
Vin=10
Vac=-Vin
Vacc=Vac
Vap=-Vout
Ic=-22.9673m
D1=125m
Vc=250m

k1=Fsw*L1*Vc*Vap*Vac/(Ic*(L1*Se+Ri*Vacc)^2)
k2=Fsw*L1*Vc^2*Vap/(2*Ic*(L1*Se+Ri*Vacc)^2)
k3=-Fsw*L1*Vc^2*Vap*Vac/(2*Ic^2*(L1*Se+Ri*Vacc)^2)
k4=Fsw*L1*Vc^2*Vac/(2*Ic*(L1*Se+Ri*Vacc)^2)
k5=-Fsw*L1*Vc^2*Vac*Vap*Ri/(Ic*(L1*Se+Ri*Vacc)^3)
k6=Fsw*L1*Vc*Vac/(L1*Se+Ri*Vacc)^2
k7=Fsw*L1*Vc^2/(2*(L1*Se+Ri*Vacc)^2)
k8=-Fsw*L1*Vc^2*Ri*Vac/(L1*Se+Ri*Vacc)^3

Figure 3.153 This is the simplified small-signal circuit diagram for studying the input impedance of the DCM CM boost converter.

As nodes c and cc are at potential V_T, we can rewrite the voltage and current sources expression to reflect this change and simplify the whole thing. The first equation defines voltage V_T by

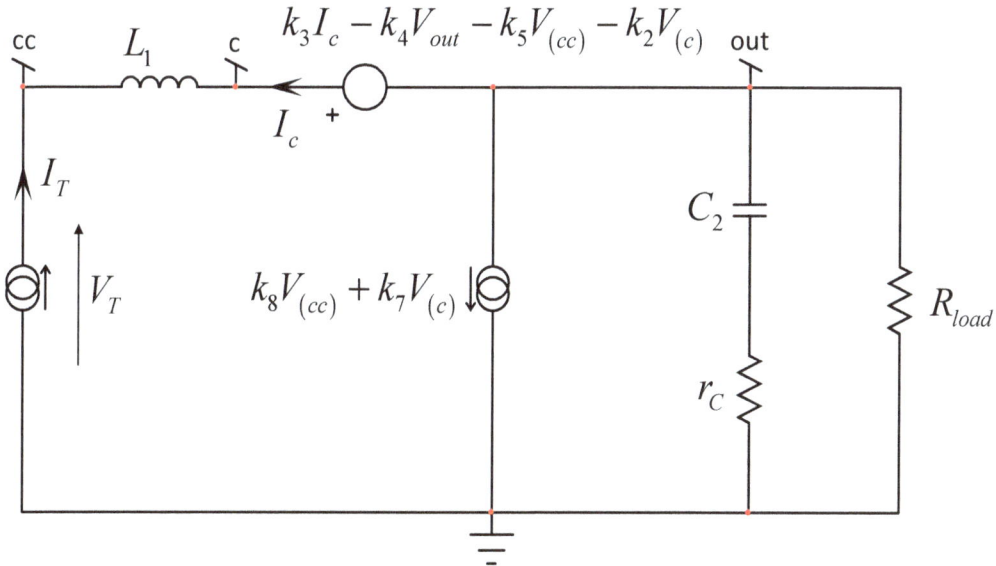

Figure 3.154 This is the small-signal circuit diagram used to determine the input impedance.

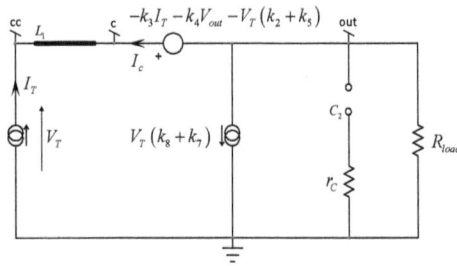

Figure 3.155 The inductor is shorted and the capacitor opened for determining the incremental input resistance.

$$V_T = -k_3 I_T - k_4 V_{out} - k_2 V_T - k_5 V_T + V_{out} \tag{3.360}$$

The output voltage is the current flowing in the load resistance multiplied by that resistance:

$$V_{out} = \left[I_T - V_T \left(k_8 + k_7 \right) \right] R_{load} \tag{3.361}$$

Substitute this definition in (3.360), solve for V_T, rearrange and factor. You obtain the incremental resistance R_0:

$$R_0 = \frac{R_{load}\left(1-k_4\right)-k_3}{R_{load}\left[k_7 + k_8 - k_4\left(k_7 + k_8\right)\right] + k_2 + k_5 + 1} \tag{3.362}$$

In this configuration, nulling the excitation means the left-side of L_1 is no longer connected to the ground and the circuit differs from the natural structure studied in the previous lines: we can cannot reuse the denominator already determined and must find the new one pertaining to this circuit:

$$D(s) = 1 + b_1 s + b_2 s^2 = 1 + s\left(\tau_1 + \tau_2\right) + s^2 \tau_1 \tau_2^1 \tag{3.363}$$

Intuitively, we can see a current source driving an inductance. As the generator fixes the inductor state variable, the denominator loses a degree and $D(s)$ will be of 1st order.

If you determine the resistance driving L_1 as shown in Figure 3.156, you can see that it is infinite, leading to a zero time constant:

$$\tau_1 = \frac{L_1}{\infty} = 0 \tag{3.364}$$

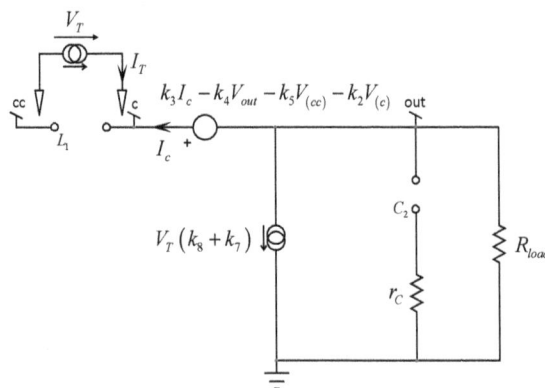

Figure 3.156 The inductor left-side is floating when the excitation is reduced to zero.

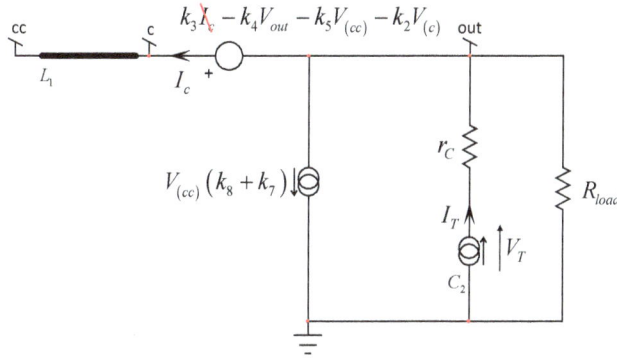

Figure 3.157 The shorted inductor allows us to determine the resistance driving capacitor C₂.

The resistance driving capacitor C_2 implies that L_1 is replaced by a short circuit (Figure 3.157). r_C can be temporarily disconnected and brought back later in series with the intermediate result. Current I_c is 0 A and V_T appears at node *out*. We can express the voltage at node *cc* with that of node *c* also:

$$V_{(cc)} = V_T - k_4 V_T - k_5 V_{(cc)} - k_2 V_{(cc)} \tag{3.365}$$

Extracting $V_{(cc)}$ leads to:

$$V_{(cc)} = \frac{V_T(1-k_4)}{1+k_2+k_5} \tag{3.366}$$

The voltage across the load resistance is:

$$V_T = -\left[V_{(cc)}(k_8+k_7) - I_T\right] R_{load} \tag{3.367}$$

Now substituting (3.366) in the above equation, solving for V_T rearranging and factoring gives the second time constant:

$$\tau_2 = C_2 \left[r_C + \frac{R_{load}}{1 + \frac{R_{load}(1-k_4)(k_7+k_8)}{1+k_2+k_5}} \right] \tag{3.368}$$

We have coefficient b_1 equal to a single term since τ_1 is zero:

$$b_1 = \tau_1 + \tau_2 = \tau_2 = C_2 \left[r_C + \frac{R_{load}}{1 + \frac{R_{load}(1-k_4)(k_7+k_8)}{1+k_2+k_5}} \right] \tag{3.369}$$

We are left to determine the second-order term, b_2, obtained by setting L_1 in its high-frequency state (an open circuit) and determining the resistance driving C_2 in this mode:

$$b_2 = \tau_1 \tau_2^1 \tag{3.370}$$

This configuration will bring a non-zero value which multiplied by τ_1 leads to:

$$b_2 = 0 \cdot \tau_2^1 = 0 \tag{3.371}$$

We have the denominator we want and it is defined as:

$$D(s) = 1 + s\tau_2 = 1 + \frac{s}{\omega_p} \tag{3.372}$$

with:

$$\omega_p = \cfrac{1}{C_2 \left[r_C + \cfrac{R_{load}}{1 + \cfrac{R_{load}(1-k_4)(k_7+k_8)}{1+k_2+k_5}} \right]} \tag{3.373}$$

To determine the numerator, we have to null the response which is the voltage across the current source I_T. We know that nulling the voltage across a current source is a degenerate case [1] and the element can be replaced by a short circuit as Figure 3.158 illustrates. You recognize the circuit returning in its natural state meaning that all time constants in this mode are those already determined in (3.315). We can either use the complete definition of $D(s)$ or use the approximate definition as in (3.319) :

$$N(s) \approx \left(1 + \frac{s}{\omega_{z_1}}\right)\left(1 + \frac{s}{\omega_{z_2}}\right) \tag{3.374}$$

In which:

$$\omega_{z_1} \approx \frac{1}{R_{load}C_2} \frac{2M-1}{M-1} \tag{3.375}$$

$$\omega_{z_2} \approx 2F_{sw} \left(\frac{1 - \frac{1}{M}}{D}\right)^2 \tag{3.376}$$

The input impedance of the DCM-operated CM boost converter can now be approximated as:

$$Z_{in}(s) \approx R_0 \frac{\left(1 + \frac{s}{\omega_{z_1}}\right)\left(1 + \frac{s}{\omega_{z_2}}\right)}{1 + \frac{s}{\omega_{p_1}}} \tag{3.377}$$

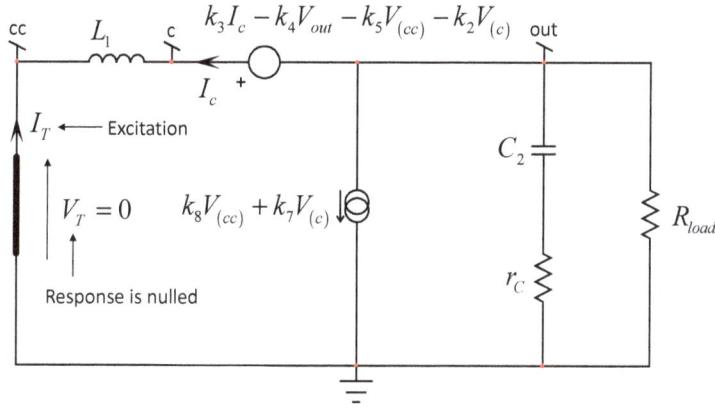

Figure 3.158 The response is nulled while the excitation is brought back in the circuit.

We can plot both expressions – the complete one and its approximation – and the results appear in Figure 3.159. The low-frequency part shows a good agreement between expressions but a deviation takes place in the upper section.

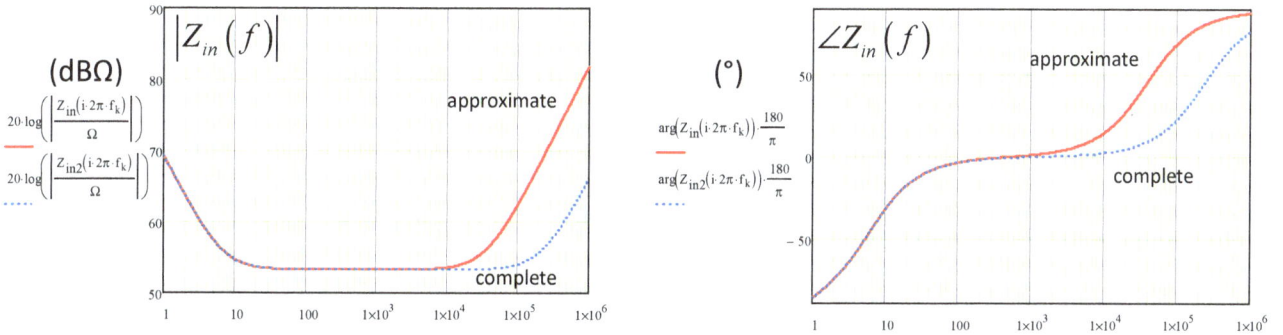

Figure 3.159 The curves deviates from each other in the upper frequency section.

We have simulated the large-signal model from Figure 3.152 and Figure 3.160 confirms our analysis is correct. We could also try a SIMPLIS® simulation configured to plot the input impedance. The schematic diagram is shown in Figure 3.161. Please note the particular configuration to measure the input impedance.

The results are given in Figure 3.162 and show a good agreement with our calculations.

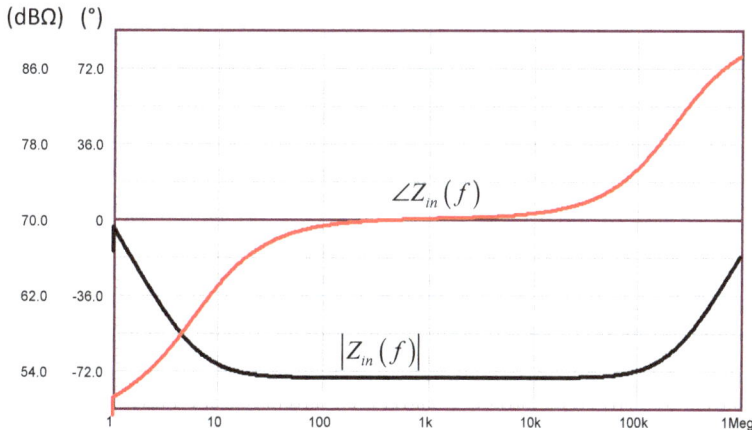

Figure 3.160 The SPICE simulations confirm our analysis is correct.

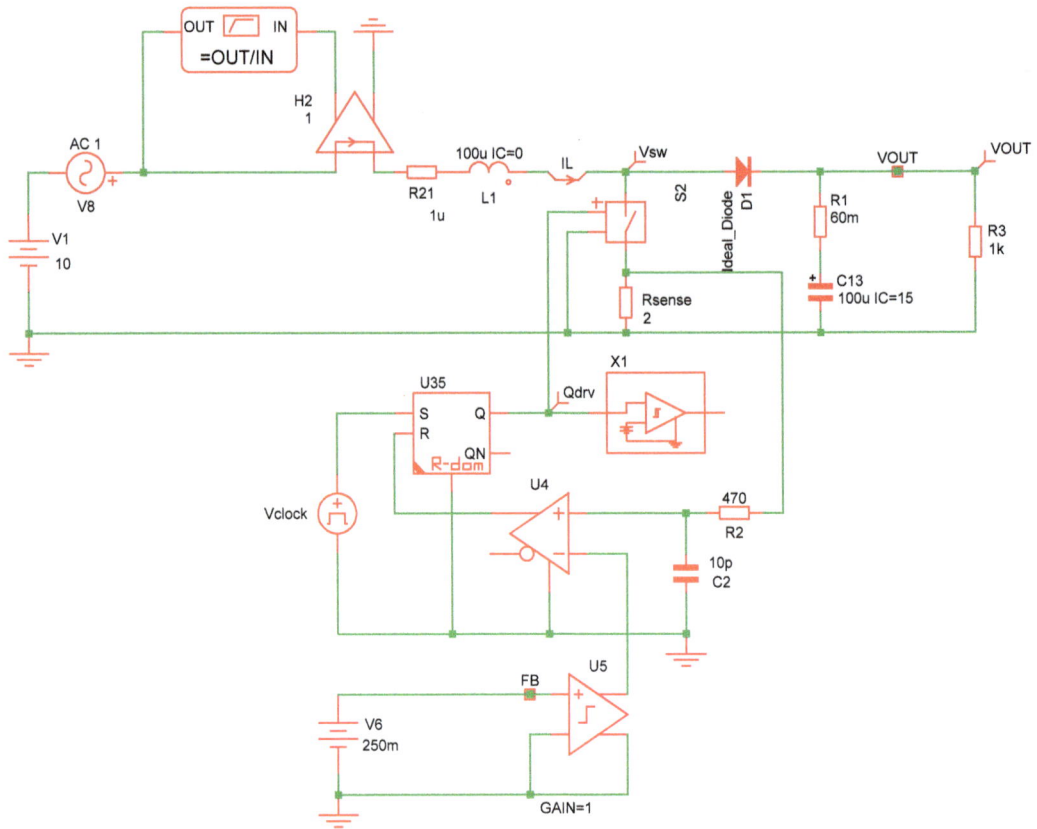

Figure 3.161 SIMPLIS® lets you simulate the input impedance by inserting the stimulus in series with the input voltage.

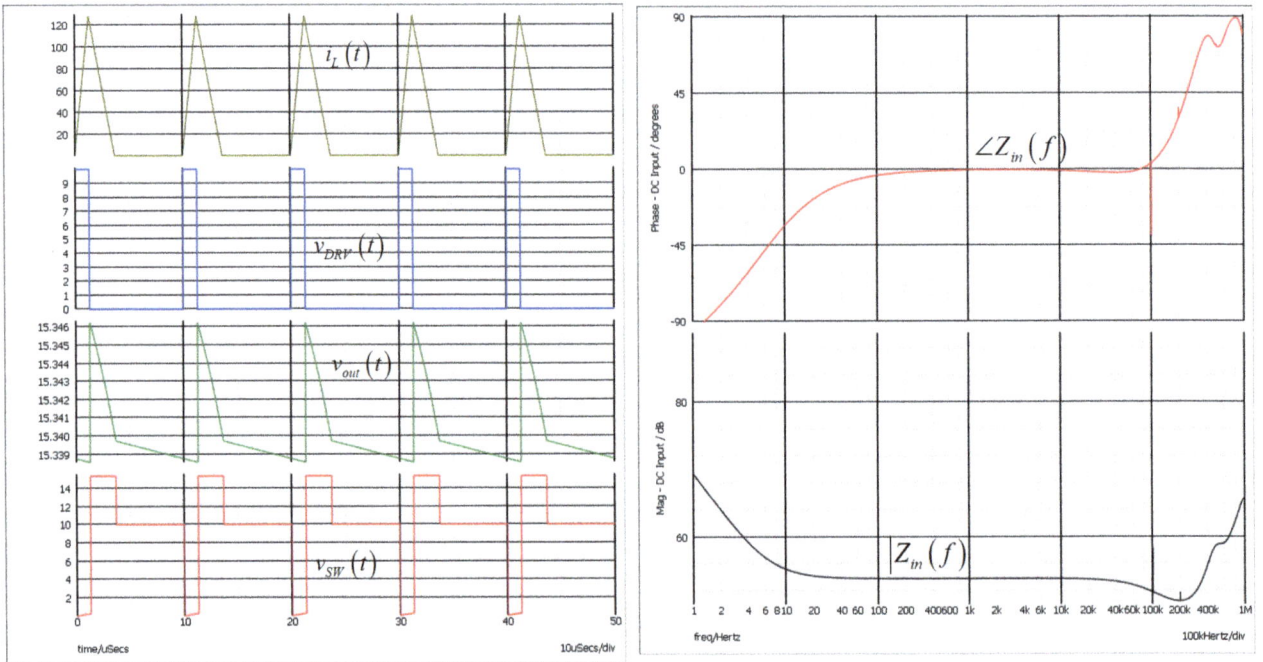

Figure 3.162 The input impedance plots are in good agreement with our derivation.

We have determined the four transfer function of the current-mode-controlled boost converter operated in DCM. These transfer functions are summarized in Figure 3.163.

$\dfrac{V_{out}(s)}{V_{err}(s)}$ Control to Output	$H_0 \dfrac{\left(1+\frac{s}{\omega_{z_1}}\right)\left(1-\frac{s}{\omega_{z_2}}\right)}{\left(1+\frac{s}{\omega_{p_1}}\right)\left(1+\frac{s}{\omega_{p_2}}\right)}$	$\omega_{z_1}=\dfrac{1}{r_C C_2}$ $\omega_{z_2}\approx\dfrac{R_{load}}{L_1 M^2}$	$\omega_{p_1}\approx\dfrac{1}{R_{load}C_2}\dfrac{2M-1}{M-1}$ $\omega_{p_2}\approx 2F_{sw}\dfrac{\left(1-\frac{1}{M}\right)^2}{D}$	$H_0 \approx V_{in}\sqrt{\dfrac{2M}{\tau_L(M-1)}}\dfrac{M-1}{2M-1}\dfrac{1}{S_n m_c T_{sw}}$
$\dfrac{V_{out}(s)}{V_{in}(s)}$ Input to Output	$H_0 \dfrac{\left(1+\frac{s}{\omega_{z_1}}\right)\left(1-\frac{s}{\omega_{z_2}}\right)}{\left(1+\frac{s}{\omega_{p_1}}\right)\left(1+\frac{s}{\omega_{p_2}}\right)}$	$\omega_{z_1}=\dfrac{1}{r_C C_2}$ $\omega_{z_2}=\dfrac{1+k_3(k_8+k_7)+k_2+k_5}{L_1\left[k_8(1+k_2)-k_3 k_7\right]}$	$\omega_{p_1}\approx\dfrac{1}{R_{load}C_2}\dfrac{2M-1}{M-1}$ $\omega_{p_2}\approx 2F_{sw}\dfrac{\left(1-\frac{1}{M}\right)^2}{D}$	$H_0=\dfrac{1+k_3(k_7+k_8)+k_5+k_2}{1-k_4-\dfrac{k_3}{R_{load}}}$
$Z_{in}(s)$ Input impedance	$R_0 \dfrac{\left(1+\frac{s}{\omega_{z_1}}\right)\left(1+\frac{s}{\omega_{z_2}}\right)}{1+\frac{s}{\omega_p}}$	$\omega_{z_1}\approx\dfrac{1}{R_{load}C_2}\dfrac{2M-1}{M-1}$ $\omega_{z_2}\approx 2F_{sw}\dfrac{\left(1-\frac{1}{M}\right)^2}{D}$	$\omega_p=\dfrac{1}{C_2\left[r_C+\dfrac{R_{load}}{1+\dfrac{R_{load}(1-k_4)(k_7+k_8)}{1+k_2+k_5}}\right]}$	$R_0=\dfrac{R_{load}(1-k_4)-k_3}{R_{load}\left[k_7+k_8-k_4(k_7+k_8)\right]+k_2+k_5+1}$
$Z_{out}(s)$ Output impedance	$R_0 \dfrac{1+\frac{s}{\omega_z}}{1+\frac{s}{\omega_p}}$	$\omega_{z_1}=\dfrac{1}{r_C C_2}$	$\omega_{p_1}\approx\dfrac{1}{R_{load}C_2}\dfrac{2M-1}{M-1}$	$R_0=\dfrac{k_3}{k_4-1}\parallel R_{load}$

C_2 is the output capacitor, L_1 the inductor. $m_c=1+\dfrac{S_e}{S_n}$ M is V_{out}/V_{in} and $\tau_L=\dfrac{L_2}{R_{load}T_{sw}}$

Figure 3.163 The four transfer functions of the DCM CM boost converter operated in current-mode control are conveniently gathered in this picture.

3.5 Tapped Boost Transfer Functions in CCM – Voltage-Mode Control

In the dc transfer function of the boost converter derived in Chapter 1.2.7, we have seen the deleterious effects of various losses on the boosting ratio. If one tries to push the conversion ratio too high by increasing the duty ratio D, a latch up effect appears and the output voltage starts diminishing rather than going up. A way to overcome this issue is to choose a tapped inductor and assemble a boost converter as suggested by Figure 3.164.

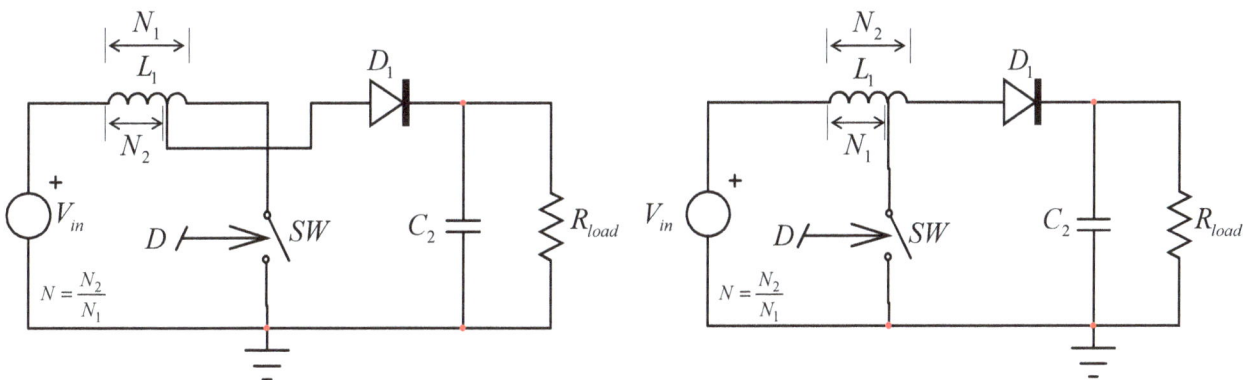

Figure 3.164 The inductor is tapped to connect the freewheel diode or the switch.

Following the recommendations in [3], it is possible to configure the large-signal voltage-mode PWM switch and make it suit both configurations. This is what Figure 3.165 illustrates. In this mode, the current and voltage of the diode in the PWM switch model are different than those of the original configuration.

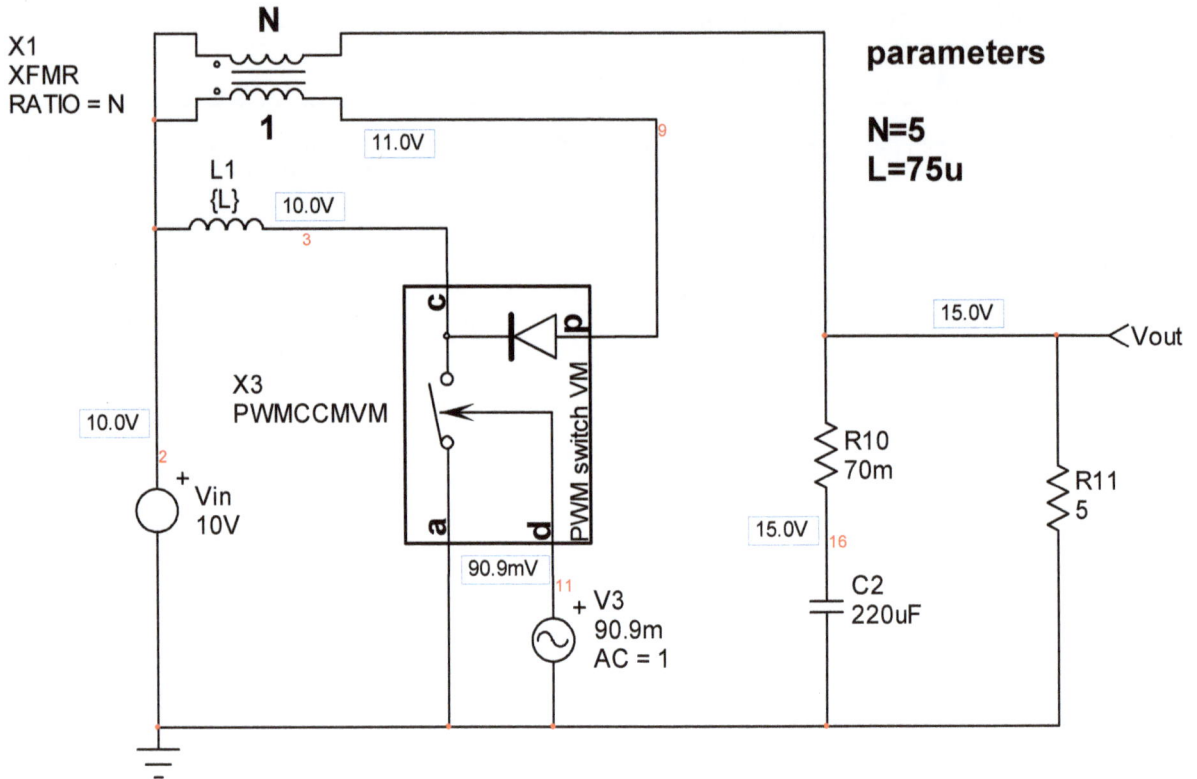

Figure 3.165 A duty ratio below 10% is sufficient to deliver 15 V from the 10-V source.

We start the analysis by determining the dc transfer function. After opening the capacitor, shorting the inductor and simplifying sources equations with $V_{(a)} = 0$, we have the schematic diagram shown in Figure 3.166. We write a first equation defining the voltage at terminal p with N the turns ratio defined as $N = N_2/N_1$.

$$V_{(p)} = V_{in} + \frac{V_{out} - V_{in}}{N} \qquad (3.378)$$

We can also determine this voltage via the controlled source:

$$V_{(p)} - V_{in} = DV_{(p)} \qquad (3.379)$$

It leads to:

$$V_{(p)} = \frac{V_{in}}{1 - D} \qquad (3.380)$$

Equating (3.378) and (3.380) then factoring gives the dc transfer function of the tapped boost:

$$M = \frac{V_{out}}{V_{in}} = 1 + \frac{ND}{1 - D} \qquad (3.381)$$

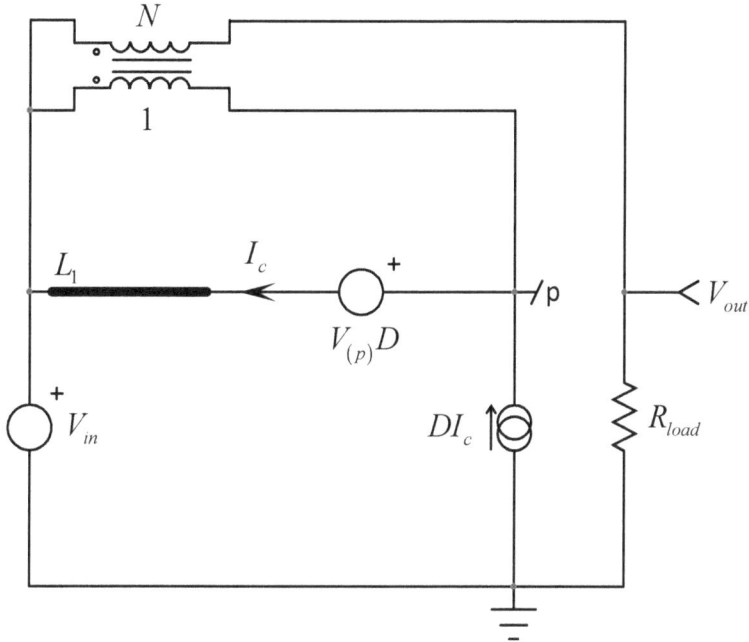

Figure 3.166 The inductor is short circuited to determine the dc transfer function.

With this definition on hand, we can immediately determine the quasi-static gain H_0:

$$H_0 = \frac{d}{dD}V_{out}(D) = \frac{NV_{in}}{(1-D)^2} \tag{3.382}$$

which is the quasi-static gain of the classical voltage-mode boost converter scaled by the turns ratio N. We now need to determine the time constants involving L_1 and C_2. The circuit is updated with the small-signal PWM model as shown in Figure 3.167. If you carry the analysis as we did many times in the previous examples, then you should find:

$$\tau_1 = \frac{L_1 N^2}{R_{load}(1-D)^2} \tag{3.383}$$

$$\tau_2 = r_C C_2 \tag{3.384}$$

$$\tau_2^1 = C_2\left(r_C + R_{load}\right) \tag{3.385}$$

Assembling these elements together leads to the following denominator expression:

$$D(s) = 1 + s\left(\tau_1 + \tau_2\right) + s^2 \tau_1 \tau_2^1 = 1 + s\left[\frac{L_1 N^2}{R_{load}(1-D)^2} + r_C C_2\right] + s^2 L_1 C_2 \frac{N^2\left(r_C + R_{load}\right)}{R_{load}(1-D)^2} \tag{3.386}$$

We can factor this formula in the well-known 2nd-order canonical form:

$$D(s) = 1 + \frac{s}{\omega_0 Q} + \left(\frac{s}{\omega_0}\right)^2 \tag{3.387}$$

with:

$$\omega_0 = \frac{1-D}{N\sqrt{L_1 C_2}}\sqrt{\frac{R_{load}}{R_{load} + r_C}} \tag{3.388}$$

$$Q = \frac{1}{\omega_0\left[\dfrac{L_1 N^2}{R_{load}(1-D)^2} + r_C C_2\right]} \tag{3.389}$$

The zeroes are obtained by nulling the response V_{out} while the excitation $D(s)$ is back in place. We have the usual suspects r_C and the output capacitor:

$$\omega_{z_1} = \frac{1}{r_C C_2} \tag{3.390}$$

While a RHP zero contributed by L_1 shows up:

$$\omega_{z_2} = \frac{R_{load}(1-D)}{M \cdot L_1 \cdot N} \tag{3.391}$$

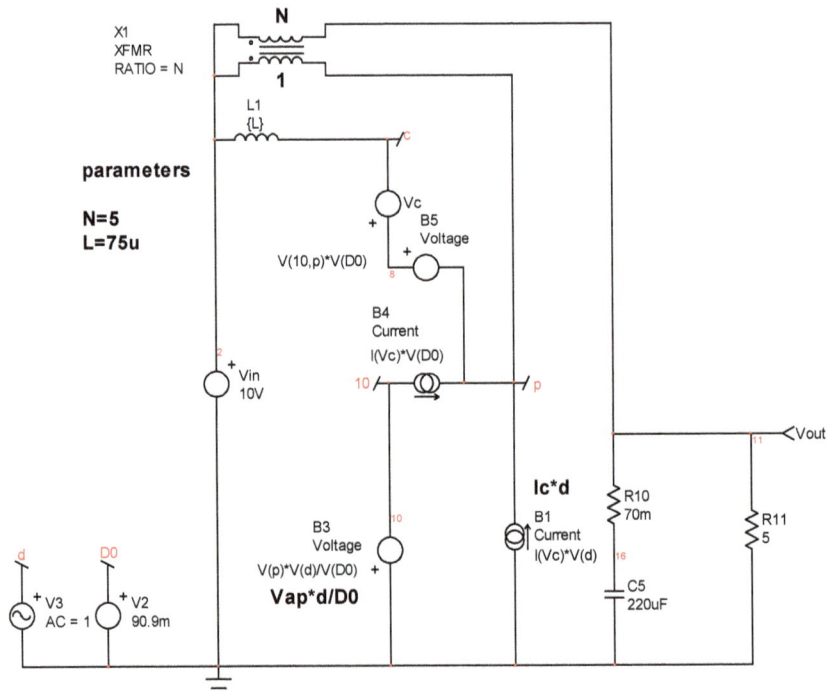

Figure 3.167 The small-signal model is in place and we can compute the various time constants.

The complete control-to-output transfer function is thus given by:

$$H(s) = H_0 \frac{\left(1 + \dfrac{s}{\omega_{z_1}}\right)\left(1 - \dfrac{s}{\omega_{z_2}}\right)}{1 + \dfrac{s}{\omega_0 Q} + \left(\dfrac{s}{\omega_0}\right)^2} \tag{3.392}$$

We have compared the dynamic response delivered by our circuit from Figure 3.165 and that of the Mathcad® sheet using (3.392). The results appear in Figure 3.168 and show a perfect matching.

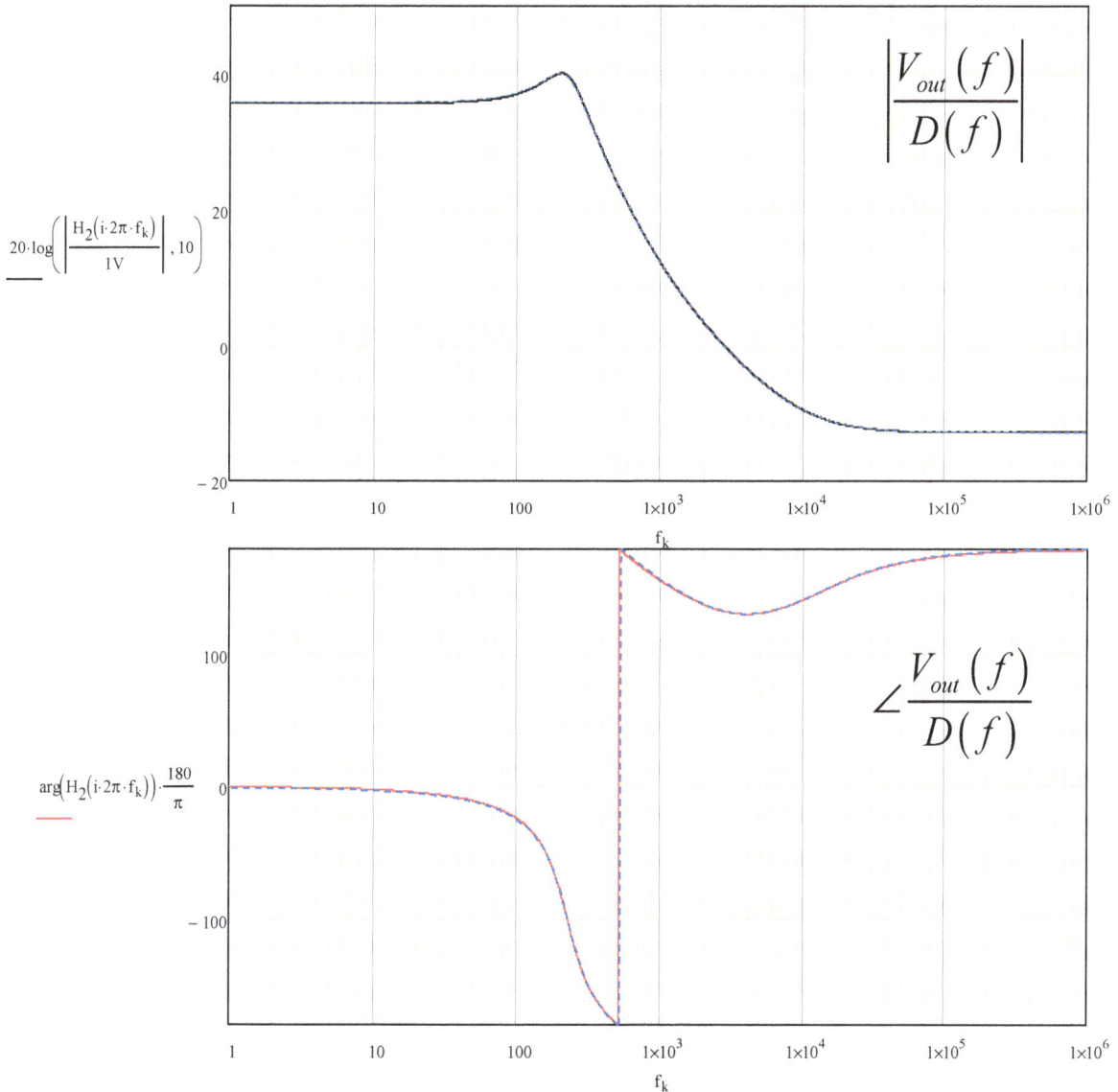

Figure 3.168 Excellent matching between the large-signal model and the transfer function we have derived.

To further confirm our analysis is correct, we have captured the schematic diagram of a tapped boost converter in SIMPLIS® (Figure 3.169).

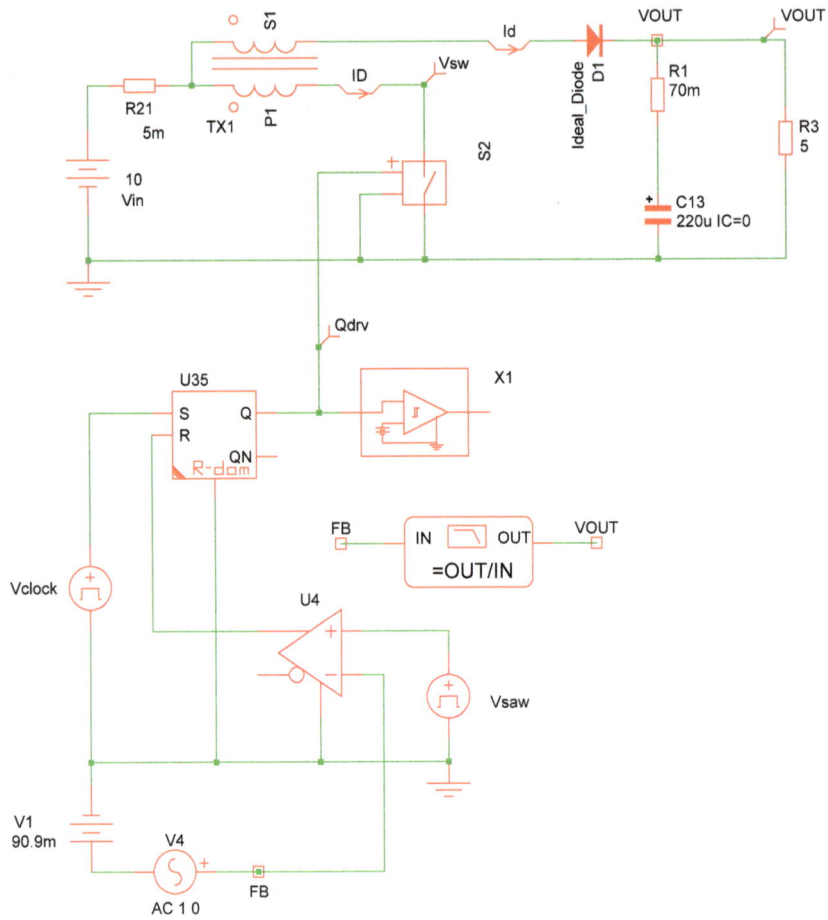

Figure 3.169 SIMPLIS® will give us the small-signal response from this switching circuit.

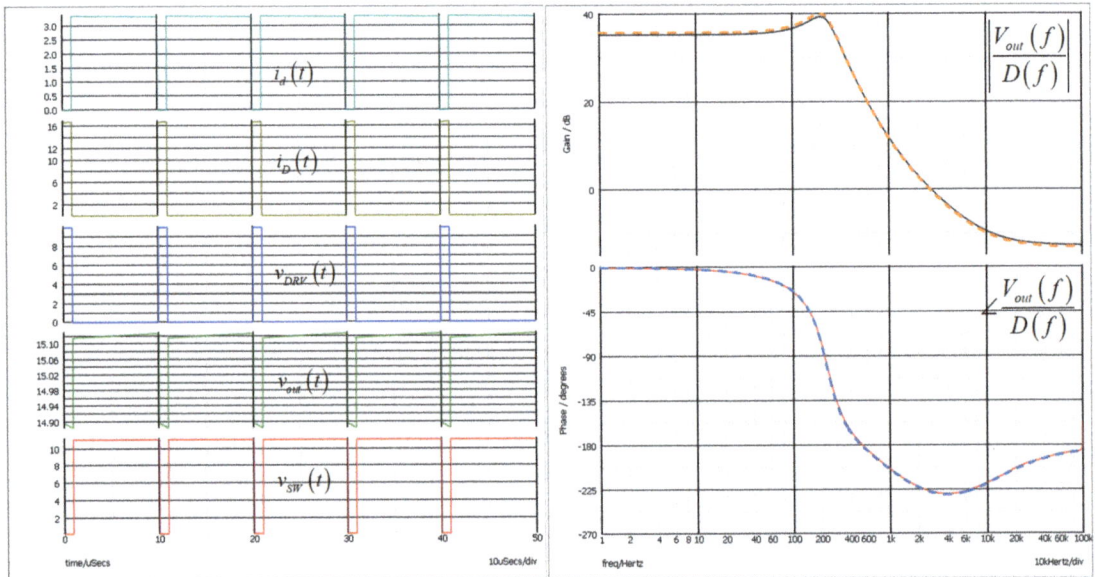

Figure 3.170 Responses from SIMPLIS® and SPICE are in perfect agreement.

As shown in Figure 3.170, the results are excellent and confirm the approach with the PWM switch model.

3.6 Fixed Frequency Multiphase Boost Transfer Function – Voltage- and Current-Mode Control

The boost converter can also be designed as a multiphase power supply to increase its output power capability and efficiency. It is the case, for instance, in battery-powered applications delivering a large output current. The simulation circuit appears in Figure 3.171 with a 2-phase example. The 50% maximum duty ratio brought by cross-connecting the 200-kHz clock signals to the reset inputs as in the buck example is replaced by independent generators which set the limit to 60% in this example.

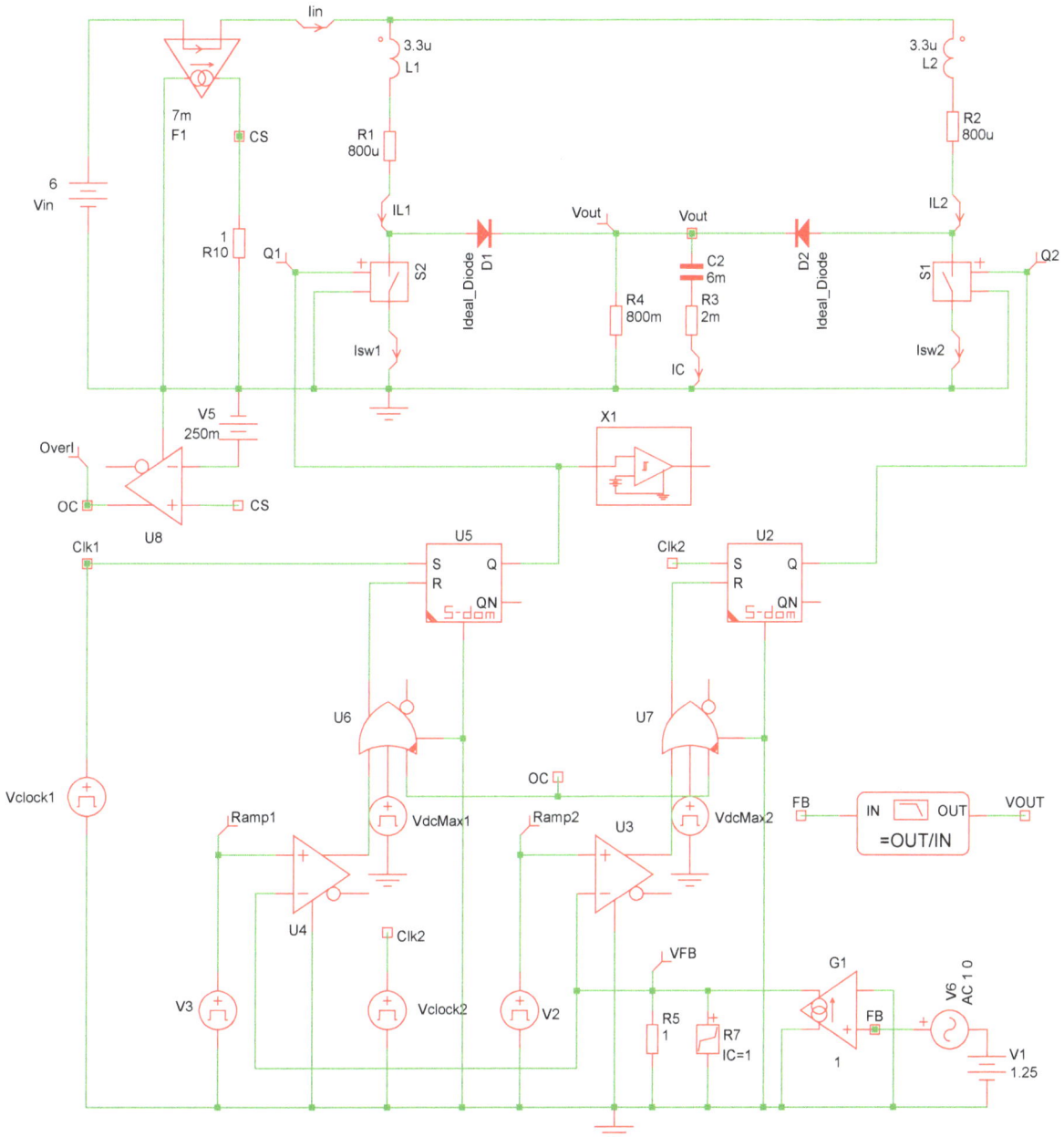

Figure 3.171 This is the two-phase version of the boost converter simulated with SIMPLIS®.

411

The converter is supplied from a 6-V source to deliver 12 V at a 15-A nominal current. The simulation results are given in Figure 3.172 and show a duty ratio slightly above 50% for a 12-V output. There is a slight overlap between the phases at this low input voltage.

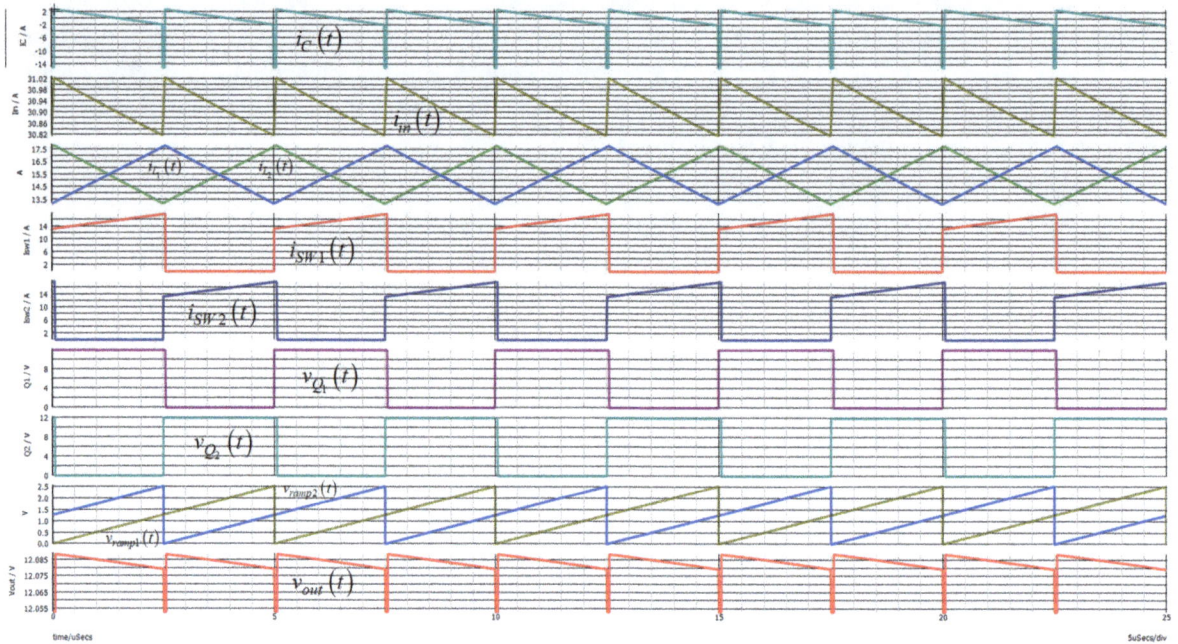

Figure 3.172 The inductor current is well interleaved in this 2-phase boost converter.

As stated in Chapter 2, the equivalent single-inductance power stage small-signal model features an inductance L_{eq} equal to the inductance L per phase divided by the number of phases n:

$$L_{eq} = \frac{L}{n} = \frac{3.3u}{2} = 1.65\ \mu\text{H} \tag{3.393}$$

The rest of the elements in the small-signal formulas from the CCM voltage-mode fixed-frequency analysis remain of identical values. With such an inductance and a 51% duty ratio at a 6-V input voltage, we can expect a resonant frequency tuned at:

$$f_0 = \frac{1-D}{2\pi\sqrt{L_{eq}C_{out}}} = \frac{1-0.51}{6.28 \times \sqrt{1.65u \times 6m}} \approx 784\ \text{Hz} \tag{3.394}$$

The ac simulation of Figure 3.173 confirms a resonant frequency occurring at the above value.

A current-mode version of this 2-phase converter appears in Figure 3.174. Considering the duty ratio beyond 50%, some slope compensation is added via controlled sources G_2 and G_3. The 50m value corresponds to a 10-kV/s artificial ramp S_e added to the current sense signal. The setpoint is adjusted by a divide-by-20 block (G_1) which ensures a comfortable control voltage V_c despite a 7-mΩ sense resistance.

The operating waveforms are given in Figure 3.175 and are very similar to those of the voltage-mode version at steady state. The input current is around 31 A while only half of it circulates in each inductor, clearly easing thermal management per branch.

Like what we have stated for the multiphase buck converter operated in current-mode control, the component values feeding the original boost formulas need an update as follows. The equivalent inductance for the single-inductor model is the same as in the multiphase voltage-mode version given in (3.393).

The sense resistance also needs an update: the single-inductance equivalent small-signal model has a current loop gain amplified by a ratio depending on the number of phases n.

Therefore, we have:

$$R_{iequ} = \frac{R_i}{n} = \frac{7 \text{ m}\Omega}{2} = 3.5 \text{ m}\Omega \tag{3.395}$$

The rest of the contributors in the small-signal formulas previously derived for the boost converter operated in CCM and current-mode control remain the same.

Using these values, we determine the quasi-static gain to be:

$$H_0 = \frac{R_{load}}{Div \cdot R_{ieq}} \frac{1}{2M + \frac{R_{load}T_{sw}}{L_{eq}M^2}\left(\frac{1}{2} + \frac{S_e}{S_n}\right)} = \frac{800m}{20 \times 3.5m} \frac{1}{2 \times 2 + \frac{800m \times 5u}{1.65u \times 2^2}\left(0.5 + \frac{10k}{12.7k}\right)} \approx 2.4 \text{ or } 7.5 \text{ dB} \tag{3.396}$$

Div in this expression represents the divide-by-20 block and M is 2.

The simulation shows a 0.2-dB difference probably explained by the various losses in the circuit.

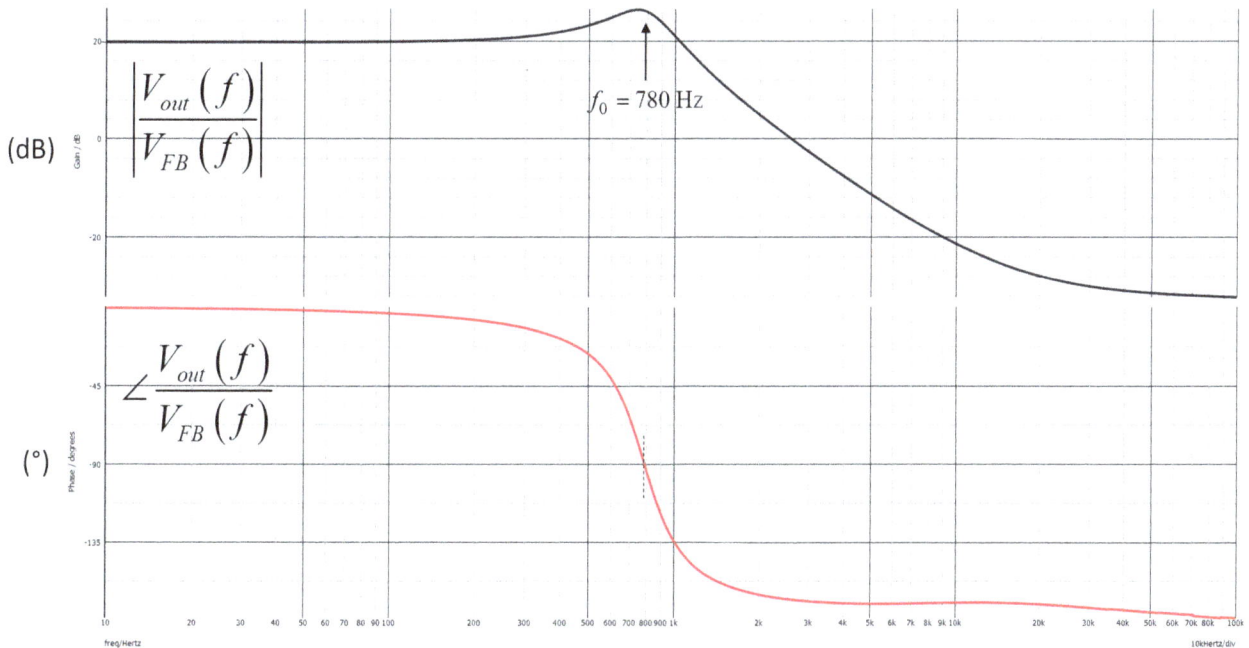

Figure 3.173 SIMPLIS® confirms a resonant frequency around 780 Hz.

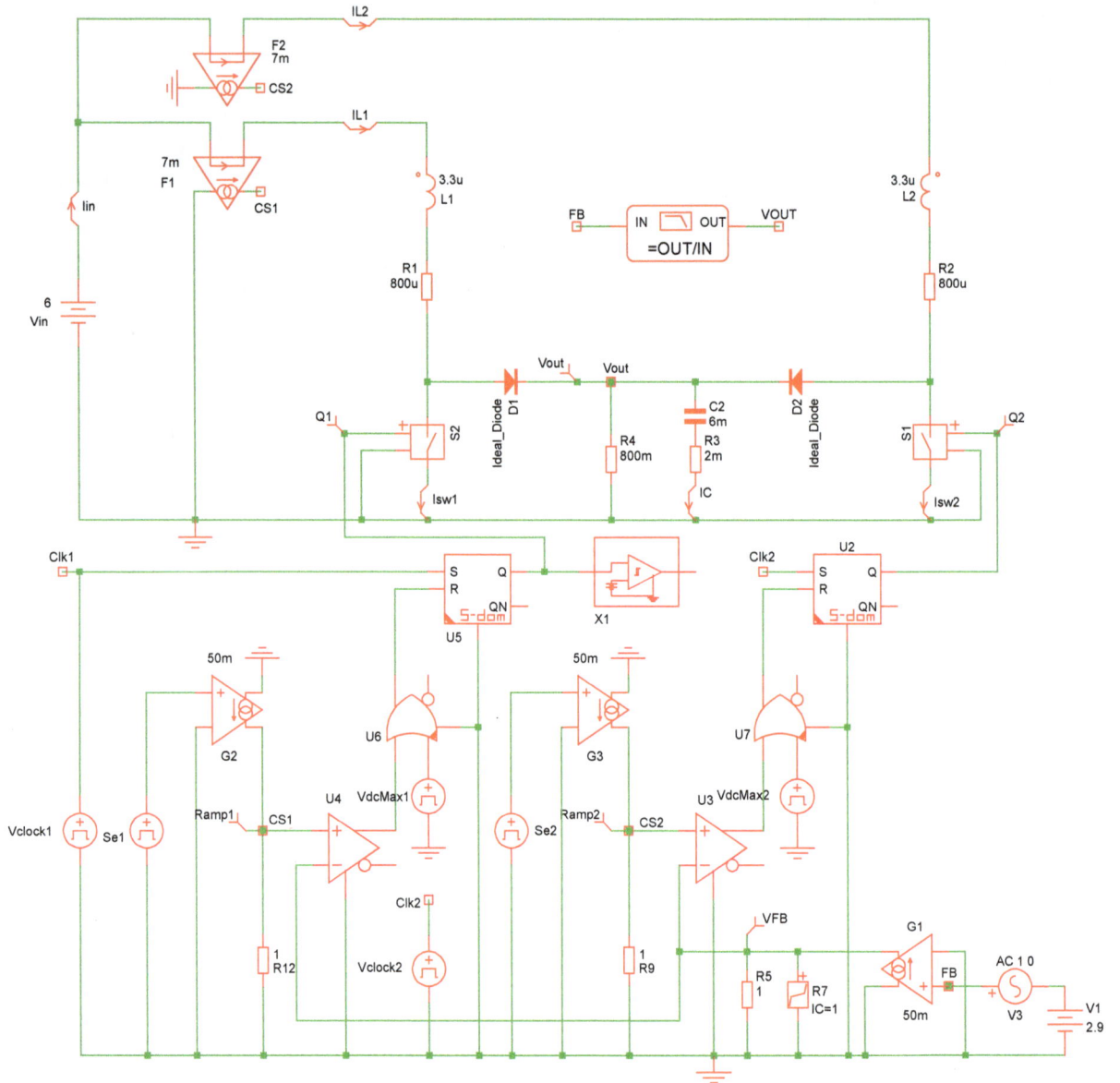

Figure 3.174 In current-mode control, some slope compensation is necessary to fight subharmonic oscillations.

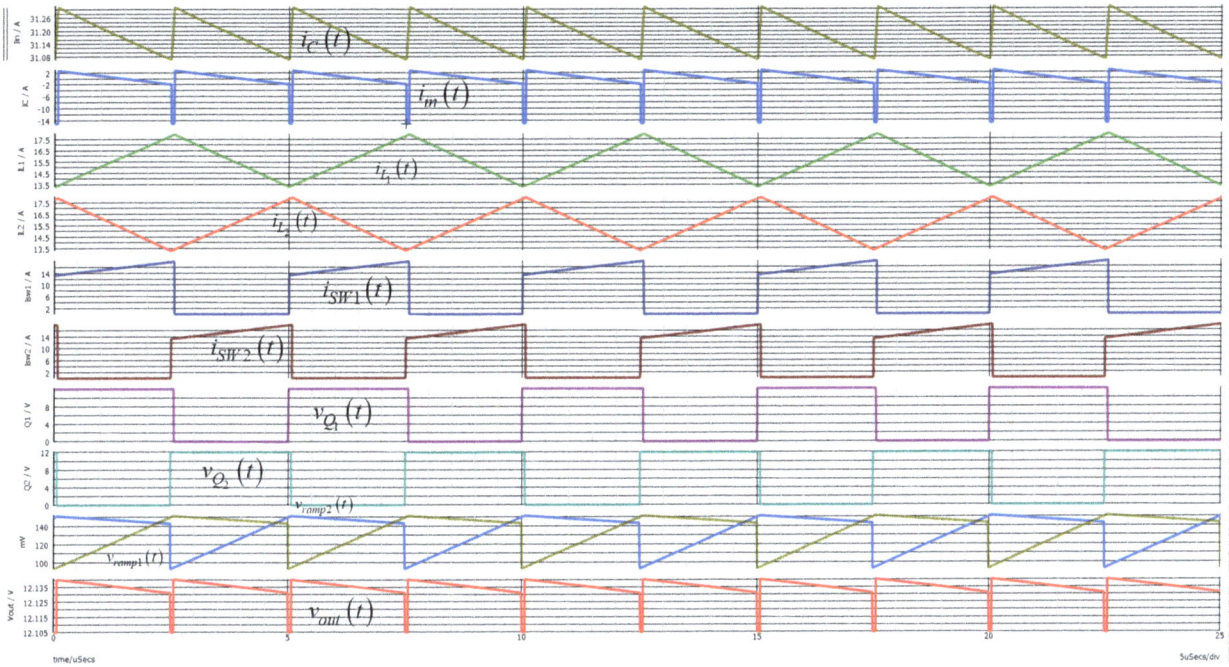

Figure 3.175 The steady-state waveforms show a smooth input current of 31 A.

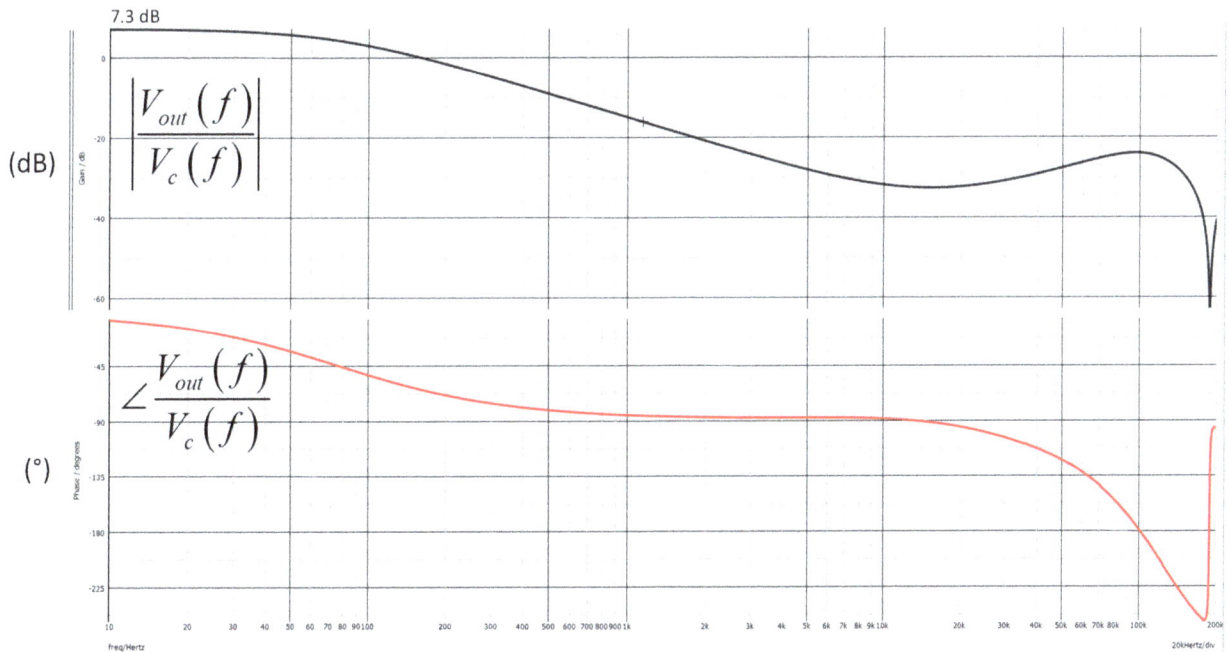

Figure 3.176 The response from SIMPLIS® shows a quasi-static gain of 7.3 dB for this 2-phase CM boost converter.

3.7 Quasi-Resonant Boost Transfer Function – Current-Mode Control

The boost converter can be advantageously used in the quasi-square-wave resonant mode also called quasi-resonant or QR. In this mode, zero-voltage switching is feasible and the stress linked to CCM disappears. Synchronous rectification for low output voltage is less of an issue as shoot-through has naturally gone with the discontinuous operation. The control-to-output transfer function can be determined using the small-signal model derived in Chapter 1. This is what Figure 3.177 shows in which two ports have been added to display the switching frequency F_{sw} and the on-time.

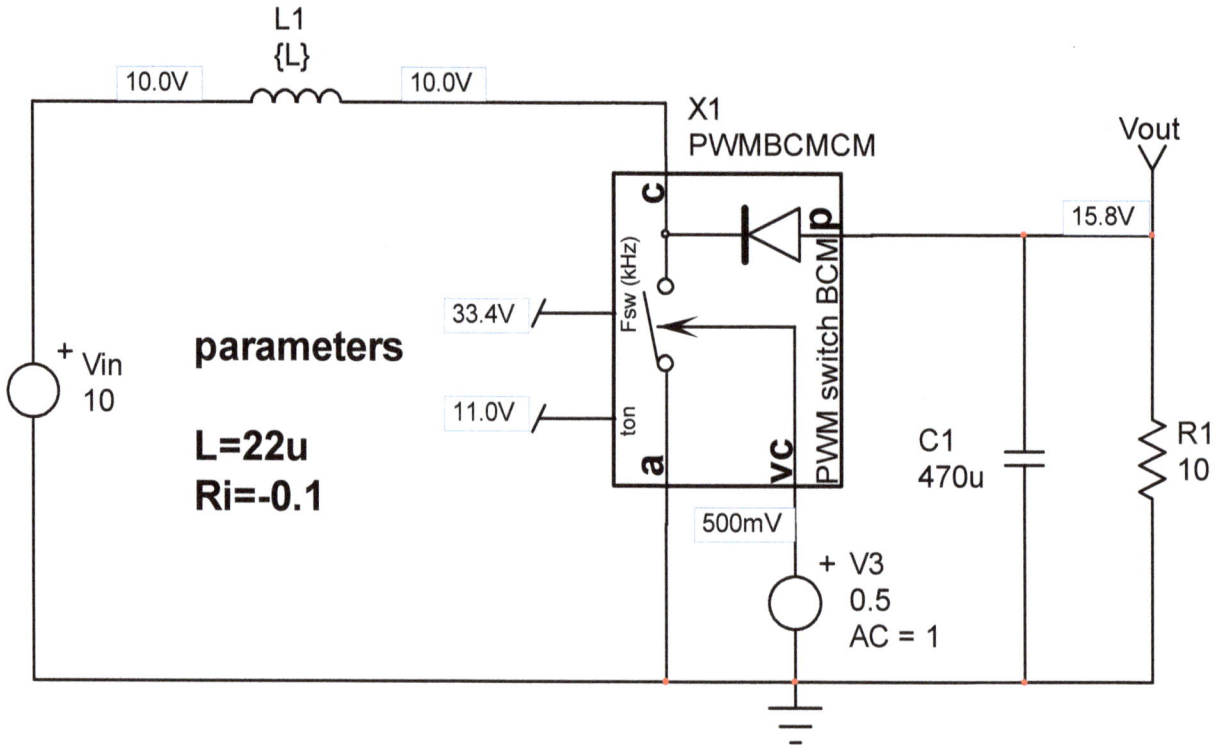

Figure 3.177 The PWM switch in its QR version lends itself well to analyzing this boost converter.

The control voltage is 500 mV and imposes a peak current of 5 A considering the 100-mΩ sense resistance. Please note that this resistance is negative here as, by convention, the I_c current must leave terminal c while it enters it in the boost structure. Once the small-signal version is plugged in, we have the simulation circuit of Figure 3.178. We have verified that the simulation of both templates leads to the exact same Bode plots.

The simplified version is shown in Figure 3.179. You can see two energy-storing elements and believe it could be a second-order circuit. However, the current source drives the inductor current and fixes its state variable: a degree is lost in the denominator and this is a first-order network.

Coefficients k_{cp}, k_{ac} and k_{ic} are determined from variables affecting the three terminals a, c and p. We can advantageously rework them for this specific application, leading to considerable simplifications and ease of manipulation later on. By observing the circuit, we can infer the following static definitions:

$$V_{cp} = V_{in} - V_{out} \tag{3.397}$$

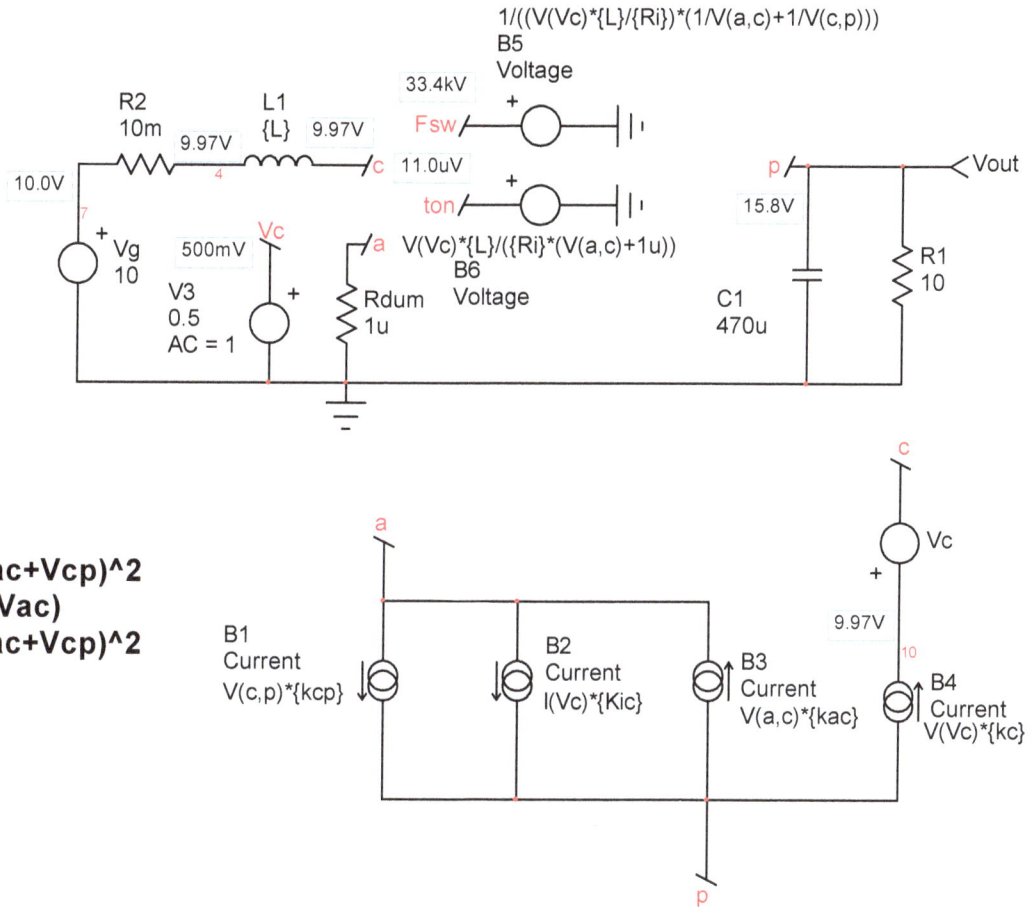

parameters

L=22u
Ri=-0.1

Vin=10
Vout=15.6
M=Vout/Vin
Ic=-2.5
Vac=-10
Vap=-Vout
Vcp=Vin-Vout
kcp=Ic*Vac/(Vac+Vcp)^2
kic=Vcp/(Vcp+Vac)
kac=Vcp*Ic/(Vac+Vcp)^2
kc=1/(2*Ri)

R2
10m

L1
{L}

33.4kV

1/((V(Vc)*{L}/{Ri})*(1/V(a,c)+1/V(c,p)))
B5
Voltage

Fsw

ton

V(Vc)*{L}/({Ri}*(V(a,c)+1u))
B6
Voltage

Vg
10

V3
0.5
AC = 1

Rdum
1u

C1
470u

R1
10

Vout

B1
Current
V(c,p)*{kcp}

B2
Current
I(Vc)*{Kic}

B3
Current
V(a,c)*{kac}

B4
Current
V(Vc)*{kc}

Vc

Figure 3.178 Once the small-signal model is connected, we can start working on transfer functions of our choice.

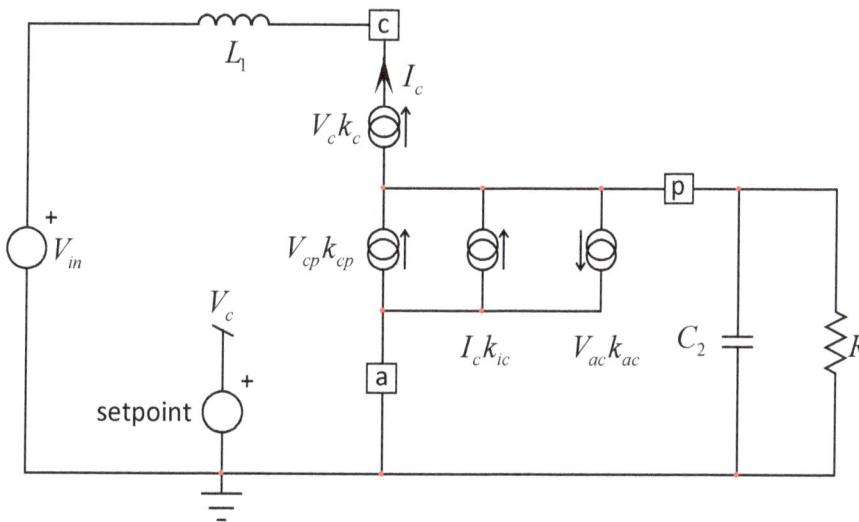

Figure 3.179 Four linear current sources form the small-signal model of the QR CM-PWM switch model.

$$I_c = -\frac{I_{out}V_{out}}{V_{in}} \qquad (3.398)$$

$$V_{ac} = -V_{in} \qquad (3.399)$$

If we now plug these expressions in the coefficient's raw expressions, we have:

$$k_{cp} = \frac{I_{out}V_{out}}{V_{in}} \frac{V_{in}}{(-V_{in}+V_{in}-V_{out})^2} = \frac{1}{R_{load}} \qquad (3.400)$$

$$k_{ac} = -\frac{I_{out}V_{out}}{V_{in}} \frac{(V_{in}-V_{out})}{(-V_{in}+V_{in}-V_{out})^2} = \frac{M}{R_{load}} - \frac{1}{R_{load}} \qquad (3.401)$$

$$k_{ic} = \frac{V_{in}-V_{out}}{V_{in}-V_{out}-V_{in}} = 1 - \frac{1}{M} \qquad (3.402)$$

$$k_c = \frac{1}{2R_i} \qquad (3.403)$$

We have everything we need to start the analysis by determining the quasi-static gain H_0 of the power stage. The inductor is shorted while the capacitor is open-circuited.

This is what Figure 3.180 represents in which the current source $V_c k_c$ has been purposely reversed to account for the direction in I_c: R_i is now a positive value.

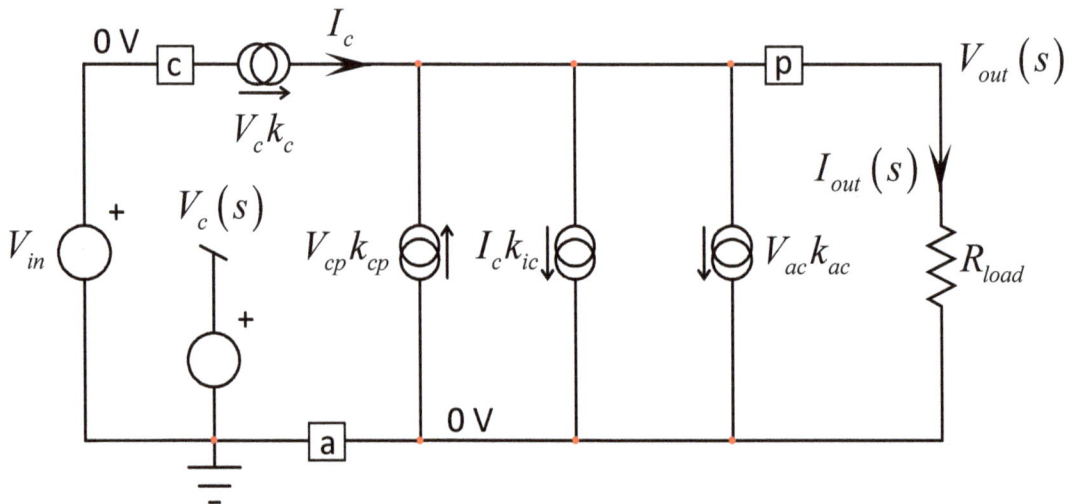

Figure 3.180 The quasi-static gain is classically determined by shorting the inductor and opening the output capacitor.

In this circuit, the input source V_{in} is actually 0 V in ac and terminal a is grounded.

The equation expressing the output current is:

$$I_{out}(s) = V_c(s)k_c + \left(V_{(c)} - V_{(p)}\right)k_{cp} - V_c(s)k_c k_{ic} - \left(V_{(a)} - V_{(c)}\right)k_{ac} \qquad (3.404)$$

Updating this expression with sources input and output names, we have:

$$I_{out}(s) = V_c(s)k_c - V_{out}(s)k_{cp} - V_c(s)k_ck_{ic} = V_c(s)k_c(1-k_{ic}) - V_{out}(s)k_{cp} \qquad (3.405)$$

The output current is simply:

$$I_{out}(s) = \frac{V_{out}(s)}{R_{load}} \qquad (3.406)$$

If you substitute this definition in (3.405) and rearrange the fraction V_{out}/V_c, you must find:

$$\frac{V_{out}(s)}{V_c(s)} = \frac{k_c(1-k_{ic})}{k_{cp} + \dfrac{1}{R_{load}}} \qquad (3.407)$$

Now substitute the coefficients definitions in the above expression, simplify and obtain:

$$H_0 = \frac{R_{load}}{4MR_i} \qquad (3.408)$$

The first time constant involves inductor L_1 and is determined by setting the excitation to zero: set V_c to 0 V and update Figure 3.179 to determine the resistance driving the inductor.

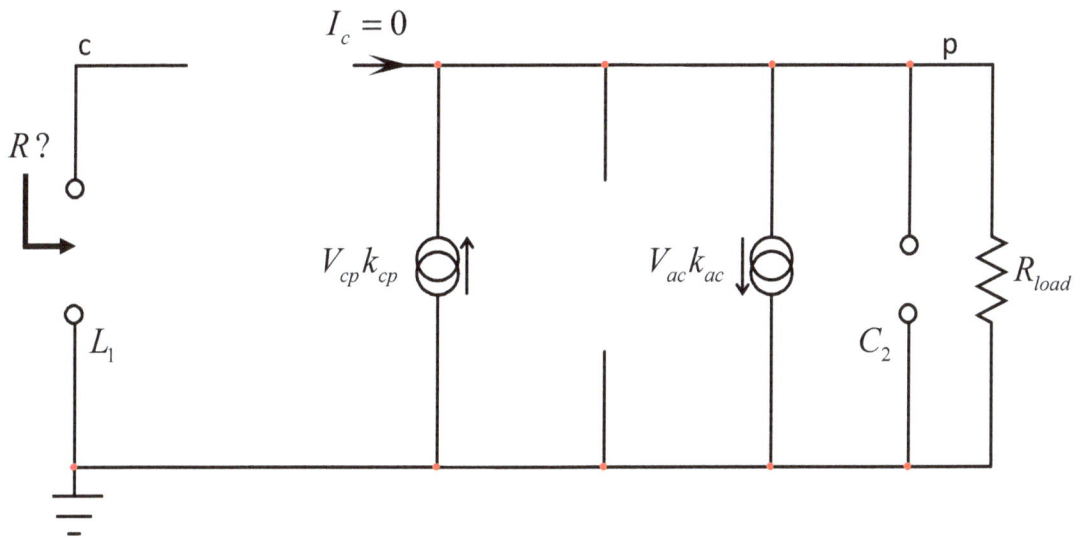

Figure 3.181 Reducing the excitation to 0 V brings current I_c to 0 A.

In this mode, the inductor has a non-connected terminal and its time constant is zero:

$$\tau_1 = \frac{L_1}{\infty} = 0 \qquad (3.409)$$

For the capacitor, bring the inductor in its dc state – a short circuit – and determine the resistance seen from the capacitor terminals. Because terminals c and a are at 0 V, the source involving V_{ac} is turned off. The current source $V_{cp}k_{cp}$ can also be rearranged to further simplify the analysis:

$$V_{(cp)}k_{cp} = k_{cp}\left[V_{(c)} - V_{(p)}\right] = -V_{(p)}k_{cp} = -\frac{V_{out}}{R_{load}}$$

(3.410)

Replace the source by a resistance R_{load} and solve the circuit shown in Figure 3.182.

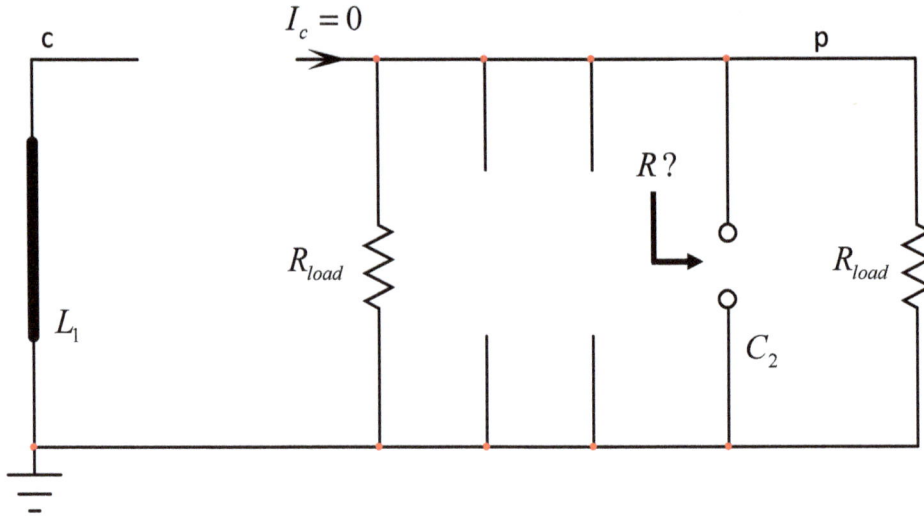

Figure 3.182 Once the circuit is properly rearranged, it is a child's play to determine the second time constant.

By inspection, you immediately determine the second time constant:

$$\tau_2 = C_2 \frac{R_{load}}{2}$$

(3.411)

With these results, we can express b_1 in the denominator:

$$b_1 = \tau_1 + \tau_2 = C_2 \frac{R_{load}}{2}$$

(3.412)

For the second-order term, $\tau_2 \tau_1^2$, C_2 is set in its high-frequency state (a short circuit) while you determine the resistance seen from L_1's terminals. It is what has been done in Figure 3.181 except that C_2 is now shorted. It does not change the result as L_1's right terminal is floating. Therefore, we have a finite value multiplied by zero which leads $\tau_2 \tau_1^2$ to be zero also:

$$b_2 = \tau_2 \tau_1^2 = C_2 \frac{R_{load}}{2} \cdot \frac{L_2}{\infty} = 0$$

(3.413)

The denominator is thus defined as:

$$D(s) = 1 + sb_1 = 1 + \frac{s}{\omega_p}$$

(3.414)

with:

$$\omega_p = \frac{2}{R_{load}C_2} \tag{3.415}$$

Now that we have the pole position, we can null the response and determine the zero. The excitation V_c is brought back in the circuit as illustrated in Figure 3.183.

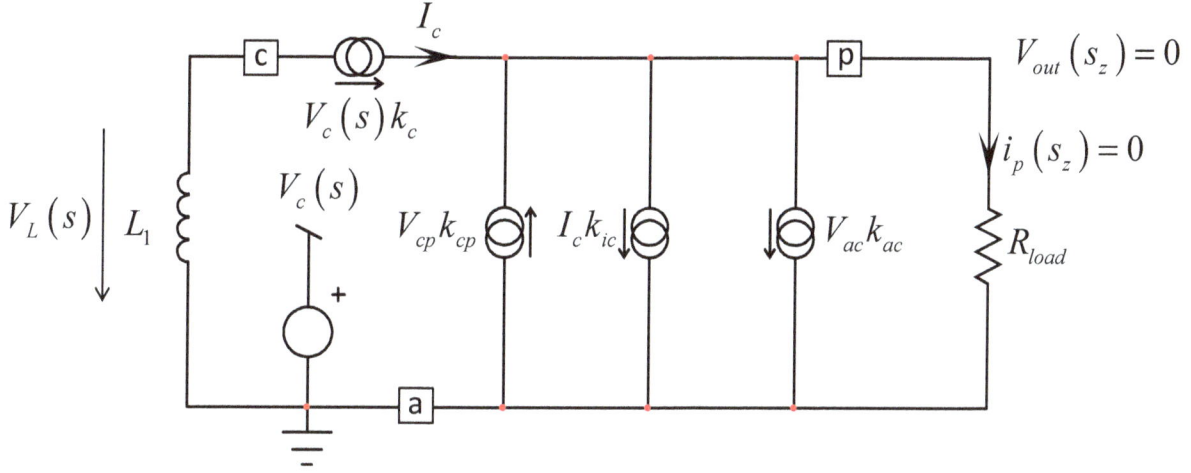

Figure 3.183 The response is nulled for a certain value of s.

In this configuration, there is no current flowing in the output resistance R_{load} if all the inductor current added to the first left-side current source is absorbed by the two remaining sources:

$$V_c(s)k_c + \left[V_{(c)}(s) - V_{(p)}(s)\right]k_{cp} = V_c(s)k_c k_{ic} + \left[V_{(a)}(s) - V_{(c)}(s)\right]k_{ac} \tag{3.416}$$

The voltage at node c is defined as:

$$V_{(c)}(s) = -V_L(s) = -I_c(s)sL_1 = -V_c(s)k_c sL_1 \tag{3.417}$$

Now substitute this definition in (3.416), simplify by V_c and k_c then solve the equation:

$$1 - k_{ic} = sL_1\left(k_{ac} + k_{cp}\right) \tag{3.418}$$

It leads to a right-half-plane root located at:

$$s_z = \frac{1 - k_{ic}}{L_1\left(k_{ac} + k_{cp}\right)} \tag{3.419}$$

After substituting the coefficients definitions and rearranging the result, we have:

$$\omega_{z_2} = \frac{R_{load}}{L_1 M^2} \tag{3.420}$$

If we now add the output capacitor equivalent series resistance r_C, we know that it contributes another zero located at:

$$\omega_{z_1} = \frac{1}{r_C C_2} \tag{3.421}$$

The complete transfer function can now be expressed as:

$$H(s) = \frac{V_{out}(s)}{V_c(s)} = H_0 \frac{\left(1 + \dfrac{s}{\omega_{z_1}}\right)\left(1 - \dfrac{s}{\omega_{z_2}}\right)}{1 + \dfrac{s}{\omega_p}} \tag{3.422}$$

The gain and the pole/zero are summarized in Figure 3.184.

$\dfrac{V_{out}(s)}{V_{err}(s)}$ Control to Output	$H_0 \dfrac{\left(1 + \dfrac{s}{\omega_{z_1}}\right)\left(1 - \dfrac{s}{\omega_{z_2}}\right)}{1 + \dfrac{s}{\omega_p}}$	$\omega_{z_1} = \dfrac{1}{r_C C_2}$ $\omega_{z_2} \approx \dfrac{R_{load}}{L_1 M^2}$	$\omega_p = \dfrac{2}{R_{load} C_2}$	$H_0 = \dfrac{R_{load}}{4 M R_i}$

C_2 is the output capacitor, L_1 the inductor. M is V_{out}/V_{in}

Figure 3.184 The QR-operated boost converter features a RHP zero despite discontinuous mode operation.

If we now gather the response of Figure 3.177 circuit and that of the Mathcad® sheet, we obtain a perfect match as shown in Figure 3.185.

To check if the whole approach is correct, I have built a SIMPLIS® model and it appears in Figure 3.186. The demagnetization is detected by observing the drain suddenly deviating from the voltage temporarily stored across C_5. A delay brought by $R_2 C_1$ lets us turn the switch on in the drain-source valley. The simulation results are given in Figure 3.187. The left-side simulation in Figure 3.187 confirms an output voltage of 15.7 V, close to what the average model delivers (15.8 V) and the switching frequency is 32.6 kHz (33.4 kHz for the model). The dynamic response confirms the dc gain and the pole position at 68 Hz. The RHPZ bends the phase response further as the magnitude gets flatter.

As the stimulus frequency approaches the switching frequency, we can observe some magnitude peaking which is normal.

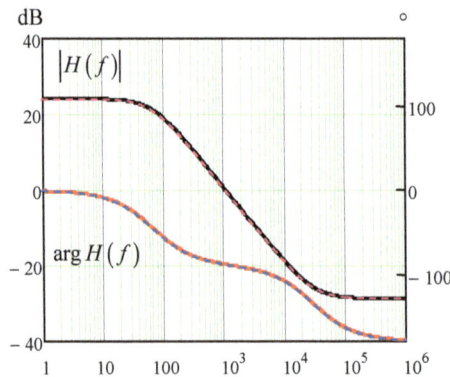

Figure 3.185 The curves obtained with the large-signal model perfectly match those obtained with Mathcad®.

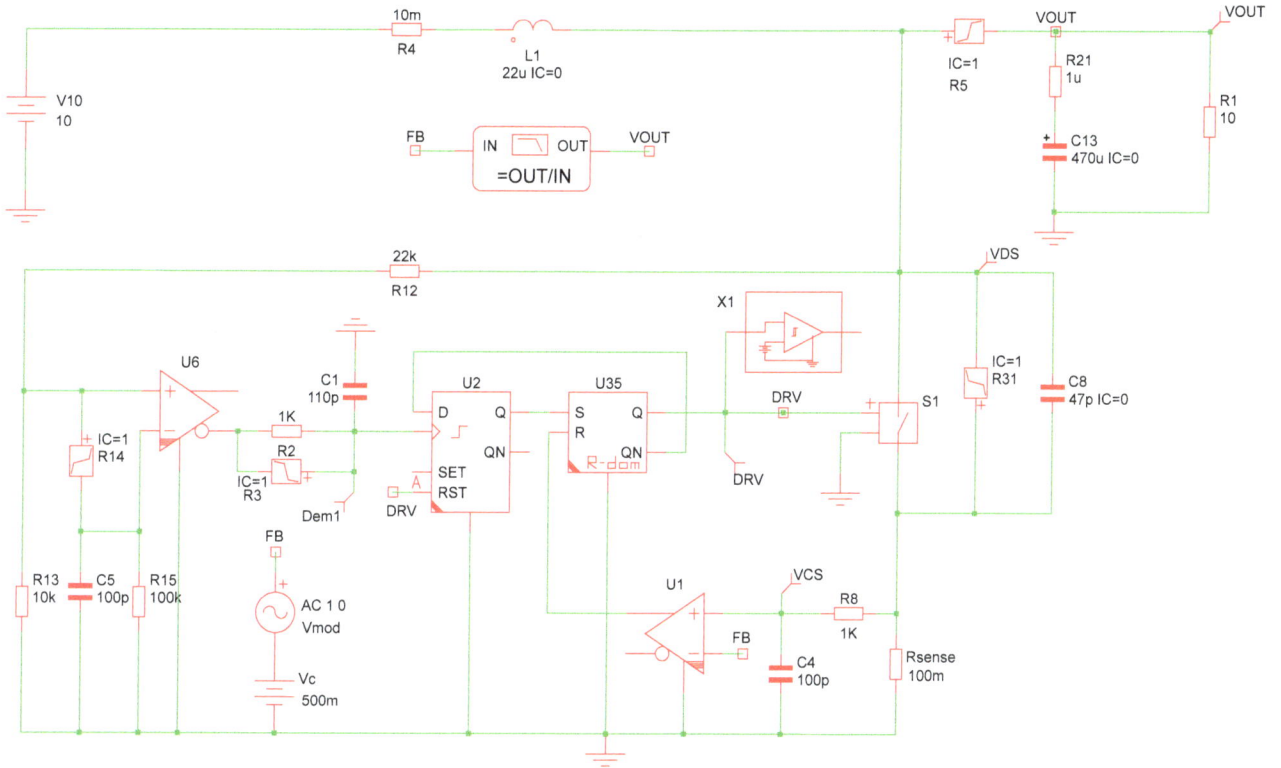

Figure 3.186 A SIMPLIS® switching circuit will give us the small-signal response in a few seconds.

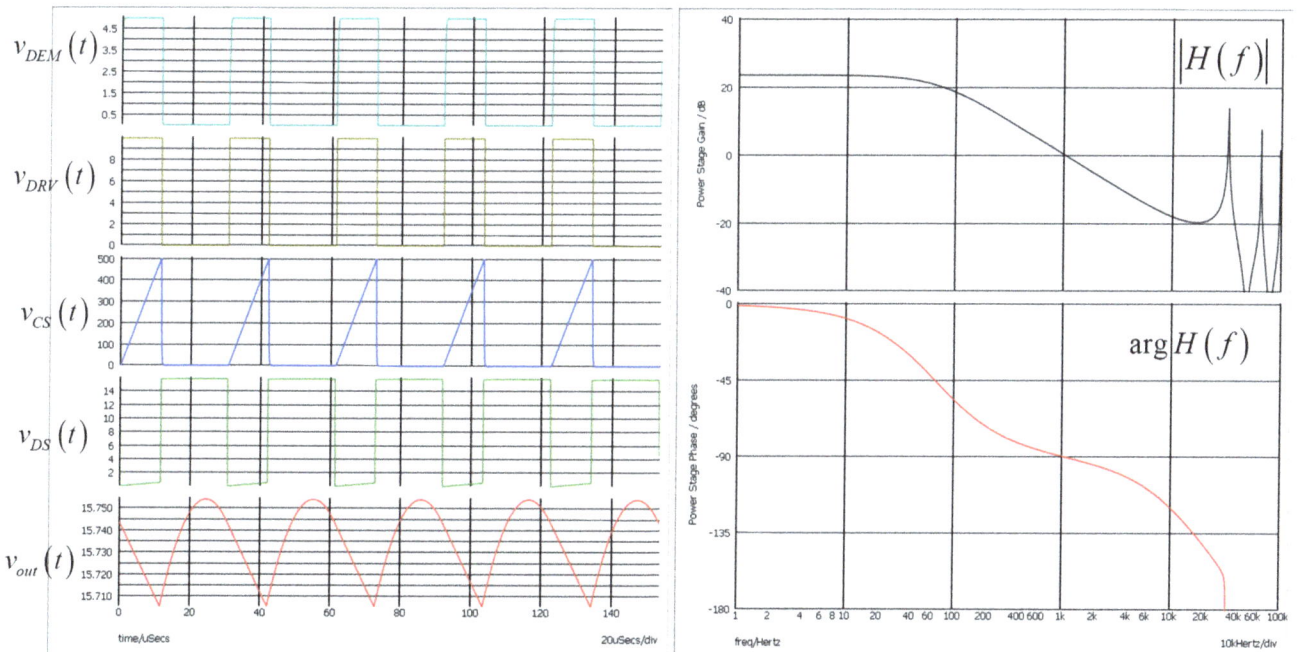

Figure 3.187 The simulation confirms the overall small-signal response and the presence of the RHPZ in particular.

3.8 Boost in Power Factor Correction – Boundary Mode Voltage–Mode Control

Boost converters are often employed as power factor correction circuits (PFC). Placed before a high-voltage dc-dc converter – they are designated as *pre-converters* – they force the absorption of a sinusoidal current from the electric grid as a purely resistive load would do. A popular scheme is the free-running QR boost converter as we described in the above lines. Used in constant on-time configuration, it can be operated in voltage- or current-mode control. In the literature, you find it under the name *borderline* or *boundary* current-mode boost converter (BCM) but also *critical* conduction mode (CrM). Many variations exist around the boost core and these structures have been thoroughly documented in [4]. Unlike a classical switching converter in which you need a certain crossover frequency for stability and performance reasons, the bandwidth of a PFC is naturally low considering the large ripple voltage at twice the mains frequency present on the output. As you do not want the control loop to react on this ripple for regulation but also for input current distortion purposes, the adopted crossover frequency is usually low, around a few Hz at low line (85 V rms). A model is thus necessary to predict the magnitude and phase responses of such a power stage so that you can determine the compensation elements adequately.

Rather than using the classical PWM switch model, we will consider the boost converter as a power-controlled current source driving the bulk capacitor and its load, the downstream converter. This approach is described in [5] and we will apply it to a voltage-mode circuit first. The principle is shown in Figure 3.188 where you see a current source delivering current to an output network. This source represents the power delivered over a complete mains cycle with 100% efficiency. We are going to determine the large-signal expression describing this source and linearize it to form the transfer function we want. This simple approach considers a power transfer without "incidents" such as right-half-plane zeroes or sub-harmonic oscillations if any. However, as these components usually manifest themselves at high frequencies, they can safely be ignored in these low-bandwidth systems. Same for the zero contributed by r_C, it will not be of any phase boost help below 100 Hz and can be neglected as we will see.

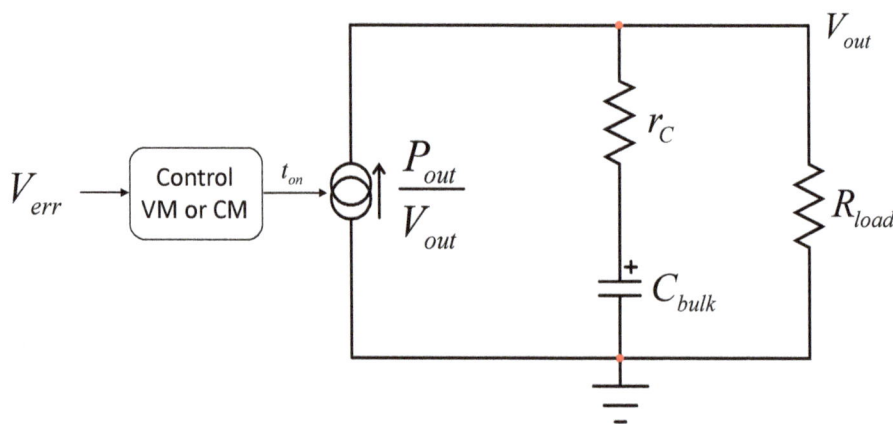

Figure 3.188 The PFC is modeled as a current source feeding the output network.

A typical voltage-mode PFC schematic diagram captured in SIMPLIS® appears in Figure 3.189. The PWM block around U_1 generates a t_{on} whose value is proportional to the delivered power. It remains constant along the mains cycle as we will see later on. The transconductance amplifier G_1 and its clamping network generate the error voltage controlling the on-time. This strategy forces the inductor peak current to follow a haversine envelope and ensures the power factor correction action we want as illustrated in Figure 3.190. Critical conduction is obtained by monitoring via U_3 the voltage at the auxiliary winding S_1 added to the main inductor (250 µH in P_1 for this simulation).

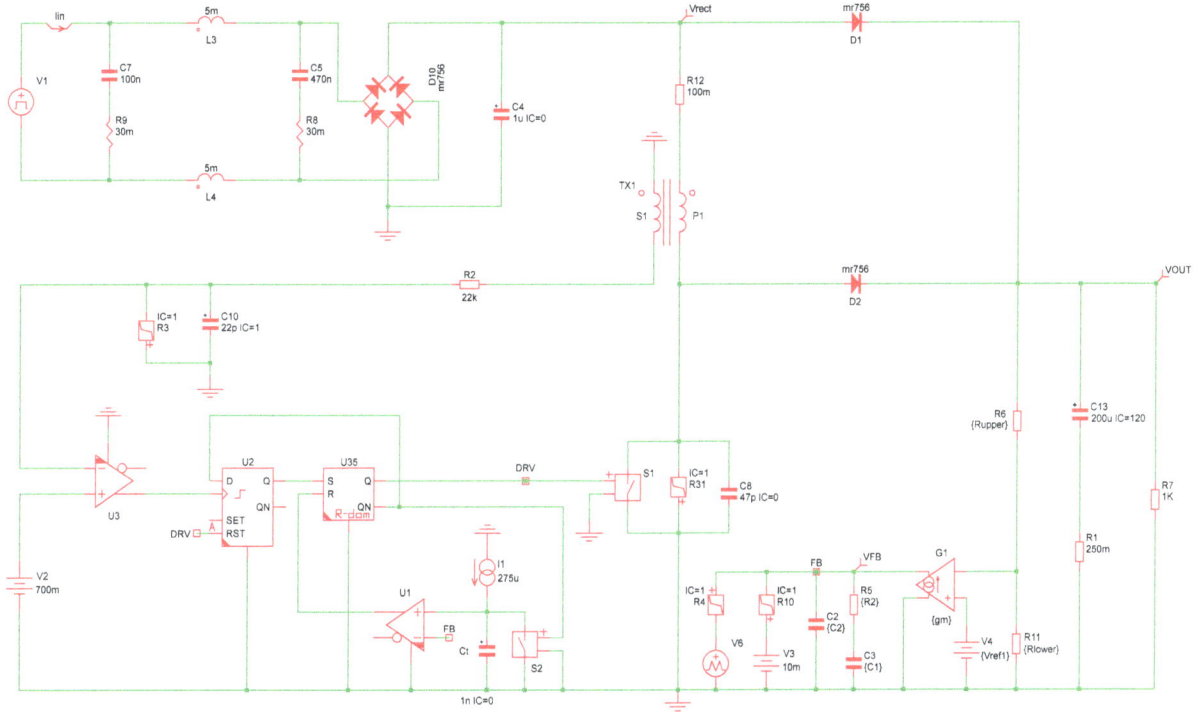

Figure 3.189 The BCM PFC operated in voltage-mode control maintains a constant on-time via the PWM block.

In borderline operation, the inductor average current is linked to its peak by a ratio of two:

$$\langle i_L(t) \rangle = \frac{I_{peak}}{2} \tag{3.423}$$

This is true as long as there is no deadtime and the power transistor is turned on immediately after the inductor current has hit 0 A. In practice, designers always insert a small delay to let the drain-source voltage fall and restart at its minimum value. This relationship linking average and peak inductor currents is illustrated in Figure 3.191 which zooms in the inductor current pattern.

To implement Figure 3.188 approach, we need a definition for the average power delivered to the load. We can start by defining its instantaneous value and later integrate it along a mains cycle.

Capitalizing on (3.423), we can write:

$$p_{in}(t) = v_{in}(t) \frac{i_{L,peak}(t)}{2} = \frac{v_{in}^2(t)}{2L} t_{on}(t) \tag{3.424}$$

This expression describes the cycle-by-cycle discrete input power points distributed along the sinusoidal input voltage.

To pursue our analysis, we can compute the input current averaged along a switching cycle.

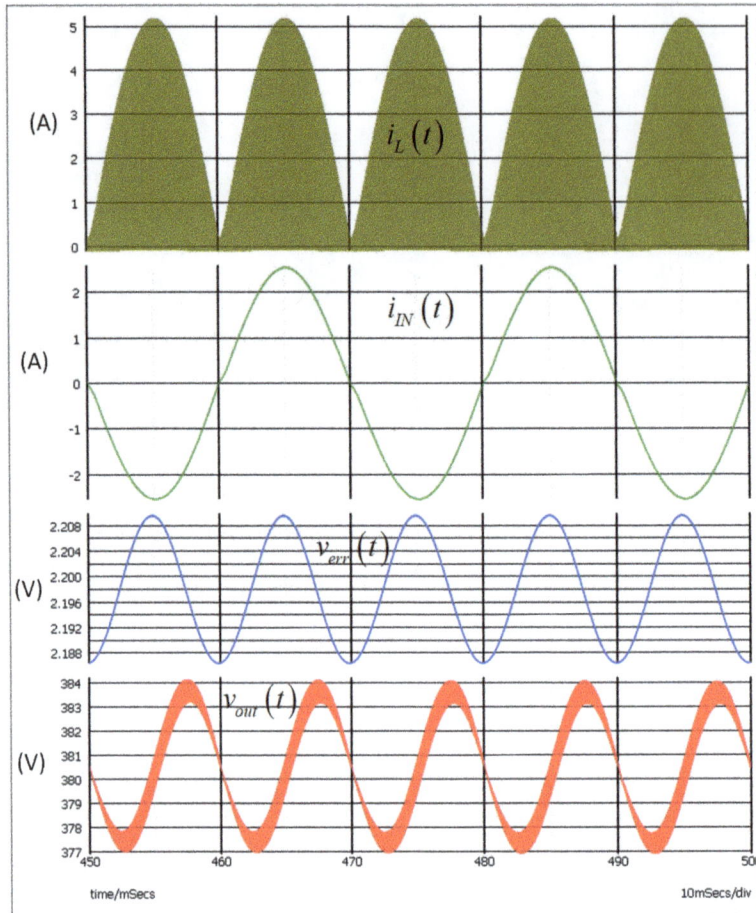

Figure 3.190 A PFC forces the input current to be a nice sinusoidal waveshape. As a consequence of its operation, some ripple shows up on the output.

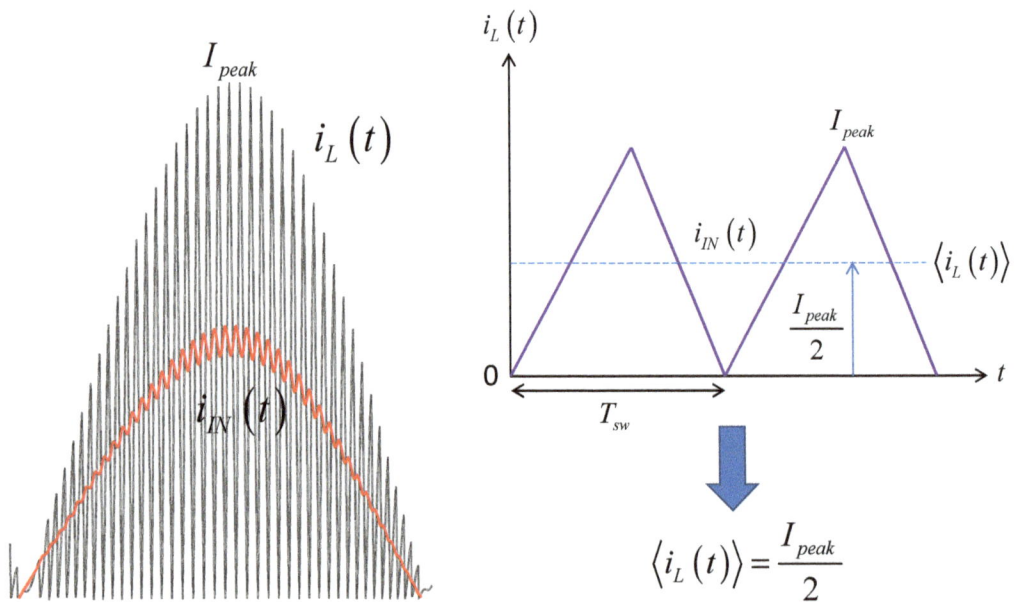

Figure 3.191 The average inductor current is the peak divided by two (no deadtime).

$$\left\langle i_L \left(t \right) \right\rangle = \frac{I_{peak}}{2}$$

$$\left\langle i_{in}\left(t\right)\right\rangle_{T_{sw}} = \frac{i_{L,peak}\left(t\right)}{2} = \frac{v_{in}\left(t\right)}{2L}t_{on}\left(t\right) \tag{3.425}$$

This input current can be linked to the input or output power as we consider a 100%-efficient converter:

$$\left\langle i_{in}\left(t\right)\right\rangle_{T_{sw}} = \frac{v_{in}\left(t\right)}{R_{in}} = \frac{v_{in}\left(t\right)}{\dfrac{V_{ac}^{\,2}}{P_{in}}} \tag{3.426}$$

Equating (3.425) and (3.426) gives us an expression for the on-time:

$$t_{on} = \frac{2LP_{in}}{V_{ac}^{\,2}} \tag{3.427}$$

In this formula, V_{ac} is the rms input voltage – 100 or 230 V for instance – and is constant. The rest of the terms are also constant, implying that t_{on} is theoretically constant along the input sinusoidal voltage. Averaged models show this to be true except when the input line approaches the 0-V area where a discontinuity exists. Having the on-time definition on hand, we can update (3.424) to define the instantaneous power as a function of t_{on}:

$$p_{in}\left(t\right) = \frac{v_{in}^{\,2}\left(t\right)}{2L}t_{on} \tag{3.428}$$

The large-signal equation is then obtained by averaging the above expression over half a line cycle:

$$\left\langle p_{in}\left(t\right)\right\rangle_{T_{line}} = \frac{t_{on}}{2L}2F_{line}\int_{0}^{1/2F_{line}} v_{in}^{\,2}\left(t\right)dt \tag{3.429}$$

$$\left\langle p_{in}\left(t\right)\right\rangle_{T_{line}} = P_{in} = \frac{t_{on}}{2L}2F_{line}\int_{0}^{1/2F_{line}}\left[\sqrt{2}V_{ac}\sin\left(2\pi F_{line}t\right)\right]^{2}\left(t\right)dt = \frac{V_{ac}^{\,2}}{2L}t_{on} \tag{3.430}$$

The delivered power depends on L and the on-time but also on the rms input voltage squared. For a maximum on-time (safely bounded by the controller), the BCM boost converter will deliver more power at high line than at low line. With a 100-230-V rms range, this ratio amounts to $2.3^2 = 5.3$. Safety can be at stake if over-power protection is not implemented. This is the reason why in some controllers, e.g. NCP1611, the maximum on-time limit is reduced at high line and avoids output power runaway. It also limits the bandwidth variation between low- and high-line operations and it contributes to the ripple reduction on the control signal.

As shown in Figure 3.189, the on-time is determined by the error voltage feeding the PWM block. A close-up of this block appears in Figure 3.192. Capacitor C_t is charged at a constant current but the setpoint at which the capacitor is discharged depends on the error voltage, effectively modulating the generated pulse width. Using the classical formula $\Delta V \cdot C = i \cdot \Delta t$, we can immediately determine the on-time:

$$t_{on} = \frac{V_{err}C_t}{I_C} \tag{3.431}$$

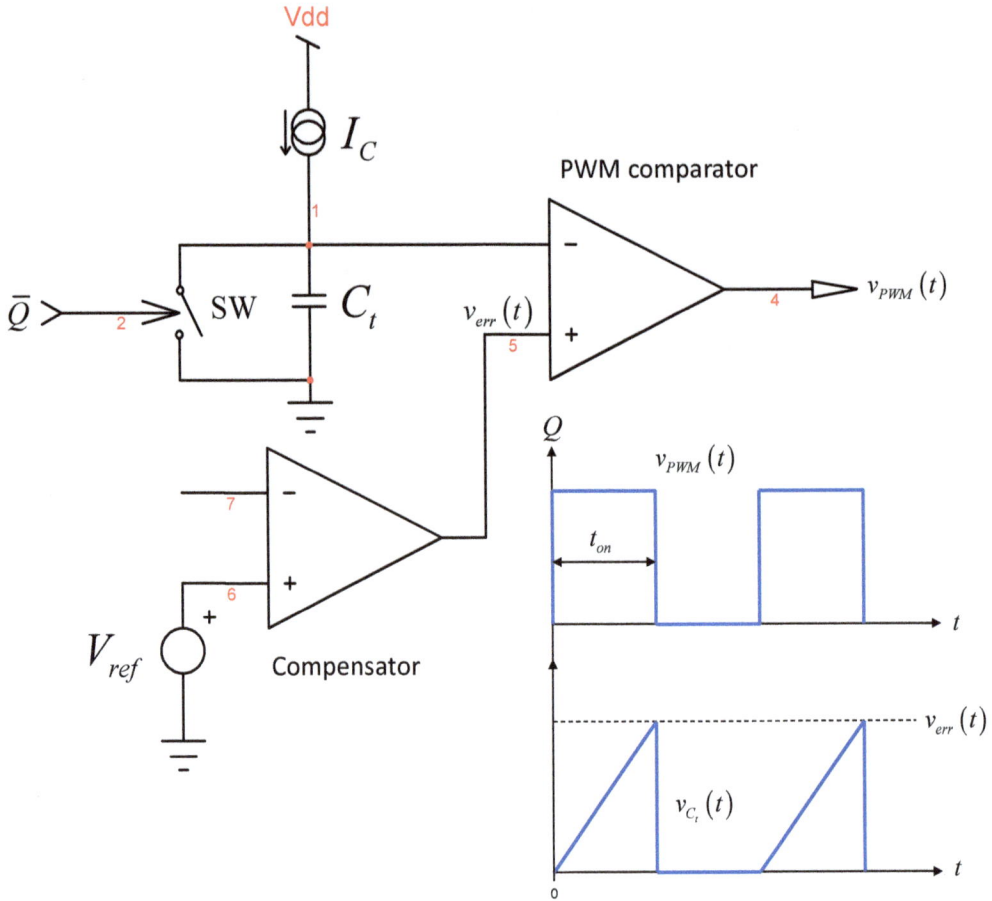

Figure 3.192 The pulse-width modulator is made around a capacitor charged at a constant current.

We can now substitute this definition in (3.430) and obtain the complete expression describing how the error voltage controls the input power:

$$P_{in} = \frac{V_{ac}^{\,2}}{2L}\frac{C_t}{I_C}V_{err} = \frac{V_{ac}^{\,2}}{2L}G_{PWM}V_{err} \tag{3.432}$$

In this expression, the PWM gain G_{PWM} is C_t/I_C .

Considering a 100% efficiency, the output current we need for fitting Figure 3.188 architecture is obtained by dividing P_{out} by V_{out}:

$$I_{out} = \frac{V_{ac}^{\,2}}{2L}G_{PWM}V_{err}\frac{1}{V_{out}} \tag{3.433}$$

This large-signal (nonlinear) expression is a function of two variables, V_{err} and V_{out}. To linearize it, we will compute the partial differentiation coefficients as follows:

$$\hat{i}_{out} = \frac{\partial}{\partial V_{out}}\left(\frac{V_{ac}^{\,2}}{2L}\frac{G_{PWM}V_{err}}{V_{out}}\right)\Bigg|_{\hat{v}_{err}=0}\hat{v}_{out} + \frac{\partial}{\partial V_{err}}\left(\frac{V_{ac}^{\,2}}{2L}\frac{G_{PWM}V_{err}}{V_{out}}\right)\Bigg|_{\hat{v}_{out}=0}\hat{v}_{err} \tag{3.434}$$

We obtain:

$$\hat{i}_{out} = \frac{V_{ac}^2}{2L}\frac{G_{PWM}}{V_{out}}\hat{v}_{err} - \frac{G_{PWM}V_{ac}^2 V_{err}}{2LV_{out}^2}\hat{v}_{out} \tag{3.435}$$

The right-end negative coefficient is actually the output power defined by (3.432) which further divided by V_{out}^2 becomes $1/R_{load}$. The small-signal output current can thus be updated as:

$$\hat{i}_{out} = \frac{V_{ac}^2}{2L}\frac{G_{PWM}}{V_{out}}\hat{v}_{err} - \frac{P_{out}}{V_{out}^2}\hat{v}_{out} = \frac{V_{ac}^2}{2L}\frac{G_{PWM}}{V_{out}}\hat{v}_{err} - \frac{1}{R_{load}}\hat{v}_{out} \tag{3.436}$$

The original large-signal circuit diagram of Figure 3.188 can now be transformed into a linear version as shown in Figure 3.193.

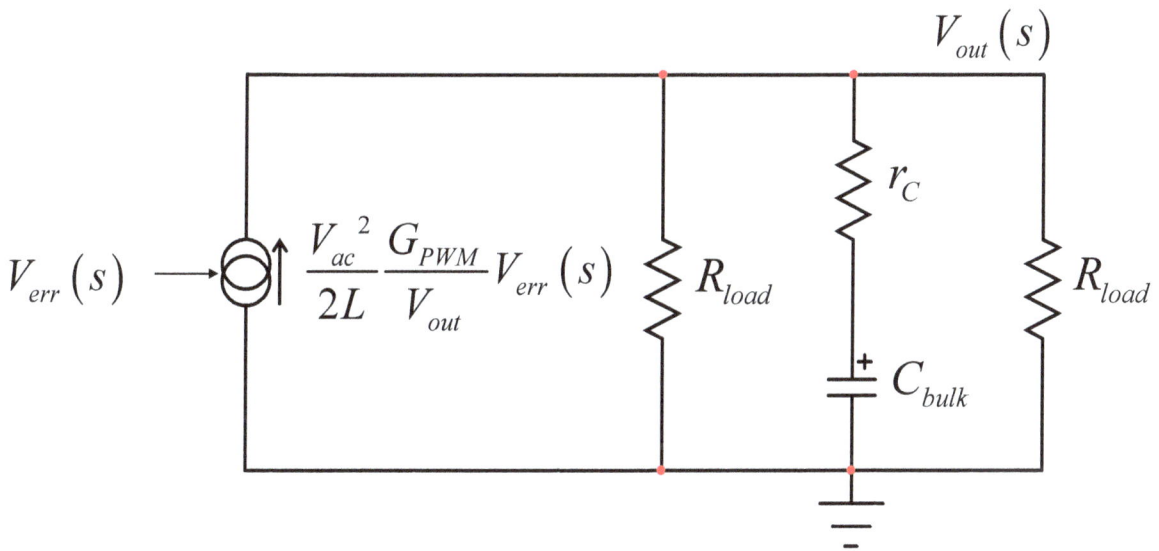

Figure 3.193 The small-signal model is that of a 1st-order circuit easy to analyze.

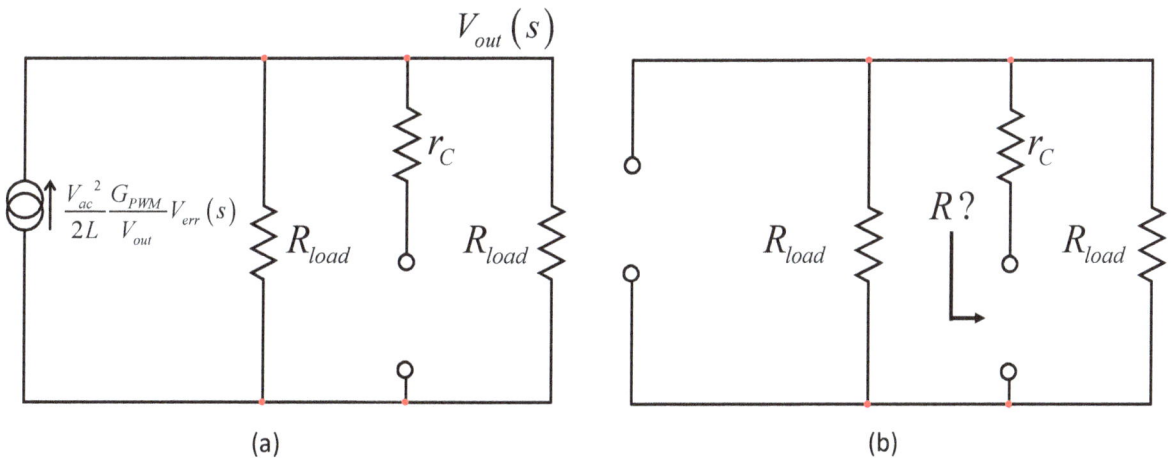

Figure 3.194 For the quasi-static gain, the output capacitor is open-circuited.

Using Figure 3.194 left-side sketch, we immediately have the quasi-static gain H_0:

$$H_0 = \frac{V_{ac}^2}{2L}\frac{G_{PWM}}{V_{out}}\frac{R_{load}}{2} = \frac{V_{ac}^2}{4L}\frac{R_{load}G_{PWM}}{V_{out}} \tag{3.437}$$

The pole and the zero are found by inspecting the right-side circuit. The resistance driving the output capacitor at zero excitation is r_C in series with the paralleled load resistors. The pole is the inverse of the time constant and equal to:

$$\omega_p = \frac{1}{C_{bulk}\left(\dfrac{R_{load}}{2} + r_C\right)} \approx \frac{2}{C_{bulk}R_{load}} \tag{3.438}$$

For the zero, it is when the series combination of r_C and C_{bulk} form a transformed short circuit:

$$r_C + \frac{1}{sC_{bulk}} = 0 \tag{3.439}$$

It gives a zero located at:

$$\omega_z = \frac{1}{r_C C_{bulk}} \tag{3.440}$$

The complete control-to-output transfer function of the voltage-mode boost converter operated in borderline conduction mode is expressed as:

$$H(s) = H_0 \frac{1+\dfrac{s}{\omega_z}}{1+\dfrac{s}{\omega_p}} = \frac{V_{ac}^2}{4L}\frac{R_{load}G_{PWM}}{V_{out}}\frac{1+sr_C C_{bulk}}{1+sC_{bulk}\left(r_C+\dfrac{R_{load}}{2}\right)} \tag{3.441}$$

The crossover frequency being in the low-frequency range, the zero brought by the bulk capacitor ESR can be ignored as it is mainly affecting the upper side of the frequency spectrum.

The transfer function can thus be simplified as:

$$H(s) \approx \frac{V_{ac}^2}{4L}\frac{R_{load}G_{PWM}}{V_{out}}\frac{1}{1+sC_{bulk}\dfrac{R_{load}}{2}} = H_0\frac{1}{1+\dfrac{s}{\omega_p}} \tag{3.442}$$

Let's assume the following component values:

$$I_C = 275 \ \mu A$$
$$C_t = 1 \ nF$$
$$L = 250 \ \mu H$$
$$C_{bulk} = 200 \ \mu F$$
$$R_{load} = 1 \ k\Omega$$
$$V_{out} = 380 \ V$$
$$V_{in} = V_{ac} = 120 \ V \ rms$$

We can first compute the resulting on-time to deliver 144.4-W of output power:

$$t_{on} = \frac{2LP_{out}}{V_{ac}^2} = \frac{2 \times 250u \times 144.4}{120^2} \approx 5 \ \mu s \qquad (3.443)$$

The PWM gain is:

$$G_{PWM} = \frac{C_t}{I_C} = \frac{1n}{275u} = 3.6 \ \mu s/V \qquad (3.444)$$

The error voltage will thus stabilize to $t_{on}/G_{PWM} = 5/3.6 \approx 1.4 \ V$ dc. In reality, it is likely to be above this value as the loop will compensate for the various losses generated in the converter. The quasi-static gain is calculated using (3.437):

$$H_0 = \frac{V_{ac}^2}{4L} \frac{R_{load}G_{PWM}}{V_{out}} = \frac{120^2}{4 \times 250u} \frac{1k \times 3.6u}{380} \approx 138 \ or \approx 43 \ dB \qquad (3.445)$$

The low-frequency pole is located at:

$$f_p = \frac{1}{\pi R_{load}C_{bulk}} = \frac{1}{3.14 \times 1k \times 200u} \approx 1.6 \ Hz \qquad (3.446)$$

A simple Mathcad® sheet is used to plot the frequency response that appears in Figure 3.195.

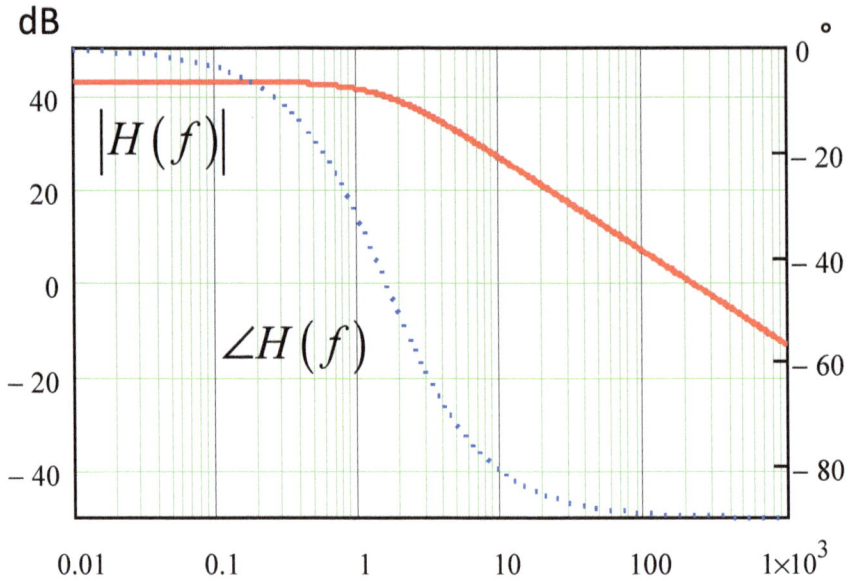

Figure 3.195 The response is that of a first-order system with a high dc gain.

To verify these results, Figure 3.196 shows the PFC boost converter powered from a dc source rather than a sinusoidal waveform as in Figure 3.189. It was shown in [6] that the small-signal response of a PFC powered from an ac source of a given rms voltage – 120 V rms for instance – is similar to the response obtained from the same dc-dc converter supplied by a dc source whose value is the selected rms level or 120 V dc in this case.

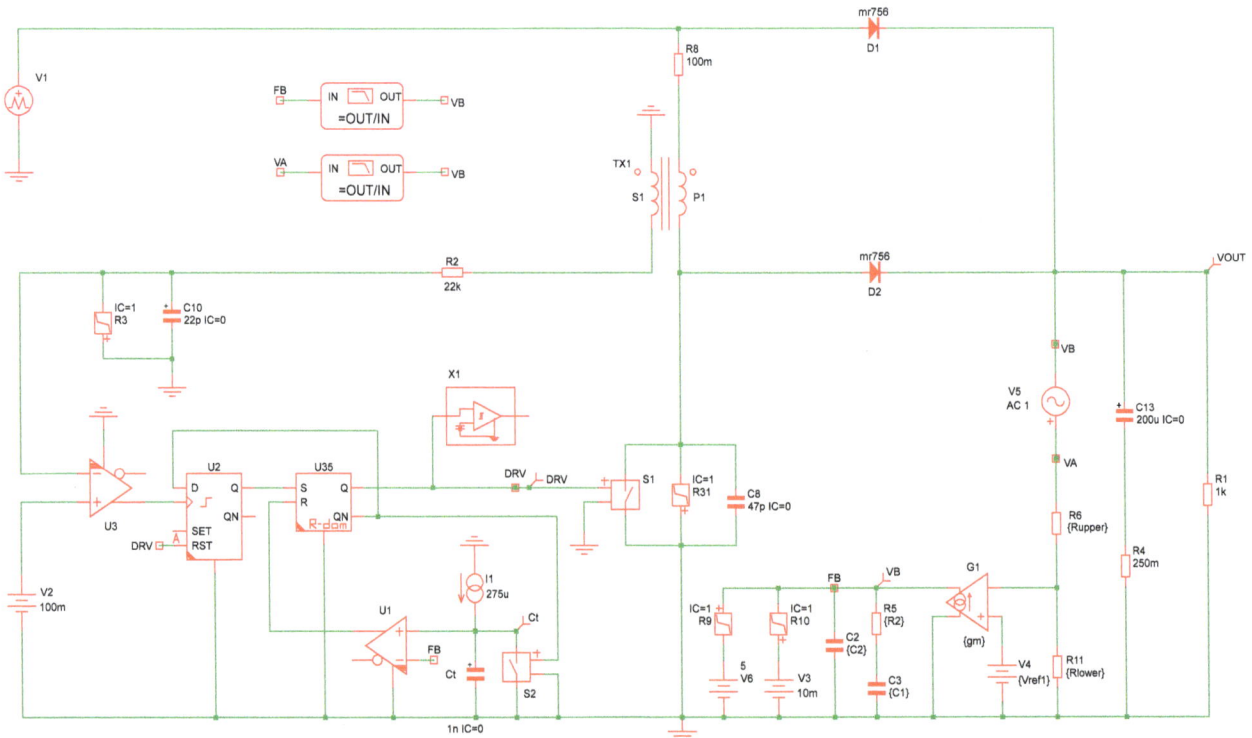

Figure 3.196 The PFC is powered from a dc source whose value is equal to the rms voltage of the ac input supply.

The simulation results appear in Figure 3.197 and confirm the results obtained with our simplified small-signal analysis. The input source V_1 delivers 120 V to the dc-dc converter which powers the 1-kΩ load with a 380-V bias.

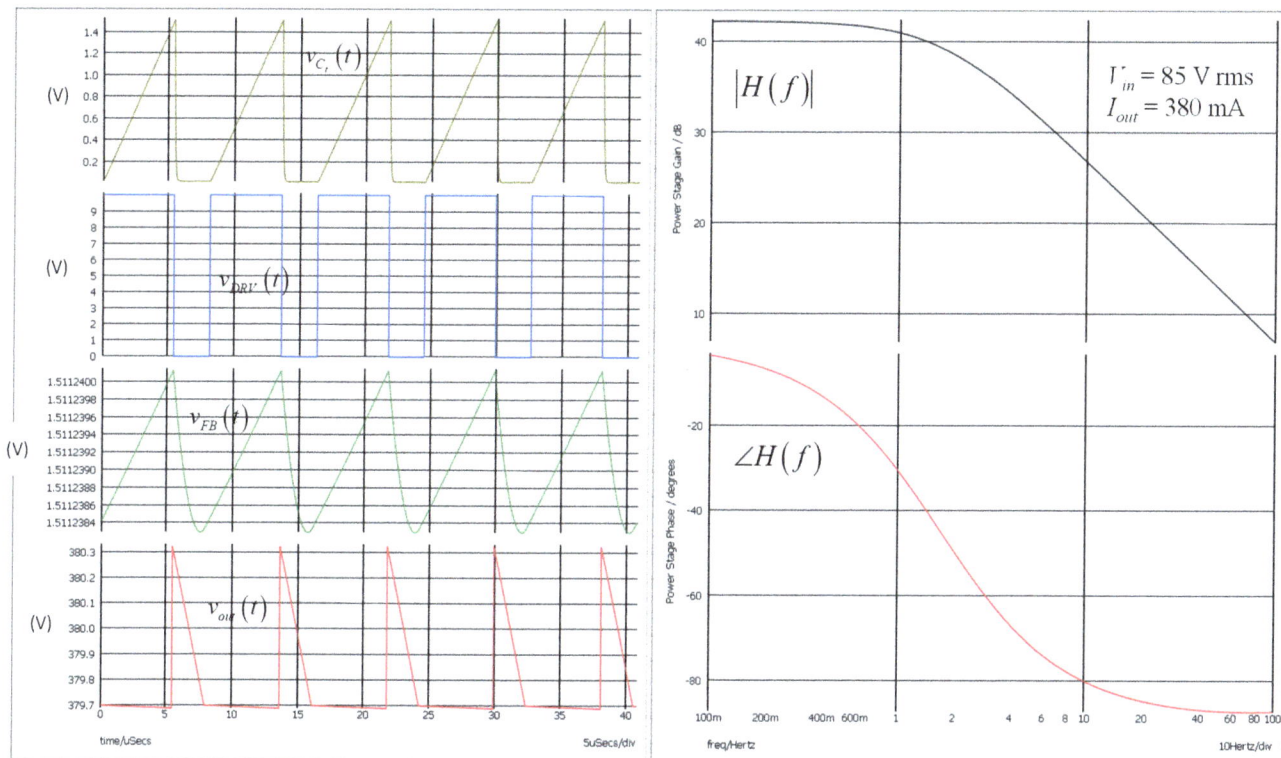

Figure 3.197 SIMPLIS® quickly delivers the switching waveforms and the dynamic power stage response.

As previously stated, a PFC is a pre-converter meaning that it is inserted between a high-voltage dc-dc converter and the mains. The ideal dc-dc converter operates in closed loop and maintains a constant output power P_{out} regardless of the input voltage value (we assume 100% efficiency and an infinite bandwidth):

$$P_{out} = I_{out}V_{out} = \text{constant} \tag{3.447}$$

If the power is constant, then differentiating the above equation should return zero. The differentiation of a product uv is defined as follows:

$$d(u \cdot v) = u \cdot dv + v \cdot du \tag{3.448}$$

Applying this concept to (3.447) leads to the following small-signal expression:

$$dP_{out} = 0 = I_{out}dV_{out} + V_{out}dI_{out} \tag{3.449}$$

Please note that dV_{out} and dI_{out} could respectively be replaced by $\hat{v}_{out}$ and $\hat{i}_{out}$.

433

Rearranging this result and factoring, we can define the incremental resistance R of a closed-loop-operated dc-dc converter powered by our PFC circuit:

$$R = \frac{dV_{out}}{dI_{out}} = -\frac{V_{out}}{I_{out}} = -R_{load} \qquad (3.450)$$

If we now update the small-signal model of Figure 3.193 with this negative resistance, the circuit evolves to that of Figure 3.198. Paralleling two resistors of equal value with one of them being negative leads to an infinite resistance: the two R_{load} resistances disappear from the picture. The system becomes a simple integrator whose transfer function is of the form:

$$H(s) = \frac{1}{\dfrac{s}{\omega_{po}}} \qquad (3.451)$$

If we express the output voltage from Figure 3.198, we have:

$$V_{out}(s) = I_{out}(s) Z_{C_{bulk}}(s) = V_{err}(s) \frac{V_{ac}^{2}}{2L} \frac{G_{PWM}}{V_{out}} \frac{1}{sC_{bulk}} \qquad (3.452)$$

We may rearrange this expression to fit the format of (3.451).

$$H(s) = \frac{1}{sC_{bulk} \dfrac{2LV_{out}}{V_{ac}^{2} G_{PWM}}} = \frac{1}{\dfrac{s}{\omega_{po}}} \qquad (3.453)$$

In which the 0-dB crossover pole is defined as:

$$\omega_{po} = \frac{1}{C_{bulk}} \frac{V_{ac}^{2} G_{PWM}}{2LV_{out}} \qquad (3.454)$$

or:

$$f_{po} = \frac{V_{ac}^{2} G_{PWM}}{4\pi L V_{out} C_{bulk}} \qquad (3.455)$$

The magnitude of (3.453) at crossover f_c is simply:

$$\left| H(f_c) \right| = \frac{f_{po}}{f_c} \qquad (3.456)$$

while the phase is a permanent 90° lag.

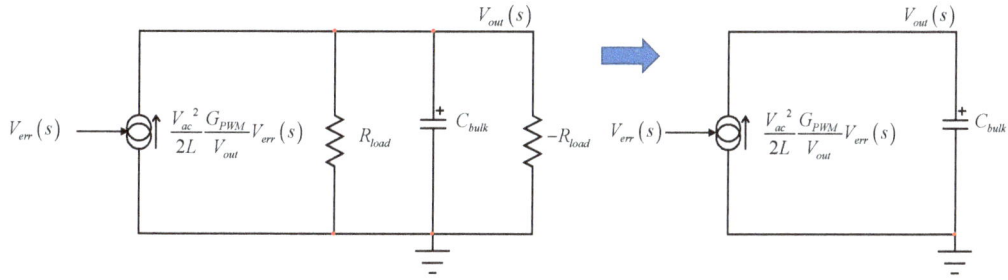

Figure 3.198 The negative resistance paralleled with the positive one cancels the association and the resulting resistance approaches infinity.

The new dynamic response is given in Figure 3.199 and confirms the disappearance of the quasi-static gain.

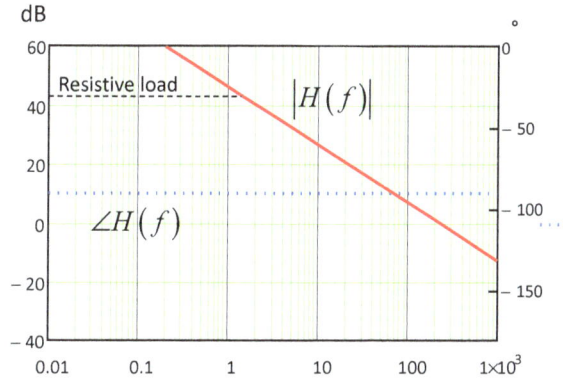

Figure 3.199 The dc plateau has disappeared because of the negative resistance now loading the PFC.

3.9 Boost in Power Factor Correction – Boundary Mode Current-Mode Control

A PFC pre-converter can also be operated in current-mode control. In this case, the inductor peak current setpoint follows a sinusoidal waveshape built with a multiplier. The principle of operation appears in Figure 3.200.

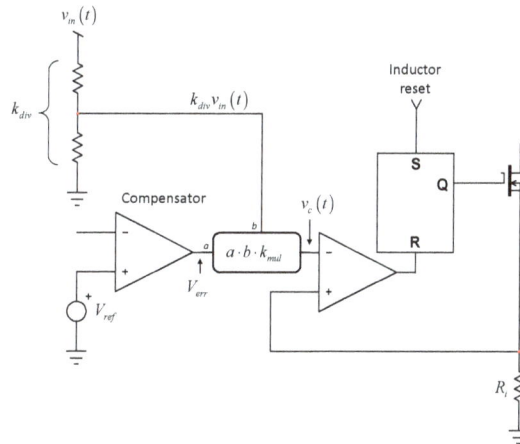

Figure 3.200 The BCM PFC operated in current mode needs a multiplier for setting the inductor peak current.

This structure was popularized with the MC33361 and MC33262 from Motorola Semiconductors in the 90's: a portion of the rectified voltage enters a multiplier (input *b*) while the error voltage feeds the second input *a*. That way, the inductor peak current setpoint follows a sinusoidal waveshape whose amplitude is scaled by the error voltage.

The circuit operates in a free-running mode and constant on-time. Unlike the voltage-mode version, you need to sense the input voltage and suffer additional losses. We have implemented the circuit in a SIMPLIS® simulation template as shown in Figure 3.201.

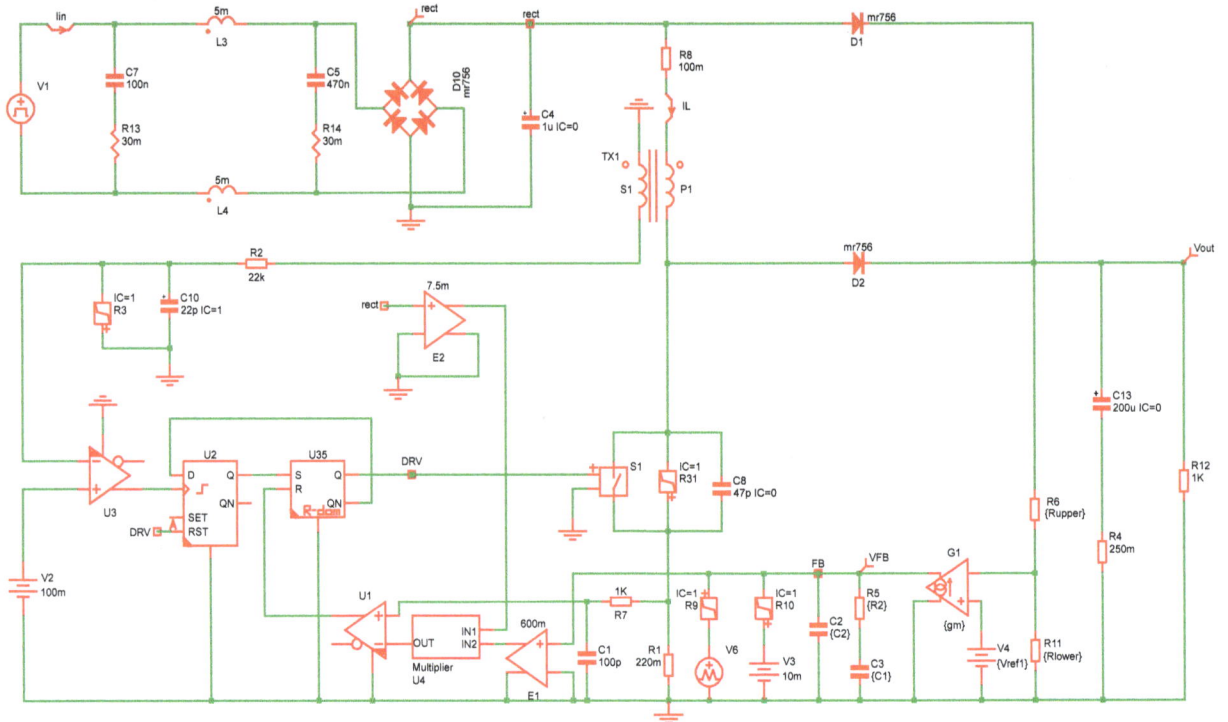

Figure 3.201 The BCM PFC operated in current mode needs a multiplier for setting the inductor peak current.

The simulation runs fast and results appear in Figure 3.202. The circuit delivers 380 V to the 1-kΩ load while the absorbed current is a nice sinusoidal waveform.

The inductor peak current can be classically determined as:

$$i_{L,peak} = \frac{v_c(t)}{R_i} \tag{3.457}$$

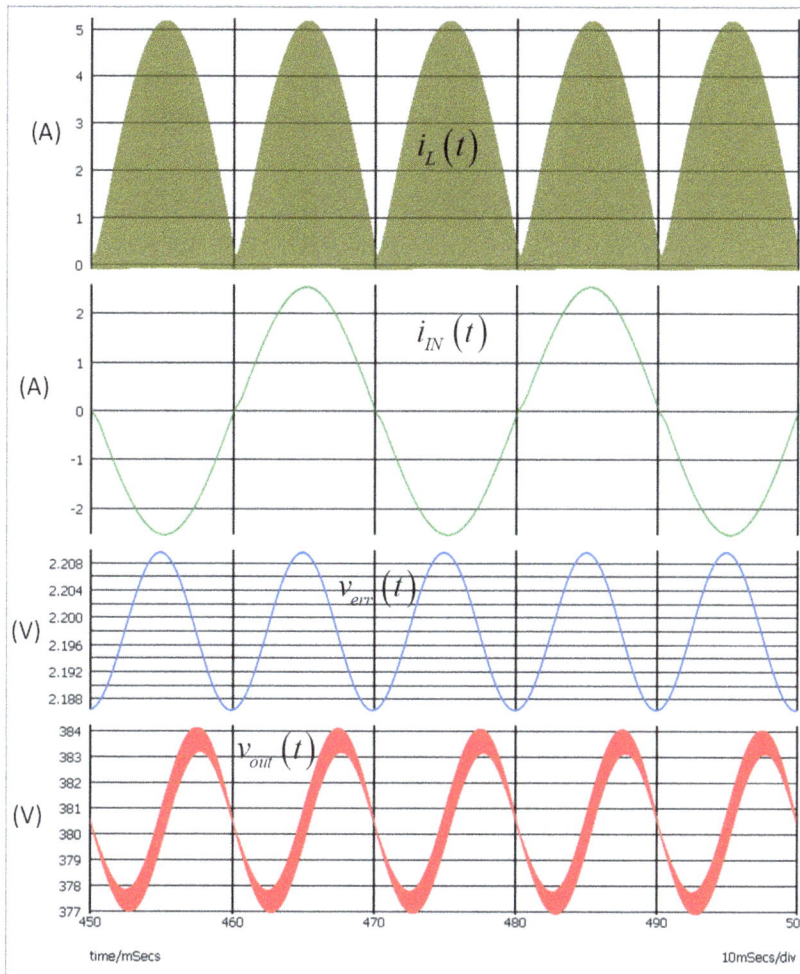

Figure 3.202 The input current is a nice sinusoidal waveshape while ripple shows up on the output as expected.

The same peak current can also be inferred by computing the inductor on-slope:

$$i_{L,peak}\left(t\right) = \frac{v_{in}\left(t\right)}{L} t_{on}\left(t\right) \tag{3.458}$$

$t_{on}(t)$, in the above expression represents the discrete on-time event at a given point of the sinusoidal input voltage. When equating both expressions, t_{on} comes out easily:

$$t_{on}\left(t\right) = \frac{v_c\left(t\right)L}{R_i v_{in}\left(t\right)} \tag{3.459}$$

In this formula, $v_c(t)$ designates the instantaneous multiplier output driving the inductor peak current setpoint. It is made of the scaled-down input voltage $k_{div} v_{in}\left(t\right)$ and the dc error voltage V_{err}. Its expression is simply the multiplication of both variables, weighted by the multiplier gain k_{mul}:

$$v_c\left(t\right) = v_{in}\left(t\right) k_{div} k_{mul} V_{err} \tag{3.460}$$

437

If we substitute (3.460) in (3.459) then we complete the definition of the on-time in this current-mode structure:

$$t_{on}(t) = \frac{L v_{in}(t) k_{div} k_{mul} V_{err}}{R_i v_{in}(t)} = \frac{L k_{div} k_{mul}}{R_i} V_{err} \tag{3.461}$$

In this expression, all the terms are constant and so is t_{on}. There is no difference in the operated mode compared with the voltage-mode architecture previously studied. The output power expression still obeys (3.430). We can update t_{on} by its definition from (3.461) and obtain the final large-signal expression we need:

$$P_{out} = \frac{V_{ac}^2}{2L} \frac{L k_{div} k_{mul} V_{err}}{R_i} = \frac{V_{ac}^2}{2} \frac{k_{div} k_{mul} V_{err}}{R_i} \tag{3.462}$$

As you can see in this expression, the inductor does not play a role in the power expression. The large-signal output current source is obtained by dividing the formula by V_{out}:

$$I_{out} = \frac{V_{ac}^2}{2} \frac{k_{div} k_{mul} V_{err}}{R_i} \frac{1}{V_{out}} \tag{3.463}$$

We will now run a partial differentiation and define the coefficients driving the two variables, V_{out} and V_{err}:

$$\hat{i}_{out} = \frac{\partial}{\partial V_{out}} \left(\frac{V_{ac}^2}{2} \frac{k_{div} k_{mul} V_{err}}{R_i} \frac{1}{V_{out}} \right)\Bigg|_{\hat{v}_{err}=0} \hat{v}_{out} + \frac{\partial}{\partial V_{err}} \left(\frac{V_{ac}^2}{2} \frac{k_{div} k_{mul} V_{err}}{R_i} \frac{1}{V_{out}} \right)\Bigg|_{\hat{v}_{out}=0} \hat{v}_{err} \tag{3.464}$$

Once computed by Mathcad®, these coefficients are:

$$\hat{i}_{out} = \frac{V_{ac}^2 k_{div} k_{mul}}{2 R_i V_{out}} \hat{v}_{err} - \frac{V_{ac}^2 V_{err} k_{div} k_{mul}}{2 R_i V_{out}} \frac{1}{V_{out}} \hat{v}_{out} = \frac{V_{ac}^2 k_{div} k_{mul}}{2 R_i V_{out}} \hat{v}_{err} - I_{out} \frac{1}{V_{out}} \hat{v}_{out} \tag{3.465}$$

In this formula, resistance R_{load} also appears and the final expression can be rewritten in the following shape:

$$\hat{i}_{out} = \frac{V_{ac}^2 k_{div} k_{mul}}{2 R_i V_{out}} \hat{v}_{err} - \frac{1}{R_{load}} \hat{v}_{out} \tag{3.466}$$

Combining these small-signal coefficients in an electrical diagram leads to Figure 3.203.

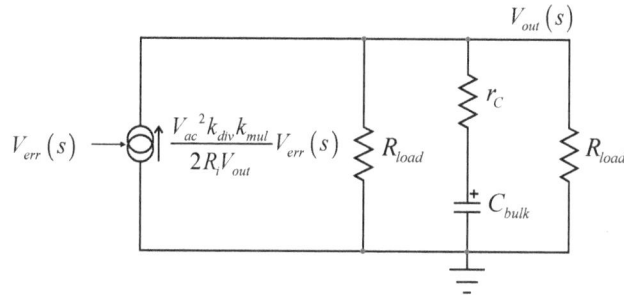

Figure 3.203 The current-mode small-signal model does not really differ from that of the voltage-mode approach except for the current-source mathematical expression.

To analyze this circuit, we can start with the quasi-static gain, determined for $s = 0$: open the capacitor and express V_{out}:

$$V_{out}(s) = I_{out}(s)\frac{R_{load}}{2} = \frac{V_{ac}{}^2 k_{div} k_{mul}}{2R_i V_{out}} \frac{R_{load}}{2} V_{err}(s)$$

(3.467)

The gain H_0 we want is simply the ratio of V_{out} over V_{err}:

$$H_0 = \frac{V_{ac}{}^2 k_{div} k_{mul}}{4V_{out}} \frac{R_{load}}{R_i}$$

(3.468)

The pole position is similar to what we have determined for the voltage-mode control, considering the two R_{load} in parallel. If we neglect the bulk capacitor ESR, then the control-to-output transfer function can be expressed as:

$$H(s) = \frac{V_{out}(s)}{V_{err}(s)} \approx H_0 \frac{1}{1 + \dfrac{s}{\omega_p}} = \frac{V_{ac}{}^2 k_{div} k_{mul}}{4V_{out}} \frac{R_{load}}{R_i} \frac{1}{1 + sC_{bulk}\dfrac{R_{load}}{2}}$$

(3.469)

We can now plot our pre-converter small-signal response when adopting the following values:

$$k_{div} = 0.0078$$
$$k_{mul} = 0.6$$
$$R_i = 0.24\ \Omega$$
$$L = 350\ \mu H$$
$$C_{bulk} = 180\ \mu F$$
$$R_{load} = 1\ k\Omega$$
$$V_{err} = 1.647\ V$$
$$V_{out} = 400\ V$$
$$V_{in} = V_{ac} = 100\ V\ rms$$

Considering a 1.647-V control voltage in this line/load configuration, what is the inductor peak current?

$$I_{L,peak} = \frac{k_{mul}k_{div}\sqrt{2}V_{ac}V_{err}}{R_i} = \frac{0.0078\times0.6\times100\times1.414\times1.647}{0.24} = 4.54 \text{ A} \tag{3.470}$$

The t_{on} duration is found using (3.461):

$$t_{on} = \frac{Lk_{div}k_{mul}}{R_i}V_{err} = \frac{350u\times0.0078\times0.6\times1.647}{0.24} = 11.24 \text{ μs} \tag{3.471}$$

And finally, as a sanity check, the output power is obtained using (3.462):

$$P_{out} = \frac{V_{ac}^{2}}{2}\frac{k_{div}k_{mul}V_{err}}{R_i} = \frac{100^2\times0.0078\times0.6\times1.647}{2\times0.24} = 160.6 \text{ W} \tag{3.472}$$

The quasi-static gain H_0 is found to be:

$$H_0 = \frac{V_{ac}^{2}k_{div}k_{mul}}{4V_{out}}\frac{R_{load}}{R_i} = \frac{100^2\times0.0078\times0.6\times1k}{4\times400\times0.24} = 121.88 \text{ or } 41.7 \text{ dB} \tag{3.473}$$

The pole is located at low frequency:

$$f_p = \frac{1}{\pi R_{load}C_{bulk}} = \frac{1}{3.14\times1k\times180u} \approx 1.8 \text{ Hz} \tag{3.474}$$

The complete transfer function has been plotted with Mathcad® and appears in Figure 3.204. It is very close in shape to the response offered by a voltage-mode version.

When you connect a constant-power downstream converter, Figure 3.204 simplifies as the two resistors go away considering one of them to be negative. We can show that the new transfer function becomes:

$$H(s) = \frac{1}{sC_{bulk}\dfrac{2V_{out}R_i}{V_{ac}^{2}k_{div}k_{mul}}} = \frac{1}{\dfrac{s}{\omega_{po}}} \tag{3.475}$$

In which the 0-dB crossover pole is defined as:

$$\omega_{po} = \frac{1}{C_{bulk}}\frac{V_{ac}^{2}k_{div}k_{mul}}{2V_{out}R_i} \tag{3.476}$$

or:

$$f_{po} = \frac{1}{C_{bulk}}\frac{V_{ac}^{2}k_{div}k_{mul}}{4\pi V_{out}R_i} \tag{3.477}$$

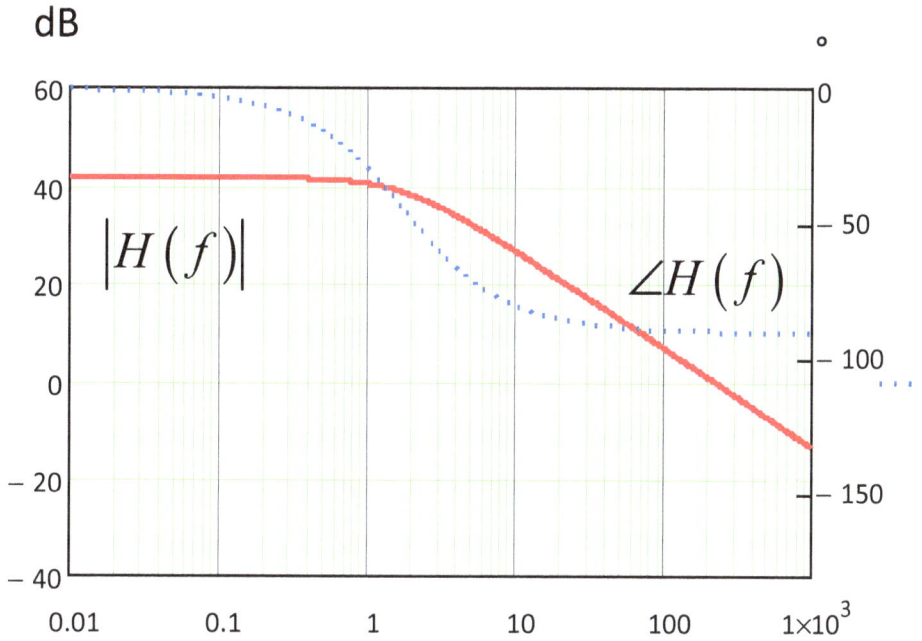

Figure 3.204 The current-mode small-signal response is similar in shape to its voltage-mode counterpart.

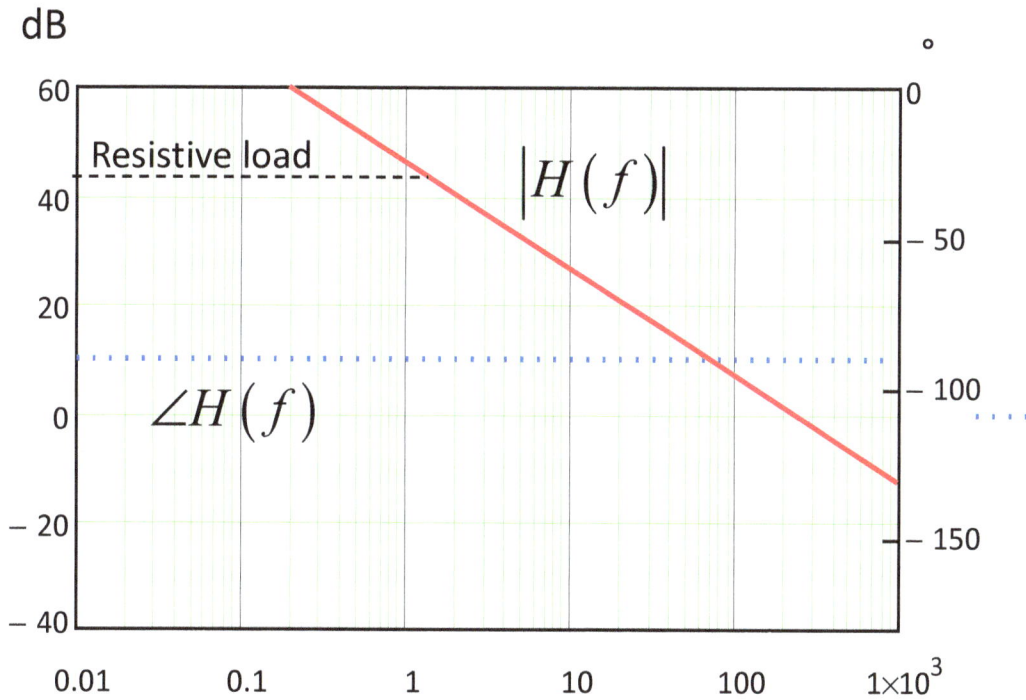

Figure 3.205 The gain plateau disappears and the response looks like that of an integrator.

Figure 3.205 confirms the disappearance of the quasi-static gain.

3.10 What Should I Retain from this Chapter?

In this third chapter, we have learned a great deal about the small-signal response of the boost converter that is summarized below:

1. A CCM boost converter operated in voltage-mode control is a second-order system affected by a right-half-plane zero. When transitioning to DCM, this converter still exhibits a second-order dynamic response but well damped. The RHPZ is still present in DCM though relegated in high frequencies.

2. In fixed-frequency current-mode control, the CCM boost converter becomes a third-order converter with a pole dominating the low-frequency response and two subharmonic poles located at half the switching frequency. Beside the left-half-plane zero contributed by the output capacitor and its ESR, there is a right-half-plane zero whose position is the same as in voltage mode. The subharmonic poles must be damped by some additional ramp as the duty ratio approaches 50%.

3. The incremental input resistance of an open-loop boost converter operated in voltage-mode is positive. On the other hand, when the same converter is operated in current-mode control, the incremental resistance is negative: if the input voltage increases, the output current reduces and so does the input current.

4. The boost converter can be operated in quasi-resonant mode for the benefit of lower turn-on losses if a delay is inserted to switch right in the drain-source valley.

5. The tapped boost version offers a higher boosting ratio together with a lower voltage stress on the power switch.

6. Boost converters lend themselves well to implementing power factor correction. A different average approach was introduced here which models the power stage as a large-signal current source. The transfer function in voltage- or current-mode control shows a low-frequency first-order response affected by a pole. When the pre-converter is loaded by a constant-power converter such as a closed-loop switching converter, the negative incremental resistance loading the PFC cancels the quasi-static plateau and brings an integrator-type frequency response.

3.11 References

1. C. Basso, Linear Circuit Transfer Functions – An Introduction to Fast Analytical Techniques, Wiley, 2016.

2. V. Vorpérian, *Simplified Analysis of PWM Converters using Model of PWM Switch, parts I and II*, IEEE Transactions on Aerospace and Electronic Systems, Vol. 26, NO. 3, 1990

3. V. Vorpérian, *Analytical Methods in Power Electronics*, In-House Power Electronics Class, Toulouse, France, 2004

4. C. Basso, Switch-Mode Power Supplies: SPICE Simulations and Practical Designs, McGraw Hill, second edition, 2014

5. J. Turchi, *Compensating a PFC Stage*, ON Semiconductor Application Note AND8321, https://www.onsemi.com/pub/Collateral/AND8321-D.PDF

6. *Running Ac Analyses on PFC Converters*, SIMPLIS® Exhibitor Seminar, APEC 2017, http://www.simplistechnologies.com/webinar/2016/11/17/ac-analysis-pfc

4 The Buck-Boost Converter and its Derivatives

THIS CHAPTER WILL detail the derivation of the four buck-boost converter transfer functions when operated in fixed-frequency CCM/DCM voltage- and current-mode control. When extended to its isolated version – the flyback converter – we will look at the quasi-square-wave resonant version as this structure is popular in ac-dc adapters for the consumer market.

Power factor correction (PFC) won't be forgotten as the single-stage flyback converter finds new usage in the lighting field for instance. The methodology follows what has already been exposed for the previous converters: the fast analytical techniques (FACTs) together with SPICE let us quickly determine the various time constants while final expressions are tested in a Mathcad® sheet and checked against a SIMPLIS® simulation. Let's begin with the popular voltage-mode control.

4.1 Buck-Boost Transfer Functions in CCM – Fixed-Frequency Voltage-Mode Control

We start the analysis by identifying the PWM switch in the buck-boost converter of Figure 4.1. Our first transfer function is the control-to-output relationship linking $D(s)$ to $V_{out}(s)$. To obtain a first reference plot, we have implemented the large-signal voltage-mode model already used in the previous chapters in the simple open-loop simulation template of Figure 4.2.

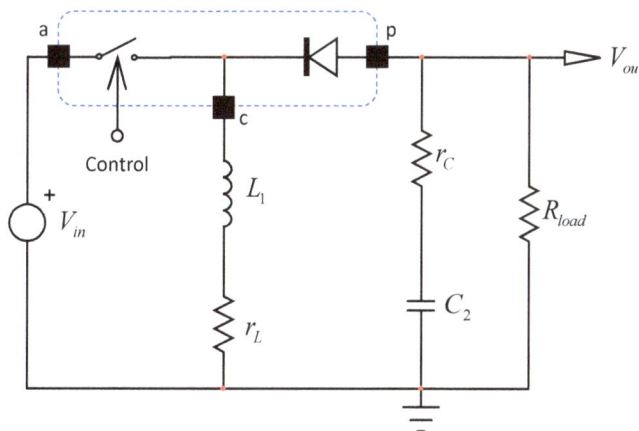

Figure 4.1 The buck-boost converter can increase or decrease the input voltage.

The calculated bias points show a -17.9-V output obtained from the 12-V source. The duty ratio is set by the V_{ctrl} source and imposes a 600-mV input bias corresponding to a 60% duty ratio. The dynamic response appears in Figure 4.3 and shows a flat quasi-static gain until the resonance appears. Beyond the peak, the magnitude falls with a -2-slope (-40 dB per decade) considering the second-order LC filter. The phase starts at 180° owing to the negative output voltage then quickly decreases down to 0° but continues to further lag despite a break in the magnitude curve.

This behavior confirms the presence of a right-half-plane zero. This RHPZ is typical of the buck-boost converter and models the delay brought by the 2-step conversion process: store energy in the inductor during the on-time first then transfer it to the capacitor during the off-time as a second step. If a sudden power demand occurs, you must first increase the energy stored in the inductor before answering the demand. If the duty ratio changes too fast for the given on-time inductor voltage, the current cannot grow quickly enough and the output voltage momentarily drops.

We have illustrated this phenomenon in Chapter 1.

After 10 kHz, the phase abandons its negative slope and goes up while the magnitude becomes flatter. It confirms the presence of the second left-half-plane zero incurred to the output capacitor ESR. This plot is our reference curve and we will compare it to the response obtained by the small-signal equivalent circuit. It is our sanity check to ensure our linear model is correct before carrying the analysis work.

The linear model derived in Chapter 1.3.1 is inserted in the converter as shown in Figure 4.4. All operating points are identical to those of the large-signal model and the ac response is also similar.

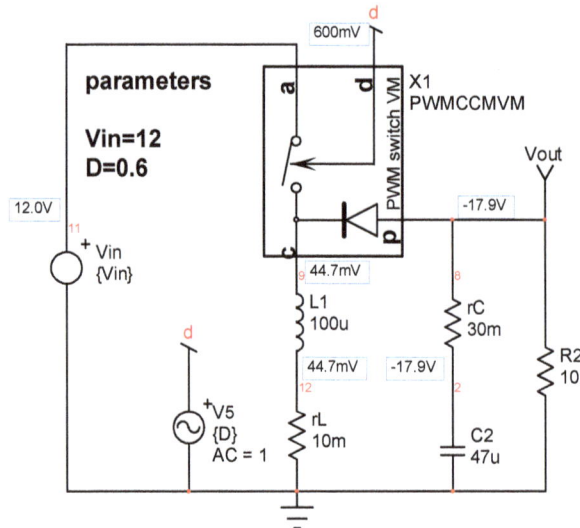

Figure 4.2 The large-signal VM-PWM switch model is easily implemented in this buck-boost simulation template.

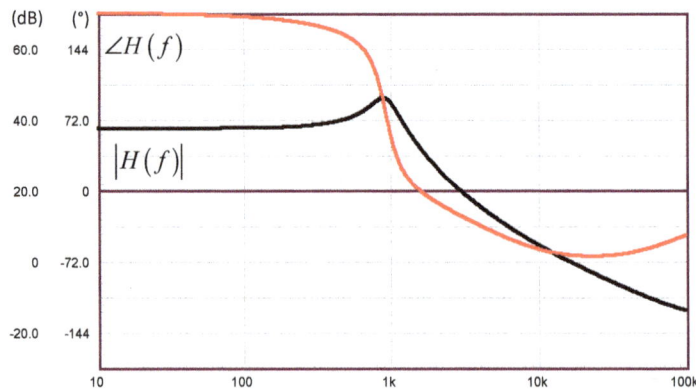

Figure 4.3 The power stage response shows the presence of a RHPZ.

As we did in previous studies, we can modifiy the circuit by rearranging the various components and make the circuit simpler to analyze. This is what Figure 4.5 details. Simulating this circuit matches exactly what has already been obtained in Figure 4.3. Please note the presence of the 0-V source V_c only included to measure the current flowing out of terminal c in these SPICE simulations.

We can start the analysis by considering $s = 0$, shorting the inductor and opening the capacitor as shown in Figure 4.6. We start by defining the current in terminal c:

$$I_c(s) = \frac{V_{out}(s) + V_{ap}D(s) - V_{out}(s)D_0}{r_L} \tag{4.1}$$

Current I_2 is current I_1 without I_c:

$$I_2(s) = I_c D(s) + I_c(s)D_0 - I_c(s) \tag{4.2}$$

Finally, the output voltage is obtained by combining I_2 and the load resistance:

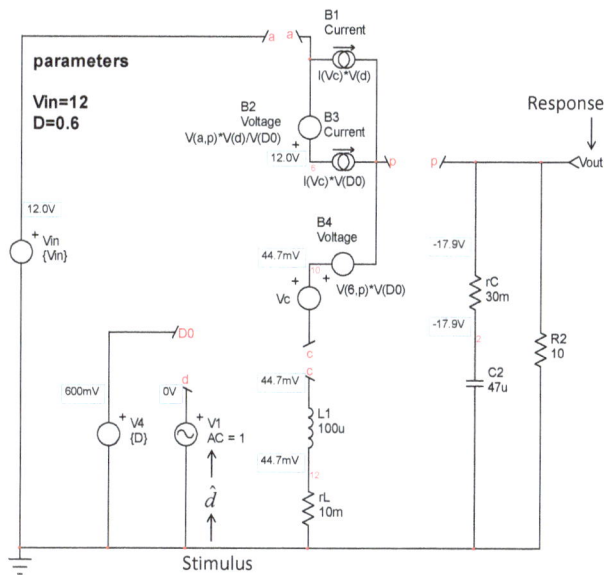

Figure 4.4 The small-signal model confirms the operating points obtained with the large-signal PWM switch model.

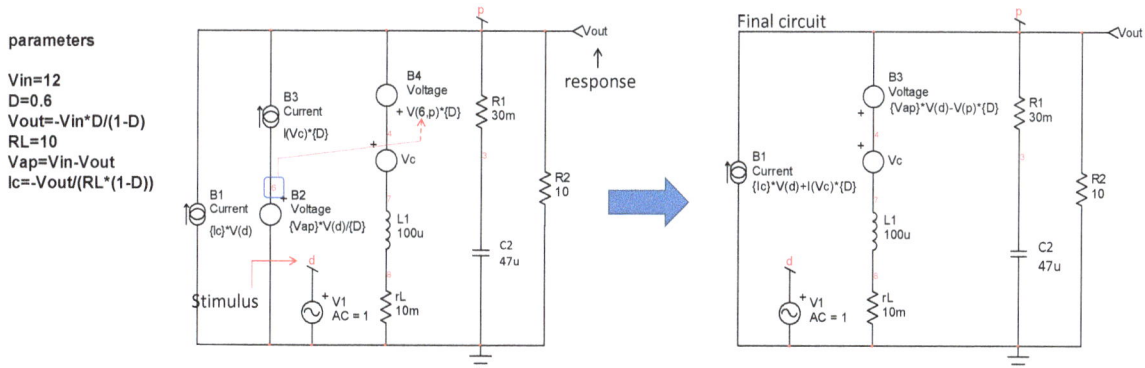

Figure 4.5 The input voltage at terminal a is 0 V in ac meaning the node can be grounded while expressions refering to it can be simplified.

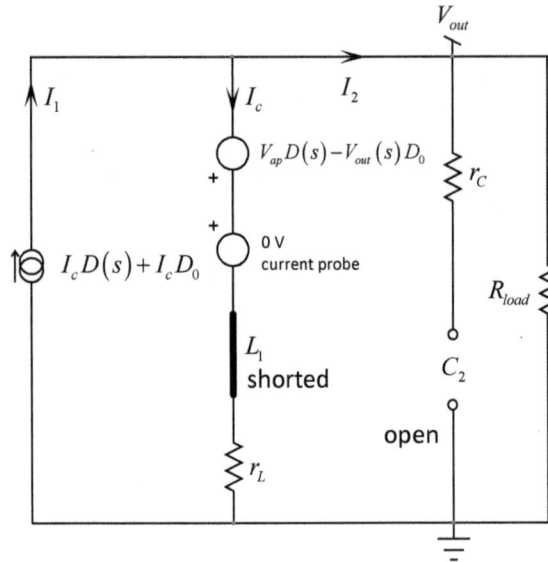

Figure 4.6 In dc analysis, the inductor is shorted while the capacitor is open-circuited.

$$V_{out}(s) = I_2(s) R_{load} \qquad (4.3)$$

If you substitute (4.1) in (4.2) then rearrange (4.3), you should find:

$$H_0 = -\left(\frac{1}{1-D_0}\right)\frac{V_{out}r_L + R_{load}(1-D_0)^2(V_{in}-V_{out})}{r_L + R_{load}(1-D_0)^2} \qquad (4.4)$$

If we neglect the inductor ohmic loss r_L and replace V_{out} by:

$$V_{out} = -V_{in}\frac{D_0}{1-D_0} \qquad (4.5)$$

then this formula simplifies to:

$$H_0 = -\frac{V_{in}}{(1-D_0)^2} \qquad (4.6)$$

We could also have derived the quasi-static gain H_0 by differentiating (4.5) with respect to the duty ratio:

$$H_0 = \frac{dV_{out}(D)}{dD} = -\frac{V_{in}}{(1-D_0)^2} \qquad (4.7)$$

Having this result on hand, we can now determine the time constants associated with this buck-boost converter. These time constants will lead us to the denominator $D(s)$ of the transfer function we want. The excitation being the duty ratio, we set it to zero. The circuit to examine is now that of Figure 4.7. We start with the resistance driving the inductor.

A test generator I_T is installed across its terminals and we must determine the voltage V_T developed by the current source.

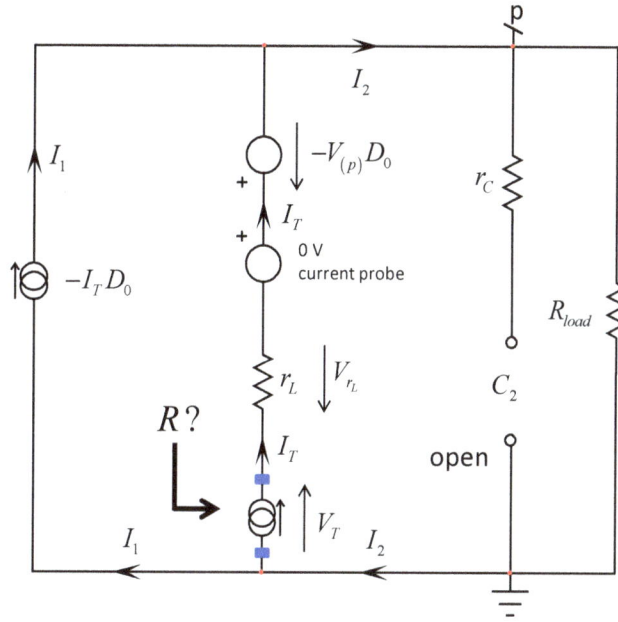

Figure 4.7 The duty ratio has been set to zero and we can determine the natural time constants of the converter.

The voltage at node p is obtained by involving the load resistance R_{load}:

$$V_{(p)} = I_2 R_{load} = (I_1 + I_T) R_{load} \tag{4.8}$$

Substituting the definition of I_1 in this expression leads to:

$$V_{(p)} = I_T (1 - D_0) R_{load} \tag{4.9}$$

Finally, the voltage V_T is obtained with the following formula:

$$V_T = V_{(p)} - V_{(p)} D_0 + r_L I_T \tag{4.10}$$

Substituting (4.9) in the above expression and rearranging gives us the resistance we want:

$$R = \frac{V_T}{I_T} = r_L + (1 - D_0)^2 R_{load} \tag{4.11}$$

Our first time constant involving L_1 is:

$$\tau_1 = \frac{L_1}{r_L + (1 - D_0)^2 R_{load}} \tag{4.12}$$

The inductor is now set in its dc state (a short circuit) and we can determine the resistance driving capacitor C_2 by installing a test generator across its terminals a shown in Figure 4.8.

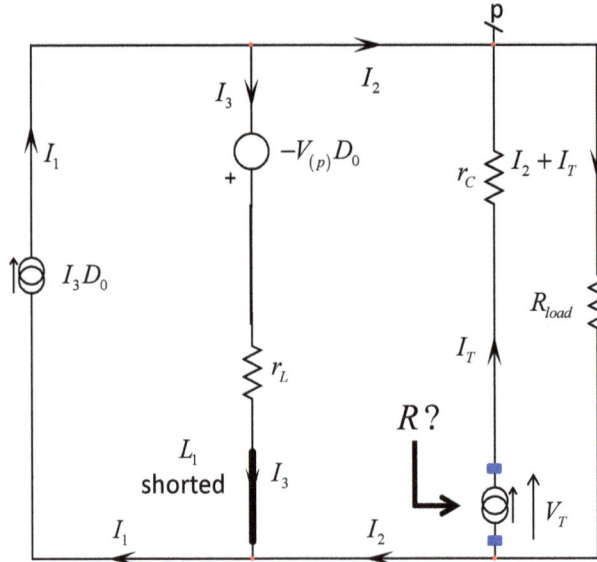

Figure 4.8 The inductor is replaced by a short circuit for this second time constant determination.

We start with current I_2:

$$I_2 = I_1 - I_3 = I_3 (D_0 - 1) \tag{4.13}$$

Current I_3 depends on the value of resistance r_L:

$$I_3 = \frac{V_{(p)} - V_{(p)} D_0}{r_L} = \frac{V_{(p)} (1 - D_0)}{r_L} \tag{4.14}$$

The voltage at terminal p is given by:

$$V_{(p)} = R_{load} (I_2 + I_T) \tag{4.15}$$

Finally, the voltage V_T across the test generator is determined by combining the voltage at terminal p and the voltage drop across the capacitor ESR:

$$R = \frac{V_T}{I_T} = r_C + \frac{R_{load} r_L}{R_{load} (1 - D_0)^2 + r_L} \tag{4.16}$$

The second time constant is thus defined as:

$$\tau_2 = C_2 \left(r_C + \frac{R_{load} r_L}{R_{load} (1 - D_0)^2 + r_L} \right) \tag{4.17}$$

The first-order term b_1 is obtained by summing the above time constants:

$$b_1 = \tau_1 + \tau_2 = \frac{L_1}{r_L + (1-D_0)^2 R_{load}} + C_2\left(r_C + \frac{R_{load}\, r_L}{R_{load}(1-D_0)^2 + r_L}\right) \qquad (4.18)$$

For the second-order term b_2, we need to set one of the energy-storing elements in its high-frequency state. We choose L_1 as it simplifies the circuit ($I_3 = 0$) as shown in Figure 4.9.

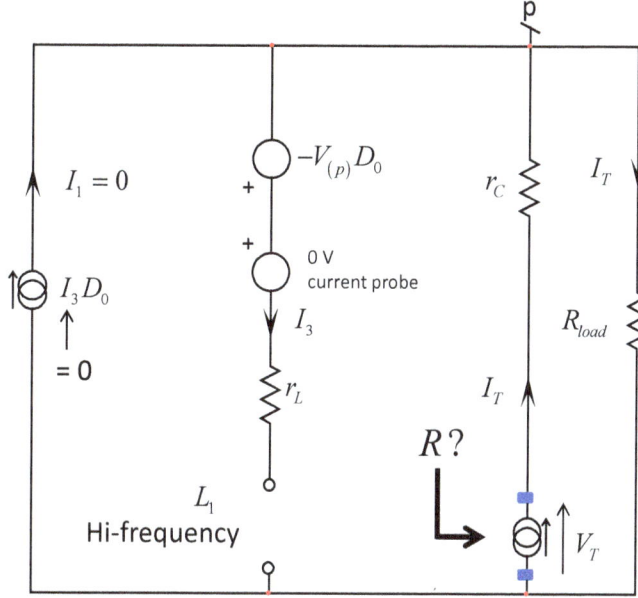

Figure 4.9 The inductor is now open-circuited for the last time constant we need.

The time constant is immediate:

$$\tau_2^1 = C_2\left(r_C + R_{load}\right) \qquad (4.19)$$

We can now form the second-order term by combining the above expression and τ_1:

$$b_2 = \tau_1 \tau_2^1 = \frac{L_1}{r_L + (1-D_0)^2 R_{load}} C_2\left(r_C + R_{load}\right) \qquad (4.20)$$

We have the complete denominator expressed as:

$$D(s) = 1 + s b_1 + s^2 b_2 = 1 + s(\tau_1 + \tau_2) + s^2 \tau_1 \tau_2^1$$

$$= 1 + \left[\frac{L_1}{r_L + (1-D_0)^2 R_{load}} + C_2\left(r_C + \frac{R_{load}\, r_L}{R_{load}(1-D_0)^2 + r_L}\right)\right] s + \left[\frac{L_1}{r_L + (1-D_0)^2 R_{load}} C_2\left(r_C + R_{load}\right)\right] s^2 \qquad (4.21)$$

We have obtained the quasi-static gain, the denominator but we are missing the zeroes. To determine them, we must identify a specific impedance combination in the circuit which prevents the excitation from generating a response.

In other words, when the stimulus $D(s)$ (the duty ratio) is back in place as in Figure 4.10, what condition would create an output null: $V_{out}(s) = 0$ V?

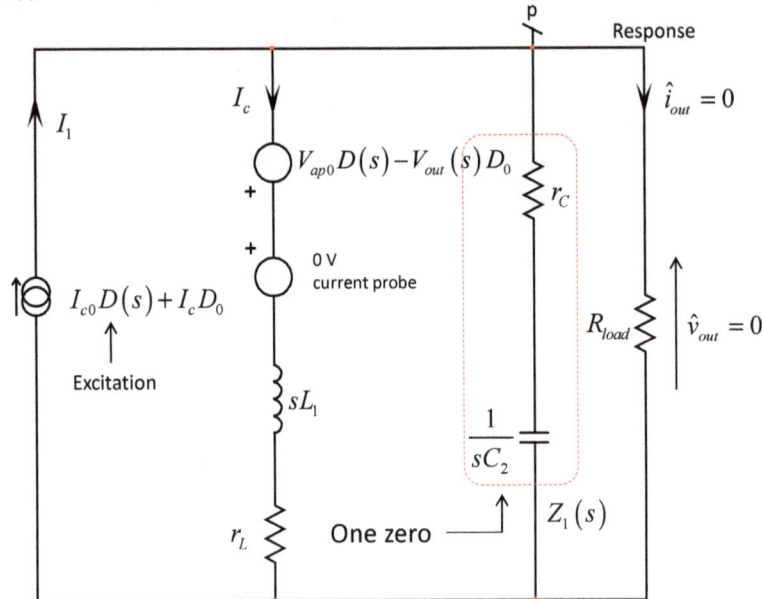

Figure 4.10 What condition in this circuit would bring a null to the output?

The obvious one is the series connection of r_C and C_2. When these two become a transformed short circuit, the response is zeroed:

$$Z_1(s) = r_C + \frac{1}{sC_2} = 0 \tag{4.22}$$

The zero is immediate and defined as:

$$\omega_{z_1} = \frac{1}{r_C C_2} \tag{4.23}$$

The second condition is when all of current I_1 is absorbed by I_c in the branch involving L_1:

$$I_1 = I_c \tag{4.24}$$

Let's narrow down the circuit to that of Figure 4.11. Current I_c is defined as follows:

$$I_c(s) = \frac{V_{ap0}D(s)}{sL_1 + r_L} \tag{4.25}$$

The equation we need to solve is the following one, obtained after replacing I_1 by its definition:

$$I_{c0}D(s) + \frac{V_{ap0}D(s)}{r_L + sL_1}D_0 - \frac{V_{ap0}D(s)}{r_L + sL_1} = 0 \tag{4.26}$$

450

Solving for the root gives:

$$S_{z_2} = -\frac{D_0 V_{ap0} - V_{ap0} + I_{c0} r_L}{I_{c0} L_1}$$

(4.27)

We now consider the following operating points determined in Chapter 1:

$$V_{ap0} = V_{in} - V_{out}$$

(4.28)

$$I_{c0} = -\frac{V_{out}}{R_L (1 - D_0)}$$

(4.29)

$$V_{out} = -\frac{D_0}{1 - D_0} V_{in}$$

(4.30)

If you substitute these variables in (4.27) and rearrange the expression, the root becomes:

$$S_{z_2} = \frac{(1 - D_0)^2 R_{load} - r_L D_0}{D_0 L_1}$$

(4.31)

This root is positive, meaning the zero lies in the right half-plane. As explained many times, its presence is due to the delay in responding to a sudden output power demand which requires to first store more energy in the inductor before feeding the output capacitor and the load.

Neglecting the inductor ohmic loss r_L, this result leads to the following RHP zero:

$$\omega_{z_2} \approx \frac{(1 - D_0)^2 R_{load}}{D_0 L_1}$$

(4.32)

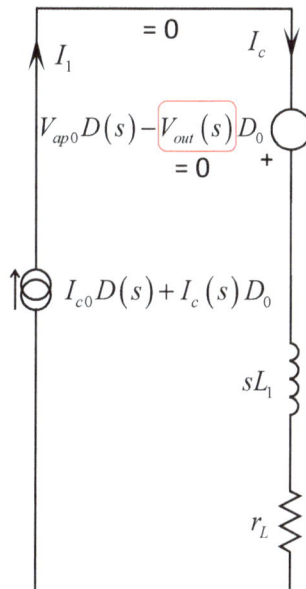

Figure 4.11 If both currents are equal, the response is nulled. The condition leading to this fact is the zero we want.

451

We have determined the complete control-to-output transfer function of the buck-boost converter operated in voltage-mode control. The denominator can be arranged in a second-order polynomial form unveiling a resonant frequency ω_0 and a quality factor Q.

These two elements can easily be determined using equations described in [1]:

$$H(s) = \frac{V_{out}(s)}{D(s)} = H_0 \frac{\left(1+\dfrac{s}{\omega_{z_1}}\right)\left(1-\dfrac{s}{\omega_{z_2}}\right)}{1+\dfrac{s}{\omega_0 Q}+\left(\dfrac{s}{\omega_0}\right)^2} \tag{4.33}$$

with:

$$H_0 \approx -\frac{V_{in}}{(1-D)^2} \tag{4.34}$$

$$\omega_{z_1} = \frac{1}{r_C C_2} \tag{4.35}$$

$$\omega_{z_2} \approx \frac{(1-D)^2 R_{load}}{D L_1} \tag{4.36}$$

$$\omega_0 = \frac{1}{\sqrt{b_2}} = \frac{1}{\sqrt{\dfrac{L_1}{r_L+(1-D)^2 R_{load}}C_2(r_C+R_{load})}} = \frac{1}{\sqrt{L_1 C_2}\sqrt{\dfrac{r_C+R_{load}}{r_L+(1-D)^2 R_{load}}}} \approx \frac{1-D}{\sqrt{L_1 C_2}} \tag{4.37}$$

$$Q = \frac{\sqrt{b_2}}{b_1} = \frac{\sqrt{\dfrac{L_1}{r_L+(1-D)^2 R_{load}}C_2(r_C+R_{load})}}{\dfrac{L_1}{r_L+(1-D)^2 R_{load}}+C_2\left(r_C+\dfrac{R_{load} r_L}{R_{load}(1-D)^2+r_L}\right)} \approx (1-D) R_{load}\sqrt{\frac{C_2}{L_1}} \tag{4.38}$$

We can now compare the dynamic response delivered by a SPICE simulation with that of our equations with Mathcad®. We use the large-signal PWM switch model shown in the buck-boost converter of Figure 4.12.

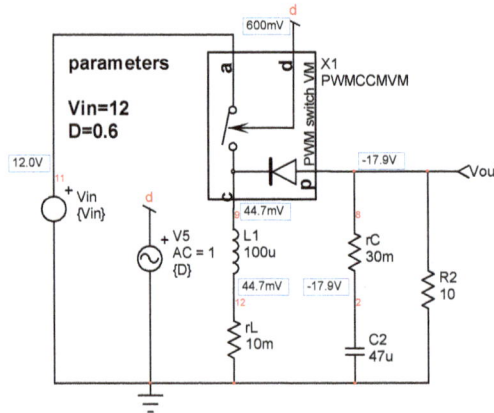

Figure 4.12 The model predicts an output voltage of –17.9 V with a duty ratio of 60%.

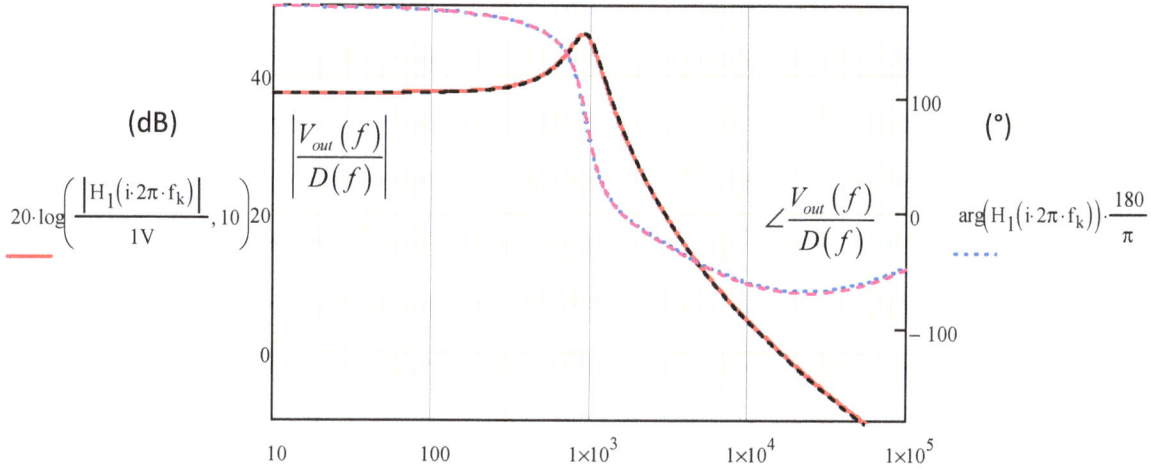

Figure 4.13 Mathcad® and the SPICE model deliver the exact same response.

Figure 4.13 confirms that our derivation is correct as both magnitude and phase responses perfectly superimpose.

We can also run a simple cycle-by-cycle simulation with SIMPLIS® as illustrated in Figure 4.14. The circuit is fairly simple and switches at a 100-kHz frequency. Simulation waveforms are obtained in a few seconds as well as the dynamic response. As confirmed by Figure 4.15, the results are in agreement with our equation-based analysis.

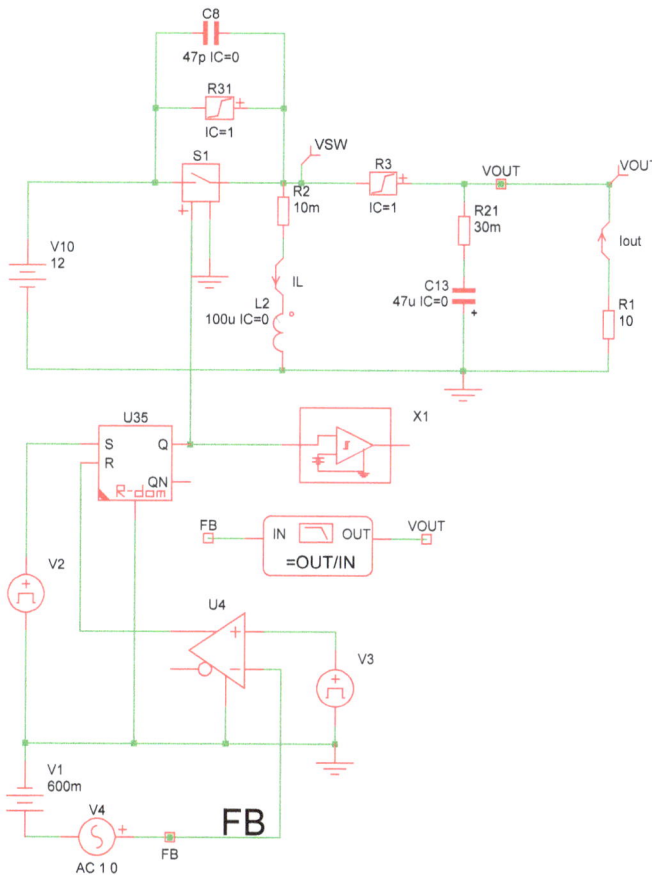

Figure 4.14 SIMPLIS® lets us simulate a switching buck-boost converter and extract its control-to-output dynamic response.

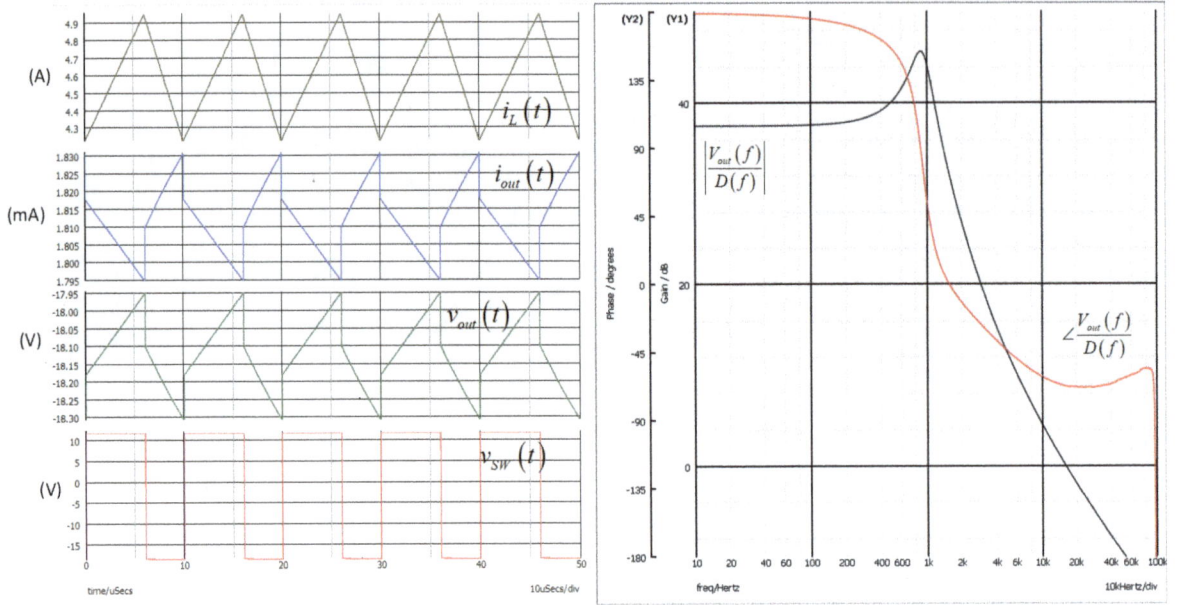

Figure 4.15 This Bode plot matches what we have analytically derived with the PWM switch model.

4.1.1 Input to Output

In this analysis, the duty ratio becomes a fixed bias, $D(s) = 0$, while the input source becomes the stimulus, $V_{in}(s)$. The new circuit appears in Figure 4.16. When V_{in} reduces to 0 V, node a is grounded and the circuit returns to the state already analyzed in Figure 4.5: time constants are similar and we can reuse the denominator $D(s)$ already determined in (4.21).

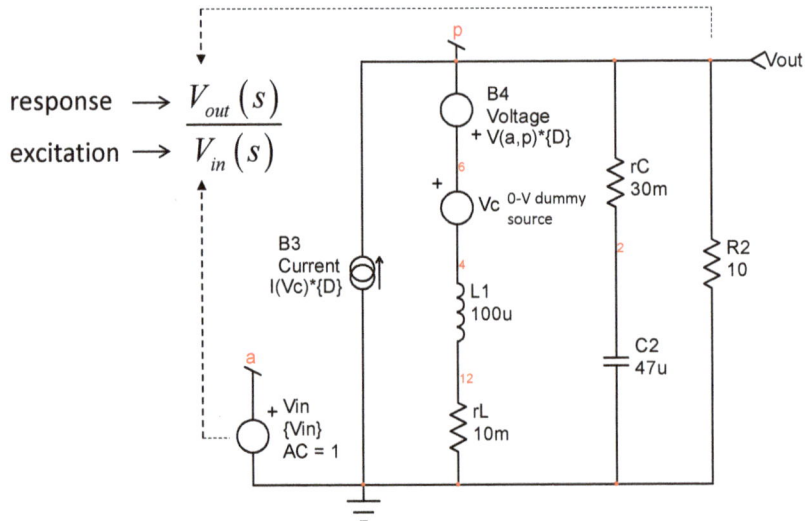

Figure 4.16 When the stimulus reduces to 0 V, node a is grounded and the circuit returns to its natural state.

The dc gain is obtained by shorting the inductor and opening the capacitor as shown in Figure 4.17.

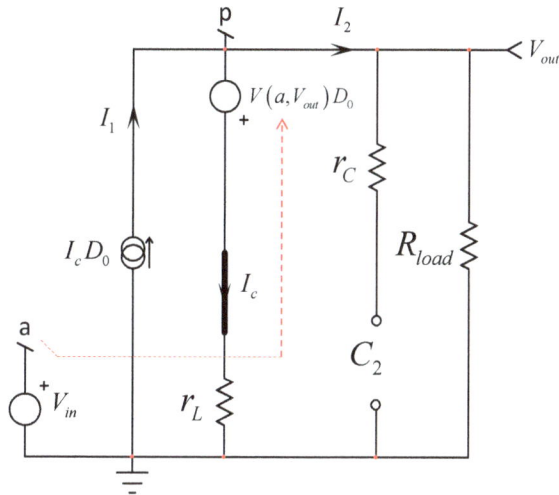

Figure 4.17 The quasi-static gain is obtained by shorting the inductor and open-circuiting the capacitor.

Current I_c can be defined as follows:

$$I_c = \frac{V_{out} + V_{in}D_0 - V_{out}D_0}{r_L} \tag{4.39}$$

This definition can be immediately substituted in I_1 definition:

$$I_1 = I_c D_0 = \left(\frac{V_{out} + V_{in}D_0 - V_{out}D_0}{r_L} \right) D_0 \tag{4.40}$$

The output voltage depends on the load and both of the above currents:

$$\frac{V_{out}}{V_{in}} = H_0 = -\frac{D_0(1-D_0)R_{load}}{(1-D_0)^2 R_{load} + r_L} \approx -\frac{D_0}{1-D_0} \tag{4.41}$$

To determine the zeroes, we bring the excitation stimulus back – V_{in} – and we observe the electrical conditions in this circuit which could null the output response (Figure 4.18).

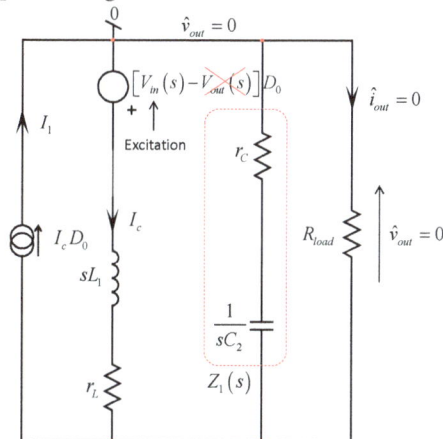

Figure 4.18 The zero is contributed by the ESR and the output capacitance.

The usual suspects are capacitance C_2 and its ESR. They contribute a zero located at:

$$\omega_z = \frac{1}{r_C C_2}$$

(4.42)

Now, does L_1 contribute a zero? To verify this fact, set the inductor in its high-frequency state [1] and check if the stimulus can make its way through the circuit in this condition to produce a response. If there is a response then you must determine the zero position.

If there is no response, then there is no zero associated with the selected energy-storing element. In that case, no need to further analyze the circuit. When L_1 is open-circuited, current I_c goes to zero which also zeroes I_1: no response is possible and there is one single zero in this circuit. The transfer function is now determined and expressed as:

$$\frac{V_{out}(s)}{V_{in}(s)} \approx H_0 \frac{1+\dfrac{s}{\omega_z}}{1+\dfrac{s}{\omega_0 Q}+\left(\dfrac{s}{\omega_0}\right)^2}$$

(4.43)

with:

$$H_0 \approx -\frac{D}{1-D}$$

(4.44)

$$\omega_0 \approx \frac{1-D}{\sqrt{L_1 C_2}}$$

(4.45)

$$Q \approx (1-D) R_{load} \sqrt{\frac{C_2}{L_1}}$$

(4.46)

and ω_z defined by (4.42). We have tested the SPICE response (Figure 4.19) versus that of the Mathcad® sheet and Figure 4.20 confirm they all agree well.

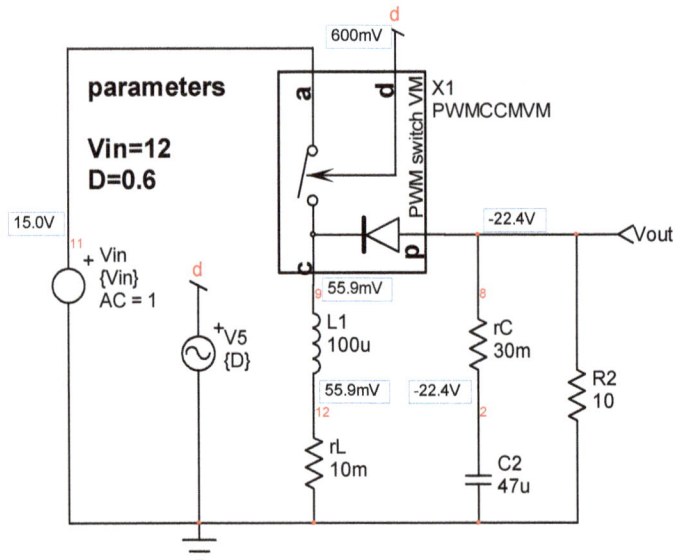

Figure 4.19 The large-signal VM PWM switch model is used to plot the input-to-output transfer function.

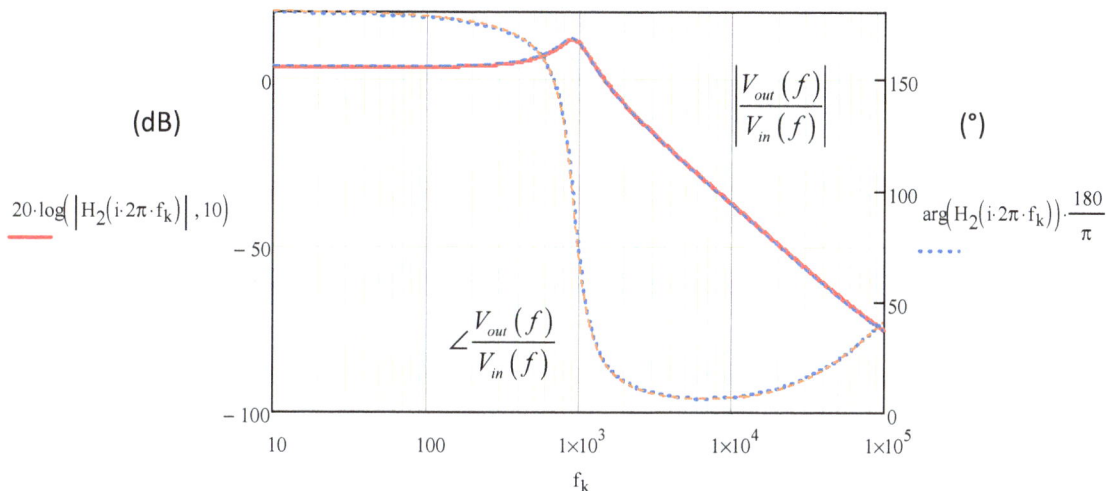

Figure 4.20 Mathcad® and SPICE plots are identical.

4.1.2 Output Impedance

The output impedance is obtained by installing a 1-A ac current source across the loading resistor. This is what Figure 4.21 illustrates. Considering the 1-A value, the voltage obtained at the Z_{out} node is the direct image of the impedance we want. To determine the natural time constants of this circuit, we turn the excitation off and open-circuit the current source (Figure 4.22). This arrangement is similar to that of Figure 4.6 implying that all time constants remain the same: we can reuse the denominator $D(s)$ already defined with (4.21).

To determine the dc resistance R_0, we install a test generator as in Figure 4.23. We start by defining the current flowing in terminal c:

$$I_c = \frac{V_{out} - V_{out}D_0}{r_L} = \frac{V_{out}(1-D_0)}{r_L} \tag{4.47}$$

We can also define currents I_1 and I_2 as:

$$I_1 = \left(\frac{V_T(1-D_0)}{r_L}\right)D_0 \tag{4.48}$$

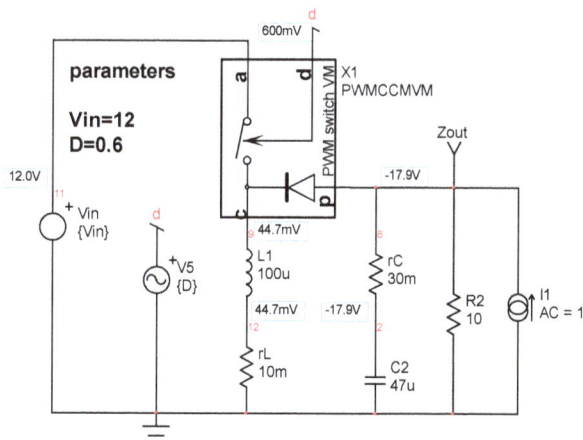

Figure 4.21 The 1-A ac current source will let you plot the output impedance.

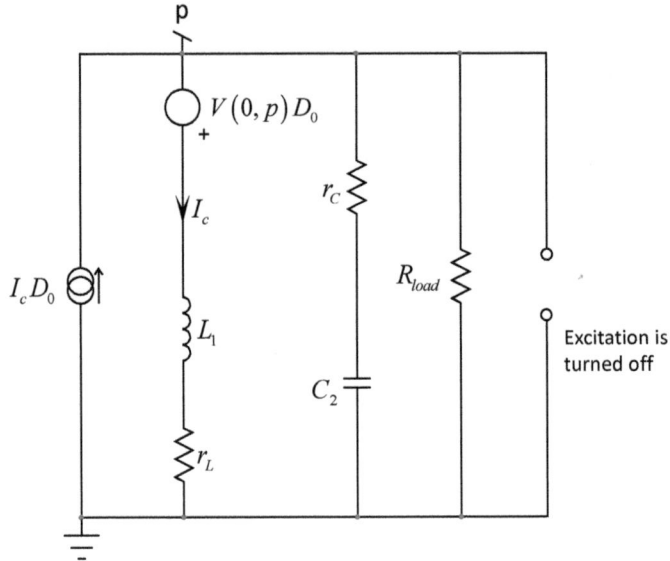

Figure 4.22 When the excitation is turned off, the circuit returs to its natural state.

$$I_2 = \frac{V_T}{R_{load}} \tag{4.49}$$

The test current I_T is a combination of these three variables:

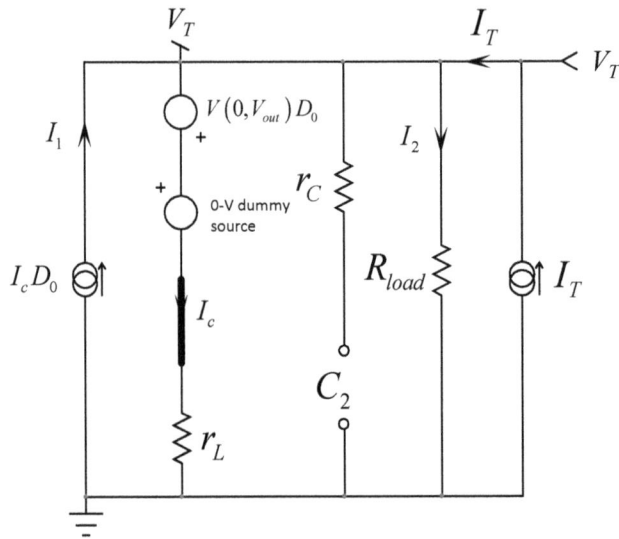

Figure 4.23 The test generator I_T lets us determine R_0, the dc term of the output impedance.

$$I_T = I_c + I_2 - I_1 \tag{4.50}$$

The resistance we want is thus defined by:

$$R_0 = \frac{V_T}{I_T} = \frac{r_L}{\left(1 - D_0\right)^2 + \dfrac{r_L}{R_{load}}} \tag{4.51}$$

Which if $r_L << R_{load}$ simplifies to:

$$R_0 \approx \frac{r_L}{\left(1-D_0\right)^2} \qquad (4.52)$$

To determine the zeroes, we bring the excitation back and find conditions which null the output: Figure 4.24 illustrates the exercise.

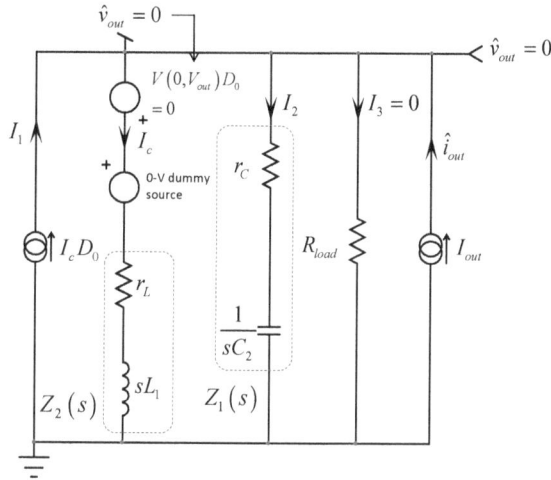

Figure 4.24 We have to find conditions which null the output to determine the zeroes of the transfer function.

Considering a null output voltage, the source $V(0,V_{out})D_0$ in series with r_L is reduced to zero and the whole circuit simplifies to that of Figure 4.25. The output is nulled if:

$$Z_1(s) = r_C + \frac{1}{sC_2} = 0 \qquad (4.53)$$

and:

$$Z_2(s) = r_L + sL_1 = 0 \qquad (4.54)$$

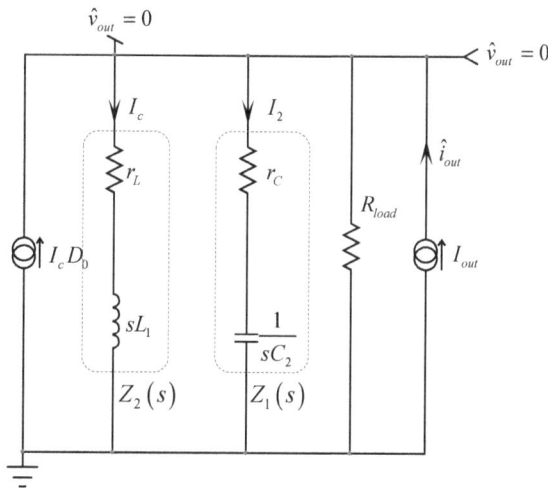

Figure 4.25 There are two conditions for which the output is nulled: one involves the inductor while the second the output capacitor.

459

These conditions are satisfied for the following angular frequencies:

$$\omega_{z_1} = \frac{1}{r_C C_2} \tag{4.55}$$

$$\omega_{z_2} = \frac{r_L}{L_1} \tag{4.56}$$

These results immediately lead to the following transfer function describing the output impedance:

$$Z_{out}(s) \approx R_0 \frac{\left(1 + \dfrac{s}{\omega_{z_1}}\right)\left(1 + \dfrac{s}{\omega_{z_2}}\right)}{1 + \dfrac{s}{\omega_0 Q} + \left(\dfrac{s}{\omega_0}\right)^2} \tag{4.57}$$

The zeroes and R_0 are given in the above lines while Q and ω_0 are respectively defined by (4.46) and (4.45). We can check how Mathcad® plots these expressions versus a SPICE simulation using the large-signal model. The test circuit appears in Figure 4.26. Magnitude and phase curves from either source perfectly superimpose, confirming our derivation is correct Figure 4.27).

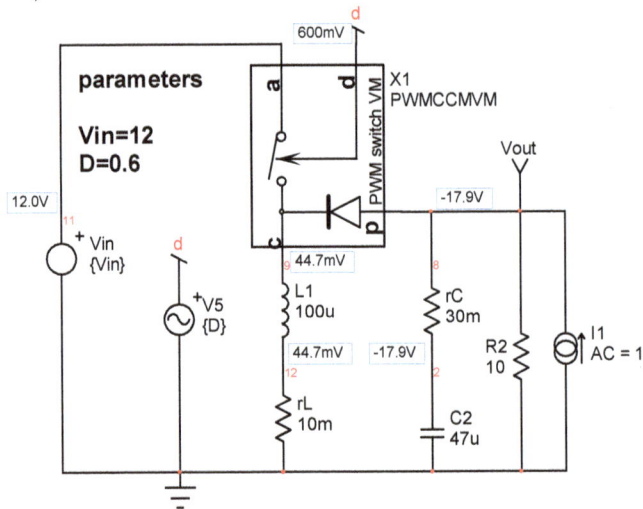

Figure 4.26 By adding a 1-A ac current source across the output load, it becomes possible to sweep the output impedance.

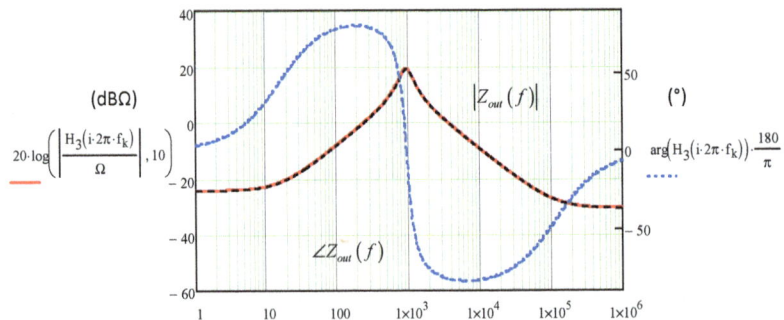

Figure 4.4.27 Results delivered by Mathcad® and SPICE are rigorously identical.

4.1.3 Input Impedance

The input impedance is the final transfer function we want to determine for this CCM buck-boost converter operated in voltage-mode control. You are now familiar with the application circuit from Figure 4.28: the operating point is ensured with the input source ($V_{in} = 12$ V) feeding the circuit via a large-value inductor.

This inductor is shorted during bias point calculations and does not bother operations. In ac analysis, it isolates the converter from the 0-Ω-resistance V_{in} source and lets you sweep the input converter impedance via the added current source. Considering a 1-A value, plotting the voltage probe VZ_{in} collected across the current source gives the magnitude and phase of Z_{in}. We have verified that this small-signal model used for analysis delivered the exact same results as those obtained in similar conditions with the large-signal model. It confirms the model is good for our analysis.

We can start with the incremental resistance R_0 obtained when shorting the inductor and opening the output capacitor. This is what Figure 4.29 shows. We can write a few equations to start with:

$$I_c D_0 = I_c + I_2 \tag{4.58}$$

From which you have:

$$I_2 = I_c \left(D_0 - 1\right) \tag{4.59}$$

Then you see that:

$$V_{(p)} = I_2 R_{load} \tag{4.60}$$

which leads to:

$$V_{(p)} = I_c \left(D_0 - 1\right) R_{load} \tag{4.61}$$

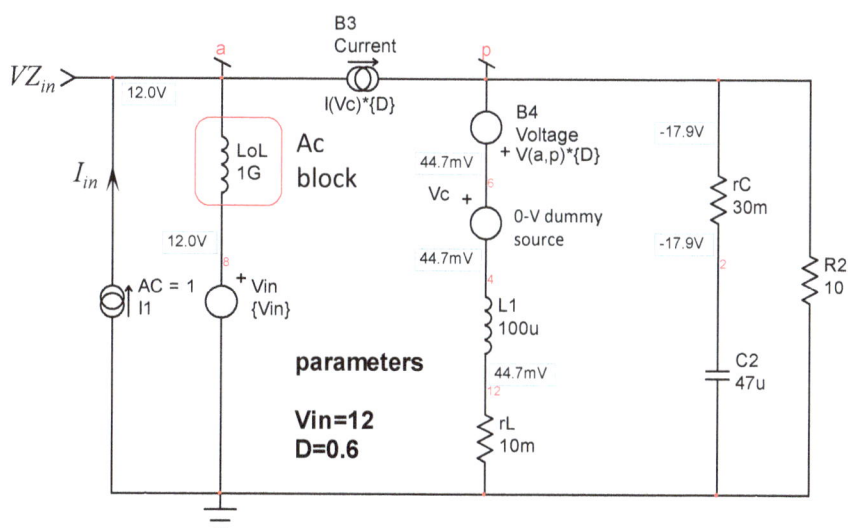

Figure 4.28 A 1-A ac current source now sweeps the circuit's input impedance.

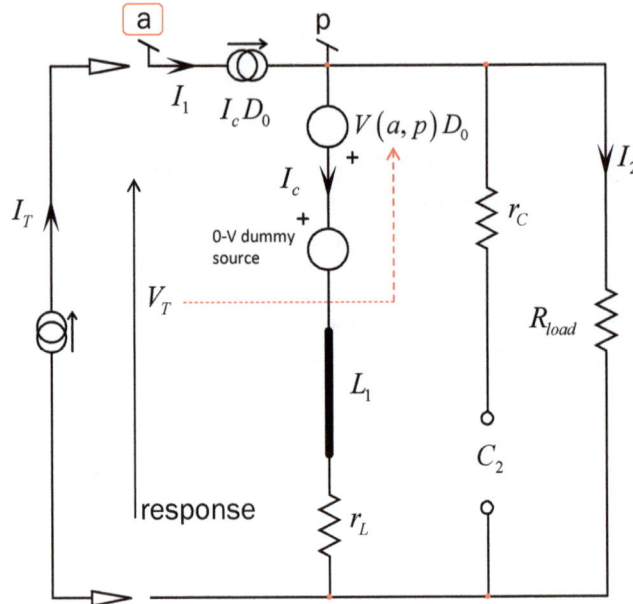

Figure 4.29 Shorting the inductor and open-circuiting the capacitor lets us determine R_0.

Current I_c is the voltage across resistor r_L divided by its resistance:

$$I_c = \frac{V_{(p)} + V_T D_0 - V_{(p)} D_0}{r_L} \tag{4.62}$$

Substitute (4.61) in the above equation and you have a new definition for current I_c:

$$I_c = \frac{D_0 V_T}{R_{load} D_0^2 - 2 R_{load} D_0 + R_{load} + r_L} \tag{4.63}$$

From the schematic diagram, you see that:

$$I_T = D_0 I_c \tag{4.64}$$

which combined with (4.63) gives you the dc input resistance R_0:

$$R_0 = \frac{V_T}{I_T} = \frac{(1 - D_0)^2 R_{load} + r_L}{D_0^2} \approx \left(\frac{V_{in}}{V_{out}}\right)^2 R_{load} \tag{4.65}$$

To determine the poles, we turn the excitation off which means open-circuiting the left-side test generator I_1 in Figure 4.28. Doing so makes node a without a return path and this changes the circuit natural structure. We cannot reuse the denominator $D(s)$ we have previously determined with (4.21) and we must identify the new time constants.

As it can create indeterminacies, we fix the situation by adding a finite resistance R_{inf} as in Figure 4.30.

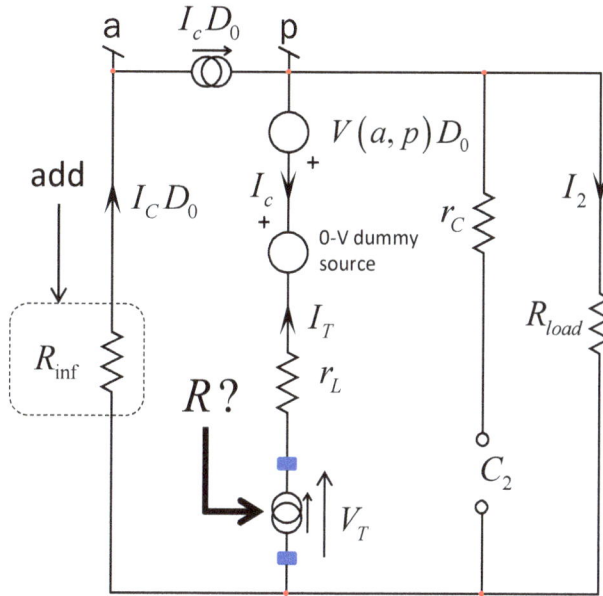

Figure 4.30 The dummy resistance R_{dum} closes the dc path at node a and conveniently avoids indeterminacies.

What we want is the time constant involving L_1 so we install a test generator I_T across its terminals. The voltage at node a is given by:

$$V_{(a)} = D_0 I_T R_{inf} \tag{4.66}$$

From the circuit, we see:

$$I_2 = I_c D_0 + I_T \tag{4.67}$$

But we know that $I_c = -I_T$ then:

$$I_2 = I_T (1 - D_0) \tag{4.68}$$

We continue with the voltage at node p:

$$V_{(p)} = I_2 R_{load} = I_T (1 - D_0) R_{load} \tag{4.69}$$

The voltage V_T is equal to:

$$V_T = I_T r_L + \left[V_{(a)} - V_{(p)} \right] D_0 + V_{(p)} = I_T \left(r_L + R_{inf} D_0^2 - (1 - D_0) R_{load} D_0 + (1 - D_0) R_{load} \right) \tag{4.70}$$

Now rearrange V_T / I_T and define the time constant involving L_1:

$$\tau_1 = \frac{L_1}{r_L + (1 - D_0) \left(R_{inf} \dfrac{D_0^2}{1 - D_0} + R_{load} (1 - D_0) \right)} \tag{4.71}$$

If you bring R_{inf} to infinity, you have:

$$\tau_1\big|_{R_{inf}\to\infty} = \frac{L_1}{\infty} = 0 \tag{4.72}$$

For the second time constant, the inductor is shorted and we install the test generator across C_2's connecting terminals (Figure 4.31). The first equation defines the voltage at terminal a:

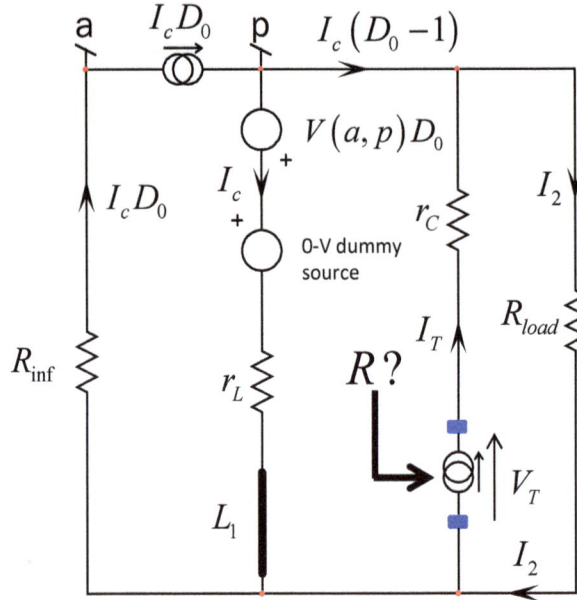

Figure 4.31 The resistance R_{inf} is left in place to offer a dc path at node a.

$$V_{(a)} = -D_0 I_c R_{inf} \tag{4.73}$$

Then, the voltage at terminal p is obtained by:

$$V_{(p)} = \left[I_c\left(D_0 - 1\right) + I_T\right]R_{load} \tag{4.74}$$

But also via:

$$V_{(p)} = V_T - I_T r_C \tag{4.75}$$

Current I_c is determined by dividing the voltage across r_L by the resistance value:

$$I_c = \frac{V_{(p)} + V_{(a)}D_0 - V_{(p)}D_0}{r_L} \tag{4.76}$$

Substitute (4.74) and (4.73) in the above expression and rearrange:

$$I_c = \frac{I_T R_{load}\left(1 - D_0\right)}{R_{load} + r_L - 2D_0 R_{load} + D_0^{\,2}\left(R_{inf} + R_{load}\right)} \tag{4.77}$$

464

Then equate (4.75) with (4.74) while substituting (4.77) for I_c. Rearrange to obtain the below equation:

$$R = \frac{V_T}{I_T} = r_C + \frac{R_{inf}\left(R_{load}D_0^2 + \frac{R_{load}r_L}{R_{inf}}\right)}{R_{inf}\left(\frac{R_{load} + r_L - 2D_0 R_{load} + D_0^2 R_{load}}{R_{inf}} + D_0^2\right)}$$ (4.78)

When R_{inf} approaches infinity, the second time constant equals

$$\tau_2 \approx \left(r_C + R_{load}\right)C_2$$ (4.79)

We can now form the first coefficient b_1 as:

$$b_1 = \tau_1 + \tau_2 = \left(r_C + R_{load}\right)C_2$$ (4.80)

For the high-frequency term, set inductor L_1 in its high-frequency state (open-circuit it) and determine the resistance seen from C_2's connecting terminals (Figure 4.32). As L_1 is open, current I_c is 0 A and therefore

$$\tau_2^1 = C_2\left(r_C + R_{load}\right)$$ (4.81)

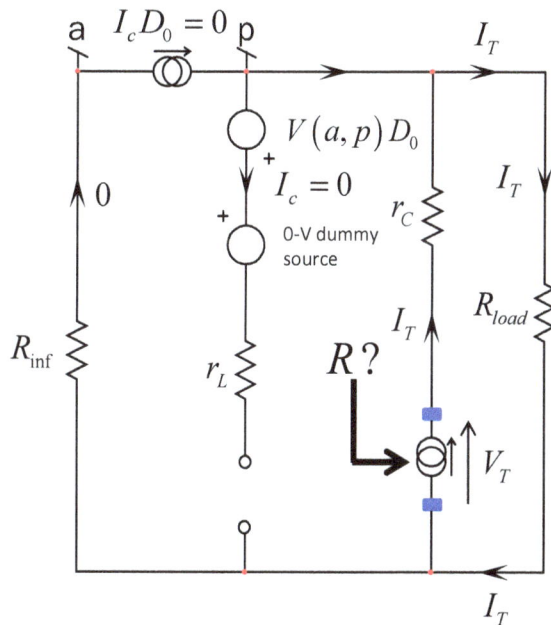

Figure 4.32 The inductor is open-circuited for the second-order term determination.

We can form b_2 as:

$$b_2 = \tau_1 \tau_2^1 = \frac{L_1}{\infty} \cdot C_2\left(r_C + R_{load}\right) = 0$$ (4.82)

The denominator is a first-order expression defined by

$$D(s) = 1 + b_1 s + b_2 s^2 = 1 + s\left(r_C + R_{load}\right)C_2$$ (4.83)

To obtain the zeroes, we null the response which is nothing else than the voltage V_T generated across the test current source. We know this is a degenerate case [1] and we can replace the current source by a short circuit. From Figure 4.33, we recognize the circuit has returned to its natural state as already described by Figure 4.7 and (4.21).

Therefore:

$$N(s) = 1 + \frac{s}{\omega_0 Q} + \left(\frac{s}{\omega_0}\right)^2 \tag{4.84}$$

with:

$$\omega_0 \approx \frac{1-D}{\sqrt{L_1 C_2}} \tag{4.85}$$

and:

$$Q \approx (1-D) R_{load} \sqrt{\frac{C_2}{L_1}} \tag{4.86}$$

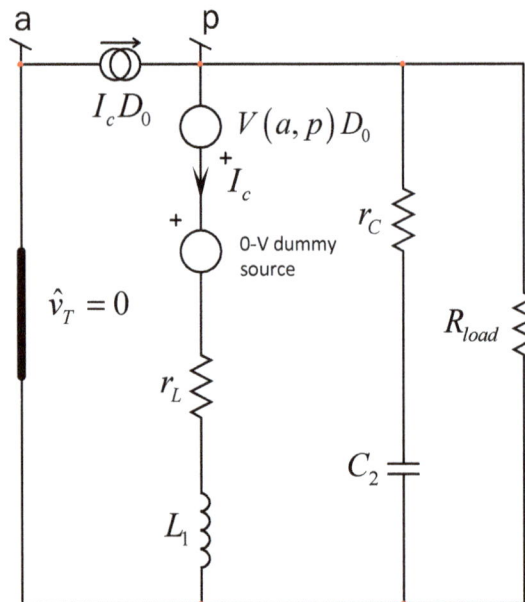

Figure 4.33 The numerator is obtained by shorting the response V_T across the test current source.

The input impedance transfer function is immediate and expressed as:

$$Z_{in}(s) \approx R_0 \frac{1 + \frac{s}{\omega_0 Q} + \left(\frac{s}{\omega_0}\right)^2}{1 + \frac{s}{\omega_p}} \tag{4.87}$$

With ω_0 and Q defined in the above lines and R_0 determined by (4.65).

The pole is the inverse of the time constant found with (4.80):

$$\omega_p = \frac{1}{\left(r_C + R_{load}\right)C_2} \tag{4.88}$$

It is time to test our equations versus the response delivered by the large-signal model shown in Figure 4.34. Figure 4.35 confirms that SPICE and Mathcad® perfectly agree on the magnitude and phase responses.

This is the last transfer function for the buck-boost converter operated in voltage-mode control and CCM. We have conveniently gathered all these expressions in the table given in Figure 4.36. In this figure, V_p is the artificial ramp peak voltage of the pulse-width modulator.

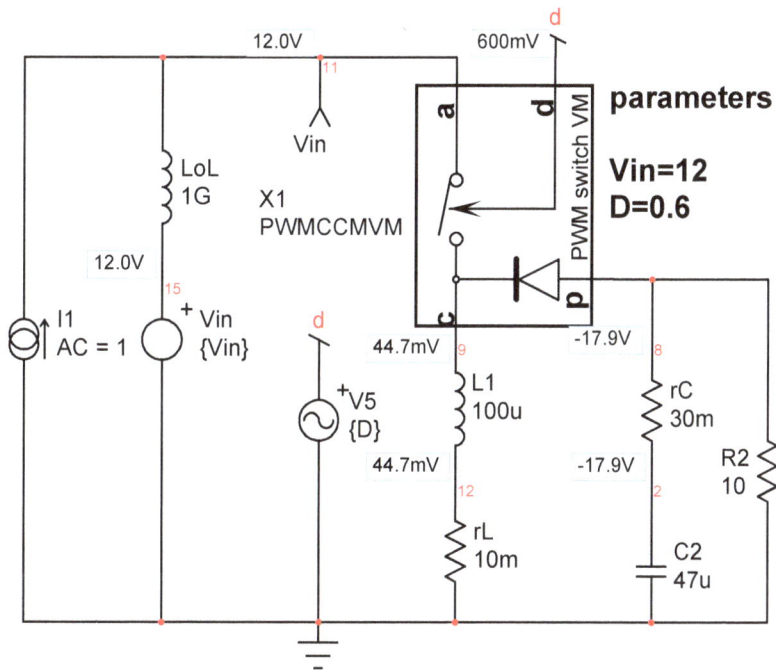

Figure 4.34 SPICE linearizes the large-signal model and delivers the input impedance immediately.

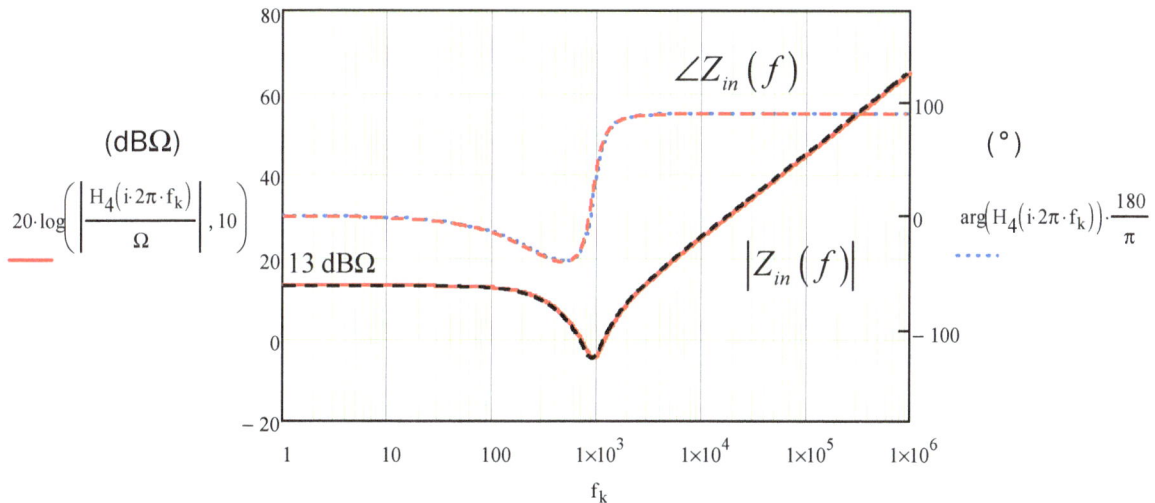

Figure 4.35 SPICE and the Mathcad® sheet deliver the exact set of magnitude-phase curves.

$\dfrac{V_{out}(s)}{V_{err}(s)}$ Control to Output	$H_0 \dfrac{\left(1+\dfrac{s}{\omega_{z_1}}\right)\left(1-\dfrac{s}{\omega_{z_2}}\right)}{1+\dfrac{s}{\omega_0 Q}+\left(\dfrac{s}{\omega_0}\right)^2}$	$\omega_{z_1}=\dfrac{1}{r_C C_2}$ $\omega_{z_2}=\dfrac{(1-D)^2 R_{load}}{DL_1}$	$\omega_0 \approx \dfrac{1-D}{\sqrt{L_1 C_2}}$ $Q \approx (1-D)R_{load}\sqrt{\dfrac{C_2}{L_1}}$	$H_0 \approx -\dfrac{V_{in}}{V_p(1-D)^2}$
$\dfrac{V_{out}(s)}{V_{in}(s)}$ Input to Output	$H_0 \dfrac{1+\dfrac{s}{\omega_z}}{1+\dfrac{s}{\omega_0 Q}+\left(\dfrac{s}{\omega_0}\right)^2}$	$\omega_z=\dfrac{1}{r_C C_2}$	$\omega_0 \approx \dfrac{1-D}{\sqrt{L_1 C_2}}$ $Q \approx (1-D)R_{load}\sqrt{\dfrac{C_2}{L_1}}$	$H_0 \approx -\dfrac{D}{1-D}$
$Z_{in}(s)$ Input impedance	$R_0 \dfrac{1+\dfrac{s}{\omega_0 Q}+\left(\dfrac{s}{\omega_0}\right)^2}{1+\dfrac{s}{\omega_p}}$	$\omega_0 \approx \dfrac{1-D}{\sqrt{L_1 C_2}}$ $Q \approx (1-D)R_{load}\sqrt{\dfrac{C_2}{L_1}}$	$\omega_p=\dfrac{1}{(r_C+R_{Load})C_2}$	$R_0 \approx \left(\dfrac{1-D}{D}\right)^2 R_{load}$
$Z_{out}(s)$ Output impedance	$R_0 \dfrac{(1+s/\omega_{z_1})(1+s/\omega_{z_2})}{1+\dfrac{s}{\omega_0 Q}+\left(\dfrac{s}{\omega_0}\right)^2}$	$\omega_{z_1}=\dfrac{r_L}{L_1}$ $\omega_{z_2}=\dfrac{1}{r_C C_2}$	$\omega_0 \approx \dfrac{1-D}{\sqrt{L_1 C_2}}$ $Q \approx (1-D)R_{load}\sqrt{\dfrac{C_2}{L_1}}$	$R_0 \approx \dfrac{r_L}{(1-D)^2}$

V_p is the PWM sawtooth

Figure 4.36 All transfer functions of the CCM-operated buck-boost converter.

4.2 Buck-Boost Transfer Functions in DCM – Fixed-Frequency Voltage-Mode Control

We will now reduce the load current and force the converter to enter the discontinuous conduction mode. To determine the load value leading to this operating condition, we can calculate the critical resistance using the expression given in Chapter 1. Please note that inductor L_1's value is arbitrarily selected to 47 µH in this example:

$$R_{critical}=\frac{2F_{sw}L_1}{(1-D)^2}=\frac{2\times 47u\times 100k}{(1-0.44)^2}\approx 30\ \Omega \tag{4.89}$$

By selecting a 100-Ω resistance, we know the converter from Figure 4.37 will operate in DCM. The large-signal model appears in the simulation template of Figure 4.37.

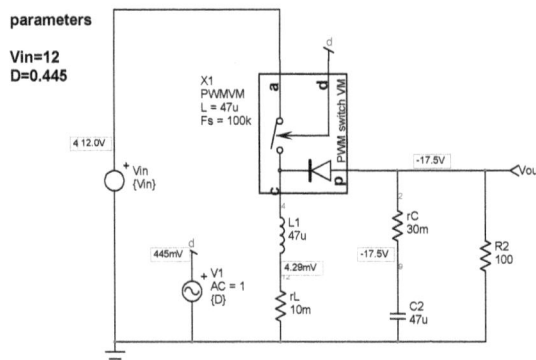

Figure 4.37 The converter now operates in the discontinuous conduction mode or DCM.

The exercise now consists of inserting the small-signal model already used in the DCM voltage-mode buck and boost converters analyses as shown in Figure 4.38. Before we carry on, we need to verify that both circuits deliver the same magnitude and phase responses. It is important to verify the sanity of this equivalent linear circuit before spending time to analyze it.

parameters

d1=445m
Vin=12
Vac=Vin
Vap=Vin-Vout
Fsw=100k
L=47u
RL=100
TauL=L*Fsw/RL
M=-d1/sqrt(2*TauL)
Vout=M*Vin
Ic=Vin*(d1^2-2*M*TauL)/(2*RL*TauL)

k1=Vac*d1/(Fsw*L)
k2=d1^2/(2*Fsw*L)
k3=Vac*Vap*d1/(Fsw*L*Ic)
k4=Vac*d1^2/(2*Fsw*L*Ic)
k5=-Vac*Vap*d1^2/(2*Fsw*Ic^2*L)
k6=Vap*d1^2/(2*Fsw*Ic*L)

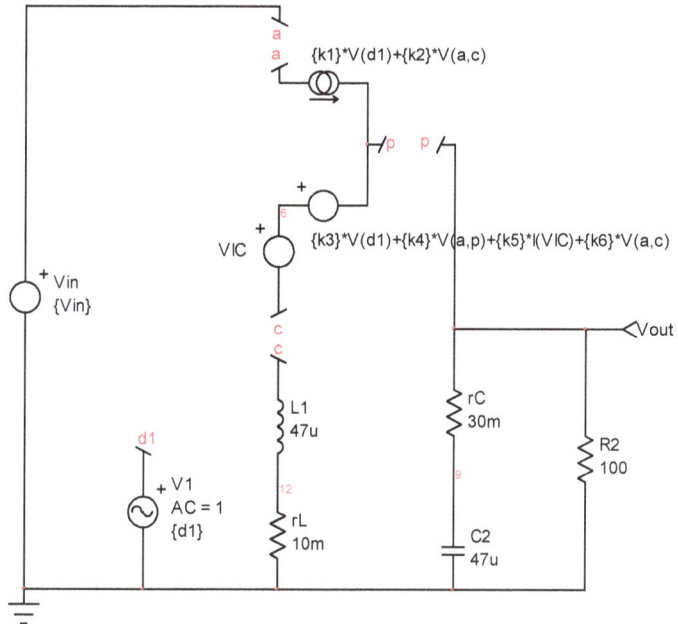

Figure 4.38 The circuit is analyzed with the small-signal DCM model in place.

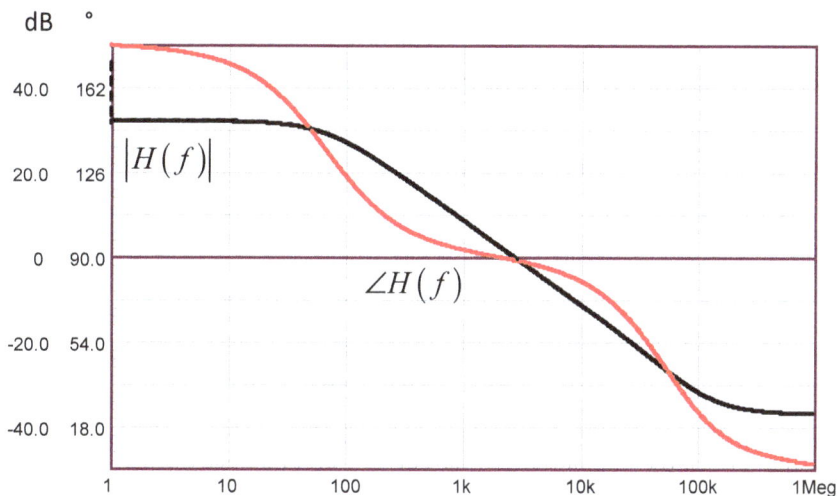

Figure 4.39 Both circuits exactly lead to the same response.

As confirmed by Figure 4.39, both responses perfectly superimpose and validate the small-signal circuit. We can start the analysis by rearranging Figure 4.38 into a circuit easier to analyze (after validating its ac response of course) proposed in Figure 4.40.

We start by shorting the inductor and opening the capacitor to determine the dc transfer function M. The input source V_{in} is present for this dc analysis as shown in Figure 4.41.

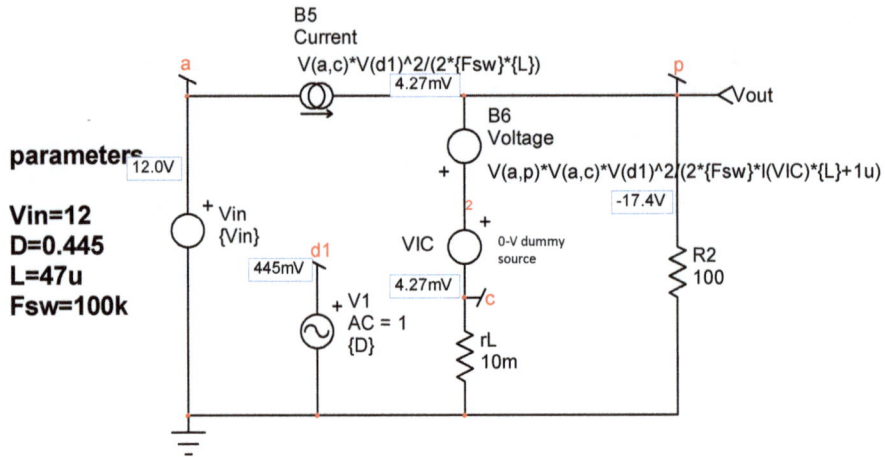

Figure 4.40 Once rearranged, the circuit looks simpler to analyze.

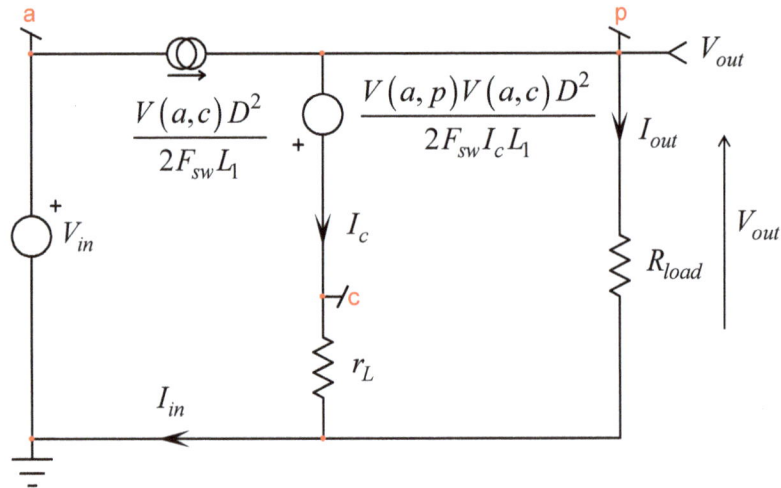

Figure 4.41 The inductor is shorted while the capacitor is open-circuited for this analysis.

We start by observing that the input current I_{in} is fixed by the upper-left-side current source:

$$I_{in} = I_c + I_{out} = \frac{D^2 V(a,c)}{2F_{sw}L} = \frac{D^2\left[V_{in} - V_{(c)}\right]}{2F_{sw}L_1} \tag{4.90}$$

We can update this definition using the output current I_{out} definition:

$$I_c = -\frac{V_{out}}{R_{load}} + \frac{D^2\left[V_{in} - V_{(c)}\right]}{2F_{sw}L_1} \tag{4.91}$$

This expression can be advantageously rewritten with the normalized time constant already defined in the previous chapters:

$$\tau_L = \frac{L_1 F_{sw}}{R_{load}} \tag{4.92}$$

Hence:

$$I_c = -\frac{V_{out}}{R_{load}} + \frac{D^2\left[V_{in} - V_{(c)}\right]}{2R_{load}\tau_L} \tag{4.93}$$

The I_c current also depends on the voltage at terminal c divided by r_L:

$$I_c = \frac{V_{(c)}}{r_L} \tag{4.94}$$

Capitalizing on (4.91), we can write:

$$\frac{V_{(c)}}{r_L} = -\frac{V_{out}}{R_{load}} + \frac{D^2\left(V_{in} - V_{(c)}\right)}{2R_{load}\tau_L} \tag{4.95}$$

Now solve for $V_{(c)}$ and you find:

$$V_{(c)} = \frac{V_{in}D^2 r_L - 2V_{out}r_L\tau_L}{r_L D^2 + 2R_{load}\tau_L} \tag{4.96}$$

The voltage at node c can also be computed involving the output voltage V_{out}:

$$V_{(c)} = V_{out} + \frac{\left(V_{in} - V_{out}\right)\left[V_{in} - V_{(c)}\right]D^2}{2\tau_L R_{load}\dfrac{V_{(c)}}{r_L}} \tag{4.97}$$

From this expression, extract V_{out} as:

$$V_{out} = \frac{2R_{load}\tau_L V_{(c)}^2 + r_L V_{(c)}V_{in}D^2 - r_L V_{in}^2 D^2}{2R_{load}V_{(c)}\tau_L + V_{(c)}D^2 r_L - V_{in}D^2 r_L} \tag{4.98}$$

Substitute (4.96) in the above, rearrange to express V_{out} again and you should find:

$$M = \frac{V_{out}}{V_{in}} = -D\frac{\left(r_L D^2 + 2R_{load}\tau_L\right)\sqrt{8\tau_L R_{load}\left(R_{load} + r_L\right) + D^2 r_L^2} - r_L D_0\left(D^2 r_L + 6R_{load}\tau_L\right)}{8\left(R_{load}^2\tau_L^2 + R_{load}D^2 r_L\tau_L + R_{load}r_L\tau_L^2 + \dfrac{D^4 r_L^2}{4}\right)} \tag{4.99}$$

If we neglect r_L, this expression simplifies to:

$$M = \frac{V_{out}}{V_{in}} = -\frac{D}{\sqrt{2\tau_L}} \tag{4.100}$$

Please note that M is a negative value in this definition considering the positive input voltage and the negative output.

To determine the quasi static gain H_0, we could resort to the small-signal model where V_{in} would now be reduced to 0 V and the stimulus is $D(s)$ but it is much faster to obtain H_0 by differentiating V_{out} with respect to D:

$$H_0 = \frac{d}{dD}V_{out}(D) = \frac{d}{dD}V_{in}\frac{D}{\sqrt{2\tau_L}} = \frac{V_{in}}{\sqrt{2\tau_L}} \tag{4.101}$$

To determine the denominator, we need the time constants of this circuit. We are going to obtain them by setting the excitation signal – $D(s)$ – to zero and express the resistances driving the inductor and the capacitor.

The corresponding circuit is given in Figure 4.42.

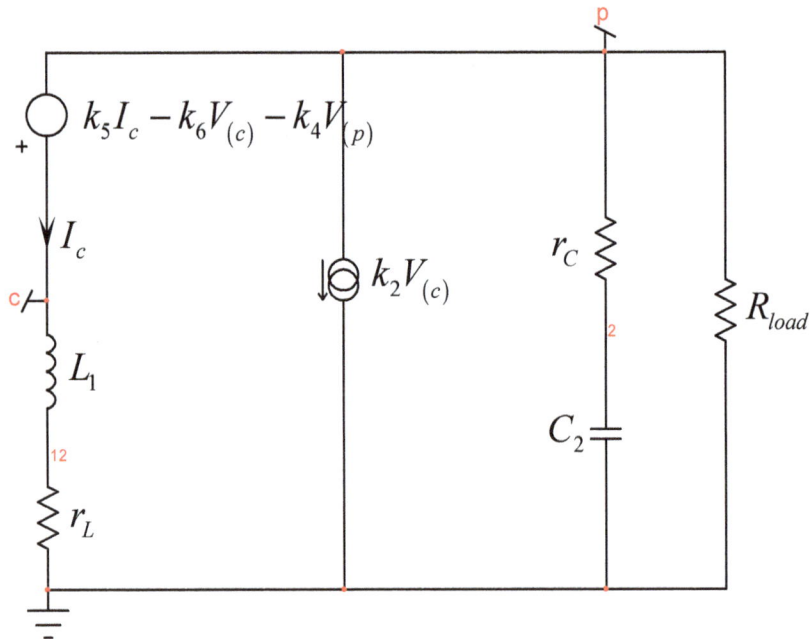

Figure 4.42 When the duty ratio is set to zero, the small-signal circuit is similar to that of the boost converter operated in DCM.

Now, the really nice thing is that this structure has already been studied in Chapter 3, Figure 4.42: it is similar to that of the DCM boost converter operated in voltage-mode control.

As a result, since time constants are similar, we can reuse the denominator $D(s)$ already determined in Chapter 3, section 3.2.

The poles are expressed with the same k coefficients given in Chapter 1 section 1.3.3 but they take on different values for the buck-boost converter (see Figure 4.43).

472

$L_1 := 47\mu H$ $F_{sw} := 100kHz$ $C_2 := 47\mu F$ $V_{in} := 12V$ $r_L := 0.01\Omega$

$\tau_L := \dfrac{L_1}{R_L} \cdot F_{sw}$ $d_1 := 0.445$ $R_L := 100\Omega$ $r_C := 0.03\Omega$

$\dfrac{2 \cdot F_{sw} \cdot L_1}{(1-d_1)^2} = 30.517\Omega$

If rL = 0 $M := -\dfrac{d_1}{\sqrt{2 \cdot \tau_L}} = -1.45143$ $V_{out} := M \cdot V_{in} = -17.41716V$

dc gain for small inductive loss (rL = 0):

$H_0 := \dfrac{V_{in}}{\sqrt{2 \cdot \tau_L}} = 39.13968V$ $20 \cdot \log\left(\dfrac{H_0}{V}\right) = 31.85235$

Small signal coefficients calculations:

$V_{ac} := V_{in} = 12V$ $V_{ap} := V_{in} - V_{out} = 29.41716V$ $I_c := \dfrac{V_{in} \cdot d_1^2}{2 \cdot F_{sw} \cdot L_1} - \dfrac{V_{out}}{R_L} = 0.42697A$ $k_{1a} := \dfrac{V_{in} \cdot d_1}{\tau_L \cdot R_L} = 1.13617A$ $k_{2a} := \dfrac{d_1^2}{2 \cdot (\tau_L \cdot R_L)} = 0.02107\dfrac{1}{\Omega}$

$k_1 := \dfrac{V_{ac} \cdot d_1}{F_{sw} \cdot L_1} = 1.13617A$ $k_2 := \dfrac{d_1^2}{2 \cdot F_{sw} \cdot L_1} = 0.02107\dfrac{1}{\Omega}$ $k_{3a} := \dfrac{2 \cdot R_L \cdot d_1 \cdot \tau_L \cdot (V_{in} - V_{out})}{F_{sw} \cdot L_1 \cdot (d_1^2 - 2 \cdot M \cdot \tau_L)} = 78.27937V$ $k_{4a} := \dfrac{R_L \cdot d_1^2 \cdot \tau_L}{F_{sw} \cdot L_1 \cdot (d_1^2 - 2 \cdot M \cdot \tau_L)} = 0.59207$

$k_3 := \dfrac{V_{ac} \cdot V_{ap} \cdot d_1}{F_{sw} \cdot I_c \cdot L_1} = 78.27937V$ $k_4 := \dfrac{V_{ac} \cdot d_1^2}{2 \cdot F_{sw} \cdot I_c \cdot L_1} = 0.59207$ $k_5 := -\dfrac{V_{ac} \cdot V_{ap} \cdot d_1^2}{2 \cdot F_{sw} \cdot I_c^2 \cdot L_1} = -40.79252\Omega$ $k_{6a} := \dfrac{R_L \cdot V_{in} \cdot d_1^2 \cdot \tau_L - R_L \cdot V_{out} \cdot d_1^2 \cdot \tau_L}{F_{sw} \cdot L_1 \cdot V_{in} \cdot d_1^2 - 2 \cdot F_{sw} \cdot L_1 \cdot M \cdot V_{in} \cdot \tau_L} = 1.45143$ $d_{2a} := -\dfrac{2 \cdot M \cdot \tau_L}{d_1} = 0.30659$

$k_6 := \dfrac{V_{ap} \cdot d_1^2}{2 \cdot F_{sw} \cdot I_c \cdot L_1} = 1.45143$ $d_2 := \dfrac{2 \cdot L_1 \cdot F_{sw} \cdot I_c}{d_1 \cdot V_{ac}} - d_1 = 0.30659$ $k_{5a} := -\dfrac{2 \cdot R_L^2 \cdot d_1^2 \cdot \tau_L^2 \cdot (V_{in} - V_{out})}{F_{sw} \cdot L_1 \cdot V_{in} \cdot (d_1^2 - 2 \cdot M \cdot \tau_L)^2} = -40.79252\Omega$ $I_a := I_c \cdot \dfrac{d_1}{d_1 + d_2} = 0.2528A$

Figure 4.43 The *k* coefficients are computed with the static values of the buck-boost converter. Here is a Mathcad® sheet with all computed results.

Rather than expressing the long formulas again, we will reproduce what is given in a simpler form by [2]:

$$D(s) \approx \left(1 + \dfrac{s}{\omega_{p_1}}\right)\left(1 + \dfrac{s}{\omega_{p_2}}\right) \qquad (4.102)$$

In this expression, a pole dominates the low-frequency response:

$$\omega_{p_1} \approx \dfrac{2}{R_{load} C_2} \qquad (4.103)$$

while the second one appears in the upper portion of the frequency spectrum:

$$\omega_{p_2} \approx 2F_{sw}\left[\dfrac{1/D}{1 + \dfrac{1}{|M|}}\right]^2 \qquad (4.104)$$

For the zeroes, we need to find the conditions nulling the output despite the presence of the stimulus $D(s)$. The structure of Figure 4.44 is similar to that of the DCM boost converter with a nulled output then the zeroes definitions still apply but lead to different results considering Figure 4.43 values.

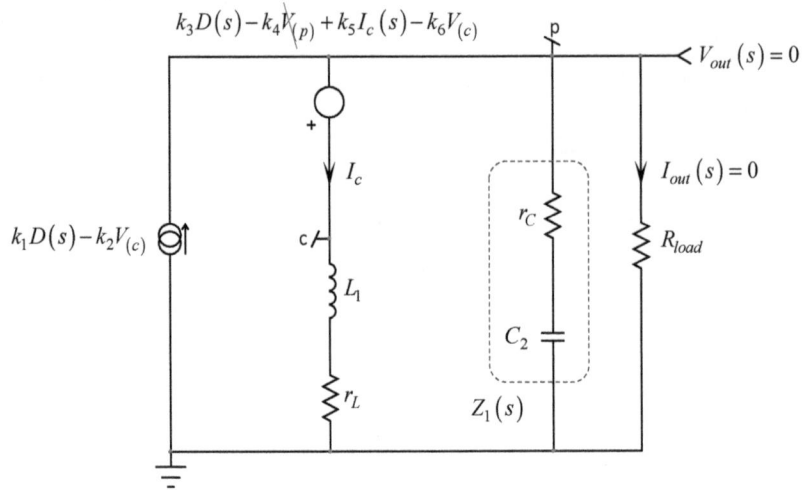

Figure 4.44 The zeroes are defined as with the DCM boost conveter but are using different coefficients definitions.

The usual suspects r_C and C_2 form a LHP zero:

$$\omega_{z_1} = \frac{1}{r_C C_2} \tag{4.105}$$

The second zero is obtain when all the current provided by the left-side source is absorbed by the L_1-r_L branch. Please look at section 3.2 in Chapter 3 to find that

$$\omega_{z_2} = \frac{\dfrac{k_3 + k_1 k_5}{k_1 (1 + k_6) - k_2 k_3}}{L_1} \tag{4.106}$$

If you now substitute the coefficients definitions described in Chapter 1 section 1.3.3, you should find

$$\omega_{z_2} = \frac{4 M R_{load} \tau_L^2 (M - 1)}{L_1 \left(D^2 - 2M\tau_L \right)^2} \approx \frac{1}{|M|(1 + |M|)} \frac{R_{load}}{L_1} \tag{4.107}$$

This is a positive root making this second zero lie in the right half-plane. We now have all the elements to assemble the control-to-output transfer function:

$$H(s) = \frac{V_{out}(s)}{V_{err}(s)} = \frac{V_{out}(s)}{D(s)} \frac{1}{V_p} = \frac{H_0}{V_p} \frac{\left(1 + \dfrac{s}{\omega_{z_1}}\right)\left(1 - \dfrac{s}{\omega_{z_2}}\right)}{\left(1 + \dfrac{s}{\omega_{p_1}}\right)\left(1 + \dfrac{s}{\omega_{p_2}}\right)} \tag{4.108}$$

In which H_0 is defined by (4.101). V_p is the peak voltage of the PWM sawtooth and the poles/zeroes are defined by equations (4.103), (4.104), (4.105) and (4.107).

We can compare the dynamic responses of the large-signal SPICE model from Figure 4.37 and that of (4.108). The result appears in Figure 4.45 and confirms our derivation is correct.

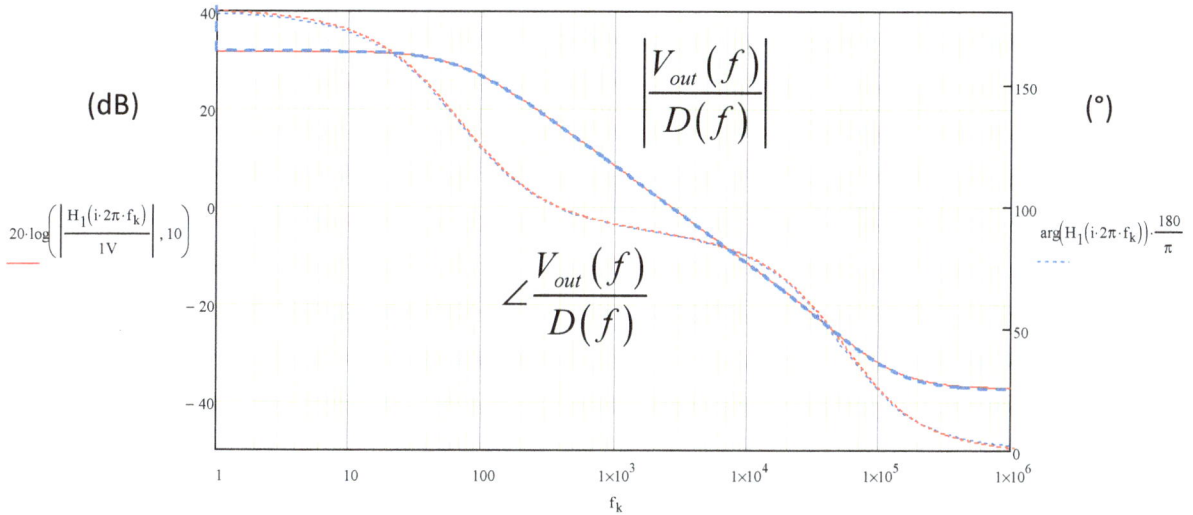

Figure 4.45 Small-signal responses of the SPICE simulation and the Mathcad® sheet.

We will now use the SIMPLIS® template from Figure 4.46 and increase the load resistance to 100 Ω. Figure 4.47 confirms the converter is now operating in DCM and delivers -17.3 V.

The Bode plot is very close to that of Figure 4.45.

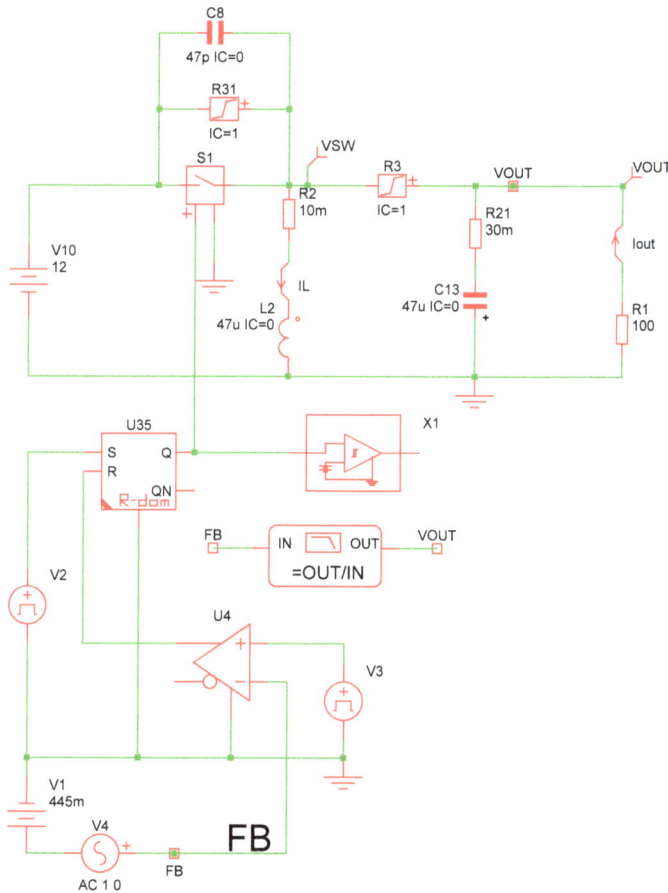

Figure 4.46 The inductance has been decreased to 47 µH and the load is now 100 Ω.

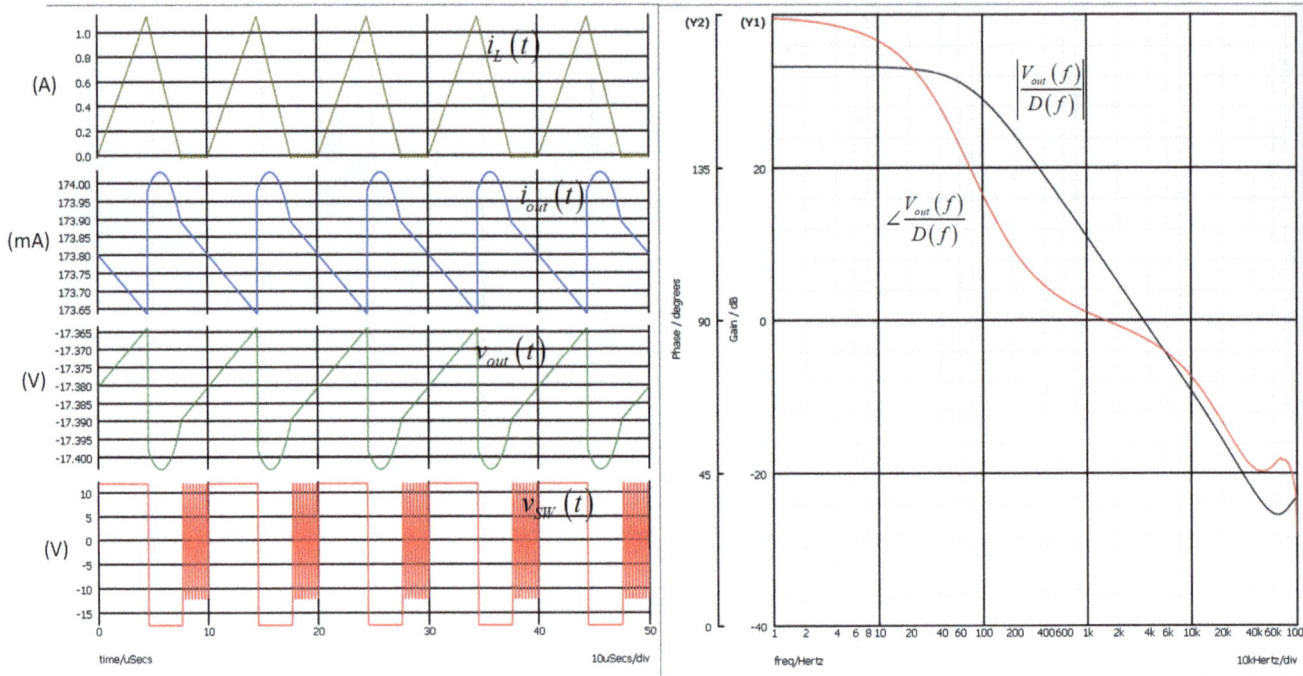

Figure 4.47 SIMPLIS® immediately delivers the new response when the converter has entered DCM.

4.2.1 Input to Output

The input-to-output transfer function is obtained by ac-stimulating the input source while the duty ratio input is kept at its dc bias. To determine the time constants, we reduce the excitation to zero volt as shown in the small-signal model of Figure 4.48.

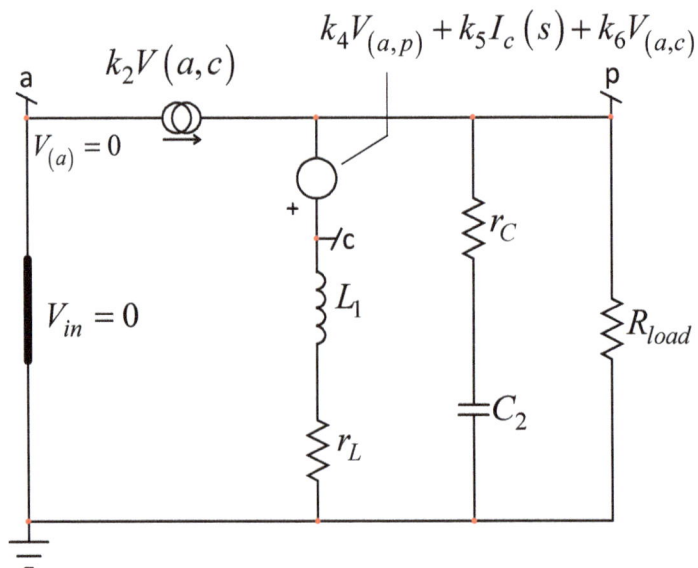

Figure 4.48 The converter's input voltage is first zeroed to determine the time constants.

If we compare this drawing with that of Figure 4.42, they are identical and we can reuse the denominator $D(s)$ already determined with (4.102).

The dc gain H_0 is obtained immediately by reusing (4.100):

$$V_{out}(V_{in}) = M \cdot V_{in} \qquad (4.109)$$

and:

$$H_0 = \frac{dV_{out}(V_{in})}{dV_{in}} = -\frac{D_0}{\sqrt{2\tau_L}} \qquad (4.110)$$

To determine the zeroes, we need to look at the condition for which the output is nulled despite the presence of an excitation. This is what Figure 4.49 shows.

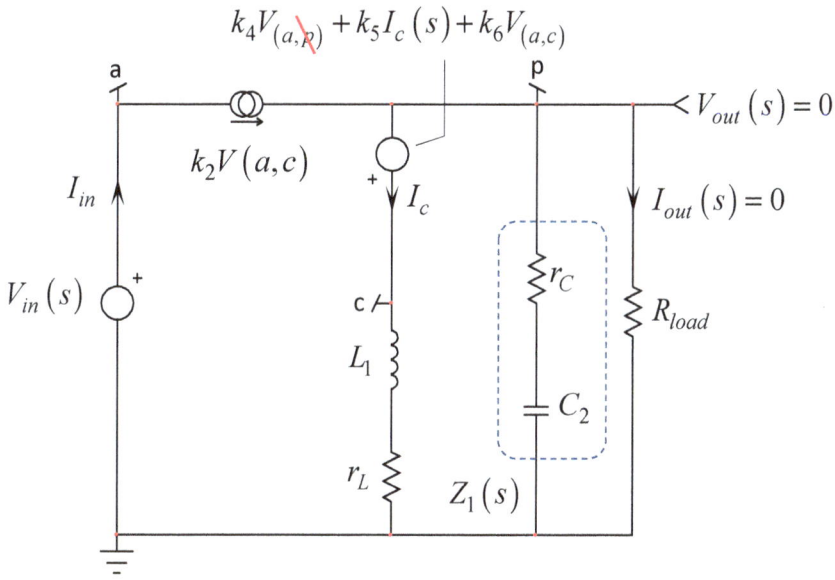

Figure 4.49 The zeroes are found by studying the conditions for which the output is nulled.

The first zero is organized around the output capacitor C_2 and its ESR r_C when their series connection brings a 0-Ω impedance:

$$Z_1(s) = r_C + \frac{1}{sC_2} = 0 \qquad (4.111)$$

Which implies a zero located at:

$$\omega_{z_1} = \frac{1}{r_C C_2} \qquad (4.112)$$

The second zero requires some lines of algebra to determine it. We start with the input current I_{in} which is determined accounting for a zeroed output current I_{out} (the output is nulled):

$$I_{in}(s) = I_c(s) = k_2 V_{in}(s) - k_2 V_{(c)}(s) \qquad (4.113)$$

The current flowing in terminal c is obtained by:

$$I_c(s) = \frac{V_{(c)}(s)}{r_L + sL_1} \qquad (4.114)$$

Equating the above expressions helps us to obtain a first definition for $V_{(c)}$:

$$V_{(c)}(s) = \frac{V_{in}(s)k_2}{k_2 + \dfrac{1}{r_L + sL_1}}$$

(4.115)

We can obtain a second definition via the voltage at node c:

$$V_{(c)}(s) = k_4 V_{in}(s) + k_5 I_c(s) + k_6 V_{in}(s) - k_6 V_{(c)}(s)$$

(4.116)

Substituting (4.114) in (4.116) and solving for the voltage at node c gives us:

$$V_{(c)}(s) = \frac{V_{in}(s)[k_4 + k_6]}{k_6 - \dfrac{k_5}{r_L + sL_1} + 1}$$

(4.117)

We can now equate (4.117) with (4.115) and solve for the s value satisfying this condition. This is the second zero we want:

$$\omega_{z_2} = \frac{k_4 + k_6 + k_2 k_5 - k_2 r_L (1 - k_4)}{L_1 k_2 (1 - k_4)}$$

(4.118)

When r_L reduces to zero and after rearranging the coefficients from Figure 4.43, we have:

$$\omega_{z_2} \approx \frac{R_{load}\left(2\tau_L M^2 - 4\tau_L M + D_0^2\right)}{L_1 M\left(2M\tau_L - D_0^2\right)}$$

(4.119)

and it is a positive root (a right-half-plane zero). We can now gather all these elements and form the transfer function linking the output to the input:

$$H(s) = \frac{V_{out}(s)}{V_{in}(s)} = H_0 \frac{\left(1 + \dfrac{s}{\omega_{z_1}}\right)\left(1 - \dfrac{s}{\omega_{z_2}}\right)}{\left(1 + \dfrac{s}{\omega_{p_1}}\right)\left(1 + \dfrac{s}{\omega_{p_2}}\right)}$$

(4.120)

$$H_0 = -\frac{D}{\sqrt{2\tau_L}}$$

(4.121)

$$\omega_{p_1} \approx \frac{2}{R_{load}C_2}$$

(4.122)

$$\omega_{p_2} \approx 2F_{sw}\left[\frac{1/D}{1 + \dfrac{1}{M}}\right]^2$$

(4.123)

$$\omega_{z_1} = \frac{1}{r_C C_2} \tag{4.124}$$

$$\omega_{z_2} \approx \frac{R_{load}\left(2\tau_L M^2 - 4\tau_L M + D^2\right)}{L_1 M \left(2M\tau_L - D^2\right)} \tag{4.125}$$

If we plot this transfer function with the Mathcad® sheet and compare the response to that of the large-signal SPICE model from Figure 4.50, there are no difference between the magnitude/phase curves—as confirmed in Figure 4.51.

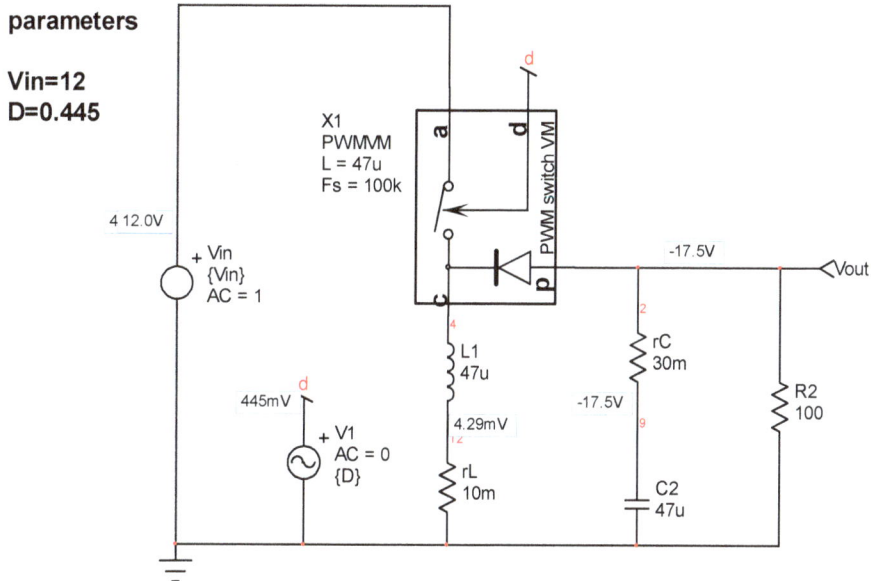

Figure 4.50 The input voltage now ac-sweeps this large-signal model.

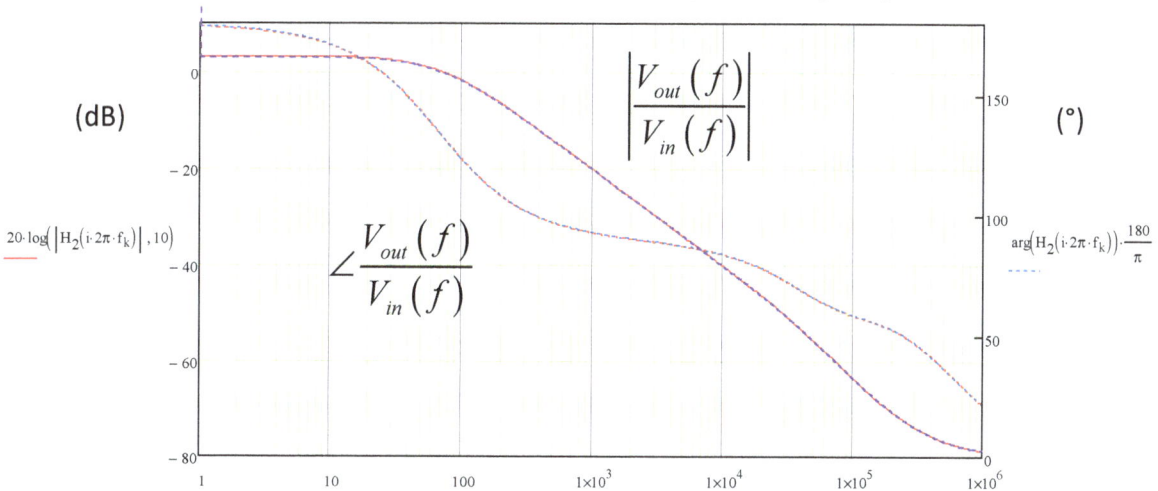

Figure 4.51 Magnitude and phase plots from Mathcad® and SPICE are identical.

4.2.2 Output Impedance

When adding a 1-A ac current source across the output load as in Figure 4.52, it becomes possible to plot the open-loop output impedance of the buck-boost converter operated in voltage-mode control and in discontinuous conduction mode.

We start by determining the resistance R_0 obtained for $s = 0$: short L_1 and open-circuit C_2 as shown in Figure 4.53 in which the test generator I_T is installed. As we did in previous examples, we can temporarily remove R_{load} and bring it back later in parallel with the intermediate result we will now determine. The current flowing out of terminal c is determined as follows:

$$I_c = I_T - k_2 V_{(c)} \tag{4.126}$$

From which we extract the value of $V_{(c)}$:

$$V_{(c)} = \frac{I_T - I_c}{k_2} \tag{4.127}$$

Knowing that:

$$V_{(c)} = I_c r_L \tag{4.128}$$

By equating the two above equations, we can solve for I_c:

$$I_c = \frac{I_T}{k_2 r_L + 1} \tag{4.129}$$

The voltage at terminal c is obtained by determining the voltage drop across r_L:

$$I_c r_L = V_T + k_5 I_c - k_6 V_{(c)} - k_4 V_T \tag{4.130}$$

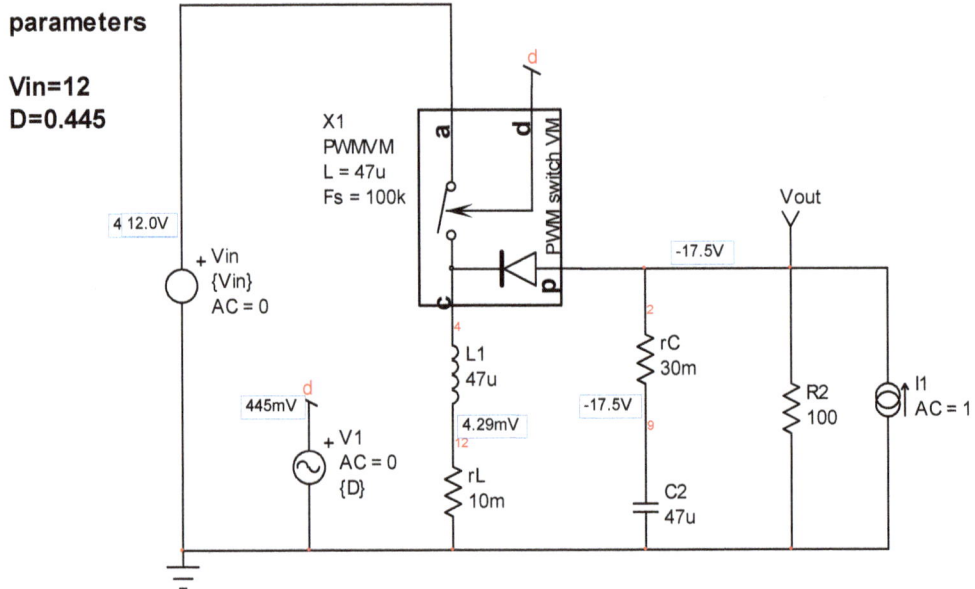

Figure 4.52 Placing a 1-A current source across the load lets us plot the open-loop ouput impedance.

If you now substitute (4.128) then (4.129) in (4.130) and solve for V_T, you have:

$$\frac{V_T}{I_T} = \frac{k_5 - r_L (1 + k_6)}{k_4 - 1 + k_2 r_L (k_4 - 1)} \tag{4.131}$$

Now bring R_{load} back in parallel and update the expression with coefficients from Figure 4.43. We have:

$$R_0 = R_{load} \,\|\, \left(\frac{V_T}{I_T}\right) = R_{load} \,\|\, \frac{R_{load}D_0^2 (M-1)}{M\left(D_0^2 - 2M\tau_L\right)} \tag{4.132}$$

Having R_0 on hand, reduce the excitation to zero ampere and look at the time constants. As shown in Figure 4.54, the circuit is similar to that of Figure 4.41 meaning we can reuse the denominator $D(s)$ already determined in (4.102). To determine the zeroes, we need to determine the conditions for which the response — V_T — is nulled despite the excitation brought back in place.

The circuit to analyze is that of Figure 4.55 where we have the usual suspects consisting of capacitor C_2 and its ESR r_C:

$$Z_1(s) = r_C + \frac{1}{r_C C_2} = 0 \tag{4.133}$$

Leading to the following zero position:

$$\omega_{z_1} = \frac{1}{r_C C_2} \tag{4.134}$$

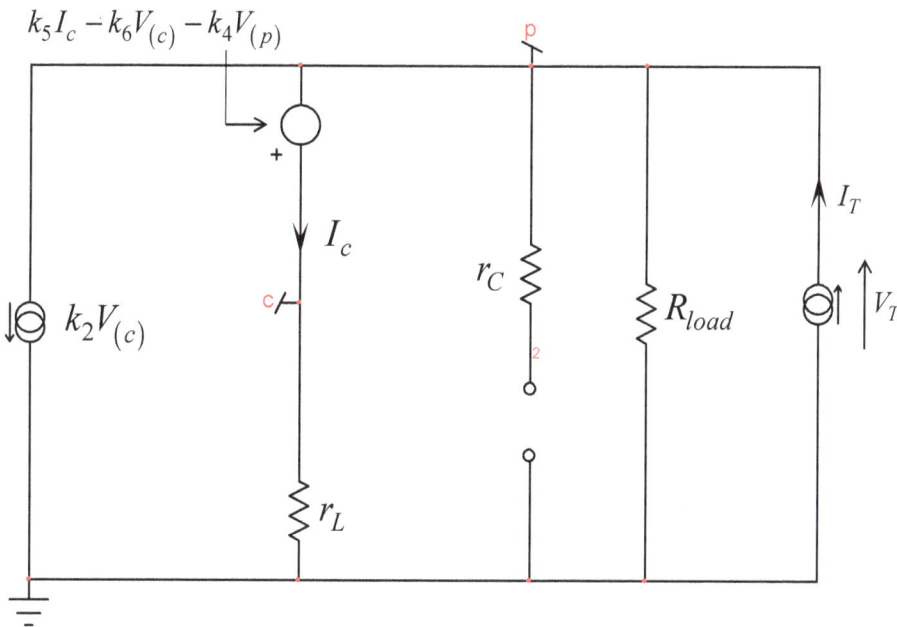

Figure 4.53 The dc output resistance is determined with L_1 shorted and C_2 open-circuited.

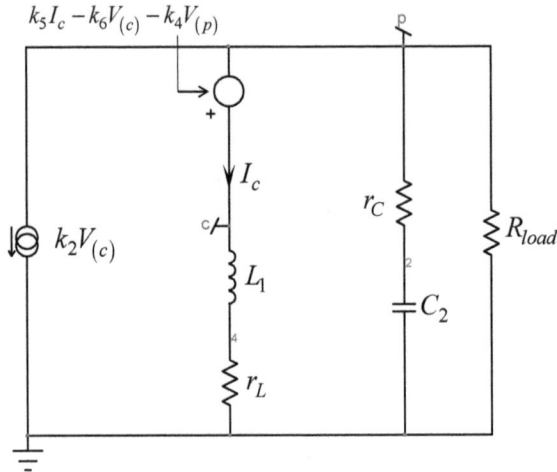

Figure 4.54 With a zeroed excitation, the circuit reduces to its natural state: we can reuse the denominator already determined.

The second zero implies that all the excitation current I_T is absorbed by I_c and I_1. The updated circuit to determine this zero is given in Figure 4.56. The response being nulled, the voltage at terminal p is zero and the voltage source simplifies. Actually, if you carefully look at the schematic diagram, it is that of the boost converter from Chapter 3 operated in VM and discontinuous mode shown in Figure 3.59. It means all equations are valid except that all coefficients values are those of Figure 4.43. We will not go through the analysis again as it has already been done:

$$\omega_{z_2} = \frac{r_L - \dfrac{k_5}{1+k_6}}{L_1} \tag{4.135}$$

If you develop this expression with the k values when considering $r_L = 0$, you have:

$$\omega_{z_2} \approx \frac{2R_{load}D_0{}^2\tau_L\left(V_{out}-V_{in}\right)}{L_1\left(D_0{}^2-2M\tau_L\right)\left(V_{out}D_0{}^2-2V_{in}D_0{}^2+2MV_{in}\tau_L\right)} \tag{4.136}$$

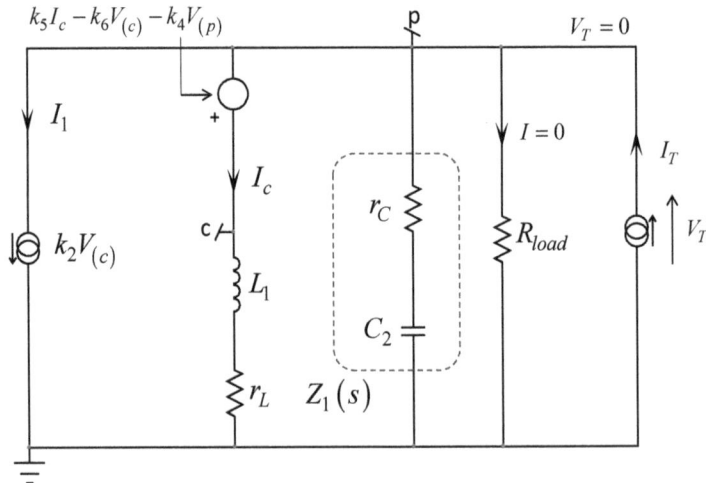

Figure 4.55 What conditions in this circuit contribute to null the response?

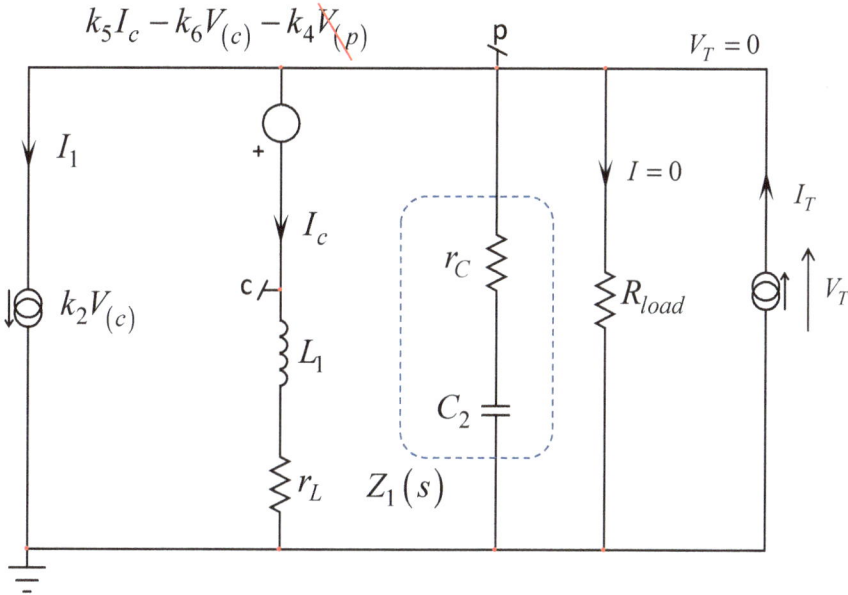

Figure 4.56 If all the excitation current splits between I_1 and I_c, there is no current flowing in R_{load} naturally bringing a null in the response.

The output impedance of the buck-boost converter operating in DCM is thus given by:

$$Z_{out}(s) = R_0 \frac{\left(1 + \dfrac{s}{\omega_{z_1}}\right)\left(1 + \dfrac{s}{\omega_{z_2}}\right)}{\left(1 + \dfrac{s}{\omega_{p_1}}\right)\left(1 + \dfrac{s}{\omega_{p_2}}\right)} \tag{4.137}$$

In which we have:

$$R_0 \approx \frac{R_{load} V_{in} D^2}{V_{in} D^2 + 2 V_{out} R_{load}} \parallel R_{load} \tag{4.138}$$

$$\omega_{z_1} = \frac{1}{r_C C_2} \tag{4.139}$$

$$\omega_{z_2} \approx \frac{2 R_{load} D^2 \tau_L (V_{out} - V_{in})}{L_1 (D^2 - 2M\tau_L)(V_{out} D^2 - 2 V_{in} D^2 + 2 M V_{in} \tau_L)} \tag{4.140}$$

$$\omega_{p_1} \approx \frac{2}{R_{load} C_2} \tag{4.141}$$

$$\omega_{p_2} \approx 2 F_{sw} \left[\frac{1/D}{1 + \dfrac{1}{M}}\right]^2 \tag{4.142}$$

483

If we consider that the second zero and pole occur in the upper-frequency range, this expression can be approximated as:

$$Z_{out}(s) \approx R_0 \frac{1 + \dfrac{s}{\omega_{z_1}}}{1 + \dfrac{s}{\omega_{p_1}}}$$

(4.143)

We gathered magnitude/phase curves from Mathcad® and SPICE in Figure 4.57 and results are identical.

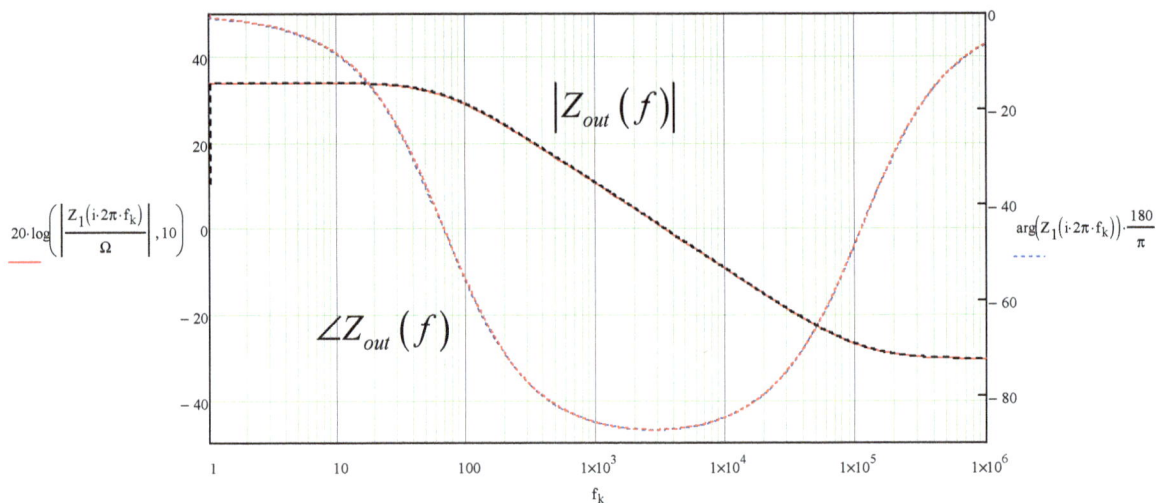

Figure 4.57 SPICE and Mathcad® results match perfectly.

4.2.3 Input Impedance

To determine the input impedance of the converter operated in DCM, we will classically sweep the input voltage source via a 1-A ac current source. The high-value inductor *LoL* blocks the ac stimulus and isolates the measurement from the zero-output impedance voltage source. The voltage collected across the current source is thus the image of the input impedance we want. The circuit diagram appears in Figure 4.58.

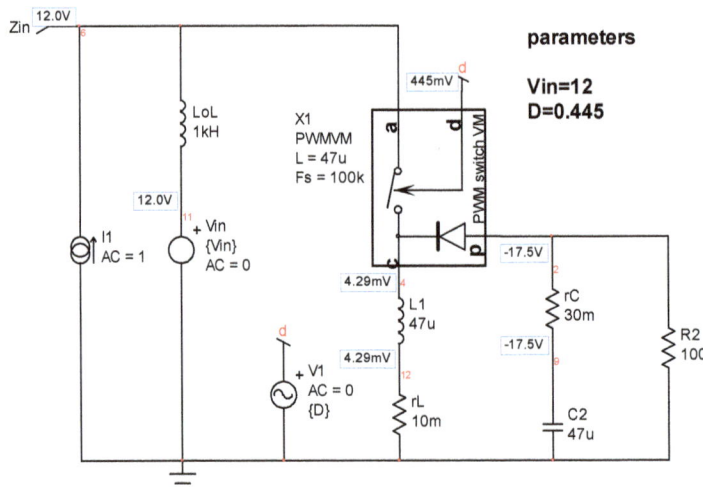

Figure 4.58 The input impedance is obtained by ac-sweeping the input port of the large-signal PWM switch model.

The small-signal model to study is that of Figure 4.59. The analysis starts with the static input resistance R_0. It is determined by solving the circuit equations when L_1 is shorted and C_2 open-circuited. The updated circuit appears in Figure 4.60.

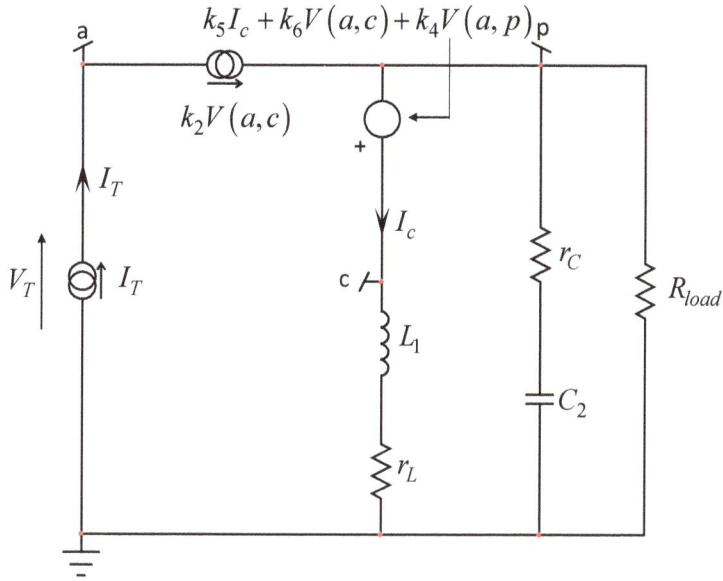

Figure 4.59 This is the small-signal circuit diagram used to determine the input impedance Z_{in}.

We can express the stimulus current I_T as follows if we consider the voltage at terminal a to be V_T:

$$I_T = k_2 V_T - k_2 V_{(c)} \qquad (4.144)$$

From which we extract $V_{(c)}$:

$$V_{(c)} = -\frac{I_T - V_T k_2}{k_2} \qquad (4.145)$$

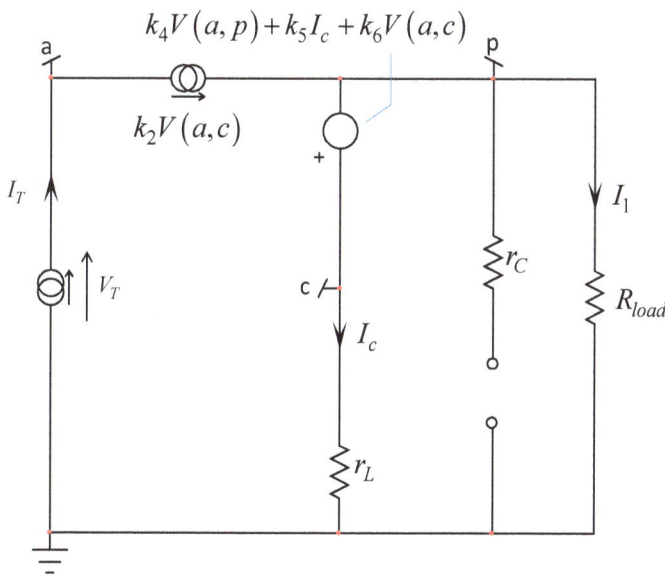

Figure 4.60 The inductor is shorted and the capacitor open-circuited to find input resistance R_0.

The voltage at node c is obtained by summing the voltage at node p with the upper-side voltage source:

$$V_{(c)} = V_p + k_4 V(V_T, p) + k_5 I_c + k_6 V(V_T, c) \qquad (4.146)$$

We know that the voltage at node p is the output voltage:

$$V_{(p)} = (I_T - I_c) R_{load} \qquad (4.147)$$

While the current I_c flowing out of terminal c is nothing else than

$$I_c = \frac{V_{(c)}}{r_L} \qquad (4.148)$$

If you now substitute (4.147) and (4.148) in (4.146) then solve for V_T, rearrange and factor, you should obtain:

$$\frac{V_T}{I_T} = \frac{r_L(1 + k_6) - R_{load}(k_4 - k_2 r_L + k_2 k_4 r_L - 1) - k_5}{R_{load}(k_2 - k_2 k_4) + k_2 r_L\left(1 - k_4 - \dfrac{k_5}{r_L}\right)} \qquad (4.149)$$

If you replace the k coefficients by their definitions from Figure 4.43 and neglect r_L, you have:

$$R_0 \approx \frac{2 R_{load} \tau_L}{D_0{}^2} \qquad (4.150)$$

We have learned from the previous examples that the input impedance analysis circuit is different compared to the one commonly used with the three other transfer functions. This is because if you turn the excitation off – I_T is zeroed – the circuit does not return to its natural state.

In this example, when the current source is off, the left-side of the upper current source is dangling and does not have a path to ground. It clearly differs from the circuit of Figure 4.41. Therefore, we cannot reuse the denominator we have previously determined. We need to find all the time constants pertinent to this circuit.

The first time constant involves L_1 and the circuit we have is shown in Figure 4.61. We see the current source I_T in series with r_L. As usual, we can temporarily disconnect r_L and bring it back later in series with the intermediate result. Therefore, we have:

$$V_T = V_{(c)} \qquad (4.151)$$

In this analysis, we will provide a temporary dc path from node a to ground with a large-valued resistance R_{inf}. Having that resistance connected, we can write:

$$V_{(a)} = R_{inf}\left(k_2 V_T - k_2 V_{(a)}\right) \qquad (4.152)$$

Solving for $V_{(a)}$ leads to:

$$V_{(a)} = \frac{R_{inf} V_T k_2}{1 + R_{inf} k_2} \qquad (4.153)$$

If R_{inf} is really large, then we have:

$$V_{(a)} \approx V_T \qquad (4.154)$$

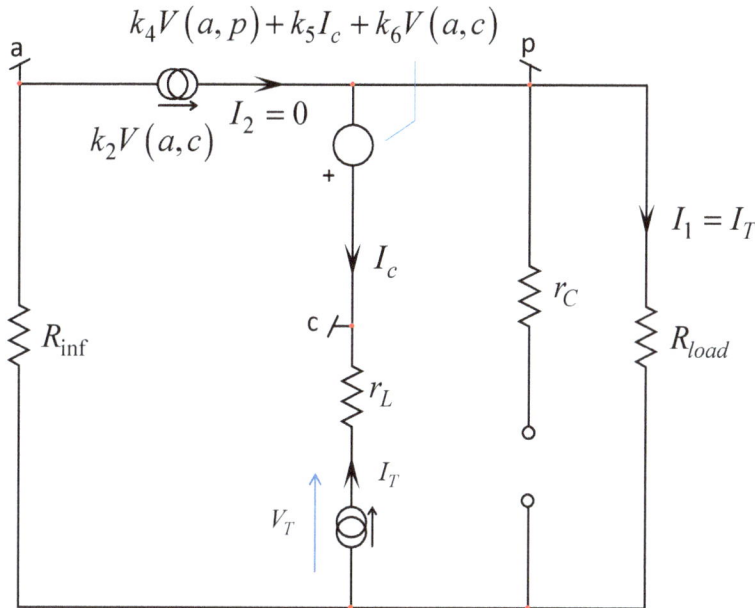

Figure 4.61 We determine the resistance driving the inductor by temporarily disconnecting r_L.

If we conveniently consider the absence of current flowing through R_{inf}, then:

$$I_c = -I_T \qquad (4.155)$$

The voltage V_T is thus the sum of the voltage across the load resistance plus the controlled voltage-source:

$$V_T = I_T R_{load} + k_4 V_T - k_4 I_T R_{load} - k_5 I_T \qquad (4.156)$$

Solve for V_T, rearrange and divide by I_T then add the series resistance r_L to obtain:

$$\tau_1 = \frac{L_1}{\dfrac{V_T}{I_T} + r_L} = \frac{L_1}{r_L + \dfrac{k_5 - R_{load}(1 - k_4)}{k_4 - 1}} \qquad (4.157)$$

Now replace the k coefficients with their values of Figure 4.43, and neglecting r_L, you should find:

$$\tau_1 \approx \frac{L_1}{\dfrac{R_{load}\left(2\tau_L M^2 - 2MD_0^2 + D_0^2\right)}{M\left(2M\tau_L - D_0^2\right)}} \qquad (4.158)$$

We now set the inductor in its dc state (a short circuit) and we install a test generator I_T across capacitor C_2. The new circuit diagram appears in Figure 4.62.

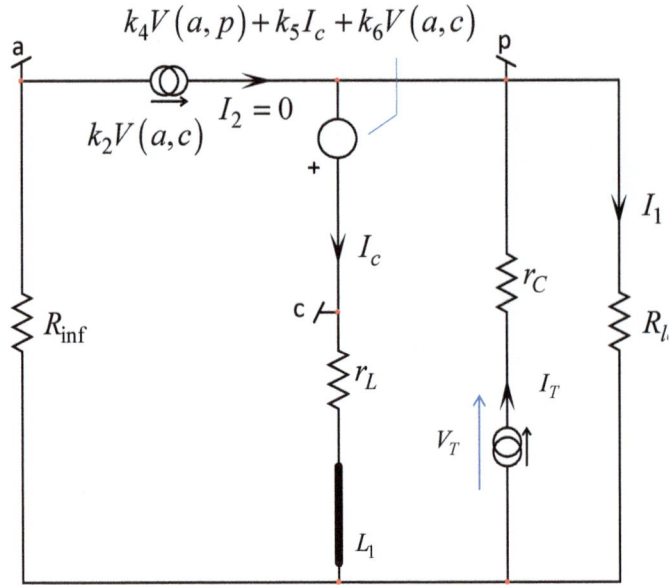

Figure 4.62 The inductor is set in its dc state to determine the time constant involving C_2.

To simplify the analysis, r_C is temporarily removed and will be brought back at the end in series with the intermediate resistive result. Our first expression is thus:

$$V_{(p)} = V_T \tag{4.159}$$

The current flowing through the load is:

$$I_1 = \frac{V_T}{R_{load}} \tag{4.160}$$

We can also write using KCL:

$$I_1 = I_T - I_c = I_T - \frac{V_{(c)}}{r_L} \tag{4.161}$$

Equating (4.160) and (4.161) leads to determining the voltage at node c:

$$V_{(c)} = -\frac{r_L \left(V_T - I_T R_{load} \right)}{R_{load}} \tag{4.162}$$

If we consider R_{inf} approaching infinity, then current I_2 is zero ampere, implying that the source involving coefficient k_2 is also zero:

$$k_2 V(a,c) = 0 \tag{4.163}$$

Which implies:

$$V_{(a)} = V_{(c)} \tag{4.164}$$

We can find an alternate way to express the voltage at node c via the upper voltage source:

$$V_{(c)} = V_{(p)} + k_4 V_{(a)} - k_4 V_{(p)} + k_5 I_c + k_6 V_{(a)} - k_6 V_{(c)} \tag{4.165}$$

Because node a and c share a similar potential, then the above expression simplifies to:

$$V_{(c)} = V_T + k_4 V_{(c)} - k_4 V_T + k_5 \frac{V_{(c)}}{r_L} \tag{4.166}$$

We can now substitute the definition from (4.162) in the above and factor the V_T / I_T under the below form:

$$\frac{V_T}{I_T} = \frac{R_{load} \left(k_5 - r_L + k_4 r_L \right)}{k_5 - r_L + R_{load} \left(k_4 - 1 \right) + k_4 r_L} \tag{4.167}$$

Bringing r_C back in place with C_2, we have the second time constant we wanted:

$$\tau_2 = \left(\frac{R_{load} \left(k_5 - r_L + k_4 r_L \right)}{k_5 - r_L + R_{load} \left(k_4 - 1 \right) + k_4 r_L} + r_C \right) C_2 \tag{4.168}$$

We can replace the k coefficients with their values from Figure 4.43 and neglect r_l to obtain a simpler expression:

$$\tau_2 \approx \left(\frac{R_{load} D_0^{\,2} \left(1 - M \right)}{2M \left(\tau_L M - D_0^{\,2} \right) + D_0^{\,2}} + r_C \right) C_2 \tag{4.169}$$

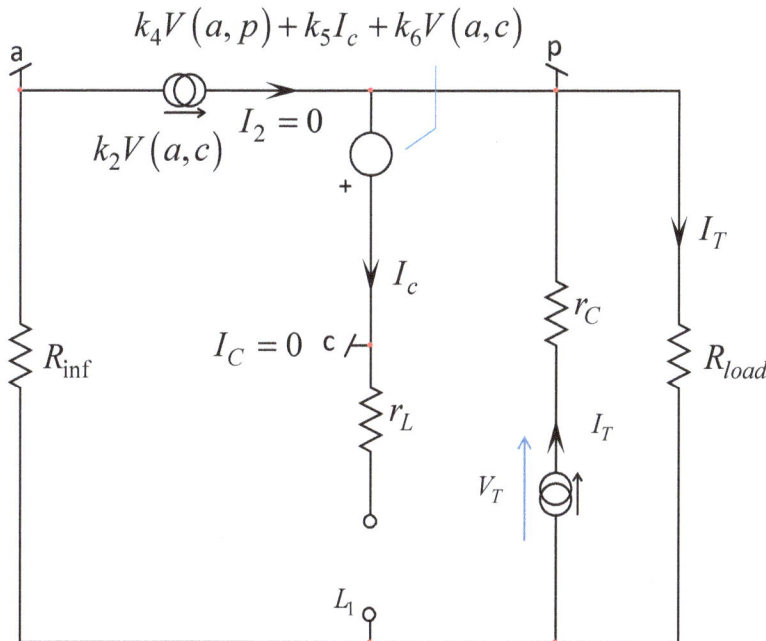

Figure 4.63 The last configuration implies inductor L_1 set in its high-frequency state which naturally zeroes current I_c and simplifies the analysis.

Christophe Basso

The first-order coefficient b_1 is the sum of the two time constants:

$$b_1 = \tau_1 + \tau_2 = \frac{L_1}{\dfrac{R_{load}\left(2\tau_L M^2 - 2MD_0^2 + D_0^2\right)}{M\left(2M\tau_L - D_0^2\right)}} + \left(\frac{R_{load}D_0^2\left(1-M\right)}{2M\left(\tau_L M - D_0^2\right) + D_0^2} + r_C\right)C_2 \tag{4.170}$$

The final time constant we need is obtained by setting L_1 in its high-frequency state as it naturally zeroes the current I_c. The circuit is really simple as show in Figure 4.63 and the time constant is immediate:

$$\tau_2^1 = \left(r_C + R_{load}\right)C_2 \tag{4.171}$$

Leading to:

$$b_2 = \tau_1\tau_2^1 = \frac{L_1}{\dfrac{R_{load}\left(2\tau_L M^2 - 2MD_0^2 + D_0^2\right)}{M\left(2M\tau_L - D_0^2\right)}}\left(r_C + R_{load}\right)C_2 \tag{4.172}$$

We are done with the denominator which is expressed by combining b_1 and b_2 terms:

$$D(s) = 1 + b_1 s + b_2 s^2 = 1 + \frac{s}{Q\omega_0} + \left(\frac{s}{\omega_0}\right)^2 \tag{4.173}$$

In which we have:

$$\omega_0 \approx \frac{1}{\sqrt{L_1 C_2}\sqrt{\dfrac{D^2\left(M-1\right)}{2\tau_L M^2 + D^2\left(1-2M\right)} + 1}} \tag{4.174}$$

$$Q \approx \frac{R_{load}\sqrt{L_1 C_2}\left(2MD^2 - D^2 - 2\tau_L M^2\right)\sqrt{\dfrac{M\left(2M\tau_L - D^2\right)}{2\tau_L M^2 - 2MD^2 + D^2}}}{C_2 R_{load}^2 D^2\left(M-1\right) + L_1 M\left(D^2 - 2\tau_L M\right)} \tag{4.175}$$

Assuming a low-Q value, we can define two cascaded poles using the low-Q approximation:

$$\omega_{p_1} = \omega_0 Q \approx \frac{R_{load}\left(2MD^2 - D^2 - 2\tau_L M^2\right)}{C_2 R_{load}^2 D^2\left(M-1\right) + L_1 M\left(D^2 - 2\tau_L M\right)} \tag{4.176}$$

490

The second pole is occurring at a high frequency value:

$$\omega_{p_2} = \frac{\omega_0}{Q} \approx \frac{C_2 R_{load}{}^2 D^2 (M-1) + L_1 M \left(D^2 - 2\tau_L M \right)}{C_2 L_1 M R_{load} \left(D^2 - 2M\tau_L \right)}$$

(4.177)

Now that we have our denominator, we can null the stimulus source in Figure 4.59. A null voltage across a current is a well-known degenerate case meaning we can replace the current source by a short circuit. If we do it, then our circuit returns to the natural state of Figure 4.42 (terminal a is grounded) and the denominator determined in (4.102) is our numerator for this input impedance transfer function:

$$N(s) \approx \left(1 + \frac{s}{\omega_{z_1}} \right) \left(1 + \frac{s}{\omega_{z_2}} \right)$$

(4.178)

In this expression, the first zero dominates the low-frequency response:

$$\omega_{z_1} \approx \frac{2}{R_{load} C_2}$$

(4.179)

while the second one appears in the upper portion of the frequency spectrum:

$$\omega_{z_2} \approx 2 F_{sw} \left[\frac{1/D}{1 + \frac{1}{M}} \right]^2$$

(4.180)

We now have all needed terms to assemble the final transfer function:

$$Z_{in}(s) = R_0 \frac{\left(1 + \frac{s}{\omega_{z_1}} \right) \left(1 + \frac{s}{\omega_{z_2}} \right)}{\left(1 + \frac{s}{\omega_{p_1}} \right) \left(1 + \frac{s}{\omega_{p_2}} \right)}$$

(4.181)

We have R_0 defined in (4.150), the poles determined in (4.176) (4.177) and finally, the two zeroes are given in the above expressions (4.179) and (4.180). We can now test the response of this transfer function versus the simulation of Figure 4.58.

The results appear in Figure 4.64 and confirm the perfect matching between the Mathcad® sheet and our SPICE simulation.

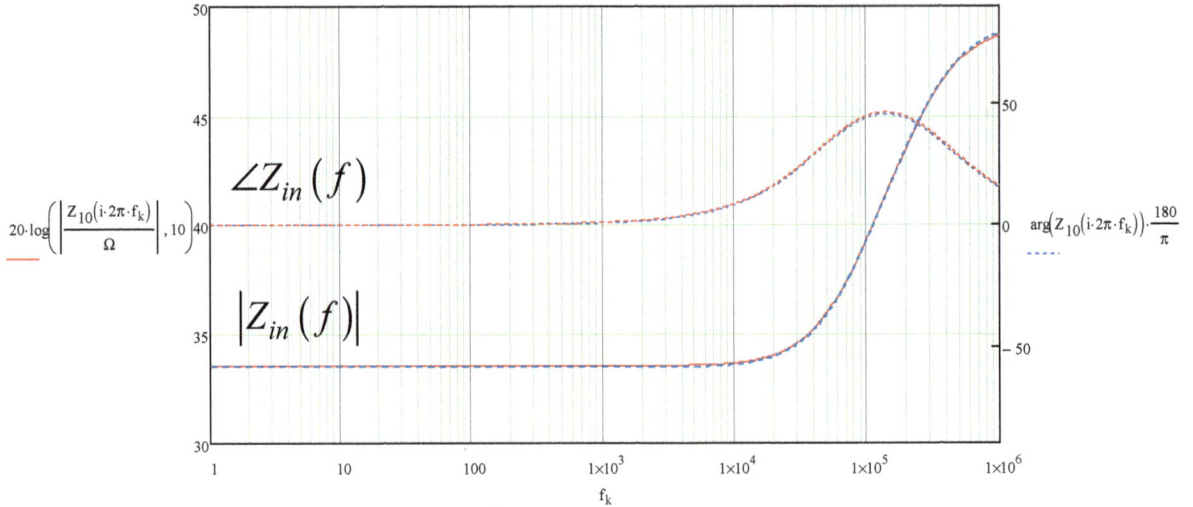

$20 \cdot \log \left(\left| \dfrac{Z_{10}(i \cdot 2\pi \cdot f_k)}{\Omega} \right|, 10 \right)$ $\arg \left(Z_{10}(i \cdot 2\pi \cdot f_k) \right) \cdot \dfrac{180}{\pi}$

Figure 4.64 The matching between the Mathcad® sheet and the simulation is perfect.

The four transfer functions of the voltage-mode buck-boost converter operated in DCM are shown in Figure 4.65.

$\dfrac{V_{out}(s)}{V_{err}(s)}$ Control to Output	$H_0 \dfrac{\left(1+\dfrac{s}{\omega_{z_1}}\right)\left(1-\dfrac{s}{\omega_{z_2}}\right)}{\left(1+\dfrac{s}{\omega_{p_1}}\right)\left(1+\dfrac{s}{\omega_{p_2}}\right)}$	$\omega_{z_1} = \dfrac{1}{r_C C_2}$ $\omega_{z_2} = \dfrac{1}{|M|(1+|M|)} \dfrac{R_{load}}{L_1}$	$\omega_{p_1} \approx \dfrac{2}{R_{load} C_2}$ $\omega_{p_2} \approx 2F_{sw} \left(\dfrac{1/D}{1+\dfrac{1}{|M|}}\right)^2$	$H_0 \approx -\dfrac{V_{in}}{V_p \sqrt{2\tau_L}}$
$\dfrac{V_{out}(s)}{V_{in}(s)}$ Input to Output	$H_0 \dfrac{\left(1+\dfrac{s}{\omega_{z_1}}\right)\left(1-\dfrac{s}{\omega_{z_2}}\right)}{\left(1+\dfrac{s}{\omega_{p_1}}\right)\left(1+\dfrac{s}{\omega_{p_2}}\right)}$	$\omega_{z_1} = \dfrac{1}{r_C C_2}$ $\omega_{z_2} \approx \dfrac{R_{load}\left(2\tau_L M^2 - 4\tau_L M + D^2\right)}{L_1 M \left(2M\tau_L - D^2\right)}$	$\omega_{p_1} \approx \dfrac{2}{R_{load} C_2}$ $\omega_{p_2} \approx 2F_{sw} \left(\dfrac{1/D}{1+\dfrac{1}{|M|}}\right)^2$	$H_0 \approx -\dfrac{D}{\sqrt{2\tau_L}}$
$Z_{in}(s)$ Input impedance	$R_0 \dfrac{\left(1+\dfrac{s}{\omega_{z_1}}\right)\left(1+\dfrac{s}{\omega_{z_2}}\right)}{\left(1+\dfrac{s}{\omega_{p_1}}\right)\left(1+\dfrac{s}{\omega_{p_2}}\right)}$	$\omega_{z_1} \approx \dfrac{2}{R_{load} C_2}$ $\omega_{z_2} \approx 2F_{sw} \left(\dfrac{1/D}{1+\dfrac{1}{|M|}}\right)^2$	$\omega_{p_1} \approx \dfrac{R_{load}\left(2MD^2 - D^2 - 2\tau_L M^2\right)}{C_2 R_{load}^2 D^2 (M-1) + L_1 M\left(D^2 - 2\tau_L M\right)}$ $\omega_{p_2} \approx \dfrac{C_2 R_{load}^2 D^2 (M-1) + L_1 M\left(D^2 - 2\tau_L M\right)}{C_2 L_1 M R_{load}\left(D^2 - 2M\tau_L\right)}$	$R_0 \approx \dfrac{2 R_{load} \tau_L}{D^2}$
$Z_{out}(s)$ Output impedance	$R_0 \dfrac{1+\dfrac{s}{\omega_{z_1}}}{1+\dfrac{s}{\omega_{p_1}}}$	$\omega_{z_1} = \dfrac{1}{r_C C_2}$	$\omega_{p_1} \approx \dfrac{2}{R_{load} C_2}$	$R_0 \approx R_{load} \| \dfrac{R_{load} D^2 (M-1)}{M\left(D^2 - 2M\tau_L\right)}$

V_p is the PWM sawtooth $M = -\dfrac{V_{out}}{V_{in}}$ $\tau_L = \dfrac{L_1}{R_{load}} F_{sw}$

Figure 4.65 Here are the definitions of the four transfer functions describing a buck-boost converter operated in DCM.

4.3 Buck-Boost Transfer Functions in CCM – Fixed-Frequency Current-Mode Control

The buck-boost converter operated in current-mode control is analyzed in a similar same way as we already did with the voltage-mode case: identify the switch/diode couple and plug the small-signal model already used in the case of the current-mode buck and boost converters. The large-signal model used for the reference curves is shown in Figure 4.66. This is a -12-V converter operated at a 100-kHz switching frequency. It delivers 1.5 A to the load.

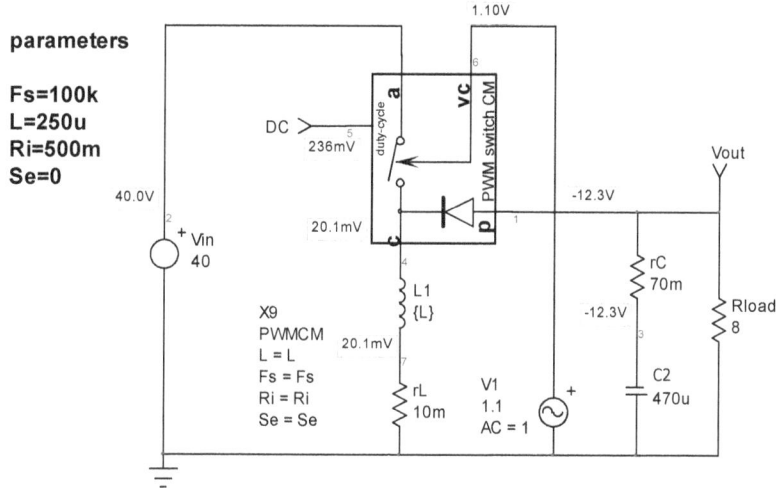

Figure 4.66 The buck-boost converter is now operated in current-mode control.

The small-signal model is then inserted in this application circuit as illustrated by Figure 4.67. Before we consider carrying on with the analysis, we need to make sure this small-signal template is sound. Figure 4.68 confirms the configurations have identical ac-responses.

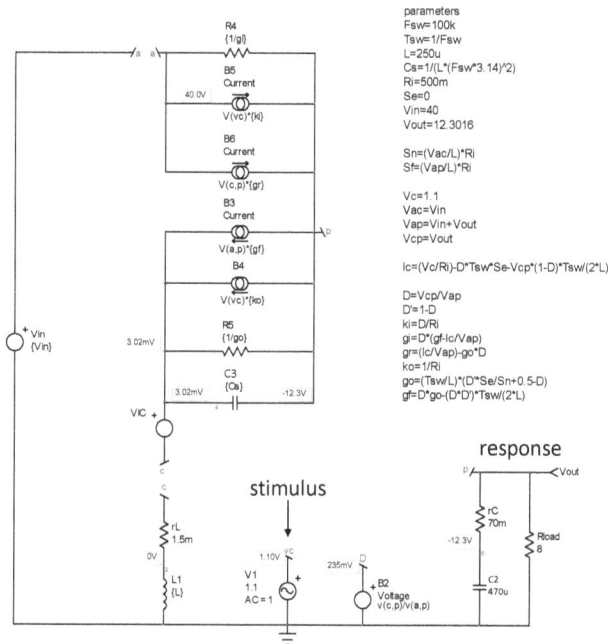

Figure 4.67 The small-signal model is now plugged in the application circuit.

493

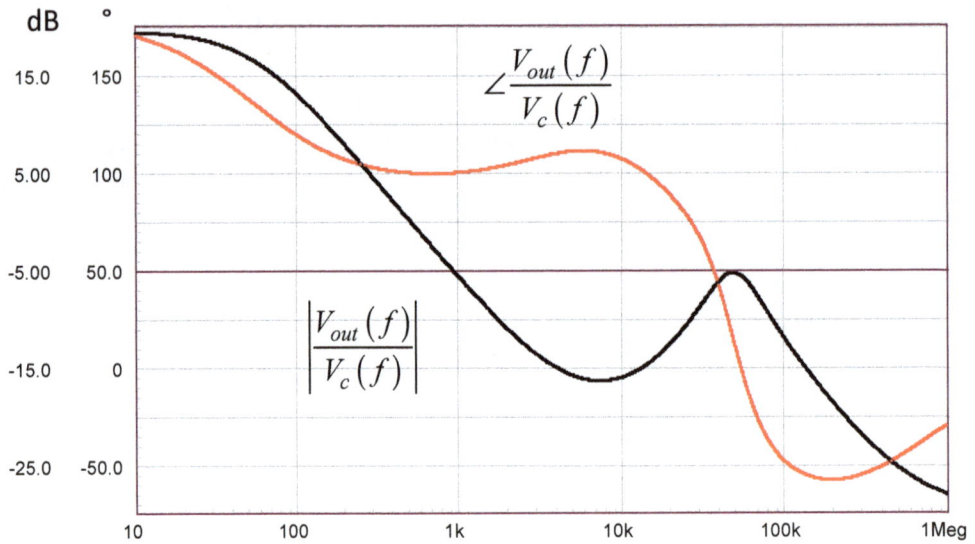

Figure 4.68 Ac responses from the large- and small-signal models are perfectly identical.

After redrawing and rearranging all elements in a convenient way, the circuit we study is that of Figure 4.69. For the quasi-static gain H_0 linking V_{out} to V_c (the control voltage) capacitors are open-circuited while the inductor is shorted. The first equation describes the voltage at node p:

$$V_{out} = V_{(p)} = I_3 R_{load} \tag{4.182}$$

Current I_2 depends on the left-side resistance and the two current sources:

$$I_2 = V_c k_i + V_{(c)} g_r - V_{out} g_r - V_{out} g_i \tag{4.183}$$

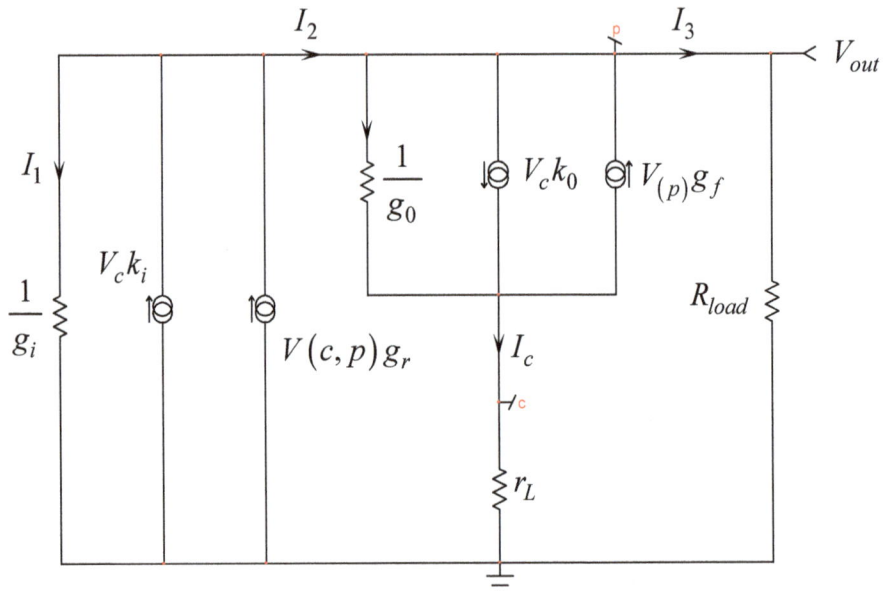

Figure 4.69 The inductor is shorted and capacitors are open-circuited for the dc analysis.

Then, using KCL, we have:

$$I_3 = I_2 - I_c \tag{4.184}$$

Therefore:

$$V_{out} = (I_2 - I_c) R_{load} \tag{4.185}$$

Current I_c can be defined with the voltage at node c and r_L:

$$\frac{V_{(c)}}{r_L} = \left[V_{out} - V_{(c)} \right] g_o + V_c k_o - V_{out} g_f \tag{4.186}$$

From which we can extract the voltage at node c as:

$$V_{(c)} = \frac{V_{out} g_o - V_{out} g_f + V_c k_o}{g_o + \dfrac{1}{r_L}} \tag{4.187}$$

Now substitute (4.187) into (4.183) and you have:

$$I_2 = V_c k_i + \frac{V_{out} g_o - V_{out} g_f + V_c k_o}{g_o + \dfrac{1}{r_L}} g_r - V_{out} g_r - V_{out} g_i \tag{4.188}$$

We now plug (4.187) divided by r_L and (4.188) in (4.185) then solve for V_{out} while factoring V_c. We obtain the quasi-static gain H_0 we want:

$$H_0 = \frac{V_{out}}{V_c} = \frac{k_i - k_o + r_L (g_o k_i + g_r k_o)}{g_o \left(\dfrac{r_L}{R_{load}} + g_i r_L + 1 \right) + g_f (g_r r_L - 1) + g_i + g_r + \dfrac{1}{R_{load}}} \tag{4.189}$$

The above gain is negative, accounting for the negative output voltage of this converter. It is possible to show [3] that this expression can be factored in the following simpler form:

$$H_0 \approx -\frac{R_{load}}{R_i} \frac{1}{\dfrac{(1-D)^2}{2\tau_L} \left(1 + 2\dfrac{S_e}{S_n} \right) + 2M + 1} \tag{4.190}$$

In this expression, S_e is the external artificial slope in [V/s] for damping the subharmonic poles and R_i represents the sense resistance.

S_n is the inductor current on-slope scaled to a voltage by R_i and is defined as:

$$S_n = \frac{V_{in}}{L_1} R_i \tag{4.191}$$

M is a positive value this time and equal to:

$$M = \frac{|V_{out}|}{V_{in}} \qquad (4.192)$$

The normalized time constant τ_L does not change and is expressed as:

$$\tau_L = \frac{L_1}{R_{load}} F_{sw} \qquad (4.193)$$

Now that the quasi-static gain is obtained, let's begin the analysis with the time constants once the excitation V_c is zeroed. We start by opening capacitors C_2 and C_3 while we install a test generator across L_1 as shown in Figure 4.70.

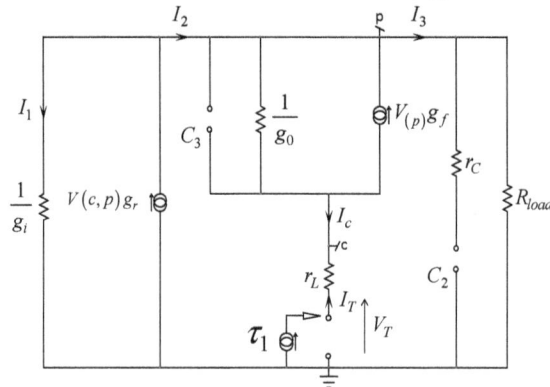

Figure 4.70 Capacitors are open to determine the time constant involving L_1.

The voltage at terminal c depends on V_T and the voltage drop across r_L:

$$V_{(c)} = V_T - r_L I_T \qquad (4.194)$$

From this expression, we extract the value of I_T:

$$I_T = \frac{V_T - V_{(c)}}{r_L} \qquad (4.195)$$

This current can also be expressed the following way:

$$I_T = \left[V_{(c)} - V_{(p)} \right] g_o + V_{(p)} g_f \qquad (4.196)$$

Equating both expressions and solving for V_T leads to:

$$V_T = V_{(c)} + V_{(c)} g_o r_L + V_{(p)} g_f r_L - V_{(p)} g_o r_L \qquad (4.197)$$

The voltage at node p is obtained via the expression of I_2:

$$I_2 = V_{(c)}g_r - V_{(p)}g_r - V_{(p)}g_i \qquad (4.198)$$

If you substitute (4.194) in this expression and solve for $V_{(p)}$, then you find:

$$V_{(p)} = \frac{R_{load}\left(I_T + V_{(c)}g_r\right)}{R_{load}\left(g_i + g_r\right)+1} \qquad (4.199)$$

Now substitute (4.199) and (4.194) in (4.197) and rearrange the ratio of V_T over I_T to obtain:

$$R_{tau1} = \frac{V_T}{I_T} = \frac{R_{load}\left(g_i - g_f + g_o + g_r + g_f g_r r_L + g_i g_o r_L\right)+ g_o r_L +1}{g_o + R_{load}\left(g_f g_r + g_i g_o\right)} \qquad (4.200)$$

Leading to the following time constant:

$$\tau_1 = \frac{L_1}{R_{tau1}} \qquad (4.201)$$

L_1 is now a short circuit and we want the resistance driving capacitor C_2.
The new circuit is in Figure 4.71.

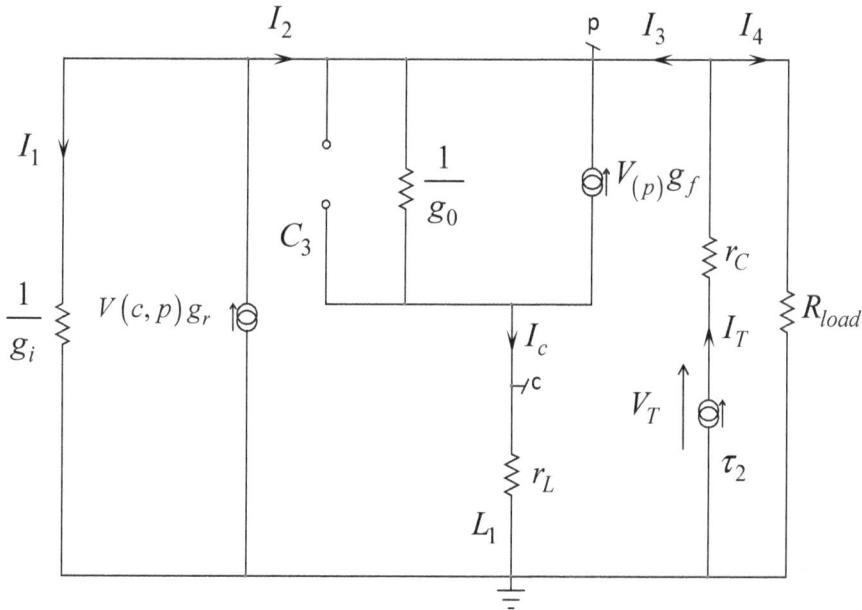

Figure 4.71 L_1 is now a short circuit and we want the time constant associated with C_2.

The current in terminal c is defined as:

$$I_c = \frac{V_{(c)}}{r_L} \qquad (4.202)$$

497

But it is also equal to:

$$I_c = \left[V_{(p)} - V_{(c)} \right] g_o - V_{(p)} g_f \tag{4.203}$$

Equating both equations gives:

$$V_{(c)} = -\frac{V_{(p)} g_f - V_{(p)} g_o}{g_o + \dfrac{1}{r_L}} \tag{4.204}$$

Current I_2 and I_3 are respectively expressed as:

$$I_2 = V_{(c)} g_r - V_{(p)} g_r - V_{(p)} g_i \tag{4.205}$$

$$I_3 = I_T - I_4 = I_T - \frac{V_{(p)}}{R_{load}} \tag{4.206}$$

Current I_c is the sum of the above currents:

$$I_c = I_2 + I_3 \tag{4.207}$$

Which leads to equating (4.202) with (4.207):

$$V_{(c)} g_r - V_{(p)} g_r - V_{(p)} g_i + \left(I_T - \frac{V_{(p)}}{R_{load}} \right) = \frac{V_{(c)}}{r_L} \tag{4.208}$$

Solving the voltage at node c gives:

$$V_{(c)} = \frac{\dfrac{V_T - I_T r_C}{R_{load}} - I_T + \left(V_T - I_T r_C \right) g_i + g_r \left(V_T - I_T r_C \right)}{g_r - \dfrac{1}{r_L}} \tag{4.209}$$

Now equate this new equation with (4.204) and rearrange the ratio of V_T over I_T to obtain:

$$R_{tau2} = \frac{V_T}{I_T} = \frac{\dfrac{r_C \left(g_f - g_o \right)}{g_o + \dfrac{1}{r_L}} + \dfrac{r_C \left(\dfrac{1}{R_{load}} + g_i + g_r \right) + 1}{g_r - \dfrac{1}{r_L}}}{\dfrac{g_i + g_r + \dfrac{1}{R_{load}}}{g_r - \dfrac{1}{r_L}} + \dfrac{g_f - g_o}{g_o + \dfrac{1}{r_L}}} \tag{4.210}$$

The second time constant is:

$$\tau_2 = R_{tau2}C_2 \tag{4.211}$$

Having the two first time constants on hand, we now can study the third one around C_3 (Figure 4.72). The voltage at node p is equal to:

$$V_{(p)} = V_T + V_{(c)} \tag{4.212}$$

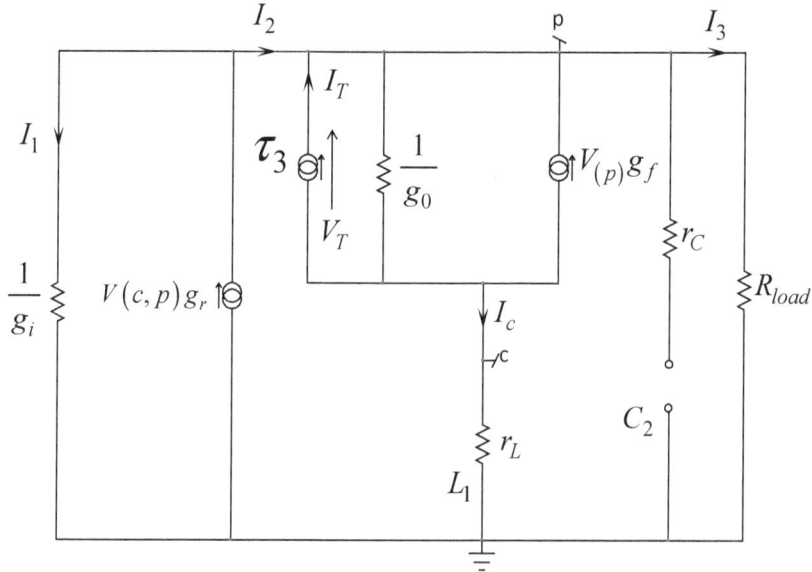

Figure 4.72 Capacitor C_3 is now subjected to the test generator I_T.

Therefore, the current I_c is expressed as:

$$I_c = V_T g_o - I_T - V_{(p)}g_f \tag{4.213}$$

Substituting (4.212) in the above gives:

$$I_c = V_T g_o - I_T - V_T g_f - V_{(c)}g_f \tag{4.214}$$

Equating this current definition with (4.202) and solving for $V_{(c)}$ leads to:

$$V_{(c)} = -\frac{I_T + V_T g_f - V_T g_o}{g_f + \dfrac{1}{r_L}} \tag{4.215}$$

Current I_2 can be defined as follows:

$$I_2 = V_{(c)}g_r - V_{(p)}g_r - V_{(p)}g_i \tag{4.216}$$

Current I_c can be also expressed as the difference between I_2 and I_3:

$$I_c = I_2 - I_3 \tag{4.217}$$

Which gives:

$$\frac{V_{(c)}}{r_L} = V_{(c)}g_r - V_{(p)}g_r - V_{(p)}g_i - \frac{V_{(p)}}{R_{load}}$$ (4.218)

Now substitute (4.212) in the above expression and extract $V_{(c)}$:

$$V_{(c)} = -\frac{V_T r_L + R_{load}V_T g_i r_L + R_{load}V_T g_r r_L}{R_{load} + r_L + R_{load}g_i r_L}$$ (4.219)

Finally, equate (4.219) and (4.215) then solve for V_T while factoring I_T. You should obtain:

$$R_{tau3} = \frac{V_T}{I_T} = \frac{1 + \dfrac{r_L}{R_{load}} + g_i r_L}{\left(\dfrac{r_L}{R_{load}} + g_i r_L + 1\right)g_o + g_f\left(g_r r_L - 1\right) + g_i + g_r + \dfrac{1}{R_{load}}}$$ (4.220)

This leads to the third time constant of this circuit:

$$\tau_3 = R_{tau3}C_3$$ (4.221)

We can now form the first-order coefficient of our denominator:

$$b_1 = \tau_1 + \tau_2 + \tau_3$$ (4.222)

The second coefficient, b_2, is obtained by adding time constants products in which one of the energy-storing elements has been set in its high-frequency state. We start with L_1 set in its high-frequency state (an open circuit) while we determine the resistance driving C_2 in this mode. This is τ_2^1. C_3 is set in its dc state (open). The circuit is that of Figure 4.73.

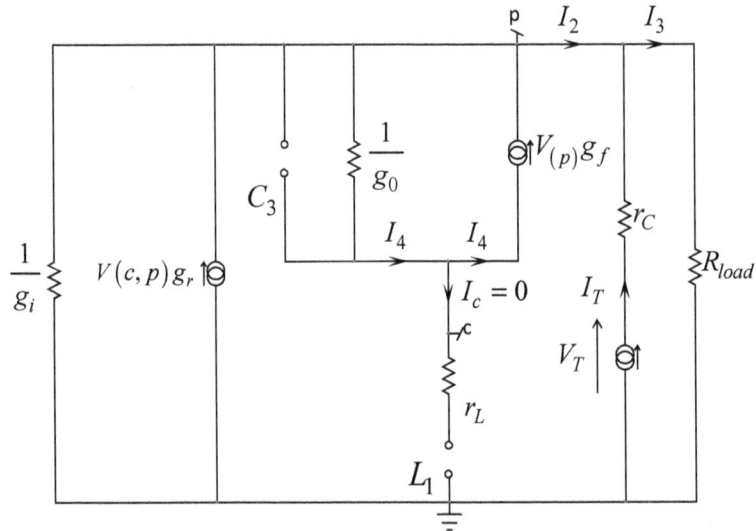

Figure 4.73 This circuit helps determining the resistance driving C_2 while L_1 is set in its high-frequency state and C_3 is open-circuited.

Because inductor L_1 is open, I_c is equal to zero. Current I_2 is thus defined as:

$$I_2 = V_{(c)}g_r - V_{(p)}g_r - V_{(p)}g_i \tag{4.223}$$

Then, we can write:

$$\left[V_{(p)} - V_{(c)} \right] g_o = V_{(p)}g_f \tag{4.224}$$

From which we extract $V_{(c)}$:

$$V_{(c)} = -\frac{V_{(p)}g_f - V_{(p)}g_o}{g_o} \tag{4.225}$$

We can substitute this definition in (4.223) and obtain a new definition for I_2:

$$I_2 = -\frac{V_{(p)}\left(g_f g_r + g_i g_o \right)}{g_o} \tag{4.226}$$

The voltage at terminal p is defined as follows:

$$V_{(p)} = V_T - I_T r_C \tag{4.227}$$

If we substitute (4.227) in (4.226), we obtain:

$$I_2 = -\frac{\left(g_f g_r + g_i g_o \right)\left(V_T - I_T r_C \right)}{g_o} \tag{4.228}$$

Then, we can write the following relationship:

$$\left(I_2 + I_T \right) R_{load} = V_T - I_T r_C \tag{4.229}$$

From which we obtain a second definition for I_2:

$$I_2 = -\frac{I_T R_{load} - V_T + I_T r_C}{R_{load}} \tag{4.230}$$

If we equate (4.228) and (4.230) then solve for V_T after factoring I_T, we have determined the resistance we want:

$$R_{tau12} = \frac{V_T}{I_T} = \frac{R_{load}\left(g_o + g_f g_r r_C + g_i g_o r_C \right) + g_o r_C}{g_o + R_{load}\left(g_f g_r + g_i g_o \right)} \tag{4.231}$$

Leading to:

$$\tau_2^1 = C_2 R_{tau12} \tag{4.232}$$

We carry on by keeping L_1 in its high-impedance state while determining the resistance driving C_3 (Figure 4.74).
We want to determine τ_3^1.

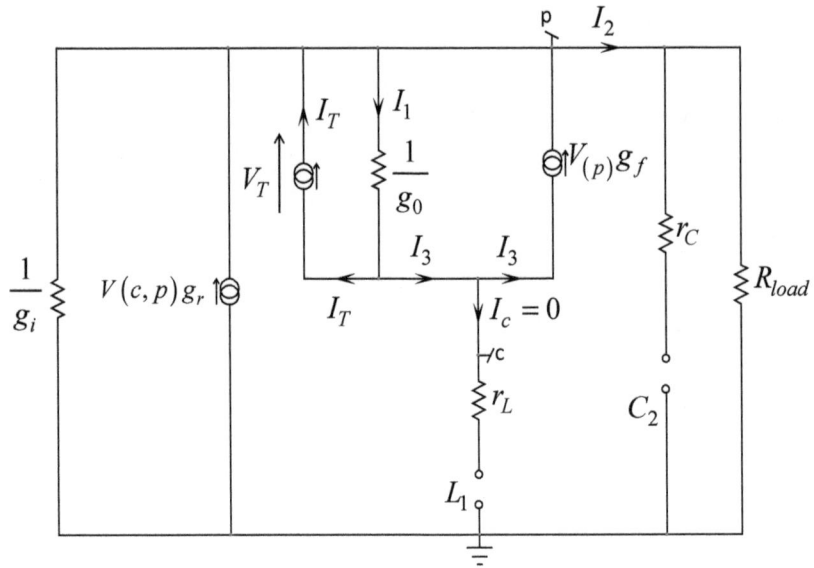

Figure 4.74 What resistance drives capacitor C_3 while L_1 is open-circuited?

Current I_2 is determined by:

$$I_2 = V_{(c)}g_r - V_{(p)}g_r - V_{(p)}g_i \tag{4.233}$$

The voltage at node c is:

$$V_{(c)} = V_{(p)} - V_T \tag{4.234}$$

Thus:

$$I_2 = \left(V_{(p)} - V_T\right)g_r - V_{(p)}g_r - V_{(p)}g_i \tag{4.235}$$

The voltage at node p depends on the load resistance R_{load}:

$$V_{(p)} = I_2 R_{load} \tag{4.236}$$

Now substitute this expression in the definition of I_2 above and update its definition:

$$I_2 = -\frac{V_T g_r}{1 + R_{load} g_i} \tag{4.237}$$

The definition for the I_T current is the following:

$$I_T = V_T g_o - V_{(p)}g_f \tag{4.238}$$

We extract the voltage at terminal p:

$$V_{(p)} = -\frac{I_T - V_T g_o}{g_f} \tag{4.239}$$

Capitalizing on (4.236), we obtain another definition for I_2:

$$I_2 = -\frac{I_T - V_T g_o}{g_f R_{load}} \tag{4.240}$$

Now equate (4.240) with (4.237) and solve for V_T while factoring I_T:

$$R_{tau13} = \frac{1 + R_{load} g_i}{g_o + R_{load}\left(g_f g_r + g_i g_o\right)} \tag{4.241}$$

Which leads to:

$$\tau_3^1 = R_{tau13} C_3 \tag{4.242}$$

Finally, we set C_2 in its high-frequency state (a short circuit) while L_1 is in its dc state, also a short circuit. We want the resistance driving C_3 in this mode: what value is τ_3^2? Figure 4.75 shows the circuit we need to solve. We start with current I_T:

$$I_T = I_1 - I_3 = I_1 - \left(I_4 + I_c\right) = V_T g_o - \left(V_{(p)} g_f + \frac{V_{(c)}}{r_L}\right) \tag{4.243}$$

We can extract the voltage at terminal c:

$$V_{(c)} = -r_L\left(I_T - V_T g_o + V_{(p)} g_f\right) \tag{4.244}$$

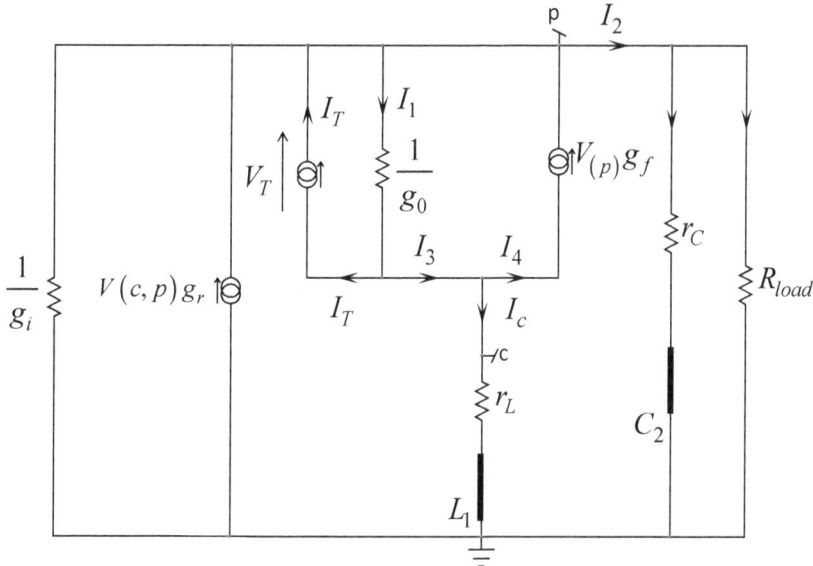

Figure 4.75 We want the resistance driving C_3 while C_2 and L_1 are shorted.

We write the following relationship linking $V_{(p)}$ and V_T:

$$V_{(p)} - V_T = I_c r_L = r_L\left(V_T g_o - I_T - V_{(p)} g_f\right) \tag{4.245}$$

From this expression, we can extract the voltage at terminal p:

$$V_{(p)} = \frac{V_T - I_T r_L + V_T g_o r_L}{1 + g_f r_L} \qquad (4.246)$$

The definition for $V_{(p)}$ and $V_{(c)}$ can now be plugged in the I_2 current expression while I_c is derived from (4.245):

$$I_2 = V_{(c)} g_r - V_{(p)} g_r - V_{(p)} g_i - I_c \qquad (4.247)$$

If you solve for I_2, you should find:

$$I_2 = -\frac{V_T g_i - V_T g_f - I_T + V_T g_o + V_T g_r - I_T g_i r_L + V_T g_f g_r r_L + V_T g_i g_o r_L}{1 + g_f r_L} \qquad (4.248)$$

Observing the circuit, we see that the voltage at terminal p is also equal to current I_2 scaled by the parallel combination of r_C and R_{load}:

$$V_{(p)} = I_2 \left(r_C \parallel R_{load} \right) \qquad (4.249)$$

If we substitute (4.246) in this expression and solve for I_2, we find:

$$I_2 = \left[V_T - \frac{r_L \left(I_T + V_T g_f - V_T g_o \right)}{1 + r_L g_f} \right] \frac{R_{load} + r_C}{R_{load} r_c} \qquad (4.250)$$

The exercise now consists of equating (4.248) with (4.250) and solve for V_T while factoring I_T:

$$R_{tau23} = \frac{V_T}{I_T} = \frac{R_{load} \left(r_C + r_L + g_i r_C r_L \right) + r_C r_L}{r_C + R_{load} \left(g_i r_C - g_f r_C + g_o r_C + g_r r_C + g_o r_L + g_f g_r r_C r_L + g_i g_o r_C r_L + 1 \right) + g_o r_C r_L} \qquad (4.251)$$

Which leads to:

$$\tau_3^2 = R_{tau23} C_3 \qquad (4.252)$$

The second-order coefficient b_2 will be obtained by assembling the time constants as follows:

$$b_2 = \tau_1 \tau_2^1 + \tau_1 \tau_3^1 + \tau_2 \tau_3^2 \qquad (4.253)$$

Almost there! We need to identify the last time constant, determined when L_1 and C_3 are set in their high-frequency state. The circuit appears in Figure 4.76 for finding τ_2^{13}.

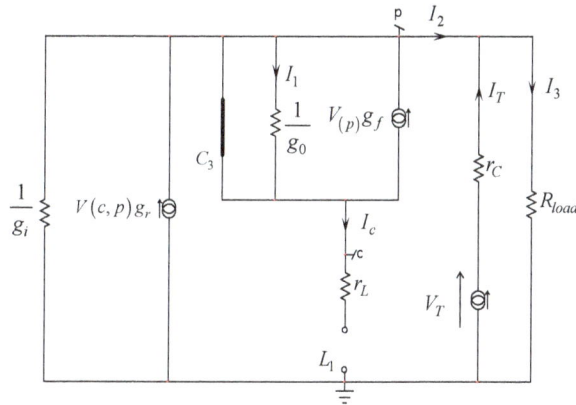

Figure 4.76 This is the last time constant determination, τ_2^{13}.

To simplify the analysis, we can redraw the schematic diagram and temporarily remove r_C to bring it back later in series with the intermediate result. The new electrical circuit is shown in Figure 4.77 where simplifications have occurred since both terminals c and p share a common potential owing to C_3 being replaced by a short circuit.

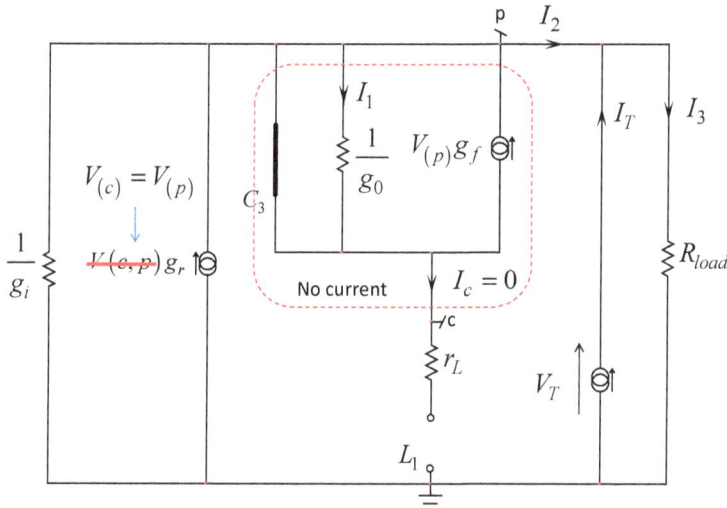

Figure 4.77 The last time constant can be found by inspection.

In this mode, the current source I_T "sees" the parallel combination of R_{load} and the conductance g_i. Finally, the last time constant is defined as:

$$\tau_2^{13} = C_2\left(\frac{1}{g_i} \| R_{load} + r_C\right) \tag{4.254}$$

b_3 can now be determined by:

$$b_3 = \tau_1 \tau_3^1 \tau_2^{13} \tag{4.255}$$

And the denominator is given by:

$$D(s) = 1 + sb_1 + s^2 b_2 + s^3 b_3 \tag{4.256}$$

According to [1], this third-order polynomial can be expressed by combining a dominating low-frequency pole and two subharmonic poles located at half the switching frequency:

$$1 + b_1 s + b_2 s^2 + b_3 s^3 \approx (1 + b_1 s)\left(1 + \frac{b_2}{b_1} s + \frac{b_3}{b_1} s^2\right)$$

(4.257)

[3] offers compact expressions compared to the raw definitions we have determined so far:

$$(1 + b_1 s)\left(1 + \frac{b_2}{b_1} s + \frac{b_3}{b_1} s^2\right) \approx \left(1 + \frac{s}{\omega_{p_1}}\right)\left[1 + \frac{s}{\omega_n Q_p} + \left(\frac{s}{\omega_n}\right)^2\right]$$

(4.258)

Where:

$$\omega_{p_1} \approx \frac{\dfrac{(1-D)^3}{2\tau_L}\left(1 + 2\dfrac{S_e}{S_n}\right) + 1 + D}{R_{load} C_2}$$

(4.259)

$$Q_p = \frac{1}{\pi(m_c D' - 0.5)}$$

(4.260)

$$\omega_n = \frac{\pi}{T_{sw}}$$

(4.261)

$$m_c = 1 + \frac{S_e}{S_n}$$

(4.262)

S_e and S_n have been defined in the quasi-static gain H_0 determination.

Now that we have the denominator, we can determine where the zeroes are located. The excitation V_c is brought back in the analysis but cannot propagate to form a response: $V_{out}(s) = 0$ and the exercise consists of identifying the impedance conditions in the circuit which bring this output null.

The circuit to analyze is that of Figure 4.78.

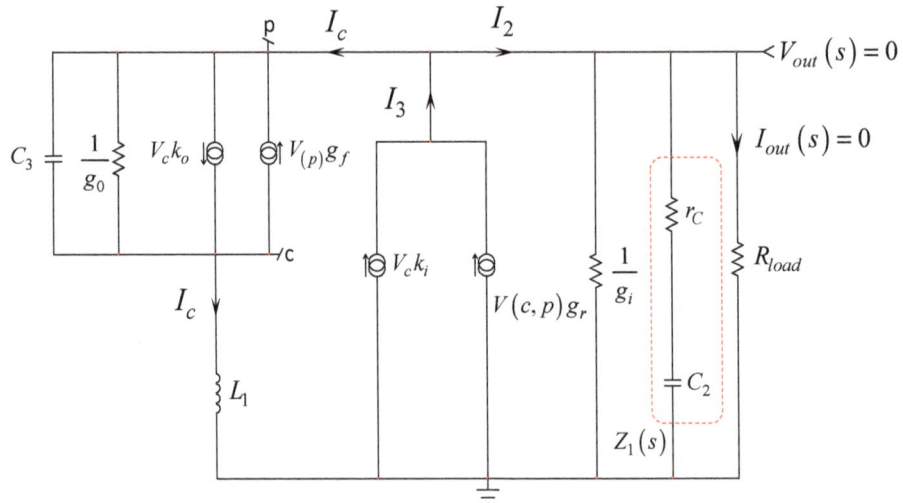

Figure 4.78 The output voltage is nulled and we need to find the conditions creating this state.

The first zero is provided by the usual suspects r_C and C_2:

$$Z_1(s) = r_C + \frac{1}{sC_2} = 0 \tag{4.263}$$

It implies a zero located at:

$$\omega_{z_1} = \frac{1}{r_C C_2} \tag{4.264}$$

The second condition bringing an output null is when I_2 equals zero. This occurs if all the I_3 current is absorbed by the left-side network in which I_c flows. Because of the null, the voltage at terminal p is zero and we can simplify the circuit a bit as shown in Figure 4.79. The first equation defines current I_3 which is in fact current I_c:

$$V_{(c)}g_r + V_c k_i = I_c \tag{4.265}$$

But I_c is also defined by:

$$I_c = \frac{V_{(c)}}{sL_1} = V_c k_o - V_{(c)}g_o - V_{(c)}sC_3 \tag{4.266}$$

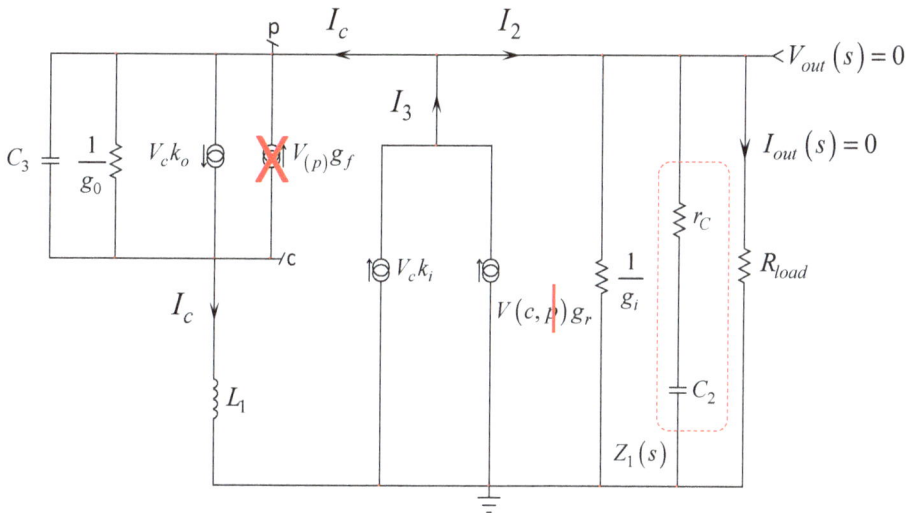

Figure 4.79 The circuit is simplified by propagating the nulled output voltage to some expressions.

From (4.266), we extract the value of $V_{(c)}$:

$$V_{(c)} = \frac{V_c k_o}{\frac{1}{sL_1} + g_o + sC_3} \tag{4.267}$$

Now equate (4.265) with the right part of (4.266)

$$V_{(c)}g_r + V_c k_i = V_c k_o - V_{(c)}g_o - V_{(c)}sC_3 \tag{4.268}$$

Then, substitute (4.267) in the expression and study the following equality:

$$\frac{V_c\left(k_i + L_1 g_o k_i s + L_1 g_r k_o s + C_3 L_1 k_i s^2\right)}{1 + sL_1 g_o + s^2 C_3 L_1} = \frac{V_c k_o}{1 + sL_1 g_o + s^2 C_3 L_1} \tag{4.269}$$

As denominators are similar, numerators must be equal:

$$k_i + L_1 g_o k_i s + L_1 g_r k_o s + C_3 L_1 k_i s^2 = k_o \tag{4.270}$$

Bring k_o on the left side and factor $k_i - k_o$:

$$\left(k_i - k_o\right)\left(1 + sL_1\left(\frac{g_o k_i + g_r k_o}{k_i - k_o}\right) + s^2 \frac{C_3 L_1 k_i}{k_i - k_o}\right) = 0 \tag{4.271}$$

What we want are the roots of the expression containing the s:

$$1 + sL_1\left(\frac{g_o k_i + g_r k_o}{k_i - k_o}\right) + s^2 \frac{C_3 L_1 k_i}{k_i - k_o} = 0 \tag{4.272}$$

We could either leave this expression for the numerator or try to apply the low-Q approximation to obtain a simpler formula. We know from [1]:

$$1 + a_1 s + a_2 s^2 \approx \left(1 + a_1 s\right)\left(1 + \frac{a_2}{a_1} s\right) \tag{4.273}$$

Therefore:

$$1 + sL_1\left(\frac{g_o k_i + g_r k_o}{k_i - k_o}\right) + s^2 \frac{C_3 L_1 k_i}{k_i - k_o} \approx \left[1 + sL_1\left(\frac{g_o k_i + g_r k_o}{k_i - k_o}\right)\right]\left[1 + s \frac{C_3 L_1 k_i}{g_o k_i - g_r k_o}\right] \tag{4.274}$$

This equation can be advantageously factored as:

$$\left[1 + sL_1\left(\frac{g_o k_i + g_r k_o}{k_i - k_o}\right)\right]\left[1 + s^2 \frac{C_3 L_1 k_i}{g_o k_i - g_r k_o}\right] = \left(1 - \frac{s}{\omega_{z_2}}\right)\left(1 + \frac{s}{\omega_{z_3}}\right) \tag{4.275}$$

with:

$$\omega_{z_2} = \frac{k_i - k_o}{L_1\left(g_o k_i + g_r k_o\right)} \approx \frac{(1-D)^2 R_{load}}{D \cdot L_1} \tag{4.276}$$

and:

$$\omega_{z_3} = \frac{g_o k_i + g_r k_o}{C_3 k_i} \tag{4.277}$$

The zero defined by (4.276) is located in the right half-plane hence the negative sign in (4.275). The third zero given by ω_{z3} occurs in high frequency and can be neglected.

Finally, the denominator can be formed by associating the three zeroes we have determined:

$$N(s) = \left(1 + \frac{s}{\omega_{z_1}}\right)\left(1 - \frac{s}{\omega_{z_2}}\right)\left(1 + \frac{s}{\omega_{z_3}}\right) \approx \left(1 + \frac{s}{\omega_{z_1}}\right)\left(1 - \frac{s}{\omega_{z_2}}\right) \tag{4.278}$$

The control-to-output transfer function of the CM CCM-operated buck-boost converter is thus defined as:

$$H(s) = \frac{V_{out}(s)}{V_c(s)} = -H_0 \cdot \frac{\left(1 + \frac{s}{\omega_{z_1}}\right)\left(1 - \frac{s}{\omega_{z_2}}\right)}{\left(1 + \frac{s}{\omega_p}\right)} \cdot \frac{1}{1 + \frac{s}{\omega_n Q_p} + \left(\frac{s}{\omega_n}\right)^2} \tag{4.279}$$

In which we have:

$$H_0 \approx \frac{R_{load}}{R_i} \frac{1}{\frac{(1-D)^2}{2\tau_L}\left(1 + 2\frac{S_e}{S_n}\right) + 2M + 1} \tag{4.280}$$

$$\omega_{z_1} = \frac{1}{r_C C_2} \tag{4.281}$$

$$\omega_{z_2} \approx \frac{(1-D)^2 R_{load}}{D \cdot L_1} \tag{4.282}$$

$$\omega_p \approx 1 + s\frac{R_{load}C_2}{\frac{(1-D)^3}{2\tau_L}\left(1 + 2\frac{S_e}{S_n}\right) + 1 + D} \tag{4.283}$$

$$Q_p = \frac{1}{\pi(m_c D' - 0.5)} \tag{4.284}$$

$$\omega_n = \frac{\pi}{T_{sw}} \tag{4.285}$$

$$m_c = 1 + \frac{S_e}{S_n} \tag{4.286}$$

As usual, we have run a SPICE simulation of Figure 4.66 and compared the results with that of the Mathcad® sheet using the raw-coefficients expressions. As confirmed by Figure 4.80, results are in perfect agreement. Separated test confirm that the magnitude and phase curves delivered by the approximated poles-zeroes are also very close to non-simplified formulas.

We will now simulate the CM buck-boost converter with SIMPLIS® shown in Figure 4.81. The circuit operates in open-loop conditions and should deliver -12 V. Simulation results in transient and ac analyses are given in Figure 4.82, confirming the good operation of the converter. The inductor current shows a CCM condition with a duty ratio at 23%, matching the value obtained from the average model in Figure 4.66. The ac response is also in good agreement with the results obtained from SPICE or Mathcad® and confirms the presence of sub-harmonic poles at 50 kHz.

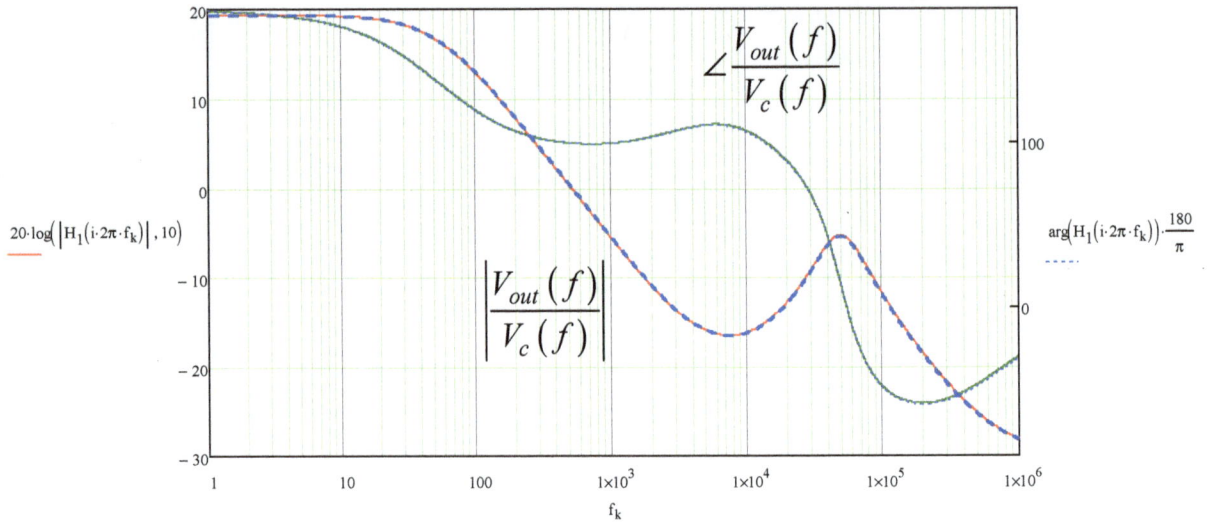

Figure 4.80 This graph gathers the SPICE simulation of the CM buck-boost converter and what our Mathcad® sheet delivers.

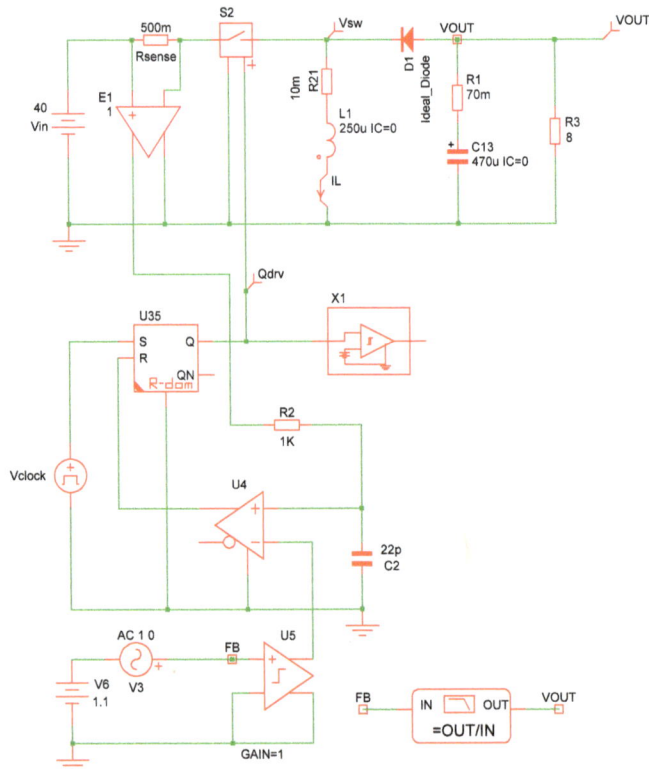

Figure 4.81 A SIMPLIS® simulation lets us verify if our analysis is correct with a cycle-by-cycle model.

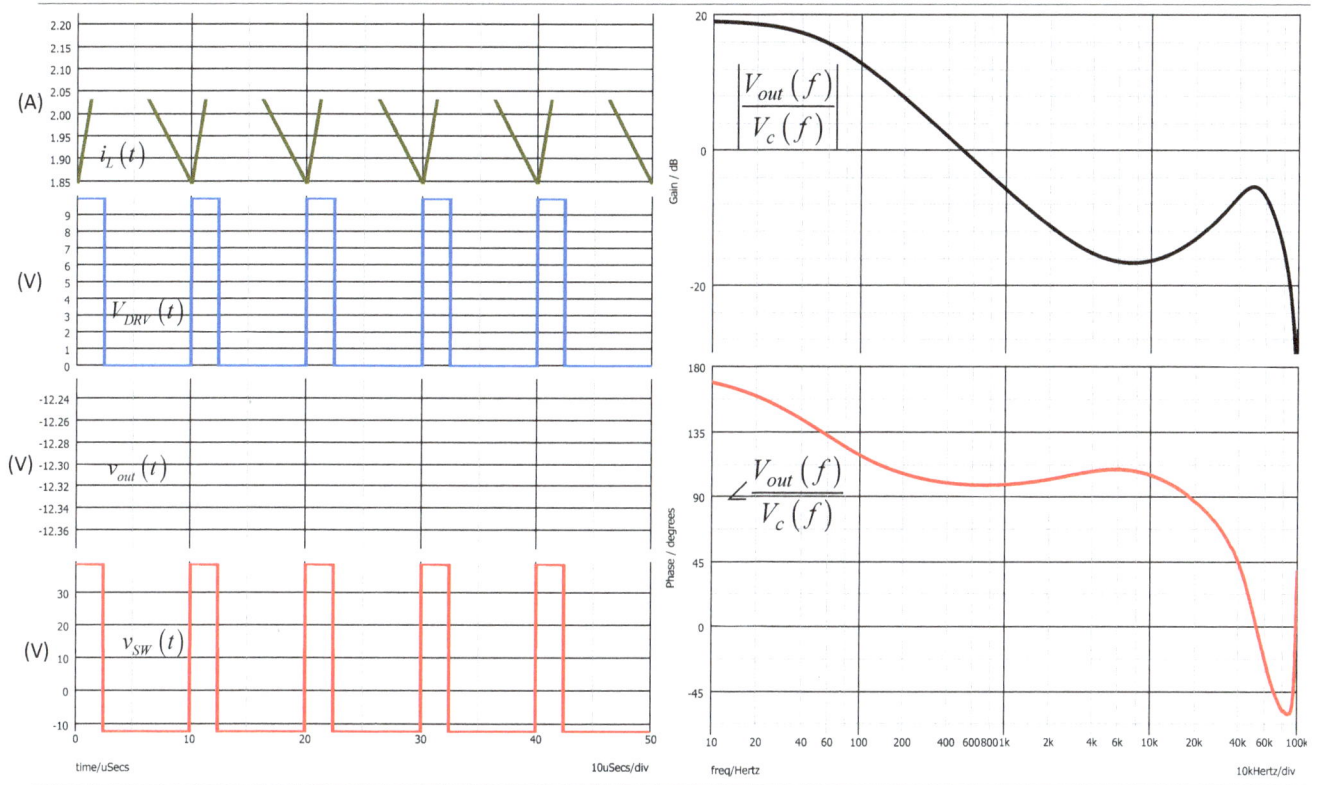

Figure 4.82 Operating points are correct as well as the ac analysis which confirms the dc gain and the subharmonic poles.

4.3.1 Input to Output

Our next task is to determine the input-to-output transfer function. In this mode, the control voltage V_c becomes ac-silent and the stimulus changes to the input source V_{in}. The large-signal configuration appears in Figure 4.83.

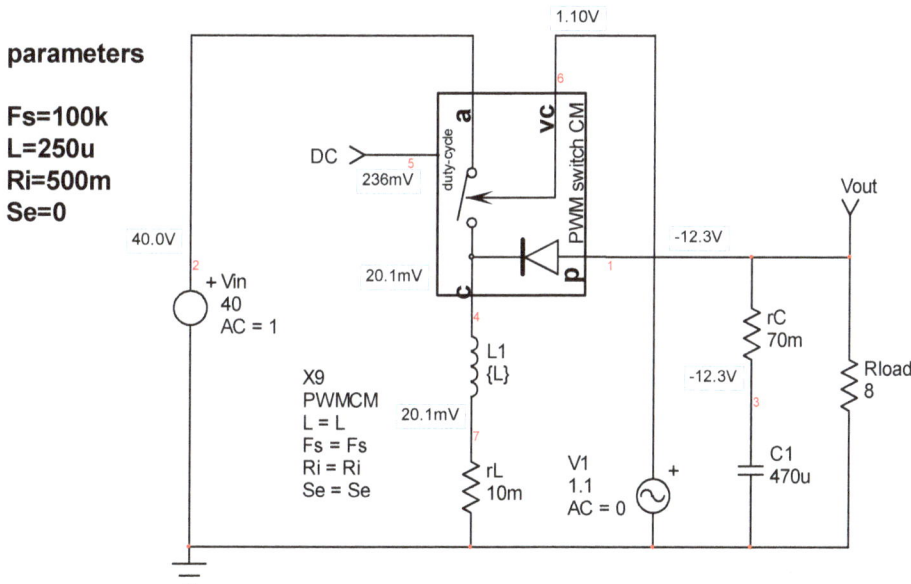

Figure 4.83 The large-signal model is configured to analyze the input-to-output transfer function.

To determine this transfer function, we insert the small-signal model as recommended in Figure 4.84. Terminal *a* is no longer grounded compared to Figure 4.67 and connects to the stimulus. As we did in the other examples, we must check that ac responses between large- and small-signal circuits are identical. This is what Figure 4.85 confirms. We can now redraw the circuit in a way that is easier to analyze and it appears in Figure 4.86.

We start with the quasi-static gain H_0. The inductor is shorted and all capacitors are open. To simplify the analysis, r_L is neglected bringing terminal *c* to ground and naturally simplifying the circuit (Figure 4.87). Our first equation links V_{out} to R_{load}:

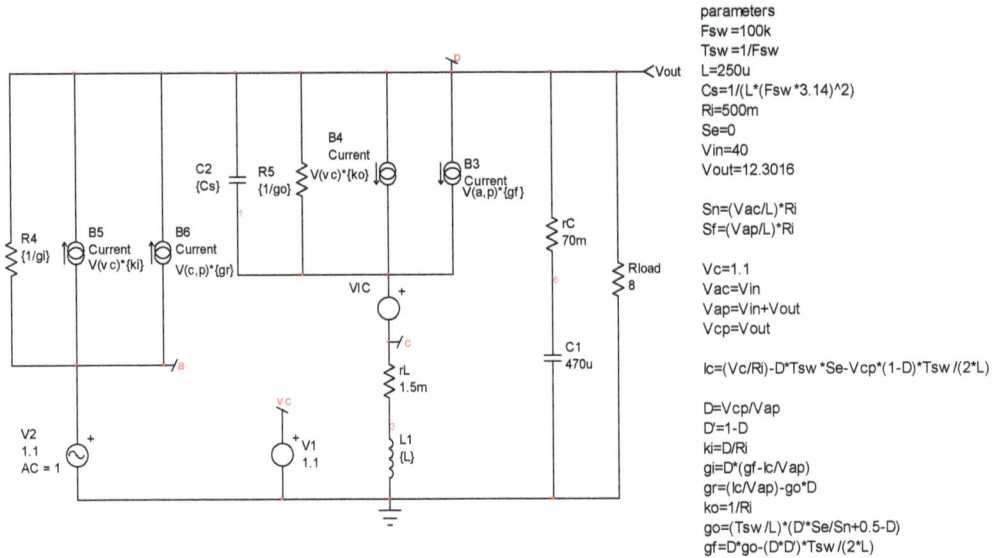

Figure 4.84 The small-signal model is necessary to determine the transfer function we want.

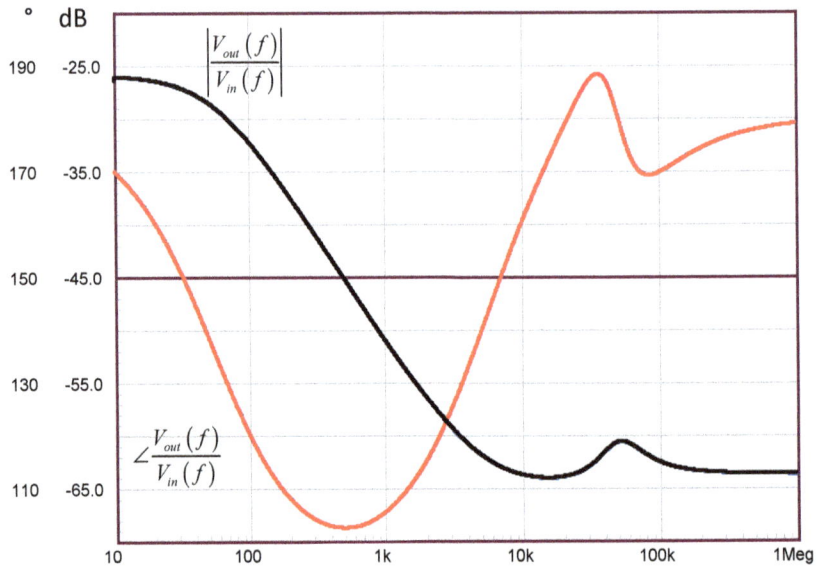

Figure 4.85 Responses between large- and small-signal models are identical.

$$V_{out} = I_1 R_{load} \tag{4.287}$$

Current I_2 is expressed by:

$$I_2 = -V_{out}g_r - V_{out}g_i - V_{in}g_i \tag{4.288}$$

The Buck-Boost Converter and its Derivatives

While current I_c also depends on V_{in} and V_{out}:

$$I_c = -V_{out}g_o + V_{in}g_f + V_{out}g_f \qquad (4.289)$$

The sum of the above two currents gives I_1:

$$I_1 = I_c + I_2 \qquad (4.290)$$

Following (4.287) this expression leads to:

$$V_{out} = R_{load}\left[-V_{out}g_r - V_{out}g_i - V_{in}g_i - V_{out}g_o + V_{in}g_f + V_{out}g_f\right] \qquad (4.291)$$

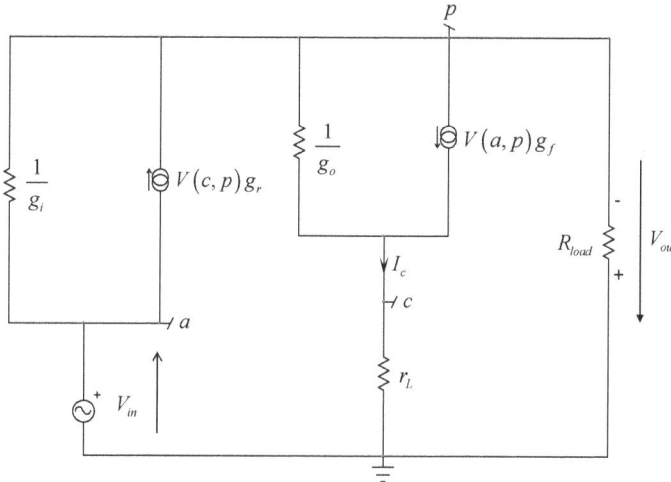

Figure 4.86 Once rearranged, the circuit looks simpler to study.

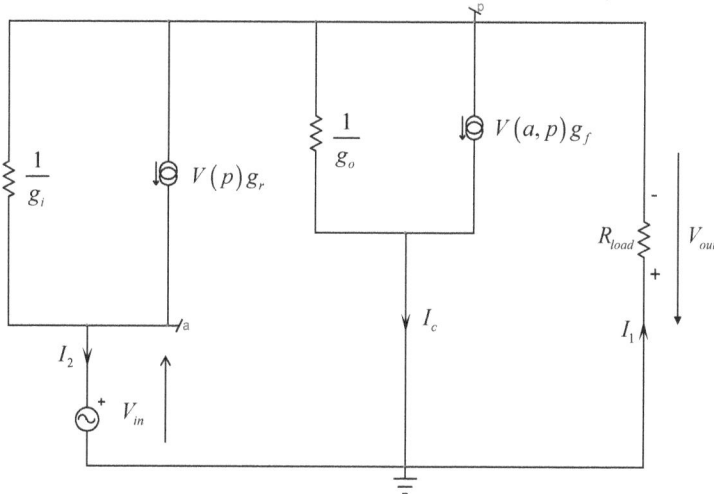

Figure 4.87 Bringing r_L to zero further simplifies the circuit.

Extract V_{out} while factoring V_{in} and rearrange. You should find the gain H_0 we want and it is equal to:

$$H_0 = \frac{R_{load}\left(g_f - g_i\right)}{R_{load}\left(g_i - g_f + g_o + g_r\right) + 1} \qquad (4.292)$$

Now, if you reduce the stimulus to zero, the circuit returns to its natural structure as observed in all of the transfer functions we have derived so far (except Z_{in}). We can advantageously reuse $D(s)$ already determined in (4.258) and the gain in time is tremendous.

We can now concentrate on the zeroes of this circuit. Zeroes manifest themselves in such a way that the response is nulled for certain stimulus frequencies. The circuit we are now looking at is that of Figure 4.88. In this drawing, we immediately see the series connection of r_C and C_2 potentially realizing a transformed short circuit at a certain frequency.

$$Z_2(s) = r_C + \frac{1}{sC_2} = 0 \tag{4.293}$$

The first zero is immediate:

$$\omega_{z_1} = \frac{1}{r_C C_2} \tag{4.294}$$

The second zero occurs if current I_2 entirely flows between terminals p and c:

$$I_2 = I_c \tag{4.295}$$

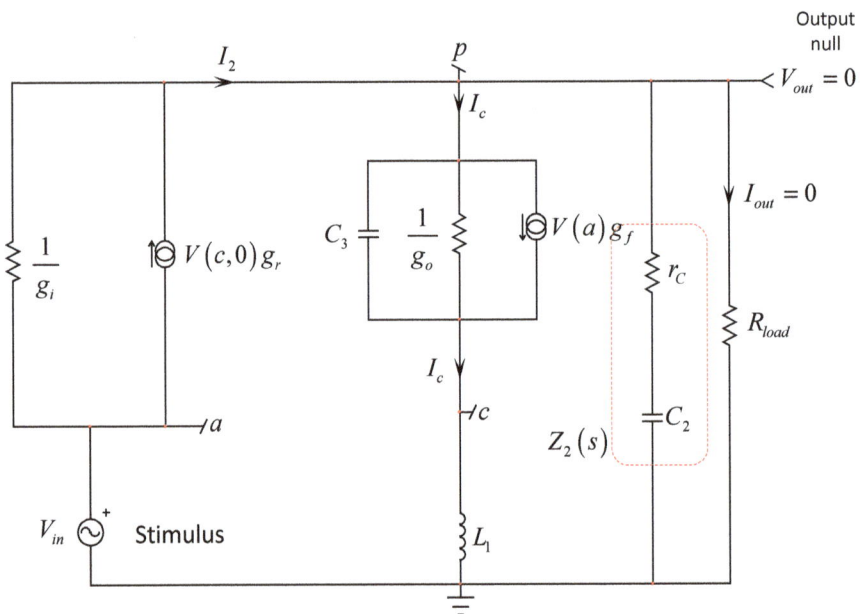

Figure 4.88 Despite the presence of a stimulus, the response is nulled. What impedance conditions bring the circuit is this condition?

Two equations define the current flowing out terminal c:

$$I_c = \frac{V_{(c)}}{sL_1} \tag{4.296}$$

and:

$$I_c = -\frac{V_{(c)}}{\frac{1}{sC_3}} - V_{(c)}g_o + V_{in}g_f \tag{4.297}$$

From these two expressions, extract the voltage at potential c:

$$V_{(c)} = \frac{V_{in} g_f}{g_o + \dfrac{1}{sL_1} + sC_3}$$

(4.298)

Because terminal p is biased at a 0-V potential, we can write:

$$\frac{V_{(c)}}{sL_1} = V_{in} g_i + V_{(c)} g_r$$

(4.299)

Now plug (4.298) in the above equation and obtain the following equality:

$$\frac{g_f}{1 + sL_1 g_o + s^2 L_1 C_3} = \frac{g_i + sL_1 \left(g_f g_r + g_i g_o \right) + s^2 L_1 C_3 g_i}{1 + sL_1 g_o + s^2 L_1 C_3}$$

(4.300)

Considering equal denominators, we can restrict the analysis to the numerators. After rearranging this expression, we obtain:

$$\left(g_i - g_f \right) \left[1 + sL_1 \left(\frac{g_f g_r + g_i g_o}{g_i - g_f} \right) + s^2 \frac{L_1 C_3 g_i}{g_i - g_f} \right] = 0$$

(4.301)

The zeroes are located in the right-side term and let us form the numerator we need:

$$N(s) = 1 + \frac{s}{Q_N \omega_{0N}} + \left(\frac{s}{\omega_{0N}} \right)^2$$

(4.302)

with:

$$\omega_{0N} = \frac{1}{\sqrt{\dfrac{L_1 C_3 g_i}{g_i - g_f}}}$$

(4.303)

$$Q_N = \frac{\left(g_i - g_f \right) \sqrt{\dfrac{L_1 C_3 g_i}{g_i - g_f}}}{L_1 \left(g_f g_r + g_i g_o \right)}$$

(4.304)

The complete transfer function can now be written as:

$$\frac{V_{out}(s)}{V_{in}(s)} = -H_0 \frac{\left(1 + \dfrac{s}{\omega_{z_1}} \right) \left[1 + \dfrac{s}{Q_N \omega_{0N}} + \left(\dfrac{s}{\omega_{0N}} \right)^2 \right]}{\left(1 + \dfrac{s}{\omega_{p_1}} \right) \dfrac{1}{1 + \dfrac{s}{\omega_n Q_p} + \left(\dfrac{s}{\omega_n} \right)^2}}$$

(4.305)

In this expression, H_0 and the numerator have been defined in the above lines while the denominator comes from (4.258).

$$\omega_{p_1} \approx \frac{(1-D)\frac{k_c}{k}\left(1+2\frac{S_e}{S_n}\right)+1+D}{R_{load}C_2} \tag{4.306}$$

$$Q_p = \frac{1}{\pi\left(m_c D'-0.5\right)} \tag{4.307}$$

$$\omega_n = \frac{\pi}{T_{sw}} \tag{4.308}$$

$$m_c = 1+\frac{S_e}{S_n} \tag{4.309}$$

S_e and S_n are respectively the external stabilizing ramp and the inductor on-slope. S_n is defined in (4.191).

We now compare the response delivered by Figure 4.83 with that from the Mathcad® sheet plotting (4.305). The curves appear in Figure 4.89 and confirm our analysis is correct.

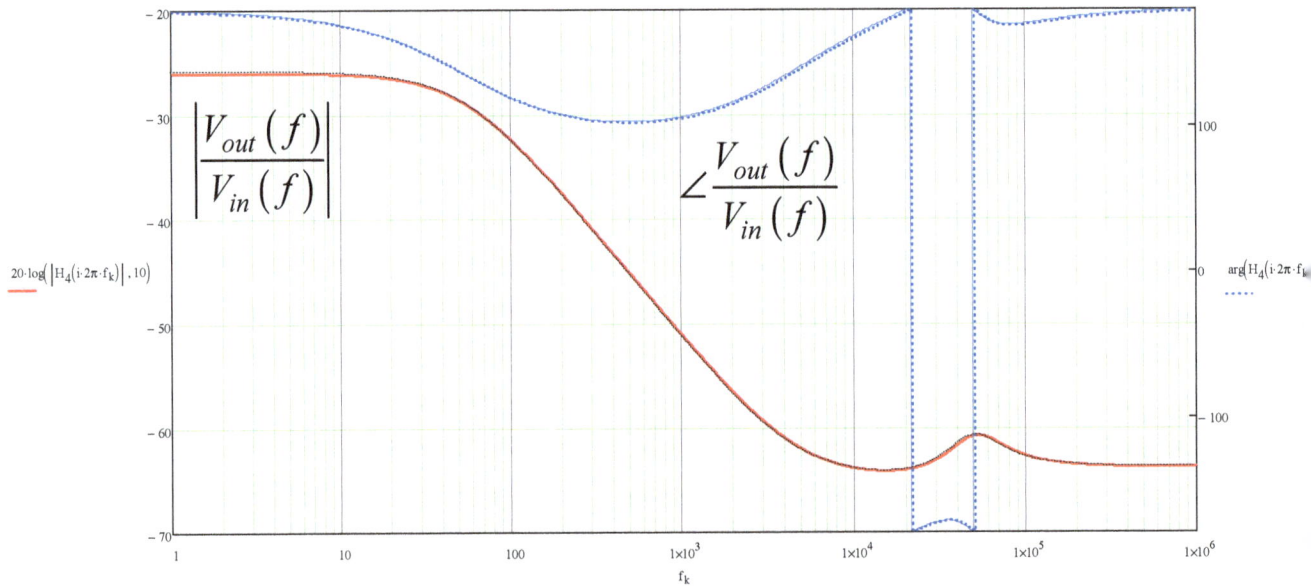

Figure 4.89 The analytical expression and the SPICE simulation perfectly agree.

4.3.2 Output Impedance

The output impedance is classically determined by installing a 1-A ac source across the converter load. This is what is shown in Figure 4.90 around the large-signal model.

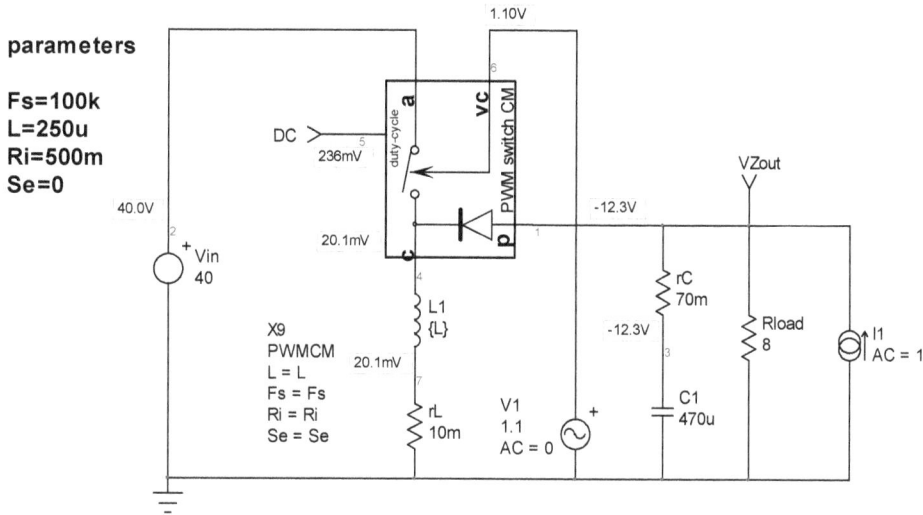

Figure 4.90 A 1-A ac current source helps determining the converter's output impedance.

The small-signal circuit with zeroed input voltage and V_c is shown in Figure 4.91 and we will perform a sanity check before attempting to study it. Results appear in Figure 4.92 and confirm both circuits lead to identical results. We can carry on with the rearranged circuit of Figure 4.93 in which the inductor has been shorted (its resistance r_l ignored) and the capacitors open-circuited: we want to determine the quasi-static output resistance R_0 that is a few equations away. First, we temporarily disconnect R_{load} and the two conductances g_i and g_o.

We will bring them back in parallel with the intermediate result. Current I_T is defined by:

$$I_T = V_{(p)}g_r - V_{(p)}g_f \tag{4.310}$$

Since $V_{(p)}$ is also V_T, I_T can also be expressed by:

$$I_T = V_T\left(g_r - g_f\right) \tag{4.311}$$

Then solving for V_T while factoring I_T, the following equivalent circuit results:

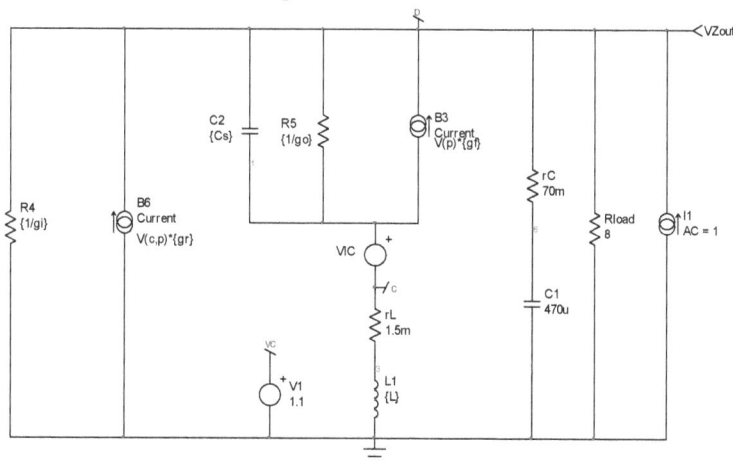

Figure 4.91 The small-signal version is tested before used for analysis.

517

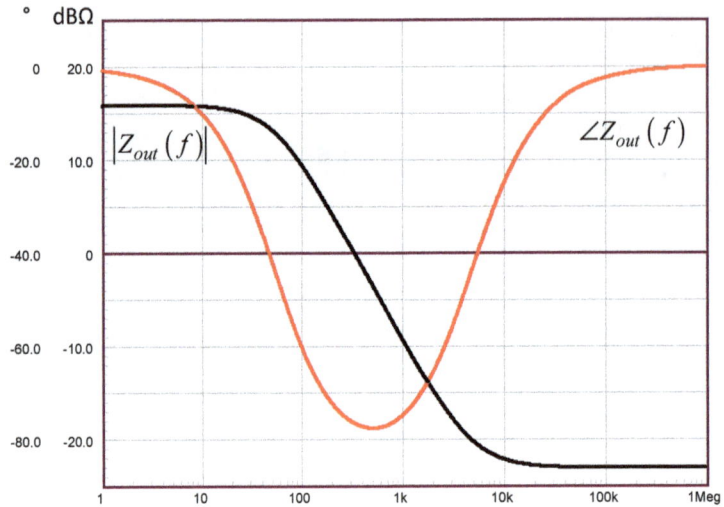

Figure 4.92 Large- and small-signal simulation circuits lead to identical results.

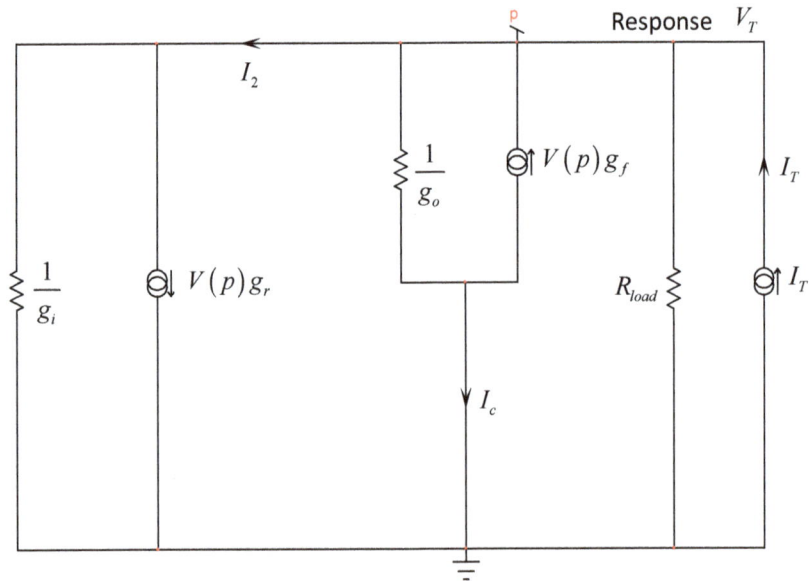

Figure 4.93 This is the circuit we will use for our quasi-static output resistance analysis.

$$\frac{V_T}{I_T} = \frac{1}{g_r - g_f} \tag{4.312}$$

Finally, if we bring back R_{load} and the two conductances, we have:

$$R_0 = \frac{1}{g_r - g_f} \parallel \frac{1}{g_i} \parallel \frac{1}{g_r} \parallel R_{load} \tag{4.313}$$

Now, if the stimulus I_T is turned off, the circuit returns to its natural state and we can advantageously reuse $D(s)$ already determined in (4.258).

Let's look at the zeroes where the impedance conditions give a null response. In other words, what impedance combination could create a transformed short across the I_T current source? The circuit appears in Figure 4.94.

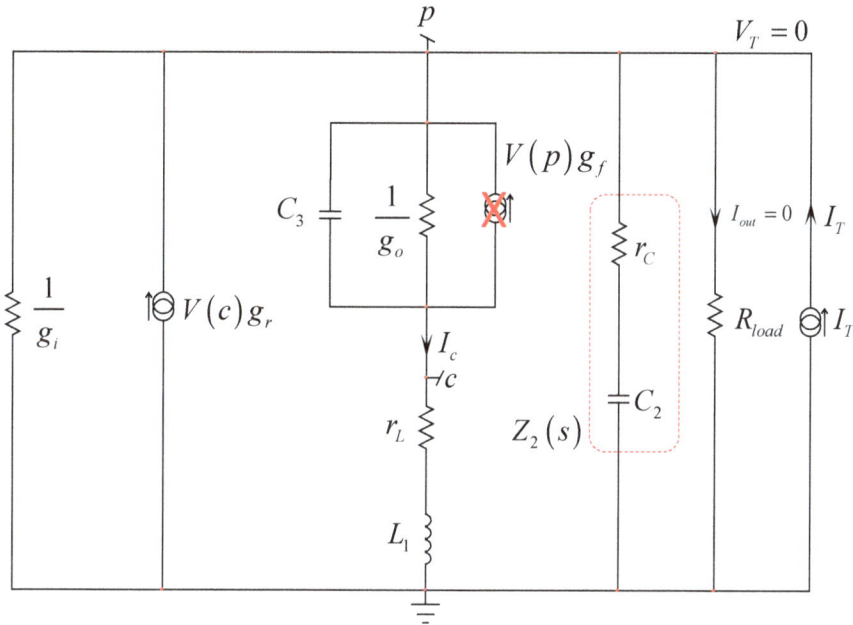

Figure 4.94 The response is nulled by two transformed short circuits across current source I_T.

You recognize the usual suspects r_C and C_2 which, together, provide the first zero:

$$Z_2(s) = r_C + \frac{1}{sC_2} = 0 \tag{4.314}$$

Which leads to:

$$\omega_{z_1} = \frac{1}{r_C C_2} \tag{4.315}$$

For the other zeroes, we will replace the current source I_T by a short circuit (a degenerate case as documented in [1]) and determine the resistances driving C_3 and L_1. Figure 4.95 gathers the sketches needed to determine these resistances.

Considering the simple configurations, inspection works well and is extremely fast. From the first sketch, we determine τ_{1N}:

$$\tau_{1N} = \frac{L_1}{r_L + \dfrac{1}{g_o}} \tag{4.316}$$

Then:

$$\tau_{3N} = C_3 \left(r_L \,\|\, \frac{1}{g_o} \right) \tag{4.317}$$

Christophe Basso

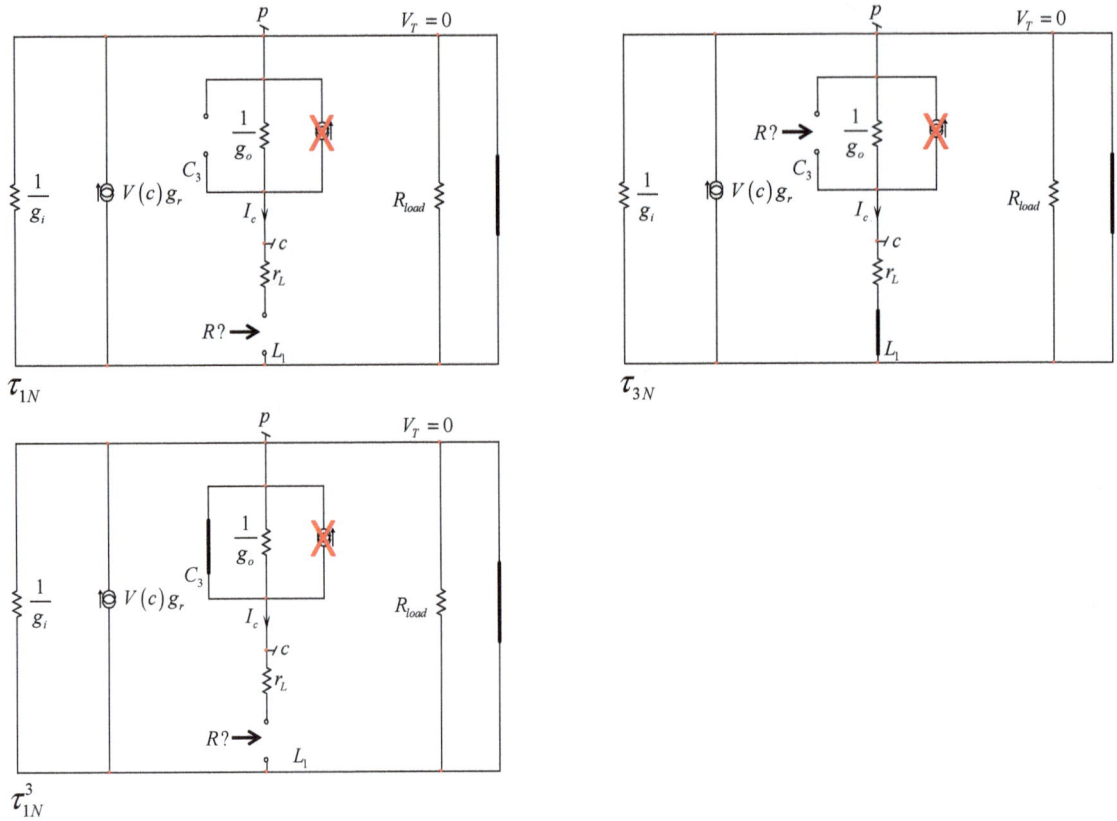

Figure 4.95 The additional zeroes are found by replacing the current source by a short circuit while "looking" through the energy-storing elements's connections.

The last one is obtained when C_3 is shorted which cancels the I_c current:

$$\tau_{1N}^3 = \frac{L_1}{r_L} \tag{4.318}$$

This is it, we have the intermediate denominator obtained by combining the above time constants as follows:

$$N_{int}(s) = 1 + s(\tau_{1N} + \tau_{3N}) + s^2(\tau_{3N}\tau_{1N}^3) \tag{4.319}$$

It can be formalized under the classical polynomial form:

$$N_{int}(s) = 1 + \frac{s}{\omega_{0N}Q_N} + \left(\frac{s}{\omega_{0N}}\right)^2 \tag{4.320}$$

with:

$$\omega_{0N} \approx \frac{1}{\sqrt{L_1 C_3}} \tag{4.321}$$

$$Q_N \approx \sqrt{\frac{C_3}{L_1}}\frac{1}{g_o} \tag{4.322}$$

We can now assemble the final transfer function describing the output impedance of the CCM CM buck-boost converter:

$$Z_{out}\left(s\right) = R_0 \frac{1+\dfrac{s}{\omega_{z_1}} \; 1+\dfrac{s}{\omega_{0N}Q_N}+\left(\dfrac{s}{\omega_{0N}}\right)^2}{1+\dfrac{s}{\omega_{p_1}} \; 1+\dfrac{s}{\omega_n Q_p}+\left(\dfrac{s}{\omega_n}\right)^2} \qquad (4.323)$$

The two values ω_{0N} and Q_N are extremely close to the parameters ω_n and Q_p characterizing the resonant poles. A simplification is therefore possible:

$$Z_{out}\left(s\right) \approx R_0 \frac{1+\dfrac{s}{\omega_{z_1}}}{1+\dfrac{s}{\omega_{p_1}}} \qquad (4.324)$$

with:

$$R_0 = \frac{1}{g_r - g_f} \left\| \frac{1}{g_i} \right\| \frac{1}{g_r} \left\| R_{load} \right. \qquad (4.325)$$

$$\omega_{p_1} \approx \frac{\dfrac{\left(1-D\right)^3}{2\tau_L}\left(1+2\dfrac{S_e}{S_n}\right)+1+D}{R_{load}C_2} \qquad (4.326)$$

and:

$$\omega_{z_1} = \frac{1}{r_C C_2} \qquad (4.327)$$

We have now gathered on a common graph the magnitude/phase given by (4.323) and the SPICE simulation of the large-signal model of Figure 4.90. Results are perfect as shown in Figure 4.96.

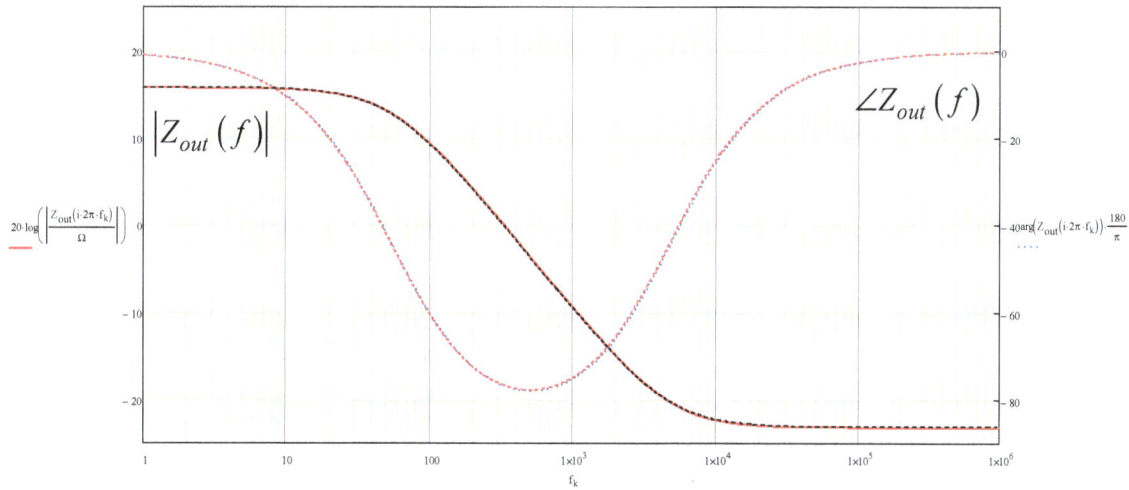

Figure 4.96 The figure confirms the excellent agreement between Mathcad® and SPICE.

4.3.3 Input Impedance

The simulation of the input impedance is performed by adding a large-value inductor in series with the input source. We have used this approach along the previous chapters; during bias point analysis, the inductor is a short circuit and SPICE computes the static voltage and current values according to input and output conditions. When the ac analysis starts, the big inductor isolates the converter from the 0-Ω source while the 1-A current source modulates the input. The voltage collected across the stimulus in this mode is thus a variable representative of the input impedance.

The circuit is given in Figure 4.97. The equivalent small-signal model is obtained by plugging-in the model you are now familiar with and the new circuit appears in Figure 4.98. Before we carry on, we verify that both circuits deliver the exact same response. This sanity check is important to ensure we are going to spend time on a valid circuit. Figure 4.99 confirms we are good to go.

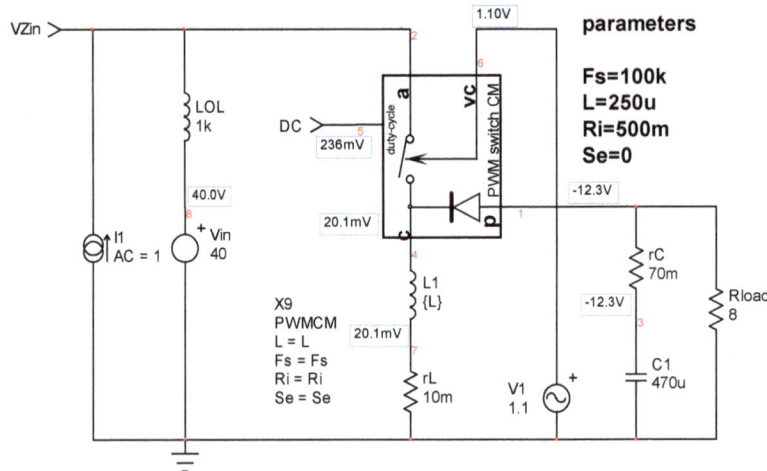

Figure 4.97 The large-signal CM model is used to determine the converter open-loop input impedance.

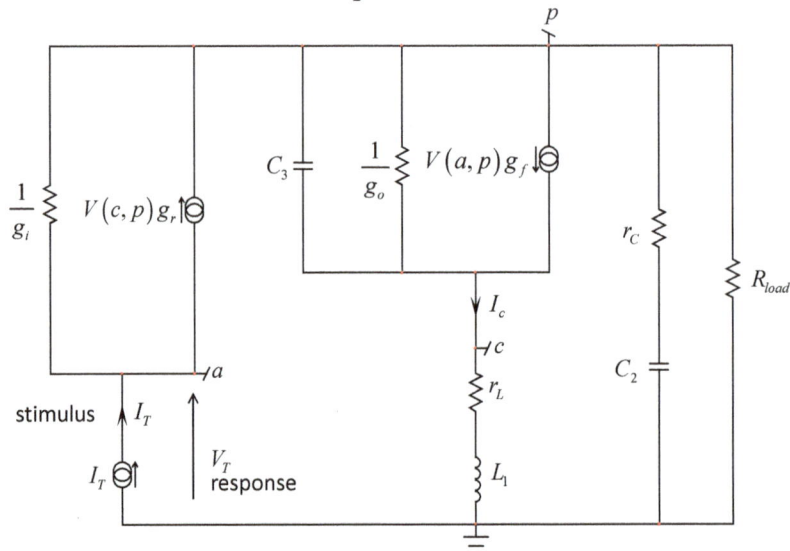

Figure 4.98 This is the small-signal model used to determine the input impedance.

We now start by determining the input resistance R_0 for which we short inductor L_1 (we ignore r_L) and open-circuit capacitors C_2 and C_3. This is what Figure 4.100 describes. If we associate g_o and R_{load} in parallel, the circuit further simplifies to that of Figure 4.101.

The first equation defines current I_T:

$$I_T = -V_{(p)}g_r + g_i \left(V_T - V_{(p)}\right)$$ (4.328)

From which we extract the voltage at node p:

$$V_{(p)} = -\frac{I_T - V_T g_i}{g_i + g_r}$$ (4.329)

A second definition for I_T is:

$$I_T = \frac{V_{(p)}}{R_{load} \| \dfrac{1}{g_o}} + V_T g_f - V_{(p)}g_f$$ (4.330)

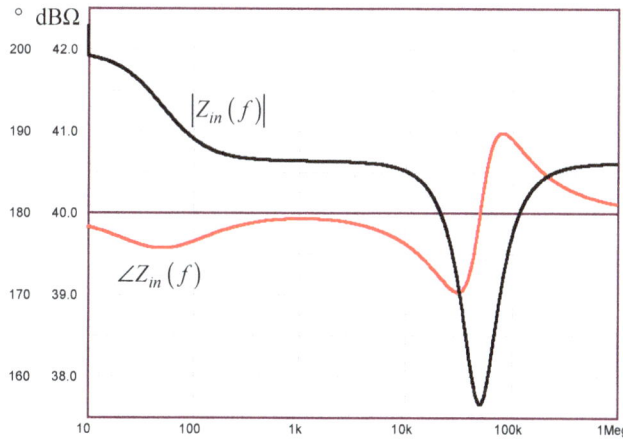

Figure 4.99 The two models deliver the exact same same magnitude and phase responses.

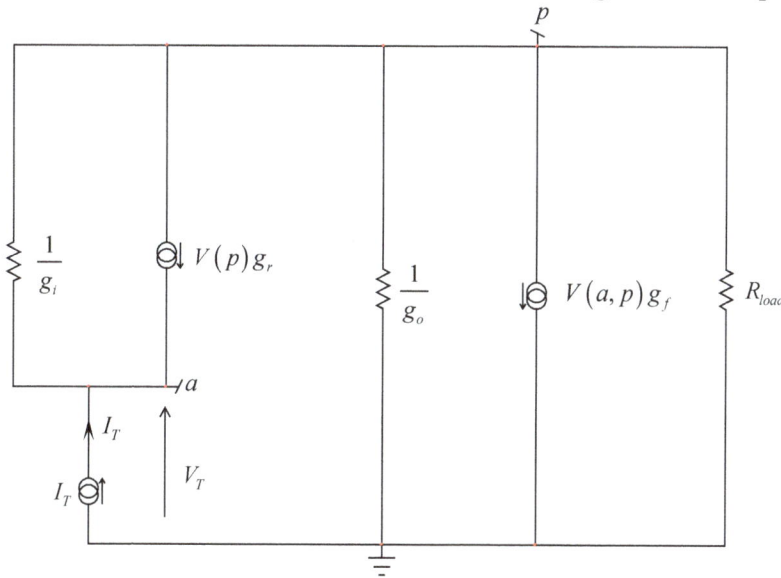

Figure 4.100 The circuit is set in its dc state to determine R_0.

Now substitute (4.329) in this definition and solve for V_T while factoring I_T. You should find:

$$R_0 = \frac{\left(g_i - g_f + g_o + g_r\right)R_{load} + 1}{\left(g_f g_r + g_i g_o\right)R_{load} + g_i}$$

(4.331)

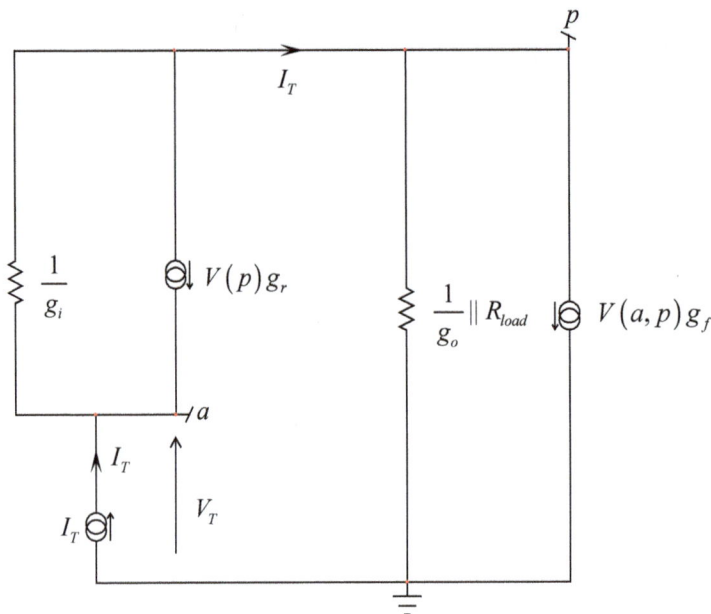

Figure 4.101 If we combine g_o and R_{load} in a single element, the circuit simplifies.

If we apply numerical value to this expression, the result is negative: the incremental open-loop input resistance of the CM CCM buck-boost converter is negative. Figure 4.102 shows the inductor current and the input current at two different line levels. The phenomenon is similar to what we have seen with the other CM converters: in high-line conditions and a fixed switching frequency operation, the on-time necessary to reach the peak current setpoint is smaller in high line than in low line. As a result, the time available to demagnetize the inductor expands and the inductor depletes longer, leading to a valley current smaller at high line than low line. This translates to the input current by a greater average current when the input source reduces.

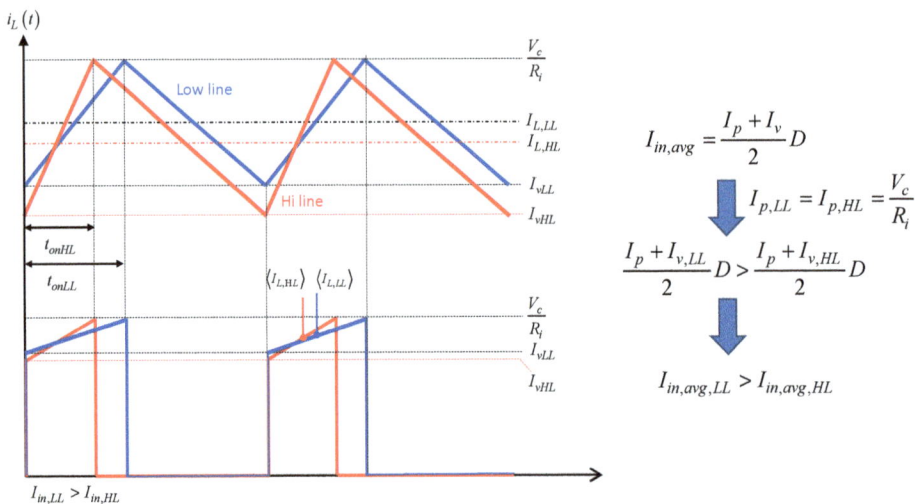

Figure 4.102 The input current decreases when the input line increases.

Now that R_0 has been determined, we can look at the time constants for a zeroed excitation. The new circuit is that of Figure 4.103 and you can see that having node a open does not bring the circuit back to its natural structure: we cannot reuse the denominator $D(s)$ determined by (4.258). We have to determine the time constants in this mode.

We start with τ_1 involving L_1 (Figure 4.104).

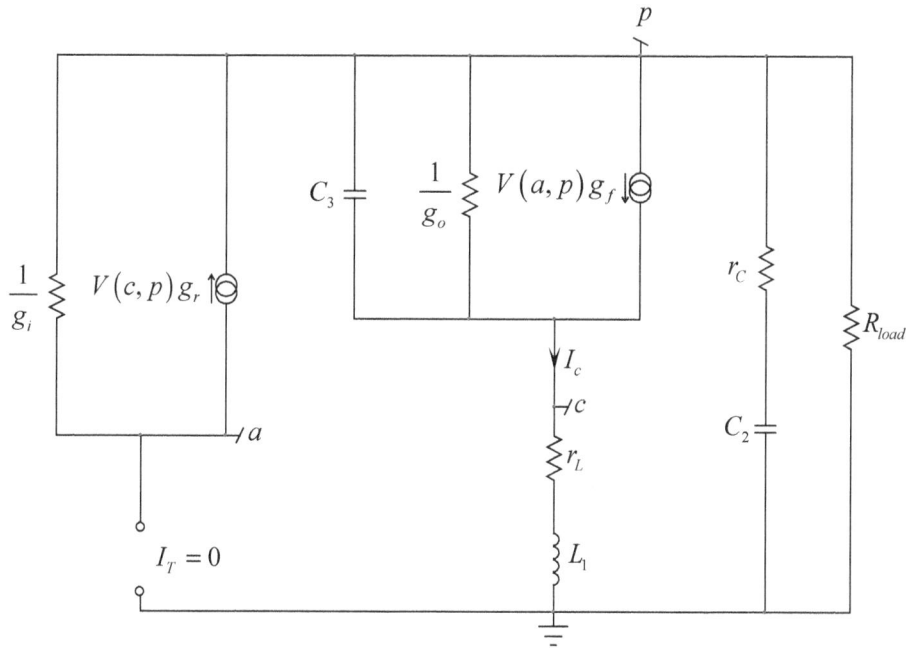

Figure 4.103 The circuit does not return to its natural state when node a is open by the zeroed stimulus.

A current source is installed across L_1's connecting terminals and we need to determine V_T. We can temporarily disconnect r_L and bring it back later with the intermediate result.

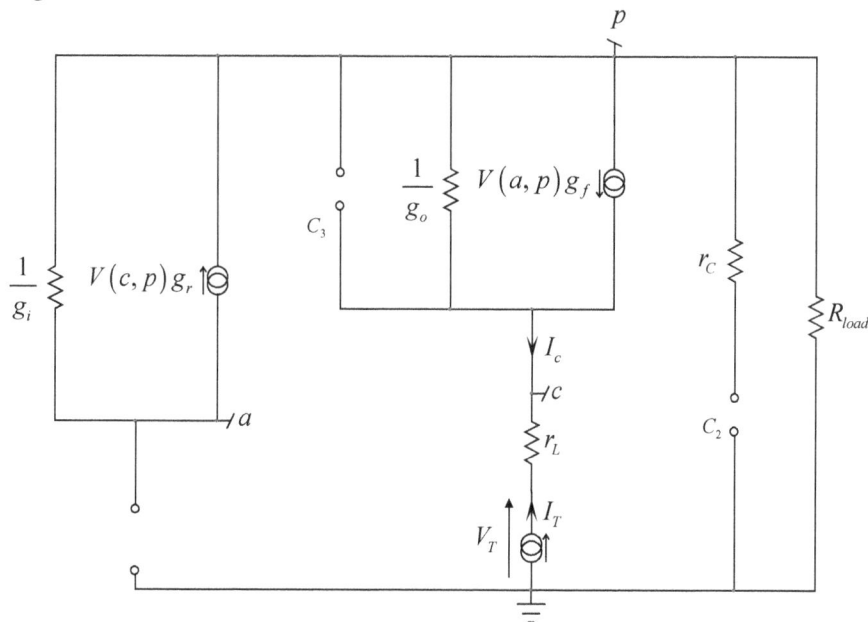

Figure 4.104 All capacitors are open-circuited for this first time constant determination.

525

We can express the voltage at node a:

$$V_{(a)} = -\left[V_{(c)} - V_{(p)}\right]\frac{g_r}{g_i} + V_{(p)} \tag{4.332}$$

The voltage V_T can be defined as:

$$V_T = V(c, p) + V_{(p)} \tag{4.333}$$

Then, $V(c,p)$ is given by:

$$V(c, p) = \left[I_T + g_f\left(V_{(a)} - V_{(p)}\right)\right]\frac{1}{g_o} \tag{4.334}$$

From the circuit, we can see that:

$$V_{(p)} = I_T R_{load} \tag{4.335}$$

and:

$$V_{(c)} = V_T \tag{4.336}$$

Now, if you substitute (4.335), (4.336) in (4.332) then you have:

$$V_{(a)} = -\left[V_T - I_T R_{load}\right]\frac{g_r}{g_i} + I_T R_{load} \tag{4.337}$$

Plug this last equation in (4.334) together with (4.335) and solve for V_T while factoring I_T. If you now bring r_L back, you should find:

$$R_{tau1} = \frac{V_T}{I_T} = \frac{R_{load}\left(g_f g_r + g_i g_o\right) + g_i}{g_f g_r + g_i g_o} + r_L \tag{4.338}$$

The first time constant involving L_1 is defined by:

$$\tau_1 = \frac{L_1}{R_{tau1}} \tag{4.339}$$

We now replace L_1 by a short circuit and we determine the resistance "seen" from C_2's connecting terminals. The new circuit updates to that of Figure 4.105.

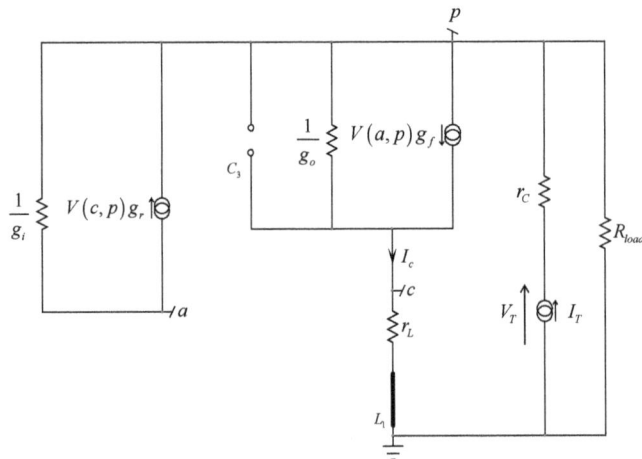

Figure 4.105 The inductor is now shorted to determine the resistance driving C_2.

To further simplify the analysis, r_l is neglected implying a grounded c node. r_C is temporarily disconnected to be brought back later on. The simplified circuit is shown in Figure 4.106.

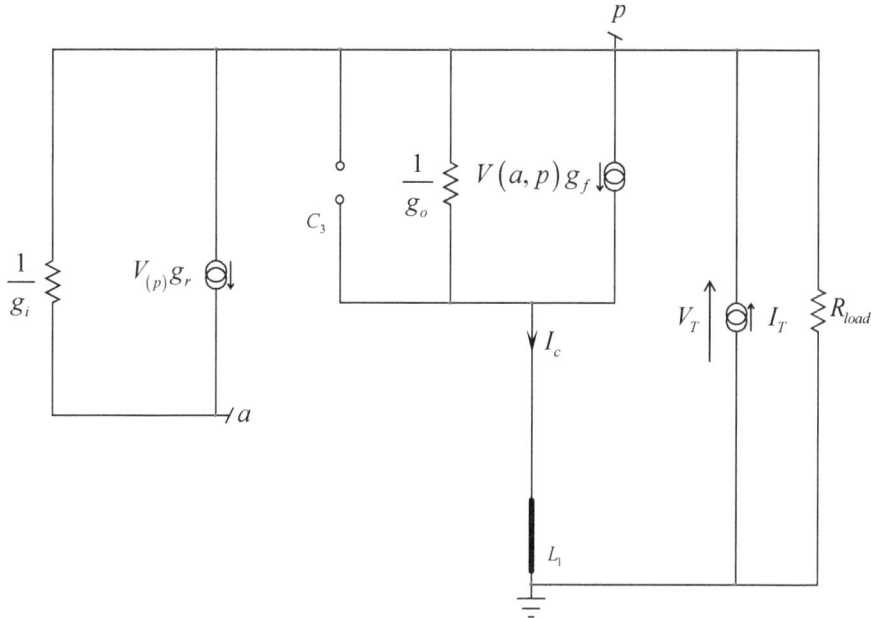

Figure 4.106 The analysis is easier with a grounded c node.

The voltage at node a is defined by:

$$V_{(a)} = V_{(p)} \frac{g_r}{g_i} + V_{(p)}$$

(4.340)

While the voltage at node p involves the combination of g_o and R_{load}:

$$V_{(p)} = \left(\frac{1}{g_o} \parallel R_{load} \right) \left[I_T - g_f \left(V_{(a)} - V_{(p)} \right) \right]$$

(4.341)

Observing the circuit reveals that the voltage at node p is actually V_T. Extracting V_T from the above expression then factoring I_T while bringing r_C back gives us the resistance we want:

$$R_{tau2} = \frac{R_{load} g_i}{R_{load} \left(g_f g_r + g_i g_o \right) + g_i} + r_C$$

(4.342)

The time constant involving C_2 is therefore:

$$\tau_2 = R_{tau2} C_2 = \left(\frac{R_{load} g_i}{R_{load} \left(g_f g_r + g_i g_o \right) + g_i} + r_C \right) C_2$$

(4.343)

For this third time constant determination, capacitor C_2 is open-circuited while L_1 is still a short circuit (Figure 4.107).

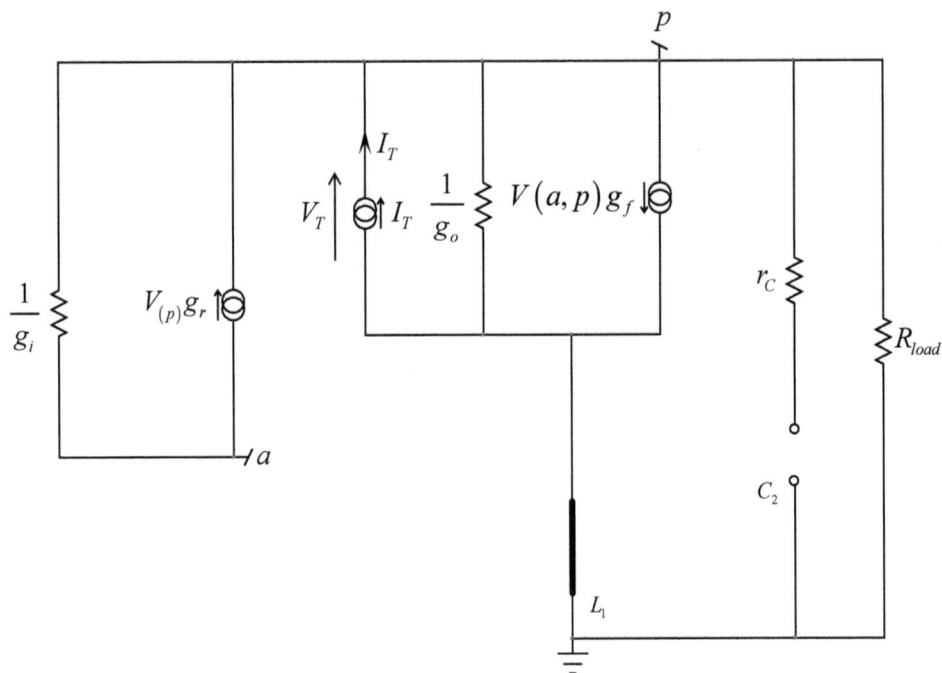

Figure 4.107 For the third time constant, L_1 is still shorted while we look through C_3's connection terminals.

Again, rearranging this sketch in a simpler way is key to determining the time constant we want. Figure 4.108 shows the new drawing. If you compare it with Figure 4.106, you can see both figures are identical.

The resistance driving C_3 is thus the same as that driving C_2 without r_C:

$$R_{tau3} = \frac{R_{load}\,g_i}{R_{load}\left(g_f g_r + g_i g_o\right) + g_i} \tag{4.344}$$

Which leads to:

$$\tau_3 = R_{tau3} C_3 \tag{4.345}$$

With these three time constants on hand, we can form the first-order coefficient b_1:

$$b_1 = \tau_1 + \tau_2 + \tau_3 = \frac{L_1}{\dfrac{R_{load}\left(g_f g_r + g_i g_o\right) + g_i}{g_f g_r + g_i g_o} + r_L} + \left(\frac{R_{load}\,g_i}{R_{load}\left(g_f g_r + g_i g_o\right) + g_i} + r_C\right)C_2 + \left(\frac{R_{load}\,g_i}{R_{load}\left(g_f g_r + g_i g_o\right) + g_i}\right)C_3, \tag{4.346}$$

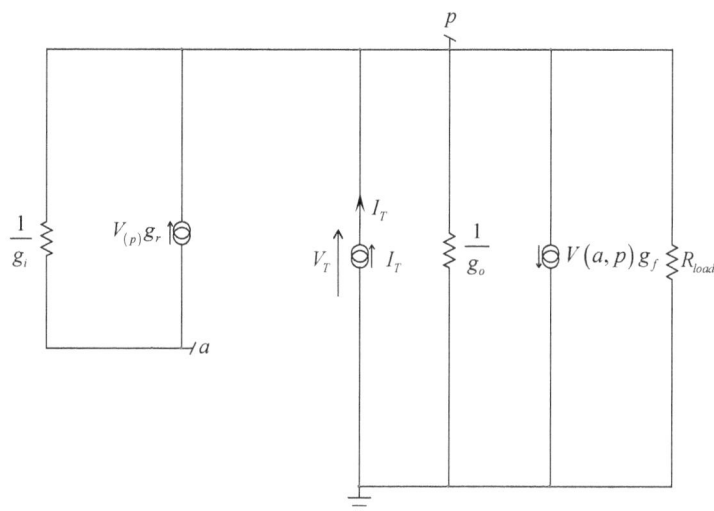

Figure 4.108 The new configuration is similar to that of Figure 4.106

The next step is to determine the second-order coefficient b_2. We start by setting L_1 in its high-frequency state as shown in Figure 4.109. Because I_c is zero in this mode, then the only path I_T can take is R_{load}. The time constant is immediate:

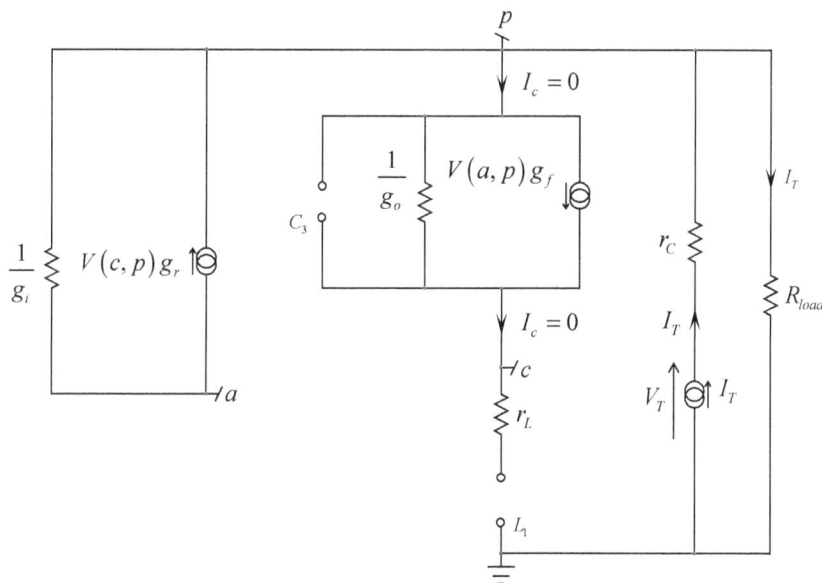

Figure 4.109 L_1 being open-circuited, the resistance we want is determined by inspection.

$$\tau_2^1 = (r_C + R_{load})C_2 \tag{4.347}$$

For the second term, L_1 is still open-circuited but we now "look" through C_3's connecting terminals as drawn in Figure 4.110. We can see that conductance g_o appears in parallel with the resistance we want. We can temporarily disconnect it and bring it back later.

We start by determining the voltage at node a:

$$V_{(a)} = -\left[V_{(c)} - V_{(p)}\right]\frac{g_r}{g_i} + V_{(p)} \tag{4.348}$$

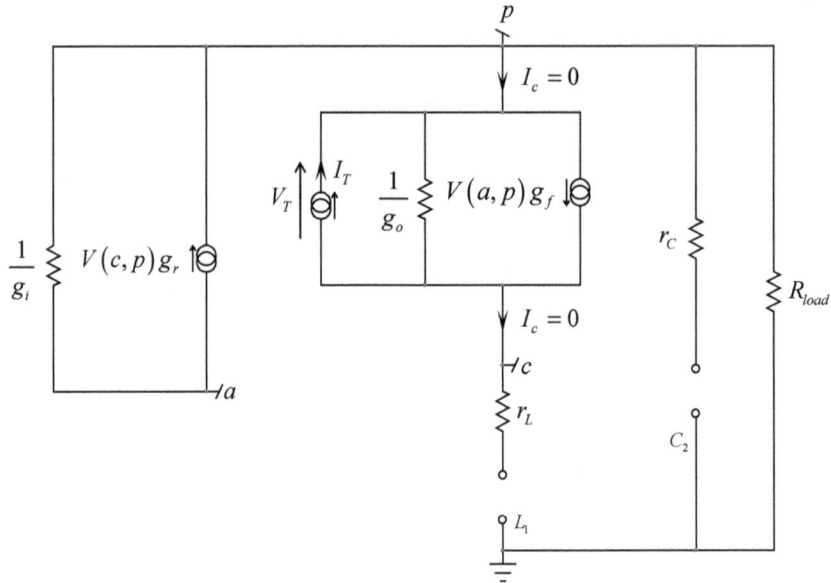

Figure 4.110 We now determine the resistance "seen" from C_3's connecting terminals while L_1 is still open-circuited.

Because there is no current flowing in R_{load} (L_1 is open-circuited and node a has no path to ground), the voltage at node p is 0 V:

$$V_{(p)} = 0 \qquad (4.349)$$

Observing the circuit also leads to write:

$$V_{(c)} = -V_T \qquad (4.350)$$

Finally, current I_T depends on node a and conductance g_f:

$$I_T = V_{(a)} g_f = \left(-[-V_T - 0]\frac{g_r}{g_i} + 0 \right) g_f \qquad (4.351)$$

If you solve for V_T and factor I_T then bring g_o back, you should find:

$$R_{tau13} = \frac{V_T}{I_T} = \left(\frac{g_i}{g_f g_r} \,\middle\|\, \frac{1}{g_o} \right) \qquad (4.352)$$

Which leads to:

$$\tau_3^1 = \left(\frac{g_i}{g_f g_r} \,\middle\|\, \frac{1}{g_o} \right) C_3 \qquad (4.353)$$

The next coefficient implies that capacitor C_2 is replaced by a short circuit as well as L_1 (dc state). The schematic diagram is that of Figure 4.112 that we have further simplified and rearranged in Figure 4.113 by neglecting r_L.

An equivalent resistance R_{eq} can be formed by assembling r_C, R_{load} and conductance g_o in parallel:

$$R_{eq} = \frac{1}{g_o} \,\|\, R_{load} \,\|\, r_C \qquad (4.354)$$

Then the voltage at node a is determined as:

$$V_{(a)} = -\left[V_{(c)} - V_{(p)}\right]\frac{g_r}{g_i} + V_{(p)} \qquad (4.355)$$

$V_{(c)}$ being equal to 0 V and $V_{(p)}$ equals V_T, we can write:

$$V_{(a)} = -\left[0 - V_{(p)}\right]\frac{g_r}{g_i} + V_{(p)} = V_T\left(\frac{g_i + g_r}{g_i}\right) \qquad (4.356)$$

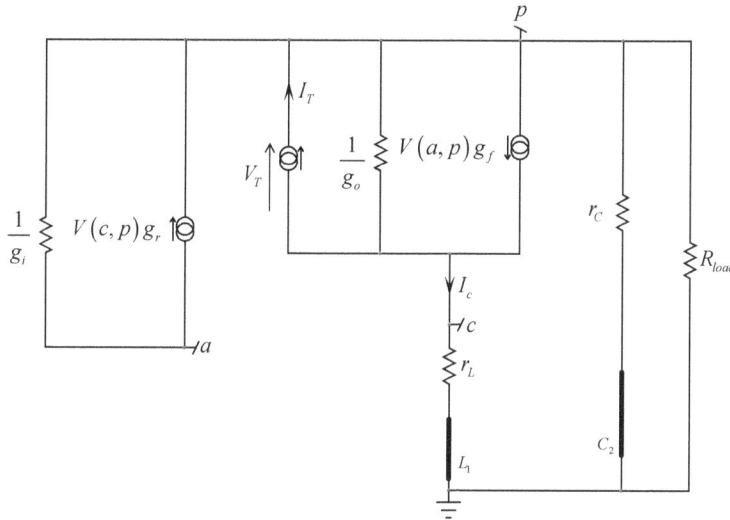

Figure 4.111 L_1 and C_2 are now replaced by short circuits in this new time constant to determine.

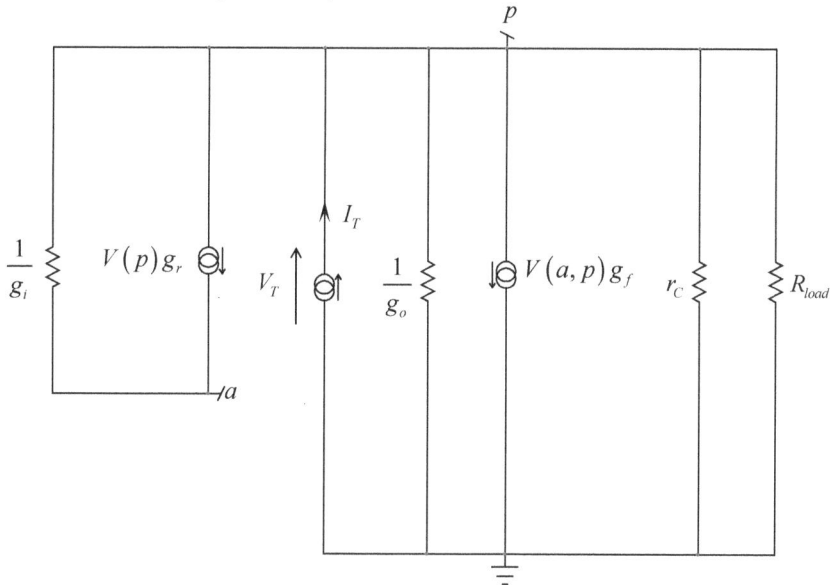

Figure 4.112 Neglecting r_L lets us ground node c which further simplifies the circuitry.

Voltage V_T can also be defined as:

$$V_T = R_{eq}\left(I_T - V_{(a)}g_f - V_T g_f\right) \qquad (4.357)$$

531

Now substitute (4.356) in the above equation and solve for V_T while factoring I_T:

$$R_{tau\,23} = \frac{V_T}{I_T} = \frac{R_{eq}g_i}{g_i\left(2R_{eq}g_f+1\right)+R_{eq}g_f g_r} \tag{4.358}$$

This resistance combined with C_3 gives a time constant equal to:

$$\tau_3^2 = \frac{R_{eq}g_i}{g_i\left(2R_{eq}g_f+1\right)+R_{eq}g_f g_r}C_3 \tag{4.359}$$

These new time constants let us express the second-order term b_2:

$$b_2 = \tau_1\tau_2^1 + \tau_1\tau_3^1 + \tau_2\tau_3^2 = \frac{L_1}{\dfrac{R_{load}\left(g_f g_r + g_i g_o\right)+g_i}{g_f g_r + g_i g_o}+r_L}\left(r_C + R_{load}\right)C_2$$

$$+ \frac{L_1}{\dfrac{R_{load}\left(g_f g_r + g_i g_o\right)+g_i}{g_f g_r + g_i g_o}+r_L}\left(\frac{g_i}{g_f g_r}\parallel\frac{1}{g_o}\right)C_3 \tag{4.360}$$

$$+ \left(\frac{R_{load}g_i}{R_{load}\left(g_f g_r + g_i g_o\right)+g_i}+r_C\right)C_2\,\frac{R_{eq}g_i}{g_i\left(2R_{eq}g_f+1\right)+R_{eq}g_f g_r}C_3$$

For the last step, L_1 and C_2 are set in their high-frequency state (L_1 is open-circuited and C_2 is a short circuit). Figure 4.113 illustrates this last circuit which is similar to that of Figure 4.110: whether C_2 is open or shorted does not change anything as the voltage at node p is 0 V.

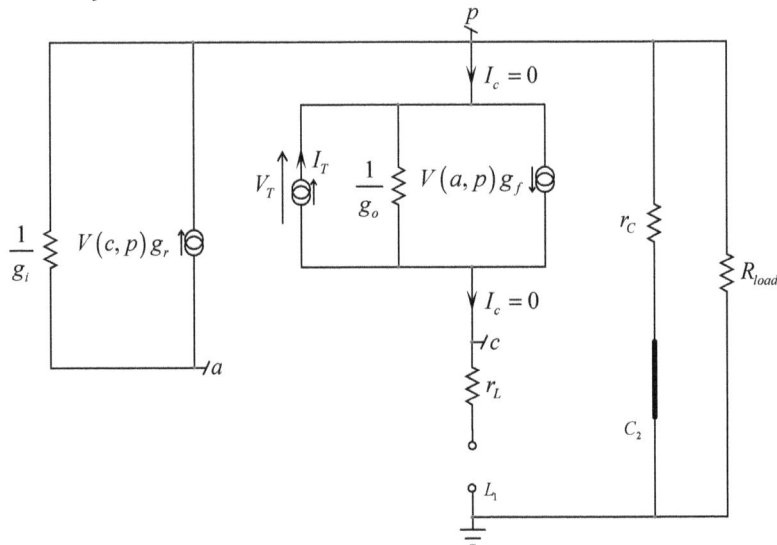

Figure 4.113 Opening L_1 nulls the I_c current which helps for the analysis.

In this case, the last time constant is simply:

$$\tau_3^{12} = \left(\frac{g_i}{g_f g_r}\parallel\frac{1}{g_o}\right)C_3 \tag{4.361}$$

The last third-order term is obtained by combining the previous time constants as follows:

$$b_3 = \tau_1 \tau_2^1 \tau_3^{12} = \frac{L_1}{\dfrac{R_{load}\left(g_f g_r + g_i g_o\right) + g_i}{g_f g_r + g_i g_o} + r_L} - \left(r_C + R_{load}\right)C_2\left(\frac{g_i}{g_f g_r} \parallel \frac{1}{g_o}\right)C_3 \tag{4.362}$$

The denominator of the input impedance transfer function is now determined and equal to:

$$D(s) = 1 + sb_1 + s^2 b_2 + s^3 b_3 \tag{4.363}$$

If we now null the response in Figure 4.98, the circuit returns to its natural state and the denominator determined in (4.258) now becomes our numerator:

$$N(s) \approx \left(1 + \frac{s}{\omega_{z_1}}\right)\left[1 + \frac{s}{\omega_N Q_N} + \left(\frac{s}{\omega_N}\right)^2\right] \tag{4.364}$$

where:

$$\omega_{z_1} \approx \frac{\dfrac{(1-D)^3}{2\tau_L}\left(1 + 2\dfrac{S_e}{S_n}\right) + 1 + D}{R_{load}C_2} \tag{4.365}$$

$$Q_N = \frac{1}{\pi(m_c D' - 0.5)} \tag{4.366}$$

$$\omega_N = \frac{\pi}{T_{sw}} \tag{4.367}$$

$$m_c = 1 + \frac{S_e}{S_n} \tag{4.368}$$

The input impedance of the CCM CM buck-boost converter has now been entirely determined:

$$Z_{in}(s) = R_0 \frac{N(s)}{D(s)} \tag{4.369}$$

In which R_0 is defined by (4.331).

For the final lap, we will test the SPICE simulation results of Figure 4.83 against the Mathcad® plots of (4.369). As Figure 4.114 shows, the matching is excellent confirming our analysis of the input impedance.

This exercise ends the analysis of the four transfer functions for the buck-boost converter operated in current-mode control and I have gathered all the formulas in the table of Figure 4.115. Figure 4.116 reproduces the small-signal parameters of the CM PWM switch model.

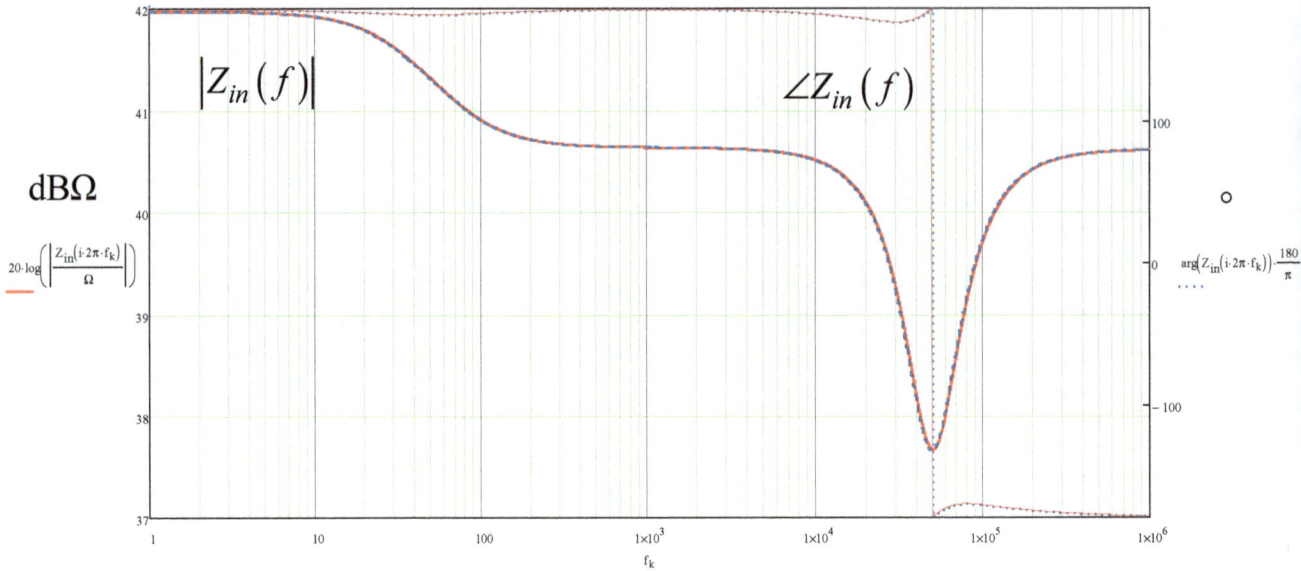

Figure 4.114 SPICE simulations and Mathcad® curves are perfectly matching.

$\dfrac{V_{out}(s)}{V_{err}(s)}$ Control to Output	$H_0\dfrac{\left(1+\dfrac{s}{\omega_{z_1}}\right)\left(1-\dfrac{s}{\omega_{z_2}}\right)}{\left(1+\dfrac{s}{\omega_{p_1}}\right)\dfrac{1}{1+\dfrac{s}{\omega_n Q_p}+\left(\dfrac{s}{\omega_n}\right)^2}}$	$\omega_{z_1}=\dfrac{1}{r_C C_2}$ $\omega_{z_2}\approx\dfrac{(1-D)^2 R_{load}}{D\cdot L_1}$	$\omega_{p_1}\approx\dfrac{\dfrac{(1-D)^3}{2\tau_L}\left(1+2\dfrac{S_e}{S_n}\right)+1+D}{R_{load}C_2}$ $Q_p=\dfrac{1}{\pi(m_c D'-0.5)}\quad \omega_n=\dfrac{\pi}{T_{sw}}$	$H_0\approx-\dfrac{R_{load}}{R_i}\dfrac{1}{\dfrac{(1-D)^2}{2\tau_L}\left(1+2\dfrac{S_e}{S_n}\right)+2M+1}$
$\dfrac{V_{out}(s)}{V_{in}(s)}$ Input to Output	$H_0\dfrac{\left(1+\dfrac{s}{\omega_{z_1}}\right)\left[1+\dfrac{s}{Q_N\omega_{0N}}+\left(\dfrac{s}{\omega_{0N}}\right)^2\right]}{\left(1+\dfrac{s}{\omega_{p_1}}\right)\dfrac{1}{1+\dfrac{s}{\omega_n Q_p}+\left(\dfrac{s}{\omega_n}\right)^2}}$	$\omega_{z_1}=\dfrac{1}{r_C C_2}\quad \omega_{0N}=\dfrac{1}{\sqrt{\dfrac{L_1 C_3 g_i}{g_i-g_f}}}$ $Q_N=\dfrac{(g_i-g_f)\sqrt{\dfrac{L_1 C_3 g_i}{g_i-g_f}}}{L_1(g_f g_r+g_i g_o)}$	$\omega_{p_1}\approx\dfrac{\dfrac{(1-D)^3}{2\tau_L}\left(1+2\dfrac{S_e}{S_n}\right)+1+D}{R_{load}C_2}$ $Q_p=\dfrac{1}{\pi(m_c D'-0.5)}\quad \omega_n=\dfrac{\pi}{T_{sw}}$	$H_0=-\dfrac{R_{load}(g_f-g_i)}{R_{load}(g_i-g_f+g_0+g_r)+1}$
$Z_{in}(s)$ Input impedance	$R_0\dfrac{\left(1+\dfrac{s}{\omega_{z_1}}\right)\dfrac{1}{1+\dfrac{s}{\omega_N Q_N}+\left(\dfrac{s}{\omega_N}\right)^2}}{D(s)}$	$\omega_{z_1}\approx\dfrac{\dfrac{(1-D)^3}{2\tau_L}\left(1+2\dfrac{S_e}{S_n}\right)+1+D}{R_{load}C_2}$ $Q_N=\dfrac{1}{\pi(m_c D'-0.5)}\quad \omega_N=\dfrac{\pi}{T_{sw}}$	See text for coefficients values $D(s)=1+sb_1+s^2 b_2+s^3 b_3$	$R_0=\dfrac{(g_i-g_f+g_o+g_r)R_{load}+1}{(g_f g_r+g_i g_o)R_{load}+g_i}$
$Z_{out}(s)$ Output impedance	$R_0\dfrac{1+\dfrac{s}{\omega_{z_1}}}{1+\dfrac{s}{\omega_{p_1}}}$	$\omega_{z_1}=\dfrac{1}{r_C C_2}$	$\omega_{p_1}\approx\dfrac{\dfrac{(1-D)^3}{2\tau_L}\left(1+2\dfrac{S_e}{S_n}\right)+1+D}{R_{load}C_2}$ $Q_p=\dfrac{1}{\pi(m_c D'-0.5)}\quad \omega_n=\dfrac{\pi}{T_{sw}}$	$R_0=\dfrac{1}{g_r-g_f}\left\|\dfrac{1}{g_i}\right\|\dfrac{1}{g_r}\|R_{load}$

$$m_c=1+\frac{S_e}{S_n}\qquad S_e \text{ in V/s}\qquad S_n=\frac{V_{in}}{L_1}R_i\qquad M=\left|\frac{V_{out}}{V_{in}}\right|\qquad \tau_L=\frac{L_1}{R_{load}}F_{sw}$$

Figure 4.115 This table conveniently gathers the four transfer functions of the CM CCM buck-boost converter.

$V_{in} := 40V$ $V_{out} := 12.3016V$ $R_i := 0.5\Omega$ $C_2 := 470\mu F$ $r_C := 0.07\Omega$

$F_{sw} := 100kHz$ $T_{sw} := \frac{1}{F_{sw}}$ $r_L := 0.01\Omega$ $L_1 := 250\mu H$ $R_L := 8\Omega$

$V_{ap} := V_{in} + V_{out}$ $V_{cp} := V_{out}$ $V_{ac} := V_{in}$ $D := \frac{V_{cp}}{V_{ap}} = 0.23521$

$V_c := 1.1V$ $S_a := 0\frac{V}{s}$ $D_p := 1 - D = 0.76479$ $D = 0.23521$ $M := \frac{V_{out}}{V_{in}} = 0.30754$

$S_n := \frac{V_{ac}}{L_1}\cdot R_i = 80\cdot\frac{kV}{s}$ $S_f := \frac{V_{ap}}{L_1}\cdot R_i = 104.6032\frac{kV}{s}$ $C_s := \frac{1}{L_1\cdot(F_{sw}\cdot\pi)^2} = 40.52847nF$

$I_c := \frac{V_c}{R_i} - V_{cp}\cdot\left(1 - \frac{V_{cp}}{V_{ap}}\right)\frac{T_{sw}}{2\cdot L_1} - \frac{S_a}{R_i}\cdot\frac{V_{cp}}{V_{ap}}\cdot T_{sw} = 2.01184A$ $m_c := 1 + \frac{S_a}{S_n} = 1$

$I_a := \frac{V_{cp}}{V_{ap}}\left[\frac{V_c}{R_i} - V_{cp}\cdot(1 - D)\cdot\frac{T_{sw}}{2\cdot L_1} - \frac{S_a}{R_i}\cdot D\cdot T_{sw}\right] = 0.47319A$ $\parallel(x,y) := \frac{x\cdot y}{x+y}$

$\omega_n := \frac{\pi}{T_{sw}}$ $f_n := \frac{\omega_n}{2\cdot\pi} = 50\cdot kHz$ $Q_p := \frac{1}{\pi\cdot[m_c\cdot(1-D)-0.5]} = 1.2021$ $\tau_L := \frac{L_1}{R_L}\cdot F_{sw}$

$g_{oo} := \frac{T_{sw}\left(\frac{V_{cp}}{V_{ap}}-1\right)}{2\cdot L_1} + \frac{T_{sw}\cdot V_{cp}}{2\cdot L_1\cdot V_{ap}} - \frac{S_a\cdot T_{sw}}{R_i\cdot V_{ap}} = -0.01059\frac{1}{\Omega}$

$k_{oo} := \frac{1}{R_i} = 2\frac{1}{\Omega}$

$g_{ff} := \frac{S_a\cdot T_{sw}\cdot V_{cp}}{R_i\cdot V_{ap}^2} - \frac{T_{sw}\cdot V_{cp}^2}{2\cdot L_1\cdot V_{ap}^2} = -1.10643\times10^{-3}\frac{1}{\Omega}$

$k_{ii} := \frac{D}{R_i} = 0.47041\frac{1}{\Omega}$

$g_{ii} := D\cdot\left(\frac{S_a\cdot T_{sw}\cdot D}{R_i\cdot V_{ap}} - \frac{T_{sw}\cdot D^2}{2\cdot L_1}\right) - \frac{D\cdot I_c}{V_{ap}} = -9.30765\times10^{-3}\frac{1}{\Omega}$

$g_{rr} := \frac{I_c}{V_{ap}} + D\cdot\left[\frac{T_{sw}\left(\frac{V_{cp}}{V_{ap}}-1\right)}{2\cdot L_1} + \frac{T_{sw}\cdot V_{cp}}{2\cdot L_1\cdot V_{ap}} - \frac{S_a\cdot T_{sw}}{R_i\cdot V_{ap}}\right] = 0.03597\frac{1}{\Omega}$

$g_o := \left[(1-D)\cdot\frac{S_a}{S_n} + \frac{1}{2} - D\right]\frac{T_{sw}}{L_1} = 0.01059\frac{1}{\Omega}$

$k_o := \frac{1}{R_i} = 2\frac{1}{\Omega}$

$g_f := D\cdot g_o - \frac{D\cdot D_p\cdot T_{sw}}{2\cdot L_1} = -1.10643\times10^{-3}\frac{1}{\Omega}$

$k_i := \frac{D}{R_i} = 0.47041\frac{1}{\Omega}$

$g_i := D\cdot\left(g_f - \frac{I_c}{V_{ap}}\right) = -9.30765\times10^{-3}\frac{1}{\Omega}$

$g_r := \frac{I_c}{V_{ap}} - g_o\cdot D = 0.03597\frac{1}{\Omega}$

Figure 4.116 These are the CM PWM switch small-signal parameters used in the above examples.

4.4 Buck-Boost Transfer Functions in DCM – Fixed-Frequency Current-Mode Control

The load resistance has now been increased to 100 ohms and the converter enters DCM. The duty ratio is driven by a dedicated source as shown in Figure 4.117 which represents the large-signal version of the DCM VM PWM switch model derived in Chapter 1.4.2. B_1 determines the duty ratio based on the inductor on-slope and the compensation ramp S_e. For a 700-mV control voltage, the converter delivers -7.8 V as indicated by the bias points on the schematic diagram.

The linear version of this circuit consists of replacing the PWM switch model by its small-signal version as drawn in Figure 4.118.

For simplicity reason, the inductor ohmic loss r_L has been neglected. It is important to verify that magnitude and phase responses are identical before carrying on with the analysis. The plots of Figure 4.119 confirm the approach is correct. As a side note, verifying the control-to-output transfer function only is sometimes not enough to confirm the validity of the small-signal model and surprises can come from other transfer functions. It is thus a good practice to give a quick look at the other configurations (Z_{in}, Z_{out} and V_{out}/V_{in}) before concluding the model is correct.

It is now time to simplify and rearrange the circuit in a more friendly way. Realizing that nodes a and cc are at a 0-V potential (V_{in} is zeroed in this analysis), some of the sources can just disappear while other see their polarity reversed.

The final circuit to study the control-to-output transfer function appears in Figure 4.121 and we have made sure its response is identical to that reproduced in Figure 4.119.

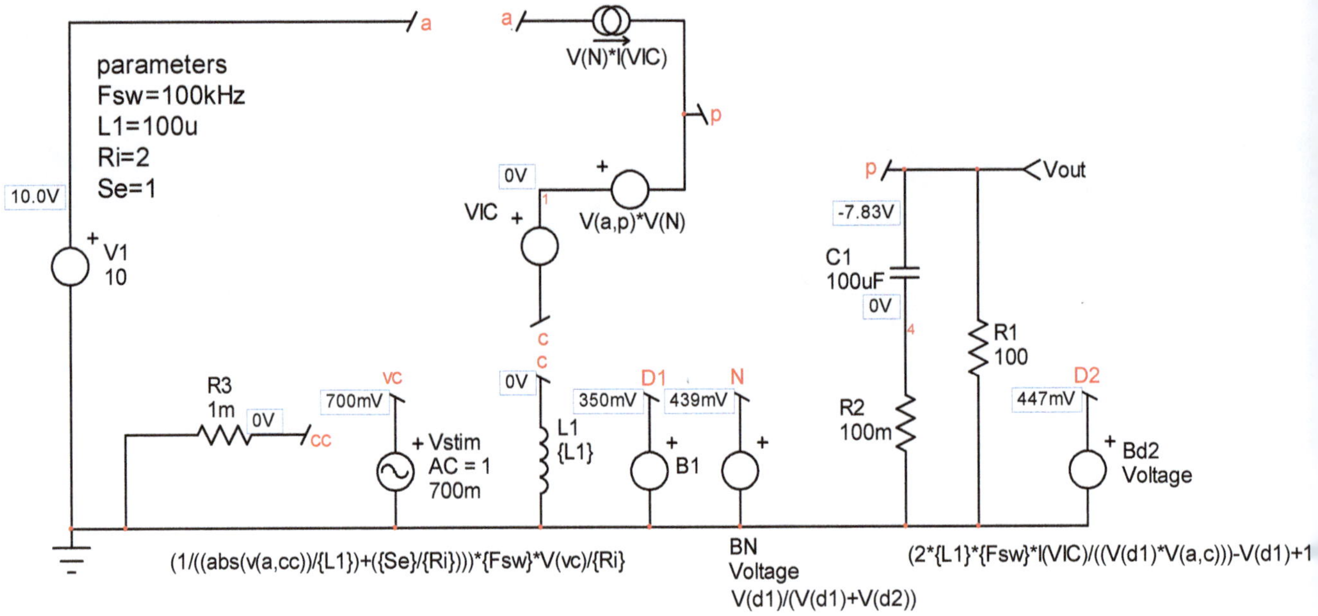

Figure 4.117

parameters
Fsw=100kHz
L1=100u
Ri=2
Se=1

10.0V

V1
10

V(N)*I(VIC)

a

a

p

0V

VIC +

V(a,p)*V(N)

p

Vout

-7.83V

C1
100uF

0V

R1
100

R3
1m

0V

VC

700mV

CC

0V

c

c

D1

350mV

N

439mV

R2
100m

D2

447mV

+ Vstim
AC = 1
700m

L1
{L1}

+
B1

+

+ Bd2
Voltage

$$(1/((abs(v(a,cc))/\{L1\})+(\{Se\}/\{Ri\})))*\{Fsw\}*V(vc)/\{Ri\}$$

BN
Voltage
V(d1)/(V(d1)+V(d2))

$$(2*\{L1\}*\{Fsw\}*I(VIC)/((V(d1)*V(a,c)))-V(d1)+1$$

Figure 4.117 The converter now operates in the discontinuous conduction mode or DCM.

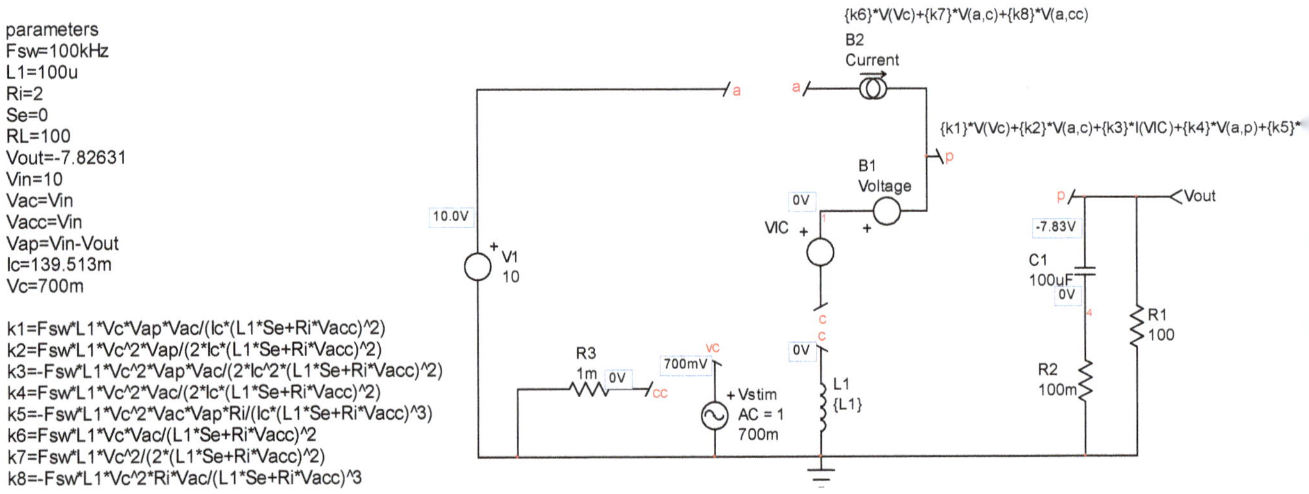

parameters
Fsw=100kHz
L1=100u
Ri=2
Se=0
RL=100
Vout=-7.82631
Vin=10
Vac=Vin
Vacc=Vin
Vap=Vin-Vout
Ic=139.513m
Vc=700m

$k1=Fsw*L1*Vc*Vap*Vac/(Ic*(L1*Se+Ri*Vacc)^2)$
$k2=Fsw*L1*Vc^2*Vap/(2*Ic*(L1*Se+Ri*Vacc)^2)$
$k3=-Fsw*L1*Vc^2*Vac/(2*Ic^2*(L1*Se+Ri*Vacc)^2)$
$k4=Fsw*L1*Vc^2*Vac/(2*Ic*(L1*Se+Ri*Vacc)^2)$
$k5=-Fsw*L1*Vc^2*Vac*Vap*Ri/(Ic*(L1*Se+Ri*Vacc)^3)$
$k6=Fsw*L1*Vc*Vac/(L1*Se+Ri*Vacc)^2$
$k7=Fsw*L1*Vc^2/(2*(L1*Se+Ri*Vacc)^2)$
$k8=-Fsw*L1*Vc^2*Ri*Vac/(L1*Se+Ri*Vacc)^3$

$\{k6\}*V(Vc)+\{k7\}*V(a,c)+\{k8\}*V(a,cc)$
B2
Current

$\{k1\}*V(Vc)+\{k2\}*V(a,c)+\{k3\}*I(VIC)+\{k4\}*V(a,p)+\{k5\}*$

a

a

p

0V

B1
Voltage

+

10.0V

V1
10

VIC +

p

Vout

-7.83V

C1
100uF

0V

R1
100

R3
1m

0V

VC

700mV

CC

0V

c

c

+ Vstim
AC = 1
700m

L1
{L1}

R2
100m

Figure 4.118 The small-signal version of the VM DCM PWM switch model is now installed in the circuit.

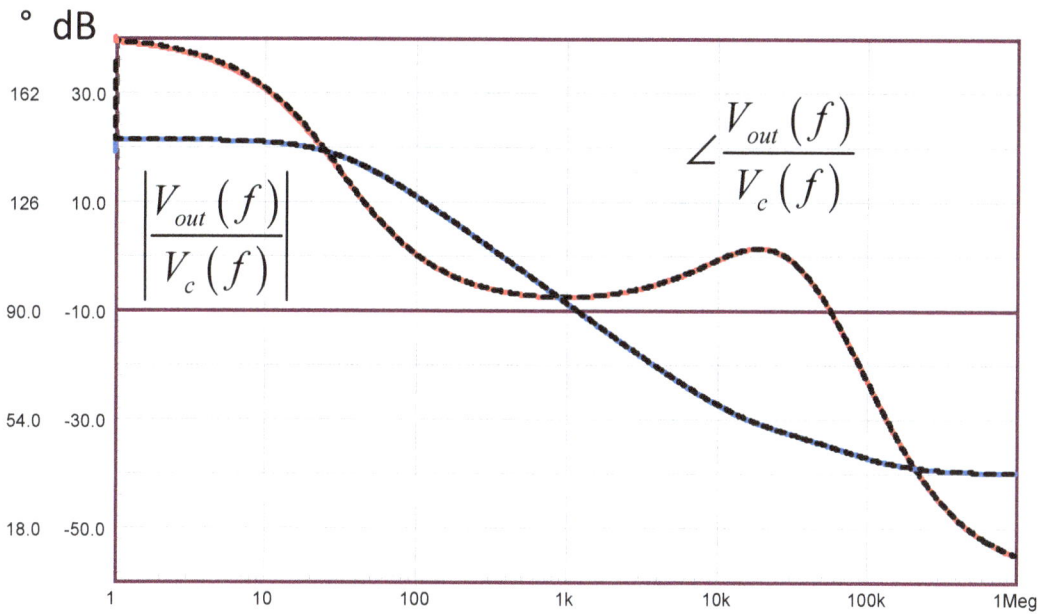

Figure 4.119 Magnitude and phase responses of the large- and small-signal versions of the DCM model are identical.

As I did in the previous paragraphs and chapters, I included below all the small-signal coefficients for the CM PWM switch model operated in discontinuous conduction mode (Figure 4.120).

$V_{in} := 10V$ $V_{out} := -7.8263\,\text{IV}$ $R_i := 2\Omega$ $C_{out} := 100\mu F$ $r_C := 0.1\Omega$ $R_{inf} := 10^{10}\Omega$

$$k_1 := \frac{F_{sw} \cdot L_1 \cdot V_c \cdot V_{ac} \cdot V_{ap}}{I_c \cdot (L_1 \cdot S_e + R_i \cdot V_{acc})^2} = 22.36065 \qquad k_6 := \frac{F_{sw} \cdot L_1 \cdot V_c \cdot V_{ac}}{(L_1 \cdot S_e + R_i \cdot V_{acc})^2} = 0.175\frac{1}{\Omega}$$

$F_{sw} := 100\text{kHz}$ $T_{sw} := \dfrac{1}{F_{sw}}$ $r_L := 0.001\text{S}\Omega$ $L_1 := 100\,\mu H$ $R_L := 100\Omega$

$V_{ap} := V_{in} - V_{out}$ $V_{cp} := -V_{out}$ $V_{ac} := V_{in}$ $V_{acc} := V_{ac}$ positive on-slope always

$$k_2 := \frac{F_{sw} \cdot L_1 \cdot V_c^2 \cdot V_{ap}}{2 \cdot I_c \cdot (L_1 \cdot S_e + R_i \cdot V_{acc})^2} = 0.78262 \qquad k_7 := \frac{F_{sw} \cdot L_1 \cdot V_c^2}{2 \cdot (L_1 \cdot S_e + R_i \cdot V_{acc})^2} = 6.125 \times 10^{-3}\frac{1}{\Omega}$$

$V_c := 700\text{mV}$ $S_e := 0\dfrac{kV}{s}$ $M := \dfrac{V_{out}}{V_{in}} = -0.78263$ $\|(x,y) := \dfrac{x \cdot y}{x+y}$

$$k_3 := \frac{F_{sw} \cdot L_1 \cdot V_c^2 \cdot V_{ac} \cdot V_{ap}}{2 \cdot I_c^2 \cdot (L_1 \cdot S_e + R_i \cdot V_{acc})^2} = -56.09673\Omega \qquad k_8 := \frac{F_{sw} \cdot L_1 \cdot R_i \cdot V_c^2 \cdot V_{ac}}{(L_1 \cdot S_e + R_i \cdot V_{acc})^3} = -0.01225\frac{1}{\Omega}$$

$S_n := \dfrac{V_{ac}}{L_1} \cdot R_i = 200\dfrac{kV}{s}$ $S_f := \dfrac{V_{ap}}{L_1} \cdot R_i = 356.5262\dfrac{kV}{s}$

$$k_4 := \frac{F_{sw} \cdot L_1 \cdot V_c^2 \cdot V_{ac}}{2 \cdot I_c \cdot (L_1 \cdot S_e + R_i \cdot V_{acc})^2} = 0.43903$$

$D_1 := \dfrac{F_{sw} \cdot V_c}{R_i} \cdot \dfrac{1}{\dfrac{V_{ac}}{L_1} + \dfrac{S_e}{R_i}} = 35\cdot\%$ $\tau_L := \dfrac{L_1}{R_L \cdot T_{sw}} = 0.1$ $m_c := 1 + \dfrac{S_e}{S_n} = 1$

$$k_5 := \frac{F_{sw} \cdot L_1 \cdot R_i \cdot V_c^2 \cdot V_{ac} \cdot V_{ap}}{I_c \cdot (L_1 \cdot S_e + R_i \cdot V_{acc})^3} = -1.56525$$

$I_c := \dfrac{V_{in} \cdot D_1^2}{2 \cdot F_{sw} \cdot L_1} - \dfrac{V_{out}}{R_L} = 139.5131\text{mA}$ $I_a := D_1 \cdot I_c = 48.82958\text{mA}$

Figure 4.120 These are all the *k* coefficients for the small-signal model of the CM PWM switch model used in DCM.

Our first step targets the quasi-static gain H_0 linking the control voltage V_c to V_{out}.

We start with the updated electrical schematic diagram of Figure 4.122 which turns into that of Figure 4.123 where the inductor is replaced by a short circuit and the output capacitor is open-circuited.

parameters
Fsw=100kHz
L1=100u
Ri=2
Se=0
RL=100
Vout=-7.82631
Vin=10
Vac=Vin
Vacc=Vin
Vap=Vin-Vout
Ic=139.513m
Vc=700m

k1=Fsw*L1*Vc*Vap*Vac/(Ic*(L1*Se+Ri*Vacc)^2)
k2=Fsw*L1*Vc^2*Vap/(2*Ic*(L1*Se+Ri*Vacc)^2)
k3=-Fsw*L1*Vc^2*Vap*Vac/(2*Ic^2*(L1*Se+Ri*Vacc)^2)
k4=Fsw*L1*Vc^2*Vac/(2*Ic*(L1*Se+Ri*Vacc)^2)
k5=-Fsw*L1*Vc^2*Vac*Vap*Ri/(Ic*(L1*Se+Ri*Vacc)^3)
k6=Fsw*L1*Vc*Vac/(L1*Se+Ri*Vacc)^2
k7=Fsw*L1*Vc^2/(2*(L1*Se+Ri*Vacc)^2)
k8=-Fsw*L1*Vc^2*Ri*Vac/(L1*Se+Ri*Vacc)^3

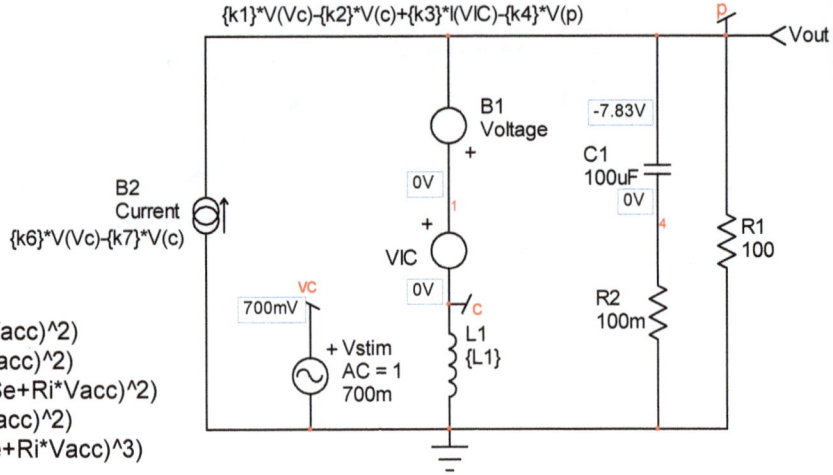

Figure 4.121 The circuit now looks simpler to analyse.

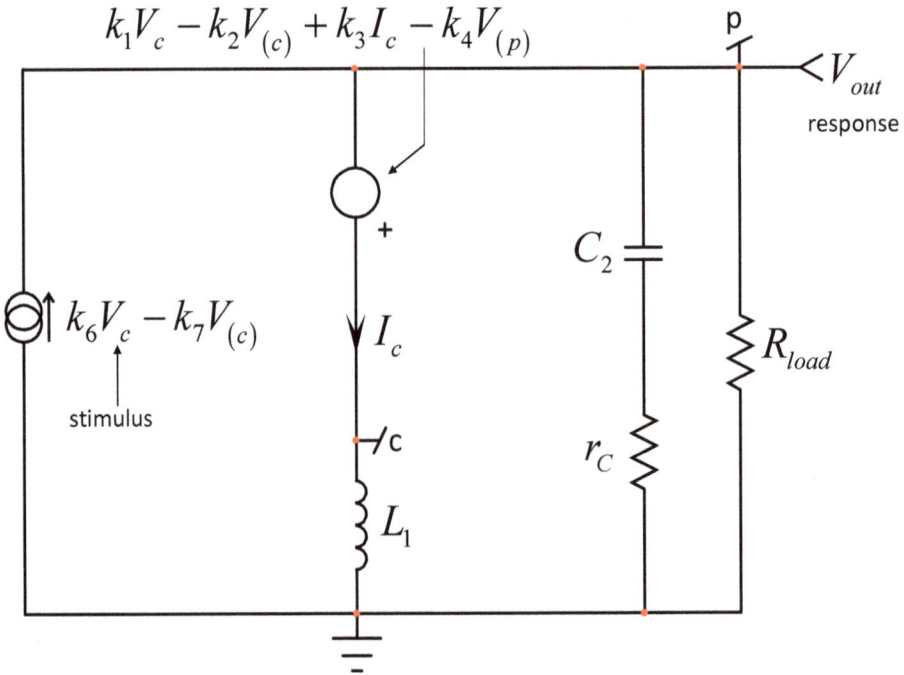

Figure 4.122 We will study the control-to-output transfer function with this second-order circuit.

We start with a first equation linking V_{out} to R_{load}:

$$-V_{out} = I_1 R_{load} \tag{4.370}$$

In which:

$$I_1 = I_c - k_6 V_c \tag{4.371}$$

We know that node p is biased at the output voltage:

$$V_{(p)} = V_{out} \qquad (4.372)$$

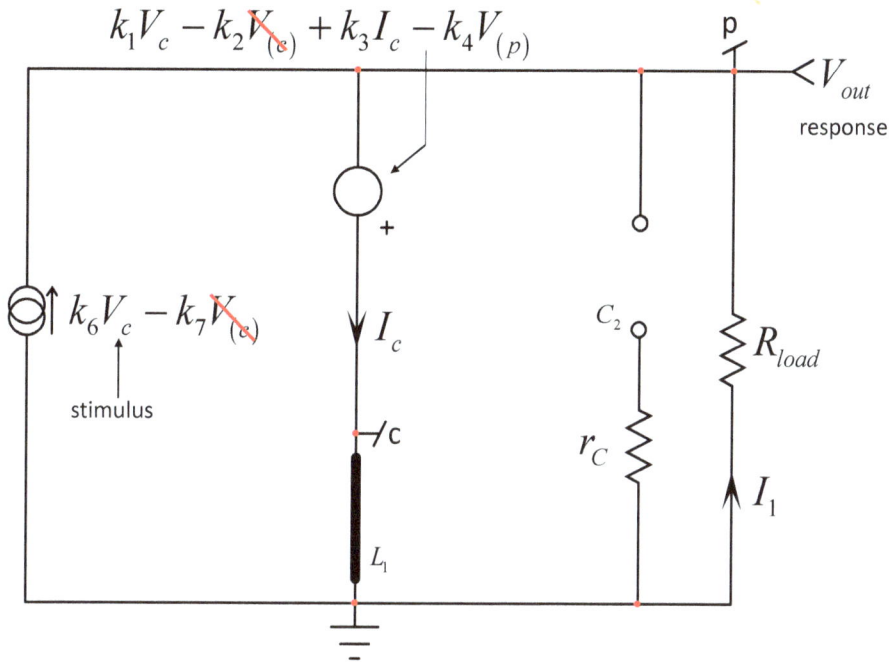

Figure 4.123 The shorted inductor grounds node c which simplifies some of the sources.

We can therefore write the following equality:

$$-V_{out} = k_1 V_c + k_3 I_c - k_4 V_{(p)} = k_1 V_c + k_3 I_c - k_4 V_{out} \qquad (4.373)$$

From which we extract current I_c:

$$I_c = -\frac{V_{out} + V_c k_1 - V_{out} k_4}{k_3} \qquad (4.374)$$

If we substitute this new definition in (4.371) then, using the result in (4.370) we obtain this intermediate result:

$$-V_{out} = \left(-\frac{V_{out} + V_c k_1 - V_{out} k_4}{k_3} + k_6 V_c \right) R_{load} \qquad (4.375)$$

Extracting V_{out} from the above expression while factoring V_c, we obtain the gain we want:

$$H_0 = \frac{V_{out}}{V_c} = \frac{R_{load} (k_1 + k_3 k_6)}{k_3 - R_{load} (1 - k_4)} \qquad (4.376)$$

In the literature [4], this expression has been elegantly reworked to become:

$$H_0 \approx -\frac{1}{S_n m_c T_{sw}} \frac{V_{in}}{\sqrt{2 \tau_L}} \qquad (4.377)$$

where:

$$S_n = \frac{V_{in}}{L_1} R_i \tag{4.378}$$

$$m_c = 1 + \frac{S_e}{S_n} \tag{4.379}$$

S_e is the external compensation ramp expressed in volts per second. The normalized time constant τ_L is defined as:

$$\tau_L = \frac{L_1}{R_{load}} F_{sw} \tag{4.380}$$

Now that we have the quasi-static gain, we can carry on with the first time constant involving L_1. To determine the resistances seen from L_1 and C_2's connection, we reduce the excitation V_c to zero and simplify sources hosting this term.

The updated schematic for τ_1 is shown in Figure 4.124.

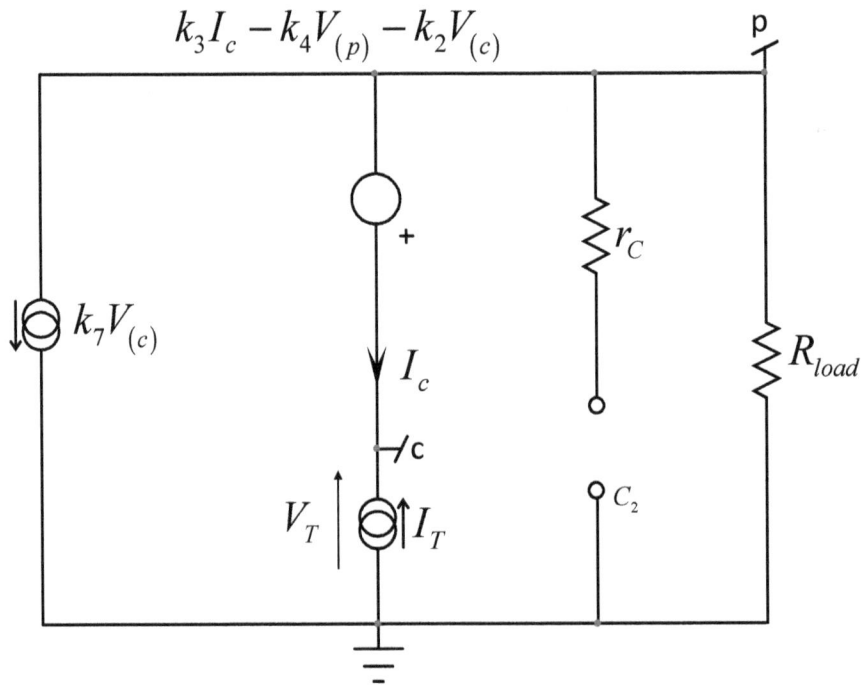

Figure 4.124 A test current installed across the inductor connections will let us determine the resistance driving L_1.

Inspecting the schematic, we recognize that:

$$V_{(c)} = V_T \tag{4.381}$$

and:

$$I_c = -I_T \tag{4.382}$$

Capitalizing on these definitions, we can express the voltage at terminal p as follows:

$$V_{(p)} = R_{load}\left(I_T - k_7 V_T\right) \tag{4.383}$$

The voltage at this node can also be defined by writing:

$$V_{(p)} - k_3 I_T - k_4 V_{(p)} - k_2 V_T = V_T \tag{4.384}$$

Extracting $V_{(p)}$ from this formula leads to:

$$V_{(p)} = -\frac{V_T + I_T k_3 + V_T k_2}{k_4 - 1} \tag{4.385}$$

Now equating (4.383) with (4.385) then solving for V_T while factoring I_T gives us the resistance driving L_1:

$$R_{tau1} = \frac{V_T}{I_T} = \frac{R_{load}\left(1-k_4\right)-k_3}{1+k_2+R_{load}k_7\left(1-k_4\right)} \tag{4.386}$$

Which leads to τ_1, the time constant involving L_1:

$$\tau_1 = \frac{L_1}{R_{tau1}} = \frac{L_1}{\dfrac{R_{load}\left(1-k_4\right)-k_3}{1+k_2+R_{load}k_7\left(1-k_4\right)}} \tag{4.387}$$

The second time constant is determined while L_1 is replaced by a short circuit as shown in Figure 4.125.

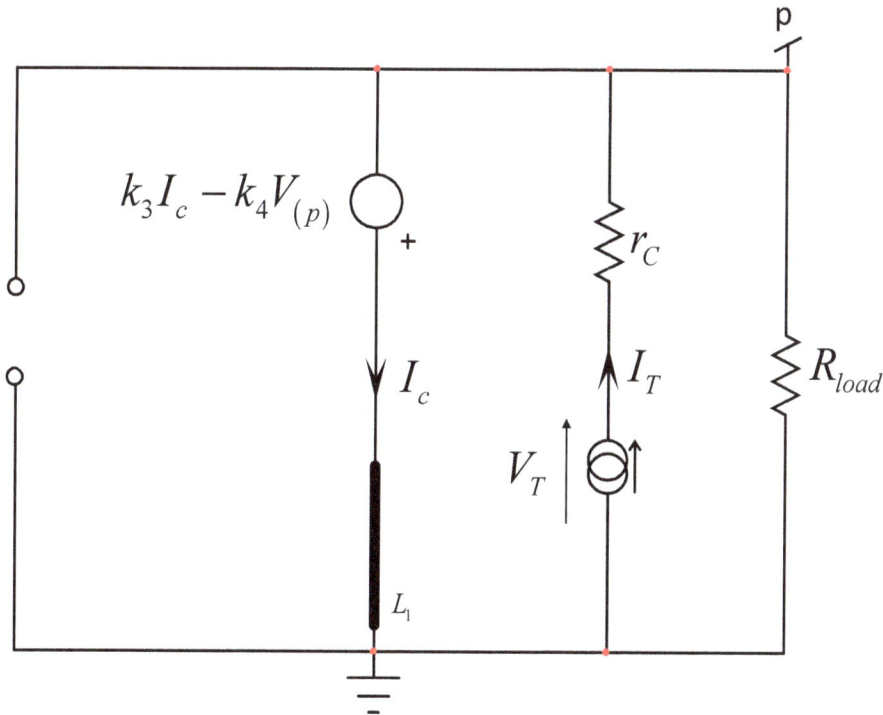

Figure 4.125 L_1 is now replaced by a short circuit for the determination of τ_2.

Because node c is grounded, $V_{(c)}$ equals zero and the left-side current source is turned off thus the voltage source expression also simplifies. We are left with a circuit in which we can temporarily disconnect r_C and bring it back later in series with the intermediate result. Doing so brings node p to V_T while I_c equals I_T.

As R_{load} appears in parallel with this intermediate resistance we want, we can also temporarily remove it and solve the simple circuit of Figure 4.126.

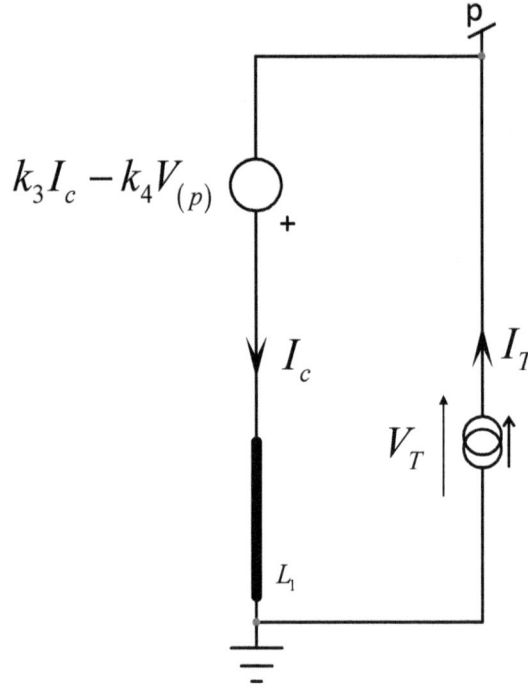

Figure 4.126 The circuit simplifies further if we remove the load resistance.

The voltage V_T can be formulated as follows:

$$V_T = -\left(k_3 I_T - k_4 V_T\right) \tag{4.388}$$

Solving for V_T, factoring I_T and rearranging gives:

$$R = \frac{V_T}{I_T} = \frac{k_3}{k_4 - 1} \tag{4.389}$$

We can bring r_C and R_{load} back to form R_{au2}:

$$R_{tau2} = \left[r_C + \left(\frac{k_3}{k_4 - 1} \right) \middle\| R_{load} \right] C_2 \tag{4.390}$$

The second time constant τ_2 is thus:

$$\tau_2 = R_{tau2} C_2 = \left[r_C + \left(\frac{k_3}{k_4 - 1} \right) \middle\| R_{load} \right] C_2 \tag{4.391}$$

The first-order term b_1 can be assembled by adding τ_1 and τ_2:

$$b_1 = \tau_1 + \tau_2 = \frac{L_1}{\dfrac{R_{load}(1-k_4)-k_3}{1+k_2+R_{load}k_7(1-k_4)}} + \left[r_C + \left(\frac{k_3}{k_4-1}\right) \| R_{load}\right]C_2 \qquad (4.392)$$

For the second-order coefficient, we will consider C_2 in its high-frequency state (a short circuit) and we determine the resistance driving L_1 in this configuration.

The circuit is that of Figure 4.127.

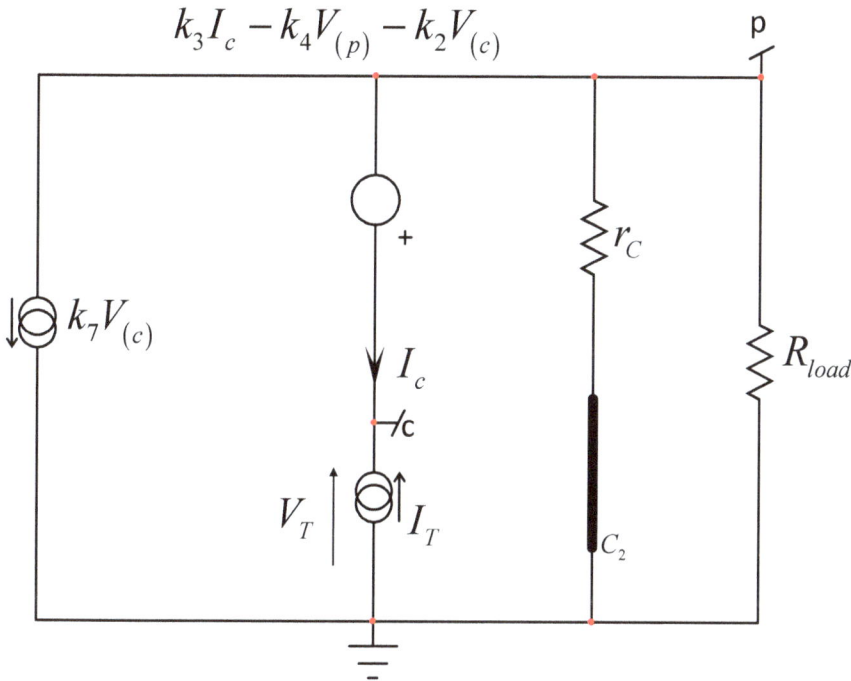

Figure 4.127 C_2 is now a short circuit and r_C comes in parallel with R_{load}.

It is similar to that of Figure 4.124 except that r_C is paralleled with R_{load}. Therefore, the new time constant is immediately obtained by updating (4.387).

$$\tau_1^2 = \frac{L_1}{\dfrac{(R_{load} \| r_C)(1-k_4)-k_3}{1+k_2+k_7(R_{load} \| r_C)(1-k_4)}} \qquad (4.393)$$

The second-order term in the denominator is obtained by involving τ_2 as follows:

$$b_2 = \tau_2\tau_1^2 = \left[r_C + \left(\frac{k_3}{k_4-1}\right) \| R_{load}\right]C_2 \frac{L_1}{\dfrac{(R_{load} \| r_C)(1-k_4)-k_3}{1+k_2+k_7(R_{load} \| r_C)(1-k_4)}} \qquad (4.394)$$

The denominator $D(s)$ can be formulated by associating b_1 and b_2 under the following normalized form:

$$D(s) = 1 + sb_1 + s^2 b_2 = 1 + s \left[\frac{L_1}{\dfrac{R_{load}(1-k_4)-k_3}{1+k_2+R_{load}k_7(1-k_4)}} + \left[r_C + \left(\frac{k_3}{k_4-1} \right) \| R_{load} \right] C_2 \right]$$

$$+ s^2 \left[r_C + \left(\frac{k_3}{k_4-1} \right) \| R_{load} \right] C_2 \frac{L_1}{\dfrac{(R_{load} \| r_C)(1-k_4)-k_3}{1+k_2+k_7(R_{load} \| r_C)(1-k_4)}} \tag{4.395}$$

Numerical application shows an extremely low Q factor and, therefore, we can apply the low-Q approximation to factor the denominator as two cascaded poles:

$$D(s) \approx \left(1 + \frac{s}{\omega_{p_1}} \right)\left(1 + \frac{s}{\omega_{p_2}} \right) \tag{4.396}$$

In this mode, the first low-frequency pole is defined as:

$$\omega_{p_1} \approx \frac{1}{b_1} \tag{4.397}$$

while:

$$\omega_{p2} \approx \frac{b_1}{b_2} \tag{4.398}$$

Fortunately, [4] offers an amazingly simpler expression than manipulating b_1 and b_2. Numerical application shows that both paths lead to very close values for the poles:

$$\omega_{p_1} = \frac{2}{R_{load}C_2} \tag{4.399}$$

and:

$$\omega_{p_2} \approx 2F_{sw} \left(\frac{1/D}{1+\dfrac{1}{|M|}} \right)^2 \tag{4.400}$$

We have the quasi-static gain and the poles, we can now consider the zeroes by nulling the output. The new circuit to consider is that of Figure 4.128 in which the output is nulled: $V_{(p)}$ now being equal to zero volt, some current and voltage sources can be simplified.

The couple r_C and C_2 forms the usual contributor for the first zero:

$$Z_1(s) = r_C + \frac{1}{sC_2} \tag{4.401}$$

Which brings a zero located at:

$$\omega_{z_1} = \frac{1}{r_C C_2}$$

(4.402)

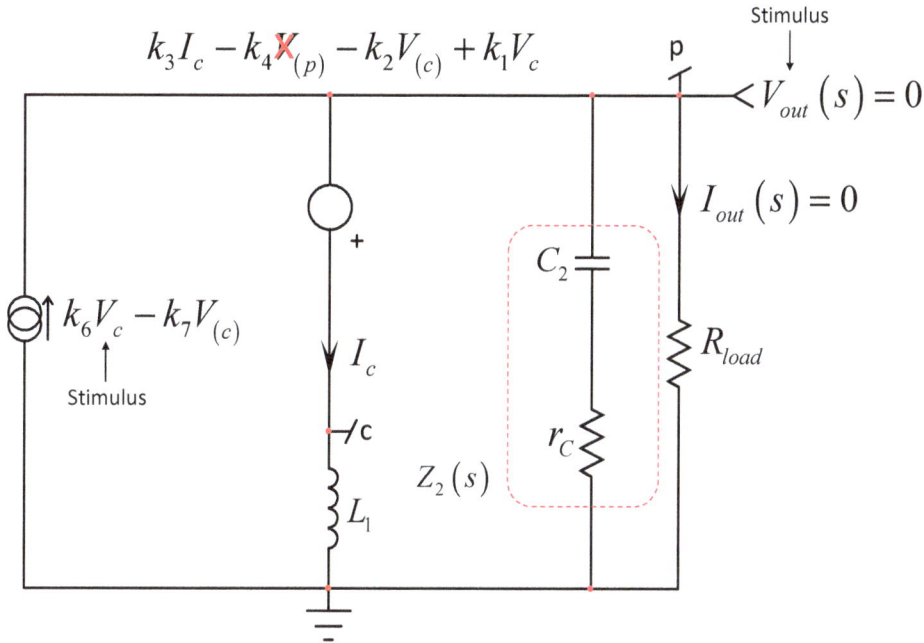

Figure 4.128 Nulling the response with stimulus to determine zeroes.

The second zero is determined considering that all the current sourced by the left-side generator is absorbed by the mesh involving L_1. In other terms, we need to solve:

$$k_6 V_c - k_7 V_{(c)} = \frac{V_{(c)}}{sL_1}$$

(4.403)

The voltage across the inductance is that of node p (which is nulled) plus the voltage delivered by the voltage source:

$$V_{(c)} = k_3 I_c - k_2 V_{(c)} + k_1 V_c$$

(4.404)

Consider:

$$I_c = \frac{V_{(c)}}{sL_1}$$

(4.405)

Substitute it in (4.404) then solve for $V_{(c)}$:

$$V_{(c)} = \frac{V_c k_1}{k_2 - \frac{k_3}{sL_1} + 1}$$

(4.406)

Now, substitute the above definition in (4.403) and you should obtain:

$$-\frac{V_c \left(k_3 k_6 - L_1 k_6 s + L_1 k_1 k_7 s - L_1 k_2 k_6 s \right)}{sL_1 \left(1 + k_2 \right) - k_3} = \frac{V_c k_1}{sL_1 \left(1 + k_2 \right) - k_3}$$

(4.407)

If now look for the s expression which ensures the numerator equality, you should find the second zero we want:

$$s_{z_2} = \frac{k_1 + k_3 k_6}{L_1 \left(k_6 - k_1 k_7 + k_2 k_6\right)} \tag{4.408}$$

Numerical application shows this is a positive root meaning we have a right-half-plane zero in the current-mode buck-boost converter operated in DCM. This zero is located at:

$$\omega_{z_2} = \left|s_{z_2}\right| = \frac{k_1 + k_3 k_6}{L_1 \left(k_6 - k_1 k_7 + k_2 k_6\right)} \tag{4.409}$$

Again, [4] offers an extremely compact expression whose value is almost identical to that returned from (4.409)

$$\omega_{z_2} \approx \frac{R_{load}}{|M|\left(1 + |M|\right) L_1} \tag{4.410}$$

This is it—we have determined the control-to-output transfer function of the CM buck-boost converter operated in DCM:

$$H(s) = \frac{V_{out}(s)}{V_c(s)} = H_0 \frac{\left(1 + \dfrac{s}{\omega_{z_1}}\right)\left(1 - \dfrac{s}{\omega_{z_2}}\right)}{\left(1 + \dfrac{s}{\omega_{p_1}}\right)\left(1 + \dfrac{s}{\omega_{p_2}}\right)} \tag{4.411}$$

where:

$$H_0 \approx -\frac{1}{S_n m_c T_{sw}} \frac{V_{in}}{\sqrt{2\tau_L}} \tag{4.412}$$

$$\omega_{p_1} = \frac{2}{R_{load} C_2} \tag{4.413}$$

$$\omega_{p_2} \approx 2F_{sw} \left(\frac{1/D}{1 + \dfrac{1}{|M|}}\right)^2 \tag{4.414}$$

$$\omega_{z_1} = \frac{1}{r_C C_2} \tag{4.415}$$

$$\omega_{z_2} = \frac{R_{load}}{|M|\left(1 + |M|\right) L_1} \tag{4.416}$$

We can verify if our analysis is correct by confronting the Mathcad® plots with the magnitude/phase response obtained by SPICE-simulating the circuit of Figure 4.118. Figure 4.119 confirms our approach is correct.

We can also run a SIMPLIS® simulation to further check our work. The application circuit shows up in Figure 4.130 while simulation results are given in Figure 4.131.

Overall curves match well up to half the switching frequency and confirm the second-order response predicted by the PWM switch model.

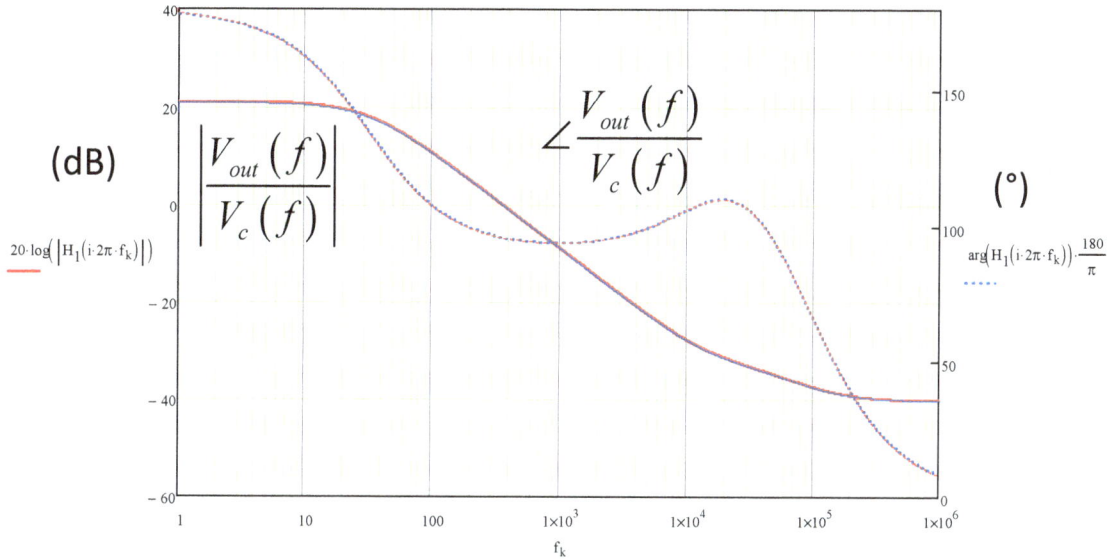

Figure 4.129 Mathcad® and SPICE agree on the magnitude/phase values.

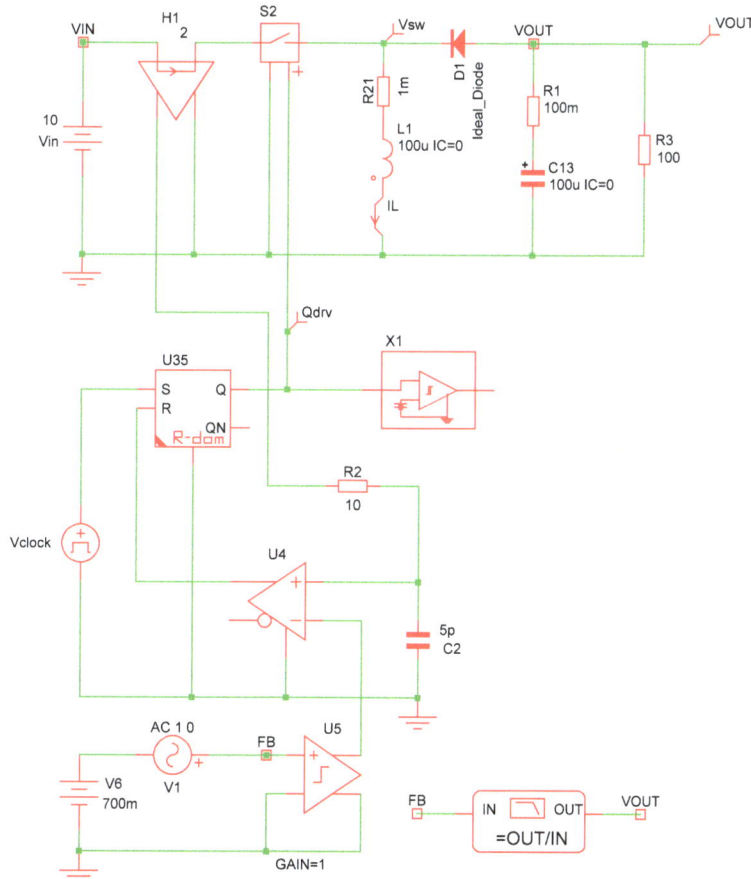

Figure 4.130 SIMPLIS® delivers the cycle-by-cycle waveforms and the small-signal response.

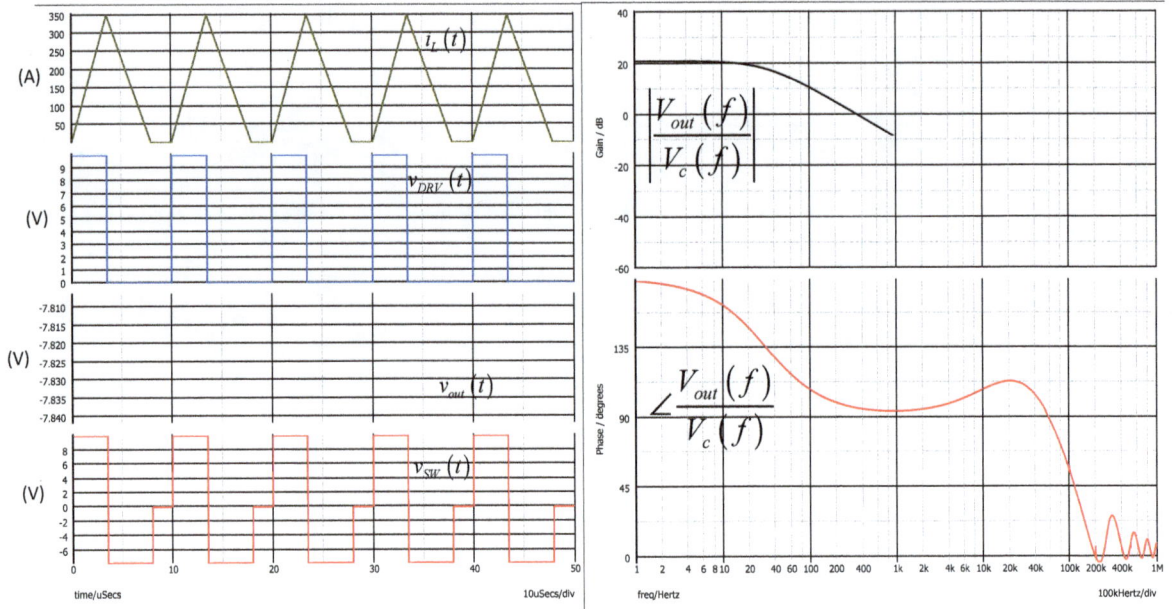

Figure 4.131 SIMPLIS® and SPICE agree quite well in magnitude and phase.

4.4.1 Input to Output

The control voltage V_c is now maintained constant (ac value is zero) and the input source is ac-modulated. The circuit becomes that of Figure 4.132. The circuit has gained in complexity as node a is no longer grounded and becomes the stimulus input.

We start by the quasi-static gain determination in which the inductor is shorted (node c is at a 0-V potential) and the capacitor open-circuited. The circuit to study in this case appears in Figure 4.133.

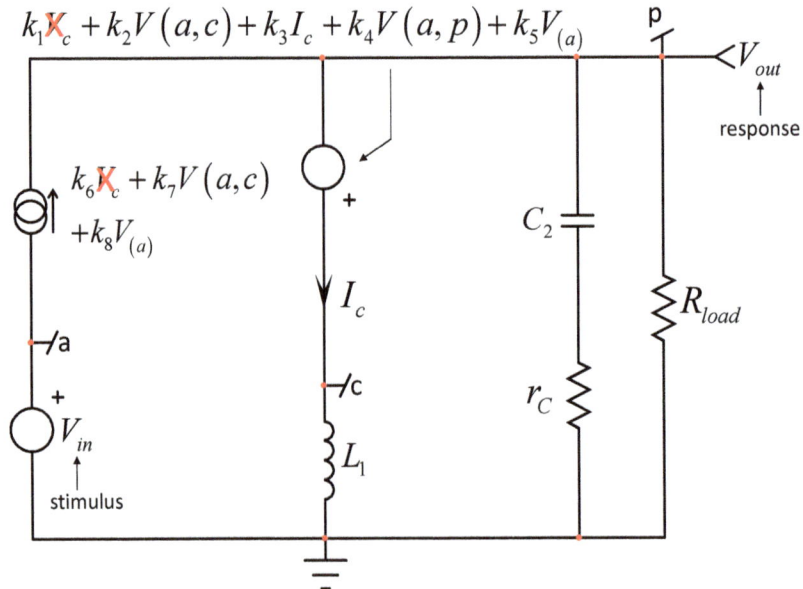

Figure 4.132 The stimulus is now the input voltage as the control voltage V_c is zeroed.

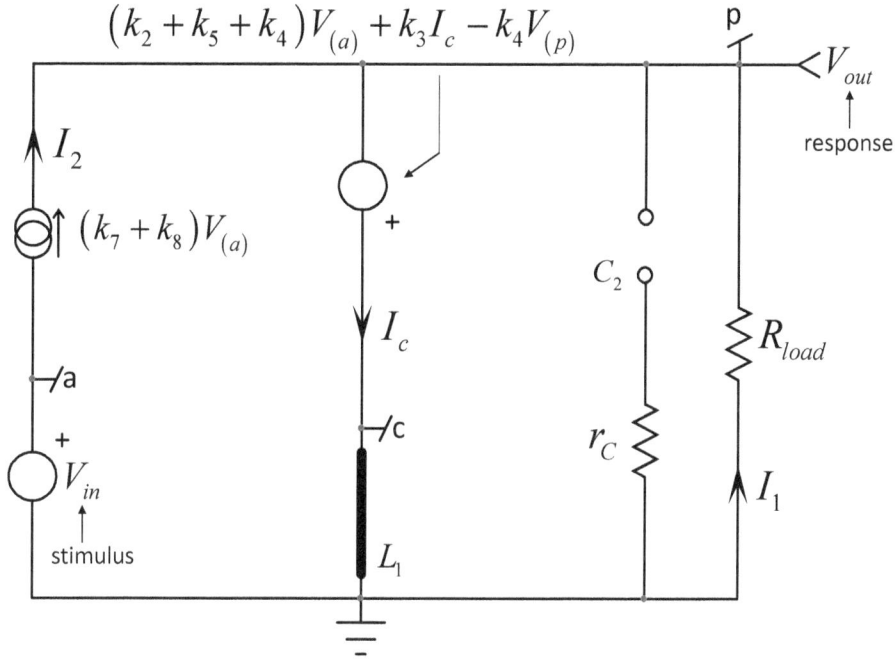

Figure 4.133 The dc gain H_0 is derived with a shorted inductor and an open-circuited capacitor.

We start by defining the output voltage in relationship with current I_1:

$$-V_{out} = I_1 R_{load} \tag{4.417}$$

The voltage at node a is our stimulus V_{in} while the voltage at node p is V_{out}. Capitalizing on these remarks, we can write:

$$-V_{out} = k_3 I_c - k_4 V_{out} + (k_2 + k_5 + k_4) V_{in} \tag{4.418}$$

From which we can extract the current I_c:

$$I_c = -\frac{V_{out} - V_{out} k_4 + V_{in} (k_2 + k_4 + k_5)}{k_3} \tag{4.419}$$

Then current I_1 is defined as:

$$I_1 = I_c - V_{in} (k_7 + k_8) \tag{4.420}$$

If this expression is plugged in (4.417) while (4.419) is substituted in this intermediate result, we can extract V_{out} while factoring V_{in} to obtain H_0:

$$H_0 = \frac{R_{load} \left[(k_7 + k_8) + \dfrac{k_2 + k_4 + k_5}{k_3} \right]}{\dfrac{R_{load} (k_4 - 1)}{k_3} + 1} \tag{4.421}$$

Now, if we turn the excitation off ($V_{in} = 0$ V) node a becomes grounded again as in Figure 4.122 when V_c is reduced to 0 V. As such, the time constants of the circuit are unchanged and we can reuse the denominator already defined with (4.396).

For the zeroes, we need to null the response as shown in Figure 4.134. The first zero, obviously, is brought by the series combination of r_C and C_2 satisfying:

$$Z_2(s) = r_C + \frac{1}{sC_2} = 0 \tag{4.422}$$

Solving this equation leads to the left-half-plane zero classically located at

$$\omega_{z_1} = \frac{1}{r_C C_2} \tag{4.423}$$

For the second zero, we will install a current generator I_T across L_1's connections as illustrated by Figure 4.135. This is a null Double Injection or NDI: there are two stimuli, V_{in} and V_T while the output is nulled. What is the resistance "seen" from the inductor connecting terminals?

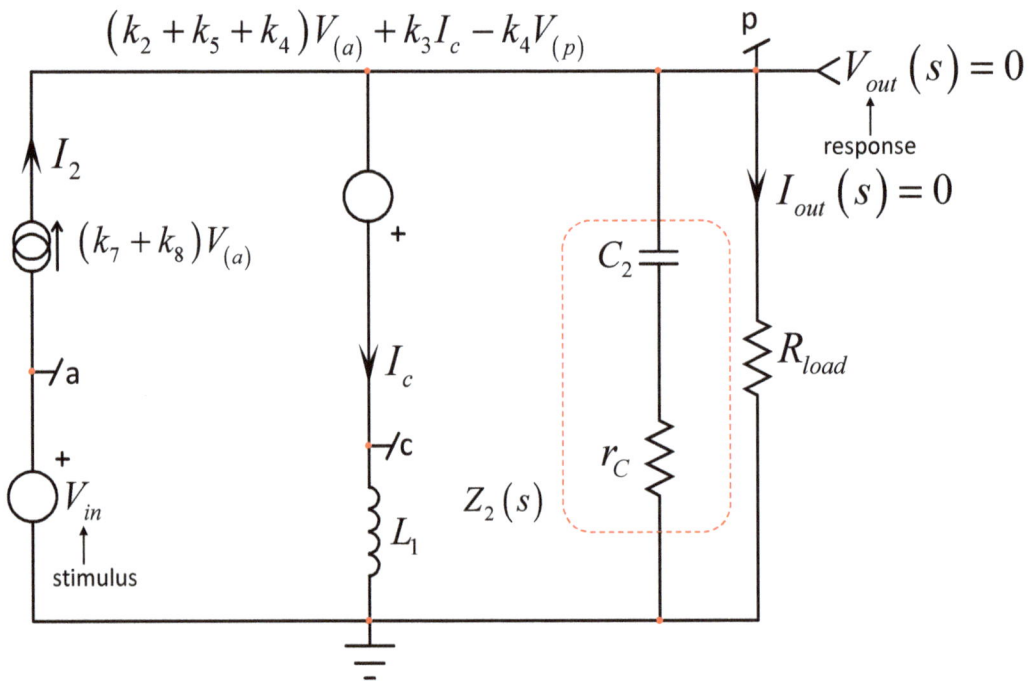

Figure 4.134 Nulling the output while the stimulus is brought back in place is the way to determine the zeroes.

As the I_T current is of reverse polarity compared with I_c and realizing that node c is biased at V_T, we can write:

$$(k_7 V_{in} + k_8 V_{in} - k_7 V_T) = -I_T \tag{4.424}$$

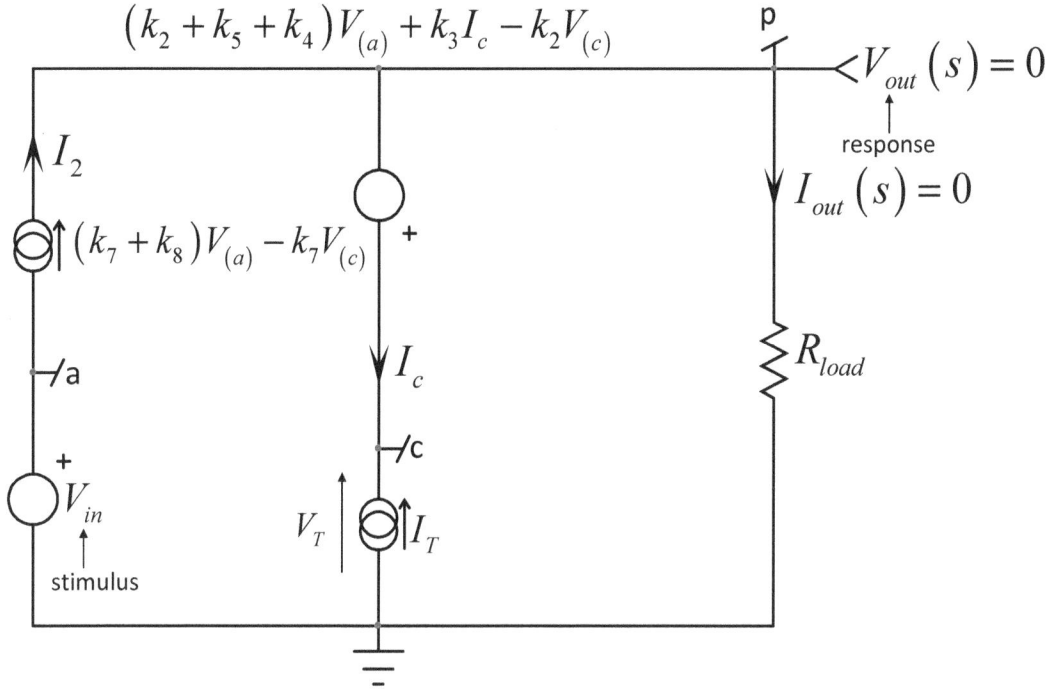

$$(k_2 + k_5 + k_4)V_{(a)} + k_3 I_c - k_2 V_{(c)}$$

$$(k_7 + k_8)V_{(a)} - k_7 V_{(c)}$$

Figure 4.135 The NDI will help us determine the position of the second zero.

From that equation, we extract the definition of V_{in}:

$$V_{in} = -\frac{I_T - V_T k_7}{k_7 + k_8} \qquad (4.425)$$

The second equation describes the voltage at node V_T imposed by the upper voltage source whose minus terminal is actually at a 0-V bias:

$$V_T = (k_2 + k_4 + k_5)V_{in} - k_3 I_T - k_2 V_T \qquad (4.426)$$

Now substitute the definition of V_{in} in the above expression and solve for V_T while factoring I_T. You should find:

$$R_{tau1N} = \frac{V_T}{I_T} = -\frac{k_2 + k_4 + k_5 + k_3(k_7 + k_8)}{k_7(1 - k_4 - k_5) + k_8(1 + k_2)} \qquad (4.427)$$

This expression leads to a second zero located in the right half-plane:

$$\omega_{z_2} = \frac{R_{tauN1}}{L_1} = \frac{\dfrac{k_2 + k_4 + k_5 + k_3(k_7 + k_8)}{k_7(1 - k_4 - k_5) + k_8(1 + k_2)}}{L_1} \qquad (4.428)$$

We now have the whole transfer function linking V_{in} to V_{out}:

$$H(s) = \frac{V_{out}(s)}{V_{in}(s)} = H_0 \frac{\left(1 + \frac{s}{\omega_{z_1}}\right)\left(1 - \frac{s}{\omega_{z_2}}\right)}{\left(1 + \frac{s}{\omega_{p_1}}\right)\left(1 + \frac{s}{\omega_{p_2}}\right)} \tag{4.429}$$

where:

$$H_0 = \frac{R_{load}\left[(k_7 + k_8) + \frac{k_2 + k_4 + k_5}{k_3}\right]}{\frac{R_{load}(k_4 - 1)}{k_3} + 1} \tag{4.430}$$

$$\omega_{p_1} = \frac{2}{R_{load}C_2} \tag{4.431}$$

$$\omega_{p_2} \approx 2F_{sw}\left(\frac{1/D}{1 + \frac{1}{|M|}}\right)^2 \tag{4.432}$$

$$\omega_{z_1} = \frac{1}{r_C C_2} \tag{4.433}$$

$$\omega_{z_2} = \frac{\frac{k_2 + k_4 + k_5 + k_3(k_7 + k_8)}{k_7(1 - k_4 - k_5) + k_8(1 + k_2)}}{L_1} \tag{4.434}$$

We can check our analysis by simulating the circuit of Figure 4.132 and comparing its response to what Mathcad® plots from (4.429). The results are given in Figure 4.136. As you can see, the quasi-static gain in this particular configuration is -110 dB which means an almost infinite rejection of the input source.

However, practically, we do not observe as high an input rejection.

Let's have a closer look with a SIMPLIS® simulation of this DCM CM buck-boost converter shown in Figure 4.137.

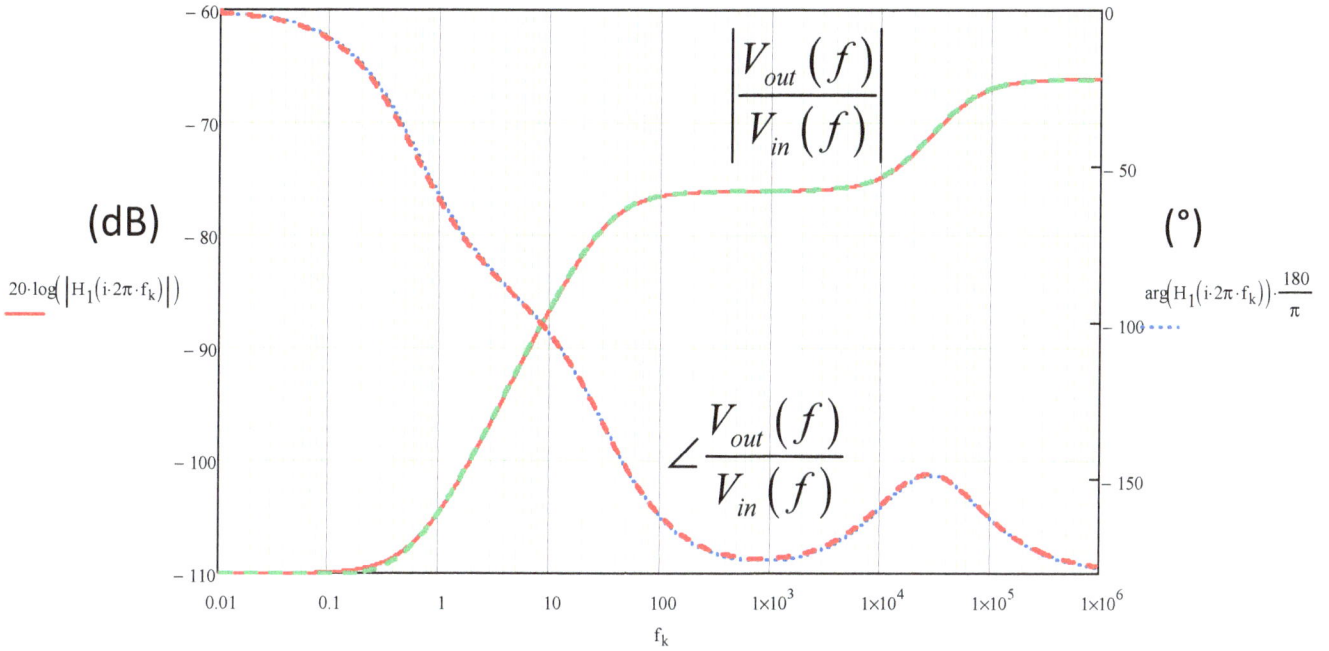

Figure 4.136 Curves from SPICE and Mathcad® agree well and confirm an excellent line rejection at dc.

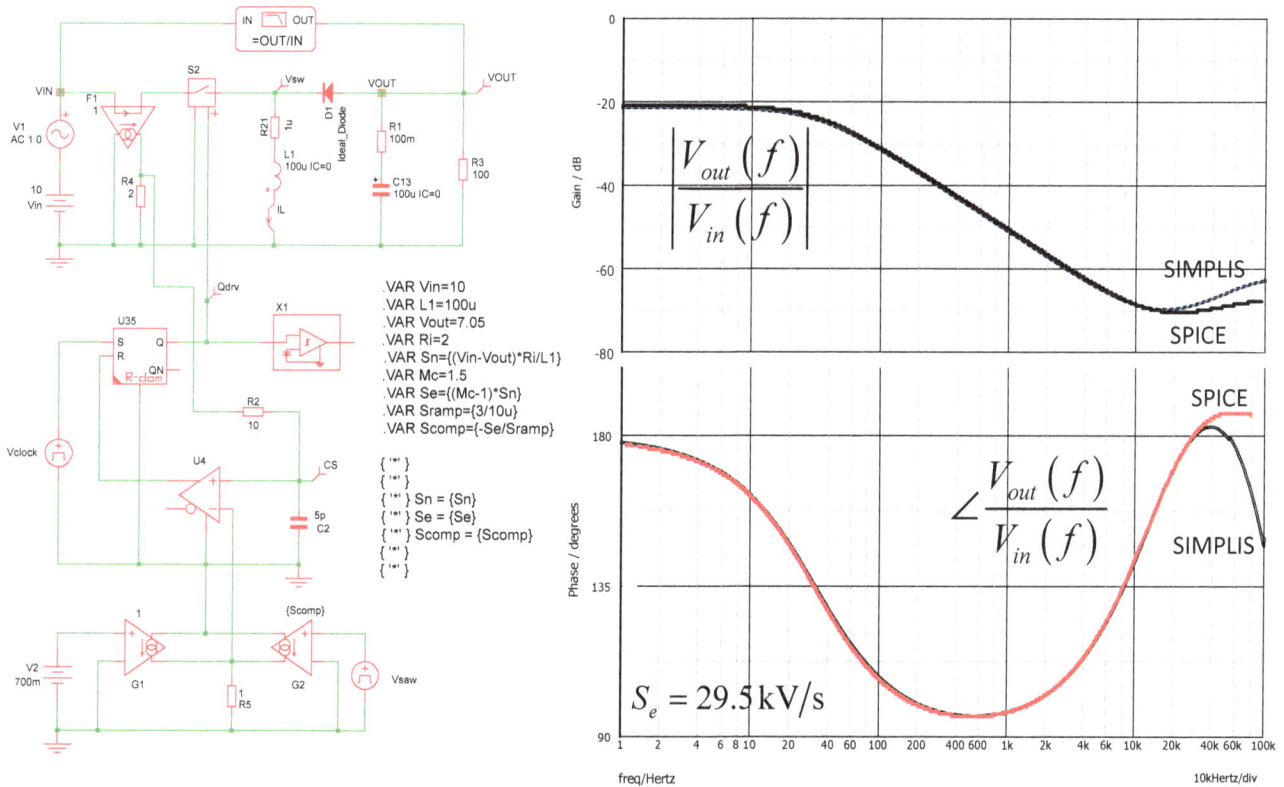

Figure 4.137 SIMPLIS® lends itself well to checking the input-to-output transfer function at a given operating point.

We have added some compensation ramp as if the converter was transitioning from CCM to DCM while the ramp is still present. As the figure confirms, agreement is good with a slight divergence when approaching half the switching frequency.

There is one contributor which is not showing up here and it is the total propagation delay. In the above analysis, it is considered zero: the inductor current is immediately interrupted when it exactly reaches the 350-mA setpoint (700 mV over the 2-Ω sensing resistance).

Therefore, the input source solely affects the slope S_n and not the peak value. In this perfect world, the input rejection is almost infinite at dc. Now, if you introduce some reaction time like a slower current-sense comparator, a stronger RC filter at the comparator input or even non-zero driver output resistance, then the inductor current will overshoot a little bit.

The overshoot amount depends on S_n and the propagation delay t_p as follows:

$$I_p = \frac{V_c}{R_i} + S_n t_p = \frac{V_c}{R_i} + \frac{V_{in}}{L_1} t_p \tag{4.435}$$

Without propagation delay, the inductor peak current setpoint is insensitive to V_{in}. With some propagation delay, the above equation shows how the door opens to the input source. This contribution is unfortunately detrimental to the input rejection ratio. As shown in Figure 4.138, the rejection, without compensation and a 1-ps-propagation delay, hits a theoretical -100 dB which is almost an infinite rejection as predicted.

As soon as a delay is included, the quasi-static gain deteriorates significantly.

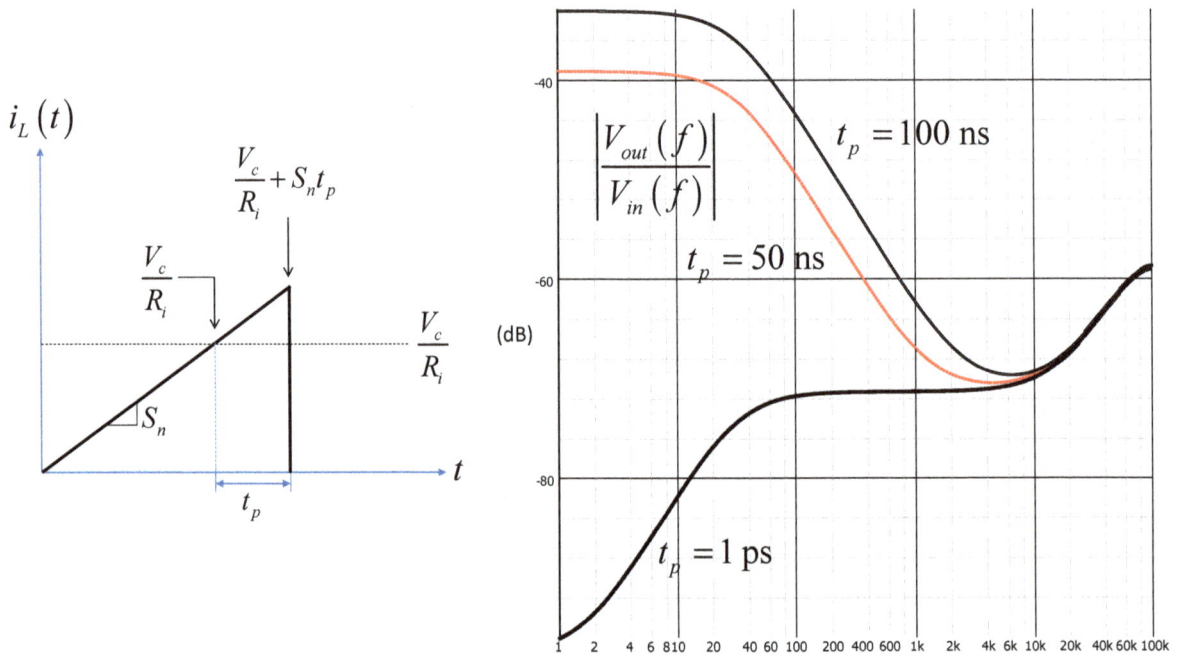

Figure 4.138 When propagation delay enters the picture, the audio susceptibility suffers significantly.

4.4.2 Output Impedance

The output impedance is obtained by sweeping the converter's output with a 1-A ac current source installed across the load resistance as shown in Figure 4.139.

parameters
Fsw=100kHz
L1=100u
Ri=2
Se=0
RL=100
Vout=-7.82631
Vin=10
Vac=Vin
Vacc=Vin
Vap=Vin-Vout
Ic=139.513m
Vc=700m

k1=Fsw*L1*Vc*Vap*Vac/(Ic*(L1*Se+Ri*Vacc)^2)
k2=Fsw*L1*Vc^2*Vap/(2*Ic*(L1*Se+Ri*Vacc)^2)
k3=-Fsw*L1*Vc^2*Vap*Vac/(2*Ic^2*(L1*Se+Ri*Vacc)^2)
k4=Fsw*L1*Vc^2*Vac/(2*Ic*(L1*Se+Ri*Vacc)^2)
k5=-Fsw*L1*Vc^2*Vac*Vap*Ri/(Ic*(L1*Se+Ri*Vacc)^3)
k6=Fsw*L1*Vc*Vac/(L1*Se+Ri*Vacc)^2
k7=Fsw*L1*Vc^2/(2*(L1*Se+Ri*Vacc)^2)
k8=-Fsw*L1*Vc^2*Ri*Vac/(L1*Se+Ri*Vacc)^3

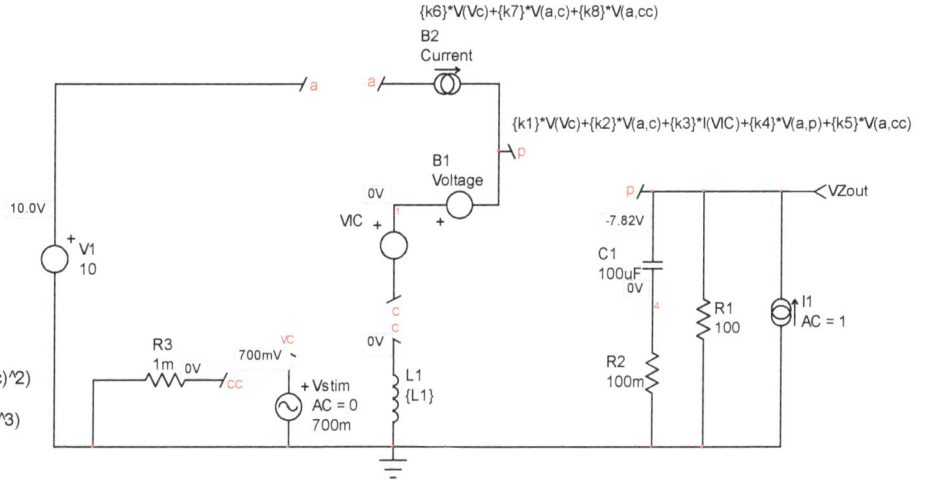

Figure 4.139 A 1-A ac source connected across the load will unveil the output impedance.

In this mode, the control voltage V_c is zeroed in ac as well as the input source V_{in}. The corresponding potential being nulled, the various sources can be simplified and the circuit becomes that of Figure 4.140.

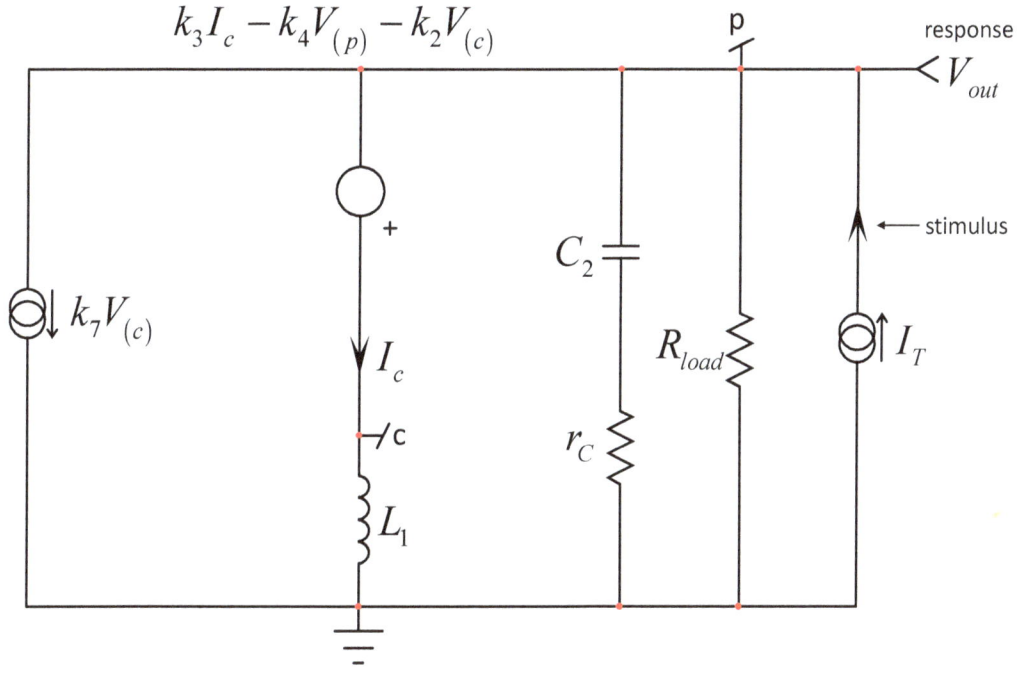

Figure 4.140 The stimulus is now the current source while the voltage collected across the load represents the response.

To determine R_0, we short the inductor and open the capacitor. We can also temporarily disconnect the load and bring it back later on in parallel with the intermediate result. Considering node c grounded by the shorted inductor, the left-side current source is turned off and the voltage loses one term.

The new circuit appears in Figure 4.141.

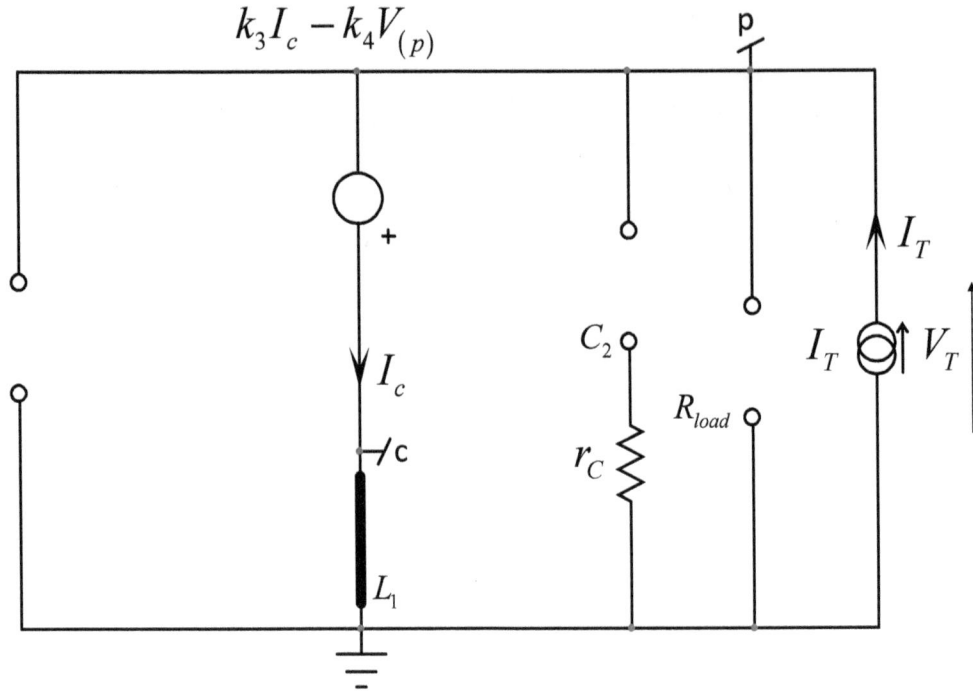

Figure 4.141 The inductor is replaced by a short while the output capacitor is open-circuited.

In this circuit, current I_c is I_T while the voltage at node p is V_T. Therefore:

$$V_T = -\left(k_3 I_T - k_4 V_T\right) \tag{4.436}$$

Extracting V_T while factoring I_T and bringing R_{load} back gives:

$$R_0 = \frac{V_T}{I_T} \| R_{load} = \frac{k_3}{k_4 - 1} \| R_{load} \tag{4.437}$$

If the stimulus I_T is now turned off, the circuit returns to the state described in Figure 4.122 where V_c is zeroed. The natural time constants of this circuit are therefore those already determined for the two previous transfer functions and we can reuse the denominator described by (4.396).

For the zeroes, we will determine them using a null double injection or NDI. We start with inductor L_1 that we bias with a current source I_T while the excitation is back in place with current source I_2 and its response is nulled: $V_{(p)} = 0$ V. The circuit we study is that of Figure 4.142.

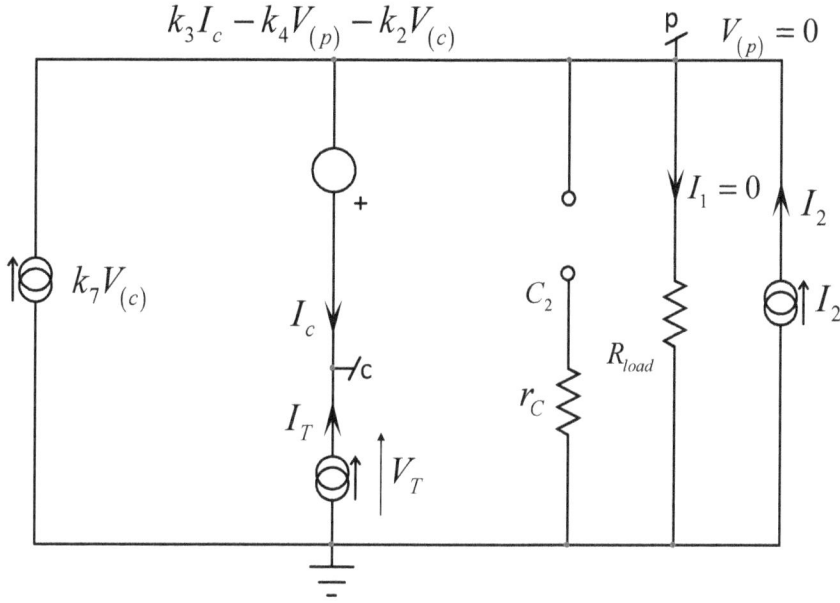

Figure 4.142 What resistance drives inductor L_1 while the response is nulled?

As the p node is at a 0-V bias and current $I_c = -I_T$, the voltage V_T we want is simply equal to the upper source:

$$V_T = V_{(c)} = -k_3 I_T - k_2 V_T \qquad (4.438)$$

Extracting V_T from this equation and factoring I_T gives us the resistance we need:

$$R_{tau1} = \frac{V_T}{I_T} = -\frac{k_3}{1+k_2} \qquad (4.439)$$

And the first time constant involving L_1 is thus equal to:

$$\tau_{1N} = \frac{L_1}{-\dfrac{k_3}{1+k_2}} \qquad (4.440)$$

The second time constant implying C_2 is quickly obtained by looking at Figure 4.143 in which L_1 is set in its dc state, a short circuit. In this mode, the only resistance "seen" by the current source is r_C because node p is at 0 V. Therefore:

$$\tau_{2N} = r_C C_2 \qquad (4.441)$$

We have our 1st order term for the numerator:

$$a_1 = \tau_{1N} + \tau_{2N} = \frac{L_1}{-\dfrac{k_3}{1+k_2}} + r_C C_2 \qquad (4.442)$$

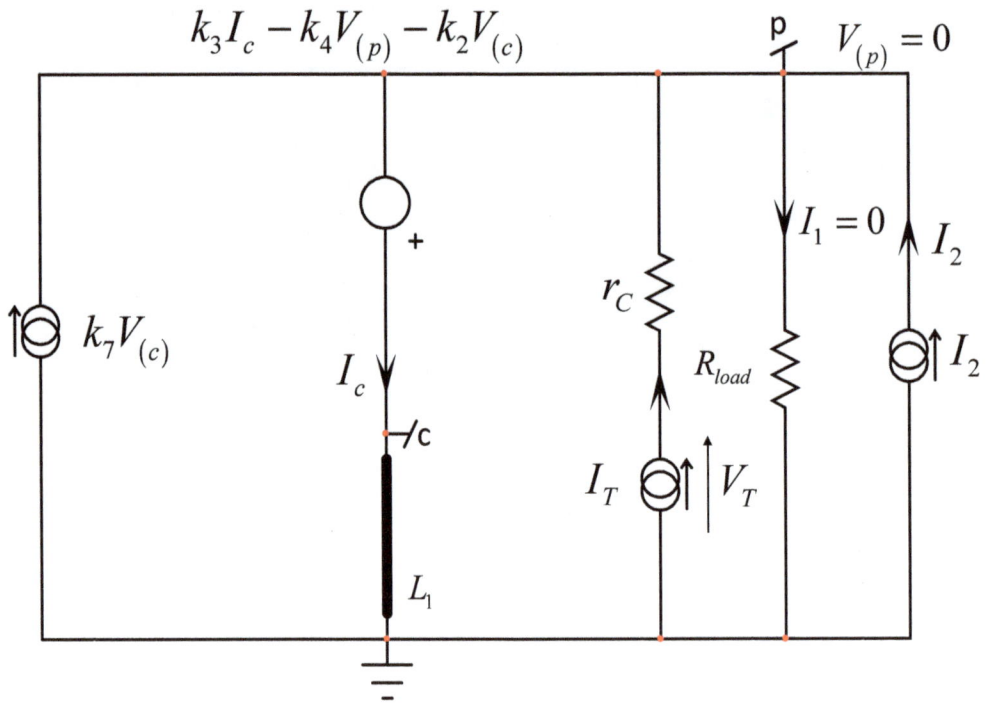

Figure 4.143 The time constant involving C_2 is immediate considering a null at node p.

The second high-frequency coefficient is determined by setting L_1 in its high-frequency state or an open circuit. It is described by Figure 4.144.

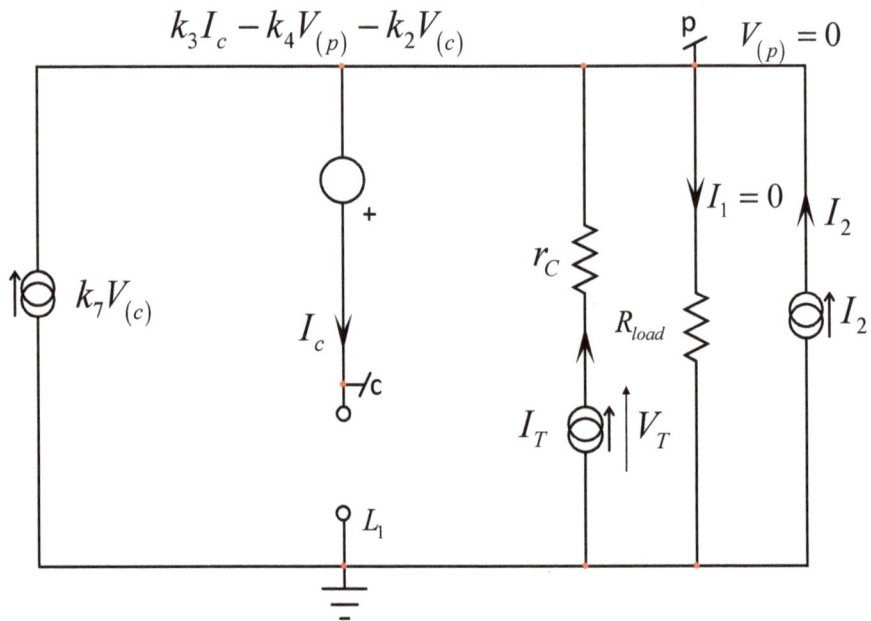

Figure 4.144 The inductor is open-circuited for this second-order time constant determination.

Again, similarly to what is described in Figure 4.143, the time constant is:

$$\tau_{2N}^1 = r_C C_2 \tag{4.443}$$

We can form the a_2 coefficient as follows:

$$a_2 = \tau_{1N} \tau_{2N}^1 = \frac{L_1}{-\dfrac{k_3}{1+k_2}} r_C C_2 \tag{4.444}$$

Which leads to expressing the numerator in the following format:

$$N(s) = 1 + a_1 s + a_2 s^2 = 1 + s \left(\frac{L_1}{-\dfrac{k_3}{1+k_2}} + r_C C_2 \right) + s^2 \frac{L_1}{-\dfrac{k_3}{1+k_2}} r_C C_2 \tag{4.445}$$

This equation follows the normalized formula:

$$1 + s(a+b) + s^2 ab = (1 + a \cdot s)(1 + b \cdot s) \tag{4.446}$$

Therefore, we can advantageously factor the denominator under a more familiar form:

$$N(s) = \left(1 + \frac{s}{\omega_{z_1}} \right) \left(1 + \frac{s}{\omega_{z_2}} \right) \tag{4.447}$$

with:

$$\omega_{z_1} = \frac{1}{r_C C_2} \tag{4.448}$$

and:

$$\omega_{z_2} = -\frac{k_3}{(1+k_2) L_1} \tag{4.449}$$

It is a high-frequency position and can be neglected in the final expression. The output impedance of the DCM CM buck-boost converter can thus be approximated considering one zero and one pole:

$$Z_{out}(s) = R_0 \frac{\left(1 + \dfrac{s}{\omega_{z_1}} \right) \left(1 + \dfrac{s}{\omega_{z_2}} \right)}{\left(1 + \dfrac{s}{\omega_{p_1}} \right) \left(1 + \dfrac{s}{\omega_{p_2}} \right)} \approx R_0 \frac{1 + \dfrac{s}{\omega_z}}{1 + \dfrac{s}{\omega_p}} \tag{4.450}$$

where:

$$R_0 = \frac{k_3}{k_4 - 1} \parallel R_{load} \tag{4.451}$$

$$\omega_p = \frac{2}{R_{load} C_2} \tag{4.452}$$

and:

$$\omega_z = \frac{1}{r_C C_2} \qquad (4.453)$$

Now that we have the complete transfer function, we can check its response versus that of the SPICE simulation of Figure 4.139. The results are given in Figure 4.145 and confirm our equations are correct.

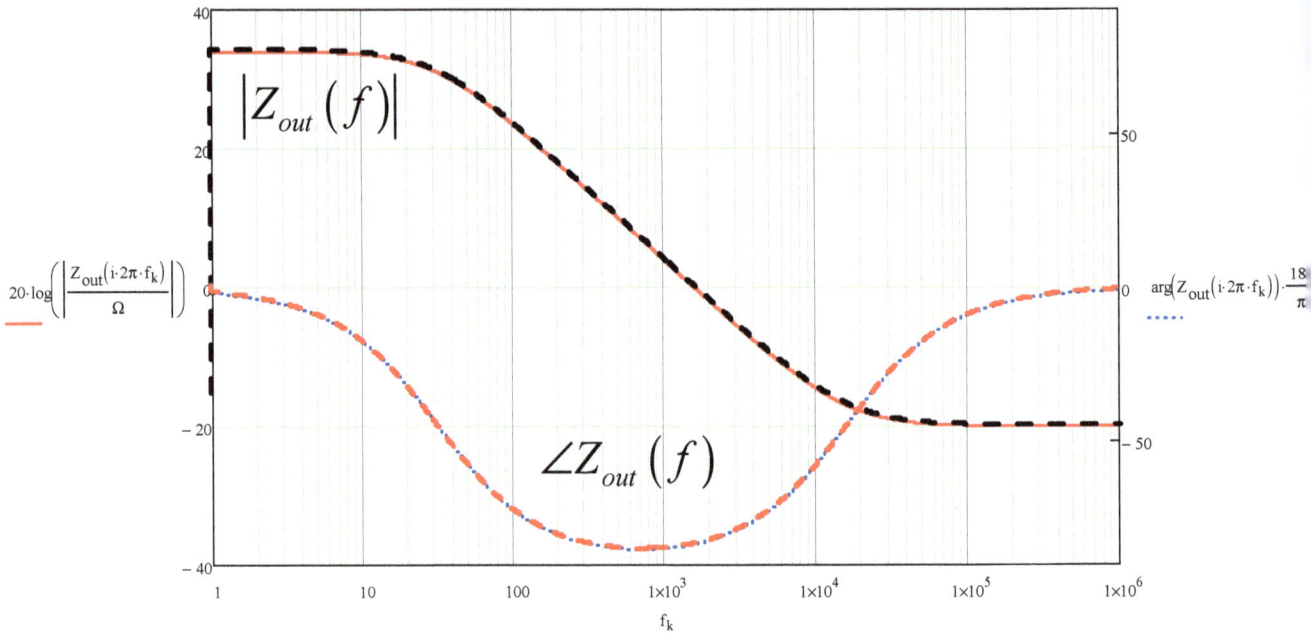

Figure 4.145 SPICE and Mathcad® curves agree very well with each other.

4.4.3 Input Impedance

The input impedance is classically obtained by connecting a 1-A ac current source at the converter's input as shown in Figure 4.146. A large-valued inductor isolates the dc source from the circuit during the ac analysis but lets the simulator compute the correct bias point.

The voltage collected across the current source during the ac sweep represents the voltage image of the small-signal input impedance we want.

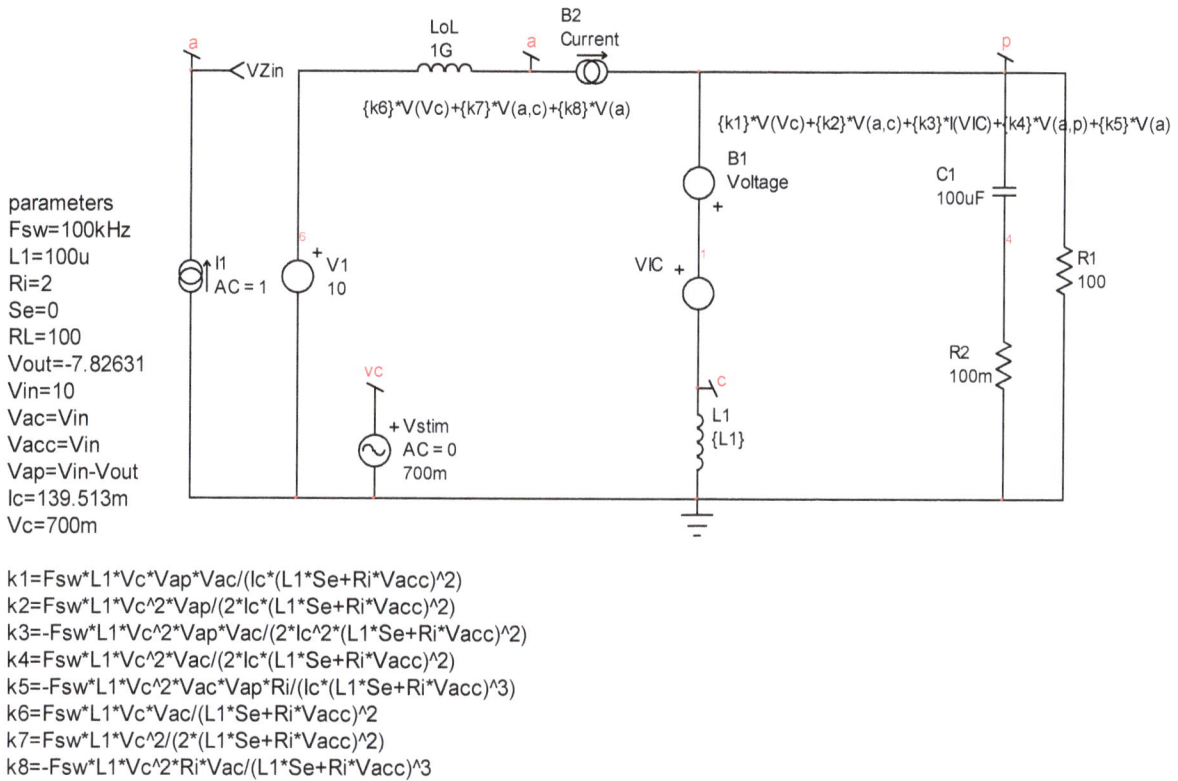

parameters
Fsw=100kHz
L1=100u
Ri=2
Se=0
RL=100
Vout=-7.82631
Vin=10
Vac=Vin
Vacc=Vin
Vap=Vin-Vout
Ic=139.513m
Vc=700m

k1=Fsw*L1*Vc*Vap*Vac/(Ic*(L1*Se+Ri*Vacc)^2)
k2=Fsw*L1*Vc^2*Vap/(2*Ic*(L1*Se+Ri*Vacc)^2)
k3=-Fsw*L1*Vc^2*Vap*Vac/(2*Ic^2*(L1*Se+Ri*Vacc)^2)
k4=Fsw*L1*Vc^2*Vac/(2*Ic*(L1*Se+Ri*Vacc)^2)
k5=-Fsw*L1*Vc^2*Vac*Vap*Ri/(Ic*(L1*Se+Ri*Vacc)^3)
k6=Fsw*L1*Vc*Vac/(L1*Se+Ri*Vacc)^2
k7=Fsw*L1*Vc^2/(2*(L1*Se+Ri*Vacc)^2)
k8=-Fsw*L1*Vc^2*Ri*Vac/(L1*Se+Ri*Vacc)^3

Figure 4.146 The input source biases the circuit for the dc point but the 1-A ac source sweeps the circuit to obtain $Z_{in}(s)$.

Now considering a zeroed control voltage V_c in the ac analysis, the circuit reduces to that of Figure 4.147 which is simpler to analyze. We start with the incremental input resistance R_0 determined when the inductor is shorted and the capacitor open-circuited. In this mode, node c is grounded and we can simplify the current and voltage sources. Figure 4.148 shows the circuit to solve.

We can see that the I_T current is imposed by the series current source which depends on $V_{(a)}$ biased at V_T:

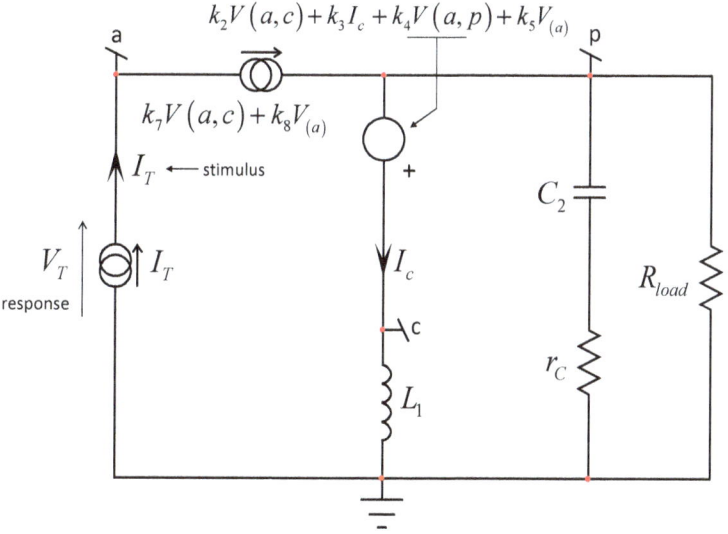

Figure 4.147 Considering a zeroed control voltage in this ac analysis, the circuit greatly simplifies.

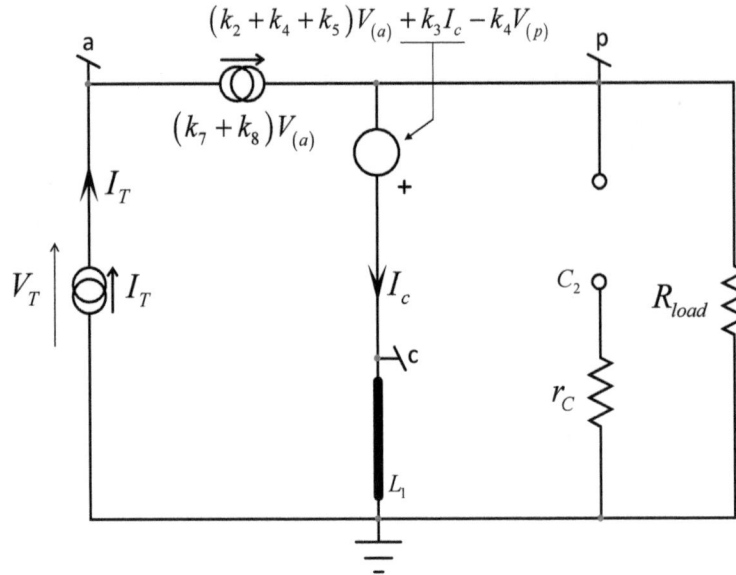

Figure 4.148 The incremental resistance R_0 is obtained with L_1 shorted and C_2 open-circuited.

$$I_T = V_T \left(k_7 + k_8 \right) \tag{4.454}$$

No need to write more equation, we immediately have the incremental resistance we want:

$$R_0 = \frac{V_T}{I_T} = \frac{1}{k_7 + k_8} \tag{4.455}$$

The numerical application gives a value of -163 Ω or 44.25 dBΩ.

We can now turn the excitation off to explore the various time constants. Turning the excitation off makes node a floating: the circuit does not return to its natural state already studied and we cannot reuse the denominator $D(s)$ we determined before. We need to determine all individual time constants starting with L_1 in Figure 4.149. Because node a is floating when the excitation is turned off, I have connected a large-value resistance R_{inf} which provides a dc path to ground. Observing that $I_T = -I_c$ and $V_T = V_{(c)}$, we can write:

$$V_{(p)} = V_T - V_{(a)} \left(k_2 + k_4 + k_5 \right) + k_3 I_T + k_4 V_{(p)} + k_2 V_T \tag{4.456}$$

Solving for $V_{(p)}$ gives:

$$V_{(p)} = -\frac{V_T + I_T k_3 + V_T k_2 - V_{(a)} \left(k_2 + k_4 + k_5 \right)}{k_4 - 1} \tag{4.457}$$

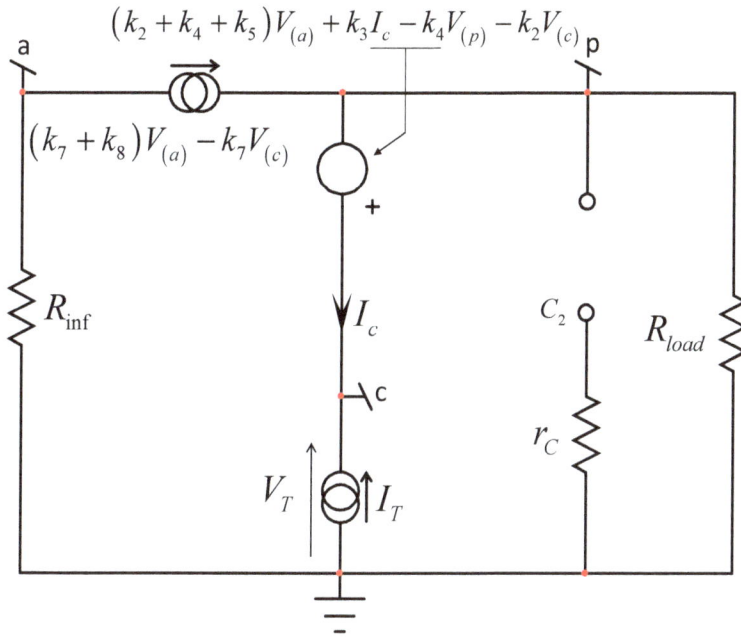

Figure 4.149 The stimulus is turned off and we look at the resistance "seen" from L_1's terminals.

The voltage at node p is also defined by:

$$V_{(p)} = \left[I_T + (k_7 + k_8)V_{(a)} - k_7 V_T \right] R_{load}$$ (4.458)

We need the voltage at node a defined as:

$$V_{(a)} = -R_{inf}\left[(k_7 + k_8)V_{(a)} - k_7 V_T \right]$$ (4.459)

Solving for $V_{(a)}$ gives:

$$V_{(a)} = \frac{R_{inf} V_T k_7}{R_{inf}(k_7 + k_8) + 1} \approx \frac{V_T k_7}{k_7 + k_8}$$ (4.460)

We can now equate (4.458) with (4.457) while replacing $V_{(a)}$ by the above definition. If we solve for V_T while factoring I_T, it leads to the resistance we want:

$$\frac{V_T}{I_T} = \frac{(k_4 k_7 - k_8 - k_7 + k_4 k_8)R_{load} + k_3(k_7 + k_8)}{k_7(k_5 + k_4 - 1) - k_8(1 + k_2)}$$ (4.461)

The time constant involving L_1 is thus:

$$\tau_1 = \frac{L_1}{\frac{(k_4 k_7 - k_8 - k_7 + k_4 k_8)R_{load} + k_3(k_7 + k_8)}{k_7(k_5 + k_4 - 1) - k_8(1 + k_2)}}$$ (4.462)

For the second time constant, inductor L_1 is shorted and we look through C_2's connections to determine the resistance we need. r_C can be temporarily disconnected and brought back later in series with the intermediate result. The circuit to study is that of Figure 4.150.

The voltage at node a, across the high-value resistance, is defined as follows:

$$V_{(a)} = -R_{\text{inf}}\left[V_{(a)}\left(k_7 + k_8\right) - k_7 V_{(c)}\right] \tag{4.463}$$

Which gives:

$$V_{(a)} = \frac{R_{\text{inf}} V_{(c)} k_7}{1 + R_{\text{inf}}\left(k_7 + k_8\right)} \approx V_{(c)} \frac{k_7}{k_7 + k_8} \tag{4.464}$$

In this expression, $V_{(c)}$ is 0 V owing to L_1 in dc state thus, $V_{(a)} = 0$ V. The circuit now nicely simplifies considering node p biased at V_T.

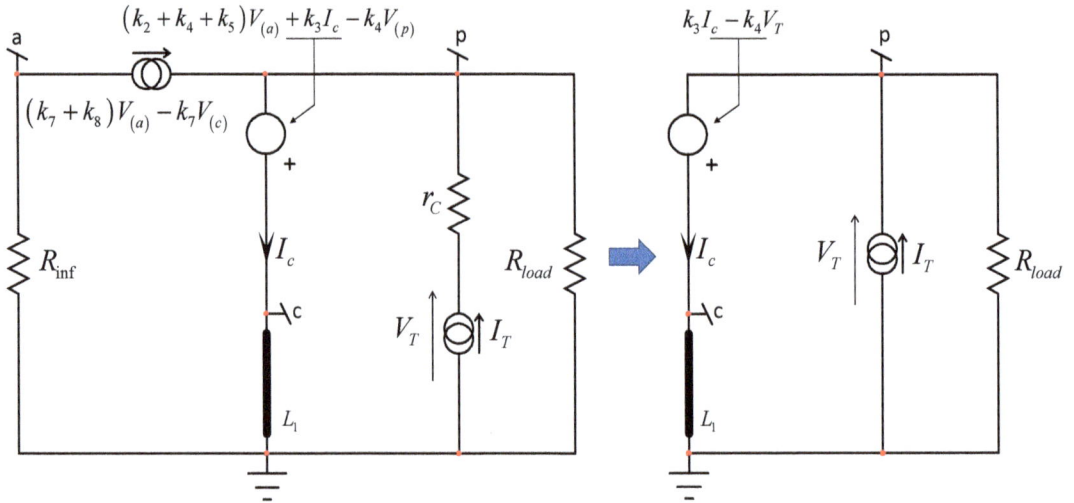

Figure 4.150 L_1 **is now replaced by a short circuit and we want to determine the second time constant involving** C_2.

We can even temporarily remove R_{load} and bring it back later in parallel with the intermediate result. Without R_{load}, we have:

$$V_T = -\left(k_3 I_c - k_4 V_T\right) \tag{4.465}$$

We can see $I_c = I_T$ thus, the above equation becomes:

$$V_T = -\left(k_3 I_T - k_4 V_T\right) \tag{4.466}$$

From which we extract V_T and factor I_T:

$$\frac{V_T}{I_T} = \frac{k_3}{k_4 - 1} \tag{4.467}$$

Now bring R_{load} back in parallel and add r_C, we have the second time constant:

$$\tau_2 = C_2\left[\frac{k_3}{k_4 - 1} \,\|\, R_{load} + r_C\right] \tag{4.468}$$

We can form the first-order coefficient b_1 of the denominator:

$$b_1 = \tau_1 + \tau_2 = \frac{L_1}{\dfrac{\left(k_4 k_7 - k_8 - k_7 + k_4 k_8\right) R_{load} + k_3\left(k_7 + k_8\right)}{k_7\left(k_5 + k_4 - 1\right) - k_8\left(1 + k_2\right)}} + C_2\left[\frac{k_3}{k_4 - 1} \,\|\, R_{load} + r_C\right] \tag{4.469}$$

The second-order term is determined by setting L_1 in its high-frequency state (an open circuit) as shown in Figure 4.151. In this mode, nodes a and c are unconnected and current in both branches is zero. The voltage source is thus non-connected and we are left with r_C in series with R_{load}:

$$\tau_2^1 = C_2 \left(r_C + R_{load} \right) \tag{4.470}$$

The second-order term b_2 is determined as follows:

$$b_2 = \tau_1 \tau_2^1 = \frac{L_1}{\dfrac{\left(k_4 k_7 - k_8 - k_7 + k_4 k_8 \right) R_{load} + k_3 \left(k_7 + k_8 \right)}{k_7 \left(k_5 + k_4 - 1 \right) - k_8 \left(1 + k_2 \right)}} C_2 \left(r_C + R_{load} \right) \tag{4.471}$$

The denominator is now completely determined and equal to:

$$D(s) = 1 + s b_1 + s^2 b_2 = 1 + s \left(\frac{L_1}{\dfrac{\left(k_4 k_7 - k_8 - k_7 + k_4 k_8 \right) R_{load} + k_3 \left(k_7 + k_8 \right)}{k_7 \left(k_5 + k_4 - 1 \right) - k_8 \left(1 + k_2 \right)}} + C_2 \left[\frac{k_3}{k_4 - 1} \| R_{load} + r_C \right] \right)$$

$$+ s^2 \frac{L_1}{\dfrac{\left(k_4 k_7 - k_8 - k_7 + k_4 k_8 \right) R_{load} + k_3 \left(k_7 + k_8 \right)}{k_7 \left(k_5 + k_4 - 1 \right) - k_8 \left(1 + k_2 \right)}} C_2 \left(r_C + R_{load} \right) \tag{4.472}$$

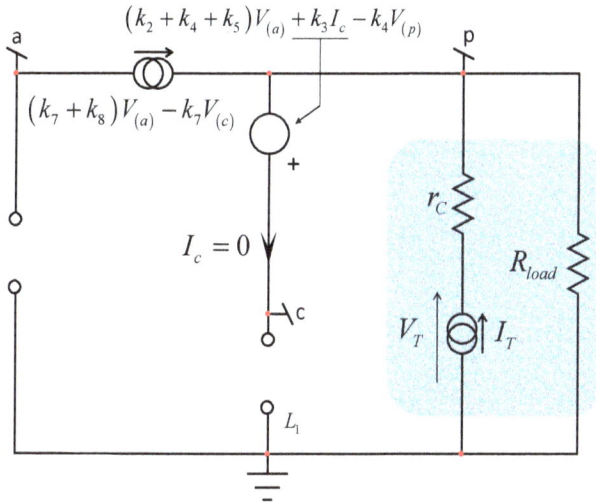

Figure 4.151 Having L_1 in an open-circuit state cancels the current I_c and simplifies the whole thing.

The zeroes of the transfer function are determined by nulling the response V_T across the current source in Figure 4.147. We know from [1] that zero volt across a current source represents a degenerate case in which the current source can be replaced by a short circuit. If we short the left-side generator I_T in Figure 4.147, the circuit returns to the state of Figure 4.122 with a zeroed V_c. The time constants are those already determined before and the poles now become the zeroes of our transfer function. $N(s)$ in $Z_{in}(s)$ is defined by:

$$N(s) = \left(1 + \frac{s}{\omega_{z_1}} \right) \left(1 + \frac{s}{\omega_{z_2}} \right) \tag{4.473}$$

with the following zeroes:

$$\omega_{z_1} = \frac{2}{R_{load} C_2}$$

(4.474)

and:

$$\omega_{z_2} \approx 2 F_{sw} \left(\frac{1/D}{1 + \frac{1}{|M|}} \right)^2$$

(4.475)

The complete transfer function is therefore defined as:

$$Z_{in}(s) = R_0 \frac{\left(1 + \frac{s}{\omega_{z_1}}\right)\left(1 + \frac{s}{\omega_{z_2}}\right)}{1 + s b_1 + s^2 b_2}$$

(4.476)

In this expression, R_0, the incremental input resistance is defined by (4.455) while b_1 and b_2 are respectively given by (4.469) and (4.471). The two zeroes have just been unveiled in the above lines. It is now time to test the transfer function we have derived versus the SPICE simulation of Figure 4.146. Results appear in Figure 4.152 and confirm that our approach is correct.

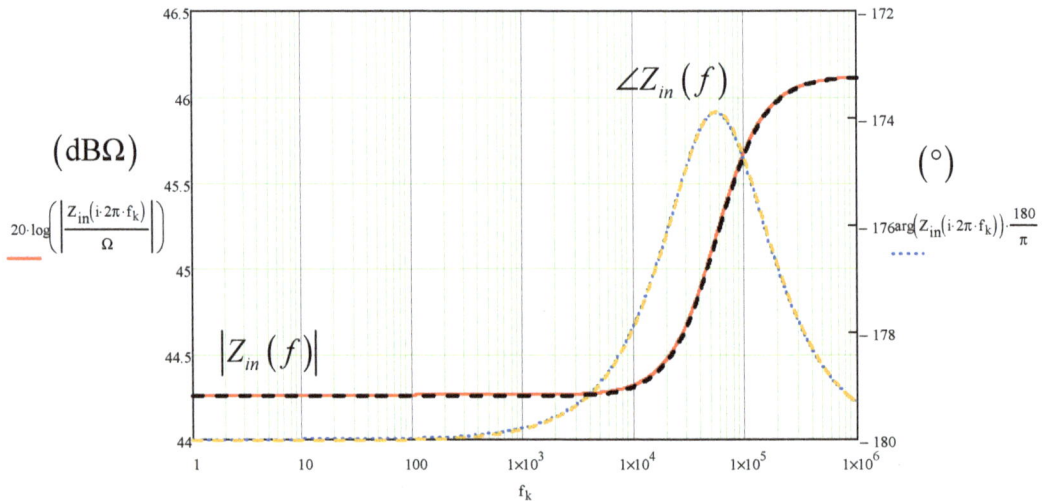

Figure 4.152 SPICE and Mathcad® agree very well in magnitude and phase.

This last transfer function ends the study of the CM buck-boost converter operated in the discontinuous conduction mode. We have conveniently gathered all the transfer functions in Figure 4.153.

		ω_z	ω_p	H_0 / R_0
$\dfrac{V_{out}(s)}{V_{err}(s)}$ Control to Output	$H_0\dfrac{\left(1+\frac{s}{\omega_{z_1}}\right)\left(1-\frac{s}{\omega_{z_2}}\right)}{\left(1+\frac{s}{\omega_{p_1}}\right)\left(1+\frac{s}{\omega_{p_2}}\right)}$	$\omega_{z_1}=\dfrac{1}{r_C C_2}$ $\omega_{z_2}=\dfrac{R_{load}}{|M|(1+|M|)L_1}$	$\omega_{p_1}=\dfrac{2}{R_{load}C_2}$ $\omega_{p_2}\approx 2F_{sw}\left(\dfrac{1/D}{1+\frac{1}{|M|}}\right)^2$	$H_0\approx-\dfrac{1}{S_n m_c T_{sw}}\dfrac{V_{in}}{\sqrt{2\tau_L}}$
$\dfrac{V_{out}(s)}{V_{in}(s)}$ Input to Output	$H_0\dfrac{\left(1+\frac{s}{\omega_{z_1}}\right)\left(1-\frac{s}{\omega_{z_2}}\right)}{\left(1+\frac{s}{\omega_{p_1}}\right)\left(1+\frac{s}{\omega_{p_2}}\right)}$	$\omega_{z_1}=\dfrac{1}{r_C C_2}$ $\omega_{z_2}=\dfrac{\frac{k_2+k_4+k_5+k_3(k_7+k_8)}{k_7(1-k_4-k_5)+k_8(1+k_2)}}{L_1}$	$\omega_{p_1}=\dfrac{2}{R_{load}C_2}$ $\omega_{p_2}\approx 2F_{sw}\left(\dfrac{1/D}{1+\frac{1}{|M|}}\right)^2$	$H_0=\dfrac{R_{load}\left[(k_7+k_8)+\frac{k_2+k_4+k_5}{k_3}\right]}{\frac{R_{load}(k_4-1)}{k_3}+1}$
$Z_{in}(s)$ Input impedance	$R_0\dfrac{\left(1+\frac{s}{\omega_{z_1}}\right)\left(1+\frac{s}{\omega_{z_2}}\right)}{1+sb_1+s^2b_2}$	$\omega_{z_1}=\dfrac{2}{R_{load}C_2}$ $\omega_{z_2}\approx 2F_{sw}\left(\dfrac{1/D}{1+\frac{1}{|M|}}\right)^2$	See text for coefficients values $D(s)=1+sb_1+s^2b_2$	$R_0=\dfrac{1}{k_7+k_8}$
$Z_{out}(s)$ Output impedance	$R_0\dfrac{1+\frac{s}{\omega_{z_1}}}{1+\frac{s}{\omega_{p_1}}}$	$\omega_{z_1}=\dfrac{1}{r_C C_2}$	$\omega_{p_1}=\dfrac{2}{R_{load}C_2}$	$R_0=\dfrac{k_3}{k_4-1}\,\|\,R_{load}$

$$m_c=1+\frac{S_e}{S_n}\quad S_e\text{ in V/s}\quad S_n=\frac{V_{in}}{L_1}R_i\quad M=\left|\frac{V_{out}}{V_{in}}\right|\quad \tau_L=\frac{L_1}{R_{load}}F_{sw}$$

Figure 4.153 The four transfer functions of the buck-boost converter operated in DCM and current mode are given here.

4.5 The Flyback Converter Operated in Voltage-Mode Control

The flyback converter represents the isolated version of the buck-boost converter. As shown in Figure 4.154, by connecting the power switch in the low side, the output voltage becomes referenced to the input source.

Insert a transformer in this intermediate circuit and you have the flyback converter with its secondary ground fully isolated from the primary side. It is an extremely popular converter, especially in the consumer world where you find it in ac-dc adapters for notebooks, CD and DVD players, auxiliary power supplies in TVs, high-power converters and so on.

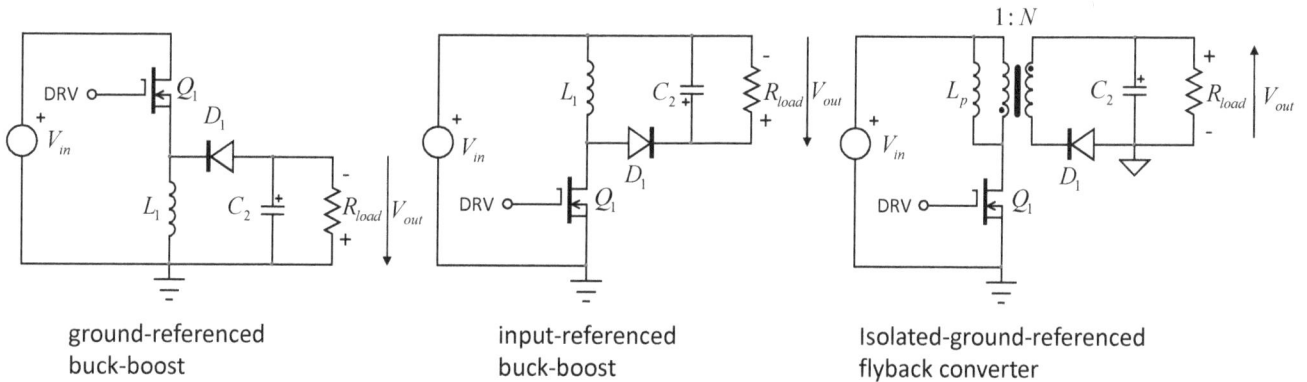

Figure 4.154 The flyback converter derives from the buck-boost converter to which a transformer has been added.

The transformer shown in Figure 4.155 is affected by several variables:

- L_p is the primary inductance. It defines the amount of magnetizing current flowing in the primary side and thus how much energy is stored during the on-time duration. When the switch turns off this energy starts transferring to the secondary side and L_p depletes.
- l_l represents the leakage inductance. It symbolizes the bad coupling between the primary and the secondary windings. In a two-winding transformer, there are two leakage inductances but one usually only considers the term in the primary side. The leakage inductance plays a marginal role in the small-signal response of the flyback converter by providing some damping in the response. It also truncates the duty ratio in CCM operation [4].
- N is the transformer turns ratio. It is referenced to the primary or the secondary side depending on authors. I reference it to the secondary side in the equations. If we have N_p turns in the primary side and N_s in the secondary then the turns ratio is $1:N$ with $N=N_s/N_p$.

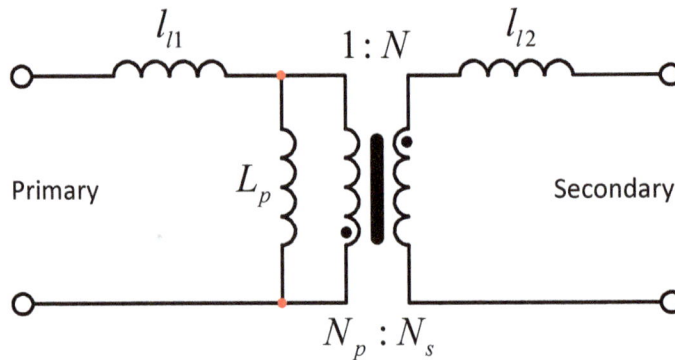

Figure 4.155 In this equivalent transformer PI-model, there are two leakage inductances.

The control-to-output transfer functions of the flyback converter operated in CCM and DCM are those of the buck-boost converter operated in similar conditions but with an inductor scaled by the transformer turns ratio squared:

$$L_1 = L_p N^2 \tag{4.477}$$

The input voltage V_{in} also needs to be scaled by N and affects the dc transfer function:

$$M = \frac{V_{out}}{NV_{in}} = \frac{D}{1-D} \tag{4.478}$$

Based on these new expressions, the transfer function for the CCM flyback converter operated in voltage-mode control is the following:

$$H(s) = \frac{V_{out}(s)}{V_{err}(s)} = \frac{H_0}{V_p} \frac{\left(1+\dfrac{s}{\omega_{z_1}}\right)\left(1-\dfrac{s}{\omega_{z_2}}\right)}{1+\dfrac{s}{\omega_0 Q}+\left(\dfrac{s}{\omega_0}\right)^2} \tag{4.479}$$

with:

$$H_0 \approx \frac{NV_{in}}{(1-D)^2} \tag{4.480}$$

$$\omega_{z_1} = \frac{1}{r_C C_2} \tag{4.481}$$

$$\omega_{z_2} \approx \frac{(1-D)^2 R_{load}}{DL_p N^2} \tag{4.482}$$

$$\omega_0 \approx \frac{1-D}{N\sqrt{L_p C_2}} \tag{4.483}$$

$$Q \approx \left(\frac{1-D}{N}\right) R_{load} \sqrt{\frac{C_2}{L_p}} \tag{4.484}$$

We have assembled a simple SPICE simulation circuit using the VM PWM switch model and a transformer. The numerical application predicts an output voltage around 20 V which is confirmed by the bias point. We have then compared the SPICE magnitude/phase response with that delivered by the above equations and results are given in Figure 4.156. The modulator is not present as we assume a sawtooth amplitude V_p of 1 V. There is a slight discrepancy in the peaking as (4.484) is approximate. Using the full definition of (4.38) instead – in which L_1 is replaced by $N^2 L_p$ – all curves perfectly superimpose.

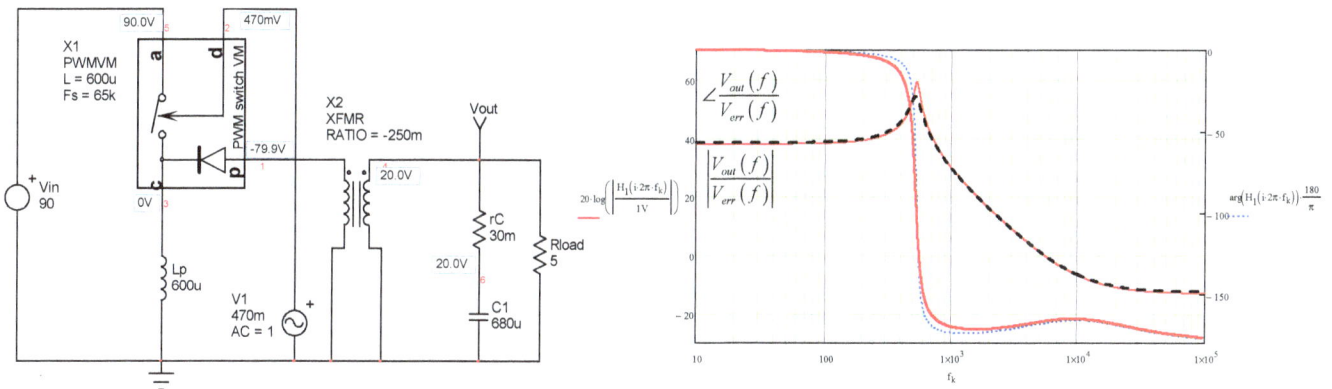

Figure 4.156 The CCM flyback converter operated in voltage-mode control peaks at the resonance.

If the load now increases to 30 Ω, the converter toggles to the DCM operation.

The transfer function changes and the system remains of 2nd-order but is heavily damped.

Using (4.100) from which the minus sign has been removed now considering a positive output, we can determine the output voltage:

$$V_{out} = M \cdot N \cdot V_{in} = V_{in} N \frac{D}{\sqrt{2\tau_L}} = V_{in} N \frac{D}{\sqrt{2\dfrac{L_p N^2}{R_{load}} F_{sw}}} = V_{in} D \sqrt{\frac{R_{load}}{2 F_{sw} L_p}} \tag{4.485}$$

569

In our simulation, to maintain a 20-V output with a 65-kHz operating frequency and a 30-Ω load, the duty ratio must be reduced to:

$$D = \frac{V_{out}}{V_{in}} \sqrt{\frac{2F_{sw}L_p}{R_{load}}} = \frac{20}{90} \times \sqrt{\frac{2 \times 65k \times 600u}{30}} \approx 36\% \tag{4.486}$$

The complete transfer function is derived from the DCM buck-boost converter tweaked to account for the transformer presence as highlighted with (4.477) and (4.478):

$$H(s) = \frac{V_{out}(s)}{V_{err}(s)} = \frac{H_0}{V_p} \frac{\left(1 + \dfrac{s}{\omega_{z_1}}\right)\left(1 - \dfrac{s}{\omega_{z_2}}\right)}{\left(1 + \dfrac{s}{\omega_{p_1}}\right)\left(1 + \dfrac{s}{\omega_{p_2}}\right)} \tag{4.487}$$

where:

$$H_0 = V_{in} \sqrt{\frac{R_{load}}{2L_p F_{sw}}} \tag{4.488}$$

$$\omega_{p_1} \approx \frac{2}{R_{load}C_2} \tag{4.489}$$

$$\omega_{p_2} \approx 2F_{sw} \left[\frac{1/D}{1 + \dfrac{1}{M}}\right]^2 \tag{4.490}$$

$$\omega_{z_1} = \frac{1}{r_C C_2} \tag{4.491}$$

$$\omega_{z_2} \approx \frac{R_{load}}{M(1+M)L_p N^2} \tag{4.492}$$

The flyback converter circuit has been updated with the 30-Ω load and a new bias point has been calculated with an updated dynamic response. Results comparing the SPICE simulation and the Mathcad® sheet are given in Figure 4.157 and show excellent agreement.

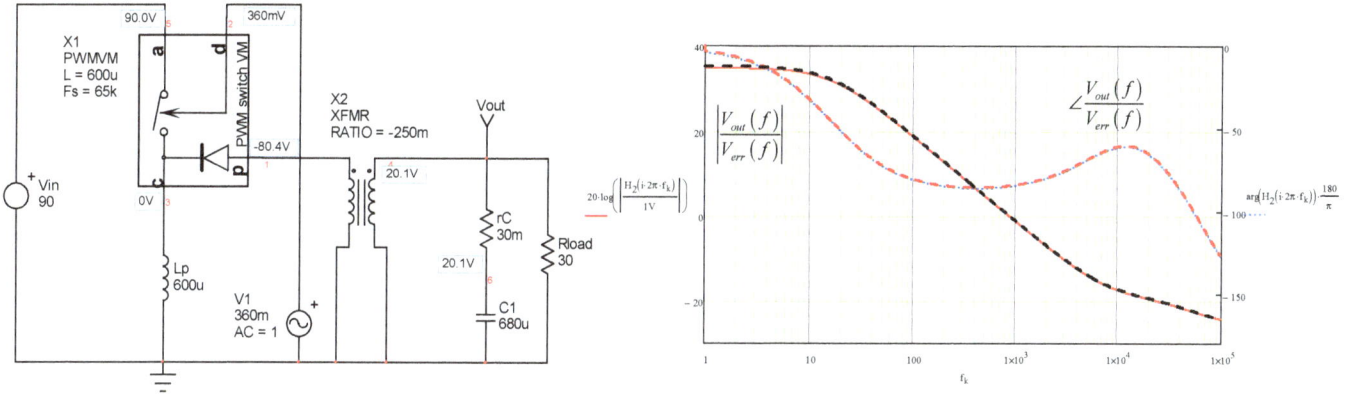

Figure 4.157 When operated in DCM, the flyback converter remains a damped 2nd-order system.

$\dfrac{V_{out}(s)}{V_{err}(s)}$ CCM	$H_0\dfrac{\left(1+\dfrac{s}{\omega_{z_1}}\right)\left(1-\dfrac{s}{\omega_{z_2}}\right)}{1+\dfrac{s}{\omega_0 Q}+\left(\dfrac{s}{\omega_0}\right)^2}$	$\omega_{z_1}=\dfrac{1}{r_C C_2}$ $\omega_{z_2}\approx\dfrac{(1-D)^2 R_{load}}{DL_p N^2}$	Double poles	$H_0\approx\dfrac{NV_{in}}{V_p(1-D)^2}$

$$\omega_0\approx\frac{1-D}{N\sqrt{L_p C_2}}\qquad Q\approx\left(\frac{1-D}{N}\right)R_{load}\sqrt{\frac{C_2}{L_p}}\qquad M=\frac{V_{out}}{NV_{in}}=\frac{D}{1-D}\quad\text{Turns ratio }1{:}N$$

$\dfrac{V_{out}(s)}{V_{err}(s)}$ DCM	$H_0\dfrac{1+\dfrac{s}{\omega_z}}{\left(1+\dfrac{s}{\omega_{p1}}\right)\left(1+\dfrac{s}{\omega_{p2}}\right)}$	$\omega_z=\dfrac{1}{r_C C_2}$	$\omega_{p1}\approx\dfrac{2}{R_{load}C_2}$ $\omega_{p2}\approx 2F_{sw}\left[\dfrac{1/D}{1+\dfrac{1}{M}}\right]^2$	$H_0=\dfrac{V_{in}}{V_p}\sqrt{\dfrac{R_{load}}{2L_p F_{sw}}}$

$$M=\frac{V_{out}}{NV_{in}}=\frac{D}{\sqrt{2\tau_L}}=\frac{D}{\sqrt{2\dfrac{L_p N^2}{R_{load}}F_{sw}}}\qquad \tau_L=\frac{L_1}{R_{load}}F_{sw}\qquad D=\frac{V_{out}}{V_{in}}\sqrt{\frac{2F_{sw}L_p}{R_{load}}}$$

Figure 4.158 This table gathers all the terms to plot the control-to-output transfer function of the voltage-mode flyback converter operated in CCM or DCM.

To let you select and apply the transfer function of your choice for the flyback converter, I have gathered the main parameters for the CCM and DCM cases in Figure 4.158. The numerical application for both CCM and DCM examples appear in Figure 4.159.

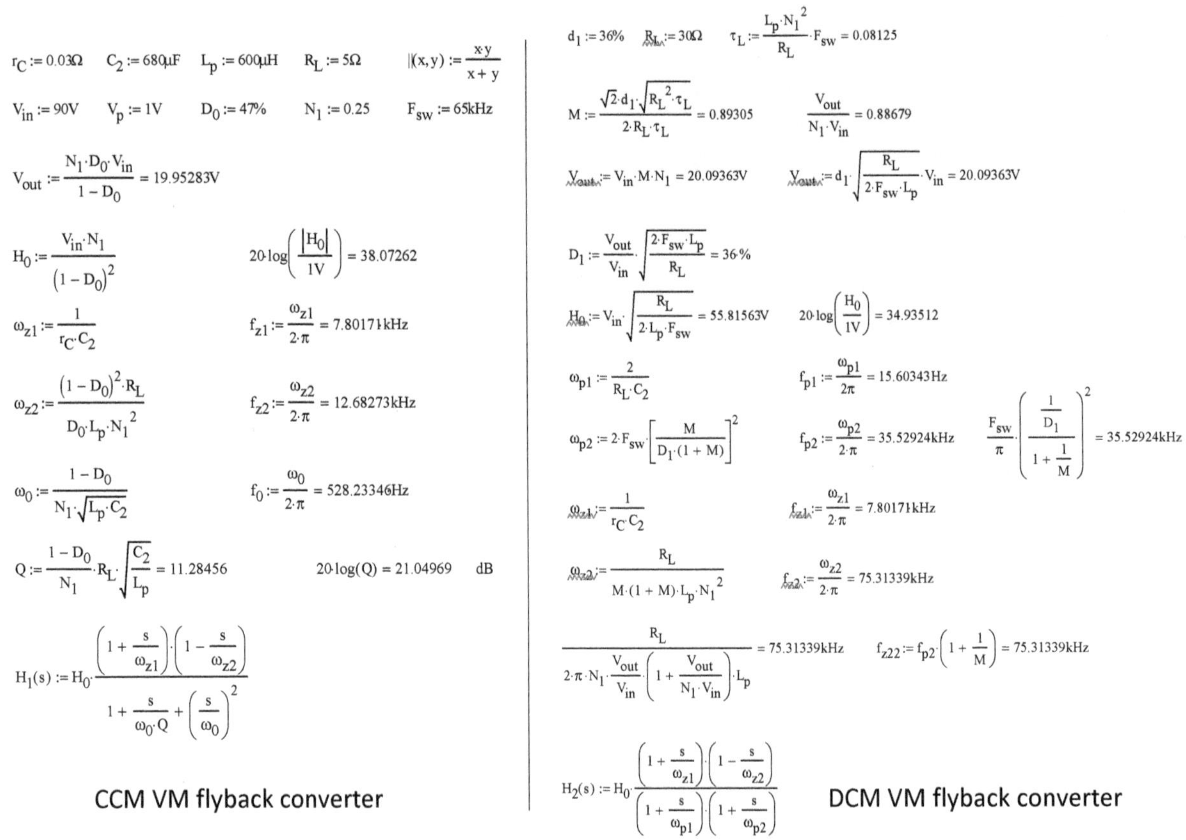

$r_C := 0.03\Omega \quad C_2 := 680\mu F \quad L_p := 600\mu H \quad R_L := 5\Omega \quad \|(x,y) := \frac{x \cdot y}{x+y}$

$d_1 := 36\% \quad R_L := 30\Omega \quad \tau_L := \frac{L_p \cdot N_1^2}{R_L} \cdot F_{sw} = 0.08125$

$V_{in} := 90V \quad V_p := 1V \quad D_0 := 47\% \quad N_1 := 0.25 \quad F_{sw} := 65kHz$

$M := \frac{\sqrt{2} \cdot d_1 \cdot \sqrt{R_L^2 \cdot \tau_L}}{2 \cdot R_L \cdot \tau_L} = 0.89305 \qquad \frac{V_{out}}{N_1 \cdot V_{in}} = 0.88679$

$V_{out} := \frac{N_1 \cdot D_0 \cdot V_{in}}{1 - D_0} = 19.95283V$

$V_{out} := V_{in} \cdot M \cdot N_1 = 20.09363V \qquad V_{out} := d_1 \sqrt{\frac{R_L}{2 \cdot F_{sw} \cdot L_p}} \cdot V_{in} = 20.09363V$

$H_0 := \frac{V_{in} \cdot N_1}{(1-D_0)^2} \qquad 20 \log\left(\frac{|H_0|}{1V}\right) = 38.07262$

$D_1 := \frac{V_{out}}{V_{in}} \sqrt{\frac{2 \cdot F_{sw} \cdot L_p}{R_L}} = 36\%$

$\omega_{z1} := \frac{1}{r_C \cdot C_2} \qquad f_{z1} := \frac{\omega_{z1}}{2\pi} = 7.80171 kHz$

$H_0 := V_{in} \sqrt{\frac{R_L}{2 \cdot L_p \cdot F_{sw}}} = 55.81563V \qquad 20 \log\left(\frac{H_0}{1V}\right) = 34.93512$

$\omega_{z2} := \frac{(1-D_0)^2 \cdot R_L}{D_0 \cdot L_p \cdot N_1^2} \qquad f_{z2} := \frac{\omega_{z2}}{2\pi} = 12.68273 kHz$

$\omega_{p1} := \frac{2}{R_L \cdot C_2} \qquad f_{p1} := \frac{\omega_{p1}}{2\pi} = 15.60343 Hz$

$\omega_{p2} := 2 \cdot F_{sw}\left[\frac{M}{D_1 \cdot (1+M)}\right]^2 \qquad f_{p2} := \frac{\omega_{p2}}{2\pi} = 35.52924 kHz \qquad \frac{F_{sw}}{\pi}\left(\frac{\frac{1}{D_1}}{1 + \frac{1}{M}}\right)^2 = 35.52924 kHz$

$\omega_0 := \frac{1-D_0}{N_1 \sqrt{L_p \cdot C_2}} \qquad f_0 := \frac{\omega_0}{2\pi} = 528.23346 Hz$

$\omega_{z1} := \frac{1}{r_C \cdot C_2} \qquad f_{z1} := \frac{\omega_{z1}}{2\pi} = 7.80171 kHz$

$Q := \frac{1-D_0}{N_1} \cdot R_L \sqrt{\frac{C_2}{L_p}} = 11.28456 \qquad 20 \log(Q) = 21.04969 \quad dB$

$\omega_{z2} := \frac{R_L}{M \cdot (1+M) \cdot L_p \cdot N_1^2} \qquad f_{z2} := \frac{\omega_{z2}}{2\pi} = 75.31339 kHz$

$\frac{R_L}{2 \pi \cdot N_1 \cdot \frac{V_{out}}{V_{in}}\left(1 + \frac{V_{out}}{N_1 \cdot V_{in}}\right) \cdot L_p} = 75.31339 kHz \qquad f_{z22} := f_{p2}\left(1 + \frac{1}{M}\right) = 75.31339 kHz$

$H_1(s) := H_0 \cdot \frac{\left(1 + \frac{s}{\omega_{z1}}\right)\left(1 - \frac{s}{\omega_{z2}}\right)}{1 + \frac{s}{\omega_0 \cdot Q} + \left(\frac{s}{\omega_0}\right)^2}$

$H_2(s) := H_0 \cdot \frac{\left(1 + \frac{s}{\omega_{z1}}\right)\left(1 - \frac{s}{\omega_{z2}}\right)}{\left(1 + \frac{s}{\omega_{p1}}\right)\left(1 + \frac{s}{\omega_{p2}}\right)}$

CCM VM flyback converter

DCM VM flyback converter

Figure 4.159 A Mathcad® sheet immediately delivers the transfer function gain, poles and zeroes.

4.6 The Flyback Converter Operated in Current-Mode Control

In current-mode control, we know from sampled-data analysis that the double poles present at half the switching frequency will make the magnitude peak at this point. Adding some compensation ramp will damp these poles and ensure a stable operation across line and output range. The control-to-output transfer function of the flyback converter operated in current-mode control and CCM is defined as follows:

$$H(s) = H_0 \frac{\left(1 + \frac{s}{\omega_{z_1}}\right)\left(1 - \frac{s}{\omega_{z_2}}\right)}{1 + \frac{s}{\omega_p}} \cdot \frac{1}{1 + \frac{s}{\omega_n Q_p} + \left(\frac{s}{\omega_n}\right)^2} \qquad (4.493)$$

with:

$$H_0 \approx \frac{R_{load}}{NR_i} \frac{1}{\frac{(1-D)^2}{2\tau_L}\left(1 + 2\frac{S_e}{S_n}\right) + 2M + 1} \qquad (4.494)$$

$$\omega_{z_1} = \frac{1}{r_C C_2} \qquad (4.495)$$

$$\omega_{z_2} \approx \frac{(1-D)^2 R_{load}}{DL_p N^2} \tag{4.496}$$

$$\omega_p \approx \frac{\dfrac{(1-D)^3}{2\tau_L}\left(1+2\dfrac{S_e}{S_n}\right)+1+D}{R_{load}C_2} \tag{4.497}$$

$$Q_p = \frac{1}{\pi\left(m_c D'-0.5\right)} \tag{4.498}$$

$$\omega_n = \frac{\pi}{T_{sw}} \tag{4.499}$$

$$m_c = 1+\frac{S_e}{S_n} \tag{4.500}$$

$$\tau_L = \frac{L_p N^2}{R_{load} T_{sw}} \tag{4.501}$$

$$S_n = \frac{V_{in}}{L_p}R_i \tag{4.502}$$

$$M = \frac{V_{out}}{NV_{in}} \tag{4.503}$$

To check our calculations, we will run a SPICE simulation as shown in Figure 4.160. Before we proceed, it is interesting to derive the setpoint for V_c, the control voltage. We want a 20-V output from a 90-V input. The ratio M is defined as:

$$M = \frac{V_{out}}{NV_{in}} = \frac{D}{1-D} \tag{4.504}$$

From which we extract the duty ratio D:

$$D = \frac{V_{out}}{V_{out}+NV_{in}} \tag{4.505}$$

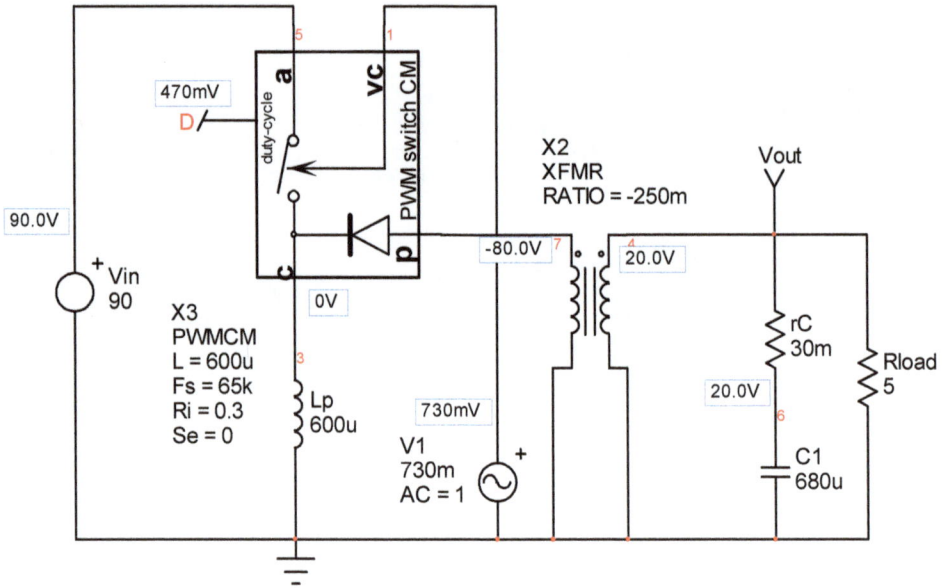

Figure 4.160 The flyback converter now operates in current-mode control and delivers the same 20-V output to a 5-Ω load.

To determine the control voltage, we can use an equation derived in the first chapter which describes the average current flowing out of terminal c:

$$I_c = \frac{V_c}{R_i} - \frac{S_e}{R_i} DT_{sw} - \frac{V_{in}}{2L_p} DT_{sw}$$

(4.506)

The average inductor current I_c is obtained by observing Figure 4.161. The average input current I_{in} is linked to the input power by the following equation when we consider a 100% efficiency:

$$I_{in} = \frac{P_{in}}{V_{in}} = \frac{P_{out}}{V_{in}}$$

(4.507)

From Figure 4.161 we can see that the instantaneous input current $i_{in}(t)$ is the average current I_c during DT_{sw}, we can write:

$$I_{in} = DI_c$$

(4.508)

From the input current definition, we then have:

$$I_c = \frac{I_{in}}{D} = \frac{P_{out}}{DV_{in}}$$

(4.509)

The output power depends on the load resistance:

$$P_{out} = \frac{V_{out}^2}{R_{load}} = \frac{(M \cdot N \cdot V_{in})^2}{R_{load}}$$

(4.510)

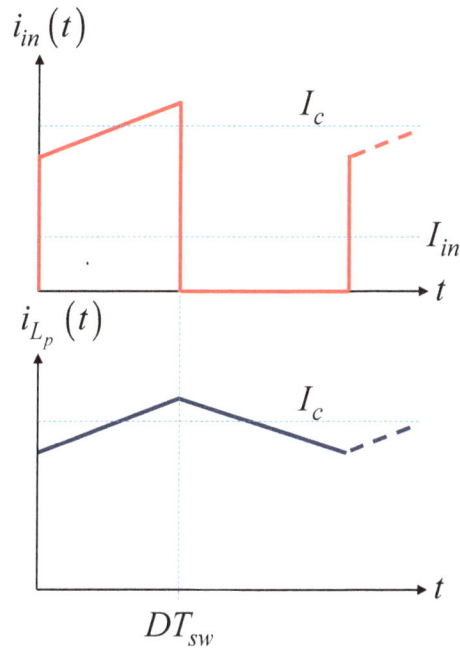

Figure 4.161 The average inductor current is obtained by averaging the input current.

Now substitute (4.510) in (4.509) and you have:

$$I_c = \frac{(M \cdot N)^2 V_{in}}{R_{load} D}$$

(4.511)

The voltage V_c must now be extracted from the following equation:

$$\frac{(M \cdot N)^2 V_{in}}{R_{load} D} = \frac{V_c}{R_i} - \frac{S_e}{R_i} DT_{sw} - \frac{V_{in}}{2L_p} DT_{sw}$$

(4.512)

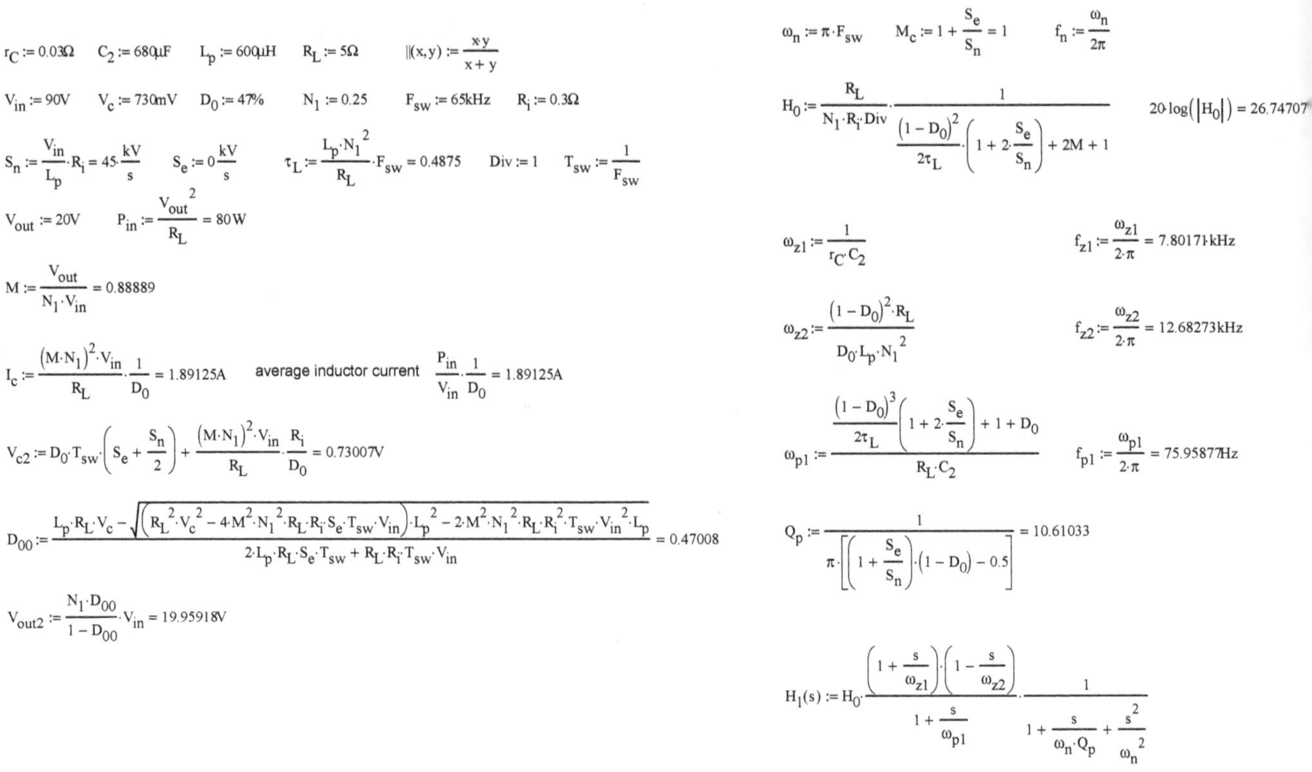

$r_C := 0.03\Omega$ $\quad C_2 := 680\mu F$ $\quad L_p := 600\mu H$ $\quad R_L := 5\Omega$ $\quad \|(x,y) := \dfrac{x \cdot y}{x+y}$

$V_{in} := 90V$ $\quad V_c := 730mV$ $\quad D_0 := 47\%$ $\quad N_1 := 0.25$ $\quad F_{sw} := 65kHz$ $\quad R_i := 0.3\Omega$

$S_n := \dfrac{V_{in}}{L_p} \cdot R_i = 45\dfrac{kV}{s}$ $\quad S_e := 0\dfrac{kV}{s}$ $\quad \tau_L := \dfrac{L_p \cdot N_1^2}{R_L} \cdot F_{sw} = 0.4875$ $\quad Div := 1$ $\quad T_{sw} := \dfrac{1}{F_{sw}}$

$V_{out} := 20V$ $\quad P_{in} := \dfrac{V_{out}^2}{R_L} = 80W$

$M := \dfrac{V_{out}}{N_1 \cdot V_{in}} = 0.88889$

$I_c := \dfrac{(M \cdot N_1)^2 \cdot V_{in}}{R_L} \cdot \dfrac{1}{D_0} = 1.89125A$ $\qquad$ average inductor current $\qquad \dfrac{P_{in}}{V_{in}} \cdot \dfrac{1}{D_0} = 1.89125A$

$V_{c2} := D_0 \cdot T_{sw} \cdot \left(S_e + \dfrac{S_n}{2}\right) + \dfrac{(M \cdot N_1)^2 \cdot V_{in}}{R_L} \cdot \dfrac{R_i}{D_0} = 0.73007V$

$D_{00} := \dfrac{L_p \cdot R_L \cdot V_c - \sqrt{\left(R_L^2 \cdot V_c^2 - 4 \cdot M^2 \cdot N_1^2 \cdot R_L \cdot R_i \cdot S_e \cdot T_{sw} \cdot V_{in}\right) \cdot L_p^2 - 2 \cdot M^2 \cdot N_1^2 \cdot R_L \cdot R_i^2 \cdot T_{sw} \cdot V_{in}^2 \cdot L_p}}{2 \cdot L_p \cdot R_L \cdot S_e \cdot T_{sw} + R_L \cdot R_i \cdot T_{sw} \cdot V_{in}} = 0.47008$

$V_{out2} := \dfrac{N_1 \cdot D_{00}}{1 - D_{00}} \cdot V_{in} = 19.95918V$

$\omega_n := \pi \cdot F_{sw}$ $\quad M_c := 1 + \dfrac{S_e}{S_n} = 1$ $\quad f_n := \dfrac{\omega_n}{2\pi}$

$H_0 := \dfrac{R_L}{N_1 \cdot R_i \cdot Div} \cdot \dfrac{1}{\dfrac{(1-D_0)^2}{2\tau_L}\left(1 + 2 \cdot \dfrac{S_e}{S_n}\right) + 2M + 1}$ $\qquad 20 \cdot \log\left(|H_0|\right) = 26.74707$

$\omega_{z1} := \dfrac{1}{r_C \cdot C_2}$ $\qquad f_{z1} := \dfrac{\omega_{z1}}{2 \cdot \pi} = 7.80171 kHz$

$\omega_{z2} := \dfrac{(1 - D_0)^2 \cdot R_L}{D_0 \cdot L_p \cdot N_1^2}$ $\qquad f_{z2} := \dfrac{\omega_{z2}}{2 \cdot \pi} = 12.68273 kHz$

$\omega_{p1} := \dfrac{\dfrac{(1-D_0)^3}{2\tau_L}\left(1 + 2 \cdot \dfrac{S_e}{S_n}\right) + 1 + D_0}{R_L \cdot C_2}$ $\qquad f_{p1} := \dfrac{\omega_{p1}}{2 \cdot \pi} = 75.95877 Hz$

$Q_p := \dfrac{1}{\pi \cdot \left[\left(1 + \dfrac{S_e}{S_n}\right) \cdot (1 - D_0) - 0.5\right]} = 10.61033$

$H_1(s) := H_0 \cdot \dfrac{\left(1 + \dfrac{s}{\omega_{z1}}\right)\left(1 - \dfrac{s}{\omega_{z2}}\right)}{1 + \dfrac{s}{\omega_{p1}}} \cdot \dfrac{1}{1 + \dfrac{s}{\omega_n \cdot Q_p} + \dfrac{s^2}{\omega_n^2}}$

Figure 4.162 The Mathcad® sheet computes the operating points but also the poles and zeroes locations.

It leads to:

$$V_c = DT_{sw}\left(S_e + \frac{S_n}{2}\right) + \frac{(M \cdot N)^2 R_i V_{in}}{DR_{load}} \tag{4.513}$$

From this equation, we can also infer the raw duty ratio if you know the control voltage:

$$D = \frac{L_p R_{load} V_c - \sqrt{\left(R_{load}^2 V_c^2 - 4M^2 N^2 R_{load} R_i S_e T_{sw} V_{in}\right) L_p^2 - 2M^2 N^2 R_{load} R_i^2 T_{sw} V_{in}^2 L_p}}{2L_p R_{load} S_e T_{sw} + R_{load} R_i T_{sw} V_{in}} \tag{4.514}$$

We have captured all these equations in a Mathcad® sheet and the computed results appear in Figure 4.162. For a 20-V output, the control voltage V_c must be set to 730 mV and this is confirmed by the reflected bias points in Figure 4.160 (node D is biased to 470 mV which is 47%). Finally, we have compared the simulation results of Figure 4.160 with what Mathcad® delivers and results are identical as confirmed by Figure 4.163.

As a final check, we can run a SIMPLIS® simulation with the circuit of Figure 4.164 and check the response. This is what is done in Figure 4.165 and the agreement is excellent despite a 2.5-dB deviation for the double poles Q_p.

If we reduce the load to 30 Ω, the converter enters DCM as we have seen in voltage mode. The simulation circuit in this case is that of Figure 4.166. We have used the CCM VM-PWM switch model that we feed with the voltage at node N. It does exactly what was shown in Figure 4.117 but in a more compact form.

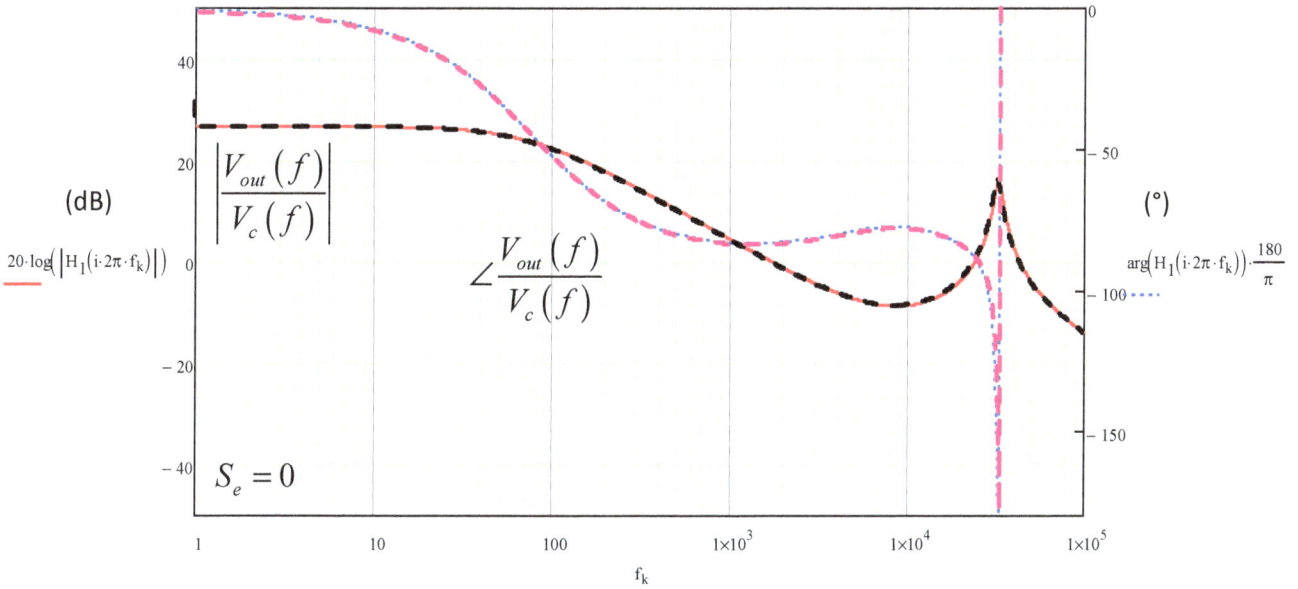

Figure 4.163 The agreement between equations and simulation is excellent.

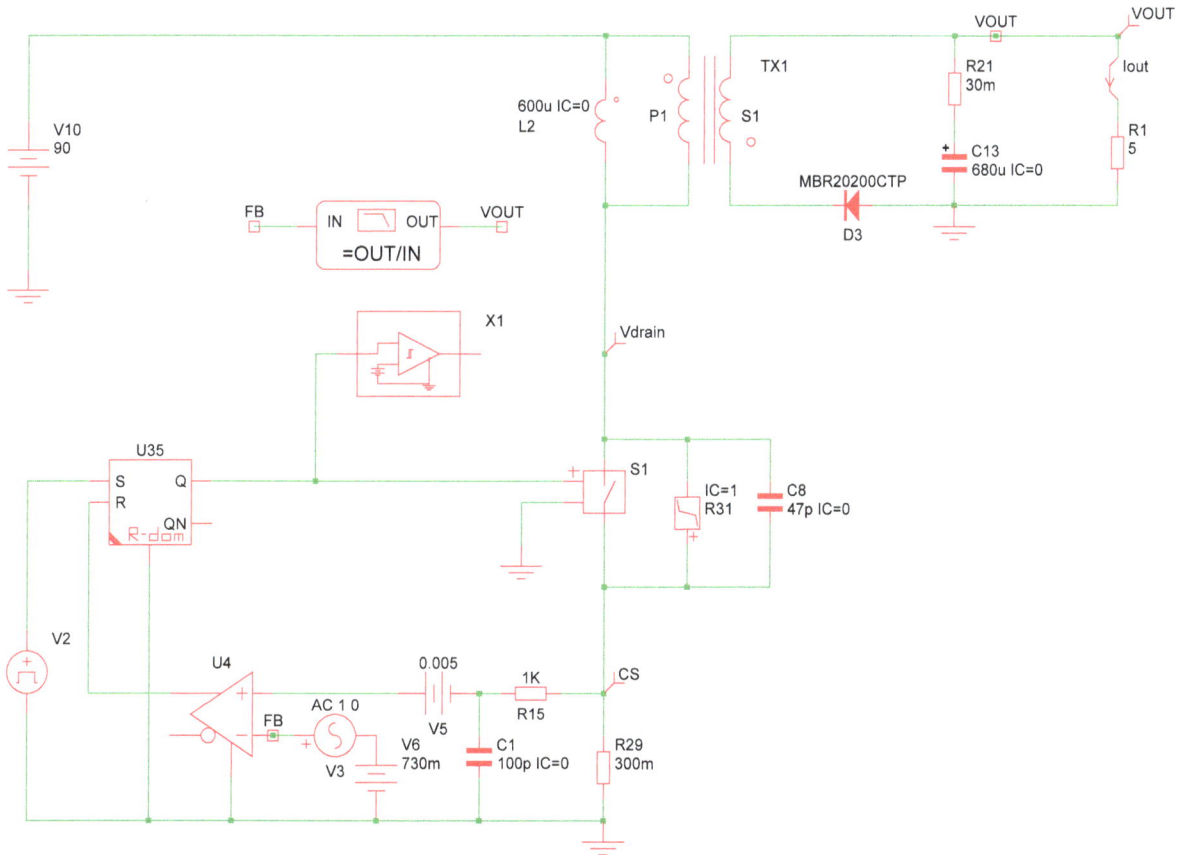

Figure 4.164 SIMPLIS® lets us quickly verify if our analysis is correct.

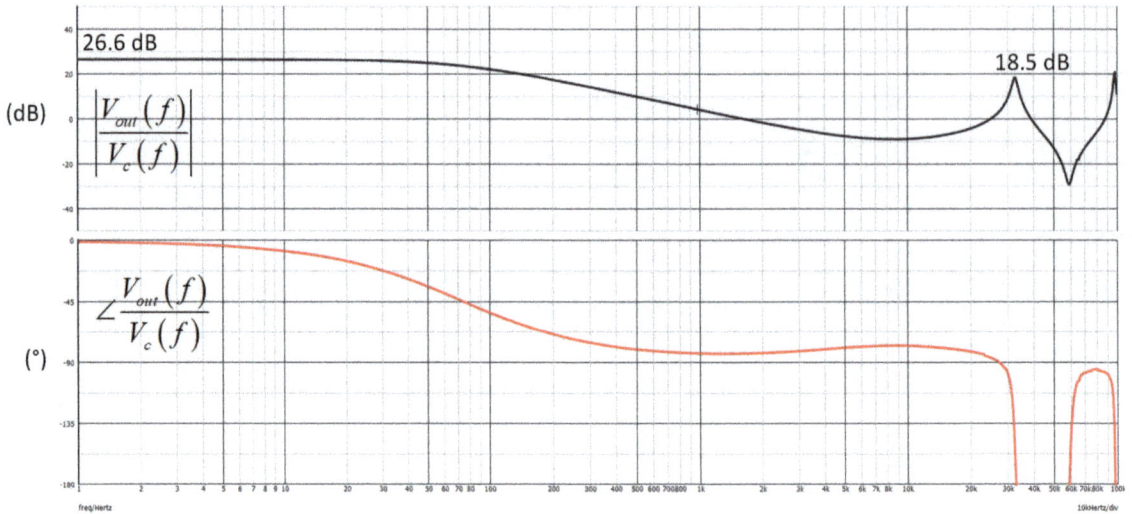

Figure 4.165 The magnitude and phase responses from SIMPLIS® are very close to the theoretical analysis.

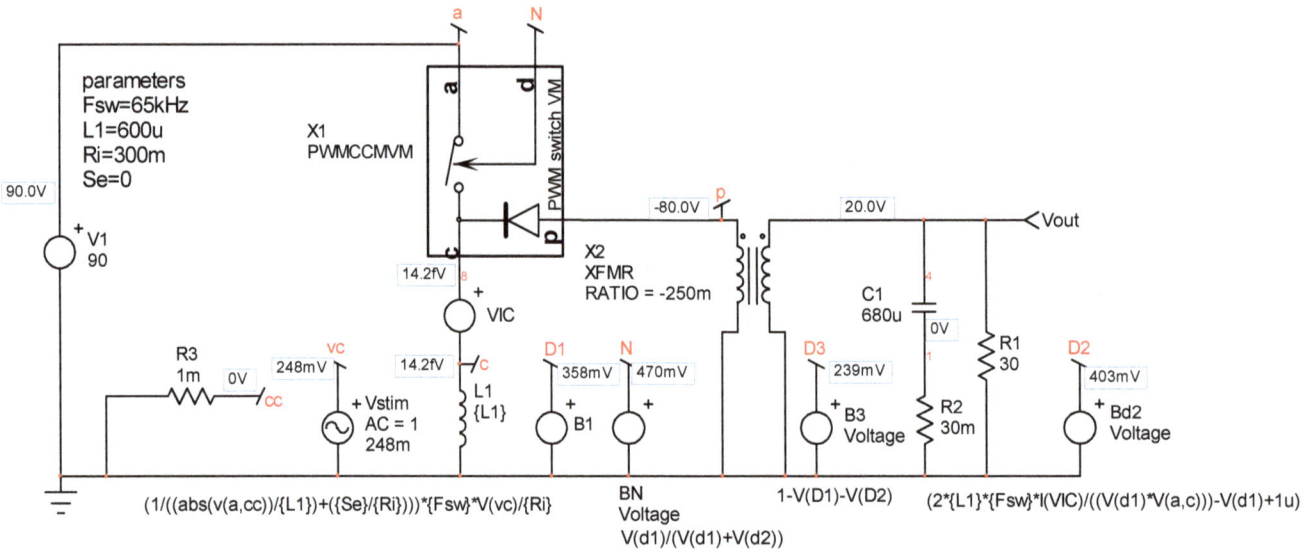

Figure 4.166 The converter now operates in DCM with a deadtime of 24% (source B_3).

The equations describing the control-to-output transfer function of the CM DCM flyback converter use those derived for the buck-boost converter but account for the transformer turns ratio scaling:

$$H(s) = \frac{V_{out}(s)}{V_c(s)} = H_0 \frac{\left(1 + \frac{s}{\omega_{z_1}}\right)\left(1 - \frac{s}{\omega_{z_2}}\right)}{\left(1 + \frac{s}{\omega_{p_1}}\right)\left(1 + \frac{s}{\omega_{p_2}}\right)} \tag{4.515}$$

where:

$$H_0 \approx \frac{1}{S_n m_c T_{sw}} \frac{N V_{in}}{\sqrt{2\tau_L}} \tag{4.516}$$

$$\omega_{p_1} = \frac{2}{R_{load} C_2} \tag{4.517}$$

$$\omega_{p_2} \approx 2F_{sw} \left(\frac{1/D}{1 + \frac{1}{M}} \right)^2 \tag{4.518}$$

$$\omega_{z_1} = \frac{1}{r_C C_2} \tag{4.519}$$

$$\omega_{z_2} = \frac{R_{load}}{M(1+M)N^2 L_p} \tag{4.520}$$

$$m_c = 1 + \frac{S_e}{S_n} \tag{4.521}$$

$$\tau_L = \frac{L_p N^2}{R_{load} T_{sw}} \tag{4.522}$$

$$S_n = \frac{V_{in}}{L_p} R_i \tag{4.523}$$

$$M = \frac{V_{out}}{N V_{in}} \tag{4.524}$$

The individual values are computed in the Mathcad® sheet shown in Figure 4.167.

$R_L := 30\Omega \qquad \tau_L := \frac{L_p \cdot N_1^{\,2}}{R_L} \cdot F_{sw} = 0.08125 \qquad V_{out} = 20\ V \qquad V_c := 248mV$

$D_1 := \frac{F_{sw} \cdot V_c}{R_i} \cdot \frac{1}{\frac{V_{in}}{L_p} + \frac{S_e}{R_i}} = 35.82222\ \%$

$M := \frac{\sqrt{2} \cdot D_1 \cdot \sqrt{R_L^{\,2} \cdot \tau_L}}{2 \cdot R_L \cdot \tau_L} = 0.88864 \qquad V_{out} := D_1 \cdot \sqrt{\frac{R_L}{2 \cdot F_{sw} \cdot L_p}} \cdot V_{in} = 19.9944\ V \qquad M_1 := \frac{D_1}{\sqrt{2 \cdot \tau_L}} = 0.88864$

$V_{out} := M \cdot V_{in} \cdot N_1 = 19.9944\ V$

$H_0 := V_{in} \cdot \sqrt{\frac{R_L \cdot F_{sw}}{2 \cdot L_p}} \cdot \frac{1}{S_e + S_n} = 80.62258 \qquad 20 \cdot \log(H_0) = 38.12913 \qquad \frac{1}{S_n \cdot M_c \cdot T_{sw}} \cdot \frac{V_{in} \cdot N_1}{\sqrt{2 \cdot \tau_L}} = 80.62258$

$\omega_{p1} := \frac{2}{R_L \cdot C_2} \qquad\qquad f_{p1} := \frac{\omega_{p1}}{2\pi} = 15.60343\ \cdot Hz$

$\omega_{p2} := 2 \cdot F_{sw} \cdot \left[\frac{M}{D_1 \cdot (1+M)} \right]^2 \qquad f_{p2} := \frac{\omega_{p2}}{2 \cdot \pi} = 35.69536\ \cdot kHz \qquad \frac{F_{sw}}{\pi} \cdot \left(\frac{\frac{1}{D_1}}{1 + \frac{1}{M}} \right)^2 = 35.69536\ \cdot kHz$

$\omega_{z1} := \frac{1}{r_C \cdot C_2} \qquad\qquad f_{z1} := \frac{\omega_{z1}}{2 \cdot \pi} = 7.80171\ \cdot kHz$

$\omega_{z2} := \frac{R_L}{M \cdot (1+M) \cdot L_p \cdot N_1^{\,2}} \qquad f_{z2} := \frac{\omega_{z2}}{2 \cdot \pi} = 75.86389\ \cdot kHz$

$\dfrac{R_L}{2 \cdot \pi \cdot N_1 \cdot \frac{V_{out}}{V_{in}} \cdot \left(1 + \frac{V_{out}}{N_1 \cdot V_{in}} \right) \cdot L_p} = 75.86389\ \cdot kHz \qquad f_{z22} := f_{p2} \cdot \left(1 + \frac{1}{M} \right) = 75.86389\ \cdot kHz$

$H_2(s) := H_0 \cdot \dfrac{\left(1 + \frac{s}{\omega_{z1}} \right) \cdot \left(1 - \frac{s}{\omega_{z2}} \right)}{\left(1 + \frac{s}{\omega_{p1}} \right) \cdot \left(1 + \frac{s}{\omega_{p2}} \right)}$

Figure 4.167 These expressions determine the control-to-output transfer function poles and zeroes.

Finally, Figure 4.168 compares the SPICE results with the Mathcad® plots and results are identical. We have conveniently gathered the control-to-output transfer functions of the CM flyback converters in both operating modes in Figure 4.169.

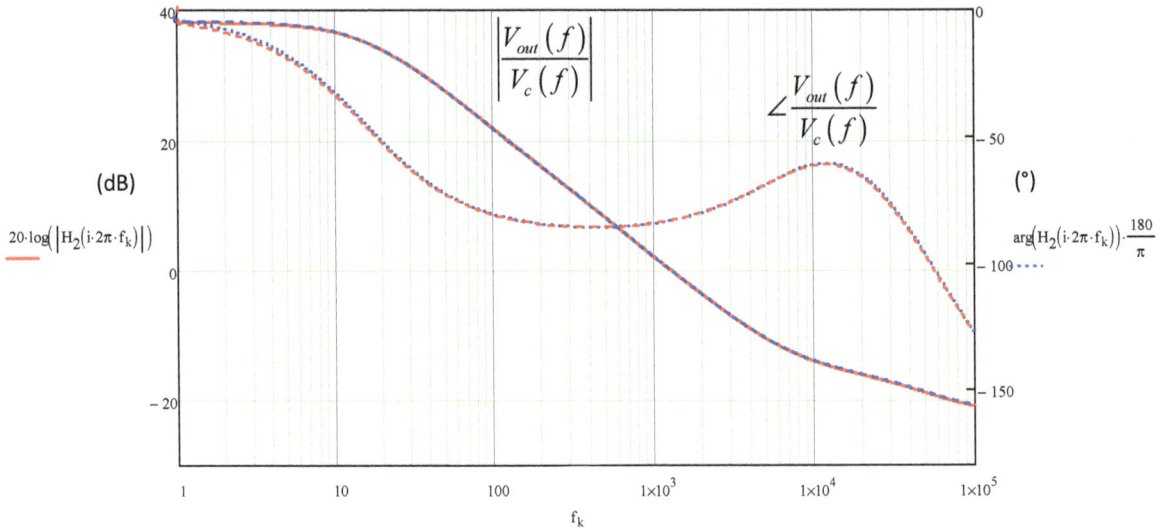

Figure 4.168 Mathcad® and SPICE deliver similar magnitude and phase curves.

$\dfrac{V_{out}(s)}{V_c(s)}$ CCM	$H_0\dfrac{\left(1+\dfrac{s}{\omega_{z_1}}\right)\left(1-\dfrac{s}{\omega_{z_2}}\right)}{\left(1+\dfrac{s}{\omega_{p_1}}\right)\dfrac{1}{1+\dfrac{s}{\omega_n Q_p}+\left(\dfrac{s}{\omega_n}\right)^2}}$	$\omega_{z_1}=\dfrac{1}{r_C C_2}$ $\omega_{z_2}\approx\dfrac{(1-D)^2 R_{load}}{D L_p N^2}$	$\omega_{p_1}\approx\dfrac{\dfrac{(1-D)^3}{2\tau_L}\left(1+2\dfrac{S_e}{S_n}\right)+1+D}{R_{load}C_2}$ $Q_p=\dfrac{1}{\pi(m_c D'-0.5)}$ $\omega_n=\dfrac{\pi}{T_{sw}}$	$H_0\approx\dfrac{R_{load}}{NR_i}\dfrac{1}{\dfrac{(1-D)^2}{2\tau_L}\left(1+2\dfrac{S_e}{S_n}\right)+2M+}$

$$m_c=1+\frac{S_e}{S_n}\quad S_e\text{ in V/s}\quad S_n=\frac{V_{in}}{L_p}R_i\quad M=\frac{V_{out}}{NV_{in}}=\frac{D}{1-D}\quad\text{Turns ratio }1:N\quad\tau_L=\frac{L_p N^2}{R_{load}}F_{sw}$$

$\dfrac{V_{out}(s)}{V_c(s)}$ DCM	$H_0\dfrac{\left(1+\dfrac{s}{\omega_{z_1}}\right)\left(1-\dfrac{s}{\omega_{z_2}}\right)}{\left(1+\dfrac{s}{\omega_{p_1}}\right)\left(1+\dfrac{s}{\omega_{p_2}}\right)}$	$\omega_{z_1}=\dfrac{1}{r_C C_2}$ $\omega_{z_2}=\dfrac{R_{load}}{M(1+M)L_p N^2}$	$\omega_{p_1}\approx\dfrac{2}{R_{load}C_2}$ $\omega_{p_2}\approx 2F_{sw}\left[\dfrac{1/D}{1+\dfrac{1}{M}}\right]^2$	$H_0\approx\dfrac{1}{S_n m_c T_{sw}}\dfrac{NV_{in}}{\sqrt{2\tau_L}}$

$$m_c=1+\frac{S_e}{S_n}\quad S_e\text{ in V/s}\quad S_n=\frac{V_{in}}{L_p}R_i\quad M=\frac{V_{out}}{NV_{in}}=\frac{D}{\sqrt{2\tau_L}}\quad\text{Turns ratio }1:N\quad\tau_L=\frac{L_p N^2}{R_{load}}F_{sw}$$

Figure 4.169 These are the transfer functions of the CM flyback converter when operated in CCM and DCM.

4.7 The Flyback Converter Operated in Quasi-Square-Wave Resonant Mode

The flyback converter lends itself very well to operating in the so-called quasi-resonant mode. In this mode, there is no internal clock and the system is self-relaxing: a first pulse turns the power MOSFET on and turns it off when the peak current has reached the target imposed by the feedback loop. Energy is then transferred from the primary to secondary side and the transformer starts demagnetizing. When the primary inductance is completely depleted, a circuit detects this event via an auxiliary winding and turns the power switch back on again for a new cycle. Operation then repeats with a frequency and a peak current depending on the input/output conditions.

I have derived the switching frequency and the small-signal model in [5]. A different approach based on a different model (the loss-free model) was proposed in [6] and both approaches lead to identical results. The QR PWM switch model proposed in Chapter 1 can be implemented to simulate this flyback converter operated in current mode. As explained, a delay can be inserted to make sure the power switch turns back on in the valley of the drain-source voltage so as to minimize turn-on losses. It is possible to show that this extra delay affects the operating frequency by inserting a deadtime and slightly impacts the dc gain. However, it can safely be neglected. The circuit featuring the average model appears in Figure 4.170. The voltage is still 20 V and the converter delivers 80 W to a 5-Ω load. The model computes the switching frequency (no deadtime) which is 57.6 kHz in this example. The ac response is given in Figure 4.171 and shows a first-order behavior however affected by a right-half-plane zero. I have reproduced below the details of the control-to-output transfer function and it starts with the operating frequency:

$$F_{sw} = \frac{4}{\left(\sqrt{4DT + \frac{2L_p P_{out} \left(V_f + V_{out} + NV_{in}\right)^2}{\eta V_{in}^2 \left(V_{out} + V_f\right)^2}} + \frac{\sqrt{2}L_p \left(V_f + V_{out} + NV_{in}\right)\sqrt{\frac{P_{out}}{\eta L_p}}}{V_{in}\left(V_{out} + V_f\right)} \right)^2}$$

(4.525)

In this expression, DT is the deadtime which depends on the capacitance lumped at the drain. It represents the total capacitance seen at this node such as the MOSFET C_{oss}, the transformer parasitic capacitance and so on.

$$DT = \pi \sqrt{L_p C_{lump}}$$

(4.526)

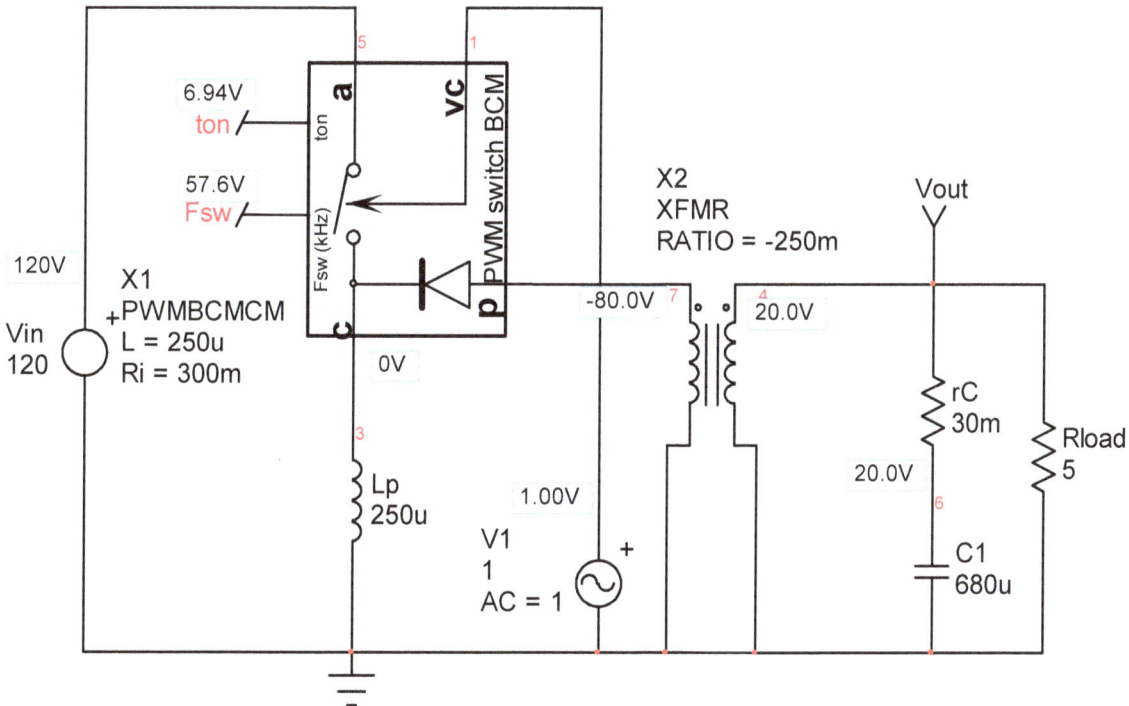

Figure 4.170 This average model lets you simulate a quasi-resonant converter such as a current-mode flyback.

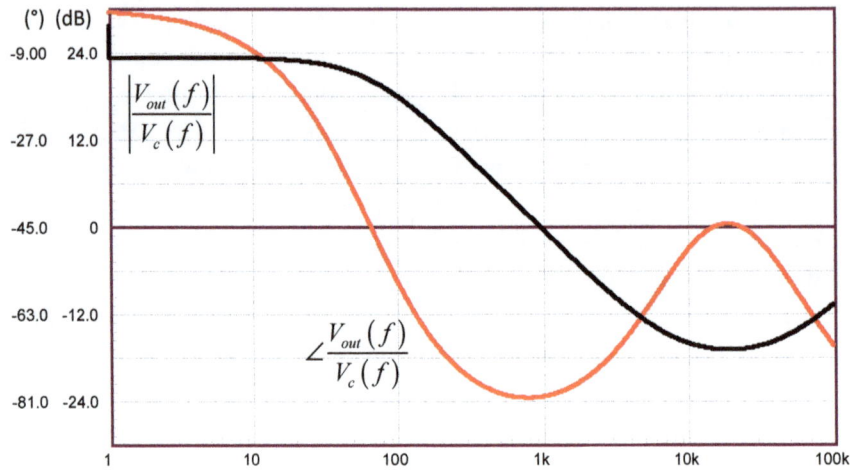

Figure 4.171 We will study the control-to-output transfer function with this first-order circuit.

V_f is the output diode forward drop and η is the power stage efficiency. The rest are known parameters. The transfer function of this converter is described by the following expression:

$$H(s) = \frac{V_{out}(s)}{V_c(s)} = H_0 \frac{\left(1 + \dfrac{s}{\omega_{z_1}}\right)\left(1 - \dfrac{s}{\omega_{z_2}}\right)}{1 + \dfrac{s}{\omega_p}} \tag{4.527}$$

where [6]:

$$H_0 \approx \frac{R_{load}}{2NR_i(1 + 2M)} \tag{4.528}$$

$$\omega_p = \frac{1}{R_{load}C_2} \frac{1 + 2M}{1 + M} \tag{4.529}$$

$$\omega_{z_1} = \frac{1}{r_C C_2} \tag{4.530}$$

$$\omega_{z_2} = \frac{R_{load}}{N^2 L_p} \frac{1}{M(1 + M)} \tag{4.531}$$

$$M = \frac{V_{out}}{NV_{in}} \tag{4.532}$$

When these equations are entered in the Mathcad® sheet of Figure 4.172, you obtain the plot of Figure 4.173. It is identical to that of Figure 4.171 therefore confirming the given expression.

$L_p := 250 \mu H$ $N_1 = 0.25$ $C_{lump} := 1pF$ $V_{out} := 20V$ $V_f := 0V$ $R_L := 5\Omega$ $V_{in} := 120V$

$DT := \pi \sqrt{L_p \cdot C_{lump}} = 49.67294 ns$ $P_{out} := \dfrac{V_{out}^2}{R_L} = 80W$ $\eta := 100\%$ $M := \dfrac{V_{out}}{N_1 \cdot V_{in}}$ $V_c := 1V$

$$F_{sw} := \dfrac{4}{\left[\sqrt{4 \cdot DT + \dfrac{2 \cdot L_p \cdot P_{out} \cdot (V_f + V_{out} + N_1 \cdot V_{in})^2}{\eta \cdot V_{in}^2 \cdot (V_{out} + V_f)^2}} + \dfrac{\sqrt{2} \cdot L_p \cdot (V_f + V_{out} + N_1 \cdot V_{in}) \cdot \sqrt{\dfrac{P_{out}}{\eta \cdot L_p}}}{V_{in} \cdot (V_{out} + V_f)}\right]^2} = 57.27273 kHz$$

$H_0 := \dfrac{R_L}{2 \cdot N_1 \cdot R_i \cdot (1 + 2 \cdot M)} = 14.28571$ $20 \cdot \log(H_0) = 23.09804$ dB

$k_{cp} := \dfrac{V_{in} \cdot V_c \cdot N_1^2}{2 \cdot R_i \cdot (V_{out} + N_1 \cdot V_{in})^2} = 5 \times 10^{-3} \dfrac{1}{\Omega}$ $k_{ic} := \dfrac{V_{out}}{V_{out} + N_1 \cdot V_{in}} = 0.4$

$k_{ac} := \dfrac{V_{out} \cdot V_c \cdot N_1}{2 \cdot R_i \cdot (V_{out} + N_1 \cdot V_{in})^2} = 3.33333 \times 10^{-3} \dfrac{1}{\Omega}$ $k_c := \dfrac{1}{2 \cdot R_i} = 1.66667 \dfrac{1}{\Omega}$

$R_{eq} := \dfrac{R_L}{R_L \cdot k_{cp} + N_1^2} = 57.14286 \Omega$

$H_{00} := \dfrac{N_1 \cdot k_c \cdot (1 - k_{ic})}{k_{cp} + \dfrac{N_1^2}{R_L}} = 14.28571$ $20 \cdot \log(H_{00}) = 23.09804$ dB

$\omega_{p1} := \dfrac{1}{R_L \cdot C_2} \cdot \dfrac{2 \cdot M + 1}{M + 1}$ $f_{p1} := \dfrac{\omega_{p1}}{2 \cdot \pi} = 65.53439 Hz$

$\omega_{p11} := \dfrac{1}{C_2 \left(\dfrac{N_1^2 + \dfrac{N_1^2 \cdot r_C}{R_L} + k_{cp} \cdot r_C}{k_{cp} + \dfrac{N_1^2}{R_L}}\right)}$ $f_{p11} := \dfrac{\omega_{p11}}{2 \cdot \pi} = 64.98849 Hz$

$\omega_{z1} := \dfrac{1}{r_C \cdot C_2}$ $f_{z1} := \dfrac{\omega_{z1}}{2 \cdot \pi} = 7.80171 kHz$

$\omega_{z22} := \dfrac{2 \cdot R_i \cdot V_{in}}{L_p \cdot V_c}$ $f_{z22} := \dfrac{\omega_{z22}}{2 \cdot \pi} = 45.83662 kHz$

$\omega_{z2} := \dfrac{R_L}{N_1^2 \cdot L_p} \cdot \dfrac{1}{M \cdot (1 + M)}$ $f_{z2} := \dfrac{\omega_{z2}}{2 \cdot \pi} = 45.83662 kHz$

$$H_2(s) := H_0 \cdot \dfrac{\left(1 + \dfrac{s}{\omega_{z1}}\right)\left(1 - \dfrac{s}{\omega_{z2}}\right)}{1 + \dfrac{s}{\omega_{p1}}}$$

Figure 4.172 This sheet computes the operating frequency for a 1-V control voltage.

To further test these results, I have assembled a CM QR-converter in SIMPLIS® and simulated the converter. The schematic diagram appears in Figure 4.174. When switch S_1 turns on, the 250-μH primary inductance is biased by the input voltage. The current grows until the voltage across the sense resistor R_{29} reaches the 1-V setpoint. At this moment, S_1 turns off and energy is transferred to the secondary side. L_1 depletes and the auxiliary winding reflects the output voltage scaled by the turns ratio. When the voltage across C_1 passes below the 45-mV threshold, the converter detects the complete demagnetization – the current in L_1 is zero – and a new cycle takes place. The time constant $R_2 C_1$ inserts a small delay which ensures valley switching operation (S_1 is turned back on when the voltage across its drain-source terminals has gone through a minimum or a valley) and minimizes switching losses. Transient results are reproduced in Figure 4.175.

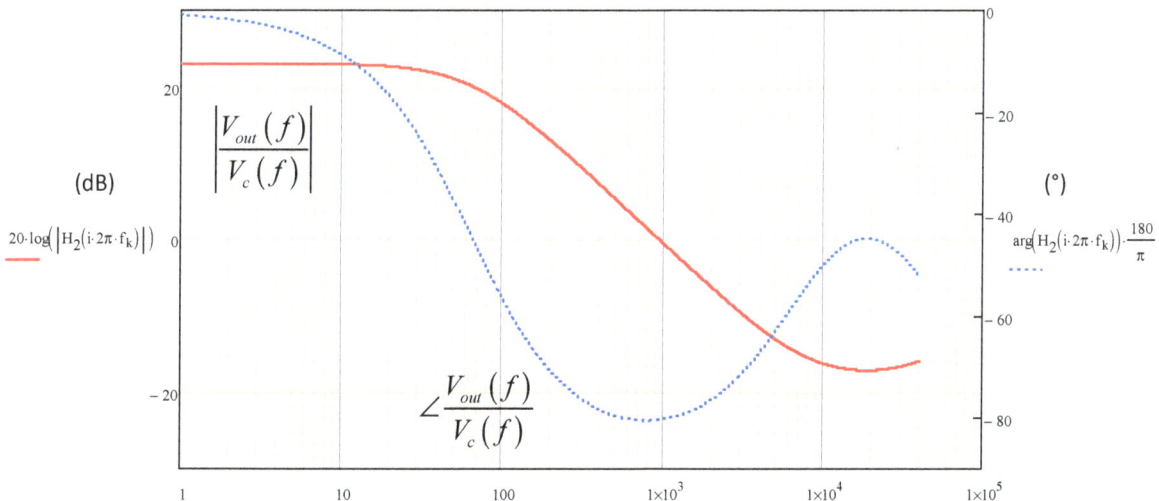

Figure 4.173 The transfer function shows a 1st-order response with a high-frequency RHP zero.

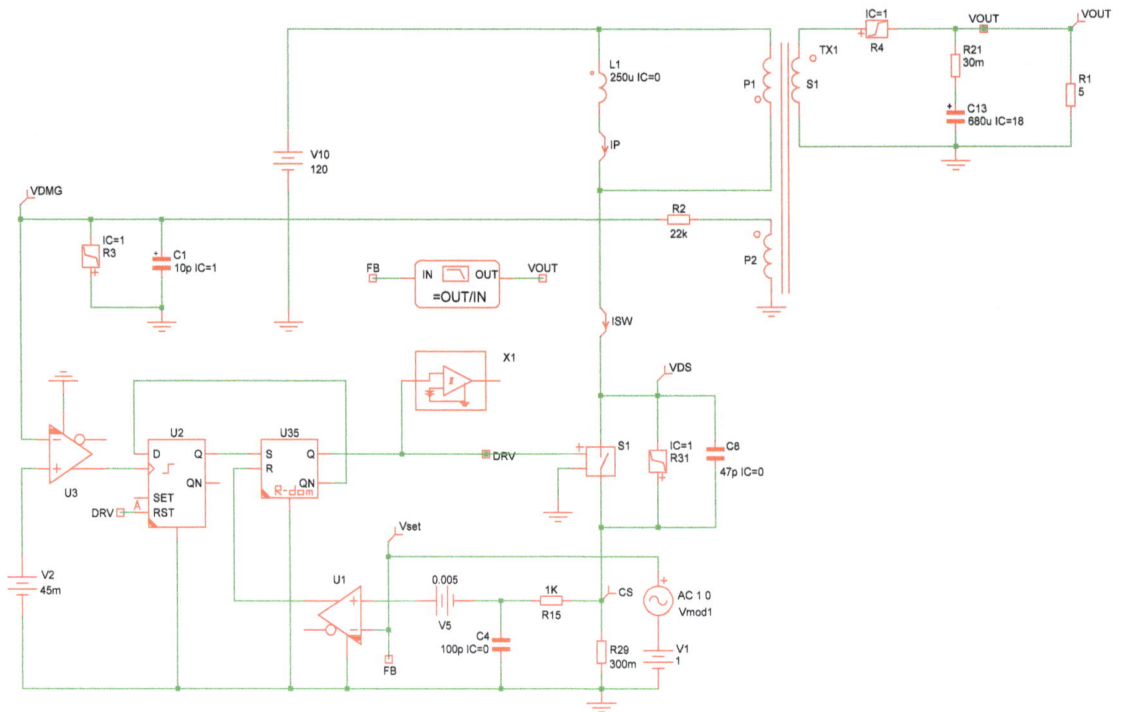

Figure 4.174 SIMPLIS® easily simulates this QR converter delivering 20 V to the 5-Ω load.

The circuit does not feature parasitics such as the leakage inductance and that is why the drain-source waveform is non-ringing at the switch opening. However, you can clearly see the minimum voltage at which the power switch is turned on again. The small-signal response is obtained by inserting an ac source in series with the 1-V setpoint. The control-to-output transfer function is shown in Figure 4.176 and confirms the results reproduced in Figure 4.173.

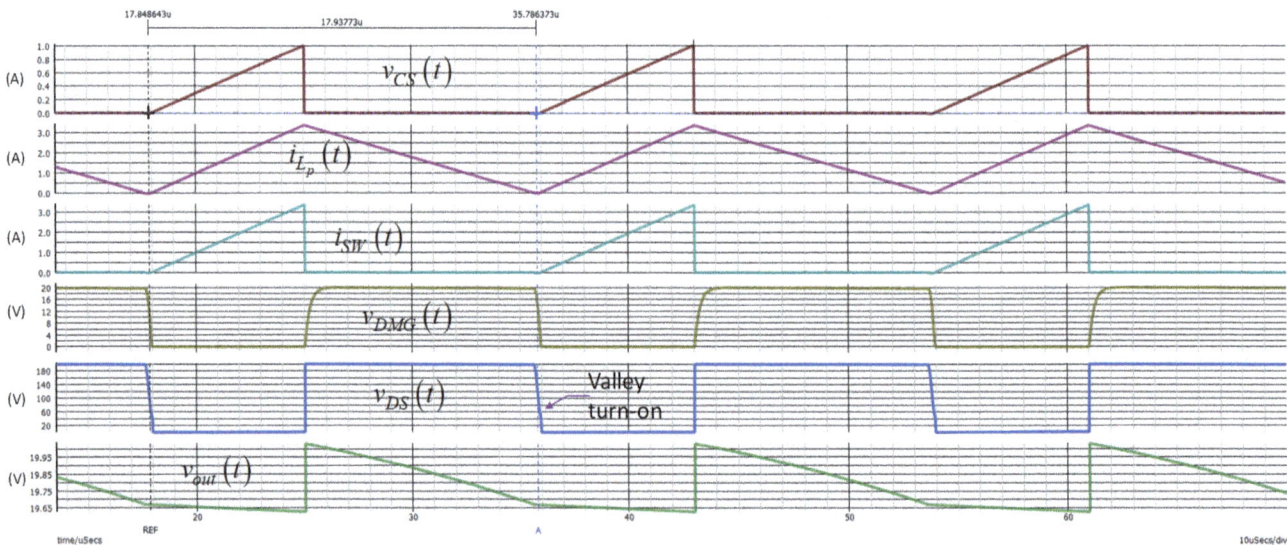

Figure 4.175 The MOSFET is turned on in when the drain-source voltage goes through a minimum.

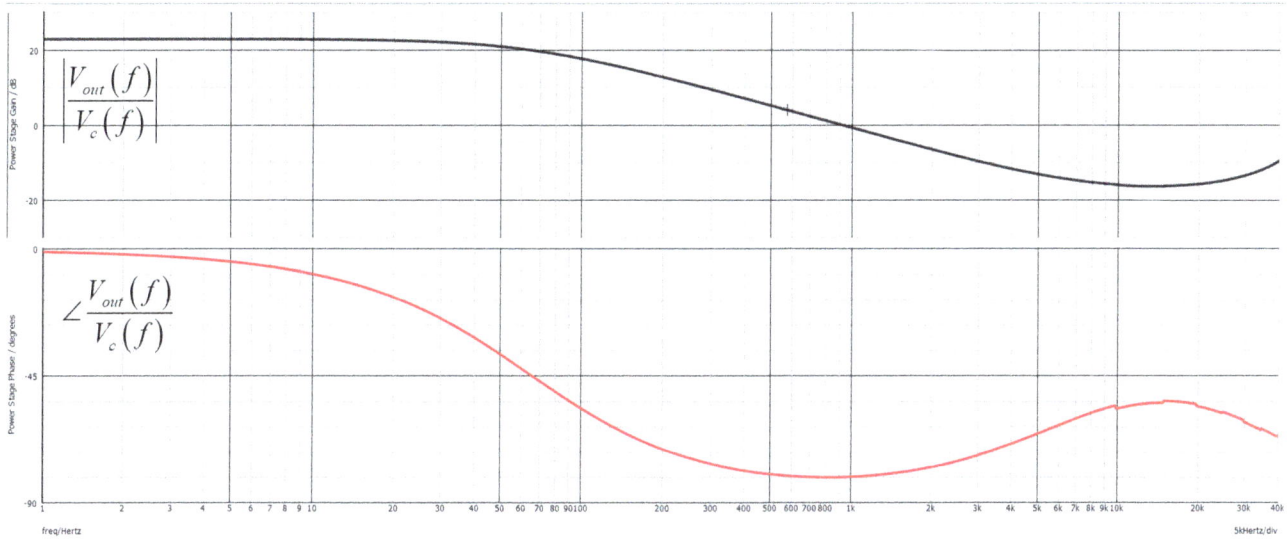

Figure 4.176 The control-to-output transfer function is of first-order type with a right-half-plane zero.

4.8 The Flyback in Single-Stage Power Factor Correction – Current Mode

The flyback converter can be used in power factor correction circuits (PFC). In this case, we talk about *single-stage* operation as you combine a PFC and an isolated converter. The single-stage approach offers several advantages over the non-isolated boost converter: galvanic isolation and low-voltage output capacitance for instance. However, compared to a cascaded structure in which you associate a boost converter with a flyback converter, the output ripple is quite large and the response time is slow. Furthermore, this structure provides no high-voltage reservoir – as with a classical bulk capacitor – and cannot provide any hold-up time capability. Nevertheless, the single-stage flyback gains popularity in lighting applications.

There are plenty of possible controllers to choose from when building a single-stage flyback converter. We will select the popular MC33262 introduced by Motorola in the 80's. This borderline conduction mode (BCM) current-mode controller hosts a multiplier which combines an image of the scaled-down input voltage $v_{in}(t)$ with the control voltage $v_c(t)$ delivered by the loop. The resulting level drives the current setpoint of the current-sense comparator and shapes the primary-side peak current setpoint following a sinusoidal waveshape imposed by the rectified mains – $v_{in}(t)$ – and modulated in amplitude by v_c according to the power demand. A small offset is added to the current setpoint via a divider k_{off}. This offset is there to artificially increase the on-time duration around the zero-volt region of the input sinewave to improve crossover distortion and reduce switching frequency runaway in this area.

The section of the circuit we are interested in appears in Figure 4.177. In this figure, an image of the rectified voltage $v_{in}(t)$ undergoes a reduction via coefficient k_m. It biases one multiplier input while the second input receives the dc control voltage V_c coming from the feedback loop. The multiplier gain is internally set by k_{mult} and equals 0.65. Then a scaled-down version of V_c takes a parallel path and ends up being added to the multiplier output to form the final peak current setpoint. The time-dependent peak current setpoint can be defined as follows:

$$i_p(t) = \frac{V_{FB} - 1.9}{R_i} \times \left[k_m k_{mult} v_{in}(t) + k_{off} \right] \qquad (4.533)$$

In this expression, V_{FB} represents the dc error voltage generated by the feedback loop to which a fixed 1.9-V offset is subtracted before becoming V_c. We will consider the setpoint at node V_c for this small-signal analysis.

585

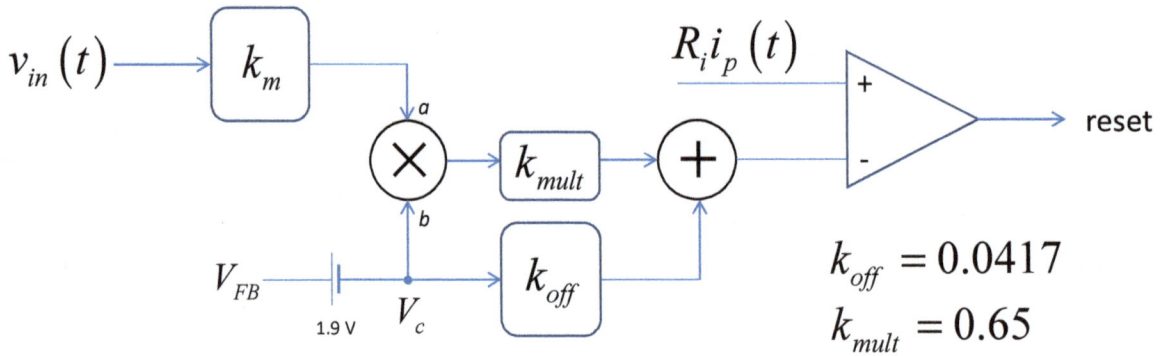

Figure 4.177 The MC33262 internals include an offset to reduce frequency excursion around zero volts.

The typical waveforms of this borderline-operated single-stage flyback converter appear in Figure 4.178. From these sketches, we can write several equations to determine the free-running switching frequency.

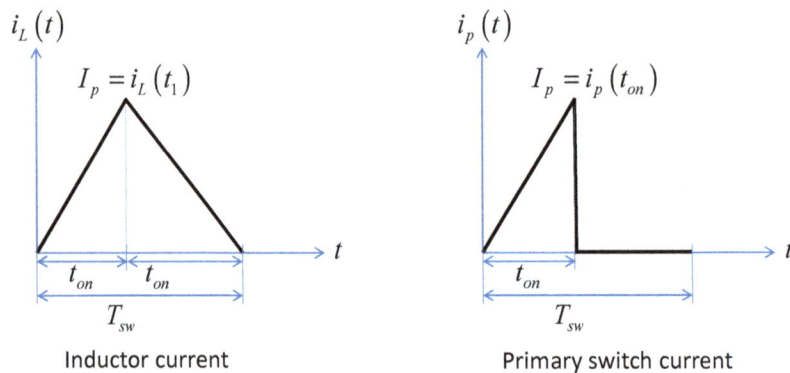

Inductor current   Primary switch current

Figure 4.178 This circuit operates in critical conduction mode also known as boundary or borderline conduction mode in the litterature.

The peak current setpoint is determined by V_c as follows:

$$i_p(t) = \frac{V_c}{R_i}\left[k_m k_{mult} v_{in}(t) + k_{off} \right] \tag{4.534}$$

But it is also expressed as:

$$i_p(t) = \frac{v_{in}(t)}{L_p} t_{on}(t) \tag{4.535}$$

Combining both equations gives us the definition of the on-time as a function of t:

$$t_{on}(t) = \frac{L_p V_c}{R_i v_{in}(t)}\left[k_m k_{mult} v_{in}(t) + k_{off} \right] \tag{4.536}$$

If we neglect the contribution of k_{off}, then this expression simplifies to:

586

$$t_{on} \approx \frac{L_p V_c}{R_i} k_m k_{mult} \tag{4.537}$$

Which confirms the constant-t_{on} operation of this circuit outside of the zero-volt input region. The peak current can also be determined via the off-time duration as the inductor current reduces to zero before a new cycle takes place:

$$i_p(t) = \frac{V_{out}}{NL_p} t_{off}(t) \tag{4.538}$$

If you extract t_{off} and substitute (4.534) in the result, we have:

$$t_{off}(t) = \frac{L_p N V_c}{R_i V_{out}} \left[k_m k_{mult} v_{in}(t) + k_{off} \right] \tag{4.539}$$

Finally, the switching frequency is obtained by summing t_{on} and t_{off}:

$$f_{sw}(t) = \frac{R_i v_{in}(t) V_{out}}{L_p V_c \left[V_{out} + N \cdot v_{in}(t) \right] \left[k_{off} + k_m k_{mult} v_{in}(t) \right]} \tag{4.540}$$

In this expression, V_{out} represents the output voltage (20 V in our example), L_p is the transformer primary inductance (250 μH), N is the transformer turns ratio 1:0.25, R_i the sense resistance (120 mΩ) and v_{in} is the mains instantaneous input voltage:

$$v_{in}(t) = V_p \sin \omega t \tag{4.541}$$

With these elements on hand, we can check how variables evolve along a mains cycle. Figure 4.179 gathers these curves generated with and without an added offset. As you can see, the offset limits the maximum frequency excursion around the 0-V area which benefits efficiency.

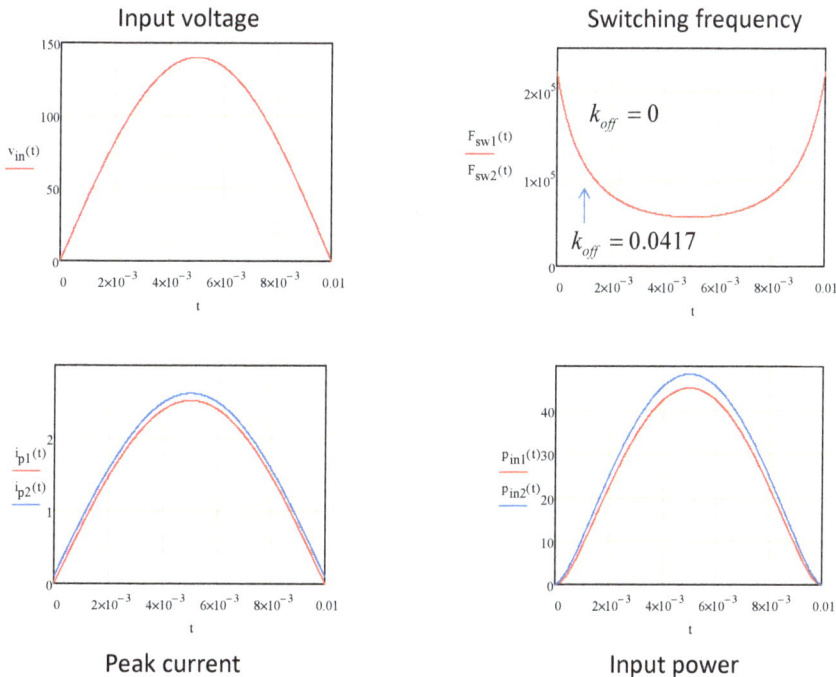

Figure 4.179 The added offset naturally expands the on-time duration around the 0-V input and keeps frequency under control.

For a 100%-efficient discontinuous conduction mode flyback converter, the instantaneous input power can be expressed as:

$$p_{in}(t) = \frac{1}{2} L_p i_p^{\,2}(t) f_{sw}(t)$$

(4.542)

If we replace i_p by its definition from (4.534) and f_{sw} by the expression we obtained a few lines above, then the instantaneous power becomes:

$$p_{in}(t) = \frac{V_c V_{out} v_{in}(t)\left[k_{off} + k_m k_{mult} v_{in}(t)\right]}{2R_i\left[V_{out} + N \cdot v_{in}(t)\right]}$$

(4.543)

Now, we want the power transmitted over an entire line cycle whose frequency, once rectified, becomes twice the line frequency (100 Hz for a 50-Hz grid and 120 Hz for the 60-Hz mains). To obtain this average power, we need to integrate the instantaneous value along the rectified sinewave:

$$P_{in} = 2F_{line} \int_0^{\frac{1}{2F_{line}}} \frac{V_c V_{out} V_p \sin(\omega t)\left[k_{off} + k_m k_{mult} V_p \sin(\omega t)\right]}{2R_i\left[V_{out} + N \cdot V_p \sin(\omega t)\right]} \cdot dt$$

(4.544)

A symbolic answer could not be derived from this expression. On average, across a line sinusoidal cycle, the power absorbed by the emulated resistive load is equivalent to having the same resistance powered by a dc source whose value is the rms voltage of the instantaneous line cycle. Consequently, we can conveniently replace (4.541) in the above expression by V_{ac}, the rms value of the input source [7, 8]. The average power expression thus greatly simplifies to:

$$P_{in} \approx \frac{V_c V_{out} V_{ac}\left[k_{off} + k_m k_{mult} V_{ac}\right]}{2R_i\left[V_{out} + N \cdot V_{ac}\right]}$$

(4.545)

in which V_{ac} is the rms value of the input voltage.

In this large-signal formula, there are two variables we are interested in, V_c and V_{out}. Our model combines a current source with the load and the output capacitor. Considering a 100% efficiency, the large-signal output current is obtained by dividing P_{out} by V_{out}:

$$I_{out} \approx \frac{V_c V_{ac}\left[k_{off} + k_m k_{mult} V_{ac}\right]}{2R_i\left[V_{out} + N \cdot V_{ac}\right]}$$

(4.546)

To linearize the expression, we can run partial differentiations with respects to the two variables as detailed below:

$$g_{vo} = \left.\frac{\partial}{\partial V_{out}}\right|_{\hat{v}_c = 0} I_{out}(V_{out}) = -\frac{V_{ac} V_c\left(k_{off} + V_{ac} k_{mult} k_m\right)}{2R_i\left(V_{out} + NV_{ac}\right)^2}$$

(4.547)

$$g_{vc} = \left.\frac{\partial}{\partial V_c}\right|_{\hat{v}_{out} = 0} I_{out}(V_c) = \frac{V_{ac}\left(k_{off} + V_{ac} k_m k_{mult}\right)}{2R_i\left(V_{out} + NV_{ac}\right)}$$

(4.548)

The small-signal output current is expressed as:

$$I_{out}(s) = g_{vo} V_{out}(s) + g_{vc} V_c(s)$$

(4.549)

Coefficient g_{vo} is a negative conductance whose current depends on the voltage across its terminals: it can be modeled as a resistance whose value is the inverse of the definition (Figure 4.180)

$$V_c = 1.34V \quad V_{ac} = 120.208V \quad k_{off} = 0.042 \quad k_m = 3.398 \times 10^{-3} \quad k_{mult} = 0.65\frac{1}{V} \quad R_i = 0.12\Omega \quad N_1 = 0.25$$

$$g_{vo} := -\frac{V_c \cdot V_{ac} \cdot (k_{off} + V_{ac} \cdot k_m k_{mult})}{2 \cdot R_i \cdot (V_{out} + N_1 \cdot V_{ac})^2} = -0.082\frac{1}{\Omega}$$

Neg. sign

$$\hat{v}_{out} \quad R = \frac{1}{g_{vo}}$$

Figure 4.180 The conductance g_{vo} is modeled as a resistance biased by the output voltage.

We can now assemble the current source and its conductance g_{vo} driving R_{load} and the output capacitor C_{out}. This is what is proposed in Figure 4.181. To determine the control-to-output transfer function, we start with $s = 0$ and open-circuit the capacitor as illustrated in Figure 4.182. The quasi-static gain H_0 linking the internal control voltage V_c to V_{out} is immediately determined as:

$$H_0 = \frac{V_{ac}\left(k_{off} + V_{ac}k_m k_{mult}\right)}{2R_i\left(V_{out} + NV_{ac}\right)}\left(\frac{1}{g_{vo}} \| R_{load}\right) \tag{4.550}$$

The time constant of this 1st-order circuit is obtained by turning the excitation off. The circuit simplifies and becomes that of Figure 4.183. By inspection, we can determine the time constant involving a series-parallel combination of resistors around capacitor C_{out}. The inverse of the result forms the pole we want:

$$\omega_p = \frac{1}{\left(r_C + \frac{1}{g_{vo}} \| R_{load}\right)C_{out}} \tag{4.551}$$

Neglecting the ESR contribution, this expression simplifies to:

$$\omega_p \approx \frac{1}{\left(\frac{1}{g_{vo}} \| R_{load}\right)C_{out}} \tag{4.552}$$

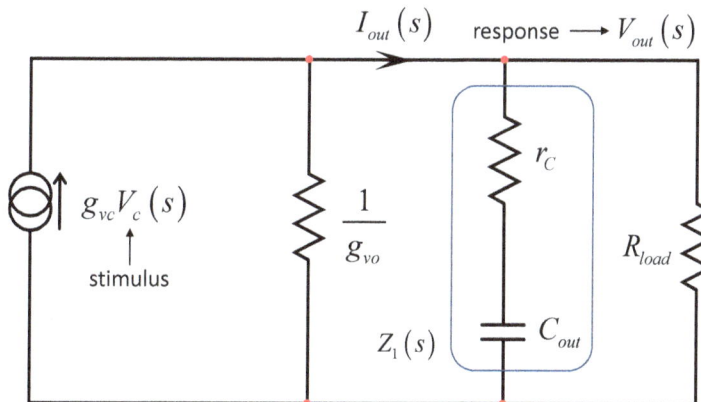

Figure 4.181 This is the small-signal circuit of the single-stage flyback converter operated in current-mode control.

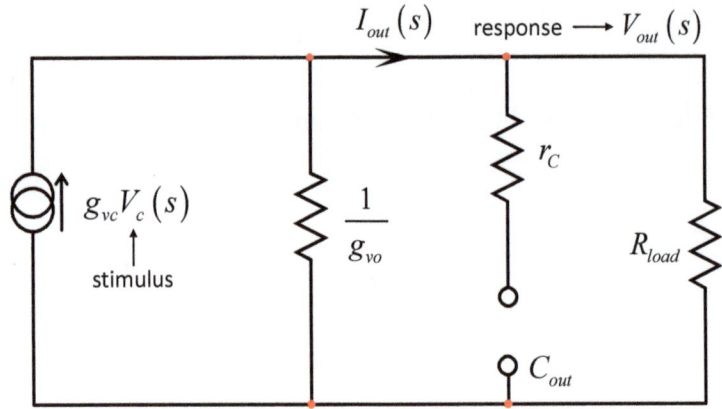

Figure 4.182 For s = 0, the capacitor is open-circuited and the gain H_0 is immediately determined.

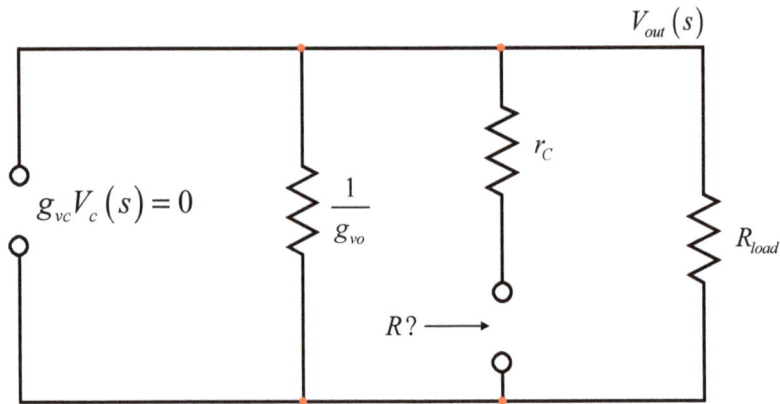

Figure 4.183 The time constant is obtained by inspecting the circuit.

There is a zero in Figure 4.184 network and it occurs when the series impedance made of r_C in series with C_{out} becomes zero:

$$Z_1(s) = r_C + \frac{1}{sC_{out}} = 0 \tag{4.553}$$

It leads to zero classically located at:

$$\omega_z = \frac{1}{r_C C_{out}} \tag{4.554}$$

We now have all the elements we want to write the transfer function of the single-stage flyback converter operated in current-mode control. It appears below:

$$H(s) = \frac{V_{out}(s)}{V_c(s)} = H_0 \frac{1 + \frac{s}{\omega_z}}{1 + \frac{s}{\omega_p}} \tag{4.555}$$

As the zero occurs in high frequency and crossover is usually a few hertz, the transfer function simplifies to:

$$H(s) \approx H_0 \frac{1}{1 + \frac{s}{\omega_p}} \tag{4.556}$$

with:

$$H_0 = \frac{V_{ac}\left(k_{off} + V_{ac}k_m k_{mult}\right)}{2R_i\left(V_{out} + NV_{ac}\right)}\left(\frac{1}{g_{vo}} \parallel R_{load}\right)$$ (4.557)

$$\omega_p \approx \frac{1}{\left(\dfrac{1}{g_{vo}} \parallel R_{load}\right)C_{out}}$$ (4.558)

We can plot this transfer function with Mathcad® and the graphs appear in Figure 4.184.

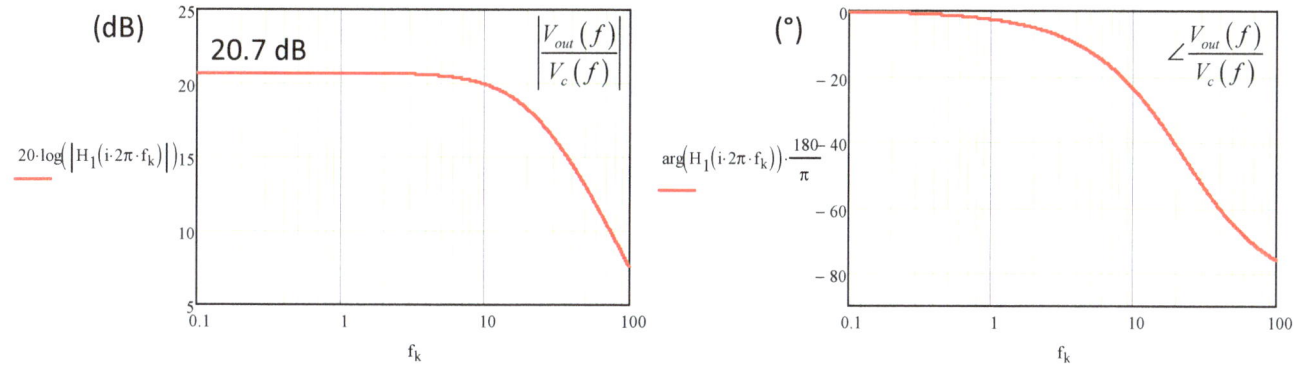

Figure 4.184 Equations reveal a dc gain of 20.7 dB.

To verify if our calculations are correct, I have captured a complete single-stage dc-dc in SIMPLIS®. The schematic diagram is given in Figure 4.185. This is the complete circuit built around the MC33262 PFC controller. You should see the multiplier section and the added offset via the transconductance amplifier G_3. The circuit operates in quasi-resonant mode via the auxiliary winding and the zero-crossing detector made by U_3 and V_2. The simulation results given in Figure 4.186 confirm that the circuit regulates and delivers 20-V across the 5-Ω load. Please note the left-side text zone which automates the calculation around the compensator for 6-Hz crossover frequency.

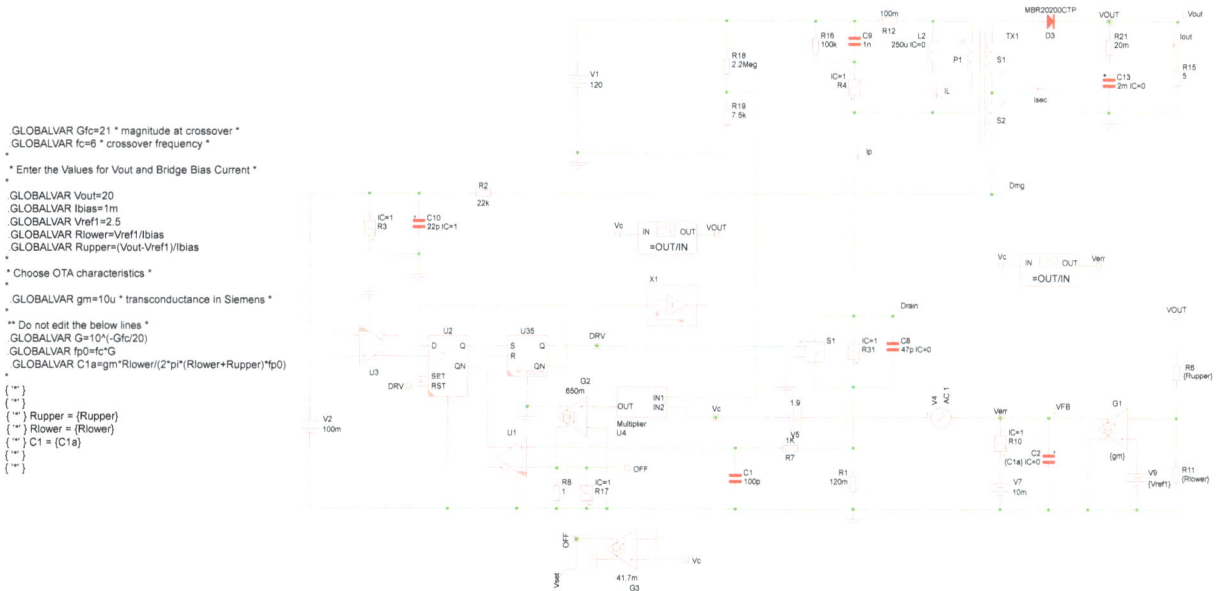

Figure 4.185 A SIMPLIS® simulation will quickly tell us if our approach is correct.

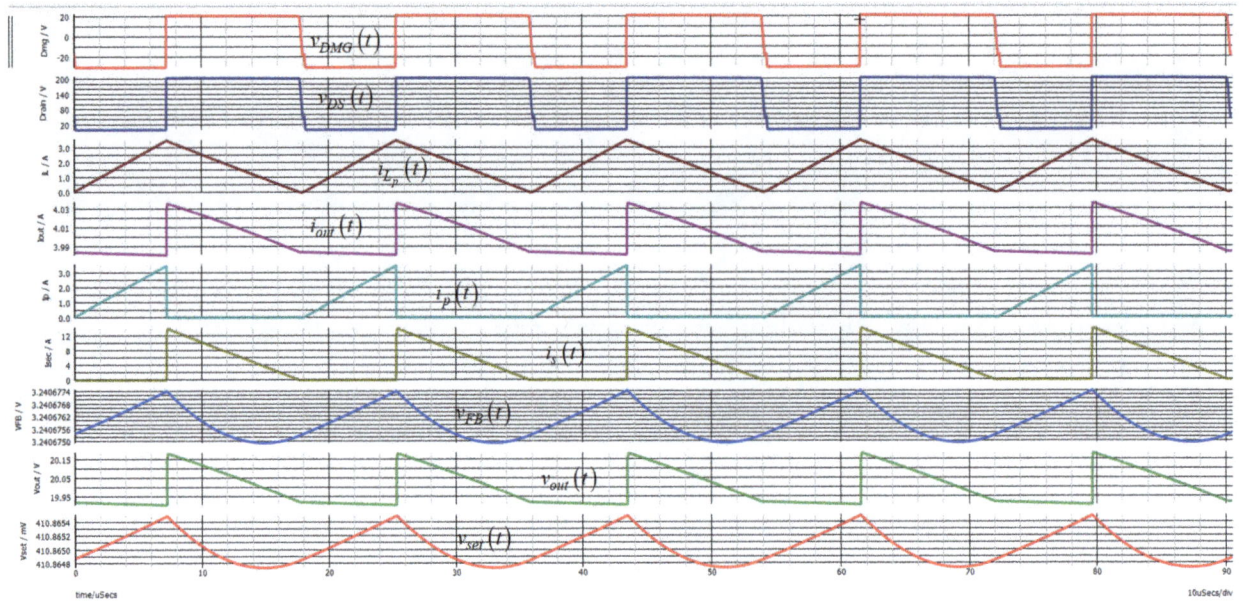

Figure 4.186 Operating waveforms are those of a 80-W flyback converter operating in quasi-resonance.

The ac response can then be extracted from this circuit and it is given in Figure 4.187. The quasi-static gain is slightly less than 21 dB, very close to what Mathcad® predicted. The -3-dB response shows a pole located at 22 Hz or so, again close to our equation prediction.

Finally, we can check the compensated open-loop gain and it appears in Figure 4.188 confirming the 6-Hz crossover frequency with a comfortable phase margin.

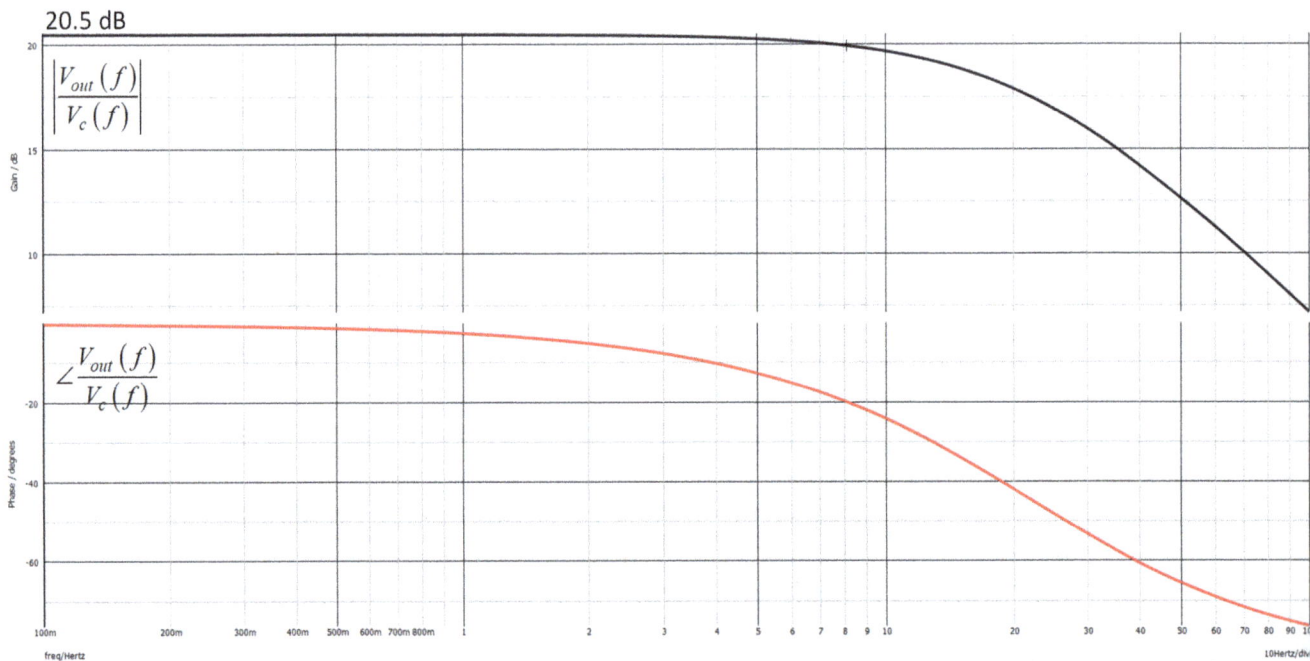

Figure 4.187 The ac simulation confirms the dc gain and the pole as predicted by the equations we derived.

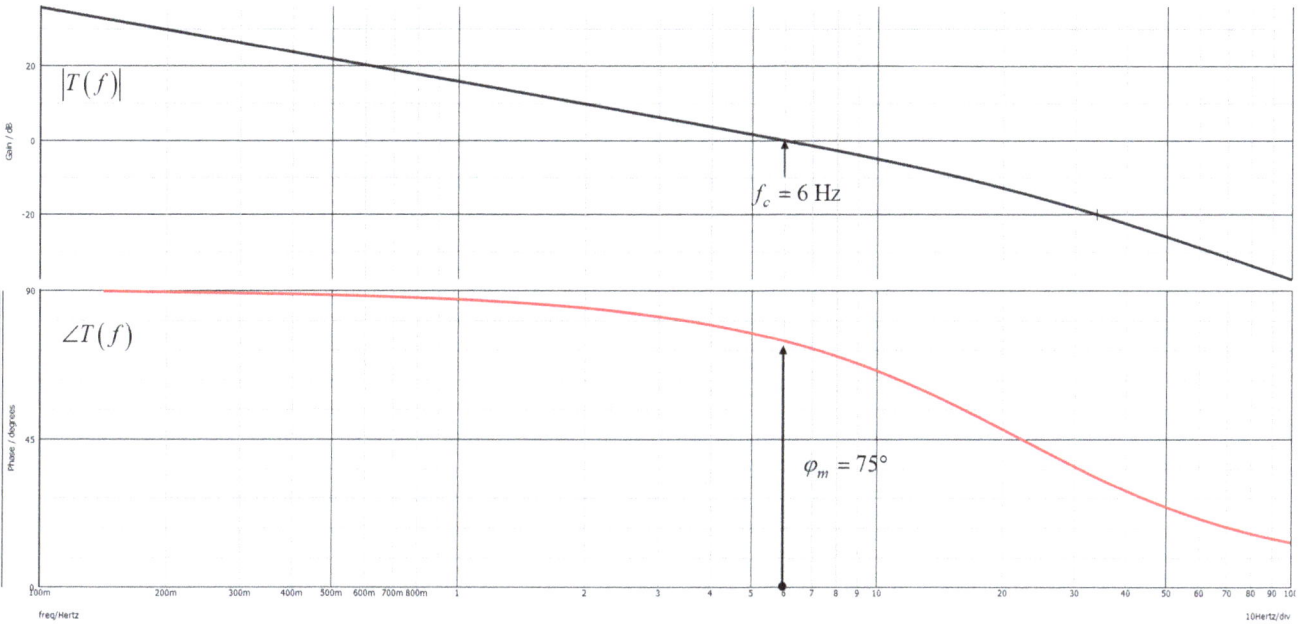

Figure 4.188 The compensated open-loop gain shows a 6-Hz crossover frequency with a good phase margin.

In Figure 4.189, I have added a front-end filter and an ac source to this PFC circuit to check the cycle-by-cycle waveforms. As confirmed by Figure 4.190, the circuit performs power factor correction and delivers a 20-V output affected by a 4-V ripple. The input current is far from being sinusoidal and is typical of a single-stage approach. Please note the oscillations on the input current coming from an interaction with the input filter [9].

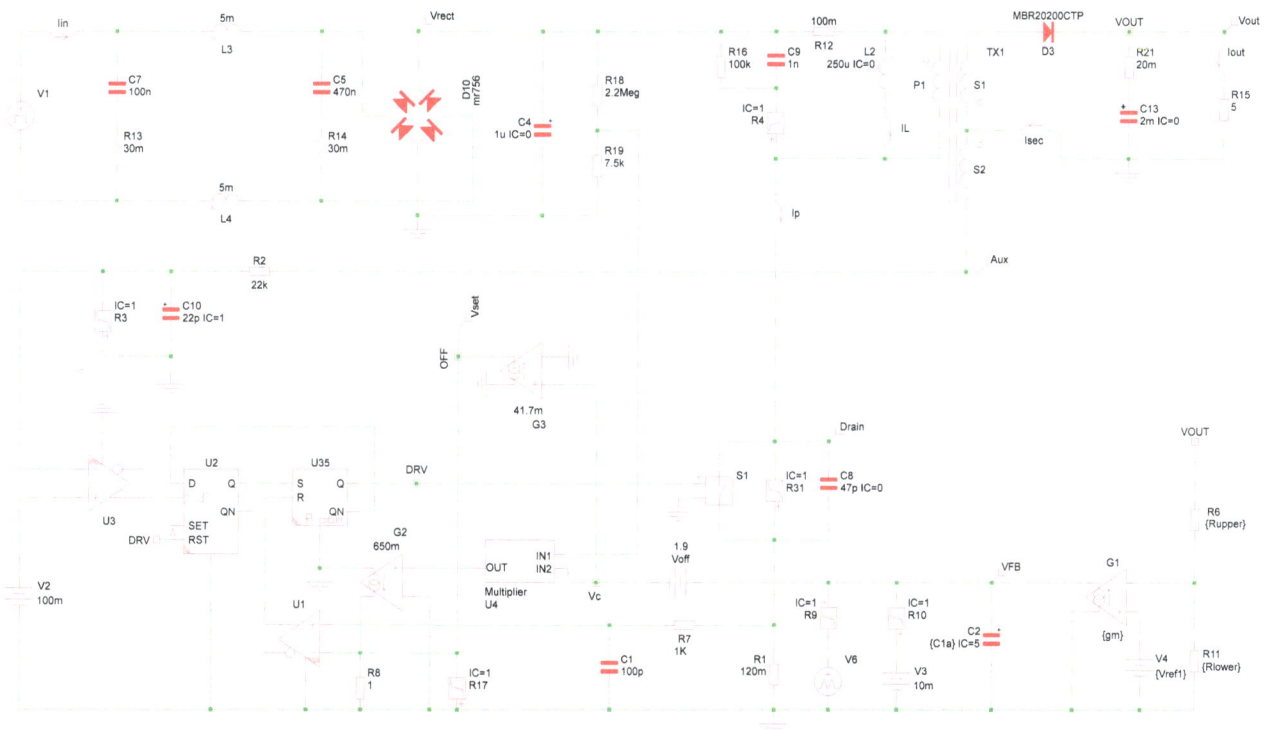

Figure 4.189 We can now test the PFC operation with an ac input.

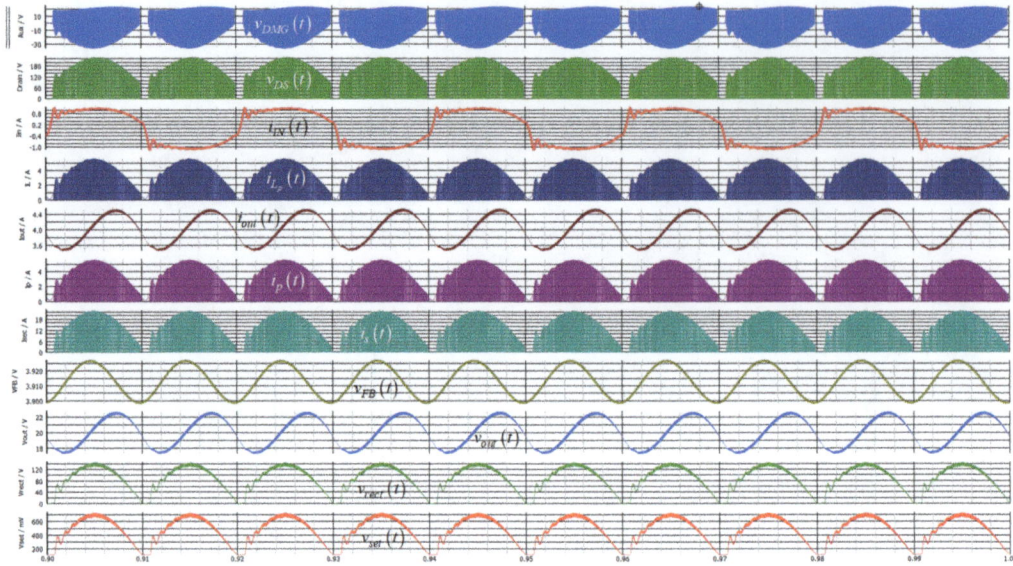

Figure 4.190 Operating waveforms show the typical signature of a single-stage circuit.

4.9 The Flyback in Single-Stage Power Factor Correction – Voltage Mode

In voltage-mode control, the error voltage drives the on-time duration via a dedicated modulator. The operating waveforms do not change compared to current-mode control as confirmed by Figure 4.191 which describes the modulator detail of the NCP1608, a dedicated voltage-mode PFC controller.

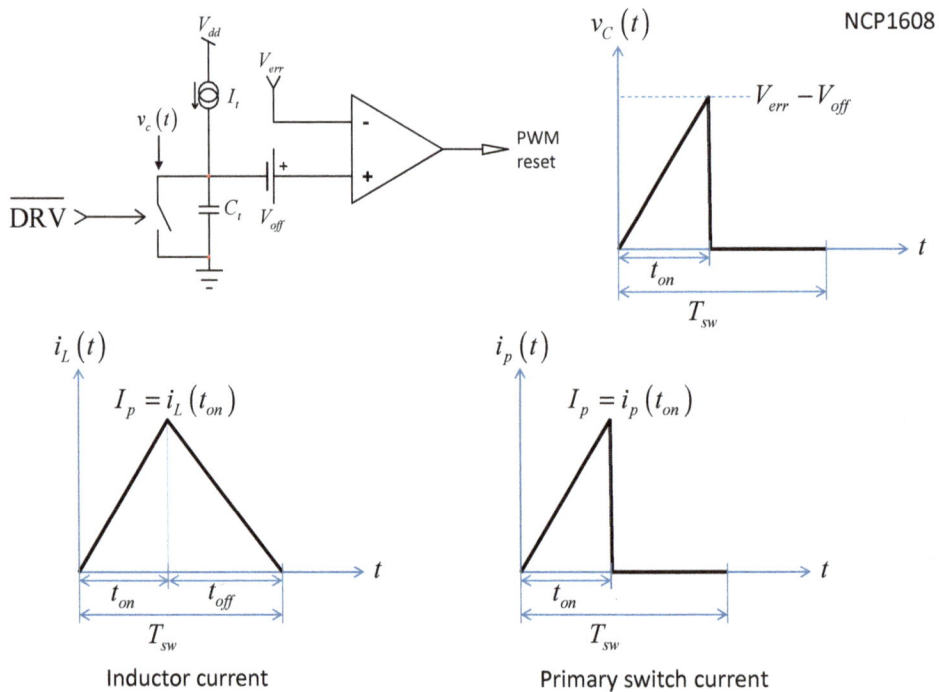

Figure 4.191 The NCP1608 from ON Semiconductor hosts an on-time modulator and operates in voltage-mode control.

In this mode, the on-time no longer depends on the inductor peak current setpoint but is imposed by the modulator via the following formula:

$$t_{on} = \frac{\left(V_{err} - V_{off}\right)C_t}{I_t} \tag{4.559}$$

I_t represents the charging current while the control voltage V_{err} minus an internal offset V_{off} fixes the peak value. The timing capacitor C_t is usually user-selectable and lets you adjust the maximum allowable on-time and thus the maximum transmitted power. The time-dependent peak current is classically expressed as:

$$i_p(t) = \frac{v_{in}(t)}{L_p} t_{on} \tag{4.560}$$

Combining both equations gives us the definition of the peak current as a function of t:

$$i_p(t) = \frac{v_{in}(t)\left(V_{err} - V_{off}\right)C_t}{L_p I_t} \tag{4.561}$$

The peak current together with the inductor current downslope fix the demagnetization time or t_{off}:

$$i_p(t) = \frac{V_{out}}{NL_p} t_{off}(t) \tag{4.562}$$

The two expressions above define the peak current, therefore:

$$\frac{v_{in}(t)\left(V_{err} - V_{off}\right)C_t}{L_p I_t} = \frac{V_{out}}{NL_p} t_{off}(t) \tag{4.563}$$

Solving for t_{off} tells us how the off-time varies in relationship with the input voltage v_{in}:

$$t_{off}(t) = \frac{NC_t\left(V_{err} - V_{off}\right)v_{in}(t)}{I_t V_{out}} \tag{4.564}$$

We have t_{on} and t_{off} and we can express the switching period t_{sw} by summing these two terms:

$$t_{sw}(t) = t_{on} + t_{off}(t) = \frac{C_t\left[V_{err} - V_{off}\right]\left[V_{out} + N \cdot v_{in}(t)\right]}{I_t V_{out}} \tag{4.565}$$

We can plot the variations of these variables in relationship to the input voltage. Figure 4.192 confirms the frequency runaway around the 0-V zone, typical of a free-running topology. More sophisticated techniques now exist which fold frequency back and limit switching losses.

The transmitted power is determined using the formula for a DCM-operated flyback converter:

$$p_{in}(t) = \frac{1}{2} L_p i_p^2(t) f_{sw}(t) = \frac{1}{2} L_p \frac{\left[\dfrac{v_{in}(t)\left(V_{err} - V_{off}\right)C_t}{L_p I_t}\right]^2}{\dfrac{C_t\left[V_{err} - V_{off}\right]\left[V_{out} + N \cdot v_{in}(t)\right]}{I_t V_{out}}} \tag{4.566}$$

595

If we develop this expression and rearrange it, we find:

$$p_{in}(t) = \frac{C_t\left(V_{err} - V_{off}\right)V_{out}\left[v_{in}(t)\right]^2}{2I_t L_p\left[V_{out} + N \cdot v_{in}(t)\right]} \tag{4.567}$$

The average power is the instantaneous power averaged across a half a line cycle:

$$P_{in} = 2F_{line}\int_0^{\frac{1}{2F_{line}}}\frac{\left(V_{err} - V_{off}\right)C_t V_{out}\left[V_p\sin\left(\omega t\right)\right]^2}{2I_t L_p\left[V_{out} + N\cdot V_p\sin\left(\omega t\right)\right]}\cdot dt \tag{4.568}$$

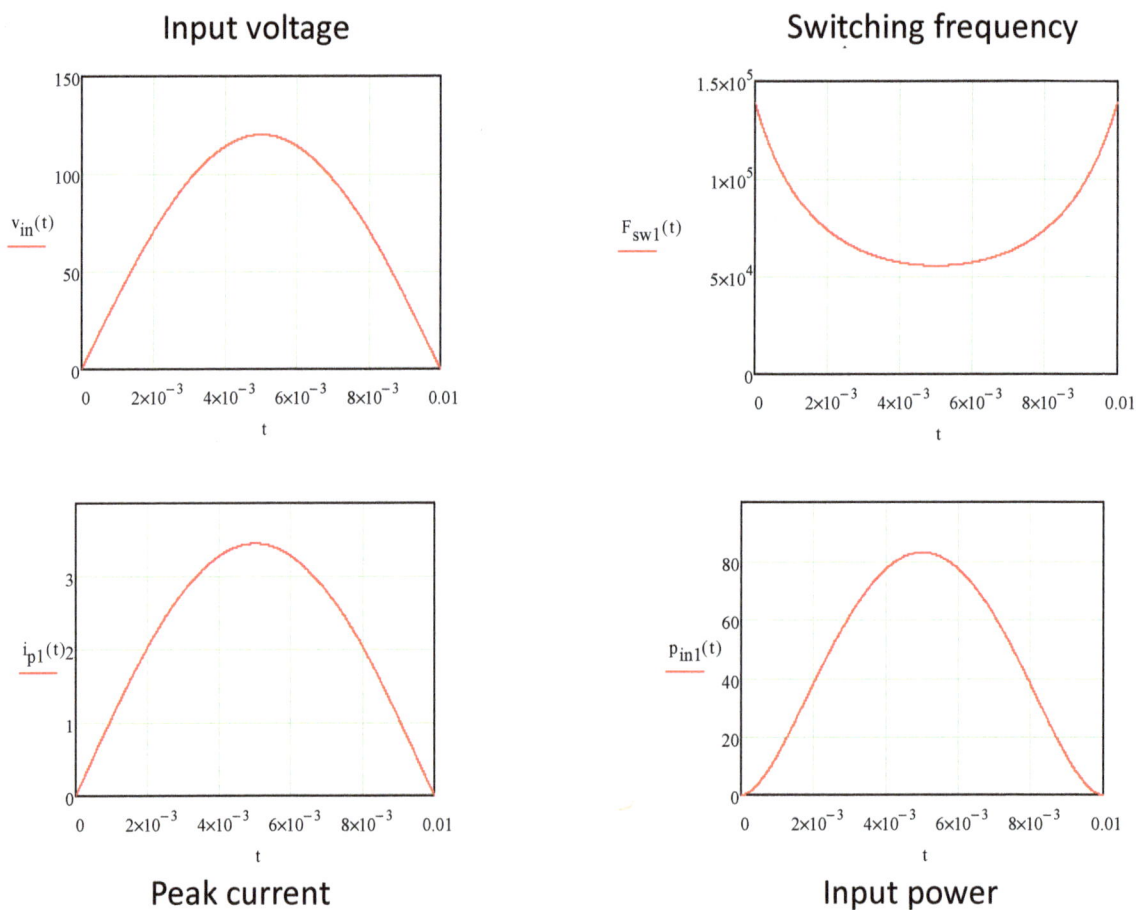

Input voltage

Switching frequency

Peak current

Input power

Figure 4.192 In absence of offset as with the MC33262, the frequency runs away around the 0-V area.

Applying the same strategy as with the current-mode case, we can conveniently approximate the input power as:

$$P_{in} \approx \frac{C_t\left(V_{err} - V_{off}\right)V_{out}V_{ac}^2}{2I_t L_p\left(V_{out} + NV_{ac}\right)} \tag{4.569}$$

in which V_{ac} is the rms value of the input voltage.

596

In this large-signal formula, there are two variables we are interested in, V_{err} and V_{out}. Our model combines a current source with the load and the output capacitor.

Considering a 100% efficiency, the large-signal output current is obtained by dividing P_{out} by V_{out}:

$$I_{out} \approx \frac{C_t \left(V_{err} - V_{off} \right) V_{ac}^{\;2}}{2 I_t L_p \left(V_{out} + N V_{ac} \right)}$$

(4.570)

To linearize the expression, we can run partial differentiations with respects to the two variables V_{out} and V_{err} as detailed below:

$$g_{vo} = \frac{\partial}{\partial V_{out}} \Big|_{\hat{v}_{err}=0} I_{out} \left(V_{out} \right) = -\frac{C_t V_{ac}^{\;2} \left(V_{err} - V_{off} \right)}{2 I_t L_p \left(V_{out} + N V_{ac} \right)^2}$$

(4.571)

$$g_{vc} = \frac{\partial}{\partial V_{err}} \Big|_{\hat{v}_{out}=0} I_{out} \left(V_{err} \right) = \frac{C_t V_{ac}^{\;2}}{2 I_t L_p \left(V_{out} + N V_{ac} \right)}$$

(4.572)

The small-signal output current is expressed as:

$$I_{out} \left(s \right) = g_{vo} V_{out} \left(s \right) + g_{vc} V_{err} \left(s \right)$$

(4.573)

Coefficient g_{vo} represents a negative conductance whose current depends on the voltage across its terminals: it can be modeled as a resistance whose value is the inverse of the definition (Figure 4.193).

$$V_{err} := V_{FB} = 3.5\,\text{V} \qquad V_{off} = 0.65\,\text{V} \qquad V_{ac} = 120.208\,\text{V} \qquad C_t = 680\,\text{pF} \qquad I_t = 270\,\mu\text{A} \qquad L_p = 250\,\mu\text{H} \qquad V_{out} = 20\,\text{V} \qquad N_1 = 0.25$$

$$g_{vo} := -\frac{C_t \cdot \left(V_{err} - V_{off} \right) \cdot V_{ac}^{\;2}}{2 \cdot I_t \cdot L_p \cdot \left(V_{out} + N_1 \cdot V_{ac} \right)^2} = -0.083 \frac{1}{\Omega} \qquad \Rightarrow \qquad \hat{v}_{out} \qquad R = \frac{1}{g_{vo}}$$

Neg. sign

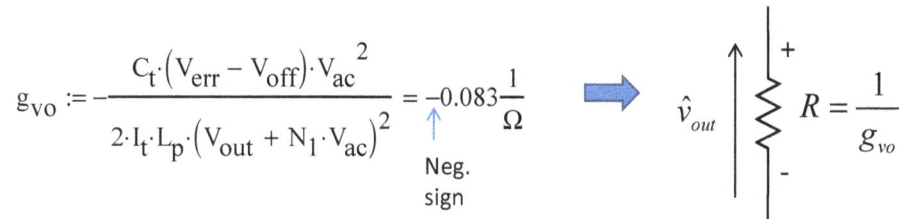

Figure 4.193 The conductance g_{vo} is modeled as a resistance biased by the output voltage.

The second equation is simply the small-signal gain linking the output current $\hat{i}_{out}$ to the control voltage $\hat{v}_{err}$.

We can now gather these sources with R_{load} and the output capacitor C_{out}. This is what is proposed in Figure 4.194. To determine the control-to-output transfer function, we start with $s = 0$ and open-circuit the capacitor as illustrated Figure 4.195.

The quasi-static gain H_0 linking the error voltage V_{err} to V_{out} is immediately determined as:

$$H_0 = \frac{C_t V_{ac}^{\;2}}{2 I_t L_p \left(V_{out} + N V_{ac} \right)} \left(\frac{1}{g_{vo}} \| R_{load} \right)$$

(4.574)

The time constant of this 1st-order circuit is obtained by turning the excitation off. The circuit simplifies and becomes that of Figure 4.196 since the left-side current source is zeroed.

By inspection, we can determine the time constant involving a series-parallel combination of resistors around capacitor C_{out}.

The inverse of the result forms the pole:

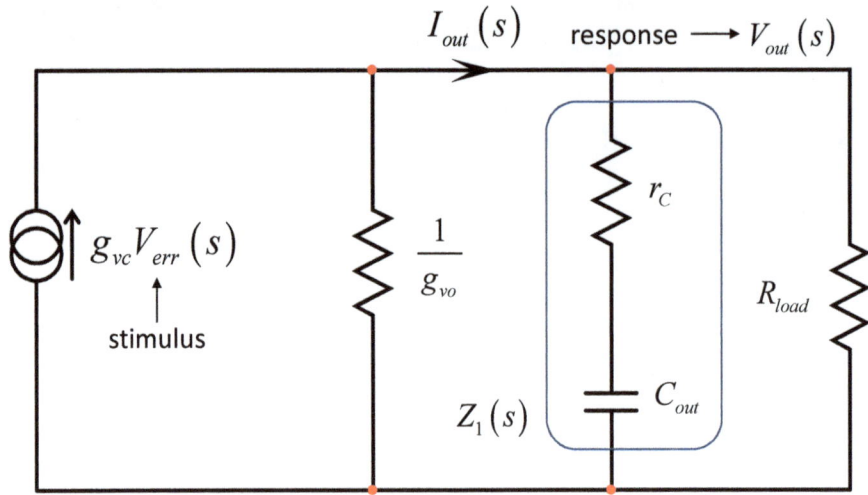

Figure 4.194 The model is similar to that of the MC33262 except that some coefficients change.

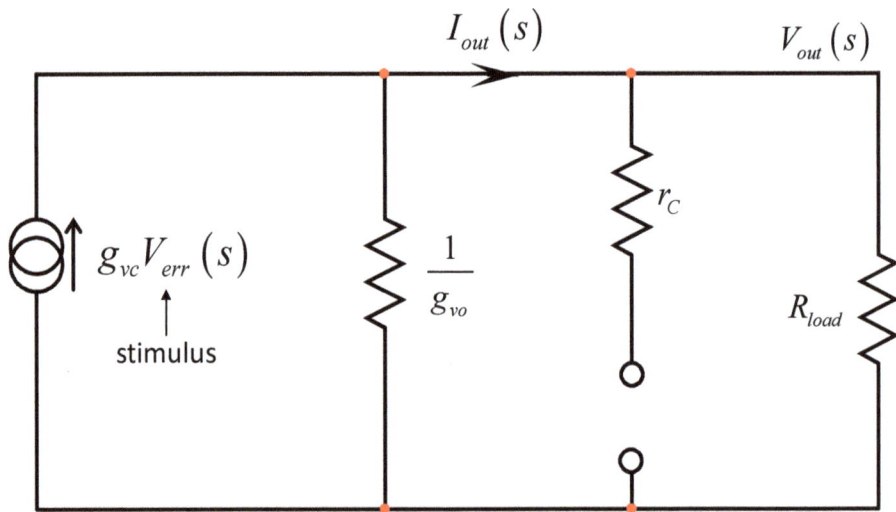

Figure 4.195 The quasi-static gain is obtained when C_{out} is open-circuited.

$$\omega_p = \frac{1}{\left(r_C + \dfrac{1}{g_{vo}} \| R_{load}\right)C_{out}}$$ (4.575)

Neglecting the ESR contribution, this expression simplifies to:

$$\omega_p \approx \frac{1}{\left(\dfrac{1}{g_{vo}} \| R_{load}\right)C_{out}}$$ (4.576)

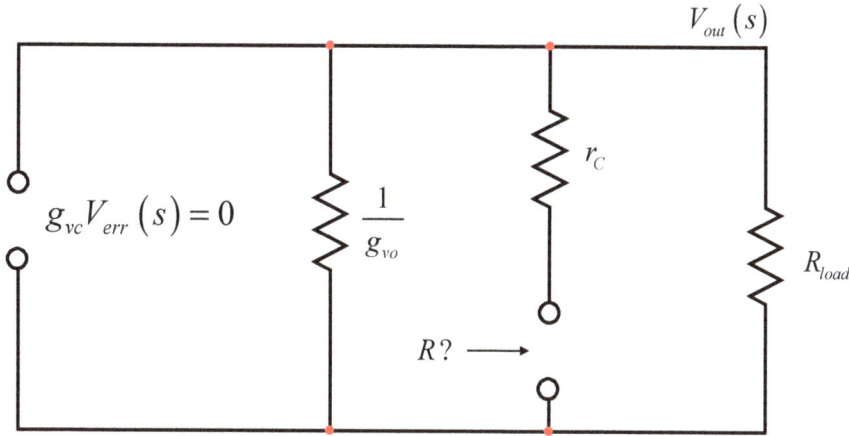

Figure 4.196 The time constant is obtained by inspecting this simple circuit.

here is a zero in Figure 4.194 network and it occurs when the series impedance made of r_C in series with C_{out} becomes ero:

$$Z_1(s) = r_C + \frac{1}{sC_{OUT}} = 0 \tag{4.577}$$

leads to a zero classically located at:

$$\omega_z = \frac{1}{r_C C_{out}} \tag{4.578}$$

Ve now have all the elements we want to write the transfer function of the single-stage flyback converter operated voltage-mode control. It appears below:

$$H(s) = \frac{V_{out}(s)}{V_{err}(s)} = H_0 \frac{1 + \frac{s}{\omega_z}}{1 + \frac{s}{\omega_p}} \tag{4.579}$$

s the zero occurs in high frequency and crossover is usually a few hertz, the transfer function simplifies to:

$$H(s) \approx H_0 \frac{1}{1 + \frac{s}{\omega_p}} \tag{4.580}$$

ith:

$$H_0 = \frac{C_t V_{ac}^2}{2 I_t L_p (V_{out} + N V_{ac})} \left(\frac{1}{g_{vo}} \parallel R_{load} \right) \tag{4.581}$$

$$\omega_p \approx \frac{1}{\left(\frac{1}{g_{vo}} \parallel R_{load} \right) C_{out}} \tag{4.582}$$

Ve can plot this transfer function with Mathcad® and the graphs appear in Figure 4.197. The dc gain in this application is around 14 dB. To verify our calculations, we have assembled an equivalent model of the NCP1608 in a SIMPLIS schematic diagram shown in Figure 4.198.

The on-time modulator is using the same architecture described in Figure 4.191. The C_t capacitor is selected to limit the maximum on-time and thus the maximum power this converter can deliver.

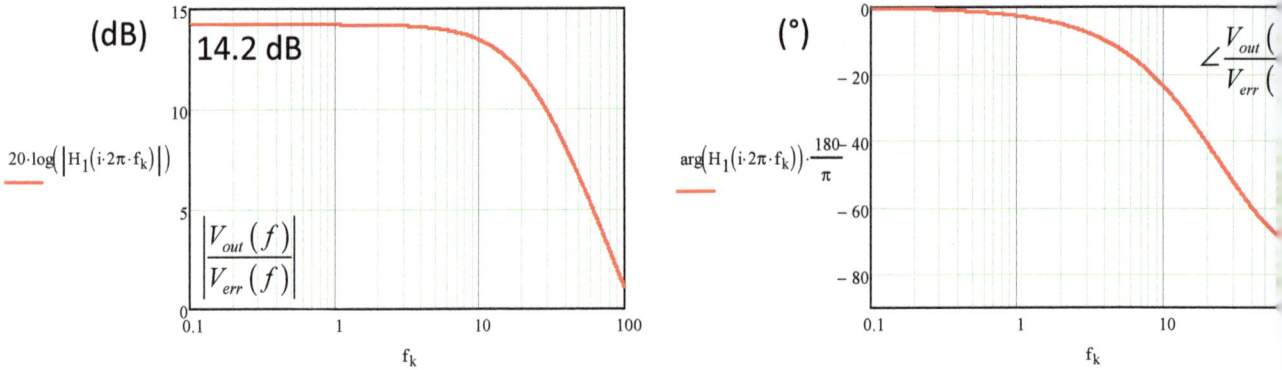

Figure 4.197 The control-to-output transfer function reveals a 14.2-dB gain in dc.

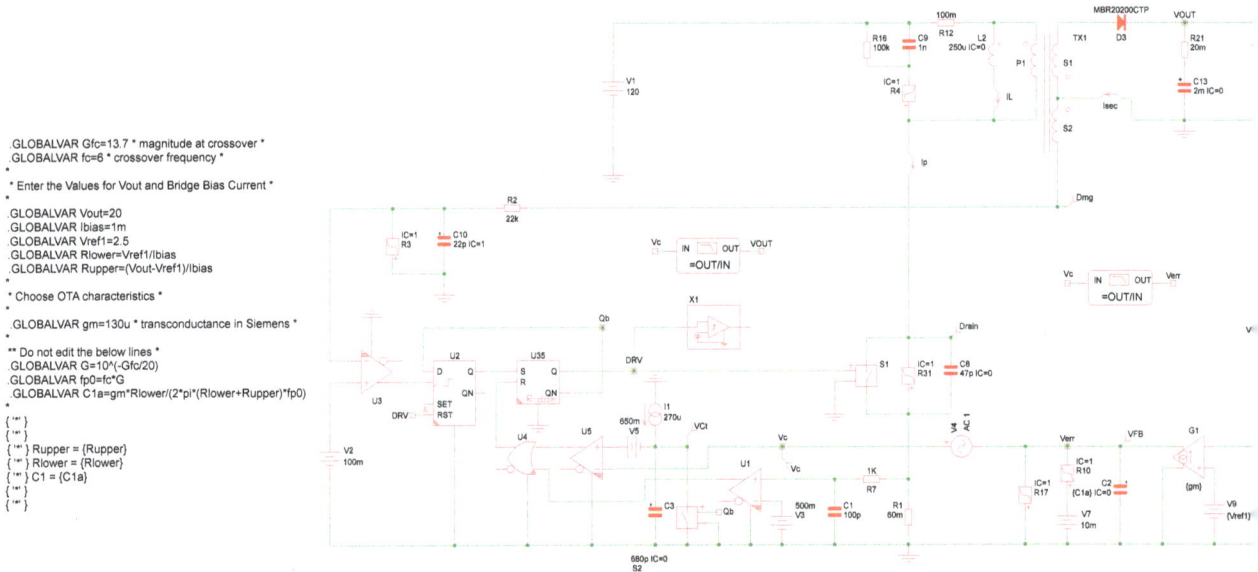

Figure 4.198 SIMPLIS® simulates this constant-on-time QR flyback converter in a PFC application.

As confirmed by Figure 4.199 simulation results, the circuit delivers 80W as expected and the power switch turns back on at the minimum of the drain-source waveform.

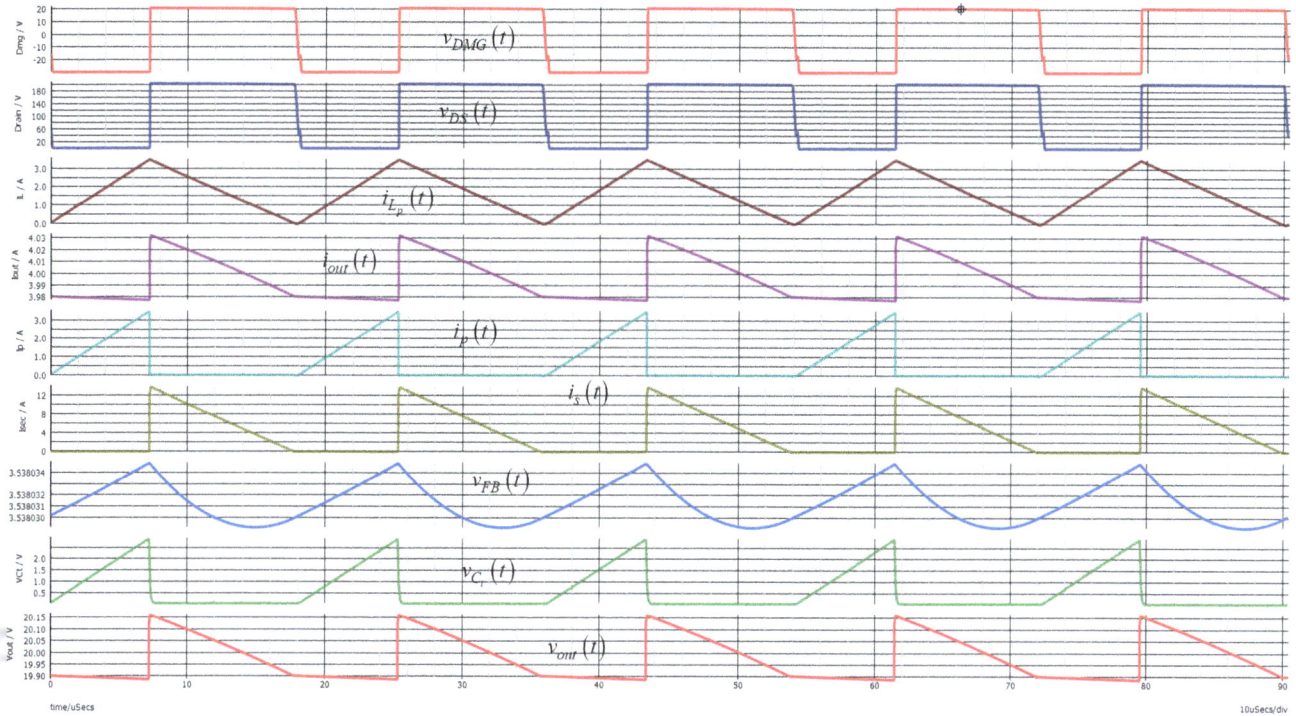

Figure 4.199 With the appropriate delay in the demagnetization path, the switch turns on in the minimum of the drain-source voltage, minimizing switching losses.

ith this circuit, we ran an ac analysis to determine the control-to-output transfer function and check our lculations. The power stage response appears in Figure 4.200 and confirms the 14-dB dc gain. The pole is also cated around 22 Hz per our calculation.

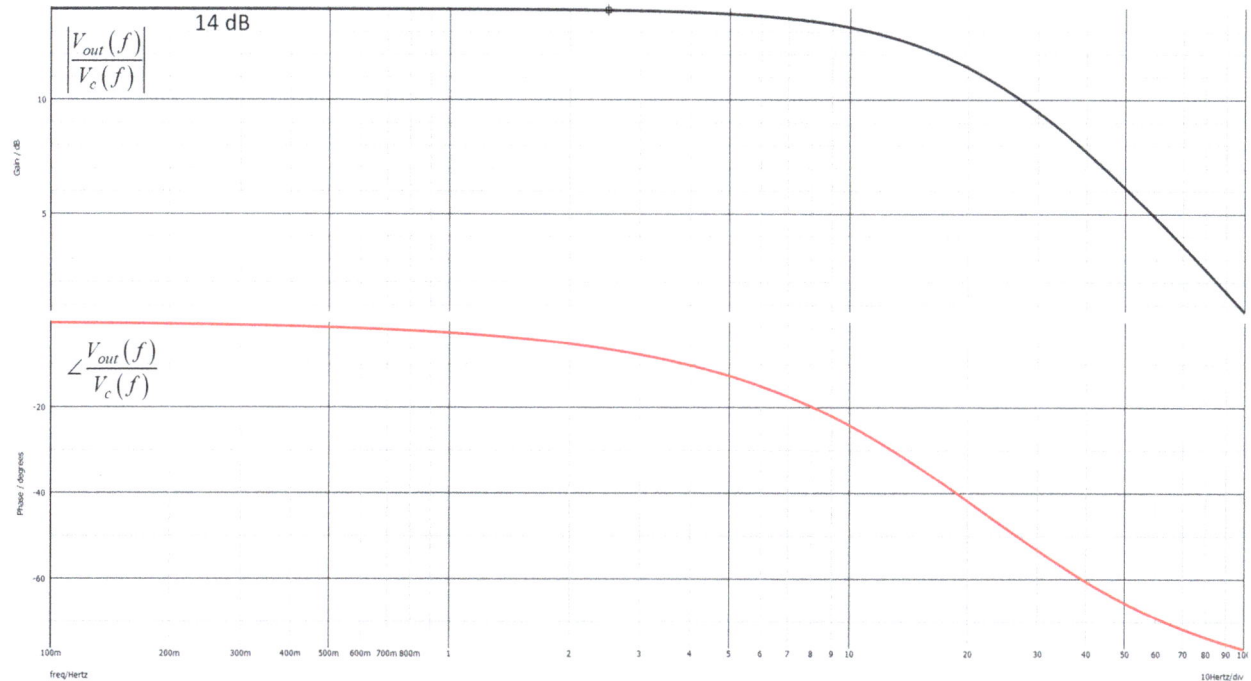

Figure 4.200 The response is exactly that of what is given in Figure 4.199: a 14-dB quasi-static gain and a pole at 22 Hz.

Once properly compensated with the type 1 filter, the crossover is 6 Hz and the phase margin quite comfortab
(Figure 4.201).

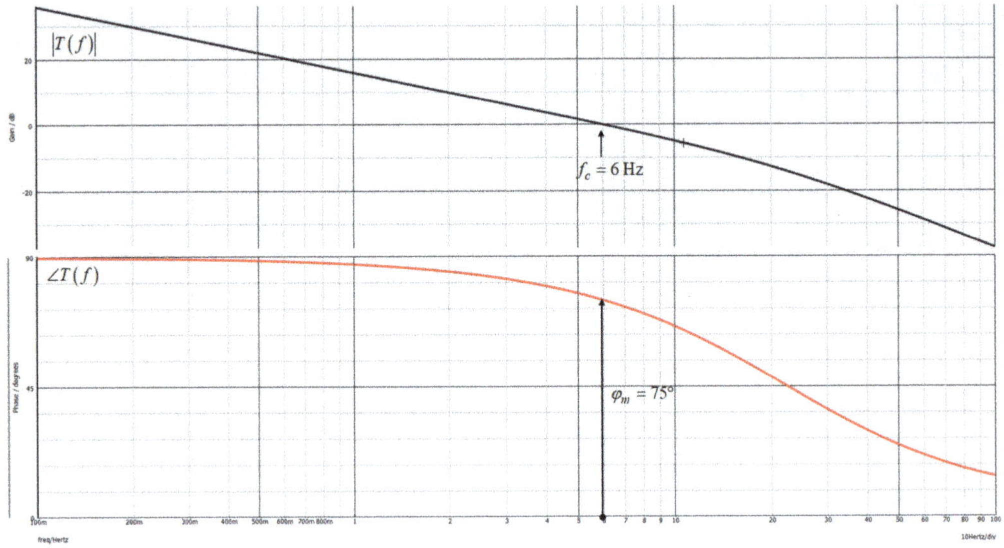

Figure 4.201 The compensator ensures a 6-Hz crossover frequency with a 75° phase margin.

We can now replace the dc input voltage by a sinewave and check the power factor correction function. The circu
to simulate appears in Figure 4.202. The electromagnetic interference (EMI) filter is in place. Care must be taken
limit the interaction between this filter and the converter but this is another story [9]. After a few minutes (which
extremely short for a 0.5-s simulation run), waveforms appear in Figure 4.203. They confirm the proper operatic
of the circuit which forces a quasi-square current but without oscillations this time unlike current-mode control.

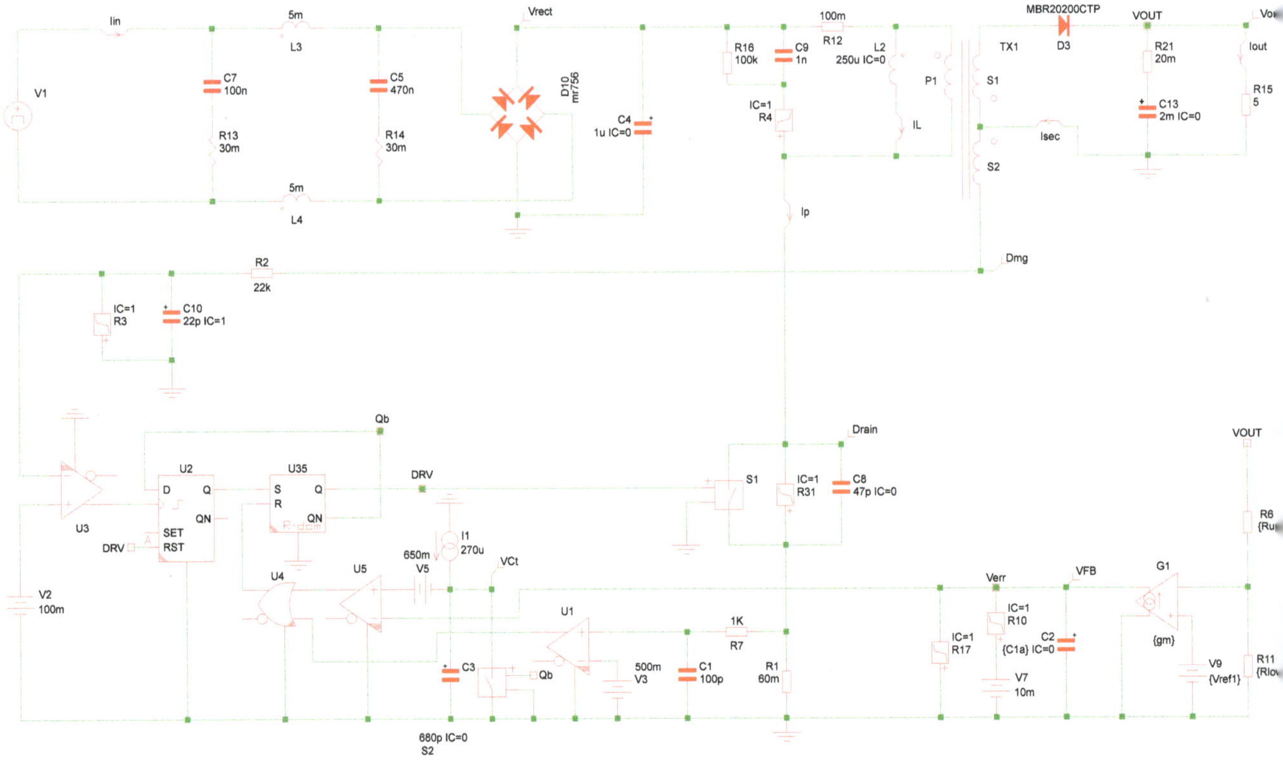

Figure 4.202 The dc input voltage is now replaced by a sinewave.

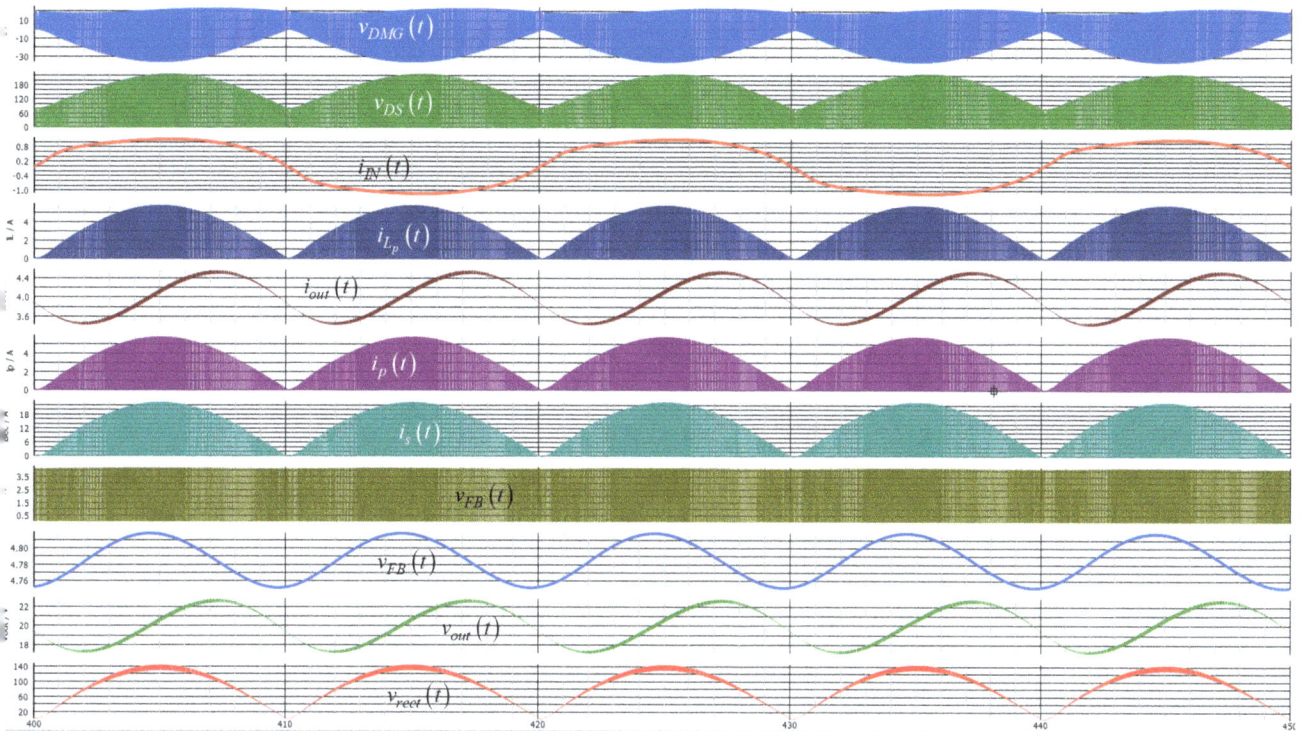

Figure 4.203 The input current is of acceptable shape for this voltage-mode version of the single-stage flyback PFC.

4.10 What Should I Retain from this Chapter?

In this fourth chapter, we have learned many information about the small-signal response of the buck-boost converter and its isolated version, the flyback. The summary is below:

1. A CCM buck-boost converter operated in voltage-mode control is a second-order system affected by a right-half-plane zero. When transitioning to DCM, this converter still exhibits a second-order dynamic response but well damped. The RHPZ is still present in DCM though relegated in high frequencies.

2. In fixed-frequency current-mode control, the CCM buck-boost converter becomes a third-order converter with a pole dominating the low-frequency response and two subharmonic poles located at half the switching frequency. Beside the left-half-plane zero contributed by the output capacitor and its ESR, there is a right-half-plane zero whose position is the same as in voltage mode. The subharmonic poles must be damped by some additional ramp as the duty ratio approaches 50%.

3. The incremental input resistance of an open-loop buck-boost converter operated in voltage-mode is positive. On the other hand, when the same converter is operated in current-mode control, the incremental resistance is negative: if the input voltage increases, the output current reduces and so does the input current.

4. The buck-boost converter can be operated in quasi-resonant mode for the benefit of lower turn-on losses if a delay is inserted to switch right in the drain-source valley. Flyback converters operated in this way are popular in high-efficiency notebook adapters.

5. Flyback converters lend themselves well to implementing power factor correction. The topology combining

603

both functions is called a single-stage flyback. The transfer function of the single-stage topology operated in voltage- or current-mode control shows a low-frequency first-order response affected by a pole. The input current in this mode is squarish but good enough for a lot of low-power applications like in the lighting market.

4.11 References

1. C. Basso, *Linear Circuit Transfer Functions – An Introduction to Fast Analytical Techniques*, Wiley, 2016.
2. V. Vorpérian, *Simplified Analysis of PWM Converters using Model of PWM Switch, parts I and II*, IEEE Transactions on Aerospace and Electronic Systems, Vol. 26, NO. 3, 1990
3. V. Vorpérian, *Analytical Methods in Power Electronics, In-House Power Electronics Class*, Toulouse, France, 2004.
4. C. Basso, *Modeling the Effects of Leakage Inductance on Flyback Converters*, How2Power http://www.how2power.com/pdf_view.php?url=/newsletters/1511/articles/H2PToday1511_design_ONSemi.pdf
5. C. Basso, *Switch-Mode Power Supplies: SPICE Simulations and Practical Designs*, 2nd edition, McGraw-Hill, New York, 2014
6. J. Chen, B. Erickson, D. Masksimović, *Average Switch Modeling of Boundary Conduction Mode Dc-to-Dc Converters*, Proc. IEEE Industrial Electronics Society Annual Conference (IECON 01), Nov. 2001, vol. 2, pp. 842-849
7. J. Turchi, *Compensating a PFC Stage*, ON Semiconductor Application Note AND8321 https://www.onsemi.com/pub/Collateral/AND8321-D.PDF
8. *Running Ac Analyses on PFC Converters*, SIMPLIS® Exhibitor Seminar, APEC 2017 http://www.simplistechnologies.com/webinar/2016/11/17/ac-analysis-pfc
9. C. Basso, *Input Filter Interactions with Switching Regulators*, APEC Professional Education Seminars, Tampa Fl, 2017

5 High-Order Converters

IN THIS LAST chapter, we will take a look at the transfer functions of less common converters such as Ćuk's, SEPIC, Zeta and LLC. However, considering the complexity of these high-order structures, I will not analyze them in detail as I did in the previous chapters but rather express their control-to-output transfer functions from known publications or personal work with working examples.

5.1 The Ćuk Converter

This converter was patented by Dr. Ćuk (pronounced *Chook*) and Dr. Middlebrook when they were professors at the California Institute of Technology (Caltech) in 1977 [1]. Different versions with integrated magnetics and imbalanced transformer appeared some years later [2, 3]. The converter in its simple two-inductor and two-capacitor version can be seen as a front-end boost converter coupled via a capacitor to a buck converter (Figure 5.1) delivering the output voltage.

Considering the presence of an inductor in the input and output of the circuit, this 4th-order converter offers non-pulsating input and output currents.

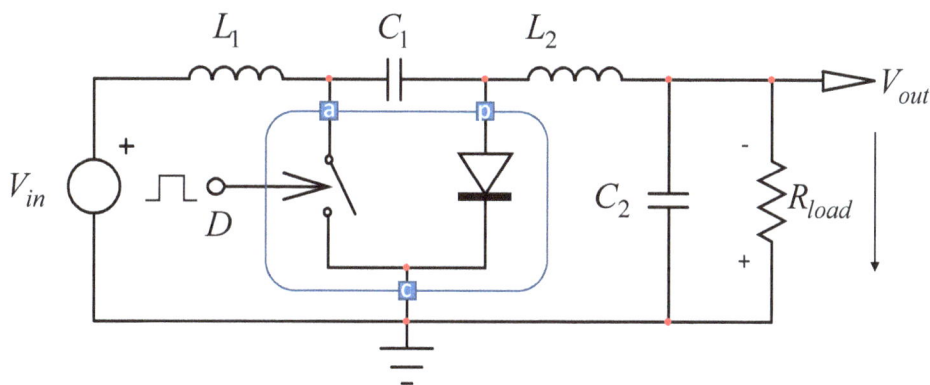

Figure 5.1 The Ćuk converter in its simplest form in which the PWM switch model appears.

This converter has been the object of many publications describing its operating details [4] but the most advanced small-signal analysis comes from a paper written by Dr. Vorpérian in 1996 [5].

In this document, the author analyzed the effect of the magnetizing inductance in an isolated version – a real analytical *tour de force* for a 6th-order system – and offered very compact factored forms derived for the non-isolated un-coupled and coupled versions we will look at.

In its simplest form shown in the above picture and without coupled inductors, the dc transfer function linking the output voltage to the source is:

$$M = -\frac{D}{1-D} \tag{5.1}$$

As such, the converter can increase or decrease the input voltage but delivers a negative voltage. The analysis of this converter can be carried by inserting the CCM VM PWM switch model as shown in Figure 5.2. The 60% duty ratio imposes a –15-V output voltage across the 5-Ω resistance with a 10-V input voltage. It is possible to show that the raw control-to-output transfer function is expressed as follows:

$$H(s) = -H_0 \frac{1 - a_1 s + a_2 s^2}{1 + b_1 s + b_2 s^2 + b_3 s^3 + b_4 s^4} \tag{5.2}$$

in which:

$$H_0 = \frac{V_{in}}{(1-D)^2} \tag{5.3}$$

$$a_1 = \frac{D^2 L_1}{R_{load}(1-D)^2} \tag{5.4}$$

$$a_2 = \frac{L_1 C_1}{1-D} \tag{5.5}$$

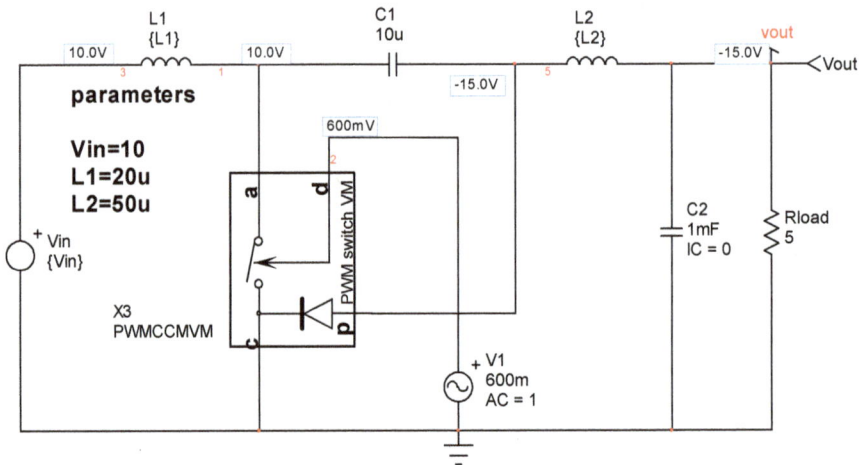

Figure 5.2 The PWM switch lends itself perfectly for simulating the control-to-output transfer function of the Ćuk converter.

$$b_1 = \frac{L_2}{R_{load}} + \frac{L_1}{R_{load}}\left(\frac{D}{1-D}\right)^2 \tag{5.6}$$

$$b_2 = \frac{L_1 C_1}{(1-D)^2} + L_2 C_2 + \left(\frac{D}{1-D}\right)^2 L_1 C_2 \tag{5.7}$$

$$b_3 = \frac{L_1 C_1 L_2}{(1-D)^2 R_{load}} \tag{5.8}$$

$$b_4 = \frac{L_1 L_2 C_1 C_2}{(1-D)^2} \tag{5.9}$$

If we plot this expression with Mathcad® against the magnitude and phase responses from the PWM switch model in Figure 5.2, graphs match with each other very well as shown in Figure 5.3.

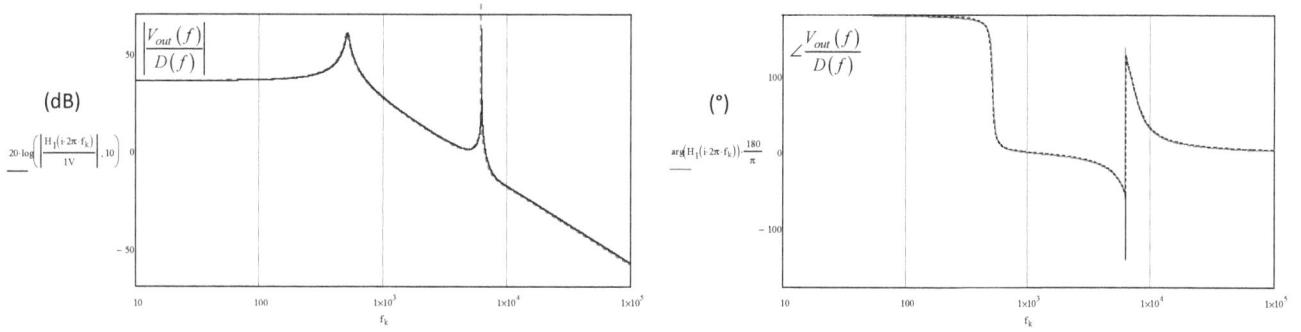

Figure 5.3 The responses from the PWM switch and the Mathcad® curves are in excellent agreement.

If we consider well-separated resonances, it is possible to factor the denominator as follows:

$$D(s) \approx \left[1 + \frac{s}{\omega_{01} Q_1} + \left(\frac{s}{\omega_{01}}\right)^2\right]\left[1 + \frac{s}{\omega_{02} Q_2} + \left(\frac{s}{\omega_{02}}\right)^2\right] \tag{5.10}$$

To determine the quality factors and the omega values, you must solve the below system of equations obtained by expanding (5.10):

$$b_1 \approx \frac{1}{\omega_{01} Q_1} \tag{5.11}$$

$$b_2 \approx \frac{1}{\omega_{01}^2} \tag{5.12}$$

$$b_3 = \frac{1}{\omega_{01} Q_1 \omega_{02}^2} + \frac{1}{\omega_{02} Q_2 \omega_{01}^2} \tag{5.13}$$

$$b_4 = \frac{1}{\omega_{01}^2 \omega_{02}^2} \tag{5.14}$$

Christophe Basso

From these expressions, you obtain the following results:

$$\omega_{01} = \frac{1}{\sqrt{b_2}} \tag{5.15}$$

$$\omega_{02} = \frac{1}{\omega_{01}\sqrt{b_4}} \tag{5.16}$$

$$Q_1 = \frac{1}{b_1\omega_{01}} \tag{5.17}$$

$$Q_2 = \frac{\omega_{02}}{\dfrac{b_3}{b_4} - \dfrac{\omega_{01}}{Q_1}} \tag{5.18}$$

The numerator can also be factored in a more convenient way:

$$N(s) = 1 - \frac{s}{\omega_{0N}Q_N} + \left(\frac{s}{\omega_{0N}}\right)^2 \tag{5.19}$$

In which [7]:

$$\omega_{0N} = \frac{1}{\sqrt{a_2}} \tag{5.20}$$

$$Q_N = \frac{\sqrt{a_2}}{a_1} \tag{5.21}$$

The transfer function can now be expressed as a compact formula showing how poles and zeroes shape the response:

$$H(s) \approx -H_0 \frac{1 - \dfrac{s}{\omega_{0N}Q_N} + \left(\dfrac{s}{\omega_{0N}}\right)^2}{\left[1 + \dfrac{s}{\omega_{01}Q_1} + \left(\dfrac{s}{\omega_{01}}\right)^2\right]\left[1 + \dfrac{s}{\omega_{02}Q_2} + \left(\dfrac{s}{\omega_{02}}\right)^2\right]} \tag{5.22}$$

A quick simulation with SIMPLIS® as shown in Figure 5.4 confirms the operating point in Figure 5.5. The ac response is given in Figure 5.6 and shows a slightly more damped response for SIMPLIS® but an excellent agreement otherwise.

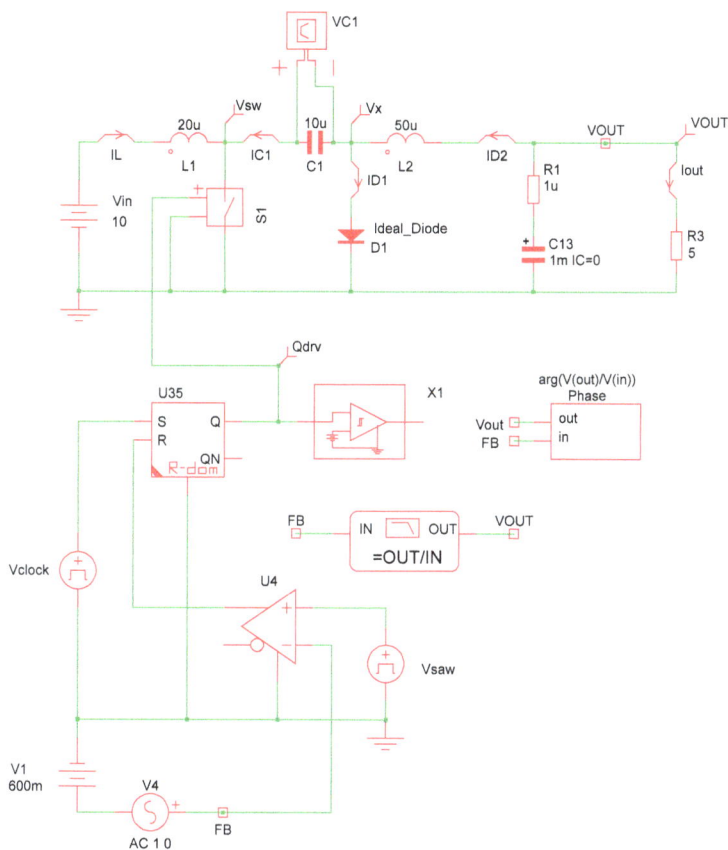

Figure 5.4 SIMPLIS® delivers the ac response in a few seconds after producing the cycle-by-cycle waveforms.

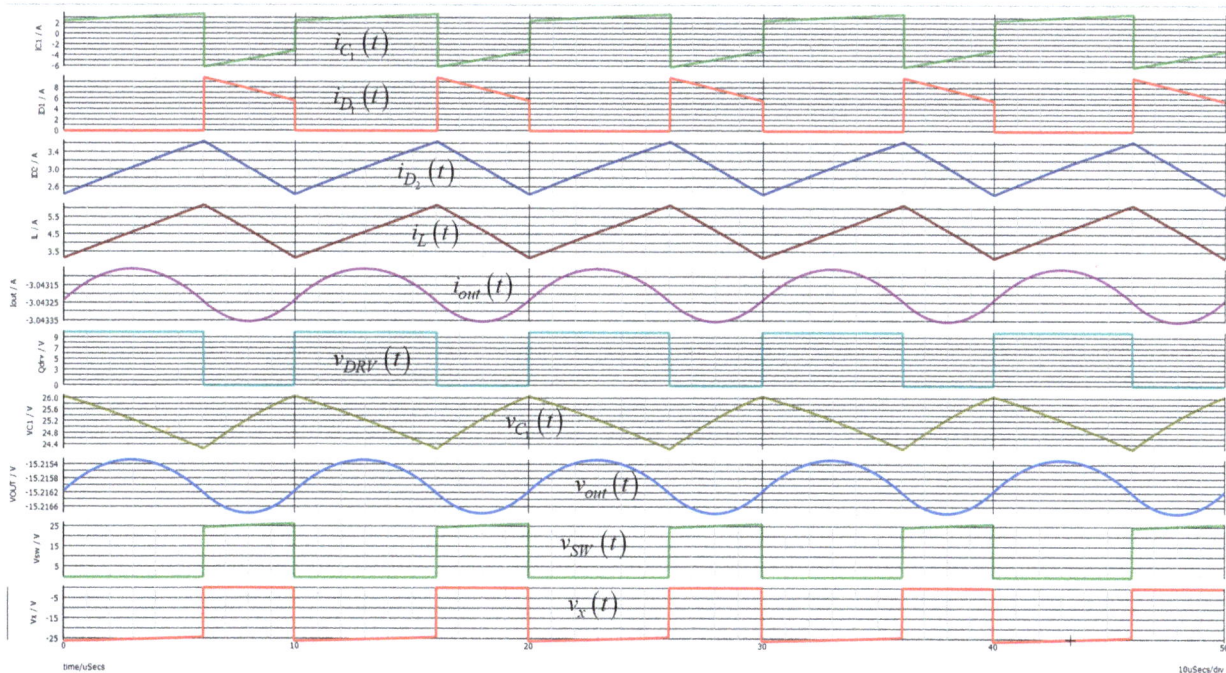

Figure 5.5 SIMPLIS® confirms the operating point of −15 V for a 60% duty ratio.

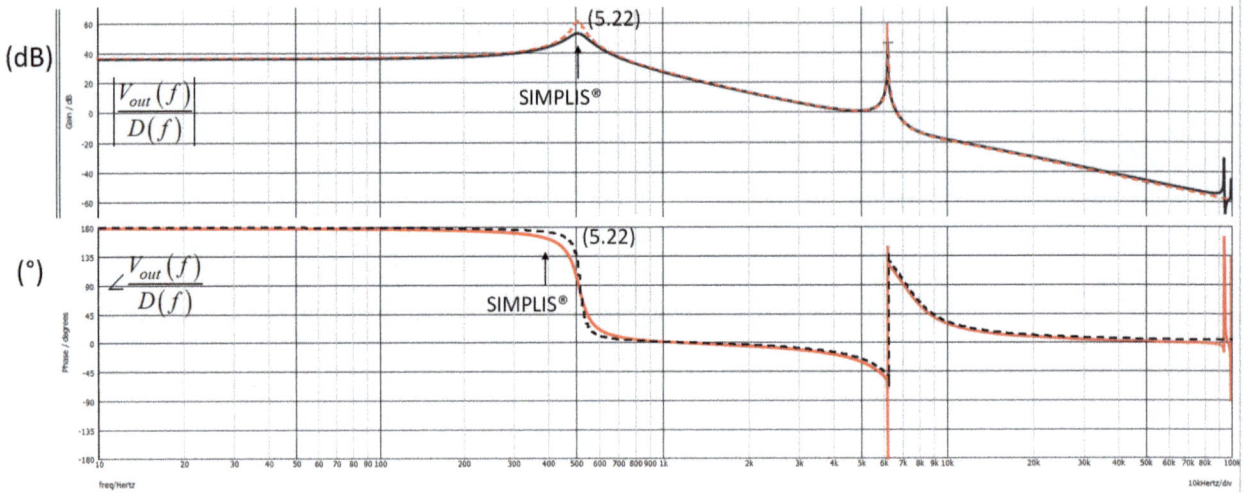

Figure 5.6 The magnitude and phase plots are very close to what the factored equation delivers for this particular configuration.

5.1.1 Coupled Inductors

As explained in details in [6], coupling inductors L_1 and L_2 can help reduce the current ripple either in the input or the output side but not simultaneously with this configuration. The output ripple is made zero or very small practically speaking if the below equality showing a coupling coefficient equal to the turns ratio but less than 1 is respected:

$$k = \sqrt{\frac{L_1}{L_2}} < 1 \qquad (5.23)$$

...in which k represents the transformer coupling coefficient between the two windings. In other terms, if L_1 is made smaller than L_2 and the coupling coefficient is adjusted accordingly, then conditions exist to cancel the ripple in L_2.

On the opposite, if one wants to cancel the input current ripple, then L_2 must be made smaller than L_1 and k set accordingly:

$$k = \sqrt{\frac{L_2}{L_1}} < 1 \qquad (5.24)$$

To update Figure 5.2 simulation schematic to a coupled-inductor version just add a SPICE coupling coefficient as shown in Figure 5.7 in which (5.23) is satisfied.

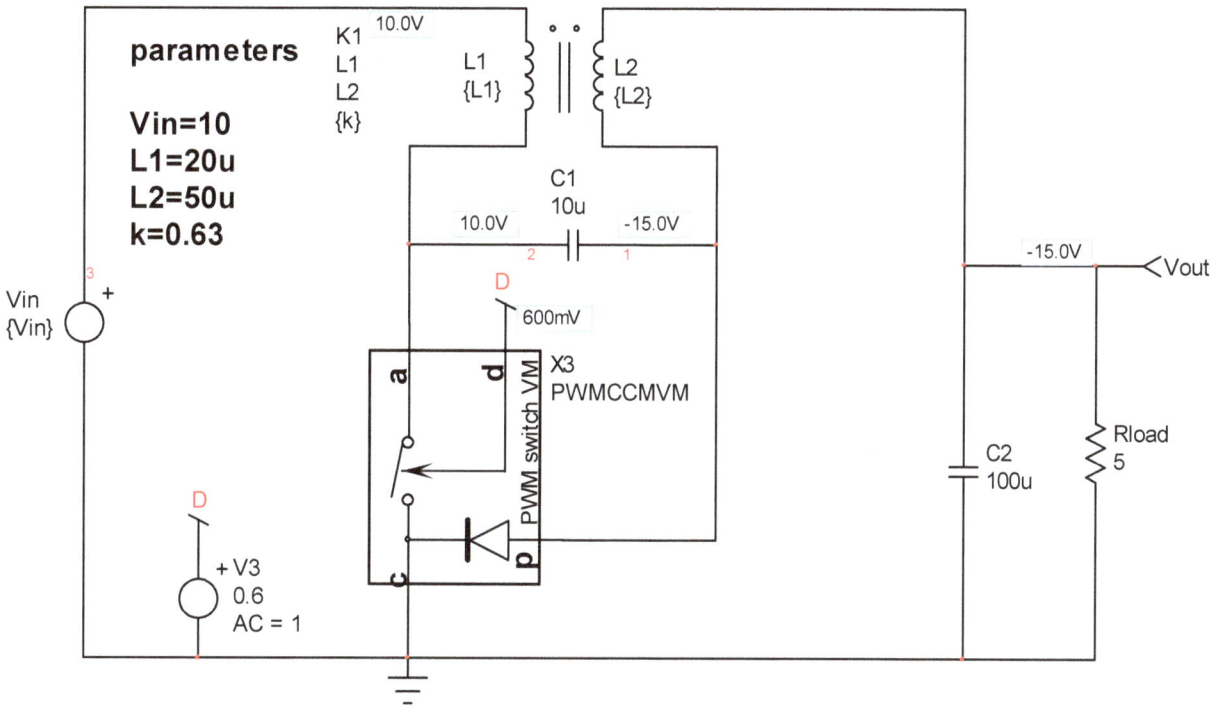

Figure 5.7 Add a simple SPICE coupling coefficient to associate L_1 and L_2 and you can simulate this new version with the PWM switch.

However, analyzing the above circuit with k in place is not practical and you have to resort to an equivalent transformer model. This model features a leakage inductance and a magnetizing inductance. The transformer turns ratio N needs to be accordingly scaled as detailed in Figure 5.8.

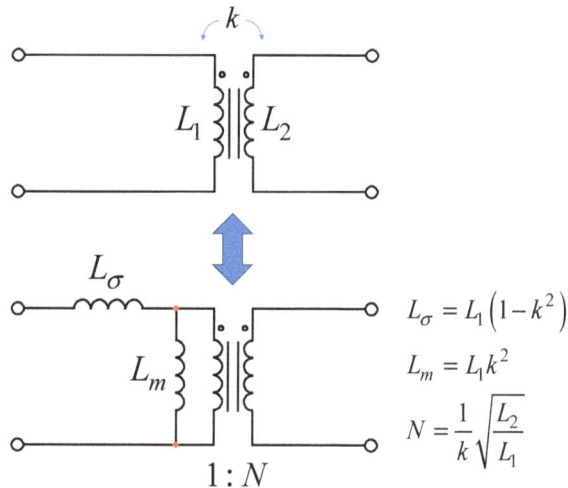

$$L_\sigma = L_1\left(1-k^2\right)$$

$$L_m = L_1 k^2$$

$$N = \frac{1}{k}\sqrt{\frac{L_2}{L_1}}$$

Figure 5.8 Coupled inductors can be advantageously replaced for the analysis by a transformer affected by a leakage term L_σ and a magnetizing inductance L_m.

You can now update the large-signal model with this equivalent transformer as proposed in Figure 5.9.

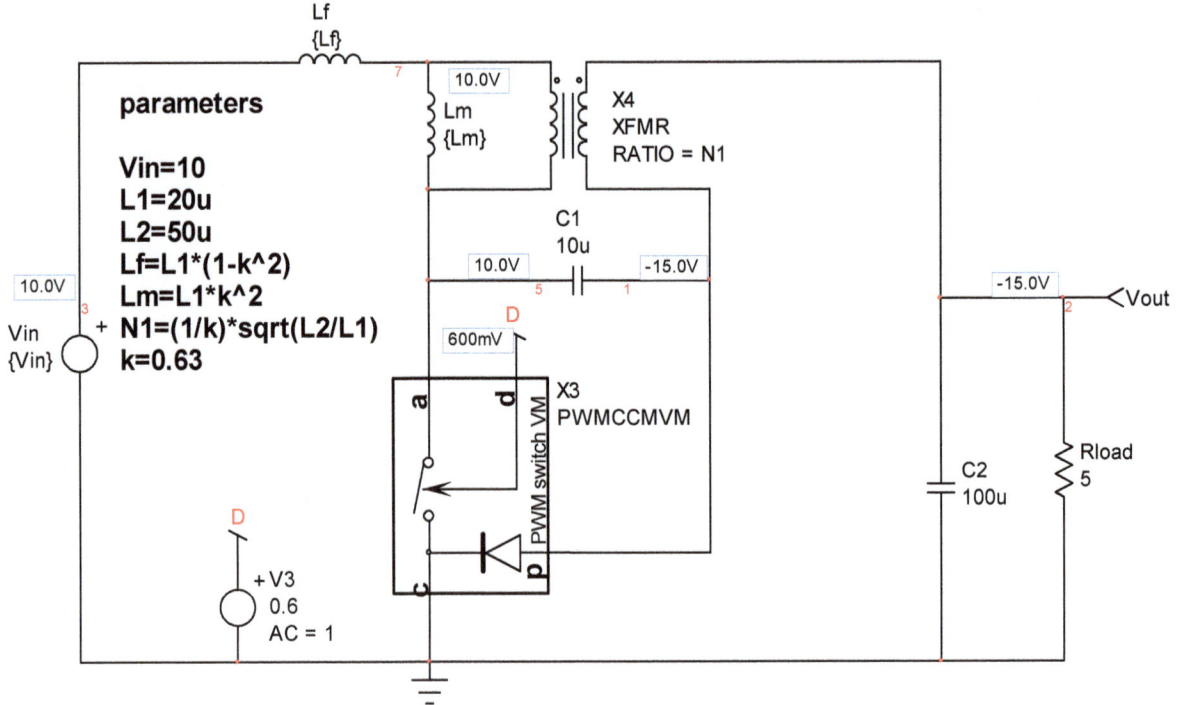

Figure 5.9 The new transformer in place lets you analyze the Ćuk converter which turns out to be a 4th-order converter.

The small-signal response of the converter can be extracted using the small-signal VM-CCM model of the PWM switch. Once plugged in the circuit and rearranged for the control-to-output transfer function, you end-up with the schematic diagram from Figure 5.10. I have of course verified that the magnitude and phase responses of the three diagrams (Figure 5.7, Figure 5.9 and Figure 5.10) are rigorously identical. If you feel like it, you can proceed with the analysis by determining all time constants of the 4th-order switching converter. Then you need to rearrange the expression in a factored form with two polynomials of second order in the denominator. Fortunately, it has been done in [5] and the transfer function is here:

$$H(s) = \frac{V_{out}(s)}{D(s)} \approx -H_0 \frac{1 + \frac{s}{\omega_N Q_N} + \left(\frac{s}{\omega_N}\right)^2}{\left[1 + \frac{s}{\omega_L Q_L} + \left(\frac{s}{\omega_L}\right)^2\right]\left[1 + \frac{s}{\omega_H Q_H} + \left(\frac{s}{\omega_H}\right)^2\right]} \tag{5.25}$$

$$L_\sigma = L_1\left(1 - k^2\right) \tag{5.26}$$

$$L_e = L_1\left(\frac{D}{1-D}\right)^2 + L_2 + 2k\sqrt{L_1 L_2}\frac{D}{1-D} \tag{5.27}$$

parameters

Vin=10
D=0.6
R=5
L1=20u
L2=50u
Lf=L1*(1-k^2)
Lm=L1*k^2
N1=(1/k)*sqrt(L2/L1)
k=0.63
Vap=Vin/(1-D)
Ic=(Vin/R)*(D/(1-D)^2)

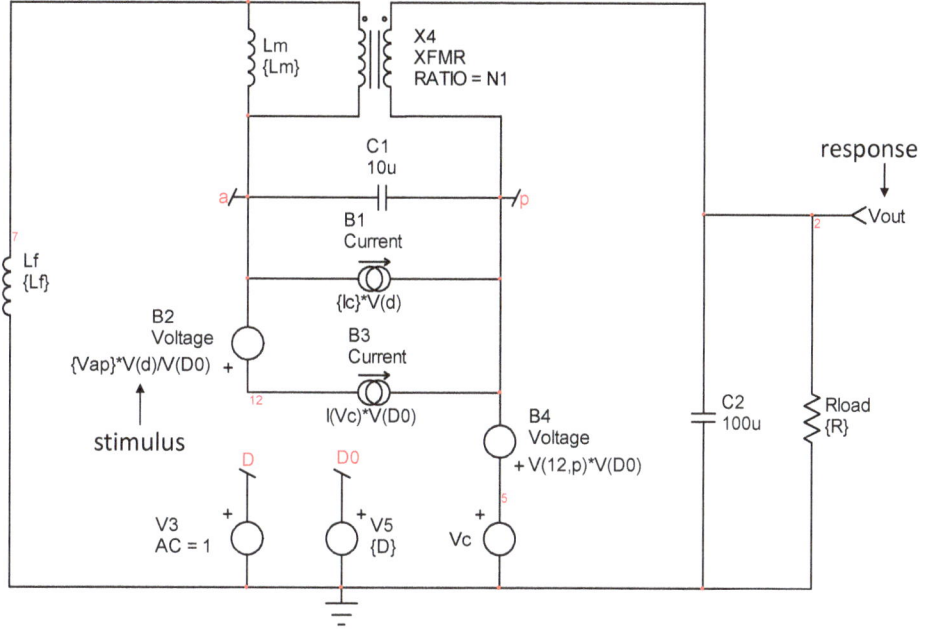

Lm {Lm}

X4
XFMR
RATIO = N1

C1
10u

B1
Current
{Ic}*V(d)

Lf {Lf}

B2
Voltage
{Vap}*V(d)/V(D0) +

B3
Current
I(Vc)*V(D0)

B4
Voltage
+ V(12,p)*V(D0)

C2
100u

Rload
{R}

response

Vout

stimulus

V3
AC = 1

V5
{D}

Vc

Figure 5.10 This is the small-signal equivalent circuit of the Ćuk converter with coupled inductors.

$$\alpha = k\sqrt{\frac{L_2}{L_1}} \tag{5.28}$$

$$a_1 = -\frac{L_1}{R_{load}}\left(\alpha + \frac{D}{1-D}\right)\frac{D}{1-D} \tag{5.29}$$

$$a_2 = L_1 C_1\left(\frac{1-\alpha}{1-D}\right) \tag{5.30}$$

$$\omega_N = \frac{1}{\sqrt{a_2}} \tag{5.31}$$

$$Q_N = \frac{\sqrt{a_2}}{a_1} \tag{5.32}$$

$$\omega_L = \frac{1}{\sqrt{L_1\frac{C_1}{(1-D)^2} + C_2 L_e}} \tag{5.33}$$

$$Q_L = \frac{R_{load}}{\omega_L L_e} \tag{5.34}$$

$$\omega_H = \sqrt{\frac{L_1}{L_2 L_\sigma C_2} + \frac{L_e}{L_2 L_\sigma} \frac{(1-D)^2}{C_1}} \qquad (5.35)$$

$$Q_L = \omega_H C_2 R_{load} \qquad (5.36)$$

We need to verify this expression versus the magnitude and phase response delivered by Figure 5.7. The result appears in Figure 5.11 and confirms the excellent matching between the formula and the large-signal model.

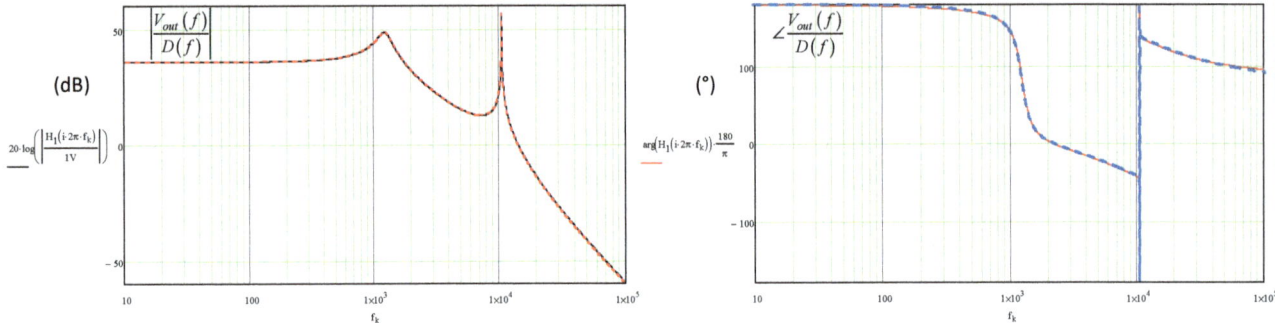

Figure 5.11 SPICE simulations and analytical expressions deliver exact same responses.

To check the effects of output current ripple cancelation, SIMPLIS® can help us confirm if respecting (5.23) brings the expected results. The circuit is proposed in Figure 5.12.

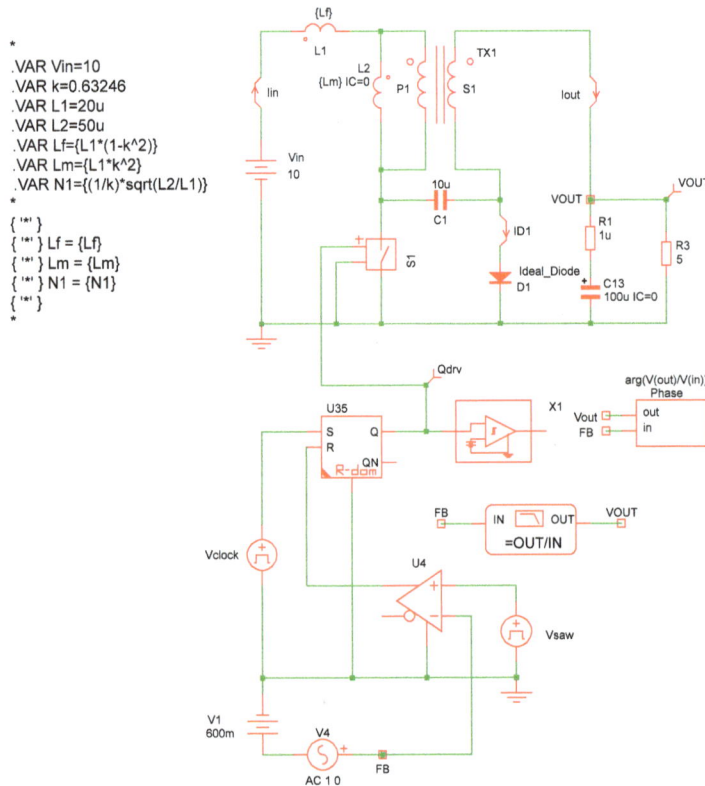

Figure 5.12 In this example, inductors and coupling coefficient *k* are sized to cancel the ripple current in the output inductor.

Once simulation is run, displaying the output current as we did in Figure 5.13 confirms the reduction to a small level of the current circulating in L_2. The small-signal simulation results in Figure 5.14 also confirm the validity of the response already obtained in Figure 5.11.

If we now make L_2 smaller than L_1 (we swapped the values) to cancel the input ripple, SIMPLIS® confirms the results in Figure 5.15 where the ripple in L_1 is brought to an extremely low level.

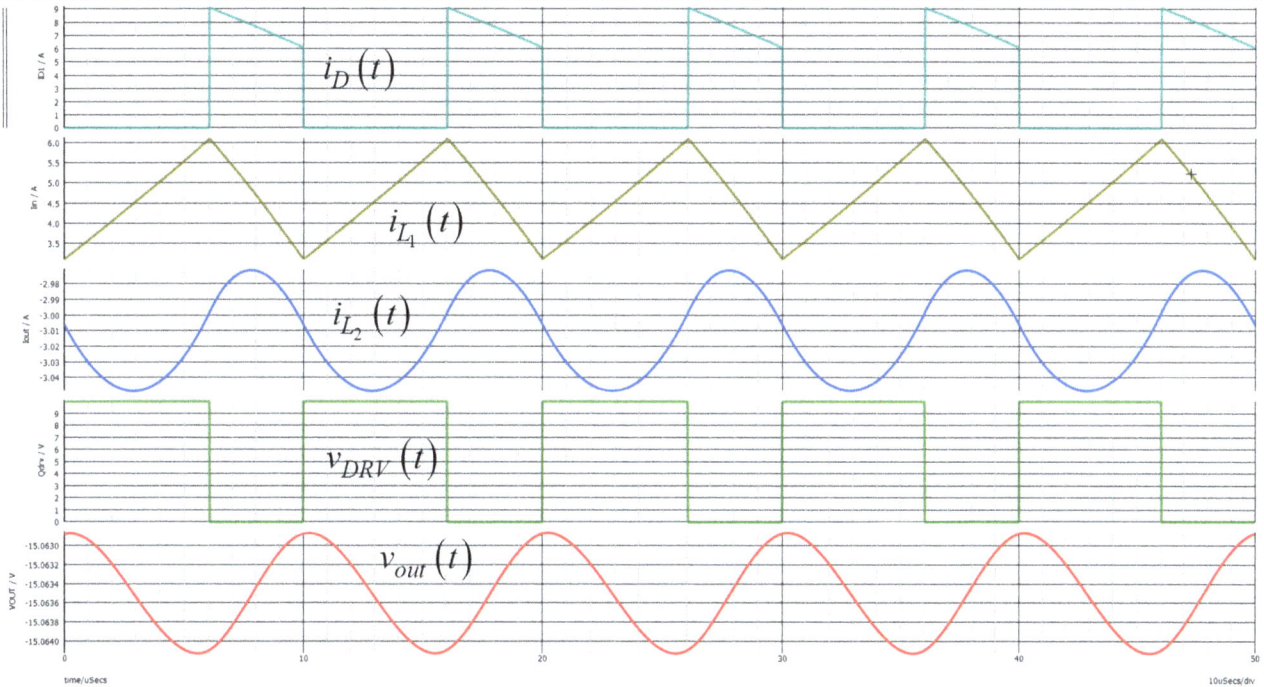

Figure 5.13 As expected, the output ripple i_{L2} is nicely reduced when choosing L_1 smaller than L_2.

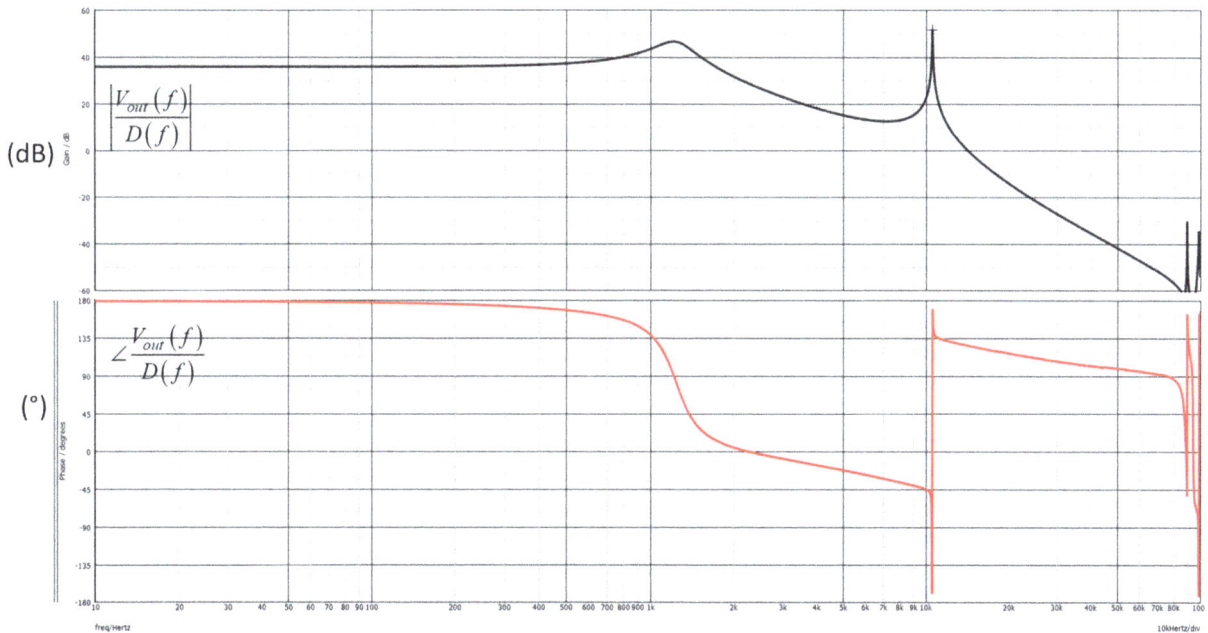

Figure 5.14 SIMPLIS® ac response is identical to that delivered by (5.25).

615

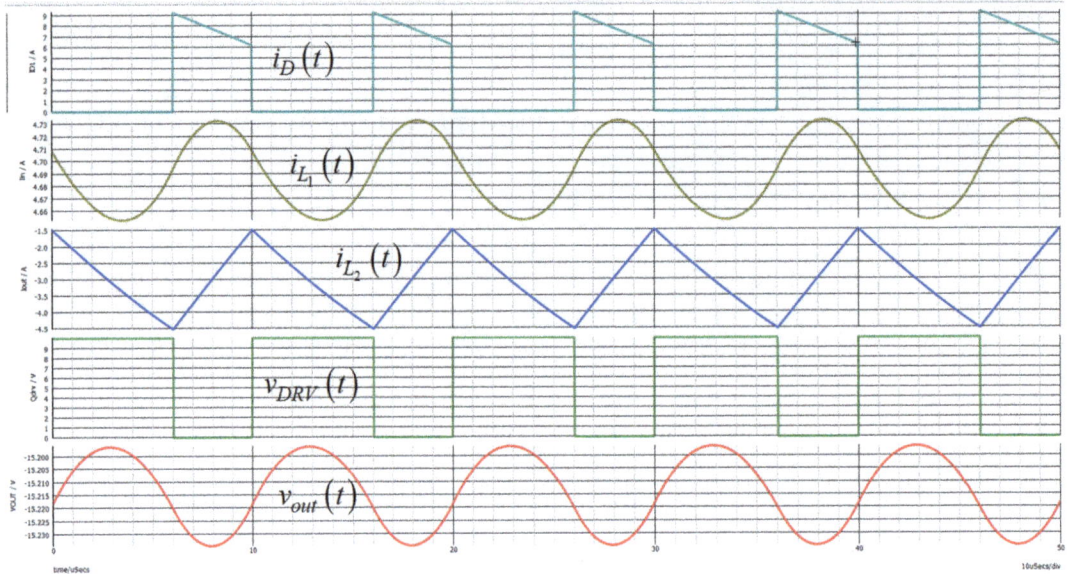

Figure 5.15 By making L_2 smaller than L_1, the input ripple current in L_1 is effectively reduced to a small amplitude.

5.1.2 Isolated Version

It is possible to isolate the output of the Ćuk converter and deliver a positive voltage. The circuit in Figure 5.16 depicts the converter in which a transformer has been inserted. Please note the windings polarity for the transformer primary and secondary sides as well as the coupled inductors L_1 and L_{2s}. The subscript s designates a component placed in the secondary side.

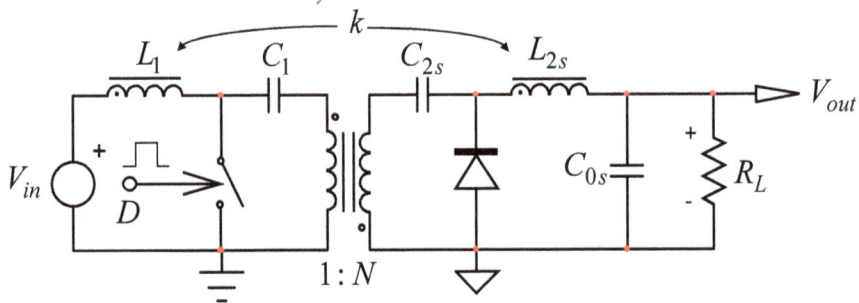

Figure 5.16 The Ćuk converter can be isolated by inserting a transformer with another coupling capacitor.

The small-signal analysis is obtained by revealing the PWM switch through several steps detailed in [5]. These steps consist of reflecting the components values to the transformer primary side. If we consider a 1:N turns ratio, then scale these components according to the below expressions:

$$C_2 = C_{2s} N^2 \tag{5.37}$$

$$C_0 = C_{0s} N^2 \tag{5.38}$$

$$L_2 = \frac{L_{2s}}{N^2} \tag{5.39}$$

616

$$R = \frac{R_L}{N^2} \tag{5.40}$$

The coupling coefficient k needs be replaced by its equivalent transformer model given in Figure 5.8:

$$L_\sigma = L_1\left(1-k^2\right) \tag{5.41}$$

$$L_M = L_1 k^2 \tag{5.42}$$

$$n = \frac{1}{k}\sqrt{\frac{L_2}{L_1}} \tag{5.43}$$

Once this is done, the equivalent circuit given in the paper appears in the Figure 5.17. You can see the appearance of the transformer magnetizing inductance L_m whose presence and impact were analyzed for the first time by Dr. Vorpérian. If you count the energy-storing elements, there are six making the converter a 6th-order system. I have left the output capacitor and the load with the isolation transformer in the right side so that the correct dc output voltage is obtained. They will be brought back to the left-side during the analysis.

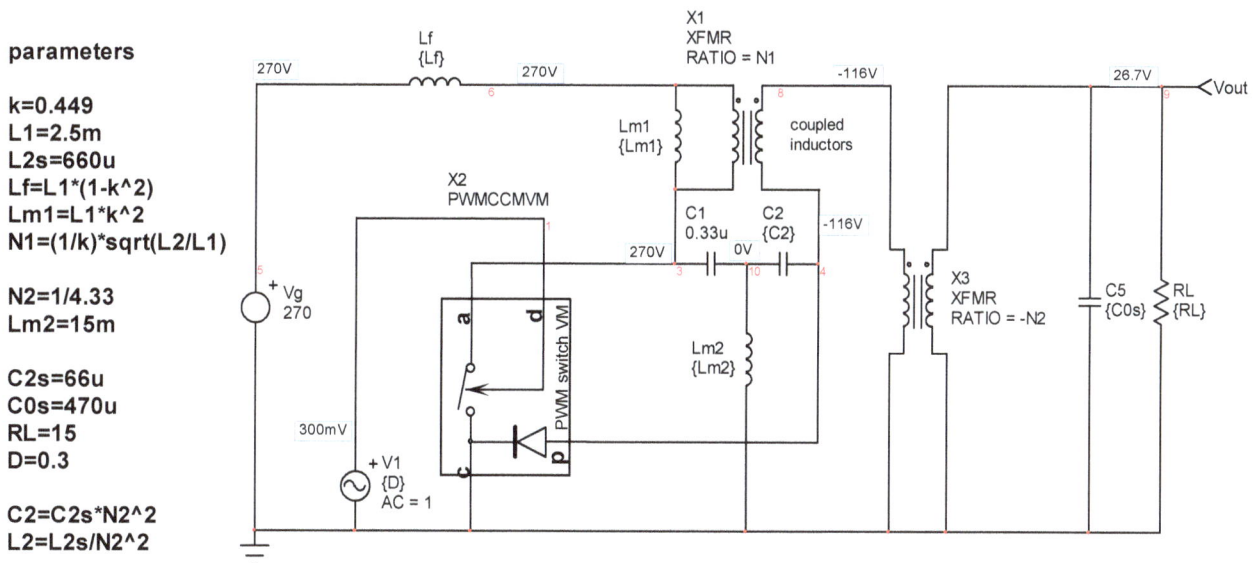

Figure 5.17 Components scaling to the primary side is necessary to build this equivalent circuit.

The dc analysis obtained when all capacitors are open and inductors are shorted gives the following transfer function:

$$M = \frac{ND}{1-D} \tag{5.44}$$

It implies a 26.7-V output voltage from a 270-V source with a 30% duty ratio ($N = 0.23$). All the components values in this application circuits are coming from the example given in [5] and correspond to a practical experiment carried in [6]. To start the analysis, replace the large-signal PWM switch by its small-signal version and have fun solving the new circuit given in Figure 5.18. Following the steps in [5], the control-to-output transfer function is given below where no parasitic elements have been accounted for:

$$H(s) = \frac{V_{out}(s)}{D(s)} = H_0 \frac{1 + a_1 s + a_2 s^2 + a_3 s^3 + a_4 s^4}{1 + b_1 s + b_2 s^2 + b_3 s^3 + b_4 s^4 + b_5 s^5 + b_6 s^6} \qquad (5.45)$$

where:

$$H_0 = \frac{N V_{in}}{(1-D)^2} \qquad (5.46)$$

$$a_1 = -\frac{L_1}{R}\left(\alpha + \frac{D}{1-D}\right)\frac{D}{1-D} \qquad (5.47)$$

$$a_2 = L_1\left(C_1 - \alpha\frac{D}{1-D}C_2\right) + L_m\left(C_1 + C_2\right) \qquad (5.48)$$

parameters

k=0.449
L1=2.5m
L2s=660u
Lf=L1*(1-k^2)
Lm1=L1*k^2
N1=(1/k)*sqrt(L2/L1)

N2=1/4.33
Lm2=15m

C2s=66u
C0s=470u
RL=15
D=0.3

C2=C2s*N2^2
C0=C0s*N2^2
L2=L2s/N2^2
R=RL/N2^2

Vg=270
Vap=Vg/(1-D)
Ic=(Vg/R)*D/(1-D)^2

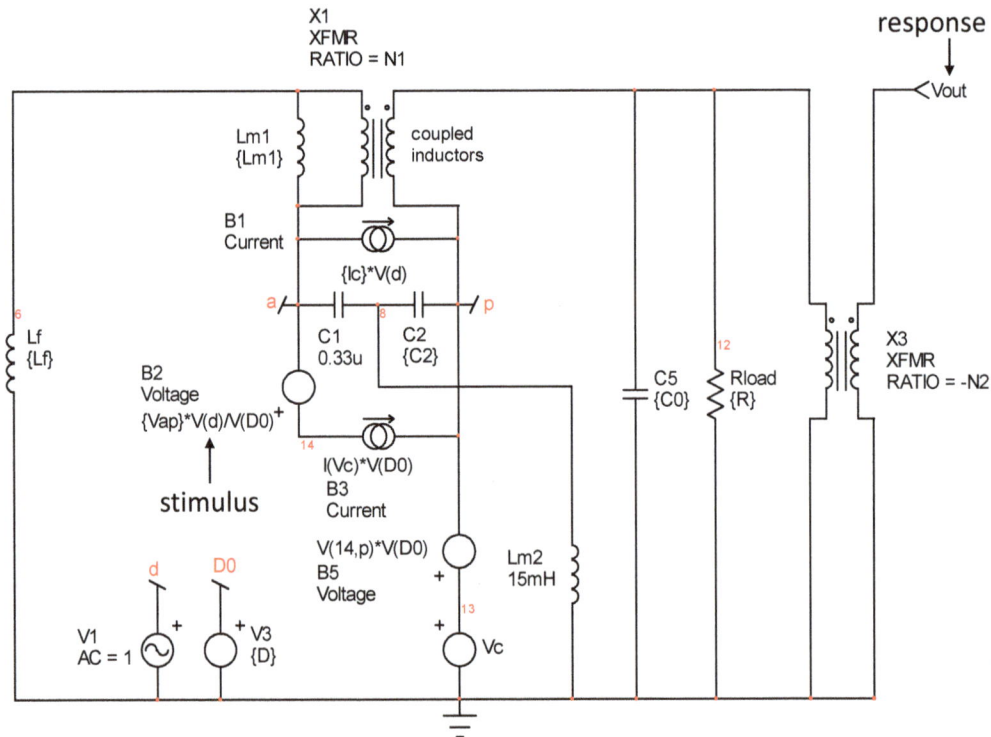

Figure 5.18 This is the setup to determine the control-to-output transfer function of the isolated Ćuk converter.

$$a_3 = -\frac{L_1}{R}L_m\left(C_1 + C_2\right)\left(\alpha + \frac{D}{1-D}\right)\frac{D}{1-D} \qquad (5.49)$$

$$a_4 = L_1 L_m C_1 C_2 \frac{1-\alpha}{1-D} \qquad (5.50)$$

in which:

$$\alpha = \frac{L_M}{L_M + L_\sigma} \quad n = k\sqrt{\frac{L_2}{L_1}} \tag{5.51}$$

The denominator $D(s)$ uses the below coefficients:

$$b_1 = \frac{L_e}{R} \tag{5.52}$$

$$b_2 = L_1 C_e + L_m \left(C_1 + C_2 \right) + L_e C_0 \tag{5.53}$$

$$b_3 = L_m \left(C_1 + C_2 \right) \frac{L_e}{R} + L_2 C_e \frac{L_\sigma}{R} \tag{5.54}$$

$$b_4 = L_1 C_1 L_m \frac{C_2}{\left(1-D\right)^2} + L_m \left(C_1 + C_2 \right) C_0 L_e + L_\sigma C_e L_2 C_0 \tag{5.55}$$

$$b_5 = \frac{L_2}{R} L_\sigma C_1 L_m \frac{C_2}{\left(1-D\right)^2} \tag{5.56}$$

$$b_6 = L_2 C_0 L_\sigma C_1 L_m \frac{C_2}{\left(1-D\right)^2} \tag{5.57}$$

where:

$$C_e = C_1 + C_2 \left(\frac{D}{1-D} \right)^2 \tag{5.58}$$

$$L_e = L_1 \left(\frac{D}{1-D} \right)^2 + L_2 + 2k\sqrt{L_1 L_2} \left(\frac{D}{1-D} \right) \tag{5.59}$$

In the above expressions, L_M is the equivalent primary inductance in the coupling coefficient equivalent transformer while L_m represents the isolation transformer primary inductance. In the referenced paper, these raw coefficients have been further assembled to form more *low-entropy* types of numerator and denominator with cascaded second-order polynomials not detailed here. As usual, it is interesting to confront the small-signal response described by (5.45) with the ac response delivered by Figure 5.17. This is what Figure 5.19 shows with amazingly-matching results.

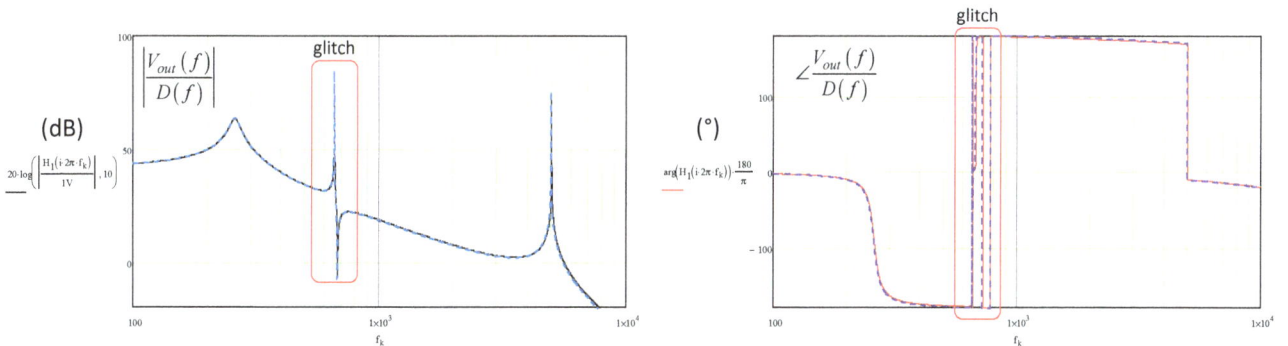

Figure 5.19 The response shows a glitch incurred to the magnetizing inductance L_m.

The glitch observed in the response is due to the isolation transformer magnetizing inductance. Dr. Vorpérian was the first to identify how the reduction of C_2 could cancel its effect. If C_2 follows the below relationship, the glitch disappears for the given input-output conditions:

$$C_2 = C_1 \frac{1-D}{D} \left[\frac{1 + D(1-\alpha)}{D + \alpha(1-D)} \right]$$

(5.60)

Applying the components values to this equation leads to a capacitor value of 14.45 μF. As confirmed by Figure 5.20, the glitch is gone when C_2 takes on the recommended value.

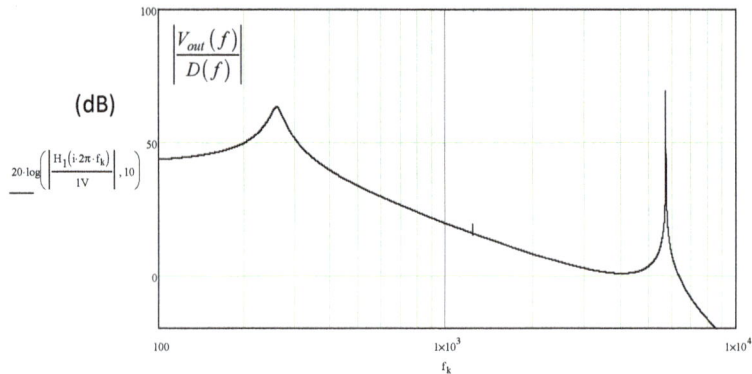

Figure 5.20 The glitch can disappear if capacitor C_2 is reduced to a certain value.

To verify all these expressions with SIMPLIS®, I have simulated the converter in Figure 5.21. The PWM circuit features a 1-V sawtooth amplitude and the switching frequency is 50 kHz. Figure 5.22 confirms the 27-V output for the 30% duty ratio.

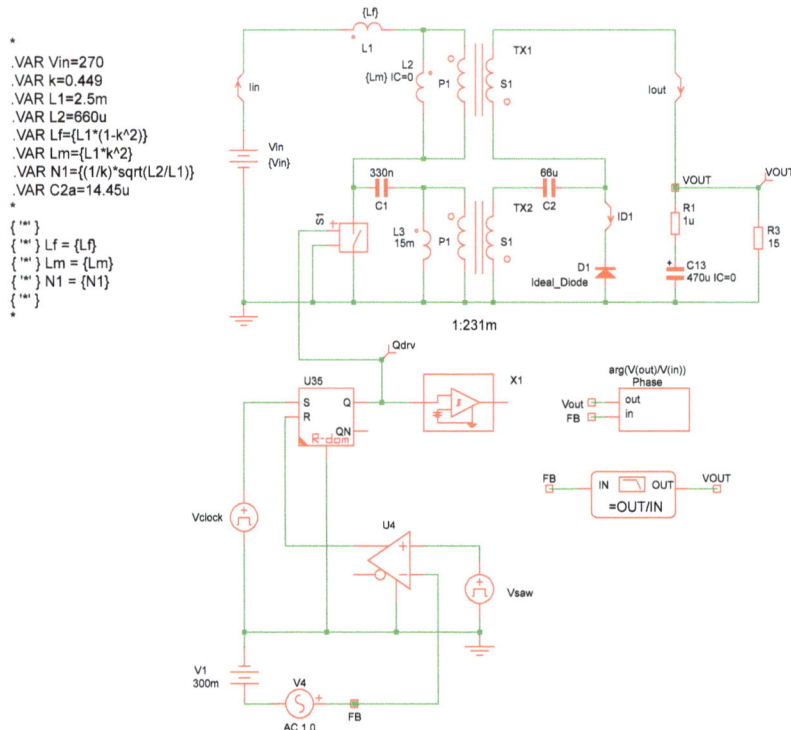

Figure 5.21 SIMPLIS® represents an excellent tool to verify the response and operating point of such a complex structure.

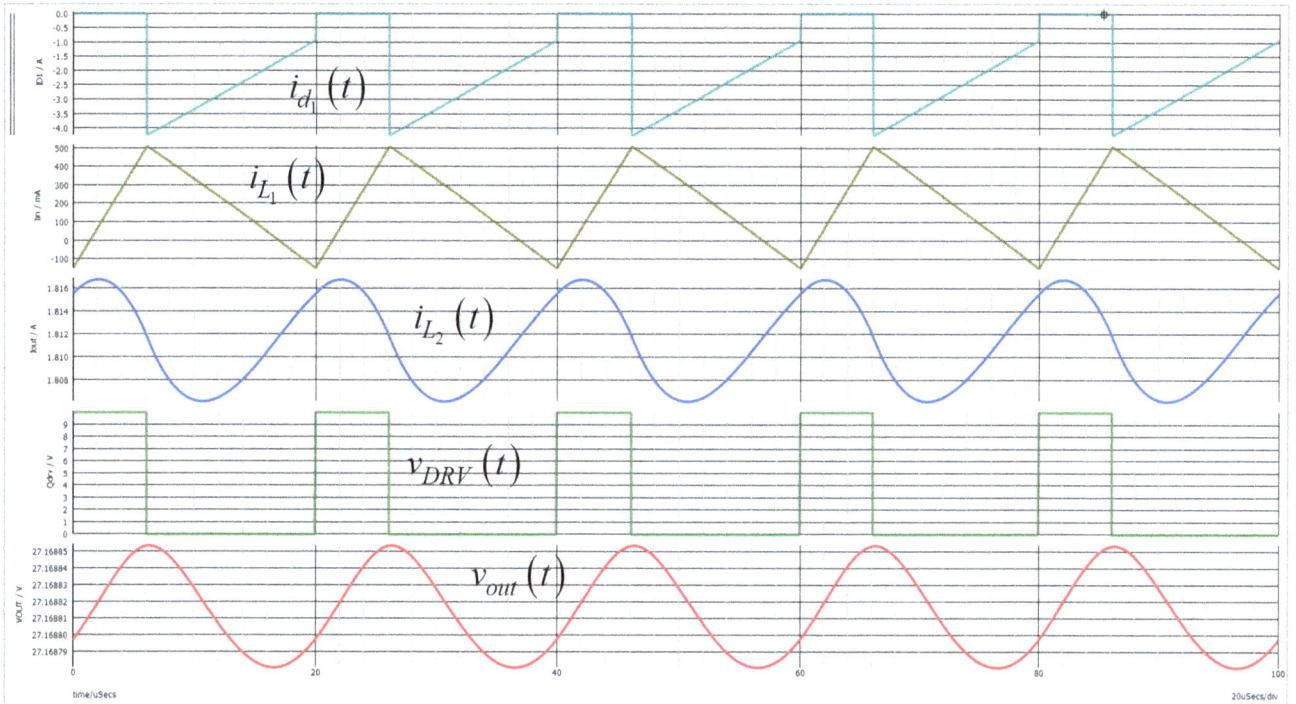

Figure 5.22 The cycle-by-cycle waveforms show a correct operating point and a 27-V output voltage.

Finally, Figure 5.23 displays the magnitude and phase responses which show to be very close to the small-signal response of Figure 5.19. This ends the quick description of the Ćuk converter which Dr. Vorpérian beautifully analyzed in great details with the help of the fast analytical circuits techniques.

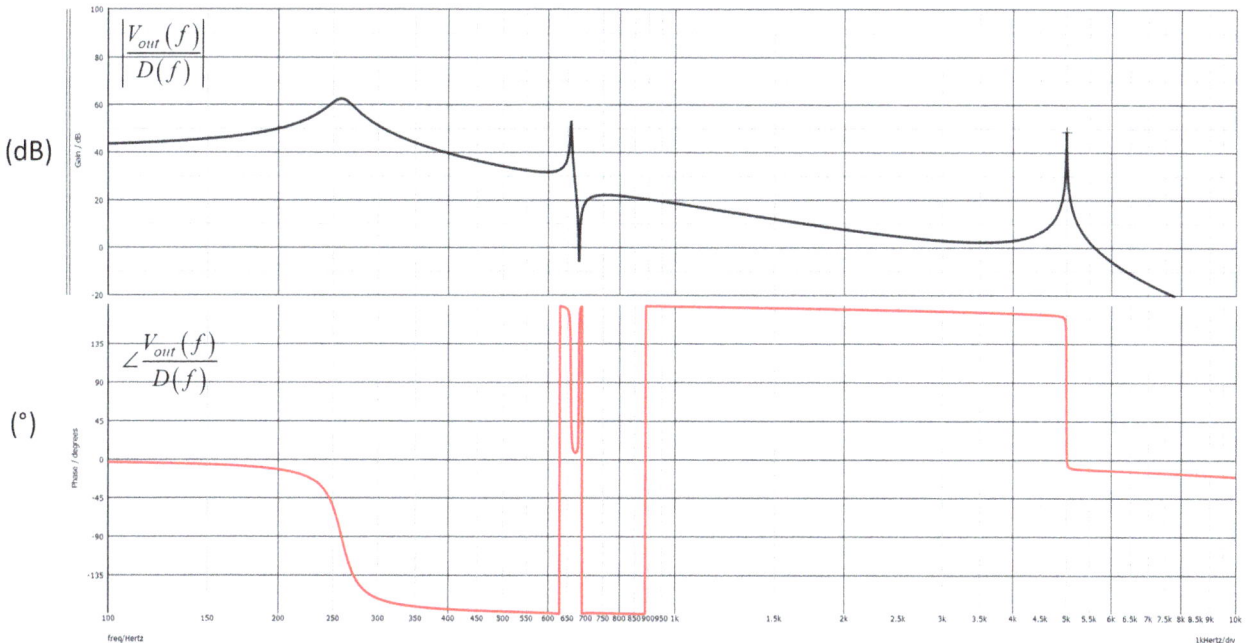

Figure 5.23 The SIMPLIS® response matches SPICE's and analytical expression quite well.

5.2 The SEPIC

The singled-ended primary-inductor converter or SEPIC can be described as a boost front-end converter (like in the Ćuk converter) but coupled to an inverted buck-boost this time. The output voltage is positive and you can couple the inductors to reduce the input ripple current as with the Ćuk converter. However, the current in the output capacitor of the SEPIC is still pulsating as with a classical boost or buck-boost converter. The circuit in its simplest form is shown in Figure 5.24.

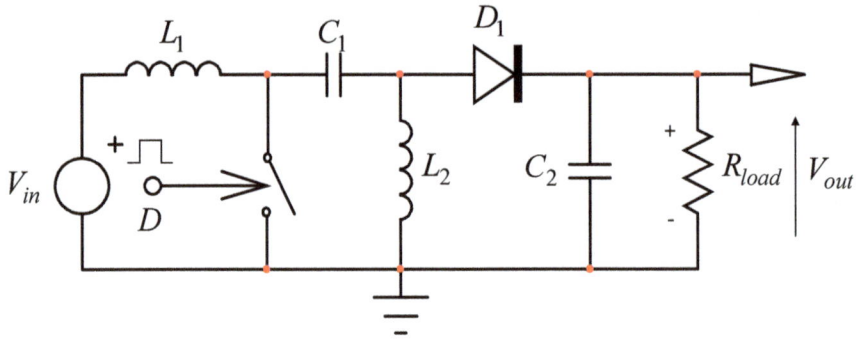

Figure 5.24 The SEPIC delivers a positive output voltage which can be greater or smaller than the input source.

You can see diode D_1 is not directly connected to the power switch. As shown in Figure 5.25 sketch (a), slide the diode to the right along the ground and the cathode now touches the switch. Then reconnect inductor L_2 as shown in sketch (b) and, there you go, you see the PWM switch connections. Please note that the input source and the load no longer share a common ground but that is not a problem.

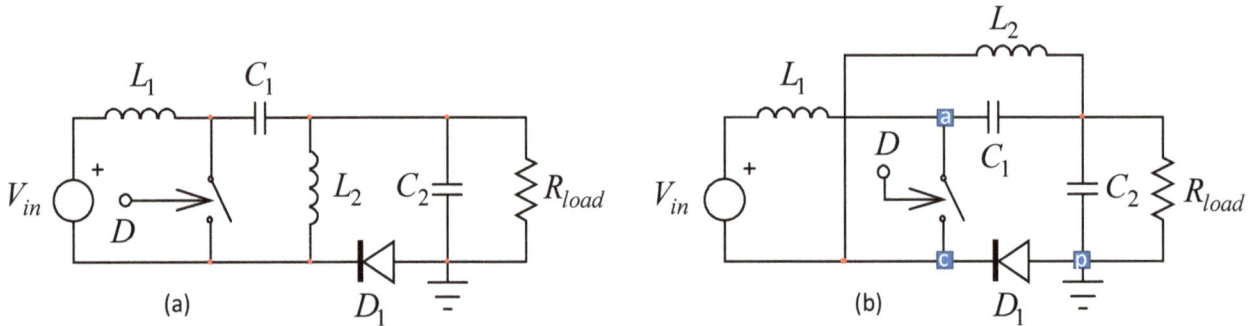

Figure 5.25 By simple manipulations, it is possible to reveal the PWM switch model.

The complete simulation model appears in Figure 5.26 and shows an output voltage of 15 V. Using this model for a dc analysis (shorted inductors and opened capacitors), it is possible to show that the conversation ratio M follows that of a classical non-inverting buck-boost converter (no parasitic elements):

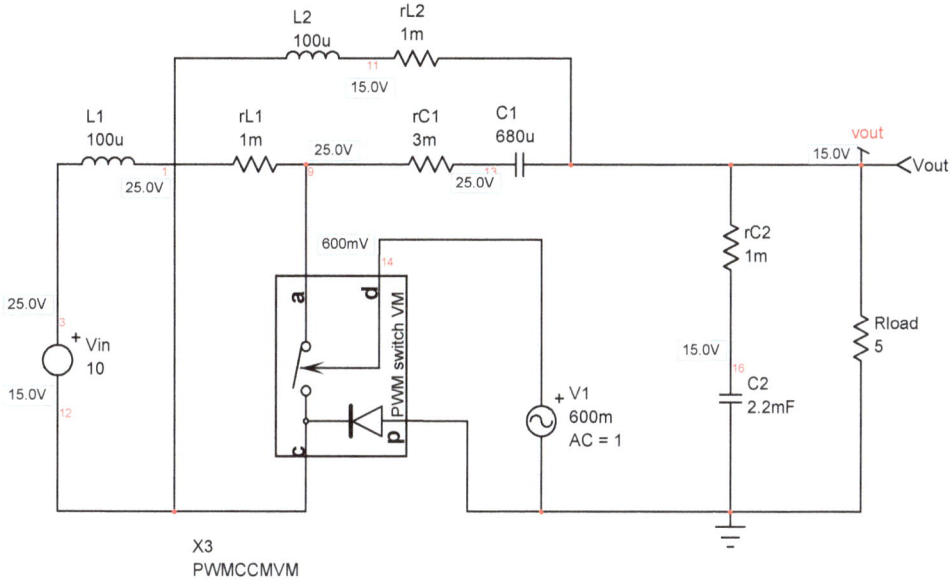

Figure 5.26 The PWM switch lends itself well to the analysis of this converter.

$$M = \frac{D}{1-D} \tag{5.61}$$

It confirms the 15-V output from a 10-V input source for a 60% duty ratio. To check this configuration, we can also implement the CoPEC model which does not require further manipulation considering the separated diode and switch approach. The circuit appears in Figure 5.27.

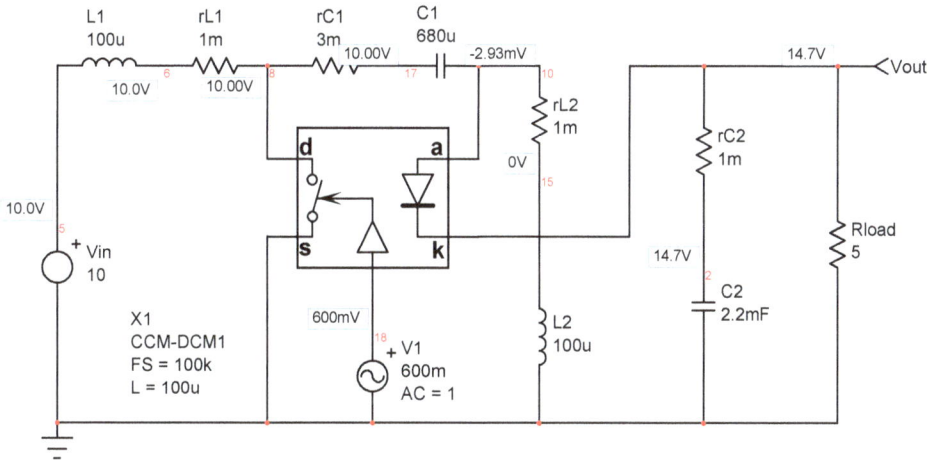

Figure 5.27 The CoPEC model is wired in a simple way to build a SEPIC.

The frequency responses of both models are similar in magnitude and phase as confirmed by Figure 5.28. You can observe a peak and a glitch which bring severe phase distortion. To determine the control-to-output transfer function in CCM, replace the PWM switch by its small-signal model as shown in Figure 5.29 after rearranging all sources in a compact way. The sanity check confirms the response of this circuit is identical to that of Figure 5.28. The complete study of the CCM SEPIC with uncoupled inductors has been carried by Dr. Vorpérian in a paper available from the web [8]. It is a 4th-order system and the author made extensive use of the fast analytical circuits techniques to unveil the time constants with a zeroed excitation and in a nulled output condition.

According to the paper, the system can be described by the following raw expression:

$$\frac{V_{out}(s)}{D(s)} = H_0 \frac{1 - a_1 s + a_2 s^2 - a_3 s^3}{1 + b_1 s + b_2 s^2 + b_3 s^3 + b_4 s^4} \tag{5.62}$$

In this expression, the quasi-static gain H_0 is:

$$H_0 = \frac{\partial V_{out}(D)}{\partial D} = \frac{V_{in}}{(1-D)^2} \tag{5.63}$$

The various coefficients are reproduced below:

$$a_1 = \frac{L_1}{R_{load}} \left(\frac{D}{1-D} \right)^2 \tag{5.64}$$

$$a_2 = C_1 (L_1 + L_2) \tag{5.65}$$

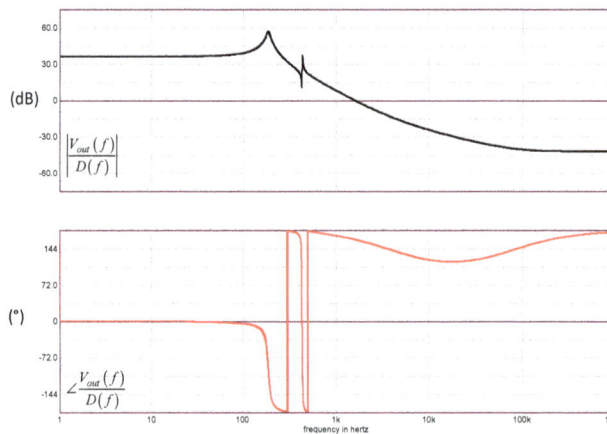

Figure 5.28 The models deliver similar magnitude/phase responses.

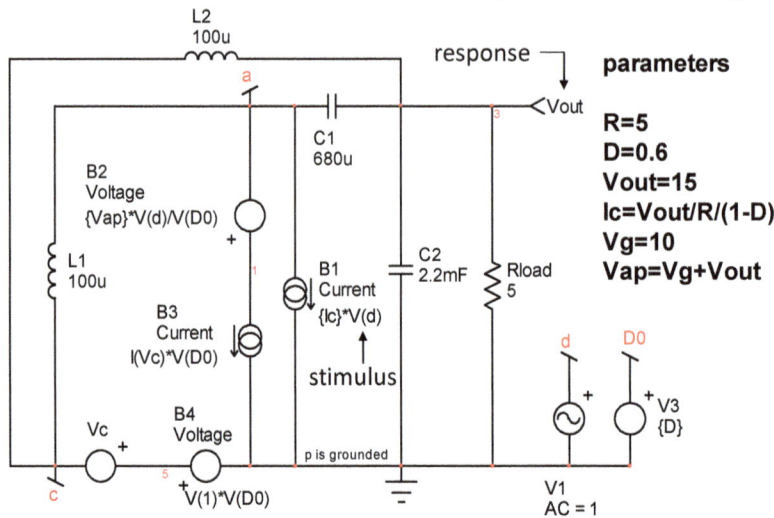

Figure 5.29 The SEPIC can be analyzed with the small-signal PWM switch model.

$$a_3 = \frac{C_1 L_1 L_2}{R_{load}} \frac{D}{(1-D)^2} \tag{5.66}$$

624

$$b_1 = \frac{L_1}{\left(\dfrac{1-D}{D}\right)^2 R_{load}} + \frac{L_2}{R_{load}} \tag{5.67}$$

$$b_2 = L_1\left[C_1 + \left(\frac{D}{1-D}\right)^2 C_2\right] + L_2\left(C_1 + C_2\right) \tag{5.68}$$

$$b_3 = L_1 L_2 C_1 \frac{1}{R_{load}\left(1-D\right)^2} \tag{5.69}$$

$$b_4 = \frac{L_1 L_2 C_1 C_2}{\left(1-D\right)^2} \tag{5.70}$$

If you compare the response given by (5.62) in a Mathcad® sheet with results delivered by Figure 5.29, they are remarkably close as represented in Figure 5.30.

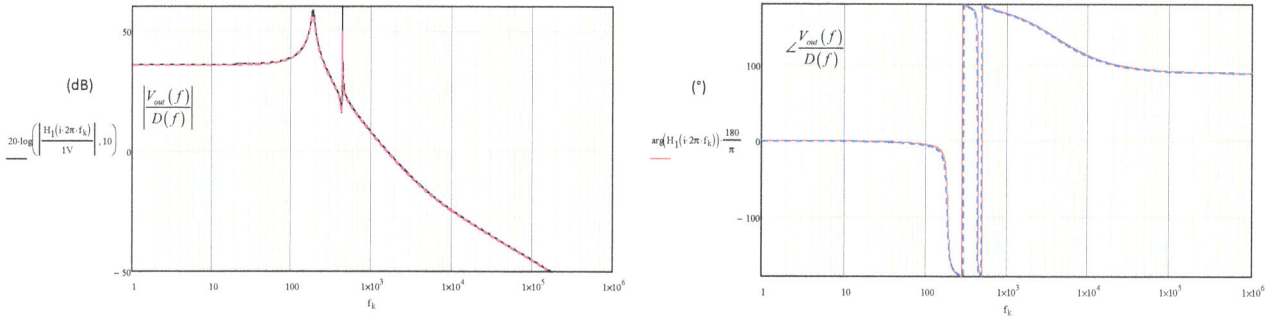

Figure 5.30 SPICE simulations and analytical expressions deliver exact same responses for the control-to-output transfer function of the SEPIC.

It is possible to factor the raw formula in a combination of poles and zeroes; however, the arrangement depends on the various time constants and load values. For full load conditions, [8] proposes the following factorization for the numerator.

It works well for $C_1 << C_2$:

$$N(s) \approx \left(1 - a_1 s\right)\left(1 - s\frac{a_2}{a_1} + s^2\frac{a_3}{a_1}\right) = \left(1 - \frac{s}{\omega_z}\right)\left[1 - \frac{s}{\omega_{0N}Q_N} + \left(\frac{s}{\omega_{0N}}\right)^2\right] \tag{5.71}$$

In which:

$$\omega_z = \frac{1}{a_1} = \left(\frac{1-D}{D}\right)^2 \frac{R_{load}}{L_1} \tag{5.72}$$

It is a RHP zero. The resonant frequency is defined as:

$$\omega_{0N} = \sqrt{\frac{a_1}{a_3}} = \sqrt{\frac{D}{L_2 C_1}} \tag{5.73}$$

$$Q_N = \frac{a_1 a_3 \sqrt{\frac{a_1}{a_3}}}{a_1 a_2 - a_3} = \left(\frac{D}{1-D}\right)^2 \frac{L_1 \sqrt{\frac{C_1 L_2}{D}}}{C_1 R_{load}\left(L_1 - \frac{L_2}{D} + L_2\right)} \tag{5.74}$$

In this arrangement, we assume that the zero defined in **Error! Reference source not found.** dominates the low-frequency response while a zero pair occurs later on. However, calculations show that the components values of Figure 5.29 do not lend themselves to the factorization in **Error! Reference source not found.**. This is because factoring a 3rd-order polynomial can be processed in different ways depending on where the resonance is located [9] and how a first zero (or a pole) dominates the low-frequency response. Assuming a choice of the component values from Figure 5.25 in which C_1 is not far from C_2, a different combination works better and is considered for light-load operation in [8]:

$$N(s) \approx \left(1 - \frac{a_3}{a_2}s\right)\left(1 - a_1 s + a_2 s^2\right) = \left(1 - \frac{s}{\omega_z}\right)\left[1 - \frac{s}{\omega_{0N} Q_N} + \left(\frac{s}{\omega_{0N}}\right)^2\right] \tag{5.75}$$

$$\omega_z = \frac{a_2}{a_3} = \frac{(1-D)^2}{D} \frac{R_{load}}{L_1 \| L_2} \tag{5.76}$$

It is also a RHP zero and the resonant frequency is defined as:

$$\omega_{0N} = \frac{1}{\sqrt{a_2}} = \frac{1}{\sqrt{C_1(L_1 + L_2)}} \tag{5.77}$$

$$Q_N = \frac{\sqrt{a_2}}{a_1} = \frac{R_{load}}{L_1}\sqrt{C_1(L_1 + L_2)}\left(\frac{1-D}{D}\right)^2 \tag{5.78}$$

I have plotted these factored expressions versus the raw numerator of **Error! Reference source not found.** which once expanded is equal to:

$$N(s) = 1 - s\frac{L_1}{R_{load}}\left(\frac{D}{1-D}\right)^2 + s^2 C_1(L_1 + L_2) - s^3 \frac{L_1 L_2 C_1}{R_{load}}\frac{D}{(1-D)^2} \tag{5.79}$$

The results appear in Figure 5.31 and show good results for (5.75) in this particular case. Should you severely reduce C_1 from its original value then (5.71) would give a better response while (5.75) would diverge from the raw expression.

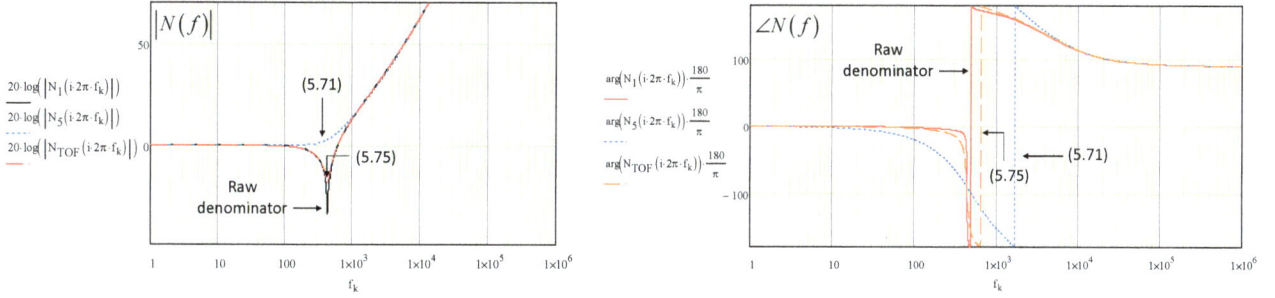

Figure 5.31 The proposed factorization is closed to the raw numerator values.

The 4th-order denominator can also be factored considering two well-separated resonances.

Following [8], we have:

$$1 + b_1 s + b_2 s^2 + b_3 s^3 + b_4 s^4 \approx \left[1 + \frac{s}{\omega_{01} Q_1} + \left(\frac{s}{\omega_{01}} \right)^2 \right] \left[1 + \frac{s}{\omega_{02} Q_2} + \left(\frac{s}{\omega_{02}} \right)^2 \right] \tag{5.80}$$

where:

$$\omega_{01} = \frac{1}{\sqrt{L_1 \left[C_2 \left(\frac{D}{1-D} \right)^2 + C_1 \right] + L_2 (C_1 + C_2)}} \tag{5.81}$$

$$Q_1 = \frac{R_{load}}{\omega_{01} \left[L_1 \left(\frac{D}{1-D} \right)^2 + L_2 \right]} \tag{5.82}$$

$$\omega_{02} = \sqrt{\frac{1}{L_2 \left[\frac{C_1}{D^2} \| \frac{C_2}{(1-D)^2} \right]} + \frac{1}{L_1 (C_1 \| C_2)}} \tag{5.83}$$

$$Q_2 = \frac{R_{load}}{\omega_{02} (L_1 + L_2) \frac{C_1}{C_2} \left(\frac{\omega_{01}}{\omega_{02}} \right)^2} \tag{5.84}$$

However, when plotting (5.80) versus the denominator in (5.62), I observed a mismatch in one of the resonant frequencies ω_{02}. I have derived another expression for this particular value that is given here:

$$\omega_{02} = (1-D) \sqrt{\frac{D^2}{C_1 \left[L_1 D^2 \| L_2 (1-D)^2 \right]} + \frac{1}{C_2 (L_1 \| L_2)}} \tag{5.85}$$

The complete factored control-to-output transfer function for the CCM voltage-mode-operated SEPIC is expressed as:

$$H(s) \approx H_0 \frac{\left(1 - \dfrac{s}{\omega_z}\right)\left[1 - \dfrac{s}{\omega_{0N}Q_N} + \left(\dfrac{s}{\omega_{0N}}\right)^2\right]}{\left[1 + \dfrac{s}{\omega_{01}Q_1} + \left(\dfrac{s}{\omega_{01}}\right)^2\right]\left[1 + \dfrac{s}{\omega_{02}Q_2} + \left(\dfrac{s}{\omega_{02}}\right)^2\right]} \tag{5.86}$$

We can now test the response of this expression – with (5.85) – and Figure 5.29 values with the SIMPLIS® simulation setup of Figure 5.32.

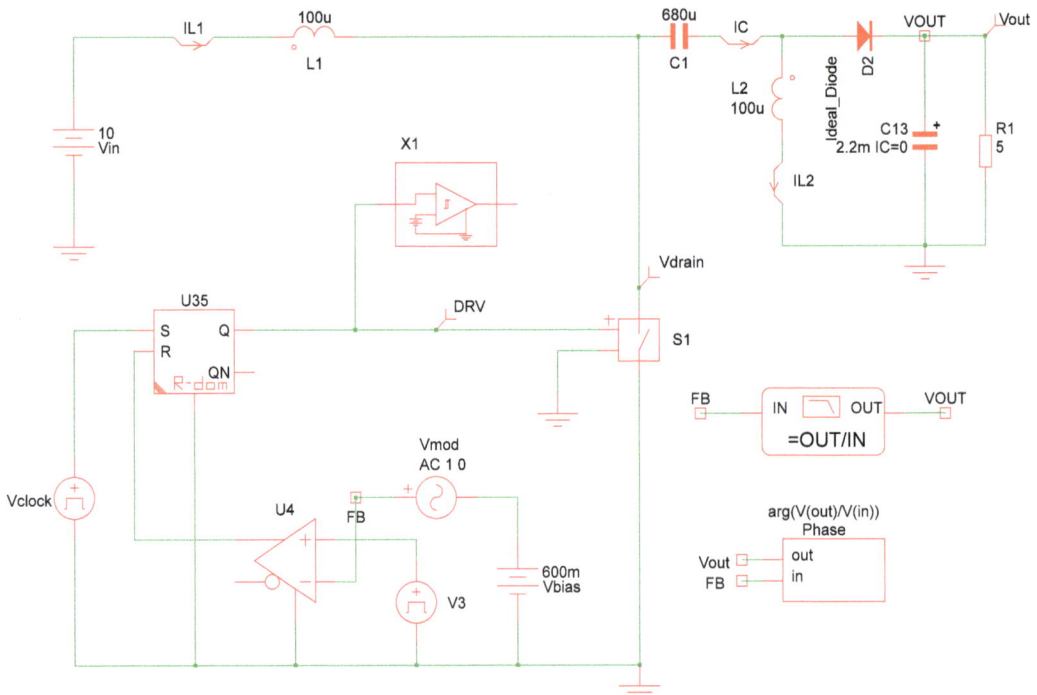

Figure 5.32 This SIMPLIS® setup lets us verify the expressions derived for the uncoupled version.

The operating waveforms in Figure 5.33 confirm a 15-V output voltage from the 10-V source. It is now interesting to compare the full-blown expression given in (5.62) with the factored version in (5.86) and the SIMPLIS® response. As shown in Figure 5.34, the complete expression fits the simulated response quite faithfully while the factored version deviates slightly.

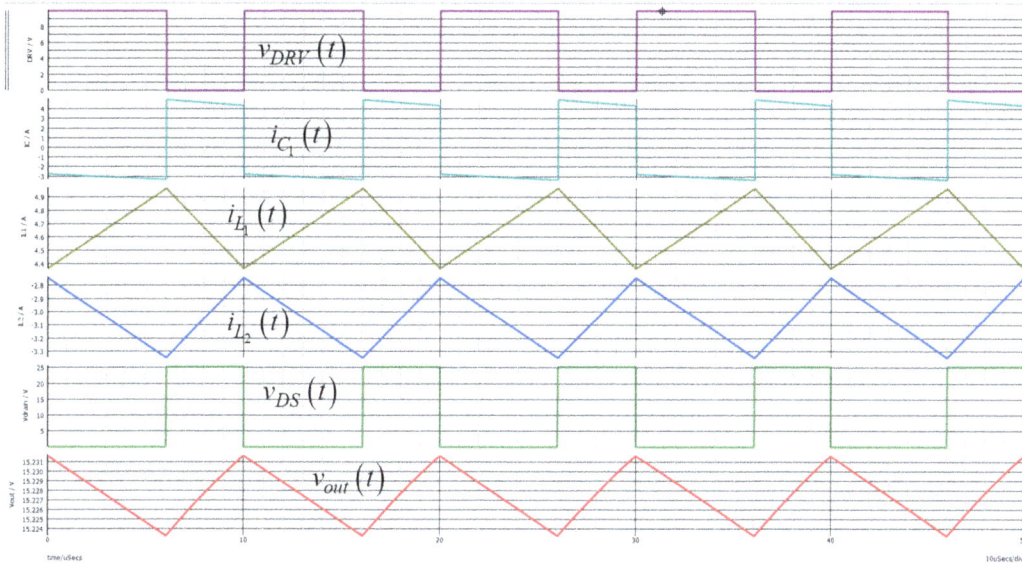

Figure 5.33 Simulations show a 15-V output voltage from the selected 60% duty ratio.

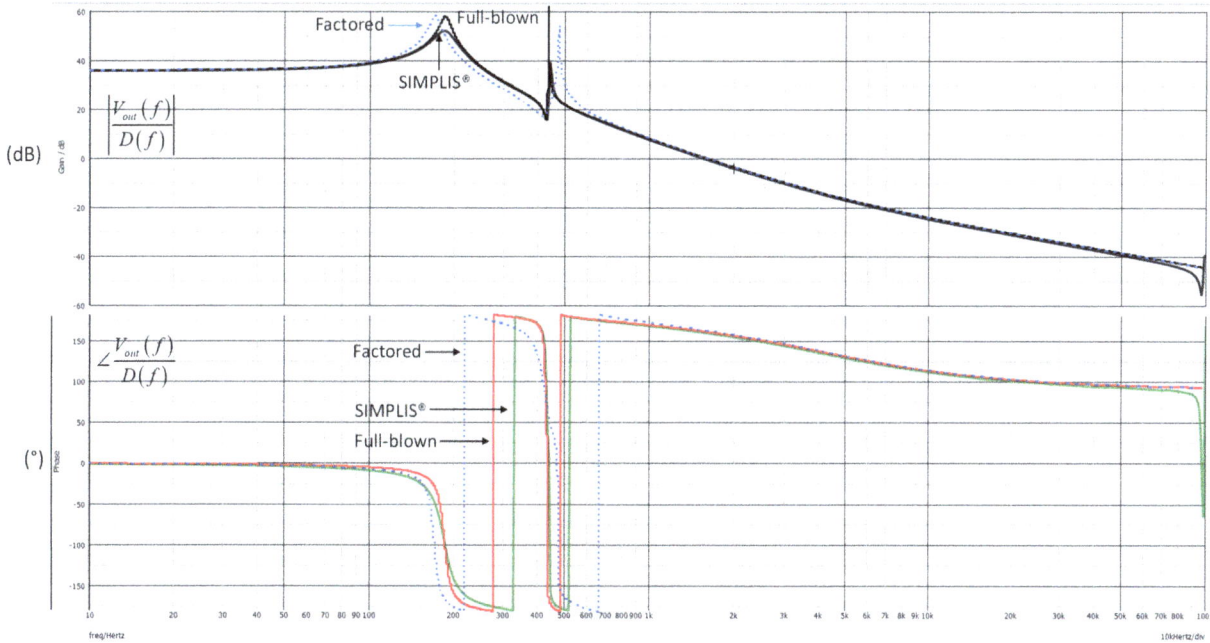

Figure 5.34 The simulation results match the PWM switch model very well while the factored version of the transfer function is not far.

You must however keep in mind that this transfer function, in reality, is extremely dependent on parasitics which exhibit temperature and production variability. This topic is well described in a series of articles published by Dr. Ridley in [10].

5.2.1 Coupled Inductors

Like with the Ćuk converter, it is possible to couple the inductors in a SEPIC. Coupling the inductors helps reduce the input ripple current and magnetics occupy less place on the printed circuit board (PCB).

629

A debate exists whether the coupling should be tight or loose [11] but we won't open it here as we are interested in the control-to-output transfer function only. I have modified the circuit from Figure 5.26 to which I added a transformer featuring a leakage inductance term as described in Figure 5.8. Changing the coupling coefficient k lets you explore loose or tight coupling conditions.

In this application circuit, the transformer must have a 1:1 turns ratio to satisfy volt-second balance while operating. The circuit is shown in Figure 5.35 and its response in Figure 5.36. With a tight coupling, e.g. k is close to 0.99, the ac response is that of a 2nd-order circuit with a phase extending beyond -180° considering the presence of a RHP zero. When you decouple the inductors with a coupling coefficient of 0.91 for instance, a glitch appears and locally distorts the phase. It is due to a resonance between the leakage inductance and the coupling capacitor C_c.

However, it is unlikely that you even notice its presence during a prototype frequency-response analysis as the various ohmic losses will damp the circuit and make the resonance disappear [10].

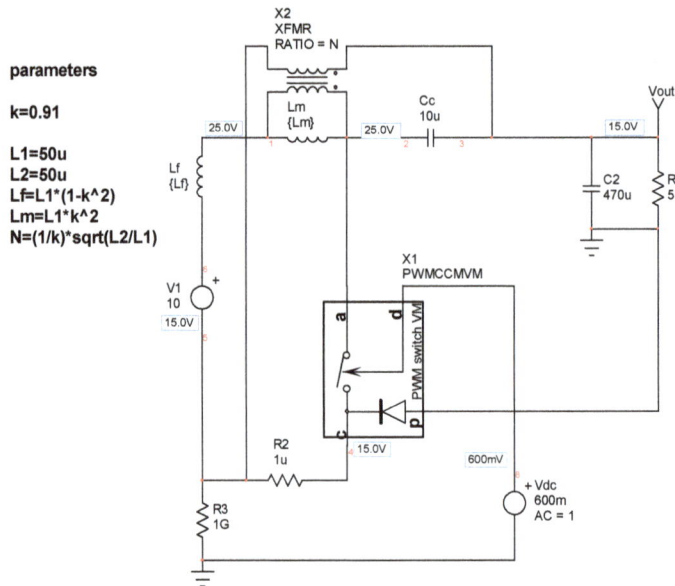

Figure 5.35 The complete large-signal model of the coupled-inductor SEPIC includes the leakage inductance contribution.

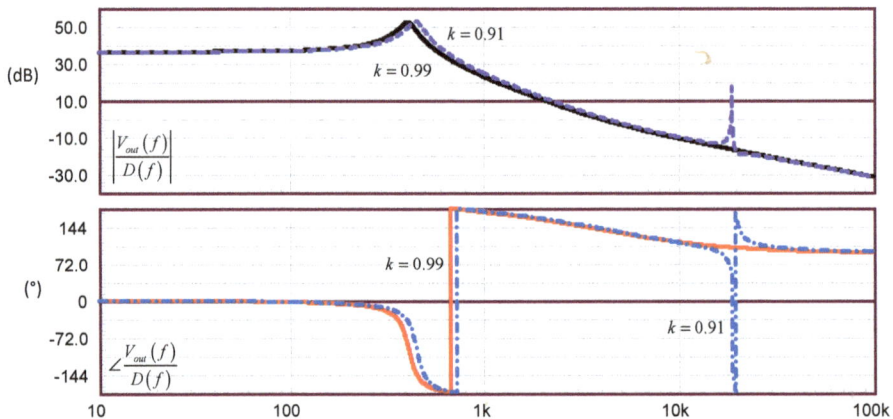

Figure 5.36 With a tight coupling factor, the response is well behaved and is that of a 2nd-order system. When a leakage inductance is added, a resonance involving the coupling capacitor is noticeable.

630

The simulated transient waveforms obtained from Figure 5.37 SIMPLIS® template are given in Figure 5.38. Despite a configuration looking very much like a flyback, the drain voltage of the SEPIC does not show ringing or extra voltage stress due to the leakage inductance presence. The input current ripple is quite smooth and will naturally limit the burden on the input capacitors. The high rms current circulated in the coupling capacitor represents the biggest constraint and a good quality type must be selected for this function. A similar remark holds for the output capacitor as its current is pulsated as with a boost or buck-boost converter.

The complete small-signal model built with the VM PWM switch model appears in Figure 5.39 and includes the leakage inductance term. If you count the energy-storing elements, it makes the whole thing a 4th-order converter and determining the control-to-output transfer function is tedious and long. However, if you reduce the leakage inductance to zero ($k = 0.999$), you see that both upper transformer connections are shorted ($L_f = 0$). In ac, and considering the 1-to-1 turns ratio, it means the voltage across the coupling capacitor C_c is also zero, removing its contribution to a 2nd-order term in the transfer function denominator: despite three energy-storing elements, the converter is of 2nd-order type.

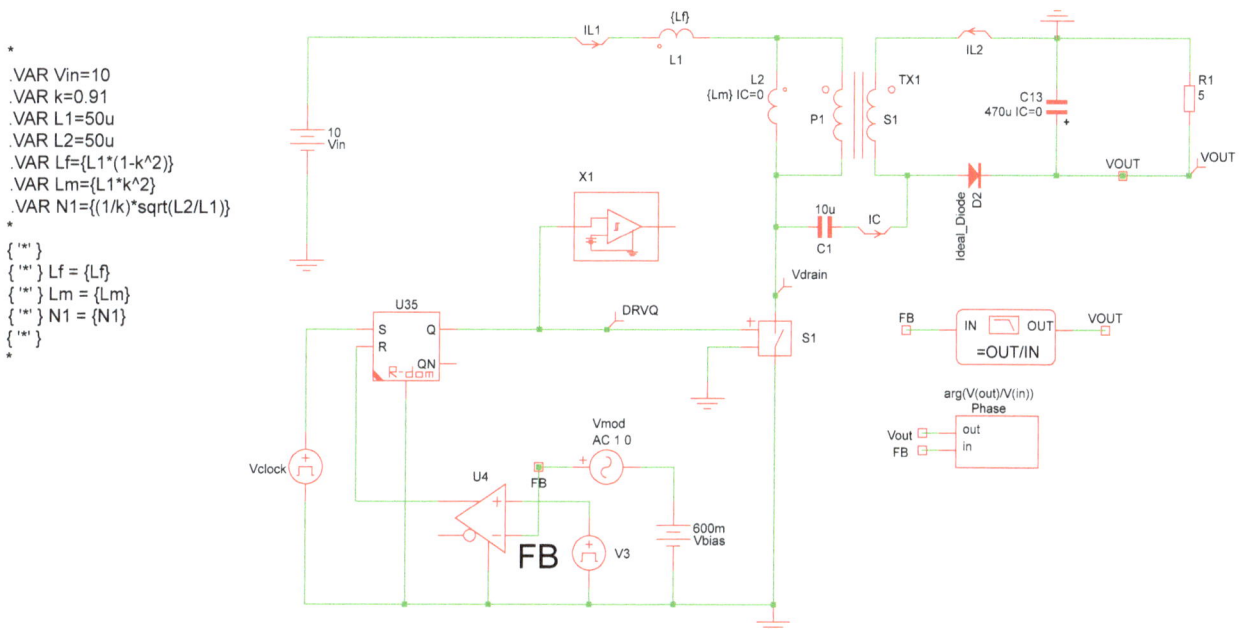

Figure 5.37 You can automate the leakage inductance calculation by adjusting the coupling coefficient in this SIMPLIS® template.

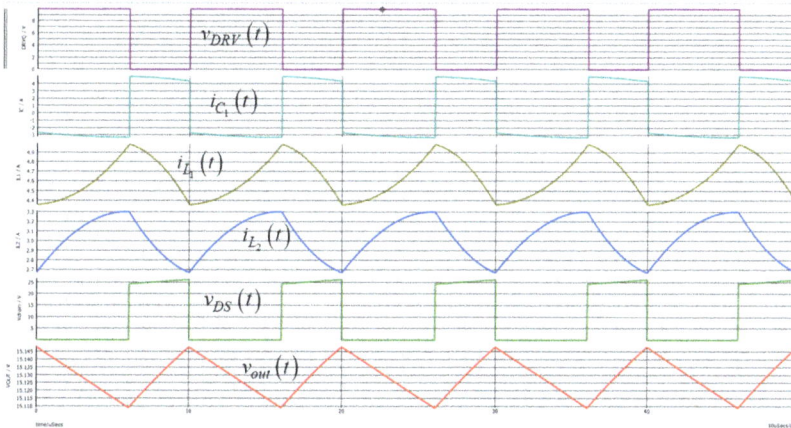

Figure 5.38 The transient simulation shows a low input ripple current inherent to the SEPIC.

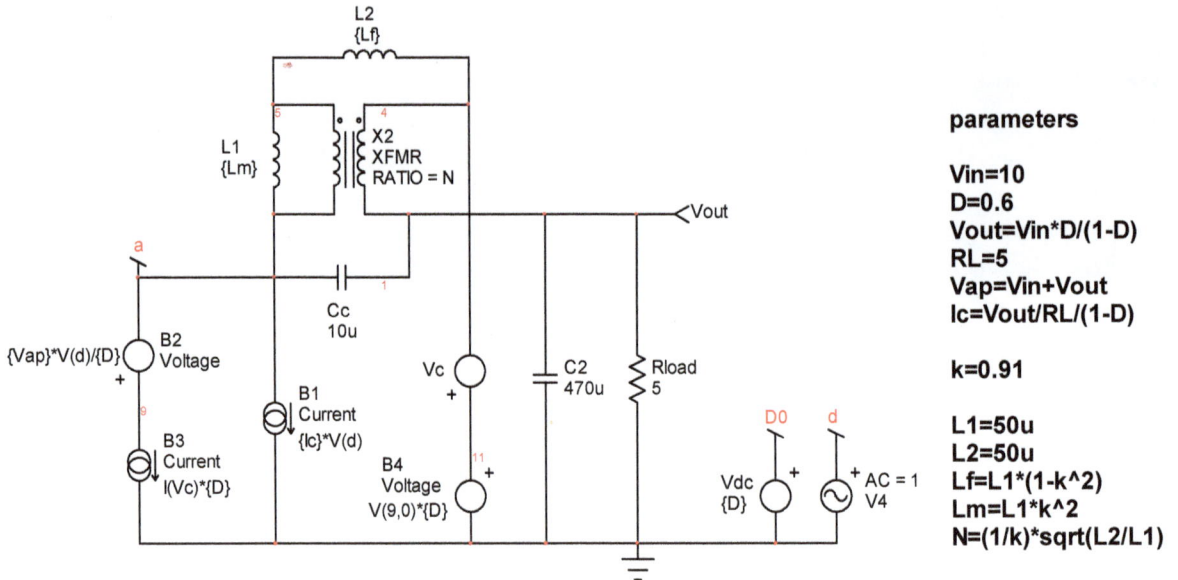

Figure 5.39 When the leakage inductance goes to zero, the response is that of a 2nd-order system in which C_c plays no role.

I have derived the control-to-output transfer function considering a perfect coupling but did not detail the steps here. The transfer function is as follows:

$$H(s) = H_0 \frac{\left(1+\frac{s}{\omega_{z_1}}\right)\left(1-\frac{s}{\omega_{z_2}}\right)}{1+\frac{s}{\omega_0 Q}+\left(\frac{s}{\omega_0}\right)^2} \tag{5.87}$$

where:

$$H_0 = \frac{V_{in}}{(1-D)^2} \tag{5.88}$$

$$Q = R_{load}(1-D)\sqrt{\frac{C_2}{L_1}} \tag{5.89}$$

$$\omega_0 = \frac{1-D}{\sqrt{L_1 C_2}} \tag{5.90}$$

$$\omega_{z_1} = \frac{1}{r_C C_2} \tag{5.91}$$

$$\omega_{z_2} = \frac{R_{load}}{L_1}\frac{(1-D)^2}{D} \tag{5.92}$$

The comparison between the Mathcad® equations and the SIMPLIS® simulations are given in Figure 5.40. The agreement is excellent.

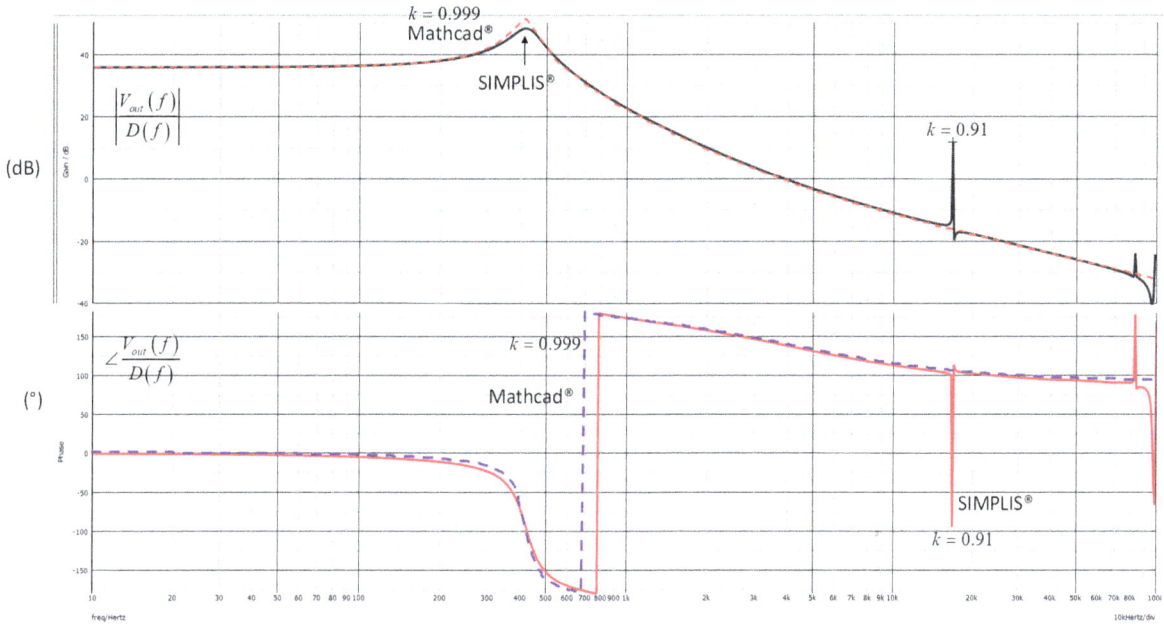

Figure 5.40 The derived expression is very close to the response obtained from SIMPLIS® with a loose coupling.

If the load is increased to 100 Ω, the converter from Figure 5.37 enters the discontinuous conduction mode of operation as confirmed by the operating waveforms from Figure 5.41. The control-to-output transfer function in CCM with a perfect coupling changes to the following DCM one:

$$H(s) = H_0 \frac{\left(1 + \dfrac{s}{\omega_{z_1}}\right)\left(1 - \dfrac{s}{\omega_{z_2}}\right)}{\left(1 + \dfrac{s}{\omega_{p_1}}\right)\left(1 + \dfrac{s}{\omega_{p_2}}\right)} \tag{5.93}$$

where:

$$H_0 = V_{in}\sqrt{\frac{1}{2\tau_L}} \tag{5.94}$$

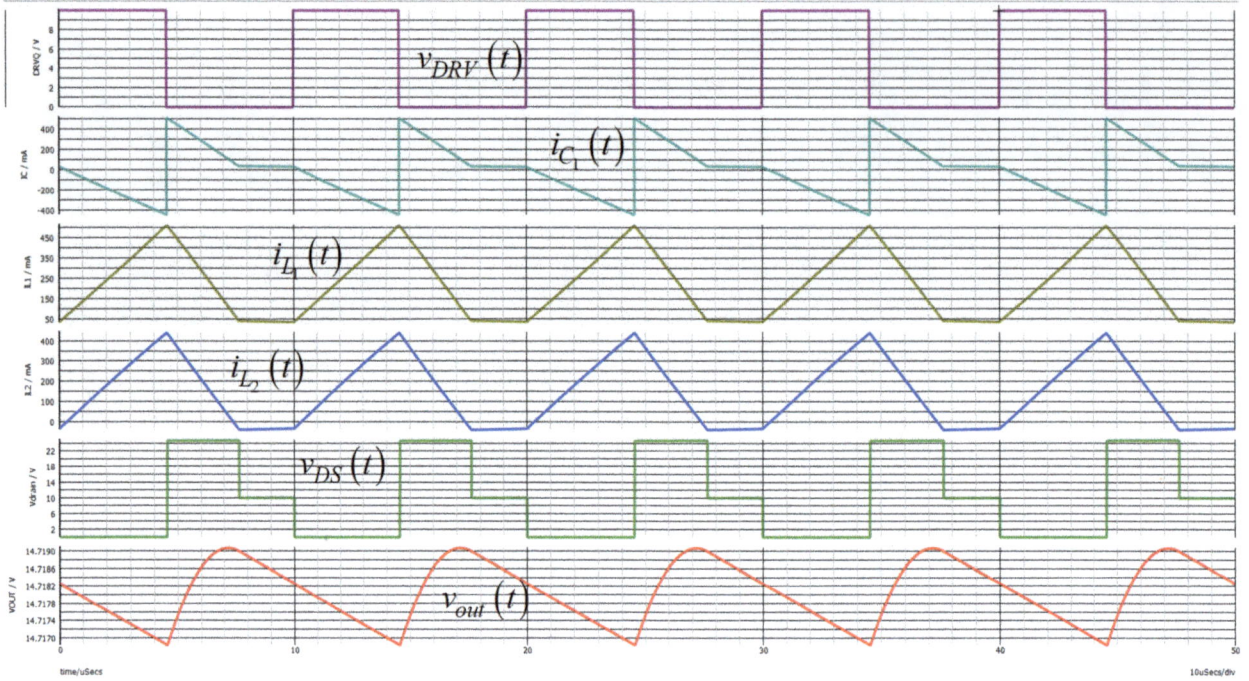

Figure 5.41 When the load is reduced to 100 Ω, the converter enters DCM operation.

$$\omega_{p_1} = \cfrac{R_{load}}{L_1\left(M^2 + M + 0.5\right) + \cfrac{C_2 R_{load}^{\,2}}{2}} \tag{5.95}$$

$$\omega_{p_2} = \cfrac{2M\left(1+M\right) + \cfrac{C_2}{L_1} R_{load}^{\,2} + 1}{C_2 R_{load}\left(1+M\right)^2} \tag{5.96}$$

$$\omega_{z_1} = \frac{1}{r_C C_2} \tag{5.97}$$

$$\omega_{z_2} = \frac{R_{load}}{L_1} \frac{\left(1-D\right)^2}{D} \tag{5.98}$$

with:

$$\tau_L = \frac{L_1}{R_{load} T_{sw}} \tag{5.99}$$

$$M = D\sqrt{\frac{1}{2\tau_L}} \tag{5.100}$$

A comparison between the plots generated by these equations and SIMPLIS® appears in Figure 5.42 and confirms the approach is correct.

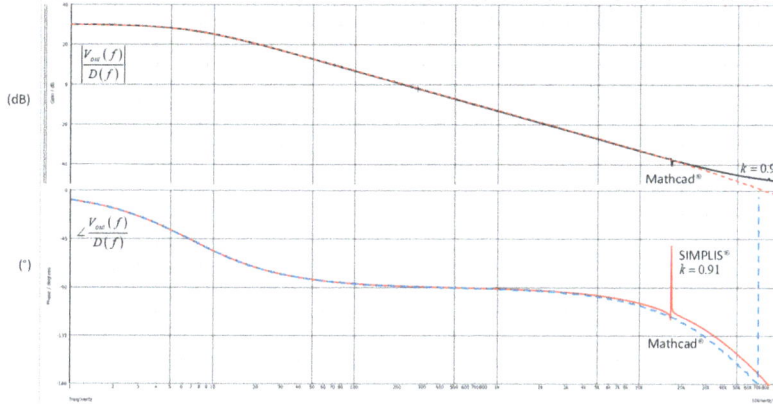

Figure 5.42 In DCM, the SEPIC exhibits the response of a damped second-order system.

5.2.2 Current Mode Control

The SEPIC can be operated in peak-current-mode control as shown in the SIMPLIS® setup of Figure 5.43. A little bit of compensation ramp is added with I_1 and R_3 at the current sense node, across C_2, to stabilize the simulation. A 1-mA sawtooth source together with the 50-Ω resistance do the job well. A 1-V control voltage imposes a 8.8-A peak current in the power switch for a 45-W output. The steady-state waveforms are given in Figure 5.44 and if you look at the primary-side current slope in $i_D(t)$, it is actually equal to the sum of the inductive slopes of L_1 and L_2. The primary slope S_n is therefore computed as follows:

$$S_n = \frac{V_{in}}{L_{eq}} R_i \tag{5.101}$$

With R_i the sense resistance (125 mΩ in this simulation) and L_{eq} defined as:

$$L_{eq} = L_1 \| L_2 \tag{5.102}$$

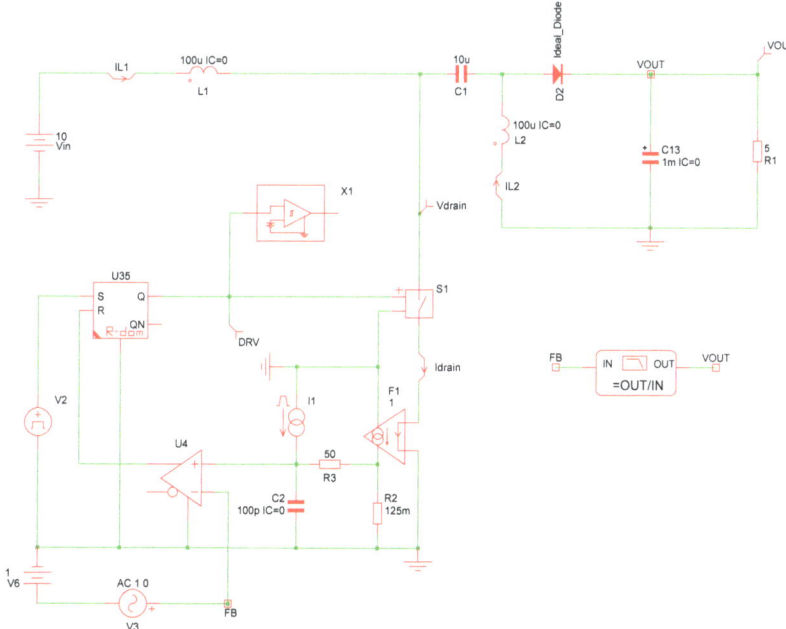

Figure 5.43 A current comparator observes the switch current and turns the SEPIC in a current-mode converter.

The coupling capacitor has been reduced to 10 μF as well as the output capacitor which is now 1000 μF. The small-signal response appears in Figure 5.45. The low-frequency resonance has gone but there is still a notch around 3 kHz confirming the presence of the RHP zeroes as with the voltage-mode-control case.

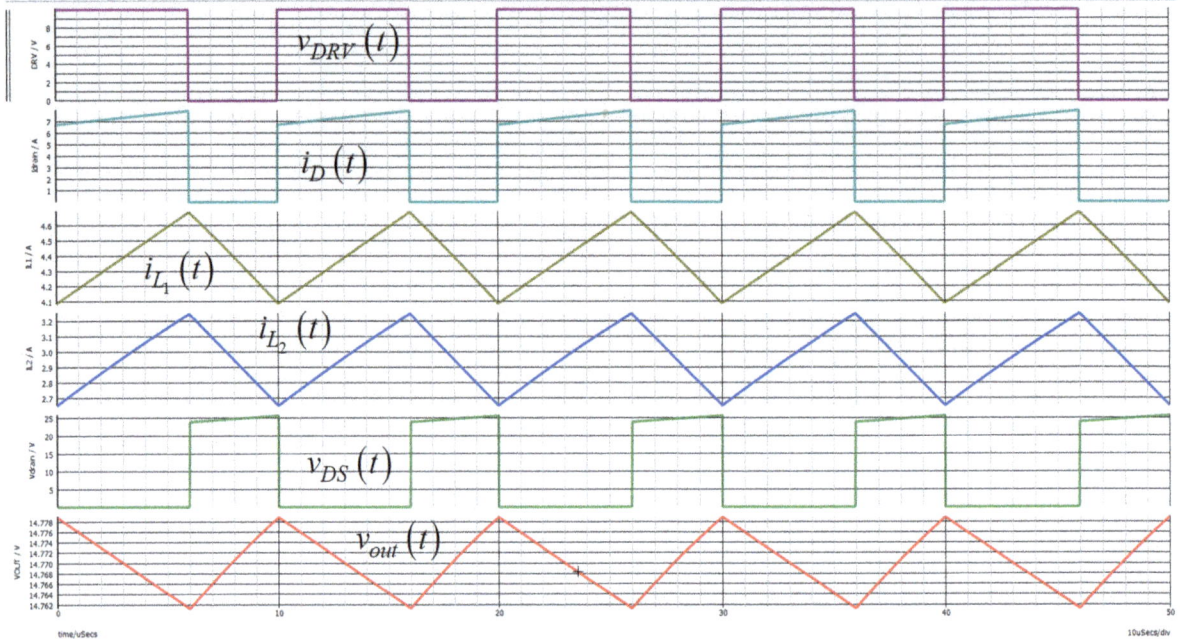

Figure 5.44 The current in the switch peaks to 8.8 A during the on-time duration and sums both inductive contributions.

The CM PWM switch model can be used to determine the control-to-output transfer function of the uncoupled-inductor SEPIC operated in current-mode control. It is shown in Figure 5.46 with the large-signal model. Considering the resonant capacitor C_s, we deal with a 5th-order converter and the small-signal circuit is given in Figure 5.47 with all computed parameters.

Both circuits lead to a very close magnitude/phase response which confirms the SIMPLIS® graph nicely.

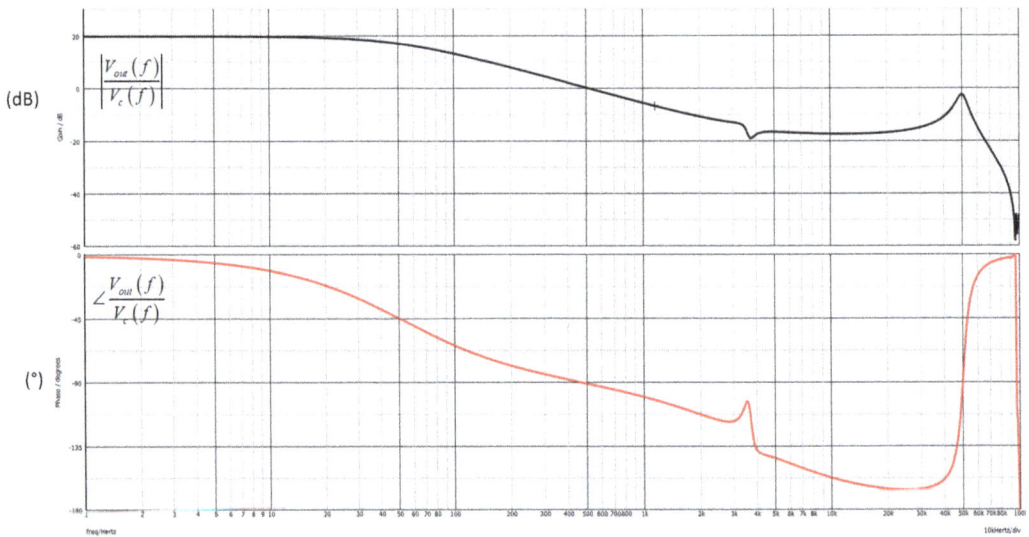

Figure 5.45 The small-signal response of the uncoupled-inductor SEPIC offers a smoother response compared to the voltage-mode case.

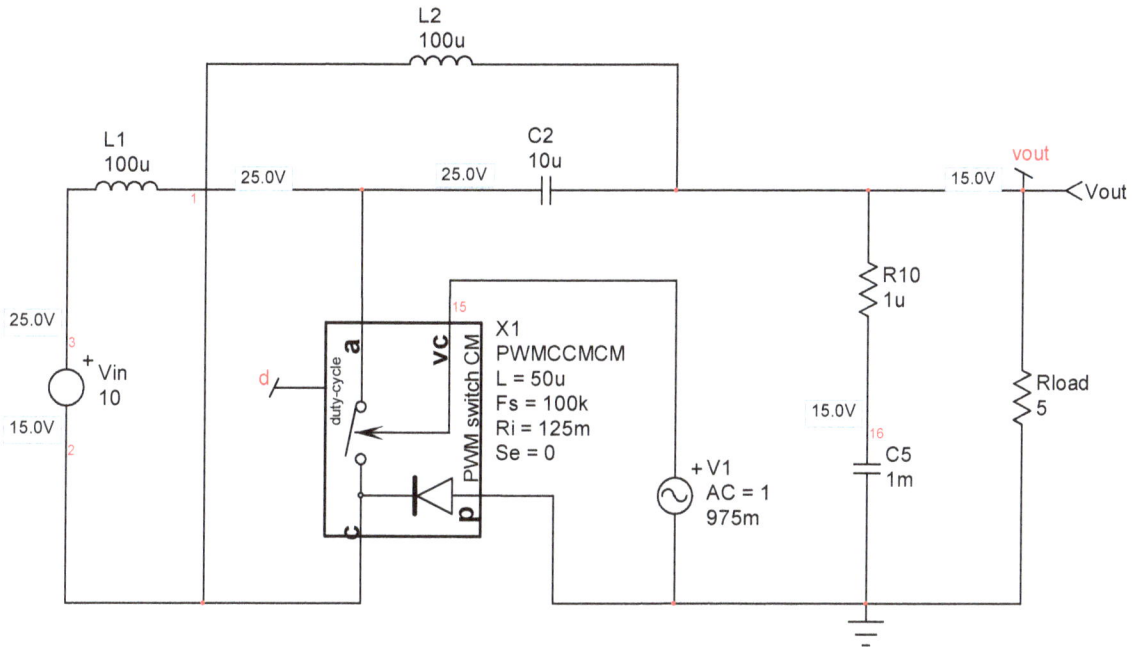

Figure 5.46 The SEPIC in current-mode control can be analyzed with the CM PWM switch model.

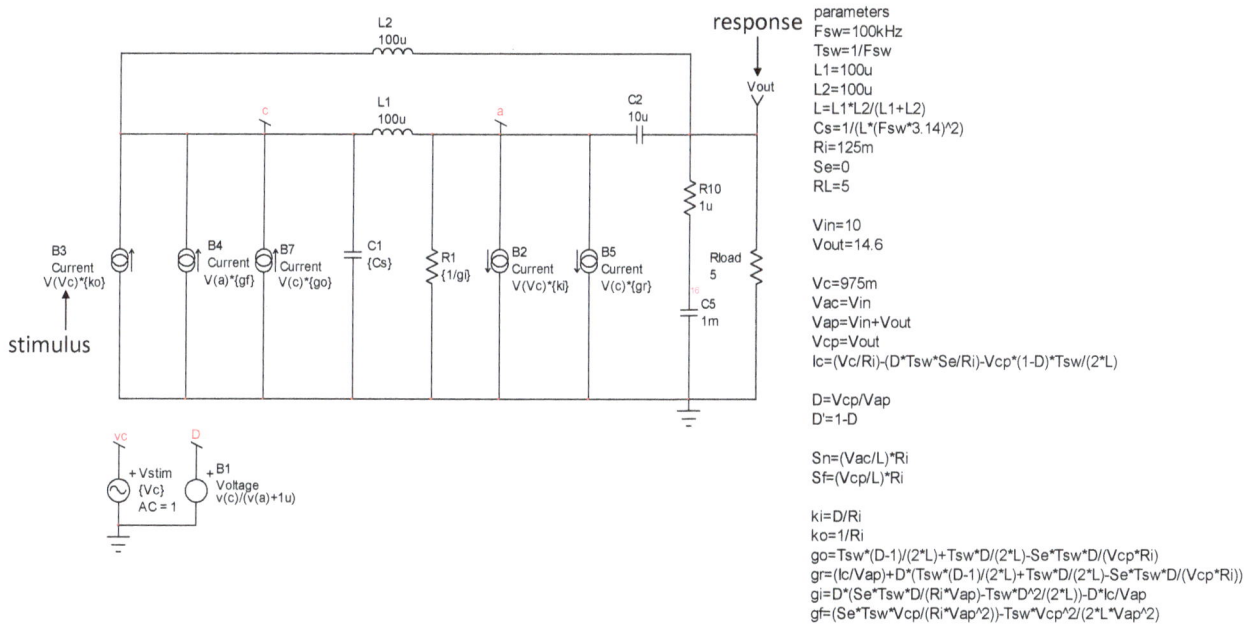

Figure 5.47 The small-signal model of the SEPIC operated in current-mode control is a 5th-order system.

To determine the control-to-output transfer function of this 5th-order system, you would start by shorting inductors and opening capacitors to check the dc gain. Then turn the excitation off: the two current sources featuring V_c are removed and you determine the constants associated with each energy-storing element as we did in the previous chapters. Nothing unsurmountable but I recommend to carefully check all intermediate results with a SPICE simulation then confirm the resistances derived with a Mathcad® sheet or equivalent.

Reference [9] gives all details to check poles and zeroes with SPICE.

5.2.3 Coupled Inductors

Just like with the voltage-mode case, it is possible to couple inductors and change the output response of the converter. If the coupling approaches 1, then the coupling capacitor no longer plays a role in ac and the response is close to that of a CM flyback converter. Figure 5.48 describes the SIMPLIS® schematic diagram used to test the CM coupled inductors while the same converter modeled with the CM PWM switch is given in Figure 5.49.

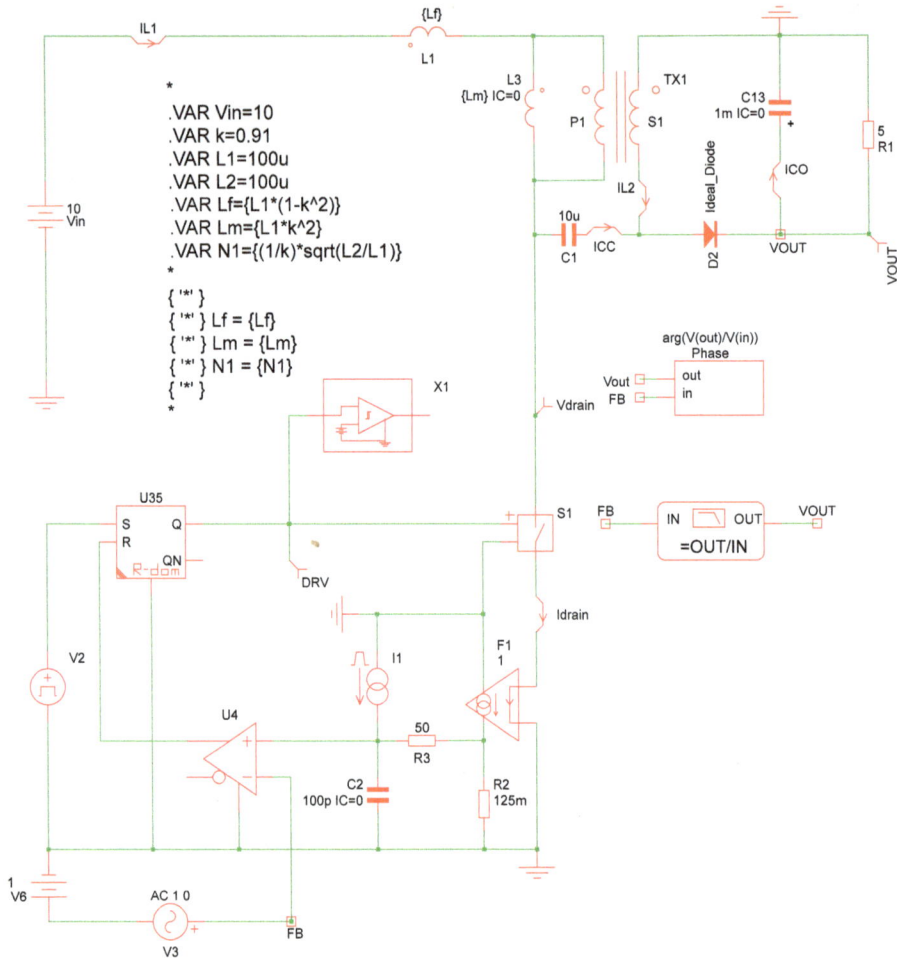

Figure 5.48 It is simple to simulate a coupled-inductor version in a SEPIC operated in current mode.

Rather than using the equivalent transformer to simulate the coupling of the two inductors, for instance if you want to confirm the computed values, you could resort to the classical coupling factor k in SPICE as in Figure 5.7. In SIMPLIS®, however, k is not supported and you have to use the mutual inductance M. You determine it as follows:

$$M = k\sqrt{L_1 L_2} \tag{5.103}$$

You then add the line in the netlist $M-L1-L2\ 91u$ for instance with 100-μH inductors and a coupling factor of 0.91. Our simulations confirm the both SPICE and SIMPLIS® deliver similar responses with k, M or the transformer.

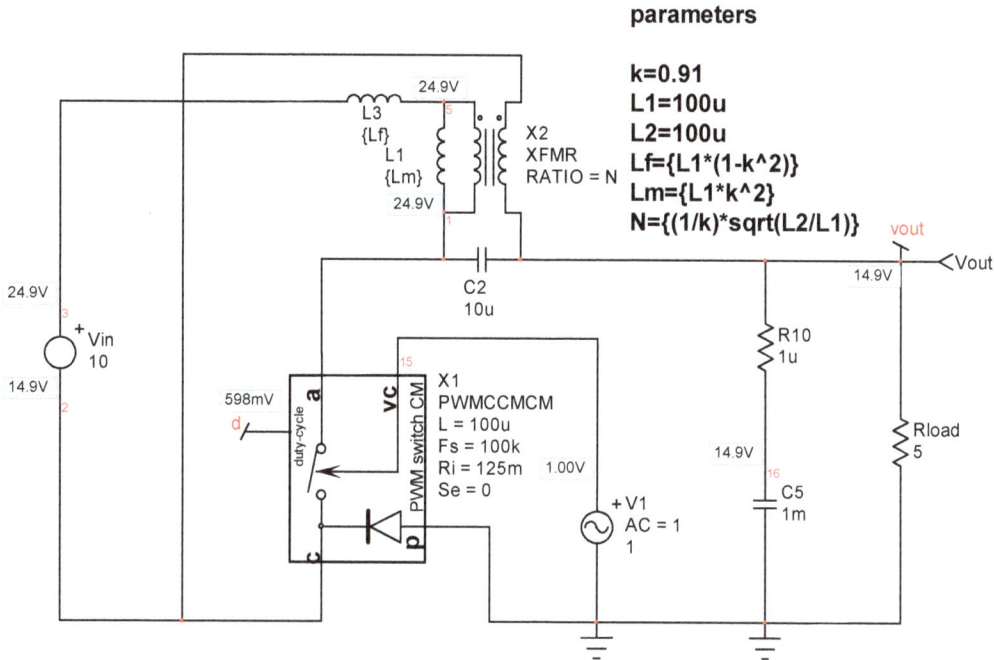

Figure 5.49 The CM PWM switch can be used to determine the control-to-output transfer function of a CM SEPIC with coupled inductors.

The small-signal response of the coupled SEPIC operated in current-mode control is classically obtained by replacing the large-signal model of Figure 5.49 by the small-signal model derived in chapter 1. This is what you see in Figure 5.50 after rearranging all sources and components in a convenient way. Please note that considering the leakage inductance and the resonating capacitor as in the uncoupled version makes the circuit a 5th-order system. You can analyze it using the fast analytical circuits techniques but we won't do it here.

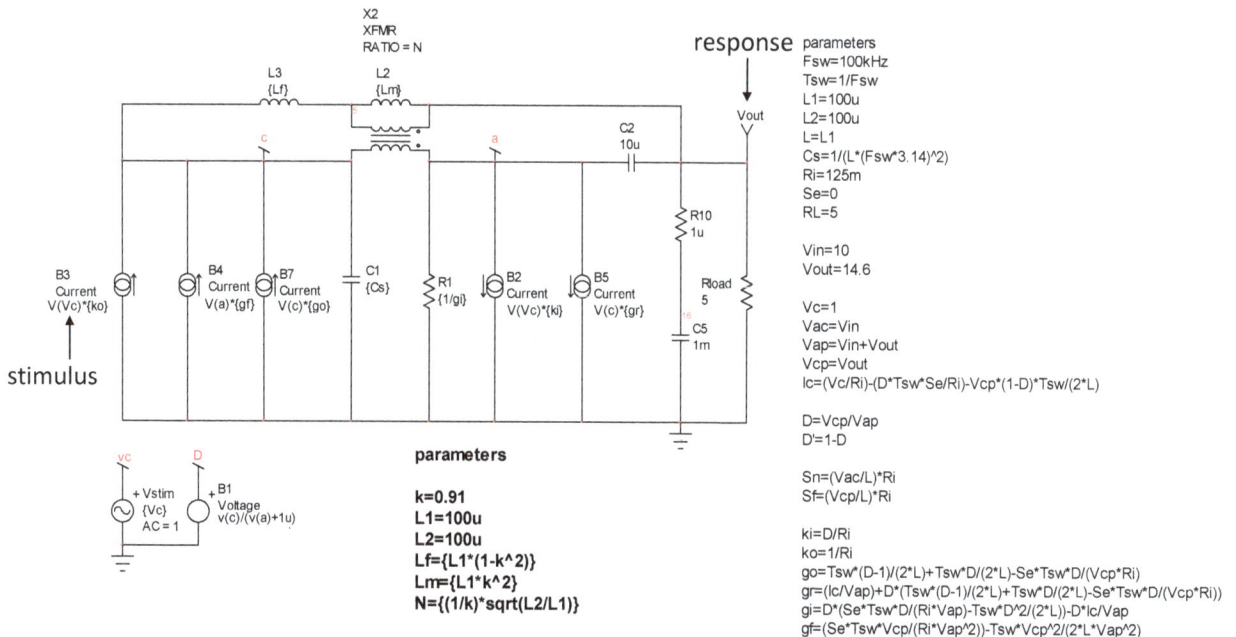

Figure 5.50 The small-signal circuit of the CM SEPIC with coupled inductors is a 5th-order system.

After we ran the cycle-by-cycle circuit of Figure 5.48, we obtain the transient waveforms shown in Figure 5.51. As expected, the input current i_{L1} is reduced compared to that in Figure 5.44 (250 mA peak to peak versus 600 mA).

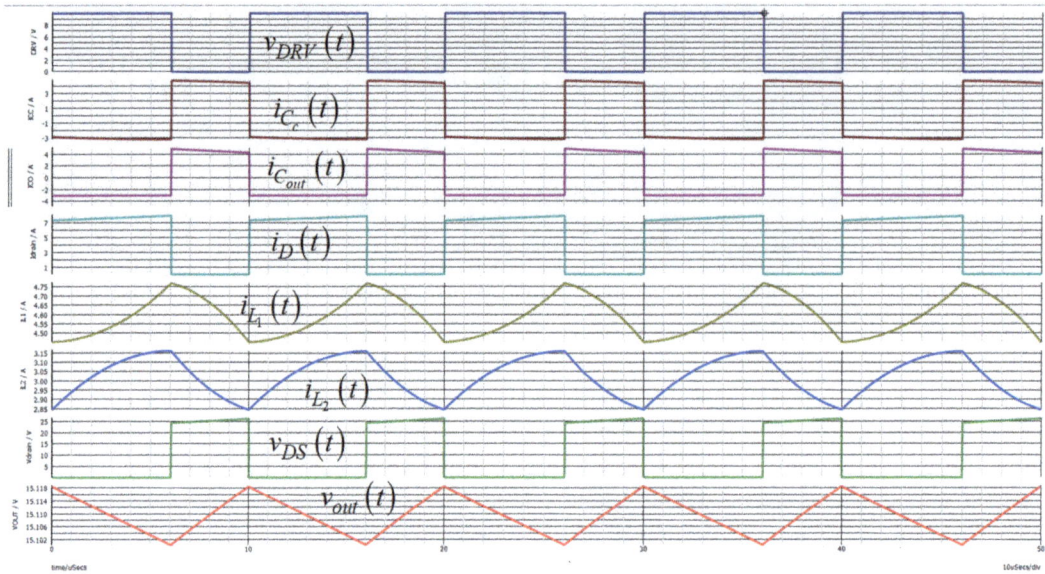

Figure 5.51 The input ripple current is reasonably low as well as in the secondary inductor. The currents circulating in the coupling and output capacitors are highly pulsating.

Finally, the control-to-output transfer function simulated by SIMPLIS® appears in Figure 5.52. We have superimposed the response given by the CM PWM switch in Figure 5.50 and the curves are very close to each other. There is a little more peaking in SIMPLIS® which explains why the phase moves faster at high frequency. You can see the glitch incurred to the leakage inductance and the coupling capacitor above 10 kHz. Again, considering ohmic losses in the prototype, you may not notice its presence during the loop measurement. If you do, a classical way to reduce this glitch consists of inserting a resistance in series with the coupling capacitor to locally damp the circuit (efficiency can suffer) or add a parallel RC damper across the capacitor.

Finally, I have assembled on the same graphs, the responses of the CM SEPIC with non-coupled inductors, tightly and loosely coupled ones. All curves are gathered in Figure 5.53.

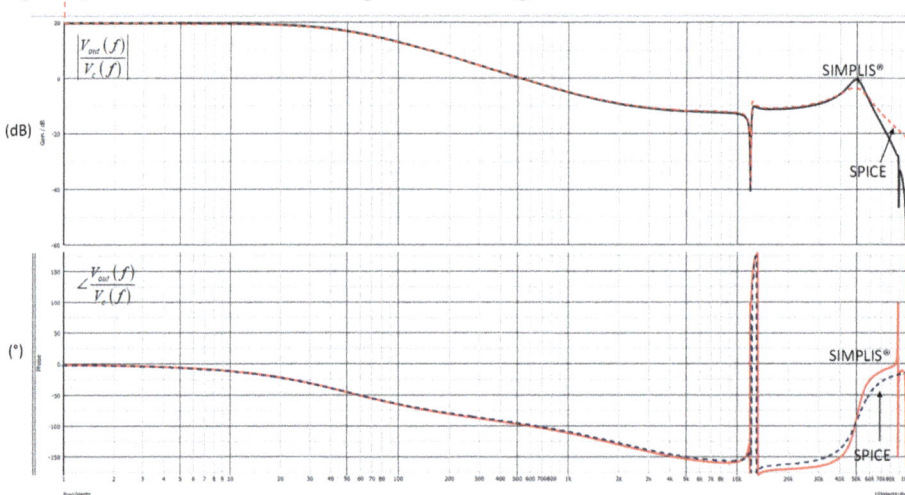

Figure 5.52 SIMPLIS® simulations and the small-signal model deliver very close responses, naturally validating the model.

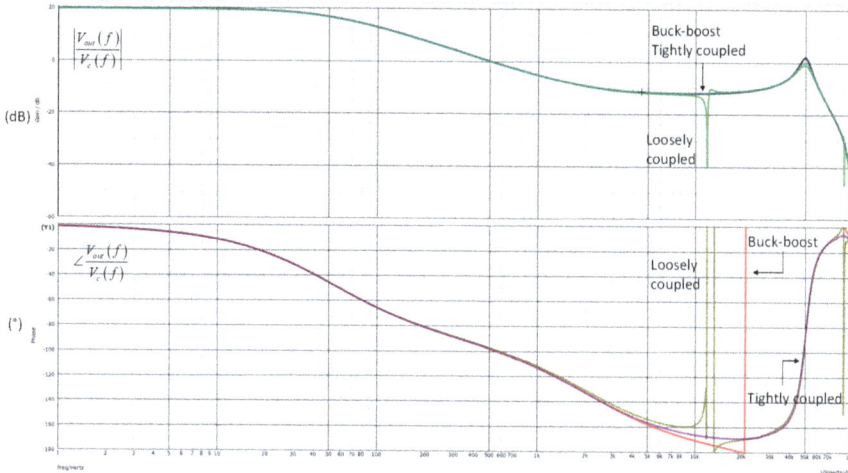

Figure 5.53 These are the magnitude/phase responses of the CM SEPIC with different magnetics.

The low-frequency magnitude/phase responses are very similar but divergences appear in the higher frequency portion especially for the loosely-coupled version. With tightly-coupled magnetics, the response approaches that of the CM buck-boost or flyback converter and you can use the expressions derived in Chapter 4, current-mode flyback section.

5.3 The Zeta

The Zeta converter represents another arrangement of a 2-switch and 4 energy-storing elements that appear in Figure 5.54. The picture offers a comparison between the three structures: the Ćuk presents non-pulsating input and output currents but delivers a negative voltage. You have to resort to an isolated version to obtain a positive voltage. The SEPIC produces a positive voltage but the output capacitor current is highly pulsating. Its input current is smooth as that of classical boost converter.

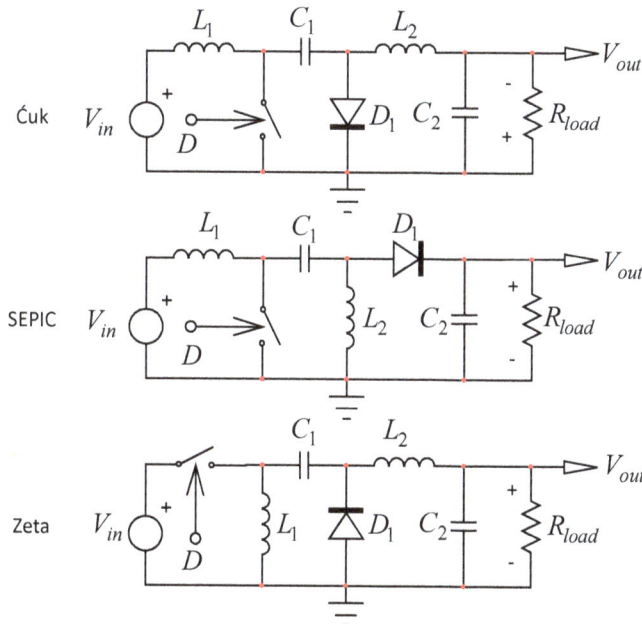

Figure 5.54 The Zeta offers a non-pulsating output current.

As an alternative, the Zeta brings the pulsating input current of a buck-boost converter but the current flowing in the output capacitor is that of a classical buck, i.e. a smooth waveform. The PWM switch can be revealed by sliding the upper-side switch along the source ground connection until it joins the diode anode forming the c connection. Reconnect inductor L_1 to the ground and you have the Zeta converter in Figure 5.55.

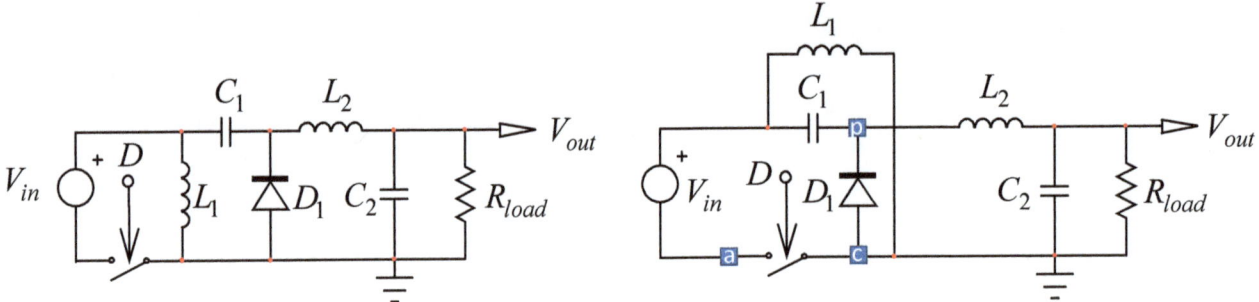

Figure 5.55 Revealing the PWM switch is easy with the Zeta converter.

The SPICE version using the large-signal VM model in an uncoupled version is given in Figure 5.56 where parasitics have been ignored. If you replace the PWM switch by its large-signal sources as in Figure 5.57 then you can short inductors, open capacitors and determine the dc transfer function which is similar to that of the SEPIC operating in voltage-mode CCM. You should find:

$$\frac{V_{out}}{V_{in}} = \frac{D}{1-D} \tag{5.104}$$

To check the response of the PWM switch model, we can run a SIMPLIS® simulation as proposed in Figure 5.58. There is no coupling between the inductors so far but the addition of (5.103) in the netlist will provide the right coupling with inductors L_1 and L_2 already fitting the right orientation.

The transient waveforms appear in Figure 5.59 and confirm a pulsating input (i_D) current and a smooth current in the output capacitor.

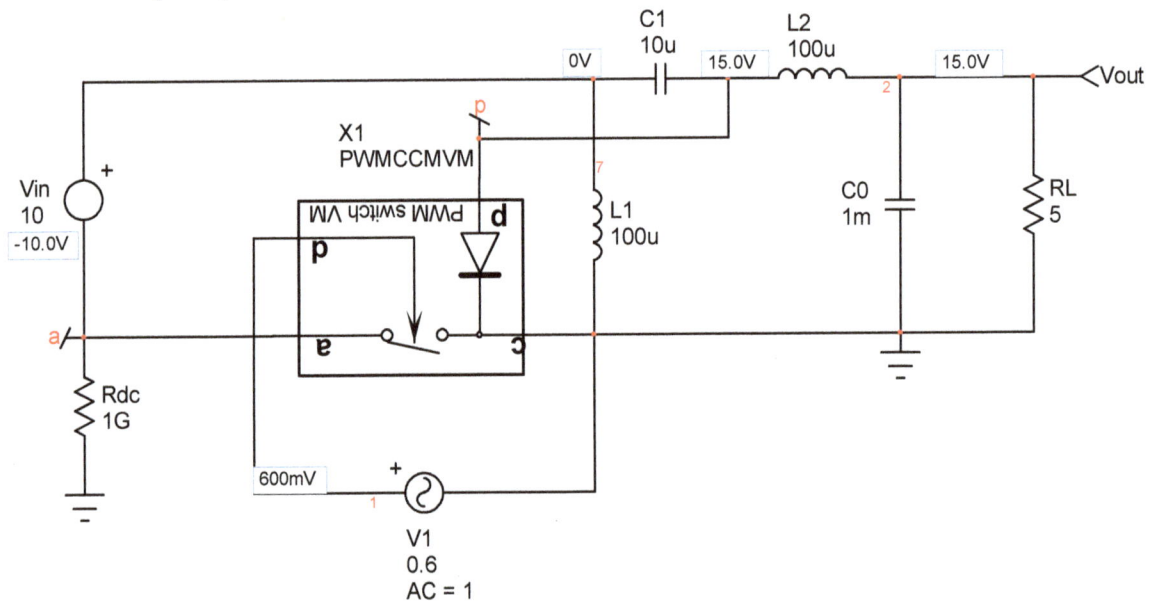

Figure 5.56 This is the Zeta converter operated in open-loop with uncoupled inductors.

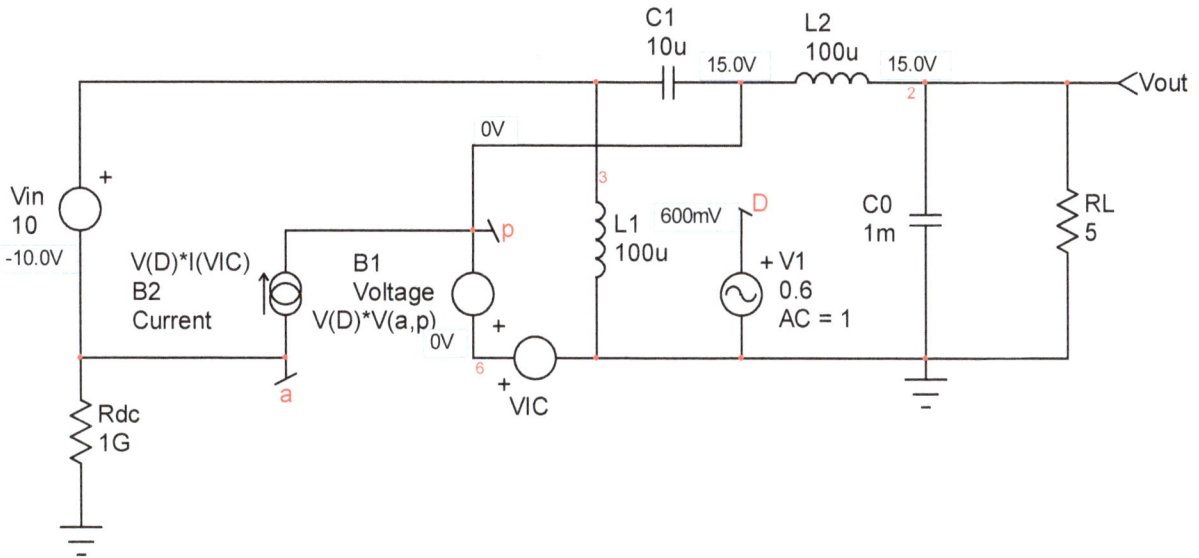

Figure 5.57 The PWM switch lends itself well to the dc analysis of the Zeta.

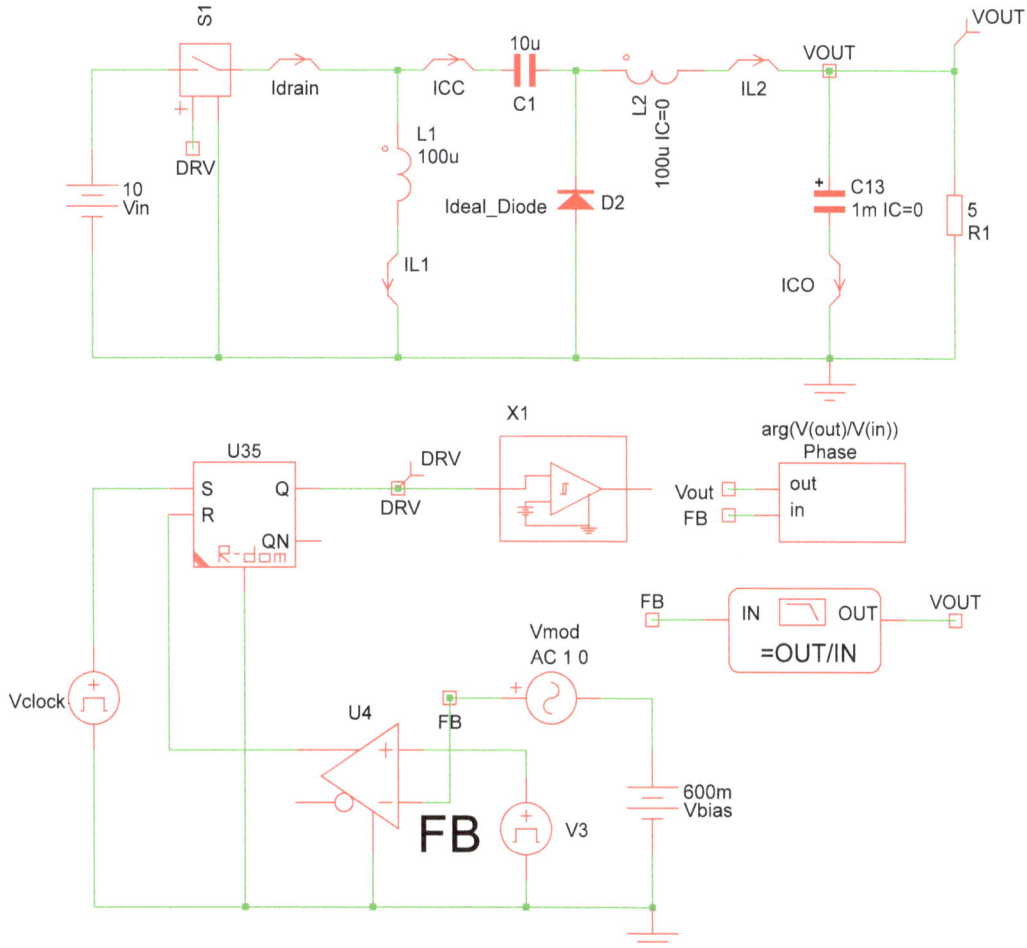

Figure 5.58 SIMPLIS® lets you easily simulate the uncoupled version. Please note the correct dot orientation ready for a coupling coefficient later on.

Further to the transient simulation, the small-signal response is delivered in a few seconds and is given in Figure 5.60.

In this graph, we have gathered the magnitude and phase response of the small-signal version of Figure 5.57 which is given in Figure 5.61.

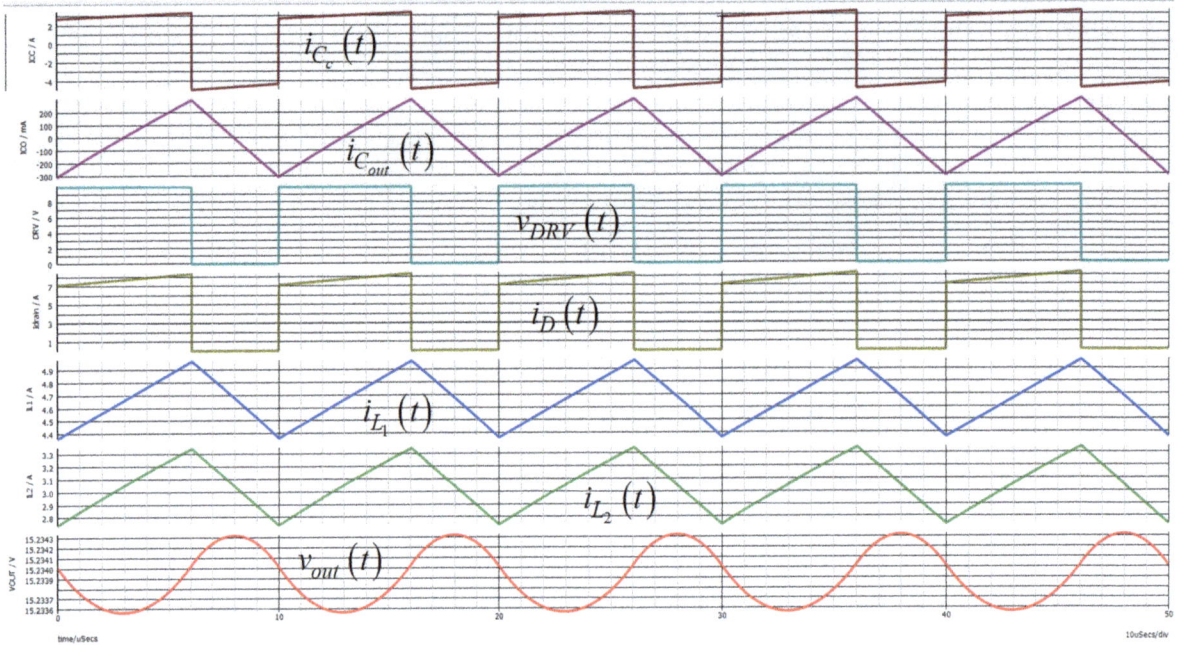

Figure 5.59 The current in the output capacitor is not pulsating unlike the SEPIC.

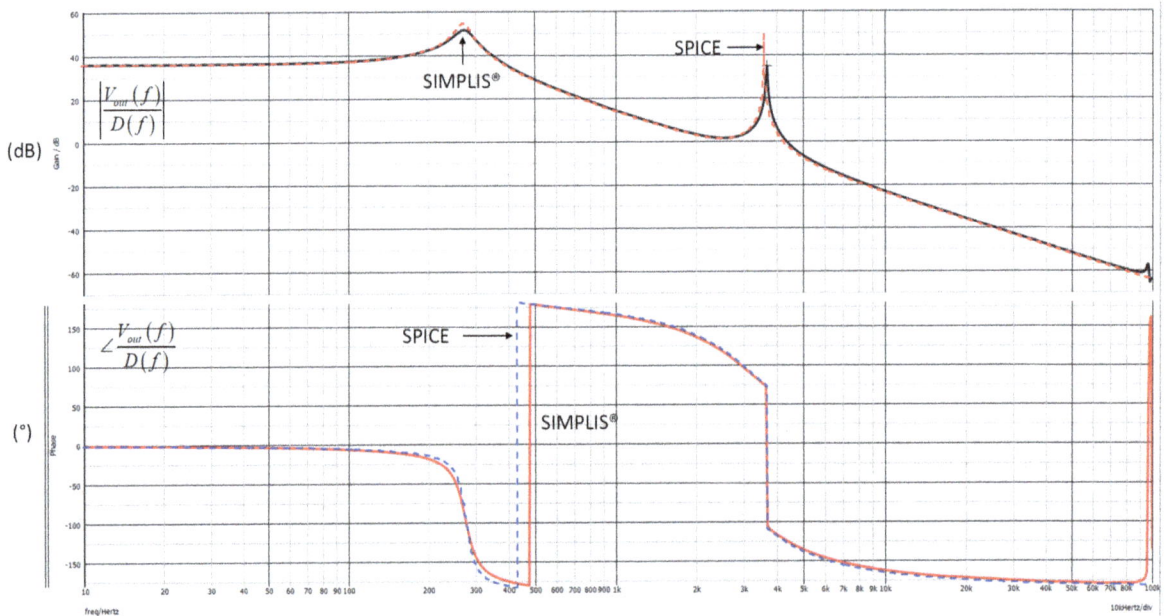

Figure 5.60 The control-to-output transfer function shows two resonances.

As you can see, the responses are very close to each other despite a slight mismatch in the two quality factors. The small-signal model can be analyzed with the FACTs and you are now familiar with the technique. I have found in [11] the transfer function for an isolated version that I reworked and tested for the uncoupled version. The transfer function sticks to the following raw format:

$$H(s) = H_0 \frac{1 + a_1 s + a_2 s^2}{1 + b_1 s + b_2 s^2 + b_3 s^3 + b_4 s^4} \tag{5.105}$$

in which:

$$H_0 = \frac{V_{in}}{(1-D)^2} \tag{5.106}$$

$$a_1 = -\frac{D^2 L_1}{R_{load}(1-D)^2} \tag{5.107}$$

$$a_2 = \frac{L_1 C_1}{1-D} \tag{5.108}$$

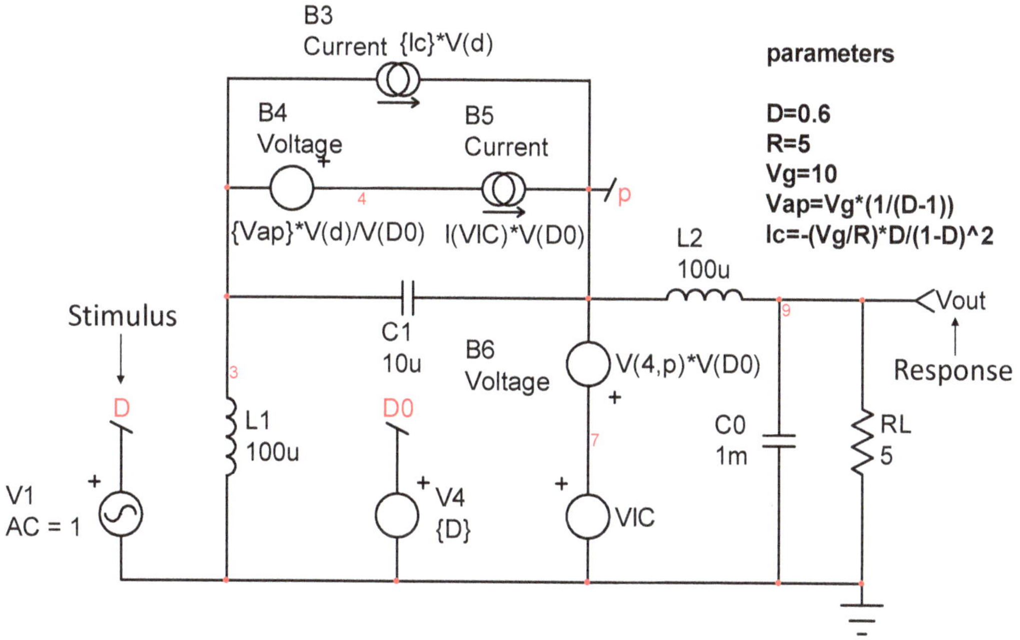

Figure 5.61 The small-signal circuit of the Zeta converter reveals a 4th-order structure.

$$b_1 = \frac{L_2(1-D)^2 + DL_1}{R_{load}(1-D)^2} \tag{5.109}$$

$$b_2 = \frac{L_2 C_2(1-D)^2 + D^2 L_1 C_2 + L_1 C_1}{(1-D)^2} \tag{5.110}$$

$$b_3 = \frac{L_1 L_2 C_1}{R_{load}(1-D)^2} \tag{5.111}$$

$$b_4 = \frac{L_1 L_2 C_1 C_2}{(1-D)^2} \tag{5.112}$$

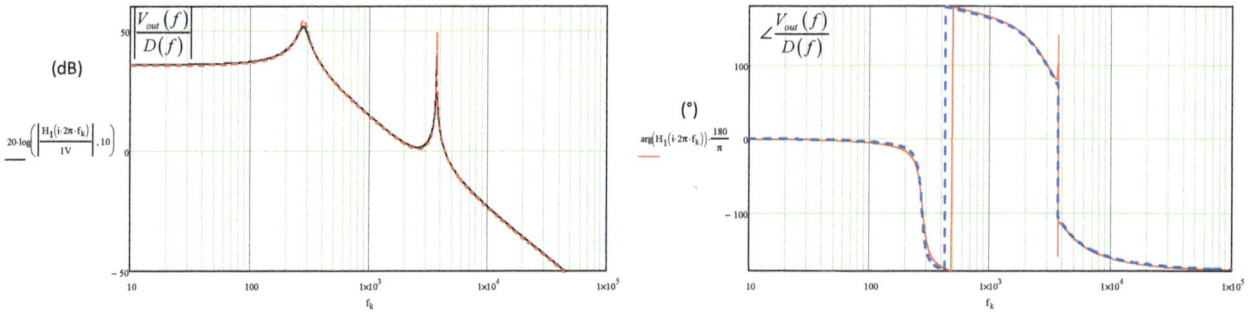

Figure 5.62 SPICE simulations and analytical expressions deliver very close responses.

Figure 5.62 compares the equation (5.105) magnitude/phase response with that of the SPICE simulation and they are extremely close. The complete control-to-output transfer function of the Zeta converter operated at full load can be approximated as follows:

$$H(s) \approx H_0 \frac{1 - \dfrac{s}{\omega_{0N} Q_N} + \left(\dfrac{s}{\omega_{0N}}\right)^2}{1 + b_1 s + b_2 s^2 + b_3 s^3 + b_4 s^4} \tag{5.113}$$

In which:

$$Q_N = \frac{R_{load}}{\omega_{0N} L_1} \left(\frac{1 - D}{D}\right)^2 \tag{5.114}$$

$$\omega_{0N} = \sqrt{\frac{1 - D}{L_1 C_1}} \tag{5.115}$$

This expression reveals a pair of RHP zeroes induced by the resonance of L_1 and C_1. Unfortunately, I could not satisfactorily rework the denominator as in the SEPIC case considering the time constants with this numerical application.

5.3.1 Coupled Inductors

Like with the other converters, it is possible to couple the inductors and reduce inductance values as detailed in [12]. The new large-signal model uses the transformer representation in Figure 5.8 and appears in Figure 5.63. Counting the energy-storing elements, you have a 4th-order converter if you account for the leakage element.

Slightly uncoupling inductors will reduce the volt-second imbalance and contribute to reducing circulating currents. It is possible to damp the corresponding resonance by inserting a resistance in series with the coupling capacitor.

If you want to study the small-signal response of this converter, the complete small-signal circuit for the control-to-output transfer function is given in Figure 5.64.

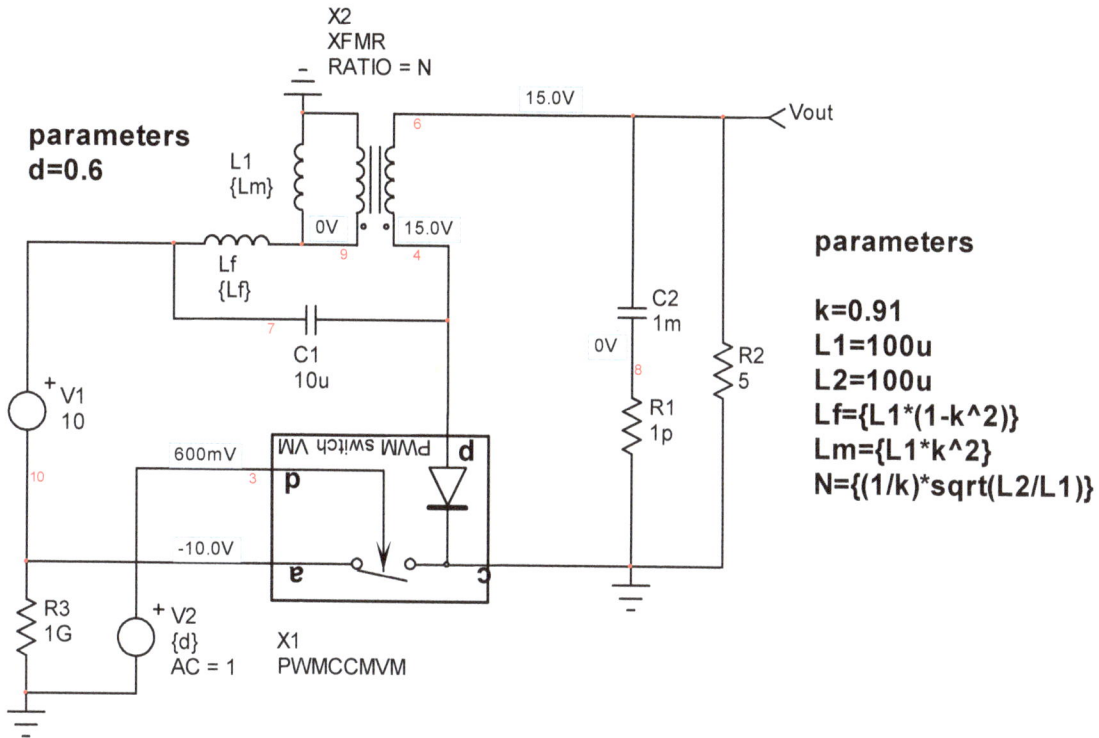

Figure 5.63 You need to resort to the equivalent transformer model to model the Zeta with coupled inductors.

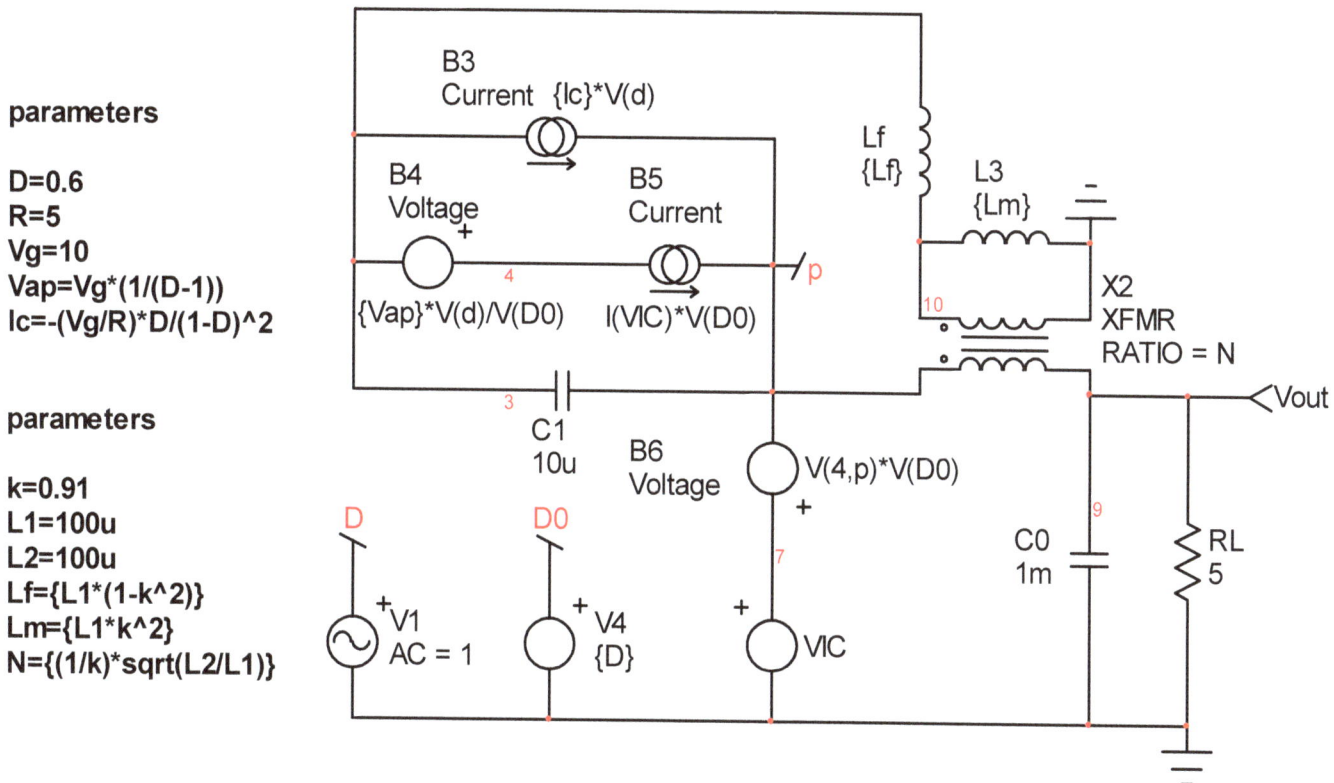

Figure 5.64 This is the small-signal model of the CCM Zeta converter operated in voltage-mode control and coupled inductors.

I have run a simulation with and without a damping resistor in series with C_1. Even though the added resistance certainly reduces the quality factor in the second resonance, the phase distortion brought by the whole system does not make it an easy circuit to stabilize (Figure 5.65). A simulation in SIMPLIS® shows how coupling with a little leakage inductance helps reduce the currents circulating in the inductors as in the SEPIC case.

The simulation schematic appears in Figure 5.66 while simulation results are given in Figure 5.67.

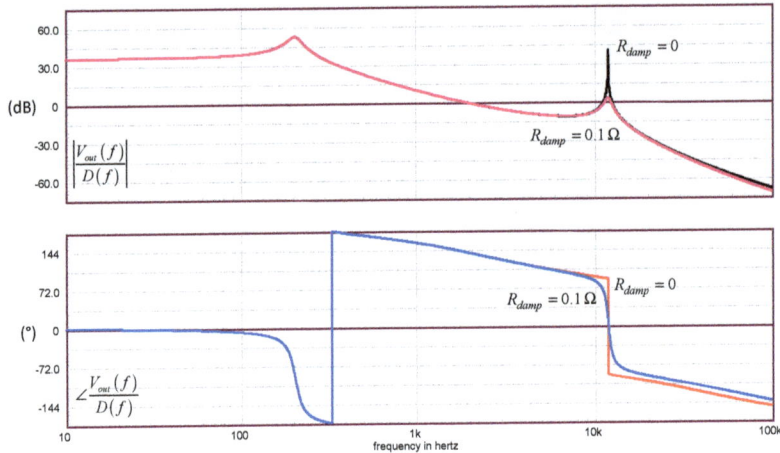

Figure 5.65 The Zeta with coupled inductors reveal a very distorted phase response. Damping the coupling capacitor does not really help to make it look better.

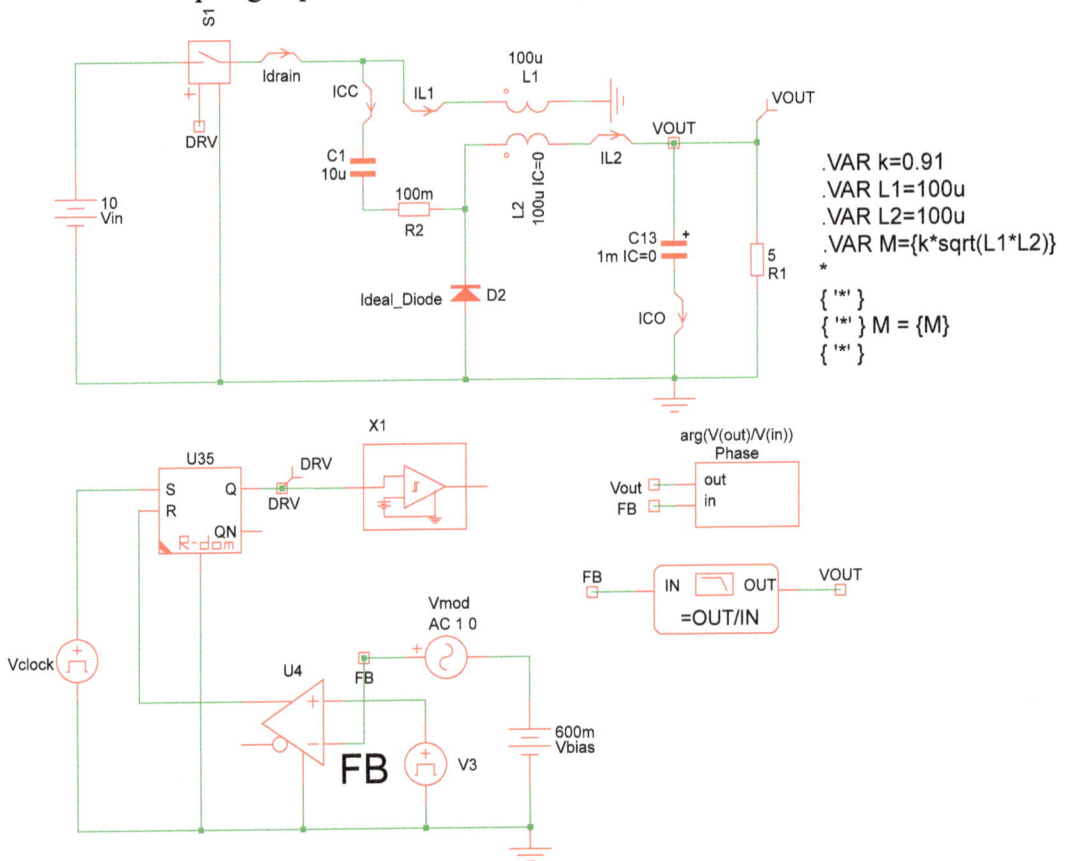

Figure 5.66 SIMPLIS® quickly shows the effect of having a little bit of leakage inductance.

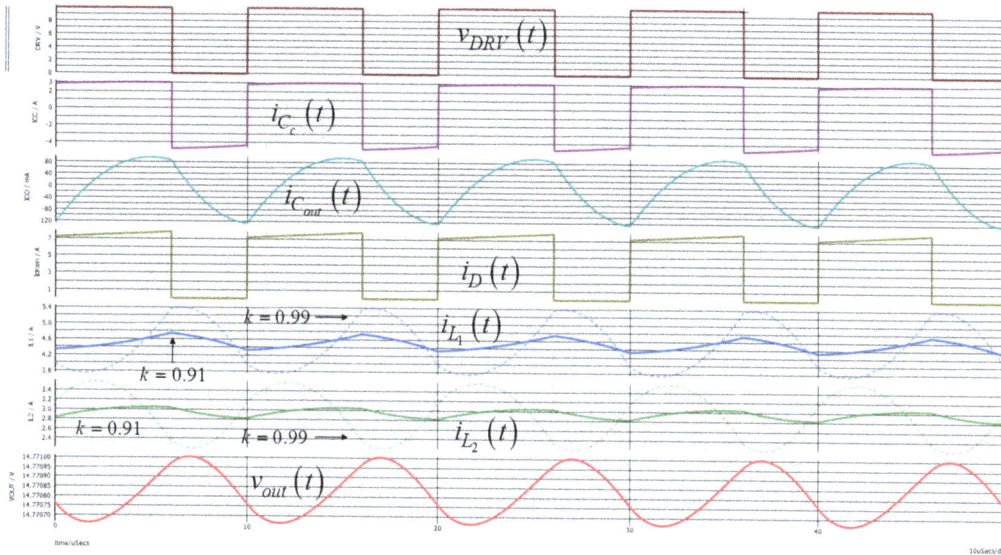

Figure 5.67 The currents circulating in the inductors are greatly reduced with slightly uncoupled inductors.

5.3.2 Current Mode Control

As in the SEPIC case, the Zeta can be operated in peak-current-mode control. The coupled-inductor version can be simulated by the CM PWM switch model described in Figure 5.68. Please note the negative resistance value considering the current entering terminal c in this configuration. You can analyze the circuit by inserting a transformer as we have done in the previous examples. The complete small-signal model useful to analyze the control-to-output transfer function of the CM Zeta appears in Figure 5.69. It is a 5th-order system if we consider the resonating capacitor C_s. The magnitude-phase responses of the large- and small-signal circuits are identical and you can use the latter in case you want to determine the transfer function linking V_{out} to V_c. Ac simulations from Figure 5.70 show a phase hitting -180° at 10 kHz versus the distorted curve from Figure 5.65 which intersected -180° at 350 Hz or so. The insertion of a damping resistance helps lowering the peaking at 10 kHz or so.

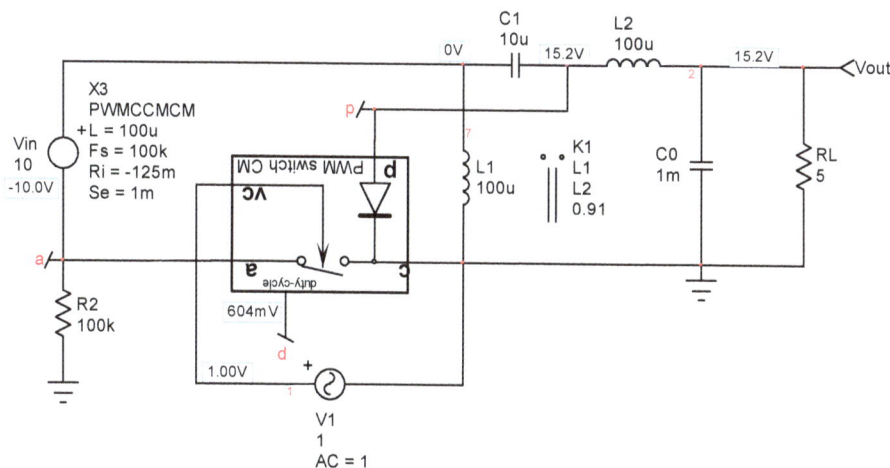

Figure 5.68 The large-signal coupled-inductor Zeta operated in current mode can be simulated with the CM PWM switch mode.

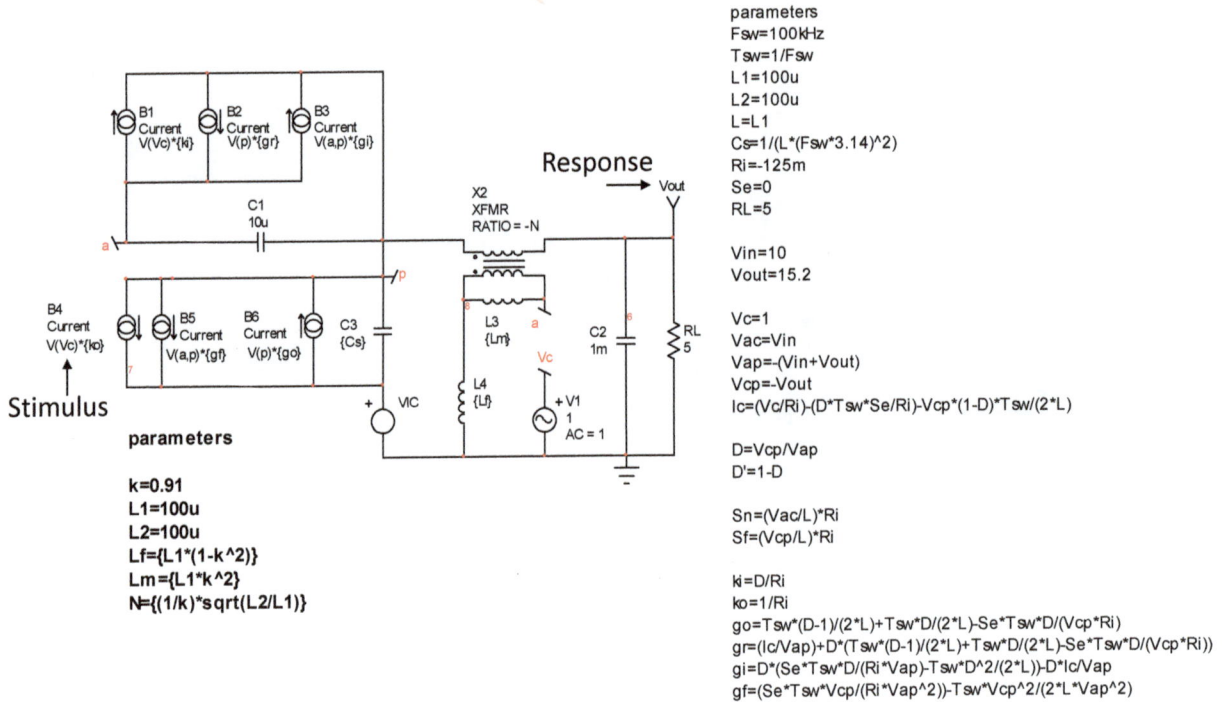

Figure 5.69 The small-signal circuit of the CM Zeta reveals a 5th-order system.

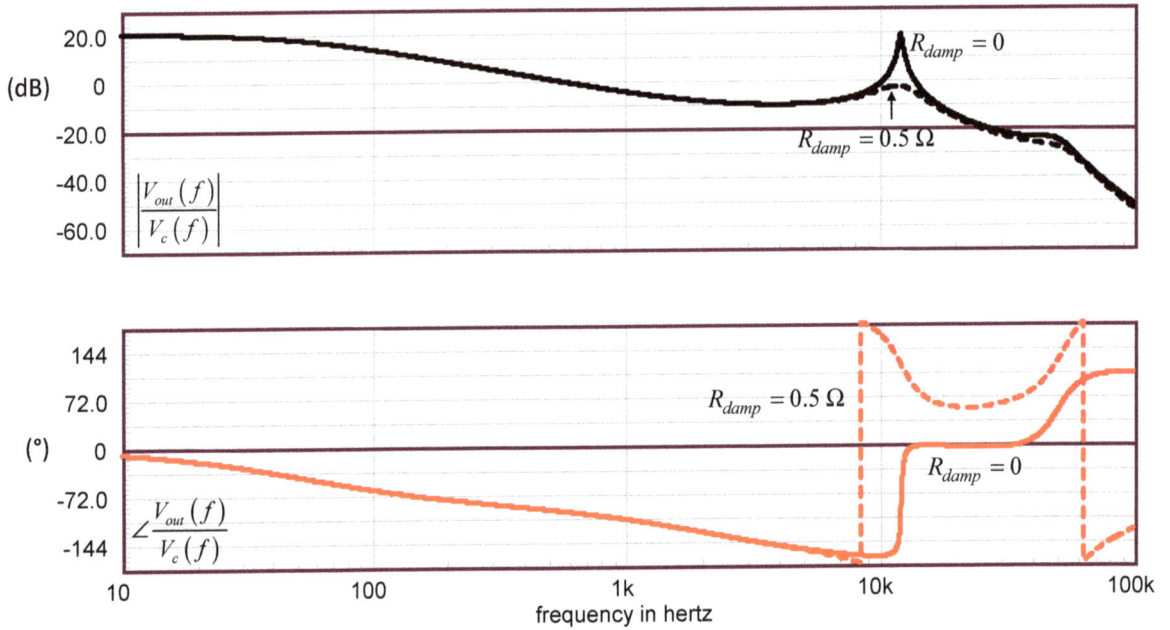

Figure 5.70 Inserting a resistance with the coupling capacitor dampens the circuit.

We have captured a schematic diagram in SIMPLIS® which reproduces current-mode control (Figure 5.71). A little bit of slope compensation is necessary to stabilize the converter which operates at a 60% duty ratio. The transient simulation results appear in Figure 5.72. The peak current is set to 8 A in this operating point.

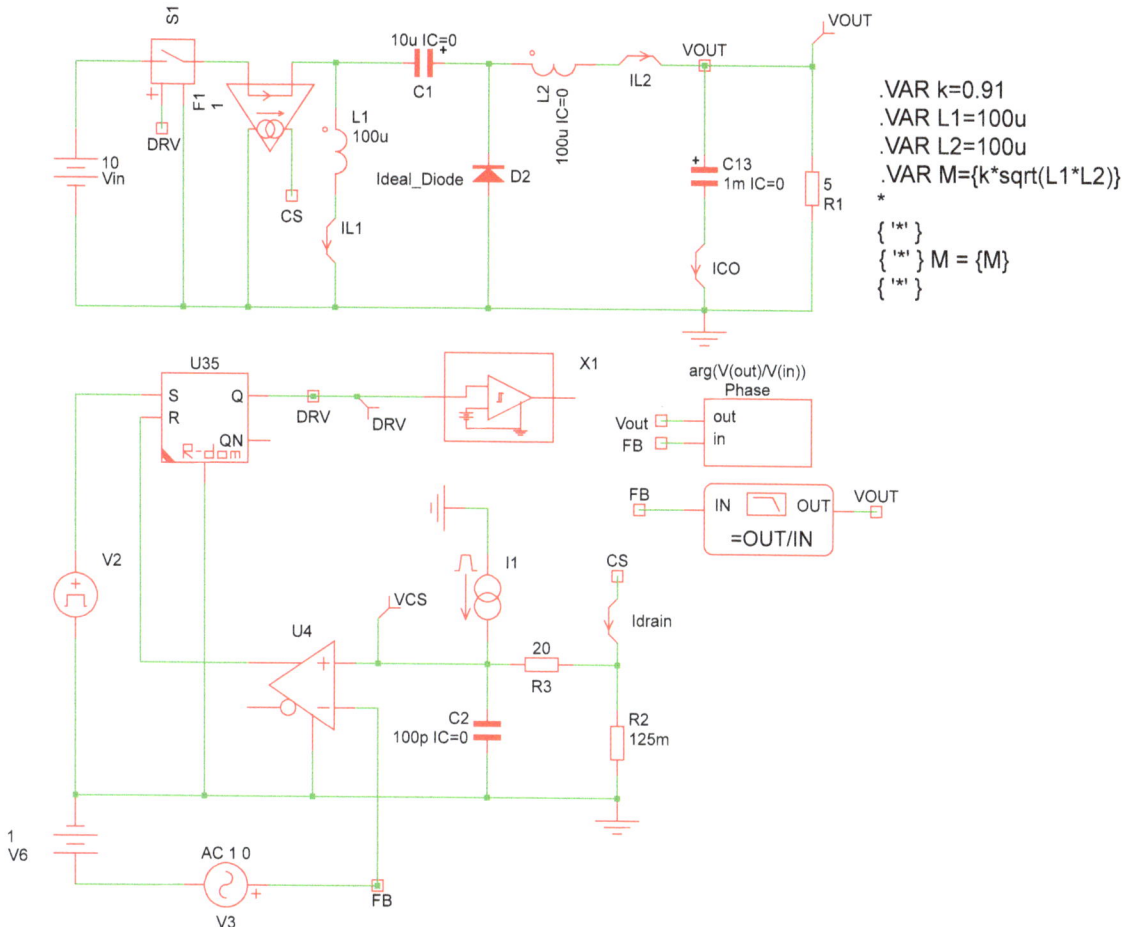

Figure 5.71 SIMPLIS® will deliver the transient response in a few seconds. You can change the value of *k* to explore different coupling conditions.

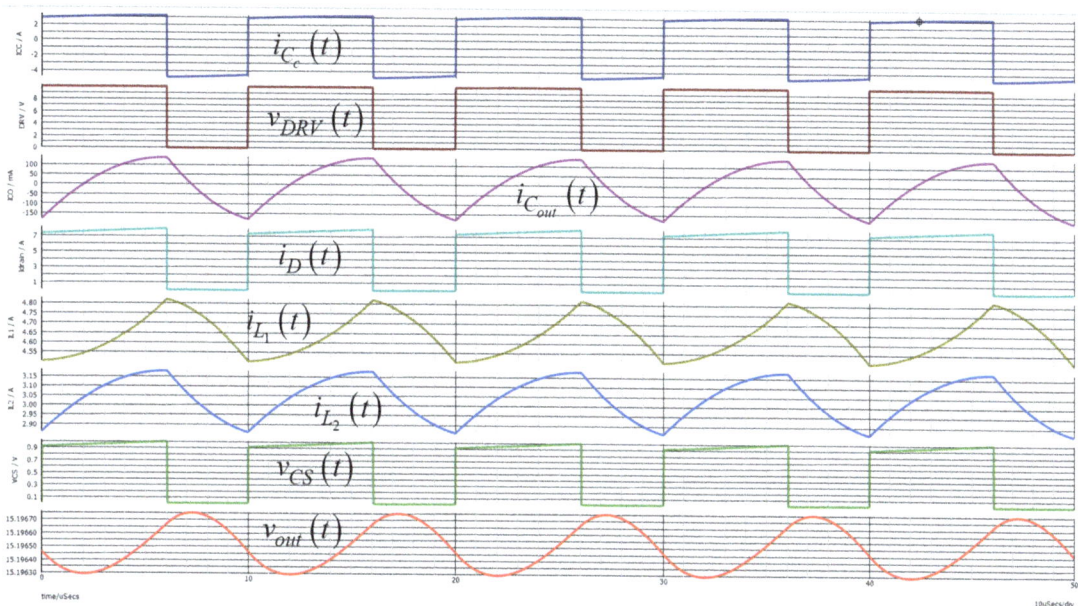

Figure 5.72 Transient waveforms for the 15-V current-mode control Zeta converter switching at 100 kHz.

The ac source modulates the control voltage V_c whose signal propagates through the circuit to affect V_{out}. The control-to-output response appears in Figure 5.73 and confirms the SPICE simulation from Figure 5.70 quite well. There is no damping resistance in this test.

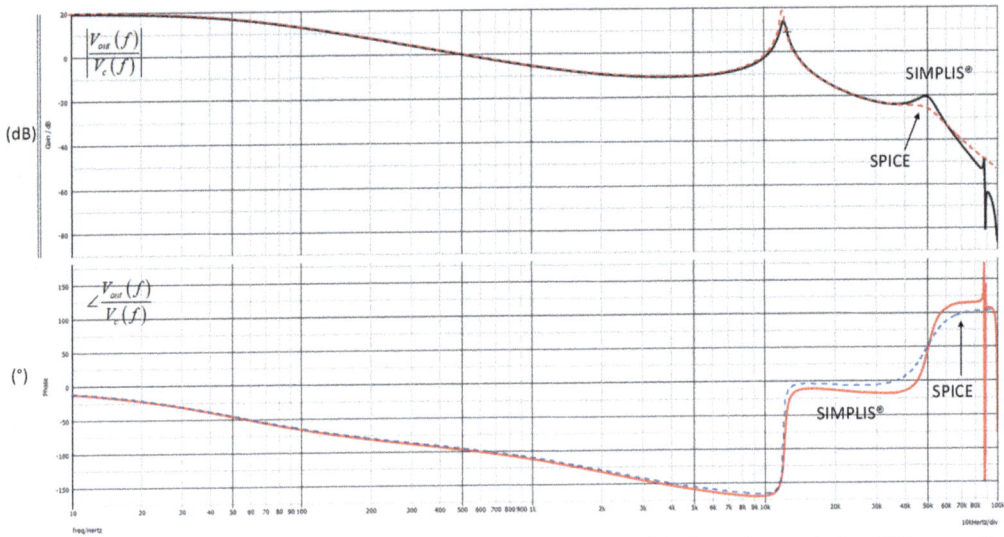

Figure 5.73 The SIMPLIS® response confirms the small-signal model of Figure 5.69.

5.4 The LLC Converter

A LLC converter, as its name implies, associates two inductors (*LL*) and one capacitor (*C*). The three energy-storing elements form a resonant tank driven by a half-bridge configuration as shown in Figure 5.74. With a constant 50% duty ratio high-voltage square waveform delivered at the HB node, the loop adjusts the switching frequency for regulating the output voltage in a constant voltage application. A deadtime is inserted in the signals driving the MOSFETs to a) avoid shoot-through currents and b) ensure zero-voltage switching (ZVS) during transitions, naturally increasing the efficiency in high-frequency applications. Detailing how this converter operates is beyond the scope of this chapter but a document such as [13] offers a smooth introduction to the subject. Literature from various sources abounds in the web, covering the subject in great detail.

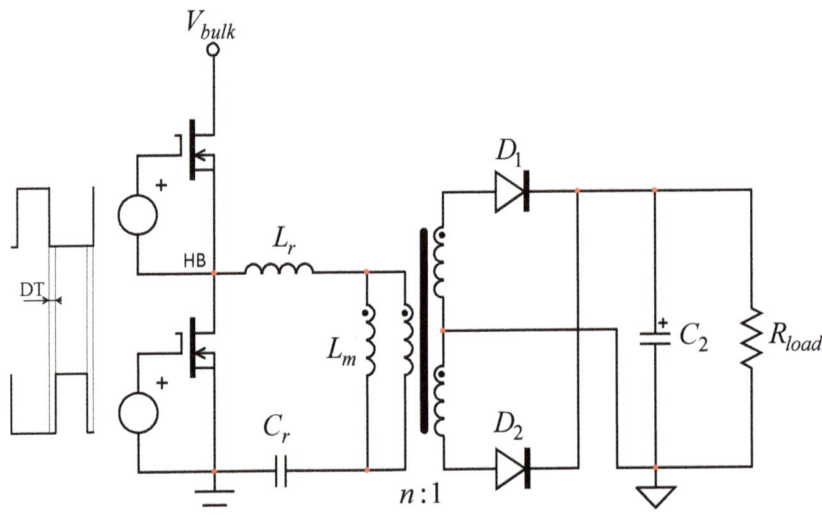

Figure 5.74 A low-impedance square-wave generator whose frequency is adjusted drives a resonating tank made of two inductors and a capacitor.

A typical LLC controller such as the NCP1397 from ON Semiconductor includes a voltage-controlled oscillator (VCO) and a high-voltage half-bridge driver. During regulation, the error amplifier adjusts the frequency to keep a constant output voltage despite load and line variations. In ac-dc applications involving a LLC converter, the input voltage is, most of the time, ensured by a Power Factor Correction (PFC) circuit and line rejection is usually not a problem. For properly compensating this converter, a transfer function is needed. Unfortunately, unlike other fixed- or variable-frequency converters, the small-signal study is not as straightforward as with the PWM switch model. In a fixed-switching frequency voltage- or current-mode-controlled converter, the switching component disappears because it is assumed that the transmitted energy depends on average values. This is the principle behind current and voltage averaging described in chapter 1. In a series-resonant converter like the LLC, there is no dc component in the transformer and the energy is conveyed by the ac components of the signal, e.g. the fundamental and its harmonics. This principle leads to complicated operating modes and gives rises to the so-called beat frequency phenomenon which describes switching frequency interaction with the different resonating modes found in a LLC converter. There have been many attempts to model the LLC control-to-output transfer function accounting for the presence of the beat frequency double poles. Until now, there was no simple model beside the extended describing functions approach used by E. Yang in [14] which lead to a faithful but uneasy to handle model. Recent work in [15] by S. Tian managed to simplify the expressions derived by Yang to form a 3rd-order model describing the frequency response of the LLC converter in a compact and efficient polynomial form.

Before we look at the power stage frequency response with SIMPLIS®, we need a modulator subcircuit. It will be designed to set the minimum and the maximum switching frequency as implemented with the selected controller for a given voltage range. For instance, if we look at NCP1397, the frequency is maximum when the feedback pin is biased to 5 V and minimum when it passes below 0.8 V. The voltage swing is thus 4.2 V from F_{min} to F_{max}. Regarding switching frequency selection, it all depends on the way you design your resonant tank and how you set F_{sw} compared to f_0, the resonating frequency when the LLC converter is loaded.

In Figure 5.75 VCO circuit, we have chosen to sweep from 120 kHz (no feedback voltage) to 350 kHz for the upper value. Considering the digital divider around the D flip-flop which delivers complementary driving signals for the power MOSFETs, the actual frequency on the bridge will vary from 60 kHz to 175 kHz in this example when V_{FB} changes from 0 to 5 V.

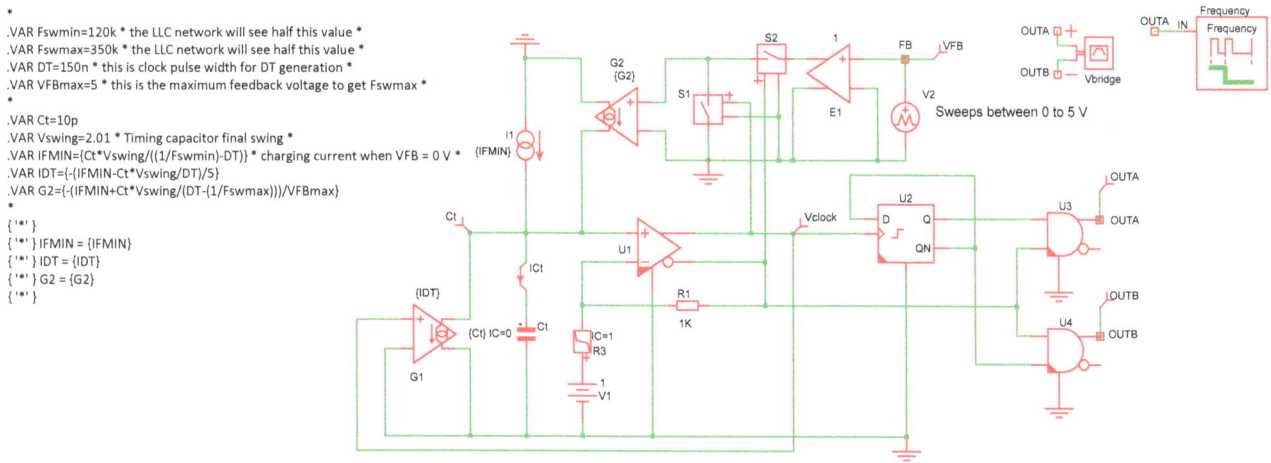

Figure 5.75 You can automate components value calculation with the left-side text block.

In absence of feedback, e.g. during the start-up sequence or a severe overload situation, the timing capacitor C_t

is charged by I_1 and discharged at a constant rate set by the I_{DT} G-source which sets the clock pulse width. This small duty ratio square-wave signal is used to generate the deadtime inserted between the two power switches conduction events via two NAND gates. Then, an extra voltage-controlled current source G_2 injects an additional charging current proportional to the feedback voltage.

To suppress this current during the discharge time a series of switches impose a 0-V control voltage during the small pulse width duration thus keeping it constant. The switching frequency as a function of V_{FB} is given below:

$$F_{sw}(V_{FB}) = \frac{1}{\frac{V_{swing}C_t}{I_{Fmin}+V_{FB}G_2}+DT} \tag{5.116}$$

Where V_{swing} is capacitor C_t peak-to-peak voltage (2 V for instance), DT represents the wanted deadtime (the clock pulse width) while I_{Fmin} and G_2 are defined below:

$$I_{Fmin} = \frac{C_t V_{swing}}{\frac{1}{F_{swmin}}-DT} \tag{5.117}$$

$$G_2 = -\frac{I_{Fmin}+\frac{C_t V_{swing}}{DT-\frac{1}{F_{max}}}}{V_{FB}} \tag{5.118}$$

A Mathcad® sheet will tell us if the frequency variation with a feedback voltage ranging from 0 V to 5 V gives the correct frequency sweep.

As shown in Figure 5.76, it nicely does.

$F_{swMIN} := 120\text{kHz}$ Select the min frequency for a 0-V feedback

$DT := 150\text{ns}$ Select the deadtime you need between the MOSFETs

$F_{max} := 350\text{kHz}$ Select the maximum switching frequency for a VFB level

$V_{FB} := 5\text{V}$ Select the maximum feedback voltage

$V_{swing} := 2.03\text{V}$

$C_t := 10\text{pF}$

$I_{FMIN} := \frac{C_t \cdot V_{swing}}{\frac{1}{F_{swMIN}}-DT} = 2.481\,\mu\text{A}$ minimum charging current to set Fswmin and DT

$G_2 := -\frac{I_{FMIN}+\frac{C_t \cdot V_{swing}}{DT-\frac{1}{F_{max}}}}{V_{FB}} = 1.004\times 10^{-6}\cdot\text{S}$ G2 transconductance value for the max Fsw

$F_{sw}(V_{FB}) := \frac{1}{\frac{V_{swing}\cdot C_t}{I_{FMIN}+V_{FB}\cdot G_2}+DT}$ Switching frequency versus feedback voltage

$V_{FB} := 0\text{V}, 100\text{mV}.. 5\text{V}$

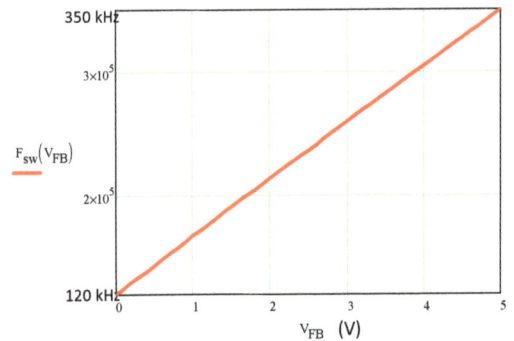

Figure 5.76 The frequency variation correctly spans between 120 and 350 kHz as expected.

If we run the VCO circuit of Figure 5.75, the simulator delivers the operating waveforms reproduced in Figure 5.77. Using the special function probe menu, plotting the frequency response of the circuit also shows a linear curve following that of Figure 5.76.

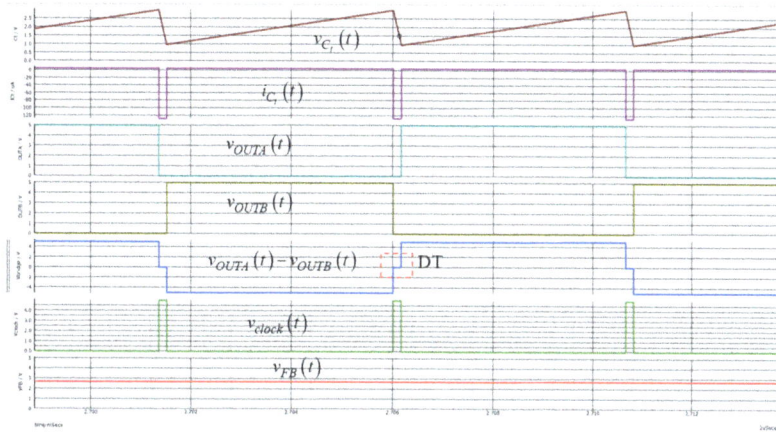

Figure 5.77 Transient waveforms of the voltage-controlled oscillator for a feedback voltage below 3 V.

The gain of this modulator can be obtained by either differentiating (5.106) with respect to V_{FB}. However, considering a fully-linear system along the input voltage span (5 V), the gain is simply:

$$G_{VCO} = \frac{F_{max} - F_{min}}{2 \cdot \Delta V_{FB}} = \frac{350k - 120k}{2 \times 5} = 23 \text{ kHz/V} \tag{5.119}$$

In this expression, the 2 is due to the D flip-flop which divides the final bridge frequency by two, e.g. from 60 to 175 kHz in this example. Now that the VCO is functional, we can associate it with the LLC power stage as shown in Figure 5.78.

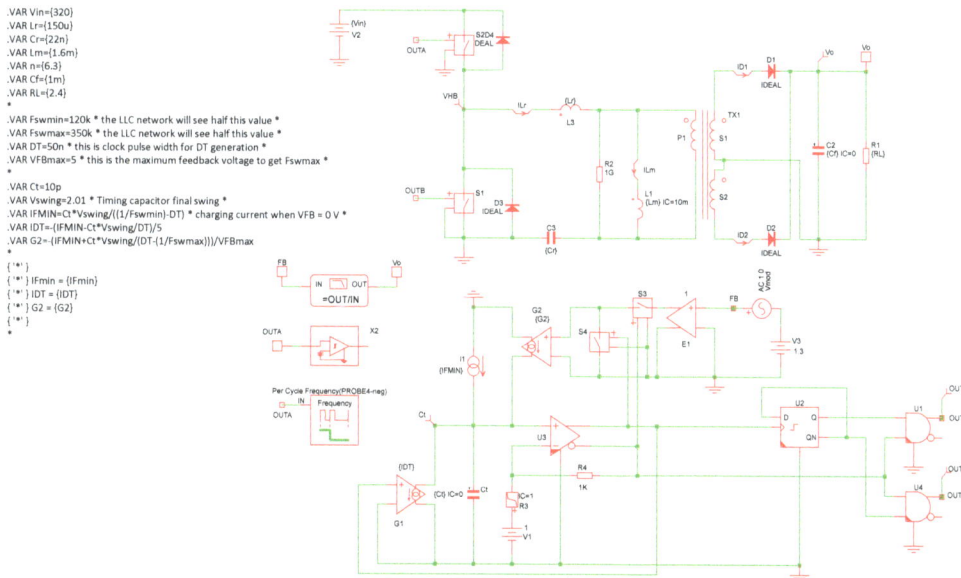

Figure 5.78 The LLC converter uses the VCO to drive a half-bridge configuration. You may need to remove gates U_1 and U_4 then drive OUTA and OUTB directly from U_2 outputs to simulate this circuit with Elements, the free SIMPLIS® demonstration version.

This converter delivers 24 V to a 2.4-Ω load from a 320-V dc source. The operating point is set by the dc source V_2. The resonant frequency at full load depends on L_r and C_r because L_m is clamped by the refelected output voltage.

The frequency is defined as follows:

$$F_0 = \frac{1}{2\pi\sqrt{L_r C_r}} = 87.6 \text{ kHz} \tag{5.120}$$

If we run the simulation, we obtain the curves reproduced in Figure 5.80. The switching frequency is at 88.8 kHz, slightly above the resonant frequency.

The primary current is nicely sinusoidal in this case. The circuit runs on the free SIMPLIS® demonstration version Elements. A simpler version of this circuit exists which runs with a primitive VCO block and uses less surrounding components (Figure 5.79). The VCO block accepts a gain parameter and you fix the nominal frequency through source V_1 set to 88.8k.

V_5 in series ensures the modulation for the ac sweep.

Simulation data in transient and ac are similar to that of Figure 5.78 circuit.

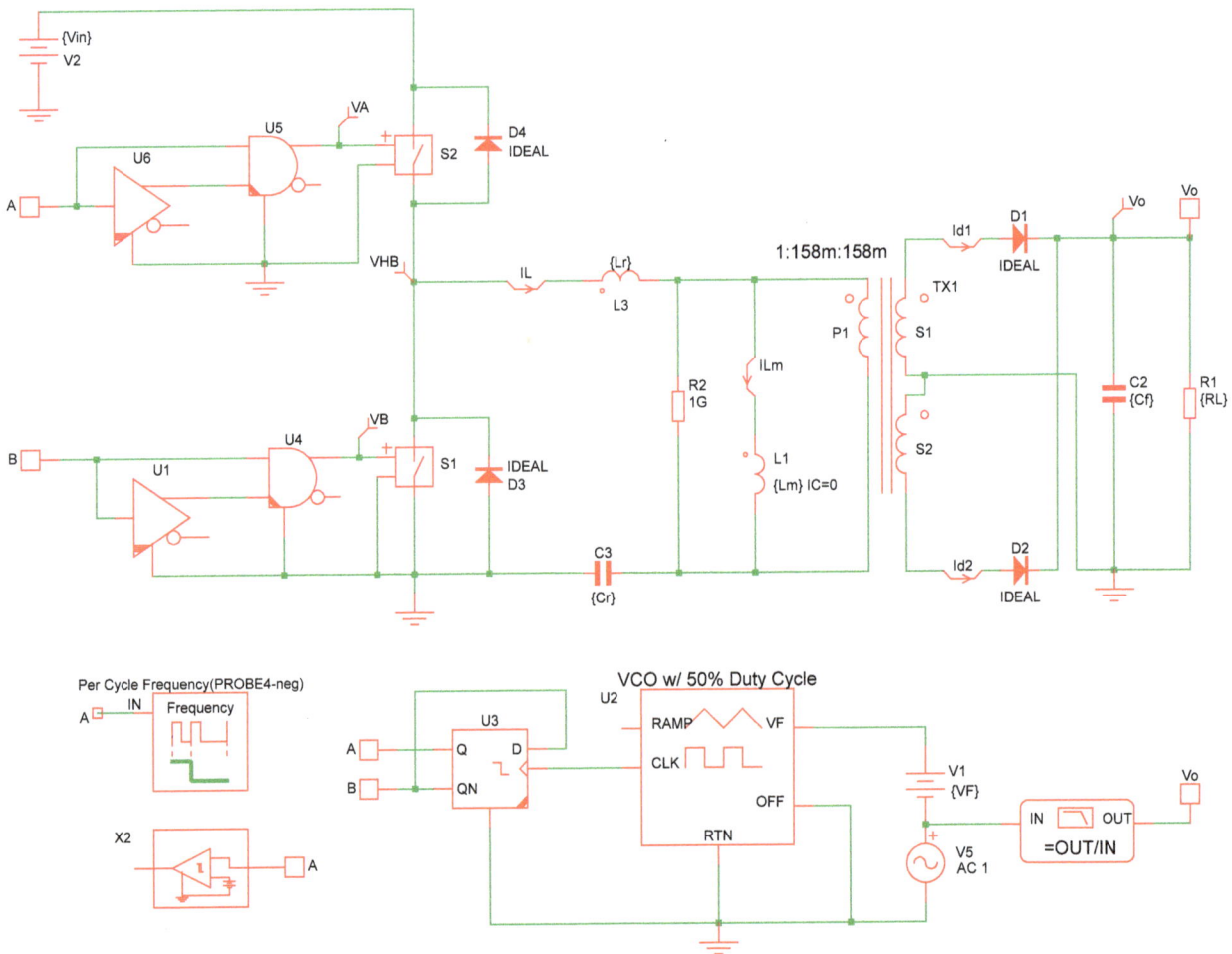

Figure 5.79 This version of the LLC converter uses a SIMPLIS® VCO block to drive the two switches.

The frequency response of this circuit appears in Figure 5.81 and confirms the presence of double poles located around 1.5 kHz. As explained in [15], the beat frequency only exists for $F_{sw} \geq F_0$. The double poles located at low frequency are the result of a high-frequency poles pair splitting: one pole goes lower and combines with the low-pass filter pole involving the output capacitor C_2 while the second pole goes higher in frequency. It is possible to show that these poles are located at:

$$\omega_{P_{1,2}} = \sqrt{\dfrac{1}{L_e \dfrac{\pi^2}{8n^2} C_2}} \tag{5.121}$$

and peak with a quality factor Q defined as:

$$Q_p = \frac{8n}{\pi^2} R_{load} \sqrt{\frac{C_2}{L_e}} \tag{5.122}$$

where:

$$L_e = \left(1 + \frac{\omega_0^2}{\omega_s^2}\right) L_r \tag{5.123}$$

$$\omega_0 = \frac{1}{\sqrt{L_r C_r}} \tag{5.124}$$

And n is the transformer turns ratio as defined in Figure 5.74.

In these equations, ω_s represents the switching angular frequency at the nominal condition. The simulation results appear in Figure 5.81 and show peaking at the frequency suggested by (5.121). Please note that there are absolutely no ohmic losses in the simulation circuits (all perfect 0-Ω types of switches) and it explains the peaky response. Any series resistance in the transformer or diodes will damp it.

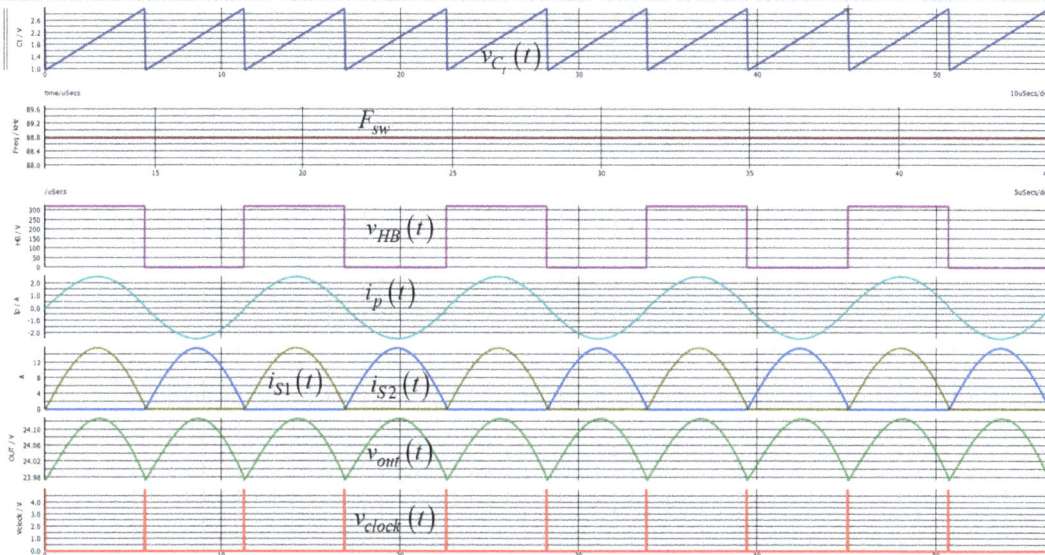

Figure 5.80 The LLC is operated slightly above the resonant frequency. It delivers 24 V to the 2.4-Ω load.

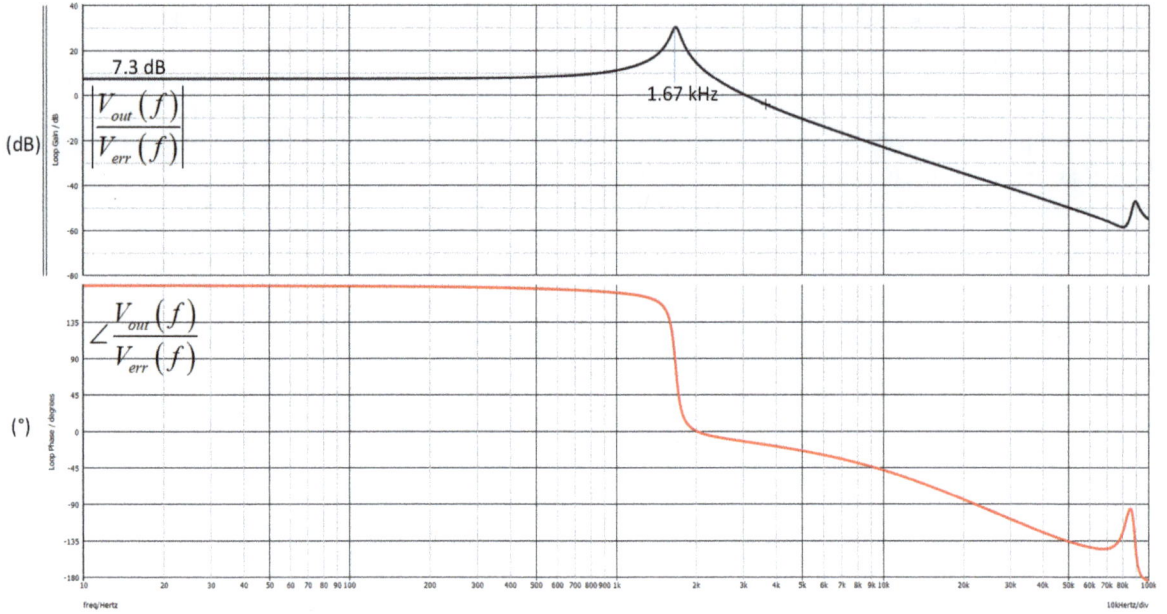

Figure 5.81 The frequency response of the LLC shows some peaking depending on operating conditions.

The small-signal transfer function proposed in [15] describes via two distinct equations a LLC converter operated at/above F_0 and below F_0. Only the first case will be studied here as it corresponds to the majority of practical cases.

We can start with the output voltage defined as:

$$V_{out} = \frac{V_{in}M}{2n} \tag{5.125}$$

With M expressed as:

$$M = \frac{\omega_n L_n}{\sqrt{\left[\omega_n\left(L_n + 1 - \frac{1}{\omega_n^2}\right)\right]^2 + \left[(1-\omega_n)^2\frac{\pi^2}{8}L_n Q_r\right]^2}} \tag{5.126}$$

The transfer function requires to know the dc gain H_0:

$$H_0 = 2\pi G_{VCO}\frac{V_{in}}{2n}\frac{L_n}{\omega_0\omega_n}\frac{\left(\frac{1}{\omega_n^2} - \omega_n^2\right)\left(\frac{\pi^2}{8}Q_r L_n\right)^2 - \left(L_n + 1 - \frac{1}{\omega_n^2}\right)\left(\frac{2}{\omega_n^2}\right)}{\left[\sqrt{\left(L_n + 1 - \frac{1}{\omega_n^2}\right)^2 + \left[\left(\frac{1}{\omega_n} - \omega_n\right)\frac{\pi^2}{8}Q_r L_n\right]^2}\right]^3} \tag{5.127}$$

Where:

$$L_n = \frac{L_m}{L_r} \tag{5.128}$$

$$Q_r = \frac{\sqrt{\dfrac{L_r}{C_r}}}{n^2 R_{load}} \tag{5.129}$$

$$\omega_n = \frac{\omega_s}{\omega_0} \tag{5.130}$$

While ω_0 and L_e are respectively defined in (5.124) and (5.123).

The complete control-to-output transfer function is determined using the below formula:

$$H(s) = H_0 \frac{\left(X_{eq}^{\ 2} + R_{eq}^{\ 2} \right)}{\left(s^2 L_e^{\ 2} + sL_e R_{eq} + X_{eq}^{\ 2} \right)\left(1 + sR_{load}C_2 \right) + R_{eq}\left(sL_e + R_{eq} \right)} \tag{5.131}$$

where:

$$X_{eq} = \omega_s L_r - \frac{1}{\omega_s C_r} \tag{5.132}$$

$$R_{eq} = \frac{8}{\pi^2} n^2 R_{load} \tag{5.133}$$

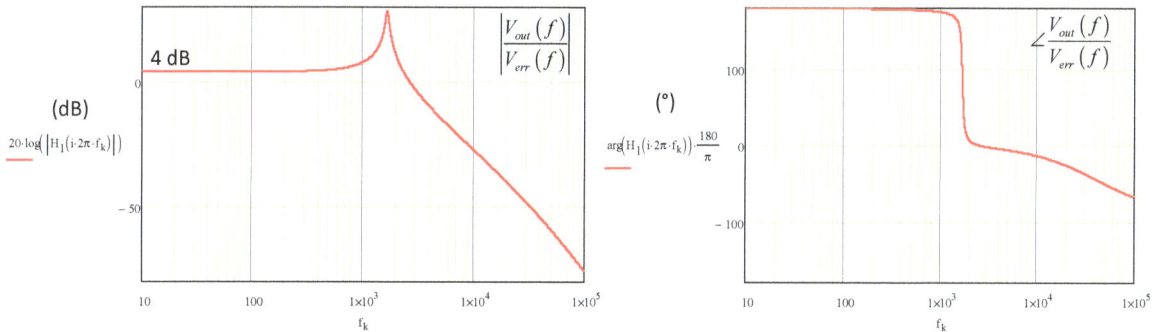

Figure 5.82 The calculated frequency response of the LLC is in good overall agreement with the SIMPLIS® simulation despite a small dc-gain error and a phase deviation in the upper section of the sprectrum.

Figure 5.82 shows the ac response as calculated with the above equations. Despite a slight gain error in dc, the overall agreement is good up to moderate frequency and certainly enough to let you think of a compensation strategy.

Before transitioning to the current-mode version, I want to thank Dr. Simon Tian for his advices and assistance when tackling the small-signal analysis of this chapter.

5.4.1 Current Mode Control

It is possible to drive the LLC converter in current-mode control like in the recently-introduced NCP1399 controller from ON Semiconductor. The operation differs from the classical VCO-based circuit as the error voltage now controls the peak current in the resonating tank. The controller turns the upper-side MOSFET on until the resonant current reaches the setpoint. At this moment, the upper-side switch turns off and the on-time duration is recorded. The lower-side MOSFET now turns on for the exact same t_{on} duration so as to create a perfect 50% duty ratio, including some deadtime between transitions. Beside the digital core, there is no clock pacing the switching frequency in the circuit and the operating period depends on the load current and the peak current setpoint. A simplified view of operating waveforms excerpted from NCP1399 technical documents is given in Figure 5.83.

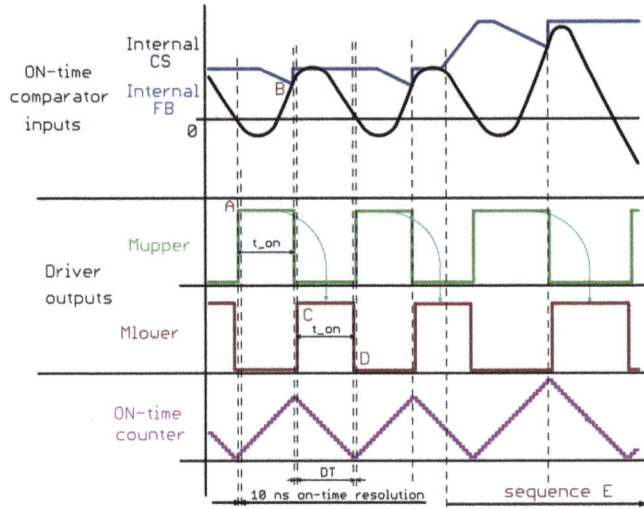

Figure 5.83 The resonant current peaks to the setpoint imposed by the loop during t_{on}. The on-time duration is then replicated during t_{off} to provide an exact 50% duty ratio in all operating conditions.

The modulator inside the controller involves high-resolution logic counters that I could not use in SIMPLIS® for the sake of model simplicity. A possible alternative analogue implementation is given in Figure 5.84. In this modulator, a capacitor is charged at a constant current rate until the peak current comparator stops the progression: this is the on-time duration. Then, the capacitor undergoes a discharge at the same current towards ground. When the voltage is 0 V on capacitor C_t, the off-time event is over and a next cycle takes place.

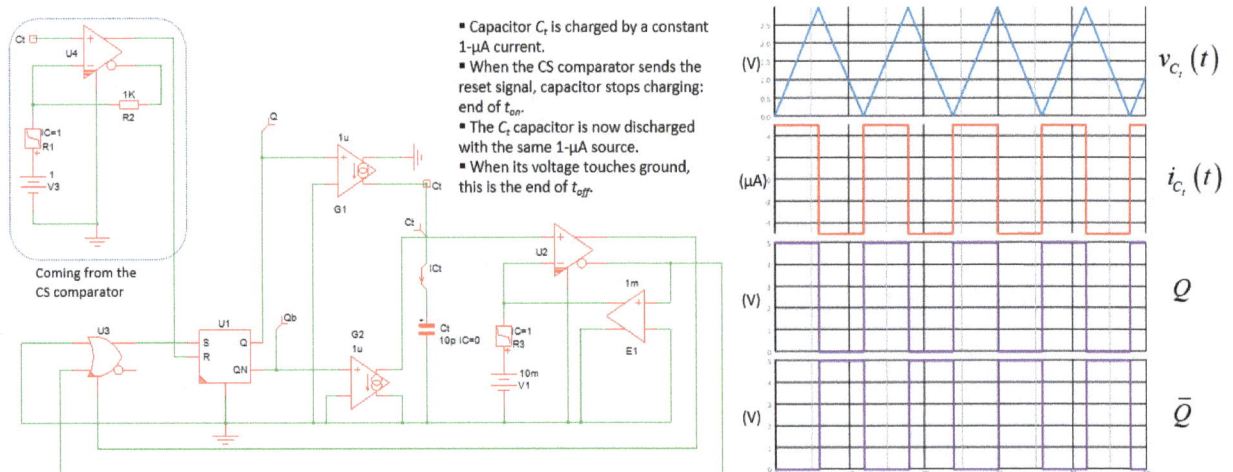

Figure 5.84 The modulator involves a timing capacitor charged and discharged at similar current for an exact 50% operation.

With this circuit on hand, I have assembled a complete 24-V/10-A LLC current-mode-controlled converter which is proposed in Figure 5.85. The voltage across resonant capacitor C_1 is scaled and filtered then routed to a current-sense comparator blanked for a small period of time.

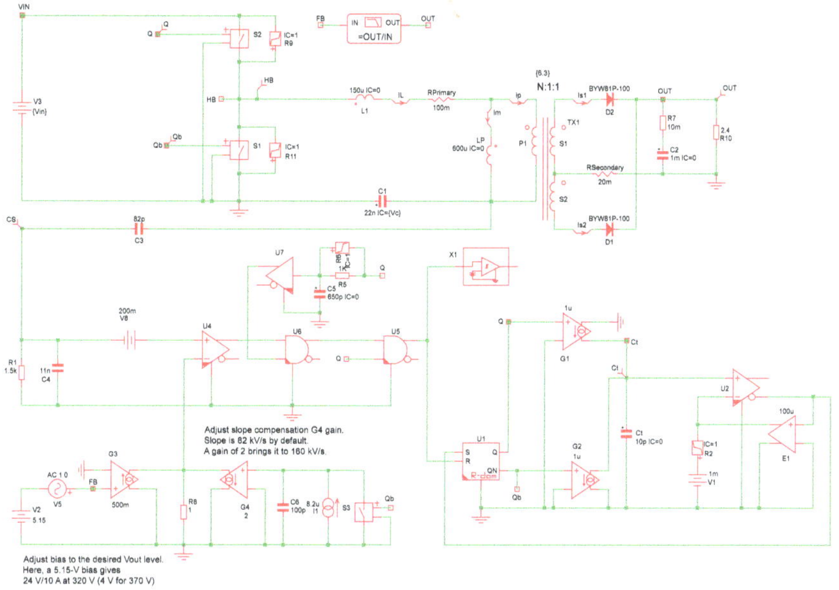

Figure 5.85 This converter operates at an exact 50% duty ratio without an internal clock owing to the current-mode operation.

Ramp compensation is necessary for a proper operation and is implemented via the voltage-controlled current source G_4. The peak current setpoint set to 5.15 V by V_2 and transformed in current via source G_3 to be further associated with the ramp signal through R_8. The input voltage is 320 V for this simulation. The cycle-by-cycle waveforms are given in Figure 5.86 and confirm an operation close to the resonant frequency with a 24-V output for a 2.4-Ω load. The duty ratio is precisely 50% and can be measured by using a dedicated per-cycle-measurement probe. The small-signal response is obtained by stimulating the setpoint with the ac source V_5.

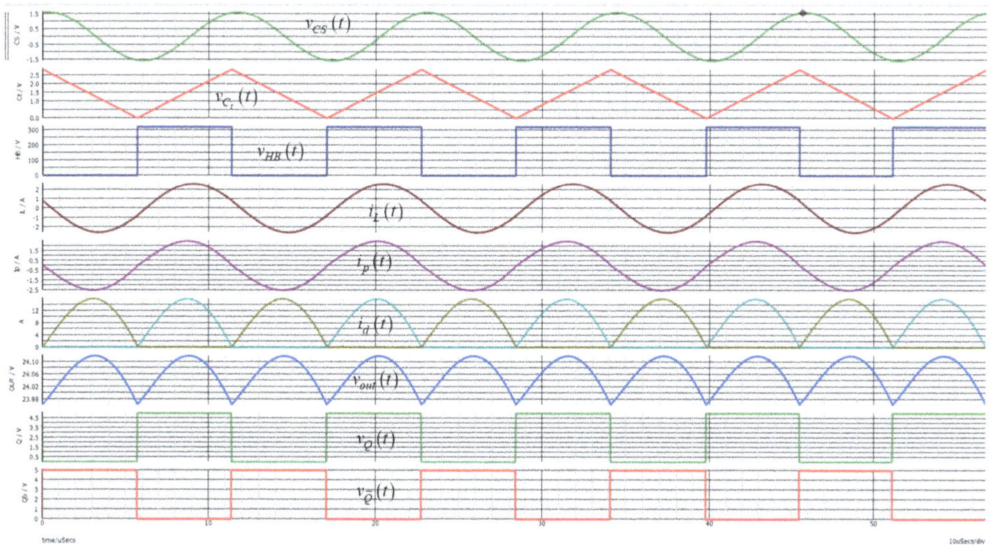

Figure 5.86 The half-bridge output node is exactly at a 50% duty ratio.

The control-to-output transfer function is shown in Figure 5.87 and confirms a small dependency to the input voltage variations. This is obviously an advantage of the current-mode structure compared to its VCO-based counterpart. I have not found a study mathematically describing the transfer function of the CM LLC converter however SIMPLIS® will be of great help backed up by measurements on a bench prototype.

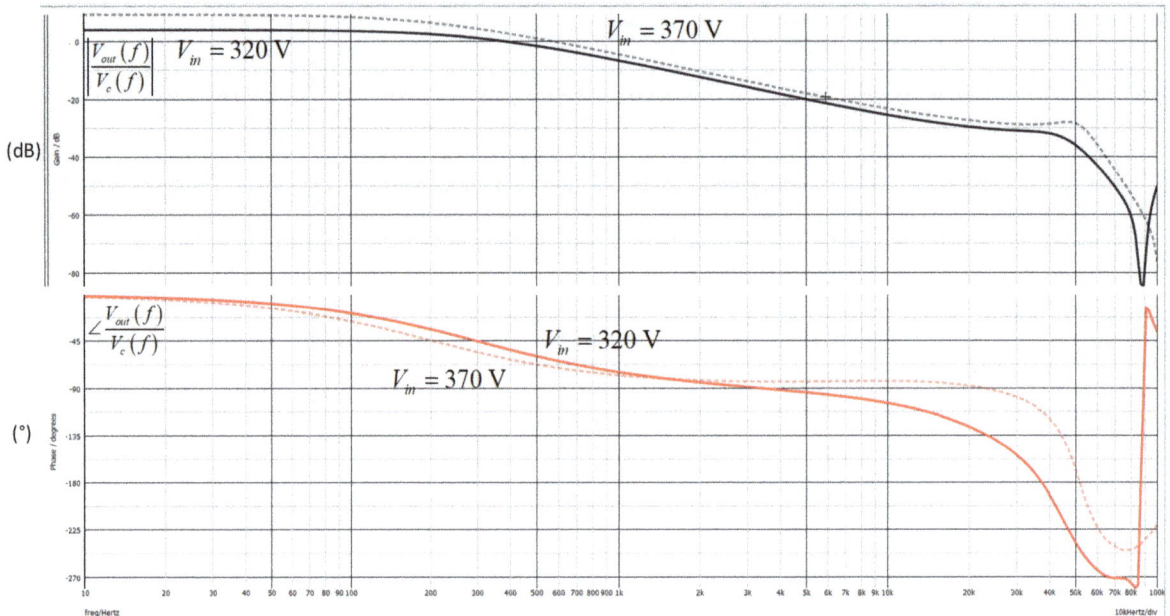

Figure 5.87 The dynamic response does not change much despite input voltage variations.

This converter ends the coverage of more complex structures described in this fifth chapter.

I thank my colleague Roman Stuler from ON Semiconductor for reviewing this section on current-mode control.

5.5 What Should I Retain from this Chapter?

In this fifth and last chapter, we have studied higher-order converters:

1. The Ćuk converter can increase or decrease the input voltage but with a negative polarity as a buck-boost converter would do. However, it is possible to reduce the input or output ripple current by coupling inductors. An isolated version exists and its mathematical 6th-order analytic transfer function was exercised versus SPICE and SIMPLIS® models.

2. The SEPIC also offers the possibility to boost or reduce the input voltage but it delivers a positive output. Unlike the Ćuk converter, the output current is pulsating just like a regular buck-boost converter. Coupling inductors and operating the converter in current-mode control ease the stabilization procedure for this high-order system.

3. The Zeta converter too delivers a positive voltage above or below the input line. Unlike the SEPIC, its input current is pulsating but its output current is smooth. This high-order converter can be analyzed with the PWM swich model with coupled or uncoupled inductors.

4. The LLC converter is a resonant structure offering many interesting characteristics such as a good

efficiency at high switching frequency and ease of control via a VCO. Until recently, there was no simple small-signal model describing its dynamic behavior but recent work from Virginia Tech lead to expressing its dynamic response through a series of transfer functions presented in this chapter.

5.6 References

1. Ćuk et al., *Dc to Dc Switching Converter*, US Patent 4,184,197, filed in September 1977.
2. Ćuk, *Dc to Dc Switching Converter with Zero Input Output Current Ripple and Integrated Magnetics Circuits*, US Patent 4,257,087, filed in April 1979.
3. Ćuk et al., *Dc to Dc Switching Converter Having Reduced Ripple without Need for Adjustment*, US Patent 4,274,133, filed in June 1979.
4. R. Erickson, D. Maksimovic, *Fundamentals of Power Electronics*, Springer, 2nd edition 2001
5. V. Vorpérian, *The Effect of the Magnetizing Inductance on the Small-Signal Dynamics of the Isolated Ćuk Converter*, IEEE Transactions of Aerospace and Electronic Systems, Vol. 32, No. 3, July 1996
6. S. Ćuk, L. Kajouke, *Electric and Magnetic Circuit Interactions in Switching Converters*, Proceedings of the IEEE Applied Power Electronics Conference, Mar. 2-6, 1987, San-Diego, CA.
7. V. Vorpérian, *Fast Analytical Techniques for Electrical and Electronic Circuits*, Cambridge University Press, 2002.
8. R. Ridley, *Analyzing the SEPIC Converter*, Power Systems Design Europe, November 2006
9. C. Basso, *Linear Circuit Transfer Functions – An Introduction to Fast Analytical Techniques*, Wiley, 2016.
10. R. Ridley, *SEPIC Converter Measurements – Parts I, II and III*, Ridley Engineering Design Center, http://www.ridleyengineering.com/design-center-ridley-engineering
11. P. Kochcha, S. Sujitjorn, *Isolated Zeta Converter: Principle of Operation and Design in Continuous Conduction Mode*, WSEAS Transactions on Circuits and Systems, Issue 7, Vol. 9, July 2010
12. J. Falin, *Designing Dc-Dc Converters Based on Zeta Topology*, Texas Instruments Application Note, SLYT372
13. C. Basso, *Understanding the LLC Structure in Resonant Applications*, AN8311/D, www.onsemi.com
14. E. X. Yang, *Extended describing function method for small-signal modeling of resonant and multi-resonant converters*, Ph.D. dissertation, Virginia Tech., VA, Feb. 1994
15. Tian, *Equivalent Circuit Model of High-Frequency PWM and Resonant Converters*, Ph.D. dissertation, Virginia Tech., VA, Aug. 2015

5.7 Conclusion

In the beginning of the project, my goal was to simply determine the control-to-output transfer function of the three classical dc-dc cells. At first I judged it was enough for the loop closing exercise but later on realized that the three remaining expressions – namely input-to-output, Z_{in} and Z_{out} – were also useful for a thorough study. I decided to extend the analysis to the four transfer functions of each cell, exploring buck- or buck-boost-derived topologies, including different operating modes, voltage- and current-mode controls, quasi-resonance, constant on-time etc. I realized too late it was a Trojan work considering the numerous available topologies, all operated in different modes and control strategies. I wrote for more than three years, deriving all expressions myself and testing them with Mathcad®, SPICE and SIMPLIS®.

What motivated me to pursue was the lack of comprehensive work available on the subject. Of course you can find papers, books and publications giving away formulas or expressions but, very often, formulas are either an assembly of matrixes or *high-entropy* expressions you cannot exploit for practical purposes. I wanted a different approach, with results you could capture in a solver sheet and obtain the response of your choice. And here we are with this new book, after a long writing period and many pages of equations and graphs. There has been ups and downs, as you can imagine, especially when you have the final equation diverging from what the model gives. However, I managed to get most of them right so that you, the power supply designer, could apply the formulas and ensure the reliable operation of your converter. Needless to say the fast analytical techniques were the tool I needed and, without them, I could not have carried the exercise. I want to thank and recognize my friend Vatché Vorpérian for paving the way with his papers, seminar and book, naturally exciting my curiosity to learn FACTs and modestly spread them to the technical community.

I hope you will appreciate the flow I adopted in this book and be able to successfully use it in your engineering tasks. Happy reading and, once again, please send your comments or the typos you may have identified – I apologise in advance but they are unavoidable despite thorough reviews – to cbasso@wanadoo.fr. I will reply with pleasure.

—Christophe Basso
Toulouse, February 2020

Appendix A – Fast Analytical Circuits Techniques or FACTs

TO SOLVE TRANSFER functions in a swift and efficient manner, nothing can beat the FACTs in terms of simplicity and ease of application. Applying the divide-and-conquer technique – you split a complicated schematic into small individual sketches that you independently solve – the FACTs naturally lead to the so-called *low-entropy* expression implying a factored form in which you immediately distinguish gains, poles and zeros, if any.

The term was forged by Dr. Middlebrook in his founding papers [1], [2] where he showed that applying brute-force analysis to high-order circuits could quickly lead to algebraic paralysis. The FACTs, on the other hand, help you build on what you have learned in the university and extend the reach to drastically simplify analyses. By using FACTs, you not only gain in execution speed but the final result appears in a well-ordered polynomial form often without the need for further factoring efforts.

This appendix offers a quick introduction to the FACTs as I use them extensively in this book to determine the transfer functions of linearized dc-dc converters. The subject is vast and a few-page appendix cannot replace some books dedicated to the subject such as [3] and [4].

Whether you are an engineer or a student, I encourage you to acquire this skill which will prove of extraordinary help when confronted with a complicated circuit whose transfer function must be determined.

1. A Quick Introduction to Fast Analytical Techniques

The basic principle behind these FACTs lies in the determination of the circuit time constants – $\tau = RC$ or $\tau = L/R$ – when the network under study is observed in two different conditions: when the excitation signal is reduced to zero and when the response is *nulled*. By using this technique, you will appreciate how quickly and intuitive it is to determine a particular transfer function.

Analysis techniques based on this method date from several decades as documented in [5] and [6].

A transfer function is a mathematical relationship linking an excitation signal, the *stimulus*, to a *response* signal resulting from that excitation.

If we consider a linear time-invariant (LTI) system without any delay and exhibiting a quasi-static gain H_0 – for instance the linearized ideal power stage of a switching converter – its control-to-output transfer function H linking V_{err} (the stimulus) to its output V_{out} (the response) can be expressed in the following form:

$$H(s) = \frac{V_{out}(s)}{V_{err}(s)} = H_0 \frac{N(s)}{D(s)} \qquad (A.1)$$

The *leading term* H_0 represents the gain or attenuation exhibited by the system evaluated for $s = 0$. This term would carry the transfer function unit (or dimension) if any.

If both the response and the excitation are expressed in volts, V_{err} and V_{out} in our case, H is unitless.

2. Zeros of the Network

The numerator $N(s)$ hosts the *zeros* of the transfer function. Mathematically, zeros are the roots for which the function magnitude is zero. With the FACTs, we use a mathematical abstraction to let us easily unveil these zeros. Rather than solely considering the vertical axis in the s-plane as we normally do in harmonic analysis ($s = j\omega$), we will cover the entire plane allowing for complex roots featuring a negative real component ($s = \sigma + j\omega$).

As such, if present in the circuit, a zero will manifest itself by the *nulling* of the output response when the input signal is tuned to the zero angular frequency s_z. The output *null* happens because some impedance in the *transformed* circuit blocks the signal propagation despite the presence of an excitation source: a series impedance in the signal path becomes infinite or a branch shunts the stimulus to ground when the transformed circuit is excited at $s = s_z$.

Please note that this convenient mathematical abstraction offers tremendous help in finding the zeros by *inspection*, often without writing a line of algebra in passive networks.

Figure 1Figure 1 offers a simple flow chart which details the procedure to unveil zeros. You see a circuit in which the energy-storing elements are replaced by their respective impedance expression: this is the *transformed network*.

Keep the excitation signal - the stimulus - in place

↓

Consider a null in the output: $V_{out}(s) = 0$ V or $\hat{v}_{out} = 0$ V

↓

Identify in the *transformed* network, one or several impedances combinations that could block the stimulus propagation and create the null:

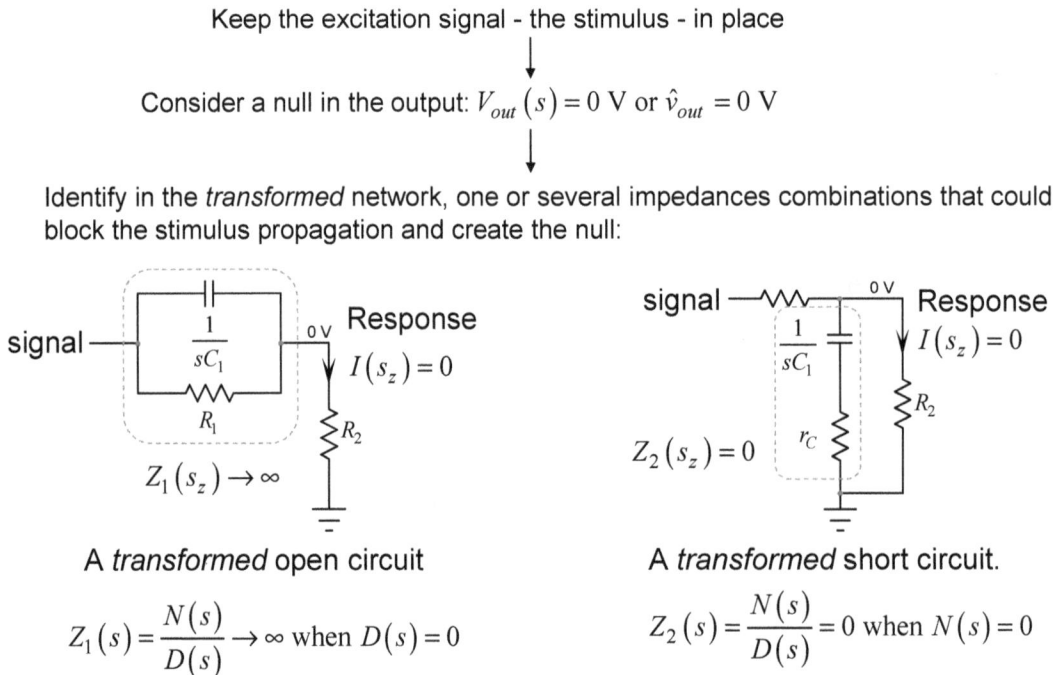

A *transformed* open circuit

$$Z_1(s) = \frac{N(s)}{D(s)} \to \infty \text{ when } D(s) = 0$$

A *transformed* short circuit.

$$Z_2(s) = \frac{N(s)}{D(s)} = 0 \text{ when } N(s) = 0$$

Figure 1: This simple flow chart guides you in determining zeros in the quickest way. When inspection does not work, you will need to go for a null double injection or NDI.

In the upper left section, you see an impedance Z_1 inserted in the signal path. Is there a specific combination for which the magnitude of this impedance could become infinite and bring a null in the response?

Express the impedance as:

$$Z_1(s) = R_1 \| \frac{1}{sC_1} = \frac{R_1 \frac{1}{sC_1}}{R_1 + \frac{1}{sC_1}} = R_1 \frac{1}{1 + sR_1C_1} \tag{A.2}$$

The magnitude of this impedance becomes infinite if the denominator $D(s)$ equals zero. In other words, the root of this expression:

$$D(s) = 0 \rightarrow s_p = -\frac{1}{R_1 C_1} \tag{A.3}$$

is the *zero* of the entire network we are looking for:

$$\omega_z = \frac{1}{R_1 C_1} \tag{A.4}$$

In the right side of the picture, what are the conditions for which the series combination of r_C and C_1 could become a transformed short circuit and shunt the stimulus to ground?

$$Z_2(s) = r_C + \frac{1}{sC_1} = \frac{1 + sr_C C_1}{sC_1} \tag{A.5}$$

The numerator equals zero when the stimulus is tuned to:

$$1 + sr_C C_1 = 0 \rightarrow s_z = -\frac{1}{r_C C_1} \tag{A.6}$$

and it implies a zero located at:

$$\omega_z = \frac{1}{r_C C_1} \tag{A.7}$$

Inspection is a very convenient way to determine zeros in a network. A simple trick [4] lets you immediately check if you have one (or several) zeros in a network, even if inspection did not reveal it at first glance: place one energy-storing element in its high-frequency state (replace the capacitor by a short circuit and open circuit the inductor) and check if, in this mode, the excitation signal gives a response.

If yes, then the considered energy-storing element contributes a zero.

If not, there is no zero associated with it. Let's exercise our skill with the 4 networks shown in Figure 2. In sketch (a), if I replace C by a short circuit in my head, then the stimulus V_{in} can propagate and produces a response V_{out}: I have a zero when r_C and C form a transformed short circuit – see (A.5).

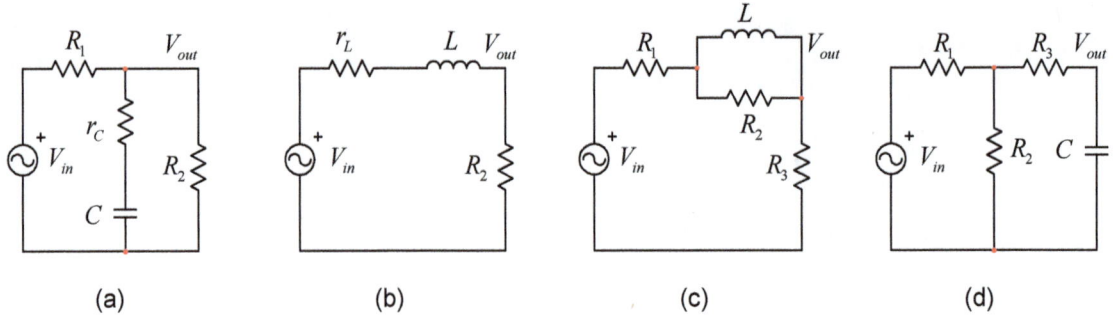

(a)   (b)   (c)   (d)

Figure 2: Some of these network feature one or several zeros, will you find them by inspection?

In (b), what if I open-circuit inductor L? The signal path is broken and there is not response in V_{out}: this network does not host a zero. In (c), when L is open circuited, because of R_2, there is a response in V_{out} and L contributes a zero. If you calculate the condition for which the parallel combination of L and R_2 brings an infinite impedance, you will identify the pole of this particular arrangement. This is the zero of our network and equals R_2/L. In (d), replace C by a short circuit and you will see that there is not response in this mode. This network does not feature a zero. Add a small resistance in series with C and you have a zero described by (A.7).

When inspection does not work, you will have to resort to a null-double injection or NDI. This term implies that you determine the time constant involving the energy-storing element while you create a null in the response with the stimulus in place. To determine the resistance driving the capacitor or the inductor and thus the time constant you want, you can install a current source I_T biasing the connecting terminals of C or L temporarily removed from the network. By tweaking the current source value, you will create a null in the output response: the double injection (the stimulus and the current source) nulls the output response when a zero exists. An example is given in Figure 3.

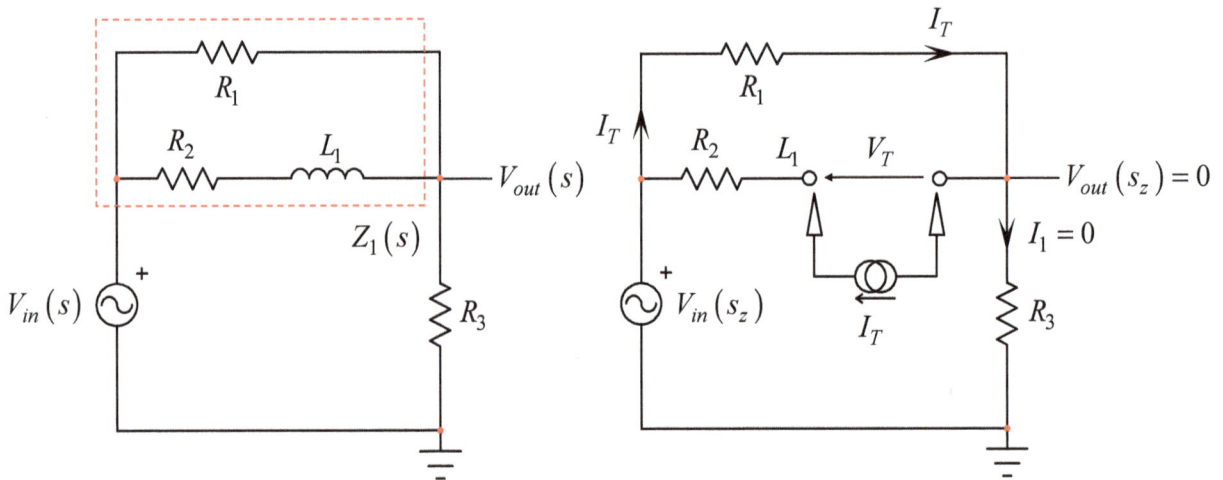

Figure 3: Tweaking the current source I_T to null the output will give you the resistance involved with L_1 to form the time constant associated with the zero.

Here, you see that placing L_1 in a high-frequency state, the stimulus V_{in} still produces a response thus implying the presence of a zero. Then, temporarily remove L_1 and install a test generator I_T across the inductor terminals as illustrated in the right side of the picture. If you null the output, the voltage across R_3 becomes 0 V meaning that no current flows in it. This is not to be confused with a short circuit but see it as a virtual ground connected across R_3.

When you inject the I_T current, as it cannot flow in R_3 because of the null (0 V across a resistance means no current flows in it) the only path it can find is via the series connection of R_1 and R_2:

$$V_T = I_T \left(R_1 + R_2 \right) \tag{A.8}$$

Also, by inspection, without installing a test generator in this simple example, the resistance you "see" through L_1's terminals in Figure 4 while the output is 0 V is the sum of R_1 and R_2. It leads to a time constant equal to

$$\tau_1 = \frac{L_1}{R_1 + R_2} \tag{A.9}$$

The pole in a 1st-order network is the inverse of its time constant:

$$Z_1 \left(s \right) = \frac{N \left(s \right)}{1 + s\tau_1} \tag{A.10}$$

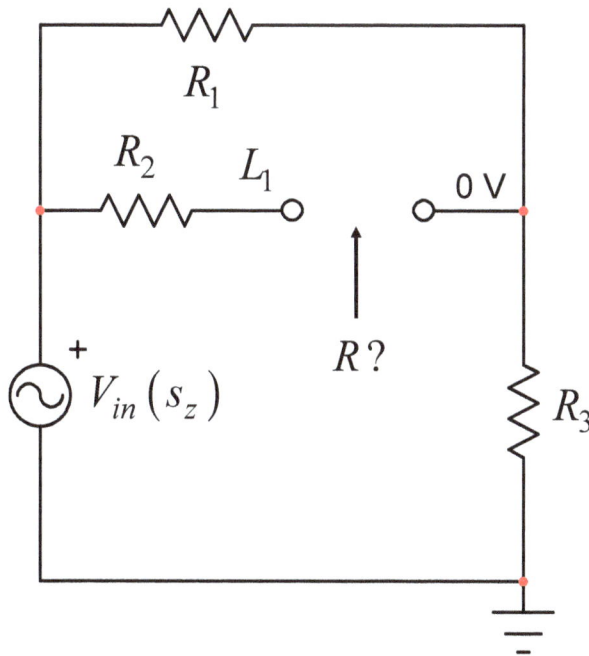

Figure 4: The pole of this network is the zero of our transfer function.

Z_1's magnitude approaches infinity and thus blocks the stimulus propagation, when:

$$s_p = -\frac{1}{\tau} = -\frac{R_1 + R_2}{L_1} \tag{A.11}$$

Leading to the zero we want and defined as:

$$\omega_z = \frac{R_1 + R_2}{L_1} \tag{A.12}$$

The NDI is physically meaningful as illustrated in Figure 5. As shown in [4], the voltage-controlled current source G_1 injects the exact amount of current to null the output. When done, you can observe a dc bias point across R_3 of an extremely low value. If you compute V_T divided by the injected current (1 mA), you obtain the value of 4 kΩ which corresponds to the series connection of R_1 and R_2.

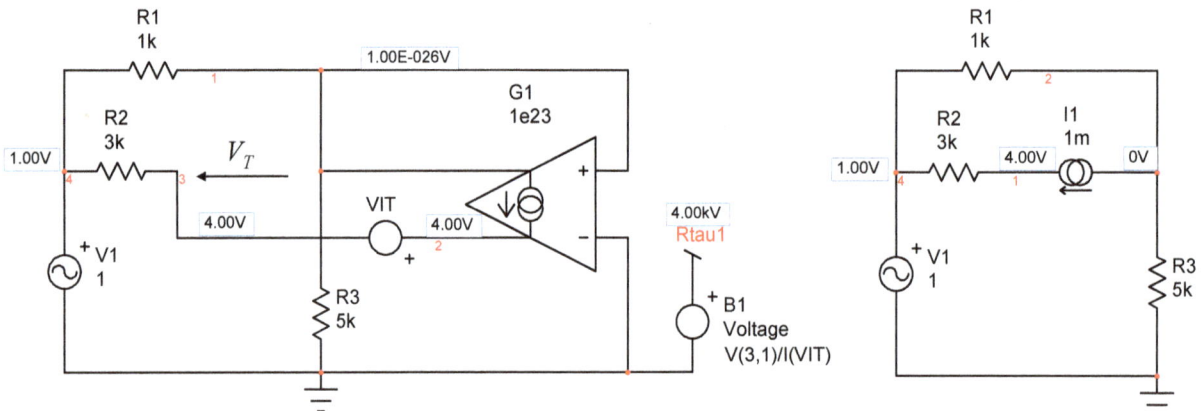

Figure 5: The NDI truly nulls the output voltage when the I_T generator is well adjusted.

In the right-side of the figure, if you build the circuit and inject a 1-mA current into R_2, then the bias point calculation confirms a 0-V potential across R_3.

3. Poles in the Network

The denominator $D(s)$ is formed by associating together the circuit *natural* time constants. These time constants are obtained by setting the stimulus signal to zero and determining the resistance "seen" from the considered capacitor or inductor terminals when temporarily removed from the circuit. By "seeing", you imagine placing an ohm-meter across the pads of the removed energy-storing element (C or L) and read the resistance it displays.

This is a quite simple exercise actually as detailed by the flow chart from Figure 6.

Count energy-storing elements with independent state variables

↓

Assume there are two energy-storing element, L_1 and C_2

↓

The denominator follows the form $D(s) = 1 + b_1 s + b_2 s^2$

H_0 ⇨ Open the capacitor, short the inductor, determine the dc gain H_0 if it exists

↓

Reduce the excitation to 0 and determine time constants for b_1 and b_2

↓

b_1 ⇨ Determine the resistance R_i driving L_1 while C_2 is open circuited: $\tau_1 = L_1/R_i$

Determine the resistance R_j driving C_2 while L_1 is short circuited: $\tau_2 = R_j C_2$

Sum the time constants: $b_1 = \tau_1 + \tau_2$

↓

b_2 ⇨ Determine the resistance R_k driving L_1 while C_2 is short circuited: $\tau_1^2 = L_1/R_k$

Determine the resistance R_l driving C_2 while L_1 is open circuited: $\tau_2^1 = C_2 R_l$

Choose the simplest combination: $b_2 = \tau_1 \tau_2^1$ or $b_2 = \tau_2 \tau_1^2$

⇩

$$D(s) = 1 + s(\tau_1 + \tau_2) + s^2 \left(\tau_1 \tau_2^1\right)$$

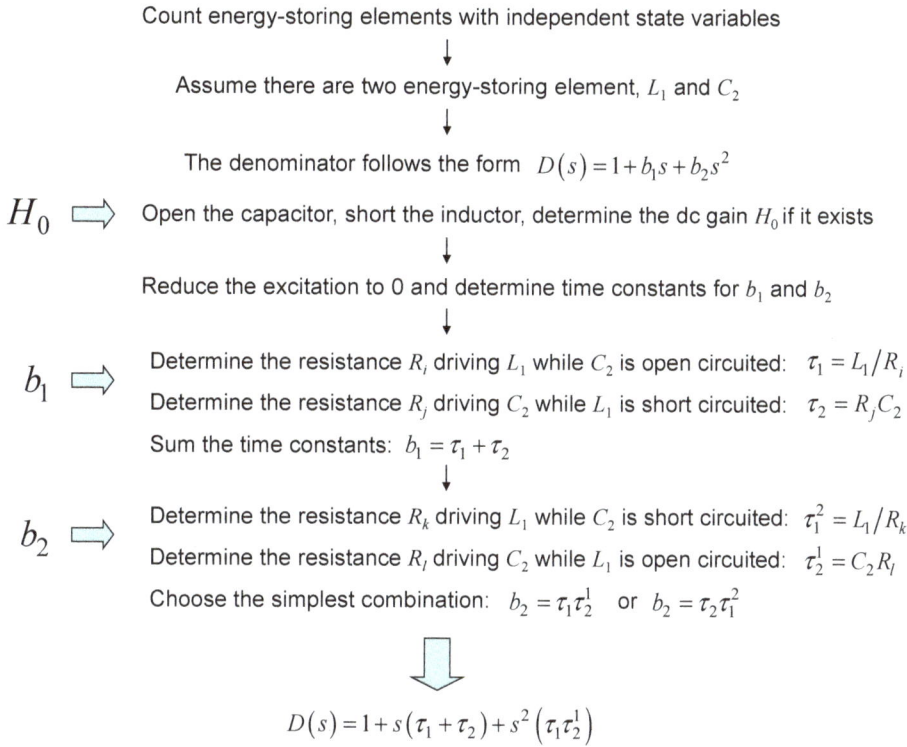

Figure 6: This flow chart explains the methodology used to determine the network time constants.

Look at Figure 7 which describes a 1st-order passive circuit involving an injection source – the stimulus – biasing the left-side of the network. The input signal V_{in} propagates through meshes and nodes to form the response V_{out} observed across resistor R_3. We are interested in deriving the transfer function G linking V_{out} to V_{in}.

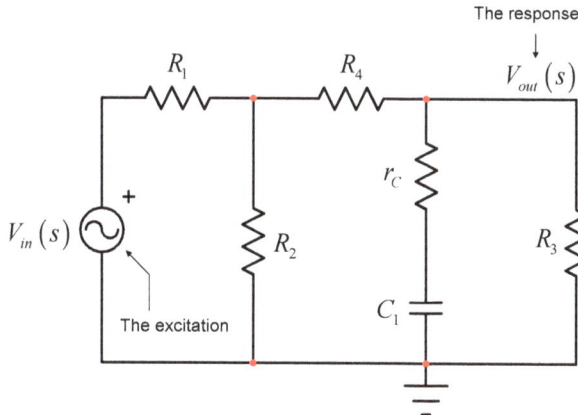

Figure 7: To determine the time constant of a circuit, set the excitation to zero and "look" at the resistance offered by the energy-storing elements terminals when temporarily removed.

To determine the time constant of this example circuit, we will set the excitation to zero (a 0-V voltage source is replaced by a short circuit while a 0-A current source would be replaced by an open circuit) and temporarily remove the capacitor. Then, we connect (in our head) an ohm-meter to determine the resistance offered by the capacitor terminals.

Figure 8 guides you in these steps.

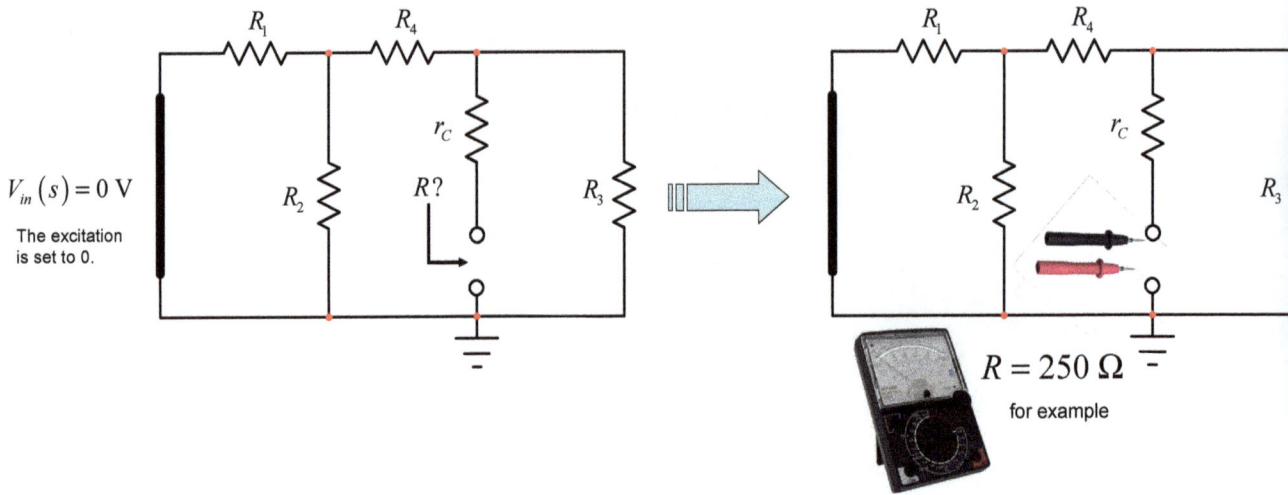

Figure 8: After replacing the 0-V source by a short circuit, you determine the resistance seen from the capacitor terminals.

If you run the exercise in Figure 8, you "see" r_C in series with the parallel combination of R_3 with the series-parallel arrangement of R_4 and R_1-R_2. The time constant of this circuit is simply the product of R and C_1:

$$\tau_1 = \left[r_C + \left(R_4 + R_1 \parallel R_2 \right) \parallel R_3 \right] C_1 \qquad (A.13)$$

We can show that the pole of a 1st-order system is the inverse of its time constant. Thus:

$$\omega_p = \frac{1}{\tau_1} = \frac{1}{\left[r_C + \left(R_4 + R_1 \parallel R_2 \right) \parallel R_3 \right] C_1} \qquad (A.14)$$

Now, what is the quasi-static gain of this circuit for $s = 0$? In dc conditions, a capacitor becomes an open circuit while an inductor becomes a short circuit. Apply this concept to Figure 7 circuit and redraw it as shown in Figure 9. In your head, you cut the connection before R_4 and you see a resistive divider involving R_1 and R_2. The Thévenin voltage across R_2 is:

$$V_{th} = V_{in} \frac{R_2}{R_1 + R_2} \qquad (A.15)$$

The output resistance R_{th} is R_1 paralleled with R_2.

The complete transfer function thus involves the resistive divider made of R_4 in series with R_{th} and loaded by R_3. r_C is off picture since capacitor C_1 is removed in this dc analysis.

You can thus write:

$$G_0 = \frac{V_{out}}{V_{in}} = \frac{R_2}{R_2 + R_1} \frac{R_3}{R_4 + R_3 + R_1 \| R_2} \tag{A.16}$$

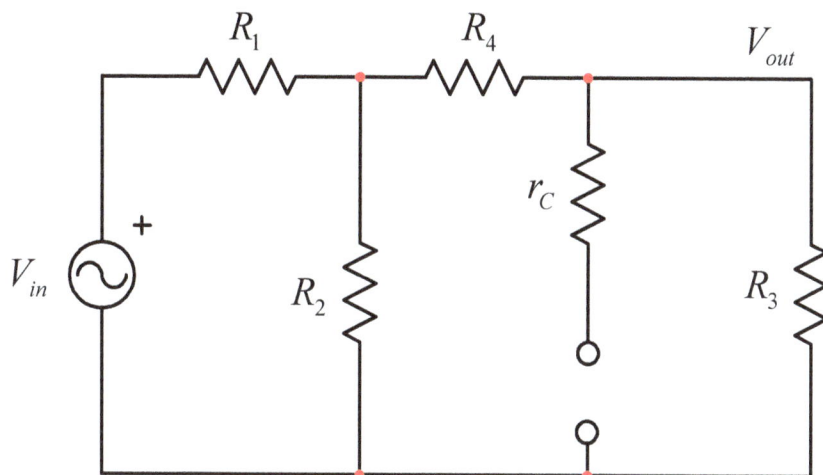

Figure 9: You open the capacitor in dc and calculate the transfer function of this simple resistive arrangement.

We are almost there and are missing the zeros. We said in preamble that a zero manifests itself in a circuit by blocking the propagation of the excitation signal and creating an output null (see Figure 1). If we consider a *transformed* circuit – in which C_1 is replaced by $1/sC_1$ – as shown in Figure 10, what particular condition would imply a nulled response when a stimulus biases the network? Having a nulled response simply means that the current circulating in R_3 is 0 A.

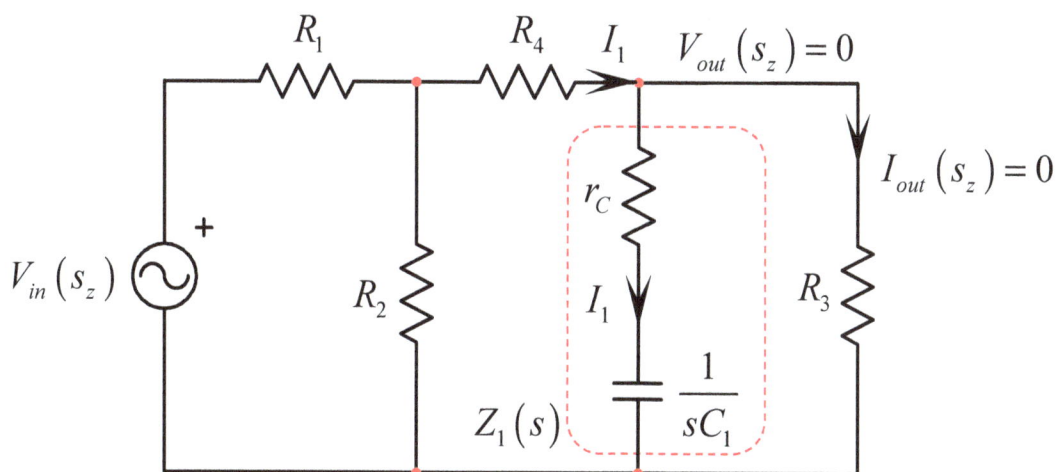

Figure 10: In this transformed circuit, when the series connection of r_C and C_1 becomes a transformed short circuit, the response disappears and no current flows in R3.

If we have no current in R_3, then the series connection of r_C and $1/sC_1$ creates a transformed short circuit:

$$Z_1(s_z) = r_C + \frac{1}{s_z C_1} = 0 \tag{A.17}$$

The root s_z is the zero location we want:

$$s_z = -\frac{1}{r_C C_1} \tag{A.18}$$

Leading to:

$$\omega_z = \frac{1}{r_C C_1} \tag{A.19}$$

We can now assemble all these results to form the final transfer function characterizing Figure 7 circuit:

$$G(s) = \frac{R_2}{R_2 + R_1} \frac{R_3}{R_4 + R_3 + R_1 \| R_2} \frac{1 + s r_C C_1}{1 + s \left[r_C + (R_4 + R_1 \| R_2) \| R_3 \right] C_1} = G_0 \frac{1 + \dfrac{s}{\omega_z}}{1 + \dfrac{s}{\omega_p}} \tag{A.20}$$

This is what is called a *low-entropy* expression in which you can immediately distinguish a quasi-static gain G_0, a pole ω_p and a zero ω_z. A *high-entropy* expression would be that obtained by applying the brute-force approach to the original circuit when considering an impedance divider for instance:

$$G(s) = \frac{R_2}{R_2 + R_1} \frac{R_3 \| \left(r_C + \dfrac{1}{sC_1} \right)}{R_3 \| \left(r_C + \dfrac{1}{sC_1} \right) + R_4 + R_1 \| R_2} \tag{A.21}$$

You can now capture these expressions into a mathematical solver such as Mathcad® and compare the dynamic responses of (A.20) and (A.21). If these expressions are identical, curves must perfectly superimpose as in Figure 11. To refine the comparison, you can also plot the difference in magnitude and phase of both expressions. The result should be the minimum value the solver can display.

$R_1 := 100\Omega$ $R_2 := 2k\Omega$ $R_3 := 470\Omega$ $R_4 := 2k\Omega$ $C_1 := 0.47\mu F$ $r_C := 50\Omega$

$\parallel(x,y) := \dfrac{x \cdot y}{x + y}$ $\tau_1 := C_1 \cdot \left[r_C + \left(R_4 + R_1 \parallel R_2 \right) \parallel R_3 \right] = 203.927\mu s$

$G_0 := \dfrac{R_2}{R_2 + R_1} \cdot \dfrac{R_3}{R_3 + R_4 + R_1 \parallel R_2} = 0.174$ $20 \cdot \log(G_0) = -15.164$ dB

$\omega_p := \dfrac{1}{\tau_1}$ $f_p := \dfrac{\omega_p}{2\pi} = 780.45 \, Hz$

$\omega_z := \dfrac{1}{r_C \cdot C_1}$ $f_z := \dfrac{\omega_z}{2\pi} = 6.773 \, kHz$

$G_1(s) := G_0 \cdot \dfrac{1 + \dfrac{s}{\omega_z}}{1 + \dfrac{s}{\omega_p}}$ $G_{ref}(s) := \dfrac{R_2}{R_2 + R_1} \cdot \dfrac{R_3 \parallel \left(r_C + \dfrac{1}{s \cdot C_1} \right)}{R_3 \parallel \left(r_C + \dfrac{1}{s \cdot C_1} \right) + R_4 + R_1 \parallel R_2}$

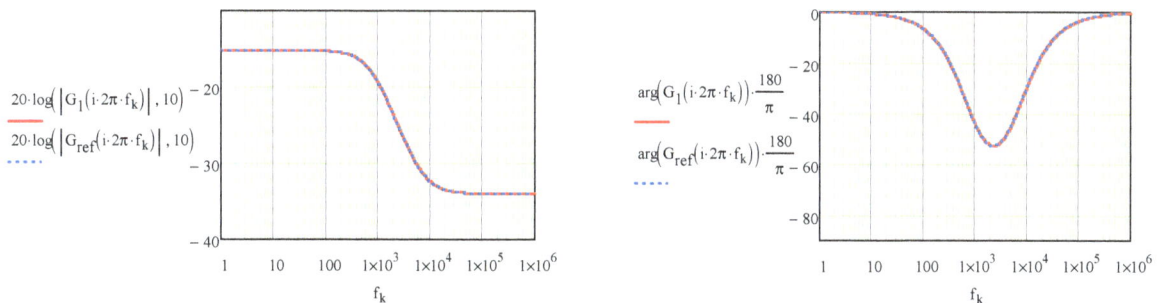

$20 \cdot \log\left(\left| G_1(i \cdot 2\pi \cdot f_k) \right|, 10 \right)$
$20 \cdot \log\left(\left| G_{ref}(i \cdot 2\pi \cdot f_k) \right|, 10 \right)$

$\arg\left(G_1(i \cdot 2\pi \cdot f_k) \right) \cdot \dfrac{180}{\pi}$
$\arg\left(G_{ref}(i \cdot 2\pi \cdot f_k) \right) \cdot \dfrac{180}{\pi}$

Figure 11: A Mathcad® sheet immediately tells you if the expression you have derived is wrong when comparing magnitude and phase response of (A.20) with that of (A.21).

Not only you could make mistakes in deriving the expression but formatting the result in something like in (A.20) would require more energy should you start from (A.21). Also, please note that in this particular example, we did not write a single line of algebra when writing (A.20). Should we later identify a mistake, then it is easy to come back to one of the individual drawings and fix it separately. The correction in (A.20) would then be simple. Try to run the same correction in (A.21) and you will probably restart from scratch.

4. A Common Denominator

When you study the different transfer functions of an electrical circuit, you change the stimulus location: a current source connects to the output node for the output impedance determination or a voltage source sweeps the input to express a gain for instance.

If you turn the stimulus off, you reveal the network natural structure. It is the original linear circuit without an excitation signal.

Now, assume you want to express two transfer functions of a particular circuit. When turning the excitation off in both cases, if the natural circuits you obtain are similar, then the transfer functions share a common denominator $D(s)$. It means that once the denominator coefficients have been determined in one configuration, as long as the stimulus insertion for the next transfer function does not change the network when it is turned off, you can reuse the denominator already on hand. This is a major gain of time as you can imagine. Let's put this observation at work with a quick example.

If you look at Figure 12, you see a 1st-order network built around inductor L_1. We want to express the transfer function H linking V_{out} to V_{in}. We begin with the denominator, setting the excitation to 0: this is a voltage source and we replace it by a short circuit. We now have the natural structure from which we can find the time constant: temporarily remove the inductor and "look" through its connecting terminals to determine the resistance which, combined with the inductor, forms the time constant τ_1 we want. This is what Figure 13a shows you. By inspection, the time constant is immediate and equal to:

$$\tau_1 = \frac{L_1}{r_L + R_1 \| R_2} \tag{A.22}$$

The denominator of this transfer function is thus:

$$D(s) = 1 + s\tau_1 = 1 + \frac{s}{\omega_p} \tag{A.23}$$

with:

$$\omega_p = \frac{r_L + R_1 \| R_2}{L_1} \tag{A.24}$$

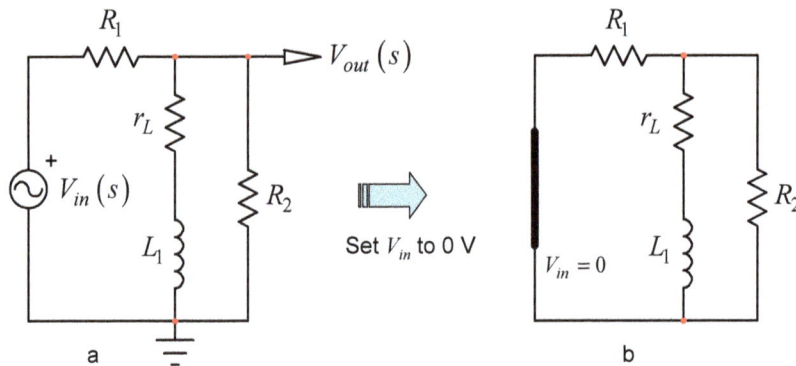

Figure 12: This 1st-order network natural structure is revealed by turning the excitation off.

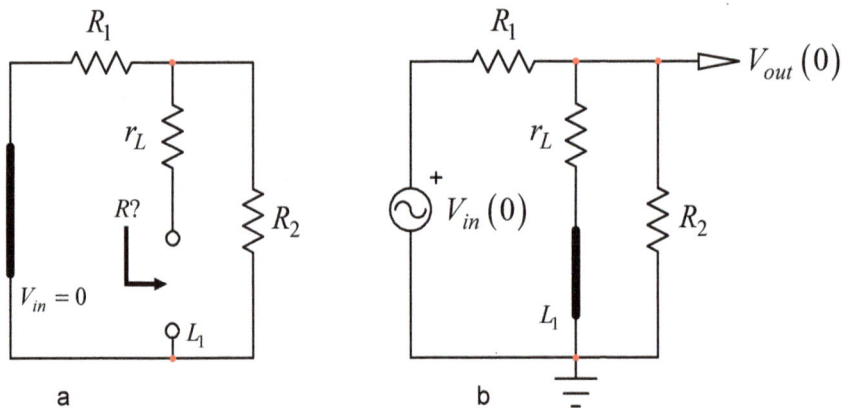

Figure 13: The resistance "seen" from the inductor terminals is the component we need to calculate the time constant.

The dc gain ($s = 0$) is obtained by shorting the inductor as shown in Figure 13b. You see a simple resistive divider leading to:

$$H_0 = \frac{R_2 \parallel r_L}{R_2 \parallel r_L + R_1} \qquad \text{(A.25)}$$

The zero is also determined by inspection when observing Figure 14.

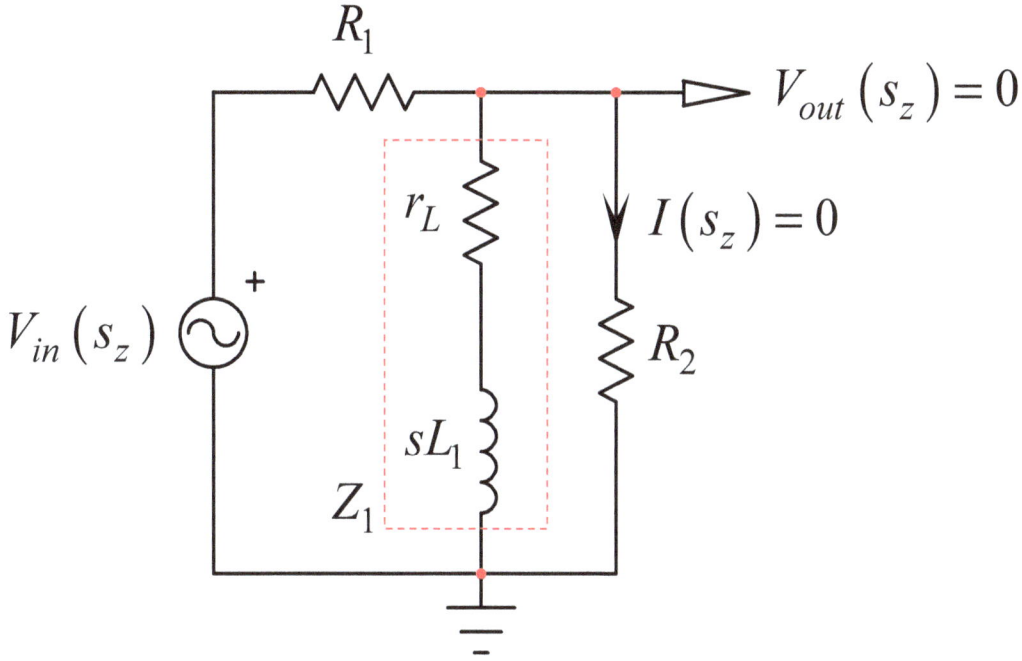

Figure 14: The null in the output occurs when Z1 becomes a transformed short.

The output null is obtained if no current circulates in R_2 implying a transformed short circuit brought by the series combination of r_L and L_1:

$$Z_1(s) = r_L + sL_1 = 0 \qquad \text{(A.26)}$$

which reveals a root located at :

$$s_z = -\frac{r_L}{L_1} \qquad \text{(A.27)}$$

and a zero positioned at:

$$\omega_z = \frac{r_L}{L_1} \qquad \text{(A.28)}$$

The complete transfer function is obtained by combining (A.24), (A.25) and (A.28):

$$H(s) = \frac{R_2 \| r_L}{R_2 \| r_L + R_1} \frac{1 + s\dfrac{L_1}{r_L}}{1 + s\dfrac{L_1}{r_L + R_1 \| R_2}} = H_0 \frac{1 + \dfrac{s}{\omega_z}}{1 + \dfrac{s}{\omega_p}} \tag{A.29}$$

Now, let's determine a second transfer function, the output impedance Z_{out}. To determine an impedance (or a resistance) at a given node, the classical solution resorts to installing a current test generator I_T which biases the identified point while you determine the voltage V_T developed across the current source.

The impedance (or the resistance) is thus equal to:

$$Z(s) = \frac{V_T(s)}{I_T(s)} \tag{A.30}$$

Figure 15a illustrates this principle at work with our 1st-order network. The output impedance is evaluated while the input bias, V_{in} in this case, is set to 0 V. The stimulus is now the injected current I_T while the response is the voltage V_T across the current source terminals. Turning the excitation off to determine the time constant and thus the network pole, means that the current source is open circuited, leading to sketch (b) in Figure 15.

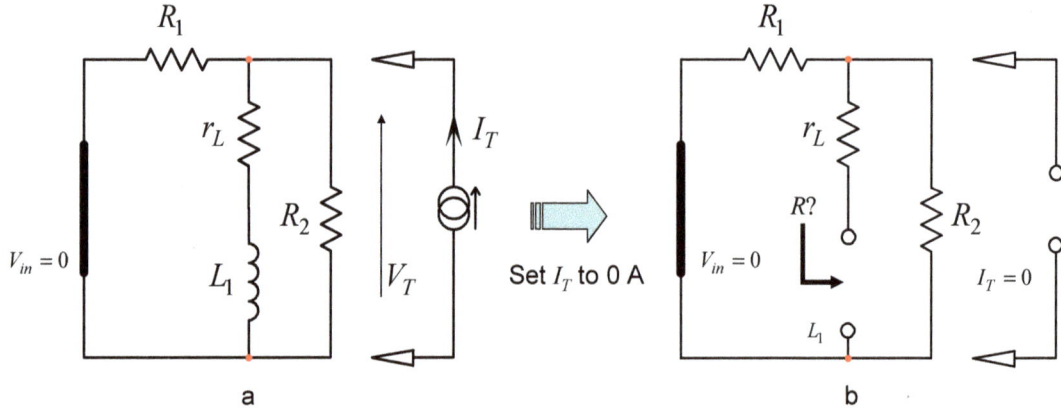

Figure 15: For the output impedance determination, you install a test generator.

You can immediately see that the network has returned to the state already depicted by Figure 13 sketch (a). This implies an identical time constant and a denominator similar to what (A.23) has already defined: no need to derive it again!

$$Z_{out}(s) = R_0 \frac{N(s)}{1 + s\dfrac{L_1}{r_L + R_1 \| R_2}} \tag{A.31}$$

Without detailing the steps (set L_1 in its dc state, a short circuit), the dc resistance is equal to:

$$R_0 = R_1 \parallel r_L \parallel R_2 \qquad (A.32)$$

and the zero is also incurred to the transformed short circuit brought by r_L and L_1 since the response is still across R_2. The complete transfer function is thus expressed as:

$$Z_{out}(s) = R_0 \frac{1 + \dfrac{s}{\omega_z}}{1 + \dfrac{s}{\omega_p}} \qquad (A.33)$$

with ω_p and ω_z respectively defined by (A.24) and (A.28).

Reusing a common denominator when determining the several transfer functions of a given circuit represents a tremendous time-saving characteristic of the FACTs, especially with n^{th}-order systems as we will see in the book. However, you have to make sure the network always returns in its natural state when turning the excitation off for each of the transfer functions you study.

Let's look at the input impedance Z_{in} now. The principle remains the same, we install a test generator I_T and express the voltage across its terminal to apply (A.30). This is what Figure 16 suggests. To determine the time constant, the stimulus is reduced to 0 A and the new circuit appears in sketch (b) of Figure 16.

If you now compare this drawing to that of Figure 13 sketch (a), you have changed the network structure meaning a different time constant. Indeed, the resistance "seen" from L_1's terminals is $r_L \parallel R_2$, leading to a new time constant equal to:

$$\tau_1 = \frac{L_1}{r_L \parallel R_2} \qquad (A.34)$$

The denominator is defined as:

$$D(s) = 1 + s\tau_1 = 1 + \frac{s}{\omega_p} \qquad (A.35)$$

with a pole located at:

$$\omega_p = \frac{r_L \parallel R_2}{L_1} \qquad (A.36)$$

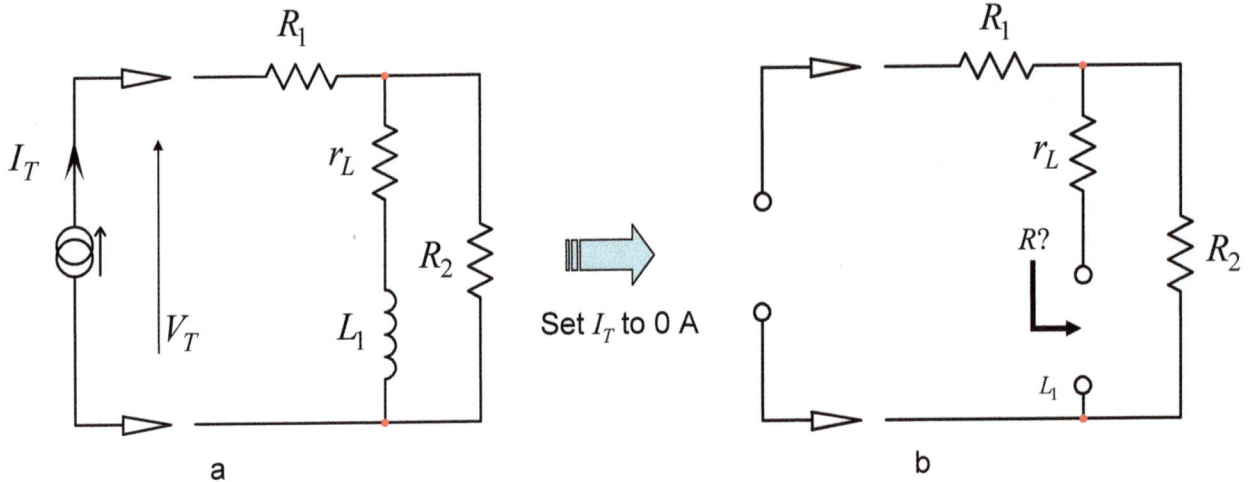

Figure 16: The input impedance is determined by opening the left-side connection of R1.

In this input impedance case, nulling the response implies that the voltage V_T across the current generator is 0 V. A current source with 0-V across its terminals, as shown in [4], represents what is called a *degenerate* case: you can replace the current source by a short circuit, it won't affect the current which flows through it. Applying this principle to input impedance measurement leads to the drawing of Figure 17.

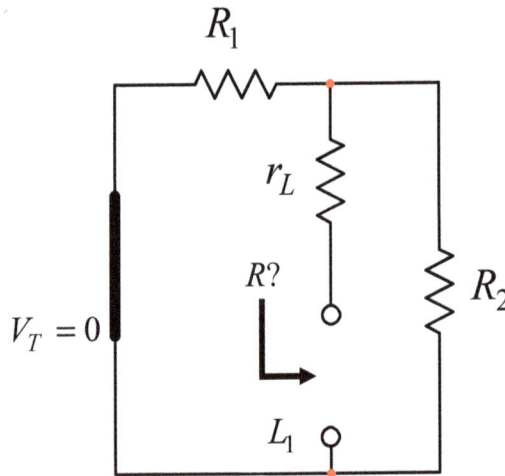

Figure 17: Nulling the response means replacing the current source by a short circuit in this input impedance measurement.

You recognize the same circuit already studied in sketch (a) of Figure 13 meaning that the numerator of our input impedance is the denominator already determined in (A.23). And it makes sense: assuming that rather than sweeping the input impedance with a current source as described by (A.30), you decide to install a voltage source and determine the current it generates. The stimulus becomes the voltage and the response a current: you want to express an admittance Y and the denominator is that of (A.23). To obtain the impedance Z, take the inverse of Y and the denominator becomes the numerator. This is an extremely useful observation when determining multiple transfer functions of a circuit.

Finally, the input impedance is defined as:

$$Z_{in}(s) = [R_1 \| r_L \| R_2] \frac{1 + s\dfrac{L_1}{r_L + R_1 \| R_2}}{1 + s\dfrac{L_1}{r_L \| R_2}} = R_0 \frac{1 + \dfrac{s}{\omega_z}}{1 + \dfrac{s}{\omega_p}} \tag{A.37}$$

With ω_p and ω_z defined as:

$$\omega_p = \frac{r_L \| R_2}{L_1} \tag{A.38}$$

$$\omega_z = \frac{r_L + R_1 \| R_2}{L_1} \tag{A.39}$$

5. The FACTs Applied to a Second-Order System

FACTs work equally well for n^{th}-order passive or active circuits. You determine the order of a circuit by counting the number of energy-storing elements whose state variables are *independent*. If we consider a second-order system H featuring a finite quasi-static gain H_0, its transfer function can be expressed the following way:

$$H(s) = H_0 \frac{1 + a_1 s + a_2 s^2}{1 + b_1 s + b_2 s^2} \tag{A.40}$$

As H_0 carries the unit of the transfer function then the ratio made of N over D is unitless. This implies that the unit for a_1 and b_1 is time [s]. You sum up the circuit time constants determined when the response is nulled for a_1 and when the excitation is zeroed for b_1. For the second-order coefficients, a_2 or b_2, the dimension is time squared [s²] and you combine time constants in a product. However, in this time constants product, you reuse one of the time constant already determined for a_1 or b_1 while the second time constant determination requires a different notation:

$$\tau_2^1 \text{ or } \tau_1^2 \tag{A.41}$$

In this definition, you set the energy-storing element whose label appears in the "exponent" in its high-frequency state: a capacitor is replaced by a short circuit while an inductor would be replaced by an open circuit. You then determine the resistance "seen" from the second element terminals when it is temporarily removed from the circuit (subscripted reference). For higher-order circuits, energy-storing elements whose labels are not involved in the superscript are left in their dc state during the analysis (capacitors are open circuited and inductors are shorted). You carry this exercise for a nulled output when a_2 must be obtained and when the excitation is reduced to 0 for b_2. Of course, when inspection works, it is always the fastest and most efficient way to obtain $N(s)$. A bit mysterious at first sight but nothing insurmountable as we will see in a few lines.

Figure 18 depicts a classical second-order filter involved in the determination of the output impedance of a voltage-mode buck converter operated in the continuous conduction mode (CCM). An impedance is a transfer function linking an excitation signal I_{out} to a response signal V_{out}. Here, I_{out} is the test generator we have installed while V_{out} is the resulting voltage produced across its terminals. To determine the various coefficients from (A.40), we can follow Figure 6 flow chart and start with $s = 0$: short the inductor and open the capacitor as shown in the picture.

The circuit is simple and the resistance R_0 seen from the current source is simply the parallel combination of r_L and R_{load}:

$$R_0 = r_L \parallel R_{load} \qquad (A.42)$$

Do we have zeros in this circuit? Let's have a look at the transformed circuit shown in Figure 19. We can check what component combinations would bring the response V_{out} to zero when the excitation current I_{out} is tuned at a zero angular frequency s_z. We can identify two transformed short circuits involving r_L-L_1 and r_C-C_2.

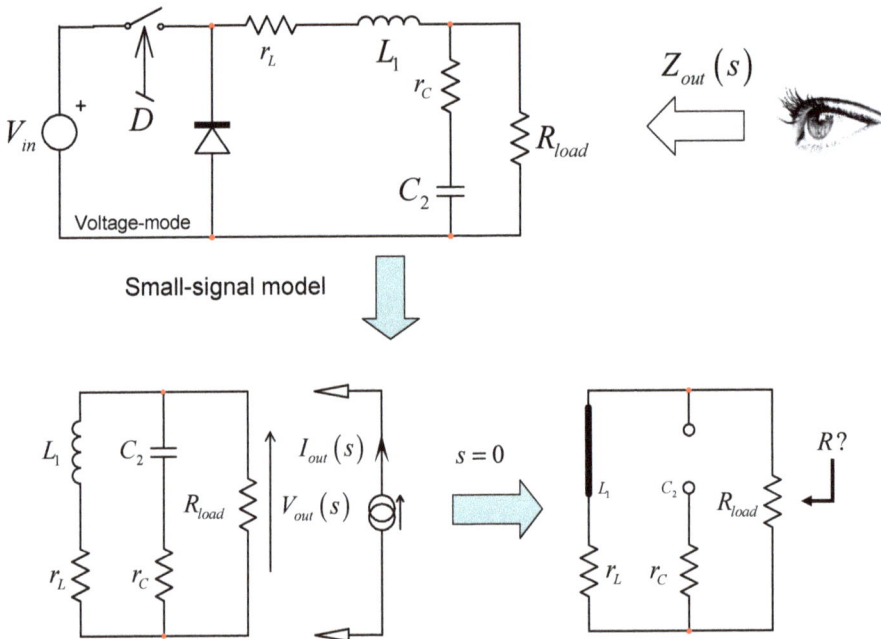

Figure 18: The determination of the CCM-operated buck converter output impedance is a good example showing how FACTs simplify analyses.

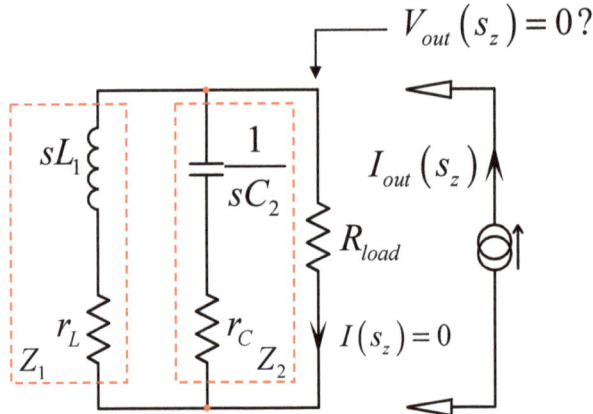

Figure 19: If impedances Z_1 or Z_2 become transformed shorts, the response Vout is nulled.

The roots for these two impedances are immediately determined:

$$r_L + sL_1 = 0 \rightarrow s_{z_1} = -\frac{r_L}{L_1} \tag{A.43}$$

$$r_C + \frac{1}{sC_2} = 0 \rightarrow s_{z_2} = -\frac{1}{r_C C_2} \tag{A.44}$$

The denominator $N(s)$ is thus expressed by:

$$N(s) = \left(1 + s\frac{L_1}{r_L}\right)(1 + sr_C C_2) \tag{A.45}$$

The first coefficient b_1 of the denominator $D(s)$ is obtained by looking at the resistance offered by L_1's terminals while C_2 is in its dc state (open): you have τ_1. Then you look at the resistance driving C_2 while L_1 is set in its dc state (short circuit): you obtain τ_2. As illustrated by Figure 20, the sketch immediately leads to the definition of b_1:

$$b_1 = \tau_1 + \tau_2 = \frac{L_1}{r_L + R_{load}} + C_2\left[(r_L \parallel R_{load}) + r_C\right] \tag{A.46}$$

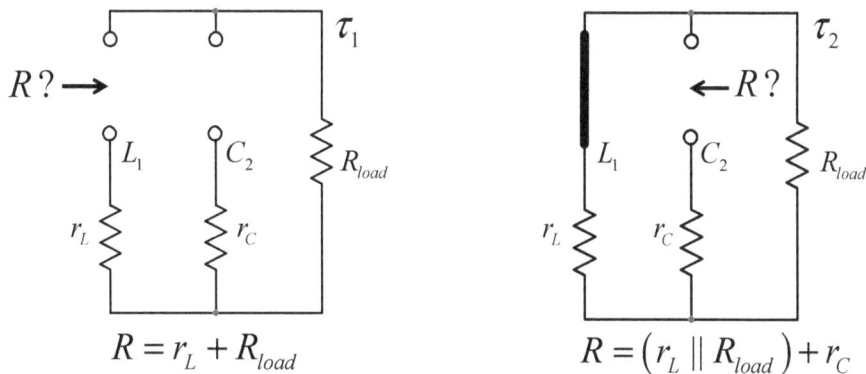

Figure 20: What resistance do you "see" between the selected component terminals while the second is set in its dc state?

The second-order coefficient b_2 is determined using the notation introduced in (A.41). Either L_1 is set in its high-frequency state (open circuit) and you look at the resistance driving C_2 to obtain τ_2^1 or C_2 is put in its high-frequency state (short circuit) and you look at the resistance driving L_1 for τ_1^2. Figure 21 shows the two possible arrangements. You usually select the one leading to the simplest expression or the one avoiding a product indeterminacy if any ($\infty \times 0$ or ∞/∞ for instance).

The below two definitions for b_2 are identical and you see that the upper one is the simplest:

$$b_2 = \tau_1 \tau_2^1 = \frac{L_1}{r_L + R_{load}} C_2 (r_c + R_{load})$$

$$b_2 = \tau_2 \tau_1^2 = C_2 [r_L \| R_{load} + r_C] \frac{L_1}{r_L + R_{load} \| r_C}$$

(A.47)

We now have all ingredients to assemble the final transfer function which is defined as:

$$Z_{out}(s) = (r_L \| R_{load}) \frac{\left(1 + s\frac{L_1}{r_L}\right)(1 + sr_C C_2)}{1 + s\left(\frac{L_1}{r_L + R_{load}} + C_2 [r_L \| R_{load} + r_C]\right) + s^2\left(L_1 C_2 \frac{r_C + R_{load}}{r_L + R_{load}}\right)}$$

(A.48)

We have determined this transfer function without writing a line of algebra, just by splitting the circuit in several simple sketches individually solved. Furthermore, as expected, (A.48) is already in a canonical form and you can easily see the presence of a quasi-static resistance (its unit is ohm), two zeros and a second-order denominator you could further rearrange with a resonant term ω_0 and a quality factor Q.

There is no way we could have obtained this result that quickly considering the parallel combination of Z_1, Z_2 and R_{load}.

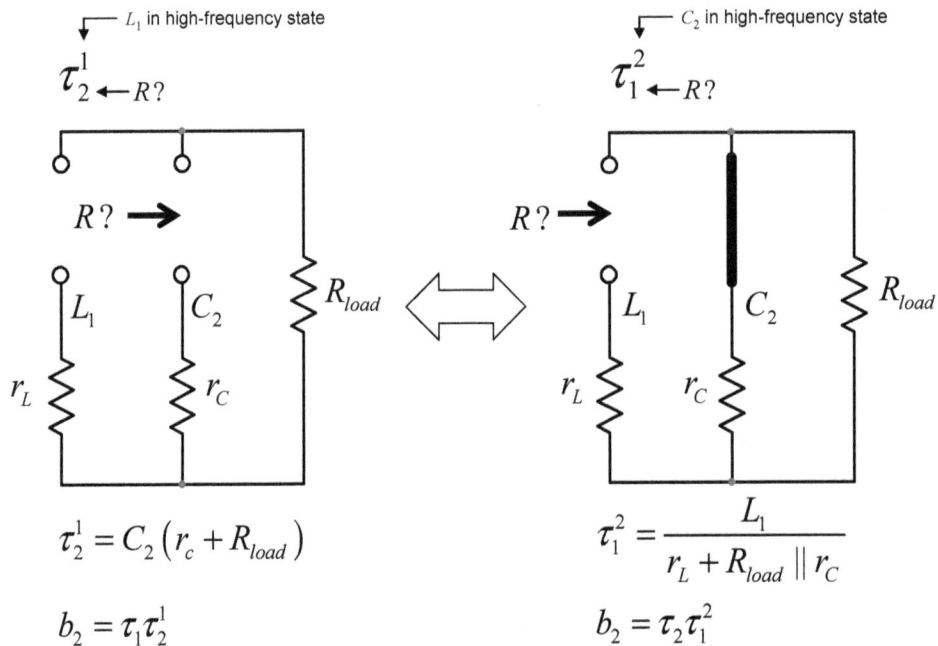

$$\tau_2^1 = C_2 (r_c + R_{load})$$

$$\tau_1^2 = \frac{L_1}{r_L + R_{load} \| r_C}$$

$$b_2 = \tau_1 \tau_2^1$$

$$b_2 = \tau_2 \tau_1^2$$

Figure 21: What resistance do you see between the selected component terminals while the second is set in its high-frequency state?

Deriving transfer functions by inspection is a possibility offered by the FACTs in particular with passive networks. As the circuit complicates and includes voltage- or current-controlled sources, inspection becomes less obvious and you need to resort to classical mesh and node analysis. The FACTs offer several advantages however: as you split the circuit into small individual sketches used to determine the coefficients of the final polynomial form, you can always come back to a particular drawing and individually correct it in case you have found a mistake in the final expression. Also, as you determine the terms associated with the a_i and b_i of the transfer function, you naturally end-up with a polynomial form without investing further energy to collect and rearrange the terms.

Finally, as shown in [4], SPICE can be of great help to verify your individual poles and zeros calculations in the presence of complicated passive and active circuits.

6. References

1. R. D. Middlebrook, *Methods of Design-Oriented Analysis: Low-Entropy Expressions*, Frontiers in Education Conference, Twenty-First Annual conference, Santa-Barbara, 1992.
2. R. D. Middlebrook, *Null Double Injection and the Extra Element Theorem*, IEEE Transactions on Education, Vol. 32, NO. 3, August 1989.
3. V. Vorpérian, *Fast Analytical Techniques for Electrical and Electronic Circuits*, Cambridge University Press, 2002.
C. Basso, Linear Circuit Transfer Functions – An Introduction to Fast Analytical Techniques, Wiley, 2016.
4. D. Feucht, *Design-Oriented Circuit Dynamics*, http://www.edn.com/electronics-blogs/outside-the-box-/4404226/Design-oriented-circuit-dynamics
5. D. Peter, *We Can do Better: A Proven, Intuitive, Efficient and Practical Design-Oriented Circuit Analysis Paradigm is Available, so why aren't we using it to teach our Students?*,
http://www.icee.usm.edu/ICEE/conferences/asee2007/papers/1362_WE_CAN_DO_BETTER__A_PROVEN__INTUITIVE__E.pdf

INDEX